Solutions Manual

Robert K. Wismer
Millersville University

GENERAL CHEMISTRY

Principles and Modern Applications

Seventh Edition

RALPH H. PETRUCCI
WILLIAM S. HARWOOD

PRENTICE HALL Upper Saddle River, NJ 07458

Production Editor: *James Buckley*
Associate Editor: *Mary Hornby*
Special Projects Manager: *Barbara Murray*
Manufacturing Buyer: *Ben Smith*
Supplement Cover Manager: *Paul Gourhan*
Cover Art: ©*Richard Megna, Fundamental Photographs, NYC*

Printed in the United States of America

10 9 8 7 6 5 4 3 2 1

ISBN 0-13-271347-0

Prentice-Hall International (UK) Limited, *London*
Prentice-Hall of Australia Pty. Limited, *Sydney*
Prentice-Hall Canada Inc., *Toronto*
Prentice-Hall Hispanoamericana, S.A., *Mexico*
Prentice-Hall of India Private Limited, *New Delhi*
Prentice-Hall of Japan, Inc., *Tokyo*
Simon & Schuster Asia Pte. Ltd., *Singapore*
Editora Prentice-Hall do Brasil, Ltda., *Rio de Janeiro*

Contents

To the Student

One of the main benefits you will obtain from your study of chemistry is the ability to explain a large number of phenomena by learning and applying a relatively small number of fundamental principles. But it is very difficult to learn these principles and how to apply them by simply reading about chemistry. Instead you must practice using them, both in the laboratory and by solving problems.

The purpose of this supplement is to help you master the use of many chemical principles through problem solving. Here are the solutions to all the Practice Examples, Review Questions, Exercises, and Feature Problems in the companion textbook: Petrucci and Harwood, General Chemistry, Principles and Modern Application, 7th ed., Prentice-Hall, Upper Saddle River, NJ, 1997. Below are given a few suggestions to help you use this book so that you obtain the maximum benefit from it.

You should attempt to solve a problem only after you have carefully studied the text in the chapter. If you try to short circuit this process, by attempting to solve the problems before you have studied the text, you will find yourself constantly returning to the chapter, paging through it for ideas that you need to approach the problem. And this search will be inefficient, because you are not familiar with the chapter. Worse, you may simply look back for an example that is similar to the problem you are attempting to work. This latter technique does not help you learn how to solve problems; it simply develops your ability to copy someone else's solution.

You should attempt to solve the Practice Exercises, Review Questions, and Exercises with a pencil and paper (and probably an electronic calculator as well). After you study an Example in the text, you should attempt to solve one of the Practice Examples. If that one comes easily, you can proceed with confidence. If it does not, you should study the Example again and try the other Practice Example. Chemistry is a vertical discipline; it builds on what has come before. Often you may not be able to understand what comes later in a chapter if you do not understand the beginning. Then for each chapter, you should attempt to solve all of the Review Questions and a representative sampling of the Exercises, so that you know you have mastered the principles presented in that chapter.

Only after you have made a determined effort on each problem should you turn to these solutions. If you simply look at the statement of the problem, think about it briefly, and then look at its solution, you will have fooled yourself into believing that you would have come up with that solution on your own. You will never develop the ability to solve problems. And you will miss something else: arriving at your own method of solving the problem. There are many ways of solving problems and, in some of these solutions, some of these alternate solutions are suggested. Some of the solutions given in these pages are slightly different thant those of the text. This will help you realize that there are other ways of arriving at a solution. Just as no two people will drive a car to a destination in exactly the same fashion, so no two people will produce the same solution to a problem. Probably your worse mistake is to continually try to copy the solution style of another person.

Be aware that you are *not* trying to "get the right answer." The correct answers to these problems are known. Rather, you are trying to learn how to solve problems, so that when you are confronted by new problems (either on an examination or in another course—or your job—that requires chemistry) you will know how to approach them. Sometimes you will arrive at a slightly wrong answer with a process that is correctly based on chemical principles with perhaps one small error, such as misplacing a decimal point.

There is, of course, a lot more to chemistry than solving mathematical problems. And many problems of a quantitative nature are presented in the textbook and solved here. These problems involve correctly defining terms; contrasting two similiar but distinct concepts; explaining chemical phenomena; predicting the products of chemical reactions; and representing chemical entities—atoms, ions, and molecules—through names, formula, and sketches; and so on. Remember to work on this aspect of your chemical education as well.

But what if you make a mistake? First, be aware that if your answer agrees with the one given here in only the last digit, you have probably not made a mistake at all; the discrepancy is simply due to rounding. The procedure followed in solving these problems is to round the result at each step in the calculation where an intermediate answer has been written down. Where an extra significant figure has been retained until the end of the calculation, that digit is written as a subscript (such as 1.5_2, rather than the misleading 1.52). Second, your answer may differ simply by a power of ten or by a simple factor (such as 2 or 3). Many times these discrepancies are the result of not pushing the correct buttons on the calculator, or due to some small error in logic. Finally, you may have a serious error. If you wrote down your technique for solving the problem in detail, you can compare it with the solution given here and see where you went wrong. If your technique differs substantially from that of the solution, you may be able to get some ideas of what you did wrong from reading the solution carefully. Or your technique may be radically different from the one presented here and still be correct, except for a small flaw. At this point you should ask your a classmate or your instructor for assistance.

The important point is to work on problem solving yourself. You are the one who is trying to learn chemistry, not your instructor and not the chemistry major down the hall who seems to know all the answers. If you make the mistakes, and you fix them, you will learn chemistry much better than if someone else shows you the right answer. And, of course, you will have the satisfaction of having done it yourself.

Millersville, Pennsylvania

Robert K. Wismer

This book is dedicated to learners in all fields, especially the three most active learners in my family: Michael, Mary, and Karen.

ACKNOWLEDGEMENTS

Any book is the product of many individuals and this one is also. Many improvements have resulted from their efforts; the errors that remain are mine. I owe special thanks to Professor Ralph H. Petrucci for his frequent comments throughout the writing of this edition, and for checking many of the solutions, especially during the process of producing the Answer Key for the textbook.

The students of The Science of Chemistry and Physical Chemistry I at Millersville University during the fall semester of 1996 have shown great patience with a professor preoccupied with writing. This is also true of my colleagues on the faculty of the Chemistry Department. I appreciate their support and encouragement.

Mary Hornby, supplements editor at Prentice-Hall Publishing Company, supervised this book. Elisabeth Belfer of college production at Macmillan taught me how to write for publication with her advice and encouragement while producing previous versions. Any beauty in these printed pages is largely the result of her guidance and of the careful readers she employed.

This book was written with Microsoft Word on a Macintosh Quadra 650 computer, using SuperPaint for most pictures, WingZ for spreadsheet calculations, Cricket Graph for graphs, and Chemintosh for chemical structures. Millersville Physical Chemistry students have helped me learn the intracacies of these programs.

Four accuracy checkers worked on this book: Michael Pearson checked the Review Problems and Exercises in the first twelve chapters, Julie Grundman checked the Practice Exercises throughout, and Scott Gable checked the Review Problems and Exercises in Chapters 22 through 25. I owe a personal debt of gratitude to Michael Wismer who gave up his other summer plans to work through every Review Problem and Exercise.

But I still owe my largest debt of gratitude to my wife Debbie and our children Michael, Mary, and Karen. They have understood a husband and father who always is thinking of and working on "a book," and have constantly loved, supported, and encouraged him.

Robert K. Wismer

1 MATTER—ITS PROPERTIES AND MEASUREMENT

PRACTICE EXAMPLES

1A Convert the Fahrenheit temperature to Celsius and compare. $C = (F - 32)\frac{5}{9} = (103\ °F - 32)\frac{5}{9} = 39.4\ °C$. The temperature of 41 °C in New Delhi is higher than the predicted high of 103 °F = 39.4 °C in Phoenix.

1B We convert the Fahrenheit temperature to Celsius. $C = (F - 32)\frac{5}{9} = (-15\ °F - 32)\frac{5}{9} = -26.1\ °C$. The antifreeze only protects to –22 °C and will not offer protection to temperatures as low as –15 °F = –26.1 °C.

2A The mass is the difference between the mass of the full and empty flask.

$$\text{density} = \frac{291.4\ \text{g} - 108.6\ \text{g}}{125\ \text{mL}} = 1.46\ \text{g/mL}$$

2B The volume of the stone is the difference between the level in the graduated cylinder with the stone present and with it absent.

$$\text{density} = \frac{\text{mass}}{\text{volume}} = \frac{28.4\ \text{g rock}}{44.1\ \text{mL rock\&water} - 33.8\ \text{mL water}} = 2.76\ \text{g/mL}$$

3A Use density as a conversion factor.

$$\text{solution mass} = 125\ \text{mL soln} \times \frac{1.081\ \text{g soln}}{1\ \text{mL soln}} = 135\ \text{g soln.}$$

3B $\text{ethanol volume} = 50.0\ \text{kg ethanol} \times \frac{1000\ \text{g}}{1\ \text{kg}} \times \frac{1\ \text{mL ethanol}}{0.789\ \text{g ethanol}} \times \frac{1\ \text{L}}{1000\ \text{mL}} = 63.4\ \text{L ethanol}$

Notice how setting up the calculation as a chain of conversions makes it easy to see that we do not need to multiply by 1000 because a later division by 1000 cancels that multiplication.

4A We use the density first to determine the mass of gasohol.

$$\text{ethanol mass} = 25\ \text{L gasohol} \times \frac{1000\ \text{mL}}{1\ \text{L}} \times \frac{0.71\ \text{g gasohol}}{1\ \text{mL}} \times \frac{10\ \text{g ethanol}}{100\ \text{g gasohol}} \times \frac{1\ \text{kg ethanol}}{1000\ \text{g ethanol}}$$

$$= 1.8\ \text{kg ethanol}$$

4B We use the percent to determine the mass of the 25.0-mL sample.

$$\text{rubbing alcohol mass} = 15.0\ \text{g 2-propanol} \times \frac{100.0\ \text{g rubbing alcohol}}{70.0\ \text{g 2-propanol}} = 21.4\ \text{g rubbing alcohol}$$

$$\text{rubbing alcohol density} = \frac{21.4\ \text{g}}{25.0\ \text{mL}} = 0.856\ \text{g/mL}$$

5A The factor 0.00456 has three significant figures.

$$\frac{62.356}{0.000456 \times 6.422 \times 10^3} = 21.3$$

5B The factor 1.3×10^{-3} determines the number of significant figures.

$$\frac{8.21 \times 10^4 \times 1.3 \times 10^{-3}}{0.00236 \times 4.071 \times 10^{-2}} = 1.1 \times 10^6$$

6A The last term has one digit to the right of the decimal. $0.236 + 128.55 - 102.1 = 26.7$

6B This is easier to visualize if the numbers are not in scientific notation.

$$\frac{(1.302 \times 10^3) + 952.7}{(1.57 \times 10^2) - 12.22} = \frac{1302 + 952.7}{157 - 12.22} = \frac{2255}{145} = 15.6$$

SUMMARIZING EXAMPLE CALCULATIONS

1. trip length $= 808\text{ km} \times \frac{1000\text{ m}}{1\text{ km}} \times \frac{100\text{ cm}}{1\text{ m}} \times \frac{1\text{ in.}}{2.54\text{ cm}} \times \frac{1\text{ ft}}{12\text{ in.}} \times \frac{1\text{ mi}}{5280\text{ ft}} = 502\text{ mi}$

2. volume of fuel consumed $= 502\text{ mi} \times \frac{1\text{ gal}}{25.5\text{ mi}} = 19.7\text{ gal}$

3. fuel mass $= 19.7\text{ gal} \times \frac{4\text{ qt}}{1\text{ gal}} \times \frac{0.9464\text{ L}}{1\text{ qt}} \times \frac{1000\text{ mL}}{1\text{ L}} \times \frac{0.775\text{ g}}{1\text{ mL}} = 5.78 \times 10^4\text{ g}$

4. methanol mass $= 5.78 \times 10^4\text{ g fuel} \times \frac{85.0\text{ g methanol}}{100.0\text{ g fuel}} = 4.91 \times 10^4\text{ g methanol} = 49.1\text{ kg methanol}$

REVIEW QUESTIONS

1. **(a)** A $\mathbf{m^3}$ is a cubic meter. As a regular solid, it is a cube one meter on a side, in volume very crudely equal to one cubic yard.
(b) **% by mass** is read "percent by mass." It is the mass in grams of one portion of a sample present in precisely 100 grams of that sample.
(c) **°C** is the temperature of substance expressed on a scale (the Celsius scale) where the freezing point of water has a value of "zero" and the boiling point of water has a value of "one hundred."
(d) **Density** is the concentration of the mass of a material. It is calculated as the mass of the material (in grams) divided by its volume (in mL or cm^3).
(e) An **element** is a substance that cannot be altered or decomposed chemically. Each element has a definite name and a specific position on the periodic table.

2. **(a)** The seven **SI base units** are those from which all other units are derived. Among them are the meter for length, the kilogram for mass, the kelvin for temperature, the second for time, and the mole for amount of substance.
(b) **Significant figures** are those digits in a number that are the result of experimental measurement, or are derived from such a measurement.
(c) A **natural law** is a summary of experimental results or of observations.

3. **(a)** The **mass** of an object is a measure of the amount of material in that object. Its **weight**, on the other hand, is the force that the object exerts due to gravitational attraction.
(b) An **extensive property** is one that depends on the quantity of material present; an **intensive** property does not so depend.
(c) A **substance** is a pure form of matter; it is either an element or a compound. A **mixture** is a blend of two or more substances, in no particular proportion.
(d) **Precision** refers to the reproducibility of an experimental measurement; **accuracy** describes the agreement between the measurement and an accepted value of the same property.
(e) A **hypothesis** is a tentative explantation of a natural law. A **theory** is a hypothesis that has survived the test of repeated experiments.

4. **(a)** $1.55\text{ kg} \times \frac{1000\text{ g}}{1\text{ kg}} = 1.55 \times 10^3\text{ g}$ **(b)** $642\text{ g} \times \frac{1\text{ kg}}{1000\text{ g}} = 0.642\text{ kg}$
(c) $2896\text{ mm} \times \frac{1\text{ cm}}{10\text{ mm}} = 289.6\text{ cm}$ **(d)** $0.086\text{ cm} \times \frac{10\text{ mm}}{1\text{ cm}} = 0.86\text{ mm}$

5. **(a)** $0.127\text{ L} \times \frac{1000\text{ mL}}{1\text{ L}} = 127\text{ mL}$ **(b)** $15.8\text{ mL} \times \frac{1\text{ L}}{1000\text{ mL}} = 0.0158\text{ L}$
(c) $981\text{ cm}^3 \times \frac{1\text{ L}}{1000\text{ cm}^3} = 0.981\text{ L}$ **(d)** $2.65\text{ m}^3 \times \left(\frac{100\text{ cm}}{1\text{ m}}\right)^3 = 2.65 \times 10^6\text{ cm}^3$

6. **(a)** $68.4\text{ in} \times \frac{2.54\text{ cm}}{1\text{ in.}} = 173\text{ cm}$ **(b)** $94\text{ ft} \times \frac{12\text{ in.}}{1\text{ ft}} \times \frac{2.54\text{ cm}}{1\text{ in.}} \times \frac{1\text{ m}}{100\text{ cm}} = 29\text{ m}$
(c) $1.42\text{ lb} \times \frac{453.6\text{ g}}{1\text{ lb}} = 644\text{ g}$ **(d)** $248\text{ lb} \times \frac{0.4536\text{ kg}}{1\text{ lb}} = 112\text{ kg}$
(e) $1.85\text{ gal} \times \frac{4\text{ qt}}{1\text{ gal}} \times \frac{0.9464\text{ L}}{1\text{ qt}} = 7.00\text{ L}$ **(f)** $3.72\text{ qt} \times \frac{0.9464\text{ L}}{1\text{ qt}} \times \frac{1000\text{ mL}}{1\text{ L}} = 3.52 \times 10^3\text{ mL}$

7. **(a)** $1.00\text{ km}^2 \times \left(\frac{1000\text{ m}}{1\text{ km}}\right)^2 = 1.00 \times 10^6\text{ m}^2$ **(b)** $1.00\text{ m}^2 \times \left(\frac{100\text{ cm}}{1\text{ m}}\right)^2 = 1.00 \times 10^4\text{ cm}^2$

(c) $1.00 \text{ mi}^2 \times \left(\frac{5280 \text{ ft}}{1 \text{ mi}} \times \frac{12 \text{ in.}}{1 \text{ ft}} \times \frac{2.54 \text{ cm}}{1 \text{ in.}} \times \frac{1 \text{ m}}{100 \text{ cm}}\right)^2 = 2.59 \times 10^6 \text{ m}^2$

8. The boiling point of water can serve as our reference. 204 °F is below the 212 °F boiling point of water, while 102 °C is above the 100 °C boiling point of water. Thus, 102 °C is the higher temperature.

9. The 80.0 g ethanol seems least massive. The 100.0 mL of benzene, with a density less than 1 g/mL, must have a mass less than 100.0 g (it is actually 87 g). On the other hand, 90.0 mL of carbon disulfide, with a density of 1.26 g/mL, should have a mass somewhat in excess of 100.0 g (it is actually 113 g). Thus 90.0 mL of carbon disulfide is the most massive.

10. Butyric acid density $= \frac{\text{mass}}{\text{volume}} = \frac{2088 \text{ g}}{2.18 \text{ L}} \times \frac{1 \text{ L}}{1000 \text{ mL}} = 0.958 \text{ g/mL}$

11. Mercury density $= \frac{\text{mass}}{\text{volume}} = \frac{5.23 \text{ kg}}{385 \text{ mL}} \times \frac{1000 \text{ g}}{1 \text{ kg}} = 13.6 \text{ g/mL}$

12. **(a)** mass $= 452 \text{ mL} \times \frac{1.11 \text{ g}}{1 \text{ mL}} = 502$ g ethylene glycol

(b) mass $= 18.6 \text{ L} \times \frac{1000 \text{ mL}}{1 \text{ L}} \times \frac{1.11 \text{ g}}{1 \text{ mL}} \times \frac{1 \text{ kg}}{1000 \text{ g}} = 20.6 \text{ kg}$

(c) volume $= 65.0 \text{ g} \times \frac{1 \text{ mL}}{1.11 \text{ g}} = 58.6$ mL ethylene gylcol

(d) volume $= 23.9 \text{ kg} \times \frac{1000 \text{ g}}{1 \text{ kg}} \times \frac{1 \text{ mL}}{1.11 \text{ g}} \times \frac{1 \text{ L}}{1000 \text{ mL}} = 21.5$ L ethylene glycol

13. Acetone mass $= 7.50 \text{ L antifreeze} \times \frac{1000 \text{ mL}}{1\text{L}} \times \frac{0.9867 \text{ g antifreeze}}{1 \text{ mL antifreeze}} \times \frac{8.50 \text{ g acetone}}{100.0 \text{ g antifreeze}} \times \frac{1 \text{ kg}}{1000 \text{ g}}$
$= 0.629$ kg acetone

14. Acetic acid mass $= 1.00 \text{ lb vinegar} \times \frac{453.6 \text{ g}}{1 \text{ lb}} \times \frac{5.4 \text{ g acetic acid}}{100.0 \text{ g vinegar}} = 24$ g acetic acid

15. Solution mass $= 1.00 \text{ kg sucrose} \times \frac{1000 \text{ g}}{1 \text{ kg}} \times \frac{100.00 \text{ g solution}}{12.62 \text{ g sucrose}} = 7.92 \times 10^3$ g solution

16. Fertilizer mass $= 775 \text{ g nitrogen} \times \frac{1 \text{ kg N}}{1000 \text{ g N}} \times \frac{100 \text{ kg fertilizer}}{21 \text{ kg N}} = 3.69$ kg fertilizer

17. **(a)** $8950. = 8.950 \times 10^3$ **(b)** $10{,}700. = 1.0700 \times 10^4$ **(c)** $0.0240 = 2.40 \times 10^{-2}$
(d) $0.0047 = 4.7 \times 10^{-3}$ **(e)** $938.3 = 9.383 \times 10^2$ **(f)** $275{,}482 = 2.75482 \times 10^5$

18. **(a)** $3.21 \times 10^{-2} = 0.0321$ **(b)** $5.08 \times 10^{-4} = 0.000508$ **(c)** $121.9 \times 10^{-5} = 0.001219$
(d) $16.2 \times 10^{-2} = 0.162$

19. **(a)** 450 has two or three significant figures; trailing zeros left of the decimal are indeterminate.
(b) 98.6 has three significant figures; non-zero digits are significant.
(c) 0.0033 has two significant digits; leading zeros are not significant.
(d) 902.10 has five significant digits; trailing zeros to the right of the decimal point are significant.
(e) 0.02173 has four significant digits; leading zeros are not significant.
(f) 7000 has one to four significant figures; trailing zeros left of the decimal are indeterminate.
(g) 7.02 has three significant figures; zeros surrounded by non-zero digits are significant.
(h) 67,000,000 may have two to eight significant figures; there is not way to determine which, if any, of the zeros are significant.

20. Each of the following is expressed with four significant figures.
(a) $3984.6 \approx 3985$ **(b)** $422.04 \approx 422.0$ **(c)** $186{,}000 = 1.860 \times 10^5$
(d) $33{,}900 \approx 3.390 \times 10^4$ **(e)** 6.321×10^4 is correct **(f)** $5.0472 \times 10^{-4} \approx 5.047 \times 10^{-4}$

21. **(a)** $0.406 \times 0.0023 = 9.3 \times 10^{-4}$ **(b)** $0.1357 \times 16.80 \times 0.096 = 2.2 \times 10^{-1}$
(c) $0.458 + 0.12 - 0.037 = 5.4 \times 10^{-1}$ **(d)** $32.18 + 0.055 - 1.652 = 3.058 \times 10^1$

22. (a) $\dfrac{320 \times 24.9}{0.080} = \dfrac{3.2 \times 10^2 \times 2.49 \times 10^1}{8.0 \times 10^{-2}} = 1.0 \times 10^5$

(b) $\dfrac{432.7 \times 6.5 \times 0.002300}{62 \times 0.103} = \dfrac{4.327 \times 10^2 \times 6.5 \times 2.300 \times 10^{-3}}{6.2 \times 10^1 \times 1.03 \times 10^{-1}} = 1.0$

(c) $\dfrac{32.44 + 4.9 - 0.304}{82.94} = \dfrac{3.244 \times 10^1 + 4.9 - 3.04 \times 10^{-1}}{8.294 \times 10^1} = 4.46 \times 10^{-1}$

(d) $\dfrac{8.002 + 0.3040}{13.4 - 0.066 + 1.02} = \dfrac{8.002 + 3.040 \times 10^{-1}}{1.34 \times 10^1 - 6.6 \times 10^{-2} + 1.02} = 5.79 \times 10^{-1}$

23. The calculated volume of the block is converted to its mass with the density of iron.

Mass = $52.8 \text{ cm} \times 6.74 \text{ cm} \times 3.73 \text{ cm} \times 7.86 \dfrac{\text{g}}{\text{cm}^3} = 1.04 \times 10^4$ g iron

24. The calculated volume of the cylinder is converted to its mass with the density of steel.

Mass = $\pi r^2 h \times d = 3.14159\ (1.88 \text{ cm})^2\ 18.35 \text{ cm} \times 7.75 \dfrac{\text{g}}{\text{cm}^3} = 1.58 \times 10^3$ g steel

EXERCISES

Scientific Method

25. No. The greater the number of experiments that conform to the predictions of the law, the more confidence we have in the law. There is no point at which the law is ever verified with certainty.

26. One theory is preferred to another if it can predict a wider range of phenomena and if it has fewer assumptions.

27. A given set of conditions, a cause, is expected to produce a certain result, an effect. Although these cause-and-effect relationships may be difficult to establish at times ("God is subtle"), they nevertheless do exist ("he is not malicious").

28. The scientific method requires that *all* observations, the results of *all* experiments, be consistent with the predictions of a theory ("the rule"). Even one exception is sufficient reason to challenge a theory and to search for a modification of the theory to explain that exception.

Properties and Classification of Matter

29. An object displaying a physical property retains its basic chemical identity. Display of a chemical property is accompanied by a change in composition.
   (a) Physical: The iron nail is not changed in any significant way when it is attracted to a magnet. Its basic chemical identity is unchanged.
   (b) Chemical: The liquid lighter fluid is converted into a gas (carbon dioxide) and water vapor, along with the evolution of considerable energy.
   (c) Chemical: The green patina is the result of the combination of water, oxygen, and carbon dioxide with the copper in the bronze to produce basic copper carbonate.
   (d) Physical: Neither the block of wood nor the water has changed its identity.

30. An object displaying a physical property retains its basic chemical identity. Display of a chemical property is accompanied by a change in composition.
   (a) Chemical: The change in color of the apple indicates that a new substance (oxidized apple) has formed by reaction with air.
   (b) Physical: The marble slab is not changed into another substance by feeling it.
   (c) Physical: The sapphire retains its identity as it displays its color.
   (d) Chemical: After firing, the properties of the clay have changed from soft and pliable to rigid and brittle. New substances have formed. (Many of the changes involve driving off water and slightly melting the silicates that remain. These molten substances cool and harden when removed from the kiln.)

31. **(a)** Heterogeneous mixture: At the least, we can see the air cells within the solid matrix. On close examination, we can distinguish different kinds of solids by their colors.
(b) Homogeneous mixture: Modern inks are solutions of dyes in water. Older inks often were heterogeneous mixtures: suspensions of particles of carbon black (soot) in water.
(c) Substance: This assumes that there are no gasses dissolved in the water.
(d) Heterogeneous mixture: The pieces of cocoa can be seen through a microscope. Most "cloudy" liquids are heterogeneous mixtures; the small particles impede the transmission of light.

32. **(a)** Homogeneous mixture: Air is a mixture of nitrogen, oxygen, argon, and traces of other gases. By "fresh," we assume no particles of smoke, pollen, etc. are present. They would produce a heterogeneous mixture.
(b) Homogeneous mixture: Most brass is a solid solution of copper and zinc. (Older brass contains zones of slightly different compositions. These show up once the surface is etched, which occurs with repeated handling on items such as doorknobs.)
(c) Heterogeneous mixture: Pieces of garlic can be distinguished from those of salt by careful examination.
(d) Substance: Ice is simply solid water (assuming no air bubbles).

33. **(a)** Physical: With sufficient care, one can scrape the cookie crumbs off each chocolate chip
(b) Chemical: Oxygen needs to be removed from the compound, iron oxide.
(c) Chemical: Sulfur needs to be removed from the compound, sulfuric acid.
(d) Physical: Seawater is a solution of various substances dissolved in water.

34. **(a)** If a magnet is drawn through the mixture, the iron filings will be attracted to the magnet and the wood will be left behind.
(b) When the sugar-sand mixture is mixed with water, the sugar will dissolve but the sand will not. The water then can be evaporated from the solution to produce pure sugar.
(c) Water and gasoline do not mix with each other. Hence, simply draw off the gasoline, which floats on top, perhaps with an eyedropper.
(d) The gold flakes will settle to the bottom if the mixture is left undisturbed. The water then can be decanted, that is, carefully poured off.

Exponential Arithmetic (see Appendix A)

35. **(a)** 34,000 centimeters/second = 3.4×10^4 cm/s
(b) six thousand three hundred seventy eight kilometers = 6378 km = 6.378×10^3 km
(c) $$\frac{(2.2 \times 10^3) + (4.7 \times 10^2)}{5.8 \times 10^{-3}} = \frac{2.7 \times 10^3}{5.8 \times 10^{-3}} = 4.7 \times 10^5$$

36. **(a)** 173 thousand trillion watts = 173,000,000,000,000,000 W = 1.73×10^{17} W
(b) ten millionths of a meter = $10 \times 0.000\,001$ m = 1×10^{-5} m = 10 μm
(c) $$\frac{5.07 \times 10^4 \times (1.8 \times 10^{-3})^2}{0.065 + (3.3 \times 10^{-2})} = \frac{0.16}{0.098} = 1.6$$

Significant figures

37. **(a)** An exact number—24 soda cans in a case.
(b) Pouring the milk into the jug is a process that is subject to error; there can be slightly more or slightly less than one gallon of milk in the jug.
(c) The distance between any pair of planetary bodies can only be determined through certain measured quantities. These measurements are subject to error.

38. **(a)** The number of pages in the text is determined by counting; the result is an exact number.
(d) An exact number. Although the number of days can vary from one month to another (say from January to February), the month of January always has 31 days.
(e) The area is determined by calculations based on measurements. These measurements are subject to error.

39. **(a)** $38.4 \times 10^{-3} \times 6.36 \times 10^{5} = 2.44 \times 10^{4}$ **(b)** $\dfrac{1.45 \times 10^{2} \times 8.76 \times 10^{-4}}{(9.2 \times 10^{-3})^{2}} = 1.5 \times 10^{3}$

(c) $24.6 + 18.35 - 2.98 = 40.0$

(d) $(1.646 \times 10^{3}) - (2.18 \times 10^{2}) + (1.36 \times 10^{4} \times 5.17 \times 10^{-2}) = 2.131 \times 10^{3}$

40. **(a)** $4.65 \times 10^{4} \times 2.95 \times 10^{-2} \times 6.663 \times 10^{-3} \times 8.2 \times 10^{-3} = 7.5 \times 10^{-2}$

(b) $\dfrac{1912 \times (0.0077 \times 10^{4}) \times 3.12 \times 10^{-3}}{(4.18 \times 10^{-4})^{3}} = 6.3 \times 10^{12}$

(c) $(3.46 \times 10^{3}) \times 0.087 \times 15.26 \times 1.0023 = 4.6 \times 10^{3}$

(d) $\dfrac{(4.505 \times 10^{-2})^{2} \times 1.080 \times 1545.9}{(0.03203 \times 10^{3})} = 1.058 \times 10^{-1}$

41. **(a)** The average speed is obtained by dividing the distance travelled (in miles) by the elapsed time (in hours). First, we need to obtain the elapsed time, in hours.

$9 \text{ days} \times \dfrac{24 \text{ h}}{1 \text{ d}} = 216.000 \text{ h}$ $\quad 3 \text{ min} \times \dfrac{1 \text{ h}}{60 \text{ min}} = 0.050 \text{ h}$ $\quad 44 \text{ s} \times \dfrac{1 \text{ h}}{3600 \text{ s}} = 0.012 \text{ h}$

Total time = 216.000 + 0.050 h + 0.012 h = 216.062 h

$\text{average speed} = \dfrac{25{,}012 \text{ mi}}{216.062 \text{ h}} = 115.76 \text{ mi/h}$

(b) First compute the mass of fuel remaining

$\text{mass} = 14 \text{ gal} \times \dfrac{4 \text{ qt}}{1 \text{ gal}} \times \dfrac{0.9464 \text{ L}}{1 \text{ qt}} \times \dfrac{1000 \text{ mL}}{1 \text{ L}} \times \dfrac{0.70 \text{ g}}{1 \text{ mL}} \times \dfrac{1 \text{ lb}}{453.6 \text{ g}} = 82 \text{ lb}$

Then determine the mass of fuel used, and finally, the fuel consumption. Notice that we know the initial quantity of fuel quite imprecisely, perhaps at best to the nearest 10 lb, certainly ("nearly 9000 lb") not to the nearest pound.

mass of fuel used = 9000 lb – 82 lb = 8920 lb $\quad$ $\text{fuel consumption} = \dfrac{25{,}012 \text{ mi}}{8920 \text{ lb}} = 2.80 \text{ mi/lb}$

42. If the proved reserve truly was an estimate, rather than an actual measurement, it would have been difficult to estimate it to the nearest trillion cubic feet. A statement such as 2,911,000 trillion cubic feet (or 2.911×10^{18} ft^3) would have more accurately reflected the precision with which the proved reserve was known.

Units of Measurement

43. Express both masses in the same units for comparison. $2172 \ \mu g \times \dfrac{1 \text{ g}}{10^{6} \ \mu g} \times \dfrac{10^{3} \text{ mg}}{1 \text{ g}} = 2.172 \text{ mg}$, which is larger than 0.00515 mg.

44. Express both masses in the same units for comparison. $0.00475 \text{ kg} \times \dfrac{1000 \text{ g}}{1 \text{ kg}} = 4.75 \text{ g}$ larger than $3257 \text{ mg} \times \dfrac{1 \text{ g}}{10^{3} \text{ mg}} = 3.257 \text{ g}$.

45. $\text{height} = 15 \text{ hands} \times \dfrac{4 \text{ in.}}{1 \text{ hand}} \times \dfrac{2.54 \text{ cm}}{1 \text{ in.}} \times \dfrac{1 \text{ m}}{100 \text{ cm}} = 1.5 \text{ m}$

46. We do know the length of a mile in feet (5280 ft = 1 mi) and can use that as a conversion factor.

$1.00 \text{ link} \times \dfrac{1 \text{ chain}}{100 \text{ links}} \times \dfrac{1 \text{ furlong}}{10 \text{ chains}} \times \dfrac{1 \text{ mile}}{8 \text{ furlongs}} \times \dfrac{5280 \text{ ft}}{1 \text{ mi}} \times \dfrac{12 \text{ in.}}{1 \text{ ft}} = 7.92 \text{ in.}$

47. **(a)** We use the speed as a conversion factor, but need to convert yards into meters.

$\text{time} = 100.0 \text{ m} \times \dfrac{9.3 \text{ s}}{100 \text{ yd}} \times \dfrac{1 \text{ yd}}{36 \text{ in.}} \times \dfrac{39.37 \text{ in.}}{1 \text{ m}} = 10._{2} \text{ s}$ $\quad$ We can keep two significant figures.

(b) We need to convert yards to meters. (Keep a third significant figure for the next part of the problem.)

$\text{speed} = \dfrac{100 \text{ yd}}{9.3 \text{ s}} \times \dfrac{36 \text{ in.}}{1 \text{ yd}} \times \dfrac{2.54 \text{ cm}}{1 \text{ in.}} \times \dfrac{1 \text{ m}}{100 \text{ cm}} = 9.8_{3} \text{ m/s}$

(c) The speed is used as a conversion factor.

$\text{time} = 1.45 \text{ km} \times \dfrac{1000 \text{ m}}{1 \text{ km}} \times \dfrac{1 \text{ s}}{9.8_{3} \text{ m}} = 1.5 \times 10^{2} \text{ s} \times \dfrac{1 \text{ min}}{60 \text{ s}} = 2.5 \text{ min}$

48. **(a)** $\text{mass (mg)} = 2 \text{ tablets} \times \dfrac{5.0 \text{ gr}}{1 \text{ tablet}} \times \dfrac{1.0 \text{ g}}{15 \text{ gr}} \times \dfrac{1000 \text{ mg}}{1 \text{ g}} = 6.7 \times 10^{2} \text{ mg}$

(b) dosage rate $= \dfrac{6.7\times10^2\text{ mg}}{155\text{ lb}}\times\dfrac{1\text{ lb}}{453.6\text{ g}}\times\dfrac{1000\text{ g}}{1\text{ kg}} = 9.5$ mg aspirin/kg body weight

(c) time $= 1.0\text{ lb}\times\dfrac{453.6\text{ g}}{1\text{ lb}}\times\dfrac{2\text{ tablets}}{0.67\text{ g}}\times\dfrac{1\text{ day}}{2\text{ tablets}} = 6.8\times10^2$ days

49. $1\text{ hectare} = 1\text{ hm}^2\times\left(\dfrac{100\text{ m}}{1\text{ hm}}\times\dfrac{100\text{ cm}}{1\text{ m}}\times\dfrac{1\text{ in.}}{2.54\text{ cm}}\times\dfrac{1\text{ ft}}{12\text{ in.}}\times\dfrac{1\text{ mi}}{5280\text{ ft}}\right)^2\times\dfrac{640\text{ acres}}{1\text{ mi}^2}$
$= 2.47$ acres

50. density $= \dfrac{0.292\text{ lb}}{1\text{ in.}^3}\times\dfrac{453.6\text{ g}}{1\text{ lb}}\times\left(\dfrac{1\text{ in}}{2.54\text{ cm}}\right)^3 = 8.08\text{ g/cm}^3$

51. pressure $= \dfrac{32\text{ lb}}{1\text{ in.}^2}\times\dfrac{453.6\text{ g}}{1\text{ lb}}\times\left(\dfrac{1\text{ in}}{2.54\text{ cm}}\right)^2 = 2.25\times10^3\text{ g/cm}^2$

pressure $= \dfrac{2.25\times10^3\text{ g}}{1\text{ cm}^2}\times\dfrac{1\text{ kg}}{1000\text{ g}}\times\left(\dfrac{100\text{ cm}}{1\text{ m}}\right)^2 = 2.25\times10^4\text{ kg/m}^2$

52. speed = Mach $1.38\times\dfrac{1130\text{ ft/s}}{\text{Mach }1}\times\dfrac{3600\text{ s}}{1\text{ h}}\times\dfrac{12\text{ in.}}{1\text{ ft}}\times\dfrac{2.54\text{ cm}}{1\text{ in.}}\times\dfrac{1\text{ m}}{100\text{ cm}}\times\dfrac{1\text{ km}}{1000\text{ m}} = 1.71\times10^3$ km/h

Temperature Scales

53. high: $°\text{C} = \tfrac{5}{9}(°\text{F} - 32) = \tfrac{5}{9}(118\ °\text{F} - 32) = 47.8\ °\text{C}$
low: $°\text{C} = \tfrac{5}{9}(°\text{F} - 32) = \tfrac{5}{9}(17\ °\text{F} - 32) = -8.3\ °\text{C}$

54. low: $°\text{F} = \tfrac{9}{5}\,°\text{C} + 32 = \tfrac{9}{5}(-15\ °\text{C}) + 32 = 5\ °\text{F}$
high: $°\text{F} = \tfrac{9}{5}\,°\text{C} + 32 = \tfrac{9}{5}(60\ °\text{C}) + 32 = 140\ °\text{F}$

55. Determine the Celsius temperature that corresponds to the highest Fahrenheit temperature, 240 °F.
$°\text{C} = \tfrac{5}{9}(°\text{F} - 32) = \tfrac{5}{9}(240°\text{F} - 32) = 116\ °\text{C}$ Because 116 °C is above the range of the thermometer, this thermometer cannot be used in this candy making assignment.

56. Let us determine the Fahrenheit equivalent of absolute zero.
$°\text{F} = (\tfrac{9}{5})°\text{C} + 32 = (\tfrac{9}{5}\times(-273.15)) + 32 = -459.67\ °\text{F}$
A temperature of −465 °F cannot be achieved because it is below absolute zero.

Density

57. The mass of acetone is the difference in masses between empty and filled masses.
Density $= \dfrac{437.5\text{ lb} - 75.0\text{ lb}}{55.0\text{ gal}}\times\dfrac{453.6\text{ g}}{1\text{ lb}}\times\dfrac{1\text{ gal}}{3.785\text{ L}}\times\dfrac{1\text{ L}}{1000\text{ mL}} = 0.790$ g/mL

58. Density is a conversion factor. volume = (283.2 g filled − 121.3 g empty) $\times\dfrac{1\text{ mL}}{1.59\text{ g}} = 102$ mL

59. Determine the mass of each item.
(1) mass of iron = $(81.5\text{ cm}\times2.1\text{ cm}\times1.6\text{ cm})\times7.86\text{ g/cm}^3 = 2.2\times10^3$ g iron
(2) mass of aluminum $= (12.12\text{ m}\times3.62\text{ m}\times0.003\text{ cm})\times\left(\dfrac{100\text{ cm}}{1\text{ m}}\right)^2\times2.70\text{ g/cm}^3 = 3._6\times10^3$ g aluminum
(3) mass of water $= 4.051\text{ L}\times\dfrac{1000\text{ cm}^3}{1\text{ L}}\times0.998\text{ g/cm}^3 = 4.04\times10^3$ g water
In order of increasing mass, the items are: iron bar < aluminum foil < water Realize, however, that the rules for significant figures do not allow us to distinguish between the masses of aluminum and water.

60. First determine the volume of the aluminum foil. Then determine its area. Finally, determine its thickness.
volume $= 2.568\text{ g}\times\dfrac{1\text{ cm}^3}{2.70\text{ g}} = 0.951\text{ cm}^3$ area $= \left(9.0\text{ in.}\times\dfrac{2.54\text{ cm}}{\text{in.}}\right)^2 = 5.2\times10^2\text{ cm}^2$

thickness $= \dfrac{\text{volume}}{\text{area}} = \dfrac{0.951\text{ cm}^3}{5.2\times10^2\text{ cm}^2}\times\dfrac{10\text{ mm}}{1\text{ cm}} = 1.8\times10^{-2}$ mm

61. Total volume of 125 pieces of shot
$V = 8.9\text{ mL} - 8.4\text{ mL} = 0.5\text{ mL}$

$$\frac{\text{mass}}{\text{shot}} = \frac{0.5\text{ mL}}{125\text{ shot}} \times \frac{8.92\text{ g}}{1\text{ cm}^3} = 0.04\text{ g/shot}$$

62. The vertical piece of steel has a volume $= 12.78\text{ cm} \times 1.35\text{ cm} \times 2.75\text{ cm} = 47.4\text{ cm}^3$
The horizontal piece of steel has a volume $= 10.26\text{ cm} \times 1.35\text{ cm} \times 2.75\text{ cm} = 38.1\text{ cm}^3$
Thus, the total volume $= 47.4\text{ cm}^3 + 38.1\text{ cm}^3 = 85.5\text{ cm}^3$
mass $= 85.5\text{ cm}^3 \times 7.78\text{ g/cm}^3 = 665$ g of steel

Percent Composition

63. The percent of students with each grade is obtained by dividing the number of students with that grade by the total number of students.

$$\%A = \frac{9\text{ As}}{76\text{ students}} \times 100\% = 12\ \%\ A$$

$$\%B = \frac{21\text{ Bs}}{76\text{ students}} \times 100\% = 28\ \%\ B \qquad \%C = \frac{36\text{ Cs}}{76\text{ students}} \times 100\% = 47\ \%\ C$$

$$\%D = \frac{8\text{ Ds}}{76\text{ students}} \times 100\% = 11\ \%\ D \qquad \%F = \frac{2\text{ Fs}}{76\text{ students}} \times 100\% = 3\ \%\ F$$

64. The number of students with a certain grade is determined by multiplying the total number of students by the fraction of students who earned that grade.

$$\text{no. A} = 84\text{ students} \times \frac{18\text{ As}}{100\text{ students}} = 15\text{ As}$$

$$\text{no. B} = 84\text{ students} \times \frac{25\text{ Bs}}{100\text{ students}} = 21\text{ Bs} \qquad \text{no. C} = 84\text{ students} \times \frac{32\text{ Cs}}{100\text{ students}} = 27\text{ Cs}$$

$$\text{no. D} = 84\text{ students} \times \frac{13\text{ Ds}}{100\text{ students}} = 11\text{ Ds} \qquad \text{no. F} = 84\text{ students} \times \frac{12\text{ Fs}}{100\text{ students}} = 10\text{ Fs}$$

65. Use the percent composition as a conversion factor.

$$\text{mass of sucrose} = 2.75\text{ L} \times \frac{1000\text{ mL}}{1\text{ L}} \times \frac{1.1175\text{ g soln}}{1\text{ mL}} \times \frac{28.0\text{ g sucrose}}{100\text{ g soln}} = 8.60 \times 10^2\text{ g sucrose}$$

66. Again, percent composition is used as a conversion factor. We are careful to label both numerator and denominator of each factor.

$$\text{soln volume, L} = 3.50\text{ kg sodium hydroxide} \times \frac{1000\text{ g}}{1\text{ kg}} \times \frac{100.0\text{ g soln}}{12.0\text{ g sodium hydroxide}}$$

$$\times \frac{1\text{ mL}}{1.131\text{ g soln}} = 2.58 \times 10^4\text{ mL soln} \times \frac{1\text{ L}}{1000\text{ mL}} = 25.8\text{ L soln}$$

FEATURE PROBLEMS

A. All of the pennies minted before 1982 weigh more than 3.00 g, while all of those minted after 1982 weigh less than 2.60 g. One might infer that the composition of a penny changed in 1982. In fact, pennies minted prior to 1982 are composed of almost pure copper (about 96% pure). Those minted after 1982 are composed of zinc with a thin copper cladding. Some pennies of each type were minted in 1982.

B. After sitting in a bathtub that was nearly ful and observing the water splashing over the side, Archimedes realized that the crown—when submerged in water—would displace a volume of water equal to its volume. Once Archimedes determined the volume in this way and determined the mass of the crown with a balance, he was able to calculate the crown's density. Since the gold-silver alloy has a different density (it is lower) than pure gold, Archimedes could tell that the crown was not pure gold.

C. Notice that the liquid does not fill each of the floating glass balls. The quantity of liquid placed in each glass ball is sufficient to give each ball a slightly different density. Note that the density of the glass ball is determined by the density of the liquid, the density of the glass (greater than the liquid's density), and the density of the air. Since the density of the liquid in the cylinder varies slightly with temperature—the liquid's volume increases as temperature goes up, but its mass does not change—different balls will be bouyant at different temperatures.

D. The density of the canoe is determined by the density of the concrete and the density of the empty space inside the canoe, where the passengers sit. That empty space is air and makes the density of the canoe less than water (1.0 g/cm³). If the concrete canoe fills with water, it will sink to the bottom, unlike a wooden canoe.

2 ATOMS AND THE ATOMIC THEORY

PRACTICE EXAMPLES

1A The total mass must be the same before and after reaction.
mass before reaction = 0.382 g magnesium + 2.652 g nitrogen = 3.034 g
mass after reaction = magnesium nitride mass + 2.505 g nitrogen = 3.034 g
magnesium nitride mass = 3.034 g – 2.505 g = 0.529 g magnesium nitride

1B Again, the total mass is the same before and after the reaction.
mass before reaction = 7.12 g magnesium + 1.80 g bromine = 8.92 g
mass after reaction = 2.07 g magnesium bromide + magnesium mass = 8.92 g
magnesium mass = 8.92 g – 2.07 g = 6.85 g magnesium

2A In Example 2-2 we are told that 0.100 g magnesium produces 0.166 g magnesium oxide.

$$\text{magnesium mass} = 0.500 \text{ g magnesium oxide} \times \frac{0.100 \text{ g magnesium}}{0.166 \text{ g magnesium oxide}} = 0.301 \text{ g magnesium}$$

2B In Example 2-2 we are told that 0.100 g magnesium forms 0.166 g magnesium oxide. With this information, we can determine the mass of magnesium needed to form 2.00 g magnesium oxide.

$$\text{mass of magnesium} = 2.00 \text{ g magnesium oxide} \times \frac{0.100 \text{ g magnesium}}{0.166 \text{ g magnesium oxide}} = 1.20 \text{ g magnesium}$$

The remainder of the 2.00 g of magnesium oxide is the mass of oxygen
mass of oxygen = 2.00 g magnesium oxide – 1.20 g magnesium = 0.80 g oxygen

3A The number of protons and electrons are equal, and thus the species has no charge. The mass number is the sum of the atomic number and the number of neutrons: 47 p + 61 n = A = 108. The atomic number of 47 is that of the element silver. The symbol thus is $^{108}_{47}\text{Ag}$.

3B The element sulfur has an atomic number of 16 and thus has 16 protons. A charge of –2 indicates two more electrons than protons; there are 16 + 2 = 18 electrons. The number of neutrons is the mass number minus the number of protons; there are 35 – 16 = 19 neutrons.

4A We know that the mass of ^{16}O = 15.9949 u and that mass of ^{16}O = 1.06632 × mass of ^{15}N. We combine these two equations and solve the resulting expression.

$$15.9949 \text{ u} = 1.06632 \times \text{mass of } ^{15}\text{N} \qquad \text{mass of } ^{15}\text{N} = \frac{15.9949 \text{ u}}{1.06632} = 15.0001 \text{ u}$$

4B We know the isotopic mass of ^{12}C is 12.000000 u. Thus, the mass ratio is found by substitution.

$$\frac{^{202}\text{Hg}}{^{12}\text{C}} = \frac{201.970617}{12.000000} = 16.830884$$

Note that the number of significant figures in the result is determined by the precision of the mass of ^{202}Hg, because the mass of ^{12}C is established by definition as an exact number.

5A The average atomic mass of boron is 10.811, which is closer to 11.009305 than to 10.012937. Thus, boron-11 is the isotope present in greater abundance.

5B We let x be the fractional abundance of lithium-6.

$$6.941 \text{ u} = [x \times 6.01513 \text{ u}] + [(1-x) \times 7.01601 \text{ u}] = 6.01513\,x \text{ u} + 7.01601 \text{ u} - 7.01601\,x \text{ u}$$

$$6.941 \text{ u} - 7.01601 \text{ u} = 6.01513\,x \text{ u} - 7.01601\,x \text{ u} = -1.00088\,x \text{ u}$$

$x = \dfrac{6.941\text{ u} - 7.01601\text{ u}}{-1.00088\text{ u}} = 0.074_9$ Percent abundances: 7.5% lithium-6 and 92.5% lithium-7

6A We assume that atoms lose or gain relatively few electrons to become ions. Thus, elements that will form cations will be on the left-hand side of the periodic table, while elements that will form anions will be on the right-hand side. The number of electrons "lost" when a cation forms is the periodic group number; the number added when an anion forms is eight minus the group number.
Li is in family 1A; it should form a cation by losing one electron: Li^+.
S is in family 6A; it should form an anion by adding two electrons: S^{2-}.
Ra is in family 2A; it should form a cation by losing two electrons: Ra^{2+}.
F and I both are in family 7A; they each should form an anion by adding one electron: F^- and I^-.
Al is in family 3A; it should form a cation by losing three electrons: Al^{3+}.

6B Main group elements are in the "A" families, while transition elements are in the "B" families. Metals, nonmetals, metalloids, and noble gases are color coded in the periodic table inside the front cover.
Na is a main-group metal in family 1A.
S is a main-group nonmetal in family 6A.
Kr is a noble gas in family 8A.
U is a transition nonmetal, an actinide.
B is a main-group nonmetal in family 3A.
As is a main-group metalloid in family 5A.
Re is a transition metal in family 7A.
I is a main-group nonmetal in family 7A.
Mg is a main-group metal in family 2A.
Si is a main-group metalloid in family 4A.
Al is a main-group metal in family 3A.
H is a main-group nonmetal in family 1A.

7A Avogadro's number serves as a conversion factor.

$$\text{no. Au atoms} = 5.07 \times 10^{-3}\text{ mol Au} \times \frac{6.022 \times 10^{23}\text{ Au atoms}}{1\text{ mol Au}} = 3.05 \times 10^{21}\text{ Au atoms}$$

7B Of all lead atoms, 24.1% are lead-25, 241 ^{206}Pb atoms in every 1000 lead atoms

$$\text{number of }^{206}\text{Mg atoms} = 8.27 \times 10^{-3}\text{ mol Pb} \times \frac{6.02214 \times 10^{23}\text{ Pb atoms}}{1\text{ mol Pb}} \times \frac{241\ ^{206}\text{Pb atoms}}{1000\text{ Pb atoms}}$$
$$= 1.20 \times 10^{21}\ ^{206}\text{Pb atoms}$$

8B $\text{number of He atoms} = 22.6\text{ g He} \times \dfrac{1\text{ mol He}}{4.0026\text{ g He}} \times \dfrac{6.022 \times 10^{23}\text{ He atoms}}{1\text{ mol He}} = 3.40 \times 10^{24}\text{ He atoms}$

8A This is the inverse of Practice Example 2-8A.

$$\text{Cu mass} = 2.35 \times 10^{24}\text{ Cu atoms} \times \frac{1\text{ mol Cu}}{6.022 \times 10^{23}\text{ atoms}} \times \frac{63.546\text{ g Cu}}{1\text{ mol Cu}} = 248\text{ g Cu}$$

9A Both the density and the molar mass of Pb serve as conversion factors.

$$\text{no. Pb atoms} = 0.105\text{ cm}^3\text{ Pb} \times \frac{11.34\text{ g}}{1\text{ cm}^3} \times \frac{1\text{ mol Pb}}{207.2\text{ g}} \times \frac{6.022 \times 10^{23}\text{ Pb atoms}}{1\text{ mol Pb}} = 3.46 \times 10^{21}\text{ Pb atoms}$$

9B First we find the number of rhenium atoms in 0.100 mg of the element.

$$\text{no. Re atoms} = 0.100\text{ mg} \times \frac{1\text{ g}}{1000\text{ mg}} \times \frac{1\text{ mol Re}}{186.2\text{ g Re}} \times \frac{6.022 \times 10^{23}\text{ atoms}}{1\text{ mol Re}} = 3.23 \times 10^{17}\text{ Re atoms}$$

$$\%\ ^{187}\text{Re} = \frac{2.02 \times 10^{17}\text{ atoms }^{187}\text{Re}}{3.23 \times 10^{17}\text{ Re atoms}} \times 100\% = 62.5\%$$

SUMMARIZING EXAMPLE CALCULATIONS

1. $\text{volume of ball bearing} = \frac{4}{3}\pi r^3 = \frac{4}{3} \times 3.14159 \times \left(6.35\text{ mm} \times \frac{1\text{ cm}}{10\text{ mm}}\right)^3 = 1.07\text{ cm}^3$
2. $\text{mass of carbon} = 1.07\text{ cm}^3\text{ steel} \times \dfrac{7.75\text{ g steel}}{1\text{ cm}^3} \times \dfrac{0.25\text{ g C}}{100.00\text{ g steel}} = 0.021\text{ g C}$
3. $\text{number of C atoms} = 0.021\text{ g C} \times \dfrac{1\text{ mol C}}{12.01\text{ g C}} \times \dfrac{6.02214 \times 10^{23}\text{ C atoms}}{1\text{ mol C}} = 1.1 \times 10^{21}\text{ C atoms}$
4. $\text{number of }^{13}\text{C atoms} = 1.1 \times 10^{21}\text{ C atoms} \times \dfrac{1.108\ ^{13}\text{C atoms}}{100\text{ C atoms}} = 1.2 \times 10^{19}\ ^{13}\text{C atoms}$

REVIEW QUESTIONS

1. **(a)** ${}^{A}_{Z}E$ is the symbol for a nuclide. E represents the symbol of the element; Z is the atomic number, the number of protons in the nucleus; and A is the mass number, the number of protons plus neutrons.

(b) A β particle refers to an electron ejected by the nucleus, and is one of the three forms of natural radioactivity.

(c) An isotope is one of at least two forms of an atom of an element which have the same number of protons in the nucleus, but different numbers of neutrons.

(d) ${}^{16}O$ is the symbol for the isotope of oxygen that has 16 nucleons in its nucleus: 8 protons (characteristic of the element oxygen) and 8 neutrons.

(e) Molar mass is the mass of a quantity of an element (or a compound) that contains Avogadro's number (6.022×10^{23}) of atoms (or formula units).

2. **(a)** The law of conservation of mass states that there is no gain or loss of mass during a chemical reaction.

(b) The atom as described by Rutherford consists of a very small (approximately 10^{-13} cm diameter), positively charged, and massive (more than 99.5% of the mass) nucleus; surrounded by a quite large (approximately 10^{-8} cm diameter), tenuous (less than 0.5% of the mass), and negatively charged cloud of electrons.

(c) The atomic mass that appears in the periodic table for each element is a weighted average, with contributions from each isotope of the element, each weighted by the relative abundance of that isotope.

(d) Radioactivity refers to the spontaneous emission from the nucleus of an atom of energy (γ radiation) or particles (α or β particles).

3. **(a)** Cathode rays are electrons moving through the evacuated volume within a glass tube as a result of a current flowing through that volume. X-rays are high-energy photons emitted when these cathode rays strike the anode within the glass tube.

(b) Protons and neutrons are both particles in the nucleus of the atom, and both have a mass of approximately 1 u. However, protons are positively charged, while neutrons have no electric charge.

(c) The nuclear charge of an atom is a positive charge equal to the number of protons in the nucleus. The ionic charge equals the nuclear charge minus the number of electrons; as a consequence, the ionic charge may be negative.

(d) A period is a horizontal row in the periodic table. A group is a vertical column, containing elements of similar chemical and physical properties.

(e) A metal is an element that has a lustre, is malleable, and conducts electricity and heat well. Also, metal atoms tend to form cations in chemical reactions. A nonmetal does not conduct heat or electricity well, and solid nonmetals typically are dull and brittle. Nonmetal atoms tend to form anions in chemical reactions.

(f) A main-group element is an element within one of the "A" families. Atoms within a main group tend to form ions of just one charge in chemical reactions, while ions of several different charges may be produced by a transition element.

(g) Avogadro's constant is equal to the number of particles of any type that are present in a mole.

4. By the law of conservation of mass, all of the magnesium initially present and all of the oxygen that reacted are present in the product. Thus, the mass of oxygen that has reacted is obtained by difference.

mass of oxygen = 0.674 g magnesium oxide – 0.406 g magnesium
= 0.268 g oxygen

5. Again we use the law of conservation of mass. The mass of the starting materials equals the mass of substances present after the reaction is complete.

mass of potassium + mass of chlorine = mass of potassium chloride + mass of unreacted chlorine
1.205 g potassium + 6.815 g chlorine = mass of potassium chloride + 3.300 g unreacted chlorine
mass of potassium chloride = (1.205 g + 6.815 g) – 3.300 g = 4.720 g potassium chloride

6. No solid residue is produced when (solid) sulfur burns because the product of combustion is sulfur dioxide gas. The law of conservation of mass is satisfied because the mass of the sulfur dioxide equals the sum of the masses of sulfur and oxygen that react.

7. If the two elements combine in the ratio 1:1, there will be one atom of sodium present for each atom of chlorine. To determine the mass percent chlorine, we simply convert these atomic quantities to masses (in u) and convert the resulting ratio to a percent.

$$\text{percent Na} = \frac{1 \text{ Na atom} \times \dfrac{22.99 \text{ u Na}}{1 \text{ Na atom}}}{\left(1 \text{ Na atom} \times \dfrac{22.99 \text{ u Na}}{1 \text{ Na atom}}\right) + \left(1 \text{ Cl atom} \times \dfrac{35.45 \text{ u Cl}}{1 \text{ Cl atom}}\right)} \times 100\% = 39.34\% \text{ Na}$$

8. (a) The mass of oxygen present in 0.166 g magnesium oxide is the remainder when the 0.100 g magnesium is deducted, or 0.066 g oxygen. Hence, there is 0.066 g oxygen/0.166 g magnesium oxide.

(b) From the numbers we have already obtained, we see that there is 0.066 g oxygen/0.100 g magnesium, or 0.66 g oxygen/1.00 g magnesium.

(c) $\% \text{ Mg, by mass} = \dfrac{0.100 \text{ g Mg}}{0.166 \text{ g magnesium oxide}} \times 100\% = 60.2\ \% \text{ Mg}$

9. (a) We can determine whether carbon dioxide has a fixed composition by determining the % C in each sample. (In the calculations below, the abbreviation "cmpd" is short for "compound.")

$\%\text{C} = \dfrac{3.62 \text{ g C}}{13.26 \text{ g cmpd}} \times 100\% = 27.3\% \text{ C}$ $\qquad \%\text{C} = \dfrac{5.91 \text{ g C}}{21.66 \text{ g cmpd}} \times 100\% = 27.3\ \% \text{ C}$

$\%\text{C} = \dfrac{7.07 \text{ g C}}{25.91 \text{ g cmpd}} \times 100\% = 27.3\% \text{ C}$

Since all three samples have the same percent of carbon, these data do establish that carbon dioxide has a fixed composition.

(b) Carbon dioxide contains only carbon and oxygen. The percent of oxygen in carbon dioxide is obtained by difference. $\qquad \%\text{ O} = 100.0\% - 27.3\% \text{ C} = 72.7\% \text{ O}$

10. By knowing that all of the 4.15 g of magnesium reacts, producing only magnesium bromide and leaving excess bromine unreacted, we are unable at this point to calculate the mass of magnesium bromide produced. In order to perform this calculation, we need to know how many moles of bromine are combined with each mole of magnesium in the compound.

11.

Name	Symbol	Number protons	Number electrons	Number neutrons	Mass number
sodium	^{23}Na	11	11	12	23
silicon	^{28}Si	14	14[a]	14	28
rubidium	^{85}Rb	37	37[a]	48	85
potassium	^{40}K	19	19	21	40
arsenic[a]	^{75}As	33[a]	33	42	75
neon	$^{20}Ne^{2+}$	10	8	10	20
bromine[b]	^{80}Br	35	35	45	80
lead[b]	^{208}Pb	82	82	126	208

[a]This result assumes that a neutral atom is involved.
[b]The information given is not enough to characterize a specific nuclide; several possibilities exist.

The minimum information needed is the atomic number (or some way to obtain it: the name or the symbol of the element involved), the number of electrons (or some way to obtain it, such as the charge on the species), and the mass number (or the number of neutrons).

12. (a) Since all of these species are neutral atoms, the number of electrons are the atomic numbers, the subscript numbers. The symbols must be arranged in order of increasing value of these subscripts.

$^{40}_{18}Ar < ^{39}_{19}K < ^{58}_{27}Co < ^{59}_{29}Cu < ^{120}_{48}Cd < ^{112}_{50}Sn < ^{122}_{52}Te$

(b) The number of neutrons is given by the difference between the mass number and the atomic number, $A - Z$. This is the difference between superscript and subscript, given in parentheses after each element in the following list.

$^{39}_{19}K(20) < ^{40}_{18}Ar(22) < ^{59}_{29}Cu(30) < ^{58}_{27}Co(31) < ^{112}_{50}Sn(62) < ^{122}_{52}Te(70) < ^{120}_{48}Cd(72)$

(c) Here the nuclides are arranged by increasing mass number, given by the superscripts.
$^{39}_{19}K < ^{40}_{18}Ar < ^{58}_{27}Co < ^{59}_{29}Cu < ^{112}_{50}Sn < ^{120}_{48}Cd < ^{122}_{52}Te$

13. **(a)** cobalt-60 $^{60}_{27}Co$ **(b)** phosphorus-32 $^{32}_{15}P$
(c) iodine-131 $^{131}_{53}I$ **(d)** sulfur-35 $^{35}_{16}S$

14. The nucleus of $^{138}_{56}Ba$ contains 56 protons and (138 – 56 =) 82 neutrons. Thus, the percent of nucleons that are neutrons is given by $\%\ \text{neutrons} = \frac{82\ \text{neutrons}}{138\ \text{nucleons}} \times 100 = 59\%\ \text{neutrons}$

15. The weighted-average atomic mass of the element iridium is just slightly more than 192 u. The mass of the first isotope is a bit less than 191 u. Hence, the mass of the second isotope must more than 192 u; that isotope must be ^{193}Ir.

16. If we let x represent the number of protons, then $x + 2$ is the number of neutrons. The mass number is the sum of the number of protons and the number of neutrons: $38 = x + (x + 2) = 2x + 2$. We solve this expression for x, and obtain $x = 18$. This is the number of protons of the nuclide and equals the atomic number. Reference to the periodic table indicates that 18 is the atomic number of the element argon.

17. Each of the listed isotopic masses is divided by the isotopic mass of ^{12}C, 12.00000 u, an exact number.
(a) $^{35}Cl \div ^{12}C = 34.96885\ u \div 12.00000\ u = 2.914071$
(b) $^{26}Mg \div ^{12}C = 25.98259\ u \div 12.00000\ u = 2.165216$
(c) $^{222}Rn \div ^{12}C = 222.0175\ u \div 12.00000\ u = 18.50146$

18. We need to work through the mass ratios in sequence to determine the mass of ^{81}Br.
mass of ^{19}F = mass of $^{12}C \times 1.5832 = 12.00000\ u \times 1.5832 = 18.998\ u$
mass of ^{35}Cl = mass of $^{19}F \times 1.8406 = 18.998\ u \times 1.8406 = 34.968\ u$
mass of ^{81}Br = mass of $^{35}Cl \times 2.3140 = 34.968\ u \times 2.3140 = 80.916\ u$

19. Each of the isotopic masses is multiplied by its fractional abundance. The resulting products are summed to obtain the average atomic mass.
contribution from $^{40}Ar = 39.9624\ u \times 0.99600 = 39.803\ u$
contribution from $^{36}Ar = 35.96755\ u \times 0.00337 = 0.121\ u$
contribution from $^{38}Ar = 37.96272\ u \times 0.00063 = 0.024\ u$
average atomic mass of uranium = 39.803 u + 0.121 u + 0.024 u = 39.948 u
Of course, this calculation can be performed in one step.
$(39.9624\ u \times 0.99600) + (35.96755\ u \times 0.00337) + (37.96272\ u \times 0.00063) = 39.948\ u$

20. **(a)** In is in group 3A and in the fifth period.
(b) Another element in family 6A is similar to S: O, Se, Te. Other elements are unlike S, but particularly metals such as Na, K, Rb.
(c) The alkali metal in the sixth period is in family 1A, Cs.
(d) The halogen (family 7A) in the fifth period is I.
(e) The element with atomic number 18 is Ar, a noble gas. Xe is a noble gas with atomic number (54) greater than 50.
(f) If an element forms an anion with charge –3, it is in family 5A. (Recall that anion charge = family number – 8.)
(g) If an element forms a cation with charge +2, it is in family 2A.

21. If the seventh period of the periodic table is 32 members long, it will be the same length as the sixth period. Elements in the same family will have atomic numbers 32 units higher. The noble gas following radon will have atomic number = 86 + 32 = 118. The alkali metal following francium will have atomic number = 87 + 32 = 119.

22. One mole of any element contains 6.022×10^{23} atoms, the Avogadro constant.
(a) $12.7\ \text{mol Ca} \times \frac{6.022 \times 10^{23}\ \text{Ca atoms}}{1\ \text{mol Ca}} = 7.65 \times 10^{24}\ \text{Ca atoms}$

(b) $0.00361 \text{ mol Ne} \times \dfrac{6.022 \times 10^{23} \text{ Ne atoms}}{1 \text{ mol Ne}} = 2.17 \times 10^{21} \text{ Ne atoms}$

(c) $1.8 \times 10^{-12} \text{ mol Pu} \times \dfrac{6.022 \times 10^{23} \text{ Pu atoms}}{1 \text{ mol Pu}} = 1.1 \times 10^{12} \text{ Pu atoms}$

23. In these problems we use the Avogadro constant and the fact that one mole of atoms of an element has a weight in grams equal to its atomic weight.

(a) $\text{no. moles Fe} = 2.18 \times 10^{26} \text{ Fe atoms} \times \dfrac{1 \text{ mol Fe}}{6.022 \times 10^{23} \text{ Fe atoms}} = 362 \text{ mol Fe}$

(b) $\text{mass of Kr, g} = 7.71 \text{ mol Kr} \times \dfrac{83.80 \text{ g Kr}}{1 \text{ mol Kr}} = 646 \text{ g Kr}$

(c) $\text{mass of Au, mg} = 6.15 \times 10^{19} \text{ Au atoms} \times \dfrac{1 \text{ mol Au}}{6.022 \times 10^{23} \text{ atoms}} \times \dfrac{196.97 \text{ g Au}}{1 \text{ mol Au}} \times \dfrac{1000 \text{ mg}}{1 \text{ g}}$

$= 20.1 \text{ mg Au}$

(d) $\text{no. Fe atoms} = 112 \text{ cm}^3 \text{ Fe} \times \dfrac{7.86 \text{ g}}{1 \text{ cm}^3} \times \dfrac{1 \text{ mol Fe}}{55.85 \text{ g Fe}} \times \dfrac{6.022 \times 10^{23} \text{ atoms}}{1 \text{ mol Fe}}$

$= 9.49 \times 10^{24} \text{ Fe atoms}$

24. Since the molar mass of nitrogen is 14.0 g/mol, 25.0 g N is almost two moles (1.79 mol N), while 6.02×10^{23} Ni atoms is almost precisely a mole, and 52.0 g Cr (52.00 g/mol Cr) is also almost exactly a mole. Finally, 10.0 cm^3 Fe (55.85 g/mol Fe) has a mass of about 79 g, and contains about 1.5 mole of atoms. Thus, 25.0 g N contains the greatest number of atoms.

25. We first determine the number of Pb atoms of all types in 1.57 g of Pb, and then use the percent abundance to determine the number of ^{204}Pb atoms present.

$\text{no. atoms } ^{204}\text{Pb} = 215 \text{ mg Pb} \times \dfrac{1 \text{ g}}{1000 \text{ mg}} \times \dfrac{1 \text{ mol Pb}}{207.2 \text{ g Pb}} \times \dfrac{6.022 \times 10^{23} \text{ atoms}}{1 \text{ mol Pb}} \times \dfrac{14\ ^{204}\text{Pb atoms}}{1000 \text{ Pb atoms}}$

$= 8.75 \times 10^{18} \text{ atoms } ^{204}\text{Pb}$

26. $\text{mass of alloy} = 6.50 \times 10^{23} \text{ Cd atoms} \times \dfrac{1 \text{ mol Cd}}{6.022 \times 10^{23} \text{ Cd atoms}} \times \dfrac{112.4 \text{ g Cd}}{1 \text{ mol Cd}} \times \dfrac{100.0 \text{ g alloy}}{8.0 \text{ g Cd}}$

$= 152 \text{ g alloy}$

EXERCISES

Law of Conservation of Mass

27. The observations cited do not necessarily violate the law of conservation of mass. The oxide formed when iron rusts is a solid and remains with the solid iron, increasing the mass of the solid by the mass of the oxygen that combined. The oxide formed when a match burns is a gas and will not remain with the solid product (the ash); the mass of the ash thus is less than that of the reactants. We would have to collect all reactants and all products and weigh them to determine if the law of conservation of mass is obeyed or violated.

28. The magnesium that is burned in air combines with some of the oxygen in the air and this oxygen (which, of course, was not weighed when the magnesium metal was weighed) adds its mass to the mass of the magnesium, making the magnesium oxide product weigh more than did the original magnesium. When this same reaction is carried out in a photoflash bulb, the oxygen (in fact, some excess oxygen) that will combine with the magnesium is already present in the bulb before the reaction. Consequently, the product contains no unweighed oxygen.

29. We compare the mass before reaction with that after reaction.
mass before reaction = 10.500 g calcium hydroxide + 11.125 g ammonium chloride = 21.625 g
mass after reaction = 14.336 g solid residue + (69.605 – 62.316) g gases = 14.336 g + 7.289 g = 21.625 g
These data support the law of conservation of mass. Note that the gain in the mass of water is due to the gases that it absorbs.

30. We compute the mass of the reactants and compare that with the mass of the products.

reactant mass = mass of calcium carbonate + mass of hydrochloric acid solution

$$= 10.00 \text{ g calcium carbonate} + 100.0 \text{ mL soln} \times \frac{1.148 \text{ g}}{1 \text{ mL soln}}$$

$$= 10.00 \text{ g calcium carbonate} + 114.8 \text{ g solution}$$

$$= 124.8 \text{ g reactants}$$

product mass = mass of solution + mass of carbon dioxide

$$= 120.40 \text{ g soln} + 2.22 \text{ L gas} \times \frac{1.9769 \text{ g}}{1 \text{ L gas}}$$

$$= 120.40 \text{ g soln} + 4.39 \text{ g carbon dioxide}$$

$$= 124.79 \text{ g products}$$

We see that the mass of the products agrees with the mass of the reactants within experimental error. The law of conservation of mass was obeyed in this case.

Law of Constant Composition

31. In the first experiment, 2.18 g of sodium produces 5.54 g of sodium chloride. In the second experiment, 2.10 g of chlorine produces 3.46 g of sodium chloride. The amount of sodium contained in this second sample of sodium chloride is given by

mass of sodium = 3.46 g sodium chloride – 2.10 g chlorine = 1.36 g sodium

We now have sufficient information to determine the % Na in each of the samples of sodium chloride.

$$\%\text{Na} = \frac{2.18 \text{ g Na}}{5.54 \text{ g cmpd}} \times 100\% = 39.4\% \text{ Na} \qquad \%\text{Na} = \frac{1.36 \text{ g Na}}{3.46 \text{ g cmpd}} \times 100\% = 39.3\% \text{ Na}$$

Thus, the two samples of sodium chloride have the same composition. Recognize that, according to the interpretation of numbers based on significant figures, each percent has an uncertainty of ±0.1%.

32. If the two samples of water have the same %H, the law of constant composition is demonstrated. Notice that, in the second experiment, the mass of the compound is equal to the sum of the masses of the elements produced from it.

$$\%\text{ H} = \frac{3.06 \text{ g}}{27.35 \text{ g } H_2O} \times 100\% = 11.2\% \text{ H} \qquad \%\text{H} = \frac{1.45 \text{ g H}}{(1.45 + 11.51)\text{g } H_2O} \times 100\% = 11.2\% \text{ H}$$

The results are consistent with the law of constant composition.

<u>33</u>. The mass of sulfur (0.312 g) needed to produce 0.623 g sulfur dioxide provides the information for the conversion factor.

$$\text{sulfur mass} = 0.842 \text{ g sulfur dioxide} \times \frac{0.312 \text{ g sulfur}}{0.623 \text{ g sulfur dioxide}} = 0.422 \text{ g sulfur}$$

34. **(a)** From the first experiment we see that 1.16 g of compound is produced per gram of Hg. These masses enable us to determine the mass of compound produced from 1.50 g Hg.

$$\text{mass of cmpd} = 1.50 \text{ g Hg} \times \frac{1.16 \text{ g cmpd}}{1.00 \text{ g Hg}} = 1.74 \text{ g cmpd}$$

(b) Since the compound weighs 0.24 g more than the mass of mercury (1.50 g) that was used, 0.24 g of sulfur must have reacted. Thus, the unreacted sulfur has a mass of 0.76 g (=1.00 g initially present – 0.24 g reacted).

Fundamental Particles

35. A fundamental particle would be expected to be found in all samples of matter. This is demonstrated with electrons by the fact that cathode rays have the same properties no matter how they are prepared. These properties are independent of the material that was used to construct the cathode ray tube, of the gas that filled the tube when it was constructed (and has since been pumped out), and of the method used to generate electricity.

36. The detection and characterization of electrons was based on the fact that they are charged particles. For instance, in Thompson's experiment, a beam of electrons is made to curve by the force between a magnetic field and a beam of charged particles. A neutron, however, does not have a charge and thus cannot be detected or characterized by the methods used for electrons.

Fundamental Charges and Mass-to-Charge Ratios

37. We can calculate the charge on each drop, express each in terms of 10^{-19} C, and finally express each in terms of $e = 1.6 \times 10^{-19}$ C.

drop 1:	1.28×10^{-18} C	$= 12.8 \times 10^{-19}$ C	$= 8\,e$
drops 2 & 3:	$1.28 \times 10^{-18} \div 2 = 0.640 \times 10^{-18}$ C	$= 6.40 \times 10^{-19}$ C	$= 4\,e$
drop 4:	$1.28 \times 10^{-18} \div 8 = 0.160 \times 10^{-18}$ C	$= 1.60 \times 10^{-19}$ C	$= 1\,e$
drop 5:	$1.28 \times 10^{-18} \times 4 = 5.12 \times 10^{-18}$ C	$= 51.2 \times 10^{-19}$ C	$= 32\,e$

We see that these values are consistent with the charge that Millikan found for that of the electron, and he could have inferred the correct charge from these data, since they are all multiples of *e*.

38. We calculate each drop's charge, express each in terms of 10^{-19} C, and then in terms of $e = 1.6 \times 10^{-19}$ C.

drop 1:	6.41×10^{-19} C	$= 6.41 \times 10^{-19}$ C	$= 4\,e$
drop 2:	$6.41 \times 10^{-19} \div 2 = 3.21 \times 10^{-19}$ C	$= 3.21 \times 10^{-19}$ C	$= 2\,e$
drop 3:	$6.41 \times 10^{-19} \times 2 = 1.28 \times 10^{-18}$ C	$= 12.8 \times 10^{-19}$ C	$= 8\,e$
drop 4:	1.44×10^{-18}	$= 14.4 \times 10^{-19}$ C	$= 9\,e$
drop 5:	$1.44 \times 10^{-18} \div 3 = 4.8 \times 10^{-19}$	$= 4.8 \times 10^{-19}$ C	$= 3\,e$

We see that these values are consistent with the charge that Millikan found for that of the electron. He could have inferred the correct charge from these values, since they are all multiples of *e*, and have no other common factor.

39. **(a)** Determine the ratio of the mass of a hydrogen atom to that of an electron. We use the mass of a proton plus that of an electron for the mass of a hydrogen atom.

$$\frac{\text{mass of proton + mass of electron}}{\text{mass of electron}} = \frac{1.0073\text{ u} + 0.00055\text{ u}}{0.00055\text{ u}} = 1.8 \times 10^{3}$$

or $$\frac{\text{mass of electron}}{\text{mass of proton + mass of electron}} = \frac{1}{1.8 \times 10^{3}} = 5.6 \times 10^{-4}$$

(b) The only two mass-to-charge ratios that we can determine from the data of Table 2-1 are those for the proton, a hydrogen ion, H^+ (given first below); and that for the electron (given second).

For the proton: $$\frac{\text{mass}}{\text{charge}} = \frac{1.673 \times 10^{-24}\text{ g}}{1.602 \times 10^{-19}\text{ C}} = 1.044 \times 10^{-5}\text{ g/C}$$

For the electron: $$\frac{\text{mass}}{\text{charge}} = \frac{9.109 \times 10^{-28}\text{ g}}{1.602 \times 10^{-19}\text{ C}} = 5.686 \times 10^{-9}\text{ g/C}$$

The hydrogen ion is the lightest positive ion available. We see that the mass-to-charge ratio for a positive particle is considerably larger than that for an electron.

40. We do not have the precise isotopic masses of the two ions, and thus the values of the mass-to-charge ratios are only approximate. As a consequence, we have used a three-significant figure mass for a nucleon, rather than the more precisely known masses of proton and neutron. (Recall that the term "nucleon" refers to a nuclear particle—either a proton or a neutron.)

$^{127}I^-$ $$\frac{m}{e} = \frac{127\text{ nucleons}}{1\text{ electron}} \times \frac{1\text{ e}}{1.602 \times 10^{-19}\text{ C}} \times \frac{1.67 \times 10^{-24}\text{ g}}{1\text{ nucleon}} = 1.32 \times 10^{-3}\text{ g/C}$$

$^{32}S^{2-}$ $$\frac{m}{e} = \frac{32\text{ nucleons}}{2\text{ electrons}} \times \frac{1\text{ e}}{1.602 \times 10^{-19}\text{ C}} \times \frac{1.67 \times 10^{-24}\text{ g}}{1\text{ nucleon}} = 1.67 \times 10^{-4}\text{ g/C}$$

Atomic Number, Mass Number, and Isotopes

41. **(a)** An atom of ^{108}Pd contains 46 protons, and thus also 46 electrons, since an atom is neutral and there is no difference between the number of protons and electrons. Since there are 108 nucleons in the nucleus, the number of neutrons is 62 (= 108 nucleons – 46 protons).

(b) The ratio of the two masses is determined as follows. $$\frac{^{108}\text{Pd}}{^{12}\text{C}} = \frac{107.90389\text{ u}}{12.00000\text{ u}} = 8.9919908$$

42. **(a)** The atomic number of Ra is 88 and equals the number of protons in the nucleus. The ion's charge is +2 and thus there are two more protons than electrons: no. protons = no. electrons + 2 = 88; no. electrons = 88 – 2 = 86. The mass number (228) is the sum of the atomic number and the number of neutrons: 228 = 88 + no. neutrons; no. neutrons = 228 – 88 = 140 neutrons.

(b) The mass of ^{16}O is 15.9949 u. ratio $= \dfrac{\text{mass of isotope}}{\text{mass of } ^{16}O} = \dfrac{228.030 \text{ u}}{15.9949 \text{ u}} = 14.2564$

43. The mass of ^{16}O is 15.9949 u. isotopic mass = 15.9949 u × 6.68374 (times ^{16}O) = 106.906 u

44. The mass of ^{16}O is 15.9949 u.
mass of heavier isotope = 15.9949 u × 7.1838 = 114.90 u = mass of ^{115}In
mass of lighter isotope $= \dfrac{114.90 \text{ u}}{1.0177} = 112.90$ u = mass of ^{113}In

45. **(a)** Species with equal numbers of protons and neutrons will also have a mass number that is twice the atomic number. The following species are approximately suitable (with numbers of protons and neutrons in parentheses). $^{24}_{12}Mg^{2+}$ (12 p, 12 n), $^{47}_{24}Cr$ (24 p, 23 n), $^{60}_{27}Co^{3+}$ (27 p, 32 n), and $^{35}_{17}Cl^{-}$ (17 p, 18 n). Of these four nuclides, only $^{24}_{12}Mg^{2+}$ has just as many protons as neutrons.

(b) A species in which protons have more than 50% of the mass must have a mass number smaller than twice the atomic number. Of these species, only in $^{47}_{24}Cr$ is more than 50% of the mass contributed by the protons.

(c) A species with 50% more neutrons than protons will have a mass number equal to 2.5 times the atomic number. Only two species might qualify: $^{124}_{50}Sn^{2+}$ (74 n, 50 p) and $^{226}_{90}Th$ (136 n, 90 p). $^{226}_{90}Th$ has more than 50% more neutrons than protons.

46. To answer these questions, we determine the number of protons, neutrons, and electrons of each species.

species:	$^{24}_{12}Mg^{2+}$	$^{47}_{24}Cr$	$^{60}_{27}Co^{3+}$	$^{35}_{17}Cl^{-}$	$^{124}_{50}Sn^{2+}$	$^{226}_{90}Th$	$^{90}_{38}Sr$
no. protons	12	24	27	17	50	90	38
no. neutrons	12	23	33	18	74	136	52
no. electrons	10	24	24	18	48	90	38

(a) The number of neutrons and electrons is equal for $^{35}_{17}Cl^{-}$.

(b) The species $^{60}_{27}Co^{3+}$ has protons (27), neutrons (33), and electrons (24) in the ratio 9:11:8.

(c) The species $^{124}_{50}Sn^{2+}$ has a number of neutrons (74) equal to its number of protons (50) plus one-half its number of electrons (48÷2 = 24).

Atomic Mass Units, Atomic Masses

47. It is exceedingly unlikely that another nuclide would have an exact integral mass. The mass of carbon-12 is *defined* as precisely 12.00000 u. Each nuclidic mass is close to integral, but none that we have encountered in this chapter are precisely integral. The reason is that each nuclide is composed of protons, neutrons, and electrons, none of which have integral masses, and there is a small quantity of the mass of each nucleon (nuclear particle) lost in the binding energy holding the nuclide together. It would be highly unlikely that all of these effects would add up to a precisely integral mass.

48. There are no copper atoms that have a mass of 63.546 u. The masses of individual atoms are close to integers and this mass (63.546 u) is about midway between two integers. It is an average atomic mass, the result of averaging two (or more) isotopic masses, each weighted by its natural abundance.

49. To determine the average atomic mass, we use the following expression (where Σ indicates the sum of the individual products). average atomic mass = Σ (isotopic mass × fractional natural abundance)
Each of the three percents given is converted to a fractional abundance by dividing it by 100.
Mg atomic mass = (23.985042 u × 0.7899) + (24.985837 u × 0.1000) + (25.982593 u × 0.1101)
= 18.95 u + 2.499 u + 2.861 u = 24.31 u

50. To determine the average atomic mass, we use the following expression (where Σ indicates the sum of the individual products). average atomic mass = Σ (isotopic mass × fractional natural abundance)
Each of the three percents given is converted to a fractional abundance by dividing it by 100.

Cr atomic mass = (49.9461 × 0.0435) + (51.9405 × 0.8379) + (52.9407 × 0.0950) + (53.9389 × 0.0236)
= 2.17 u + 43.52 u + 5.03 u + 1.27 u = 51.99 u

51. We use the expression for determining the weighted-average atomic mass.
107.8682 u = (106.905092 u × 0.5184) + (^{109}Ag × 0.4816) = 55.42 u + 0.4816 ^{109}Ag
107.8682 u – 55.42 u = 0.4816 ^{109}Ag = 52.45 u $\qquad$ $^{109}\text{Ag} = \dfrac{52.45\text{ u}}{0.4816} = 108.9\text{ u}$

52. The percent abundances of the two isotopes must add to 100.00%, since there are only two naturally occurring isotopes of bromine. Thus, we can determine the percent natural abundance of the second isotope by difference. % second isotope = 100.00% – 50.69% = 49.31%
From the periodic table, we see that the weighted-average atomic mass of bromine is 79.904 u. We use this value in the expression for determining the weighted-average atomic mass, along with the isotopic mass of ^{79}Br and the fractional abundances of the two isotopes (the percent abundances divided by 100).
79.904 u = (0.5069 × 78.918336 u) + (0.4931 × other isotope) = 40.00 u + (0.4931 × other isotope)

$$\text{other isotope} = \frac{79.904\text{ u} - 40.00\text{ u}}{0.4931} = 80.92\text{ u} = \text{mass of } ^{81}\text{Br, the other isotope}$$

53. Since the three percent abundances total 100%, the percent abundance of ^{40}K is found by difference.
% ^{40}K = 100.0000% – 93.2581% – 6.7302% = 0.0117%
Then the expression for the weighted-average atomic mass is used, with the percent abundances converted to fractional abundances by dividing by 100. The average atomic mass of potassium is 39.0983 u.
39.0983 u = (0.932581 × 38.963707 u) + (0.000117 × 39.963999 u) + (0.067302 × ^{41}K)
= 36.3368 u + 0.00468 u + (0.067302 × ^{41}K)

$$\text{mass of } ^{41}\text{K} = \frac{39.0983\text{ u} - (36.3368\text{ u} + 0.00468\text{ u})}{0.067302} = 40.962\text{ u}$$

54. We use the expression for determining the weighted-average atomic mass, allowing x to represent the fractional abundance of ^{10}B and $(1 - x)$ the fractional abundance of ^{11}B
10.811 u = (10.012937 u × x) + [11.009305 × $(1 - x)$] = 10.012937x + 11.009305 – 11.009305x
10.811 – 11.009305 = –0.198 = 10.012937x – 11.009305x = –0.996368x

$$x = \frac{0.198}{0.996368} = 0.199 \qquad 19.9\%\ ^{10}\text{B} \qquad (100.0 - 19.9) = 80.1\%\ ^{11}\text{B}$$

Mass spectrometry

55. **(a)**

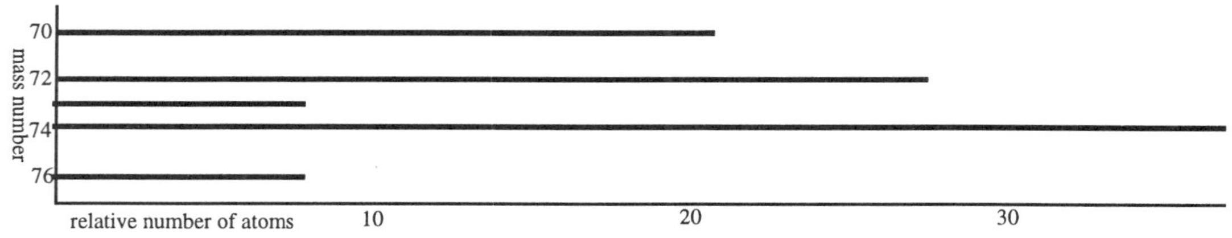

(b) As before, we multiply each isotopic mass by its fractional abundance; we sum these products to obtain the (average) atomic mass.
(0.205 × 70) + (0.274 × 72) + (0.078 × 73) + (0.365 × 74) + (0.078 atoms × 76)
14 + 20. + 5.7 + 27 + 5.9 = $72._6$ = average atomic weight of germanium
The result is only approximately correct because the isotopic masses are given to only two significant figures. Thus, only a two-significant-figure result can be obtained.

56. **(a)** There are six possible types of molecule: ^{1}H^{35}Cl, ^{2}H^{35}Cl, ^{3}H^{35}Cl, ^{1}H^{37}Cl, ^{2}H^{37}Cl, and ^{3}H^{37}Cl
The mass numbers of the six different possible types of molecules are obtained by summing the mass numbers of the two atoms in each molecule:
^{1}H^{35}Cl has A = 36 $\qquad$ ^{2}H^{35}Cl has A = 37 $\qquad$ ^{3}H^{35}Cl has A = 38
^{1}H^{37}Cl has A = 38 $\qquad$ ^{2}H^{37}Cl has A = 39 $\qquad$ ^{3}H^{37}Cl has A = 40

(b) The most abundant molecule contains the most abundant of each element's isotope. It is ^{1}H^{35}Cl. The second most abundant molecule is ^{1}H^{37}Cl.

The relative abundance of each type of molecule is determined by multiplying together the fractional abundances of the two isotopes that are present in that type. Because ^3H is so scarce, it will be difficult to detect. Relative abundances of the molecules are as follows.

$^1H^{35}Cl$: 75.52%	$^1H^{37}Cl$: 24.47%
$^2H^{35}Cl$: 0.011%	$^2H^{37}Cl$: 0.004%
$^3H^{35}Cl$: < 0.001%	$^3H^{37}Cl$: < 0.001%

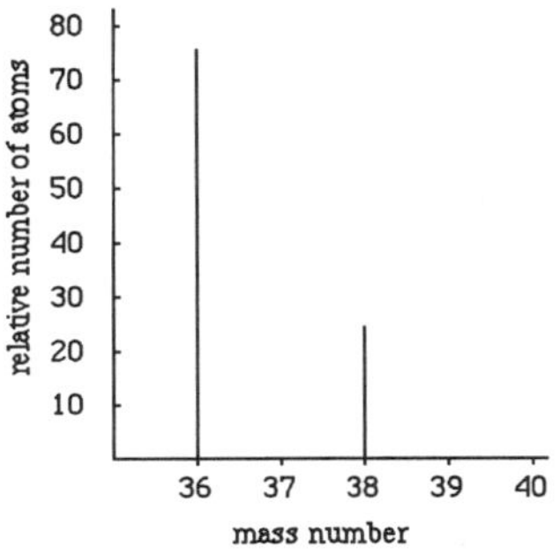

The Avogadro Constant and the Mole

57. Each of these calculations employs the average atomic mass as a conversion factor.

(a) $\text{amount of Rb} = 167.0 \text{ g Rb} \times \dfrac{1 \text{ mol Rb}}{85.468 \text{ g Rb}} = 1.954 \text{ mol Rb}$

(b) $\text{number of Fe atoms} = 363.2 \text{ kg Fe} \times \dfrac{1000 \text{ g}}{1 \text{ kg}} \times \dfrac{1 \text{ mol Fe}}{55.847 \text{ g}} \times \dfrac{6.022 \times 10^{23} \text{ Fe atoms}}{1 \text{ mol Fe}}$

$= 3.916 \times 10^{27} \text{ Fe atoms}$

(c) $\text{Ag mass} = 1.0 \times 10^{12} \text{ Ag atoms} \times \dfrac{1 \text{ mol Ag}}{6.022 \times 10^{23} \text{ Ag atoms}} \times \dfrac{107.87 \text{ g Ag}}{1 \text{ mol Ag}}$

$= 1.8 \times 10^{-10} \text{ g Ag} = 0.18 \text{ ng Ag}$

58. Each of these calculations employs the average atomic mass as a conversion factor.

(a) $\text{number of Ar atoms} = 5.25 \text{ mg Ar} \times \dfrac{1 \text{ g}}{1000 \text{ mg}} \times \dfrac{1 \text{ mol Ar}}{39.948 \text{ g}} \times \dfrac{6.022 \times 10^{23} \text{ Ar atoms}}{1 \text{ mol Ar}}$

$= 7.91 \times 10^{19} \text{ Ar atoms}$

(b) $\text{molar mass} = \dfrac{4.24 \text{ g}}{2.80 \times 10^{22} \text{ atoms}} \times \dfrac{6.022 \times 10^{23} \text{ atoms}}{1 \text{ mole}} = 91.2 \text{ g/mol}$

(c) $\text{Al mass} = 35.55 \text{ g Zn} \times \dfrac{1 \text{ mol Zn}}{65.39 \text{ g Zn}} \times \dfrac{1 \text{ mol Al}}{1 \text{ mol Zn}} \times \dfrac{26.98 \text{ g Al}}{1 \text{ mol Al}} = 14.67 \text{ g Al}$

Quantities with the same number of moles have the same number of atoms.

59. We determine the mass of Ag in the piece of jewelry and then the number of Ag atoms.

$\text{no. Ag atoms} = 38.7 \text{ g sterling} \times \dfrac{92.5 \text{ g Ag}}{100.0 \text{ g sterling}} \times \dfrac{1 \text{ mol Ag}}{107.9 \text{ g Ag}} \times \dfrac{6.022 \times 10^{23} \text{ atoms}}{1 \text{ mol Ag}} = 2.00 \times 10^{23} \text{ Ag atoms}$

60. We first determine the amount in moles of each metal.

$\text{amount of Pb} = 75.0 \text{ cm}^3 \times \dfrac{9.4 \text{ g}}{1 \text{ cm}^3} \times \dfrac{67 \text{ g Pb}}{100 \text{ g solder}} \times \dfrac{1 \text{ mol Pb}}{207.2 \text{ g Pb}} = 2.28 \text{ mol Pb}$

$\text{amount of Sn} = 75.0 \text{ cm}^3 \times \dfrac{9.4 \text{ g}}{1 \text{ cm}^3} \times \dfrac{33 \text{ g Sn}}{100 \text{ g solder}} \times \dfrac{1 \text{ mol Sn}}{118.7 \text{ g Sn}} = 1.96 \text{ mol Sn}$

$\text{total atoms} = (2.28 \text{ mol Pb} + 1.96 \text{ mol Sn}) \times \dfrac{6.022 \times 10^{23} \text{ atoms}}{1 \text{ mol}} = 2.55 \times 10^{24} \text{ atoms}$

61. We use the average atomic mass of lead, 207.2 g/mol.

(a) $\dfrac{30 \text{ µg Pb}}{1 \text{ dL}} \times \dfrac{1 \text{ dL}}{0.1 \text{ L}} \times \dfrac{1 \text{ g Pb}}{10^6 \text{ µg Pb}} \times \dfrac{1 \text{ mol Pb}}{207.2 \text{ g}} = 1.4 \times 10^{-6} \text{ mol Pb/L}$

(b) $\dfrac{1.4 \times 10^{-6} \text{ mol Pb}}{\text{L}} \times \dfrac{1 \text{ L}}{1000 \text{ mL}} \times \dfrac{6.022 \times 10^{23} \text{ atoms}}{1 \text{ mol}} = 8.4 \times 10^{14} \text{ Pb atoms/mL}$

62. The concentration of Pb in air provides the principal conversion factor. Other conversion factors are needed to convert to and from its units, beginning with the 0.500-L volume, and ending with the number of atoms.

$\text{no. Pb atoms} = 0.500 \text{ L} \times \dfrac{1 \text{ m}^3}{1000 \text{ L}} \times \dfrac{3.01 \text{ µg Pb}}{1 \text{ m}^3} \times \dfrac{1 \text{ g Pb}}{10^6 \text{ µg Pb}} \times \dfrac{1 \text{ mol Pb}}{207.2 \text{ g Pb}} \times \dfrac{6.022 \times 10^{23} \text{ Pb atoms}}{1 \text{ mol Pb}}$

$= 4.37 \times 10^{12} \text{ Pb atoms}$

FEATURE PROBLEMS

A. The product mass differs from that of the reactants by (5.62 – 2.50 =) 3.12 grains. In order to determine the percent gain in mass, we need to convert the reactant mass entirely to grains.

$$13 \text{ onces} \times \frac{8 \text{ gros}}{1 \text{ once}} = 104 \text{ gros} \qquad (104 + 2) \text{ gros} \times \frac{72 \text{ grains}}{1 \text{ gros}} = 7632 \text{ grains}$$

$$\% \text{ mass increase} = \frac{3.12 \text{ grains increase}}{(7632 + 2.50) \text{ grains original}} \times 100\% = 0.0409\% \text{ mass increase}$$

The sensitivity of Lavoisier's balance can be as little as 0.01 grain, which seems to be the limit of the readability of the balance; or it can be as large as 3.12 grains, which assumes that all of the error in the experiment is due to the (in)sensitivity of the balance. Let us convert 0.01 grains to a mass in grams.

$$\text{minimum error} = 0.01 \text{ gr} \times \frac{1 \text{ gros}}{72 \text{ gr}} \times \frac{1 \text{ once}}{8 \text{ gros}} \times \frac{1 \text{ livre}}{16 \text{ once}} \times \frac{30.59 \text{ g}}{1 \text{ livre}} = 3 \times 10^{-5} \text{ g} = 0.03 \text{ mg}$$

$$\text{maximum error} = 3.12 \text{ gr} \times \frac{3 \times 10^{-5} \text{ g}}{0.01 \text{ g}} = 9 \times 10^{-3} \text{ g} = 9 \text{ mg}$$

The maximum error is close to that of a somewhat common modern laboratory balance, which typically has a sensitivity of 1 mg. The minimum error is approximated by a good quality analytical balance.

B. One way to determine the common factor of which all 13 numbers are multiples is to first divide all of them by the smallest. The ratios thus obtained may either be integers or they may be rational numbers whose decimal equivalents are easy to recognize.

Obs.	1	2	3	4	5	6	7	8	9	10	11	12	13
Quan.	19.66	24.60	29.62	34.47	39.38	44.42	49.41	53.91	59.12	63.68	68.65	78.34	83.22
Ratio	1.000	1.251	1.507	1.753	2.003	2.259	2.513	2.742	3.007	3.239	3.492	3.984	4.233
Mult.	4.000	5.005	6.026	7.013	8.012	9.038	10.05	10.97	12.03	12.96	13.97	15.94	16.93
Int.	4	5	6	7	8	9	10	11	12	13	14	16	17

In the row labeled "Mult." we have given the multiplier of the common factor (4.915) of each measured quantity. In the row labeled "Int." we give the integer closest to each of these multipliers. It is obvious that each of the 13 measurements is exceedingly close to an integer multiple of a common quantity.

C. In a 50-year-old chemistry textbook the value printed in the periodic table for oxygen would be 16.000..., because chemists assigned precisely 16 as the atomic weight of the naturally occuring mixture of oxygen isotopes. This value is slightly higher than the value of 15.9994 in modern chemistry textbooks. Thus, we would expect all other atomic weights in the 50-year-old textbook to be slightly higher as well.

D. The grade-point average is defined as the quotient of two sums. The sum in the numerator is that of the product of the number of units for a course times the points corresponding to the grade earned. The denominator is the total number of units taken during the term.

$$\text{grade-point average} = \frac{[(\text{units, course } \alpha) \times (\text{points, course } \alpha)] + [\text{units}_\beta \times \text{points}_\beta] + [\text{units}_\gamma \times \text{points}_\gamma] +}{(\text{units, course } \alpha) + \text{units}_\beta + \text{units}_\gamma +}$$

Of course, the ratio of number of units for a course to total number of units equals the fraction of total units allocated to that course (fraction, course α = $\text{units}_\alpha \div$ total units). So our general formula becomes:

$$\text{grade-point average} = [(\text{fraction, course } \alpha) \times (\text{points, course } \alpha) + [\text{fraction}_\beta \times \text{points}_\beta] + [\text{fraction}_\gamma \times \text{points}_\gamma] +$$

We apply our formula to this specific case, where total units = 5 + 3 + 4 = 12 units.

$$\text{grade-point average} = \left(\frac{5}{12} \times 4.0\right) + \left(\frac{3}{12} \times 3.0\right) + \left(\frac{4}{12} \times 2.0\right) = 1.67 + 0.75 + 0.67 = 3.09$$

E. We begin with the amount of reparations and obtain the volume in cubic kilometers with a series of conversion factors.

$$\text{volume} = \$28.8 \times 10^9 \times \frac{1 \text{ troy oz}}{\$21.25} \times \frac{1 \text{ lb}}{12 \text{ troy oz}} \times \frac{453.6 \text{ g Au}}{1 \text{ lb}} \times \frac{1 \text{ mol Au}}{196.97 \text{ g}} \times \frac{6.022 \times 10^{23} \text{ atoms}}{1 \text{ mol Au}}$$
$$\times \frac{1 \text{ ton sea water}}{4.67 \times 10^{17} \text{ Au atoms}} \times \frac{2000 \text{ lb}}{1 \text{ ton}} \times \frac{453.6 \text{ g}}{1 \text{ lb seawater}} \times \frac{1 \text{ cm}^3}{1.03 \text{ g}} \times \left(\frac{1 \text{ m}}{100 \text{ cm}} \times \frac{1 \text{ km}}{1000 \text{ m}}\right)^3$$
$$= 2.95 \times 10^5 \text{ km}^3$$

3 CHEMICAL COMPOUNDS

PRACTICE EXAMPLES

1A For one conversion factor we need the molar mass of ZnO.

$$\mathcal{M} = \left(\frac{1\text{ mol Zn}}{1\text{ mol ZnO}} \times \frac{65.39\text{ g Zn}}{1\text{ mol Zn}}\right) + \left(\frac{1\text{ mol O}}{1\text{ mol ZnO}} \times \frac{16.00\text{ g O}}{1\text{ mol O}}\right) = \frac{81.39\text{ g ZnO}}{1\text{ mol ZnO}}$$

Then determine the number of ions in 1.0 g of ZnO. Note that each mole of ZnO contains two moles of ions: 1 mole of Zn^{2+} ions, and 1 mole of O^{2-} ions.

$$?\text{ ions} = 1.0\text{ g ZnO} \times \frac{1\text{ mol ZnO}}{81.39\text{ g ZnO}} \times \frac{2\text{ mol ions}}{1\text{ mol ZnO}} \times \frac{6.022 \times 10^{23}\text{ ions}}{1\text{ mol ions}} = 1.5 \times 10^{22}\text{ ions}$$

1B For one conversion factor we need the molar mass of $MgCl_2$.

$$\mathcal{M} = \left(\frac{1\text{ mol Mg}}{1\text{ mol MgCl}_2} \times \frac{24.305\text{ g Mg}}{1\text{ mol Mg}}\right) + \left(\frac{2\text{ mol Cl}}{1\text{ mol MgCl}_2} \times \frac{35.453\text{ g Cl}}{1\text{ mol Cl}}\right) = \frac{95.211\text{ g MgCl}_2}{1\text{ mol MgCl}_2}$$

Then convert the number of chloride ions to the mass of $MgCl_2$.

$$\text{MgCl}_2\text{ mass} = 5.0 \times 10^{23}\text{ Cl}^-\text{ ions} \times \frac{1\text{ f.u. MgCl}_2}{2\text{ Cl}^-\text{ ions}} \times \frac{1\text{ mol MgCl}_2}{6.022 \times 10^{23}\text{ f.u.}} \times \frac{95.211\text{ g MgCl}_2}{1\text{ mol MgCl}_2}$$

$$= 4.0 \times 10^1\text{ g MgCl}_2$$

2A The volume of gold is converted to its mass and then to the amount in moles.

$$?\text{ Au atoms} = (2.50\text{ cm})^2 \times \left(0.100\text{ mm} \times \frac{1\text{ cm}}{10\text{ mm}}\right) \times \frac{19.32\text{ g}}{1\text{ cm}^3} \times \frac{1\text{ mol Au}}{196.97\text{ g Au}} \times \frac{6.022 \times 10^{23}\text{ atoms}}{1\text{ mol Au}}$$

$$= 3.69 \times 10^{21}\text{ Au atoms}$$

2B We need the molar mass of ethyl mercaptan for one conversion factor.

$\mathcal{M} = (2 \times 12.011\text{ g C}) + (6 \times 1.008\text{ g H}) + (1 \times 32.066\text{ g S}) = 62.136\text{ g/mol C}_2\text{H}_6\text{S}$

$$\text{C}_2\text{H}_6\text{S conc'n} = \frac{1.0\ \mu\text{L C}_2\text{H}_6\text{S}}{1500\text{ m}^3} \times \frac{1\text{ L}}{1 \times 10^6\ \mu\text{L}} \times \frac{1000\text{ mL}}{1\text{ L}} \times \frac{0.84\text{ g}}{1\text{ mL}} \times \frac{1\text{ mol C}_2\text{H}_6\text{S}}{62.136\text{ g}} \times \frac{10^6\ \mu\text{mol}}{1\text{ mol}}$$

$$= 9 \times 10^{-3}\ \mu\text{mol/m}^3 > 0.9 \times 10^{-3}\ \mu\text{mol/m}^3 = \text{the detectable limit}$$

Thus, the vapor will be detectable.

3A The mole weight of halothane is given in Example 3-3 *in the text* as 197.4 g/mol. The rest of the solution uses conversion factors to change units.

$$\text{C mass} = 75.0\text{ mL C}_2\text{HBrClF}_3 \times \frac{1.871\text{ g}}{1\text{ mL}} \times \frac{1\text{ mol halothane}}{197.4\text{ g}} \times \frac{2\text{ mol C}}{1\text{ mol C}_2\text{HBrClF}_3} \times \frac{12.01\text{ g C}}{1\text{ mol C}}$$

$$= 17.1\text{ g C}$$

3B Again, the mole weight of halothane is given in Example 3-3 *in the text* as 197.4 g/mol.

$$\text{Halothane volume} = 100.0\text{ g Br} \times \frac{1\text{ mol Br}}{79.904\text{ g Br}} \times \frac{1\text{ mol C}_2\text{HBrClF}_3}{1\text{ mol Br}} \times \frac{197.4\text{ g C}_2\text{HBrClF}_3}{1\text{ mol C}_2\text{HBrClF}_3} \times \frac{1\text{ mL}}{1.871\text{ g}}$$

$$= 132.0\text{ mL C}_2\text{HBrClF}_3$$

4A The molecular formula of acetic acid is $C_2H_4O_2$. We first determine the mole weight of acetic acid.

$$\mathcal{M} = \left(\frac{2\text{ mol C}}{1\text{ mol acid}} \times \frac{12.011\text{ g C}}{1\text{ mol C}}\right) + \left(\frac{4\text{ mol H}}{1\text{ mol acid}} \times \frac{1.008\text{ g H}}{1\text{ mol H}}\right) + \left(\frac{2\text{ mol O}}{1\text{ mol acid}} \times \frac{15.9994\text{ g O}}{1\text{ mol O}}\right)$$

$$\mathcal{M} = \frac{24.022 \text{ g C}}{1 \text{ mol } C_2H_4O_2} + \frac{4.032 \text{ g H}}{1 \text{ mol } C_2H_4O_2} + \frac{31.9988 \text{ g O}}{1 \text{ mol } C_2H_4O_2} = \frac{60.053 \text{ g } C_2H_4O_2}{1 \text{ mol } C_2H_4O_2}$$

The percent of each element is determined by comparing the mass of that element present in a mole of the compound with the mass of one mole. $\%C = \frac{24.022 \text{ g C/1 mol acid}}{60.053 \text{ g acetic acid/1 mol acid}} \times 100\% = 40.001\% \text{ C}$

$\%H = \frac{4.032 \text{ gH}}{60.053 \text{ g } C_2H_4O_2} \times 100\% = 6.714\% \text{ H}$ $\quad \%O = \frac{31.9988 \text{ g O}}{60.053 \text{ g } C_2H_4O_2} \times 100\% = 53.284\% \text{ O}$

Note that the three percents sum to 100%, within the limits of significant figures, as they should.
40.001% C + 6.714 % H + 53.284% O = 99.999% total

4B We use the same technique as before: determine the mass of each element in a mole of the compound. Their sum is the molar mass of the compound. The percent composition then is determined by comparing the mass of each element with the molar mass of the compound.

$\mathcal{M} = (10 \times 12.011 \text{ g C}) + (11 \times 1.008 \text{ g H}) + (5 \times 14.01 \text{ g N}) + (3 \times 30.97 \text{ g P}) + (13 \times 15.999 \text{ g O})$
$= 120.11 \text{ g C} + 11.09 \text{ g H} + 70.05 \text{ g N} + 92.91 \text{ g P} + 207.99 \text{ g O} = 502.15 \text{ g ATP/mol}$

$\%C = \frac{120.11 \text{ g C}}{502.15 \text{ g ATP}} \times 100\% = 23.919\% \text{ C}$ $\quad \%H = \frac{11.09 \text{ g H}}{502.15 \text{ g ATP}} \times 100\% = 2.208\% \text{ H}$

$\%N = \frac{70.05 \text{ g N}}{502.15 \text{ g ATP}} \times 100\% = 13.95\% \text{ N}$ $\quad \%P = \frac{92.91 \text{ g P}}{502.15 \text{ g ATP}} \times 100\% = 18.50\% \text{ P}$

$\%O = \frac{207.99 \text{ g O}}{502.15 \text{ g ATP}} \times 100\% = 41.420\% \text{ O}$

5A We begin by using a sample of 100.00 g of the compound. In this way, each elemental mass in grams is numerically equal to its percent. We convert each mass to an amount in moles, and then determine the simplest integer set of molar amounts. This determination begins by dividing all three molar amounts by the smallest.

$55.37 \text{ g C} \times \frac{1 \text{ mol C}}{12.011 \text{ g C}} = 4.610 \text{ mol C} \quad \div 2.305 = 2.000 \text{ mol C} \times 3.000 = 6.000 \text{ mol C}$

$7.75 \text{ g H} \times \frac{1 \text{ mol H}}{1.008 \text{ g H}} = 7.69 \text{ mol H} \quad \div 2.305 = 3.34 \text{ mol H} \times 3.000 = 10.02 \text{ mol H}$

$36.88 \text{ g O} \times \frac{1 \text{ mol O}}{15.9994 \text{ g O}} = 2.305 \text{ mol O} \quad \div 2.305 = 1.000 \text{ mol O} \times 3.000 = 3.000 \text{ mol O}$

Thus, the empirical formula of the compound is $C_6H_{10}O_3$. The empirical weight of this compound is:
$(6 \times 12.01 \text{ g C}) + (10 \times 1.008 \text{ g H}) + (3 \times 16.00 \text{ g O}) = 72.06 + 10.08 + 48.00 = 130.14 \text{ g/mol}$
The empirical weight is almost precisely one half the reported molar mass, leading to the conclusion that the molecular formula must be twice the empirical formula in order to double the molar mass. Thus, the molecular formula is $C_{12}H_{20}O_6$.

5B We begin by using a sample of 100.00 g of the compound. In this way, each elemental mass in grams is numerically equal to its percent. We convert each mass to an amount in moles, and then determine the simplest integer set of molar amounts. This determination begins by dividing all three molar amounts by the smallest.

$39.56 \text{ g C} \times \frac{1 \text{ mol C}}{12.011 \text{ g C}} = 3.294 \text{ mol C} \quad \div 3.294 = 1.000 \text{ mol C} \times 3.000 = 3.000 \text{ mol C}$

$7.74 \text{ g H} \times \frac{1 \text{ mol H}}{1.008 \text{ g H}} = 7.68 \text{ mol H} \quad \div 3.294 = 2.33 \text{ mol H} \times 3.000 = 6.99 \text{ mol H}$

$52.70 \text{ g O} \times \frac{1 \text{ mol O}}{15.9994 \text{ g O}} = 3.294 \text{ mol O} \quad \div 3.294 = 1.000 \text{ mol O} \times 3.000 = 3.000 \text{ mol O}$

Thus, the empirical formula of the compound is $C_3H_7O_3$. The empirical weight of this compound is:
$(3 \times 12.01 \text{ g C}) + (7 \times 1.008 \text{ g H}) + (3 \times 16.00 \text{ g O}) = 36.03 + 7.056 + 48.00 = 91.09 \text{ g/mol}$
The empirical weight is almost precisely one half the reported molar mass, leading to the conclusion that the molecular formula must be twice the empirical formula in order to double the molar mass. Thus, the molecular formula is $C_6H_{14}O_6$.

6A We calculate the amount in moles of each element in the sample (determining the mass of oxygen by difference) and transform these molar amounts to the simplest integral amounts, by first dividing all three by the smallest.

$2.726 \text{ g } CO_2 \times \frac{1 \text{ mol } CO_2}{44.010 \text{ g } CO_2} \times \frac{1 \text{ mol C}}{1 \text{ mol } CO_2} = 0.06194 \text{ mol C} \times \frac{12.011 \text{ g C}}{1 \text{ mol C}} = 0.7440 \text{ g C}$

$$1.116 \text{ g H}_2\text{O} \times \frac{1 \text{ mol H}_2\text{O}}{18.015 \text{ g H}_2\text{O}} \times \frac{2 \text{ mol H}}{1 \text{ mol H}_2\text{O}} = 0.1239 \text{ mol H} \times \frac{1.008 \text{ g H}}{1 \text{ mol H}} = 0.1249 \text{ g H}$$

$$(1.152 \text{ g cmpd} - 0.7440 \text{ g C} - 0.1249 \text{ g H}) = 0.283 \text{ g O} \times \frac{1 \text{ mol O}}{16.00 \text{ g O}} = 0.0177 \text{ mol O}$$

0.06194 mol C ÷ 0.0177 = 3.499 0.1239 mol H ÷ 0.0177 = 7.00 0.0177 mol O ÷ 0.0177 = 1.00

All of these amounts in moles are multiplied by 2 to make them integral. Thus, the empirical formula of isobutyl propionate is $C_7H_{14}O_2$.

6B Notice that we do not have to obtain the mass of any element in this compound by difference; there is no oxygen present in the compound. We calculate the amount in mole of each element in the sample and transform these molar amounts to the simplest integral amounts, by first dividing all three by the smallest.

$$3.149 \text{ g CO}_2 \times \frac{1 \text{ mol CO}_2}{44.010 \text{ g CO}_2} \times \frac{1 \text{ mol C}}{1 \text{ mol CO}_2} = 0.07155 \text{ mol C} \quad \div 0.01789 = 3.999 \text{ mol C}$$

$$0.645 \text{ g H}_2\text{O} \times \frac{1 \text{ mol H}_2\text{O}}{18.015 \text{ g H}_2\text{O}} \times \frac{2 \text{ mol H}}{1 \text{ mol H}_2\text{O}} = 0.0716 \text{ mol H} \quad \div 0.01789 = 4.00 \text{ mol H}$$

$$1.146 \text{ g SO}_2 \times \frac{1 \text{ mol SO}_2}{64.065 \text{ g SO}_2} \times \frac{1 \text{ mol S}}{1 \text{ mol SO}_2} = 0.01789 \text{ mol S} \quad \div 0.01789 = 1.000 \text{ mol S}$$

Thus, the empirical formula of thiophene is C_4H_4S.

7A $\underline{S}_8$ For an atom of a free element, the oxidation state is 0 (rule 1).

$\underline{Cr}_2O_7^{2-}$ The total of all the oxidation numbers in the ion is –2 (rule 2). The O.S. of each oxygen is –2 (rule 6). Thus, the total for all seven oxygens is –14. The total for both chromiums must be +12. Thus, each Cr has an O.S. = +6.

$\underline{Cl}_2O$ The sum of all oxidation numbers in the compound is 0 (rule 2). The O.S. of oxygen is –2 (rule 6). The total for the two chlorines must be +2. Thus, each chlorine must have O.S. = +1.

$K\underline{O}_2$ The total for all the oxidation numbers in the compound is 0 (rule 2). The O.S. of potassium is +1 (rule 3). The sum of the oxidation numbers of the two oxygens must be –1. Thus, each oxygen must have O.S. = $-^1/_2$.

7B $\underline{S}_2O_3^{2-}$ The total of all the oxidation numbers in the ion is –2 (rule 2). The O.S. of oxygen is –2 (rule 6). Thus, the total for three oxygens must be –6. The total for both sulfurs must be +4. Thus, each S has an O.S. = +2.

$\underline{Hg}_2Cl_2$ The O.S. of each Cl is –1 (rule 7). The total of all O.S. is 0 (rule 2). Thus the total for two Hg is +2 and each Hg has O.S. = +1.

$K\underline{Mn}O_4$ The O.S. of each O is –2 (rule 6). Thus, the total for 4 oxygens must be –8. The K has O.S. = +1 (rule 3). The total of all O.S. is 0 (rule 2). Thus, the O.S. of Mn is +7.

$H_2\underline{C}O$ The O.S. of each H is +1 (rule 5), producing a total for both hydrogens of +2. The O.S. of O is –2 (rule 6). Thus, the O.S. of C is 0, because total of all O.S. is 0 (rule 2).

8A In each case, we determine the formula *with its accompanying charge* of each ion in the compound. We then produce a formula for the compound in which the total positive charge equals the total negative charge.

lithium oxide	Li^+ and O^{2-}	*two* Li^+ and *one* O^{2-}	Li_2O
tin(II) fluoride	Sn^{2+} and F^-	*one* Sn^{2+} and *two* F^-	SnF_2
lithium nitride	Li^+ and N^{3-}	*three* Li^+ and *one* N^{3-}	Li_3N

8B In each case, we determine the formula *with its accompanying charge* of each ion in the compound. We then produce a formula for the compound in which the total positive charge equals the total negative charge.

aluminum sulfide	Al^{3+} and S^{2-}	*two* Al^{3+} and *two* S^{2-}	Al_2S_3
magnesium nitride	Mg^{2+} and N^{3-}	*three* Mg^{2+} and *two* N^{3-}	Mg_3N_2
vanadium(III) oxide	V^{3+} and O^{2-}	*two* V^{3+} and *three* O^{2-}	V_2O_3

9A The name of each of these ionic compounds is the name of the cation followed by that of the anion. Each anion name is a modified (with the ending "ide") version of the name of the element. Each cation name is the name of the metal, with the oxidation state appended in Roman numerals in parentheses if there is more than one type of cation for that metal.

CsI cesium iodide CaF_2 calcium fluoride

FeO The O.S. of O = –2 (rule 6). Thus, the O.S. of Fe = +2 (rule 2). The cation is iron(II). The name of the compound is iron(II) oxide.

$CrCl_3$ The O.S. of Cl = –1 (rule 7). Thus, the O.S. of Cr = +3 (rule 2). The cation is chromium(III). The name of the compound is chromium(III) chloride.

9B The name of each of these ionic compounds is the name of the cation followed by that of the anion. Each anion name is a modified (with the ending "ide") version of the name of the element. Each cation name is the name of the metal, with the oxidation state appended in Roman numerals in parentheses if there is more than one type of cation for that metal.

CaH_2 calcium hydride Ag_2S silver sulfide

In the next two compounds, the oxidation state of chlorine is –1 (rule 7) and thus the oxidation state of the metal in each cation must be +1 (rule 2).

$CuCl$ copper(I) chloride Hg_2Cl_2 mercury(I) chloride

10A SF_6 Both S and F are nonmetals. This is a binary molecular compound: sulfur hexafluoride.

HNO_2 The NO_2^- ion is the nitrite ion. Its acid is nitrous acid.

$Ca(HCO_3)_2$ HCO_3^- is the bicarbonate ion or the hydrogen carbonate ion. This compound is calcium bicarbonate or calcium hydrogen carbonate.

$FeSO_4$ The SO_4^{2-} ion is the sulfate ion. The cation is Fe^{2+}, iron(II). This compound is iron(II) sulfate.

10B NH_4NO_3 The cation is NH_4^+, ammonium ion. The anion is NO_3^-, nitrate ion. This compound is ammonium nitrate.

PCl_3 Both P and Cl are nonmetals. This is a binary molecular compound: phosphorus trichloride.

$AgClO_4$ The anion is perchlorate ion, ClO_4^-. The compound is silver perchlorate.

$Fe_2(SO_4)_3$ The SO_4^{2-} ion is the sulfate ion. The cation is Fe^{3+}, iron(III). This compound is iron(III) sulfate.

11A boron trifluoride Both elements are nonmetals. This is a binary molecular compound: BF_3

potassium dichromate The potassium ion is K^+, while the dichromate ion is $Cr_2O_7^{2-}$. This is $K_2Cr_2O_7$.

sulfuric acid The anion is the sulfate anion, SO_4^{2-}. There must be two H^+'s. This is H_2SO_4.

calcium chloride The ions are Ca^{2+} and Cl^-. There must be *one* Ca^{2+} and *two* Cl^-'s: $CaCl_2$.

11B aluminum nitrate The aluminum ion is Al^{3+}; the nitrate ion is NO_3^-. This is $Al(NO_3)_3$.

tetraphosphorus decoxide Both elements are nonmetals. This is a binary molecular compound: P_4O_{10}

chromium(III) hydroxide The chromium(III) ion is Cr^{3+}; the hydroxide ion is OH^-. This is $Cr(OH)_3$.

iodic acid The "ic" acid of the halogens has halogen in a +5 oxidation state. This is HIO_3.

SUMMARIZING EXAMPLE CALCULATIONS

1. mass of water in the original sample = mass of original sample - mass of heated sample = 2.574 g hydrate – 1.647 g $CuSO_4$ = 0.927 g H_2O

2. amount of H_2O $= 0.927\text{ g H}_2\text{O} \times \frac{1\text{ mol H}_2\text{O}}{18.02\text{ g H}_2\text{O}} = 0.0514\text{ mol H}_2\text{O}$

amount of $CuSO_4$ $= 1.647\text{ g CuSO}_4 \times \frac{1\text{ mol CuSO}_4}{159.61\text{ g}} = 0.01032\text{ mol CuSO}_4$

3. $\frac{\text{amount of H}_2\text{O}}{\text{amount of CuSO}_4} = \frac{0.0514\text{ mol H}_2\text{O}}{0.01032\text{ mol CuSO}_4} = \frac{4.98\text{ mol H}_2\text{O}}{1\text{ mol CuSO}_4}$

The formula of the compound is $CuSO_4 \cdot 5H_2O$

REVIEW QUESTIONS

1. (a) The formula unit of a compound is a group of atoms that has atoms of the same type and number as they are present in the formula of that compound. For example, if "Na_2" appears in the formula of the compound, then there will be two sodium atoms in the formula unit of that compound.
   (b) S_8 is a molecule of elemental sulfur. In the same manner as several other elements (including H_2, F_2, Cl_2, Br_2, I_2, N_2, O_2, and P_4), elemental sulfur exists as molecules rather than as isolated atoms.
   (c) An ionic compound is one that is composed of (positively charged) cations and (negatively charged) anions. Most binary ionic compounds are composed of a metal (which becomes the cation) and a nonmetal (which becomes the anion).

(d) An oxoacid is an acid that contains the element oxygen, in addition to some other element and the element hydrogen. H_2SO_4, HNO_3, and $HClO_4$ all are oxoacids, in which the "other element" is S, N, and Cl, respectively.

(e) A hydrate is a compound that contains water, rather loosely bound. Usually this water of hydration can be driven off by mild heating of the compound.

2. **(a)** A molecule of an element refers to the smallest independent particle of that element. Usually this is an atom, but in some cases (notably H_2, F_2, Cl_2, Br_2, I_2, N_2, O_2, S_8, and P_4) it is a grouping of two or more atoms.

(b) The structural formula of a compound not only indicates which atoms are present in the formula unit, but also how they are joined together; it represents the bonding in the molecule, for instance.

(c) The oxidation state of an element in a compound is an indication of how many electrons each atom of that element has lost (positive oxidation state) or gained (negative). Since oxidation state is determined by a set of rules, rather than by experiment, its connection to the number of electrons actually transferred is rather tenuous. Still, the concept is useful in naming compounds and balancing some types of chemical equations.

(d) The determination of the carbon–hydrogen–oxygen content of a compound by combustion analysis involves realizing that all of the carbon has formed carbon dioxide, all of the hydrogen has formed water, and the amount of oxygen present in the original compound must be determined by difference.

3. **(a)** A chemical symbol is the symbol for one element, while a chemical formula is the "word" made up by several symbols (and numbers) and indicates the composition of a compound.

(b) An empirical formula indicates the simplest grouping of atoms that has the same ratio of elements as are present in the compound. The molecular formula indicates the actual number of atoms of each type present in the molecule. The molecular formula is an integral multiple of the empirical formula.

(c) The systematic name for a compound is based on the elements present in it and gives an indication of its composition. The trivial or common name is simply a label for the substance.

(d) A binary acid consists of hydrogen and one other element. A ternary acid consists of hydrogen, the other element, and the element oxygen: three elements in all.

4. **(a)** The atomic mass of oxygen is the mass of one (average) atom, 15.9994 u.

(b) The molecular mass of oxygen is the mass of one (average) molecule of O_2, 31.9988 u.

(c) The molar mass of molecular oxygen is the mass of one mole of oxygen molecules, 31.9988 g. That of atomic oxygen is the mass of one mole of oxygen atoms, 15.9994 g.

5. **(a)** A nitroglycerine molecule, $C_3H_5(NO_3)_3$, contains 3 C atoms, 5 H atoms, 3 N atoms, and 3×3 = 9 O atoms, for a total of (3 + 5 + 3 + 9 =) 20 atoms.

(b) Each molecule of C_2H_6 contains 6 H atoms and 2 C atoms, 8 atoms total.

$$\text{number of atoms} = 0.00102 \text{ mol } C_2H_6 \times \frac{6.022 \times 10^{23} \text{ } C_2H_6 \text{ molecules}}{1 \text{ mol } C_2H_6} \times \frac{8 \text{ atoms}}{1 \text{ } C_2H_6 \text{ molecule}}$$
$$= 4.91 \times 10^{21} \text{ atoms}$$

(c) $$\text{number of F atoms} = 12.15 \text{ mol } C_2HBrClF_3 \times \frac{3 \text{ mol F}}{1 \text{ mol } C_2HBrClF_3} \times \frac{6.022 \times 10^{23} \text{ F atoms}}{1\text{mol F atoms}}$$
$$= 2.195 \times 10^{25} \text{ F atoms}$$

6. **(a)** To convert amount in moles to mass, we need the molar mass of N_2O_4.

$$\text{molar mass } N_2O_4 = \left(2 \text{ mol N} \times \frac{14.01 \text{ g N}}{1 \text{ mol N}}\right) + \left(4 \text{ mol O} \times \frac{16.00 \text{ g O}}{1 \text{ mol O}}\right) = 92.02 \text{ g/mol } N_2O_4$$

$$\text{mass } N_2O_4 = 7.34 \text{ mol } N_2O_4 \times \frac{92.02 \text{ g } N_2O_4}{1 \text{ mol}} = 675 \text{ g } N_2O_4$$

(b) $$\text{mass of } O_2 = 3.16 \times 10^{24} \text{ } O_2 \text{ molecules} \times \frac{1 \text{ mol } O_2}{6.022 \times 10^{23} \text{ molecules}} \times \frac{32.00 \text{ g } O_2}{1 \text{ mol } O_2} = 168 \text{ g } O_2$$

(c) $$\text{molar mass } CuSO_4 \cdot 5H_2O = 63.5 \text{ g Cu} + 32.1 \text{ g S} + (9 \times 16.0 \text{ g O}) + (10 \times 1.01 \text{ g H})$$
$$= 249.7 \text{ g/mol } CuSO_4 \cdot 5H_2O$$

$$\text{mass of } CuSO_4 \cdot 5H_2O = 18.6 \text{ mol} \times \frac{249.7 \text{ g } CuSO_4 \cdot 5H_2O}{1 \text{ mol}} = 4.64 \times 10^3 \text{ g } CuSO_4 \cdot 5H_2O$$

(d) $$\text{molar mass } C_2H_4(OH)_2 = (2 \times 12.01 \text{ g C}) + (6 \times 1.01 \text{ g H}) + (2 \times 16.00 \text{ g O}) = 62.08 \text{ g/mol}$$

$$\text{mass of } C_2H_4(OH)_2 = 4.18 \times 10^{24} \text{ molecules} \times \frac{1 \text{ mol}}{6.022 \times 10^{23} \text{ molec.}} \times \frac{62.08 \text{ g}}{1 \text{ mol } C_2H_4(OH)_2}$$
$$= 431 \text{ g } C_2H_4(OH)_2$$

7. (a) $\text{amount of } Br_2 = 8.08 \times 10^{22} \text{ } Br_2 \text{ molecules} \times \frac{1 \text{ mole } Br_2}{6.022 \times 10^{23} \text{ } Br_2 \text{ molecules}} = 0.134 \text{ mol } Br_2$

(b) $\text{amount of } Br_2 = 2.17 \times 10^{24} \text{ Br atoms} \times \frac{1 \text{ } Br_2 \text{ molecule}}{2 \text{ Br atoms}} \times \frac{1 \text{ mole } Br_2}{6.022 \times 10^{23} \text{ } Br_2 \text{ molecules}}$

$= 1.80 \text{ mol } Br_2$

(c) $\text{amount of } Br_2 = 11.3 \text{ kg } Br_2 \times \frac{1000 \text{ g}}{1 \text{ kg}} \times \frac{1 \text{ mol } Br_2}{159.8 \text{ g } Br_2} = 70.7 \text{ mol } Br_2$

(d) $\text{amount of } Br_2 = 2.65 \text{ L } Br_2 \times \frac{1000 \text{ mL}}{1 \text{ L}} \times \frac{3.10 \text{ g } Br_2}{1 \text{ mL } Br_2} \times \frac{1 \text{ mol } Br_2}{159.8 \text{ g } Br_2} = 51.4 \text{ mol } Br_2$

8. (a) $\text{molec.ms. } C_5H_{11}NO_2S = (5 \times 12.0 \text{ u C}) + (11 \times 1.01 \text{ u H}) + 14.0 \text{ u N} + (2 \times 16.0 \text{ u O}) + 32.1 \text{ u S}$
$= 149.2 \text{ u}/C_5H_{11}NO_2S \text{ molecule}$

(b) Since there are 11 H atoms in each $C_5H_{11}NO_2S$ molecule, there are 11 moles of H atoms in each mole of $C_5H_{11}NO_2S$ molecules

(c) $\text{mass C} = 1 \text{ mol } C_5H_{11}NO_2S \times \frac{5 \text{ mol C}}{1 \text{ mol } C_5H_{11}NO_2S} \times \frac{12.011 \text{ g C}}{1 \text{ mol C}} = 60.055 \text{ g C}$

(d) $\text{no. C atoms} = 9.07 \text{ mol } C_5H_{11}NO_2S \times \frac{5 \text{ mol C}}{1 \text{ mol } C_5H_{11}NO_2S} \times \frac{6.022 \times 10^{23} \text{ atoms}}{1 \text{ mol C}}$

$= 2.73 \times 10^{25} \text{ C atoms}$

9. The information obtained in the course of calculating the molar mass is used to determine the mass percent of H in decane.

$$\text{molar mass } C_{10}H_{22} = \left(\frac{10 \text{ mol C}}{1 \text{ mol } C_{10}H_{22}} \times \frac{12.011 \text{ g C}}{1 \text{ mol C}}\right) + \left(\frac{22 \text{ mol H}}{1 \text{ mol } C_{10}H_{22}} \times \frac{1.00794 \text{ g H}}{1 \text{ mol H}}\right)$$
$$= \frac{120.11 \text{ g C}}{1 \text{ mol } C_{10}H_{22}} + \frac{22.1747 \text{ g H}}{1 \text{ mol } C_{10}H_{22}} = \frac{142.28 \text{ g}}{1 \text{ mol } C_{10}H_{22}}$$
$$\%H = \frac{22.1747 \text{ g H/mol octane}}{142.28 \text{ g } C_{10}H_{22}\text{/mol octane}} \times 100\% = 15.585\% \text{ H} = 15.59\% \text{ H (to two decimal places)}$$

10. We first determine the mass of O in each mol of $Cu_2(OH)_2CO_3$ and the molar mass of the compound.

$$\text{mass O/mol } Cu_2(OH)_2CO_3 = \frac{5 \text{ mol O}}{1 \text{ mol } Cu_2(OH)_2CO_3} \times \frac{16.00 \text{ g O}}{1 \text{ mol O}} = 80.00 \text{ g O/mol } Cu_2(OH)_2CO_3$$
$$\text{molar mass } Cu_2(OH)_2CO_3 = (2 \times 63.55 \text{ g Cu}) + (5 \times 16.00 \text{ g O}) + (2 \times 1.01 \text{ g H}) + 12.01 \text{ g C}$$
$$= 221.13 \text{ g/mol } Cu_2(OH)_2CO_3$$
$$\%\text{ O} = \frac{80.00\text{g O/mol malachite}}{221.13 \text{ g } Cu_2(OH)_2CO_3\text{/mol malachite}} \times 100\% = 36.18\% \text{ O}$$

11. Determine the molar mass of $Cr(NO_3)_3{\cdot}9H_2O$, and then the mass of water per mole of $Cr(NO_3)_3{\cdot}9H_2O$.

$$\text{molar mass } Cr(NO_3)_3{\cdot}9H_2O = 52.00 \text{ g Cr} + (3 \times 14.01 \text{ g N}) + (18 \times 16.00 \text{ g O}) + (18 \times 1.01 \text{ g H})$$
$$= 400.2 \text{ g/mol } Cr(NO_3)_3{\cdot}9H_2O$$
$$\text{mass } H_2O = \frac{9 \text{ mol } H_2O}{1 \text{ mol } Cr(NO_3)_3{\cdot}9H_2O} \times \frac{18.02 \text{ g } H_2O}{1 \text{ mol } H_2O} = 162.2 \text{ g } H_2O\text{/mol } Cr(NO_3)_3{\cdot}9H_2O$$
$$\frac{162.2 \text{ g } H_2O\text{/mol } Cr(NO_3)_3{\cdot}9H_2O}{400.2 \text{ g/mol } Cr(NO_3)_3{\cdot}9H_2O} \times 100\% = 40.53\% \text{ } H_2O$$

12. In each case, we first determine the molar mass of the compound, and then the mass of the indicated element in one mole of the compound. Finally, we determine the percent by mass of the indicated element to four significant figures.

(a) $\text{molar mass } Pb(C_2H_5)_4 = 207.2 \text{ g Pb} + (8 \times 12.01 \text{ g C}) + (20 \times 1.008 \text{ g H}) = 323.4 \text{ g/mol } Pb(C_2H_5)_4$

$$\text{mass Pb/mol } Pb(C_2H_5)_4 = \frac{1 \text{ mol Pb}}{1 \text{ mol } Pb(C_2H_5)_4} \times \frac{207.2 \text{ g Pb}}{1 \text{ mol Pb}} = 207.2 \text{ g Pb/mol } Pb(C_2H_5)_4$$
$$\%\text{ Pb} = \frac{207.2 \text{ g Pb}}{323.4 \text{ g } Pb(C_2H_5)_4} \times 100\% = 64.07\% \text{ Pb}$$

(b) $\text{molar mass } Fe_4[Fe(CN)_6]_3 = (7 \times 55.85 \text{ g Fe}) + (18 \times 12.01 \text{ g C}) + (18 \times 14.01 \text{ g N})$

$= 859.3 \text{ g/mol } Fe_4[Fe(CN)_6]_3$

$$\text{mass Fe/mol } Fe_4[Fe(CN)_6]_3 = \frac{7 \text{ mol Fe}}{1 \text{ mol } Fe_4[Fe(CN)_6]_3} \times \frac{55.85 \text{ g}}{1 \text{ mol Fe}} = 391.0 \text{ g Fe/mol } Fe_4[Fe(CN)_6]_3$$

$$\% \text{ Fe} = \frac{391.0 \text{ g Fe}}{859.3 \text{ g } Fe_4[Fe(CN)_6]_3} \times 100\% = 45.50\% \text{ Fe}$$

13. For SO_2 and Na_2S, a mole of each contains a mole of S and two moles of another element; in the case of SO_2 the other element has a smaller mole weight, causing SO_2 to have a higher percent sulfur. For S_2Cl_2 and $Na_2S_2O_3$, a mole of each contains two moles of S; for S_2Cl_2 the rest of the mole weighs 71.0 g, while for Na_2SO_3 it weighs $(2 \times 23) + (3 \times 16) = 94$ g. The S makes up the greater proportion of the mass in S_2Cl_2, giving it the larger percent of S. Now we compare SO_2 and S_2Cl_2; S_2O_4 has the same mass proportions as does SO_2 but also has the two moles S, as does S_2Cl_2. In S_2Cl_2 the remainder of a mole weighs 71.0 g, while in S_2O_4 the remainder of a mole weighs $4 \times 16.0 = 64.0$ g; SO_2 has the highest percent of S of the four compounds listed.

14. Express each percent as a mass in grams, convert each mass to moles of that element, then determine the simplest ratio of numbers of moles by dividing these mole numbers by the smallest.

$$\text{no. mol Cr} = 68.42 \text{ g Cr} \times \frac{1 \text{ mol Cr}}{52.00 \text{ g Cr}} = 1.316 \text{ mol Cr}$$

$$\text{no. mol O} = 31.58 \text{ g O} \times \frac{1 \text{ mol O}}{16.00 \text{ g O}} = 1.974 \text{ mol O}$$

Divide each number by 1.316 to obtain 1.000 mol Cr and 1.500 mol O. Empirical formula is Cr_2O_3.

15. Determine the percent oxygen by difference first. % O = 100.00% – 45.27% C – 9.50% H = 45.23% O

$$\text{no. mol O} = 45.23 \text{ g} \times \frac{1 \text{ mol O}}{16.00 \text{ g O}} = 2.827 \text{ mol O} \qquad \div 2.827 \longrightarrow 1.000 \text{ mol O}$$

$$\text{no mol C} = 45.27 \text{ g C} \times \frac{1 \text{ mol C}}{12.01 \text{ g C}} = 3.769 \text{ mol C} \qquad \div 2.827 \longrightarrow 1.333 \text{ mol C}$$

$$\text{no. mol H} = 9.50 \text{ g H} \times \frac{1 \text{ mol H}}{1.008 \text{ g H}} = 9.42 \text{ mol H} \qquad \div 2.827 \longrightarrow 3.33 \text{ mol H}$$

Multiply all amounts by 3 to obtain integers. Empirical formula is $C_4H_{10}O_3$.

16. We base our calculation on 100.0 g of monosodium glutamate.

$$13.6 \text{ g Na} \times \frac{1 \text{ mol Na}}{22.99 \text{ g Na}} = 0.592 \text{ mol Na} \qquad \div 0.592 \longrightarrow 1.00 \text{ mol Na}$$

$$35.5 \text{ g C} \times \frac{1 \text{ mol Cl}}{12.01 \text{ g C}} = 2.96 \text{ mol C} \qquad \div 0.592 \longrightarrow 5.00 \text{ mol C}$$

$$4.8 \text{ g H} \times \frac{1 \text{ mol H}}{1.01 \text{ g H}} = 4.8 \text{ mol H} \qquad \div 0.592 \longrightarrow 8.1 \text{ mol H}$$

$$8.3 \text{ g N} \times \frac{1 \text{ mol N}}{14.01 \text{ g N}} = 0.59 \text{ mol N} \qquad \div 0.592 \longrightarrow 1.0 \text{ mol N}$$

$$37.8 \text{ g O} \times \frac{1 \text{ mol O}}{16.00 \text{ g O}} = 2.36 \text{ mol O} \qquad \div 0.592 \longrightarrow 3.99 \text{ mol O}$$

Empirical formula: $NaC_5H_8NO_4$

17. First determine the empirical formula. Begin by determining the percent oxygen by difference.
% O = 100% – 57.83% C – 3.64% H = 38.53% O

$$\text{no. mol O} = 38.53 \text{ g} \times \frac{1 \text{ mol O}}{16.00 \text{ g O}} = 2.408 \text{ mol O} \qquad \div 2.408 \longrightarrow 1.000 \text{ mol O}$$

$$\text{no. mol C} = 57.83 \text{ g C} \times \frac{1 \text{ mol C}}{12.01 \text{ g C}} = 4.815 \text{ mol C} \qquad \div 2.408 \longrightarrow 2.000 \text{ mol C}$$

$$\text{no. mol H} = 3.64 \text{ g H} \times \frac{1 \text{ mol H}}{1.008 \text{ g H}} = 3.61 \text{ mol H} \qquad \div 2.408 \longrightarrow 1.50 \text{ mol H}$$

Empirical formula is $C_4H_3O_2$. The mass of an empirical formula unit is
empirical molecular mass = (4 × 12.0 u C) + (3 × 1.0 u H) + (2 × 16.0 u O) = 83.0 u
This empirical molecular mass is one-half of the measured molecular mass. Thus, the molecular formula of terephthalic acid is twice the empirical formula. Molecular formula: $C_8H_6O_4$

18. **(a)** First determine the masses of carbon and hydrogen in the original sample.

$$\text{mass C} = 6.029 \text{ g } CO_2 \times \frac{1 \text{ mol } CO_2}{44.010 \text{ g}} \times \frac{1 \text{ mol C}}{1 \text{ mol } CO_2} = 0.1370 \text{ mol C} \times \frac{12.011 \text{ g C}}{1 \text{ mol C}} = 1.646 \text{ g C}$$

$$\text{mass H} = 1.709 \text{ g } H_2O \times \frac{1 \text{ mol } H_2O}{18.02 \text{ g } H_2O} \times \frac{2 \text{ mol H}}{1 \text{ mol } H_2O} = 0.1897 \text{ mol H} \times \frac{1.008 \text{ g H}}{1 \text{ mol H}} = 0.1912 \text{ g H}$$

Then the percents of the two elements in the compound are computed.

$\% \text{ C} = \frac{1.646 \text{ g C}}{2.174 \text{ g cmpd}} \times 100\% = 75.71\% \text{ C}$ $\qquad \% \text{ H} = \frac{0.1912 \text{ g H}}{2.174 \text{ g cmpd}} \times 100\% = 8.795\% \text{ H}$

The % O is determined by difference. % O = 100% – 75.71% C – 8.795% H = 15.50% O

(b) In part (a), we determined the number of moles of C and H in the original sample of the compound. We can determine the mass of oxygen in that sample by difference, and then the number of moles of oxygen in that sample. We divide each of these numbers of moles by the smallest number to determine the empirical formula.

mass O = 2.174 g cmpd – 1.646 g C – 0.1912 g H = 0.337 g O

$\text{mol O} = 0.337 \text{ g O} \times \frac{1 \text{ mol O}}{16.00 \text{ g O}} = 0.0211 \text{ mol O} \qquad \div 0.0211 \longrightarrow 1.00 \text{ mol O}$

$0.1897 \text{ mol H} \qquad \div 0.0211 \longrightarrow 8.99 \text{ mol H}$

$0.1370 \text{ mol C} \qquad \div 0.0211 \longrightarrow 6.49 \text{ mol C}$

Multiply all amounts by 2 to obtain integers; the empirical formula of ibuprofen is $C_{13}H_{18}O_2$.

19. The element chromium has an atomic weight of 52.0 u. Thus, there can only be one chromium atom per formula unit of the compound. (Two atoms of chromium weigh 104 u, more than the formula weight of the compound.) The remaining three of the four atoms in the formula unit must be oxygen. Thus, the oxide is CrO_3, chromium(VI) oxide.

20.

(a)	Pb^{2+}	lead(II)	(b)	Co^{3+}	cobalt(III)	(c) Ba^{2+}	barium
(d)	Cr^{2+}	chromium(II)	(e)	IO_4^-	periodate	(f) ClO_2^-	chlorite
(g)	Au^{3+}	gold(III)	(h)	HSO_3^-	bisulfite *or* hydrogen sulfite		
(i)	HCO_3^-	hydrogen carbonate	(j)	CN^-	cyanide		

21.

(a)	KBr	potassium bromide	(b)	$SrCl_2$	strontium chloride
(c)	ClF_3	chlorine trifluoride	(d)	N_2O_4	dinitrogen tetroxide
(e)	PCl_5	phosphorus pentachloride			

22.

(a)	KCN	potassium cyanide	(b)	HClO	hypochlorous acid
(c)	$(NH_4)_2SO_4$	ammonium sulfate	(d)	KIO_3	potassium iodate

23. The desired oxidation state is given first, followed by the method of assigning that oxidation state.

(a)	Zn = 0	The oxidation state (O. S.) of an uncombined element is 0.
(b)	S = –2 in BaS	The O. S. of Ba in its compounds is +2. Oxidation states in a compound must sum to zero.
(c)	N = +4 in NO_2	The O. S. of O in its compounds is –2.
(d)	N = +3 in HNO_2	The O. S. of H in its compounds is +1; that of O is –2.
(e)	V = +4 in VO^{2+}	The O. S. of O in its compounds is –2. O. S. = ionic charge.
(f)	P = +5 in $H_2PO_4^-$	The O. S. of H in its compounds is +1; that of O is –2. O.S. in a polyatomic ion must sum to the charge on that ion.

24.

(a)	$MgBr_2$	magnesium bromide	(b)	BaO	barium oxide
(c)	$Hg(C_2H_3O_2)_2$	mercury(II) acetate	(d)	$Fe_2(C_2O_4)_3$	iron(III) oxalate
(e)	$Sr(ClO_4)_2$	strontium perchlorate	(f)	$KHSO_4$	potassium hydrogen sulfate
(g)	NCl_3	nitrogen trichloride	(h)	BrF_5	bromine pentafluoride

25.

(a)	$HClO_2$	chlorous acid	(b)	H_2SO_3	sulfurous acid
(c)	H_2Se	hydroselenic acid	(d)	HNO_2	nitrous acid

26.

(a)	HI	hydroiodic acid	(b)	HNO_3	nitric acid
(c)	H_3PO_4	phosphoric acid	(d)	H_2SO_4	sulfuric acid

EXERCISES

The Avogadro Constant and the Mole

27. The greatest number of S atoms is contained in the compound with the greatest number of moles of S.

The solid sulfur contains 8 × 0.12 = 0.96 mol S atoms.

There are 0.50 × 2 mol S atoms in 0.50 mol S_2O.

There is a bit more than 1 mole (64.1 g) of SO_2 in 65 g, and thus a bit more than 1 mole of S atoms.

The molar mass of thiophene is (4 mol C × 12.0 g C) + (4 g H × 1.0 g H) + (1 mol S × 32.1 g S) = 84.1 g; 75 mL weighs 79.5 g and thus contains less than 1 mole of S.

So 65 g SO_2 has the greatest number of S atoms.

28. The greatest number of N atoms is contained in the compound with the greatest number of moles of N.

The molar mass of N_2O is (2 mol N × 14.0 g N) + (1 mol O × 16.0 g O) = 44.0 g/mol N_2O. Thus, 50.0 g N_2O is slightly more than 1 mole of N_2O, and contains slightly more than 2 moles of N.

Each mole of N_2 contains 2 moles of N.

The molar mass of NH_3 is 17.0 g. Thus, there is 1 mole of NH_3 present, which contains 1 mole of N.

The molar mass of pyridine is (5 mol C × 12.0 g C) + (5 mol H × 1.01 g H) + 14.0 g N = 79.1 g/mol. Because each mole of pyridine contains 1 mole of N, we need slightly more than 2 moles of pyridine to have more N than in the N_2O. But that would be a mass of about 158 g pyridine, and 150 mL has a mass of less than 150 g.

Thus, the greatest number of N atoms is present in 50.0 g N_2O.

29. **(a)** $\text{P mass} = 6.25 \times 10^{-2}\ \text{mol P}_4 \times \frac{4\ \text{mol P}}{1\ \text{mol P}_4} \times \frac{30.97\ \text{g P}}{1\ \text{mol P}} = 7.74\ \text{g P}$

(b) First we need the molar mass of $C_{18}H_{36}O_2$, stearic acid:

molar mass = (18 mol C × 12.01 g C) + (36 mol H × 1.01 g H) + (2 mol O × 16.00 g O) = 284.5 g/mol

$$\text{Stearic acid mass} = 4.03 \times 10^{24}\ \text{molecules} \times \frac{1\ \text{mole}}{6.022 \times 10^{23}\ \text{molecules}} \times \frac{284.5\ \text{g}}{1\ \text{mole C}_{18}\text{H}_{36}\text{O}_2}$$

$= 1.90 \times 10^3$ g stearic acid.

(c) molar mass = (6 mol C × 12.01 g C) + (14 mol H × 1.01 g H) + (2 mol N × 14.00 g N) + (2 mol O × 16.00 g O) = 146.2 g/mol

$$\text{lysine mass} = 1.15\ \text{mol N} \times \frac{1\ \text{mol C}_6\text{H}_{14}\text{N}_2\text{O}_2}{2\ \text{mol N}} \times \frac{146.2\ \text{g lysine}}{1\ \text{mol lysine}} = 84.1\ \text{g lysine}$$

30. **(a)** molar mass = (2 mol N × 14.01 g N) + (4 mol O × 16.00 g O) = 92.02 g/mol

$$\text{amount N}_2\text{O}_4 = 82.5\ \text{g N}_2\text{O}_4 \times \frac{1\ \text{mol N}_2\text{O}_4}{92.02\ \text{g N}_2\text{O}_4} = 0.897\ \text{mol N}_2\text{O}_4$$

(b) molar mass = (1 mol Mg × 24.31 g Mg) + (2 mol N × 14.01 g N) + (6 mol O × 16.00 g O) = 148.33 g/mol

$$\text{amount N} = 106\ \text{g Mg(NO}_3)_2 \times \frac{1\ \text{mol Mg(NO}_3)_2}{148.33\ \text{g}} \times \frac{2\ \text{mol N}}{1\ \text{mol Mg(NO}_3)_2} = 1.43\ \text{mol N atoms}$$

(c) molar mass = (6 mol C × 12.01 g C) + (12 g H × 1.008 g H) + (6 mol O × 16.00 g O) = 180.16 g/mol $C_6H_{12}O_6$

$$\text{amt N} = \frac{56.5\ \text{g}}{\text{C}_6\text{H}_{12}\text{O}_6} \times \frac{1\ \text{mol C}_6\text{H}_{12}\text{O}_6}{180.16\ \text{g}} \times \frac{6\ \text{mol O}}{1\ \text{mol C}_6\text{H}_{12}\text{O}_6} \times \frac{1\ \text{mol C}_7\text{H}_5(\text{NO}_2)_3}{6\ \text{mol O}} \times \frac{2\ \text{mol N}}{1\ \text{mol C}_7\text{H}_5(\text{NO}_2)_3}$$

= 0.627 mol N

31. **(a)** $\text{amount S}_8 = 0.568\ \text{mm}^3 \times \frac{1\ \text{cm}^3}{1000\ \text{mm}^3} \times \frac{2.07\ \text{g}}{1\ \text{cm}^3} \times \frac{1\ \text{mol S}}{32.07\ \text{g}} \times \frac{1\ \text{mol S}_8}{8\ \text{mol S}} = 4.58 \times 10^{-6}\ \text{mol S}_8$

(b) $\text{no. S atoms} = 4.58 \times 10^{-6}\ \text{mol S}_8 \times \frac{8\ \text{mol S}}{1\ \text{mol S}_8} \times \frac{6.022 \times 10^{23}\ \text{atoms}}{1\ \text{mol S}} = 2.21 \times 10^{19}\ \text{S atoms}$

32. $$\text{no. Fe atoms} = 6\ \text{L blood} \times \frac{1000\ \text{mL}}{1\ \text{L}} \times \frac{15.5\ \text{g Hb}}{100\ \text{mL blood}} \times \frac{1\ \text{mol Hb}}{64{,}500\ \text{g}} \times \frac{4\ \text{mol Fe}}{1\ \text{mol Hb}} \times \frac{6.022 \times 10^{23}\ \text{atoms}}{1\ \text{mol Fe}}$$

$= 3 \times 10^{22}$ Fe atoms

Chemical Formulas

33. For glucose (blood sugar), $C_6H_{12}O_6$,

(1) FALSE The percentages by mass of C and O are *different* than in CO. For one thing, CO contains no hydrogen.

(2) TRUE In dihydroxyacetone, $(CH_2OH)_2CO$ or $C_3H_6O_3$, the ratio of C : H : O = 3 : 6 : 3 = 1 : 2 : 1. In glucose, this ratio is C : H : O = 6 : 12 : 6 = 1 : 2 : 1. Thus, the ratios are the *same*.

(3) FALSE The proportions, by number of atoms, of C and O are the same in glucose. Since, however, C and O have different mole weights, their proportions by mass must be *different*.

(4) FALSE Each mole of glucose contains $(12 \times 1.01 =)$ 12.1 g H. But each mole also contains 72.0 g C and 96.0 g O. Thus, the highest percentage, by mass, is that of O. The highest percentage, by number of atoms, is that of H.

34. For sorbic acid, $C_6H_8O_2$,

(1) FALSE The C:H:O mole ratio is 3:4:1, but the mass ratio differs because moles of different elements have different masses.

(2) TRUE Since the two compounds have the same empirical formula, they have the same mass percent composition.

(3) TRUE Aspidinol, $C_{12}H_{16}O_4$, and sorbic acid have the same empirical formula, C_3H_4O

(4) TRUE The ratio of H atoms to O atoms is 8:2 = 4:1.

Thus, the mass ratio is $(4 \text{ mol H} \times 1 \text{ g H}):(1 \text{ mol O} \times 16.0 \text{ g O}) = 4 \text{ g H}:16 \text{ g O} = 1 \text{ g H}: 4 \text{ g O}$.

35. (a) A formula unit of $C_2HBrClF_3$ contains:

2 C atoms 1 H atom 1 Br atom 1 Cl atom 3 F atoms

For a total of 2 + 1 + 1 + 1 + 3 = 8 atoms

(b) $\dfrac{\text{no. F atoms}}{\text{no. C atoms}} = \dfrac{3 \text{ F atoms}}{2 \text{ C atoms}}$

(c) $\dfrac{\text{mass Br}}{\text{mass F}} = \dfrac{1 \text{ mol Br}}{3 \text{ mol F}} \times \dfrac{79.90 \text{ g Br}}{1 \text{ mol Br}} \times \dfrac{1 \text{ mol F}}{19.00 \text{ g F}} = 1.402 \text{ g Br/g F}$

(d) The element present in greatest mass percent is the one with the greatest mass in one mole. To determine that, we determine molar mass.

$$\text{molar mass} = (2 \text{ mol C} \times 12.01 \text{ g C}) + (1 \text{ mol H} \times 1.008 \text{ g H}) + (1 \text{ mol Br} \times 79.90 \text{ g Br}) + (1 \text{ mol Cl} \times 35.45 \text{ g Cl}) + (3 \text{ mol F} \times 19.00 \text{ g F})$$
$$= 24.02 \text{ g C} + 1.008 \text{ g H} + 79.09 \text{ g Br} + 35.45 \text{ g Cl} + 57.00 \text{ g F} = 196.57 \text{ g/mol}$$

Bromine is present in greatest mass percent.

(e) $\text{compound mass} = 1.00 \text{ g F} \times \dfrac{1 \text{ mol F}}{19.00 \text{ g F}} \times \dfrac{1 \text{ mol cmpd}}{3 \text{ mol F}} \times \dfrac{196.57 \text{ g cmpd}}{1 \text{ mol cmpd}} = 3.45 \text{ g cmpd}$

36. (a) A formula unit of $Ge[S(CH_2)_4CH_3]_4$ contains:

1 Ge atom 4 S atoms 4 (4 + 1) = 20 C atoms 4[4(2) + 3] = 44 H atoms

For a total of 1 + 4 + 20 + 44 = 69 atoms per formula unit

(b) $\dfrac{\text{no. C atoms}}{\text{no. H atoms}} = \dfrac{20 \text{ C atoms}}{44 \text{ H atoms}} = \dfrac{5 \text{ C atoms}}{11 \text{ H atoms}} = 0.454 \text{ C atom/H atom}$

(c) $\dfrac{\text{mass Ge}}{\text{mass Se}} = \dfrac{1 \text{ mol Ge} \times \dfrac{72.6 \text{ g Ge}}{1 \text{ mol Ge}}}{4 \text{ mol S} \times \dfrac{32.07 \text{ g S}}{1 \text{ mol S}}} = \dfrac{72.6 \text{ g Ge}}{128.3 \text{ g S}} = 0.566 \text{ g Ge/g S}$

(d) $\text{mol wt} = 72.6 \text{ g Ge} + (4 \times 32.1 \text{ g S}) + (20 \times 12.01 \text{ g C}) + (44 \times 1.01 \text{ g H}) = 485.6 \text{ g/mol}$

$\text{mass of S} = 1 \text{ mol Ge[S(CH}_2)_4\text{CH}_3]_4 \times \dfrac{4 \text{ mol S}}{1 \text{ mol Ge[S(CH}_2)_4\text{CH}_3]_4} \times \dfrac{32.07 \text{ g S}}{1 \text{ mol S}} = 128.3 \text{ g S}$

(e) $\text{no. C atoms} = 33.10 \text{ g cmpd} \times \dfrac{1 \text{ mol cmpd}}{485.6 \text{ g cmpd}} \times \dfrac{20 \text{ mol C}}{1 \text{ mol cmpd}} \times \dfrac{6.022 \times 10^{23} \text{ C atoms}}{1 \text{ mol C}}$

$= 8.210 \times 10^{23} \text{ C atoms}$

Percent Composition of Compounds

37. The information obtained in the course of calculating the molar mass is used to determine the mass percent of each element in stearic acid, $C_{18}H_{36}O_2$, abbreviated as SA below.

molar mass $C_8H_{18} = \mathcal{M}$

$$\mathcal{M} = \left(\frac{18 \text{ mol C}}{1 \text{ mol SA}} \times \frac{12.011 \text{ g C}}{1 \text{ mol C}}\right) + \left(\frac{36 \text{ mol H}}{1 \text{ mol SA}} \times \frac{1.00794 \text{ g H}}{1 \text{ mol H}}\right) + \left(\frac{2 \text{ mol O}}{1 \text{ mol SA}} \times \frac{15.9994 \text{ g O}}{1 \text{ mol O}}\right)$$

$$\mathcal{M} = \frac{216.20 \text{ g C}}{1 \text{ mol SA}} + \frac{36.2858 \text{ g H}}{1 \text{ mol SA}} + \frac{31.9988 \text{ g O}}{1 \text{ mol SA}} = \frac{284.48 \text{ g}}{1 \text{ mol SA}}$$

$$\%C = \frac{216.20 \text{ g C/mol SA}}{284.48 \text{ g SA/mol SA}} \times 100\% = 75.998\% \text{ C}; \ \%H = \frac{36.2858 \text{ g H/mol SA}}{284.48 \text{ g SA/mol SA}} \times 100\% = 12.755\% \text{ H}$$

$$\%O = \frac{31.9988 \text{ g O/mol SA}}{284.48 \text{ g SA/mol SA}} \times 100\% = 11.248\% \text{ O}$$

38. The method of solution of Exercise 37 is abbreviated in the following problem solution.

$$\text{molar mass} = (20 \text{ mol C} \times 12.011 \text{ g C}) + (24 \text{ mol H} \times 1.00794 \text{ g H}) + (2 \text{ mol N} \times 14.0067 \text{ g N}) + (2 \text{ mol O} \times 15.9994 \text{ g O})$$

$$= 240.22 \text{ g C} + 24.1906 \text{ g H} + 28.0134 \text{ g N} + 31.9988 \text{ g O} = 324.42 \text{ g/mol}$$

$$\%C = \frac{240.22}{324.42} \times 100\% = 74.046\ \%C \qquad \%H = \frac{24.1906}{324.42} \times 100\% = 7.4566\ \%H$$

$$\%N = \frac{28.0134}{324.42} \times 100\% = 8.6349\ \%N \qquad \%O = \frac{31.9988}{324.42} \times 100\% = 9.8634\ \%O$$

39. **(a)** $\%Zr = \frac{1 \text{ mol Zr}}{1 \text{ mol ZrSiO}_4} \times \frac{1 \text{ mol ZrSiO}_4}{183.31 \text{ g ZrSiO}_4} \times \frac{91.224 \text{ g Zr}}{1 \text{ mol Zr}} \times 100\% = 49.765\% \text{ Zr}$

(b) $\%Be = \frac{3 \text{ mol Be}}{1 \text{ mol Be}_3\text{Al}_2\text{Si}_6\text{O}_{18}} \times \frac{1 \text{ mol Be}_3\text{Al}_2\text{Si}_6\text{O}_{18}}{537.502 \text{ g Be}_3\text{Al}_2\text{Si}_6\text{O}_{18}} \times \frac{9.01218 \text{ g Be}}{1 \text{ mol Be}} \times 100\% = 5.0300\% \text{ Be}$

(c) $\%Fe = \frac{3 \text{ mol Fe}}{1 \text{ mol Fe}_3\text{Al}_2\text{Si}_3\text{O}_{12}} \times \frac{1 \text{ mol Fe}_3\text{Al}_2\text{Si}_3\text{O}_{12}}{497.753 \text{ g Fe}_3\text{Al}_2\text{Si}_3\text{O}_{12}} \times \frac{55.847 \text{ g Fe}}{1 \text{ mol Fe}} \times 100\% = 33.659\% \text{ Fe}$

(d) $\%S = \frac{1 \text{ mol S}}{1 \text{ mol Na}_4\text{SSi}_3\text{Al}_3\text{O}_{12}} \times \frac{1 \text{ mol Na}_4\text{SSi}_3\text{Al}_3\text{O}_{12}}{481.219 \text{ g Na}_4\text{SSi}_3\text{Al}_3\text{O}_{12}} \times \frac{32.066 \text{ g S}}{1 \text{ mol S}} \times 100\% = 6.6635\% \text{ S}$

40. **(a)** $\%K = \frac{1 \text{ mol K}}{1 \text{ mol C}_{16}\text{H}_{17}\text{KN}_2\text{O}_4\text{S}} \times \frac{1 \text{ mol C}_{16}\text{H}_{17}\text{KN}_2\text{O}_4\text{S}}{372.47 \text{ g C}_{16}\text{H}_{17}\text{KN}_2\text{O}_4} \times \frac{39.098 \text{ g K}}{1 \text{ mol K}} \times 100\% = 10.497\% \text{ K}$

(b) $\%N = \frac{3 \text{ mol N}}{1 \text{ mol C}_{14}\text{H}_{21}\text{N}_3\text{O}_6\text{S}} \times \frac{1 \text{ mol C}_{14}\text{H}_{21}\text{N}_3\text{O}_6\text{S}}{359.40 \text{ g C}_{14}\text{H}_{21}\text{N}_3\text{O}_6\text{S}} \times \frac{14.0067 \text{ g N}}{1 \text{ mol N}} \times 100\% = 11.6917\% \text{ N}$

(c) $\%S = \frac{2 \text{ mol S}}{1 \text{ mol C}_{14}\text{H}_{18}\text{ClKN}_2\text{O}_4\text{S}_2} \times \frac{1 \text{ mol C}_{14}\text{H}_{18}\text{ClKN}_2\text{O}_4\text{S}_2}{417.00 \text{ g C}_{14}\text{H}_{18}\text{ClKN}_2\text{O}_4\text{S}_2} \times \frac{32.066 \text{ g S}}{1 \text{ mol S}} \times 100\% = 15.379\%\text{S}$

(d) $\%C = \frac{32 \text{ mol C}}{1 \text{ mol Ca(C}_{16}\text{H}_{17}\text{N}_2\text{O}_4\text{S)}_2} \times \frac{1 \text{ mol Ca(C}_{16}\text{H}_{17}\text{N}_2\text{O}_4\text{S)}_2}{706.84 \text{ g Ca(C}_{16}\text{H}_{17}\text{N}_2\text{O}_4\text{S)}_2} \times \frac{12.011 \text{ g C}}{1 \text{ mol C}} \times 100\% = 54.376\%\text{C}$

41. The oxide with the largest %Cr will have the largest number of moles of Cr per mole of oxygen.

$CrO \quad \frac{1 \text{ mol Cr}}{1 \text{ mol O}} = 1 \text{ mol Cr/mol O} \qquad Cr_2O_3 \quad \frac{2 \text{ mol Cr}}{3 \text{ mol O}} = 0.667 \text{ mol Cr/mol O}$

$CrO_2 \quad \frac{1 \text{ mol Cr}}{2 \text{ mol O}} = 0.500 \text{ mol Cr/mol O} \qquad CrO_3 \quad \frac{1 \text{ mol Cr}}{3 \text{ mol O}} = 0.333 \text{ mol Cr/mol O}$

Arranged in order of increasing %Cr: $CrO_3 < CrO_2 < Cr_2O_3 < CrO$

42. In each case, a P atom is associated with four O atoms. The species with the largest percent phosphorus will have the smallest mass (per P atom) of atoms other than P and O.

In H_3PO_4, the mass of those other atoms is (3 H atoms × 1 u/H atom) = 3 u

In Na_3PO_4, their mass is (3 Na atoms × 23 u/Na atom) = 69 u

In $(NH_4)_2HPO_4$, their mass is (2 N atoms × 14 u/N atom) + (9 H atoms × 1 u/H atom) = 37 u

In $Ca(H_2PO_4)_2$, their mass is ($^1/_2$ Ca atom × 40 u/Ca atom) + (2 H atoms × 1 u/H atom) = 22 u

Thus, arranged in order of increasing %P: $Na_3PO_4 < (NH_4)_2HPO_4 < Ca(H_2PO_4)_2 < H_3PO_4$

Chemical Formulas from Percent Composition

43. Convert each of the percentages into the mass in 100.00 g, and then to the amount in moles of that element.

$93.71 \text{ g C} \times \frac{1 \text{ mol C}}{12.01 \text{ g C}} = 7.803 \text{ mol C} \qquad \div 6.23 \longrightarrow 1.25 \text{ mol C}$

$6.29 \text{ g H} \times \frac{1 \text{ mol H}}{1.01 \text{ g H}} = 6.23 \text{ mol H} \qquad \div 6.23 \longrightarrow 1.00 \text{ mol H}$

The empirical formula is C_5H_4, which has an empirical molar mass of [(5 × 12.0 g C) + (4 × 1.0 g H)] = 64.0 g/mol. Since this empirical molar mass is one-half of the 128 g/mol correct molar mass, the molecular formula must be twice the empirical formula. Molecular formula: $C_{10}H_8$

44. The percent of selenium in each oxide is found by difference.
First oxide: % Se = 100.0% – 28.8% O = 71.2% Se

$$28.8\%\ \text{O} \times \frac{1\ \text{mol O}}{16.0\ \text{g O}} = 1.80\ \text{mol O} \qquad \div 0.901 \longrightarrow 2.00\ \text{mol O}$$

$$71.2\%\ \text{Se} \times \frac{1\ \text{mol Se}}{79.0\ \text{g Se}} = 0.901\ \text{mol Se} \qquad \div 0.901 \longrightarrow 1.00\ \text{mol Se}$$

The empirical formula is SeO_2. An appropriate name is selenium dioxide.
Second oxide: % Se = 100.0% – 37.8% O = 62.2% Se

$$37.8\%\ \text{O} \times \frac{1\ \text{mol O}}{16.0\ \text{g O}} = 2.36\ \text{mol O} \qquad \div 0.787 \longrightarrow 3.00\ \text{mol O}$$

$$62.2\%\ \text{Se} \times \frac{1\ \text{mol Se}}{79.0\ \text{g Se}} = 0.787\ \text{mol Se} \qquad \div 0.787 \longrightarrow 1.00\ \text{mol Se}$$

The empirical formula is SeO_3. An appropriate name is selenium trioxide.

45. **(a)**

$$74.01\ \text{g C} \times \frac{1\ \text{mol C}}{12.01\ \text{g C}} = 6.162\ \text{mol C} \qquad \div 1.298 \longrightarrow 4.747\ \text{mol C}$$

$$5.23\ \text{g H} \times \frac{1\ \text{mol H}}{1.01\ \text{g H}} = 5.18\ \text{mol H} \qquad \div 1.298 \longrightarrow 3.99\ \text{mol H}$$

$$20.76\ \text{g O} \times \frac{1\ \text{mol O}}{16.00\ \text{g O}} = 1.298\ \text{mol O} \qquad \div 1.298 \longrightarrow 1.00\ \text{mol O}$$

Now multiply each of the mole numbers by 4 to obtain an empirical formula of $C_{19}H_{16}O_4$.

(b)

$$30.20\ \text{g C} \times \frac{1\ \text{mol C}}{12.01\ \text{g C}} = 2.515\ \text{mol C} \qquad \div 0.6288 \longrightarrow 4.000\ \text{mol C}$$

$$5.07\ \text{g H} \times \frac{1\ \text{mol H}}{1.01\ \text{g H}} = 5.02\ \text{mol H} \qquad \div 0.6288 \longrightarrow 7.98\ \text{mol H}$$

$$44.58\ \text{g Cl} \times \frac{1\ \text{mol Cl}}{35.45\ \text{g Cl}} = 1.258\ \text{mol Cl} \qquad \div 0.6288 \longrightarrow 2.001\ \text{mol Cl}$$

$$20.16\ \text{g S} \times \frac{1\ \text{mol S}}{32.06\ \text{g H}} = 0.6288\ \text{mol S} \qquad \div 0.6288 \longrightarrow 1.000\ \text{mol S}$$

The empirical formula is $C_4H_8Cl_2S$.

46. **(a)**

$$95.21\ \text{g C} \times \frac{1\ \text{mol C}}{12.01\ \text{g C}} = 7.928\ \text{mol C} \qquad \div 4.74 \longrightarrow 1.67\ \text{mol C}$$

$$4.79\ \text{g H} \times \frac{1\ \text{mol H}}{1.01\ \text{g H}} = 4.74\ \text{mol H} \qquad \div 4.74 \longrightarrow 1.00\ \text{mol H}$$

Multiply each of the mole numbers by 3 to obtain an empirical formula of C_5H_3

(b) Each percent is numerically equal to the mass of that element present in 100.00 g of the compound. These masses then are converted to amounts of the elements, in moles.

$$\text{amount C} = 38.37\ \text{g C} \times \frac{1\ \text{mol C}}{12.01\ \text{g C}} = 3.195\ \text{mol C} \qquad \div 0.491 \longrightarrow 6.51\ \text{mol C}$$

$$\text{amount H} = 1.49\ \text{g H} \times \frac{1\ \text{mol H}}{1.01\ \text{g H}} = 1.48\ \text{mol H} \qquad \div 0.491 \longrightarrow 3.01\ \text{mol H}$$

$$\text{amount Cl} = 52.28\ \text{g Cl} \times \frac{1\ \text{mol Cl}}{35.45\ \text{g Cl}} = 1.475\ \text{mol Cl} \qquad \div 0.491 \longrightarrow 3.004\ \text{mol Cl}$$

$$\text{amount O} + 7.86\ \text{g O} \times \frac{1\ \text{mol O}}{16.0\ \text{g O}} = 0.491\ \text{mol O} \qquad \div 0.491 \longrightarrow 1.00\ \text{mol O}$$

Multiply each of these numbers of moles by 2 to obtain the empirical formula: $C_{13}H_6Cl_6O_2$.

47. Determine the mass of oxygen by difference. Then convert all masses to amounts in moles.
oxygen mass = 100.00 g – 73.27 g C – 3.84 g H – 10.68 g N = 12.21 g O

$$\text{amount C} = 73.27\ \text{g C} \times \frac{1\ \text{mol C}}{12.011\ \text{g C}} = 6.100\ \text{mol C} \qquad \div 0.7625 = 8.000\ \text{mol C}$$

$$\text{amount H} = 3.84\ \text{g H} \times \frac{1\ \text{mol H}}{1.008\ \text{g H}} = 3.81\ \text{mol H} \qquad \div 0.7625 = 5.00\ \text{mol H}$$

$$\text{amount N} = 10.68\ \text{g N} \times \frac{1\ \text{mol N}}{14.007\ \text{g N}} = 0.7625\ \text{mol N} \qquad \div 0.7625 = 1.000\ \text{mol N}$$

$$\text{amount O} = 12.21\ \text{g O} \times \frac{1\ \text{mol O}}{15.999\ \text{g O}} = 0.7632\ \text{mol O} \qquad \div 0.7625 = 1.001\ \text{mol O}$$

The empirical formula is C_8H_5NO, which has an empirical mass of
(8 mol C × 12.0 u C) + (5 mol H × 1.0 u H) + (1 mol N × 14.0 u N) + (1 mol O × 16.0 u O)

= 131 u. This is almost exactly half the molecular mass of 262.3 u. Thus, the molecular formula is twice the empirical formula and is $C_{16}H_{10}N_2O_2$.

48. β-Carotene contains only carbon and hydrogen.

$$\text{amount C} = 89.49 \text{ g C} \times \frac{1 \text{ mol C}}{12.011 \text{ g C}} = 7.451 \text{ mol C} \qquad \div 7.451 = 1.000 \text{ mol C}$$

$$\text{amount H} = 10.51 \text{ g H} \times \frac{1 \text{ mol H}}{1.0079 \text{ g H}} = 10.43 \text{ mol H} \qquad \div 7.451 = 1.400 \text{ mol H}$$

The empirical formula, C_5H_7, has an empirical mass of (5 mol C × 12.01 u C) + (7 mol H × 1.01 u H) = 67.12 u. The molecular weight is 536.9/67.12 = 7.999 = eight times the empirical weight. Thus, the molecular formula is eight times the empirical formula and is $C_{40}H_{56}$.

49. In each 100 g of the compound there are 65 g of F and 35 g of X. The number of moles of X is given by

$$\text{no. mol X} = 65 \text{ g F} \times \frac{1 \text{ mol F}}{19.0 \text{ g}} \times \frac{1 \text{ mol X}}{3 \text{ mol F}} = 1.14 \text{ mol X}$$

$$\text{Thus, the molar mass of X} = \frac{35 \text{ g X}}{1.14 \text{ mol X}} = 31 \text{ g/mol X}$$

50. The atomic weight of element X has the units of grams per mole. We can determine the amount, in moles of Cl, and convert that to the amount of X, equivalent to 25.0 g of X.

$$\text{atomic weight} = \frac{25.0 \text{ g X}}{75.0 \text{ g Cl}} \times \frac{35.453 \text{ g Cl}}{1 \text{ mol Cl}} \times \frac{4 \text{ mol Cl}}{1 \text{ mol X}} = \frac{47.3 \text{ g X}}{1 \text{ mol X}}$$

This atomic weight is close to that of the element titanium, which therefore is identified as element X.

Combustion Analysis

51. **(a)** Determine the mass of carbon and of hydrogen present in the sample. A hydrocarbon only contains the two elements hydrogen and carbon.

$$0.8661 \text{ g } CO_2 \times \frac{1 \text{ mol } CO_2}{44.010 \text{ g}} \times \frac{1 \text{ mol C}}{1 \text{ mol } CO_2} = 0.01968 \text{ mol C} \times \frac{12.011 \text{ g C}}{1 \text{ mol C}} = 0.2364 \text{ g C}$$

$$0.2216 \text{ g } H_2O \times \frac{1 \text{ mol } H_2O}{18.015 \text{ g}} \times \frac{2 \text{ mol H}}{1 \text{ mol } H_2O} = 0.02460 \text{ mol H} \times \frac{1.0079 \text{ g H}}{1 \text{ mol H}} = 0.02479 \text{ g H}$$

The % C and % H then are found.

$$\% \text{ C} = \frac{0.2364 \text{ g C}}{0.2612 \text{ g cmpd}} \times 100\% = 90.51\% \text{ C} \qquad \% \text{ H} = \frac{0.02479 \text{ g H}}{0.2612 \text{ g cmpd}} \times 100\% = 9.491\% \text{ H}$$

(b) Use the moles of C and H from part (a), and divide both by the smallest.

0.01968 mol C ÷ 0.01968 = 1.000 mol C   0.02460 mol H ÷ 0.01968 = 1.250 mol H

The empirical formula is obtained by multiplying these mole numbers by 4. It is C_4H_5.

(c) The empirical formula C_4H_5 has an empirical molar mass of [(4 × 12.0 g C) + (5 × 1.0 g H) =] 53.0 g/mol. This value is one-half of the actual molar mass. Hence, the molecular formula is twice the empirical formula. Molecular formula: C_8H_{10}.

52. Determine the mass of carbon and of hydrogen present in the sample.

$$1.0420 \text{ g } CO_2 \times \frac{1 \text{ mol } CO_2}{44.01 \text{ g}} \times \frac{1 \text{ mol C}}{1 \text{ mol } CO_2} = 0.02368 \text{ mol C} \times \frac{12.01 \text{ g C}}{1 \text{ mol C}} = 0.2844 \text{ g C}$$

$$0.2437 \text{ g } H_2O \times \frac{1 \text{ mol } H_2O}{18.02 \text{ g}} \times \frac{2 \text{ mol H}}{1 \text{ mol } H_2O} = 0.02705 \text{ mol H} \times \frac{1.008 \text{ g H}}{1 \text{ mol H}} = 0.02726 \text{ g H}$$

(b) To find the empirical formula determine the mass of oxygen by difference, and its amount in moles.

mass O = 0.3654 g cmpd – 0.2844 g C – 0.02726 g H = 0.0537 g O

$$0.0537 \text{ g O} \times \frac{1 \text{ mol O}}{16.0 \text{ g O}} = 0.00336 \text{ mol O} \div 0.00336 \longrightarrow 1.00 \text{ mol O}$$

$$0.02368 \text{ mol C} \div 0.00336 \longrightarrow 7.05 \text{ mol C}$$

$$0.02705 \text{ mol H} \div 0.00336 \longrightarrow 8.05 \text{ mol H}$$

This gives an empirical formula of C_7H_8O.

(c) The molecular formula is found by realizing that a mole of empirical units has a mass of (7 × 12.0 g C + 8 × 1.0 g H + 16.0 g O =) 107.0 g. Since this agrees with the molecular weight, the molecular formula is the same as the empirical formula: C_7H_8O.

(a) The percent composition can be determined any time after the masses of C and H are determined.

$$\% \text{ C} = \frac{0.2844 \text{ g C}}{0.3654 \text{ g cmpd}} \times 100\% = 77.83\% \text{ C} \qquad \% \text{ H} = \frac{0.02726 \text{ g H}}{0.3654 \text{ g cmpd}} \times 100\% = 6.460\% \text{ H}$$

%O = 100.00% – 77.83% C – 6.460% H = 15.71% O

Of course, these percents can be used as input in determining the empirical formula if one wishes.

53. First determine the mass of carbon and hydrogen present in the sample.

$$0.741\text{ g }CO_2 \times \frac{1\text{ mol }CO_2}{44.01\text{ g}} \times \frac{1\text{ mol C}}{1\text{ mol }CO_2} = 0.0168\text{ mol C} \times \frac{12.01\text{ g C}}{1\text{ mol C}} = 0.202\text{ g C}$$

$$0.605\text{ g }H_2O \times \frac{1\text{ mol }H_2O}{18.02\text{ g}} \times \frac{2\text{ mol H}}{1\text{ mol }H_2O} = 0.0671\text{ mol H} \times \frac{1.008\text{ g H}}{1\text{ mol H}} = 0.0677\text{ g H}$$

Then the mass of N that this sample would have produced is determined. (Note that this is also the mass of N_2 produced in the reaction.) $0.226\text{ g N} \times \frac{0.505\text{ g 1st sample}}{0.486\text{ g 2nd sample}} = 0.235\text{ g N}$

We also can determine this mass of N by difference. 0.505 g cmpd – 0.202 g C – 0.0677 g H = 0.235 g N

Then we can calculate the relative number of moles of each element.

$$0.235\text{ g N} \times \frac{1\text{ mol N}}{14.01\text{ g N}} = 0.0168\text{ mol N} \quad \div 0.0168 \longrightarrow 1.00\text{ mol N}$$

$$0.0168\text{ mol C} \quad \div 0.0168 \longrightarrow 1.00\text{ mol C}$$

$$0.0671\text{ mol H} \quad \div 0.0168 \longrightarrow 3.99\text{ mol H}$$

Thus, the empirical formula is CH_4N.

54. Thiophene contains just carbon, hydrogen, and sulfur, so there is no need to determine the mass of oxygen by difference. We simply determine the amount of each element from the mass of its combustion product.

$$2.272\text{ g }CO_2 \times \frac{1\text{ mol }CO_2}{44.010\text{ g }CO_2} \times \frac{1\text{ mol C}}{1\text{ mol }CO_2} = 0.05162\text{ mol C} \quad \div 0.0129 \longrightarrow 4.00\text{ mol C}$$

$$0.465\text{ g }H_2O \times \frac{1\text{ mol }H_2O}{18.02\text{ g }H_2} \times \frac{2\text{ mol H}}{1\text{ mol }H_2O} = 0.0516\text{ mol H} \quad \div 0.0129 \longrightarrow 4.00\text{ mol H}$$

$$0.827\text{ g }SO_2 \times \frac{1\text{ mol }SO_2}{64.06\text{ g }SO_2} \times \frac{1\text{ mol S}}{1\text{ mol }SO_2} = 0.0129\text{ mol S} \quad \div 0.0129 \longrightarrow 1.00\text{ mol S}$$

The empirical formula of thiophene is C_4H_4S.

55. Each mole of CO_2 is produced from a mole of C. Therefore, the compound with the largest number of moles of C per mole of the compound will produce the largest amount of CO_2 and thus also the largest mass of CO_2. Of the compounds listed—CH_4, C_2H_5OH, $C_{10}H_8$, and C_6H_5OH—$C_{10}H_8$ has the largest number of moles of C per mole of the compound. $C_{10}H_8$ thus will produce the greatest mass of CO_2 per mole on complete combustion.

56. The compound that produces the largest mass of water per gram of the compound will have the largest amount of hydrogen per gram of the compound. Thus, we need to compare the ratios of amount of hydrogen per mole to the mole weight for each compound.

Note that C_2H_5OH has as much H per mole as does C_6H_5OH, but C_6H_5OH has a higher mole weight. Thus, C_2H_5OH produces more H_2O per gram than will C_6H_5OH. Notice also that CH_4 has 4 H's per C, while $C_{10}H_8$ has 8 H's per 10 C's or 0.8 H per C. Thus CH_4 will produce more H_2O than will $C_{10}H_8$.

Thus, we are left with comparing CH_4 to C_2H_5OH. The O in the second compound has about the same mass (16 u) as does C (12 u). Thus, in CH_4 there are 4 H's per C, while in C_2H_5OH there are about 2 H's per C. CH_4 will produce more water per gram, on combustion, the most of all four compounds.

Oxidation States

57. The oxidation state (O. S.) is given first, followed by the explanation for its assignment.

(a) C = –4 in CH_4 — H has an oxidation state of +1 in its compounds (Remember that the sum of the oxidation states in a compound equals 0.)

(b) S = +4 in SF_4 — F has O. S. = –1 in its compounds.

(c) O = –1 in Na_2O_2 — Na has O. S. = +1 in its compounds.

(d) C = 0 in $C_2H_3O_2^-$ — H has O. S. = +1 in its compounds; that of O = –2. (Remember that the sum of the oxidation states in a polyatomic ion equals the charge on that ion.)

(e) Fe = +6 in FeO_4^{2-} — O has O. S. = –2 in its compounds.

58. The oxidation state of sulfur in each species is determined below. Remember that the oxidation state of O is –2 in its compounds. And the total of the oxidation states in an ion equals the charge on that ion.

(a) S = +4 in SO_3^{2-} **(b)** S = +2 in $S_2O_3^{2-}$ **(c)** S = +7 in $S_2O_8^{2-}$

(d) S = +6 in HSO_4^- **(e)** S = –2 in HS^-

59. Remember that the oxidation state of oxygen is –2 in its compounds.
Cr^{3+} and O^{2-} form Cr_2O_3, chromium(III) oxide. Cr^{4+} and O^{2-} form CrO_2, chromium (IV) oxide.
Cr^{6+} and O^{2-} form CrO_3, chromium(VI) oxide.

60. Remember that oxygen has an oxidation state of –2 in its compounds.
N = +1 in N_2O, nitrous oxide or dinitrogen monoxide N = +2 in NO, nitric oxide or nitrogen monoxide
N = +3 in N_2O_3, dinitrogen trioxide N = +4 in NO_2, nitrogen dioxide
N = +5 in N_2O_5, dinitrogen pentoxide

Nomenclature

61. (a) SrO strontium oxide (b) ZnS zinc sulfide
(c) K_2CrO_4 potassium chromate (d) Cs_2SO_4 cesium sulfate
(e) Cr_2O_3 chromium(III) oxide (f) $Fe_2(SO_4)_3$ iron(III) sulfate
(g) $Mg(HCO_3)_2$ magnesium hydrogen carbonate (h) $(NH_4)_2HPO_4$ ammonium hydrogen phosphate
(i) $Ca(HSO_3)_2$ calcium hydrogen sulfite (j) $Cu(OH)_2$ copper(II) hydroxide
(k) HNO_3 nitric acid (l) $KClO_4$ potassium perchlorate
(m) $HBrO_3$ bromic acid (n) H_3PO_3 phosphorous acid

62. (a) $Ba(NO_3)_2$ barium nitrate (b) HNO_2 nitrous acid
(c) CrO_2 chromium(IV) oxide (d) KIO_3 potassium iodate
(e) LiCN lithium cyanide (f) KIO potassium hypoiodite
(g) $Fe(OH)_2$ iron(II) hydroxide (h) $Ca(H_2PO_4)_2$ calcium dihydrogen phosphate
(i) H_3PO_4 phosphoric acid (j) $NaHSO_4$ sodium hydrogen sulfate
(k) $Na_2Cr_2O_7$ sodium dichromate (l) $NH_4C_2H_3O_2$ ammonium acetate
(m) MgC_2O_4 magnesium oxalate (n) $Na_2C_2O_4$ sodium oxalate

63. (a) CS_2 carbon disulfide (b) SiF_4 silicon tetrafluoride
(c) ClF_5 chlorine pentafluoride (d) N_2O_5 dinitrogen pentoxide
(e) SF_6 sulfur hexafluoride (f) I_2Cl_6 diiodine hexachloride

64. (a) ICl iodine monochloride (b) ClF_3 chlorine trifluoride
(c) SF_4 sulfur tetrafluoride (d) BrF_5 bromine pentafluoride
(e) N_2O_4 dinitrogen tetroxide (g) S_4N_4 tetrasulfur tetranitride

65. (a) $Al_2(SO_4)_3$ aluminum sulfate (b) $(NH_4)_2Cr_2O_7$ ammonium dichromate
(c) SiF_4 silicon tetrafluoride (d) Fe_2O_3 iron(III) oxide
(e) C_3S_2 tricarbon disulfide (f) $Co(NO_3)_2$ cobalt(II) nitrate
(g) $Sr(NO_2)_2$ strontium nitrite (h) HBr hydrobromic acid
(i) HIO_3 iodic acid (j) PCl_2F_3 phosphorus dichloride trifluoride

66. (a) $Mg(ClO_4)_2$ magnesium perchlorate (b) $Pb(C_2H_3O_2)_2$ lead(II) acetate
(c) SnO_2 tin(IV) oxide (d) HI hydroiodic acid
(e) $HClO_2$ chlorous acid (f) $NaHSO_3$ sodium hydrogen sulfite
(g) $Ca(H_2PO_4)_2$ calcium dihydrogen phosphate (h) $AlPO_4$ aluminum phosphate
(i) N_2O_4 dinitrogen tetroxide (j) S_2Cl_2 disulfur dichloride

67. (a) Ti^{4+} and Cl^- produce $TiCl_4$ (b) Fe^{3+} and SO_4^{2-} produce $Fe_2(SO_4)_3$
(c) $Cl^{(7+)}$ and $O^{(2-)}$ produce Cl_2O_7 (d) $S^{(7+)}$ and $O^{(2-)}$ produce $S_2O_8^{2-}$

68. (a) $N^{(5+)}$ and $O^{(2-)}$ produce N_2O_5 (b) $N^{(3+)}$ and $O^{(2-)}$ produce N_2O_3
(c) $C^{(+4/3)}$ and $O^{(2-)}$ produce C_3O_2 (d) $S^{(+2.5)}$ and $O^{(2-)}$ produce $S_4O_6^{2-}$

Hydrates

69. The hydrates with the greatest % H_2O will be the ones with the highest ratio of the molar amount of water (in a mole of the compound) to the mass of one mole of the anhydrous compound.

$$\frac{5\ H_2O}{CuSO_4} = \frac{5\ \text{mol}\ H_2O}{159.6\ \text{g}} = 0.03133 \qquad \frac{6\ H_2O}{MgCl_2} = \frac{6\ \text{mol}\ H_2O}{95.3\ \text{g}} = 0.0630$$

$\frac{18\ H_2O}{Cr_2(SO_4)_3} = \frac{18\ \text{mol}\ H_2O}{392.3\ \text{g}} = 0.04588$ $\frac{2\ H_2O}{LiC_2H_3O_2} = \frac{2\ \text{mol}\ H_2O}{65.9\ \text{g}} = 0.0303$

The hydrate with the greatest % H_2O therefore is $MgCl_2{\cdot}6\ H_2O$

70. A mole of this hydrate will contain about the same mass of H_2O and of Na_2SO_3.

molar mass $Na_2SO_3 = (2 \times 23.0\ \text{g Na}) + 32.1\ \text{g S} + (3 \times 16.0\ \text{g O}) = 126.1\ \text{g/mol}$

$$\text{no. mol } H_2O = 126.1\ \text{g} \times \frac{1\ \text{mol}\ H_2O}{18.0\ \text{g}\ H_2O} = 7.01\ \text{mol}\ H_2O$$

Thus, the formula of the hydrate is $Na_2SO_3{\cdot}7H_2O$.

71. molar mass $CuSO_4 = 63.5\ \text{g Cu} + 32.1\ \text{g S} + (4 \times 16.0\ \text{g O}) = 159.6\ \text{g}\ CuSO_4\text{/mol}$

The mass of this solid needed to combine with 8.5 mL of water is depends on the absorption of 5 moles of water by one mole of the solid, based on the formula $CuSO_4{\cdot}5H_2O$.

$$\text{mass } CuSO_4 = 8.5\ \text{mL}\ H_2O \times \frac{1.00\ \text{g}\ H_2O}{1\ \text{mL}\ H_2O} \times \frac{1\ \text{mol}\ H_2O}{18.0\ \text{g}\ H_2O} \times \frac{1\ \text{mol}\ CuSO_4}{5\ \text{mol}\ H_2O} \times \frac{159.6\ \text{g}\ CuSO_4}{1\ \text{mol}\ CuSO_4}$$

$= 15\ \text{g}\ CuSO_4$ required to remove all the water

72. The increase in mass of the solid is related to each mole of the solid absorbing ten moles of water.

$$\text{increase in mass} = 3.50\ \text{g}\ Na_2SO_4 \times \frac{1\ \text{mol}\ Na_2SO_4}{142.0\ \text{g}\ Na_2SO_4} \times \frac{10\ \text{mol}\ H_2O\ \text{added}}{1\ \text{mol}\ Na_2SO_4} \times \frac{18.015\ \text{g}\ H_2O}{1\ \text{mol}\ H_2O}$$

$= 4.44\ \text{g}\ H_2O$ added

FEATURE PROBLEMS

A. $$N_A = \frac{96{,}485\ \text{C}}{1\ \text{mol Ag}} \times \frac{1\ \text{mol Ag}}{1\ \text{mol Ag}^+} \times \frac{1\ \text{mol Ag}^+}{1\ \text{mol e}^-} \times \frac{1\ \text{electron}}{1.602 \times 10^{-19}\ \text{C}} = 6.023 \times 10^{23}\ \frac{\text{electrons}}{\text{mole}}$$

B. To verify the law of multiple proportions, we determine the mass of nitrogen that combines with 1.00 g of oxygen in each compound. In NO_2: $\frac{14.0\ \text{g N}}{2 \times 16.0\ \text{g O}} = 0.4375\ \text{g N/g O}$

In NO: $\frac{14.0\ \text{g N}}{16.0\ \text{g O}} = 0.875\ \text{g N/g O}$ In N_2O: $\frac{2 \times 14.0\ \text{g N}}{16.0\ \text{g O}} = 1.75\ \text{g N/g O}$

We demonstrate that these three ratios are related as small whole numbers by dividing each of them by the smallest, 0.4375. $0.4375 : 0.875 : 1.75 \div 0.4375 \longrightarrow 1 : 2 : 4$ (a series of small whole numbers)

To determine if there can be a compound with 26.40% N, we attempt to determine the empirical formula.

$$26.40\ \text{g N} \times \frac{1\ \text{mol N}}{14.007\ \text{g N}} = 1.885\ \text{mol N} \quad \div 1.885 \longrightarrow 1.000\ \text{mol N}$$

$$73.60\ \text{g O} \times \frac{1\ \text{mol O}}{15.999\ \text{g O}} = 4.600\ \text{mol O} \quad \div 1.885 \longrightarrow 2.440\ \text{mol O}$$

The simplest empirical formula is $N_{25}O_{61}$. We conclude that no nitrogen oxide has 26.40% N.

C. **1.** "5-10-5" fertilizer contains 5 g N (that is, 5% N), 10 g P_2O_5, and 5 g K_2O in 100 g fertilizer. We convert the last two numbers to masses of the two elements.

(a) $$\%P = 10\%\ P_2O_5 \times \frac{1\ \text{mol}\ P_2O_5}{141.94\ \text{g}\ P_2O_5} \times \frac{2\ \text{mol P}}{1\ \text{mol}\ P_2O_5} \times \frac{30.97\ \text{g P}}{1\ \text{mol P}} = 4.4\%\ \text{P}$$

(b) $$\%K = 5\%\ K_2O \times \frac{1\ \text{mol}\ K_2O}{94.20\ \text{g}\ K_2O} \times \frac{2\ \text{mol K}}{1\ \text{mol}\ K_2O} \times \frac{39.10\ \text{g K}}{1\ \text{mol K}} = 4.2\%\ \text{K}$$

2. First we determine %P and then convert it to $\%P_2O_5$, given that 10.0% P_2O_5 is equivalent to 4.37% P.

(a) $$\%P_2O_5 = \frac{2\ \text{mol P}}{1\ \text{mol}\ Ca(H_2PO_4)_2} \times \frac{30.97\ \text{g P}}{1\ \text{mol P}} \times \frac{1\ \text{mol}\ Ca(H_2PO_4)_2}{234.05\ \text{g}\ Ca(H_2PO_4)_2} \times 100\% \times \frac{10.0\%\ P_2O_5}{4.37\%\ \text{P}}$$

$= 60.6\%\ P_2O_5$

(b) $$\%P_2O_5 = \frac{1\ \text{mol P}}{1\ \text{mol}\ (NH_4)_2HPO_4} \times \frac{30.97\ \text{g P}}{1\ \text{mol P}} \times \frac{1\ \text{mol}\ (NH_4)_2HPO_4}{132.06\ \text{g}\ (NH_4)_2HPO_4} \times 100\% \times \frac{10.0\%\ P_2O_5}{4.37\%\ \text{P}}$$

$= 53.7\%\ P_2O_5$

4 CHEMICAL REACTIONS

PRACTICE EXAMPLES

1A When combusted (burned in oxygen) a carbon–hydrogen compound produces $CO_2(g)$ and $H_2O(l)$.

$$C_2H_4(g) + O_2(g) \longrightarrow CO_2(g) + H_2O(l)$$

The formula expression then is balanced with integer coefficients. First balance C and H.

$$C_2H_4(g) + O_2(g) \longrightarrow 2\ CO_2(g) + 2\ H_2O(l)$$

Then balance O. $C_2H_4(g) + 3\ O_2(g) \longrightarrow 2\ CO_2(g) + 2\ H_2O(l)$

Self	Left:	2 C	4 H	$(3 \times 2) = 6$ O
Check	Right:	2 C	$(2 \times 2) = 4$ H	$[(2 \times 2) + 2] = 6$ O

1B When combusted (burned in oxygen) C, H, and S produce $CO_2(g)$, $H_2O(l)$, and $SO_2(g)$, respectively.

$$C_4H_4S + O_2(g) \longrightarrow CO_2(g) + H_2O(l) + SO_2(g) \quad \text{(not balanced)}$$

The expression obtained then is balanced with integer coefficients. First balance C, H, and S.

$$C_4H_4S + O_2(g) \longrightarrow 4\ CO_2(g) + 2\ H_2O(l) + SO_2(g)$$

Then balance O. $C_4H_4S + 6\ O_2(g) \longrightarrow 4\ CO_2(g) + 2\ H_2O(l) + SO_2(g)$

Self	Left:	4 C	4 H	1 S	$(6 \times 2) = 12$ O
Check	Right:	4 C	$(2 \times 2) = 4$ H	1 S	$[(4 \times 2) + 2 + 2] = 12$ O

2A The combustion of a carbon–hydrogen–oxygen compound produces $CO_2(g)$ and $H_2O(l)$.

$$C_3H_8O_3(l) + O_2(g) \longrightarrow CO_2(g) + H_2O(l)$$

The formula expression then is balanced with integer coefficients. First balance C and H.

$$C_3H_8O_3(l) + O_2(g) \longrightarrow 3\ CO_2(g) + 4\ H_2O(l)$$

Then balance O. $C_3H_8O_3(l) + \frac{7}{2}\ O_2(g) \longrightarrow 3\ CO_2(g) + 4\ H_2O(l)$

Obtain integral coefficients. $2\ C_3H_8O_3(l) + 7\ O_2(g) \longrightarrow 6\ CO_2(g) + 8\ H_2O(l)$

Self	Left:	$(2 \times 3) = 6$ C	$(2 \times 8) = 16$ H	$(3 \times 2) + (7 \times 2) = 20$ O
Check	Right:	6 C	$(8 \times 2) = 16$ H	$[(6 \times 2) + 8] = 20$ O

2B The solution proceeds in the same way as the previous Practice Example.

$$C_7H_6O_2S + O_2(g) \longrightarrow CO_2(g) + H_2O(l) + SO_2(g) \quad \text{(not balanced)}$$

The expression obtained then is balanced with integer coefficients. First balance C, H, and S.

$$C_7H_6O_2S + O_2(g) \longrightarrow 7\ CO_2(g) + 3\ H_2O(l) + 1\ SO_2(g)$$

Then balance O and multiply the equation by 2 to obtain all coefficients as integers.

$$2\ C_7H_6O_2S + 17\ O_2(g) \longrightarrow 14\ CO_2(g) + 6\ H_2O(l) + 2\ SO_2(g)$$

Self	Left:	$(2 \times 7) = 14$ C	$(2 \times 6) = 12$ H	2 S	$[(2 \times 2) + (17 \times 2)] = 38$ O
Check	Right:	14 C	$(6 \times 2) = 12$ H	2 S	$[(14 \times 2) + 6 + (2 \times 2)] = 38$ O

3A The balanced chemical equation provides the factor needed to convert from moles $KClO_3$ to moles O_2.

$$\text{amount } O_2 = 1.76 \text{ mol } KClO_3 \times \frac{3 \text{ mol } O_2}{2 \text{ mol } KClO_3} = 2.64 \text{ mol } O_2$$

3B First, find the mole weight of Ag_2O. $(2 \text{ mol Ag} \times 107.87 \text{ g Ag}) + 16.00 \text{ g O} = 231.74 \text{ g } Ag_2O/\text{mol}$

$$\text{amount Ag} = 1.00 \text{ kg } Ag_2O \times \frac{1000 \text{ g}}{1.00 \text{ kg}} \times \frac{1 \text{ mol } Ag_2O}{231.74 \text{ g } Ag_2O} \times \frac{2 \text{ mol Ag}}{1 \text{ mol } Ag_2O} = 8.63 \text{ mol Ag}$$

4A The balanced chemical equation provides the factor to convert from amount of Mg to amount of Mg_3N_2. First, we determine the molar mass of Mg_3N_2.

molar mass = (3 mol Mg × 24.305 g Mg) + (2 mol N × 14.007 g N) = 100.93 g Mg_3N_2

$$\text{mass } Mg_3N_2 = 3.82 \text{ g Mg} \times \frac{1 \text{ mol Mg}}{24.31 \text{ g Mg}} \times \frac{1 \text{ mol } Mg_3N_2}{3 \text{ mol Mg}} \times \frac{100.93 \text{ g } Mg_3N_2}{1 \text{ mol } Mg_3N_2} = 5.29 \text{ g } Mg_3N_2$$

4B The pivotal conversion is from $H_2(g)$ to CH_3OH. For this we use the balanced equations, which requires that we use the amounts in moles of both substances. The result of the solution involves converting to and from amounts, principally with molar masses.

$$\text{mass } H_2(g) = 1.00 \text{ kg } CH_3OH \times \frac{1000 \text{ g}}{1 \text{ kg}} \times \frac{1 \text{ mol } CH_3OH}{32.04 \text{ g } CH_3OH} \times \frac{2 \text{ mol } H_2}{1 \text{ mol } CH_3OH} \times \frac{2.016 \text{ g } H_2}{1 \text{ mol } H_2} = 126 \text{ g } H_2$$

5A The equation for the cited reaction is: $2\,H_2(g) + O_2(g) \longrightarrow 2\,H_2O(l)$ The pivotal conversion is from one substance to another, in moles with the balanced chemical equation providing the conversion factor.

$$\text{mass } H_2(g) = 1.00 \text{ g } O_2(g) \times \frac{1 \text{ mol } O_2}{32.00 \text{ g } H_2} \times \frac{2 \text{ mol } H_2}{1 \text{ mol } O_2} \times \frac{2.016 \text{ g } H_2}{1 \text{ mol } H_2} = 0.126 \text{ g } H_2$$

5B The equation for the combustion reaction is: $C_8H_{18}(l) + \frac{25}{2}\,O_2(g) \longrightarrow 8\,CO_2(g) + 9\,H_2O(l)$

$$\text{mass } O_2 = 1.00 \text{ g } C_8H_{18} \times \frac{1 \text{ mol } C_8H_{18}}{114.23 \text{ g } C_8H_{18}} \times \frac{12.5 \text{ mol } O_2}{1 \text{ mol } C_8H_{18}} \times \frac{32.00 \text{ g } O_2}{1 \text{ mol } O_2} = 3.50 \text{ g } O_2(g)$$

6A If we are permitted to use the result of Example 4-6, the solution of this problem is particularly simple. That result is that 0.208 g H_2 is produced from 0.691 cm^3 of the alloy. The solution follows.

$$\text{volume of alloy} = 1.000 \text{ g } H_2 \times \frac{0.691 \text{ cm}^3 \text{ alloy}}{0.208 \text{ g } H_2} = 3.32 \text{ cm}^3 \text{ alloy}$$

If, instead, we work out the problem as if we did not know the result of Example 4-6, we must convert mass $H_2 \longrightarrow$ amount of $H_2 \longrightarrow$ amount of Al $\longrightarrow$ mass of Al $\longrightarrow$ mass of alloy $\longrightarrow$ volume of alloy. The calculation is performed as follows; each arrow in the preceding sentence requires a conversion factor.

$$\text{volume of alloy} = 1.000 \text{ g } H_2 \times \frac{1 \text{ mol } H_2}{2.0159 \text{ g } H_2} \times \frac{2 \text{ mol Al}}{3 \text{ mol } H_2} \times \frac{26.982 \text{ g Al}}{1 \text{ mol Al}} \times \frac{100.0 \text{ g alloy}}{93.7 \text{ g Al}} \times \frac{1 \text{ cm}^3}{2.85 \text{ g}}$$

$= 3.34 \text{ cm}^3$ alloy The difference in the two answers is a consequence of rounding off each intermediate answer during the solution of Example 4-6.

6B In the example, 0.208 g H_2 is collected from 1.97 g alloy; the alloy is 6.3% Cu by mass. This information provides the conversion factors we need.

$$\text{mass Cu} = 1.31 \text{ g } H_2 \times \frac{1.97 \text{ g alloy}}{0.208 \text{ g } H_2} \times \frac{6.3 \text{ g Cu}}{100.0 \text{ g alloy}} = 0.78 \text{ g Cu}$$

Notice that we do not have to reproduce the entire progress of a calculation. We can simply use values produced in the course of the calculation as conversion factors.

7A The cited reaction is $2\,Al(s) + 6\,HCl(aq) \longrightarrow 2\,AlCl_3(aq) + 3\,H_2(g)$. The HCl(aq) solution has a density of 1.14 g/mL and contains 28.0% HCl. We need to convert between the substances HCl and H_2; the important conversion factor comes from the balanced chemical equation. The sequence of conversions is volume of HCl(aq) $\longrightarrow$ mass of HCl(aq) $\longrightarrow$ mass of pure HCl $\longrightarrow$ amount of HCl $\longrightarrow$ amount of H_2 $\longrightarrow$ mass of H_2. In the calculation below, each arrow in the sequence is replaced by a conversion factor.

$$\text{mass } H_2 = 0.05 \text{ mL HCl(aq)} \times \frac{1.14 \text{ g}}{1 \text{ mL soln}} \times \frac{28.0 \text{ g HCl}}{100.0 \text{ g soln}} \times \frac{1 \text{ mol HCl}}{36.46 \text{ g HCl}} \times \frac{3 \text{ mol } H_2}{6 \text{ mol HCl}} \times \frac{2.016 \text{ g}}{1 \text{ mol } H_2}$$

mass $H_2 = 4 \times 10^{-4}$ g $H_2(g)$ = 0.4 mg $H_2(g)$

7B The density is necessary to determine the mass of the vinegar, and then the mass of acetic acid.

$$\text{mass } CO_2 = 5.00 \text{ mL vinegar} \times \frac{1.01 \text{ g}}{1 \text{ mL}} \times \frac{0.040 \text{ g acid}}{1 \text{ g vinegar}} \times \frac{1 \text{ mol } HC_2H_3O_2}{60.05 \text{ g } HC_2H_3O_2} \times \frac{1 \text{ mol } CO_2}{1 \text{ mol } HC_2H_3O_2} \times \frac{44.01 \text{ g } CO_2}{1 \text{ mol } CO_2}$$

$= 0.15 \text{ g } CO_2$

8A We need to determine the amount in moles of acetone and the volume in liters of the solution.

$$\text{molarity of acetone} = \frac{22.3 \text{ g } (CH_3)_2CO \times \frac{1 \text{ mol } (CH_3)_2CO}{58.08 \text{ g } (CH_3)_2CO}}{1.25 \text{ L soln}} = 0.307 \text{ M}$$

8B The molar mass of acetic acid, $HC_2H_3O_2$, is 60.05 g/mol. We begin with the quantity of acetic acid in the numerator and that of the solution in the denominator, and transform to the appropriate units for each.

$$\text{molarity} = \frac{15.0\text{ mL }HC_2H_3O_2}{500.0\text{ mL soln}} \times \frac{1000\text{ mL}}{1\text{ L soln}} \times \frac{1.048\text{ g }HC_2H_3O_2}{1\text{ mL }HC_2H_3O_2} \times \frac{1\text{ mol }HC_2H_3O_2}{60.05\text{ g }HC_2H_3O_2} = \frac{0.262\text{ mol}}{0.500\text{ L}}$$
$$= 0.524\text{ M}$$

9A The molar mass of $NaNO_3$ is 84.99 g/mol. We recall that "M" stands for "mol $NaNO_3$/L soln."

$$\text{mass }NaNO_3 = 125\text{ mL soln} \times \frac{1\text{ L}}{1000\text{ mL}} \times \frac{10.8\text{ mol }NaNO_3}{1\text{ L soln}} \times \frac{84.99\text{ g }NaNO_3}{1\text{ mol NaCl}} = 115\text{ g }NaNO_3$$

9B We begin by determining the molar mass of $Na_2SO_4{\cdot}10H_2O$. The amount of solute needed is computed from the concentration and volume of the solution.

$$\text{molar mass} = (2 \times 22.99\text{ g Na}) + 32.07\text{ g S} + (14 \times 16.00\text{ g O}) + (20 \times 1.008\text{ g H}) = 322.21\text{ g/mol}$$
$$\text{mass }Na_2SO_4{\cdot}10H_2O = 355\text{ mL soln} \times \frac{1\text{ L}}{1000\text{ mL}} \times \frac{0.445\text{ mol }Na_2SO_4}{1\text{ L soln}} \times \frac{1\text{ mol }Na_2SO_4{\cdot}10H_2O}{1\text{ mol }Na_2SO_4}$$
$$\times \frac{322.21\text{ g }Na_2SO_4{\cdot}10H_2O}{1\text{ mol }Na_2SO_4{\cdot}10H_2O} = 50.9\text{ g }Na_2SO_4{\cdot}10H_2O$$

10A The amount of solute in the concentrated solution is still present when the solution is diluted. We take advantage of an alternate definition of molarity: millimoles of solute/milliliter of solution.

$$\text{amount }K_2CrO_4 = 15.00\text{ mL} \times \frac{0.450\text{ mmol }K_2CrO_4}{1\text{ mL soln}} = 6.75\text{ mmol }K_2CrO_4$$
$$K_2CrO_4\text{ molarity, dilute solution} = \frac{6.75\text{ mmol }K_2CrO_4}{100.00\text{ mL soln}} = 0.0675\text{ M}$$

10B We know the initial concentration (0.105 M) and volume (275 mL) of the solution, along with its final volume (237 mL). The final concentration equals the initial concentration times a ratio of the two volumes.

$$c_f = c_i \times \frac{V_i}{V_f} = 0.105\text{ M} \times \frac{275\text{ mL}}{237\text{ mL}} = 0.122\text{ M}$$

11A The balanced equation is $K_2CrO_4(aq) + 2\,AgNO_3(aq) \longrightarrow Ag_2CrO_4(s) + 2\,KNO_3(aq)$. The molar mass of Ag_2CrO_4 is 331.73 g/mol. The conversions needed are mass $Ag_2CrO_4 \longrightarrow$ amount $Ag_2CrO_4 \longrightarrow$ amount $K_2CrO_4 \longrightarrow$ volume $K_2CrO_4(aq)$.

$$\text{volume }K_2CrO_4 = 1.50\text{ g }Ag_2CrO_4 \times \frac{1\text{ mol }Ag_2CrO_4}{331.73\text{ g }Ag_2CrO_4} \times \frac{1\text{ mol }K_2CrO_4}{1\text{ mol }Ag_2CrO_4} \times \frac{1\text{ L soln}}{0.250\text{ mol }K_2CrO_4} \times \frac{1000\text{ mL}}{1\text{ L}}$$
$$= 18.1\text{ mL soln}$$

11B The balanced chemical equation is $K_2CrO_4(aq) + 2\,AgNO_3(aq) \longrightarrow Ag_2CrO_4(s) + 2\,KNO_3(aq)$. The conversions needed are mass $Ag_2CrO_4 \longrightarrow$ amount $Ag_2CrO_4 \longrightarrow$ amount $AgNO_3 \longrightarrow$ volume $AgNO_3(aq)$. In the calculation below, each arrow in the previous sentence gives rise to a conversion factor.

$$\text{volume }AgNO_3(aq) = 1.00\text{ g }Ag_2CrO_4 \times \frac{1\text{ mol }Ag_2CrO_4}{331.7\text{ g }Ag_2CrO_4} \times \frac{2\text{ mol }AgNO_3}{1\text{ mol }Ag_2CrO_4} \times \frac{1\text{ L soln}}{0.150\text{ mol }AgNO_3}$$
$$\times \frac{1000\text{ mL }AgNO_3(aq)}{1\text{ L}} = 40.2\text{ mL }AgNO_3(aq)$$

12A The balanced equation is $P_4(l) + 6\,Cl_2(g) \longrightarrow 4\,PCl_3(l)$. We calculate the mass of PCl_3 produced from each reactant.

$$\text{mass }PCl_3 = 215\text{ g }P_4 \times \frac{1\text{ mol }P_4}{123.90\text{ g }P_4} \times \frac{4\text{ mol }PCl_3}{1\text{ mol }P_4} \times \frac{137.33\text{ g }PCl_3}{1\text{ mol }PCl_3} = 953\text{ g }PCl_3$$
$$\text{mass }PCl_3 = 725\text{ g }Cl_2 \times \frac{1\text{ mol }Cl_2}{70.91\text{ g }Cl_2} \times \frac{4\text{ mol }PCl_3}{6\text{ mol }Cl_2} \times \frac{137.33\text{ g }PCl_3}{1\text{ mol }PCl_3} = 936\text{ g }PCl_3$$

Thus, 936 g PCl_3 are produced; there is not enough Cl_2 to produce any more.

12B Since data are supplied and the answer is requested in kilograms (thousands of grams), we can use kilomoles (thousands of moles) to solve the problem. We calculate the amount in kilomoles of $POCl_3$ that would be produced if each of the reactants were completely converted to product. The smallest of these amounts is that actually produced; there is not enough of the limiting reactant to produce any more.

$$\text{amount }POCl_3 = 1.00\text{ kg }PCl_3 \times \frac{1\text{ kmol }PCl_3}{137.33\text{ kg }PCl_3} \times \frac{10\text{ kmol }POCl_3}{6\text{ kmol }PCl_3} = 0.0121_4\text{ kmol }POCl_3$$
$$\text{amount }POCl_3 = 1.00\text{ kg }Cl_2 \times \frac{1\text{ kmol }Cl_2}{70.905\text{ kg }Cl_2} \times \frac{10\text{ kmol }POCl_3}{6\text{ kmol }Cl_2} = 0.0235_1\text{ kmol }POCl_3$$

$$\text{amount } POCl_3 = 1.00 \text{ kg } P_4O_{10} \times \frac{1 \text{ kmol } P_4O_{10}}{283.89 \text{ kg } P_4O_{10}} \times \frac{10 \text{ kmol } POCl_3}{1 \text{ kmol } P_4O_{10}} = 0.0352_2 \text{ kmol } POCl_3$$

Thus, 0.0121_4 kmol $POCl_3$ is produced. We determine the mass of the product.

$$\text{mass } POCl_3 = 0.0121_4 \text{ kmol } POCl_3 \times \frac{153.33 \text{ kg } POCl_3}{1 \text{ kmol } POCl_3} = 1.86 \text{ kg } POCl_3$$

13A The 725 g Cl_2 limits the mass of product formed. The $P_4(l)$ therefore is the reactant in excess. The quantity of excess reactant is sufficient to form the excess product: 953 g PCl_3 – 936 g PCl_3 = 17 g PCl_3. We calculate how much P_4 this is, both in the traditional way and by using the initial (215 g P_4) and final (953 g PCl_3) values of the previous calculation.

$$\text{mass } P_4 = 17 \text{ g } PCl_3 \times \frac{1 \text{ mol } PCl_3}{137.33 \text{ g } PCl_3} \times \frac{1 \text{ mol } P_4}{4 \text{ mol } PCl_3} \times \frac{123.90 \text{ g } P_4}{1 \text{ mol } P_4} = 3.8 \text{ g } P_4$$

$$\text{mass } P_4 = 17 \text{ g } PCl_3 \times \frac{215 \text{ g } P_4}{953 \text{ g } PCl_3} = 3.8 \text{ g } P_4$$

13B We compute the amount of $H_2O(l)$ formed from each mass of reactant, to determine the limiting reactant.

$$\text{amount } H_2O = 12.2 \text{ g } H_2 \times \frac{1 \text{ mol } H_2}{2.016 \text{ g } H_2} \times \frac{1 \text{ mol } H_2O}{2 \text{ mol } H_2} = 3.02_6 \text{ mol } H_2O$$

$$\text{amount } H_2O = 154 \text{ g } O_2 \times \frac{1 \text{ mol } O_2}{32.00 \text{ g } O_2} \times \frac{1 \text{ mol } H_2O}{1 \text{ mol } O_2} = 4.81_3 \text{ mol } H_2O$$

Since H_2 is the limiting reactant, we compute the mass of O_2 needed to react with all of the H_2.

$$\text{mass } O_2 \text{ reacting} = 3.02_6 \text{ mol } H_2O \text{ produced} \times \frac{1 \text{ mol } O_2}{1 \text{ mol } H_2O} \times \frac{32.00 \text{ g } O_2}{1 \text{ mol } O_2} = 96.8 \text{ g } O_2 \text{ reacting}$$

mass O_2 remaining = 154 g originally present – 96.8 g O_2 reacting = 57 g O_2 remaining

14A **(a)** The theoretical yield is the product mass we predict by calculation.

$$\text{mass } CH_2O(g) = 1.00 \text{ mol } CH_3OH \times \frac{1 \text{ mol } CH_2O}{1 \text{ mol } CH_3OH} \times \frac{30.03 \text{ g } CH_2O}{1 \text{ mol } CH_2O} = 30.03 \text{ g } CH_2O$$

(b) The actual yield is what is obtained experimentally: 25.7 g $CH_2O(g)$.

(c) The percent yield is the ratio of actual to theoretical yields:

$$\% \text{ yield} = \frac{25.7 \text{ g } CH_2O \text{ produced}}{30.03 \text{ g } CH_2O \text{ calculated}} \times 100\% = 85.6\% \text{ yield}$$

14B We determine the mass of product formed by each reactant, assuming that all of each yields product.

$$\text{mass } PCl_3 = 25.0 \text{ g } P_4 \times \frac{1 \text{ mol } P_4}{123.90 \text{ g } P_4} \times \frac{4 \text{ mol } PCl_3}{1 \text{ mol } P_4} \times \frac{137.33 \text{ g } PCl_3}{1 \text{ mol } PCl_3} = 111 \text{ g } PCl_3$$

$$\text{mass } PCl_3 = 91.5 \text{ g } Cl_2 \times \frac{1 \text{ mol } Cl_2}{70.91 \text{ g } Cl_2} \times \frac{4 \text{ mol } PCl_3}{6 \text{ mol } Cl_2} \times \frac{137.33 \text{ g } PCl_3}{1 \text{ mol } PCl_3} = 118 \text{ g } PCl_3$$

The limiting reactant is P_4, and 111 g PCl_3 should be produced. This is the theoretical yield. The actual yield is 104 g PCl_3. Thus, the percent yield of the reaction is $\frac{104 \text{ g } PCl_3 \text{ produced}}{111 \text{ g } PCl_3 \text{ calculated}} \times 100\% = 93.7\%$ yield

15A The balanced equation is $2\,NH_3 + CO_2 \longrightarrow CO(NH_2)_2 + H_2O$. We need to distinguish between mass of urea produced (actual yield) and mass of urea predicted (theoretical yield).

$$\text{mass } CO_2 = \frac{50.0 \text{ g } CO(NH_2)_2}{\text{produced}} \times \frac{100.0 \text{ g predicted}}{87.5 \text{ g produced}} \times \frac{1 \text{ mol } CO(NH_2)_2}{60.1 \text{ g } CO(NH_2)_2} \times \frac{1 \text{ mol } CO_2}{1 \text{ mol } CO(NH_2)_2} \times \frac{40.01 \text{ g } CO_2}{1 \text{ mol } CO_2} = 38.0 \text{ g } CO_2 \text{ needed}$$

15B Careful labeling of each part of each conversion factor is important when working problems dealing with percent purity. Note that you cannot calculate the mole weight of an impure material, only of a compound or an element. Also notice how the use of the labels "produced" and "calculated" enables one to properly apply the percent yield of the reaction.

$$\text{mass } C_6H_{11}OH = 45.0 \text{ g } C_6H_{10} \text{ produced} \times \frac{100.0 \text{ g } C_6H_{10} \text{ calc'd}}{86.2 \text{ g } C_6H_{10} \text{ prodc'd}} \times \frac{1 \text{ mol } C_6H_{10}}{82.1 \text{ g } C_6H_{10}} \times \frac{1 \text{ mol } C_6H_{11}OH}{1 \text{ mol } C_6H_{10}} \times \frac{100.2 \text{ g pure } C_6H_{11}OH}{1 \text{ mol } C_6H_{11}OH} \times \frac{100.0 \text{ g impure } C_6H_{11}OH}{92.3 \text{ g pure } C_6H_{11}OH} = 69.0 \text{ g impure } C_6H_{11}OH$$

16A We can trace the nitrogen through the sequence of reactions. We notice that 4 moles of N (as 4 mol NH_3) is consumed in the first reaction, and 4 moles of N (as 4 mole NO) is produced. In the second reaction, 2 moles of N (as 2 mol NO) is consumed and 2 moles of N (as 2 mol NO_2) is produced. In the last reaction, 3 moles of N (as 3 mol NO_2) is consumed and just 2 moles of N (as 2 mol HNO_3) is consumed.

$$\text{mass HNO}_3 = 1.00 \text{ kg NH}_3 \times \frac{1000 \text{ g NH}_3}{1 \text{ kg NH}_3} \times \frac{1 \text{ mol NH}_3}{17.03 \text{ g NH}_3} \times \frac{4 \text{ mol NO}}{4 \text{ mol NH}_3} \times \frac{2 \text{ mol NO}_2}{2 \text{ mol NO}}$$
$$\times \frac{2 \text{ mol HNO}_3}{3 \text{ mol NO}_2} \times \frac{63.01 \text{ g HNO}_3}{1 \text{ mol HNO}_3} = 2.47 \times 10^3 \text{ g HNO}_3$$

If we assume that the NO by-product in the last reaction is recycled (which is the case industrially) and all the nitrogen in NH_3 is converted to nitrogen in HNO_3, we obtain a different result.

$$\text{mass HNO}_3 = 1.00 \text{ kg NH}_3 \times \frac{1000 \text{ g NH}_3}{1 \text{ kg NH}_3} \times \frac{1 \text{ mol NH}_3}{17.03 \text{ g NH}_3} \times \frac{1 \text{ mol N}}{1 \text{ mol NH}_3} \times \frac{1 \text{ mol HNO}_3}{1 \text{ mol N}} \times \frac{63.01 \text{ g HNO}_3}{1 \text{ mol HNO}_3}$$
$$= 3.70 \times 10^3 \text{ g HNO}_3$$

16B $$\text{mass H}_2\text{(Al)} = 0.710 \text{ g alloy} \times \frac{0.700 \text{ g Al}}{1.000 \text{ g alloy}} \times \frac{1 \text{ mol Al}}{26.98 \text{ g Al}} \times \frac{3 \text{ mol H}_2}{2 \text{ mol Al}} \times \frac{2.016 \text{ g H}_2}{1 \text{ mol H}_2} = 0.0557 \text{ g}$$
$$\text{mass H}_2\text{(Mg)} = 0.710 \text{ g alloy} \times \frac{0.300 \text{ g Mg}}{1.000 \text{ g alloy}} \times \frac{1 \text{ mol Mg}}{24.31 \text{ g Mg}} \times \frac{1 \text{ mol H}_2}{1 \text{ mol Mg}} \times \frac{2.016 \text{ g H}_2}{1 \text{ mol H}_2} = 0.0177 \text{ g}$$

total mass of H_2 = 0.0557 g H_2 from Al + 0.0177 g H_2 from Mg = 0.0734 g H_2

SUMMARIZING EXAMPLE CALCULATIONS

1. Formula expression: $Na_2CO_3(aq) + NO(g) + O_2(g) \longrightarrow NaNO_2(aq) + CO_2(g)$
Balance Na: $Na_2CO_3(aq) + NO(g) + O_2(g) \longrightarrow 2\,NaNO_2(aq) + CO_2(g)$
Balance N: $Na_2CO_3(aq) + 2\,NO(g) + O_2(g) \longrightarrow 2\,NaNO_2(aq) + CO_2(g)$
Balance O and × 2: $2\,Na_2CO_3(aq) + 4\,NO(g) + O_2(g) \longrightarrow 4\,NaNO_2(aq) + 2\,CO_2(g)$

2. Determine the theoretical amount produced from each limited reactant.

$$\text{amount NaNO}_2 = 225 \text{ mL Na}_2\text{CO}_3\text{(aq)} \times \frac{1 \text{ L}}{1000 \text{ mL}} \times \frac{1.50 \text{ mol Na}_2\text{CO}_3}{1 \text{ L soln}} \times \frac{4 \text{ mol NaNO}_2}{2 \text{ mol Na}_2\text{CO}_3}$$
$$= 0.675 \text{ mol NaNO}_2$$
$$\text{amount NaNO}_2 = 22.1 \text{ g NO} \times \frac{1 \text{ mol NO}}{30.01 \text{ g NO}} \times \frac{4 \text{ mol NaNO}_2}{4 \text{ mol NO}} = 0.736 \text{ mol NaNO}_2$$

Na_2CO_3 is the limiting reactant. Determine the theoretical yield of $NaNO_2$.

$$\text{mass NaNO}_2 = 0.675 \text{ mol NaNO}_2 \times \frac{69.00 \text{ g NaNO}_2}{1 \text{ mol NaNO}_2} = 46.6 \text{ g NaNO}_2$$

3. Now determine the actual yield of the reaction.

$$\text{product mass} = 46.6 \text{ g NaNO}_2 \text{ calculated} \times \frac{95.0 \text{ g NaNO}_2 \text{ produced}}{100.0 \text{ g NaNO}_2 \text{ calculated}} = 44.3 \text{ g NaNO}_2 \text{ produced}$$

REVIEW QUESTIONS

1. **(a)** The symbol "$\xrightarrow{\Delta}$" indicates that a reaction mixture is heated to produce the reaction.
(b) The symbol "(aq)" indicates that the species preceeding this symbol is dissolved in aqueous solution, that is, it indicates a solution with water as the solvent.
(c) The stoichiometric coefficient is the number that appears in a chemical equation immediately before the chemical formula of a species.
(d) The net equation is the chemical equation that remains after species that appear on both sides of an equation are "cancelled." The term also is used to describe an equation that summarizes the overall result of a process consisting of several reactions.

2. **(a)** One "balances a chemical equation" by inserting stoichiometric coefficients into the formula expression, so that the resulting equation has the same number and type of atoms on each side.
(b) When one "prepares a solution by dilution" one begins with a more concentrated solution (a homogeneous mixture with a larger concentration of solute) and adds solvent, thus producing a less concentrated (or more dilute) solution.
(c) One "determines the limiting reactant in a reaction" by discovering which reactant will produce the smallest quantity of product. That reactant will limit the quantity of product that can be formed from the other reactants, and also limit the quantity that will be consumed of the other reactants.

3. **(a)** A chemical formula is a short-hand representation of a chemical species: atom, ion, or molecule. A chemical equation is a written representation of a chemical reaction; it involves two or more species. Whereas a chemical formula is rather analogous to a "word," chemical equations parallel "sentences."
(b) A decomposition reaction is one in which a compound is broken down into simpler substances. A synthesis reaction is one in which two or more substances combine to form a third.

(c) The solute is the substance that is dispersed in a solution. The solvent is the substance that does the dispersing. Usually, a solution is of the same physical state (solid, liquid, or gas) as the solvent, and the solvent is the component present in larger amount.

(d) The actual yield of a chemical reaction is the quantity of product that actually was formed. The percent yield relates the actual yield to the quantity of product that was calculated to be produced, assuming that all reactants produced only one set of products and the reaction continued until one reactant was exhausted.

4. (a) $Na_2SO_4(s) + 4\ C(s) \longrightarrow Na_2S(s) + 4\ CO(g)$ (b) $4\ HCl(g) + O_2(g) \longrightarrow 2\ H_2O(l) + 2\ Cl_2(g)$
(c) $PCl_5(l) + 4\ H_2O(l) \longrightarrow H_3PO_4(aq) + 5\ HCl(aq)$
(d) $3\ PbO(s) + 2\ NH_3(g) \longrightarrow 3\ Pb(s) + N_2(g) + 3\ H_2O(l)$
(e) $Mg_3N_2(s) + 6\ H_2O(l) \longrightarrow 3\ Mg(OH)_2(s) + 2\ NH_3(g)$

5. (a) $2\ Mg(s) + O_2(g) \longrightarrow 2\ MgO(s)$ (b) $2\ NO(g) + O_2(g) \longrightarrow 2\ NO_2(g)$
(c) $2\ C_2H_6(g) + 7\ O_2(g) \longrightarrow 4\ CO_2(g) + 6\ H_2O(l)$
(d) $Ag_2SO_4(aq) + BaI_2(aq) \longrightarrow BaSO_4(s) + 2\ AgI(s)$

6. (a) $C_7H_{16}(l) + 11\ O_2(g) \longrightarrow 7\ CO_2(g) + 8\ H_2O(l)$
(b) $C_4H_9OH(l) + 6\ O_2(g) \longrightarrow 4\ CO_2(g) + 5\ H_2O(l)$
(c) $2\ HI(aq) + Na_2CO_3(aq) \longrightarrow 2\ NaI(aq) + H_2O(l) + CO_2(g)$
(d) $3\ NaOH(aq) + FeCl_3(aq) \longrightarrow Fe(OH)_3(s) + 3\ NaCl(aq)$

7. Expression **(c)** is incorrect because KClO is potassium hypochlorite, but the stated product is potassium chloride, KCl. Expression **(a)** and **(b)** are incorrect because O(g) is not normally produced in chemical reactions; $O_2(g)$ is the more stable species. The correct equation is $2\ KClO_3(s) \longrightarrow 2\ KCl(s) + 3\ O_2(g)$.

8. For the reaction $2\ H_2S + SO_2 \longrightarrow 3\ S + 2\ H_2O$
(1) FALSE 3 moles of S are produced per *two* moles of H_2S.
(2) FALSE 3 *moles* of S are produced for every *mole* of SO_2 consumed.
(3) TRUE 1 mole of H_2O *is* produced per mole of H_2S consumed.
(4) TRUE Two-thirds of the S produced *does* come from the H_2S.
(5) FALSE There are *five* moles of products and *three* moles of reactants.

9. The conversion factor is obtained from the balanced chemical equation.

$$\text{no. mol FeCl}_3 = 7.26 \text{ mol Cl}_2 \times \frac{2 \text{ mol FeCl}_3}{3 \text{ mol Cl}_2} = 4.84 \text{ mol FeCl}_3$$

10. The pivotal conversion factor, from the balanced equation, enables one to related the amounts of O_2 and $KClO_3$.

$$\text{mass O}_2 = 43.4 \text{ g KClO}_3 \times \frac{1 \text{ mol KClO}_3}{122.5 \text{ g KClO}_3} \times \frac{3 \text{ mol O}_2}{2 \text{ mol KClO}_3} \times \frac{32.00 \text{ g O}_2}{1 \text{ mol O}_2} = 17.0 \text{ g O}_2$$

11. Each calculation uses the stoichiometric coefficients from the balanced chemical equation and the mole weight of the reactant.

$$\text{mass Cl}_2 = 0.337 \text{ mol PCl}_3 \times \frac{6 \text{ mol Cl}_2}{4 \text{ mol PCl}_3} \times \frac{70.91 \text{ g Cl}_2}{1 \text{ mol Cl}_2} = 35.8 \text{ g Cl}_2$$

$$\text{mass P}_4 = 0.337 \text{ mol PCl}_3 \times \frac{1 \text{ mol P}_4}{4 \text{ mol PCl}_3} \times \frac{123.9 \text{ g P}_4}{1 \text{ mol P}_4} = 10.4 \text{ g P}_4$$

12. (a) $$\text{amount O}_2 = 156 \text{ g CO}_2 \times \frac{1 \text{ mol CO}_2}{44.01 \text{ g CO}_2} \times \frac{3 \text{ mol O}_2}{2 \text{ mol CO}_2} = 5.32 \text{ mol O}_2$$

(b) $$\text{mass KO}_2 = 100.0 \text{ g CO}_2 \times \frac{1 \text{ mol CO}_2}{44.01 \text{ g CO}_2} \times \frac{4 \text{ mol KO}_2}{2 \text{ mol CO}_2} \times \frac{71.10 \text{ g KO}_2}{1 \text{ mol KO}_2} = 323.1 \text{ g KO}_2$$

(c) $$\text{no. O}_2 \text{ molecules} = 1.00 \text{ mg KO}_2 \times \frac{1 \text{ g KO}_2}{1000 \text{ mg}} \times \frac{1 \text{ mol KO}_2}{71.10 \text{ g KO}_2} \times \frac{3 \text{ mol O}_2}{4 \text{ mol KO}_2} \times \frac{6.022 \times 10^{23} \text{ molecules}}{1 \text{ mol O}_2}$$

$$= 6.35 \times 10^{18} \text{ O}_2 \text{ molecules}$$

13. (a) $$\text{CH}_3\text{OH molarity} = \frac{2.92 \text{ mol CH}_3\text{OH}}{7.16 \text{ L}} = 0.408 \text{ M}$$

(b) C_2H_5OH molarity $= \dfrac{7.69 \text{ mmol } C_2H_5OH}{50.00 \text{ mL}} = 0.154 \text{ M}$

(c) $CO(NH_2)_2$ molarity $= \dfrac{25.2 \text{ g } CO(NH_2)_2}{275 \text{ mL}} \times \dfrac{1 \text{ mol } CO(NH_2)_2}{60.06 \text{ g } CO(NH_2)_2} \times \dfrac{1000 \text{ mL}}{1 \text{ L}} = 1.53 \text{ M}$

(d) molarity $= \dfrac{18.5 \text{ mL } C_3H_5(OH)_3}{375 \text{ mL soln}} \times \dfrac{1.26 \text{ g}}{1 \text{ cm}^3} \times \dfrac{1 \text{ mol } C_3H_5(OH)_3}{92.0 \text{ g } C_3H_5(OH)_3} \times \dfrac{1000 \text{ mL}}{1 \text{ L}} = 0.676 \text{ M}$

14. **(a)** no. mol NaI $= 2.55 \times 10^3 \text{ L} \times \dfrac{0.125 \text{ mol NaI}}{1 \text{ L soln}} = 319 \text{ mol NaI}$

(b) mass $Na_2CO_3 = 475 \text{ mL} \times \dfrac{1 \text{ L}}{1000 \text{ mL}} \times \dfrac{0.398 \text{ mol } Na_2CO_3}{1 \text{ L}} \times \dfrac{106.0 \text{ g } Na_2CO_3}{1 \text{ mol } Na_2CO_3} = 20.0 \text{ g } Na_2CO_3$

(c) no. mg $CaCl_2 = 1.00 \text{ mL} \times \dfrac{0.148 \text{ mmol } CaCl_2}{1 \text{ mL}} \times \dfrac{111.0 \text{ mg } CaCl_2}{1 \text{ mmol } CaCl_2} = 16.4 \text{ mg } CaCl_2$

15. A 1.00 M KCl solution contains 1 mol KCl per liter of solution. The molar mass of KCl is 74.6 g. Thus, a 1.00 M KCl solution contains 74.6 g KCl per liter of solution.

"1.00 L … containing 100 g" is incorrect; 1.00 L should contain 74.6 g.

"500 mL … containing 74.6 g" also is incorrect; 74.6 g should be contained in 1000 mL.

"a solution … 7.46 mg KCl/mL" is incorrect; there should be 74.6 mg.

5.00 L of 1.00 M KCl contains five times the mass of solute as does 1.00 L of this solution, 373 g.

The last description is correct.

16. volume conc'd $AgNO_3$ soln $= 250.0 \text{ mL dilute soln} \times \dfrac{0.423 \text{ mmol } AgNO_3}{1 \text{ mL dilute soln}} \times \dfrac{1 \text{ mL conc'd soln}}{0.625 \text{ mmol } AgNO_3}$

$= 163 \text{ mL conc'd } AgNO_3 \text{ soln}$

17. The same amount in moles of K_2SO_4 is present in both solutions. That amount is given in the numerator of the following expression.

$$K_2SO_4 \text{ molarity} = \dfrac{135 \text{ mL} \times \dfrac{0.188 \text{ mmol } K_2SO_4}{1 \text{ mL}}}{105 \text{ mL}} = 0.242 \text{ M}$$

18. The balanced chemical equation provides a conversion factor between the two compounds.

mass $CuCO_3 = 415 \text{ mL} \times \dfrac{1 \text{ L}}{1000 \text{ mL}} \times \dfrac{0.275 \text{ mol } Cu(NO_3)_2}{1 \text{ L}} \times \dfrac{1 \text{ mol } CuCO_3}{1 \text{ mol } Cu(NO_3)_2} \times \dfrac{123.6 \text{ g } CuCO_3}{1 \text{ mol } CuCO_3}$

$= 14.1 \text{ g } CuCO_3$

19. After determining the amount of $CaCO_3$ (100.09 g/mol), we find the volume of HCl.

volume HCl(aq) $= 1.75 \text{ g } CaCO_3 \times \dfrac{1 \text{ mol } CaCO_3}{100.09 \text{ g}} \times \dfrac{2 \text{ mol HCl}}{1 \text{ mol } CaCO_3} \times \dfrac{1 \text{ L soln}}{2.35 \text{ mol HCl}} \times \dfrac{1000 \text{ mL}}{1 \text{ L}}$

$= 14.9 \text{ mL HCl(aq) solution}$

20. In this situation, since 2 mol H_2O are required per mol CaH_2, 1.52 mol H_2O is the limiting reactant. Thus, the amount in moles of H_2 can be computed as follows.

amount $H_2 = 1.52 \text{ mol } H_2O \times \dfrac{2 \text{ mol } H_2}{2 \text{ mol } H_2O} = 1.52 \text{ mol } H_2$

21. Determine the number of moles of NO produced by each reactant. The one producing the smaller amount of NO is the limiting reactant.

no. mol NO $= 0.696 \text{ mol Cu} \times \dfrac{2 \text{ mol NO}}{3 \text{ mol Cu}} = 0.464 \text{ mol NO}$

no. mol NO $= 136 \text{ mL } HNO_3(aq) \times \dfrac{1 \text{ L}}{1000 \text{ mL}} \times \dfrac{6.0 \text{ mol } HNO_3}{1 \text{ L}} \times \dfrac{2 \text{ mol NO}}{8 \text{ mol } HNO_3} = 0.204 \text{ mol NO}$

Since $HNO_3(aq)$ is the limiting reactant, it will be completely consumed, leaving some Cu unreacted.

22. **(a)** Since the stoichiometry indicates that 1 mole CCl_2F_2 is produced per mole CCl_4, the use of 1.80 mole CCl_4 will produce 1.80 mole CCl_2F_2. This is the theoretical yield of the reaction.

(b) The actual yield of the reaction is the amount actually produced, 1.55 mol CCl_2F_2.

(c) % yield $= \dfrac{1.55 \text{ mol } CCl_2F_2 \text{ obtained}}{1.80 \text{ mol } CCl_2F_2 \text{ calculated}} \times 100\% = 86.1\%$ yield

23. **(a)** mass $C_6H_{10} = 100.0 \text{ g } C_6H_{11}OH \times \dfrac{1 \text{ mol } C_6H_{11}OH}{100.16 \text{ g } C_6H_{11}OH} \times \dfrac{1 \text{ mol } C_6H_{10}}{1 \text{ mol } C_6H_{11}OH} \times \dfrac{82.146 \text{ g } C_6H_{10}}{1 \text{ mol } C_6H_{10}}$

$= 82.01 \text{ g } C_6H_{10} = \text{theoretical yield}$

(b) percent yield $= \dfrac{64.0\text{ g } C_6H_{10}\text{ produced}}{82.01\text{ g } C_6H_{10}\text{ calculated}} \times 100\% = 78.0\%$ yield

(c) mass $C_6H_{11}OH = 100.0\text{ g } C_6H_{10}\text{ produced} \times \dfrac{1.000\text{ g calculated}}{0.780\text{ g produced}} \times \dfrac{1\text{ mol } C_6H_{10}}{82.15\text{ g } C_6H_{10}}$
$\times \dfrac{1\text{ mol } C_6H_{11}OH}{1\text{ mol } C_6H_{10}} \times \dfrac{100.2\text{ g } C_6H_{11}OH}{1\text{ mol } C_6H_{11}OH} = 156\text{ g } C_6H_{11}OH$ needed

24. mass $CaCO_3 = 0.981\text{ g } CO_2 \times \dfrac{1\text{ mol } CO_2}{44.01\text{ g } CO_2} \times \dfrac{1\text{ mol } CaCO_3}{1\text{ mol } CO_2} \times \dfrac{100.1\text{ g } CaCO_3}{1\text{ mol } CaCO_3} = 2.23\text{ g } CaCO_3$

$\%\ CaCO_3 = \dfrac{2.23\text{ g } CaCO_3}{3.28\text{ g sample}} \times 100\% = 68.0\%\ CaCO_3$

EXERCISES

Writing and Balancing Chemical Equations

25. **(a)** $Cr_2O_3(s) + 2\ Al(s) \xrightarrow{\Delta} Al_2O_3(s) + 2\ Cr(l)$
(b) $CaC_2(s) + 2\ H_2O(l) \longrightarrow Ca(OH)_2(s) + C_2H_2(g)$
(c) $3\ H_2(g) + Fe_2O_3(s) \xrightarrow{\Delta} 2\ Fe(l) + 3\ H_2O(g)$
(d) $NCl_3(g) + 3\ H_2O(l) \longrightarrow NH_3(g) + 3\ HOCl(aq)$

26. **(a)** $(NH_4)_2Cr_2O_7(s) \xrightarrow{\Delta} Cr_2O_3(s) + N_2(g) + 4\ H_2O(g)$
(b) $3\ NO_2(g) + H_2O(l) \longrightarrow 2\ HNO_3(aq) + NO(g)$
(c) $2\ H_2S(g) + SO_2(g) \longrightarrow 3\ S(s) + 2\ H_2O(g)$
(d) $SO_2Cl_2 + 8\ HI \longrightarrow H_2S + 2\ H_2O + 2\ HCl + 4\ I_2$

27. **(a)** $2\ C_4H_{10} + 13\ O_2(g) \longrightarrow 8\ CO_2(g) + 10\ H_2O(l)$
(b) $2\ C_3H_7OH + 9\ O_2(g) \longrightarrow 6\ CO_2(g) + 8\ H_2O(l)$
(c) $HC_3H_5O_3(s) + 3\ O_2(g) \longrightarrow 3\ CO_2(g) + 3\ H_2O(l)$

28. **(a)** $2\ C_3H_6(g) + 9\ O_2(g) \longrightarrow 6\ CO_2(g) + 6\ H_2O(l)$
(b) $2\ CH_2OHCHOHCH_2OH(l) + 7\ O_2(g) \longrightarrow 6\ CO_2(g) + 8\ H_2O(l)$
(c) $C_6H_5COSH(s) + 9\ O_2(g) \longrightarrow 7\ CO_2(g) + 3\ H_2O(l) + SO_2(g)$

29. **(a)** $NH_4NO_3(s) \xrightarrow{\Delta} N_2O(g) + 2\ H_2O(g)$
(b) $Na_2CO_3(aq) + 2\ HCl(aq) \longrightarrow 2\ NaCl(aq) + H_2O(l) + CO_2(g)$
(c) $2\ CH_4(g) + 2\ NH_3(g) + 3\ O_2(g) \longrightarrow 2\ HCN(g) + 6\ H_2O(g)$

30. **(a)** $2\ SO_2(g) + O_2(g) \longrightarrow 2\ SO_3(g)$
(b) $CaCO_3(s) + H_2O(l) + CO_2(aq) \longrightarrow Ca(HCO_3)_2(aq)$
(c) $4\ NH_3(g) + 6\ NO(g) \longrightarrow 5\ N_2(g) + 6\ H_2O(l)$

Stoichiometry of Chemical Reactions

31. **(a)** no. mol $O_2 = 32.8\text{ g } KClO_3 \times \dfrac{1\text{ mol } KClO_3}{122.6\text{ g } KClO_3} \times \dfrac{3\text{ mol } O_2}{2\text{ mol } KClO_3} = 0.401\text{ mol } O_2$

(b) mass $KClO_3 = 50.0\text{ g } O_2 \times \dfrac{1\text{ mol } O_2}{32.00\text{ g } O_2} \times \dfrac{2\text{ mol } KClO_3}{3\text{ mol } O_2} \times \dfrac{122.6\text{ g } KClO_3}{1\text{ mol } KClO_3} = 128\text{ g } KClO_3$

(c) mass $KCl = 28.3\text{ g } O_2 \times \dfrac{1\text{ mol } O_2}{32.00\text{ g } O_2} \times \dfrac{2\text{ mol } KCl}{3\text{ mol } O_2} \times \dfrac{74.55\text{ g } KCl}{1\text{ mol } KCl} = 44.0\text{ g } KCl$

32. **(a)** amount $H_2 = 42.7\text{ g Fe} \times \dfrac{1\text{ mol Fe}}{55.85\text{ g Fe}} \times \dfrac{4\text{ mol } H_2}{3\text{ mol Fe}} = 1.02\text{ mol } H_2$

(b) mass $H_2O = 63.5\text{ g Fe} \times \dfrac{1\text{ mol Fe}}{55.85\text{ g Fe}} \times \dfrac{4\text{ mol } H_2O}{3\text{ mol Fe}} \times \dfrac{18.02\text{ g } H_2O}{1\text{ mol } H_2O} = 27.3\text{ g } H_2O$

(c) mass $Fe_3O_4 = 7.36\text{ mol } H_2 \times \dfrac{1\text{ mol } Fe_3O_4}{4\text{ mol } H_2} \times \dfrac{231.55\text{ g } Fe_3O_4}{1\text{ mol } Fe_3O_4} = 426\text{ g } Fe_3O_4$

33. Balance the given equation, then solve the problem. $2\,Ag_2CO_3(s) \xrightarrow{\Delta} 4\,Ag(s) + 2\,CO_2(g) + O_2(g)$

$$\text{mass } Ag_2CO_3 = 75.1\text{ g Ag} \times \frac{1\text{ mol Ag}}{107.87\text{ g Ag}} \times \frac{2\text{ mol } Ag_2CO_3}{4\text{ mol Ag}} \times \frac{275.75\text{ g } Ag_2CO_3}{1\text{ mol } Ag_2CO_3} = 96.0\text{ g } Ag_2CO_3$$

34. First, balance the equation. $Ca_3(PO_4)_2 + 4\,HNO_3 \longrightarrow Ca(H_2PO_4)_2 + 2\,Ca(NO_3)_2$

$$\text{mass } HNO_3 = 125\text{ kg } Ca(H_2PO_4)_2 \times \frac{1\text{ kmol } Ca(H_2PO_4)_2}{234.05\text{ kg } Ca(H_2PO_4)_2} \times \frac{4\text{ kmol } HNO_3}{1\text{ kmol } Ca(H_2PO_4)_2} \times \frac{63.01\text{ kg } HNO_3}{1\text{ kmol } HNO_3}$$
$$= 135\text{ kg } HNO_3$$

35. The balanced equation is $Fe_2O_3(s) + 3\,C(s) \xrightarrow{\Delta} 2\,Fe(l) + 3\,CO(g)$

$$\text{mass } Fe_2O_3 = 523\text{ kg Fe} \times \frac{1\text{ kmol Fe}}{55.85\text{ kg Fe}} \times \frac{1\text{ kmol } Fe_2O_3}{2\text{ kmol Fe}} \times \frac{159.7\text{ kg } Fe_2O_3}{1\text{ kmol } Fe_2O_3} = 748\text{ kg } Fe_2O_3$$

$$\%\ Fe_2O_3 \text{ in ore} = \frac{748\text{ kg } Fe_2O_3}{938\text{ kg ore}} \times 100\% = 79.7\%\ Fe_2O_3$$

36. The following reaction occurs: $2\,Ag_2O(s) \xrightarrow{\text{heat}} 4\,Ag(s) + O_2(g)$

$$\text{mass } Ag_2O = 0.187\text{ g } O_2 \times \frac{1\text{ mol } O_2}{32.0\text{ g } O_2} \times \frac{2\text{ mol } Ag_2O}{1\text{ mol } O_2} \times \frac{231.8\text{ g } Ag_2O}{1\text{ mol } Ag_2O} = 2.71\text{ g } Ag_2O$$

$$\%\ Ag_2O = \frac{2.71\text{ g } Ag_2O}{3.13\text{ g sample}} \times 100\% = 86.6\%\ Ag_2O$$

37. $2\,Al(s) + 6\,HCl(aq) \longrightarrow 2\,AlCl_3(aq) + 3\,H_2(g)$ First determine the mass of Al in the piece of foil.

$$\text{mass Al} = (10.25\text{ cm} \times 5.50\text{ cm} \times 0.601\text{ mm}) \times \frac{1\text{ cm}}{10\text{ mm}} \times \frac{2.70\text{ g}}{1\text{ cm}^3} = 9.15\text{ g Al}$$

$$\text{mass } H_2 = 9.15\text{ g Al} \times \frac{1\text{ mol Al}}{26.98\text{ g Al}} \times \frac{3\text{ mol } H_2}{2\text{ mol Al}} \times \frac{2.016\text{ g } H_2}{1\text{ mol } H_2} = 1.03\text{ g } H_2$$

38. $2\,Al(s) + 6\,HCl(aq) \longrightarrow 2\,AlCl_3(aq) + 3\,H_2(g)$

$$\text{mass } H_2 = 225\text{ mL soln} \times \frac{1.088\text{ g}}{1\text{ mL}} \times \frac{18.0\text{ g HCl}}{100.0\text{ g soln}} \times \frac{1\text{ mol HCl}}{36.46\text{ g HCl}} \times \frac{3\text{ mol } H_2}{6\text{ mol HCl}} \times \frac{2.016\text{ g } H_2}{1\text{ mol } H_2}$$
$$= 1.22\text{ g } H_2$$

39. In each balanced reaction, one mole of $O_2(g)$ is produced from two moles of solid reactant. Thus, the reaction that produces the most $O_2(g)$ per gram of reactant is the one with the smallest molar mass of reactant. NH_4NO_3 is 80.04 g/mol; Ag_2O is 231.74 g/mol; HgO is 216.59 g/mol; and $Pb(NO_3)_2$ is 331.2 g/mol. Thus, NH_4NO_3 (reaction 1) produces the most oxygen per gram of reactant.

40. First write the balanced chemical equation for each reaction.

$2\,Na(s) + 2\,HCl(aq) \longrightarrow 2\,NaCl(aq) + H_2(g)$ $\qquad Mg(s) + 2\,HCl(aq) \longrightarrow MgCl_2(aq) + H_2(g)$

$2\,Al(s) + 6\,HCl(aq) \longrightarrow 2\,AlCl_3(aq) + 3\,H_2(g)$ $\qquad Zn(s) + 2\,HCl(aq) \longrightarrow ZnCl_2(aq) + H_2(g)$

Three of the reactions—those of Na, Mg, and Zn—produce 1 mole of $H_2(g)$. The one of these three that produces the most hydrogen per gram of metal is the one for which the metal's atomic weight is the smallest, remembering to compare twice the atomic weight for Na. The atomic weights are: 2×23 u for Na, 24.3 u for Mg, and 65.4 u for Zn. Thus, among these three Mg produces the most H_2 per gram of metal, specifically 1 mol H_2 per 24.3 g Mg. In the case of Al, 3 moles of H_2 are produced by 2 moles of the metal, or 54 g Al. This reduces as follows: 3 mol H_2/54 g Al = 1 mol H_2/18 g Al.
Thus, Al produces the largest amount of H_2 per gram of metal.

41. Determine the amount of HCl needed to react with each component of the mixture.

$Mg(OH)_2 + 2\,HCl \longrightarrow MgCl_2 + 2\,H_2O$ $\qquad MgCO_3 + 2\,HCl \longrightarrow MgCl_2 + H_2O + CO_2$

$$\text{no. mol HCl} = 425\text{ g mixt.} \times \frac{35.2\text{ g } MgCO_3}{100.0\text{ g mixt.}} \times \frac{1\text{ mol } MgCO_3}{84.3\text{ g } MgCO_3} \times \frac{2\text{ mol HCl}}{1\text{ mol } MgCO_3} = 3.55\text{ mol HCl}$$

$$\text{no. mol HCl} = 425\text{ g mixt.} \times \frac{64.8\text{ g } Mg(OH)_2}{100.0\text{ g mixt.}} \times \frac{1\text{ mol } Mg(OH)_2}{58.3\text{ g } Mg(OH)_2} \times \frac{2\text{ mol HCl}}{1\text{ mol } MgCO_3} = 9.45\text{ mol HCl}$$

$$\text{mass HCl} = (3.55 + 9.45)\text{ mol HCl} \times \frac{36.45\text{ g HCl}}{1\text{ mol HCl}} = 474\text{ g HCl}$$

42. Determine the amount of CO_2 produced from each reactant.

$C_3H_8(g) + 5\,O_2(g) \longrightarrow 3\,CO_2(g) + 4\,H_2O(l)$ $\qquad 2\,C_4H_{10}(g) + 13\,O_2(g) \longrightarrow 8\,CO_2(g) + 10\,H_2O(l)$

$$\text{no. mol } CO_2 = 406 \text{ g mixt.} \times \frac{72.7 \text{ g } C_3H_8}{100.0 \text{ g mixt.}} \times \frac{1 \text{ mol } C_3H_8}{44.10 \text{ g } C_3H_8} \times \frac{3 \text{ mol } CO_2}{1 \text{ mol } C_3H_8} = 20.1 \text{ mol } CO_2$$

$$\text{no. mol } CO_2 = 406 \text{ g mixt.} \times \frac{27.3 \text{ g } C_4H_{10}}{100.0 \text{ g mixt}} \times \frac{1 \text{ mol } C_4H_{10}}{58.12 \text{ g } C_4H_{10}} \times \frac{8 \text{ mol } CO_2}{2 \text{ mol } C_4H_{10}} = 7.63 \text{ mol } CO_2$$

$$\text{mass } CO_2 = (20.1 + 7.63) \text{ mol } CO_2 \times \frac{44.01 \text{ g } CO_2}{1 \text{ mol } CO_2} = 1.22 \times 10^3 \text{ g } CO_2$$

43. The molar ratios given by the stoichiometric coefficients in the balanced chemical equations are used in the solution.

$$\text{amount } Cl_2 = 2.25 \times 10^3 \text{ g } CCl_2F_2 \times \frac{1 \text{ mol } CCl_2F_2}{120.91 \text{ g } CCl_2F_2} \times \frac{1 \text{ mol } CCl_4}{1 \text{ mol } CCl_2F_2} \times \frac{4 \text{ mol } Cl_2}{1 \text{ mol } CCl_4} = 74.4 \text{ mol } Cl_2$$

44. $\text{mass } C_2H_6 = 0.506 \text{ g } BaCO_3 \times \frac{1 \text{ mol } BaCO_3}{197.3 \text{ g } BaCO_3} \times \frac{1 \text{ mol } CO_2}{1 \text{ mol } BaCO_3} \times \frac{2 \text{ mol } C_2H_6}{4 \text{ mol } CO_2} \times \frac{30.07 \text{ g } C_2H_6}{1 \text{ mol } C_2H_6}$
$= 0.0386 \text{ g } C_2H_6$

Molarity

45. (a) $C_{12}H_{22}O_{11} \text{ concn} = \frac{150.0 \text{ g } C_{12}H_{22}O_{11}}{250.0 \text{ mL soln}} \times \frac{1000 \text{ mL}}{1 \text{ L}} \times \frac{1 \text{ mol } C_{12}H_{22}O_{11}}{342.3 \text{ g } C_{12}H_{22}O_{11}} = 1.753 \text{ M}$

(b) $CO(NH_2)_2 \text{ concn} = \frac{98.3 \text{ mg solid}}{5.00 \text{ mL soln}} \times \frac{97.9 \text{ mg } CO(NH_2)_2}{100.0 \text{ mg solid}} \times \frac{1 \text{ mmol } CO(NH_2)_2}{60.06 \text{ mg } CO(NH_2)_2} = 0.320 \text{ M}$

(c) $CH_3OH \text{ concn} = \frac{125.0 \text{ mL } CH_3OH}{15.0 \text{ L soln}} \times \frac{0.792 \text{ g}}{1 \text{ mL}} \times \frac{1 \text{ mol } CH_3OH}{32.04 \text{ g } CH_3OH} = 0.206 \text{ M}$

46. (a) $H_2C_4H_5NO_4 \text{ concn} = \frac{0.405 \text{ g } H_2C_4H_5NO_4}{100.0 \text{ mL}} \times \frac{1000 \text{ mL}}{1 \text{ L}} \times \frac{1 \text{ mol } H_2C_4H_5NO_4}{133.10 \text{ g } H_2C_4H_5NO_4} = 0.0304 \text{ M}$

(b) $C_3H_6O \text{ concn} = \frac{35.0 \text{ mL } C_3H_6O}{425 \text{ mL soln}} \times \frac{1000 \text{ mL}}{1 \text{ L}} \times \frac{0.790 \text{ g } C_3H_6O}{1 \text{ mL}} \times \frac{1 \text{ mol}}{58.08 \text{ g } C_3H_6O} = 1.12 \text{ M}$

(c) $(C_2H_5)_2O \text{ concn} = \frac{8.8 \text{ mg } (C_2H_5)_2O}{3.00 \text{ L soln}} \times \frac{1 \text{ g}}{1000 \text{ mg}} \times \frac{1 \text{ mol } (C_2H_5)_2O}{74.12 \text{ g } (C_2H_5)_2O} = 4.0 \times 10^{-5} \text{ M}$

47. (a) $\text{mass } C_6H_{12}O_6 = 75.0 \text{ mL soln} \times \frac{1 \text{ L}}{1000 \text{ mL}} \times \frac{0.350 \text{ mol } C_6H_{12}O_6}{1 \text{ L soln}} \times \frac{180.16 \text{ g } C_6H_{12}O_6}{1 \text{ mol } C_6H_{12}O_6} = 4.73 \text{ g}$

(b) $\text{volume } CH_3OH = 2.25 \text{ L soln} \times \frac{0.485 \text{ mol}}{1 \text{ L}} \times \frac{32.04 \text{ g } CH_3OH}{1 \text{ mol } CH_3OH} \times \frac{1 \text{ mL}}{0.792 \text{ g}} = 44.1 \text{ mL } CH_3OH$

48. (a) $\text{volume } C_2H_5OH = 200.0 \text{ L soln} \times \frac{1.65 \text{ mol } C_2H_5OH}{1 \text{ L}} \times \frac{46.07 \text{ g } C_2H_5OH}{1 \text{ mol } C_2H_5OH} \times \frac{1 \text{ mL}}{0.789 \text{ g}} \times \frac{1 \text{ L}}{1000 \text{ cm}^3}$
$= 19.3 \text{ L } C_2H_5OH$

(b) $\text{vol conc HCl} = 12.0 \text{ L} \times \frac{0.234 \text{ mol HCl}}{1 \text{ L}} \times \frac{36.45 \text{ g HCl}}{1 \text{ mol HCl}} \times \frac{100 \text{ g soln}}{36.0 \text{ g HCl}} \times \frac{1 \text{ mL soln}}{1.18 \text{ g}}$
$= 241 \text{ mL conc HCl}$

49. We determine the molar concentration of the 46% by mass sucrose solution.

$$C_{12}H_{22}O_{11} \text{ molarity} = \frac{46 \text{ g } C_{12}H_{22}O_{11} \times \frac{1 \text{ mol } C_{12}H_{22}O_{11}}{342.3 \text{ g } C_{12}H_{22}O_{11}}}{100 \text{ g soln} \times \frac{1 \text{ mL}}{1.21 \text{ g soln}} \times \frac{1 \text{ L}}{1000 \text{ mL}}} = 1.63 \text{ M}$$

The 46% by mass sucrose solution is the more concentrated.

50. We calculate the $[C_2H_5OH]$ in the white wine and compare it with 1.71 M C_2H_5OH, the concentration of the solution described in Example 4-8.

$$[C_2H_5OH] = \frac{11.0 \text{ g } C_2H_5OH}{100.0 \text{ g soln}} \times \frac{0.95 \text{ g soln}}{1 \text{ cm}^3} \times \frac{1000 \text{ cm}^3}{1 \text{ L}} \times \frac{1 \text{ mol } C_2H_5OH}{46.1 \text{ g } C_2H_5OH} = 2.3 \text{ M } C_2H_5OH$$

The white wine has a greater ethyl alcohol content.

51. $KNO_3 \text{ concn} = \dfrac{10.00 \text{ mL conc'd soln} \times \frac{2.05 \text{ mmol } KNO_3}{1 \text{ mL}}}{250.0 \text{ mL}} = 0.0820 \text{ M}$

52. $HCl \text{ concn} = \dfrac{500.0 \text{ mL dilute soln} \times \frac{0.085 \text{ mmol HCl}}{1 \text{ mL soln}}}{25.0 \text{ mL}} = 1.7 \text{ M}$

53. Let us compute how many mL of dilute ($_d$) solution we obtain from each mL of concentrated ($_c$) solution.
$V_c \times C_c = V_d \times C_d$ becomes $1.00 \text{ mL} \times 0.250 \text{ M} = x \text{ mL} \times 0.0125 \text{ M}$ and $x = 20$
Thus, the ratio of the volume of the volumetric flask to that of the pipet would be 20:1. We could use a 100.0-mL flask and a 5.00-mL pipet, a 1000.0-mL flask and a 50.00-mL pipet, or a 500.0-mL flask and a 25.00-mL pipet.

54. First determine the amount of solute in the final solution and then the volume of the initial, more concentrated, solution that must be used.

$$\text{volume conc'd soln} = 250.0 \text{ mL} \times \frac{0.175 \text{ mmol KCl}}{1 \text{ mL dil soln}} \times \frac{1 \text{ mL conc'd soln}}{0.496 \text{ mmol KCl}} = 88.2 \text{ mL}$$

Thus the instructions are: Place 88.2 mL of 0.496 M KCl in a 250-mL volumetric flask. Dilute to the mark with distilled water, stopping to mix thoroughly several times during the addition of water.

Chemical Reactions in Solutions

55. (a) $$\text{mass } Na_2S = 27.8 \text{ mL} \times \frac{1 \text{ L}}{1000 \text{ mL}} \times \frac{0.163 \text{ mol } AgNO_3}{1 \text{ L soln}} \times \frac{1 \text{ mol } Na_2S}{2 \text{ mol } AgNO_3} \times \frac{78.04 \text{ g } Na_2S}{1 \text{ mol } Na_2S} = 0.177 \text{ g } Na_2S$$

(b) $$\text{mass } Ag_2S = 0.177 \text{ g } Na_2S \times \frac{1 \text{ mol } Na_2S}{78.04 \text{ g } Na_2S} \times \frac{1 \text{ mol } Ag_2S}{1 \text{ mol } Na_2S} \times \frac{247.81 \text{ g } Ag_2S}{1 \text{ mol } Ag_2S} = 0.562 \text{ g } Ag_2S$$

56. (a) $$\text{mass } Ca(OH)_2 = 415 \text{ mL} \times \frac{1 \text{ L}}{1000 \text{ mL}} \times \frac{0.477 \text{ mol HCl}}{1 \text{ L soln}} \times \frac{1 \text{ mol } Ca(OH)_2}{2 \text{ mol HCl}} \times \frac{74.1 \text{ g } Ca(OH)_2}{1 \text{ mol } Ca(OH)_2} = 7.33 \text{ g } Ca(OH)_2$$

(b) $$\text{mass } Ca(OH)_2 = 324 \text{ L} \times \frac{1.12 \text{ kg}}{1 \text{ L}} \times \frac{24.28 \text{ kg HCl}}{100.00 \text{ kg soln}} \times \frac{1 \text{ kmol HCl}}{36.45 \text{ kg HCl}} \times \frac{1 \text{ kmol } Ca(OH)_2}{2 \text{ kmol HCl}} \times \frac{74.10 \text{ kg } Ca(OH)_2}{1 \text{ kmol } Ca(OH)_2} = 89.6 \text{ kg } Ca(OH)_2$$

57. (a) We know that the Al forms the $AlCl_3$.

$$\text{no. mol } AlCl_3 = 1.87 \text{ g Al} \times \frac{1 \text{ mol Al}}{26.98 \text{ g Al}} \times \frac{1 \text{ mol } AlCl_3}{1 \text{ mol Al}} = 0.0693 \text{ mol } AlCl_3$$

(b) $$AlCl_3 \text{ concn} = \frac{0.0693 \text{ mol } AlCl_3}{23.8 \text{ mL}} \times \frac{1000 \text{ mL}}{1 \text{ L}} = 2.91 \text{ M } AlCl_3$$

58. The balanced chemical equation indicates that 4 mol $NaNO_2$ are formed from 2 mol Na_2CO_3.

$$[NaNO_2] = \frac{138 \text{ g } Na_2CO_3}{1.42 \text{ L soln}} \times \frac{1 \text{ mol } Na_2CO_3}{106.0 \text{ g } Na_2CO_3} \times \frac{4 \text{ mol } NaNO_2}{2 \text{ mol } Na_2CO_3} = 1.83 \text{ M } NaNO_2$$

59. The molarity unit can be interpreted as millimoles of solute per milliliter of solution.

$$\text{volume } K_2CrO_4(aq) = 415 \text{ mL} \times \frac{0.186 \text{ mmol } AgNO_3}{1 \text{ mL soln}} \times \frac{1 \text{ mmol } K_2CrO_4}{2 \text{ mmol } AgNO_3} \times \frac{1 \text{ mL } K_2CrO_4(aq)}{0.650 \text{ mmol } K_2CrO_4} = 59.4 \text{ mL } K_2CrO_4(aq)$$

60. The volume of solution determines the amount of product.

$$\text{mass } Ag_2CrO_4 = 415 \text{ mL} \times \frac{1 \text{ L}}{1000 \text{ mL}} \times \frac{0.186 \text{ mol } AgNO_3}{1 \text{ L soln}} \times \frac{1 \text{ mol } Ag_2CrO_4}{2 \text{ mol } AgNO_3} \times \frac{331.73 \text{ g } Ag_2CrO_4}{1 \text{ mol } Ag_2CrO_4} = 12.8 \text{ g } Ag_2CrO_4$$

61. $$\text{mass Na} = 155 \text{ mL soln} \times \frac{1 \text{ L}}{1000 \text{ mL}} \times \frac{0.175 \text{ mol NaOH}}{1 \text{ L soln}} \times \frac{2 \text{ mol Na}}{2 \text{ mol NaOH}} \times \frac{22.99 \text{ g Na}}{1 \text{ mol Na}} = 0.624 \text{ g Na}$$

62. We determine the amount of HCl present initially, and the amount desired.

$$\text{amount HCl present} = 250.0 \text{ mL} \times \frac{1.023 \text{ mmol HCl}}{1 \text{ mL soln}} = 255.8 \text{ mmol HCl}$$

$$\text{amount HCl desired} = 250.0 \text{ mL} \times \frac{1.000 \text{ mmol HCl}}{1 \text{ mL soln}} = 250.0 \text{ mmol HCl}$$

$$\text{mass Mg} = (255.8 - 250.0) \text{ mmol HCl} \times \frac{1 \text{ mmol Mg}}{2 \text{ mmol HCl}} \times \frac{24.3 \text{ mg Mg}}{1 \text{ mmol Mg}} = 70. \text{ mg Mg}$$

63. The mass of oxalic acid enables us to determine the amount of NaOH in the solution.

$$[NaOH] = \frac{0.3126 \text{ g } H_2C_2O_4}{26.21 \text{ mL soln}} \times \frac{1000 \text{ mL}}{1 \text{ L soln}} \times \frac{1 \text{ mol } H_2C_2O_4}{90.04 \text{ g } H_2C_2O_4} \times \frac{2 \text{ mol NaOH}}{1 \text{ mol } H_2C_2O_4} = 0.2649 \text{ M}$$

64. The total amount of HCl present is the amount that reacted with the $CaCO_3$ plus the amount that reacted with the $Ba(OH)_2(aq)$.

$$CaCO_3 \text{ amount} = 0.1000 \text{ g } CaCO_3 \times \frac{1 \text{ mol } CaCO_3}{100.09 \text{ g } CaCO_3} \times \frac{2 \text{ mol HCl}}{1 \text{ mol } CaCO_3} \times \frac{1000 \text{ mmol}}{1 \text{ mol}} = 1.998 \text{ mmol HCl}$$

$$Ba(OH)_2 \text{ amount} = 43.82 \text{ mL} \times \frac{0.01185 \text{ mmol } Ba(OH)_2}{1 \text{ mL soln}} \times \frac{2 \text{ mmol HCl}}{1 \text{ mmol } Ba(OH)_2} = 1.039 \text{ mmol HCl}$$

The HCl molarity is this total amount divided by the volume of 25.00 mL.

$$\text{HCl concn} = \frac{(1.998 + 1.039) \text{ mmol HCl}}{25.00 \text{ mL}} = 0.1215 \text{ M}$$

Determining the Limiting Reagent

65. There are equal number of moles of each reactant present, but more O_2 is needed than NH_3. Thus, $O_2(g)$ is the limiting reactant, and all of the $O_2(g)$ is consumed. The mass of product produced from 1.00 mol $O_2(g)$ is then calculated.

$$\text{mass NO(g)} = 1.00 \text{ mol } O_2 \times \frac{4 \text{ mol NO(g)}}{5 \text{ mol } O_2(g)} \times \frac{30.01 \text{ g NO(g)}}{1 \text{ mol NO(g)}} = 24.0 \text{ g NO(g)}$$

66. Determine the mass of H_2 produced from each of the reactants. The smaller mass is that produced by the limiting reactant and is the mass actually formed.

$$\text{mass } H_2 = 1.84 \text{ g Al} \times \frac{1 \text{ mol Al}}{26.98 \text{ g Al}} \times \frac{3 \text{ mol } H_2}{2 \text{ mol Al}} \times \frac{2.016 \text{ g } H_2}{1 \text{ mol } H_2} = 0.206 \text{ g } H_2$$

$$\text{mass } H_2 = 75.0 \text{ mL} \times \frac{1 \text{ L}}{1000 \text{ mL}} \times \frac{2.95 \text{ mol HCl}}{1 \text{ L}} \times \frac{3 \text{ mol } H_2}{6 \text{ mol HCl}} \times \frac{2.016 \text{ g } H_2}{1 \text{ mol } H_2} = 0.223 \text{ g } H_2$$

Thus, 0.206 g H_2 is produced.

67. Determine the amount of Na_2CS_3 produced from each of the reactants.

$$\text{amount } Na_2CS_3 = 92.5 \text{ mL } CS_2 \times \frac{1.26 \text{ g}}{1 \text{ mL}} \times \frac{1 \text{ mol } CS_2}{76.14 \text{ g } CS_2} \times \frac{2 \text{ mol } Na_2CS_3}{3 \text{ mol } CS_2} = 1.02 \text{ mol } Na_2CS_3$$

$$\text{amount } Na_2CS_3 = 2.78 \text{ mol NaOH} \times \frac{2 \text{ mol } Na_2CS_3}{6 \text{ mol NaOH}} = 0.927 \text{ mol } Na_2CS_3$$

Thus, the mass produced is $0.927 \text{ mol } Na_2CS_3 \times \frac{154.3 \text{ g } Na_2CS_3}{1 \text{ mol } Na_2CS_3} = 143 \text{ g } Na_2CS_3$

68. Since the two reactants combined in an equimolar basis, the one present with the fewer number of moles is the limiting reactant and determines the mass of the products.

$$\text{no. mol } ZnSO_4 = 315 \text{ mL} \times \frac{1 \text{ L}}{1000 \text{ mL}} \times \frac{0.275 \text{ mol } ZnSO_4}{1 \text{ L soln}} = 0.0866 \text{ mol } ZnSO_4$$

$$\text{no. mol BaS} = 285 \text{ mL} \times \frac{1 \text{ L}}{1000 \text{ mL}} \times \frac{0.315 \text{ mol BaS}}{1 \text{ L soln}} = 0.0898 \text{ mol BaS}$$

Thus, $ZnSO_4$ is the limiting reactant and 0.0866 mol of each of the products will be produced.

$$\text{mass products} = \left(0.0866 \text{ mol } BaSO_4 \times \frac{233.4 \text{ g } BaSO_4}{1 \text{ mol } BaSO_4}\right) + \left(0.0866 \text{ mol ZnS} \times \frac{97.46 \text{ g ZnS}}{1 \text{ mol ZnS}}\right)$$

$$= 28.7 \text{ g product mixture (lithopone)}$$

69. $Ca(OH)_2(s) + 2\ NH_4Cl(s) \longrightarrow CaCl_2(aq) + 2\ H_2O + 2\ NH_3(g)$

Compute the amount of NH_3 formed from each reactant in this limiting reactant problem.

$$\text{amount } NH_3 = 33.0 \text{ g } NH_4Cl \times \frac{1 \text{ mol } NH_4Cl}{53.49 \text{ g } NH_4Cl} \times \frac{2 \text{ mol } NH_3}{2 \text{ mol } NH_4Cl} = 0.617 \text{ mol } NH_3$$

$$\text{amount } NH_3 = 33.0 \text{ g } Ca(OH)_2 \times \frac{1 \text{ mol } Ca(OH)_2}{74.09 \text{ g } Ca(OH)_2} \times \frac{2 \text{ mol } NH_3}{1 \text{ mol } Ca(OH)_2} = 0.891 \text{ mol } NH_3$$

0.617 mol NH_3 is produced. $\text{mass } NH_3 = 0.617 \text{ mol } NH_3 \times \frac{17.03 \text{ g } NH_3}{1 \text{ mol } NH_3} = 10.5 \text{ g } NH_3$

Now we determine the mass of reactant in excess, $Ca(OH)_2$.

$$\text{mass } Ca(OH)_2 \text{ used} = 0.617 \text{ mol } NH_3 \times \frac{1 \text{ mol } Ca(OH)_2}{2 \text{ mol } NH_3} \times \frac{74.09 \text{ g } Ca(OH)_2}{1 \text{ mol } Ca(OH)_2} = 22.9 \text{ g } Ca(OH)_2 \text{ used}$$

excess $Ca(OH)_2$ = 33.0 g $Ca(OH)_2$ supplied – 22.9 g $Ca(OH)_2$ used = 10.1 g excess $Ca(OH)_2$

70. We first write the balanced chemical equation. $Ca(OCl)_2 + 4\ HCl \longrightarrow CaCl_2 + 2\ Cl_2 + 2\ H_2O$

$$\text{amount } Cl_2 = 50.0 \text{ g } Ca(OCl)_2 \times \frac{1 \text{ mol } Ca(OCl)_2}{142.98 \text{ g } Ca(OCl)_2} \times \frac{2 \text{ mol } Cl_2}{1 \text{ mol } Ca(OCl)_2} = 0.699 \text{ mol } Cl_2$$

$$\text{amount } Cl_2 = 275 \text{ mL} \times \frac{1 \text{ L}}{1000 \text{ mL}} \times \frac{6.00 \text{ mol HCl}}{1 \text{ L soln}} \times \frac{2 \text{ mol } Cl_2}{4 \text{ mol HCl}} = 0.825 \text{ mol } Cl_2$$

$$\text{mass } Cl_2 \text{ produced} = 0.699 \text{ mol } Cl_2 \times \frac{70.91 \text{ g } Cl_2}{1 \text{ mol } Cl_2} = 49.6 \text{ g } Cl_2$$

The excess reactant is the one producing the most Cl_2; it is HCl(aq). The quantity of excess HCl(aq) is determined from the amount of excess $Cl_2(g)$ it produces.

$$\text{volume excess HCl} = (0.825 - 0.699) \text{ mol } Cl_2 \times \frac{4 \text{ mol HCl}}{2 \text{ mol } Cl_2} \times \frac{1000 \text{ mL}}{6.00 \text{ mol HCl}} = 42.0 \text{ mL excess HCl(aq)}$$

$$\text{mass excess HCl} = (0.825 - 0.699) \text{ mol } Cl_2 \times \frac{4 \text{ mol HCl}}{2 \text{ mol } Cl_2} \times \frac{36.46 \text{ g HCl}}{1 \text{ mol HCl}} = 9.19 \text{ g excess HCl(aq)}$$

Theoretical, Actual, and Percent Yields

71. **(a)** We first need to solve the limiting reactant problem involved here.

$$\text{no. mol } C_4H_9Br = 15.0 \text{ g } C_4H_9OH \times \frac{1 \text{ mol } C_4H_9OH}{74.12 \text{ g } C_4H_9OH} \times \frac{1 \text{ mol } C_4H_9Br}{1 \text{ mol } C_4H_9OH} = 0.202 \text{ mol } C_4H_9Br$$

$$\text{no. mol } C_4H_9Br = 22.4 \text{ g NaBr} \times \frac{1 \text{ mol NaBr}}{102.9 \text{ g NaBr}} \times \frac{1 \text{ mol } C_4H_9Br}{1 \text{ mol NaBr}} = 0.218 \text{ mol } C_4H_9Br$$

$$\text{no. mol } C_4H_9Br = 32.7 \text{ g } H_2SO_4 \times \frac{1 \text{ mol } H_2SO_4}{98.1 \text{ g } H_2SO_4} \times \frac{1 \text{ mol } C_4H_9Br}{1 \text{ mol } H_2SO_4} = 0.333 \text{ mol } C_4H_9Br$$

$$\text{Thus, mass } C_4H_9Br = 0.202 \text{ mol } C_4H_9Br \times \frac{136.9 \text{ g } C_4H_9Br}{1 \text{ mol } C_4H_9Br} = 27.7 \text{ g } C_4H_9Br = \text{theoretical yield}$$

(b) The actual yield is the mass obtained, 17.1 g C_4H_9Br.

(c) Then $$\%\text{ yield} = \frac{17.1 \text{ g } C_4H_9Br \text{ produced}}{27.7 \text{ g } C_4H_9Br \text{ expected}} \times 100\% = 61.7\% \text{ yield}$$

72. **(a)** Again, we solve the limiting reactant problem first.

$$\text{amount } (C_6H_5N)_2 = 0.10 \text{ L } C_6H_5NO_2 \times \frac{1000 \text{ mL}}{1 \text{ L}} \times \frac{1.20 \text{ g}}{1 \text{ mL}} \times \frac{1 \text{ mol } C_6H_5NO_2}{123.1 \text{ g } C_6H_5NO_2} \times \frac{1 \text{ mol} (C_6H_5N)_2}{2 \text{ mol } C_6H_5NO_2} = 0.49 \text{ mol } (C_6H_5N)_2$$

$$\text{amount } (C_6H_5N)_2 = 0.30 \text{ L } C_6H_{14}O_4 \times \frac{1000 \text{ mL}}{1 \text{ L}} \times \frac{1.12 \text{ g}}{1 \text{ mL}} \times \frac{1 \text{ mol } C_6H_{14}O_4}{150.0 \text{ g } C_6H_{14}O_4} \times \frac{1 \text{ mol } (C_6H_5N)_2}{4 \text{ mol } C_6H_{14}O_4} = 0.56 \text{ mol } (C_6H_5N)_2$$

$$\text{mass } (C_6H_5N)_2 = 0.49 \text{ mol } (C_6H_5N)_2 \times \frac{182.2 \text{ g } (C_6H_5N)_2}{1 \text{ mol } (C_6H_5N)_2} = 89 \text{ g } (C_6H_5N)_2 = \text{theoretical yield}$$

(b) actual yield = 55 g $(C_6H_5N)_2$ produced

(c) $$\text{percent yield} = \frac{55 \text{ g } (C_6H_5N)_2 \text{ produced}}{89 \text{ g } (C_6H_5N)_2 \text{ expected}} \times 100\% = 62\% \text{ yield}$$

73. Balanced equation: $3\ C_2H_4O_2 + PCl_3 \longrightarrow 3\ C_2H_3OCl + H_3PO_3$

$$\text{mass acid} = 75 \text{ g } C_2H_3OCl \times \frac{100.0 \text{ g calculated}}{78.2 \text{ g produced}} \times \frac{1 \text{ mol } C_2H_3OCl}{78.5 \text{ g } C_2H_3OCl} \times \frac{3 \text{ mol } C_2H_4O_2}{3 \text{ mol } C_2H_3OCl}$$
$$\times \frac{60.1 \text{ g pure } C_2H_4O_2}{1 \text{ mol } C_2H_4O_2} \times \frac{100 \text{ g commercial}}{97 \text{ g pure } C_2H_4O_2} = 76 \text{ g commercial } C_2H_4O_2$$

74. $$\text{mass } CH_2Cl_2 = 112 \text{ g } CH_4 \times \frac{1 \text{ mol } CH_4}{16.04 \text{ g } CH_4} \times \frac{1 \text{ mol } CH_3Cl}{1 \text{ mol } CH_4} \times \frac{0.92 \text{ mol } CH_3Cl \text{ produced}}{1.00 \text{ mol } CH_3Cl \text{ expected}}$$
$$\times \frac{1 \text{ mol } CH_2Cl_2}{1 \text{ mol } CH_3Cl} \times \frac{85.0 \text{ g } CH_2Cl_2}{1 \text{ mol } CH_2Cl_2} \times \frac{0.92 \text{ g } CH_2Cl_2 \text{ produced}}{1.00 \text{ g } CH_2Cl_2 \text{ calculated}} = 5.0 \times 10^2 \text{ g } CH_2Cl_2$$

75. A less-than-100% yield of desired product in synthesis reactions is almost always the case. This is because of side reactions that yield products other than the desired one (by-products) and because of the loss of material in various steps of the synthesis process. A main criterion for choosing a synthesis reaction is how economically it can be run. In the analysis of a compound, on the other hand, it is essential that all of the material present be detected. Therefore, a 100% yield is required; none of the material present in the sample can be lost during the analysis. Therefore analysis reactions are carefully chosen to meet this criterion; they need not be economical to run.

76. The theoretical yield is 2.07 g Ag_2CrO_4. If the mass actually obtained is less than this, it is likely that some of the pure material was lost in the reaction, perhaps stuck to the walls of the flask in which the reaction occurred, or suspended in the solution. But the maximum mass of Ag_2CrO_4 that can be produced is 2.07 g. If the precipitate weighs more than 2.07 g, the extra mass must be impurities.

FEATURE PROBLEMS

A. If the sample that was caught is representative of all fish in the lake, there are five marked fish for every 18 fish. Thus, the total number of fish in the lake is determined.

$$\text{total fish} = 100 \text{ marked fish} \times \frac{18 \text{ fish}}{5 \text{ marked fish}} = 360 \text{ fish}$$

B. **1.** The graph obtained is one of two straight lines, meeting at a peak of about 2.50 g $Pb(NO_3)_2$.

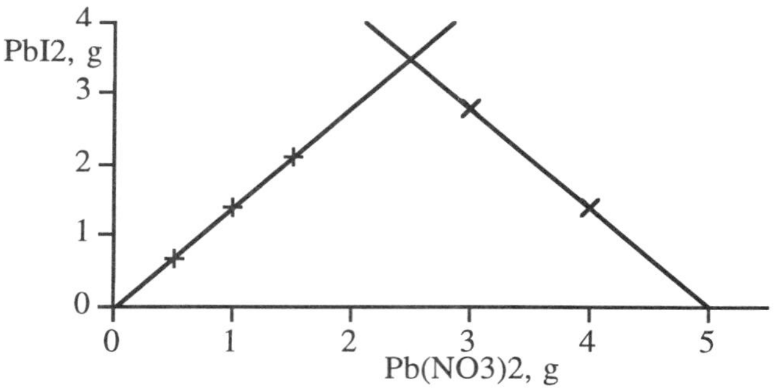

2. The total quantity of reactant is limited to 5.000 g. If either reactant is in excess, the amount in excess will be "wasted," because it cannot be used to form product. Thus, we obtain the maximum amount of product when neither reactant is in excess; there is a stoichiometric amount of each.

The balanced chemical equation for this reaction, $2\,KI + Pb(NO_3)_2 \longrightarrow 2\,KNO_3 + PbI_2$ shows that stoichiometric quantities are two moles of KI (166.00 g/mol) for each mole of $Pb(NO_3)_2$ (331.21 g/mol). If we have 5.000 g total, we can let the mass of KI equal x g, so that the mass of $Pb(NO_3)_2$ = $(5.000 - x)$ g. Then we have

$$\text{amount KI} = x \text{ g KI} \times \frac{1 \text{ mol KI}}{166.00 \text{ g}} = \frac{x}{166.00}$$

$$\text{amount } Pb(NO_3)_2 = (5.000 - x) \text{ g } Pb(NO_3)_2 \times \frac{1 \text{ mol } Pb(NO_3)_2}{331.21 \text{ g}} = \frac{5.000 - x}{331.21}$$

At the point of stoichiometric balance, amount KI = 2 × amount $Pb(NO_3)_2$

$$\frac{x}{166.00} = 2 \times \frac{5.000 - x}{331.21} \quad \text{OR} \quad 331.21x = 10.00 \times 166.00 - 332.00x$$

$$x = \frac{1660.0}{331.21 + 332.00} = 2.503 \text{ g KI} \times \frac{1 \text{ mol KI}}{166.00 \text{ g KI}} = 0.01508 \text{ mol KI}$$

$$5.000 - x = 2.497 \text{ g } Pb(NO_3)_2 \times \frac{1 \text{ mol } Pb(NO_3)_2}{331.21 \text{ g } Pb(NO_3)_2} = 0.007539 \text{ mol } Pb(NO_3)_2$$

3. To determine the proportions precisely, we used the balanced chemical equation.

$$\text{maximum } PbI_2 \text{ mass} = 2.503 \text{ g KI} \times \frac{1 \text{ mol KI}}{166.0 \text{ g KI}} \times \frac{1 \text{ mol } PbI_2}{2 \text{ mol KI}} \times \frac{461.01 \text{ g } PbI_2}{1 \text{ mol } PbI_2} = 3.476 \text{ g } PbI_2$$

C. **1.** For the balanced equation, the order is immaterial; the relative amount of each is important.

$20 \text{ rd} + 20 \text{ bl} + 30 \text{ gr} \longrightarrow 1 \text{ necklace}$

2. This is similar to a limiting reactant problem. We determine how many necklaces can be made from each quantity of beads.

$$\text{number of necklaces} = 10.0 \text{ kg beads} \times \frac{1000\text{g}}{1 \text{ kg}} \times \frac{1 \text{ rd bead}}{1.98 \text{ g}} \times \frac{1 \text{ necklace}}{20 \text{ rd beads}} = 252._5 \text{ necklaces}$$

$$\text{number of necklaces} = 10.0 \text{ kg beads} \times \frac{1000\text{g}}{1 \text{ kg}} \times \frac{1 \text{ bl bead}}{3.05 \text{ g}} \times \frac{1 \text{ necklace}}{20 \text{ bl beads}} = 163._9 \text{ necklaces}$$

$$\text{number of necklaces} = 10.0 \text{ kg beads} \times \frac{1000\text{g}}{1 \text{ kg}} \times \frac{1 \text{ gr bead}}{1.82 \text{ g}} \times \frac{1 \text{ necklace}}{30 \text{ gr beads}} = 183._1 \text{ necklaces}$$

We have expressed each result with an additional significant figure, written as a subscript, so that we can see the effect of rounding. With the beads available, we can produce 163 necklaces, since we are unable to produce a fraction of a necklace.

3. Because the mass of a bead, and the total mass available of each type of bead, both are known to just three significant figures, our results are only known that well. The best we can state is that we can make at least 163 necklaces, because 164 is uncertain by one unit. We should not be surprised if we actually made just 161 necklaces, or if we produced 165 of them. More precise masses would help.

5 INTRODUCTION TO REACTIONS IN AQUEOUS SOLUTIONS

PRACTICE EXAMPLES

1A The number of each type of ion is determined from the chemical formula.

$$[PO_4^{3-}] = \frac{0.358 \text{ mol } Na_3PO_4}{1 \text{ L soln}} \times \frac{1 \text{ mol } PO_4^{3-}}{1 \text{ mol } Na_3PO_4} = 0.358 \text{ M}$$

$$[Na^+] = \frac{0.358 \text{ mol } Na_3PO_4}{1 \text{ L soln}} \times \frac{3 \text{ mol } Na^+}{1 \text{ mol } Na_3PO_4} = 1.07 \text{ M}$$

1B In determining total $[Cl^-]$, we recall the definition of molarity: moles of solute per liter of solution.

$$\text{from NaCl, } [Cl^-] = \frac{0.438 \text{ mol NaCl}}{1 \text{ L soln}} \times \frac{1 \text{ mol } Cl^-}{1 \text{ mol NaCl}} = 0.438 \text{ M } Cl^-$$

$$\text{from } MgCl_2\text{, } [Cl^-] = \frac{0.0512 \text{ mol } MgCl_2}{1 \text{ L soln}} \times \frac{2 \text{ mol } Cl^-}{1 \text{ mol } MgCl_2} = 0.102_4 \text{ M } Cl^-$$

$[Cl^-]$ total = $[Cl^-]$ from NaCl + $[Cl^-]$ from $MgCl_2$ = 0.438 M + 0.102_4 M = 0.540 M Cl^-

2A In each case we use the solubility rules to determine whether either of the product compounds is insoluble. The ions in each product compound are determined by simply "switching the partners" of the reactant compounds. The designation "(aq)" on each reactant indicates that it is soluble.

(a) Possible products are potassium chloride, KCl, which is soluble, and aluminum hydroxide, $Al(OH)_3$, which is not. The net ionic equation is: $Al^{3+}(aq) + 3\,OH^-(aq) \longrightarrow Al(OH)_3(s)$

(b) Possible products are iron(III) sulfate, $Fe_2(SO_4)_3$, and potassium bromide, KBr, both of which are soluble. No reaction occurs.

(c) Possible products are calcium nitrate, $Ca(NO_3)_2$, which is soluble, and lead(II) iodide, PbI_2, which is insoluble. The net ionic equation is: $Pb^{2+}(aq) + 2\,I^-(aq) \longrightarrow PbI_2(s)$

2B **(a)** Possible products are sodium chloride, NaCl, which is soluble, and aluminum phsophate, $AlPO_4$, which is insoluble. The net ionic equation is: $Al^{3+}(aq) + PO_4^{3-}(aq) \longrightarrow AlPO_4(s)$

(b) Possible products are aluminum chloride, $AlCl_3$, which is soluble, and barium sulfate, $BaSO_4$, which is insoluble. The net ionic equation is: $Ba^{2+}(aq) + SO_4^{2-}(aq) \longrightarrow BaSO_4(s)$

(c) Possible products are ammonium nitrate, NH_4NO_3, which is soluble, and lead (II) carbonate, $PbCO_3$, which is insoluble. The net ionic equation is: $Pb^{2+}(aq) + CrO_3^{2-}(aq) \longrightarrow PbCrO_3(s)$

3A Propionic acid is a weak acid, not dissociated completely in aqueous solution. Ammonia is a weak base. The acid and base react to form a salt, whose name is ammonium propionate.

$NH_3(aq) + HC_3H_5O_2(aq) \longrightarrow NH_4^+(aq) + C_3H_5O_2^-(aq)$

3B Since acetic acid is a weak acid, it is not dissociated completely in aqueous solution; it is misleading to write it in ionic form. The products of this reaction are the gas carbon dioxide, the covalent compound water, and the ionic solute calcium acetate. Only the latter exists as ions in aqueous solution.

$CaCO_3(s) + 2\,HC_2H_3O_2(aq) \longrightarrow CO_2(g) + H_2O(l) + Ca^{2+}(aq) + 2\,C_2H_3O_2^-(aq)$

4A **(a)** This is a metathesis or double displacement reaction. The same ions, with elements in the same oxidation states, are present on each side of the reaction. It is not an oxidation–reduction reaction.

(b) The presence of $O_2(g)$ as a produced indicates that this is an oxidation–reduction reaction. Oxygen is oxidized from O.S. = –2 in $Pb(NO_3)_2(s)$ to O.S. = 0 in $O_2(g)$. Nitrogen is reduced from O.S. = +5 in $Pb(NO_3)_2(s)$ to O.S. = +4 in $NO_2(g)$

4B We determine the oxidation state (O.S.) of each element on each side of the equation. The O.S. of H is +1 on each side of the equation, and the O.S. of O is –2. For *vanadium*, the O.S. of V is +4 in VO^{2+}, and the O.S. of V is +5 in VO_2^+; since the oxidation state of V has increased during the reaction, VO^{2+} has been oxidized. For *manganese*, the O.S. of Mn in MnO_4^- is +7, and the O.S. of Mn in Mn^{2+} is +2; since the oxidation state of Mn has decreased during the reaction, MnO_4^- is the species reduced.

5A Aluminum is oxidized (from an O.S. of 0 to an O.S. of +3), while hydrogen is reduced (from an O.S. of +1 to an O.S. of 0).

Oxidation: $\{Al(s) \longrightarrow Al^{3+}(aq) + 3\,e^-\} \quad \times 2$

Reduction: $\{2\,H^+(aq) + 2\,e^- \longrightarrow H_2(g)\} \quad \times 3$

Net equation: $2\,Al(s) + 6\,H^+(aq) \longrightarrow 2\,Al^{3+}(aq) + 3\,H_2(g)$

5B Bromide is oxidized (from –1 to 0) while chlorine is reduced (from 0 to –1).

Oxidation: $2\,Br^-(aq) \longrightarrow Br_2(l) + 2\,e^-$

Reduction: $Cl_2(g) + 2\,e^- \longrightarrow 2\,Cl^-(aq)$

Net equation: $2\,Br^-(aq) + Cl_2(g) \longrightarrow Br_2(l) + 2\,Cl^-(aq)$

6A Step 1: Write the two skeleton half-equations.

$MnO_4^-(aq) \longrightarrow Mn^{2+}(aq)$ *and* $Fe^{2+}(aq) \longrightarrow Fe^{3+}(aq)$

Step 2: Balance each skeleton half-equation for O (with H_2O) and for H atoms (with H^+).

$MnO_4^-(aq) + 8\,H^+ \longrightarrow Mn^{2+}(aq) + 4\,H_2O$ *and* $Fe^{2+}(aq) \longrightarrow Fe^{3+}(aq)$

Step 3: Balance electric charge by adding electrons.

$MnO_4^-(aq) + 8\,H^+ + 5\,e^- \longrightarrow Mn^{2+}(aq) + 4\,H_2O$ *and* $Fe^{2+}(aq) \longrightarrow Fe^{3+}(aq) + e^-$

Step 4: Combine the two half-equations.

$\{Fe^{2+}(aq) \longrightarrow Fe^{3+}(aq) + e^-\} \times 5$

$MnO_4^-(aq) + 8\,H^+ + 5\,e^- \longrightarrow Mn^{2+}(aq) + 4\,H_2O$

$MnO_4^-(aq) + 8\,H^+ + 5\,Fe^{2+}(aq) \longrightarrow Mn^{2+}(aq) + 4\,H_2O + 5\,Fe^{3+}(aq)$

Step 5: VERIFY
left side: Mn 5 Fe 8 H 4 O $-1 + 8 + (5\times+2) = +17$ charge
right side: Mn 5 Fe $(4\times2) = 8$ H 4 O $+2 + (5\times+3) = +17$ charge

6B Step 1: Uranium is oxidized and chromium is reduced in this reaction. The "skeleton" half-equations are:

$UO^{2+}(aq) \longrightarrow UO_2^{2+}(aq)$ *and* $Cr_2O_7^{2-}(aq) \longrightarrow Cr^{3+}(aq)$

Step 2: First balance the chromium skeleton half-equation for chromium atoms:

$Cr_2O_7^{2-}(aq) \longrightarrow 2\,Cr^{3+}(aq)$

Next balance oxygen atoms with water molecules in each half-equation:

$UO^{2+}(aq) + H_2O \longrightarrow UO_2^{2+}(aq)$ *and* $Cr_2O_7^{2-}(aq) \longrightarrow 2\,Cr^{3+}(aq) + 7\,H_2O$

Then balance hydrogen atoms with hydrogen ions in each half-equation:

$UO^{2+}(aq) + H_2O \longrightarrow UO_2^{2+}(aq) + 2\,H^+(aq)$ *and* $Cr_2O_7^{2-}(aq) + 14\,H^+(aq) \longrightarrow 2\,Cr^{3+}(aq) + 7\,H_2O$

Step 3: Balance the charge of each half-equation with electrons.

$UO^{2+}(aq) + H_2O \longrightarrow UO_2^{2+}(aq) + 2\,H^+(aq) + 2\,e^-$

$Cr_2O_7^{2-}(aq) + 14\,H^+(aq) + 6\,e^- \longrightarrow 2\,Cr^{3+}(aq) + 7\,H_2O$

Step 4: Multiply the uranium half-equation by 3 and add it to the chromium half-equation.

$\{UO^{2+}(aq) + H_2O \longrightarrow UO_2^{2+}(aq) + 2\,H^+(aq) + 2\,e^-\} \times 3$

$Cr_2O_7^{2-}(aq) + 14\,H^+(aq) + 6\,e^- \longrightarrow 2\,Cr^{3+}(aq) + 7\,H_2O$

$3\,UO^{2+}(aq) + Cr_2O_7^{2-}(aq) + 14\,H^+(aq) + 3\,H_2O \longrightarrow 3\,UO_2^{2+}(aq) + 2\,Cr^{3+}(aq) + 7\,H_2O + 6\,H^+(aq)$

Step 5: SIMPLIFY. Subtract 3 H_2O and 6 $H^+(aq)$ from each side of the equation.

$3\,UO^{2+}(aq) + Cr_2O_7^{2-}(aq) + 8\,H^+(aq) \longrightarrow 3\,UO_2^{2+}(aq) + 2\,Cr^{3+}(aq) + 4\,H_2O$

Step 6: VERIFY
left side: 3 U 2 Cr 8 H $3 + 7 = 10$ O $(3\times+2) - 2 + 8 = +12$ charge
right side: 3 U 2 Cr $(4\times2) = 8$ H $(3\times2) + 4 = 10$ O $(3\times+2) + (2\times+3) = +12$ charge

7A Step 1: Write the two skeleton half-equations.

$S(s) \longrightarrow SO_3^{2-}(aq)$ *and* $OCl^-(aq) \longrightarrow Cl^-(aq)$

Step 2: Balance each skeleton half-equation for O (with H_2O) and for H atoms (with H^+).

$3\,H_2O + S(s) \longrightarrow SO_3^{2-}(aq) + 6\,H^+$ *and* $OCl^-(aq) + 2\,H^+ \longrightarrow Cl^-(aq) + H_2O$

Step 3: Balance electric charge by adding electrons.

$3\,H_2O + S(s) \longrightarrow SO_3^{2-}(aq) + 6\,H^+ + 4\,e^-$ *and* $OCl^-(aq) + 2\,H^+ + 2\,e^- \longrightarrow Cl^-(aq) + H_2O$

Step 4: Change from an acidic medium to a basic one by adding OH^- to eliminate H^+.

$\cancel{3\,H_2O} + S(s) + 6\,OH^- \longrightarrow SO_3^{2-}(aq) + 3\,\cancel{6}\,H^+ + 3\,\cancel{6}\,OH^- + 4\,e^-$

$OCl^-(aq) + 1\,\cancel{2}\,H^+ + 1\,\cancel{2}\,OH^- + 2\,e^- \longrightarrow Cl^-(aq) + \cancel{H_2O} + 2\,OH^-$

Step 5: Simplify by removing the items present on both sides (indicated by ~~crossing out~~ above) of each half-equation, and combine the half-equations to obtain the net redox equation.

$\{S(s) + 6\,OH^- \longrightarrow SO_3^{2-}(aq) + 3\,H_2O + 4\,e^-\} \times 1$

$\{OCl^-(aq) + H_2O + 2\,e^- \longrightarrow Cl^-(aq) + 2\,OH^-\} \times 2$

$S(s) + 2\,\cancel{6}\,OH^- + 2\,OCl^-(aq) + \cancel{2\,H_2O} \longrightarrow SO_3^{2-}(aq) + 1\,\cancel{3}\,H_2O + 2\,Cl^-(aq) + \cancel{4\,OH^-}$

Simplify by removing the items present on both sides (indicated by ~~crossing out~~ above).

NET EQUATION $S(s) + 2\,OH^- + 2\,OCl^-(aq) \longrightarrow SO_3^{2-}(aq) + H_2O + 2\,Cl^-(aq)$

Step 6: VERIFY	*left side:*	1 S	(2 + 2) O = 4 O	2 Cl	2 H	−2 − 2 = −4 charge
	right side:	1 S	(3 + 1) O = 4 O	2 Cl	2 H	−2 − 2 = −4 charge

7B Step 1: Write the two skeleton half-equations.

$MnO_4^-(aq) \longrightarrow MnO_2(s)$ *and* $SO_3^{2-}(aq) \longrightarrow SO_4^{2-}(aq)$

Step 2: Balance each skeleton half-equation for O (with H_2O) and for H atoms (with H^+).

$MnO_4^-(aq) + 4\,H^+(aq) \longrightarrow MnO_2(s) + 2\,H_2O$ *and* $SO_3^{2-}(aq) + H_2O \longrightarrow SO_4^{2-}(aq) + 2\,H^+(aq)$

Step 3: Balance electric charge by adding electrons.

$MnO_4^-(aq) + 4\,H^+(aq) + 3\,e^- \longrightarrow MnO_2(s) + 2\,H_2O$

and $SO_3^{2-}(aq) + H_2O \longrightarrow SO_4^{2-}(aq) + 2\,H^+(aq) + 2\,e^-$

Step 4: Change from an acidic medium to a basic one by adding OH^- to eliminate H^+.

$MnO_4^-(aq) + \cancel{4}\,2\,H^+(aq) + \cancel{4}\,2\,OH^-(aq) + 3\,e^- \longrightarrow MnO_2(s) + \cancel{2\,H_2O} + 4\,OH^-(aq)$

$SO_3^{2-}(aq) + \cancel{H_2O} + 2\,OH^-(aq) \longrightarrow SO_4^{2-}(aq) + \cancel{2}\,H^+(aq) + \cancel{2}\,OH^-(aq) + 2\,e^-$

Step 5: Simplify by removing the items present on both sides (indicated by ~~crossing out~~ above) of each half-equation, and combine the half-equations to obtain the net redox equation.

$\{MnO_4^-(aq) + 2\,H_2O + 3\,e^- \longrightarrow MnO_2(s) + 4\,OH^-(aq)\} \times 2$

$\{SO_3^{2-}(aq) + 2\,OH^-(aq) \longrightarrow SO_4^{2-}(aq) + H_2O + 2\,e^-\} \times 3$

$2\,MnO_4^-(aq) + 3\,SO_3^{2-}(aq) + \cancel{6\,OH^-(aq)} + \cancel{4}\,1\,H_2O \longrightarrow$

$2\,MnO_2(s) + 3\,SO_4^{2-}(aq) + \cancel{3\,H_2O} + \cancel{8}\,2\,OH^-(aq)$

Simplify by removing the items present on both sides (indicated by ~~crossing out~~ above).

NET EQUATION $2\,MnO_4^-(aq) + 3\,SO_3^{2-}(aq) + H_2O \longrightarrow 2\,MnO_2(s) + 3\,SO_4^{2-}(aq) + 2\,OH^-(aq)$

Step 6: VERIFY	*left side:*	2 Mn	3 S	2 H	(2×4) + (3×3) + 1 = 18 O	−2 + (3×−2) = −8 charge
	right side:	2 Mn	3 S	2 H	(2×2) + (3×4) + 2 = 18 O	(3×−2) − 2 = −8 charge

8A Since the oxidation state of H is 0 in $H_2(g)$ and is +1 in both $NH_3(g)$ and $H_2O(g)$, hydrogen is oxidized. A substance that is oxidized is itself a reducing agent. In addition, the oxidation state of N in $NO_2(g)$ is +4, while it is −3 in NH_3; the oxidation state of the element N decreases during this reaction, meaning that $NO_2(g)$ is reduced. This is another indication that $H_2(g)$ is a reducing agent.

8B Because they are present as elements on the reactant side and thus have oxidation states of zero and are combined on the product side, we first investigate Au and O_2. In $[Au(CN)_2]^-(aq)$, gold has an oxidation state of +1; Au has been oxidized and thus Au(s) is the reducing agent. In $OH^-(aq)$, oxygen has an oxidation state of −2; O has been reduced and thus $O_2(g)$ is the oxidizing agent.

9A We first determine the amount of NaOH that reacts with 0.500 g KHP.

$$\text{amount NaOH} = 0.500\text{ g KHP} \times \frac{1\text{ mol KHP}}{204.22\text{ g KHP}} \times \frac{1\text{ mol OH}^-}{1\text{ mol KHP}} \times \frac{1\text{ mol NaOH}}{1\text{ mol OH}^-} = 0.00245\text{ mol NaOH}$$

$$\text{NaOH molarity} = \frac{0.00245 \text{ mol NaOH}}{24.03 \text{ mL soln}} \times \frac{1000 \text{ mL}}{1 \text{ L}} = 0.102 \text{ M}$$

9B The net ionic equation for the reaction of solid hydroxides with a strong acid is $OH^- + H^+ \longrightarrow H_2O$. There are two sources of hydroxide ion: KOH and $Ba(OH)_2$. We compute the amount of hydroxide ion from each source and add the results.

$$\text{amount OH}^- \text{ from KOH} = 0.235 \text{ g sample} \times \frac{92.5 \text{ g KOH}}{100.0 \text{ g sample}} \times \frac{1 \text{ mol KOH}}{56.11 \text{ g KOH}} \times \frac{1 \text{ mol OH}^-}{1 \text{ mol KOH}}$$
$$= 0.00387 \text{ mol OH}^-$$

$$\text{amount OH}^- \text{ from Ba(OH)}_2 = 0.235 \text{ g sample} \times \frac{7.5 \text{ g Ba(OH)}_2}{100.0 \text{ g sample}} \times \frac{1 \text{ mol Ba(OH)}_2}{171.3 \text{ g Ba(OH)}_2} \times \frac{2 \text{ mol OH}^-}{1 \text{ mol Ba(OH)}_2}$$
$$= 0.00021 \text{ mol OH}^-$$

total amount OH = 0.00387 mol from KOH + 0.00021 mol from $Ba(OH)_2$ = 0.00408 mol OH^-

$$\text{HCl molarity} = \frac{0.00408 \text{ mol OH}^-}{45.6 \text{ mol HCl soln}} \times \frac{1 \text{ mol H}^+}{1 \text{ mol OH}^-} \times \frac{1 \text{ mol HCl}}{1 \text{ mol H}^+} \times \frac{1000 \text{ mL soln}}{1 \text{ L soln}} = 0.0895 \text{ M}$$

10A The balanced equation provides stoichiometric coefficients used in the solution.

$$\text{amount MnO}_4^- = 0.2482 \text{ g Na}_2\text{C}_2\text{O}_4 \times \frac{1 \text{ mol Na}_2\text{C}_2\text{O}_4}{134.00 \text{ g Na}_2\text{C}_2\text{O}_4} \times \frac{1 \text{ mol C}_2\text{O}_4^{2-}}{1 \text{ mol Na}_2\text{C}_2\text{O}_4} \times \frac{2 \text{ mol MnO}_4^-}{5 \text{ mol C}_2\text{O}_4^{2-}}$$
$$= 0.0007409 \text{ mol MnO}_4^-$$

$$\text{KMnO}_4 \text{ concn} = \frac{0.0007409 \text{ mol MnO}_4^-}{23.68 \text{ mL soln}} \times \frac{1000 \text{ mL}}{1 \text{ L}} \times \frac{1 \text{ mol KMnO}_4}{1 \text{ mol MnO}_4^-} = 0.03129 \text{ M}$$

10B First determine the mass of iron that has reacted as Fe^{2+} with the titrant. The balanced chemical equation provides the essential conversion factor.

$$\text{mass Fe} = \frac{41.25 \text{ mL}}{\text{titrant}} \times \frac{1 \text{ L titrant}}{1000 \text{ mL titrant}} \times \frac{0.02140 \text{ mol MnO}_4^-}{1 \text{ L titrant}} \times \frac{5 \text{ mol Fe}^{2+}}{1 \text{ mol MnO}_4^-} \times \frac{55.847 \text{ g Fe}}{1 \text{ mol Fe}^{2+}}$$
$$= 0.246 \text{ g Fe}$$

Then determine the % Fe in the ore. $\quad \% \text{ Fe} = \frac{0.246 \text{ g Fe}}{0.376 \text{ g ore}} \times 100\% = 65.4\% \text{ Fe}$

SUMMARIZING EXAMPLE CALCULATIONS

1. dithionite ion + chromate ion $\longrightarrow$ chromium(III) hydroxide + sulfite ion in basic solution

$S_2O_4^{2-}(aq) + CrO_4^{2-}(aq) \longrightarrow Cr(OH)_3(s) + SO_3^{2-}(aq)$

2. $CrO_4^{2-}(aq) \longrightarrow Cr(OH)_3(s)$ *and* $S_2O_4^{2-}(aq) \longrightarrow 2\ SO_3^{2-}(aq)$

$CrO_4^{2-}(aq) \longrightarrow Cr(OH)_3(s) + H_2O$ *and* $S_2O_4^{2-}(aq) + 2\ H_2O \longrightarrow 2\ SO_3^{2-}(aq)$

$CrO_4^{2-}(aq) + 2\ H^+(aq) \longrightarrow Cr(OH)_3(s) + H_2O$ *and* $S_2O_4^{2-}(aq) + 2\ H_2O \longrightarrow 2\ SO_3^{2-}(aq) + 4\ H^+(aq)$

$CrO_4^{2-}(aq) + 5\ H^+(aq) \longrightarrow Cr(OH)_3(s) + H_2O$ *and* $S_2O_4^{2-}(aq) + 2\ H_2O \longrightarrow 2\ SO_3^{2-}(aq) + 4\ H^+(aq)$

$CrO_4^{2-}(aq) + 5\ H^+(aq) + 3\ e^- \longrightarrow Cr(OH)_3(s) + H_2O$

and $S_2O_4^{2-}(aq) + 2\ H_2O \longrightarrow 2\ SO_3^{2-}(aq) + 4\ H^+(aq) + 2\ e^-$

$CrO_4^{2-}(aq) + \cancel{5}\ 4\ H^+(aq) + \cancel{5}\ 4\ OH^-(aq) + 3\ e^- \longrightarrow Cr(OH)_3(s) + \cancel{H_2O} + 5\ OH^-(aq)$

and $S_2O_4^{2-}(aq) + \cancel{2\ H_2O} + 4\ OH^-(aq) \longrightarrow 2\ SO_3^{2-}(aq) + \cancel{4}\ 2\ H^+(aq) + \cancel{4}\ 2\ OH^-(aq) + 2\ e^-$

$\{CrO_4^{2-}(aq) + 4\ H_2O + 3\ e^- \longrightarrow Cr(OH)_3(s) + 5\ OH^-(aq)\} \times 2$

$\{S_2O_4^{2-}(aq) + 4\ OH^-(aq) \longrightarrow 2\ SO_3^{2-}(aq) + 2\ H_2O + 2\ e^-\} \times 3$

$2\ CrO_4^{2-}(aq) + 3\ S_2O_4^{2-}(aq) + \cancel{1}2\ OH^-(aq) + \cancel{8}\ 2\ H_2O$

$\longrightarrow 2\ Cr(OH)_3(s) + 6\ SO_3^{2-}(aq) + \cancel{6\ H_2O} + \cancel{10\ OH^-(aq)}$

NET EQUATION: $2\ CrO_4^{2-}(aq) + 3\ S_2O_4^{2-}(aq) + 2\ OH^-(aq) + 2\ H_2O \longrightarrow 2\ Cr(OH)_3(s) + 6\ SO_3^{2-}(aq)$

3. $$\text{mass Na}_2\text{S}_2\text{O}_4 = 100.0 \text{ L wastewater} \times \frac{0.0148 \text{ mol CrO}_4^{2-}}{1 \text{ L}} \times \frac{3 \text{ mol S}_2\text{O}_4^{2-}}{2 \text{ mol CrO}_4^{2-}} \times \frac{1 \text{ mol Na}_2\text{S}_2\text{O}_4}{1 \text{ mol S}_2\text{O}_4^{2-}}$$
$$\times \frac{174.11 \text{ g Na}_2\text{S}_2\text{O}_4}{1 \text{ mol Na}_2\text{S}_2\text{O}_4} = 387 \text{ g Na}_2\text{S}_2\text{O}_4$$

REVIEW QUESTIONS

1. **(a)** The symbol "$\rightleftharpoons$" means that a chemical reaction reaches a point of balance or equilibrium somewhere short of the complete consumption of (the limiting) reactant.
(b) The square brackets, [ ], surrounding the formula of a species, are the symbol for the molarity of that species in solution.
(c) A "spectator" ion is one that is present in a solution in which a reaction takes place but is not indicated in the net ionic equation for that reaction because this ion is unaffected by the reaction.
(d) A weak acid is a species that produces hydrogen ion in aqueous solution (an acid), but that does not dissociate completely into its ions (a weak electrolyte).

2. **(a)** In the half-reaction method of balancing redox equations, each species that is oxidized or reduced is the basis for a balanced half-equation. These half-equations then are combined to produce the balanced net ionic equation for the redox reaction.
(b) A disproportionation reaction is one in which the same species is both oxidized and reduced.
(c) Titration is the procedure of adding a measured amount of one material to a measured amount of another, in such a way that chemically equivalent amounts of substances are present at the end of the titration. The concentration of substance in one of the two materials is known; the technique permits the determination of the other concentration.
(d) Standardization of a solution refers to the determination of its concentration by titration.

3. **(a)** A strong electrolyte is a substance that dissociates completely into its ions when it is dissolved in aqueous solution. A strong acid is a strong electrolyte that produces hydrogen ions and anions when it dissociates.
(b) An oxidizing agent is a species that causes another species to be oxidized, that is, to donate electrons and thereby have the oxidation state of one of its elements increased. A reducing agent is a species that causes another species to be reduced: to accept electrons and have the oxidation state of one of its elements decreased.
(c) A precipitation reaction is one in which an insoluble substance is formed when solutions of two soluble substances are mixed. A neutralization reaction is the reaction of an acid with a base; the normal products are water and a salt that has the same cation as the base and the same anion as the acid.
(d) A half-reaction refers to just the oxidation or just the reduction aspect of a redox reaction. A net reaction refers to the entire chemical reaction, with only spectator ions omitted (or it can be the net result of several reactions that, together, make up a process).

4. **(a)** The best electrical conductor is the solution of the strong electrolyte: 0.10 M NaCl. In each liter of this solution, there is 0.10 mol Na^+ ions and 0.10 mol Cl^- ions.
(b) The poorest electrical conductor is the solution of the nonelectrolyte: 0.10 M C_2H_5OH. In this solution, the concentration of ions is almost nonexistent.

5. **(a)** Na_2SO_4 is a *salt* of sodium hydroxide (NaOH) and sulfate ion SO_4^{2-} from sulfuric acid, H_2SO_4.
(b) $Ba(OH)_2$ is a *strong base*, one of the common strong bases listed in Table 5-1.
(c) $Ba(NO_3)_2$ is the *salt* of barium hydroxide [$Ba(OH)_2$] and nitric acid (HNO_3).
(d) H_3PO_4 is an acid, since its formula begins with hydrogen. It is not listed in Table 5-1, so it must be a *weak acid*.
(e) HBr is a *strong acid*, listed in Table 5-1.
(f) HNO_2 is an acid since its formula begins with hydrogen, but it is not listed in Table 5-1. It is a *weak acid*.
(g) NH_3 is the *weak base*, ammonia.
(h) NH_4I is the *salt* of ammonia (NH_3) and hydroiodic acid (HI).
(i) KOH is a *strong base*, one of the common ones listed in Table 5-1.

6. For all these solutes but one—$Al_2(SO_4)_3$—there is one sulfate ion per formula unit. Consequently, the concentration of the compound and the solufate ion concentration in that compound's aqueous solution will be the same. This makes 0.22 M $MgSO_4$ the solution with the highest $[SO_4^{2-}]$ among these four. But there are three sulfate ions per formula unit of $Al_2(SO_4)_3$. Thus $[SO_4^{2-}]$ in the $Al_2(SO_4)_3$ solution is three times the concentration of the solute, or $[SO_4^{2-}] = 3 \times 0.080\text{ M} = 0.24\text{ M}$, making this solution the one with the highest $[SO_4^{2-}]$.

7. (a) $[K^+] = \frac{0.238 \text{ mol } KNO_3}{1 \text{ L soln}} \times \frac{1 \text{ mol } K^+}{1 \text{ mol } KNO_3} = 0.238 \text{ M } K^+$

(b) $[NO_3^-] = \frac{0.167 \text{ mol } Ca(NO_3)_2}{1 \text{ L soln}} \times \frac{2 \text{ mol } NO_3^-}{1 \text{ mol } Ca(NO_3)_2} = 0.334 \text{ M } NO_3^-$

(c) $[Al^{3+}] = \frac{0.083 \text{ mol } Al_2(SO_4)_3}{1 \text{ L soln}} \times \frac{2 \text{ mol } Al^{3+}}{1 \text{ mol } Al_2(SO_4)_3} = 0.17 \text{ M } Al^{3+}$

(d) $[Na^+] = \frac{0.209 \text{ mol } Na_3PO_4}{1 \text{ L soln}} \times \frac{3 \text{ mol } Na^+}{1 \text{ mol } Na_3PO_4} = 0.627 \text{ M } Na^+$

8. The amount of chloride ion in each solution in millimoles is computed.

$Cl^- \text{ amount} = 200.0 \text{ mL} \times \frac{0.35 \text{ mmol NaCl}}{1 \text{ mL soln}} \times \frac{1 \text{ mmol } Cl^-}{1 \text{ mmol NaCl}} = 70. \text{ mmol } Cl^-$

$Cl^- \text{ amount} = 500.0 \text{ mL} \times \frac{0.065 \text{ mmol } MgCl_2}{1 \text{ mL soln}} \times \frac{2 \text{ mmol } Cl^-}{1 \text{ mmol } MgCl_2} = 65 \text{ mmol } Cl^-$

$Cl^- \text{ amount} = 1.00 \text{ L} \times \frac{1000 \text{ mL}}{1 \text{ L}} \times \frac{0.068 \text{ mmol HCl}}{1 \text{ mL}} \times \frac{1 \text{ mmol } Cl^-}{1 \text{ mmol HCl}} = 68 \text{ mmol } Cl^-$

The 200.0 mL of 0.035 M NaCl contains the largest amount of Cl^-.

9. $[OH^-] = \frac{0.132 \text{ g } Ba(OH)_2{\cdot}8H_2O}{275 \text{ mL soln}} \times \frac{1000 \text{ mL}}{1 \text{ L}} \times \frac{1 \text{ mol } Ba(OH)_2{\cdot}8H_2O}{315.5 \text{ } Ba(OH)_2{\cdot}8H_2O} \times \frac{2 \text{ mol } OH^-}{1 \text{ mol } Ba(OH)_2{\cdot}8 \text{ } H_2O}$

$= 3.04 \times 10^{-3} \text{ M } OH^-$

10. $[K^+] = \frac{0.126 \text{ mol KCl}}{1 \text{ L soln}} \times \frac{1 \text{ mol } K^+}{1 \text{ mol KCl}} = 0.126 \text{ M}$

$[Mg^{2+}] = \frac{0.148 \text{ mol } MgCl_2}{1 \text{ L soln}} \times \frac{1 \text{ mol } Mg^{2+}}{1 \text{ mol } MgCl_2} = 0.148 \text{ M}$

Now determine the amount of Cl^- in 1.00 L of the solution.

$\text{no. mol } Cl^- = \left(\frac{0.126 \text{ mol KCl}}{1 \text{ L soln}} \times \frac{1 \text{ mol } Cl^-}{1 \text{ mol KCl}}\right) + \left(\frac{0.148 \text{ mol } MgCl_2}{1 \text{ L soln}} \times \frac{2 \text{ mol } Cl^-}{1 \text{ mol } MgCl_2}\right)$

$= 0.126 \text{ mol } Cl^- + 0.296 \text{ mol } Cl^- = 0.422 \text{ mol } Cl^-$

$[Cl^-] = \frac{0.422 \text{ mol } Cl^-}{1 \text{ L soln}} = 0.422 \text{ M } Cl^-$

11. Determine the amount of I^- in the solution as it now exists, and the amount of I^- in the solution of the desired concentration. The different in these two amounts is the amount of I^- that must be added. Convert this amount to a mass of MgI_2 in grams.

$I^- \text{ amount in final solution} = 250.0 \text{ mL} \times \frac{1 \text{ L}}{1000 \text{ mL}} \times \frac{0.1000 \text{ mol } I^-}{1 \text{ L soln}} = 0.02500 \text{ mol } I^-$

$I^- \text{ amount in KI solution} = 250.0 \text{ mL} \times \frac{1 \text{ L}}{1000 \text{ mL}} \times \frac{0.0876 \text{ mol KI}}{1 \text{ L soln}} \times \frac{1 \text{ mol } I^-}{1 \text{ mol KI}} = 0.0219 \text{ mol } I^-$

$\text{mass } MgI_2 \text{ to be added} = (0.02500 - 0.0219) \text{ mol } I^- \times \frac{1 \text{ mol } MgI_2}{2 \text{ mol } I^-} \times \frac{278.11 \text{ g } MgI_2}{1 \text{ mol } MgI_2} \times \frac{1000 \text{ mg}}{1 \text{ g}}$

$= 4.3 \times 10^2 \text{ mg } MgI_2$

12. Nitrates, acetates, and alkali metal compounds are water-soluble. $Zn(NO_3)_2$, $Pb(C_2H_3O_2)_2$, and NaI are soluble.
Most halides are soluble in water; $CuCl_2$ is soluble in water.
Although most sulfates are soluble in water, $BaSO_4(s)$ is not soluble in water.
Only a few hydroxides are soluble in water; $Al(OH)_3(s)$ is not soluble in water.

13. HCl(aq) reacts with active metals and some anions to produce a gas.
Ca is an active metal: $Ca(s) + 2\,HCl(aq) \longrightarrow CaCl_2(aq) + H_2(g)$
HSO_3^- produces a gas with an acid: $KHSO_3(s) + HCl(aq) \longrightarrow KCl(aq) + H_2O + SO_2(g)$

14. In each case, each available cation is paired with the available anions, one at a time, to determine if a compound is produced that is insoluble, based on the solubility rules of Chapter 5. Then a net ionic equation is written to summarize this information.

(a) $Pb^{2+}(aq) + 2\,Br^-(aq) \longrightarrow PbBr_2(s)$

(b) No reaction occurs.

(c) $Fe^{3+}(aq) + 3\,OH^-(aq) \longrightarrow Fe(OH)_3(s)$

(d) $Ca^{2+}(aq) + CO_3^{2-}(aq) \longrightarrow CaCO_3(s)$

(e) $Ba^{2+}(aq) + SO_4^{2-}(aq) \longrightarrow BaSO_4(s)$

(f) No reaction; CaS(s) is moderately soluble.

15. The type of reaction is given first, followed by the net ionic equation.

(a) Neutralization: $OH^-(aq) + HC_2H_3O_2(aq) \longrightarrow H_2O(l) + C_2H_3O_2^-(aq)$

(b) No reaction occurs. This is the mixing of two acids.

(c) Gas evolution: $FeS(s) + 2\ H^+(aq) \longrightarrow H_2S(g) + Fe^{2+}(aq)$

(d) Gas evolution: $HCO_3^-(aq) + H^+(aq) \longrightarrow "H_2CO_3(aq)" \longrightarrow H_2O(l) + CO_2(g)$

(e) Redox: $Mg(s) + 2\ H^+(aq) \longrightarrow Mg^{2+}(aq) + H_2(g)$

(f) No reaction occurs, based on the information in Table 5-3.

16. The problem is most easily solved with amounts in millimoles.

$$\text{volume NaOH(aq)} = 10.00\ \text{mL HCl(aq)} \times \frac{0.128\ \text{mmol HCl}}{1\ \text{mL HCl(aq)}} \times \frac{1\ \text{mmol H}^+}{1\text{mmol HCl}} \times \frac{1\ \text{mmol OH}^-}{1\ \text{mmol H}^+}$$
$$\times \frac{1\ \text{mmol NaOH}}{1\ \text{mmol OH}^-} \times \frac{1\ \text{mL NaOH(aq)}}{0.0962\ \text{mmol NaOH}} = 13.3\ \text{mL NaOH(aq) soln}$$

17. $$\text{NaOH molarity} = \frac{10.00\ \text{mL acid} \times \dfrac{0.1012\ \text{mmol H}_2\text{SO}_4}{1\ \text{mL acid}} \times \dfrac{2\ \text{mmol NaOH}}{1\ \text{mmol H}_2\text{SO}_4}}{23.31\ \text{mL base}} = 0.08683\ \text{M}$$

18. The net ionic equation for the reaction of KOH, a strong base, with HCl, a strong acid, is:

$$OH^-(aq) + H^+(aq) \longrightarrow H_2O$$

Thus, the reactant that produces the smaller amount of ions is the limiting reactant. More to the point, the difference between the larger amount of ions and the smaller amount, determines whether the resulting solution is acidic or basic. If the difference is zero, the solution is neutral.

$$\text{amount OH}^- = 23.58\ \text{mL KOH(aq)} \times \frac{0.1278\ \text{mmol KOH}}{1\ \text{mL KOH(aq)}} \times \frac{1\ \text{mmol OH}^-}{1\ \text{mmol KOH}} = 3.014\ \text{mmol OH}^-$$

$$\text{amount H}^+ = 25.13\ \text{mL HCl(aq)} \times \frac{0.1264\ \text{mmol HCl}}{1\ \text{mL HCl(aq)}} \times \frac{1\ \text{mmol H}^+}{1\ \text{mmol HCl}} = 3.176\ \text{mmol H}^+$$

excess ion = 3.176 mmol H^+ – 3.014 mmol OH^- = 0.162 mmol H^+ The solution is acidic.

19. **(a)** The O.S. of H is +1, that of O is –2, that of C is +4, and that of Mg is +2 on each side of this equation. This is not a redox equation.

(b) The O.S. of Cl_2 is 0 on the left and –1 on the right side of this equation. The O.S. of Br is –1 on the left and 0 on the right side of this equation. This is the equation of a redox reaction.

(c) The O.S. of Ag is 0 on the left and +1 on the right side of this equation. The O.S. of N is +5 on the left and +2 on the right side of this equation. This is the equation of a redox reaction.

(d) The O.S. of O is –2, that of Ag is +1, and that of Cr is +6 on each side of this equation. This is not a redox equation.

20. **(a)** The O.S. of O is –2 on both sides of this equation. The O.S. of H is 0 on the left and +1 on the right side of this equation; H is oxidized and thus NO must be an oxidizing agent. The O.S. of N is +2 on the left and –3 on the right side of this equation; N is reduced and thus H_2 must be a reducing agent.

(b) The O.S. of O is –2 and that of H is +1 on both sides of this equation. The O.S. of Cu is 0 on the left and +2 on the right side of this equation; Cu is oxidized and thus NO_3^- must be an oxidizing agent. The O.S. of N is +5 on the left and +2 on the right side of this equation; N is reduced and thus Cu must be a reducing agent.

(c) The O.S. of O is –2 and that of H is +1 on both sides of this equation. The O.S. of Cl is 0 on the left side of this equation; on the right side, the O.S. of Cl is –1 in Cl^- and it is +5 in ClO_3^-. Cl is both oxidized and reduced and Cl_2 serves as both an oxidizing agent and as a reducing agent.

21. **(a)** Reduction: $2\ SO_3^{2-}(aq) + 6\ H^+(aq) + 4\ e^- \longrightarrow S_2O_3^{2-}(aq) + 3\ H_2O$

(b) Reduction: $2\ NO_3^-(aq) + 10\ H^+(aq) + 8\ e^- \longrightarrow N_2O(g) + 5\ H_2O(l)$

(c) Oxidation: $I^-(aq) + 3\ H_2O(l) \longrightarrow IO_3^-(aq) + 6\ H^+(aq) + 6\ e^-$

(d) Oxidation: $Al(s) + 4\ OH^-(aq) \longrightarrow Al(OH)_4^-(aq) + 3\ e^-$

22. **(a)** Oxidation $Zn(s) \longrightarrow Zn^{2+}(aq) + 2\ e^-$ $\times 3$

Reduction: $NO_3^-(aq) + 4\ H^+(aq) + 3\ e^- \longrightarrow NO(g) + 2\ H_2O$ $\times 2$

Net: $3\ Zn(s) + 2\ NO_3^-(aq) + 8\ H^+(aq) \longrightarrow 3\ Zn^{2+}(aq) + 2\ NO(g) + 4\ H_2O$

(b) Oxidation: $Zn(s) \longrightarrow Zn^{2+}(aq) + 2\,e^-$ $\times 4$

Reduction: $NO_3^-(aq) + 10\,H^+(aq) + 8e^- \longrightarrow NH_4^+(aq) + 3\,H_2O(l)$ $\times 1$

Net: $4\,Zn(s) + NO_3^-(aq) + 10\,H^+(aq) \longrightarrow 4\,Zn^{2+}(aq) + NH_4^+(aq) + 3\,H_2O(l)$

(c) Oxidation: $Fe^{2+}(aq) \longrightarrow Fe^{3+}(aq) + e^-$ $\times 6$

Reduction $Cr_2O_7^{2-}(aq) + 14\,H^+(aq) + 6\,e^- \longrightarrow 2\,Cr^{3+}(aq) + 7\,H_2O$ $\times 1$

Net: $Cr_2O_7^{2-}(aq) + 14\,H^+(aq) + 6\,Fe^{2+}(aq) \longrightarrow 6\,Fe^{3+}(aq) + 2\,Cr^{3+}(aq) + 7\,H_2O$

(d) Oxidation: $H_2O_2(aq) \longrightarrow O_2(g) + 2\,H^+(aq) + 2\,e^-$ $\times 5$

Reduction: $MnO_4^-(aq) + 8\,H^+(aq) + 5\,e^- \longrightarrow Mn^{2+}(aq) + 4\,H_2O$ $\times 2$

Net: $2\,MnO_4^-(aq) + 6\,H^+(aq) + 5\,H_2O_2(aq) \longrightarrow 2\,Mn^{2+}(aq) + 8\,H_2O + 5\,O_2(g)$

<u>23</u>. **(a)** Oxidation: $MnO_2(s) + 4\,OH^-(aq) \longrightarrow MnO_4^-(aq) + 2\,H_2O + 3\,e^-$ $\times 2$

Reduction: $ClO_3^-(aq) + 3\,H_2O + 6\,e^- \longrightarrow Cl^-(aq) + 6\,OH^-(aq)$ $\times 1$

Net: $2\,MnO_2(s) + ClO_3^-(aq) + 2\,OH^-(aq) \longrightarrow 2\,MnO_4^-(aq) + Cl^-(aq) + H_2O$

(b) Oxidation: $Fe(OH)_3(s) + 5\,OH^-(aq) \longrightarrow FeO_4^{2-}(aq) + 4\,H_2O + 3\,e^-$ $\times 2$

Reduction: $OCl^-(aq) + H_2O + 2\,e^- \longrightarrow Cl^-(aq) + 2\,OH^-(aq)$ $\times 3$

Net: $2\,Fe(OH)_3(s) + 3\,OCl^-(aq) + 4\,OH^-(aq) \longrightarrow 2\,FeO_4^{2-}(aq) + 3\,Cl^-(aq) + 5\,H_2O$

(c) Oxidation: $ClO_2 + 2\,OH^-(aq) \longrightarrow ClO_3^-(aq) + H_2O(l) + e^-$ $\times 5$

Reduction: $ClO_2 + 2\,H_2O(l) + 5\,e^- \longrightarrow Cl^- + 4\,OH^-(aq)$ $\times 1$

Net: $6\,ClO_2 + 6\,OH^-(aq) \longrightarrow 5\,ClO_3^-(aq) + Cl^-(aq) + 3\,H_2O$

<u>24</u>. Oxidation: $C_2O_4^{2-}(aq) \longrightarrow 2\,CO_2(g) + 2\,e^-$ $\times 5$

Reduction: $MnO_4^-(aq) + 8\,H^+(aq) + 5\,e^- \longrightarrow Mn^{2+}(aq) + 4\,H_2O$ $\times 2$

Net: $2\,MnO_4^-(aq) + 16\,H^+(aq) + 5\,C_2O_4^{2-}(aq) \longrightarrow 2\,Mn^{2+}(aq) + 8\,H_2O + 10\,CO_2(g)$

$$[MnO_4^-] = \frac{0.2879\text{ g }Na_2C_2O_4 \times \dfrac{1\text{ mol }Na_2C_2O_4}{134.00\text{ g }Na_2C_2O_4} \times \dfrac{1\text{ mol }C_2O_4^{2-}}{1\text{ mol }Na_2C_2O_4} \times \dfrac{2\text{ mol }MnO_4^-}{5\text{ mol }C_2O_4^{2-}}}{25.12\text{ mL soln} \times \dfrac{1\text{ L soln}}{1000\text{ mL soln}}}$$

$= 0.03421$ M Thus, 0.03421 M $= KMnO_4$ molarity.

EXERCISES

Strong Electrolytes, Weak Electrolytes, and Nonelectrolytes

<u>25</u>. **(a)** Because its formula begins with hydrogen, HC_6H_5O is an acid. It is not listed in Table 5-1, so it is a weak acid. A weak acid is a *weak electrolyte*.

(b) Li_2SO_4 is an ionic compound, that is, a salt. A salt is a *strong electrolyte*.

(c) MgI_2 also is a salt, a *strong electrolyte*.

(d) $(CH_3CH_2)_2O$ is a covalent compound whose formula does not begin with H. Thus it is neither an acid nor a salt. It also is not built around nitrogen, making it not a weak base. This is a *nonelectrolyte*.

(e) $Sr(OH)_2$ is a *strong electrolyte*, one of the strong bases listed in Table 5-1.

26. $NH_3(aq)$ is a weak base; $HC_2H_3O_2(aq)$ is a weak acid. Reaction produces a solution of ammonium acetate, $NH_4C_2H_3O_2(aq)$, a salt and a strong electrolyte. $NH_3(aq) + HC_2H_3O_2(aq) \longrightarrow NH_4^+(aq) + C_2H_3O_2^-(aq)$

Ion Concentrations

<u>27</u>. **(a)** $[Ca^{2+}] = \dfrac{35.0\text{ mg }Ca^{2+}}{1\text{ L}} \times \dfrac{1\text{ g}}{1000\text{ mg}} \times \dfrac{1\text{ mol }Ca^{2+}}{40.078\text{ g }Ca^{2+}} = 8.73 \times 10^{-4}\text{ M }Ca^{2+}$

(b) $[K^+] = \frac{25.6 \text{ mg } K^+}{100 \text{ mL}} \times \frac{1 \text{ mmol } K^+}{39.098 \text{ mg } K^+} = 6.55 \times 10^{-3} \text{ M } K^+$

(c) $[Zn^{2+}] = \frac{0.168 \text{ mg } Zn^{2+}}{1 \text{ mL}} \times \frac{1 \text{ mmol } Zn^{2+}}{65.39 \text{ mg } Zn^{2+}} = 2.57 \times 10^{-3} \text{ M } Zn^{2+}$

28. $\text{NaF concn} = \frac{0.9 \text{ mg } F^-}{1 \text{ L}} \times \frac{1 \text{ g}}{1000 \text{ mg}} \times \frac{1 \text{ mol } F^-}{19.00 \text{ g } F^-} \times \frac{1 \text{ mol NaF}}{1 \text{ mol } F^-} = 5 \times 10^{-5} \text{ M NaF}$

29. Let us determine the concentration of each solution in mg Na^+/mL.

(a) $\text{mg } Na^+/\text{mL} = \frac{0.208 \text{ mmol } Na_2SO_4}{1 \text{ mL}} \times \frac{2 \text{ mmol } Na^+}{1 \text{ mmol } Na_2SO_4} \times \frac{23.0 \text{ mg } Na^+}{1 \text{ mmol } Na^+} = 9.57 \text{ mg } Na^+/\text{mL}$

(b) $\text{mg } Na^+/\text{mL} = \frac{1.05 \text{ g NaCl}}{100 \text{ mL}} \times \frac{1 \text{ mol NaCl}}{58.5 \text{ g NaCl}} \times \frac{1 \text{ mol } Na^+}{1 \text{ mol NaCl}} \times \frac{23.0 \text{ g } Na^+}{1 \text{ mol } Na^+} \times \frac{1000 \text{ mg}}{1 \text{ g}}$

$= 4.13 \text{ mg } Na^+/\text{mL}$

(c) The solution with 14.7 mg Na^+/mL has the highest concentration of Na^+.

30. NH_3 is a weak base and would have an exceedingly low $[H^+]$; the answer is not 1.00 M NH_3.
$HC_2H_3O_2$ is a very weak acid; 0.011 M $HC_2H_3O_2$ would have a low $[H^+]$.
H_2SO_4 has two ionizable protons per mole while HCl has but one.
Thus, H_2SO_4 would have the highest $[H^+]$ in a 0.010 M aqueous solution.

31. $\text{mol } Cl^- = \left(0.225 \text{ L} \times \frac{0.625 \text{ mol KCl}}{1 \text{ L soln}} \times \frac{1 \text{ mol } Cl^-}{1 \text{ mol KCl}}\right) + \left(0.615 \text{ L} \times \frac{0.385 \text{ mol } MgCl_2}{1 \text{ L soln}} \times \frac{2 \text{ mol } Cl^-}{1 \text{ mol } MgCl_2}\right)$

$= 0.141 \text{ mol } Cl^- + 0.474 \text{ mol } Cl^-$

$[Cl^-] = \frac{(0.141 + 0.474) \text{ mol } Cl^-}{0.225 \text{ L} + 0.615 \text{ L}} = 0.732 \text{ M}$

32. amount of NO_3^- ion =

$\left(0.275 \text{ L} \times \frac{0.283 \text{ mol } KNO_3}{1 \text{ L soln}} \times \frac{1 \text{ mol } NO_3^-}{1 \text{ mol } KNO_3}\right) + \left(0.328 \text{ L} \times \frac{0.421 \text{ mol } Mg(NO_3)_2}{1 \text{ L soln}} \times \frac{2 \text{ mol } NO_3^-}{1 \text{ mol } Mg(NO_3)_2}\right)$

$= 0.0778 \text{ mol } NO_3^- + 0.276 \text{ mol } NO_3^-$

$[NO_3^-] = \frac{(0.0778 + 0.276) \text{ mol } NO_3^-}{0.275 \text{ L} + 0.328 \text{ L} + 0.784 \text{ L}} = 0.255 \text{ M}$

Predicting Precipitation Reactions

33.

	Mixture	Result (net ionic equation)
(a)	$HI(aq) + Zn(NO_3)_2(aq) \rightarrow$	No reaction occurs.
(b)	$CuSO_4(aq) + Na_2CO_3(aq) \rightarrow$	$Cu^{2+}(aq) + CO_3^{2-}(aq) \longrightarrow CuCO_3(s)$
(c)	$Cu(NO_3)_2(aq) + Na_3PO_4(aq) \rightarrow$	$3\ Cu^{2+}(aq) + 2\ PO_4^{3-}(aq) \longrightarrow Cu_3(PO_4)_2(s)$

34.

	Mixture	Result (net ionic equation)
(a)	$AgNO_3(aq) + CuCl_2(aq) \rightarrow$	$Ag^+(aq) + Cl^-(aq) \longrightarrow AgCl(s)$
(b)	$Na_2S(aq) + FeCl_2(aq) \rightarrow$	$S^{2-}(aq) + Fe^{2+}(aq) \longrightarrow FeS(s)$
(c)	$Na_2CO_3(aq) + AgNO_3(aq) \rightarrow$	$CO_3^{2-}(aq) + 2\ Ag^+(aq) \longrightarrow Ag_2CO_3(s)$

35. (a) Add $K_2SO_4(aq)$; $BaSO_4(s)$ will form. $CaCl_2(s)$ will dissolve.
$BaCl_2(s) + K_2SO_4(aq) \longrightarrow BaSO_4(s) + 2\ KCl(aq)$

(b) Add $H_2O(l)$; $Na_2CO_3(s)$ will dissolve, but $MgCO_3(s)$ will not dissolve.
$Na_2CO_3(s) \xrightarrow{\text{water}} 2\ Na^+(aq) + CO_3^{2-}(aq)$

(c) Add KCl(aq); AgCl(s) will form, while $Cu(NO_3)(s)$ will dissolve.
$AgNO_3(s) + KCl(aq) \longrightarrow AgCl(s) + KNO_3(aq)$

36. (a) Add H_2O. $Cu(NO_3)_2(s)$ will dissolve, while $PbSO_4(s)$ will not.
$Cu(NO_3)_2(s) \xrightarrow{\text{water}} Cu^{2+}(aq) + 2\ NO_3^-(aq)$

(b) Add HCl(aq). $Mg(OH)_2(s)$ will dissolve, but $BaSO_4(s)$ will not.
$Mg(OH)_2(s) + 2\ HCl(aq) \longrightarrow MgCl_2(aq) + 2\ H_2O$

(c) Add HCl(aq). Both carbonates will react, but $PbSO_4(s)$ will form while $CaCl_2(aq)$ remains dissolved.
$PbCO_3(s) + 2\ HCl(aq) \longrightarrow PbCl_2(s) + H_2O + CO_2(g)$
$CaCO_3(s) + 2\ HCl(aq) \longrightarrow CaCl_2(aq) + H_2O + CO_2(g)$

37.

	Mixture	Net ionic equation
(a)	$Sr(NO_3)_2(aq) + K_2SO_4(aq) \longrightarrow$	$Sr^{2+}(aq) + SO_4^{2-}(aq) \longrightarrow SrSO_4(s)$
(b)	$Mg(NO_3)_2(aq) + NaOH(aq) \longrightarrow$	$Mg^{2+}(aq) + 2\ OH^-(aq) \longrightarrow Mg(OH)_2(s)$
(c)	$BaCl_2(aq) + K_2SO_4(aq) \longrightarrow$	$Ba^{2+}(aq) + SO_4^{2-}(aq) \longrightarrow BaSO_4(s)$ KCl(aq) solution left.

38.

	Mixture	Net ionic equation
(a)	$BaCl_2(aq) + K_2SO_4(aq) \longrightarrow$	$Ba^{2+}(aq) + SO_4^{2-}(aq) \longrightarrow BaSO_4(s)$
(b)	$NaCl(aq) + Ag_2SO_4(s) \longrightarrow$	$Ag_2SO_4(s) + 2\ Cl^-(aq) \longrightarrow 2\ AgCl(s) + SO_4^{2-}(aq)$

$BaCl_2(aq) + Ag_2SO_4(s)$ will simply give a mixed precipitate: $AgCl(s) + BaSO_4(s)$

(c)	$Sr(NO_3)_2(aq) + K_2SO_4(aq) \longrightarrow$	$Sr^{2+}(aq) + SO_4^{2-}(aq) \longrightarrow SrSO_4(s)$ $KNO_3(aq)$ soln left.

Acid–Base Reactions

39. **(a)** $NaHCO_3(s) + H^+(aq) \longrightarrow Na^+(aq) + H_2O(l) + CO_2(g)$
(b) $CaCO_3(s) + 2\ H^+(aq) \longrightarrow Ca^{2+}(aq) + H_2O(l) + CO_2(g)$
(c) $Mg(OH)_2(s) + 2\ H^+(aq) \longrightarrow Mg^{2+}(aq) + 2\ H_2O(l)$
(d) $Mg(OH)_2(s) + 2\ H^+(aq) \longrightarrow Mg^{2+}(aq) + 2\ H_2O(l)$
$Al(OH)_3(s) + 3\ H^+(aq) \longrightarrow Al^{3+}(aq) + 3\ H_2O(l)$
(e) $NaAl(OH)_2CO_3(s) + 4\ H^+(aq) \longrightarrow Al^{3+}(aq) + Na^+(aq) + 3\ H_2O(l) + CO_2(g)$

40. Dissolve $Na_2CrO_4(s)$ and $ZnSO_4(s)$ each in H_2O to produce solutions of, respectively, Na^+ and Zn^{2+}. Dissolve each of $BaCO_3(s)$ and $Al(OH)_3(s)$ in HCl(aq) to produce solutions of $Ba^{2+}(aq)$ and $Al^{3+}(aq)$.

41. As a salt: $NaHSO_4(aq) \longrightarrow Na^+(aq) + HSO_4^-(aq)$
As an acid: $HSO_4^-(aq) + OH^-(aq) \longrightarrow H_2O(l) + SO_4^{2-}(aq)$

42. Because all three compounds have an ammonium cation, all are formed by the reaction of an acid with aqueous ammonia. The identity of the anion determines which acid is used.
$2\ NH_3(aq) + H_3PO_4(aq) \longrightarrow (NH_4)_2HPO_4(aq)$ $NH_3(aq) + HNO_3(aq) \longrightarrow NH_4NO_3(aq)$
$2\ NH_3(aq) + H_2SO_4(aq) \longrightarrow (NH_4)_2SO_4(aq)$

Oxidation–Reduction (Redox) Equations

43. **(a)** In this reaction, iron is reduced from $Fe^{3+}(aq)$ to $Fe^{2+}(aq)$ *and* manganese is reduced from a +7 O.S. in $MnO_4^-(aq)$ to a +2 O.S. in $Mn^{2+}(aq)$. Thus, there are two reductions and no oxidation, an impossibility.
(b) In this reaction, chlorine is oxidized from an O.S. of 0 in $Cl_2(aq)$ to an O.S. of +1 in $ClO^-(aq)$ *and* oxygen is oxidized from an O.S. of –1 in $H_2O_2(aq)$ to an O.S. of 0 in $O_2(g)$. There thus are two oxidations and no reduction, also an impossibility.

44. The story is not accurate. The reaction between sodium hydroxide and hydrochloric acid is an acid–base reaction, $NaOH(aq) + HCl(aq) \longrightarrow NaCl(aq) + H_2O(l)$. The formation of $Cl_2(g)$ would require an oxidation–reduction reaction.

45. We first balance the equation by inspection, and then by the half-equation method.
(a) Inspection just involves placing the proper coefficients before the products, because there is only one reactant. $(NH_4)_2Cr_2O_7(s) \longrightarrow N_2(g) + Cr_2O_3(s) + 4\ H_2O$
Oxidation: $2\ NH_4^+ \longrightarrow N_2(g) + 8\ H^+(aq) + 6\ e^-$
Reduction: $Cr_2O_7^{2-} + 8\ H^+(aq) + 6\ e^- \longrightarrow Cr_2O_3(s) + 4\ H_2O$
Net: $(NH_4)_2Cr_2O_7(s) \longrightarrow N_2(g) + Cr_2O_3(s) + 4\ H_2O$

(b) We double the reaction to avoid $\frac{1}{2} O_2(g)$: $2\ Pb(NO_3)_2 \longrightarrow 2\ PbO(s) + 4\ NO_2(g) + O_2(g)$

Oxidation: $2\ H_2O \longrightarrow O_2(g) + 4\ H^+(aq) + 4\ e^-$

Reduction: $NO_3^- + 2\ H^+(aq) + e^- \longrightarrow NO_2(g) + H_2O$ $\times 4$

Net: $4\ NO_3^- + 4\ H^+(aq) \longrightarrow 4\ NO_2(g) + O_2(g) + 2\ H_2O$

Add $2\ O^{2-}$: $4\ NO_3^- +$ ~~$2\ H_2O$~~ $\longrightarrow 2\ O^{2-} + 4\ NO_2(g) + O_2(g) +$ ~~$2\ H_2O$~~

46. Each of the reactions is treated as if it takes place in acidic aqueous solution, for the purpose of balancing its redox equation.

(b) Oxidation: $NH_3(g) + H_2O \longrightarrow NO(g) + 5\ H^+ + 5\ e^-$ $\times 4$

Reduction: $O_2(g) + 4\ H^+ + 4\ e^- \longrightarrow 2\ H_2O$ $\times 5$

Net: $4\ NH_3(g) + 4\ H_2O + 5\ O_2(g) + 20\ H^+ \longrightarrow 4\ NO(g) + 20\ H^+ + 10\ H_2O$

Simplify: $4\ NH_3(g) + 5\ O_2(g) \longrightarrow 4\ NO(g) + 6\ H_2O(g)$

(b) Oxidation: $H_2(g) \longrightarrow 2\ H^+(aq) + 2\ e^-$ $\times 5$

Reduction: $2\ NO(g) + 10\ H^+(aq) + 10\ e^- \longrightarrow 2\ NH_3(g) + 2\ H_2O$

Net: $5\ H_2(g) + 2\ NO(g) \longrightarrow 2\ NH_3(g) + 2\ H_2O(g)$

47. (a) Oxidation: $2\ I^-(aq) \longrightarrow I_2(s) + 2\ e^-$ $\times 5$

Reduction: $MnO_4^-(aq) + 8\ H^+(aq) + 5\ e^- \longrightarrow Mn^{2+}(aq) + 4\ H_2O(l)$ $\times 2$

Net: $10\ I^-(aq) + 2\ MnO_4^-(aq) + 16\ H^+(aq) \longrightarrow 5\ I_2(s) + 2\ Mn^{2+}(aq) + 8\ H_2O(l)$

(b) Oxidation: $N_2H_4(l) \longrightarrow N_2(g) + 4\ H^+(aq) + 4\ e^-$ $\times 3$

Reduction: $BrO_3^-(aq) + 6\ H^+(aq) + 6\ e^- \longrightarrow Br^-(aq) + 3\ H_2O(l)$ $\times 2$

Net: $3\ N_2H_4(l) + 2\ BrO_3^-(aq) \longrightarrow 3\ N_2(g) + 2\ Br^-(aq) + 6\ H_2O(l)$

(c) Oxidation: $Fe^{2+}(aq) \longrightarrow Fe^{3+}(aq) + e^-$ $\times 1$

Reduction: $VO_4^{3-}(aq) + 6\ H^+(aq) + e^- \longrightarrow VO^{2+}(aq) + 3\ H_2O(l)$ $\times 1$

Net: $Fe^{2+}(aq) + VO_4^{3-}(aq) + 6\ H^+(aq) \longrightarrow Fe^{3+}(aq) + VO^{2+}(aq) + 3\ H_2O(l)$

(d) Oxidation: $UO^{2+}(aq) + H_2O(l) \longrightarrow UO_2^{2+}(aq) + 2\ H^+(aq) + 2\ e^-$ $\times 3$

Reduction: $NO_3^-(aq) + 4\ H^+(aq) + 3\ e^- \longrightarrow NO(g) + 2\ H_2O(l)$ $\times 2$

Net: $3\ UO^{2+}(aq) + 2\ NO_3^-(aq) + 2\ H^+(aq) \longrightarrow 3\ UO_2^{2+}(aq) + 2\ NO(g) + H_2O(l)$

48. (a) Oxidation: $P_4(s) + 16\ H_2O \longrightarrow 4\ H_2PO_4^-(aq) + 24\ H^+(aq) + 20\ e^-$ $\times 3$

Reduction: $NO_3^-(aq) + 4\ H^+(aq) + 3\ e^- \longrightarrow NO(g) + 2\ H_2O$ $\times 20$

Net: $3\ P_4(s) + 20\ NO_3^-(aq) + 8\ H_2O + 8\ H^+(aq) \longrightarrow 12\ H_2PO_4^-(aq) + 20\ NO(g)$

(b) Oxidation: $S_2O_3^{2-}(aq) + 5\ H_2O(l) \longrightarrow 2\ SO_4^{2-}(aq) + 10\ H^+(aq) + 8\ e^-$ $\times 5$

Reduction: $MnO_4^-(aq) + 8\ H^+(aq) + 5\ e^- \longrightarrow Mn^{2+}(aq) + 4\ H_2O$ $\times 8$

Net: $5\ S_2O_3^{2-}(aq) + 8\ MnO_4^-(aq) + 14\ H^+(aq) \longrightarrow 10\ SO_4^{2-}(aq) + 8\ Mn^{2+}(aq) + 7\ H_2O(l)$

(c) Oxidation: $2\ HS^-(aq) + 3\ H_2O(l) \longrightarrow S_2O_3^{2-}(aq) + 8\ H^+(aq) + 8\ e^-$ $\times 1$

Reduction: $2\ HSO_3^-(aq) + 4\ H^+(aq) + 4\ e^- \longrightarrow S_2O_3^{2-}(aq) + 3\ H_2O$ $\times 2$

Net: $2\ HS^-(aq) + 4\ HSO_3^-(aq) \longrightarrow 3\ S_2O_3^{2-}(aq) + 3\ H_2O(l)$

(d) Oxidation: $2\ NH_3OH^+(aq) \longrightarrow N_2O(g) + H_2O(l) + 6\ H^+(aq) + 4\ e^-$ $\times 1$

Reduction: $Fe^{3+}(aq) + e^- \longrightarrow Fe^{2+}(aq)$ $\times 4$

Net: $4\ Fe^{3+}(aq) + 2\ NH_3OH^+(aq) \longrightarrow 4\ Fe^{2+}(aq) + N_2O(g) + H_2O(l) + 6\ H^+(aq)$

49. (a) Oxidation: $CN^-(aq) + 2\ OH^-(aq) \longrightarrow CNO^-(aq) + H_2O + 2\ e^-$ $\times 3$

Reduction: $MnO_4^-(aq) + 2\ H_2O(l) + 3\ e^- \longrightarrow MnO_2(s) + 4\ OH^-(aq)$ $\times 2$

Net: $3\ CN^-(aq) + 2\ MnO_4^-(aq) + H_2O(l) \longrightarrow 3\ CNO^-(aq) + 2\ MnO_2(s) + 2\ OH^-(aq)$

(b) Oxidation: $N_2H_4(l) + 4\,OH^-(aq) \longrightarrow N_2(g) + 4\,H_2O + 4\,e^-$ × 1
Reduction: $[Fe(CN)_6]^{3-}(aq) + e^- \longrightarrow [Fe(CN)_6]^{4-}(aq)$ × 4

Net: $4\,[Fe(CN)_6]^{3-}(aq) + N_2H_4(l) + 4\,OH^-(aq) \longrightarrow 4\,[Fe(CN)_6]^{4-}(aq) + N_2(g) + 4\,H_2O(l)$

(c) Oxidation: $Fe(OH)_2(s) + OH^-(aq) \longrightarrow Fe(OH)_3(s) + e^-$ × 4
Reduction: $O_2(g) + 2\,H_2O(l) + 4\,e^- \longrightarrow 4\,OH^-(aq)$ × 1

Net: $4\,Fe(OH)_2(s) + O_2(g) + 2\,H_2O(l) \longrightarrow 4\,Fe(OH)_3(s)$

(d) Oxidation: $C_2H_5OH(aq) + 5\,OH^-(aq) \longrightarrow C_2H_3O_2^-(aq) + 4\,H_2O(l) + 4\,e^-$ × 3
Reduction: $MnO_4^-(aq) + 2\,H_2O(l) + 3\,e^- \longrightarrow MnO_2(s) + 4\,OH^-(aq)$ × 4

Net: $3\,C_2H_5OH(aq) + 4\,MnO_4^-(aq) \longrightarrow 3\,C_2H_3O_2^-(aq) + 4\,MnO_2(s) + OH^-(aq) + 4\,H_2O(l)$

50. (a) Reduction: $Cl_2(g) + 2\,e^- \longrightarrow 2\,Cl^-(aq)$ × 5
Oxidation: $Cl_2(g) + 12\,OH^-(aq) \longrightarrow 2\,ClO_3^-(aq) + 6\,H_2O + 10\,e^-$ × 1

Net: $6\,Cl_2(g) + 12\,OH^-(aq) \longrightarrow 10\,Cl^-(aq) + 2\,ClO_3^-(aq) + 6\,H_2O$
Or: $3\,Cl_2(g) + 6\,OH^-(aq) \longrightarrow 5\,Cl^-(aq) + ClO_3^-(aq) + 3\,H_2O$

(b) Oxidation: $S_2O_4^{2-}(aq) + 2\,H_2O \longrightarrow 2\,HSO_3^-(aq) + 2\,H^+(aq) + 2\,e^-$
Reduction: $S_2O_4^{2-}(aq) + 2\,H^+(aq) + 2\,e^- \longrightarrow S_2O_3^{2-}(aq) + H_2O$

Net: $2\,S_2O_4^{2-}(aq) + H_2O \longrightarrow 2\,HSO_3^-(aq) + S_2O_3^{2-}(aq)$

(c) Oxidation: $MnO_4^{2-}(aq) \longrightarrow MnO_4^-(aq) + e^-$ × 2
Reduction: $MnO_4^{2-}(aq) + 2\,H_2O + 2\,e^- \longrightarrow MnO_2(s) + 4\,OH^-(aq)$ × 1

Net: $3\,MnO_4^{2-}(aq) + 2\,H_2O \longrightarrow 2\,MnO_4^-(aq) + MnO_2(s) + 4\,OH^-(aq)$

(d) Oxidation: $P_4(s) + 8\,OH^-(aq) \longrightarrow 4\,H_2PO_2^-(aq) + 4\,e^-$ × 3
Reduction: $P_4(s) + 12\,H_2O + 12\,e^- \longrightarrow 4\,PH_3(g) + 12\,OH^-(aq)$ × 1

Net: $4\,P_4(s) + 12\,OH^-(aq) + 12\,H_2O \longrightarrow 12\,H_2PO_2^-(aq) + 4\,PH_3(g)$

51. (a) Oxidation: $S_2O_3^{2-}(aq) + 5\,H_2O \longrightarrow 2\,SO_4^{2-}(aq) + 10\,H^+(aq) + 8\,e^-$ × 1
Reduction: $Cl_2(g) + 2\,e^- \longrightarrow 2\,Cl^-(aq)$ × 4

Net: $S_2O_3^{2-}(aq) + 5\,H_2O(l) + 4\,Cl_2(g) \longrightarrow 2\,SO_4^{2-}(aq) + 8\,Cl^-(aq) + 10\,H^+(aq)$

(b) Oxidation: $Sn^{2+}(aq) \longrightarrow Sn^{4+}(aq) + 2e^-$ × 3
Reduction $Cr_2O_7^{2-}(aq) + 14\,H^+(aq) + 6\,e^- \longrightarrow 2\,Cr^{3+}(aq) + 7\,H_2O$ × 1

Net: $Cr_2O_7^{2-}(aq) + 14\,H^+(aq) + 3\,Sn^{2+}(aq) \longrightarrow 3\,Sn^{4+}(aq) + 2\,Cr^{3+}(aq) + 7\,H_2O$

(c) Oxidation: $S_8(s) + 24\,OH^-(aq) \longrightarrow 4\,S_2O_3^{2-}(aq) + 12\,H_2O(l) + 16\,e^-$
Reduction: $S_8(s) + 16\,e^- \longrightarrow 8\,S^{2-}(aq)$

Net: $2\,S_8(s) + 24\,OH^-(aq) \longrightarrow 8\,S^{2-}(aq) + 4\,S_2O_3^{2-}(aq) + 12\,H_2O(l)$
Halved: $S_8(s) + 12\,OH^-(aq) \longrightarrow 4\,S^{2-}(aq) + 2\,S_2O_3^{2-}(aq) + 6\,H_2O(l)$

(d) Oxidation: $As_2S_3(s) + 40\,OH^-(aq) \longrightarrow 2\,AsO_4^{3-}(aq) + 3\,SO_4^{2-}(aq) + 20\,H_2O + 28\,e^-$ ×
Reduction: $H_2O_2(aq) + 2\,e^- \longrightarrow 2\,OH^-(aq)$ × 14

Net: $As_2S_3(s) + 12\,OH^-(aq) + 14\,H_2O_2(aq) \longrightarrow 2\,AsO_4^{3-}(aq) + 3\,SO_4^{2-}(aq) + 20\,H_2O$

52. (a) Oxidation: $NO_2^-(aq) + H_2O(l) \longrightarrow NO_3^-(aq) + 2\,H^+(aq) + 2\,e^-$ × 5
Reduction: $MnO_4^-(aq) + 8\,H^+(aq) + 5\,e^- \longrightarrow Mn^{2+}(aq) + 4\,H_2O(l)$ × 2

Net: $5\,NO_2^-(aq) + 2\,MnO_4^-(aq) + 6\,H^+(aq) \longrightarrow 5\,NO_3^-(aq) + 2\,Mn^{2+}(aq) + 3\,H_2O(l)$

(b) Oxidation: $H_2S(aq) \longrightarrow S(s) + 2\,H^+(aq) + 2\,e^-$ × 2
Reduction: $SO_2(aq) + 4\,H^+(aq) + 4\,e^- \longrightarrow S(s) + 2\,H_2O(l)$ × 1

Net: $2\,H_2S(aq) + SO_2(aq) \longrightarrow 3\,S(s) + 2\,H_2O(l)$

(c) $2\ Na(s) + 2\ HI(aq) \longrightarrow 2\ NaI(aq) + H_2(g)$

(d) Oxidation: $Zn(s) \longrightarrow Zn^{2+}(aq) + 2\ e^-$ $\times 1$

Reduction: $VO^{2+}(aq) + 2\ H^+(aq) + e^- \longrightarrow V^{3+}(aq) + H_2O(l)$ $\times 2$

Net: $Zn(s) + 2\ VO^{2+}(aq) + 4\ H^+(aq) \longrightarrow Zn^{2+}(aq) + 2\ V^{3+}(aq) + 2\ H_2O(l)$

Oxidizing and Reducing Agents

53. The oxidizing agents experience a decrease in the oxidation state of one of their elements, while the reducing agents experience an increase in the oxidation state of one of their elements.

(a) $SO_3^{2-}(aq)$ is the reducing agent; the O.S. of S = +4 in SO_3^{2-} and = +6 in SO_4^{2-}.
$MnO_4^-(aq)$ is the oxidizing agent; the O.S. of Mn = +7 in MnO_4^- and = +2 in Mn^{2+}.

(b) $H_2(g)$ is the reducing agent; the O.S. of H = 0 in $H_2(g)$ and = +1 in $H_2O(g)$.
$NO_2(g)$ is the oxidizing agent; the O.S. of N = +4 in $NO_2(g)$ and = –3 in NH_3

(c) $[Fe(CN)_6]^{4-}(aq)$ is the reducing agent; the O.S. of Fe = +2 in $[Fe(CN)_6]^{4-}$ and = +3 in $[Fe(CN)_6]^{3-}$
$H_2O_2(aq)$ is the oxidizing agent; the O.S. of O = –1 in H_2O_2 and = –2 in H_2O.

54. **(a)** $2\ S_2O_3^{2-}(aq) + I_2(s) \longrightarrow S_4O_6^{2-}(aq) + 2\ I^-(aq)$

(b) $S_2O_3^{2-}(aq) + 4\ Cl_2(g) + 5\ H_2O(l) \longrightarrow 2\ HSO_4^-(aq) + 8\ Cl^-(aq) + 8\ H^+(aq)$

(c) $S_2O_3^{2-}(aq) + 4\ OCl^-(aq) + 2\ OH^-(aq) \longrightarrow 2\ SO_4^{2-}(aq) + 4\ Cl^-(aq) + H_2O(l)$

Neutralization and Acid–Base Titrations

55. $NaOH(aq) + HCl(aq) \longrightarrow NaCl(aq) + H_2O$ is the titration reaction.

$$\text{NaOH molarity} = \frac{0.02834\ \text{L} \times \frac{0.1085\ \text{mol HCl}}{1\ \text{L soln}} \times \frac{1\ \text{mol NaOH}}{1\ \text{mol HCl}}}{0.02500\ \text{L sample}} = 0.1230\ \text{M NaOH}$$

56. $$[NH_3] = \frac{28.72\ \text{mL acid} \times \frac{1.021\ \text{mmol HCl}}{1\ \text{mL acid}} \times \frac{1\text{mmol H}^+}{1\ \text{mmol HCl}} \times \frac{1\ \text{mmol NH}_3}{1\ \text{mmol H}^+}}{5.00\ \text{mL sample}} = 5.86\ \text{M NH}_3$$

57. The net reaction is $OH^-(aq) + HC_3H_5O_2(aq) \longrightarrow H_2O + C_3H_5O_2^-(aq)$

$$\text{volume base} = 25.00\ \text{mL acid} \times \frac{0.3057\ \text{mmol HC}_3\text{H}_5\text{O}_2}{1\ \text{mL acid}} \times \frac{1\ \text{mmol KOH}}{1\ \text{mmol HC}_3\text{H}_5\text{O}_2} \times \frac{1\ \text{mL base}}{0.2155\ \text{mmol KOH}}$$
$$= 35.46\ \text{mL KOH solution}$$

58. Titration reaction: $Ba(OH)_2(aq) + 2\ HNO_3(aq) \longrightarrow Ba(NO_3)_2(aq) + 2\ H_2O(l)$

$$\text{volume base} = 50.00\ \text{mL acid} \times \frac{0.0526\ \text{mol HNO}_3}{1\ \text{ml acid}} \times \frac{1\ \text{mmol Ba(OH)}_2}{2\ \text{mmol HNO}_3} \times \frac{1\ \text{mL base}}{0.0884\ \text{mmol Ba(OH)}_2}$$
$$= 14.9\ \text{mL Ba(OH)}_2\ \text{solution}$$

59. The mass of acetylsalicyclic acid is converted to the amount of NaOH, in millimoles, that will react with it.

$$[\text{NaOH}] = \frac{0.32\ \text{g HC}_9\text{H}_7\text{O}_4}{23\ \text{mL NaOH(aq)}} \times \frac{1\ \text{mol HC}_9\text{H}_7\text{O}_4}{180.1\ \text{g HC}_9\text{H}_7\text{O}_4} \times \frac{1\ \text{mol NaOH}}{1\ \text{mol HC}_9\text{H}_7\text{O}_4} \times \frac{1000\ \text{mmol NaOH}}{1\ \text{mol NaOH}}$$
$$= 0.077\ \text{M NaOH}$$

60. **(a)** $$\text{vol conc. acid} = 20.0\ \text{L} \times \frac{0.10\ \text{mol HCl}}{1\ \text{L soln}} \times \frac{36.5\ \text{g HCl}}{1\ \text{mol HCl}} \times \frac{100\ \text{g conc}}{38\ \text{g HCl}} \times \frac{1\ \text{mL}}{1.19\ \text{g conc}}$$
$$= 1.6 \times 10^2\ \text{mL conc. acid}$$

(b) The titration reaction is $HCl(aq) + NaOH(aq) \longrightarrow NaCl(aq) + H_2O(l)$

$$[\text{HCl}] = \frac{20.93\ \text{mL base} \times \frac{0.1186\ \text{mmol NaOH}}{1\ \text{mL base}} \times \frac{1\ \text{mmol HCl}}{1\ \text{mmol NaOH}}}{25.00\ \text{mL acid}} = 0.09929\ \text{M HCl}$$

(c) First of all the volume of the dilute solution (20 L) is known to a precision of at best two significant figures. Secondly, HCl is somewhat volatile (we can smell its odor above the solution) and some will likely be lost during the process of producing the solution.

61. The equation for the reaction is $HNO_3(aq) + KOH(aq) \longrightarrow KNO_3(aq) + H_2O(l)$ This equation shows that equal numbers of moles are needed for a complete reaction. We compute the amount of each reactant.

$$\text{no. mmol } HNO_3 = 25.00 \text{ mL acid} \times \frac{0.132 \text{ mmol } HNO_3}{1 \text{ mL acid}} = 3.30 \text{ mmol } HNO_3$$

$$\text{no. mmol KOH} = 10.00 \text{ mL acid} \times \frac{0.318 \text{ mmol KOH}}{1 \text{ mL base}} = 3.18 \text{ mmol KOH}$$

There is more acid present than base. Thus, the resulting solution is acidic.

62. Compute the amount of acetic acid in the vinegar and the amount of acetic acid needed to react with the sodium carbonate. If there is more than enough acid to react with the solid, the solution will remain acidic.

$$\text{acetic acid in vinegar} = 125 \text{ mL} \times \frac{0.762 \text{ mmol } HC_2H_3O_2}{1 \text{ mL vinegar}} = 95.3 \text{ mmol } HC_2H_3O_2$$

$$Na_2CO_3(s) + 2\, HC_2H_3O_2(aq) \longrightarrow 2\, NaC_2H_3O_2(aq) + H_2O + CO_2(g)$$

$$\text{acetic acid for solid} = 7.55 \text{ g } Na_2CO_3 \times \frac{1000 \text{ mmol}}{106.0 \text{ g } Na_2CO_3} \times \frac{2 \text{ mmol } HC_2H_3O_2}{1 \text{ mmol } Na_2CO_3} = 142.5 \text{ mmol } HC_2H_3O_2$$

There is not enough acetic acid present to react with all of the sodium carbonate. The resulting solution will not be acidic. In fact, it will not be a solution, since not all of the solid will react and dissolve.

63. $$\text{volume base} = 5.00 \text{ mL vinegar} \times \frac{1.01 \text{ g vinegar}}{1 \text{ mL}} \times \frac{4.0 \text{ g } HC_2H_3O_2}{100.0 \text{ g vinegar}} \times \frac{1 \text{ mol } HC_2H_3O_2}{60.0 \text{ g } HC_2H_3O_2} \times \frac{1 \text{ mol NaOH}}{1 \text{ mol } HC_2H_3O_2} \times \frac{1 \text{ L base}}{0.1000 \text{ mol NaOH}} \times \frac{1000 \text{ mL}}{1 \text{ L}} = 34 \text{ mL base}$$

64. The titration reaction is $2\, NaOH(aq) + H_2SO_4(aq) \longrightarrow Na_2SO_4(aq) + 2\, H_2O$

It is most convenient to consider molarity as millimoles per milliliter when solving this problem.

$$[H_2SO_4] = \frac{49.74 \text{ mL base} \times \frac{0.935 \text{ mmol NaOH}}{1 \text{ mL base}} \times \frac{1 \text{ mmol } H_2SO_4}{2 \text{ mmol NaOH}}}{5.00 \text{ mL battery acid}} = 4.65 \text{ M}$$

The battery acid is *not* sufficiently concentrated.

Stoichiometry of Oxidation–Reduction Equations

65. $$\text{concn} = \frac{0.1078 \text{ g } As_2O_3 \times \frac{1 \text{ mol } As_2O_3}{197.84 \text{ g } As_2O_3} \times \frac{4 \text{ mol } MnO_4^-}{5 \text{ mol } As_2O_3} \times \frac{1 \text{ mol } KMnO_4}{1 \text{ mol } MnO_4^-}}{22.15 \text{ mL} \times \frac{1 \text{ L}}{1000 \text{ mL}}}$$

$$= 0.01968 \text{ M } KMnO_4$$

66. The balanced equation: $5\, SO_3^{2-}(aq) + 2\, MnO_4^-(aq) + 6\, H^+(aq) \longrightarrow 5\, SO_4^{2-}(aq) + 2\, Mn^{2+}(aq) + 3\, H_2O(l)$

$$[SO_3^{2-}] = \frac{31.46 \text{ mL} \times \frac{0.02237 \text{ mmol } KMnO_4}{1 \text{ mL soln}} \times \frac{1 \text{ mmol } MnO_4^-}{1 \text{ mmol } KMnO_4} \times \frac{5 \text{ mmol } SO_3^{2-}}{2\, MnO_4^-}}{25.00 \text{ mL } SO_3^{2-} \text{ soln}}$$

$$= 0.07038 \text{ M } SO_3^{2-}$$

67. First determine the mass of Fe, then the percentage of iron in the ore.

$$\text{mass Fe} = 28.72 \text{ mL} \times \frac{1 \text{ L}}{1000 \text{ mL}} \times \frac{0.05051 \text{ mol } Cr_2O_7^{2-}}{1 \text{ L soln}} \times \frac{6 \text{ mol } Fe^{2+}}{1 \text{ mol } Cr_2O_7^{2-}} \times \frac{55.85 \text{ g Fe}}{1 \text{ mol } Fe^{2+}}$$

$$= 0.4861 \text{ g Fe}$$

$$\%\text{ Fe} = \frac{0.4861 \text{ g Fe}}{0.9132 \text{ g ore}} \times 100\% = 53.23\% \text{ Fe}$$

68. First balance the equation.

Oxidation: $Mn^{2+}(aq) + 4\, OH^-(aq) \longrightarrow MnO_2(s) + 2\, H_2O(l) + 2\, e^-$ $\times 3$

Reduction: $MnO_4^-(aq) + 2\, H_2O(aq) + 3\, e^- \longrightarrow MnO_2(s) + 4\, OH^-(aq)$ $\times 2$

Net: $3\, Mn^{2+}(aq) + 2\, MnO_4^-(aq) + 4\, OH^-(aq) \longrightarrow 5\, MnO_2(s) + 2\, H_2O(l)$

$$[Mn^{2+}] = \frac{37.21 \text{ mL titrant} \times \frac{0.04162 \text{ mmol } MnO_4^-}{1 \text{ mL titrant}} \times \frac{3 \text{ mmol } Mn^{2+}}{2 \text{ mmol } MnO_4^-}}{25.00 \text{ mL soln}} = 0.09292 \text{ M } Mn^{2+}$$

69. Oxidation: $C_2O_4^{2-}(aq) \longrightarrow 2\,CO_2(g) + 2\,e^-$ $\times 5$

Reduction: $MnO_4^-(aq) + 8\,H^+(aq) + 5\,e^- \longrightarrow Mn^{2+}(aq) + 4\,H_2O(l)$ $\times 2$

Net: $5\,C_2O_4^{2-}(aq) + 2\,MnO_4^-(aq) + 16\,H^+(aq) \longrightarrow 10\,CO_2(g) + 2\,Mn^{2+}(aq) + 8\,H_2O(l)$

$$\text{mass } Na_2C_2O_4 = 1.00 \text{ L satd soln} \times \frac{1000 \text{ mL}}{1 \text{ L}} \times \frac{25.8 \text{ mL}}{50.0 \text{ mL satd soln}} \times \frac{0.02140 \text{ mol } KMnO_4}{1000 \text{ mL}}$$
$$\times \frac{1 \text{ mol } MnO_4^-}{1 \text{ mol } KMnO_4} \times \frac{5 \text{ mol } C_2O_4^{2-}}{2 \text{ mol } MnO_4^-} \times \frac{1 \text{ mol } Na_2C_2O_4}{1 \text{ mol } C_2O_4^{2-}} \times \frac{134.0 \text{ g } Na_2C_2O_4}{1 \text{ mol } Na_2C_2O_4}$$
$$= 3.70 \text{ g } Na_2C_2O_4$$

70. Balanced equation: $3\,S_2O_4^{2-}(aq) + 2\,CrO_4^{2-}(aq) + 4\,H_2O(l) \longrightarrow 6\,SO_3^{2-}(aq) + 2\,Cr(OH)_3(s) + 2\,H^+(aq)$

(a) $$\text{mass } Cr(OH)_3 = 100.\text{ L soln} \times \frac{0.0126 \text{ mol } CrO_4^{2-}}{1 \text{ L soln}} \times \frac{2 \text{ mol } Cr(OH)_3}{2 \text{ mol } CrO_4^{2-}} \times \frac{103.0 \text{ g } Cr(OH)_3}{1 \text{ mol } Cr(OH)_3}$$
$$= 130.\text{ g } Cr(OH)_3$$

(b) $$\text{mass } Na_2S_2O_4 = 100.\text{ L soln} \times \frac{0.0126 \text{ mol } CrO_4^{2-}}{1 \text{ L soln}} \times \frac{3 \text{ mol } S_2O_4^{2-}}{2 \text{ mol } CrO_4^{2-}} \times \frac{1 \text{ mol } Na_2S_2O_4}{1 \text{ mol } S_2O_4^{2-}}$$
$$\times \frac{174.1 \text{ g } Na_2S_2O_4}{1 \text{ mol } Na_2S_2O_4} = 329 \text{ g } Na_2S_2O_4$$

FEATURE PROBLEMS

A. **(a)** $2\,Na(s) + 2\,H_2O(l) \longrightarrow 2\,NaOH(aq) + H_2(g)$

(b) $Fe^{3+}(aq) + 3\,OH^-(aq) \longrightarrow Fe(OH)_3(s)$

(c) $Fe(OH)_3(s) + 3\,H^+(aq) \longrightarrow Fe^{3+}(aq) + 3\,H_2O(l)$

B. From the volume of titrant we can calculate both the amount in moles of NaC_5H_5 and (through its molar mass of 88.08 g/mol) the mass of NaC_5H_5 in a sample. The remaining mass in a sample is that of C_4H_8O (72.11 g/mol) whose amount in moles we calculate. The ratio of the amount of C_4H_8O in a sample to the amount of NaC_5H_5 is the value of x.

$$\text{amount } NaC_5H_5 = 0.01492 \text{ L} \times \frac{0.1001 \text{ mol HCl}}{1 \text{ L soln}} \times \frac{1 \text{ mol NaOH}}{1 \text{ mol HCl}} \times \frac{1 \text{ mol } NaC_5H_5}{1 \text{ mol NaOH}} = \begin{array}{c}0.001493 \\ \text{mol } NaC_5H_5\end{array}$$

$$\text{mass } C_4H_8O = 0.242 \text{ g sample} - \left(0.001493 \text{ mol } NaC_5H_5 \times \frac{88.08 \text{ g } NaC_5H_5}{1 \text{ mol } NaC_5H_5}\right) = 0.110 \text{ g } C_4H_8O$$

$$x = \frac{0.110 \text{ g } C_4H_8O \times \dfrac{1 \text{ mol } C_4H_8O}{72.11 \text{ g } C_4H_8O}}{0.001493 \text{ mol } NaC_5H_5} = 1.02$$

For the second sample, parallel calculations give 0.001200 mol NaC_5H_5, 0.093 g C_4H_8, $x = 1.07$. There is a rounding error in this second calculation because it is limited to two significant figures.

C. First we balance the two equations.

Oxidation: $H_2C_2O_4(aq) \longrightarrow 2\,CO_2(g) + 2\,H^+(aq) + 2\,e^-$ $\times 1$

Reduction: $MnO_2(s) + 4\,H^+(aq) + 2\,e^- \longrightarrow Mn^{2+}(aq) + 2\,H_2O(l)$ $\times 1$

Net: $H_2C_2O_4(aq) + MnO_2(s) + 2\,H^+(aq) \longrightarrow 2\,CO_2(g) + Mn^{2+}(aq) + 2\,H_2O(l)$

Oxidation: $H_2C_2O_4(aq) \longrightarrow 2\,CO_2(g) + 2\,H^+(aq) + 2\,e^-$ $\times 5$

Reduction: $MnO_4^-(aq) + 8\,H^+(aq) + 5\,e^- \longrightarrow Mn^{2+}(aq) + 4\,H_2O(l)$ $\times 2$

Net: $5\,H_2C_2O_4(aq) + 2\,MnO_4^-(aq) + 6\,H^+(aq) \longrightarrow 10\,CO_2(g) + 2\,Mn^{2+}(aq) + 8\,H_2O(l)$

Now we determine the mass of the excess oxalic acid.

$$\text{mass } H_2C_2O_4 = 0.03006 \text{ L} \times \frac{0.1000 \text{ mol } KMnO_4}{1 \text{ L}} \times \frac{1 \text{ mol } MnO_4^-}{1 \text{ mol } KMnO_4} \times \frac{5 \text{ mol } H_2C_2O_4}{2 \text{ mol } MnO_4^-}$$
$$\times \frac{1 \text{ mol } H_2C_2O_4{\cdot}2H_2O}{1 \text{ mol } H_2C_2O_4} \times \frac{126.07 \text{ g } H_2C_2O_4{\cdot}2H_2O}{1 \text{ mol } H_2C_2O_4{\cdot}2H_2O} = 0.9474 \text{ g } H_2C_2O_4{\cdot}2H_2O$$

The mass of $H_2C_2O_4{\cdot}2H_2O$ that reacted with MnO_2 = 1.651 g − 0.9474 g = 0.704 g $H_2C_2O_4{\cdot}2H_2O$

$$\text{mass } MnO_2 = 0.704 \text{ g } H_2C_2O_4{\cdot}2H_2O \times \frac{1 \text{ mol } H_2C_2O_4}{126.07 \text{ g } H_2C_2O_4{\cdot}2H_2O} \times \frac{1 \text{ mol } MnO_2}{1 \text{ mol } H_2C_2O_4} \times \frac{86.9 \text{ g } MnO_2}{1 \text{ mol } MnO_2}$$
$$= 0.485 \text{ g}$$

$$\%\ MnO_2 = \frac{0.485 \text{ g } MnO_2}{0.533 \text{ g sample}} \times 100\% = 91.0\%\ MnO_2$$

6 GASES

PRACTICE EXAMPLES

1A The pressure measured by each liquid must be the same. They are related through $P = g\,h\,d$
Thus we have the following $g\,h_{PCE}\,d_{PCE} = g\,h_{Hg}\,d_{Hg}$ The g's cancel; we substitute known values.

$6.38\text{ m PCE} \times 1.62\text{ g/cm}^3\text{ PCE} = h_{Hg} \times 13.6\text{ g/cm}^3\text{ Hg}$ $h_{Hg} = 6.38\text{ m} \times \frac{1.62\text{ g/cm}^3}{13.6\text{ g/cm}^3} = 0.760\text{ m Hg}$

$P = 0.760\text{ m Hg} = 760\text{ mmHg}$

1B The solution is found through the expression relating density and height. $h_{Et}\,d_{Et} = h_{Hg}\,d_{Hg}$
We substitute known values and solve for ethanol density. $13.04\text{ m Et} \times d_{Et} = 757\text{ mmHg} \times 13.6\text{ g/cm}^3\text{ Hg}$

$$d_{Et} = \frac{757\text{ mm} \times 13.6\text{ g/cm}^3}{13.04\text{ m}} \times \frac{1\text{ m}}{1000\text{ mm}} = 0.790\text{ g/cm}^3\text{ ethanol}$$

2A We know that $P_{gas} = P_{bar} + \Delta P$ with $P_{bar} = 748.2$ mmHg. We are now given that $\Delta P = 7.8$ mmHg. Thus P_{gas} = 748.2 mmHg + 7.8 mmHg = 756.0 mmHg

2B The difference in pressure between the two levels must be the same, just expressed in different units. Hence, this problem is almost a repetition of Practice Example 6-1.
$h_{Hg} = 748.2\text{ mmHg} - 739.6\text{ mmHg} = 8.6\text{ mmHg}$ Again we have the following $g\,h_g\,d_g = g\,h_{Hg}\,d_{Hg}$
This becomes $8.6\text{ mmHg} \times 13.6\text{ g/cm}^3\text{ Hg} = h_g \times 1.26\text{ g/cm}^3\text{ glycerol}$

$$h_g = 8.6\text{ mmHg} \times \frac{13.6\text{ g/cm}^3\text{ Hg}}{1.26\text{ g/cm}^3\text{ glycerol}} = 93\text{ mm glycerol}$$

3A Boyle's Law relates the pressure-volume product. $P_1\,V_1 = P_2\,V_2$

$5.25\text{ atm} \times V_1 = 1.85\text{ atm} \times 12.5\text{ L}$ $V_1 = \frac{1.85\text{ atm} \times 12.5\text{ L}}{5.25\text{ atm}} = 4.40\text{ L}$

3B We use Boyle's law, solved for the final pressure. After that, the pressure is converted to mmHg.

$$P_2 = P_1 \times \frac{V_1}{V_2} = 2.25\text{ atm} \times \frac{1.50\text{ L}}{8.10\text{ L}} = 0.417\text{ atm} \times \frac{760\text{ mmHg}}{1\text{ atm}} = 317\text{ mmHg}$$

4A Charles's law states that the volume/temperature ratio is constant with the temperature in kelvins.

$$\frac{V_1}{T_1} = \frac{V_2}{T_2} = \frac{0.250\text{ L}}{(25 + 273.15)\text{ K}} = \frac{1.65\text{ L}}{T_2} \qquad T_2 = \frac{298\text{ K} \times 1.65\text{ L}}{0.250\text{ L}} = 1.97 \times 10^3\text{ K} = 1.70 \times 10^3\text{ °C}$$

4B We use another simple gas law, which states that the pressure is proportional to the absolute temperature, for a fixed amount of gas at constant volume. The two absolute (kelvin) temperatures in this problem are T_1 = 273 + 22 = 295 K and T_2 = 273 + 935 = 1208 K.

$$\frac{P_1}{T_1} = \frac{P_2}{T_2} \quad \text{This becomes} \quad P_2 = P_1 \times \frac{T_2}{T_1} = 1.82\text{ atm} \times \frac{1208\text{ K}}{295\text{ K}} = 7.45\text{ atm}$$

5A We are comparing 0.60 L $N_2(g)$ at 150 K and 0.98 atm with 1.25 L $O_2(g)$ at STP. The $N_2(g)$ temperature is slightly less than half of standard temperature of 273 K; raising temperature to 273 K will not bring the volume to 1.20 atm, twice the initial volume. And raising the pressure to 1.00 atm will compress the gas. The final volume of $N_2(g)$ is less than the 1.25 L of $O_2(g)$.

5B Case (a) is at a temperature slightly higher than STP; decreasing the temperature will cause the volume to contract below 1.20 L. It also is at a pressure slightly lower than STP; increasing the pressure also will cause the volume to contract still further below 1.20 L.

Case (b) is at a temperature slightly below STP; increasing the temperature will cause the gas to expand above 1.25 L. It also is at a pressure slightly greater than STP; decreasing the pressure will cause the volume to expand still further above 1.25 L.

As a result, at STP the volume of gas (a) will be significantly below 1.20 L, while that of gas (b) will be significantly above 1.25 L. Gas (b) will have the greater volume at STP. Note that we assumed that the simple gas laws apply to each gas. Their identities—$N_2(g)$ and $O_2(g)$—were not needed to solve this problem.

6A The STP molar volume of 22.414 L enables us to determine the amount in moles of propane, from which we find the mass with the use of the molar mass.

$$\text{mass propane} = 30.0\text{ L} \times \frac{1\text{ mol}}{22.414\text{ L}} \times \frac{44.10\text{ g } C_3H_8}{1\text{ mol } C_3H_8} = 59.0\text{ g } C_3H_8$$

6B A gas's STP molar volume is 22.414 L. In addition, we need the molar mass of $CO_2(g)$, 44.01 g/mol.

$$CO_2(g)\text{ volume} = 128\text{ g} \times \frac{1\text{ mol } CO_2}{44.01\text{ g } CO_2} \times \frac{22.414\text{ L at STP}}{1\text{ mol } CO_2} = 65.2\text{ L } CO_2(g)\text{ at STP}$$

7A The ideal gas equation is solved for volume. Conversions are made within the equation.

$$V = \frac{nRT}{P} = \frac{\left(20.2\text{ g } NH_3 \times \frac{1\text{ mol } NH_3}{17.03\text{ g } NH_3}\right) \times \frac{0.0821\text{ L·atm}}{\text{mol·K}} \times (-25 + 273)\text{ K}}{752\text{ mmHg} \times \frac{1\text{ atm}}{760\text{ mmHg}}} = 24.4\text{ L } NH_3$$

7B The amount of $Cl_2(g)$ is 0.193 mol Cl_2 and the pressure is 0.980 atm, as they are in Example 6-7. This information is substituted into the ideal gas equation after it has been solved for temperature.

$$T = \frac{PV}{nR} = \frac{0.980\text{ atm} \times 7.50\text{ L}}{0.193\text{ mol} \times 0.08206\text{ L atm mol}^{-1}\text{ K}^{-1}} = 464\text{ K}$$

8A The ideal gas equaltion is solved for amount and the quantities are substituted.

$$n = \frac{PV}{RT} = \frac{10.5\text{ atm} \times 5.00\text{ L}}{\frac{0.0821\text{ L·atm}}{\text{mol·K}} \times (30.0 + 273.2)\text{ K}} = 2.11\text{ mol He}$$

8B We use the ideal gas equation, solved for pressure.

$$P = \frac{nRT}{V} =$$

$$= \frac{\left(1.00 \times 10^{20}\text{ molecules} \times \frac{1\text{ mol}}{6.022 \times 10^{23}\text{ molecules}}\right) \times \frac{0.08206\text{ L atm}}{\text{mol K}} \times (175 + 273)\text{ K}}{3.05\text{ L}}$$

$$= 2.00 \times 10^{-3}\text{ atm} = 1.52\text{ mmHg}$$

9A The general gas equation is solved for volume, after the constant amount in moles is cancelled. Temperatures are converted to kelvins.

$$V_2 = \frac{V_1 P_1 T_2}{P_2 T_1} = \frac{1.00\text{ mL} \times 2.14\text{ atm} \times (37.8 + 273.2)\text{ K}}{1.02\text{ atm} \times (36.2 + 273.2)\text{ K}} = 2.11\text{ mL}$$

9B The flask has a volume of 1.00 L and initially contains $O_2(g)$ at STP. The mass of $O_2(g)$ that must be released is obtained from the difference in amount of $O_2(g)$ at the two temperatures, 273 K and 373 K. We also could compute the masses separately and subtract them.

$$\text{mass released} = (n_{STP} - n_{100°C}) \times \mathcal{M}_{O_2} = \left(\frac{PV}{R\,273\text{ K}} - \frac{PV}{R\,373\text{ K}}\right) \times \mathcal{M}_{O_2} = \frac{PV}{R}\left(\frac{1}{273\text{ K}} - \frac{1}{373\text{ K}}\right) \times \mathcal{M}_{O_2}$$

$$= \frac{1.00\text{ atm} \times 1.00\text{ L}}{0.08206\text{ L atm mol}^{-1}\text{ K}^{-1}}\left(\frac{1}{273\text{ K}} - \frac{1}{373\text{ K}}\right) \times \frac{32.00\text{ g}}{1\text{ mol } O_2} = 0.383\text{ g } O_2$$

10A The volume of the vessel is 0.09841 L. We substitute other values into the expression for molar mass.

$$\mathcal{M} = \frac{m\ R\ T}{P\ V} = \frac{(40.4868\text{ g} - 40.1305\text{ g}) \times \dfrac{0.08206\text{ L·atm}}{\text{mol·K}} \times (22.4 + 273.2)\text{ K}}{\left(772\text{ mmHg} \times \dfrac{1\text{ atm}}{760\text{ mmHg}}\right) \times 0.09841\text{ L}} = 86.46\text{ g/mol}$$

10B The gas's molar mass is its mass (1.27 g) divided by its amount in moles. The amount can be determined from the ideal gas equation.

$$n = \frac{PV}{RT} = \frac{\left(737\text{ mm Hg} \times \dfrac{1\text{ atm}}{760\text{ mmHg}}\right) \times 1.07\text{ L}}{\dfrac{0.08206\text{ L atm}}{\text{mol K}} \times (25 + 273)\text{K}} = 0.0424\text{ mol gas}$$

$$\mathcal{M} = \frac{1.27\text{ g}}{0.0424\text{ mol}} = 30.0\text{ g/mol}$$ This agrees very well with the molar mass of NO, 30.006 g/mol.

11A The molar mass of He is 4.003 g/mol. This is substituted into the expression for density.

$$d = \frac{\mathcal{M}P}{R\ T} = \frac{4.003\text{ g mol}^{-1} \times 0.987\text{ atm}}{0.08206\text{ L·atm mol}^{-1}\text{ K}^{-1} \times 298\text{ K}} = 0.162\text{ g/L}$$

When compared to the density of air under the same conditions (1.29 g/L) the density of He is only about one eighth as much. He is less dense ("lighter") than air.

11B The suggested solution is a simple one; merely solve for the temperature.

$$T = \frac{\mathcal{M}P}{Rd} = \frac{\dfrac{32.00\text{ g }O_2}{1\text{ mol }O_2} \times \left(745\text{ mmHg} \times \dfrac{1\text{ atm}}{760\text{ mmHg}}\right)}{\dfrac{0.08206\text{ L atm}}{\text{mol K}} \times \dfrac{1.00\text{ g}}{1\text{ L}}} = 382\text{ K}$$

However, suppose that you have forgotten the convenient formula for the density of an ideal gas? You can still solve a problem such as this one. The density of 1.00 g/L indicates a mass of 1.00 g of gas in a 1.00-L volume. Of course, the mass of a gas, given its identity (oxygen in this case) enables us to determine the amount in moles of the gas (*n* in the ideal gas equation). Then we can solve the ideal gas equation for the desired property, such as temperature, as follows.

$$T = \frac{PV}{nR} = \frac{\left(745\text{ mmHg} \times \dfrac{1\text{ atm}}{760\text{ mmHg}}\right) \times 1.00\text{ L}}{\left(1.00\text{ g }O_2 \times \dfrac{1\text{ mol }O_2}{32.00\text{ g }O_2}\right) \times \dfrac{0.08206\text{ L atm}}{\text{mol K}}} = 382\text{ K}$$

12A The balanced equation is $2\ NaN_3(s) \xrightarrow{\Delta} 2\ Na(l) + 3\ N_2(g)$

$$\text{amount } N_2 = \frac{P\ V}{R\ T} = \frac{\left(776\text{ mmHg} \times \dfrac{1\text{ atm}}{760\text{ mmHg}}\right) \times 20.0\text{ L}}{\dfrac{0.08206\text{ L·atm}}{\text{mol·K}} \times (30.0 + 273.2)\text{ K}} = 0.821\text{ mol }N_2$$

Now solve the stoichiometry problem.

$$\text{mass } NaN_3 = 0.821\text{ mol }N_2 \times \frac{2\text{ mol }NaN_3}{3\text{ mol }N_2} \times \frac{65.01\text{ g }Na_3N}{1\text{ mol }Na_3N} = 35.6\text{ g }Na_3N$$

12B Here we are not dealing with gaseous reactants; the law of combining volumes cannot be used. From the ideal gas equation we determine the amount of $N_2(g)$ per liter under the specified conditions. Then we determine the amount of Na(l) produced simultaneously, and finally the mass of that Na(l).

$$\text{mass Na(l)} = \frac{\left(751\text{ mmHg} \times \dfrac{1\text{ atm}}{760\text{ mmHg}}\right) \times 1.000\text{ L}}{\dfrac{0.08206\text{ L atm}}{\text{mol K}} \times (25 + 273)\text{ K}} \times \frac{2\text{ mol Na}}{3\text{ mol }N_2} \times \frac{22.99\text{ g Na}}{1\text{ mol Na}} = 0.619\text{ g Na(l)}$$

13A The law of combining volumes permits us to use stoichiometric coefficients for volume ratios.

$$O_2\text{ volume} = 1.00\text{ L NO(g)} \times \frac{5\text{ L }O_2}{4\text{ L NO}} = 1.25\text{ L }O_2(g)$$

13B The first task is to balance the chemical equation. There must be three moles of hydrogen for every mole of nitrogen in both products (because of the formula of NH_3) and reactants: $N_2(g) + 3\ H_2(g) \longrightarrow 2\ NH_3(g)$
Then volumes of gaseous reactants and products are related as their stoichiometric coefficients, as long as all gases are measured at the same temperature and pressure.

$$\text{volume } NH_3(g) = 225 \text{ L } H_2(g) \times \frac{2 \text{ L } NH_3(g)}{3 \text{ L } H_2(g)} = 150. \text{ L } NH_3$$

14A We can work easily with the ideal gas equation, with a new temperature of $T = (55 + 273)\text{ K} = 328\text{ K}$.
The amount of Ne added is readily computed. $n_{Ne} = 12.5 \text{ g Ne} \times \frac{1 \text{ mol Ne}}{20.18 \text{ g Ne}} = 0.619 \text{ mol Ne}$

$$P = \frac{n_{total}RT}{V} = \frac{(1.75 + 0.619)\text{ mol} \times \frac{0.08206 \text{ L atm}}{\text{mol K}} \times 328 \text{ K}}{5.0 \text{ L}} = 13 \text{ atm}$$

14B The total volume initially is $2.0\text{ L} + 8.0\text{ L} = 10.0\text{ L}$. These two mixed ideal gases then obey the general gas equation as if they were one gas.

$$P_2 = \frac{P_1 V_1 T_2}{V_2 T_1} = \frac{1.00 \text{ atm} \times 10.0 \text{ L} \times 298 \text{ K}}{2.0 \text{ L} \times 273 \text{ K}} = 5.46 \text{ atm}$$

15A The partial pressures are proportional to the mole fractions.

$$P_{H_2O} = \frac{n_{H_2O}}{n_{tot}} \times P_{tot} = \frac{0.00278 \text{ mol } H_2O}{0.197 \text{ mol } CO_2 + 0.00278 \text{ mol } H_2O} \times 2.50 \text{ atm} = 0.0348 \text{ atm } H_2O(g)$$

$$P_{CO_2} = P_{tot} - P_{H_2O} = 2.50 \text{ atm} - 0.0348 \text{ atm} = 2.47 \text{ atm } CO_2(g)$$

15B Expression (6.17) indicates that, in a mixture of gases, the mole percent equals the volume percent, which in turn equals the pressure percent. Thus, we can apply these volume percents—converted to fractions by dividing by 100—directly to the total pressure.

N_2 pressure $= 0.7808 \times 748$ mmHg $= 584$ mmHg $\qquad$ O_2 pressure $= 0.2095 \times 748$ mmHg $= 157$ mmHg
CO_2 pressure $= 0.00036 \times 748$ mmHg $= 0.27$ mmHg $\qquad$ Ar pressure $= 0.0093 \times 748$ mmHg $= 7.0$ mmHg

16A First compute amount in moles of $H_2(g)$, then use stoichiometry to convert to moles of HCl.

$$\text{amount HCl} = \frac{\left((755 - 25.2) \text{ torr} \times \frac{1 \text{ atm}}{760 \text{ mmHg}}\right) \times 0.0355 \text{ L}}{\frac{0.0821 \text{ L·atm}}{\text{mol·K}} \times (26 + 273) \text{ K}} \times \frac{6 \text{ mol HCl}}{3 \text{ mol } H_2} = 0.00278 \text{ mol HCl}$$

16B The volume occupied by the $O_2(g)$ at its partial pressure is the same as the volume occupied by the mixed gases: water vapor and $O_2(g)$. The partial pressure of $O_2(g)$ is found by difference.

O_2 pressure $= 749.2$ total pressure $- 23.8$ mmHg (H_2O pressure) $= 725.4$ mmHg

The mass of Ag_2O is related to the amount of $O_2(g)$ produced.

$$\text{mass } Ag_2O = 8.07 \text{ g sample} \times \frac{88.3 \text{ g } Ag_2O}{100.0 \text{ g sample}} \times \frac{1 \text{ mol } Ag_2O}{231.74 \text{ g } Ag_2O} \times \frac{1 \text{ mol } O_2}{2 \text{ mol } Ag_2O} = 0.0154 \text{ mol } O_2$$

The volume of $O_2(g)$ is found with the ideal gas law.

$$V = \frac{nRT}{P} = \frac{0.0154 \text{ mol } O_2 \times 0.08206 \frac{\text{L·atm}}{\text{mol·K}} \times (273 + 25) \text{ K}}{725.4 \text{ mmHg} \times \frac{1 \text{ atm}}{760 \text{ mmHg}}} = 0.395 \text{ L } O_2$$

17A The gas with the smaller molar mass, NH_3 at 17.0 g/mol, has the greater root-mean-square speed.

$$u_{rms} = \sqrt{\frac{3RT}{\mathcal{M}}} = \sqrt{\frac{3 \times 8.314 \text{ kg m}^2 \text{ s}^{-2} \text{ mol}^{-1} \text{ K}^{-1} \times 298 \text{ K}}{0.0170 \text{ kg mol}^{-1}}} = 661 \text{ m/s}$$

17B $\text{bullet speed} = \frac{2180 \text{ mi}}{1 \text{ h}} \times \frac{1 \text{ h}}{3600 \text{ s}} \times \frac{5280 \text{ ft}}{1 \text{ mi}} \times \frac{12 \text{ in.}}{1 \text{ ft}} \times \frac{2.54 \text{ cm}}{1 \text{ in.}} \times \frac{1 \text{ m}}{100 \text{ cm}} = 974.5 \text{ m/s}$

Solve the rms-speed equation [12] for temperature by first squaring both sides.

$$u_{rms}^2 = \frac{3RT}{\mathcal{M}} \qquad T = \frac{u_{rms}^2 \mathcal{M}}{3R} = \frac{\left(\frac{974.5 \text{ m}}{1 \text{ s}}\right)^2 \times \frac{2.016 \times 10^{-3} \text{ kg}}{1 \text{ mol } H_2}}{3 \times \frac{8.3145 \text{ kg m}^2}{\text{s}^2 \text{ mol K}}} = 76.75 \text{ K}$$

We expected the temperature to be lower than 298 K. Note that the speed of the bullet is about half the speed of a H_2 molecule at 298 K. To halve the speed of a molecule, its temperature mustbe divided by four.

18A The only difference is the molar mass of the gas. 2.2×10^{-4} mol N_2 effuses through the orifice in 105 s.

$$\frac{?\text{ mol } O_2}{2.2 \times 10^{-4}\text{ mol } N_2} = \sqrt{\frac{\mathcal{M}_{N2}}{\mathcal{M}_{O2}}} = \frac{28.014\text{ g/mol}}{31.999\text{ g/mol}} = 0.87546$$

amount $O_2 = 0.87546 \times 2.2 \times 10^{-4} = 1.9 \times 10^{-4}$ mol O_2

18B The two rates of effusion are related as the square root of the ratio of the molar masses of the two gases. The lighter gas, H_2, effuses faster, and thus requires a shorter time for the same amount of gas to effuse.

$$H_2 \text{ time} = N_2 \text{ time} \times \sqrt{\frac{\mathcal{M}_{H2}}{\mathcal{M}_{N2}}} = 105\text{ s} \times \sqrt{\frac{2.016\text{ g } H_2/\text{mol } H_2}{28.013\text{ g } N_2/\text{mol } N_2}} = 28.2\text{ s}$$

19A The times of effusion are related as the square root of the molar mass. It requires 87.3 s for Kr to effuse.

$$\frac{\text{unknown time}}{\text{Kr time}} = \sqrt{\frac{\mathcal{M}_{unk}}{\mathcal{M}_{Kr}}} = \frac{131.3\text{ s}}{87.3\text{ s}} = \sqrt{\frac{\mathcal{M}_{unk}}{83.80\text{ g/mol}}} = 1.504$$

$\mathcal{M}_{unk} = (1.504)^2 \times 83.80$ g/mol $= 189.6$ g/mol

19B This problem is solved in virtually the same manner as the previous Practice Example. The lighter gas is ethane, with a molar mass of 30.07 g/mol.

$$C_2H_6 \text{ time} = \text{Kr time} \times \sqrt{\frac{\mathcal{M}(C_2H_6)}{\mathcal{M}(Kr)}} = 87.3\text{ s} \times \sqrt{\frac{30.07\text{ g } C_2H_6/\text{mol } C_2H_6}{83.80\text{ g Kr/mol Kr}}} = 52.3\text{ s}$$

20A Because one mole of gas is being considered, the value of n^2a is numerically the same as the value of a, and the value of nb is numerically the same as the value of b.

$$P = \frac{nRT}{V - nb} - \frac{n^2a}{V^2} = \frac{1.00\text{ mol} \times \frac{0.08206\text{ L atm}}{\text{mol K}} \times 273\text{ K}}{(2.00 - 0.0427)\text{ L}} - \frac{3.59\text{ L}^2\text{ atm}}{(2.00\text{ L})^2} = 11.4_4\text{ atm} - 0.898\text{ atm}$$

$= 10.5$ atm $CO_2(g)$ compared with 9.9 atm for $Cl_2(g)$

$$P_{ideal} = \frac{nRT}{V} = \frac{1.00\text{ mol} \times \frac{0.08206\text{ L atm}}{\text{mol K}} \times 273\text{ K}}{2.00\text{ L}} = 11.2\text{ atm}$$

$Cl_2(g)$ shows a greater deviation from ideal gas behavior than does $CO_2(g)$.

20B Because one mole of gas is being considered, the value of n^2a is numerically the same as the value of a, and the value of nb is numerically the same as the value of b.

$$P = \frac{nRT}{V - nb} - \frac{n^2a}{V^2} = \frac{1.00\text{ mol} \times \frac{0.08206\text{ L atm}}{\text{mol K}} \times 273\text{ K}}{(2.00 - 0.0399)\text{ L}} - \frac{1.49\text{ L}^2\text{ atm}}{(2.00\text{ L})^2} = 11.4_3\text{ atm} - 0.373\text{ atm}$$

$= 11.1$ atm $CO(g)$ compared to 9.9 atm for $Cl_2(g)$, 11.2 atm for an ideal gas, and 10.5 atm for $CO_2(g)$
$Cl_2(g)$ displays the greatest deviation from ideality.

SUMMARIZING EXAMPLE CALCULATIONS

1. The relative amounts of $CO_2(g)$ and $H_2O(l)$ are established by the ratio of 1 C to four H's in CH_4
   $CH_4(g) + 2\,O_2(g) \longrightarrow CO_2(g) + 2\,H_2O(l)$

2. $$\text{amount } O_2(g) = \frac{3.55\text{ atm} \times 1.00\text{ L } CH_4}{\frac{0.08206\text{ L atm}}{\text{mol K}} \times (22 + 273)\text{ K}} \times \frac{2\text{ mol } O_2}{1\text{ mol } CH_4} = 0.293\text{ mol } O_2$$

3. Calculate the volume of $O_2(g)$ under the stated conditions.
   $$O_2 \text{ volume} = \frac{0.293\text{ mol } O_2 \times \frac{0.08206\text{ L atm}}{\text{mol K}} \times 295\text{ K}}{745\text{ mmHg} \times \frac{1\text{ atm}}{760\text{ mmHg}}} = 7.24\text{ L } O_2(g)$$

4. Then use the percentage composition of air to determine the volume of air.
   $$\text{air volume} = 7.24\text{ L } O_2(g) \times \frac{100.00\text{ L air}}{20.95\text{ L } O_2(g)} = 34.6\text{ L air}$$

REVIEW QUESTIONS

1. **(a)** "atm" is the abbreviation for "atmosphere," a unit of pressure equal to 760 mmHg, 101,325 Pa, or 14.7 lb/in.^{2}.
(b) "STP" is the abbreviation for "standard temperature and pressure:" 0 °C and 1 atm pressure.
(c) R is the symbol for the ideal gas constant. It has a value of 0.08206 L atm mol^{-1} K^{-1}.
(d) Partial pressure is the pressure that one of the gases in a mixture of gases would exert if it were present in the container by itself under the same conditions.
(e) "u_{rms}" is the abbreviation for the root mean square speed of a number of moving objects, molecules in our considerations. It is the square root of the average of the squares of the speeds. It also is the median speed: half the molecules are traveling faster than this speed, and half are traveling slower.

2. **(a)** The absolute zero of temperature, –273.15 °C, is the lowest temperature possible. All molecular motion ceases at this temperature.
(b) A gas is collected over water by bubbling the gas into an upside-down container that is filled with water. The gas rises to the top of the container, displacing the water, and is trapped in the container.
(c) Effusion of a gas refers to the very slow leakage of the gas out of a small hole in a container and into a vacuum.
(d) The law of combining volumes states that gases, when present at the same temperature and pressure, react in volumes that are related as small whole numbers. These small whole numbers turn out to be the stoichioimetric coefficients.

3. **(a)** A barometer is a device used to measure absolute pressures; it measures the difference in pressure between some gas and a vacuum. A manometer measures the difference in pressure between two gases.
(b) Celsius and Kelvin temperatures both have the same size degree, large enough that 100 degrees span the temperature range from the boiling point to the freezing point of water. The zero point of Celsius temperature is the freezing point of water, that of Kelvin is –273.15 °C, absolute zero.
(c) The ideal gas equation relates the properties of pressure, volume, temperature, and amount for one gas. The general gas equation relates the initial and final values of these four properties, or the properties for two gases.
(d) An ideal gas is one that obeys the ideal gas law. On a molecular level, the molecules of such a gas are dimensionless points that exert no forces of attraction or repulsion. The molecules of a real gas occupy some space (but not much compared to the volume of a gas container) and exert weak attractions on each other. At room temperature and pressure the differences in the properties of real and ideal gases are almost insignificant.

4. **(a)** $P = 736\text{ mmHg} \times \frac{1\text{ atm}}{760\text{ mmHg}} = 0.968\text{ atm}$ **(b)** $P = 58.2\text{ cm Hg} \times \frac{1\text{ atm}}{76\text{ cmHg}} = 0.766\text{ atm}$
(c) $P = 892\text{ torr} \times \frac{1\text{ atm}}{760\text{ torr}} = 1.17\text{ atm}$ **(d)** $P = 48\text{ psi} \times \frac{1\text{ atm}}{14.7\text{ psi}} = 3.3\text{ atm}$

5. **(a)** $h = 0.984\text{ atm} \times \frac{760\text{ mmHg}}{1\text{ atm}} = 748\text{ mmHg}$ **(b)** $h = 928\text{ torr} = 928\text{ mmHg}$
(c) $h = 142\text{ ft } H_2O \times \frac{12\text{ in.}}{1\text{ ft}} \times \frac{2.54\text{ cm}}{1\text{ in.}} \times \frac{10\text{ mmH}_2\text{O}}{1\text{ cm H}_2\text{O}} \times \frac{1\text{ mmHg}}{13.6\text{ mmH}_2\text{O}} \times \frac{1\text{ mHg}}{1000\text{ mmHg}} = 3.18\text{ mHg}$
$= 10.4\text{ ft Hg}$

6. The atmospheric pressure is less than the pressure of the gas. The difference in pressures is
$\Delta P = 16.1\text{ mmHg} - 7.6\text{ mmHg} = 8.5\text{ mmHg}$
$P_{gas} = P_{atm} + \Delta P = 744\text{ mmHg} + 8.5\text{ mmHg} = 753\text{ mmHg}$

7. **(a)** $V = 26.7\text{ L} \times \frac{762\text{ mmHg}}{385\text{ mmHg}} = 52.8\text{ L}$ **(b)** $V = 26.7\text{ L} \times \frac{762\text{ mmHg}}{3.68\text{ atm} \times \frac{760\text{ mmHg}}{1\text{ atm}}} = 7.27\text{ L}$

8. Apply Charles's law: $V = k\,T$. $T_i = 26 + 273 = 299$ K
(a) $T = 273 + 98 = 371$ K $V = 886\text{ mL} \times \frac{371\text{ K}}{299\text{ K}} = 1.10 \times 10^3\text{ mL}$

(b) $T = 273 - 20 = 253\text{ K}$ $\quad V = 886\text{ mL} \times \frac{253\text{ K}}{299\text{ K}} = 7.50 \times 10^2\text{ mL}$

9. Apply Charles's law. $T_f = (22 + 273)\text{ K} \times \frac{165\text{ mL}}{57.3\text{ mL}} = 849\text{ K} = 576\text{ °C}$

10. $V = n \times V_m = \left(49.6\text{ g } C_2H_2 \times \frac{1\text{ mol } C_2H_2}{26.04\text{ g}}\right) \times \frac{22.414\text{ L at STP}}{1\text{ mol}} = 42.7\text{ L } C_2H_2$

11. $V = n \times V_m = \left(250.0\text{ g } Cl_2 \times \frac{1\text{ mol } Cl_2}{70.905\text{ g}}\right) \times \frac{22.414\text{ L at STP}}{1\text{ mol}} = 79.03\text{ L } Cl_2$

12. The gas with the greatest density at STP is the one with the highest mole weight: Cl_2 = 70.9 g/mol, SO_3 = 80.1 g/mol, N_2O = 44.0 g/mol; and PF_3 = 88.0 g/mol. Thus, PF_3 would have the highest STP density of the four gases listed.

13. Assume that the $CO_2(g)$ behaves ideally and use the ideal gas law: $PV = nRT$

$$V = \frac{nRT}{P} = \frac{\left(89.2\text{ g} \times \frac{1\text{ mol } CO_2}{44.01\text{ g}}\right) 0.08206 \frac{\text{L atm}}{\text{mol K}} (37 + 273.2)\text{K}}{737\text{ mmHg} \times \frac{1\text{ atm}}{760\text{ mmHg}}} \times \frac{1000\text{ mL}}{1\text{ L}} = 5.32 \times 10^4\text{ mL}$$

14. $P = \frac{nRT}{V} = \frac{\left(285\text{ g} \times \frac{1\text{ mol } SO_2}{64.07\text{ g}}\right) 0.08206 \frac{\text{L atm}}{\text{mol K}} (27 + 273.2)\text{K}}{40.0\text{ L}} = 2.74\text{ atm}$

15. Use the ideal gas law to determine the amount in moles of the given quantity of gas.

$$n = \frac{PV}{RT} = \frac{\left(743\text{ mmHg} \times \frac{1\text{ atm}}{760\text{ mmHg}}\right)\left(115\text{ mL} \times \frac{1\text{ L}}{1000\text{ mL}}\right)}{0.08206 \frac{\text{L atm}}{\text{mol K}} (273.2 + 66.3)\text{K}} = 0.00404\text{ mol gas}$$

$$\mathcal{M} = \frac{0.418\text{ g}}{0.00404\text{ mol}} = 103\text{ g/mol}$$

16. Density, d (g/L) = molar mass, $\mathcal{M}$(g/mol) ÷ molar volume, V/n (L/mol)

$$V/n = \frac{RT}{P} = \frac{0.08206 \frac{\text{L atm}}{\text{mol K}} \times (32.7 + 273.2)\text{K}}{758\text{ mmHg} \times \frac{1\text{ atm}}{760\text{ mmHg}}} = 25.2\text{ L/mol} \qquad d = \frac{44.01\text{ g/mol}}{25.2\text{ L/mol}} = 1.75\text{ g/L}$$

17. Each mole of gas occupies 22.414 L at STP.

$$H_2\text{ STP volume} = 1.00000\text{ g Al} \times \frac{1\text{ mol Al}}{26.9815\text{ g Al}} \times \frac{3\text{ mol } H_2(g)}{2\text{ mol Al(s)}} \times \frac{22.4\text{ L } H_2(g)\text{ at STP}}{1\text{ mol } H_2}$$
$$= 1.25\text{ L } H_2(g)$$

18. Determine first the amount of $CO_2(g)$ that can be removed. Then use the ideal gas law.

$$\text{no. mol } CO_2 = 1.00\text{ kg LiOH} \times \frac{1000\text{ g}}{1\text{ kg}} \times \frac{1\text{ mol LiOH}}{23.95\text{ g LiOH}} \times \frac{1\text{ mol } CO_2}{2\text{ mol LiOH}} = 20.9\text{ mol } CO_2$$

$$V = \frac{nRT}{P} = \frac{20.9\text{ mol} \times 0.08206 \frac{\text{L atm}}{\text{mol K}} \times (25.9 + 273.2)\text{K}}{751\text{ mmHg} \times \frac{1\text{ atm}}{760\text{ mmHg}}} = 519\text{ L } CO_2(g)$$

19. Determine the total amount of gas; then use the ideal gas law, assuming that the gases behave ideally.

$$\text{no. moles gas} = \left(15.2\text{ g Ne} \times \frac{1\text{ mol Ne}}{20.18\text{ g Ne}}\right) + \left(34.8\text{ g Ar} \times \frac{1\text{ mol Ar}}{39.95\text{ g Ar}}\right)$$
$$= 0.753\text{ mol Ne} + 0.871\text{ mol Ar} = 1.624\text{ mol gas}$$

$$V = \frac{nRT}{P} = \frac{1.624\text{ mol} \times 0.08206 \frac{\text{L atm}}{\text{mol K}} \times (26.7 + 273.2)\text{K}}{7.15\text{ atm}} = 5.59\text{ L gas}$$

20. 2.24 L $H_2(g)$ at STP is 0.100 mol $H_2(g)$. After 0.10 mol He is added, the container holds 0.20 mol gas.

$$V = \frac{nRT}{P} = \frac{0.20 \text{ mol} \times 0.0821 \frac{\text{L atm}}{\text{mol K}} (273 + 100)\text{K}}{1.00 \text{ atm}} = 6.1 \text{ L gas}$$

21. **(a)** The total pressure is the sum of the partial pressure of $O_2(g)$ and the vapor pressure of water.
$P_{total} = P_{O2} + P_{H2O} = 756 \text{ mmHg} = P_{O2} + 19 \text{ mmHg}$ $\quad P_{O2} = (756 - 19) \text{ mmHg} = 737 \text{ mmHg}$

(b) The volume percent is equal to the pressure percent.

$$\%O_2(g) \text{ by volume} = \frac{V_{O2}}{V_{total}} \times 100\% = \frac{P_{O2}}{P_{total}} \times 100\% = \frac{737 \text{ mmHg of } O_2}{756 \text{ mmHg total}} \times 100\% = 97.5\%$$

(c) $P_{oxygen} = P_{total} - P_{water} = 741 \text{ mmHg} - 21 \text{ mmHg} = 720. \text{ mmHg}$ Determine the mass of $O_2(g)$ collected by multiplying the amount of O_2 collected, in moles, by the molar mass of $O_2(g)$.

$$\text{Mass } O_2 = \frac{PV}{RT}\mathcal{M} = \frac{\left(720. \text{ mmHg} \times \frac{1 \text{ atm}}{760 \text{ mmHg}}\right)\left(89.3 \text{ mL} \times \frac{1 \text{ L}}{1000 \text{ mL}}\right)}{0.08206 \frac{\text{L atm}}{\text{mol K}} (23 + 273)\text{K}} \times \frac{32.00 \text{ g } O_2}{1 \text{ mol } O_2} = 0.111 \text{ g } O_2$$

22. **(1)** is the true statement; average molecular kinetic energy depends only on absolute temperature, which is the same for these two gases. **(2)** is incorrect, since average molecular speed depends also on mole weight, which is different for these two gases. **(3)** also is incorrect, since volumes at the same temperature and pressure depend on the amount in moles, which is different for these two gases. And **(4)** is incorrect; the effusion rates at the same temperature and pressure depends inversely on the square root of the mole weights, which differ for these two gases.

23. Effusion time is proportional to the square root of molar mass.

$$\frac{\text{effusion time for NO}}{\text{effusion time for } Cl_2} = \sqrt{\frac{30.01 \text{ g/mol NO}}{70.91 \text{ g/mol } Cl_2}} = 0.6505 = \frac{\text{NO effusion time}}{28.6 \text{ s}}$$

NO effusion time = 0.6505 × 28.6 s = 18.6 s.

24. The best choice for ideal behavior is **(3)**, 200 °C and 0.50 atm. Gases are most nearly ideal at high temperatures where the molecules move rapidly, and low pressures where molecules are relatively far apart.

EXERCISES

Pressure and Its Measurement

25. We use: $h_{bnz}d_{bnz} = h_{Hg}d_{Hg}$ $\quad h_{bnz} = 0.970 \text{ atm} \times \frac{0.760 \text{ m Hg}}{1 \text{ atm}} \times \frac{13.6 \text{ g/cm}^3 \text{ Hg}}{0.879 \text{ g/cm}^3 \text{ benzene}} = 11.4 \text{ m benzene}$

26. We use: $h_{gly}d_{gly} = h_{CCl4}d_{CCl4}$ $\quad h_{gly} = 3.022 \text{ m } CCl_4 \times \frac{1.59 \text{ g/cm}^3 \text{ } CCl_4}{1.26 \text{ g/cm}^3 \text{ glycerol}} = 3.81 \text{ m glycerol}$

27. The mercury level difference equals the difference in pressure between the container and the atmosphere. This mercury level difference is $\Delta P = 276 \text{ mmHg} - 49 \text{ mmHg} = 227 \text{ mmHg}$. Since the mercury level in the arm open to the atmosphere is higher than in the arm connected to the container, the pressure in the container is higher than atmospheric pressure. $P = \Delta P + P_{atm} = 227 \text{ mmHg} + 749 \text{ mmHg} = 976 \text{ mmHg}$

28. The gas inside the container is at a lower pressure than the barometric pressure.

$$\Delta P = P_{bar.} - P_{gas} = 4.5 \text{ cm } H_2O \times \frac{10 \text{ mm} H_2O}{1 \text{ cm } H_2O} \times \frac{1 \text{ mmHg}}{13.6 \text{ mm } H_2O} = 3.3 \text{ mmHg}$$

$P_{gas} = 756.2 \text{ mmHg} - 3.3 \text{ mmHg} = 752.9 \text{ mmHg}$

The Simple Gas Laws

29. $$P_i = P_f \times \frac{V_f}{V_i} = \left(721 \text{ mmHg} \times \frac{35.8 \text{ L} + 1875 \text{ L}}{35.8\text{L}}\right) \times \frac{1 \text{ atm}}{760 \text{ mmHg}} = 50.7 \text{ atm}$$

30. We let P represent barometric pressure, and solve the Boyle's law expression below for P.
$P \times 42.0 \text{ mL} = (P + 85 \text{ mmHg}) \times 37.7 \text{ mL}$ $\quad 42.0\, P = 37.7\, P + 3.2_0 \times 10^3$

$$P = \frac{3.2_0 \times 10^3}{42.0 - 37.7} = 7.4 \times 10^2 \text{ mmHg}$$

31. $T_f = T_i \times \frac{P_f}{P_i} = (23.4 + 273.2)\text{K} \times \frac{1.00 \text{ atm}}{798 \text{ mmHg} \times \frac{1 \text{ atm}}{760 \text{ mmHg}}} = 282 \text{ K} = 9\ °\text{C}$

32. We can obtain the ratio of the number of moles both before and after heating, using $V_f = V_i$, and $P_f = P_i$. This will be the same as the ratio of the masses before and after, because the mole weight of the gas is the same. $\frac{n_f}{n_i} = \frac{P_fV_f/RT_f}{P_iV_i/RT_i} = \frac{T_i}{T_f} = \frac{(21 + 273)\text{K}}{(210 + 273)\text{K}} = 0.609 = \frac{m_f}{m_i} = \frac{m_f}{12.5 \text{ g}}$

$m_f = 0.609 \times 12.5 \text{ g} = 7.61 \text{ g}$ Thus, mass that escapes = 12.5 g – 7.61 g = 4.9 g of gas

33. **(a)** mass $= 27.6 \text{ mL} \times \frac{1 \text{ L}}{1000 \text{ mL}} \times \frac{1 \text{ mol}}{22.414 \text{ L STP}} = 0.00123 \text{ mol } PH_3 \times \frac{34.0 \text{ g } PH_3}{1 \text{ mol } PH_3} \times \frac{1000 \text{ mg}}{1 \text{ g}}$

$= 41.8 \text{ mg } PH_3$

(b) molec. $PH_3 = 0.00123 \text{ mol } PH_3 \times \frac{6.022 \times 10^{23} \text{ molecules}}{1 \text{ mol } PH_3} = 7.41 \times 10^{20}$ molecules

34. **(a)** mass $= 5.0 \times 10^{17} \text{ atoms} \times \frac{1 \text{ mol Rn}}{6.022 \times 10^{23} \text{ atoms}} = 8.30 \times 10^{-7} \text{ mol} \times \frac{222 \text{ g Rn}}{1 \text{ mol}} \times \frac{10^6\ \mu\text{g}}{1 \text{ g}}$

$= 1.8 \times 10^2\ \mu\text{g Rn(g)}$

(b) volume $= 8.30 \times 10^{-7} \text{ mol} \times \frac{22.414 \text{ L}}{1 \text{ mol}} \times \frac{10^6\ \mu\text{L}}{1 \text{ L}} = 19\ \mu\text{L Rn(g)}$

35. At the higher elevation of the mountains, the atmospheric pressure is lower than at the beach. However, the bag is leak proof; no gas escapes. Thus, the gas inside the bag expands in the lower pressure until the bag is filled to nearly bursting. (It would have been difficult to predict this result. The temperature in the mountains is usually lower than at the beach. The lower temperature would *decrease* the pressure of the gas.)

36. We first compute the pressure of the water at 30 m $h_{Hg} = 30 \text{ m } H_2O \times \frac{1.00 \text{ g/mL } H_2O}{13.6 \text{ g/mL Hg}} = 2.21 \text{ mHg}$

$P_{water} = 2.21 \text{ mHg} \times \frac{1000 \text{ mm}}{1 \text{ m}} \times \frac{1 \text{ atm}}{760 \text{ mmHg}} = 2.91 \text{ atm}$ To this we add the pressure of the atmosphere above the water: $P_{total} = 2.91 \text{ atm} + 1.00 \text{ atm} = 3.91 \text{ atm}$. When the diver rises to the surface, she rises to a pressure of 1.00 atm. Since the pressure is about one fourth of the pressure below the surface, the gas in her lungs attempt the expand to four times the volume of her lungs. It is possible that her lungs would burst.

Ideal Gas Equation

37. $P = \frac{nRT}{V} = \frac{\left(35.8 \text{ g } O_2 \times \frac{1 \text{ mol } O_2}{32.00 \text{ g } O_2}\right) \times \frac{0.08206 \text{ L atm}}{\text{mol K}} \times (46 + 273.2) \text{ K}}{12.8 \text{ L}} = 2.29 \text{ atm}$

38. mass $= n \times \mathcal{M} = \frac{PV}{RT}\mathcal{M} = \frac{11.2 \text{ atm} \times 18.5 \text{ L} \times 83.80 \text{ g/mol}}{0.08206 \frac{\text{L atm}}{\text{mol K}} (28.2 + 273.2)\text{K}} = 702 \text{ g Kr}$

39. $T = \frac{PV}{nR} = \frac{3.50 \text{ atm} \times 72.8 \text{ L}}{1.85 \text{ mol} \times 0.08206 \frac{\text{L atm}}{\text{mol K}}} = 1.68 \times 10^3 \text{ K}$ $t(°\text{C}) = 1.68 \times 10^3 - 273 = 1.41 \times 10^3\ °\text{C}$

40. Because the number of moles of gas does not change, $\frac{P_i \times V_i}{T_i} = n\ R = \frac{P_f \times V_f}{T_f}$ is obtained from the ideal gas equation. This expression can be rearranged as follows.

$$V_f = \frac{V_i \times P_i \times T_f}{P_f \times T_i} = \frac{4.25 \text{ L} \times 748 \text{ mmHg} \times (273.2 + 26.8)\text{K}}{742 \text{ mmHg} \times (273.2 + 25.6)\text{K}} = 4.30 \text{ L}$$

41. First determine the mass of O_2 in the cylinder under the final conditions.

$$\text{mass } O_2 = n \times \mathcal{M} = \frac{PV}{RT}\,\mathcal{M} = \frac{1.15\text{ atm} \times 34.0\text{ L} \times 32.0\text{ g/mol}}{0.08206\,\frac{\text{L atm}}{\text{mol K}}\,(22 + 273)\text{K}} = 51.7\text{ g } O_2$$

mass of O_2 to be released = 305 g – 51.7 g = 253 g O_2

42. We first compute the pressure at 35 °C as a result of the additional gas. Of course, the gas pressure increases proportionally to the increase in the mass of gas, because the mass is proportional to the number of moles of gas.

$$P = \frac{12.5\text{ g}}{10.0\text{ g}} \times 762\text{ mmHg} = 953\text{ mmHg}$$

Now we compute the pressure resulting from increasing the temperature.

$$P = \frac{(62 + 273)\text{ K}}{(25 + 273)\text{ K}} \times 953\text{ mmHg} = 1.07 \times 10^3\text{ mmHg}$$

Determining Molar Mass

43. We first determine the empirical formula of propylene.

$$\text{no. mol C} = 85.63\text{ g C} \times \frac{1\text{ mol C}}{12.01\text{ g C}} = 7.130\text{ mol C} \quad \div 7.130 \longrightarrow 1.000\text{ mol C}$$

$$\text{no. mol H} = 14.37\text{ g H} \times \frac{1\text{ mol H}}{1.008\text{ g H}} = 14.26\text{ mol H} \quad \div 7.130 \longrightarrow 2.000\text{ mol H}$$

The empirical formula is CH_2 and the empirical molar mass is 14.0 g/mol. The molar mass of propylene is 42.08 g/mol, three times the empirical molar mass. The molecular formula is C_3H_6, three times the empirical formula.

44. We first determine the molar mass of the gas.

$$T = 24.3 + 273.2 = 297.5\text{ K} \qquad P = 742\text{ mmHg} \times \frac{1\text{atm}}{760\text{ mmHg}} = 0.976\text{ atm}$$

$$\mathcal{M} = \frac{mRT}{PV} = \frac{2.650\text{ g} \times 0.08206\text{ L atm mol}^{-1}\text{ K}^{-1} \times 297.5\text{ K}}{0.976\text{ atm} \times \left(428\text{ mL} \times \frac{1\text{ L}}{1000\text{ mL}}\right)} = 155\text{ g/mol}$$

Then we determine the empirical formula of the gas, basing our calculations on a 100.0-g sample.

$$\text{amount C} = 15.5\text{ g C} \times \frac{1\text{ mol C}}{12.01\text{ g C}} = 1.29\text{ mol C} \quad \div 0.649 \longrightarrow 1.99\text{ mol C}$$

$$\text{amount Cl} = 23.0\text{ g Cl} \times \frac{1\text{ mol Cl}}{35.45\text{ g Cl}} = 0.649\text{ mol Cl} \quad \div 0.649 \longrightarrow 1.00\text{ mol Cl}$$

$$\text{amount F} = 61.5\text{ g F} \times \frac{1\text{ mol F}}{19.00\text{ g F}} = 3.24\text{ mol F} \quad \div 0.649 \longrightarrow 4.99\text{ mol F}$$

The empirical formula thus is C_2ClF_5, which has an empirical molar mass of 154.5 g/mol. This is the same as the experimentally determined molar mass; the molecular formula is C_2ClF_5.

45.

$$\mathcal{M} = \frac{mRT}{PV} = \frac{0.231\text{ g} \times 0.08206\,\frac{\text{L atm}}{\text{mol K}} \times (23 + 273)\text{K}}{\left(749\text{ mmHg} \times \frac{1\text{ atm}}{760\text{ mmHg}}\right)\left(102\text{ mL} \times \frac{1\text{ L}}{1000\text{ mL}}\right)} = 55.8\text{ g/mol}$$

The formula contains 4 atoms of carbon. (5 atoms of carbon gives a mole weight of 60—too high—and 3 C atoms gives a mole weight of 36—too low to be made up by adding H's.) To produce a mole weight of 56 with 3 carbons requires the inclusion of 8 atoms of H in the formula of the compound: C_4H_8.

46. First, we obtain the mass of acetylene, and then acetylene's molar mass.
mass of acetylene = 56.2445 g – 56.1035 g = 0.1410 g acetylene

$$\mathcal{M} = \frac{mRT}{PV} = \frac{0.1410\text{ g} \times 0.082057\,\frac{\text{L atm}}{\text{mol K}} \times (20.02 + 273.15)\text{K}}{\left(749.3\text{ mmHg} \times \frac{1\text{ atm}}{760\text{ mmHg}}\right)\left(132.1\text{ mL} \times \frac{1\text{ L}}{1000\text{ mL}}\right)} = 26.04\text{ g/mol}$$

Gas Densities

47. $d = \frac{\mathcal{M}P}{RT} \rightarrow P = \frac{dRT}{\mathcal{M}} = \frac{1.80\text{ g/L} \times 0.08206 \frac{\text{L atm}}{\text{mol K}} \times (32 + 273)\text{ K}}{28.0\text{ g/mol}} \times \frac{760\text{ mmHg}}{1\text{ atm}} = 1.22 \times 10^3\text{ mmHg}$

48. $\mathcal{M} = \frac{dRT}{P} = \frac{2.56\text{ g/L} \times 0.08206 \frac{\text{L atm}}{\text{mol K}} \times (22.8 + 273.2)\text{ K}}{756\text{ mmHg} \times \frac{1\text{ atm}}{760\text{ mmHg}}} = 62.5\text{ g/mol}$

49. (a) $d = \frac{\mathcal{M}P}{RT} = \frac{28.96\text{ g/mol} \times 1.00\text{ atm}}{0.0821 \frac{\text{L atm}}{\text{mol K}} \times (273 + 25)\text{K}} = 1.18\text{ g/L air}$

(b) $d = \frac{\mathcal{M}P}{RT} = \frac{44.0\text{ g/mol } CO_2 \times 1.00\text{ atm}}{0.08206 \frac{\text{L atm}}{\text{mol K}} \times (273 + 25)\text{K}} = 1.80\text{ g/L } CO_2$

Since this density is greater than air's, the balloon will not rise in air when filled with CO_2 at 25 °C.

50. $d = \frac{\mathcal{M}P}{RT}$ becomes $T = \frac{\mathcal{M}P}{Rd} = \frac{44.0\text{ g/mol} \times 1.00\text{ atm}}{0.08206 \frac{\text{L atm}}{\text{mol K}} \times 1.18\text{ g/L}} = 454\text{ K} = 181\text{ °C}$

51. $d = \frac{\mathcal{M}P}{RT}$ becomes $\mathcal{M} = \frac{dRT}{P} = \frac{2.64\text{ g/L} \times 0.0821 \frac{\text{L atm}}{\text{mol K}} \times (310 + 273)\text{K}}{775\text{ mmHg} \times \frac{1\text{ atm}}{760\text{ mmHg}}} = 124\text{ g/mol}$

Since the atomic mass of phosphorus is 31.0, the formula of phosphorus molecules in the vapor is P_4. (4 atoms/molecule × 31.0 = 124)

52. We first determine the molar mass of the gas, then its empirical formula. These two pieces of information are combined to obtain the molecular formula of the gas.

$\mathcal{M} = \frac{dRT}{P} = \frac{2.33\text{ g/L} \times 0.08206 \frac{\text{L atm}}{\text{mol K}} \times 296\text{ K}}{746\text{ mmHg} \times \frac{1\text{ atm}}{760\text{ mmHg}}} = 57.7\text{ g/mol}$

no. mol C = $82.7\text{ g C} \times \frac{1\text{ mol C}}{12.0\text{ g C}} = 6.89\text{ mol C} \quad \div 6.89 \longrightarrow 1.00\text{ mol C}$

no. mol H = $17.3\text{ g H} \times \frac{1\text{ mol H}}{1.01\text{ g H}} = 17.1\text{ mol H} \div 6.89 \longrightarrow 2.48\text{ mol H}$

Multiply both of these mole numbers by 2 to obtain the empirical formula, C_2H_5, which has an empirical molar mass of 29.0 g/mol. Since the molar mass (calculated as 57.7 g/mol above) is twice this empirical molar mass, twice the empirical formula is the molecular formula: C_4H_{10}.

Gases in Chemical Reactions

53. Balanced equation: $C_3H_8(g) + 5\ O_2(g) \longrightarrow 3\ CO_2(g) + 4\ H_2O(l)$

Use the law of combining volumes. O_2 volume = $75.6\text{ L } C_3H_8 \times \frac{5\text{ L } O_2}{1\text{ L } C_3H_8} = 378\text{ L } O_2$

54. We first use the law of combining volumes, and then the general gas law.

$3\ CO(g) + 7\ H_2(g) \longrightarrow C_3H_8(g) + 3\ H_2O(l) \qquad 28.5\text{ L CO(g)} \times \frac{7\text{ L } H_2(g)}{3\text{ L CO(g)}} = 66.5\text{ L } H_2(g)$

volume $H_2(g) = 66.5\text{ L } H_2(g) \times \frac{760\text{ mmHg}}{751\text{ mmHg}} \times \frac{(26 + 273)\text{K}}{273\text{ K}} = 73.7\text{ L } H_2(g)$

55. Determine the moles of $SO_2(g)$ produced and then use the ideal gas equation.

$2.7 \times 10^6\text{ lb coal} \times \frac{3.28\text{ lb S}}{100.00\text{ lb coal}} \times \frac{454\text{ g S}}{1\text{ lb S}} \times \frac{1\text{ mol S}}{32.1\text{ g S}} \times \frac{1\text{ mol } SO_2}{1\text{ mol S}} = 1.3 \times 10^6\text{ mol } SO_2$

$$V = \frac{nRT}{P} = \frac{1.3 \times 10^6 \text{ mol SO}_2 \times 0.0821 \frac{\text{L atm}}{\text{mol K}} \times 296 \text{ K}}{738 \text{ mmHg} \times \frac{1 \text{ atm}}{760 \text{ mmHg}}} = 3.3 \times 10^7 \text{ L SO}_2$$

56. Determine the amount of O_2, and then the mass of $KClO_3$ that produced this amount of O_2.

$$\text{no. mol O}_2 = \frac{\left(738 \text{ mmHg} \times \frac{1 \text{ atm}}{760 \text{ mmHg}}\right) \times \left(119 \text{ mL} \times \frac{1 \text{ L}}{1000 \text{ mL}}\right)}{0.08206 \frac{\text{L atm}}{\text{mol K}} \times (22.4 + 273.2)\text{K}} = 0.00476 \text{ mol O}_2$$

$$\text{mass KClO}_3 = 0.00476 \text{ mol O}_2 \times \frac{2 \text{ mol KClO}_3}{3 \text{ mol O}_2} \times \frac{122.6 \text{ g KClO}_3}{1 \text{ mol KClO}_3} = 0.389 \text{ g KClO}_3$$

$$\% \text{ KClO}_3 = \frac{0.389 \text{ g KClO}_3}{3.57 \text{ g sample}} \times 100\% = 10.9\% \text{ KClO}_3$$

57. Determine the amount of O_2 liberated, and then its volume. $2\,H_2O_2(aq) \longrightarrow 2\,H_2O(l) + O_2(g)$

$$\text{no. mol O}_2 = 10.0 \text{ mL soln} \times \frac{1.01 \text{ g}}{1 \text{ mL}} \times \frac{0.0300 \text{ g H}_2\text{O}_2}{1 \text{ g soln}} \times \frac{1 \text{ mol H}_2\text{O}_2}{34.0 \text{ g H}_2\text{O}_2} \times \frac{1 \text{ mol O}_2}{2 \text{ mol H}_2\text{O}_2}$$

$$= 0.00446 \text{ mol O}_2$$

$$V = \frac{0.00446 \text{ mol O}_2 \times 0.08206 \frac{\text{L atm}}{\text{mol K}} \times (22 + 273)\text{K}}{752 \text{ mmHg} \times \frac{1 \text{ atm}}{760 \text{ mmHg}}} \times \frac{1000 \text{ mL}}{1 \text{ L}} = 109 \text{ mL O}_2$$

58. **(a)** $\text{volume NH}_3 = 313 \text{ L H}_2 \times \frac{2 \text{ L NH}_3}{3 \text{ L H}_2} = 209 \text{ L NH}_3$

(b) $$\text{no. mol NH}_3 = \frac{525 \text{ atm} \times 313 \text{ L}}{0.08206 \frac{\text{L atm}}{\text{mol K}} \times (515 + 273)\text{K}} = 2.54 \times 10^3 \text{ mol H}_2 \times \frac{2 \text{ mol NH}_3}{3 \text{ mol H}_2}$$

$$= 1.69 \times 10^3 \text{ mol NH}_3$$

$$V = \frac{1.69 \times 10^3 \text{ mol NH}_3 \times 0.08206 \frac{\text{L atm}}{\text{mol K}} \times 298\text{K}}{727 \text{ mmHg} \times \frac{1 \text{ atm}}{760 \text{ mmHg}}} = 4.32 \times 10^4 \text{ L NH}_3$$

Mixtures of Gases

59. The two pressures are related as are the number of moles of $N_2(g)$ and the total number of moles.

$$\text{amount N}_2 = \frac{PV}{RT} = \frac{28.2 \text{ atm} \times 53.7 \text{ L}}{0.08206 \frac{\text{L atm}}{\text{mol K}} \times (26 + 273)\text{K}} = 61.7 \text{ mol N}_2$$

$$\text{total amount of gas} = 61.7 \text{ mol N}_2 \times \frac{75.0 \text{ atm}}{28.2 \text{ atm}} = 164 \text{ mol gas}$$

$$\text{mass Ne} = (164 \text{ mol total} - 61.7 \text{ mol N}_2) \times \frac{20.18 \text{ g Ne}}{1 \text{ mol Ne}} = 2.1 \times 10^3 \text{ g Ne}$$

60. Solve a Boyle's law problem for each gas and add the resulting partial pressures.

$$P_{H_2} = 762 \text{ mm Hg} \times \frac{2.35 \text{ L}}{5.52 \text{ L}} = 324 \text{ mmHg} \qquad P_{He} = 728 \text{ mmHg} \times \frac{3.17 \text{ L}}{5.52 \text{ L}} = 418 \text{ mmHg}$$

$$P_{total} = P_{H_2} + P_{He} = 324 \text{ mmHg} + 418 \text{ mmHg} = 742 \text{ mmHg}$$

61. **(a)** We first determine the amount of each gas, then the total amount, and the total pressure.

$$\text{amount H}_2 = 4.0 \text{ g H}_2 \times \frac{1 \text{ mol H}_2}{2.02 \text{ g H}_2} = 2.0 \text{ mol H}_2 \qquad \text{amount He} = 10.0 \text{ g He} \times \frac{1 \text{ mol He}}{4.00 \text{ g He}} = 2.50 \text{ mol He}$$

$$P = \frac{nRT}{V} = \frac{(2.0 + 2.50) \text{ mol} \times 0.0821 \frac{\text{L atm}}{\text{mol K}} \times 273 \text{ K}}{4.3 \text{ L}} = 23.5 \text{ atm}$$

(b) $P_{H_2} = 23.5 \text{ atm} \times \frac{2.0 \text{ mol H}_2}{4.5 \text{ mol total}} = 10.4 \text{ atm}$ $\qquad P_{He} = 23.5 \text{ atm} - 10.4 \text{ atm} = 13.1 \text{ atm}$

62. **(a)** $$P_{ben} = \frac{nRT}{V} = \frac{\left(0.728\text{ g} \times \frac{1\text{ mol } C_6H_6}{78.11\text{ g } C_6H_6}\right) \times 0.08206 \frac{\text{L atm}}{\text{mol K}} \times (35 + 273)}{2.00\text{ L}} \times \frac{760\text{ mmHg}}{1\text{ atm}}$$
$= 89.5$ mmHg

$P_{total} = 89.5$ mmHg $C_6H_6(g) + 752$ mmHg $Ar(g) = 842$ mmHg

(b) $P_{benzene} = 89.5$ mmHg $\qquad P_{Ar} = 752$ mmHg

63. **(a)** Density should be related to average molar mass. We expect the average molar mass of air to be between the molar masses of its two principal constituents, N_2 (28.0 g/mol) and O_2 (32.0 g/mol). The average molar mass of normal air is approximately 28.9 g/mol. Expired air would be made more dense by the presence of more CO_2 (44.0 g/mol) and less dense by the presence of more H_2O (18.0 g/mol). The change might be minimal. In fact, it is, as the following calculation shows.

$$\mathcal{M}_{exp.air} = \left(0.742\text{ mol } N_2 \times \frac{28.013\text{ g } N_2}{1\text{ mol } N_2}\right) + \left(0.152\text{ mol } O_2 \times \frac{31.999\text{ g } O_2}{1\text{ mol } O_2}\right)$$
$$+ \left(0.038\text{ mol } CO_2 \times \frac{44.01\text{ g } CO_2}{1\text{ mol } CO_2}\right) + \left(0.059\text{ mol } H_2O \times \frac{18.02\text{ g } H_2O}{1\text{ mol } H_2O}\right)$$
$$+ \left(0.009\text{ mol Ar} \times \frac{39.9\text{ g Ar}}{1\text{ mol Ar}}\right) = 28.7\text{ g/mol of expired air}$$

(b) We see from equation (6.17) that the mole ratio, the volume ratio, and the ratio of partial pressures all are equal. Thus, to determine the ratio of partial pressure, we can use the ratios of volumes. The $\%CO_2$ in ordinary air is 0.036%, while from the data of this problem, the $\%CO_2$ in expired air is 3.8%.

$$\frac{P\{CO_2\text{ expired air}\}}{P\{CO_2\text{ ordinary air}\}} = \frac{3.8\%\ CO_2}{0.036\%\ CO_2} = 1.1 \times 10^2\ CO_2\text{ (expired air to ordinary air)}$$

64. The volume percents are also equal to the mole percents and to the partial pressure percents. We use the mole percents, converted to mole fractions, to compute a molar mass of producer gas.

$$\text{molar mass} = 0.080\text{ mol } CO_2 \times \frac{44.01\text{ g } CO_2}{1\text{ mol } CO_2} + 0.232\text{ mol CO} \times \frac{28.01\text{ g CO}}{1\text{ mol CO}}$$
$$+ 0.177\text{ mol } H_2 \times \frac{2.016\text{ g } H_2}{1\text{ mol } H_2} + 0.011\text{ mol } CH_4 \times \frac{16.04\text{ g } CH_4}{1\text{ mol } CH_4} + 0.500\text{ mol } N_2 \times \frac{28.01\text{ g } N_2}{1\text{ mol } N_2}$$
$$= 3.5\text{ g } CO_2 + 6.50\text{ g CO} + 0.357\text{ g } H_2 + 0.18\text{ g } CH_4 + 14.0\text{ g } N_2$$
$$= 24.6\text{ g/mole of producer gas}$$

(a) $P = 763\text{ mmHg} \times \frac{1\text{ atm}}{760\text{ mmHg}} = 1.00_4\text{ atm} \qquad T = 23 + 273 = 296\text{ K}$

$$\text{density} = \frac{m}{V} = \frac{\mathcal{M}P}{RT} = \frac{24.6\text{ g mol}^{-1} \times 1.00_4\text{ atm}}{0.08206\text{ L atm mol}^{-1}\text{ K}^{-1} \times 296\text{ K}} = 1.01_7\text{ g/L}$$

(b) The partial pressure of CO equals its mole fraction times the total pressure.

$P_{CO} = 0.232 \times 1.00\text{ atm} = 0.232\text{ atm}$

Collecting Gases over Liquids

65. The pressure of the liberated $H_2(g)$ is 744 mmHg – 23.8 mmHg = 720. mmHg

$$V = \frac{nRT}{P} = \frac{\left(1.65\text{ g Al} \times \frac{1\text{ mol Al}}{26.98\text{ g}} \times \frac{3\text{ mol } H_2}{2\text{ mol Al}}\right) 0.08206 \frac{\text{L atm}}{\text{mol K}} (273 + 25)\text{K}}{720.\text{ mmHg} \times \frac{1\text{ atm}}{760\text{ mmHg}}} = 2.37\text{ L } H_2(g)$$

This is the total volume of both gases, each with a different partial pressure.

66. If the gas were measured at the same volume before and after passing it through H_2O, its pressure would increase by 23.8 mmHg from 748 mmHg to 772 mmHg (= 748 + 23.8 mmHg). Since it is measured at the same pressure, we can determine its new volume with Boyle's law. What we are doing is assuming that the gas is collected at the same volume but at 772 mmHg. We then reduce the pressure to 748 mmHg by expanding the volume of the gas. $\qquad \text{volume} = 367\text{ mL} \times \frac{772\text{ mmHg}}{748\text{ mmHg}} = 379\text{ mL}$

67. We first determine the pressure of the gas collected. This would be its "dry gas" pressure and, when added to 22.4 mmHg, gives the barometric pressure.

$$P=\frac{nRT}{V}=\frac{\left(1.46\ \text{g}\times\frac{1\ \text{mol O}_2}{32.0\ \text{g O}_2}\right)0.08206\ \frac{\text{L atm}}{\text{mol K}}\times 297\ \text{K}}{1.16\ \text{L}}\times\frac{760\ \text{mmHg}}{1\ \text{atm}}=729\ \text{mmHg}$$

barometric pressure = 729 mm Hg + 22.4 mmHg = 751 mmHg

68. We first determine the "dry gas" pressure of helium. This pressure, subtracted from the barometric pressure of 738.6 mmHg, gives the vapor pressure of hexane at 25 °C.

$$P=\frac{nRT}{V}=\frac{\left(1.072\ \text{g}\times\frac{1\ \text{mol He}}{4.003\ \text{g He}}\right)0.08206\ \frac{\text{L atm}}{\text{mol K}}\times 298.2\ \text{K}}{8.446\ \text{L}}\times\frac{760\ \text{mmHg}}{1\ \text{atm}}=589.7\ \text{mmHg}$$

vapor pressure = 738.6 – 589.7 = 148.9 mmHg

Kinetic molecular theory

69. $$u_{rms}=\sqrt{\frac{3RT}{\mathcal{M}}}=\sqrt{\frac{3\times 8.3145\ \frac{\text{J}}{\text{mol K}}\times 303\ \text{K}}{\frac{70.91\times 10^{-3}\ \text{kg Cl}_2}{1\ \text{mol Cl}_2}}}=326\ \text{m/s}$$

70. $u_{rms}=\sqrt{\frac{3RT}{\mathcal{M}}}=1.84\times 10^3$ m/s Solve this equation for temperature with u_{rms} doubled.

$$T=\frac{\mathcal{M}\,u_{rms}^2}{3R}=\frac{2.016\times 10^{-3}\ \text{kg/mol}\ (2\times 1.84\times 10^3\ \text{m/s})^2}{3\times 8.3145\ \frac{\text{J}}{\text{mol K}}}=1.09\times 10^3\ \text{K}$$

71. $$\mathcal{M}=\frac{3RT}{u_{rms}^2}=\frac{3\times 8.3145\ \frac{\text{J}}{\text{mol K}}\times 298\ \text{K}}{\left(2180\ \frac{\text{mi}}{\text{hr}}\times\frac{1\ \text{hr}}{3600\ \text{sec}}\times\frac{5280\ \text{ft}}{1\ \text{mi}}\times\frac{12\ \text{in.}}{1\ \text{ft}}\times\frac{1\ \text{m}}{39.37\ \text{in.}}\right)^2}=0.0078\ \text{kg/mol}$$

= 7.8 g/mol The molecular mass of the gas is 7.8.

72. A noble gas with molecules having u_{rms} at 25 °C greater than that of a rifle bullet will have a molar mass less than 7.8 g/mol. Helium is the only possibility. A noble gas with a slower u_{rms} will have a molar mass greater than 7.8 g/mol; any one of the other noble gases will have a slower u_{rms}.

73. We equate the two expressions for root mean square speed, cancel the common factors, and solve for the temperature of Ne. Note that the units of molar masses do not have to be in kg/mol in this calculation; they simply must be expressed in the same units.

$$\sqrt{\frac{3RT}{\mathcal{M}}}=\sqrt{\frac{3R\times 300\ \text{K}}{4.003}}=\sqrt{\frac{3R\times T_{Ne}}{20.18}}$$

$$\frac{300\ \text{K}}{4.003}=\frac{T_{Ne}}{20.18}\qquad T_{Ne}=300\ \text{K}\times\frac{20.18}{4.003}=1.51\times 10^3\ \text{K}$$

74. u_m, the modal speed, is the speed that occurs most often, 55 mi/h

$$\text{Average speed}=\frac{38+44+45+48+50+55+55+57+58+60}{10}=51.0\ \text{mi/h}=\overline{u}$$

$$\text{Root mean square speed}=\sqrt{\frac{38^2+44^2+45^2+48^2+50^2+55^2+55^2+57^2+58^2+60^2}{10}}$$

$$=\sqrt{\frac{1444+1936+2025+2304+2500+3025+3025+3249+3364+3600}{10}}$$

$$=\sqrt{\frac{26472}{10}}=51.5\ \text{mi/h}=u_{rms}$$

Diffusion and Effusion of Gases

75. $\frac{\text{rate (NO}_2)}{\text{rate (N}_2\text{O)}} = \sqrt{\frac{\mathcal{M}(\text{N}_2\text{O})}{\mathcal{M}(\text{NO}_2)}} = \sqrt{\frac{44.02}{46.01}} = 0.9781 = \frac{x \text{ mol NO}_2/t}{0.00484 \text{ mol N}_2\text{O}/t}$

mol NO_2 = 0.00484 mol × 0.9781 = 0.00473 mol NO_2

76. $\frac{\text{rate (N}_2)}{\text{rate (unk)}} = \frac{\text{mol (N}_2)/38 \text{ s}}{\text{mol (unk)}/64 \text{ s}} = \frac{64 \text{ s}}{38 \text{ s}} = 1.68 = \sqrt{\frac{\mathcal{M}(\text{unk})}{\mathcal{M}(\text{N}_2)}}$

$\mathcal{M}(\text{unk}) = (1.68)^2 \mathcal{M}(\text{N}_2) = (1.68)^2 (28.01 \text{ g/mol}) = 79 \text{ g/mol}$

77. (a) $\frac{\text{rate (N}_2)}{\text{rate (O}_2)} = \sqrt{\frac{\mathcal{M}(\text{O}_2)}{\mathcal{M}(\text{N}_2)}} = \sqrt{\frac{32.00}{28.01}} = 1.07$ (b) $\frac{\text{rate (H}_2\text{O)}}{\text{rate (D}_2\text{O)}} = \sqrt{\frac{\mathcal{M}(\text{D}_2\text{O})}{\mathcal{M}(\text{H}_2\text{O})}} = \sqrt{\frac{20.0}{18.02}} = 1.05$

(c) $\frac{\text{rate}(^{14}\text{CO}_2)}{\text{rate}(^{12}\text{CO}_2)} = \sqrt{\frac{\mathcal{M}(^{12}\text{CO}_2)}{\mathcal{M}(^{14}\text{CO}_2)}} = \sqrt{\frac{44.0}{46.0}} = 0.978$

(d) $\frac{\text{rate}(^{235}\text{UF}_6)}{\text{rate}(^{238}\text{UF}_6)} = \sqrt{\frac{\mathcal{M}(^{238}\text{UF}_6)}{\mathcal{M}(^{235}\text{UF}_6)}} = \sqrt{\frac{352}{349}} = 1.004$

78. The distances that the two molecules travel are related in the same way as their rates of effusion.

$\frac{\text{distance NH}_3}{\text{distance HCl}} = \sqrt{\frac{\mathcal{M}(\text{HCl})}{\mathcal{M}(\text{NH}_3)}} = \sqrt{\frac{36.5}{17.0}} = 1.46 = \frac{x}{1.00 - x}$, if we let distance $NH_3 = x$.

$1.46 - 1.46\,x = x$ $\qquad 2.46\,x = 1.46$ $\qquad x = \frac{1.46}{2.46} = 0.593$ meter

The $NH_4Cl(s)$ cloud will form about 600 cm from the NH_3 and 400 cm from the HCl.

Nonideal Gases

79. $P = \frac{nRT}{V - nb} - \frac{n^2a}{V^2} = \frac{1.50 \text{ mol} \times \frac{0.08206 \text{ L atm}}{\text{mol K}} \times 298 \text{ K}}{(5.00 - 1.50 \times 0.0564) \text{ L}} - \frac{1.50^2 \times 6.71 \text{ L}^2 \text{ atm}}{(5.00 \text{ L})^2}$

$= 7.46_2 \text{ atm} - 0.604 \text{ atm} = 6.86 \text{ atm SO}_2\text{(g)}$

$P_{\text{ideal}} = \frac{1.50 \text{ mol} \times \frac{0.08206 \text{ L atm}}{\text{mol K}} \times 298 \text{ K}}{5.00 \text{ L}} = 7.34 \text{ atm}$ The ideal pressure is 2% less than the van der Waals pressure.

80. For 1.000 mol $Cl_2(g)$, $n^2a = 6.49 \text{ L}^2$ atm and $nb = 0.0562$ L. $P_{\text{vdW}} = \frac{nRT}{V - nb} - \frac{n^2a}{V^2}$

At 0 °C, $P_{\text{vdW}} = 9.9$ atm and $P_{\text{ideal}} = 11.2$ atm, off by 1.3 atm or +13%

(a) At 100 °C $P_{\text{ideal}} = \frac{nRT}{V} = \frac{1.00 \text{ mol} \times \frac{0.08206 \text{ L atm}}{\text{mol K}} \times 373 \text{ K}}{2.00 \text{ L}} = 15.3 \text{ atm}$

$P_{\text{vdW}} = \frac{1.00 \text{ mol} \times \frac{0.08206 \text{ L atm}}{\text{mol K}} \times T}{(2.00 - 0.0562) \text{ L}} - \frac{6.49 \text{ L}^2 \text{ atm}}{(2.00 \text{ L})^2} = 0.0422_2\, T \text{ atm} - 1.62_3 \text{ atm}$

$= 0.0422_2 \times 373 \text{ K} - 1.62_3 = 14.1_3 \text{ atm}$ $\qquad P_{\text{ideal}}$ is off by 1.2 atm or +8.5%

(b) At 200 °C $P_{\text{ideal}} = \frac{nRT}{V} = \frac{1.00 \text{ mol} \times \frac{0.08206 \text{ L atm}}{\text{mol K}} \times 473 \text{ K}}{2.00 \text{ L}} = 19.4 \text{ atm}$

$P_{\text{vdW}} = 0.0422_2 \times 473 \text{ K} - 1.62_3 = 18.3_5 \text{ atm}$ $\qquad P_{\text{ideal}}$ is off by 1.0 atm or +5.7%

(c) At 400 °C $P_{\text{ideal}} = \frac{nRT}{V} = \frac{1.00 \text{ mol} \times \frac{0.08206 \text{ L atm}}{\text{mol K}} \times 673 \text{ K}}{2.00 \text{ L}} = 27.6 \text{ atm}$

$P_{\text{vdW}} = 0.0422_2 \times 673 \text{ K} - 1.62_3 = 26.7_9 \text{ atm}$ $\qquad P_{\text{ideal}}$ is off by 0.8 atm or $+3._0$%

FEATURE PROBLEMS

A. The illustrated difference in levels indicates that the pressure inside the tube is greater than atmospheric pressure. Thus $P_{tube} = P_{bar} + B$. Boyle's law is demonstrated if $A \times P_{tube}$ = constant. We tabulate the data. The column labeled 1/A is for use in Exercise 82.

A, cm	B, cm	B, mm	P_{tube}, mm	1/A, cm^{-1}	$A \times P_{tube}$, mmHg cm
30.5	0.0	0.0	739.8	0.0328	$2.25_6 \times 10^4$
27.9	7.1	71	811	0.0358	$2.26_3 \times 10^4$
25.4	15.7	157	897	0.0394	$2.27_8 \times 10^4$
22.9	25.7	257	997	0.0437	$2.28_3 \times 10^4$
20.3	38.3	383	1123	0.0493	$2.28_0 \times 10^4$
17.8	53.8	538	1278	0.0562	$2.27_5 \times 10^4$
15.2	75.4	754	1494	0.0658	$2.27_1 \times 10^4$
12.7	105.6	1056	1796	0.0787	$2.28_1 \times 10^4$
10.2	147.6	1476	2216	0.0980	$2.26_0 \times 10^4$
7.6	224.6	2246	2986	0.131_6	$2.26_9 \times 10^4$

We see that Boyle's law is indeed demonstrated by the data. In fact, we had to display an additional significant figure (as a subscript) in order to show the variation in the pressure-volume product.

B. We plot the data from the table in Feature Problem A. The slope of this line depends on the temperature and the amount of gas. After all, we are plotting $P = \frac{nRT}{V}$ vs. $1/V$. The slope of this line equals nRT. Of course, the slope also depends on the units chosen for P and V.

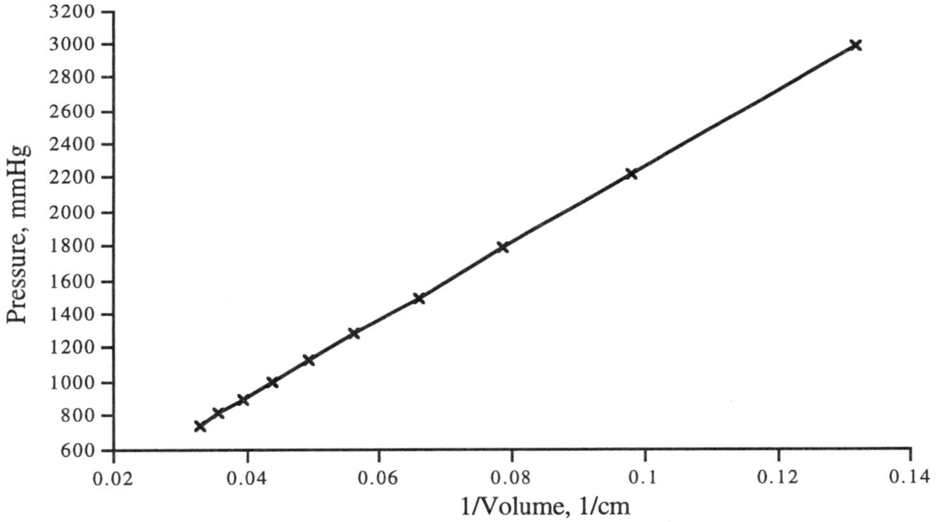

C. 1. The pressure drops to about one third of its STP value, from 760 mmHg to 250 mmHg, while temperature and amount of gas stay fixed. This means that the volume of the container about triples, the speed (indicated by the length of their "streamers") of the gas molecules and their number stay the same.

2. The temperature drops to about half of its STP value, from 273 K to 140 K, while pressure and amount of gas stay fixed. This means that the volume of the container about halves, the speed of the gas molecules is 71% ($\sqrt{0.5} = 0.71$) times the original speed, and their number stays the same.

3. The pressure halves while the temperature doubles. This means the container's volume about quadruples, the speed of the gas molecules increases 41% (it is $\sqrt{2} = 1.41$ times the original speed), but their number remains the same.

4. The temperature and the amount go up by 50% each (each multiplied by 1.5), and the pressure is multiplied by 1.5^2. This means that the size of the container remains the same, there are 50% more molecules, and they are moving 22% faster. (Speed is $\sqrt{1.50} = 1.22$ times the original speed.)

D. We calculate the values of *d*/*P* and plot against pressure.

P, mmHg	760.00	570.00	380.00	190.00
d, g/L	1.428962	1.071485	0.714154	0.356985
d/*P*	0.0018802	0.0018798	0.0018794	0.0018789

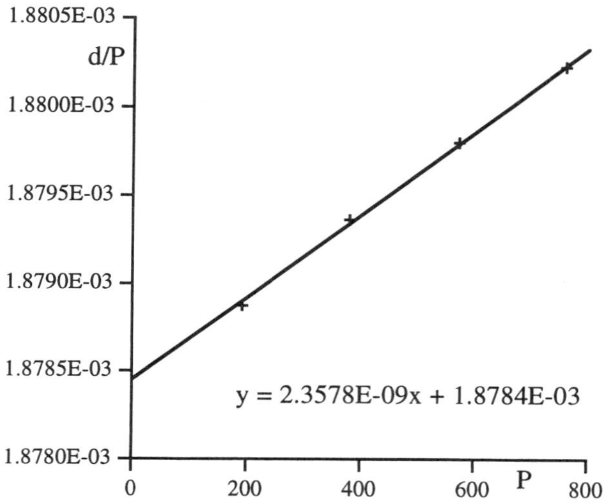

The intercept of this graph is the value of *d*/*P* at zero pressure, that is, under ideal conditions. From this ratio we can calculate the molar mass of $O_2(g)$ as an ideal gas.

$$\mathcal{M} = \frac{0.0018784 \text{ g}}{\text{L·mmHg}} \times \frac{760 \text{ mmHg}}{1 \text{ atm}} \times \frac{0.082057 \text{ L·atm}}{\text{mol·K}} \times 273.15 \text{ K} = 31.998 \text{ g/mol}$$

The molar mass of O(g) is one-half this value, or 15.999 g/mol, compared to a tabulated value of 15.9994 g/mol.

7 THERMOCHEMISTRY

PRACTICE EXAMPLES

1A The expression for constant-pressure work as a result of volume change is $w = P\Delta V$. The conversion factor between L·atm and J can be obtained as the ratio between two values of the ideal gas constant, R.

$$w = P\Delta V = 0.987 \text{ atm} \times (23.5\text{ L} - 35.0\text{ L}) \times \frac{8.3145 \text{ J mol}^{-1}\text{ K}^{-1}}{0.082058 \text{ L·atm mol}^{-1}\text{K}^{-1}} = -1.15 \times 10^3 \text{ J done by the gas}$$

work done *on* the gas = $+1.15 \times 10^3$ J

1B The ideal gas equation is used to find the initial volume; Boyle's law, to determine the final volume.

$$V = \frac{0.225 \text{ mol N}_2 \times \dfrac{0.08206 \text{ L atm}}{\text{mol K}} \times (23 + 273)\text{ K}}{2.15 \text{ atm}} = 2.54 \text{ L}$$

$$P_{\text{final}} = 746 \text{ mmHg} \times \frac{1 \text{ atm}}{760 \text{ mmHg}} = 0.982 \text{ atm} \qquad V_{\text{final}} = 2.54 \text{ L} \times \frac{2.15 \text{ atm}}{0.982 \text{ atm}} = 5.56 \text{ L}$$

$$w = P\Delta V = 0.982 \text{ atm } (5.56 \text{ L} - 2.54 \text{ L}) \times \frac{101.3 \text{ J}}{1 \text{ L atm}} = 300. \text{ J}$$

2A The heat absorbed is the product of the mass of water, its specific heat (4.18 J g^{-1} °C^{-1}), and the temperature change that occurs.

$$\text{heat energy} = 237 \text{ g} \times \frac{4.18 \text{ J}}{\text{g °C}} \times (37.0 \text{ °C} - 4.0 \text{ °C}) \times \frac{1 \text{ kJ}}{1000 \text{ J}} = 32.7 \text{ kJ of heat energy}$$

2B The heat absorbed is the product of the amount of mercury, its molar heat capacity, and the temperature change that occurs.

$$\text{heat energy} = \left(2.50 \text{ kg} \times \frac{1000 \text{ g}}{1 \text{ kg}} \times \frac{1 \text{ mol Hg}}{200.59 \text{ g Hg}}\right) \times \frac{28.0 \text{ J}}{\text{mol °C}} \times [-6.0 - (-20.0)] \text{ °C} \times \frac{1 \text{ kJ}}{1000 \text{ J}}$$

$$= 4.89 \text{ kJ of heat energy}$$

3A First calculate the quantity of heat absorbed by the lead. That heat (which actually is given off, as indicated by its negative sign) is absorbed by the water.

$$q_{\text{lead}} = 1.00 \text{ kg} \times \frac{1000 \text{ g}}{1 \text{ kg}} \times \frac{0.13 \text{ J}}{\text{g °C}} \times (35.2 \text{ °C} - 100.0 \text{ °C}) = -8.4 \times 10^3 \text{ J} = -q_{\text{water}}$$

$$8.4 \times 10^3 \text{ J} = m_{\text{water}} \times \frac{4.18 \text{ J}}{\text{g °C}} \times (35.2 \text{ °C} - 28.5 \text{ °C}) = 28\, m_{\text{water}} \qquad m_{\text{water}} = \frac{8.4 \times 10^3}{28} = 3.0 \times 10^2 \text{ g}$$

3B We use the same equation, equating the heat lost by the copper to the heat absorbed by the water, except now we solve for final temperature.

$$q_{\text{Cu}} = 100.0 \text{ g} \times \frac{0.385 \text{ J}}{\text{g °C}} \times (x \text{ °C} - 100.0 \text{ °C}) = -50.0 \text{ g} \times \frac{4.18 \text{ J}}{\text{g °C}} \times (x \text{ °C} - 26.5 \text{ °C}) = -q_{\text{water}}$$

$$38.5\,x - 3850 = -209\,x + 5539 \text{ J} \qquad 38.5\,x + 209\,x = 5539 + 3850 = 247.5\,x = 9389$$

$$x = \frac{9389}{247.5} = 37.9 \text{ °C}$$

4A The molar mass of $C_8H_8O_3$ is 152.15 g/mol. The calorimeter has a heat capacity of 4.90 kJ/°C

$$q_{\text{calor}} = \frac{4.90 \text{ kJ °C}^{-1} \times (30.09 \text{ °C} - 24.89 \text{ °C})}{1.013 \text{ g}} \times \frac{152.15 \text{ g}}{1 \text{ mol}} = 3.83 \times 10^3 \text{ kJ/mol}$$

$$\Delta H_{\text{comb}} = -q_{\text{calor}} = -3.83 \times 10^3 \text{ kJ/mol}$$

4B The heat that is liberated by the benzoic acid's combustion serves to raise the temperature of the assembly. We designate the calorimeter's heat capacity by C. $q_{\text{rxn}} = 1.176\text{ g} \times \dfrac{-26.42\text{ kJ}}{1\text{ g}} = -31.07\text{ kJ} = -q_{\text{calorim}}$

$q_{\text{calorim}} = C\,\Delta t = 31.07\text{ kJ} = C \times 4.96\text{ °C} \qquad C = \dfrac{31.07\text{ kJ}}{4.96\text{ °C}} = 6.26\text{ kJ/°C}$

5A The heat that is liberated by the reaction raises the temperature of the reaction mixture. We assume that this reaction mixture has the same density and specific heat as pure water.

$$q_{\text{rxn}} = \left(200.0\text{ mL} \times \frac{1.00\text{ g}}{1\text{ mL}}\right) \times \frac{4.18\text{ J}}{\text{g °C}} \times (30.2 - 22.4)\text{ °C} = 6.52 \times 10^3\text{ J}$$

$$\text{amount AgCl} = 100.0\text{ mL} \times \frac{1\text{ L}}{1000\text{ mL}} \times \frac{1.00\text{ M AgNO}_3}{1\text{ L}} \times \frac{1\text{ mol AgCl}}{1\text{ mol AgNO}_3} = 0.100\text{ mol AgCl}$$

$$q_{\text{rxn}} = \frac{6.52 \times 10^3\text{ J}}{0.100\text{ mol}} \times \frac{1\text{ kJ}}{1000\text{ J}} = 65.2\text{ kJ/mol}$$

5B The assumptions include no heat loss to the surroundings or to the calorimeter, a solution density of 1.00 g/mL and specific heat of 4.18 J $\text{g}^{-1}\,\text{°C}^{-1}$, and that the initial and final solution volumes are the same. The equation for the reaction that occurs is $NaOH(aq) + HCl(aq) \longrightarrow NaCl(aq) + H_2O(l)$ Since the two reactants combine in a one to one mole ratio, the limiting reactant is the one present in smaller amount.

$$\text{amount HCl} = 100.0\text{ mL} \times \frac{1.020\text{ mmol HCl}}{1\text{ mL soln}} = 102.0\text{ mmol HCl}$$

$$\text{amount NaOH} = 50.0\text{ mL} \times \frac{1.988\text{ mmol NaOH}}{1\text{ mL soln}} = 99.4\text{ mmol NaOH}$$ This is the limiting reactant.

$$q_{\text{neutr}} = 99.4\text{ mmol NaOH} \times \frac{1\text{ mmol H}_2\text{O}}{1\text{ mmol NaOH}} \times \frac{1\text{ mol H}_2\text{O}}{1000\text{ mmol H}_2\text{O}} \times \frac{-57\text{ kJ}}{1\text{ mol H}_2\text{O}} = -5.6_7\text{ kJ}$$

$$q_{\text{calorim}} = -q_{\text{neutr}} = 5.6_7\text{ kJ} = (100.0 + 50.0)\text{ mL} \times \frac{1.00\text{ g}}{1\text{ mL}} \times \frac{4.18\text{ J}}{\text{g °C}} \times \frac{1\text{ kJ}}{1000\text{ J}} \times (t - 24.52\text{ °C})$$

$$= 0.627\,t - 15.3_7 \qquad t = \frac{5.6_7 + 15.3_7}{0.627} = 33.5_6\text{ °C} = 33.6\text{ °C}$$

6A The work is $w = +355$ J. The heat is $q = -185$ J. These two are related to the energy change of the system by the first law equation: $\Delta E = q + w$ which becomes $\Delta E = +355\text{ J} - 185\text{ J} = +1.70 \times 10^2\text{ J}$

6B The internal energy change is $\Delta E = -125$ J. The heat is $q = +54$ J. These two are related to the work done on the system by the first law equation: $\Delta E = q + w$ which becomes $-125\text{ J} = +54\text{ J} + w$
The solution to this equation is $w = -125\text{ J} - 54\text{ J} = -179\text{ J}$ which means that 179 J of work is done by the system.

7A Heat that is given off has a negative sign. In addition, we use the molar mass of sucrose, 342.30 g/mol.

$$\text{sucrose mass} = -1.00 \times 10^3\text{ kJ} \times \frac{1\text{ mol C}_{12}\text{H}_{22}\text{O}_{11}}{-5.65 \times 10^3\text{ kJ}} \times \frac{342.30\text{ g C}_{12}\text{H}_{22}\text{O}_{11}}{1\text{ mol C}_{12}\text{H}_{22}\text{O}_{11}} = 60.6\text{ g C}_{12}\text{H}_{22}\text{O}_{11}$$

7B Although the equation does not say so explicitly, 56 kJ of heat is given off *per mole* of water formed. The equation then is the source of a conversion factor.

$$\text{heat} = 25.0\text{ mL} \times \frac{1\text{ L}}{1000\text{ mL}} \times \frac{0.1045\text{ mol HCl}}{1\text{ L soln}} \times \frac{1\text{ mol H}_2\text{O}}{1\text{ mol HCl}} \times \frac{56\text{ kJ evolved}}{1\text{ mol H}_2\text{O}} = 0.15\text{ kJ heat evolved}$$

8A We combine the three combustion reactions to produce the hydrogenation reaction.

$C_3H_6(g) + \frac{9}{2}O_2(g) \longrightarrow 3\,CO_2(g) + 3\,H_2O(l)$	$\Delta H_{\text{comb}} = \Delta H_1 = -2058$ kJ
$H_2(g) + \frac{1}{2}O_2(g) \longrightarrow H_2O(l)$	$\Delta H_{\text{comb}} = \Delta H_2 = -285.8$ kJ
$3\,CO_2(g) + 4\,H_2O(l) \longrightarrow C_3H_8(g) + 5\,O_2(g)$	$-\Delta H_{\text{comb}} = \Delta H_3 = +2219.1$ kJ
$C_3H_6(g) + H_2(g) \longrightarrow C_3H_8(g)$	$\Delta H_{\text{rxn}} = \Delta H_1 + \Delta H_2 + \Delta H_3 = -125$ kJ

8B The combustion reaction has propanol and $O_2(g)$ as reactants; the products are $CO_2(g)$ and $H_2O(l)$. We reverse the reaction given and combine it with the combustion reaction of $C_3H_6(g)$.

$C_3H_7OH(l) \longrightarrow C_3H_6(g) + H_2O(l)$	$\Delta H_1 = +36.8$ kJ
$C_3H_6(g) + \frac{9}{2}O_2(g) \longrightarrow 3\,CO_2(g) + 3\,H_2O(l)$	$\Delta H_2 = -2058$ kJ
$C_3H_7OH(l) + \frac{9}{2}O_2(g) \longrightarrow 3\,CO_2(g) + 4\,H_2O(l)$	$\Delta H_{\text{rxn}} = \Delta H_1 + \Delta H_2 = -2021$ kJ

9A The enthalpy of formation is the enthalpy change for the reaction in which one mole of the product, $C_6H_{13}O_2N(s)$, is produced from appropriate amounts of the most stable forms of the elements.

$$6\,C(\text{graphite}) + \tfrac{13}{2}\,H_2(g) + O_2(g) + \tfrac{1}{2}\,N_2(g) \longrightarrow C_6H_{13}O_2N(s)$$

9B The enthalpy of formation is the enthalpy change for the reaction in which one mole of the product, $NH_3(g)$, is produced from appropriate amounts of the most stable forms of the elements, in this case from 0.5 mol $N_2(g)$ and 1.5 mol $H_2(g)$, that is, for the reaction: $\tfrac{1}{2}N_2(g) + \tfrac{3}{2}H_2(g) \longrightarrow NH_3(g)$

The specified reaction is twice the reverse of the formation reaction, and its enthalpy change is twice negative of the enthalpy of formation of $NH_3(g)$: $-2 \times (-46.11\text{ kJ}) = +92.22\text{ kJ}$

10A We write the combustion reaction for each compound, and use that reaction to determine the compound's heat of combustion.

$$C_3H_8(g) + 5\,O_2(g) \longrightarrow 3\,CO_2(g) + 4\,H_2O(l)$$

$$\Delta H^\circ_{\text{comb}} = 3 \times \Delta H^\circ_f[CO_2(g)] + 4 \times \Delta H^\circ_f[H_2O(l)] - \Delta H^\circ_f[C_3H_8(g)] - 5 \times \Delta H^\circ_f[O_2(g)]$$
$$= [3 \times (-393.5\text{ kJ})] + [4 \times (-285.8\text{ kJ})] - [-104.7] - [5 \times 0.00\text{ kJ}]$$
$$= -1181\text{ kJ} - 1143\text{ kJ} + 104.7 - 0.00 = -2219\text{kJ/mol } C_3H_8$$

$$C_4H_{10}(g) + \tfrac{13}{2}\,O_2(g) \longrightarrow 4\,CO_2(g) + 5\,H_2O(l)$$

$$\Delta H^\circ_{\text{comb}} = 4 \times \Delta H^\circ_f[CO_2(g)] + 5 \times \Delta H^\circ_f[H_2O(l)] - \Delta H^\circ_f[C_4H_{10}(g)] - 6.5 \times \Delta H^\circ_f[O_2(g)]$$
$$= [4 \times (-393.5\text{ kJ})] + [5 \times (-285.8\text{ kJ})] - [-125.6] - [6.5 \times 0.00\text{ kJ}]$$
$$= -1574\text{ kJ} - 1429\text{ kJ} + 125.6 - 0.00 = -2877\text{ kJ/mol } C_4H_{10}$$

In 1.00 mole of the mixture there are 0.62 mol $C_3H_8(g)$ and 0.38 mol $C_4H_{10}(g)$.

$$\text{heat of combustion} = \left(0.62\text{ mol } C_3H_8 \times \frac{-2219\text{ kJ}}{1\text{ mol } C_3H_8}\right) + \left(0.38\text{ mol } C_4H_{10} \times \frac{-2877\text{ kJ}}{1\text{ mol } C_4H_{10}}\right)$$
$$= -1.4 \times 10^3\text{ kJ} - 1.2 \times 10^3\text{ kJ} = -2.5 \times 10^3\text{ kJ/mole of mixture}$$

10B $\Delta H^\circ_{\text{rxn}} = 12 \times \Delta H^\circ_f[CO_2(g)] + 6 \times \Delta H^\circ_f[H_2O(l)] - 2 \times \Delta H^\circ_f[C_6H_6(l)] - 15 \times \Delta H^\circ_f[O_2(g)]$

$$-6535\text{ kJ} = [12 \times (-393.5\text{ kJ})] + [6 \times (-285.8\text{ kJ})] - 2 \times \Delta H^\circ_f[C_6H_6(l)] - [15 \times 0.00\text{ kJ}]$$
$$= -4722\text{ kJ} - 1715\text{ kJ} - 2 \times \Delta H^\circ_f[C_6H_6(l)] - 0.00$$
$$\Delta H^\circ_f[C_6H_6(l)] = \frac{+6535\text{ kJ} - 4722\text{ kJ} - 1715\text{ kJ}}{2} = +49\text{ kJ/mol } C_6H_6(l)$$

11A The net ionic equation is: $Ag^+(aq) + I^-(aq) \longrightarrow AgI(s)$ and we have the following.

$$\Delta H^\circ_{\text{rxn}} = \Delta H^\circ_f[AgI(s)] - \Delta H^\circ_f[Ag^+(aq)] - \Delta H^\circ_f[I^-(aq)]$$
$$= -61.84\text{ kJ/mol} - (+105.6\text{ kJ/mol}) - (-55.19\text{ kJ/mol}) = -112.25\text{ kJ/mol}$$

11B The solubility rules of Chapter 5 indicate that $Mg(OH)_2(s)$ precipitates. The net ionic equation is:

$Mg^{2+}(aq) + 2\,OH^-(aq) \longrightarrow Mg(OH)_2(s)$ and we have the following.

$$\Delta H^\circ_{\text{rxn}} = \Delta H^\circ_f[Mg(OH)_2(s)] - \Delta H^\circ_f[Mg^{2+}(aq)] - 2\,\Delta H^\circ_f[OH^-(aq)]$$
$$= -924.5\text{ kJ/mol} - (-466.9\text{ kJ/mol}) - 2\,(-230.0\text{ kJ/mol}) = +2.4\text{ kJ/mol}$$

SUMMARIZING EXAMPLE CALCULATIONS

1. Since one mole of gas occupies 22.414 L at STP, the amount of each gas present in 1.00 L at STP is determined as follows.

$$\text{amount CO} = 1.00\text{ L synthesis gas} \times \frac{0.550\text{ L CO}}{1\text{ L synth gas}} \times \frac{1\text{ mol CO}}{22.414\text{ L CO}} = 0.0245\text{ mol CO}$$

$$\text{amount } H_2 = 1.00\text{ L synthesis gas} \times \frac{0.330\text{ L } H_2}{1\text{ L synth gas}} \times \frac{1\text{ mol } H_2}{22.414\text{ L } H_2} = 0.0147\text{ mol } H_2$$

2. Since the combustion of one mole of $H_2(g)$ forms one mole of $H_2O(l)$—note that there is but one mole of product in the combustion reaction and stable forms of the elements are the only reactants—the heat of combustion of hydrogen is equal to the heat of formation of water, –285.8 kJ/mol $H_2(g)$

For CO(g), the combustion reaction is $CO(g) + \tfrac{1}{2}O_2(g) \longrightarrow CO_2(g)$

$$\Delta H^\circ_{\text{rxn}} = \Delta H^\circ_f[CO_2(g)] - \Delta H^\circ_f[CO(g)] - \tfrac{1}{2}\,\Delta H^\circ_f[O_2(g)] = -393.5\text{ kJ} - (-110.5\text{ kJ}) - (0.5 \times 0.00) = -283.0\text{ kJ}$$

$$q_{H2} = 0.0147\text{ mol } H_2 \times \frac{-285.8\text{ kJ}}{1\text{ mol } H_2} = -4.20\text{ kJ} \qquad q_{CO} = 0.0245\text{ mol CO} \times \frac{-283.0\text{ kJ}}{1\text{ mol CO}} = -6.93\text{ kJ}$$

$$\text{Total} = -4.20\text{ kJ} - 6.93\text{ kJ} = -11.13\text{ kJ} = q_{\text{comb}} = -q_{\text{water}}$$

3. $q_{water} = m_{water} \times \text{sp ht}_{water} \times \Delta T_{water}$ $\quad 11.13 \text{ kJ} \times \frac{1000 \text{ J}}{1 \text{ kJ}} = 1.00 \text{ kg} \times \frac{1000 \text{ g}}{1 \text{ kg}} \times \frac{4.18 \text{ J}}{\text{g °C}} \times \Delta T$

$\Delta T = \frac{11.13 \times 10^3}{4.18 \times 10^3} = 2.66°\text{C}$ $\quad T_{final} = T_{initial} + \Delta T = 25.0°\text{C} + 2.66°\text{C} = 27.7°\text{C}$

REVIEW QUESTIONS

1. (a) ΔH, the enthalpy change for a process, is the quantity of heat absorbed when the process occurs at constant pressure.
   (b) $P\Delta V$, the change in volume multiplied by a constant pressure, is the expression for the pressure-volume work in a process that occurs at constant pressure.
   (c) ΔH_f°, the standard enthalpy of formation, is the heat absorbed at constant pressure when 1 mole of product is formed from the stable form of the elements, with reactants and products in their standard states.
   (d) Standard state is defined as a pressure of exactly 1 atm for a pure substance, or an aqueous solute at a 1 M concentration, each at a temperature of interest.
   (e) A fossil fuel is a material that can be burned for heat and that was produced by material that lived eons ago.

2. (a) The law of conservation of energy states that energy is neither created nor destroyed during a process.
   (b) Bomb calorimetry is the technique of running a chemical reaction in a constant-volume container and measuring the heat absorbed by the process.
   (c) A function of state is a measurable property that depends only on the initial and final conditions of a process and not on its path.
   (d) An enthalpy diagram represents the enthalpy values of reactants and products by their vertical positions, and the progress of the reaction horizontally.
   (e) Hess's law states that if several reactions can be combined to form a net reaction, the enthalpy changes of those reactions combine in the same way to produce the enthalpy change of the net reaction.

3. (a) The system is that part of the universe that we are considering, in which we are interested. The surroundings are the rest of the universe, particularly the rest that influences the system.
   (b) Heat is energy in transport that is associated with either a difference in temperature or a phase change (such as solid to liquid) of a material. Work is organized energy that has the ability to move a force through a distance.
   (c) The specific heat of a substance is the quantity of heat needed to raise the temperature of one *gram* of that substance by 1.00 °C. The (molar) heat capacity of a substance the the quantity of heat needed to raise the temperature (of one *mole*) of that substance by 1.00 °C.
   (d) An endothermic reaction is one that absorbs heat from the surroundings. An exothermic reaction is one that evolves heat to the surroundings.

4. (a) $q = 9.25 \text{ L} \times \frac{1000 \text{ cm}^3}{1 \text{ L}} \times \frac{1.00 \text{ g}}{1 \text{ cm}^3} \times \frac{1.00 \text{ cal}}{1 \text{ g °C}} \times (29.4 \text{ °C} - 22.0 \text{ °C}) \times \frac{1 \text{ kcal}}{1000 \text{ cal}} = +68.5 \text{ kcal}$
   (b) $q = 5.85 \text{ kg} \times \frac{1000 \text{ g}}{1 \text{ kg}} \times \frac{0.903 \text{ J}}{\text{g °C}} \times (-33.5 \text{ °C}) \times \frac{1 \text{ kJ}}{1000 \text{ J}} = -177 \text{ kJ}$

5. heat = mass × sp ht × Δt becomes $\Delta t = \frac{\text{heat}}{\text{mass} \times \text{sp ht}}$
   (a) $\Delta t = \frac{+875 \text{ J}}{12.6 \text{ g} \times 4.18 \text{ J g}^{-1} \text{ °C}^{-1}} = +16.6 \text{ °C}$ $\quad t_f = t_i + \Delta t = 22.9 \text{ °C} + 16.6 \text{ °C} = 39.5 \text{ °C}$
   (b) $\Delta t = \frac{-1.05 \text{ kcal} \times \frac{1000 \text{ cal}}{1 \text{ kcal}}}{\left(1.59 \text{ kg} \times \frac{1000 \text{ g}}{1 \text{ kg}}\right) 0.032 \frac{\text{cal}}{\text{g °C}}} = -21 \text{ °C}$ $\quad t_f = t_i + \Delta t = 78.2 \text{ °C} - 21 \text{ °C} = 57 \text{ °C}$

6. (a) $\text{sp.ht} = \frac{\text{heat}}{\text{mass} \times \Delta t} = \frac{186 \text{ J}}{15.0 \text{ g} \times (29.6 - 22.3)} = 1.7 \text{ J g}^{-1} \text{ °C}^{-1}$

(b) $\Delta t = \dfrac{\text{heat}}{\text{mass} \times \text{sp.ht}} = \dfrac{-2.75 \text{ kcal} \times \dfrac{1000 \text{ cal}}{1 \text{ kcal}}}{\left(2.25 \text{ kg} \times \dfrac{1000 \text{ g}}{1 \text{ kg}}\right) 1.00 \dfrac{\text{cal}}{\text{g °C}}} = -1.22 \text{ °C}$

$t_f = t_i + \Delta t = 23.1 \text{ °C} - 1.22 \text{ °C} = 21.9 \text{ °C}$

7. The heat capacities of the two substances are added and then multiplied by the temperature change.

$$\Delta H = \left(118 \text{ g Cu} \times \frac{0.385 \text{ J}}{\text{g °C}} + 197 \text{ g } H_2O \times \frac{4.18 \text{ J}}{\text{g °C}}\right)(79.2 \text{ °C} - 22.7 \text{ °C}) \times \frac{1 \text{ kJ}}{1000 \text{ J}} = +49.1 \text{ kJ}$$

8. Heat is transferred from the iron to the water.

$$q_{water} = 981 \text{ g} \times \frac{4.18 \text{ J}}{\text{g °C}} \times (34.4 - 22.1) \text{ °C} = 5.04 \times 10^4 \text{ J} = -q_{iron}$$

$$q_{iron} = -5.04 \times 10^4 \text{ J} = 1.22 \text{ kg} \times \frac{1000 \text{ g}}{1 \text{ kg}} \times \text{sp ht} \times (34.4 - 126.5) \text{ °C} = -1.12 \times 10^5 \text{ sp ht}$$

$$\text{sp ht} = \frac{-5.04 \times 10^4}{-1.12 \times 10^5} = \frac{0.450 \text{ J}}{\text{g °C}}$$

9. If the two volumes were the same, a straight average of the temperatures would be the final temperature. But that is not true, so option **(3)** [50 °C] is incorrect. In fact, there is more of the cooler water (100.0 mL) than of the warmer water (75.0 mL). The cooler side of the straight average should be somewhat favored, but not grossly so; option **(4)** [28 °C] is incorrect. In order to arrive at 40 °C, the cold water should increase by 20 °C and the warm water should decrease by 40 °C; their masses, and thus their volume should be in the inverse relationship as their temperature changes: 2:1. That is not the case, so option **(1)** [40 °C] is incorrect. The correct final temperature is likely to be 46 °C, option **(2)**.

10. **(a)** $\Delta E = q + w = 67 \text{ J heat} - 67 \text{ J work} = 0 \text{ J}$ **(b)** $\Delta E = q + w = 356 \text{ J heat} - 592 \text{ J work} = -236 \text{ J}$
(c) $\Delta E = q + w = -38 \text{ J heat} + 171 \text{ J work} = +133 \text{ J}$ **(d)** $\Delta E = q + w = 0 \text{ J heat} - 416 \text{ J work} = -416 \text{ J}$

11. **(a)** $q = \dfrac{-29.4 \text{ kJ}}{0.584 \text{ g } C_3H_8} \times \dfrac{44.10 \text{ g } C_3H_8}{1 \text{ mol } C_3H_8} = -2.22 \times 10^3 \text{ kJ/mol } C_3H_8$

(b) $q = \dfrac{-1.26 \text{ kcal}}{0.136 \text{ g } C_{10}H_{16}O} \times \dfrac{4.184 \text{ kJ}}{1 \text{ kcal}} \times \dfrac{152.24 \text{ g } C_{10}H_{16}O}{1 \text{ mol } C_{10}H_{16}O} = -5.90 \times 10^3 \text{ kJ/mol } C_{10}H_{16}O$

(c) $q = \dfrac{-58.3 \text{ kJ}}{2.35 \text{ mL } (CH_3)_2CO} \times \dfrac{1 \text{ mL}}{0.791 \text{ g}} \times \dfrac{58.08 \text{ g } (CH_3)_2CO}{1 \text{ mol } (CH_3)_2CO} = -1.82 \times 10^3 \text{ kJ/mol } (CH_3)_2CO$

12. $\text{heat capacity} = \dfrac{\text{heat absorbed}}{\Delta t} = \dfrac{5228 \text{ cal}}{4.39 \text{ °C}} \times \dfrac{4.184 \text{ J}}{1 \text{ cal}} \times \dfrac{1 \text{ kJ}}{1000 \text{ J}} = 4.98 \text{ kJ/°C}$

13. In each case, heat absorbed by calorimeter ($= -q_{comb} \times$ no. mol) = heat cap × Δt or $\Delta t = \dfrac{q_{comb} \times \text{no. mol}}{\text{heat cap}}$

(a) $\Delta t = \dfrac{\left(1014.2 \dfrac{\text{kcal}}{\text{mol}} \times 4.184 \dfrac{\text{kJ}}{\text{kcal}}\right)\left(0.3268 \text{ g} \times \dfrac{1 \text{ mol } C_8H_{10}O_2N_4}{194.19 \text{ g } C_8H_{10}O_2N_4}\right)}{5.136 \text{ kJ/°C}} = 1.390 \text{ °C}$

$t_f = t_i + \Delta t = 22.43 \text{ °C} + 1.390 \text{ °C} = 23.82 \text{ °C}$

(b) $\Delta t = \dfrac{2444 \dfrac{\text{kJ}}{\text{mol}}\left(1.35 \text{ mL} \times \dfrac{0.805 \text{ g}}{1 \text{ mL}} \times \dfrac{1 \text{ mol } C_4H_8O}{72.11 \text{ g } C_4H_8O}\right)}{5.136 \text{ kJ/°C}} = 7.17 \text{ °C}$

$t_f = 22.43 \text{ °C} + 7.17 \text{ °C} = 29.60 \text{ °C}$

14. **(a)** $\dfrac{\text{heat}}{\text{mass}} = \dfrac{\text{heat cap.} \times \Delta t}{\text{mass}} = \dfrac{4.728 \text{ kJ/°C} \times (27.19 - 23.29) \text{ °C}}{1.183 \text{ g}} = 15.6 \text{ kJ/g xylose}$

$\Delta H = \text{heat given off/g} \times \mathcal{M}(\text{g/mol}) = \dfrac{-15.6 \text{ kJ}}{1 \text{ g } C_5H_{10}O_5} \times \dfrac{150.13 \text{ g } C_5H_{10}O_5}{1 \text{ mol}}$

$= -2.34 \times 10^3 \text{ kJ/mol } C_5H_{10}O_5$

(b) $C_5H_{10}O_5(g) + 5\ O_2(g) \longrightarrow 5\ CO_2(g) + 5\ H_2O(l) \quad \Delta H = -2.34 \times 10^3 \text{ kJ/mol } C_5H_{10}O_5$

15. This is first a limiting reactant problem. There is 0.1000 L × 0.300 M = 0.0300 mol HCl and 1.82/65.39 = 0.0278 mol Zn. Stoichiometry demands 2 mol HCl for every 1 mol Zn. HCl is the limiting reactant. The reaction is exothermic. We assume that the specific heat of the solution is 4.18 J g^{-1} °C^{-1}. The enthalpy change in kJ/mol Zn is obtained from the heat absorbed per gram Zn.

$$\Delta H = -\frac{\left(5.20 \text{ g Zn} + 100.0 \text{ mL} \times \frac{1.00 \text{ g}}{1 \text{ mL}}\right)\frac{4.18 \text{ J}}{\text{g °C}}(30.5 - 20.3)\text{ °C}}{0.0300 \text{ mol HCl} \times \frac{1 \text{ mol Zn}}{2 \text{ mol HCl}}} \times \frac{1 \text{ kJ}}{1000 \text{ J}}$$

$= -299$ kJ/mol

16. **(a)** Because the temperature of the mixture decreases, the reaction (the system) must have absorbed heat from the reaction mixture (the surroundings). Consequently, the reaction must be endothermic.

(b) We assume that the specific heat of the solution is 4.18 J g^{-1} °C^{-1}. The enthalpy change in kJ/mol KCl is obtained by the heat absorbed per gram KCl.

$$\Delta H = -\frac{(0.75 + 35.0)\text{g}\,\frac{4.18 \text{ J}}{\text{g °C}}(23.6 - 24.8)\text{°C}}{0.75 \text{ g KCl}} \times \frac{1 \text{ kJ}}{1000 \text{ J}} \times \frac{74.55 \text{ g KCl}}{1 \text{ mol KCl}} = +18 \text{ kJ/mol}$$

17. As indicated by the negative sign for the enthalpy change, this is an exothermic reaction; the temperature of the system should increase.

$$q_{rxn} = 0.136 \text{ mol } KC_2H_3O_2 \times \frac{15.3 \text{ kJ}}{1 \text{ mol } KC_2H_3O_2} \times \frac{1000 \text{ J}}{1 \text{ kJ}} = 2.08 \times 10^3 \text{ J} = -q_{calorim}$$

Now we assume that the density of water is 1.00 g/mL, the specific heat of the solution in the calorimeter is 4.18 J g^{-1} °C^{-1}, and no heat is absorbed by the calorimeter.

$$q_{calorim} = 2.08 \times 10^3 \text{ J} = \left(\left(525 \text{ mL} \times \frac{1.00 \text{ g}}{1 \text{ mL}}\right) + \left(0.136 \text{ mol } KC_2H_3O_2 \times \frac{98.14 \text{ g}}{1 \text{ mol } KC_2H_3O_2}\right)\right)$$

$$\times \frac{4.18 \text{ J}}{\text{g °C}} \times \Delta t = 2.25 \times 10^3\, \Delta t \qquad \Delta t = \frac{2.08 \times 10^3}{2.25 \times 10^3} = +0.924 \text{ °C}$$

$t_{final} = t_{initial} + \Delta t = 25.1$ °C $+ 0.924$ °C $= 26.0$ °C

18. **(a)** $N_2(g) + \frac{1}{2} O_2(g) \longrightarrow N_2O(g)$ $\quad \Delta H° = +82.05$ kJ/mol

(b) $S(\text{rhombic}) + O_2(g) + Cl_2(g) \longrightarrow SO_2Cl_2(l)$ $\quad \Delta H° = -394.1$ kJ/mol

(c) $CH_3CH_2COOH(l) + \frac{7}{2} O_2(g) \longrightarrow 3\, CO_2(g) + 3\, H_2O(l)$ $\quad \Delta H° = -1527$ kJ/mol

19. **(a)** $\text{heat evolved} = 1.325 \text{ g } C_4H_{10} \times \frac{1 \text{ mol } C_4H_{10}}{58.123 \text{ g } C_4H_{10}} \times \frac{2877 \text{ kJ}}{1 \text{ mol } C_4H_{10}} = 65.59 \text{ kJ}$

(b) $\text{heat evolved} = 28.4\, L_{STP}\, C_4H_{10} \times \frac{1 \text{ mol } C_4H_{10}}{22.414\, L_{STP}\, C_4H_{10}} \times \frac{2877 \text{ kJ}}{1 \text{ mol } C_4H_{10}} = 3.65 \times 10^3 \text{ kJ}$

(c) Use the ideal gas equation to determine the amount of propane in moles and multiply this amount by 2877 kJ heat produced per mole.

$$\text{heat evolved} = \frac{\left(738 \text{ mmHg} \times \frac{1 \text{ atm}}{760 \text{ mmHg}}\right) \times 12.6 \text{ L}}{\frac{0.08206 \text{ L atm}}{\text{mol K}} \times (273.2 + 23.6) \text{ K}} \times \frac{2877 \text{ kJ}}{1 \text{ mol } C_4H_{10}} = 1.45 \times 10^3 \text{ kJ}$$

20. The formation reaction for $NH_3(g)$ is $\frac{1}{2} N_2(g) + \frac{3}{2} H_2(g) \longrightarrow NH_3(g)$. The given reaction is two-thirds the reverse of the formation reaction. The sign of the enthalpy is changed and it is multiplied by two-thirds. Thus, the enthalpy of the given reaction is $-(-46.11 \text{ kJ}) \times \frac{2}{3} = +30.74$ kJ.

21.

−(1)	$CO(g) \longrightarrow C(\text{graphite}) + \frac{1}{2} O_2(g)$	$\Delta H = +110.54$ kJ
+(2)	$C(\text{graphite}) + O_2(g) \longrightarrow CO_2(g)$	$\Delta H = -393.51$ kJ
	$CO(g) + \frac{1}{2} O_2(g) \longrightarrow CO_2(g)$	$\Delta H = -282.97$ kJ

22.

−(3)	$3\, CO_2(g) + 5\, O_2(g) \longrightarrow C_3H_8(g) + 5\, O_2(g)$	$\Delta H = +2219.1$ kJ
+(2)	$C_3H_4(g) + 4\, O_2(g) \longrightarrow 3\, CO_2(g) + 2\, H_2O(l)$	$\Delta H = -1937$ kJ
2(1)	$2\, H_2(g) + O_2(g) \longrightarrow 2\, H_2O(l)$	$\Delta H = -\ 571.6$ kJ
	$C_3H_4(g) + 2\, H_2(g) \longrightarrow C_3H_8(g)$	$\Delta H = -290$ kJ

23. The second reaction is the only one in which NO(g) appears; it must be run twice to produce 2 NO(g).

$$2\,NH_3(g) + \tfrac{5}{2}\,O_2(g) \longrightarrow 2\,NO(g) + 3\,H_2O(l) \qquad 2\,\Delta H_2$$

The first reaction is the only one that can eliminate 2 NH_3(g); it must be run twice to eliminate 2 NH_3(g).

$$N_2(g) + 3\,H_2(g) \longrightarrow 2\,NH_3(g) \qquad 2\,\Delta H_1$$

We triple and reverse the third reaction to eliminate 3 H_2(g).

$$3\,H_2O(l) \longrightarrow 3\,H_2(g) + \tfrac{3}{2}\,O_2(g) \qquad -3\,\Delta H_3$$

Result: $N_2(g) + O_2(g) \longrightarrow 2\,NO(g) \qquad \Delta H_{rxn} = 2\,\Delta H_1 + 2\,\Delta H_2 - 3\,\Delta H_3$

24. (a) $\Delta H^\circ = \Delta H^\circ_f[C_2H_6(g)] + \Delta H^\circ_f[CH_4(g)] - \Delta H^\circ_f[C_3H_8(g)] - \Delta H^\circ_f[H_2(g)]$
$= -84.68 - 74.81 - (-104.7) - 0.00 = -54.8$ kJ

(b) $\Delta H^\circ = 2\,\Delta H^\circ_f[SO_2(g)] + 2\,\Delta H^\circ_f[H_2O(l)] - 2\,\Delta H^\circ_f[H_2S(g)] - 3\,\Delta H^\circ_f[O_2(g)]$
$= 2\,(-296.8) + 2\,(-285.8) - 2\,(-20.63) - 3\,(0.00) = -1123.9$ kJ

25. $\Delta H^\circ = \Delta H^\circ_f[H_2O(l)] + \Delta H^\circ_f[NH_3(g)] - \Delta H^\circ_f[NH_4^+(aq)] - \Delta H^\circ_f[OH^-(aq)]$
$= -285.8 + (-46.11) - (-132.5) - (-230.0) = +30.6$ kJ

26. $ZnO(s) + SO_2(g) \longrightarrow ZnS(s) + \tfrac{3}{2}\,O_2(g) \qquad \Delta H^\circ = -(-878.2\text{ kJ})/2 = +439.1\text{ kJ}$

$439.1\text{ kJ} = \Delta H^\circ_f[ZnS(s)] + \tfrac{3}{2}\,\Delta H^\circ_f[O_2(g)] - \Delta H^\circ_f[ZnO(s)] - \Delta H^\circ_f[SO_2(g)]$
$= \Delta H^\circ_f[ZnS(s)] + \tfrac{3}{2}\,(0.00) - (-348.3) - (-296.8)$

$\Delta H^\circ_f[ZnS(s)] = 439.1 - 348.3 - 296.8 = -206.0$ kJ/mol

EXERCISES

Work

27. $P = 748\text{ mmHg} \times \dfrac{1\text{ atm}}{760\text{ mmHg}} = 0.984\text{ atm}$

(a) $w = P\Delta V = 0.984\text{ atm} \times 3.5\text{ L} = 3.4\text{ L atm}$

(b) $w = 3.4\text{ L atm} \times \dfrac{8.3145\text{ J}}{0.08206\text{ L atm}} = 3.4 \times 10^2\text{ J}$

(c) $w = 3.4\text{ L atm} \times \dfrac{1.9872\text{ cal}}{0.08206\text{ L atm}} = 82\text{ cal}$

28. $w = P\Delta V = 1.23\text{ atm }(5.62\text{ L} - 3.37\text{ L}) \times \dfrac{8.3145\text{ J}}{0.08206\text{ L atm}} = 2.80 \times 10^2\text{ J}$

Heat Capacity (Specific Heat)

29. heat gained by the water = heat lost by the metal; each heat = mass × sp.ht. × Δt

(a) $50.0\text{ g} \times 4.18\,\dfrac{\text{J}}{\text{g °C}}\,(38.9 - 22.0)\text{ °C} = 3.53 \times 10^3\text{ J} = -150.0\text{ g} \times \text{sp.ht.} \times (38.9 - 100.0)\text{ °C}$

$\text{sp.ht.} = \dfrac{3.53 \times 10^3\text{ J}}{150.0\text{ g} \times 61.1\text{ °C}} = 0.385\text{ J g}^{-1}\text{ C}^{\circ-1}$ for Zn

(b) $50.0\text{ g} \times 4.18\,\dfrac{\text{J}}{\text{g °C}}\,(28.8 - 22.0)\text{ °C} = 1.4 \times 10^3\text{ J} = -150.0\text{ g} \times \text{sp.ht.} \times (28.8 - 100.0)\text{ °C}$

$\text{sp.ht.} = \dfrac{1.4 \times 10^3\text{ J}}{150.0\text{ g} \times 71.2\text{ °C}} = 0.13\text{ J g}^{-1}\text{ C}^{\circ-1}$ for Pt

(c) $50.0\text{ g} \times 4.18\,\dfrac{\text{J}}{\text{g °C}}\,(52.7 - 22.0)\text{ °C} = 6.42 \times 10^3\text{ J} = -150.0\text{ g} \times \text{sp.ht.} \times (52.7 - 100.0)\text{ °C}$

$\text{sp.ht.} = \dfrac{6.42 \times 10^3\text{ J}}{150.0\text{ g} \times 47.3\text{ °C}} = 0.905\text{ J g}^{-1}\text{ C}^{\circ-1}$ for Al

30. $50.0\text{ g} \times 4.18\,\dfrac{\text{J}}{\text{g °C}}\,(27.6 - 23.2)\text{ °C} = 9.2 \times 10^2\text{ J} = -75.0\text{ g} \times \text{sp.ht.} \times (27.6 - 80.0)\text{ °C}$

$\text{sp.ht.} = \dfrac{9.2 \times 10^2\text{ J}}{75.0\text{ g} \times 52.4\text{ °C}} = 0.23\text{ J g}^{-1}\text{ C}^{\circ-1}$ for Ag

31. $q_{water} = 375 \text{ g} \times 4.18 \frac{\text{J}}{\text{g °C}} (87 - 26) \text{ °C} = 9.6 \times 10^4 \text{ J} = -q_{iron}$

$q_{iron} = -9.6 \times 10^4 \text{ J} = 465 \text{ g} \times 0.449 \frac{\text{J}}{\text{g °C}} (87 - t_i) = 1.8_2 \times 10^4 \text{ J} - 2.0_9 \times 10^2\, t_i$

$t_i = \frac{-9.6 \times 10^4 - 1.8 \times 10^4}{-2.0_9 \times 10^2} = 5.4 \times 10^2 \text{ °C}$

The number of significant figures in the final answer is limited by the two significant figures in the temperatures given.

32. heat gained by steel = –heat lost by steel = heat gained by water

$-m \times 0.50 \frac{\text{J}}{\text{g °C}} (51.5 - 183) \text{ °C} = 66\, m = 125 \text{ mL} \times \frac{1.00 \text{ g}}{1 \text{ mL}} \times 4.18 \frac{\text{J}}{\text{g °C}} (51.5 - 23.2) \text{ °C} = 1.48 \times 10^4 \text{ J}$

$m = \frac{1.48 \times 10^4}{66} = 2.2 \times 10^2$ g stainless steel.

The accuracy of this calculation is limited by the obedience of the experiment to the assumption that no heat leaks out of the system. When we deal with temperatures far above (or far below) room temperature, this assumption becomes less and less accurate.

On the other hand, the precision of the calculation is limited to two significant figures by the specific heat of the steel. If the two specific heats were known to an unlimited number of significant figures, then the temperature difference would determine the final precision of the calculation. It is unlikely that we could readily measure temperature more precisely than ±0.01 °C, without expensive equipment. The mass of steel in this case would be known to four significant figures, to ±0.1 g.

33. heat gained by Mg = –heat lost by Mg = heat gained by water

$-\left(1.00 \text{ kg Mg} \times \frac{1000 \text{ g}}{1 \text{ kg}}\right) 1.024 \frac{\text{J}}{\text{g °C}} (t_f - 40.0 \text{ °C})$

$= \left(1.00 \text{ L} \times \frac{1000 \text{ cm}^3}{1 \text{ L}} \times \frac{1.00 \text{ g}}{1 \text{ cm}^3}\right) 4.18 \frac{\text{J}}{\text{g °C}} (t_f - 20.0 \text{ °C})$

$-1.024 \times 10^3\, t_f + 4.10 \times 10^4 = 4.18 \times 10^3\, t_f - 8.36 \times 10^4$

$4.10 \times 10^4 + 8.36 \times 10^4 = (4.18 \times 10^3 + 1.024 \times 10^3)\, t_f = 12.46 \times 10^4 = 5.20 \times 10^3\, t_f$

$t_f = \frac{12.46 \times 10^4}{5.20 \times 10^3} = 24.0 \text{ °C}$

34. heat gained by the water = heat gained by the brass = –heat lost by the brass

$150.0 \text{ g} \times 4.18 \frac{\text{J}}{\text{g °C}} \times (t_f - 22.4 \text{ °C}) = -\left(15.2 \text{ cm}^3 \times \frac{8.40 \text{ g}}{1 \text{ cm}^3}\right) 0.385 \frac{\text{J}}{\text{g °C}} (t_f - 163 \text{ °C})$

$6.27 \times 10^2\, t_f - 1.40 \times 10^4 = -49.2\, t_f + 8.01 \times 10^3$ $\quad t_f = \frac{1.40 \times 10^4 + 8.01 \times 10^3}{6.27 \times 10^2 + 49.2} = 32.6 \text{ °C}$

35. heat gained by copper = –heat lost by copper = heat gained by glycerol

$-74.8 \text{ g} \times \frac{0.385 \text{ J}}{\text{g °C}} \times (31.1 \text{ °C} - 143.2 \text{ °C}) = 165 \text{ mL} \times \frac{1.26 \text{ g}}{1 \text{ mL}} \times \text{sp ht} \times (31.1 \text{ °C} - 24.8 \text{ °C})$

$3.23 \times 10^3 = 1.3 \times 10^3 \text{ sp ht}$ $\quad \text{sp ht} = \frac{3.23 \times 10^3}{1.3 \times 10^3} = 2.5 \text{ J g}^{-1} \text{ °C}^{-1}$

$\text{molar heat capacity} = 2.5 \text{ J g}^{-1} \text{ °C}^{-1} \times \frac{92.1 \text{ g}}{1 \text{ mol } C_3H_8O_3} = 2.3 \times 10^2 \text{ J mol}^{-1} \text{ °C}^{-1}$

36. The additional water simply acts as a heat transfer medium. The essential relationship is heat gained by iron = –heat lost by iron = heat gained by water (of unknown mass)

$-\left(1.23 \text{ kg} \times \frac{1000 \text{ g}}{1 \text{ kg}}\right) 0.449 \frac{\text{J}}{\text{g °C}} (25.6 - 68.5) \text{ °C} = x \text{ g } H_2O \times 4.18 \frac{\text{J}}{\text{g °C}} (25.6 - 18.5) \text{ °C}$

$2.37 \times 10^4 \text{ J} = 29.7\, x$ $\quad x = \frac{2.37 \times 10^4}{29.7} = 780 \text{ g } H_2O \times \frac{1 \text{ mL } H_2O}{1.00 \text{ g } H_2O} = 780 \text{ mL } H_2O$

Heats of reaction

37. heat $= 283 \text{ kg} \times \frac{1000 \text{ g}}{1 \text{ kg}} \times \frac{1 \text{ mol Ca(OH)}_2}{74.10 \text{ g Ca(OH)}_2} \times \frac{65.2 \text{ kJ}}{1 \text{ mol Ca(OH)}_2} = 2.49 \times 10^5$ kJ of heat evolved.

38. heat $= 1.00 \text{ gal} \times \frac{3.785 \text{ L}}{1 \text{ gal}} \times \frac{1000 \text{ mL}}{1 \text{ L}} \times \frac{0.703 \text{ g}}{1 \text{ mL}} \times \frac{1 \text{ mol C}_8\text{H}_{18}}{114.2 \text{ g C}_8\text{H}_{18}} \times \frac{5.48 \times 10^3 \text{ kJ}}{1 \text{ mol C}_8\text{H}_{18}}$

$= 1.28 \times 10^5$ kJ

39. **(a)** mass $= 2.80 \times 10^7 \text{ kJ} \times \frac{1 \text{ mol CH}_4}{890.3 \text{ kJ}} \times \frac{16.04 \text{ g CH}_4}{1 \text{ mol CH}_4} \times \frac{1 \text{ kg}}{1000 \text{ g}} = 504 \text{ kg CH}_4$

(b) First determine the moles of CH_4 present, with the ideal gas law.

$$\text{no. mol CH}_4 = \frac{\left(768 \text{ mmHg} \times \frac{1 \text{ atm}}{760 \text{ mmHg}}\right) 1.65 \times 10^4 \text{ L}}{0.08206 \frac{\text{L atm}}{\text{mol K}} \times (18.6 + 273.2) \text{ K}} = 697 \text{ mol CH}_4$$

$$\text{heat} = 697 \text{ mol CH}_4 \times \frac{-890.3 \text{ kJ}}{1 \text{ mol CH}_4} = -6.21 \times 10^5 \text{ kJ of heat}$$

(c) $$\text{volume H}_2\text{O} = \frac{6.21 \times 10^5 \text{ kJ} \times \frac{1000 \text{ J}}{1 \text{ kJ}}}{4.18 \frac{\text{J}}{\text{g °C}} (60.0 - 8.8) \text{ °C}} \times \frac{1 \text{ mL H}_2\text{O}}{1 \text{ g}} = 2.90 \times 10^6 \text{ mL} = 2.90 \times 10^3 \text{ L H}_2\text{O}$$

40. The heat of combustion of 1.00 L (STP) of synthesis gas produces 11.13 kJ of heat. The volume of synthesis gas needed to heat 40.0 gal of water is given by first determining the quantity of heat needed to raise the temperature of the water.

$$\text{heat water} = \left(40.0 \text{ gal} \times \frac{3.785 \text{ L}}{1 \text{ gal}} \times \frac{1000 \text{ mL}}{1 \text{ L}} \times \frac{1.00 \text{ g}}{1 \text{ mL}}\right) 4.18 \frac{\text{J}}{\text{g °C}} (65.0 - 15.2) \text{ °C}$$

$$= 3.15 \times 10^7 \text{ J} \times \frac{1 \text{ kJ}}{1000 \text{ J}} = 3.15 \times 10^4 \text{ kJ}$$

$$\text{gas volume} = 3.15 \times 10^4 \text{ kJ} \times \frac{1 \text{ L (STP)}}{11.13 \text{ kJ of heat}} = 283 \text{ L at STP}$$

41. Since the mole weight of H_2 (2.0 g/mol) is $\frac{1}{16}$ of the mole weight of O_2 (32.0 g/mol) and only twice as many moles of H_2 are needed as O_2, we see that $O_2(g)$ is the limiting reagent in this reaction.

$$\frac{180}{2} \text{ g O}_2 \times \frac{1 \text{ mol O}_2}{32.0 \text{ g O}_2} \times \frac{241.8 \text{ kJ heat}}{0.500 \text{ mol O}_2} = 1.4 \times 10^3 \text{ kJ heat}$$

42. The amounts of the two reactants provided are the same as their stoichiometric coefficients in the balanced equation. Thus 852 kJ of heat is given off by the reaction. We can use this quantity of heat, along with the specific heat of the mixture to determine the temperature change that will occur if all of the heat is retained in the reaction mixture.

$$\text{heat} = 852 \text{ kJ} \times \frac{1000 \text{ J}}{1 \text{ kJ}} = 8.52 \times 10^5 \text{ J} = \text{mass} \times \text{sp.ht.} \times \Delta t$$

$$= \left(\left(1 \text{ mol Al}_2\text{O}_3 \times \frac{102 \text{ g Al}_2\text{O}_3}{1 \text{ mol Al}_2\text{O}_3}\right) + \left(2 \text{ mol Fe} \times \frac{55.8 \text{ g Fe}}{1 \text{ mol Fe}}\right)\right) 0.8 \frac{\text{J}}{\text{g °C}} \times \Delta t$$

$$\Delta t = \frac{8.52 \times 10^5 \text{ J}}{214 \text{ g} \times 0.8 \text{ J g}^{-1} \text{ °C}^{-1}} = 5 \times 10^3 \text{ °C}$$

The temperature only needs to increase from 25 °C to 1530 °C or $\Delta t = 1505 \text{ °C} \approx 1.5 \times 10^3$ °C. Since the actual Δt is more than three times as large as this value, the iron indeed will melt, even if a large fraction of the heat evolved is lost to the surroundings and is not retained in the products.

43. **(a)** We first compute the heat produced by this reaction, then determine the value of ΔH in kJ/mol KOH.

$$\text{heat produced} = (0.205 + 55.9) \text{ g} \times 4.18 \frac{\text{J}}{\text{g °C}} (24.4 - 23.5) = 2 \times 10^2 \text{ J heat} = -q_{\text{absorbed}}$$

$$\Delta H = -\frac{2 \times 10^2 \text{ J} \times \frac{1 \text{ kJ}}{1000 \text{ J}}}{0.205 \text{ g} \times \frac{1 \text{ mol KOH}}{56.1 \text{ g KOH}}} = -5 \times 10^1 \text{ kJ/mol}$$

(b) The temperature change in this experiment is known to just one significant figure (0.9 °C). Doubling the amount of solute should give a temperature change known to two significant figures (1.6 °C) and using twenty times the mass of solute should give a temperature change known to three significant figures (16.0 °C). This would require just 4.10 g KOH rather than the 0.205 g KOH actually used, and would increase the precision from one part in five to one part in 500, an imprecision of about 0.1%.

44. First determine the heat absorbed by the solute during the chemical reaction, q_{rxn}. This is the negative of the heat absorbed by the solution, q_{soln}. Since the solution (water plus solute) actually gives up heat (the heat that is used by the dissolving reaction), the temperature of the solution drops.

$$\text{heat of reaction} = 150.0 \text{ mL} \times \frac{1 \text{ L}}{1000 \text{ mL}} \times \frac{2.50 \text{ mol KI}}{1 \text{ L soln}} \times \frac{20.3 \text{ kJ}}{1 \text{ mol KI}} = 7.61 \text{ kJ} = q_{rxn}$$

$$-q_{rxn} = q_{soln} = \left(150.0 \text{ mL} \times \frac{1.30 \text{ g}}{1 \text{ mL}}\right) \times \frac{2.7 \text{ J}}{\text{g °C}} \times \Delta t \qquad \Delta t = \frac{-7.61 \times 10^3 \text{ J}}{150.0 \text{ mL} \times \frac{1.30 \text{ g}}{1 \text{ mL}} \times \frac{2.7 \text{ J}}{\text{g °C}}} = -14 \text{ °C}$$

final t = initial $t + \Delta t$ = 23.5 °C – 14 °C = 10 °C

45. Let x represent the mass of NH_4Cl added to the water.

heat = mass × sp.ht. × Δt

$$x \text{ g } NH_4Cl \times \frac{1 \text{ mol } NH_4Cl}{53.49 \text{ g}} \times \frac{14.7 \text{ kJ}}{1 \text{ mol } NH_4Cl} \times \frac{1000 \text{ J}}{1 \text{ kJ}}$$

$$= -\left(\left(1400 \text{ mL} \times \frac{1.00 \text{ g}}{1 \text{ mL}}\right) + x \text{ g } NH_4Cl\right) 4.18 \frac{\text{J}}{\text{g °C}} (10. - 25) \text{ °C}$$

$$275\,x = 8.8 \times 10^4 + 63\,x \qquad x = \frac{8.8 \times 10^4}{275 - 63} = 4.2 \times 10^2 \text{ g } NH_4Cl$$

Our final value is approximate because of the assumed density (1.00 g/mL). The solution's density probably is a bit larger than 1.00 g/mL. Many aqueous solutions are somewhat more dense than water.

46. $$\text{heat} = 500 \text{ mL} \times \frac{1 \text{ L}}{1000 \text{ mL}} \times \frac{7.0 \text{ mol NaOH}}{1 \text{ L soln}} \times \frac{-44.5 \text{ kJ}}{1 \text{ mol NaOH}} = -1.6 \times 10^2 \text{ kJ}$$

= heat of reaction = –heat absorbed by solution OR $q_{rxn} = -q_{soln}$

$$\Delta t = \frac{1.6 \times 10^5 \text{ J}}{500. \text{ mL} \times \frac{1.08 \text{ g}}{1 \text{ mL}} \times \frac{4.00 \text{ J}}{\text{g °C}}} = 74 \text{ °C}$$

final t = 21 °C + 74 °C = 95 °C

47. We assume that the solution volumes are additive; that is, that 200.0 mL of solution is formed. Then we compute the heat needed to warm the solution and the cup, and then ΔH for the reaction.

$$\text{heat} = \left(200.0 \text{ mL} \times \frac{1.02 \text{ g}}{1 \text{ mL}}\right) 4.02 \frac{\text{J}}{\text{g °C}} (27.8 - 21.1) \text{ °C} + 10 \frac{\text{J}}{\text{°C}} (27.8 - 21.1) = 5.6 \times 10^3 \text{ J}$$

$$\Delta H_{neutr.} = \frac{-5.6 \times 10^3 \text{J}}{0.100 \text{ mol}} \times \frac{1 \text{ kJ}}{1000 \text{ J}} = -56 \text{ kJ/mol}$$ (–55.6 kJ/mol to three significant figures)

48. Neutralization reaction: $NaOH(aq) + HCl(aq) \longrightarrow NaCl(aq) + H_2O$

Since NaOH and HCl react in a one-to-one molar ratio, and since there is twice the volume of NaOH solution as of HCl solution, but the [HCl] is not twice the [NaOH], the HCl solution is the limiting reagent.

$$\text{heat} = 25.00 \text{ mL} \times \frac{1 \text{ L}}{1000 \text{ mL}} \times \frac{1.86 \text{ mol HCl}}{1 \text{ L}} \times \frac{1 \text{ mol } H_2O}{1 \text{ mol HCl}} \times \frac{-55.84 \text{ kJ}}{1 \text{ mol } H_2O} = -2.60 \text{ kJ}$$

= heat of reaction = –heat absorbed by solution or $q_{rxn} = -q_{soln}$

$$\Delta t = \frac{2.60 \times 10^3 \text{ J}}{75.00 \text{ mL} \times \frac{1.02 \text{ g}}{1 \text{ mL}} \times \frac{3.98 \text{ J}}{\text{g °C}}} = 8.54 \text{ °C}$$

final t = 8.54 °C + 24.72 °C = 33.26 °C

Bomb calorimetry

49. To determine the heat capacity of the calorimeter, recognize that the heat evolved by the reaction is the negative of the heat of combustion.

$$\text{heat capacity} = \frac{\text{heat evolved}}{\Delta t} = \frac{1.620 \text{ g } C_{10}H_8 \times \frac{1 \text{ mol } C_{10}H_8}{128.2 \text{ g } C_{10}H_8} \times \frac{5156.1 \text{ kJ}}{1 \text{ mol } C_{10}H_8}}{8.44 \text{ °C}} = 7.72 \text{ kJ/°C}$$

50. Note that the heat evolved is the negative of the heat absorbed.

$$\text{heat capacity} = \frac{\text{heat evolved}}{\Delta t} = \frac{1.201 \text{ g} \times \frac{1 \text{ mol } C_7H_6O_3}{138.12 \text{ g } C_7H_6O_3} \times \frac{3023 \text{ kJ}}{1 \text{ mol } C_7H_6O_3}}{(29.82 - 23.68)\ °C} = 4.28 \text{ kJ/°C}$$

51. The temperature should increase as the result of an exothermic combustion reaction.

$$\Delta T = 1.227 \text{ g } C_{12}H_{22}O_{11} \times \frac{1 \text{ mol } C_{12}H_{22}O_{11}}{342.2 \text{ g } C_{12}H_{22}O_{11}} \times \frac{5.65 \times 10^3 \text{ kJ}}{1 \text{ mol } C_{12}H_{22}O_{11}} \times \frac{1\ °C}{3.87 \text{ kJ}} = 5.23\ °C$$

52.
$$q_{comb} = \frac{-11.23\ °C \times 4.68 \text{ kJ/°C}}{1.397 \text{ g } C_{10}H_{14}O \times \frac{1 \text{ mol } C_{10}H_{14}O}{150.2 \text{ g } C_{10}H_{14}O}} = -5.65 \times 10^3 \text{ kJ/mol } C_{10}H_{14}O$$

Hess's Law

53.

$2\,HCl(g) + C_2H_4(g) + \frac{1}{2}\,O_2(g) \longrightarrow C_2H_4Cl_2(l) + H_2O(l)$	$\Delta H = -318.7$ kJ
$Cl_2(g) + H_2O(l) \longrightarrow 2\,HCl(g) + \frac{1}{2}\,O_2(g)$	$\Delta H = 0.5\,(+202.4) = +101.2$ kJ
$C_2H_4(g) + Cl_2(g) \longrightarrow C_2H_4Cl_2(l)$	$\Delta H = -217.5$ kJ

54.

$N_2H_4(l) + O_2(g) \longrightarrow N_2(g) + 2\,H_2O(l)$	$\Delta H = -622.2$ kJ
$2\,H_2O_2(l) \longrightarrow 2\,H_2(g) + 2\,O_2(g)$	$\Delta H = -2\,(-187.8 \text{ kJ}) = +375.6$ kJ
$2\,H_2(g) + O_2(g) \longrightarrow 2\,H_2O(l)$	$\Delta H = 2\,(-285.8 \text{ kJ}) = -571.6$ kJ
$N_2H_4(l) + 2\,H_2O_2(l) \longrightarrow N_2(g) + 4\,H_2O(l)$	$\Delta H = -818.2$ kJ

55.

$CO(g) + \frac{1}{2}\,O_2(g) \longrightarrow CO_2(g)$	$\Delta H = -283.0$ kJ
$3\,C(\text{graphite}) + 6\,H_2(g) \longrightarrow 3\,CH_4(g)$	$\Delta H = 3\,(-74.81) = -224.43$ kJ
$2\,H_2(g) + O_2(g) \longrightarrow 2\,H_2O(l)$	$\Delta H = 2\,(-285.8) = -571.6$ kJ
$3\,CO(g) \longrightarrow \frac{3}{2}\,O_2(g) + 3\,C(\text{graphite})$	$\Delta H = 3\,(+110.5) = +331.5$ kJ
$4\,CO(g) + 8\,H_2(g) \longrightarrow CO_2(g) + 3\,CH_4(g) + 2\,H_2O(l)$	$\Delta H = -747.5$ kJ

56.

$CS_2(l) + 3\,O_2(g) \longrightarrow CO_2(g) + 2\,SO_2(g)$	$\Delta H = -1077$ kJ
$2\,S(s) + Cl_2(g) \longrightarrow S_2Cl_2(l)$	$\Delta H = -58.2$ kJ
$C(s) + 2\,Cl_2(g) \longrightarrow CCl_4(l)$	$\Delta H = -135.4$ kJ
$2\,SO_2(g) \longrightarrow 2\,S(s) + 2\,O_2(g)$	$\Delta H = -2\,(-296.8 \text{ kJ}) = +593.6$ kJ
$CO_2(g) \longrightarrow C(s) + O_2(g)$	$\Delta H = -(-393.5 \text{ kJ}) = +393.5$ kJ
$CS_2(l) + 3\,Cl_2(g) \longrightarrow CCl_4(l) + S_2Cl_2(l)$	$\Delta H = -284$ kJ

57.

$CH_4(g) + CO_2(g) \longrightarrow 2\,CO(g) + 2\,H_2(g)$	$\Delta H = +206$ kJ
$2\,CH_4(g) + 2\,H_2O(g) \longrightarrow 2\,CO(g) + 6\,H_2(g)$	$\Delta H = 2\,(+247 \text{ kJ}) = +494$ kJ
$CH_4(g) + 2\,O_2(g) \longrightarrow CO_2(g) + 2\,H_2O(g)$	$\Delta H = -802$ kJ
$4\,CH_4(g) + 2\,O_2(g) \longrightarrow 4\,CO(g) + 8\,H_2(g)$	$\Delta H = -102$ kJ

÷ 4 produces $CH_4(g) + \frac{1}{2}\,O_2(g) \longrightarrow CO(g) + 2\,H_2(g)$ $\quad \Delta H = -25.5$ kJ

58. The thermochemical combustion reactions follow.

$C_4H_6(g) + \frac{11}{2}\,O_2(g) \longrightarrow 4\,CO_2(g) + 3\,H_2O(l)$	$\Delta H = -2540.2$ kJ
$C_4H_{10}(g) + \frac{13}{2}\,O_2(g) \longrightarrow 4\,CO_2(g) + 5\,H_2O(l)$	$\Delta H = -2877.6$ kJ
$H_2(g) + \frac{1}{2}\,O_2(g) \longrightarrow H_2O(l)$	$\Delta H = -285.85$ kJ

Then these equations are combined in the following manner.

$C_4H_6(g) + \frac{11}{2}\,O_2(g) \longrightarrow 4\,CO_2(g) + 3\,H_2O(l)$	$\Delta H = -2540.2$kJ
$4\,CO_2(g) + 5\,H_2O(l) \longrightarrow C_4H_{10}(g) + \frac{13}{2}\,O_2(g)$	$\Delta H = -(-2877.6 \text{ kJ}) = +2878.6$ kJ
$2\,H_2(g) + O_2(g) \longrightarrow 2\,H_2O(l)$	$\Delta H = 2\,(-285.8 \text{ kJ}) = -571.6$ kJ
$C_4H_6(g) + 2\,H_2(g) \longrightarrow C_4H_{10}(g)$	$\Delta H = -234.2$ kJ

Standard Enthalpies (Heats) of Formation

59. The compounds that have negative standard enthalpies of formation are more stable than the elements from which they are made. Those that have positive standard enthalpies of formation are less stable than the elements. It is unlikely that many compounds would have zero standard enthalpy of formation. Such compounds would be exactly as stable as the elements from which they are made.

60. The correct answer is (b): "The standard enthalpy of formation of $CO_2(g)$ is the standard enthalpy of combustion of C(graphite)." The formation reaction for $CO_2(g)$ is the reaction in which one mol $CO_2(g)$ is formed from 1 mole of C(graphite) and 1 mole $O_2(g)$; that is: $C(graphite) + O_2(g) \longrightarrow CO_2(g)$. This is the combustion reaction of graphite. (a) This reaction does not have an enthalpy change of zero. (c) *and* (d) Carbon monoxide is not involved at all in the formation reaction.

61. $\Delta H° = 4\,\Delta H_f°[HCl(g)] + \Delta H_f°[O_2(g)] - 2\,\Delta H_f°[Cl_2(g)] - 2\,\Delta H_f°[H_2O(l)]$
$= 4\,(-92.31) + (0.00) - 2\,(0.00) - 2\,(-285.8) = +202.4 \text{ kJ}$

62. $\Delta H° = 2\,\Delta H_f°[Fe(s)] + 3\,\Delta H_f°[CO_2(g)] - \Delta H_f°[Fe_2O_3(s)] - 3\,\Delta H_f°[CO(g)]$
$= 2\,(0.00) + 3\,(-393.5) - (-824.2) - 3\,(-110.5) = -24.8 \text{ kJ}$

63. Balanced equation: $C_2H_5OH(l) + 3\,O_2(g) \longrightarrow 2\,CO_2(g) + 3\,H_2O(l)$
$\Delta H° = 2\,\Delta H_f°[CO_2(g)] + 3\,\Delta H_f°[H_2O(l)] - \Delta H_f°[C_2H_5OH(l)] - 3\Delta H_f°[O_2(g)]$
$= 2\,(-393.5) + 3\,(-285.8) - (-277.7) - 3\,(0.00) = -1366.7 \text{ kJ}$

64. First determine the value of $\Delta H_f°$ for $C_5H_{12}(l)$.
Balanced combustion equation: $C_5H_{12}(l) + 8\,O_2(g) \longrightarrow 5\,CO_2(g) + 6\,H_2O(l)$
$\Delta H° = -3509 \text{ kJ} = 5\,\Delta H_f°[CO_2(g)] + 6\,\Delta H_f°[H_2O(l)] - \Delta H_f°[C_5H_{12}(l)] - 8\,\Delta H_f°[O_2(g)]$
$= 5\,(-393.5) + 6\cdot(-285.8) - \Delta H_f°[C_5H_{12}(l)] - 8\,(0.00) = -3683 \text{ kJ} - \Delta H_f°[C_5H_{12}(l)]$
$\Delta H_f°[C_5H_{12}(l)] = -3683 + 3509 = -173 \text{ kJ}$
Then use this value to determine the value of $\Delta H°$ of the reaction in question.
$\Delta H° = \Delta H_f°[C_5H_{12}(l)] + 5\,\Delta H_f°[H_2O(l)] - 5\,\Delta H_f°[CO(g)] - 11\,\Delta H_f°[H_2(g)]$
$= -173 + 5\,(-285.8) - 5\,(-110.5) - 11\,(0.00) = -1050 \text{ kJ}$

65. $\Delta H° = -397.3 \text{ kJ} = \Delta H_f°[CCl_4(g)] + 4\,\Delta H_f°[HCl(g)] - \Delta H_f°[CH_4(g)] - 4\,\Delta H_f°[Cl_2(g)]$
$= \Delta H_f°[CCl_4(g)] + 4\,(-92.31) - (-74.81) - 4\,(0.00) = \Delta H_f°[CCl_4(g)] - 294.4$
$\Delta H_f°[CCl_4(g)] = -397.3 + 294.4 = -102.9 \text{ kJ}$

66. $\Delta H° = -8326 \text{ kJ} = -2\,\Delta H_f°[C_6H_{14}(l)] - 19\,\Delta H_f°[O_2(g)] + 12\,\Delta H_f°[CO_2(g)] + 14\,\Delta H_f°[H_2O(g)]$
$= -2\,\Delta H_f°[C_6H_{14}(l)] - 19\,(0.00) + 12\,(-393.5) + 14\,(-285.8) = -2\,\Delta H_f°[C_6H_{14}(l)] - 8723$
$$\Delta H_f°[C_6H_{14}(l)] = \frac{+8326 - 8723}{2} = -199 \text{ kJ/mol}$$

67. $\Delta H° = \Delta H_f°[Al(OH)_3(s)] - \Delta H_f°[Al^{3+}(aq)] - 3\,\Delta H_f°[OH^-(aq)]$
$= (-1276) - (-531) - 3\,(-230.0) = -55 \text{ kJ}$

68. $\Delta H° = \Delta H_f°[Mg^{2+}(aq)] + 2\,\Delta H_f°[NH_3(g)] + 2\,\Delta H_f°[H_2O(l)] - \Delta H_f°[Mg(OH)_2(s)] - 2\,\Delta H_f°[NH_4^+(aq)]$
$= (-466.9) + 2\,(-46.11) + 2\,(-285.8) - (-924.5) - 2\,(-132.5) = +58.8 \text{ kJ}$

69. Balanced equation: $CaCO_3(s) \longrightarrow CaO(s) + CO_2(g)$
$\Delta H° = \Delta H_f°[CaO(s)] + \Delta H_f°[CO_2(g)] - \Delta H_f°[CaCO_3(s)] = -635.1 - 393.5 - (-1207) = +178 \text{ kJ}$
$$\text{heat} = 1.35 \times 10^3 \text{ kg } CaCO_3 \times \frac{1000 \text{ g}}{1 \text{ kg}} \times \frac{1 \text{ mol } CaCO_3}{100.09 \text{ g } CaCO_3} \times \frac{178 \text{ kJ}}{1 \text{ mol } CaCO_3} = 2.40 \times 10^6 \text{ kJ}$$

70. Determine the heat of combustion: $C_4H_{10}(g) + \frac{13}{2}\,O_2(g) \longrightarrow 4\,CO_2(g) + 5\,H_2O(l)$
$\Delta H° = 4\,\Delta H_f°[CO_2(g)] + 5\,\Delta H_f°[H_2O(l)] - \Delta H_f°[C_4H_{10}(g)] - \frac{13}{2}\,\Delta H_f°[O_2(g)]$
$= 4\,(-393.5) + 5\,(-285.8) - (-125.7) - 6.5\,(0.00) = -2877.3 \text{ kJ/mol}$
Now compute the volume of gas needed, with the ideal gas equation, rearranged to: $V = nRT/P$.
$$\text{volume} = \frac{\left(5.00 \times 10^4 \text{ kJ} \times \frac{1 \text{ mol } C_4H_{10}}{2877.3 \text{ kJ}}\right) 0.08206 \frac{\text{L atm}}{\text{mol K}} (24.6 + 273.2)\text{K}}{756 \text{ mmHg} \times \frac{1 \text{ atm}}{760 \text{ mmHg}}} = 427 \text{ L } CH_4$$

FEATURE PROBLEMS

A. We first determine the energy needed in Joules.

$$\text{energy} = \left(250\ \text{mL} \times \frac{1\ \text{g}}{1\ \text{mL}}\right) \times \frac{4.2\ \text{J}}{\text{g}\ ^\circ\text{C}} \times (50\ ^\circ\text{C} - 4\ ^\circ\text{C}) = 4.8 \times 10^4\ \text{J}$$

$$\text{time} = 4.8 \times 10^4\ \text{J} \times \frac{1\ \text{s}}{700\ \text{W}} = 69\ \text{s, a bit more than one minute.}$$

B. **(1)** We plot specific heat vs. the inverse of atomic weight.

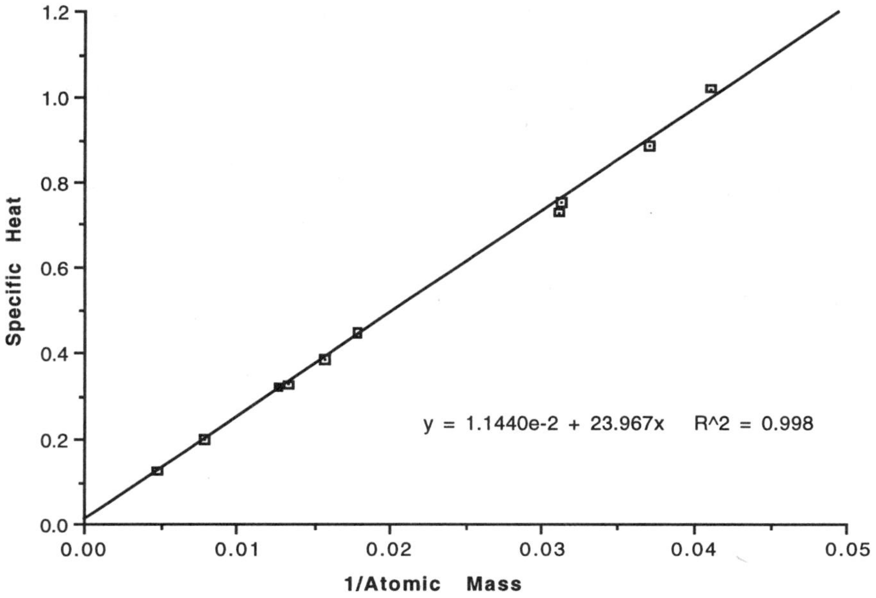

The equation is sp.ht. = 0.011440 + 23.967 ÷ (atomic mass)

(2) $0.23\ \text{J g}^{-1}\ ^\circ\text{C}^{-1} = 0.011440 + 23.967 / \text{atomic mass}$

$23.967 / \text{atomic mass} = 0.23 - 0.011440$

$$\text{atomic mass} = \frac{23.967}{0.23 - 0.01144} = 110\ \text{u}$$

Cadmium's tabulated atomic mass is 112.4 u.

(3)
$$\text{sp.ht.} = \frac{450\ \text{J}}{75.0\ \text{g} \times 15.\ ^\circ\text{C}} = 0.40\ \text{J g}^{-1}\ ^\circ\text{C}^{-1} = 0.011440 + 23.967 / \text{atomic mass}$$

$$\text{atomic mass} = \frac{23.967}{0.40 - 0.01144} = 62\ \text{u}$$

The metal most likely is Cu, 63.5 u

C. The plot's maximum is the equivalence point.

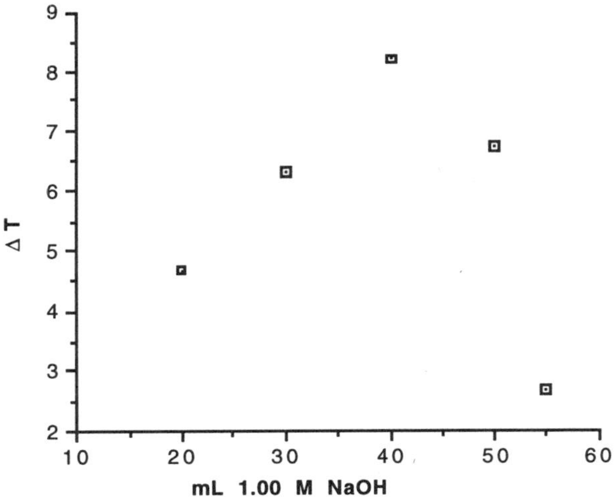

(1) The equivalence point occurs with 45.0 mL of 1.00 M NaOH(aq) [45.0 mmol NaOH] and 15.0 mL of 1.00 M citric acid [15.0 mmol citric acid].

(2) Heat is a product of reaction, just as chemical species (products) are. Products are maximized at the exact stoichiometric proportions. Since each reaction mixture has the same volume, and thus about the same mass to heat, the temperature also is a maximum at this point.

(3) $H_3C_6H_5O_7(s) + 3\ OH^-(aq) \longrightarrow 3\ H_2O + C_6H_5O_7^{3-}(aq)$

D. The reactions, and their temperature changes, are as follows.

(1st) $NH_3(\text{conc. aq}) + HCl(aq) \longrightarrow NH_4Cl(aq)$ $\Delta t = (35.8 - 23.8)\ °C = 12.0\ °C$

(2nd,a) $NH_3(\text{conc. aq}) \longrightarrow NH_3(g)$ $\Delta t = (13.2 - 19.3)\ °C = -6.1\ °C$

(2nd,b) $NH_3(g) + HCl(aq) \longrightarrow NH_4Cl(aq)$ $\Delta t = (42.9 - 23.8)\ °C = 19.1\ °C$

The sum of reactions (2nd, a) + (2nd,b) produces the same change as reaction (1st). We now compute the heat absorbed by the surroundings for each. Hess's law is demonstrated if $\Delta H_1 = \Delta H_{2a} + \Delta H_{2b}$, where in each case $\Delta H = -q$.

$q_1 = \{[(100.0\text{ mL} + 8.00\text{ mL}) \times 1.00\text{ g/mL}]\ 4.18\text{ J g}^{-1}\ °C^{-1} \times 12.0\ °C\} = 5.42 \times 10^3\text{ J} = -\Delta H_1$

$q_{2a} = [(100.0\text{ mL} \times 1.00\text{ g/mL})\ 4.18\text{ J g}^{-1}\ °C^{-1} \times (-6.1\ °C)] = -2.54 \times 10^3\text{ J} = -\Delta H_{2a}$

$q_{2b} = [(100.0\text{ mL} \times 1.00\text{ g/mL})\ 4.18\text{ J g}^{-1}\ °C^{-1} \times (19.1\ °C)] = 7.98 \times 10^3\text{ J} = -\Delta H_{2b}$

$\Delta H_{2a} + \Delta H_{2b} = +2.54 \times 10^3\text{ J} - 7.98 \times 10^3\text{ J} = -5.44 \times 10^3\text{ J} \approx -5.42 \times 10^3\text{ J} = \Delta H_1$

8 THE ATMOSPHERIC GASES AND HYDROGEN

PRACTICE EXAMPLES

1A By volume, the atmosphere is 0.934% Ar.

$$\text{volume of air} = 5.00 \text{ L Ar} \times \frac{100.00 \text{ L air}}{0.934 \text{ L Ar}} = 535 \text{ L air}$$

1B Calculate the volume of $CO_2(g)$, and then the volume of air. The atmosphere is 0.036% $CO_2(g)$ by volume.

$$\text{volume of air} = \frac{nRT}{P} = \frac{\left(5.00 \text{ g } CO_2 \times \frac{1 \text{ mol } CO_2}{44.01 \text{ g } CO_2}\right) \times 0.08206 \frac{\text{L atm}}{\text{mol K}} \times 298 \text{ K}}{1 \text{ atm}} \times \frac{100.00 \text{ L air}}{0.036 \text{ L } CO_2}$$

$= 7.7 \times 10^3$ L air

2A First substitute chemical formulas for names. $PbO_2(s) + HNO_3(aq) \longrightarrow Pb(NO_3)_2(aq) + O_2(g) + H_2O$

Balance by inspection. The oxidation and reduction reactants are in the same substance, $PbO_2(s)$.

$PbO_2(s) + 2\,H^+(aq) \longrightarrow Pb^{2+}(aq) + H_2O + \frac{1}{2}O_2(g)$

$2\,PbO_2(s) + 4\,HNO_3(aq) \longrightarrow 2\,Pb(NO_3)_2(aq) + O_2(g) + 2\,H_2O(l)$

2B First substitute chemical formulas for names. $Zn(s) + HNO_3(aq) \longrightarrow Zn(NO_3)_2(aq) + NH_4NO_3(aq) + H_2O$

Then write the two skeleton half-equations: $Zn(s) \longrightarrow Zn^{2+}(aq)$ $\quad NO_3^-(aq) \longrightarrow NH_4^+(aq)$

Balance the nitrogen half-equation for oxygen: $NO_3^-(aq) \longrightarrow NH_4^+(aq) + 3\,H_2O$

and for hydrogen: $10\,H^+(aq) + NO_3^-(aq) \longrightarrow NH_4^+(aq) + 3\,H_2O$

Balance the two half-equations for charge with electrons and then combine them to produce the net ionic equation. Finally, add in the nitrate spectator ions.

Oxidation: $Zn(s) \longrightarrow Zn^{2+}(aq) + 2\,e^-$ $\quad \times 4$

Reduction: $10\,H^+(aq) + NO_3^-(aq) + 8\,e^- \longrightarrow NH_4^+(aq) + 3\,H_2O$

Net: $4\,Zn(s) + 10\,H^+(aq) + NO_3^-(aq) \longrightarrow 4\,Zn^{2+}(aq) + NH_4^+(aq) + 3\,H_2O$

Spectators: $9\,NO_3^-(aq) \longrightarrow 9\,NO_3^-(aq)$

Final: $4\,Zn(s) + 10\,HNO_3^-(aq) \longrightarrow 4\,Zn(NO_3)_2(aq) + NH_4NO_3(aq) + 3\,H_2O$

3A As we learned in Chapter 5, half-equations for reactions in basic solution are initially balanced as if they occur in acidic solution. Both of the products in this case are composed of the elements of water. It is easiest to begin with skeleton half-equations that contain only products.

Skeletons: $\longrightarrow H_2(g)$ at cathode $\qquad \longrightarrow O_2(g)$ at anode

Balance oxygens: $\longrightarrow H_2(g)$ $\qquad 2\,H_2O \longrightarrow O_2(g)$

Balance hydrogens: $2\,H^+(aq) \longrightarrow H_2(g)$ $\qquad 2\,H_2O \longrightarrow O_2(g) + 4\,H^+(aq)$

Balance charge: $2\,H^+(aq) + 2\,e^- \longrightarrow H_2(g)$ $\qquad 2\,H_2O \longrightarrow O_2(g) + 4\,H^+(aq) + 4\,e^-$

Add OH⁻'s: $2\,OH^-(aq) + 2\,H^+(aq) + 2\,e^- \longrightarrow H_2(g) + 2\,OH^-(aq)$

$2\,H_2O + 4\,OH^-(aq) \longrightarrow O_2(g) + 4\,H^+(aq) + 4\,OH^-(aq) + 4\,e^-$

Simplify: $2\,H_2O + 2\,e^- \xrightarrow[\text{reduction}]{\text{cathode}} H_2(g) + 2\,OH^-(aq)$

$4\,OH^-(aq) \xrightarrow[\text{oxidation}]{\text{anode}} O_2(g) + 2\,H_2O + 4\,e^-$

3B First substitute chemical formulas for names. $Li_2O_2(s) + CO_2(g) \longrightarrow Li_2CO_3(s) + O_2(g)$
Simply use "2" as the stoichiometric coefficient of all species except $O_2(g)$.
$2\ Li_2O_2(s) + 2\ CO_2(g) \longrightarrow 2\ Li_2CO_3(s) + O_2(g)$ Oxygen has an oxidation state of –1 in $Li_2O_2(s)$, and –2 in $Li_2CO_3(s)$ and 0 in $O_2(g)$. Thus $Li_2O_2(s)$ is both an oxidizing agent and a reducing agent.

REVIEW QUESTIONS

1. (a) Allotropy refers to an element existing in two forms at the same temperature and pressure, such as $O_2(g)$ and $O_3(g)$.
   (b) Electrolysis refers to causing a chemical change by passing electric current through a material.
   (c) A chlorofluorocarbon is a compound, derived from a hydrocarbon, that contains the elements carbon, chlorine, and fluorine. These compounds are good to excellent refrigerants, but their release into the atmosphere is implicated in the destruction of the ozone layer in the upper atmosphere.
   (d) A nonstoichiometric compound is one in which the ratio of moles of the constituent elements is not a ratio of small whole numbers.

2. (a) A noble gas is one of the elements in periodic table family 8A: He, Ne, Ar, Kr, Xe, and Rn. These elements are called "noble" because they form few compounds with the other elements. Like "nobility," they do not associate with the other elements.
   (b) Water gas reactions are chemical reactions in which a combustible fuel is made (usually industrially) using water as one reactant.
   (c) In a hydrogenation reaction, the element hydrogen is chemically combined with another compound.
   (d) The "greenhouse" explanation of global warming is that gases such as CO_2 absorb infrared (heat) radiation coming from the Earth's surface, preventing it from escaping into space. This trapped heat will cause the temperature of the Earth to rise abnormally rapidly.

3. (a) The troposphere is the first 12 km of the Earth's atmosphere, the region in which we live and within which weather occurs. The stratosphere is the next 12 to 50 km above the Earth's surface.
   (b) An allotrope is a form of an element that differs from another form in physical and chemical properties. An isotope is an alternative version of an atom that differs in the neutron number.
   (c) A "fuel lean" mixture in an engine is one in which there is more air than is needed to completely burn the fuel. In a "fuel rich" mixture, there is an abundance of fuel and insufficient air.
   (d) A metallic hydride is composed of the hydride ion and an active metal cation; it is a stoichiometric compound. A nonmetallic hydride is composed of hydrogen and a less active metal. It actually consists of hydrogen atoms in the voids present in the metallic lattice. It is a nonstoichiometric compound.

4. (a) O_3 ozone (b) N_2O dinitrogen monoxide (nitrous oxide)
   (c) KO_2 potassium superoxide (d) CaH_2 calcium hydride
   (e) Mg_3N_2 magnesium nitride (f) K_2CO_3 potassium carbonate
   (g) $(NH_4)H_2PO_4$ ammonium dihydrogen phosphate

5. (a) ammonia: NH_3 (b) coke: almost pure C
   (c) urea: $CO(NH_2)_2$ (d) limestone: principally $CaCO_3$
   (e) synthesis gas: a mixture of CO(g) and $H_2(g)$, produced by steam reforming a hydrocarbon.

6. (a) NO_2 has N in O.S. = +4. (b) KO_2 has O in O.S. = $-1/2$.
   (c) N_2O has N in O.S. = +1. (d) CaH_2 has H in O.S. = –1. (e) CO has C in O.S. = +2.

7. (a) Small quantities of $O_2(g)$ can be prepared either by the electrolysis of water, or by the gentle heating of $KClO_3(s)$ in the presence of a $MnO_2(s)$ catalyst.
   $2\ H_2O(aq) \xrightarrow{\text{electrolysis}} 2\ H_2(g) + O_2(g)$ $2\ KClO_3(s) \xrightarrow{\text{heat, MnO2}} 2\ KCl(s) + 3\ O_2(g)$
   (b) Small quantities of $N_2O(g)$ can be produced by the thermal decomposition of ammonium nitrate.
   $NH_4NO_3(s) \xrightarrow{200\text{–}260°C} N_2O(g) + 2\ H_2O(g)$
   (c) Small quanitites of $H_2(g)$ can be produced by the reaction of a moderately active metal with a strong acid: $Zn(s) + 2\ HCl(aq) \longrightarrow ZnCl_2(aq) + H_2(g)$
   (d) Small quantities of $CO_2(g)$ can be produced by the reaction of a metal carbonate with a strong acid.
   $CaCO_3(s) + 2\ HCl(aq) \longrightarrow CaCl_2(aq) + H_2O(l) + CO_2(g)$

8. Helium occurs in the atmosphere to the extent of 0.000524 percent by volume.

$$\text{air volume} = 5.00\text{ L He} \times \frac{100.00\text{ L air}}{0.000524\text{ L He}} = 9.54 \times 10^5\text{ L air}$$

9. **(a)** $LiH(s) + H_2O \longrightarrow Li^+(aq) + OH^-(aq) + H_2(g)$ **(b)** $C(s) + H_2O(g) \xrightarrow{\Delta} CO(g) + H_2(g)$
(c) $3\ NO_2(g) + H_2O(l) \longrightarrow 2\ HNO_3(aq) + NO(g)$

10. **(a)** $Mg(s) + 2\ HCl(aq) \longrightarrow MgCl_2(aq) + H_2(g)$ **(b)** $NH_3(g) + HNO_3(aq) \longrightarrow NH_4NO_3(aq)$
(c) $MgCO_3(s) + 2\ HCl(aq) \longrightarrow MgCl_2(aq) + H_2O + CO_2(g)$
(d) $NaHCO_3(s) + HC_2H_3O_2(aq) \longrightarrow NaC_2H_3O_2(aq) + H_2O + CO_2(g)$

11. Aqueous sulfuric acid is $H_2SO_4(aq)$; aqueous ammonia is $NH_3(aq)$. The equation for their complete neutralization is: $H_2SO_4(aq) + 2\ NH_3(aq) \longrightarrow (NH_4)_2SO_4(aq)$

12. We can identify oxidizing and reducing agents by changes in oxidation state. The oxidation state of one element in an oxidizing agent is lowered, while the oxidation state of one element in a reducing agent is raised.

(8.4) $4\ NH_3(g) + 5\ O_2(g) \xrightarrow{450°C,\ Pt} 4\ NO(g) + 6\ H_2O(g)$

The oxidation state of N = –3 on the left- and = +2 on the right-hand side of this equation; the O.S. of N increases. $NH_3(g)$ is the reducing agent. The oxidation state of O = 0 on the left- and = –2 on the right-hand side of this equation; the O.S. of O decreases. $O_2(g)$ is the oxidizing agent.

(8.8) $3\ NO_2(g) + H_2O(l) \longrightarrow 2\ HNO_3(aq) + NO(g)$

The oxidation state of N = +4 on the left- and = +5 in $HNO_3(aq)$ on the right-hand side of this equation; the O.S. of N increases. $NO_2(g)$ is the reducing agent. The oxidation state of N = +4 on the left- and = +2 in NO(g) on the right-hand side of this equation; the O.S. of O decreases. $NO_2(g)$ is also the oxidizing agent. This is a disproportionation reaction.

(8.16) $C_8H_{18}(l) + \frac{25}{2}\ O_2(g) \longrightarrow 8\ CO_2(g) + 9\ H_2O(l)$

The oxidation state of C = $-\frac{9}{4}$ on the left- and = +4 on the right-hand side of this equation; the O.S. of C increases. $C_8H_{18}(l)$ is the reducing agent. The oxidation state of O = 0 on the left- and = –2 on the right-hand side of this equation; the O.S. of oxygen decreases. $O_2(g)$ is the oxidizing agent.

13. The oxide of iron with an oxidation state of +3 is $Fe_2O_3(s)$: $Fe_2O_3(s) + 3\ H_2(g) \longrightarrow 2\ Fe(s) + 3\ H_2O(g)$

14. We assume that copper is oxidized to copper(II) ion. The skeleton half-equations are as follows.

$Cu(s) \longrightarrow Cu^{2+}(aq)$ $NO_3^-(aq) \longrightarrow NO(g)$

The nitrogen-containing half-equation has its elements balanced in two steps, first for O and then for H.

$NO_3^-(aq) \longrightarrow NO(g) + 2\ H_2O$ $NO_3^-(aq) + 4\ H^+(aq) \longrightarrow NO(g) + 2\ H_2O$

The two half-equations then are balanced for charge with electrons and combined to produce the net ionic equation. Finally, nitrate spectator ions are added to produce the balanced equation.

Oxidation:	$Cu(s) \longrightarrow Cu^{2+}(aq) + 2\ e^-$	$\times 3$
Reduction:	$NO_3^-(aq) + 4\ H^+(aq) + 3\ e^- \longrightarrow NO(g) + 2\ H_2O$	$\times 2$
Net:	$3\ Cu(s) + 2\ NO_3^-(aq) + 8\ H^+(aq) \longrightarrow 3\ Cu^{2+}(aq) + 2\ NO(g) + 4\ H_2O$	

15. The gaseous thermal decomposition product is given as the last product in each of these chemical equations.

$2\ KClO_3(s) \xrightarrow{heat,\ MnO2} 2\ KCl(s) + 3\ O_2(g)$ Oxygen gas is produced.

$CaCO_3(s) \xrightarrow{\Delta} CaO(s) + CO_2(g)$ Carbon dioxide gas is formed.

$NH_4NO_3(s) \xrightarrow{200–260°C} 2\ H_2O(g) + N_2O(g)$ Dinitrogen monoxide gas is formed.

16. $\text{mass } NO_x = 75 \times 10^9\text{ gal} \times \frac{15\text{ mi}}{1\text{ gal}} \times \frac{5\text{ g } NO_x}{1\text{ mi}} \times \frac{1\text{ kg}}{1000\text{ g}} = 6 \times 10^9\text{ kg } NO_x$

17. He is found in natural gas deposits principally because alpha particles are produced during natural radioactive decay processes. These alpha particles are ^4He nuclei; they obtain two electrons from the surrounding material to become helium atoms. This gaseous helium then accumulates with the natural gas trapped beneath the earth. Although other noble gases are produced by radioactive decay—notably ^{40}Ar—they are not produced in the large quantities that helium is.

18. The correct answer is natural gas. H accounts for $4/16 = 25\%$ of the mass in CH_4, about $2/18 = 11\%$ of the mass of seawater (which is principally H_2O), about $3/17 = 18\%$ of the mass of ammonia (NH_3), and almost none of the mass of the atmosphere.

19. The reaction is $CaH_2(s) + 2\ H_2O(l) \longrightarrow Ca(OH)_2(aq) + 2\ H_2(g)$
First calculate the amount of $H_2(g)$ needed and then the mass of $CaH_2(s)$ required.

$$\text{mol } H_2 = \frac{PV}{RT} = \frac{722 \text{ mmHg} \times \frac{1 \text{ atm}}{760 \text{ mmHg}} \times 235 \text{ L}}{0.08206 \text{ L atm mol}^{-1} \text{ K}^{-1} \times (273.2 + 19.7) \text{ K}} = 9.29 \text{ mol } H_2$$

$$\text{mass } CaH_2 = 9.29 \text{ mol } H_2 \times \frac{1 \text{ mol } CaH_2}{2 \text{ mol } H_2} \times \frac{42.09 \text{ g } CaH_2}{1 \text{ mol } CaH_2} = 196 \text{ g } CaH_2$$

20. $NH_3(g) + \frac{3}{4} O_2(g) \longrightarrow \frac{1}{2} N_2(g) + \frac{3}{2} H_2O(g)$

$$\Delta H^\circ = \tfrac{1}{2} \Delta H_f^\circ[N_2(g)] + \tfrac{3}{2} \Delta H_f^\circ[H_2O(g)] - \Delta H_f^\circ[NH_3(g)] + \tfrac{3}{4} \Delta H_f^\circ[O_2(g)]$$
$$= 0.5 \times (0.00) + 1.5 \times (-241.8) - (-46.11) - 0.75 \times (0.00) = -316.6 \text{ kJ}$$

EXERCISES

The Atmosphere

21. We assume that the gases in the atmosphere obey the ideal gas law, $PV = nRT$. Then the amount in moles of a particular gas will be given by $n = \frac{PV}{RT}$. Since all of the gases in the mixture are at the same temperature and subjected to the same total pressure, their amounts in moles are proportional to their volumes, with the same proportionality constant (P/RT) for each gas in the mixture. Therefore, volume percents and mole percents will be the same.

22. The composition of air on a mass percent basis is not the same as on a volume percent basis. The easiest way to see this is to realize that volume percents and mole percents are equal (Exercise 20), and that different gases have different mole weights. The gases with the heavier mole weights—CO_2, Ar, and O_2—will have a higher percent by mass than by volume. The calculation for the four most abundant gases in air, based on 100.00 moles of air, follows.

$$\text{mass } N_2 = 78.084 \text{ mol } N_2 \times \frac{28.0135 \text{ g } N_2}{1 \text{ mol } N_2} = 2187.4 \text{ g } N_2$$

$$\text{mass } O_2 = 20.946 \text{ mol } O_2 \times \frac{31.9988 \text{ g } O_2}{1 \text{ mol } O_2} = 670.25 \text{ g } O_2$$

$$\text{mass Ar} = 0.934 \text{ mol Ar} \times \frac{39.95 \text{ g Ar}}{1 \text{ mol Ar}} = 37.3 \text{ g} \qquad \text{mass } CO_2 = 0.036 \text{ mol } CO_2 \times \frac{44.01 \text{ g } CO_2}{1 \text{ mol } CO_2} = 1.6 \text{ g}$$

$$\text{mass \% } N_2 = \frac{2187.4 \text{ g } N_2}{2187.4 \text{ g } N_2 + 670.25 \text{ g } O_2 + 37.3 \text{ g Ar} + 1.6 \text{ g } CO_2} \times 100\% = \frac{2187.4 \text{ g } N_2}{2896.6 \text{ g total}} \times 100\%$$

$$= 75.516 \% \ N_2 \qquad O_2 \text{ mass\%} = \frac{670.25 \text{ g } O_2}{2896.6 \text{ g total}} \times 100\% = 23.139\% \ O_2$$

$$\text{Ar mass\%} = \frac{37.3 \text{ g Ar}}{2896.6 \text{ g total}} \times 100\% = 1.29\% \text{ Ar} \qquad CO_2 \text{ mass\%} = \frac{1.6 \text{ g } CO_2}{2896.6 \text{ g total}} \times 100\% = 0.056\% \ CO_2$$

Nitrogen

23. **(a)** The Haber-Bosch process is the principal artificial method of fixing atmospheric nitrogen.
$N_2(g) + 3\ H_2(g) \rightleftharpoons 2\ NH_3(g)$

(b) The first step of the Ostwald process: $4\ NH_3(g) + 5\ O_2(g) \xrightarrow{450°C, Pt} 4\ NO(g) + 6\ H_2O(g)$

(c) The second and third steps of the Ostwald process: $2\ NO(g) + O_2(g) \longrightarrow 2\ NO_2(g)$
$3\ NO_2(g) + H_2O(l) \longrightarrow 2\ HNO_3(aq) + NO(g)$

24. **(a)** $2\ NH_4NO_3(s) \xrightarrow{400°C} 2\ N_2(g) + O_2(g) + 4\ H_2O(g)$ The equation was balanced by inspection, by realizing first that each mole of $NH_4NO_3(s)$ would produce 1 mol $N_2(g)$ and 2 mol $H_2O(l)$, with 1 mol O [or $1/2$ mol $O_2(g)$] remaining. Each coefficient was then doubled.

(b) $NaNO_3(s) \longrightarrow NaNO_2(s) + 1/2\ O_2(g)$ or $2\ NaNO_3(s) \longrightarrow 2\ NaNO_2(s) + O_2(g)$

(c) $Pb(NO_3)_2(s) \longrightarrow PbO(s) + 2\ NO_2(g) + {}^1/_2\ O_2(g)$
$2\ Pb(NO_3)_2(s) \longrightarrow 2\ PbO(s) + 4\ NO_2(g) + O_2(g)$

25. Decomposition of $HNO_3(l)$: $HNO_3(l) \longrightarrow N_2O_4(g) + H_2O(l) + O_2(g)$
Although this equation can be balanced by the half-reaction method, it also can be balanced by inspection. First notice that all of the N is present in $N_2O_4(g)$ in the products. This implies that there should be two $HNO_3(aq)$: $2\ HNO_3(l) \longrightarrow N_2O_4(g) + H_2O(l) + O_2(g)$ With this addition, N and H are balanced, and we need to only consider balancing oxygen. There are 6 O's on the left and 7 O's on the right, an imbalance that can be corrected either by making ${}^1/_2$ the coefficient of $O_2(g)$ or by doubling all of the other coefficients and leaving the $O_2(g)$ coefficient alone: $4\ HNO_3(l) \longrightarrow 2\ N_2O_4(g) + 2\ H_2O(l) + O_2(g)$

26. Begin with the chemical formulas of the species involved: $Na_2CO_3(aq) + O_2(g) + NO(g) \longrightarrow NaNO_2(aq)$
Oxygen is reduced and nitrogen (in NO) is oxidized. We use the ion-electron method.

two couples:	$NO(g) \longrightarrow NO_2^-(aq)$ AND	$O_2 \longrightarrow$ products
balance oxygens:	$H_2O + NO \longrightarrow NO_2^-$	$O_2 \longrightarrow 2\ H_2O$
balance hydrogens:	$H_2O + NO \longrightarrow NO_2^- + 2\ H^+$	$4\ H^+ + O_2 \longrightarrow 2\ H_2O$
balance charge:	$H_2O + NO \longrightarrow NO_2^- + 2\ H^+ + e^-$	$4\ e^- + 4\ H^+ + O_2 \longrightarrow 2\ H_2O$
combine:	$4\ H_2O + 4\ NO + 4\ H^+ + O_2 \longrightarrow 2\ H_2O + 4\ NO_2^- + 8\ H^+$	
simplify:	$2\ H_2O + 4\ NO + O_2 \longrightarrow 4\ NO_2^- + 4\ H^+$	
add spectator ions:	$4\ Na^+ + 2\ CO_3^{2-} \longrightarrow 4\ Na^+ + 2\ CO_3^{2-}$	
	$2\ Na_2CO_3 + 2\ H_2O + 4\ NO + O_2 \longrightarrow 4\ NaNO_2 + 2\ \text{“}H_2CO_3\text{”}\ (2\ H_2O + 2\ CO_2)$	
simplify:	$2\ Na_2CO_3(aq) + 4\ NO(g) + O_2(g) \longrightarrow 4\ NaNO_2(aq) + 2\ CO_2(g)$	

27. We recall that 1 mol of gas occupies 22.414 L at STP.

$$\text{mass } N_2 = 9.39 \times 10^{11}\ \text{ft}^3 \times \left(\frac{12\ \text{in.}}{1\ \text{ft}} \times \frac{2.54\ \text{cm}}{1\ \text{in.}}\right)^3 \times \frac{1\ \text{L}}{1000\ \text{cm}^3} \times \frac{1\ \text{mol } N_2}{22.414\ \text{L}} \times \frac{28.01\ \text{g } N_2}{1\ \text{mol}} \times \frac{1\ \text{kg}}{1000\ \text{g}} = 3.32 \times 10^{10}\ \text{kg } N_2$$

28.
$$\text{mass\% } HNO_3 = \frac{15\ \text{mol } HNO_3 \times \dfrac{63.01\ \text{g } HNO_3}{1\ \text{mol } HNO_3}}{1\ \text{L soln} \times \dfrac{1000\ \text{mL}}{1\ \text{L}} \times \dfrac{1.41\ \text{g}}{1\ \text{mL soln}}} \times 100\% = 67\%\ HNO_3$$

Oxygen

29. **(a)** $2\ HgO(s) \xrightarrow{\Delta} 2\ Hg(l) + O_2(g)$ **(b)** $2\ KClO_4(s) \xrightarrow{\Delta} 2\ KClO_3(s) + O_2(g)$

30. **(a)** $O_3(g) + 2\ I^-(aq) + 2\ H^+(aq) \longrightarrow I_2(aq) + H_2O + O_2(g)$
(b) $S(s) + H_2O(l) + 3\ O_3(g) \longrightarrow H_2SO_4(aq) + 3\ O_2(g)$
(c) $2\ [Fe(CN)_6]^{4-}(aq) + O_3(g) + H_2O \longrightarrow 2\ [Fe(CN)_6]^{3-}(aq) + O_2(g) + 2\ OH^-(aq)$

31. We first write the formulas of the four substances: N_2O_4, Al_2O_3, P_4O_6, CO_2. The one constant in all these substances is oxygen. If we compare amounts of substance with the same amount (in moles) of oxygen, the one with the smallest mass of the other element will have the highest percent oxygen.
3 mol N_2O_4 contains 12 mol O and 6 mol N: $6 \times 14.0 = 84.0$ g N
4 mol Al_2O_3 contains 12 mol O and 8 mol Al: $8 \times 27.0 = 216$ g Al
2 mol P_4O_6 contains 12 mol O and 8 mol P: $8 \times 31.0 = 248$ g P
6 mol CO_2 contains 12 mol O and 6 mol C: $6 \times 12.0 = 72.0$ g C
Thus, of the oxides listed, CO_2 contains the largest percent oxygen by mass.

32. $2\ NH_4NO_3(s) \xrightarrow{400°C} 2\ N_2(g) + O_2(g) + 4\ H_2O(g)$ $2\ H_2O_2(l) \longrightarrow 2\ H_2O(l) + O_2(g)$
$2\ KClO_3(s) \xrightarrow{\Delta} 2\ KCl(s) + 3\ O_2(g)$

(a) All three reactions have two moles of reactant, and the decomposition of potassium chlorate produces three moles of $O_2(g)$, compared to one mole of $O_2(g)$ for each of the others. The potassium chlorate decomposition produces the most oxygen per mole of reactant.

(b) If each reaction produced the same amount of $O_2(g)$, then the one with the lightest mass of reactant would produce the most oxygen per gram of reactant. When the first two reactions are compared, it is clear that 2 mol H_2O_2 have less mass than 2 mol NH_4NO_3. (Notice that each mol NH_4NO_3 contains the amounts of elements—2 mol H and 2 mol O—present in each mol H_2O_2 plus some more.) So now we need to compare hydrogen peroxide with potassium chlorate. Notice that 6 H_2O_2 produces the same amount of $O_2(g)$ as 2 $KClO_3$. 6 mol H_2O_2 has a mass of $34.01 \times 6 = 204.1$ g, while 2 mol $KClO_3$ has a mass of $122.2 \times 2 = 244.2$ g. Thus, H_2O_2 produces the most $O_2(g)$ per gram of reactant.

33. Recall that fraction by volume and fraction by pressure are numerically equal. Additionally, one atmosphere pressure is equivalent to 760 mmHg. We combine these two facts.

$$P[O_3] = 760 \text{ mmHg} \times \frac{0.04 \text{ mmHg of } O_3}{10^6 \text{ mmHg of atmosphere}} = 3 \times 10^{-5} \text{ mmHg}$$

34. $$P = \frac{nRT}{V} = \frac{\left(5 \times 10^{12} \text{ molecules} \times \dfrac{1 \text{ mol } O_3}{6.022 \times 10^{23} \text{ molec.}}\right) \times \dfrac{0.08206 \text{ L atm}}{\text{mol K}} \times 220 \text{ K}}{1 \text{ cm}^3 \times \dfrac{1 \text{ L}}{1000 \text{ cm}^3}}$$

$$= 1._5 \times 10^{-7} \text{ atm} \times \frac{760 \text{ mmHg}}{1 \text{ atm}} = 1._1 \times 10^{-4} \text{ mmHg}$$

35. The electrolysis reaction is $2\,H_2O(l) \xrightarrow{\text{electrolysis}} 2\,H_2(g) + O_2(g)$ In this reaction, 2 moles of $H_2(g)$ are produced for each mole of $O_2(g)$, by the law of combining volumes, we would expect the volume of hydrogen to be twice the volume of oxygen produced. (Actually the volumes are not exactly in the ratio of 2:1 because of the different solubilities of oxygen and hydrogen in water.)

36. The electrolysis reaction is $2\,H_2O(l) \xrightarrow{\text{electrolysis}} 2\,H_2(g) + O_2(g)$ In this reaction, 2 moles of water are decomposed to produce each mole of $O_2(g)$. We use the data in the problem to determine the amount of $O_2(g)$ produced, which we then convert to the mass of H_2O needed.

$$\text{mass } H_2O(l) = \frac{\left(736.7 \text{ mmHg} \times \dfrac{1 \text{ atm}}{760 \text{ mmHg}}\right) \times \left(22.83 \text{ mL} \times \dfrac{1 \text{ L}}{1000 \text{ mL}}\right)}{\dfrac{0.08206 \text{ L atm}}{\text{mol K}} \times (25.0 + 273.2) \text{ K}} \times \frac{2 \text{ mol } H_2O}{1 \text{ mol } O_2}$$

$$\times \frac{18.02 \text{ g } H_2O}{1 \text{ mol } H_2O} = 0.03259 \text{ g } H_2O = 32.59 \text{ mg } H_2O \text{ decomposed}$$

The Noble Gases

37. First we use the ideal gas law to determine the amount in moles of argon.

$$n = \frac{PV}{RT} = \frac{145 \text{ atm} \times 55 \text{ L}}{0.08206 \text{ L atm mol}^{-1} \text{ K}^{-1} \times 299 \text{ K}} = 3.3 \times 10^2 \text{ mol Ar}$$

$$\text{no. L air} = 330. \text{ mol Ar} \times \frac{22.414 \text{ L Ar STP}}{1 \text{ mol Ar}} \times \frac{100.000 \text{ L air}}{0.934 \text{ L Ar}} = 7.9 \times 10^5 \text{ L air}$$

38. First we compute the volume of 5.00 g He at STP, and then the volume of natural gas.

$$\text{natural gas volume} = 5.00 \text{ g He} \times \frac{1 \text{ mol He}}{4.003 \text{ g He}} \times \frac{22.414 \text{ L}}{1 \text{ mol He}} \times \frac{100 \text{ L air}}{8 \text{ L He}} = 3._5 \times 10^2 \text{ L air at STP}$$

39. 1 mol of the mixture at STP occupies a volume of 22.414 L. It contains 0.79 mol He and 0.21 mol O_2.

$$\text{STP density} = \frac{\text{mass}}{\text{volume}} = \frac{0.79 \text{ mol He} \times \dfrac{4.003 \text{ g He}}{1 \text{ mol He}} + 0.21 \text{ mol } O_2 \times \dfrac{32.0 \text{ g } O_2}{1 \text{ mol } O_2}}{22.414 \text{ L}} = 0.44 \text{ g/L}$$

25°C is a temperature higher than STP. This condition increases the 1.00-L volume that contains 0.44 g of the mixture at STP. We calculate the expanded volume with the combined gas law.

$$V_{\text{final}} = 1.00 \text{ L} \times \frac{(25 + 273.2) \text{ K}}{273.2 \text{ K}} = 1.09 \text{ L} \qquad \text{final density} = \frac{0.44 \text{ g}}{1.09 \text{ L}} = 0.40 \text{ g/L}$$

40. We determine the apparent molar masses of each mixture by multiplying the mole fraction (numerically equal to the volume fraction) of each gas by its molar mass, and then summing these products for all gases in the mixture.

$$\mathcal{M}_{air} = \left(0.78084 \times \frac{28.01 \text{ g } N_2}{1 \text{ mol } N_2}\right) + \left(0.20946 \times \frac{32.00 \text{ g } O_2}{1 \text{ mol } O_2}\right) + \left(0.00934 \times \frac{39.95 \text{ g Ar}}{1 \text{ mol Ar}}\right) =$$

$$= 21.87 \text{ g } N_2 + 6.703 \text{ g } O_2 + 0.373 \text{ g Ar} = 28.95 \text{ g/mol air}$$

$$\mathcal{M}_{mix} = \left(0.79 \times \frac{4.003 \text{ g He}}{1 \text{ mol He}}\right) + \left(0.21 \times \frac{32.00 \text{ g } O_2}{1 \text{ mol } O_2}\right) = 3.2 \text{ g He} + 6.7 \text{ g } O_2 = 9.9 \text{ g mixture/mol}$$

In order to give the two gases the same density, the volume of the gas of smaller molar mass must be smaller by a factor equal to the ratio of the molar masses. According to Boyle's law, this means that the pressure on the less dense gas must be larger by a factor equal to a ratio of mole weights.

$$P_{mix} = \frac{28.95}{9.9} \times 1.00 \text{ atm} = 2.9 \text{ atm}$$

Carbon

41. **(a)** $2\ C_6H_{14}(l) + 19\ O_2(g) \longrightarrow 12\ CO_2(g) + 14\ H_2O(l)$

(b) $PbO(s) + CO(g) \xrightarrow{\text{heat}} Pb(s) + CO_2(g)$

(c) $2\ KOH(aq) + CO_2(g) \longrightarrow K_2CO_3(aq) + H_2O$

(d) $MgCO_3(s) + 2\ HCl(aq) \longrightarrow MgCl_2(aq) + H_2O(l) + CO_2(g)$

42. **(a)** $HC_2H_3O_2(aq) + NaHCO_3(s) \longrightarrow NaC_2H_3O_2(aq) + H_2O + CO_2(g)$

(b) $ZnO(s) + CO(g) \longrightarrow Zn(s) + CO_2(g)$ (The heat from this reaction may melt the Zn.)

(c) $2\ CO(g) + 3\ H_2(g) \longrightarrow CH_2OHCH_2OH(l)$

43. The reaction referred to is exemplified by $CaCO_3(s) + 2\ HCl(aq) \longrightarrow CaCl_2(aq) + CO_2(g) + H_2O$ the action of hydrochloric acid on calcium carbonate. Notice that no oxidation states change in this reaction; it is a double displacement reaction followed by the decomposition of "$H_2CO_3(aq)$." Since the carbon dioxide produced in this reaction is not formed from elemental carbon or by the oxidation of a carbon-containing compound, it cannot be partially oxidized. Thus, no CO(g) will form.

44. Both carbon dioxide and chlorofluorocarbons are greenhouse gases. They absorb infrared radiation, preventing this heat from leaving the Earth. But only the chlorine released by chlorofluorocarbons plays a role in the destruction of ozone in the upper atmosphere.

45. The combustion reaction: $C_8H_{18}(l) + {}^{25}/_2 O_2(g) \longrightarrow 8\ CO_2(g) + 9\ H_2O(l)$

First compute the enthalpy change for combustion.

$$\Delta H° = 8\ \Delta H_f°[CO_2(g)] + 9\ \Delta H_f°[H_2O(l)] - \Delta H_f°[C_8H_{18}(l)] - {}^{25}/_2\ \Delta H_f°[O_2(g)]$$

$$= 8\ (-393.5) + 9\ (-285.8) - (-250.0) - {}^{25}/_2\ (0.00) = -5470.\ \text{kJ/mol } C_8H_{18}$$

Then determine the heat produced from each gallon of gasoline.

$$\text{heat produced} = 1.00 \text{ gal} \times \frac{3.785 \text{ L}}{1 \text{ gal}} \times \frac{1000 \text{ mL}}{1 \text{ L}} \times \frac{0.703 \text{ g}}{1 \text{ mL}} \times \frac{1 \text{ mol } C_8H_{18}}{114.23 \text{ g } C_8H_{18}} \times \frac{5470 \text{ kJ}}{1 \text{ mol } C_8H_{18}}$$

$$= 1.27 \times 10^5 \text{ kJ of heat} \qquad (\text{Note that } \Delta H = -1.27 \times 10^5 \text{ kJ})$$

46. Let us determine the difference between the two reactions.

combustion: $C_8H_{18}(l) + {}^{25}/_2 O_2(g) \longrightarrow 8\ CO_2(g) + 9\ H_2O(l)$

–(incomplete): $7\ CO_2(g) + CO(g) + 9\ H_2O(l) \longrightarrow C_8H_{18}(l) + 12\ O_2(g)$

difference: $CO(g) + {}^{1}/_2 O_2(g) \longrightarrow CO_2(g)$

The heat the accompanies the "difference" reaction is the difference in the heat of the two reactions. To determine the percent of the maximum possible heat this is, first compute the enthalpy change for the combustion reaction.

$$\Delta H° = 8\ \Delta H_f°[CO_2(g)] + 9\ \Delta H_f°[H_2O(l)] - \Delta H_f°[C_8H_{18}(l)] - {}^{25}/_2\ \Delta H_f°[O_2(g)]$$

$$= 8\ (-393.5) + 9\ (-285.8) - (-250.0) - {}^{25}/_2\ (0.00) = -5470.\ \text{kJ/mol } C_8H_{18}$$

Then compute the enthalpy change for the "difference" reaction.

$$\Delta H° = \Delta H_f°[CO_2(g)] - \Delta H_f°[CO(g)] - {}^{1}/_2\ \Delta H_f°[O_2(g)] = (-393.5) - (-110.5) - {}^{1}/_2\ (0.00) = -283 \text{ kJ}$$

And finally compute the percent decrease that the difference reaction represents.

$$\%\text{decrease} = \frac{-283\text{ kJ}}{-5470\text{ kJ}} \times 100\% = 5.17\%\text{ decrease}$$

Hydrogen

47. **(a)** $2\,Al(s) + 6\,HCl(aq) \longrightarrow 2\,AlCl_3(aq) + 3\,H_2(g)$

(b) $C_3H_8(g) + 3\,H_2O(g) \longrightarrow 3\,CO(g) + 7\,H_2(g)$

(c) $MnO_2(s) + 2\,H_2(g) \xrightarrow{\Delta} Mn(s) + 2\,H_2O(g)$

48. **(a)** $2\,H_2O(l) \xrightarrow{\text{electricity}} O_2(g) + 2\,H_2(g)$

(b) $2\,HI(aq) + Zn(s) \longrightarrow ZnI_2(aq) + H_2(g)$ Any moderately active metal is suitable.

(c) $Mg(s) + H_2SO_4(aq) \longrightarrow MgSO_4(aq) + H_2(g)$ Any strong acid is suitable.

(d) $CO(g) + H_2O(g) \longrightarrow CO_2(g) + H_2(g)$

49. $CaH_2(s) + 2\,H_2O \longrightarrow Ca(OH)_2(aq) + 2\,H_2(g)$ $Ca(s) + 2\,H_2O \longrightarrow Ca(OH)_2(aq) + H_2(g)$
$2\,Na(s) + 2\,H_2O \longrightarrow 2\,NaOH(aq) + H_2(g)$

(a) The one of these three that produces the largest volume of $H_2(g)$ per liter of water also produces the largest amount of hydrogen gas per mole of water used. All three reactions use two moles of water and the reaction with $CaH_2(s)$ produces the greatest amount—2 moles—of hydrogen gas.

(b) We can compare three reactions that produce the same amount of hydrogen; the one that requires the smallest mass of solid produces the greatest amount of H_2 per gram of solid. The amount of hydrogen we choose is 2 moles, which means that we compare 1 mol CaH_2 (42.09 g) with 2 mol Ca (80.16 g) with 4 mol Na (91.96 g). Clearly CaH_2 produces the greatest amount of H_2 per gram of solid.

50. The balanced equation is $C_{17}H_{33}COOH(l) + H_2(g) \longrightarrow C_{17}H_{35}COOH(s)$ One mole of oleic acid requires one mole of $H_2(g)$.

$$V = \frac{nRT}{P} = \frac{1.00\text{ mol} \times 0.08206\,\frac{\text{L atm}}{\text{mol K}} \times 298\text{ K}}{752\text{ mmHg} \times \frac{1\text{ atm}}{760\text{ mmHg}}} \times \frac{0.95\text{ L produced}}{1\text{ L calculated}} = 24\text{ L } H_2(g)$$

FEATURE PROBLEMS

A. **1.** An engine with a fuel-lean mixture will produce large proportions of nitrogen oxides since there is a large amount of air (which contains nitrogen and oxygen) to the fuel. On the other hand, an engine with a fuel-rich mixture will produce a lot of unburned fuel (RH) and carbon monoxide, because of the lack of oxygen.

2. First we balance the combustion reaction. $2\,C_8H_{18}(l) + 25\,O_2(g) \longrightarrow 16\,CO_2(g) + 18\,H_2O(l)$

$$\frac{\text{air}}{\text{fuel}} = \frac{25\text{ mol }O_2}{2\text{ mol }C_8H_{18}} \times \frac{100\text{ mol air}}{20.946\text{ mol }O_2} \times \frac{29.0\text{ g}}{1\text{ mol air}} \times \frac{1\text{ mol }C_8H_{18}}{114.2\text{ g }C_8H_{18}} = 15{:}1$$

We used the average molar mass of air from Additional Exercise 51.

B. **1.** The $N_2(g)$ extracted from liquid air has some Ar(g) mixed in. Only $O_2(g)$ was removed from liquid air.

2. Because of the presence of Ar(g) [39.95 g/mol], the $N_2(g)$ [28.01 g/mol] from liquid air will have a greater density than $N_2(g)$ from nitrogen compounds.

3. Magnesium will react with nitrogen [$3\,Mg(s) + N_2(g) \longrightarrow Mg_3N_2(s)$] but not with Ar. Thus, magnesium reacts with all the nitrogen in the mixture, but leaves the relatively inert Ar(g) unreacted.

4. The "nitrogen" remaining after oxygen is extracted from each mole of air (Rayleigh's mixture) contains (0.78084 + 0.00934 =) 0.79018 mol and has a mass of (0.78084 × 28.013 g/mol N_2) + (0.00934 × 39.948 g/mol Ar) = 21.874 g N_2 + 0.373 g Ar = 22.247 g mixture. Then, the molar mass of the mixture can be computed: 22.247 g mixture / 0.79018 mol = 28.154 g/mol. Since the STP molar volume of an ideal gas is 22.414 L, we can compute the two densities.

$$d(N_2) = \frac{28.013\text{ g/mol}}{22.414\text{ L/mol}} = 1.2498\text{ g/mol} \qquad d(\text{mixture}) = \frac{28.154\text{ g/mol}}{22.414\text{ L/mol}} = 1.2561\text{ g/mol}$$

These densities differ by 0.50%.

C. The goal is to demonstrate that the three reactions result in the decomposition of water as the net reaction:

Net: $2\ H_2O \longrightarrow 2\ H_2 + O_2$ First balance each equation.

(1) $3\ FeCl_2 + 4\ H_2O \longrightarrow Fe_3O_4 + HCl + H_2$ Balance by inspection. Notice that there are 3 Fe and 4 O on the right-hand side. Then balance Cl. $3\ FeCl_2 + 4\ H_2O \longrightarrow Fe_3O_4 + 6\ HCl + H_2$

(2) $Fe_3O_4 + HCl + Cl_2 \longrightarrow FeCl_3 + H_2O + O_2$ Try the half-equation method. $Cl_2 + 2\ e^- \longrightarrow 2\ Cl^-$

But realize that $Fe_3O_4 = Fe_2O_3 \cdot FeO$; only iron(II) needs to be oxidized. $2\ FeO \longrightarrow 2\ Fe^{3+} + O_2 + 6\ e^-$

Now combine the two half-equations. $2\ FeO + 3\ Cl_2 \longrightarrow 2\ Fe^{3+} + O_2 + 6\ Cl^-$

Add in $2\ Fe_2O_3 + 12\ H^+ \longrightarrow 4\ Fe^{3+} + 6\ H_2O$

$$2\ Fe_3O_4 + 3\ Cl_2 + 12\ H^+ \longrightarrow 6\ Fe^{3+} + O_2 + 6\ Cl^- + 6\ H_2O$$

And 12 Cl^- spectators: $2\ Fe_3O_4 + 3\ Cl_2 + 12\ HCl \longrightarrow 6\ FeCl_3 + O_2 + 6\ H_2O$

(3) $FeCl_3 \longrightarrow FeCl_2 + Cl_2$ by inspection $2\ FeCl_3 \longrightarrow 2\ FeCl_2 + Cl_2$

One strategy is to consider each of the three equations and the net equation. Only equation (1) produces hydrogen; run it twice. Only equation (2) produces oxygen; run it once. Equation (3) can balance out the Cl_2 required by equation (2); run it three times.

2 × (1)	$6\ FeCl_2 + 8\ H_2O \longrightarrow 2\ Fe_3O_4 + 12\ HCl + 2\ H_2$
1 × (2)	$2\ Fe_3O_4 + 3\ Cl_2 + 12\ HCl \longrightarrow 6\ FeCl_3 + O_2 + 6\ H_2O$
3 × (3)	$6\ FeCl_3 \longrightarrow 6\ FeCl_2 + 3\ Cl_2$
Net:	$2\ H_2O \longrightarrow 2\ H_2 + O_2$

9 ELECTRONS IN ATOMS

PRACTICE EXAMPLES

1A We solve the equation $c = \lambda\nu$ for frequency. $\nu = \dfrac{2.9979 \times 10^8 \text{ m/s}}{690 \text{ nm}} \times \dfrac{10^9 \text{ nm}}{1 \text{ m}} = 4.34 \times 10^{14}$ Hz

1B Wavelength and frequency are related through the equation $c = \lambda\nu$, which can be solved for either one.

$$\lambda = \frac{c}{\nu} = \frac{2.9979 \times 10^8 \text{ m/s}}{91.5 \times 10^6 \text{ s}^{-1}} = 3.28 \text{ m}$$ Note that Hz = s^{-1}

2A X-rays have wavelengths of about 10^{-10} m, while microwaves have wavelengths of about 1 cm = 10^{-2} m. We recall that short wavelength corresponds to high frequency. Thus, the shorter wavelength X-rays have higher frequency than microwaves.

2B The colors of these light sources are, respectively: red, green, yellow, and violet. Of the colors of the visible spectrum, red has the longest wavelength and violet the shortest. The wavelength of yellow light is longer than that of green. Thus, ranked in order of decreasing wavelength, the four colors, and light sources are:

red > yellow > green > violet "stop" > "caution" > "go" > Hg lamp

3A Into the equation $E = h\nu$ can be substituted the relationship $\nu = c/\lambda$ to obtain $E = hc/\lambda$. This energy, in J/photon can then be converted to kJ/mol.

$$E = \frac{hc}{\lambda} = \frac{6.626 \times 10^{-34} \text{ J s photon}^{-1} \times 2.998 \times 10^8 \text{ m s}^{-1}}{230 \text{ nm} \times \dfrac{1 \text{ m}}{10^9 \text{ nm}}} \times \frac{6.022 \times 10^{23} \text{ photons}}{1 \text{ mol}} \times \frac{1 \text{ kJ}}{1000 \text{ J}}$$

= 520 kJ/mol

With a similar calculation one finds that 290 nm corresponds to 410 kJ/mol. Thus, the energy range is from 400 to 520 kJ/mol.

3B The equation $E = h\nu$ is solved for frequency and the two frequencies are calculated.

$$\nu = \frac{E}{h} = \frac{3.056 \times 10^{-19} \text{ J/photon}}{6.626 \times 10^{-34} \text{ J·s/photon}} = 4.612 \times 10^{14} \text{ Hz} \qquad \nu = \frac{E}{h} = \frac{4.414 \times 10^{-19} \text{ J/photon}}{6.626 \times 10^{-34} \text{ J·s/photon}} = 6.662 \times 10^{14} \text{ Hz}$$

To determine color, we calculate the wavelength of each frequency and compare with *text* Figure 9-3.

$$\lambda = \frac{c}{\nu} = \frac{2.9979 \times 10^8 \text{ m/s}}{4.612 \times 10^{14} \text{ Hz}} \times \frac{10^9 \text{ nm}}{1 \text{ m}} = 650 \text{ nm} \quad \text{orange} \qquad \lambda = \frac{c}{\nu} = \frac{2.9979 \times 10^8 \text{ m/s}}{6.662 \times 10^{14} \text{ Hz}} \times \frac{10^9 \text{ nm}}{1 \text{ m}} = 450 \text{ nm} \quad \text{indigo}$$

The colors of the spectrum that are not absorbed are what we see when we look at a plant: blue, green, and yellow. The plant appears green.

4A We solve the Rydberg equation for n, to see if we obtain an integer.

$$n = \sqrt{n^2} = \sqrt{\frac{-R_H}{E_n}} = \sqrt{\frac{-2.179 \times 10^{-18} \text{ J}}{-2.69 \times 10^{-20} \text{ J}}} = \sqrt{81.00} = 9.00$$ Thus $E_9 = -2.69 \times 10^{-20}$ J

4B It is not likely that an atomic radius would be precisely equal to an arbitrary unit of length, but let us see how close the radii are. 1 nm = 1000 pm, so we solve the following for n.

$1000 \text{ pm} = n^2\, 53 \text{ pm}$ $\qquad n = \sqrt{\dfrac{1000}{53}} = 4.3$ $\qquad$ We see that no radius is exactly 1 nm. The closest:

$r_4 = 4^2 a_0 = 16 \times 0.053 \text{ nm} = 0.85 \text{ nm}$ $\qquad r_5 = 5^2 a_0 = 25 \times 0.053 \text{ nm} = 1.3 \text{ nm}$

5A We first determine the energy difference, and then the wavelength of light corresponding to that energy.

$$\Delta E = R_H\left(\frac{1}{2^2} - \frac{1}{4^2}\right) = 2.179 \times 10^{-18} \text{ J}\left(\frac{1}{2^2} - \frac{1}{4^2}\right) = 4.086 \times 10^{-19} \text{ J}$$

$$\lambda = \frac{hc}{\Delta E} = \frac{6.626 \times 10^{-34} \text{ J s} \times 2.998 \times 10^{8} \text{ m s}^{-1}}{4.086 \times 10^{-19} \text{ J}} = 4.862 \times 10^{-7} \text{ m} = 486.2 \text{ nm}$$

5B The longest wavelength light results from the transition that spans the smallest difference in energy. Since all Lyman series emissions end with $n_f = 1$, the smallest energy transition has $n_i = 2$. From this, we obtain the value of ΔE.

$$\Delta E = R_H\left(\frac{1}{n_i^2} - \frac{1}{n_f^2}\right) = 2.179 \times 10^{-18} \text{ J}\left(\frac{1}{2^2} - \frac{1}{1^2}\right) = -1.634 \times 10^{-18} \text{ J}$$

From this energy emitted, we can obtain the wavelength of the emitted light: $\qquad \Delta E = hc/\lambda$

$$\lambda = \frac{hc}{\Delta E} = \frac{6.626 \times 10^{-34} \text{ J s} \times 2.998 \times 10^{8} \text{ m s}^{-1}}{1.634 \times 10^{-18} \text{ J}} = 1.216 \times 10^{-7} \text{ m} = 121.6 \text{ nm}$$

Of course, we could have solved this equation for $1/\lambda$, and then inverted the result.

6A Superman's de Broglie wavelength is given by the relationship $\qquad \lambda = h/mv$

$$\lambda = \frac{h}{mv} = \frac{6.626 \times 10^{-34} \text{ J·s}}{91 \text{ kg} \times \frac{1}{5} \times 2.998 \times 10^{8} \text{ m/s}} = 1.21 \times 10^{-43} \text{ m}$$

6B The de Broglie wavelength is given by the relationship $\lambda = h/mv$ which can be solved for v.

$$v = \frac{h}{m\lambda} = \frac{6.626 \times 10^{-34} \text{ J s}}{1.673 \times 10^{-27} \text{ kg} \times 0.0100 \times 10^{-9} \text{ m}} = 3.96 \times 10^{4} \text{ m/s} = 39.6 \text{ km/s}$$

We used the facts that $\text{J} = \text{kg m}^2\text{ s}^{-2}$ $\qquad \text{nm} = 10^{-9}\text{ m}$ $\qquad$ and $\qquad \text{g} = 10^{-3}\text{ kg}$

7A The value of ℓ can range from 0 to $n - 1$; in this case, $\ell = 0$ is acceptable. The value of m_ℓ is integral and ranges from $-\ell$ to $+\ell$; in this case, $m_\ell = 0$ is acceptable. Yes, an orbital can have $n = 3$, $\ell = 0$, and $m_\ell = 0$.

7B The first restriction is that ℓ must be an integer smaller than n. This restricts ℓ to the values: 2, 1, and 0. The second restriction is that the absolute value of m_ℓ must be an integer equal to or smaller than ℓ. This further restricts ℓ, and the only allowed values are: $\ell = 2$ and $\ell = 1$.

8A The magnetic quantum number, m_ℓ, is not reflected in the orbital designation. Because $\ell = 1$, this is a p orbital. Because $n = 3$, the designation is $3p$.

8B The designation $5s$ already indicates the values of two quantum numbers. The principal quantum number, $n = 5$. And the orbital quantum number, $\ell = 0$, as indicated by the designation s. The magnetic quantum number may have only the value $m_\ell = 0$

9A We can simply sum the exponents to obtain the number of electrons in the neutral atom and thus the atomic number of the element. $\qquad Z = 2 + 2 + 6 + 2 + 6 + 2 + 2 = 22$, which is the atomic number of Ti.

9B Iodine has an atomic number of 53. The first 36 electrons have the same electron configuration as Kr: $1s^2 2s^2 2p^6 3s^2 3p^6 3d^{10} 4s^2 4p^6$. The next two electrons go into the $5s$ subshell ($5s^2$), then 10 electrons fill the $4d$ subshell ($4d^{10}$), accounting for a total of 48 electrons. The last five electrons partially fill the $5p$ subshell ($5p^5$). The electron configuration of I is therefore $1s^2 2s^2 2p^6 3s^2 3p^6 3d^{10} 4s^2 4p^6 4d^{10} 5s^2 5p^5$. Each iodine atom has ten $3d$ electrons and one unpaired electron, one of the electrons in the $5p$ subshell.

10A Iron has 26 electrons, of which 18 are accounted for by the [Ar] configuration. Beyond [Ar] there are two $4s$ electrons and six $3d$ electrons, as shown in the following orbital diagram.

Fe: [Ar] $3d$ [↑↓][↑][↑][↑][↑] $\quad 4s$ [↑↓]

10B Bismuth has 83 electrons, of which 54 are accounted for by the [Xe] configuration. Beyond [Xe] there are two $6s$ electrons, fourteen $4f$ electrons, ten $5d$ electrons, and three $6p$ electrons, as shown in the following orbital diagram.

		$4f$	$5d$	$6s$	$6p$
Bi:	[Xe]	[⇅][⇅][⇅][⇅][⇅][⇅][⇅]	[⇅][⇅][⇅][⇅][⇅]	[⇅]	[↑][↑][↑]

11A **(a)** Tin is in the fifth period, which means that five electronic shells are filled or partially filled.
(b) The $3p$ subshell was filled with Ar; there are six $3p$ electrons in an atom of Sn.
(c) The outer electron configuration of Sn is [Kr] $4d^{10}\,5s^2\,5p^2$. There are no $5d$ electrons.
(d) Both of the $5p$ electrons are unpaired; there are two unpaired electrons in each atom of Sn.

11B **(a)** The $3d$ subshell was filled at Zn; each Y atom has ten $3d$ electrons.
(b) Ge is in the $4p$ row; each germanium atom has two $4p$ electrons.
(c) We would expect each Au atom to have ten $5d$ electrons and one $6s$ electron. Each Au atom should have one unpaired electron.

SUMMARIZING EXAMPLE CALCULATIONS

1. $\nu = \dfrac{c}{\lambda} = \dfrac{2.998 \times 10^8 \text{ m s}^{-1}}{12.2 \text{ cm} \times \dfrac{1 \text{ m}}{100 \text{ cm}}} = 2.46 \times 10^9 \text{ Hz}$

2. $\Delta E = h\nu = 6.626 \times 10^{-34} \text{ J s} \times 2.46 \times 10^9 \text{ Hz} = 1.63 \times 10^{-24} \text{ J}$

3. Another solution is to attempt to find values of n_i and n_f in equation (9.6) *of the text*, equation [2] here.

$$1.63 \times 10^{-24} \text{ J} = \Delta E = R_H\left(\frac{1}{n_i^2} - \frac{1}{n_f^2}\right) = 2.179 \times 10^{-18} \text{ J}\left(\frac{1}{n_i^2} - \frac{1}{n_f^2}\right)$$

$$\frac{1}{n_i^2} - \frac{1}{n_f^2} = \frac{1.63 \times 10^{-24} \text{ J}}{2.179 \times 10^{-18} \text{ J}} = 7.48 \times 10^{-7}$$

How about we guess that the two integers are 100 and 101? The result follows.

$\dfrac{1}{100^2} - \dfrac{1}{101^2} = 1.97 \times 10^{-6}$ This is somewhat larger than the result we obtained.

How about 200 and 201? $\dfrac{1}{200^2} - \dfrac{1}{201^2} = 2.48 \times 10^{-7}$ This is somewhat smaller than the result.

Thus we know that the transition occurs between two levels whose quantum numbers fall somewhere between 100 and 200. With similar further refinement, we could possibly pin down the two levels, but we have achieved our goal: such a transition *is* possible.

REVIEW QUESTIONS

1. **(a)** λ is the symbol for wavelength, the distance between like features (two peaks, for instance) of successive waves.
(b) ν is the symbol for frequency, the number of times that a particular feature of successive waves (such as a peak) passes a fixed point in a second.
(c) h is the symbol for Planck's constant, which relates the energy and frequency of radiation: $E = h\nu$.
(d) ψ is the symbol for the wavefunction, the function that contains all the information known about a given atomic or molecular system.
(e) The principal quantum number, n, is related to both the distance of an electron from the nucleus and to the electron's energy.

2. **(a)** The atomic (line) spectrum is the result of atoms being energized and then radiating energy as they lose that energy and return to lower energy states. Since only certain energy levels are permitted for each atom, the energies radiated are definite, giving rise to light of only certain wavelengths, or lines in the spectrum.
(b) The photoelectric effect occurs when light (*photo-*) shining on a surface causes electrons (*-electric*) to be emitted from that surface.
(c) A matter wave refers to the wave properties associated with matter.
(d) The Heisenberg uncertainty principle states that it is impossible to simultaneously determine the position and linear momentum of a particle to any desired degree of precision.

(e) Electron spin is that property of electrons that makes them appear as if they were spinning, or rotating, on an axis. Electrons can spin in only one of two states: clockwise (spin up) or counterclockwise (spin down).

(f) The Pauli exclusion principle states that two electrons in an atom do not share the same four quantum numbers.

3. **(a)** The frequency of radiation is the number of oscillations that occur per second (as the radiation is observed passing a fixed point), the wavelength is the distance between two wave peaks on successive waves.

(b) Ultraviolet light is radiation that has wavelengths just a bit shorter than those of visible light. Infrared light has wavelengths just a bit longer than visible light.

(c) A continuous spectrum is visible light of all wavelengths that has been spread out by a prism or a grating. A discontinuous spectrum is light of only certain wavelengths, such as light emitted from an atomic species that has been excited.

(d) A traveling wave is one whose wave peaks move past a given point as time passes, such as waves in the ocean. A standing wave is one whose nodes remain fixed, such as sound waves in an organ pipe or the waves in a violin string.

(e) A quantum number is one of the four (principal, orbital, magnetic, spin) definite values that are used to specify the properties of an electron in an atom. An orbital is a region of space in which there is a good chance of finding an electron.

(f) *spdf* notation specifies the electron configuration of an atom by giving the principal and orbital quantum numbers of electrons, grouped into subshells. An orbital diagram places each electron of an atom, symbolized as an arrow, into a space that represents an orbital.

(g) An *s*-block element is one in Group 1A or Group 2A. A *p*-block element is one in Group 3A, 4A, 5A, 6A, 7A, or 8A.

(h) A main-group element is an element in one of the "A" Groups, that is, the *s* block or the *p* block. A transition element is one in a "B" family, that is a *d*-block element. (Sometimes *f*-block elements are included as transition elements as well.)

4. **(a)** $\text{length (nm)} = 1625\ \text{Å} \times \dfrac{1\ \text{m}}{10^{10}\ \text{Å}} \times \dfrac{10^9\ \text{nm}}{1\ \text{m}} = 162.5\ \text{nm}$

(b) $\text{length (µm)} = 3880\ \text{Å} \times \dfrac{1\ \text{m}}{10^{10}\ \text{Å}} \times \dfrac{10^6\ \text{µm}}{1\ \text{m}} = 0.388\ \text{µm}$

(c) $\text{length (nm)} = 7.27 \times 10^{-3}\ \text{m} \times \dfrac{10^9\ \text{nm}}{1\ \text{m}} = 7.27 \times 10^6\ \text{nm}$

(d) $\text{length (m)} = 546\ \text{nm} \times \dfrac{1\ \text{m}}{10^9\ \text{nm}} = 5.46 \times 10^{-7}\ \text{m}$

(e) $\text{length (nm)} = 1.12\ \text{cm} \times \dfrac{1\ \text{m}}{10^2\ \text{cm}} \times \dfrac{10^9\ \text{nm}}{1\ \text{m}} = 1.12 \times 10^7\ \text{nm}$

(f) $\text{length (cm)} = 2.6 \times 10^4\ \text{Å} \times \dfrac{1\ \text{m}}{10^{10}\ \text{Å}} \times \dfrac{10^2\ \text{cm}}{1\ \text{m}} = 2.6 \times 10^{-4}\ \text{cm}$

5. **(a)** $\lambda = \dfrac{c}{\nu} = \dfrac{3.00 \times 10^8\ \text{m/s}}{6.8 \times 10^{12}\ \text{s}^{-1}} = 4.4 \times 10^{-5}\ \text{m}$ infrared

(b) $\lambda = \dfrac{c}{\nu} = \dfrac{3.00 \times 10^8\ \text{m/s}}{9.8 \times 10^{15}\ \text{s}^{-1}} = 3.1 \times 10^{-8}\ \text{m}$ ultraviolet

(c) $\lambda = \dfrac{c}{\nu} = \dfrac{2.998 \times 10^8\ \text{m/s}}{2.54 \times 10^7\ \text{s}^{-1}} = 11.8\ \text{m}$ radio

(d) $\lambda = \dfrac{c}{\nu} = \dfrac{2.998 \times 10^8\ \text{m/s}}{1.07 \times 10^8\ \text{s}^{-1}} = 2.80\ \text{m}$ radio

6. Because frequency and wavelength are inversely related ($\nu = c/\lambda$), the shortest wavelength will correspond to the highest frequency. Because 1 mm = 0.1 cm, 5.9×10^{-4} cm is smaller than 1.13 mm. Because 1 cm = 10^{-2} m and 1 µm = 10^{-6} m, 6.92 µm is just a bit larger than 5.9×10^{-4} cm (5.9 µm). Because 1 Å = 10^{-10} m, 860 Å is smaller than 5.9×10^{-4} cm (5.9×10^4 Å). Thus, light of 860 Å has the shortest wavelength and highest frequency.

7. The wavelength is the distance between successive peaks. Thus, $4 \times 1.17\ \text{nm} = \lambda = 4.68\ \text{nm}$.

8. (a) $\nu = \dfrac{c}{\lambda} = \dfrac{3.00 \times 10^8 \text{ m/s}}{4.68 \text{ nm} \times \dfrac{1 \text{ m}}{10^9 \text{ nm}}} = 6.41 \times 10^{16} \text{ Hz}$

(b) $E = h\nu = 6.626 \times 10^{-34} \text{ J s} \times 6.41 \times 10^{16} \text{ Hz} = 4.25 \times 10^{-17} \text{ J}$

9. (a) The velocity of electromagnetic radiation in a vacuum, the speed of light, is a universal constant.
(b) The wavelength of electromagnetic radiation is inversely proportional to its frequency: $\lambda = c/\nu$.
(c) The energy per mole of electromagnetic radiation is directly proportional to its frequency: $E = N_A h\nu$.

10. (a) $\nu = 3.2881 \times 10^{15} \text{ s}^{-1}\left(\dfrac{1}{2^2} - \dfrac{1}{n^2}\right) = 3.2881 \times 10^{15} \text{ s}^{-1}\left(\dfrac{1}{2^2} - \dfrac{1}{5^2}\right) = 6.9050 \times 10^{14} \text{ s}^{-1}$

(b) $\nu = 3.2881 \times 10^{15} \text{ s}^{-1}\left(\dfrac{1}{2^2} - \dfrac{1}{7^2}\right) = 7.5492 \times 10^{14} \text{ s}^{-1}$

$$\lambda = \frac{2.9979 \times 10^8 \text{ m/s}}{7.5492 \times 10^{14} \text{ s}^{-1}} = 3.9711 \times 10^{-7} \text{ m} \times \frac{10^9 \text{ nm}}{1 \text{ m}} = 397.11 \text{ nm}$$

(c) $\nu = \dfrac{3.00 \times 10^8 \text{ m}}{380 \text{ nm} \times \dfrac{1 \text{ m}}{10^9 \text{ nm}}} = 7.89 \times 10^{14} \text{ s}^{-1} = 3.2881 \times 10^{15} \text{ s}^{-1}\left(\dfrac{1}{2^2} - \dfrac{1}{n^2}\right)$

$$0.250 - \frac{1}{n^2} = \frac{7.89 \times 10^{14} \text{ s}^{-1}}{3.2881 \times 10^{15} \text{ s}^{-1}} = 0.240 \qquad \frac{1}{n^2} = 0.250 - 0.240 = 0.010 \qquad n = 10$$

11. The frequencies of hydrogen emission lines in the infrared region of the spectrum other than the visible region would be predicted by replacing the constant "2" in the Balmer equation by the variable m, where m is an integer smaller than n: $m = 3, 4, \ldots$ The resulting equation is $\nu = 3.2881 \times 10^{15} \text{ s}^{-1}\left(\dfrac{1}{m^2} - \dfrac{1}{n^2}\right)$.

12. (a) $E = h\nu = 6.626 \times 10^{-34} \text{ J s} \times 8.62 \times 10^{15} \text{ s}^{-1} = 5.71 \times 10^{-18} \text{ J/photon}$

(b) $E_m = 6.626 \times 10^{-34} \text{ J s} \times 1.53 \times 10^{14} \text{ s}^{-1} \times \dfrac{6.022 \times 10^{23} \text{ photons}}{\text{mol}} \times \dfrac{1 \text{ kJ}}{1000 \text{ J}} = 61.1 \text{ kJ/mol}$

13. (a) $\nu = \dfrac{E}{h} = \dfrac{4.18 \times 10^{-21} \text{ J}}{6.626 \times 10^{-34} \text{ J s}} = 6.31 \times 10^{12} \text{ s}^{-1} = 6.31 \times 10^{12} \text{ Hz}$

(b) $E = h\nu = \dfrac{hc}{\lambda}$; $\lambda = \dfrac{hc}{E} = \dfrac{6.626 \times 10^{-34} \text{ J s} \times 3.00 \times 10^8 \text{ m/s}}{215 \text{ kJ/mol} \times \dfrac{1000 \text{ J}}{1 \text{ kJ}} \times \dfrac{1 \text{ mol}}{6.022 \times 10^{23} \text{ photons}}} = 5.57 \times 10^{-7} \text{ m} = 557 \text{ nm}$

14. We know that $E = h\nu$ and that $\nu = c/\lambda$, thus $E = hc/\lambda$. This means that energy and wavelength are inversely proportional. The radiation of shortest wavelength will have the greatest energy per photon; that with the longest wavelength will be the least energetic. It would be nice to have all four wavelengths in the same units for comparison; we choose μm since 1 μm = 10^{-6} m = 10^{+3} nm = 10^{-4} cm. Thus we have

(a) 662 nm = 0.662 μm	(b) 2.1×10^{-5} cm = 0.21 μm	most energetic
(c) 3.58 μm	(d) 4.1×10^{-6} m = 4.1 μm	least energetic

15. (a) Bohr atom radius = $r_n = n^2 a_0 = 4^2 \times 0.53 \text{ Å} = 8.5 \text{ Å}$

(b) We let $r_n = 4.00$ Å, and solve for n. Only if n is an integer does such an orbit exist.

$4.00 \text{ Å} = n^2 \times 0.53 \text{ Å} \qquad n = \sqrt{\dfrac{4.00}{0.53}} = 2.75$ There is no Bohr orbit with radius = 4.00 Å.

(c) $E_n = -\dfrac{R_H}{n^2} = -\dfrac{2.179 \times 10^{-18} \text{ J}}{8^2} = -3.405 \times 10^{-20} \text{ J}$

(d) We let $E_n = -5.00 \times 10^{-18}$ J, and solve for n. Only if n is an integer does such an orbit exist.

$$-5.00 \times 10^{-18} \text{ J} = -\frac{2.179 \times 10^{-18} \text{ J}}{n^2} \qquad n = \sqrt{\frac{-2.179 \times 10^{-18}}{-5.00 \times 10^{-18}}} = 0.660$$

There is no Bohr orbit with energy = -5.00×10^{-18} J.

16. We use equation (9.6) to determine the energy of the transition, followed by the Planck equation to find the frequency of the radiation.

$$\Delta E = 2.179 \times 10^{-18}\,\text{J}\left(\frac{1}{3^2} - \frac{1}{5^2}\right) = 1.5495 \times 10^{-19}\,\text{J} \qquad \nu = \frac{E}{h} = \frac{1.5495 \times 10^{-19}\,\text{J}}{6.626 \times 10^{-34}\,\text{J s}} = 2.339 \times 10^{+14}\,\text{s}^{-1}$$

17. **(a)** Light of the longest wavelength will be light of the smallest energy. Energy change is computed from the expression $\Delta E = B[(1/n^2) - (1/m^2)]$, where B is a constant. Also, light is emitted only when we proceed from a state with a higher value of n to one with a lower value; thus answers **(b)** and **(c)** are incorrect, because they refer to absorption of energy rather than to emission. We see that energy emitted is proportional to the difference between the inverses of the squares of the quantum numbers. Remember that we want the smaller value. The two possibilities are: **(a)** $[(1/3^2) - (1/4^2)] = 0.0486$, **(d)** $[(1/2^2) - (1/3^2)] = 0.1389$. Thus answer **(a)** designates the transition that produces light of the longest wavelength: from $n = 4$ to $n = 3$.

18. **(a)** deBroglie's equation ($\lambda = h/mv$) indicates that, if the velocity is constant, the wavelength is inversely proportional to the mass. Thus, the particle with the least mass will have the longest wavelength. Of the four particles given, the *electron* **(a)** is the least massive.

19. **(a)** $m_\ell = 0, \pm 1$ Because $\ell = 1$ and $|m_\ell|$ must be $\leq \ell$.
(b) $\ell = 1, 2, 3$ ℓ must be less than n (which equals 4) and also must be $\geq |m_\ell|$, and $m_\ell = -1$.
(c) $n = 2, 3, \ldots$ n must be greater than ℓ (which equals 1 in this case).

20.

(a)	4*s*	has	$n = 4$	$\ell = 0$	$m_\ell = 0$
(b)	3*p*	has	$n = 3$	$\ell = 1$	$m_\ell = -1, 0, +1$
(c)	5*f*	has	$n = 5$	$\ell = 3$	$m_\ell = -3, -2, -1, 0, +1, +2, +3$
(d)	3*d*	has	$n = 3$	$\ell = 2$	$m_\ell = -2, -1, 0, +1, +2$

21.

(a)	$n = 3$	$\ell = 2$	$m_\ell = -1$	is permitted.
(b)	$n = 2$	$\ell = 3$	$m_\ell = -1$	is not allowed, because it has $\ell > n$.
(c)	$n = 3$	$\ell = 0$	$m_\ell = +1$	is not allowed, because it has $|m_\ell| > \ell$.
(d)	$n = 6$	$\ell = 2$	$m_\ell = -1$	is permitted.
(e)	$n = 4$	$\ell = 4$	$m_\ell = +4$	is not permitted, because it has $\ell = n$.
(f)	$n = 4$	$\ell = 3$	$m_\ell = -1$	is permitted.

22. The number of orbitals in a subshell is determined by the value of the orbital quantum number, ℓ. The possible m_ℓ values range from $-\ell$ to $+\ell$ in whole number steps. Each m_ℓ value corresponds to a different orbital. Therefore, the number of orbitals in a subshell = $2\ell + 1$.
(a) Since an *s* orbital has $\ell = 0$, there is but one orbital in the 2*s* subshell.
(b) There is no 3*f* subshell. For 3*f*, $n = 3$ and $\ell = 3$, but ℓ must always be less than n.
(c) A *p* orbital has $\ell = 1$. There are three orbitals in the 4*p* subshell.
(d) A *d* orbital has $\ell = 2$. There are five orbitals in the 5*d* subshell.
(e) An *f* orbital has $\ell = 3$. There are seven orbitals in the 5*f* subshell.
(f) A *p* orbital has $\ell = 1$. There are three orbitals in the 6*p* subshell.

23. The order of subshell filling is 1*s* 2*s* 2*p* 3*s* 3*p* 4*s* 3*d* 4*p* 5*s* 4*d* 5*p* 6*s* 4*f* 5*d* 6*p* 7*s* 5*f* 6*d*. An *s* subshell can have a maximum of 2 electrons, a *p* subshell can have 6 electrons maximum, a *d* subshell can have 10, and an *f* subshell can have 14.
(a) Br has 35 electrons. $1s^22s^22p^63s^23p^64s^23d^{10}4p^5$ or $1s^22s^22p^63s^23p^63d^{10}4s^24p^5$
(b) S has 16 electrons. $1s^22s^22p^63s^23p^4$
(c) Sb has 51 electrons. $1s^22s^22p^63s^23p^64s^23d^{10}4p^65s^24d^{10}5p^3$
or $1s^22s^22p^63s^23p^63d^{10}4s^24p^64d^{10}5s^25p^3$
(d) Si has 14 electrons. $1s^22s^22p^63s^23p^2$
or $1s^22s^22p^63s^23p^63d^{10}4s^24p^64d^{10}5s^25p^4$

24. **(a)** Ca has 20 electrons, 2 more than the noble gas Ar. [Ar] 4*s* [↑↓]
(b) Sc has 21 electrons, 3 more than the noble gas Ar. [Ar] 3*d* [↑][ ][ ][ ][ ] 4*s* [↑↓]
(c) Ga has 31 electrons, 13 more than the noble gas Ar. [Ar] 3*d* [↑↓][↑↓][↑↓][↑↓][↑↓] 4*s* [↑↓] 4*p* [↑][ ][ ]
(d) Ni has 28 electrons, 10 more than the noble gas Ar. [Ar] 3*d* [↑↓][↑↓][↑↓][↑][↑] 4*s* [↑↓]
(e) Bi has 83 electrons, 29 more than Xe. [Xe] 4*f* [↑↓][↑↓][↑↓][↑↓][↑↓][↑↓][↑↓] 5*d* [↑↓][↑↓][↑↓][↑↓][↑↓] 6*s* [↑↓] 6*p* [↑][↑][↑]

25. We write the orbital diagram of each element and determine the number of unpaired electrons from that.

(a) Mg has 12 electrons, 2 more than the noble gas Ne. [Ne] $3s$ [↑↓] Zero unpaired electrons.

(b) Tl has 81 electrons, 27 more than the noble gas Xe.
[Xe] $4f$ [↑↓][↑↓][↑↓][↑↓][↑↓][↑↓][↑↓] $5d$ [↑↓][↑↓][↑↓][↑↓][↑↓] $6s$ [↑↓] $6p$ [↑][ ][ ] One unpaired electron.

(c) Te has 52 electrons, 16 more than the noble gas Kr.
[Kr] $4d$ [↑↓][↑↓][↑↓][↑↓][↑↓] $5s$ [↑↓] $4p$ [↑↓][↑][↑] Two unpaired electrons.

(d) Al has 13 electrons, 3 more than the noble gas Ne. [Ne] $3s$ [↑↓] $3p$ [↑][ ][ ] One unpaired electron.

26. (a) Main group metals include those in Groups 1A and 2A, along with Al, Ga, In, Tl, Sn, Pb, and Bi.

(b) Main group nonmetals are: H, F, Cl, Br, I, O, S, Se, N, P, C, and B.

(c) Noble gases are He, Ne, Ar, Kr, Xe, and Rn.

(d) The *d*-block elements are those in Groups 1B, 2B, 3B, 4B, 5B, 6B, 7B, and 8B.

(e) The inner transition elements include those with atomic numbers from $Z = 58$ (Ce) through $Z = 71$ (Lu) and those from $Z = 90$ (Th) through $Z = 103$ (Lr).

27. Base the prediction of each electron configuration upon the configuration of the preceding noble gas.

		prediction		from the *text* Appendix
(a)	In:	[Kr] $4d^{10}\,5s^2\,5p^1$	in family 3A	[Kr] $4d^{10}\,5s^2\,5p^1$
(b)	Cd:	[Kr] $4d^{10}\,5s^2$	in family 2B	[Kr] $4d^{10}\,5s^2$
(c)	Sb:	[Kr] $4d^{10}\,5s^2\,5p^3$	in family 3A	[Kr] $4d^{10}\,5s^2\,5p^3$
(d)	Au:	[Xe] $4f^{14}\,5d^{10}\,6s^1$	in family 1B	[Xe] $4f^{14}\,5d^{10}\,6s^1$

The predictions exactly match the given electron configurations for these four atoms.

28. (a) K is in family 1A and in the fourth period. Elements in family 1A have an s^1 outer electron configuration; those in the fourth period are filling the 4th principal quantum level. Thus K has one $4s$ electron.

(b) I is in family 7A (s^2p^5 outer electron configuration) and in the 5th period (5th principal quantum level). I has five $5p$ electrons.

(c) Zn is in family 2B [$(n-1)\,d^{10}\,n\,s^2$ outer electron configuration] and in the 4th period ($n = 4$). Zn has ten $3d$ electrons.

(d) S is in family 6A [$n\,s^2\,n\,p^4$] and in the 3rd period. The $2p^6$ electron configuration was complete with the preceding noble gas (Ne). S has six $2p$ electrons.

(e) Pb follows the lanthanide series in which fourteen $4f$ electrons were added. Pb has fourteen $4f$ electrons.

(f) Ni is in the *d*-block of elements, two from the end. It therefore has eight *d* electrons. It also is in the fourth period, so these eight *d* electrons are $3d$ electrons. Ni has eight 3d electrons.

EXERCISES

Electromagnetic Radiation

29. (a) TRUE Since frequency and wavelength are inversely related to each other, radiation of shorter wavelength has higher frequency.

(b) FALSE Light of wavelengths between 390 nm and 790 nm is visible to the eye.

(c) FALSE All electromagnetic radiation has the same speed in vacuum.

(d) TRUE The wavelength of an x ray approximates 0.1 nm.

30. (a) $$\nu = \frac{c}{\lambda} = \frac{2.9979 \times 10^8 \text{ m/s}}{418.7 \text{ nm}} \times \frac{10^9 \text{ nm}}{1 \text{ m}} = 7.160 \times 10^{14} \text{ Hz}$$

(b) Light of wavelength 418.7 nm is in the visible region of the spectrum.

(c) The light is visible, with a violet color.

31. The speed of light is used to convert the distance into an elapsed time.

$$\text{time} = 93 \times 10^6 \text{ mi} \times \frac{5280 \text{ ft}}{1 \text{ mi}} \times \frac{12 \text{ in.}}{1 \text{ ft}} \times \frac{2.54 \text{ cm}}{1 \text{ in}} \times \frac{1 \text{ s}}{3.00 \times 10^{10} \text{ cm}} \times \frac{1 \text{ min}}{60 \text{ s}} = 8.3 \text{ min}$$

32. The speed of light is used to convert the time into a distance spanned by light.

$$\text{distance} = 1\text{ y} \times \frac{365.25\text{ d}}{1\text{ y}} \times \frac{24\text{ h}}{1\text{ d}} \times \frac{3600\text{ s}}{1\text{ h}} \times \frac{2.9979 \times 10^8\text{ m}}{1\text{ s}} \times \frac{1\text{ km}}{1000\text{ m}} = 9.4607 \times 10^{12}\text{ km}$$

Atomic Spectra

33. The longest wavelength component has the lowest frequency (and the smallest energy).

$$\nu = 3.2881 \times 10^{15}\text{ s}^{-1}\left(\frac{1}{2^2} - \frac{1}{3^2}\right) = 4.5668 \times 10^{14}\text{ s}^{-1} \qquad \lambda = \frac{c}{\nu} = \frac{2.9979 \times 10^8\text{ m/s}}{4.5668 \times 10^{14}\text{ s}^{-1}} = \begin{matrix}6.5646 \times 10^{-7}\text{ m}\\ 656.46\text{ nm}\end{matrix}$$

$$\nu = 3.2881 \times 10^{15}\text{ s}^{-1}\left(\frac{1}{2^2} - \frac{1}{4^2}\right) = 6.1652 \times 10^{14}\text{ s}^{-1} \qquad \lambda = \frac{c}{\nu} = \frac{2.9979 \times 10^8\text{ m/s}}{6.1652 \times 10^{14}\text{ s}^{-1}} = \begin{matrix}4.8626 \times 10^{-7}\text{ m}\\ 486.26\text{ nm}\end{matrix}$$

$$\nu = 3.2881 \times 10^{15}\text{ s}^{-1}\left(\frac{1}{2^2} - \frac{1}{5^2}\right) = 6.9050 \times 10^{14}\text{ s}^{-1} \qquad \lambda = \frac{c}{\nu} = \frac{2.9979 \times 10^8\text{ m/s}}{6.9050 \times 10^{14}\text{ s}^{-1}} = \begin{matrix}4.3416 \times 10^{-7}\text{ m}\\ 434.16\text{ nm}\end{matrix}$$

$$\nu = 3.2881 \times 10^{15}\text{ s}^{-1}\left(\frac{1}{2^2} - \frac{1}{6^2}\right) = 7.3069 \times 10^{14}\text{ s}^{-1} \qquad \lambda = \frac{c}{\nu} = \frac{2.9979 \times 10^8\text{ m/s}}{7.3069 \times 10^{14}\text{ s}^{-1}} = \begin{matrix}4.1028 \times 10^{-7}\text{ m}\\ 410.28\text{ nm}\end{matrix}$$

34. $\lambda = 1880\text{ nm} \times \frac{1\text{ m}}{10^9\text{ nm}} = 1.88 \times 10^{-6}\text{ m}$ From Exercise 33, we see that wavelengths in the Balmer series range downward from 6.5646×10^{-7} m. Since this is less than 1.88×10^{-6} m, we conclude that light with a wavelength of 1880 nm cannot be in the Balmer series.

35. First we determine the frequency of the radiation, and then match it with the Balmer equation.

$$\nu = \frac{c}{\lambda} = \frac{2.9979 \times 10^8\text{ m s}^{-1} \times \frac{10^9\text{ nm}}{1\text{ m}}}{389\text{ nm}} = 7.71 \times 10^{14}\text{ s}^{-1} = 3.2881 \times 10^{15}\text{ s}^{-1}\left(\frac{1}{2^2} - \frac{1}{n^2}\right)$$

$$\left(\frac{1}{2^2} - \frac{1}{n^2}\right) = \frac{7.71 \times 10^{14}\text{ s}^{-1}}{3.2881 \times 10^{15}\text{ s}^{-1}} = 0.234 = 0.2500 - \frac{1}{n^2} \qquad \frac{1}{n^2} = 0.016 \qquad n = 7.9 \approx 8$$

36. (a) The maximum wavelength occurs when $n = 2$ and the minimum wavelength occurs at the series convergence limit, when n is exceedingly large (and $1/n^2 \approx 0$).

$$\nu = 3.2881 \times 10^{15}\text{ s}^{-1}\left(\frac{1}{1^2} - \frac{1}{2^2}\right) = 2.4661 \times 10^{15}\text{ s}^{-1} \qquad \lambda = \frac{c}{\nu} = \frac{2.9979 \times 10^8\text{ m/s}}{2.4661 \times 10^{15}\text{ s}^{-1}} = \begin{matrix}1.2156 \times 10^{-7}\text{ m}\\ 121.56\text{ nm}\end{matrix}$$

$$\nu = 3.2881 \times 10^{15}\text{ s}^{-1}\left(\frac{1}{1^2} - 0\right) = 3.2881 \times 10^{15}\text{ s}^{-1} \qquad \lambda = \frac{c}{\nu} = \frac{2.9979 \times 10^8\text{ m/s}}{3.2881 \times 10^{15}\text{ s}^{-1}} = \begin{matrix}9.1174 \times 10^{-8}\text{ m}\\ 91.174\text{ nm}\end{matrix}$$

(b) First determine the frequency of this spectral line, and then the value of n to which it corresponds.

$$\nu = \frac{c}{\lambda} = \frac{2.9979 \times 10^8\text{ m/s}}{95.0\text{ nm} \times \frac{1\text{ m}}{10^9\text{ nm}}} = 3.16 \times 10^{15}\text{ s}^{-1} = 3.2881 \times 10^{15}\text{ s}^{-1}\left(\frac{1}{1^2} - \frac{1}{n^2}\right)$$

$$\left(\frac{1}{1^2} - \frac{1}{n^2}\right) = \frac{3.16 \times 10^{15}\text{ s}^{-1}}{3.2881 \times 10^{15}\text{ s}^{-1}} = 0.961 \qquad \frac{1}{n^2} = 1.000 - 0.961 = 0.039 \qquad n = 5$$

(c) Let us pursue the same attack as in part (b).

$$\nu = \frac{c}{\lambda} = \frac{2.9979 \times 10^8\text{ m/s}}{108.5\text{ nm} \times \frac{1\text{ m}}{10^9\text{ nm}}} = 2.763 \times 10^{15}\text{ s}^{-1} = 3.2881 \times 10^{15}\text{ s}^{-1}\left(\frac{1}{1^2} - \frac{1}{n^2}\right)$$

$$\left(\frac{1}{1^2} - \frac{1}{n^2}\right) = \frac{2.763 \times 10^{15}\text{ s}^{-1}}{3.2881 \times 10^{15}\text{ s}^{-1}} = 0.8403 \qquad \frac{1}{n^2} = 1.000 - 0.8403 = 0.1597$$

This gives as a result $n = 2.502$. Since n is not an integer, there is no line in the Lyman spectrum with a wavelength of 108.5 nm.

Quantum Theory

37. (a) Combine $E = h\nu$ and $c = \nu\lambda$ to obtain $E = hc/\lambda$

$$E = \frac{6.626 \times 10^{-34} \text{ J s} \times 2.998 \times 10^8 \text{ m/s}}{474 \text{ nm} \times \frac{1 \text{ m}}{10^9 \text{ nm}}} = 4.19 \times 10^{-19} \text{ J/photon}$$

(b) $E_m = 4.19 \times 10^{-19} \frac{\text{J}}{\text{photon}} \times 6.022 \times 10^{23} \frac{\text{photons}}{\text{mol}} = 2.52 \times 10^5 \text{ J/mol} = 252 \text{ kJ/mol}$

38. First determine the energy of an individual photon, and then its wavelength in nm.

$$E = \frac{1799 \frac{\text{kJ}}{\text{mol}} \times \frac{1000 \text{ J}}{1 \text{ kJ}}}{6.022 \times 10^{23} \frac{\text{photons}}{\text{mol}}} = 2.987 \times 10^{-18} \frac{\text{J}}{\text{photon}} = \frac{hc}{\lambda} \quad \text{or} \quad \frac{hc}{E} = \lambda$$

$$\lambda = \frac{6.626 \times 10^{-34} \text{ J s} \times 2.998 \times 10^8 \text{ m/s}}{2.987 \times 10^{-18} \text{ J}} \times \frac{10^9 \text{ nm}}{1 \text{ m}} = 66.5 \text{ nm}$$

This is ultraviolet radiation.

39. Notice that energy and wavelength are inversely related: $E = \frac{hc}{\lambda}$. Therefore radiation that is 100 times as energetic as radiation with a wavelength of 988 nm will have a wavelength one hundredth as long: 9.88 nm. The frequency of this radiation is: $\nu = \frac{c}{\lambda} = \frac{2.998 \times 10^8 \text{ m/s}}{9.88 \text{ nm} \times \frac{1 \text{ m}}{10^9 \text{ nm}}} = 3.03 \times 10^{16} \text{ s}^{-1}$ From Figure 9-3, we see that this is ultraviolet radiation.

40. $E_1 = h\nu = \frac{hc}{\lambda} = \frac{6.62608 \times 10^{-34} \text{ J·s} \times 2.99792 \times 10^8 \text{ m s}^{-1}}{589.00 \text{ nm} \times \frac{1 \text{ m}}{10^9 \text{ nm}}} = 3.3726 \times 10^{-19} \text{ J/photon}$

$$E_2 = \frac{6.62608 \times 10^{-34} \text{ J·s} \times 2.99792 \times 10^8 \text{ m s}^{-1}}{589.59 \text{ nm} \times \frac{1 \text{ m}}{10^9 \text{ nm}}} = 3.3692 \times 10^{-19} \text{ J/photon}$$

$\Delta E = E_1 - E_2 = 3.3726 \times 10^{-19} \text{ J} - 3.3692 \times 10^{-19} \text{ J} = 0.0034 \times 10^{-19} \text{ J/photon} = 3.4 \times 10^{-22} \text{ J/photon}$

The Photoelectric Effect

41. **(a)** $E = h\nu = 6.63 \times 10^{-34} \text{ J s} \times 9.96 \times 10^{14} \text{ s}^{-1} = 6.60 \times 10^{-19} \text{ J/photon}$

(b) Indium will display a photoelectric effect when exposed to ultraviolet light since ultraviolet light has a maximum frequency of $1 \times 10^{16} \text{ s}^{-1}$, which is above the threshhold frequency of indium. It will not display a photoelectric effect when exposed to infrared light since the maximum frequency of infrared light is $4 \times 10^{14} \text{ s}^{-1}$, which is below the threshhold frequency of indium.

42. In his explanation of the photoelectric effect, Einstein stated that each photon strikes and is absorbed by only one electron (cannot kill two birds with one stone). He also, and very importantly, stated that no more than one photon could contribute its energy to a given electron (cannot kill one bird with two stones). It is this second principle which explains why photoelectrons are not produced in larger number when the intensity of (sub-threshhold frequency) light is increased.

The Bohr Atom

43. **(a)** radius $= n^2 a_0 = 6^2 \times 0.53 \text{ Å} \times \frac{1 \text{ m}}{10^{10} \text{ Å}} \times \frac{10^9 \text{ nm}}{1 \text{ m}} = 1.9 \text{ nm}$

(b) $E = -\frac{B}{n^2} = -\frac{2.179 \times 10^{-18} \text{ J}}{6^2} = -6.053 \times 10^{-20} \text{ J}$

44. **(a)** $r_1 = 1^2 \times 0.53 \text{ Å} = 0.53 \text{ Å}$ $\quad r_4 = 3^2 \times 0.53 \text{ Å} = 4.8 \text{ Å}$

increase in distance $= r_4 - r_1 = 4.8 \text{ Å} - 0.53 \text{ Å} = 4.3 \text{ Å}$

(b) $E_1 = \frac{-2.179 \times 10^{-18} \text{ J}}{1^2} = -2.179 \times 10^{-18} \text{ J}$ $\quad E_4 = \frac{-2.179 \times 10^{-18} \text{ J}}{3^2} = -2.421 \times 10^{-19} \text{ J}$

increase in energy $= -2.421 \times 10^{-19} \text{ J} - (-2.179 \times 10^{-18} \text{ J}) = 1.937 \times 10^{-18} \text{ J}$

45. (a) $\nu = \dfrac{2.179 \times 10^{-18}\ \text{J}}{6.626 \times 10^{-34}\ \text{J s}}\left(\dfrac{1}{4^2} - \dfrac{1}{7^2}\right) = 1.384 \times 10^{14}\ \text{s}^{-1}$ (c) This is infrared radiation.

(b) $\lambda = \dfrac{c}{\nu} = \dfrac{2.998 \times 10^{8}\ \text{m/s}}{1.384 \times 10^{14}\ \text{s}^{-1}} = 2.166 \times 10^{-6}\ \text{m} \times \dfrac{10^9\ \text{nm}}{1\ \text{m}} = 2166\ \text{nm}$

46. The greatest quantity of energy is absorbed in the situation where the difference between the inverses of the squares of the two quantum numbers is the largest, and the system begins with a lower quantum number than it finishes with. The second condition eliminates answer **(d)** from consideration. Now we can consider the difference of the inverses of the squares of the two quantum numbers in each case.

(a) $\left(\dfrac{1}{1^2} - \dfrac{1}{2^2}\right) = 0.75$ (b) $\left(\dfrac{1}{2^2} - \dfrac{1}{4^2}\right) = 0.1875$ (c) $\left(\dfrac{1}{3^2} - \dfrac{1}{9^2}\right) = 0.0988$

Thus, the largest amount of energy is absorbed in the transition from $n = 1$ to $n = 2$, answer **(a)**, among the four choices given.

47. If infrared light is produced, the quantum number of the final state must have a lower value (be of lower energy) than the quantum number of the initial state. First we compute the frequency of the transition being considered (from $\nu = c/\lambda$), and then solve for the final quantum number.

$$\nu = \frac{c}{\lambda} = \frac{3.00 \times 10^8\ \text{m/s}}{2170\ \text{nm} \times \dfrac{1\ \text{m}}{10^9\ \text{nm}}} = 1.38 \times 10^{14}\ \text{s}^{-1}$$

$$= \frac{2.179 \times 10^{-18}\ \text{J}}{6.626 \times 10^{-34}\ \text{J s}}\left(\frac{1}{n^2} - \frac{1}{7^2}\right) = 3.289 \times 10^{15}\ \text{s}^{-1}\left(\frac{1}{n^2} - \frac{1}{7^2}\right)$$

$$\left(\frac{1}{n^2} - \frac{1}{7^2}\right) = \frac{1.38 \times 10^{14}\ \text{s}^{-1}}{3.289 \times 10^{15}\ \text{s}^{-1}} = 0.0419_6 \qquad \frac{1}{n^2} = 0.0419_6 + \frac{1}{7^2} = 0.0623_7 \qquad n = 4$$

48. If infrared light is produced, the quantum number of the final state must have a lower value (be of lower energy) than the quantum number of the initial state. First we compute the frequency of the transition being considered (from $\nu = c/\lambda$), and then solve for the initial quantum number, n.

$$\nu = \frac{c}{\lambda} = \frac{3.00 \times 10^8\ \text{m/s}}{3740\ \text{nm} \times \dfrac{1\ \text{m}}{10^9\ \text{nm}}} = 8.02 \times 10^{13}\ \text{s}^{-1}$$

$$= \frac{2.179 \times 10^{-18}\ \text{J}}{6.626 \times 10^{-34}\ \text{J s}}\left(\frac{1}{5^2} - \frac{1}{n^2}\right) = 3.289 \times 10^{15}\ \text{s}^{-1}\left(\frac{1}{5^2} - \frac{1}{n^2}\right)$$

$$\left(\frac{1}{5^2} - \frac{1}{n^2}\right) = \frac{8.02 \times 10^{13}\ \text{s}^{-1}}{3.289 \times 10^{15}\ \text{s}^{-1}} = 0.0243_8 \qquad \frac{1}{n^2} = -0.0243_8 + \frac{1}{5^2} = 0.0156_2 \qquad n = 8$$

Wave-Particle Duality

49. The de Broglie equation is $\lambda = h/mv$. This means that, for a given wavelength to be produced, a lighter particle would have to be moving faster. Thus, electrons would have to move faster than protons to display matter waves of the same wavelength.

50. First, we rearrange the de Broglie equation, solving it for velocity: $v = h/m\lambda$. Then we obtain the velocity of the electron. From Table 2-1, the mass of an electron is 9.109×10^{-28} g.

$$v = \frac{h}{m\lambda} = \frac{6.626 \times 10^{-34}\ \text{J s}}{\left(9.109 \times 10^{-28}\ \text{g} \times \dfrac{1\ \text{kg}}{1000\ \text{g}}\right)\left(1\ \mu\text{m} \times \dfrac{1\ \text{m}}{10^6\ \mu\text{m}}\right)} = 7 \times 10^2\ \text{m/s}$$

51. $$\lambda = \frac{h}{mv} = \frac{6.626 \times 10^{-34}\ \text{J s}}{\left(145\ \text{g} \times \dfrac{1\ \text{kg}}{1000\ \text{g}}\right)\left(168\ \text{km/h} \times \dfrac{1\ \text{h}}{3600\ \text{s}} \times \dfrac{1000\ \text{m}}{1\ \text{km}}\right)} = 9.79 \times 10^{-35}\ \text{m}$$

The diameter of a nucleus approximates 10^{-15} m, far larger than this wavelength.

52. $\lambda = \frac{h}{mv} = \frac{6.626 \times 10^{-34} \text{ J s}}{1000 \text{ kg} \times 25 \text{ m/s}} = 2.7 \times 10^{-38} \text{ m}$

This wavelength is so much smaller than the smallest particle of matter known (nuclear diameter ≈ 10^{-15} m) that its experimental measurement is virtually impossible.

The Uncertainty Principle

53. The Bohr model is a determinant model of an atom. It implies that the position of the electron is exactly known at any time in the future, once that position is known at the present. The distance of the electron from the nucleus also is exactly known, as is its energy. And finally, the velocity of the electron in its orbit is exactly known. All of these exactly known quantities—position, distance from nucleus, energy, and velocity—should, according to the Heisenberg uncertainty principle, be known with some imprecision.

54. Einstein believed very strongly in the law of cause and effect, what is known as a deterministic view of the universe. He felt that the need to use probability and chance ("playing dice") in describing atomic structure resulted because a suitable theory had not yet been developed to permit accurate predictions. He believed that such a theory could be developed, and had good reason for his belief: The developments in the theory of atomic structure came very rapidly during the first thirty years of this century, and those, such as Einstein, who had lived through this period had seen the revision of a number of theories and explanations that were thought to be the final answer. Another viewpoint in this area is embodied in another famous quotation: "Nature is subtle, but not malicious." The meaning of this statement is that the causes of various effects may be quite small (subtle) but there are causes nonetheless (not malicious).

Bohr was stating that we should accept theories as they are revealed by experimentation and logic, rather than attempting to make these theories fit our preconceived notions of what we believe they should be. In other words, Bohr was telling Einstein to keep an open mind.

Wave Mechanics

55. A sketch of this situation is presented at right. We see that 2.50 waves span the space of the 42 cm. Thus, the length of each wave is obtained by equating: 2.50 λ = 42 cm, giving λ = 16.8 cm.

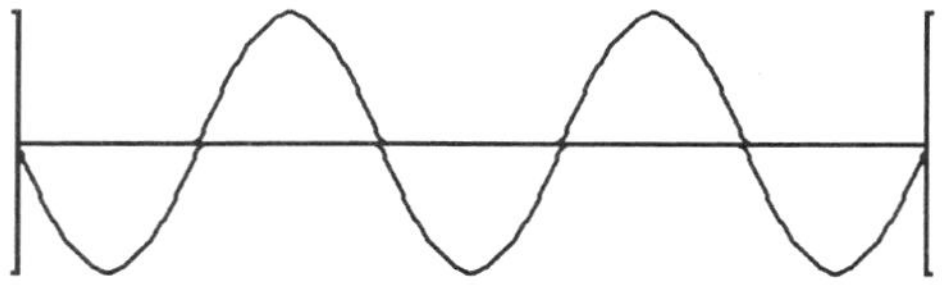

56. If there are four nodes, then there are three half-wavelengths within the string: one between the first and second node, the second half-wavelength between nodes 2 and 3, and the third between nodes 3 and 4. Therefore, $3 \times \lambda/2$ = length = 3×17 cm/2 = 25.5 cm long string.

57. The differences between Bohr orbits and wave mechanical orbitals are given below.

a. The first difference is that of shape. Bohr orbits, as originally proposed, are circular (later Sommerfeld proposed elliptical orbits). Orbitals, on the other hand, are spherical; or shaped like two tear drops or two squashed spheres; or shaped like four tear drops meeting at their points.

b. Bohr orbits are planar pathways, while orbitals are three-dimensional regions of space in which there is a high probability of finding electrons.

c. The electron in a Bohr orbit has a definite trajectory. Its position and velocity are known at all times. The electron in an orbital, however, does not have a well-known position or velocity. In fact, there is a small but definite probability that the electron may be found outside the boundaries generally drawn for the orbital.

Orbits and orbitals are similar in that the radii of Bohr orbits correspond to the distance from the nucleus in an orbital at which the electron is found with high probability.

58. We must be careful to distinguish between probability density—the chance of finding the electron within a definite volume of space—and the probability of finding the electron at a certain distance from the nucleus. The probability density—that is the probability of finding the electron within a small volume element (such as 1 pm^2)—at the nucleus may well be high, in fact higher than the probability density at a distance 0.53 Å from the nucleus. But the probability of finding the electron at a fixed distance from the nucleus is this probability density multiplied by all of the many small volume elements that are located at this distance. (Recall the dart board analogy of Figure 9-19.)

Quantum Numbers and Electron Orbitals

59. Answer (a) is incorrect because the values of m_s may be either $+\frac{1}{2}$ or $-\frac{1}{2}$. Answers (b) and (d) are incorrect because the value of ℓ may be any integer $\geq |m_\ell|$, and less than n. Thus, answer (c) is correct.

60.
- **(a)** $n = 3$, $\ell = \mathbf{2}$, $m_\ell = 2$, $m_s = +\frac{1}{2}$ — (ℓ must be smaller than n and $\geq |m_\ell|$.) This is a $3d$ orbital.
- **(b)** $n \geq \mathbf{3}$, $\ell = 2$, $m_\ell = -1$, $m_s = +\frac{1}{2}$ — (n must be larger than ℓ.) This is a $3d$, $4d$, $5d$, … orbital.
- **(c)** $n = 4$, $\ell = 2$, $m_\ell = 0$, $m_s = +\frac{1}{2}$ — (m_s can be either $+\frac{1}{2}$ or $-\frac{1}{2}$.) This is $4d$ orbital.
- **(d)** $n = \mathbf{1}$, $\ell = 0$, $m_\ell = \mathbf{0}$, $m_s = +\frac{1}{2}$ — (The principles are stated above; n can be any positive integer, m_s could also equal $-\frac{1}{2}$.) This is a $1s$ orbital.

61.
- **(a)** $n = 5$ $\ell = 1$ $m_\ell = 0$ designates a $5p$ orbital. ($\ell = 1$ for all p orbitals.)
- **(b)** $n = 4$ $\ell = 2$ $m_\ell = -2$ designates a $4d$ orbital. ($\ell = 2$ for all d orbitals.)
- **(c)** $n = 2$ $\ell = 0$ $m_\ell = 0$ designates a $2s$ orbital. ($\ell = 0$ for all s orbitals.)

62.
- **(a)** TRUE — The fourth principal shell has $n = 4$.
- **(b)** TRUE — A d orbital has $\ell = 2$. Since $n = 4$, ℓ can be $= 3, 2, 1, 0$. Since $m_\ell = -2$, ℓ can be $= 2, 3$.
- **(c)** FALSE — A p orbital has $\ell = 1$. But we demonstrated in part (b) that the only allowed values for ℓ are 2 and 3.
- **(4)** FALSE — Either $+\frac{1}{2}$ or $-\frac{1}{2}$ is permitted as a value of m_s.

63.
- **(a)** Each subshell has a different value of the orbital quantum number. When $n = 4$, the possible values of ℓ are 0, 1, 2, and 3. There are four subshells in the $n = 4$ level.
- **(b)** In the $n = 3$ level, the possible value of the orbital quantum number are $\ell = 0$, 1, and 2. These correspond to the $3s$ subshell, the $3p$ subshell, and the $3d$ subshell.
- **(c)** In any $\ell = 3$ subshell, the possible values of the magnetic quantum number are $m_\ell = -3, -2, -1, 0, 1, 2, 3$. This means that there are seven orbitals in an f ($\ell = 3$) subshell.
- **(d)** The values $n = 4$, $\ell = 3$, and $m_\ell = -2$ completely designate an orbital. Just one orbital has these three quantum numbers.
- **(e)** Within the $n = 4$ level, there is a $4s$ subshell with one orbital, a $4p$ subshell with three orbitals, a $4d$ subshell with five orbitals, and a $4f$ subshell with seven orbitals. The total number of orbitals in the $n = 4$ level is $1 + 3 + 5 + 7 = 16$ orbitals.

64.
- **(a)** Just one electron can have $n = 3$, $\ell = 2$, $m_\ell = 0$, and $s = +\frac{1}{2}$. Four quantum numbers completely designate an electron.
- **(b)** These three quantum numbers designate an orbital, which can hold two electrons. Two electrons can have the three quantum numbers $n = 3$, $\ell = 2$, $m_\ell = 0$.
- **(c)** These two quantum numbers designate the $3d$ subshell, which contains five orbitals, with the possibility of two electrons in each. Ten electrons can have the two quantum numbers $n = 3$, $\ell = 2$.
- **(d)** $n = 3$ designates the shell that contains the $3d$ subshell that has five orbitals, the $3p$ subshell with three orbitals, and the $3s$ subshell with one orbital. Each of these nine orbitals in the shell can hold two electrons, for a total of 18 electrons.
- **(e)** The first two quantum numbers designate the $3d$ subshell, which has five orbitals. Each orbital can accomodate one electron with spin up. Five electrons can have quantum numbers $n = 3$, $\ell = 2$, $s = \frac{1}{2}$.

Electron Configurations

65. Configuration **(b)** is correct for phosphorus. The reason why each other configuration is incorrect follows.
- **(a)** The two electrons in the $3s$ subshell must have opposed spins, that is different m_s quantum numbers.
- **(c)** The three $3p$ orbitals must each contain one electron, before a pair of electrons is placed in any one of these orbitals.
- **(d)** The three unpaired electrons in the $3p$ subshell must all have the same spin, either all spin up or all spin down.

66. The electron configuration of Mo is [Kr] $_{4d}$[↑][↑][↑][↑][↑] $_{5s}$[↑]
- **(a)** [Ar] $_{3d}$[↑↓][↑↓][↑↓][↑↓][↑↓] $_{3f}$[↑↓][↑↓][↑↓][↑↓][↑↓][↑↓][↑↓] — This configuration has the correct number of electrons. But it incorrectly assumes that there is a $3f$ subshell.
- **(b)** [Kr] $_{4d}$[↑][↑][↑][↑][↑] $_{5s}$[↑] — This is the correct electron configuration.

(c) [Kr] $_{4d}$[↑][↑][↑][↑][↑] $_{5s}$[↑↓] This configuration has one electron too many.

(d) [Ar] $_{3d}$[↑↓][↑↓][↑↓][↑↓][↑↓] $_{4s}$[↑↓] $_{4p}$[↑↓][↑↓][↑↓] $_{4d}$[↑↓][↑][↑][↑][↑] This configuration incorrectly assumes that electrons enter the 4*d* subshell before they enter the 5*s* subshell.

67. We write the correct electron configuration first in each case.

(a) P: $[Ne]3s^2 3p^3$ where $3p^3$ is [↑] [↑] [↑] There are 3 unpaired electrons in each P atom.

(b) Br: $[Ar]3d^{10}4s^2 4p^5$ There are ten 3*d* electrons in an atom of Br.

(c) Ge: $[Ar]3d^{10}4s^2 4p^2$ There are two 4*p* electrons in an atom of Ge.

(d) Ba: $[Xe]6s^2$ There are two 6*s* electrons in an atom of Ba.

(e) Au: $[Xe]4f^{14}5d^9 6s^2$ There are fourteen 4*f* electrons in an atom of Au.

68. **(a)** The 4*p* subshell of Br contains 5 electrons [Ar] $_{3d}$[↑↓][↑↓][↑↓][↑↓][↑↓] $_{4s}$[↑↓] $_{4p}$[↑↓][↑↓][↑]

(b) The 3*d* subshell of Co^{2+} contains 7 electrons [Ar] $_{3d}$[↑↓][↑↓][↑][↑][↑] $_{4s}$[ ]

(c) The 5*d* subshell of Pb contains 10 electrons [Xe] $_{4f}$[↑↓][↑↓][↑↓][↑↓][↑↓][↑↓][↑↓] $_{5d}$[↑↓][↑↓][↑↓][↑↓][↑↓] $_{6s}$[↑↓] $_{6p}$[↑][↑][ ]

69. **(a)** N is the third element in the *p*-block of the second period and thus has three 2*p* electrons.

(b) Rb is the first element in the *s*-block of the *fifth* period, thus Rb has one 4*s* electrons.

(c) As is in the *p*-block of the fourth period. The 3*d* subshell is filled with ten electrons, but no 4*d* electrons have been added.

(d) Au is in the *d*-block of the sixth period; the 4*f* subshell is filled. Au has fourteen 4*f* electrons.

(e) Pb is the second element in the *p*-block of the sixth period; it has two 6*p* electrons. Since these two electrons are placed in separate 6*p* orbitals, they are unpaired. There are two unpaired electrons.

(f) Group 4A of the periodic table is the group with the elements C, Si, Ge, Sn, and Pb. This group has five elements.

(g) The sixth period begins with the element Cs ($Z = 55$) and ends with the element Rn ($Z = 86$). This period is 32 elements long.

70. **(a)** Sb is in family 5A, with outer shell electron configuration $ns^2\ np^3$. Sb has five outer-shell electrons.

(b) Pt has an atomic number of $Z = 78$. The fourth principal shell fills as follows: 4*s* from $Z = 19$ (K) to $Z = 20$ (Ca); 4*p* from $Z = 31$ (Ga) to $Z = 36$ (Kr); 4*d* from $Z = 39$ (Y) to $Z = 48$ (Cd); and 4*f* from $Z = 58$ (Ce) to $Z = 71$ (Lu). Since the atomic number of Pt is greater than $Z = 71$, the entire fourth principal shell is filled, with 32 electrons.

(c) The five elements with six outer-shell electrons are those in family 6A: O, S, Se, Te, Po.

(d) The outer-shell electron configuration is $ns^2\ np^4$, giving the following as a partial orbital diagram:

s^2 [↑↓] p^4 [↑↓][↑][↑] There are two unpaired electrons in an atom of Te.

(e) The sixth period begins with Cs and ends with Rn. There are 10 outer transition elements in this period (La, and Hf through Hg) and there are 14 inner transition elements in the period (Ce through Lu). Thus, there are (10 + 14 =) 24 transition elements in the sixth period.

71. Since the periodic table is based on electron structure, two elements in the same family (Pb and element 114, Uuq) should have similar electron configurations.

(a) Pb: $[Xe]\ 4f^{14}\ 5d^{10}\ 6s^2\ 6p^2$ **(b)** Uuq: $[Rn]\ 5f^{14}\ 6d^{10}\ 7s^2\ 7p^2$

72. **(a)** A fifth period noble gas is the element in Group 8A and the fifth period, the element Xe.

(b) A sixth period element whose atoms have three unpaired electrons is an element in family 5A, which has an outer electron configuration of $ns^2\ np^3$, and thus has three unpaired *p* electrons. This is the element Bi.

(c) One *d*-block element that has one 4*s* electron is Cu: $[Ar]\ 3d^{10}\ 4s^1$. Another is Cr: $[Ar]\ 3d^5\ 4s^1$. Others are Mo, W, Ag, and Au.

(d) There are several *p*-block elements that are metals: Al, Ga, In, Tl, Sn, Pb, and Bi.

FEATURE PROBLEMS

A. By carefully observing the diagram, we note that there are no spectral lines in the area of 304 nm and 309 nm, nor at 318 nm, and 327 nm. Likewise, there are none between 435 and 440 nm. We conclude that V is absent.

There are spectral lines the correspond to each of the Cr spectral lines—between 355 nm and 362 nm, and between 425 nm and 430 nm. Cr is present.

There are no spectral lines close to 403 nm; Mn is absent.

There are spectral lines at about 344 nm, 358 nm, 372 nm, 373 nm, and 386 nm. Fe is present.

There are spectral lines at 341 nm, 344 nm to 352 nm, and 362 nm. Ni is present.

There are spectral lines between 310 and 315 nm and at about 415 nm. Another element is present.

Thus, Cr, Fe, and Ni are present. V and Mn are absent. There is an additional element present.

B. The equation of a straight line is $y = mx + b$, where m is the slope of the line and b is its y-intercept. The Balmer equation is $\nu = 3.2881 \times 10^{15}\text{ Hz}\left(\frac{1}{2^2} - \frac{1}{n^2}\right) = \frac{c}{\lambda}$. In this equation, one plots ν on the vertical axis, and $1/n^2$ on the horizontal axis. The slope is $b = -3.2881 \times 10^{15}$ Hz and the intercept is $3.2881 \times 10^{15}\text{ Hz} \div 2^2 = 8.2203 \times 10^{14}$ Hz. The plot of the data of Figure 9.9 follows.

λ	656.3 nm	486.1 nm	434.0 nm	410.1 nm
ν	4.568×10^{14} Hz	6.167×10^{14} Hz	6.908×10^{14} Hz	7.310×10^{14} Hz
n	3	4	5	6

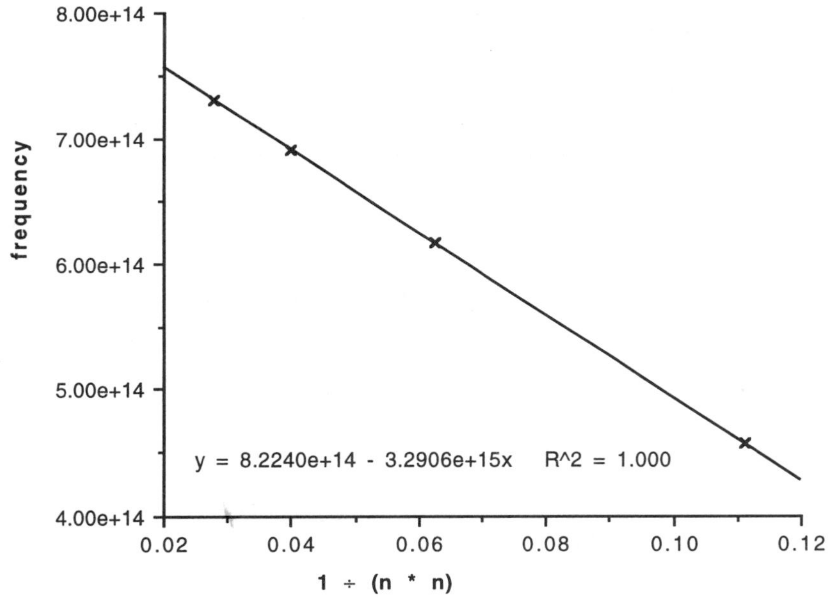

We see that the slope (-3.2906×10^{15}) and y-intercept (8.2240×10^{14}) are almost exactly what we had predicted from the Balmer equation.

C. **1.** This graph differs from the one of ψ^2 in Figure 9-18(a) because this graph factors in the volume of the thin shell. Figure 9-18(a) simply is a graph of the probability of finding an electron at a distance r from the nucleus. But the graph that accompanies this problem multiplies that "radial probability by the volume of the shell that is a distance r from the nucleus. The volume of the shell is the thickness of the shell (a very small value dr) times the area of the shell ($4\pi r^2$). Close to the nucleus, the area is very small because r is very small. Therefore, the relative probability also is very small near the nucleus; there just isn't enough volume to contain the electrons.

2. What we plot is the product of the number of darts times the circumference of the outer boundary of the scoring ring. (probability = number × circumference)

darts	200	300	400	250	200
score	"50"	"40"	"30"	"20"	"10"
radius	1.0	2.0	3.0	4.0	5.0
circumference	3.14	6.28	9.42	12.6	15.7
probability	628	1884	3768	3150	3140

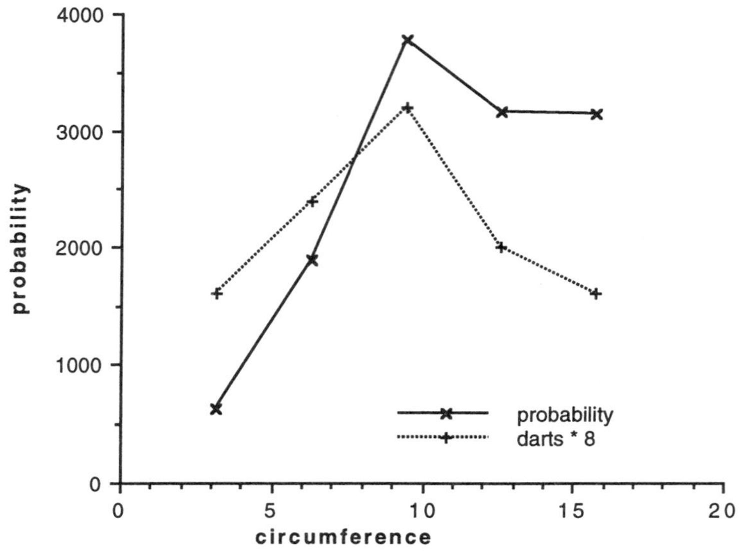

3. The graph of "probability" is close to the graph that accompanies this problem, except that the dart board is two-dimensional, while the atom is three-dimensional. This added dimension means that the volume close to the nucleus is relatively much smaller than is the area close to the center. The other difference, of course, is that it is harder for darts to get close to the center, while electrons are attracted to the nucleus.

D. We use no leaders in the answers that follow, just place the letters over the appropriate places.

1. (Mn): $1s^2\ 2s^2\ 2p^6\ 3s^2\ 3p^6\ 3d^5\ 4s^2$ — (i) over $2p^6$; <—(f)—> over $3s^2\ 3p^6\ 3d^5$; (k) over $4s^2$

(i) a p subshell (f) a principal shell
(k) an electron pair

2. (Nb): [Kr] [↑][↑][↑][↑][ ] [↑] — (j) over the first set of boxes; (d) over the last box

(j) a d subshell (d) a valence electron

3. (I): [Kr] [↑↓][↑↓][↑↓][↑↓][↑↓] [↑↓] [↑↓][↑↓][↑] — <—(b)—> over the first set of boxes; <—(c)—> over the last two sets

(b) a filled subshell (c) a valence shell

4. (U): [Rn] [↑][↑][↑][ ][ ][ ][ ] [↑][ ][ ][ ][ ] [↑↓] — 5*f* over the first set; (a) over the second set; (h) and (l) over the last box

(a) an unpaired electron (h) a d orbital
(l) an s subshell

There quite possibly are a number of other ways of assigning these labels. For instance, it is obvious that labels (l) and (k) could be interchanged above.

10 THE PERIODIC TABLE AND SOME ATOMIC PROPERTIES

PRACTICE EXAMPLES

1A Atomic size decreases left to right through a period, and bottom to top in a family. We expect the smallest elements to be in the upper right corner of the periodic table. S is the element closest to the upper right corner and should be the smallest element. S = 104 pm As = 121 pm I = 133 pm

1B From the periodic table inside the front cover, we see that Na is in the same period as Al (period 3), but in a different family from K, Ca, and Br (period 4), which might predict that Na and Al are about the same size. However, there is a substantial decrease in size as one moves from left to right in a period, perhaps substantial enough that Ca should be about the same size as Na. The sizes given in Table 10-1 are:

227 pm for K < 197 pm for Ca ≈ 184 pm for Na < 143 pm for Al < 114 pm for Br

2A Ti^{2+} and V^{3+} are isoelectronic; the one with higher positive charge should be smaller: $V^{3+} < Ti^{2+}$. Sr^{2+} and Br^- are isoelectronic; again, the one with higher positive charge should be smaller: $Sr^{2+} < Br^-$. In addition Ca^{2+} and Sr^{2+} both are ions of Group 2A; the one of lower atomic number should be smaller: $Ca^{2+} < Sr^{2+} < Br^-$. Finally, we know that size of atoms decreases left to right across a period; we expect size of like-charged ions to follow the same trend: $Ti^{2+} < Ca^{2+}$. Thus, summarized, with actual sizes in parentheses:

V^{3+} (64 pm) < Ti^{2+} (86 pm) < Ca^{2+} (100 pm) < Sr^{2+} (113 pm) < Br^- (196 pm)

2B Br^- clearly is larger than As since Br^- is an anion in the same period as As. In turn, As is larger than N since both are in the same family, with As toward the bottom of the family. As also should be larger than P, which is larger than Mg^{2+}. All that remains is to note that Cs is a truly large atom, one of the largest in the periodic table. The As atom should be in the middle. Data from Table 10-1 are:

65 pm for Mg^{2+} < 70 pm for N < 125 pm for As < 196 pm for Br^- < 265 pm for Cs

3A Ionization increases from bottom to top of a group and from left to right through a period. The first ionization energy of K is less than that of Mg and the first ionization energy of S is less than that of Cl. We would suppose also that the first ionization energy of Mg is smaller than that of S, because Mg is a metal.
K (418.8 kJ/mol) < Mg (737.7 kJ/mol) < S (1000 kJ/mol) < Cl (1251.1 kJ/mol)

3B We would expect an alkali metal (Rb) or an alkaline earth metal (Sr) to have a low first ionization energy and nonmetals (Br and As) to have relatively high first ionization energies. A metalloid (such as Sb) should have an intermediate ionization energy. Thus, the first ionization energy of Sb should be in the middle. Data that were used to produce Figure 10-9 include the following (in kJ/mol): 403 for Rb, 549 for Sr, 834 for Sb, 947 for As, and 1140 for Br.

4A Cl and Al must be paramagnetic, since they each have an odd number of electrons. The electron configurations of K^+ ([Ar]) and O^{2-} ([Ne]) are those of the nearest noble gas; because all electrons are paired, they are diamagnetic. In Zn: [Ar] $3d^{10}\,4s^2$ all electrons are paired; the atom is diamagnetic.

4B The electron configuration of Cr is [Ar] $3d^5\,4s^1$; it has six unpaired electrons. The electron configuration of Cr^{2+} is [Ar] $3d^4\,4s^0$; it has four unpaired electrons. The electron configuration of Cr^{3+} is [Ar] $3d^3\,4s^0$; it has three unpaired electrons. Thus, of the two ions, Cr^{2+} has the greater number of unpaired electrons.

5A We expect the melting point of bromine to be the average of the melting points of chlorine and iodine:
estimated melting point of $Br_2 = \dfrac{172\text{ K} + 387\text{ K}}{2} = 280\text{ K}$. The actual melting point is 266 K.

5B If the boiling point of I (458 K) is the average of the boiling points of Br (349 K) and At, then
458 K = (349 K + ?)/2 ? = 2 × 458 K – 349 K = 567 K = boiling point of astatine, about 570 K.

SUMMARIZING EXAMPLE CALCULATIONS

1. Fr is element 87 in period 7 and Group 1A.

2. The given melting points are: 174°C for Li, 97.8°C for Na, 63.7°C for K, 38.9°C for Rb, and 28.5°C for Cs. Note that the melting point decreases toward the bottom of the family, with the difference in melting points also decreasing down the family. One might expect a five-degree difference between the melting points of Cs and Fr, giving ≈24°C as the melting point of Fr.

3. Based on the trend in the atomic volumes of the alkali metals, one might estimate the atomic volume of Fr to be 80 cm³/mol. The molar masses of Rn (222) and Ra (226) give an average of 224 g/mol. The expected density of Fr then is computed. $\text{density} = \dfrac{224 \text{ g/mol}}{80 \text{ cm}^3\text{/mol}} = 2.8 \text{ g/cm}^3$

4. The atomic radius of Cs (265 pm) is about 15 pm larger than that of Rb (250 pm). We might expect the atomic radius of Fr to be 15 pm larger still, that is, 280 pm.

REVIEW QUESTIONS

1. **(a)** Two isoelectronic species (an atom and an ion, or two ions) have the same number of electrons.

(b) The valence electrons in an atom are those that have the highest principal quantum number. They also are the ones furthest from the nucleus and thus on the outside of the atom. Because they are the electrons that other atoms "see," they are most important in determining the atom's chemical behavior.

(c) A metal is an element that is relatively easily oxidized. Most elements are metals. In the periodic table, all of the transition elements are metals, as well as the lanthanides and actinides. In addition, metals are the representative elements in families 1A and 2A, as well as Al, Ga, In, Tl, Sn, Pb, and Bi.

(d) A nonmetal is an element that is relatively easily reduced. In the periodic table, nonmetals are at the right and include H, F, Cl, Br, I, O, S, Se, N, P, C, and B.

(e) A metalloid is an element that is fairly easy to oxidize or to reduce. That is, their behavior is between that of metals and of nonmetals. Metalloids include Si, Ge, As, Sb, Te, Po, and At.

2. **(a)** The periodic law is, like all natural laws, a summary of experimental observations. It states that a given property of elements varies periodically when the elements are arranged in order of increasing atomic number.

(b) Ionization energy is the quantity of energy that, when added to a gaseous atom (or ion), will remove an electron.

(c) Electron affinity is the quantity of energy absorbed by an atom (or ion) when it gains an electron. Since the electron affinity process usually is exothermic for atoms, most atomic electron affinities are negative.

(d) Paramagnetism, the result of a species having one or more unpaired electrons, is the small attraction of that species into a magnetic field. It is not nearly as strong as the more familiar property of ferromagnetism.

3. **(a)** Lanthanide elements are *f*-block elements are those in which the 4*f* subshell is filling: La through Lu. Actinide elements are those in which the 5*f* subshell is filling: Ac through Lr.

(b) A covalent radius is the radius of an atom that is bonded covalently to another. For example, one half the internuclear distance in Cl_2 is the covalent radius of chlorine. A metallic radius is one half the shortest internuclear distance in a crystal of solid metal.

(c) The atomic number of an atom equals the number of protons in the nucleus and thus the positive charge of the nucleus. The effective nuclear charge is the positive charge experienced by the outermost electrons. Much of the nuclear charge is cancelled out by intervening electrons of inner shells.

(d) Ionization energy is that needed to remove an electron from a gaseous atom (or ion). Electron affinity is the energy needed to add an electron to a gaseous atom (or ion).

(e) A paramagnetic species has one or more unpaired electrons and is attracted into a magnetic field. A diamagnetic species has all electrons paired and is slightly repelled by a magnetic field.

4. The pairs of elements that are "out of order" based on their atomic masses are presented here, together with their atomic numbers. It is clear that in each instance the elements are ordered by increasing atomic number, not increasing atomic mass. For one of these pairs there is a further explanation. Most of the Ar in the atmosphere is thought to result from the radioactive decay of ^{40}K, a nuclide of potassium that once was more plentiful than it is now.

Element	Ar	K	Te	I	Co	Ni
Z	18	19	52	53	27	28
At. ms.	39.948	39.098	127.60	126.90	58.93	58.69
Element	Th	Pa	Pu	Am	Unh	Uns
Z	90	91	94	95	106	107
At. ms.	232.038	231.0359	244	243	263	262

5. **(a)** The number of protons in the nucleus of $^{119}_{50}Sn$ equals the atomic number: 50 protons.
(b) The number of neutrons in the nucleus is the difference between the mass number and the atomic number of the nuclide: 119 – 50 = 69 neutrons.
(c) Sn is in Group 4A of the fifth period of the periodic table. The 4*d* subshell completed filling with element 47 (Ag). Thus, there are ten 4*d* electrons in an atom of Sn.
(d) The 3*s* subshell is filled. Sn has two 3*s* electrons.
(e) The outer (valence) electron configuration of Sn is $5s^2\ 5p^2$. Sn has two 5*p* electrons.
(f) There are four electrons in the shell of highest principal quantum number ($n = 5$) in an atom of Sn.

6. In the literal sense, isoelectronic means having the same number of electrons. (In another sense, not used *in the text*, it means having the same electron configuration.) We determine the total number of electrons and the electron configuration for each species and make our decisions based on this information.

Fe^{2+}	24 electrons	[Ar] $3d^6\ 4s^0$	Sc^{3+}	18 electrons	[Ar] $3d^0\ 4s^0$
Ca^{2+}	18 electrons	[Ar] $3d^0\ 4s^0$	F^-	10 electrons	[He] $2s^2\ 2p^6$
Co^{2+}	25 electrons	[Ar] $3d^7\ 4s^0$	Co^{3+}	24 electrons	[Ar] $3d^6\ 4s^0$
Sr^{2+}	36 electrons	[Ar] $3d^{10}\ 4s^2\ 4p^6$	Cu^+	28 electrons	[Ar] $3d^{10}$
Zn^{2+}	28 electrons	[Ar] $3d^{10}\ 4s^0$	Al^{3+}	10 electrons	[He] $2s^2\ 2p^6$

Species with the same number of electrons and the same electron configuration are the following.
Fe^{2+} and Co^{3+} Sc^{3+} and Ca^{2+} F^- and Al^{3+} Zn^{2+} and Cu^+

7. Because isoelectronic species must have the same number of electrons, and each element has a different atomic number, atoms of different elements cannot be isoelectronic. Two different cations may be isoelectronic, as may two different anions, or an anion and a cation. An example would be two anions (or two cations, or an anion and a cation) that have the electron configuration of a nearby noble gas, such as: O^{2-} and F^-, Na^+ and Mg^{2+}, or F^- and Na^+.

8. **(a)** The most nonmetallic element is the one farthest to the right and the top (with the exception of the noble gases). This is the element fluorine.
(b) The transition elements are those with incomplete *d* subshells, or which form ions with incomplete *d* subshells. Scandium ([Ar] $3d^1\ 4s^2$) is the transition element with the lowest atomic number.
(c) The metalloids are defined as the elements in Groups 4A and 5A that are adjacent to the stair-step diagonal line in the periodic table, that is, Si, Ge, As, Sb, and Te; along with the elements at the bottom of Groups 6A and 7A, that is, Po and At. Of these 6 elements, only Si (Z = 14) has an atomic number exactly midway between those of two noble gases, Ne (Z = 10) and Ar (Z = 18).

9. In general, atomic size in the periodic table increases from top to bottom through a family and increases from right to left through a period, as indicated in Figures 10-4 and 10-8. The larger element is indicated first, followed by the reason for making the choice.
(a) Te   Te is to the left of Br and also in the period below that of Br in the 4th period.
(b) K   K is to the left of Ca in the periodic table, Period 4.
(c) Cs   Cs is both below and to the left of Ca in the periodic table.
(d) N   N is to the left of O in the periodic table, Period 2.
(e) P   P is to the left of O and also in the period below that of O which lies in the 2nd period.
(f) Au   Au is both below and to the left of Al in the periodic table.

10. An Al atom is larger than a F atom since Al is both below and to the left of F in the periodic table. As is larger than Al, since As is below Al in the periodic table. (Even though As is to the right of Al, we would not conclude that As is smaller than Al, since increases in size down a group are more pronounced than decreases in size across—left to right—a period.) A Cs^+ ion is isoelectronic with an I^- ion, and in an isoelectronic series, anions are larger than cations; thus I^- is larger than Cs^+. I^- also is larger than As, since I is below As in the periodic table (and increases in size down a group are more pronounced than those across a period). Finally, N is larger than F, since N is to the left of F in the periodic table. Therefore, we conclude that F is the smallest species listed and I^- is the largest. In fact, with the exception of Cs^+, we can rank the species in order of decreasing size. $I^- > As > Al > N > F$ and also $I^- > Cs^+$

11. Ionization energy in the periodic table decreases from top to bottom of a family, and increases from left to right of a period, as summarized in Figure 10-9. Cs has the lowest ionization energy; it is furthest to the left and nearest to the bottom of the periodic table. Next comes Sr, then As, then S, and finally F, the most nonmetallic element in the group (and in the periodic table). Thus the elements, listed in order of increasing ionization energy are: $Cs < Sr < As < S < F$

12. **(a)** The first element in a group has the smallest atoms. In Group 4A this is C.
(b) The atom in a period with the largest atoms is furthest to the left. In the fifth period this is Rb.
(c) The bottom element in a group has atoms with the lowest first ionization energy. In Group 7A this is At (or I if we do not wish to consider radioactive elements).

13. First we convert the mass of Na given to an amount in moles of Na. Then we compute the energy needed to ionize this much Na.

$$\text{Energy} = 1.00 \text{ mg Na} \times \frac{1 \text{ g Na}}{1000 \text{ mg}} \times \frac{1 \text{ mol Na}}{22.99 \text{ g Na}} \times \frac{495.8 \text{ kJ}}{1 \text{ mol Na}} = 0.0216 \text{ kJ} \times \frac{1000 \text{ J}}{1 \text{ kJ}} = 21.6 \text{ J}$$

14. **(a)** The most metallic element is the one closest to the lower left-hand corner. This is the element Ba.
(b) The most nonmetallic element is the one closest to the upper right-hand corner, the element S.
(c) There are two distinct metals in this group, Ba and Ca; they have the lowest ionization energies.There is one distinct nonmetal, S; it has the highest ionization energy. The remaining two elements, As and Bi, are a metalloid and a metal, respectively. The metalloid has a higher ionization energy than does the metal. Thus Bi has the intermediate value of ionization energy of the five elements listed.

15. Metallic character decreases from left to right and from bottom to top in the periodic table. Thus, in order of decreasing metallic character the elements listed are: $Rb > Ca > Sc > Fe > Te > Br > O > F$ The difficulty in establishing this series is in placing the elements Te and Br. First, the metal Fe is more metallic than the nonmetal Te. Further, Te clearly is more metallic than the halogen Br. Finally, we only need to recognize that Cl and O have approximately the same nonmetallic character, and Br clearly is more metallic than is Cl.

16. Paramagnetism indicates unpaired electrons, which in turn are often associated with partially filled subshells. First we write the electron configurations of the elements, and then those of the ions. From those electron configurations, we determine whether the species is paramagnetic or diamagnetic.

K	[Ar] $4s^1$	K^+	[Ar] $4s^0$	diamagnetic, all subshells filled.
Cr	[Ar] $3d^5 4s^1$	Cr^{3+}	[Ar] $3d^3$	paramagnetic.
Zn	[Ar] $3d^{10} 4s^2$	Zn^{2+}	[Ar] $3d^{10}$	diamagnetic, all subshells filled.
Cd	[Kr] $4d^{10} 5s^2$			diamagnetic, all subshells filled.
Co	[Ar] $3d^7 4s^2$	Co^{3+}	[Ar] $3d^6$	paramagnetic.
Sn	[Kr] $4d^{10} 5s^2 5p^2$	Sn^{2+}	[Kr] $4d^{10} 5s^2 5p^0$	diamagnetic, all subshells filled.
Br	[Ar] $3d^{10} 4s^2 4p^5$			paramagnetic, an odd number of electrons.

17. **(a)** 6. Tl's electron configuration [Xe] $4f^{14} 5d^{10} 6s^2 6p^1$ has one *p* electron in its outermost shell.
(b) 8. $Z = 70$ identifies the element as Yb. Yb should have the largest radius of the elements given—but *not* of all the elements—since atomic radii generally increase from top to bottom in the periodic table and from right to left. Yb also is in the *f*-block of the periodic table, an inner transition element.
(c) 5. Ni has the electron configuration [Ar] $4s^2 3d^8$. It also is a *d*-block element.
(d) 1. An s^2 outer electron configuration, with the underlying configuration of a noble gas, is characteristic of elements of family 2A, the alkaline earth elements.

(e) 2. The element in the fifth period and Group 5A is Sb, a metalloid.

(f) 4. The element in the fourth period and Group 6A is Se, a nonmetal. (6 might also be a choice, since boron also is a nonmetal, and it has one *p* electron in the shell of highest principal quantum number, but the other elements of Group 3A are *not* nonmetals.)

18. We expect periodic properties to be functions of atomic number.

element, atomic number	He, 2		Ne, 10		Ar, 18		Kr, 36		Xe, 54	Rn, 86
boiling point, K	4.2 K		27.1 K		87.3 K		119.7 K		165 K	
Δb.p. / ΔZ		2.86		7.5		1.8		2.5		

With one notable exception, we see that the boiling point increases about 2.4 K per unit of atomic number. The atomic number increases by 32 units from Xe to Rn. We might expect the boiling point to increase by $(2.4 \times 32 =)$ 77 K to $(165 + 77 =)$ 242 K for Rn.

A simpler manner is to expect that the boiling point of xenon (165 K) is the average of the boiling points of radon and krypton (120 K).

$165 \text{ K} = (120 \text{ K} + ?)/2 \qquad ? = 2 \times 165 \text{ K} - 120 \text{ K} = 210 \text{ K}$ = boiling point of radon

Often the simplest way is best. The tabulated value is 211 K.

EXERCISES

The Periodic Law

19. Element 114 will be a metal in the same family as Pb, element 82 (18 cm^3/mol); Sn, element 50 (18 cm^3/mol); and Ge, element 32 (14 cm^3/mol). We note that the atomic volume of Pb and Sn are essentially equal, probably due to the lanthanide contraction. If there is also an actinide contraction, element 114 will have an atomic volume of 18 g/cm^3; if not, the value is probably about 22 cm^3/mol. This need to estimate atomic volume is what makes the value for density inaccurate.

$$\text{density (g/cm}^3) = \frac{298 \text{ g/mol}}{18 \text{ cm}^3\text{/mol}} = 16 \text{ g/cm}^3 \qquad \text{density (g/cm}^3) = \frac{298 \text{ g/mol}}{22 \text{ cm}^3\text{/mol}} = 14 \text{ g/cm}^3$$

20. Lanthanum has an atomic number of $Z = 57$, and thus its atomic volume is somewhat less than 25 g/cm^3. Let us assume 23 cm^3/mol.

atomic mass = density × atomic volume = 6.145 g/cm^3 × 23 cm^3/mol = 141 g/mol

This compares very well with the listed value of 139 for La.

21. The following data are plotted at right. Density clearly is a periodic property for these two periods of main group elements. It rises, falls a bit, rises again, and falls back to the axis, in both cases.

	Z	density, g/cm^3
Na	11	0.968
Mg	12	1.738
Al	13	2.699
Si	14	2.336
P	15	1.823
S	16	2.069
Cl	17	0.0032
Ar	18	0.0018
K	19	0.856
Ca	20	1.550
Ga	31	5.904
Ge	32	5.323
As	33	5.778
Se	34	4.285
Br	35	3.100
Kr	36	0.0037

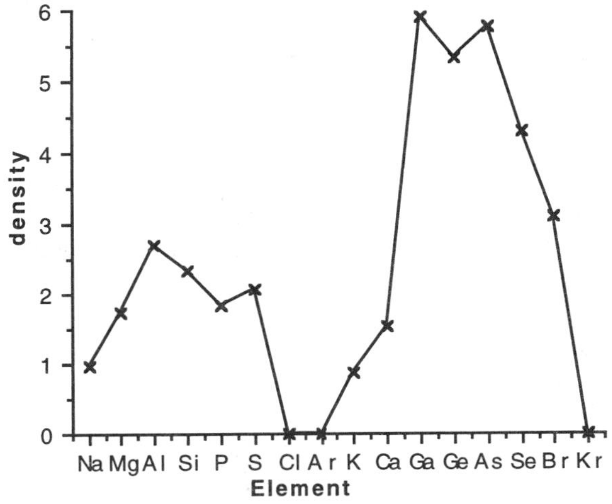

22. The following data are plotted at right below. Melting point clearly is a periodic property for these two periods. It rises to a maximum and then falls off in each case.

	Z	m.p., °C
Li	3	179
Be	4	1278
B	5	2300
C	6	3350
N	8	–210
O	7	–218
F	9	–220
Ne	10	–249
Na	11	98
Mg	12	651
Al	13	660
Si	14	1410
P	15	590
S	16	119
Cl	17	–101
Ar	18	–189

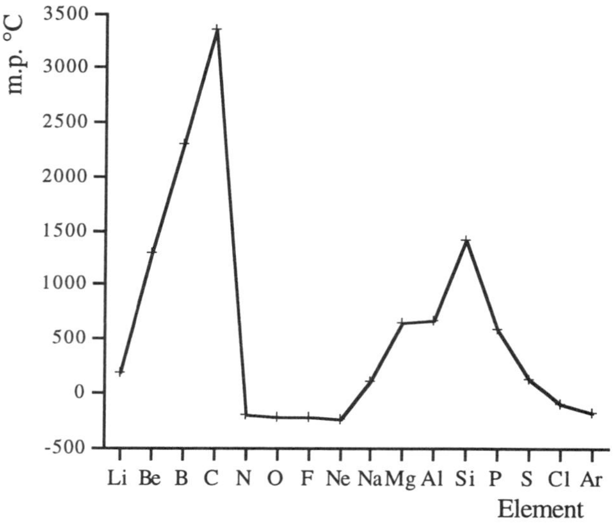

The Periodic Table

23. Mendeleev arranged elements in the periodic table in order of increasing atomic weight. Of course, fractional atomic weights are permissable. Hence, there always is room for an added element between two elements that already are present in the table. On the other hand, Moseley arranged elements in order of increasing atomic number. Only integral (whole number) values of atomic number are permitted. Thus, when elements with all possible integral values in a certain range have been discovered, no new elements are possible in that range.

24. For there to be the same number of elements in each period of the periodic table, each shell of an electron configuration would have to contain the same number of electrons. This is not the case, the shells have 2 (*K* shell), 8 (*L* shell), 18 (*M* shell), and 32 (*N* shell) electrons each. This is because each of the periods of the periodic table begins with one *s* electron beyond a noble gas electron configuration, and the noble gas electron configuration corresponds to either s^2 (He) or $s^2\,p^6$, with various other full subshells.

<u>25</u>. **(a)** The noble gas following radon (Z = 86) will have an atomic number of (86 + 32 =) 118.
(b) The alkali metal following francium (Z = 87) will have an atomic number of (87 + 32 =) 119.
(c) The mass number of radon (A = 222) is (222 ÷ 86 =) 2.58 times its atomic number. The mass number of Lr (A = 260) is (260 ÷ 103 =) 2.52 times its atomic number. Thus, we would expect the mass numbers, and hence approximate atomic masses of elements 118 and 119 to be about 2.52 times their atomic numbers, that is, $A_{118} \approx 297$ u and $A_{119} \approx 300$ u.

26. **(a)** The 6*d* subshell will be complete with an element in the same family as Hg. This is three elements beyond Une (Z = 109) and thus this element has an atomic number of Z = 112.
(b) The element in this period that will most closely resemble bismuth is six elements beyond Une (Z = 109) and thus has Z = 115.
(c) The element in this period that would be a noble gas would fall below Rn, nine elements beyond Une (Z = 109) and thus has Z = 118.

Atomic Radii

27. The reason why the sizes of atoms do not simply increase with atomic number is because electrons often are added successively to the same subshell. These electrons do not screen each other from the nuclear charge (they do not effectively get between each other and the nucleus). Consequently, as each electron is added to a subshell and the nuclear charge increases by one unit, all of the electrons in this subshell are drawn more closely into the nucleus.

28. The reason why atomic sizes are uncertain is because the electron cloud that surrounds an atom has no specific limit. It can be pictured as gradually fading away, rather like the edge of a town. In both cases we pick an arbitrary boundary.

29. **(a)** The smallest atom in Group 3A is the first: B
(b) Po is in the sixth period, and is larger than the others, which are rewritten in the following list from left to right in the fifth period, that is, from largest to smallest: Sr, In, Sb, Te. Thus, Te is the smallest of the elements given.

30. The hydrogen ion contains no electrons, only a nucleus. It is exceedingly tiny, much smaller than any other atom or electron-containing ion. Both H and H^- have a nuclear charge of +1, but H^- has two electrons to H's one, and thus is larger. Both He and H^- contain two electrons, but He has a nuclear charge of +2, while H^- has one of only +1. The smaller nuclear charge of H^- is less effective at attracting electrons than the nuclear charge of He. The only comparison left is between H and He; we expect He to be smaller since atomic size decreases left to right across a period. Thus, ordered by increasing size: $H^+ < He < H < H^-$.

31. Ions can be isoelectronic without having noble-gas electron configurations. An example is Cu^+ and Zn^{2+}, which both have the electron configuration [Ar] $3d^{10}\,4s^0$.

32. In an isoelectronic series, all of the species have the same number of electrons. The size is determined by the nuclear charge. Those species with the largest (positive) nuclear charge are the smallest. Those with smaller nuclear charges are larger in size. Thus, the more positively charged an ion is in an isoelectronic series, the smaller it will be. $Y^{3+} < Sr^{2+} < Rb^+ < Kr < Br^- < Se^{2-}$

33. Li^+ is the smallest; it not only is in the second period, but also is a cation. I^- is the largest, an anion in the 5th period. Next largest is Se in the previous (the 4th) period. We expect Br to be smaller than Se because it is both to the right of Se and in the same period. $Li^+ < Br < Se < I^-$

34. Size decreases from left to right in the periodic table; on this basis I should be smaller than Al. But size increases from top to bottom in the periodic table; on this basis I should be larger than Al. There really is no good way of resolving these conflicting predictions.

Ionization Energies; Electron Affinities

35. The second ionization energy for an atom (I_2) cannot be smaller than the first ionization energy (I_1) for the same atom. The reason is that, when the first electron is removed it is being taken away from a species with a charge of +1. On the other hand, when the second electron is removed, it is being taken from a species with a charge of +2. Since the force between two charged particles is proportional to q_+q_-/r^2 (r is the distance between the particles), the higher the charge, the more difficult it will be to remove an electron.

36. In the case of a first electron affinity, a negative electron is being added to a neutral atom. This process may be either exothermic or endothermic. In the case of an ionization potential, however, a negatively charged electron is being removed from a positively charged cation, a process that must always require energy, because unlike charges attract each other.

37. There are four third-shell electrons in each atom of silicon. In Table 10.4, the values for the ionization energies of these four electrons are listed in kJ/mol. Thus, the energy needed to ionize all four electrons from a mole of silicon atoms is given by:

$$\text{Energy} = 786.5 + 1577 + 3232 + 4356 = 9952\ \text{kJ}$$

38. Data (I_1 = 375.7 kJ/mol) are obtained from Table 10.3.

$$\text{no. } Cs^+ \text{ ions} = 1\ \text{J} \times \frac{1\ \text{kJ}}{1000\ \text{J}} \times \frac{1\ \text{mol } Cs^+}{375.7\ \text{kJ}} \times \frac{6.022 \times 10^{23}\ Cs^+ \text{ ions}}{1\ \text{mol}} = 1.603 \times 10^{18}\ Cs^+ \text{ ions}$$

39. The electron affinity of bromine is –324.6 kJ/mol (Figure 10-10). We use Hess's law to determine the heat of reaction for $Br_2(g)$ becoming $Br^-(g)$.

$Br_2(g) \longrightarrow 2\,Br(g)$ $\quad \Delta H = +\,193$ kJ

$2\,Br(g) + 2\,e^- \longrightarrow 2\,Br^-(g)$ $\quad E.A. = 2(-324.6)$ kJ

$Br_2(g) + 2\,e^- \longrightarrow 2\,Br^-(g)$ $\quad \Delta H = -456.2$ kJ $\quad$ The overall process is *exothermic*.

40. The electron affinity of fluorine is –328.0 kJ/mol (Figure 10-10) and the first and second ionization energies of Mg (Table 10.4) are 737.7 kJ/mol and 1451 kJ/mol.

$Mg(g) \longrightarrow Mg^{+}(g) + e^{-}$	$I_1 = 737.7$ kJ/mol
$Mg^{+}(g) \longrightarrow Mg^{2+}(g) + e^{-}$	$I_2 = 1451$ kJ/mol
$2\ F(g) + 2\ e^{-} \longrightarrow 2\ F^{-}(g)$	$2\ E.A. = 2(-328.0)$ kJ/mol
$Mg(g) + 2\ F(g) \longrightarrow Mg^{2+}(g) + 2\ F^{-}(g)$	$\Delta H = +1533$ kJ/mol

This reaction is endothermic.

41. The reasoning here is similar to that for the answer to Exercise 35. In both cases, an electron is being removed from a neon electron configuration. But in the case of Na^{+}, the electron is being removed from a species with a charge of +2, while in the case of Ne, the electron is being removed from a species with a charge of +1. The more highly charged the resulting species is, the more difficult it is to produce it by removing an electron.

42. While the electron affinity of Li is –59.6 kJ/mol, the smallest ionization energy listed in Table 10.3 (and, except for Fr, displayed in Figure 10-9) is that of Cs, 375.7 kJ/mol. Thus, insufficient energy is produced by the electron affinity of Li to account for the ionization of Cs. And we would predict that compounds containing the Li^{-} ion will not be stable. (The other consideration is the energy released when the positive and negative ion combine to form an ion pair. This is an exothermic process, but we have no way of assessing the value of this energy.)

Magnetic Properties

43. Three of the ions have noble gas electron configurations and thus have no unpaired electrons:
F^{-} is $1s^2\ 2s^2\ 2p^6$ Ca^{2+} and S^{2-} are [Ne] $3s^2\ 3p^6$
Only Fe^{2+} has unpaired electrons in its configuration: [Ar] $3d^6\ 4s^0$.

44. First we write the electron configuration of the element, then that of the ion, and then an orbital diagram. In each case, the number of unpaired electrons shown in the orbital diagram agrees with the data given in the statement of the problem.

(a)	Ni	[Ar] $3d^8\ 4s^2$	Ni^{2+} [Ar] $3d^8\ 4s^0$	[Ar] 3d [↑↓][↑↓][↑↓][↑][↑] 4s [ ]	two
(b)	Cu	[Ar] $3d^{10}\ 4s^1$	Cu^{2+} [Ar] $3d^9\ 4s^0$	[Ar] 3d [↑↓][↑↓][↑↓][↑↓][↑] 4s [ ]	one
(c)	Cr	[Ar] $3d^5\ 4s^1$	Cr^{3+} [Ar] $3d^3\ 4s^0$	[Ar] 3d [↑][↑][↑][ ][ ] 4s [ ]	three

45. All atoms with an odd number of electrons must be paramagnetic. There is no way to create all pairs with an odd number of electrons. Many atoms with an even number of electrons are diamagnetic, but some are paramagnetic. The one of lowest atomic number is carbon ($Z = 6$): [He] 2s [↑↓] 2p [↑][↑][ ].

46. The electron configuration of each iron ion can be determined by starting with the electron configuration of the iron atom. [Fe] = [Ar] $3d^6\ 4s^2$ = [Ar] 3d [↑↓][↑][↑][↑][↑] 4s [↑↓] Then the two ions result from the loss of both 4*s* electrons, followed by the loss of a 3*d* electron in the case of Fe^{3+}.
[Fe^{2+}] = [Ar] $3d^6\ 4s^0$ = [Ar] 3d [↑↓][↑][↑][↑][↑] 4s [ ] four unpaired electrons
[Fe^{3+}] = [Ar] $3d^5\ 4s^0$ = [Ar] 3d [↑][↑][↑][↑][↑] 4s [ ] five unpaired electrons

Predictions Based on the Periodic Table

47. **(a)** Elements that one would expect to exhibit the photoelectric effect with visible light should be ones that have a small value of their first ionization energy. Based on Figure 10-9, the alkali metals have the lowest first ionization potentials: Cs, Rb, and K are three suitable metals.

Metals that would not exhibit the photoelectric effect with visible light are those that have high values of their first ionization energy. Again from Figure 10-9, Zn, Cd, and Hg seem to be three metals that would not exhibit the photoelectric effect with visible light.

(b) From Figure 10-1, we notice that the atomic (molar) volume increases for the solid forms of the noble gases as we travel down the group (the data points just before the alkali metal peaks). But it seems to increase less rapidly than the molar mass. This means that the density should increase with atomic mass, and Rn should be the most dense solid. We expect densities of liquids to follow the same trend as densities of solids.

(c) To estimate the first ionization energy of fermium, we note in Figure 10-9 that the ionization energies of the lanthanides (following the Cs valley) are approximately the same. We expect similar behavior of the actinides, and estimate a first ionization energy of about +600 kJ/mol.

(d) We can estimate densities of solids from the information in Figure 10-1. Radium has $Z = 88$ and an approximate atomic volume of 40 cm³/mol. Then we use the atomic weight of radium to determine its density:

$$\text{density} = \frac{1\ \text{mol}}{40\ \text{cm}^3} \times \frac{226\ \text{g Ra}}{1\ \text{mol}} = 5.7\ \text{g/cm}^3$$

48. Germanium lies between silicon and tin in Group 4A. We would expect the heat of atomization of germanium to approximate the average of the heats of atomization ofSi and Sn.

$$\text{average} = \frac{452\ \text{kJ/mol Si} + 302\ \text{kJ/mol Sn}}{2} = 377\ \text{kJ/mol Ge}$$

(This is precisely the same value given by a handbook.)

49. **(a)** From Figure 10-1, the atomic (molar) volume of Al is 10. cm³/mol and that of In is 15 cm³/mol. Thus, we predict 12.5 cm³/mol as the molar volume of Ga. Then we compute the density of Ga.

$$\text{density} = \frac{1\ \text{mol Ga}}{12.5\ \text{cm}^3} \times \frac{69.7\ \text{g Ga}}{1\ \text{mol Ga}} = 5.58\ \text{g/cm}^3 \text{ or } 5.6\ \text{g/cm}^3 \text{ as closely as the graph can be read.}$$

(b) Since Ga is in family 3A (Gruppe III on Mendeleev's table), the formula of its oxide should be Ga_2O_3. We use Mendeleev's atomic weights. mol weight $= 2 \times 68$ g Ga $+ 3 \times 16$ g O $= 184$ g Ga_2O_3

$$\%\text{Ga} = \frac{2 \times 68\ \text{g Ga}}{184\ \text{g Ga}_2\text{O}_3} \times 100\% = 74\%\ \text{Ga}$$

With modern atomic weights we obtain 74.4% Ga.

50. **(a)** The boiling point increases by 52 C° from CH_4 to SiH_4, and by 22 C° from SiH_4 to GeH_4. One expects another increase in boiling point from GeH_4 to SnH_4, probably by about 15 C°. Thus the predicted boiling point of SnH_4 is –75°C. (The actual boiling point of SnH_4 is –52 °C.)

(b) The boiling point decreases by 39 C° from H_2Te to H_2Se, and by 20 C° from H_2Se to H_2S. It probably will decrease by about 10 C° to reach H_2O. Therefore, one predicts a value of the boiling point of H_2O is –71°C. Of course, the actual boiling point of water is 100°C. The prediction is seriously in error because of hydrogen bonding between water molecules, a topic that is discussed in Chapter 13.

51. **(a)** Size increases down a group and from right to left in a period. Ba is closest to the lower left corner of the periodic table and has the largest size.

(b) Ionization energy decreases down a group and from right to left in a period. Although Pb is closest to the bottom of its group, Sr is farthest left in its period (and only one period above Pb). Sr should have the lowest first ionization energy.

(c) Electron affinity becomes more negative from left to right in a period and from bottom to top in a group. Cl is closest to the upper right in the periodic table and has the most negative (smallest) electron affinity.

(d) The number of unpaired electrons can be determined from the orbital diagram for each species.

F	[He] 2s [↑↓] 2p [↑↓][↑↓][↑]	1 unpaired e⁻	N	[He] 2s [↑↓] 2p [↑][↑][↑]	3 unpaired e⁻
S^{2-}	[Ne] 3s [↑↓] 3p [↑↓][↑↓][↑↓]	0 unpaired e⁻	Mg^{2+}	[He] 2s [↑↓] 2p [↑↓][↑↓][↑↓]	0 unpaired e⁻
Sc^{3+}	[Ne] 3s [↑↓] 3p [↑↓][↑↓][↑↓]	0 unpaired e⁻	Ti^{3+}	[Ar] 4s [ ] 3d [↑][ ][ ][ ][ ]	1 unpaired e⁻

Thus, N has the largest number of unpaired electrons.

52. **(a)** $Z = 32$ 1. This is the element Ge, with an outer electron configuration of $ns^2\ np^2$. Thus, Ge has two unpaired p electrons.

(b) $Z = 8$ 1. This is the element O. Each atom has an outer electron configuration of $ns^2\ np^4$, which produces a partial (valence) orbital diagram of: 2s [↑↓] 2p [↑↓][↑][↑]. Thus, there are two unpaired p electrons.

(c) $Z = 53$ 3. This is the element I, with an electron affinity more negative than that of the adjacent atoms: Xe and Te.

(d) $Z = 38$ 4. This is the element Sr, which has two 5s electrons. It is easier to remove one 5s electron than to remove the outermost 4s electron of Ca, but harder than removing the outermost 6s electron of Cs.

(e) $Z = 48$ 2. This is the element Cd. Its outer electron configuration is s^2 and thus it is diamagnetic.

(f) $Z = 20$ 2. This is the element Ca. Since its electron configuration is [Ar] $4s^2$, all electrons are paired and it is diamagnetic.

11 Chemical Bonding I: Basic Concepts

Practice Examples

1A Mg is in Family 2A, and thus has 2 valence electrons and 2 dots in its Lewis symbol.
Ge is in Family 4A, and thus has 4 valence electrons and 4 dots in its Lewis symbol.
K is in Family 1A, and thus has 1 valence electron and 1 dot in its Lewis symbol.
Ne is in Family 8A, and thus has 8 valence electrons and 8 dots in its Lewis symbol.

·Mg·   ·Ġe·   K·   :N̈e:

1B Sn is in Family 4A, and thus has 4 electrons and 4 dots in its Lewis symbol.
Br is in Family 7A with 7 valence electrons. Adding an electron produces an ion with 8 valence electrons.
Tl is in Family 3A with 3 valence electrons. Removing an electron produces a cation with 2 valence electrons.
S is in Family 6A with 6 valence electrons. Adding 2 electrons produces an anion with 8 valence electrons.

·Ṡn· $[:\ddot{\text{Br}}:]^-$ $[\cdot\text{Tl}\cdot]^+$ $[:\ddot{\text{S}}:]^{2-}$

2A The Lewis structure for the cation, the anion, and the compound follows the explanation.

(a) Na loses one electron to form Na^+, while S gains two to form S^{2-}.

Na· – 2 e⁻ ⟶ $[Na]^+$ ·S̈· + e⁻ ⟶ $[:\ddot{\text{S}}:]^{2-}$ Lewis Structure: $[Na]^+\ [:\ddot{\text{S}}:]^{2-}\ [Na]^+$

(b) Mg loses two electrons to form Mg^{2+}, while N gains three to form N^{3-}.

·Mg· – 2 e⁻ ⟶ $[Mg]^{2+}$ ·Ṅ· + 3 e⁻ ⟶ $[:\ddot{\text{N}}:]^{3-}$ Lewis Structure: $[Mg]^{2+}\ [:\ddot{\text{N}}:]^{3-}\ [Mg]^{2+}\ [:\ddot{\text{N}}:]^{3-}\ [Mg]^{2+}$

2B Below each explanation are the Lewis structures: for the cation, for the anion, and for the compound.

(a) In order to acquire a noble-gas electron configuration, Ca loses two electrons, and I gains one, forming the ions Ca^{2+} and I^-. The formula of the compound is CaI_2.

·Ca· – 2 e⁻ ⟶ $[Ca]^{2+}$:Ï· + e⁻ ⟶ $[:\ddot{\text{I}}:]^-$ Lewis Structure: $[:\ddot{\text{I}}:]^-\ [Ca]^{2+}\ [:\ddot{\text{I}}:]^-$

(b) Ba loses two electrons and S gains two to acquire a noble-gas electron configuration, forming the ions Ba^{2+} and S^{2-}. The formula of the compound is BaS.

·Ba· – 2 e⁻ ⟶ $[Ba]^{2+}$ ·S̈· + 2 e⁻ ⟶ $[:\ddot{\text{S}}:]^{2-}$ Lewis Structure: $[Ba]^{2+}\ [:\ddot{\text{S}}:]^{2-}$

(c) Each Li loses one electron and each O gains two to attain a noble-gas electron configuration, producing the ions Li^+ and O^{2-}. The formula of the compound is Li_2O.

Li· – 2 e⁻ ⟶ $[Li]^+$ ·Ö + 2 e⁻ ⟶ $[:\ddot{\text{O}}:]^{2-}$ Lewis Structure: $[Li]^+\ [:\ddot{\text{O}}:]^{2-}\ [Li]^+$

3A The Lewis structure of CH_3Br has all single bonds. From Table 11.1 the length of a C—H bond is 110 pm. The length of a C—Br bond is not given in the table. A reasonable value is the average of the C—C and Br—Br bond lengths.

$$C\text{—}Br = \frac{C\text{—}C + Br\text{—}Br}{2} = \frac{154 \text{ pm} + 228 \text{ pm}}{2} = 191 \text{ pm}$$

$$\begin{array}{c} H \\ | \\ H\text{—}C\text{—}\overline{Br}| \\ | \\ H \end{array}$$

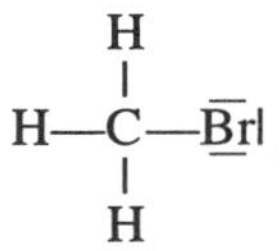

3B In Table 11-3, a C—O bond length is 143 pm, while a C=O bond length is 120 pm, much closer to the experimental bond length in CO_2. In determining the Lewis structure, we begin with 4 valence electrons from C and 6 valence electrons from each of the two O atoms, for a total of $4 + (2 \times 6) = 16$ valence electrons, or 8 pairs of electrons. Two of these eight pairs are used to attach the C atom to each O atom. The remaining 6 pairs complete the octet on each O atom. This produces this Lewis structure: $\overline{|O}\text{—}C\text{—}\overline{O}|$ This Lewis structure has two flaws: (1) There is no octet on C. (2) The carbon-to-oxygen bonds are single bonds, in contradiction to experimental evidence. If a lone pair of electrons from each O is shared with C, the following Lewis structure is produced, that overcomes both objections. $|O\text{=}C\text{=}O|$

4A The bond with the most ionic character is the one in which the two bonded atoms are the most different in their electronegativities. We find electronegativities in Table 11-2 and calculate ΔEN for each bond.
Electronegativities: H = 2.1 Br = 2.8 N = 3.0 O = 3.5 P = 2.1 Cl = 3.0
Bonds: H—Br N—H N—O P—Cl
ΔEN values: 0.7 0.9 0.5 0.9
Therefore, the P—Cl or the N—H bond are the most polar of the four bonds cited.

4B The most polar bond is the one with the greatest electronegativity difference.
Electronegativities: C = 2.5 S = 2.5 P = 2.1 O = 3.5 F = 4.0
Bonds: C—S C—P P—O O—F
ΔEN values: 0.0 0.4 1.4 0.5
Therefore, the P—O bond is the most polar of the four bonds cited.

5A **(a)** N has 5 valence electrons and each I has 7 valence electrons: $5 + (3 \times 7) = 26$ valence electrons or 13 pairs of valence electrons. We place N at the center of three I, and use three pairs to hold the molecule together, one between each I and N. We complete the octet on each I with three pairs for each I. This uses nine more pairs, for a total of twelve pairs used. The thirteenth pair is used to complete N's octet.

$$\begin{array}{c} \overline{\overline{I}}| \\ | \\ |\overline{I}\text{—}N\text{—}\overline{I}| \end{array}$$

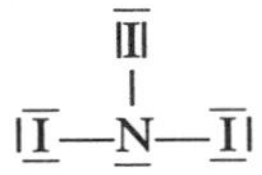

(b) C has 4 valence electrons and each S has 6 valence electrons: $4 + (2 \times 6) = 16$ valence electrons or 8 pairs of valence electrons. We place C between two S, and use two pairs to hold the molecule together, one between each C and S. We complete the octet on each S with three pairs for each S. This uses six more pairs, for a total of eight pairs used. $|\overline{S}\text{—}C\text{—}\overline{S}|$ But C does not have an octet. We correct this situation by moving one lone pair of each S into a bonding position between C and S. $\overline{S}\text{=}C\text{=}\overline{S}$

5B Each of the two N atoms contributes 5 valence electrons to the Lewis structure and each of the F atoms contributes 7 valence electrons. There are $(2 \times 5) + (2 \times 7) = 24$ valence electrons, or 12 pairs of valence electrons. The first attempt at a Lewis structure places three of these pairs of electrons between the four atoms to bond them together, three more on each of the terminal F's to complete their octets, two more pairs as lone pairs on one of the N's to complete its octet, and one pair on the remaining N. $|\overline{F}\text{—}\overline{N}\text{—}\overline{N}\text{—}\overline{F}|$ This means that this "last" N does not have an octet of electrons. This can be corrected by converting one of the other N's lone pairs into a bonding pair, forming a double bond. $|\overline{F}\text{—}\overline{N}\text{=}\overline{N}\text{—}\overline{F}|$

6A

Atom type	H—	—O=	—$\overline{O}$|	—$\overline{O}$—	N	=$\overline{O}$
Valence electrons	1	6	6	6	5	6
–Lone-pair electrons	–0	–2	–6	–4	–0	–4
–Bond-pair electrons	–1	–3	–1	–2	–4	–2
Formal charge	0	+1	–1	0	+1	0

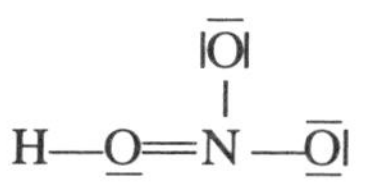

In each structure, H has zero formal charge and N has +1 formal charge. In the left-hand structure two O have –1 formal charge, the other has +1. In the right-hand structure, two O have zero formal charge, the other has –1. We prefer the right-hand structure because it has more formal charges equal to zero and it does not put a positive formal charge on an electronegative atom.

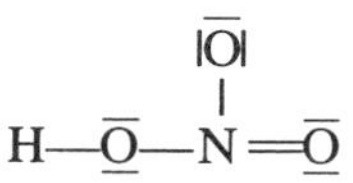

6B C has 4 valence electrons, each N has 5, and each H has one: $4 + (2 \times 5) + (2 \times 1) = 16$ valence electrons or eight pairs. We bond to Hs to one N, bond that N to C and that C to the other N. With a pair of electrons between each bonded pair of atoms, we have used four electron pairs. We complete the octet on the right-most N with three pairs, and then on the left-most N with the remaining pair. But C does not have an octet. We can solve this by creating either two double bonds (C=N), or a single bond (N—C) and a triple bond (C≡N).

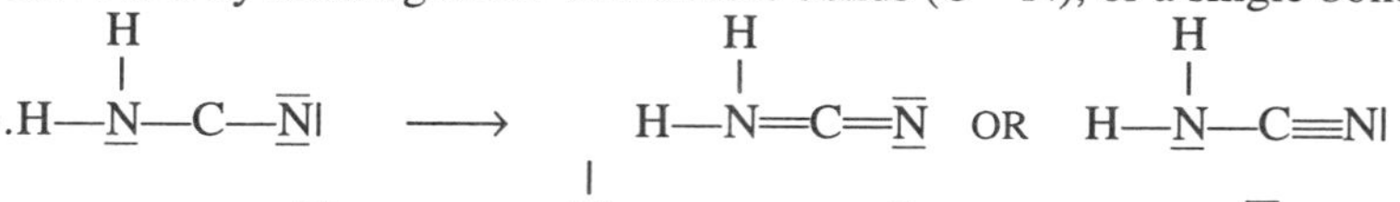

Atom type	H—	—N=	C	=N	≡N|
Valence electrons	1	5	4	5	5
–Lone-pair electrons	–0	–0	–0	–4	–2
–Bond-pair electrons	–1	–4	–4	–2	–3
Formal charge	0	+1	0	–1	0

Thus, in each structure H has zero formal charge and C has zero formal charge. In the left-hand structure the two Ns have +1 and –1 formal charge, while in the right-hand structure, each N has zero formal charge. The right-hand structure is preferred.

7A B has 3 valence electrons, each F has 7 valence electrons, and the ionic charge contributes one electron: $3 + (4 \times 7) + 1 = 32$ valence electrons, 16 electron pairs. Arrange the four Fs around the B and hold each in place with a bond pair, using four electron pairs. Complete the octet of each F, requiring three electron pairs each, or twelve electron pairs. All 16 electron pairs have been used. Formal charges are computed.

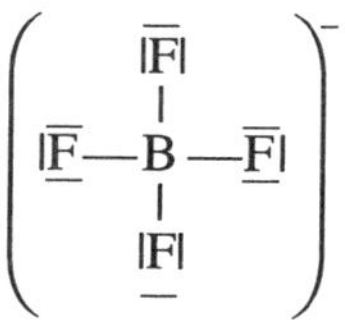

Formal charge on F = 7 valence electrons – (3×2) lone pair electrons – 1 bond pair electron = 0.
Formal charge on B = 3 valence electrons – 0 lone pair electrons – 4 bond pair electrons = –1.
It is the B that carries the negative formal charge.

7B Sodium peroxide consists of sodium cations, Na^+, and peroxide anions, O_2^{2-}. A Na atom loses one electron to attain a noble-gas electron configuration producing the ion Na^+. $Na\cdot - e^- \longrightarrow [Na]^+$
Each O atom contributes 6 valence electrons to the peroxide anion, and 2 more electrons are added to account for the –2 charge, for a total of $(2 \times 6) + 2 = 14$ valence electrons, or 7 pairs. In the Lewis structure of the peroxide anion, one pair of electrons is used to bond the O atoms together, and the remaining 6 pairs are used, 3 each, to complete the octet of each O. $Na^+ \,[|\overline{\underline{O}}—\overline{\underline{O}}|]^{2-}\, Na^+$

8A Each H has one valence electron, each C has four valence electrons, each O has six valence electrons, and the charge contributes one electron: $(3 \times 1) + (2 \times 4) + (2 \times 6) + 1 = 24$ valence electrons, 12 pairs. Attach the three Hs to one C, that C to the second C and the two Os to the second C. Place an electron pair between each pair of bonded atoms and six electron pairs have been used. To complete the octet of the two oxygens requires the remaining six electron pairs.

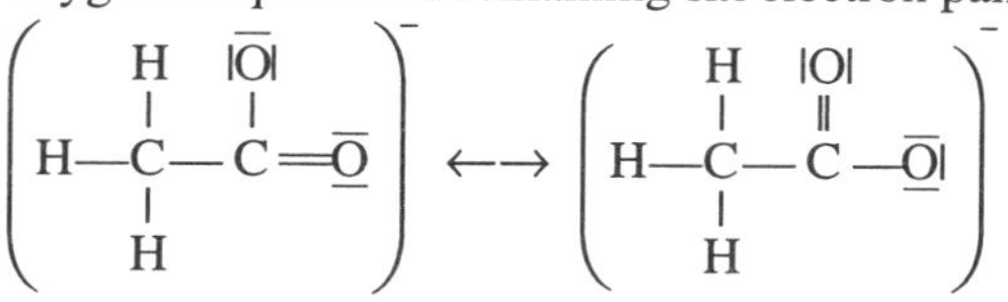

This leaves the octet incomplete on C. The double bond that completes that octet can be to either oxygen.

8B Nitrogen-to-nitrogen bond lengths are: N—N = 145 pm   N=N = 123 pm   N≡N = 109.8 pm
Thus, the resulting Lewis structure should depict each bond as part way between a double and a triple bond. In the azide ion, each N atom contributes 5 valence electrons, and 2 more electrons are added for the –1 charge. This gives a total of $(3 \times 5) + 1 = 16$ valence electrons, or 8 pairs. In a first attempt at a Lewis

structure, 2 pairs are used to join the atoms together and the remaining 6 pairs complete the octets of the terminal N's. First attempt: $[|\overline{N}—N—\overline{N}|]^-$ This leaves the central N without an octet, a deficiency that is remedied by converting two lone pairs into bonding pairs.

$$[|\overline{N}—N{\equiv}N|]^- \leftrightarrow [|N{\equiv}N—\overline{N}|]^- \leftrightarrow [|\overline{N}{=}N{=}\overline{N}|]^-$$

These three resonance forms give bond lengths approximately those of nitrogen-to-nitrogen double bonds.

9A The Lewis structure of NCl_3 has three Cl atoms bonded to N and one lone pair attached to N. These four electron groups around N produce a tetrahedral electron-group geometry. The fact that one of the electron groups is a lone pair means that the molecular geometry is trigonal pyramidal.

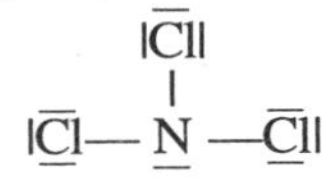

9B The Lewis structure of $POCl_3$ has three Cl atoms single bonded to P and one O atom single bonded to P. These four electron groups around P produce a tetrahedral electron-group geometry. No lone pairs are attached to P and thus the molecular geometry also is tetrahedral.

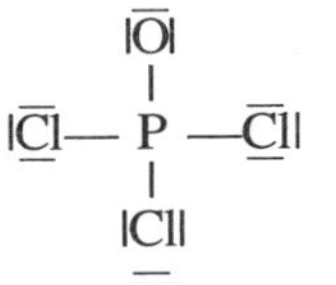

10A The Lewis structure of COS has one S double-bonded to C and an O double-bonded to C. There are no lone pairs attached to C. The electron-group and molecular geometries are the same: linear. $|\overline{S}{=}C{=}\overline{O}|$

10B Nitrous oxide, N_2O, is the familiar "laughing gas" used as an anesthetic in dentistry. Predict the shape of the N_2O molecule. (*Hint:* What is the central atom in this molecule?)

The Lewis structure of N_2O is the subject of Exercise 48 *in the text*. The result is that N is the central atom. $|N{\equiv}N—\overline{\underline{O}}|$ This gives an octet on each atom, a formal charge of +1 on the central N, and a –1 on the O atom. There are no lone pairs on the central N atom and two bonding pairs. The N_2O molecule is linear.

11A In the Lewis structure of methanol, each H atom contributes 1 valence electron, the C atom contributes 4, and the O atom contributes 6, for a total of $(4 \times 1) + 4 + 6 = 14$ valence electrons, or 7 pairs. 4 pairs are used to connect the H atoms to the C and the O, 1 pair is used to connect C to O, and the remaining 2 pairs are lone pairs on O that complete its octet.

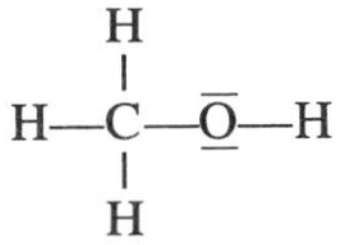

The resulting molecule has two central atoms. Around the C there are four bonding pairs, resulting in a tetrahedral electron-group geometry and molecular geometry. The H—C—H bond angles are 109.5° as are the H—C—O bond angles. Around the O there are two bonding pairs and two lone pairs, resulting in a tetrahedral electron-group geometry and a bent molecular shape around the O atom, with a C—O—H bond angle of slightly less than 109.5°.

11B The Lewis structure is drawn at right. With four electron groups surrounding each of them, the electron-group geometries of N, the central C, and the right-hand O are all tetrahedral. The H—N—H bond angle and the H—N—C bond angles are almost the tetrahedral angle of 109.5°, made a bit smaller by the lone pair.

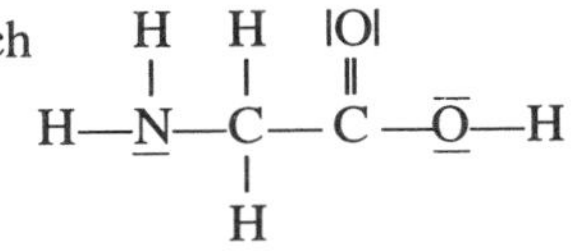

The H—C—N angles, the H—C—H angle and the H—C—C angles all are very close to 109.5°. The C—O—H bond angle is made somewhat smaller than 109.5° by the presence of two lone pairs on O. Three electron groups surround the right-hand O, making its electron-group and molecular geometries trigonal planar. The O—C—O bond angle and the O—C—C bond angles all are very close to 120°.

12A Lewis structures of the three molecules are drawn at right. Around the S in the SF_6 molecule are six bonding pairs, and no lone pairs. The molecule is octahedral; each of the S—F bond moments is cancelled by one on the other side of the molecule.

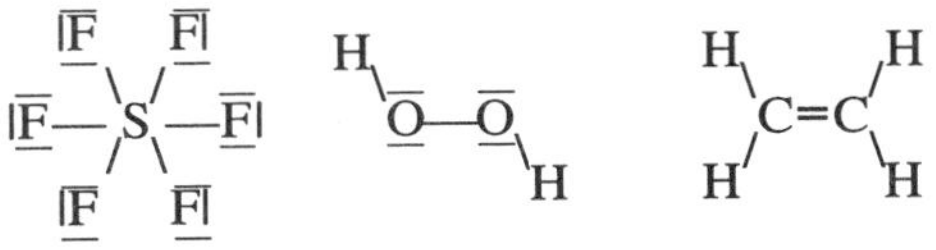

SF_6 is nonpolar. In H_2O_2, the molecular geometry around each O atom is bent; the bond moments do not cancel. H_2O_2 is polar. Around each C in C_2H_4 are three bonding pairs; the molecule is planar around each C and planar overall. The polarity of each —CH_2 group is cancelled by the polarity of the other H_2C— group. C_2H_2 is nonpolar.

12B Lewis structures of the four molecules are drawn at right. In CCl_3CH_3, we can consided the three C—H bonds and the one C—C bond to be non polar. The three C—Cl bonds are tetrahedrally oriented.

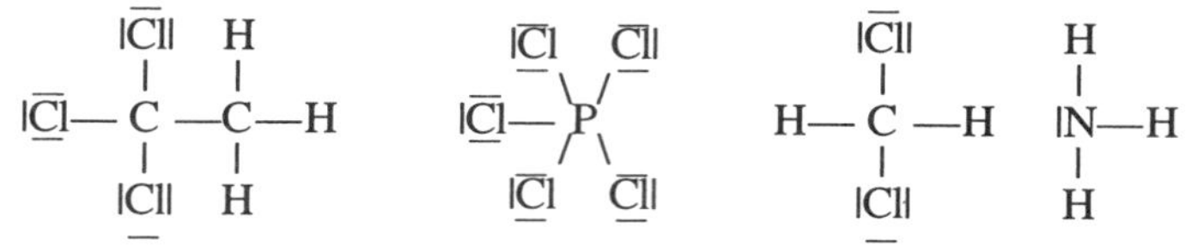

If there were a fourth C—Cl bond on the left-hand C, the bond dipoles would cancel out, producing a nonpolar molecule. Since it is not there, the molecule is polar. A similar argument is made for NH_3, where three tetrahedrally-oriented N—H polar bonds are not balanced by a fourth, and CH_2Cl_2, where two tetrahedrally oriented C—Cl bonds are not balanced by two others. This leaves nonpolar PCl_5, a symmetrical molecule in which bond dipoles cancel.

13A We first draw Lewis structures of all molecules involved.

2 H—H + O=O ⟶ 2 H—O—H

Break 1 O=O + 2 H—H = 498 kJ/mol + (2 × 436 kJ/mol) = 1370 kJ/mol absorbed

Form 4 H—O = (4 × 464 kJ/mol) = 1856 kJ/mol given off

Enthalpy change = 1370 kJ/mol – 1856 kJ/mol = –486 kJ/mol

13B The chemical equation, with Lewis structures is: $\frac{1}{2}$ |N≡N| + $\frac{3}{2}$ |H—H| ⟶ H—N(—H)—H

Energy required to break bonds = $\frac{1}{2}$ N≡N + $\frac{3}{2}$ H—H

= (0.5 × 946 kJ/mol) + (1.5 × 436 kJ/mol) = 1.13×10^3 kJ/mol

Energy realized by forming bonds = 3 N—H = 3 × 389 kJ/mol = 1.17×10^3 kJ/mol

$\Delta H = 1.13 \times 10^3$ kJ/mol – 1.17×10^3 kJ/mol = -4×10^1 kJ/mol of N_2

Thus, $\Delta H_f = -4 \times 10^1$ kJ/mol NH_3 (The value in Appendix D is $\Delta H_f = -46.11$ kJ/mol NH_3)

14A The reaction with Lewis structures is ·O—N=O + ·O· ⟶ Ṅ=O + O=O

The overall result seems to be converting a single N—O bond to a double O=O bond. Since we expect a double bond to be stronger than a single bond, we predict that the products will be more stable than the reactants and this reaction will be exothermic.

14B Double the chemical equation, as Lewis structures: 2 H—O—H + 2 |Cl—Cl| ⟶ |O=O| + 4 H—Cl

Energy required to break bonds = 2 Cl—Cl + 4 H—O

= (2 × 243 kJ/mol) + (4 × 464 kJ/mol) = 2342 kJ/mol

Energy realized by forming bonds = 1 O=O + 4 × H—Cl

= 498 kJ/mol + (4 × 431 kJ/mol) = 2222 kJ/mol

$\Delta H = \frac{1}{2}$(2342 kJ/mol –2222 kJ/mol) = +60 kJ/mol The reaction is endothermic.

SUMMARIZING EXAMPLE CALCULATIONS

1. Convert percents to masses in grams and then to amounts in moles. Divide all amounts by the smallest to find whole number ratios of amounts.

$$21.55 \text{ g N} \times \frac{1 \text{ mol N}}{14.0067 \text{ g N}} = 1.539 \text{ mol N} \quad \div 1.539 \longrightarrow \quad 1.000 \text{ mol N}$$

$$49.23 \text{ g O} \times \frac{1 \text{ mol O}}{15.9994 \text{ g O}} = 3.077 \text{ mol O} \quad \div 1.539 \longrightarrow \quad 1.999 \text{ mol O}$$

Empirical formula: NO_2F

$$29.23 \text{ g F} \times \frac{1 \text{ mol F}}{18.9984 \text{ g F}} = 1.539 \text{ mol F} \quad \div 1.539 \longrightarrow \quad 1.000 \text{ mol F}$$

2. First calculate the amount of gas in 1.00 L, and then use the mass of that liter to calculate the molar mass.

$$n = \frac{PV}{RT} = \frac{1.00 \text{ atm} \times 1.00 \text{ L}}{\frac{0.08206 \text{ L atm}}{\text{mol K}} \times (20 + 273) \text{ K}} = 0.0416 \text{ mol} \qquad m = \frac{2.7 \text{ g}}{0.0416 \text{ mol}} = 65 \text{ g/mol}$$

The molar mass of the empirical formula if 14.0 g N + (2 × 16.0 g O) + 19.0 g F = 65 g/mol. Thus, the empirical formula and the molecular formula are the same: NO_2F.

3. In NO_2F there are 5 + (2 × 6) + 7 = 24 valence electrons, or 12 pairs. N is the central atom, since it has the lowest electronegativity.

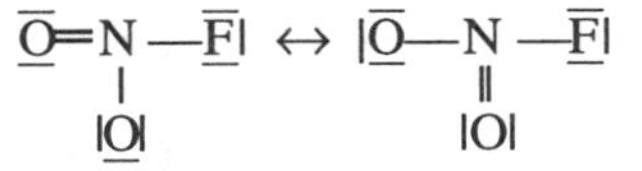

4. In the Lewis structures, there are three bonding pairs and no lone pairs attached to the central N. Thus, the electron-pair geometry and the molecular geometry both are trigonal planar, with 120° bond angles.

5. Each bond is polar from nitrogen toward either O or F. If all three bonded atoms were O, the bond moments would cancel, resulting in a nonpolar molecule. Since F is slightly more electronegative than is O, the molecule is slightly polar, toward F.

REVIEW QUESTIONS

Important Note: In this and subsequent chapters, a lone pair of electrons in a Lewis structure often is shown as a line rather than a pair of dots. Thus, the Lewis structure of Be is Bel rather than Be:

1. **(a)** Valence electrons are those of highest principal quantum number, those in the outermost shell, those furthest from the nucleus.
(b) Electronegativity is a measure of the attraction that an atom in a compound has for electrons.
(c) Bond dissociation energy is the energy needed to break a mole of bonds of a given type.
(d) A double covalent bond results when two pairs of electrons are shared by two atoms.
(e) A coordinate covalent bond is formed when both electrons of a shared pair come originally from one of the bonded atoms.

2. **(a)** Formal charge is a measure of how many valence electrons surround an atom in a covalently bonded structure, compared with the number of valence electrons of the isolated atom.
(b) Resonance occurs when the bonding in a compound cannot be completely represented by one Lewis structure, but requires the "blending" of two or more.
(c) An "expanded octet" refers to more than eight valence electrons surrounding an atom in a covalently bonded structure.
(d) Bond energy is the energy required to break a mole of bonds of a given type, averaged over all the compounds in which that type of bond appears.

3. **(a)** An ionic bond is the result of the attraction between a cation and an anion. A covalent bond results when two atoms share one or more pairs of electrons.
(b) A bonding pair of electrons is a pair that is shared between two atoms. A lone pair resides on one atom only, not shared between atoms.
(c) Electron-group geometry describes the orientation of electron groups—bonding and lone—around a central atom. Molecular geometry describes the orientation of bonding pairs around the central atom.
(d) A bond moment is the result of the unequal sharing of bonding electrons between two atoms. The resultant dipole moment of a molecule is the result of all of the bond moments and the molecule's geometry in producing a net dipole moment.
(e) A polar molecule is one that has a net dipole moment, a slight overall separation of positive and negative charge. In a nonpolar molecule, there is no such net dipole moment.

4. **(a)** $[H:]^-$ **(b)** $:\ddot{Kr}:$ **(c)** $[\cdot Sn\cdot]^{2+}$ **(d)** $[K]^+$ **(e)** $[:\ddot{Br}:]^-$

(f) $\cdot Ge\cdot$ **(g)** $:N\cdot$ **(h)** $\cdot Ca\cdot$ **(i)** $[:\ddot{Se}:]^{2-}$ **(j)** $[Sc]^{3+}$

5. **(a)** $[:\ddot{Cl}:]^-$ $[Ca]^{2+}$ $[:\ddot{Cl}:]^-$ **(b)** $[Ba]^{2+}$ $[:\ddot{S}:]^{2-}$ **(c)** $[Li]^+$ $[:\ddot{O}:]^{2-}$ $[Li]^+$

(d) $[Na]^+$ $[:\ddot{F}:]^-$ **(e)** $[Mg]^{2+}$ $[:\ddot{N}:]^{3-}$ $[Mg]^{2+}$ $[:\ddot{N}:]^{3-}$ $[Mg]^{2+}$

6. (a) $|\bar{I}—\bar{C}l|$ (b) $|\bar{B}r—\bar{B}r|$ (c) $|\bar{F}—\bar{O}—\bar{F}|$ (d) $|\bar{I}—\overset{\displaystyle}{N}—\bar{I}|$ (e) $H—\bar{S}e—H$

under (d): $\overset{|}{\underset{|\bar{I}|}{}}$

7. (a) $|\bar{S}=C=\bar{S}|$ (b) $H—\overset{H}{\underset{H}{C}}—\overset{|O|}{C}—\overset{H}{\underset{H}{C}}—H$ (c) $|\bar{C}l—\overset{|O|}{C}—\bar{C}l|$ (d) $|\bar{F}—\bar{N}=\bar{O}|$

8. (a) $H—H—\bar{N}—\bar{O}—H$ has two bonds to (four electrons around) the second hydrogen, and only three

bonds to (six electrons around) the nitrogen. A better Lewis structure is $H—\overset{H}{\underset{}{N}}—\bar{O}—H$

(b) $|\bar{O}—\bar{C}l—\bar{O}|$ has 20 valence electrons, whereas the molecule ClO_2 has 19 valence electrons. This is a proper Lewis structure for the chlorite ion, although the brackets and the minus charge are missing. A plausible Lewis structure for the molecule ClO_2 is $|\bar{O}—\overset{\cdot}{\underset{}{C}l}—\bar{O}|$

(c) $[\cdot\overset{\cdot}{C}=\bar{N}|]^-$ has only six electrons around the C atom. $[|C\equiv N|]^-$ is a more plausible Lewis structure for the cyanide ion.

(d) $Ca—\bar{O}|$ is improperly written as a covalent Lewis structure, although CaO is an ionic compound. In addition, there are only two electrons around the Ca atom. $[Ca]^{2+} [|\bar{O}|]^{2-}$ is a more plausible Lewis structure for CaO.

9. (a) $[|\bar{O}—H]^- Ca^{2+} [|\bar{O}—H]^-$ (b) $\left(H—\overset{\overset{H}{|}}{\underset{\underset{H}{|}}{N}}—H\right)^+ [|\bar{B}r|]^-$ (c) $[|\bar{O}—\bar{C}l|]^- Ca^{2+} [|\bar{O}—\bar{C}l|]^-$

10. (a) computations for:

	H—	—C≡	≡C:
number of valence electrons	1	4	4
−2 × no. lone pairs	−0	−0	−2
−no. bonds	−1	−4	−3
formal charge	0	0	−1

(b) computations for:

	=O	—O	C
number of valence electrons	6	6	4
−2 × no. lone pairs	−4	−6	−0
−no. bonds	−2	−1	−4
formal charge	0	−1	0

(c) computations for:

	—H	side C	central C
number of valence electrons	1	4	4
−2 × no. lone pairs	−0	−0	−0
−no. bonds	−1	−4	−3
formal charge	0	0	+1

(d) The formal charge on each I is 0, computed as follows:

number of valence electrons	= 7
−2 × no. lone pairs	= −6
− no. bonds	= −1
formal charge	= 0

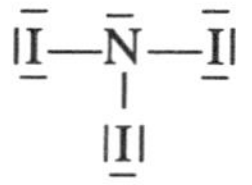

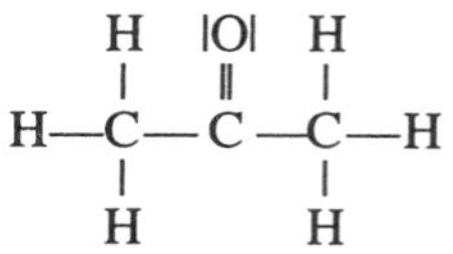

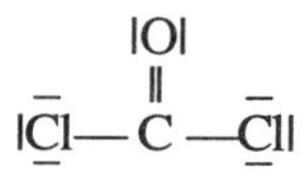

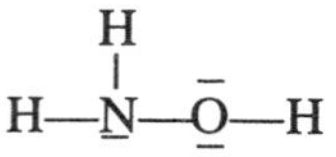

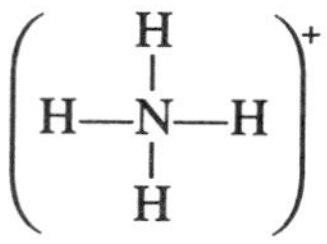

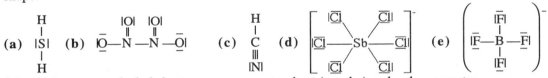

(e)	computations for:	=O	—O	=S—
	number of valence electrons	6	6	6
	$-2 \times$ no. lone pairs	-4	-6	-2
	$-$no. bonds	$\underline{-2}$	$\underline{-1}$	$\underline{-3}$
	formal charge	0	-1	$+1$
(f)	computations for:	=O	—O	N
	number of valence electrons	6	6	5
	$-2 \times$ no. lone pairs	-4	-6	-1
	$-$no. bonds	$\underline{-2}$	$\underline{-1}$	$\underline{-3}$
	formal charge	0	-1	$+1$

11. Since electrons pair up (if at all possible) in plausible Lewis structures, we can get a very good indication if a species is paramagnetic if it has an odd number of (valence) electrons.

(a)	OH^-	$6 + 1 + 1 = 8$ valence electrons	diamagnetic
(b)	OH	$6 + 1 = 7$ valence electrons	paramagnetic
(c)	NO_3	$5 + (3 \times 6) = 23$ valence electrons	paramagnetic
(d)	SO_3	$6 + (3 \times 6) = 24$ valence electrons	diamagnetic
(e)	SO_3^{2-}	$6 + (3 \times 6) + 2 = 26$ valence electrons	diamagnetic
(f)	HO_2	$1 + (2 \times 6) = 13$ valence electrons	paramagnetic

12. **(a)** 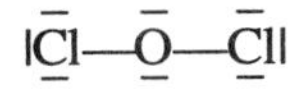 **(b)** 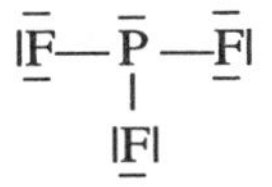 **(c)** 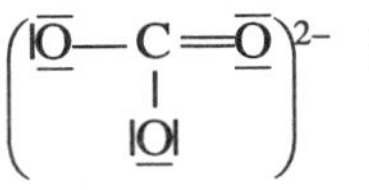 **(d)** 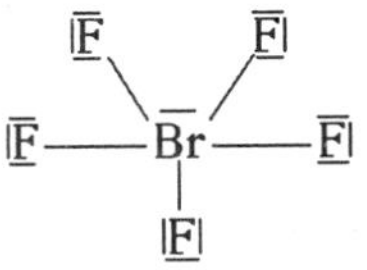

13. We first give all the Lewis structures. From each, we deduce the electron-group geometry and the molecular shape.

(a) 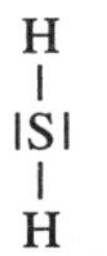 **(b)** 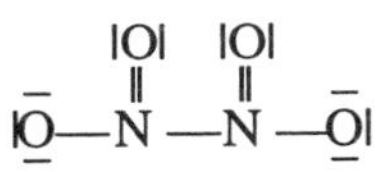 **(c)**  **(d)** 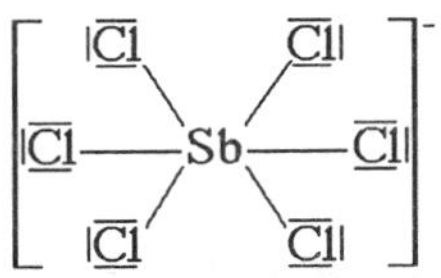 **(e)** 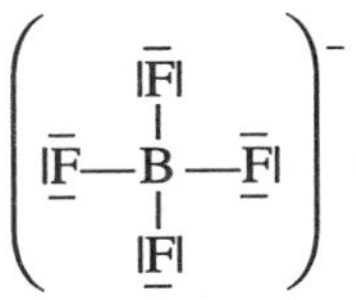

(a)	H_2S	tetrahedral electron-group geometry, bent (angular) molecular geometry
(b)	N_2O_4	trigonal planar electron-group geometry around each N, planar molecule
(c)	HCN	linear electron-group geometry, linear molecular geometry
(d)	$SbCl_6^-$	octahedral electron-group geometry, octahedral geometry
(e)	BF_4^-	tetrahedral electron-group geometry, tetrahedral molecular geometry

14. **(a)** In SO_2, there are a total of $6 + (2 \times 6) = 18$ valence electrons, or 9 pairs. A plausible Lewis structure has two resonance forms, of which one is: 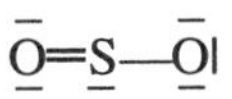 This molecule is of the type AX_2E; it has a trigonal planar electron-group geometry and a bent molecular shape.

(b) In SO_3^{2-} the total number of valence electrons is $2 + (3 \times 6) + 6 = 26$ valence electrons, or 13 pairs. A plausible Lewis structure is 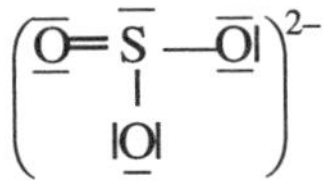 Two other resonance forms can be drawn. This molecule is of the type AX_3E; it has a tetrahedral electron-group geometry and a trigonal pyramidal molecular shape.

(c) In SO_4^{2-}, there are a total of $2 + 6 + (4 \times 6) = 32$ valence electrons, or 16 pairs. There are several possible resonance forms, but a simple Lewis structure is 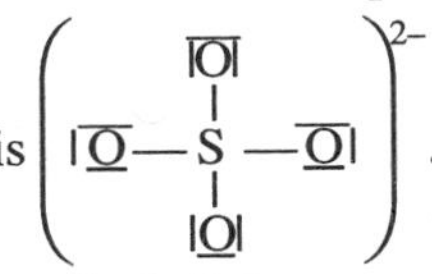. This ion is of the type AX_4; it has a tetrahedral electron-group geometry and a tetrahedral shape.

15. In each case, a plausible Lewis structure is given first, followed by the AX_nE_m notation of each species. Then comes the electron-group geometry and, finally, the molecular geometry.

	(a)	**(b)**	**(c)**	**(d)**
AX_nE_m electron-group geometry	AX_1E_1 linear	AX_4E_0 tetrahedral	AX_3E_1 tetrahedral	trigonal bipyramid
molecular geometry	linear	tetrahedral	trigonal pyramid	T-shaped

	(e)	**(f)**	**(g)**
AX_nE_m electron-group geometry	AX_5E_0 trigonal bipyramid	AX_2E_1 triangular planar	AX_6E_0 octahedral
molecular geometry	trigonal bipyramid	bent	octahedral

16.

bond	H—C	C=O	C—C	C—Cl	
(a) bond length	110	120	154	178	pm
(b) bond energy	414	736	347	339	kJ/mol

17. (c) is the longest bond. Lewis structures are of assistance, for single bonds are generally the longest. Of the two molecules with single bonds, Br_2 is expected to have longer bonds than BrCl, since Br is larger than a Cl atom.   (a) $\overline{O}=\overline{O}$ (b) $|N\equiv N|$ (c) $|\overline{Br}—\overline{Br}|$ (d) $|\overline{Br}—\overline{Cl}|$

18. The reaction $O_2 \longrightarrow 2\ O$ is an endothermic reaction since it requires the breaking of the bond between two oxygen atoms without the formation of any bonds. Since bond breakage is endothermic, the entire process must be endothermic.

19. **(a)** The net result of this reaction, per mole of $CH_4(g)$, is to break one mole of C—H bonds (which requires 414 kJ) and to form 1 mole of H—I bonds (which produces 297 kJ). Thus, this reaction is endothermic; a net infusion of energy is necessary.

(b) The net result of this reaction, per mole of $H_2(g)$, is to break 1 mol of H—H bonds (which requires 436 kJ) and 1 mol of I—I bonds (which requires 151 kJ), and to form two moles of H—I bonds (which produces $2 \times 297 = 594$ kJ). Thus, this reaction is just barely exothermic.

20. We draw Lewis structures of the reactants and the products to aid in determining the value of ΔH.

$$2\ \overset{\bullet}{\underset{}{N}}=\overline{O} + 5\ H—H \longrightarrow 2\ H—\overset{\displaystyle H}{\underset{\displaystyle |}{\overline{N}}}—H + 2\ H—\overline{O}—H$$

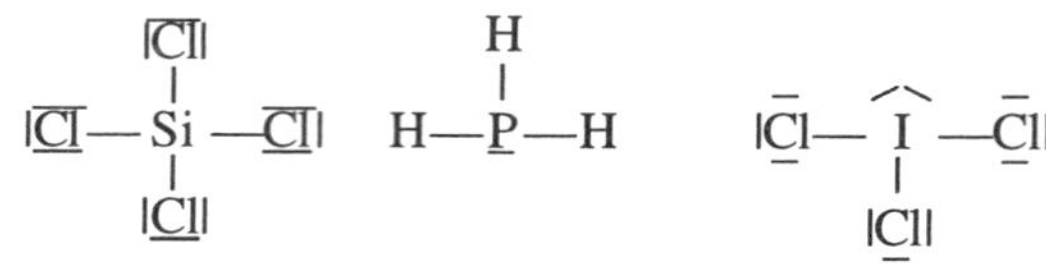

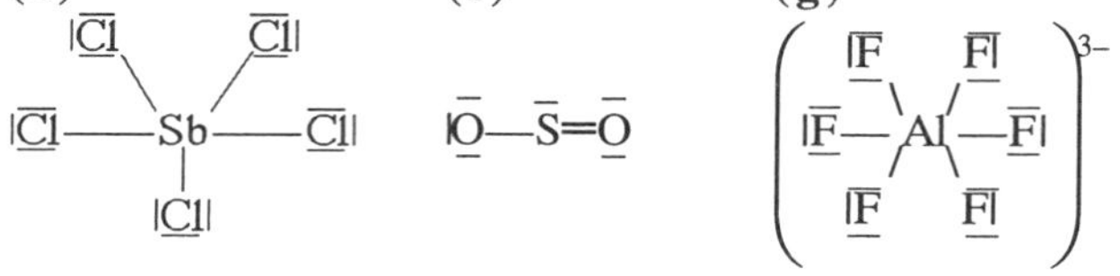

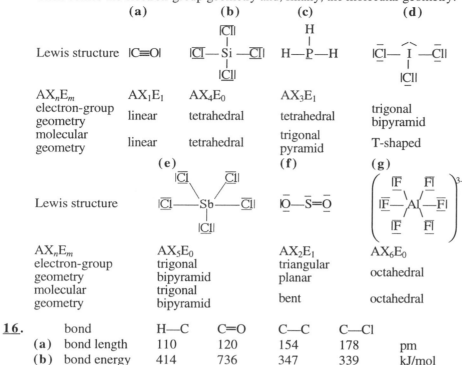

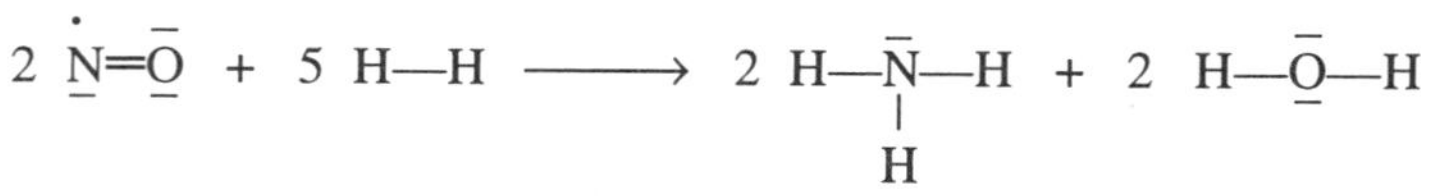

$$\Delta H = \text{energy of bonds broken} - \text{energy of bonds formed}$$
$$= [(2 \times N{=}O) + (5 \times H{—}H)] - [(6 \times N{—}H) + (4 \times H{—}O)]$$
$$= [(2 \times 590\ \text{kJ/mol}) + (5 \times 436\ \text{kJ/mol})] - [(6 \times 389\ \text{kJ/mol}) + (4 \times 464\ \text{kJ/mol})]$$
$$= (1.18 \times 10^3 + 2.18 \times 10^3) - 2.33 \times 10^3 - 1.86 \times 10^3$$
$$= -0.83 \times 10^3\ \text{kJ/mol} = -8.3 \times 10^2\ \text{kJ/mol}$$

21. Recall that electronegativity increases in the periodic table from lower left to upper right, and specifically that metals have lower electronegativities than do nonmetals, with metalloids having intermediate electronegativities. Based on these principles, the electronegativity of S is greater than that of As, which in turn is greater than that of Bi. The electronegativity of Ba is less than that of Mg. And finally, we would predict the electronegativity of Mg (a definite metal) to be less than that of Bi (somewhat metalloid in character). Thus, ranked in order of increasing electronegativity, these elements are: Ba < Mg < Bi < As < S, with Bi having the intermediate electronegativity of this group of five elements. The actual electronegativities are parenthetically indicated: Ba (0.9) < Mg (1.2) < Bi (1.9) < As (2.0) < S(2.5)

22. Na—Cl and K—F both are bonds between a metal and a nonmetal; they have the largest ionic character, with the ionic character of K—F being greater than that of Na—Cl, both because K is more metallic (closer to the lower left of the periodic table) than Na and F is more nonmetallic (closer to the upper right) than Cl. The remaining three bonds are covalent bonds to H. Since H and C have about the same electronegativity (a fact you need to memorize), the H—C bond is the most covalent (or the least ionic). Br is somewhat more electronegative than is C, while F is considerably more electronegative than C, making the F—H bond the most polar of the three covalent bonds. Thus, ranked in order of increasing ionic character, these five bonds are: C—H < Br—H < F—H < Na—Cl < K—F The actual electronegativity differences follow: C(2.5)—H(2.1) < Br(2.8)—H(2.1) < F(4.0)—H(2.1) < Na(0.9)—Cl(3.0) < K(0.8)—F(4.0)

ΔEN = 0.4 = 0.7 = 1.9 = 2.1 = 3.2

23. **(a)** F_2 cannot possess a dipole moment, since all of the atoms in the molecule are the same. This means that there is no electronegativity difference between atoms, and hence no polar bonds.

(b) |O—Ṅ=O Each nitrogen-to-oxygen bond in this molecule is polar toward oxygen, the more electronegative element. The molecule is of the AX_2E_2 category and hence is bent. Therefore the two bond dipoles do not cancel, and the molecule is polar.

(c) |F—B—F| (with |F| below B) Although each B—F bond is polar toward F, in this trigonal planar AX_3E_0 molecule these bond dipoles cancel. The molecule is nonpolar.

(d) H—Br| The H—Br bond is polar toward Br, and this molecule is polar as well.

(e) H—C—Cl| (with H above C and |Cl| below C) The H—C bonds are not polar, but the C—Cl bonds are, toward Cl. The molecular shape is tetrahedral (AX_4) and thus these two C—Cl dipoles do not cancel each other; the molecule is polar.

(f) |F—Si—F| (with |F| above and below Si) Although each Si—F bond is polar toward F, in this tetrahedral AX_4E_0 molecule these bond dipoles oppose and cancel each other; the molecule is nonpolar.

(g) O=C=S In this linear molecule, the two bonds from carbon both are polar away from carbon. But the C=O bond is more polar than the C=S bond and hence the molecule is polar.

24. **(a)** There is at least one instance in which one atom must bear a formal charge—a polyatomic ion—because the charge on the species equals the sum of the formal charges. There are other types of molecules that have formal charges. Sometimes this occurs when removing formal charges would require that there be more than one unpaired electron in the structure.

(b) Because three points define a plane, stating that a triatomic molecule is planar is not saying anything new. In fact, it is misleading; some triatomic molecules are linear. HCN is one, as is CO_2. Of course, some molecules with more than three atoms also are planar; examples are XeF_4 and $H_2C{=}CH_2$.

(c) This statement is incorrect, for in some molecules with polar bonds, the bonds are so oriented that there is no resulting molecular dipole moment. Examples are CO_2, BCl_3, CCl_4, PCl_5, and SF_6.

EXERCISES

Lewis theory

25. Hydrogen never has an octet of electrons in any of its compounds, but only a pair (or duet, if you prefer). An example is the Lewis structure of H_2O (below). In many compounds in which the central atom is from the second period or higher, there are more than eight electrons around the central atom; an example of a compound with such an "expanded octet" is ICl_3 (below). Finally, in some compounds, there simply are not eight electrons around the central atom; one such "electron deficient" compound is BF_3.

$$H—\overline{O}—H \qquad |\overline{Cl}—\overset{\frown}{I}—\overline{Cl}| \qquad |\overline{F}—B—\overline{F}|$$
$$\qquad\qquad\qquad |\underline{Cl}| \qquad\qquad\qquad |\underline{F}|$$

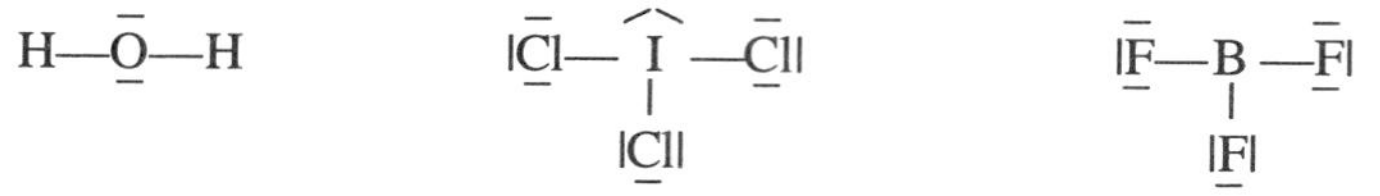

26. NH_3 $5 + (3 \times 1) = 8$ v.e. = 4 pairs BF_3 $3 + (3 \times 7) = 24$ v.e. = 12 pairs

SF_6 $6 + (6 \times 7) = 48$ v.e. = 24 pairs SO_3 $6 + (3 \times 6) = 24$ v.e. = 12 pairs

NH_4^+ $5 + (4 \times 1) - 1 = 8$ v.e. = 4 pairs SO_4^{2-} $6 + (4 \times 6) + 2 = 32$ v.e. = 16 pairs

NO_2 $5 + (2 \times 6) = 17$ v.e. = 8.5 pairs

NO_2 cannot obey the octet rule; there is no way to pair all electrons when the number of electrons is odd.

$$H—\overline{N}—H \qquad |\overline{F}—B—\overline{F}| \qquad |\overline{F}—\overset{|\overline{F}\quad\overline{F}|}{\underset{|\overline{F}\quad\overline{F}|}{S}}—\overline{F}| \qquad \overset{..}{O}=\overset{}{S}—\overline{O}| \qquad \left(H—\overset{H}{\underset{H}{N}}—H\right)^+ \qquad \left(|\overline{O}—\overset{|\overline{O}|}{\underset{|\overline{O}|}{S}}—\overline{O}|\right)^{2-}$$
$$\quad H \qquad\qquad |\overline{F}| \qquad\qquad\qquad\qquad\quad |O|$$

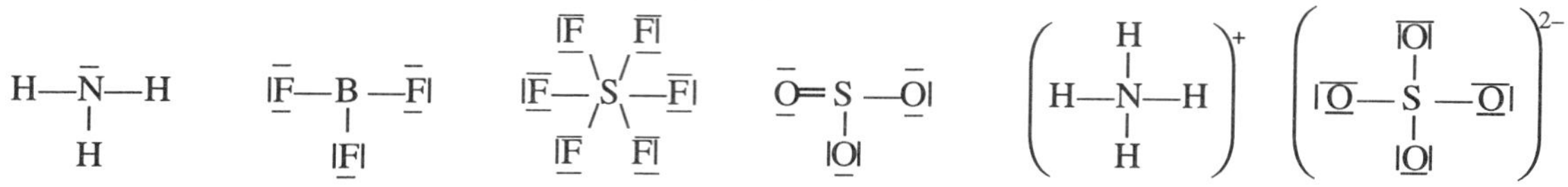

All of these Lewis structures obey the octet rule except for BF_3 which is electron deficient and SF_6, which has an expanded octet.

27. (a) $Cs^+ [:\overset{..}{\underset{..}{Br}}:]^-$ (b) $H—\overset{}{\underset{H}{\overline{Sb}}}—H$ (c) $|\overline{Cl}—\overset{}{\underset{|\underline{Cl}|}{B}}—\overline{Cl}|$

      CsBr, cesium bromide H_3Sb, hydrogen antimonide BF_3, boron trichloride

(d) $Cs^+ [:\overset{..}{\underset{..}{Cl}}:]^-$ (e) $Li^+ [:\overset{..}{\underset{..}{O}}:]^{2-} Li^+$ (f) $|\overline{I}—\overline{Cl}|$

      CsCl, cesium chloride Li_2O, lithium oxide   ICl, iodine chloride

28. (a) In an H_3 molecule, one of the H atoms would have to be a central atom, with two pairs of electrons, and one bond to each of the other H atoms. This places more than 2 electrons around this central H atom, more than the stable pair found around H in most Lewis structures.

(b) In HHe there would either be a shared triplet of electrons between the two atoms, or three electrons around the He atom H—He· Either of these is not a particularly stable situation.

(c) He_2 would have either a double bond between two He atoms   He=He   and thus four electrons around each He atom, or three electrons around each He atom   ·He—He·   Neither situation achieves the electron configuration of the nearest noble gas.

(d) $H—\overset{H}{\underset{..}{\overset{|}{O}}}—H$ has an expanded octet (9 electrons) on oxygen; expanded octets are not found on elements of the second period. Other structures place a multiple bond between O and H. Both situations are unstable.

29. (c) is the correct answer.   (a) $[\overline{O}—C=\overline{N}]^-$ does not have an octet of electrons around C.

(b) $[C\equiv Cl]^-$ does not have an octet around the left-hand C, it has only 8 valence electrons, and should have 10, and the formal charges on the two carbons are different.   (d) The total number of valence electrons in $|\overline{N}=\overline{O}|$ is incorrect; this odd-electron species should have 11 valence electrons, not 12.

30. (a) $Mg—\overline{O}|$ is written as a covalent structure even though the compound is composed of a metal (Mg) and a nonmetal (O). One expects an ionic Lewis structure. $[Mg]^{2+} [|\overline{O}|]^{2-}$

(b) $[|\overline{Cl}|]^+ [|\overline{O}|]^{2-} [\overline{Cl}|]^+$ is written as an ionic structure, even though we expect a covalent structure between nonmetallic atoms. A more plausible structure is $|\overline{Cl}—\overline{O}—\overline{Cl}|$

(c) $[|\overline{O}—\overset{.}{N}=\overline{O}|]^+$ has too many valence electrons—8.5 pairs or 19 valence electrons—it should have (2 $\times$ 6) + 5 - 1 = 16 valence electrons: 8 pairs. A plausible Lewis structure is $[|\overline{O}=N=\overline{O}|]^+$, which has +1 formal charge on N and 0 formal charge on each oxygen.

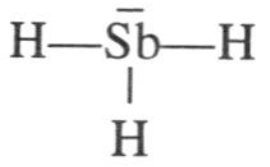

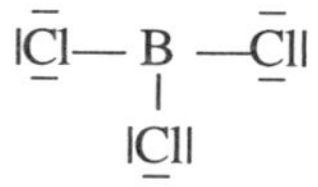

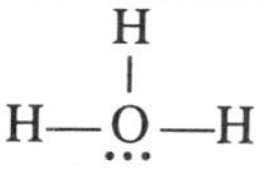

(d) In the structure $[|\overline{S}—C=\overline{N}|]^-$ there is not an octet of electrons around either S nor C. In addition, there are only 7 pairs of valence electrons in this structure, 14 valence electrons. There should be $6 + 4 + 5 + 1 = 16$ valence electrons, or 8 pairs. At least two structures are possible. $[\overline{S}=C=\overline{N}|]^-$ has a formal charge of -1 on N and is preferred over $[|\overline{S}—C\equiv N|]^-$, with its formal charge of -1 on S, which is less electronegative than N.

Ionic bonding

31. (a) Li· forms Li$^+$ cations, and :S̈: forms $[:\overset{..}{\underset{..}{S}}:]^{2-}$ anions. Thus, lithium sulfide is Li$^+$ $[:\overset{..}{\underset{..}{S}}:]^{2-}$ Li$^+$

(b) Na· forms Na$^+$ cations, and :F̈· forms $[:\overset{..}{\underset{..}{F}}:]^-$ anions. Thus, sodium fluoride is Na$^+$ $[:\overset{..}{\underset{..}{F}}:]^-$

(c) ·Ca· forms Ca^{2+} cations, and :Ï· forms $[:\overset{..}{\underset{..}{I}}:]^-$ anions. Thus, calcium iodide is $[:\overset{..}{\underset{..}{I}}:]^-$ Ca^{2+} $[:\overset{..}{\underset{..}{I}}:]^-$

(d) ·Sc· forms Sc^{3+} cations, and :C̈l: forms $[:\overset{..}{\underset{..}{Cl}}:]^-$ anions. Scandium chloride is $[:\overset{..}{\underset{..}{Cl}}:]^-$ Sc^{3+} $[:\overset{..}{\underset{..}{Cl}}:]^-$

The formulas of the compounds are Li$_2$S, NaF, CaI$_2$, and ScCl$_3$. $[:\overset{..}{\underset{..}{Cl}}:]^-$

32. The Lewis symbols are $[H|]^-$ for the hydride ion, $[|\overline{N}|]^{3-}$ for the nitride ion.
   (a) $[Li]^+$ $[H|]^-$ (b) $[H|]^-$ $[Ca]^{2+}$ $[H|]^-$ (c) $[Mg]^{2+}$ $[|\overline{N}|]^{3-}$ $[Mg]^{2+}$ $[|\overline{N}|]^{3-}$ $[Mg]^{2+}$
   lithium hydride   calcium hydride   magnesium nitride

Formal Charge

33. There are three common features of formal charge and oxidation state. Both indicate how electrons are distributed in the bonding of the compound. Second, negative formal charge (in the most plausible Lewis structure) and negative oxidation state are generally those of more electronegative atoms. And third, both numbers are determined by a set of rules, rather than being determined experimentally. Differences include some instances in which the same type of atom in a compound (which of course has the same oxidation state), has different formal charges, such as oxygen in ozone, O$_3$. Another is that formal charges are used to decide between alternative Lewis structures, while oxidation state is used in balancing equations and naming compounds. The most significant difference, though, is that whereas the oxidation state of an element in its compounds is usually not zero, its formal charge often is.

34. The most common instance in which formal charge is not kept to a minimum occurs in the case of ionic compounds. For example, in Mg—$\overline{O}|$ the formal charge on Mg is $+1$ and on O is -1, while in the ionic version $[Mg]^{2+}$ $[|\overline{O}|]^{2-}$, formal charges are $+2$ and -2, respectively. Additionally, in some resonance hybrids formal charge is not minimized. In order to have bond lengths agree with experimental results, it may not be acceptable to create multiple bonds. Yet a third instance is when double bonds are created to lower formal charge, particularly when this results in the octet rule being violated. Thus, all the Cl—O bonds in ClO$_4^-$ are single bonds, although including some double bonds would minimize formal charges.

35. Formal charge = number of valence electrons – 2 × number of lone pairs – number of bond pairs
The formal charge of the central atom is calculated below the Lewis structure of each species.

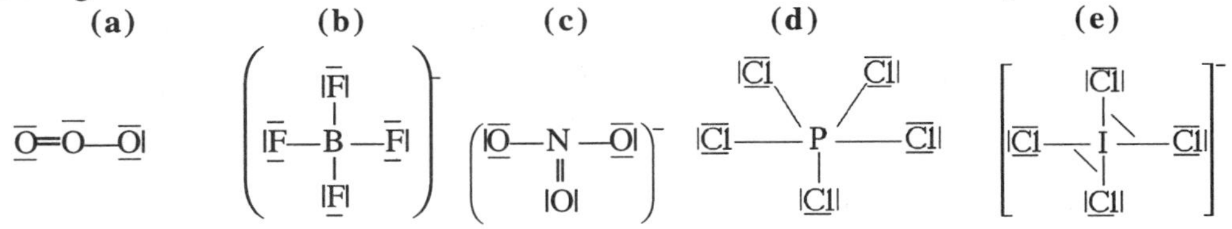

	(a)	(b)	(c)	(d)	(e)
	$\overline{O}=\overset{..}{O}—\overline{O}|$	$\left(|F—\underset{|\underline{F}|}{\overset{|\overline{F}|}{B}}—\overline{F}|\right)^-$	$\left(|\overline{O}—\underset{|O|}{N}=\overline{O}|\right)^-$	Cl···P···Cl structure	$[Cl—I—Cl]^-$ structure
val. elns.	6	3	5	5	7
–2 lone prs.	–2	–0	–0	–0	–4
–bond prs.	–3	–4	–4	–5	–4
formal charge	+1	–1	+1	0	–1

36. We calculate formal charge = no. valence electrons – no. bonds – 2 × no. lone pairs.

   (a)   H—N̄—Ō—H   f.c. of N = 5 – 3 – (1 × 2) = 0   most plausible
          | f.c. of O = 6 – 2 – (2 × 2) = 0
          H

      H—Ō—N̄—H   f.c. of N = 5 – 2 – (2 × 2) = –1
          | f.c. of O = 6 – 3 – (1 × 2) = +1
          H

   (b)   S̄=C=S̄   f.c. of C = 4 – 4 – 0 = 0   f.c. of S = 6 – 2 – (2 × 2) = 0   most plausible

       C̄=S=S̄   f.c. of C = 4 – 2 – (2 × 2) = –2   f.c. of central S = 6 – 4 – 0 = +2

                   f.c. of terminal S = 6 – 2 – (2 × 2) = 0

   (c)   N̄=Ō—F̄| f.c. of N = 5 – 2 – (2 × 2) = –1   f.c. of O = 6 – 3 – (1 × 2) = +1

                   f.c. of F = 7 – 1 – (3 × 2) = 0

       Ō=N̄—F̄| f.c. of O = 6 – 2 – (2 × 2) = 0   f.c. of N = 5 – 3 – (1 × 2) = 0

                   f.c. of F = 7 – 1 – (3 × 2) = 0   most plausible

   (d)

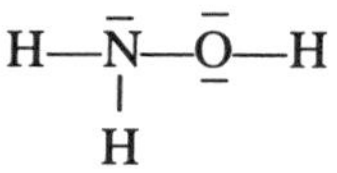

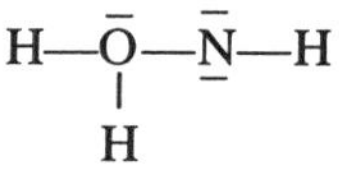

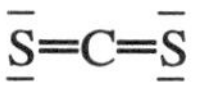

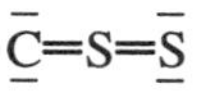

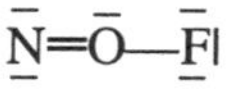

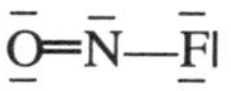

S̄=Ō—C̄l|	Ō=S̄—C̄l|	|Ō—C̄l—C̄l—S̄|
|C̄l|	|C̄l|	

     f.c. of S   6 – 2 – (2 × 2) = 0   6 – (4 × 1) – 2 = 0   6 – 1 – (3 × 2) = –1
     f.c. of O   6 – (4 × 1) – 2 = 0   6 – 2 – (2 × 2) = 0   6 – 1 – (3 × 2) = –1
     f.c. of Cl   7 –1 – (3 × 2) = 0   7 –1 – (3 × 2) = 0   7 – (2 × 2) – (2 × 1) = +1
         O's expanded octet forbidden   most plausible

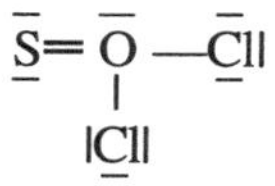

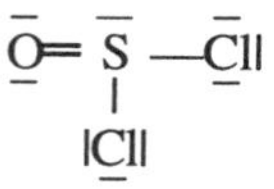

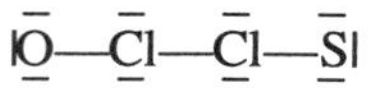

Lewis structures

37. **(a)**   In H_2NOH, N and O atoms are the central atoms and the terminal atoms are the H atoms. The number of valence electrons in the molecule totals: $(3 \times 1) + 5 + 6 = 14$ valence electrons, or seven pairs. A resonable Lewis structure is   H—N̄—Ō—H
                                                        |
                                                        H

   (b)   In $HOClO_2$, two O atoms and one H atom are terminal atoms, and the Cl and O atoms are the central atoms. The total number of valence electrons in the structure is $1 + 7 + (3 \times 6) = 26$ valence electrons, or 13 pairs. A reasonable Lewis structure is   H—Ō—C̄l—Ō̄
                                                                               |
                                                                               |O|

   (c)   In HONO, the H atom and one O atom are the terminal atoms; the other O atom and the N atom are the central atoms. The total number of valence electrons in the structure is $1 + 5 + (2 \times 6) = 18$ valence electrons, or nine (9) pairs. A plausible Lewis structure is   H—Ō—N̄=Ō

   (d)   In O_2SCl_2, S is the central atom, with Cl and O as terminal atoms. The total number of valence electrons is $(2 \times 6) + 6 + (2 \times 7) = 32$ valence electrons, 16 pairs. A reasonable Lewis structure is

              |Ō|
               |
         |C̄l— S —C̄l\|
               |
              |Ō|

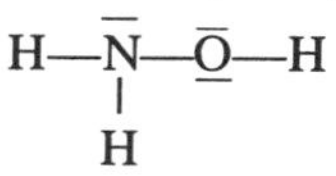

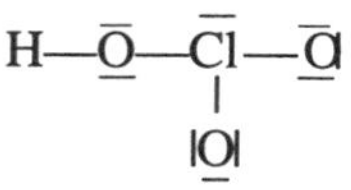

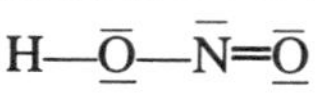

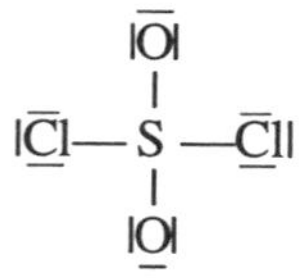

38. The total number of valence electrons is $(2 \times 7) + (2 \times 6) = 26$ valence electrons, or 13 pairs of valence electrons. It is unlikely to have F as a central atom; that would require an expanded octet on F. One structure is $|\overline{F}—\overline{S}—\overline{S}—\overline{F}|$ another is $\overline{S}{=}\overline{S}—\overline{F}|$ Both have zero formal charge on each atom, but only the first has

$$\quad\quad\quad\quad |\overline{F}|$$

just an octet around each atom. The second has an expanded octet around the central S.

39. (a) The total number of valence electrons in SO_3^{2-} is $6 + (3 \times 6) + 2 = 26$, or 13 pairs. A plausible Lewis structure is $\left(|\overline{O}—\overline{S}—\overline{O}|\right)^{2-}$ Two other resonance forms can be drawn.

$$\quad\quad\quad\quad\quad |\overline{O}|$$

(b) The total number of valence electrons in NO_2^- is $5 + (2 \times 6) + 1 = 18$, or 9 pairs. There are two resonance forms for the nitrite ion: $[\overline{O}—\overline{N}{=}\overline{O}]^- \leftrightarrow [\overline{O}{=}\overline{N}—\overline{O}]^-$

(c) The total number of valence electrons in CO_3^{2-} is $4 + (3 \times 6) + 2 = 24$, or 12 pairs. There are three resonance forms for the carbonate ion: $\left(|\overline{O}—C{=}\overline{O}\right)^{2-} \leftrightarrow \left(\overline{O}{=}C—\overline{O}|\right)^{2-} \leftrightarrow \left(|\overline{O}—C—\overline{O}|\right)^{2-}$

$$\quad\quad |\overline{O}| \quad\quad\quad\quad |\overline{O}| \quad\quad\quad\quad\quad |\overline{O}|$$

(d) The total number of valence electrons in HO_2^- is $1 + (2 \times 6) + 1 = 14$, or 7 pairs. A plausible Lewis structure is $[H—\overline{O}—\overline{O}|]^-$

40. Each of the cations has an empty valence shell as the result of ionization. The main task is to determine the Lewis structure of each anion.

(a) The total number of valence electrons in OH^- is $6 + 1 + 1 = 8$, or 4 pairs. A plausible Lewis structure for barium hydroxide is $[\overline{O}—H]^-$ $[Ba]^{2+}$ $[|\overline{O}—H]^-$

(b) The total number of valence electrons in NO_2^- is $5 + (2 \times 6) + 1 = 18$, or 9 pairs. A plausible Lewis structure for sodium nitrite is $[Na]^+$ $[\overline{O}—\overline{N}{=}\overline{O}]^- \leftrightarrow [\overline{O}{=}\overline{N}—\overline{O}|]^-$

(c) The total number of valence electrons in IO_3^- is $7 + (3 \times 6) + 1 = 26$, or 13 pairs. A plausible Lewis structure for magnesium iodate is $\left(\begin{matrix} |\overline{O}| \\ | \\ |\overline{O}—I—\overline{O}| \end{matrix}\right)^- [Mg]^{2+} \left(\begin{matrix} |\overline{O}| \\ | \\ |\overline{O}—I—\overline{O}| \end{matrix}\right)^-$

(d) The total number of valence electrons in SO_4^{2-} is $6 + (4 \times 6) + 2 = 32$, or 16 pairs. A plausible structure for aluminum sulfate is $[SO_4]^{2-}$ $[Al]^{3+}$ $[SO_4]^{2-}$ $[Al]^{3+}$ $[SO_4]^{2-}$ Because of the ability of S to expand its octet, SO_4^{2-} has several resonance forms, a few of which are:

$$\left(\begin{matrix} |\overline{O}| \\ | \\ \overline{O}—S—\overline{O}| \\ | \\ |\overline{O}| \end{matrix}\right)^{2-} \leftrightarrow \left(\begin{matrix} |\overline{O}| \\ | \\ \overline{O}{=}S—\overline{O}| \\ | \\ |\overline{O}| \end{matrix}\right)^{2-} \leftrightarrow \left(\begin{matrix} |\overline{O}| \\ | \\ |\overline{O}—S—\overline{O}| \\ || \\ |\overline{O}| \end{matrix}\right)^{2-} \leftrightarrow \left(\begin{matrix} |\overline{O}| \\ | \\ \overline{O}{=}S{=}\overline{O} \\ | \\ |\overline{O}| \end{matrix}\right)^{2-} \leftrightarrow \left(\begin{matrix} |O| \\ || \\ \overline{O}{=}S—\overline{O}| \\ | \\ |\overline{O}| \end{matrix}\right)^{2-}$$

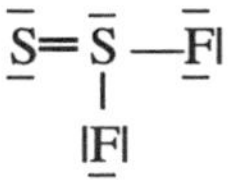

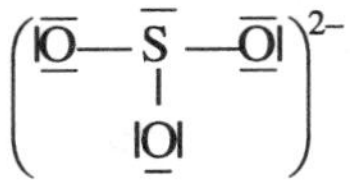

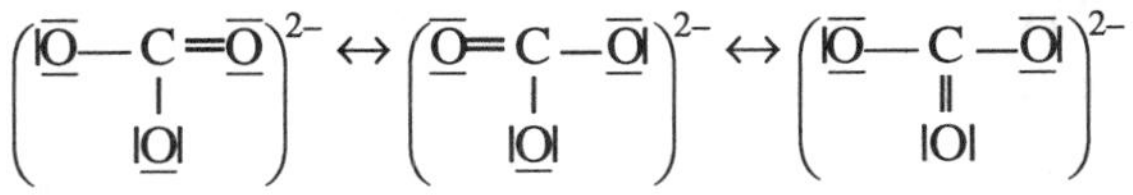

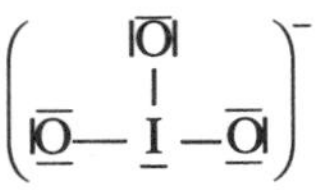

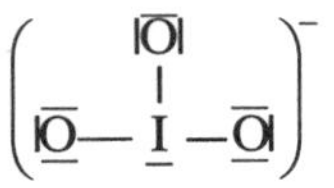

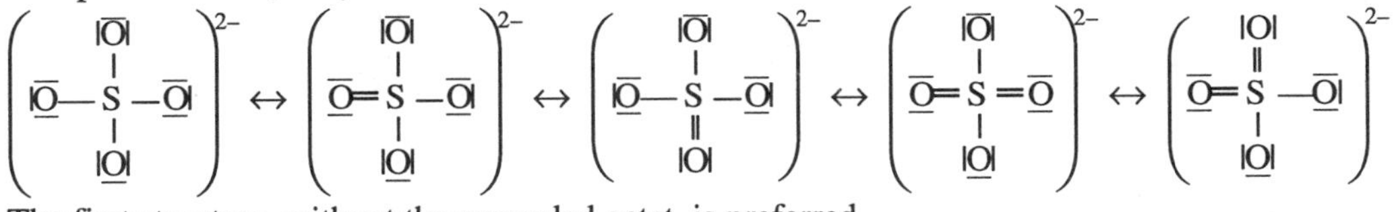

The first structure, without the expanded octet, is preferred.

41. In $CH_3CHCHCHO$ there are $(4 \times 4) + (6 \times 1) + 6 = 28$ valence electrons, or 14 pairs. We expect that the carbon atoms bond to each other. A plausible Lewis structure is

$$\begin{matrix} H & H & H & |\overline{O}| \\ | & | & | & || \\ H—C—C{=}C—C—H \\ | \\ H \end{matrix}$$

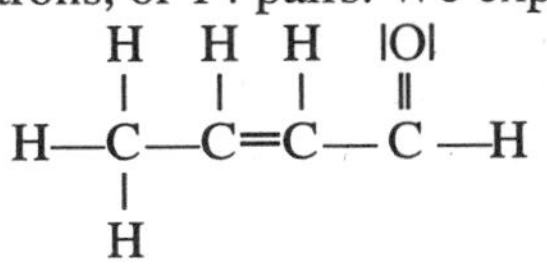

42. In C_3O_2 there are $(3 \times 4) + (2 \times 6) = 24$ valence electrons or 12 valence electron pairs. A plausible Lewis structure follows. $\overline{O}{=}C{=}C{=}C{=}\overline{O}$

Bond lengths

43. A heteronuclear bond length (one between two different atoms) is equal to the average of two homonuclear bond lengths (one between two like atoms) of the same order (both single, both double, or both triple).

(a) I—Cl bond length = [(I—I bond length) + (Cl—Cl bond length)] ÷ 2
= [266 pm + 199 pm] ÷ 2 = 233 pm

(b) O—Cl bond length = [(O—O bond length) + (Cl—Cl bond length)] ÷ 2
= [145 pm + 199 pm] ÷ 2 = 172 pm

(c) C—F bond length = [(C—C bond length) + (F—F bond length)] ÷ 2
= [154 pm + 143 pm] ÷ 2 = 149 pm

(d) C—Br bond length = [(C—C bond length) + (Br—Br bond length)] ÷ 2
= [154 pm + 228 pm] ÷ 2 = 191 pm

44. First we need to draw the Lewis structure of each of the compounds cited, so that we can determine the order, and thus the relative length, of each O-to-O bond.

(a) In H_2O_2, there are (2 × 1) + (2 × 6) = 14 valence electrons or 7 pairs. A plausible Lewis structure is H—$\overline{\underline{O}}$—$\overline{\underline{O}}$—H

(b) In O_2, the total number of valence electrons is (2 × 6) = 12 valence electrons, or 6 pairs. A plausible Lewis structure is $\overline{\underline{O}}$=$\overline{\underline{O}}$

(c) In O_3, the total number of valence electrons is (3 × 6) = 18 valence electrons, or 9 pairs. A plausible Lewis structure is $\overline{\underline{O}}$=$\overline{O}$—$\overline{\underline{O}}$| ⟵⟶ |$\overline{\underline{O}}$—$\overline{O}$=$\overline{\underline{O}}$

O_2 should have the shortest O-to-O bond, a double bond. The single O—O bond in H_2O_2 should be longest.

45. 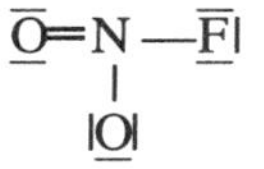 The N—F bond is a single bond. Its bond length should be the average of the N—N single bond (145 pm) and the F—F single bond (143 pm). This average is N—F bond length = (145 + 143) ÷ 2 = 144 pm

46. In H_2NOH, there are (3 × 1) + 5 + 6 = 14 valence electrons total, or 7 pairs. N and O are the two central atoms. A plausible Lewis structure has zero formal charge on each atom. 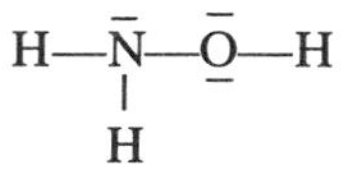 The N—H bond lengths are 100 pm, the O—H bond is 97 pm, and the N—O bond is 136 pm. All values are taken from Table 11.1. All bond angles approximate the tetrahedral bond angle of 109.5°, but are expected to be somewhat smaller, perhaps by 2° to 4° each.

Polar Covalent Bonds

47. The percent ionic character of a bond is based on the difference in electronegativity of its constituent atoms from Figure 11-5.

	(a) S(2.5)—H(2.1)	(b) O(3.5)—Cl(3.0)	(c) Al(1.5)—O(3.5)	(d) As(2.0)—O(3.5)
ΔEN	0.4	0.5	2.0	1.5
%ionic	≈4%	≈5%	≈60%	≈33%

48. 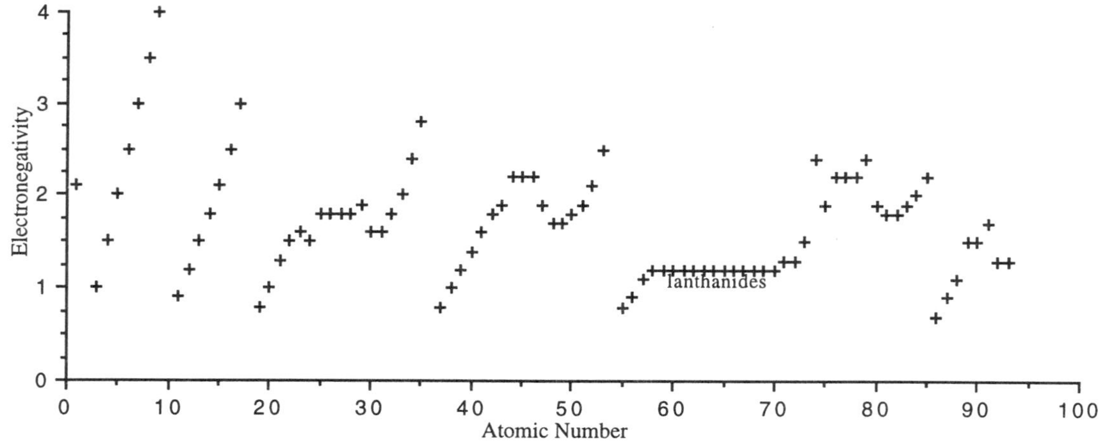

The property of electronegativity does indeed conform to the periodic law. Each of the "low points" corresponds to an alkali metal, and the end of each trend corresponds to a halogen. This is not unexpected,

Resonance

49. In NO_2^-, the total number of valence electrons is $1 + 5 + (2 \times 6) = 18$ valence electrons, or 9 pairs. N is the central atom. There are two resonance forms. $[\overline{I}O{-}\overline{N}{=}\overline{O}]^- \longleftrightarrow [\overline{O}{=}N{-}\overline{O}I]^-$

50. The only one of the four species that requires resonance forms for its adequate representation is CO_3^{2-}. In none of the others are there dissimilar electron distributions around like atoms. All four Lewis structures are drawn below.
  (a) In CO_2, there are $4 + (2 \times 6) = 16$ valence electrons, or 8 pairs.
  (b) In OCl^-, there are $6 + 7 + 1 = 14$ valence electrons, or 7 pairs,
  (c) In CO_3^{2-}, there are $4 + (3 \times 6) + 2 = 24$ valence electrons, or 12 pairs.
  (d) In OH^-, there are $6 + 1 + 1 = 8$ valence electrons, or 4 pairs.

$\overline{O}{=}C{=}\overline{O}$ $[\overline{I}\underline{C}l{-}\overline{O}I]^-$ $\left(\overline{I}O{-}C{=}\overline{O}\atop{|}\atop{IOI}\right)^{2-} \leftrightarrow \left(\overline{O}{=}C{-}\overline{O}I\atop{|}\atop{IOI}\right)^{2-} \leftrightarrow \left(\overline{I}O{-}C{-}\overline{O}I\atop{\|}\atop{IOI}\right)^{2-}$ $[\overline{I}O{-}H]^-$

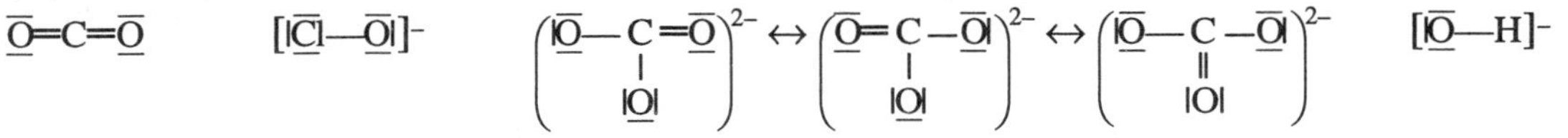

51. Bond length data from Table 11-1 follow: N≡N 109.8 pm   N=N 123 pm   N—N 145 pm
                                                          N=O 120 pm   N—O 136 pm
  The experimental N—N bond length of 113 pm approximates that of the N≡N triple bond, which appears in structure (1). The experimental N—O bond length of 119 pm approximates that of the N=O double bond, which appears in structure (2). Structure (4) is highly unlikely because it contains no nitrogen-to-nitrogen bonds, one of which is found experimentally. Structure (3) also is unlikely, because it contains a very long (145 pm) N—N single bond, which does not agree at all well with the experimental N-to-N bond length. The molecule seems best represented as a resonance hybrid of (1) and (2). $IN{\equiv}N{-}\overline{O}I \longleftrightarrow \overline{N}{=}N{=}\overline{O}$

52. We begin by drawing all three resonance forms of HNO_3 and then analyzing their distribution of formal charge to determine which is the most plausible.
  (a) $H{-}\overline{O}{-}N{-}\overline{O}I\atop{\|}\atop{IOI}$ **(b)** $H{-}\overline{O}{-}N{=}\overline{O}\atop{|}\atop{I\underline{O}I}$ **(c)** $H{-}\overline{O}{=}N{-}\overline{O}I\atop{|}\atop{I\underline{O}I}$

  In all three structures, the formal charges of N and H are the same:
    f.c. of H $= 1 - 1 - 0 = 0$ f.c. of N $= 5 - 4 - 0 = +1$
  For an oxygen that forms two bonds (either 2 single or one double), f.c. $= 6 - 2 - (2 \times 2) = 0$
  For an oxygen that forms only one bond, f.c. $= 6 - 1 - (3 \times 2) = -1$
  For the oxygen that forms three bonds (a single and a double), f.c. $= 6 - 3 - (1 \times 2) = +1$
  Thus, structures **(a)** and **(b)** are equivalent in their distributions of formal charges, zero on all atoms except $+1$ on N and -1 on one O. These are resonance forms. Structure **(c)** is quite different, with formal charges of -1 on two O's, and $+1$ on the other, and a formal charge of $+1$ on N. Structure **(c)** thus is the least plausible.

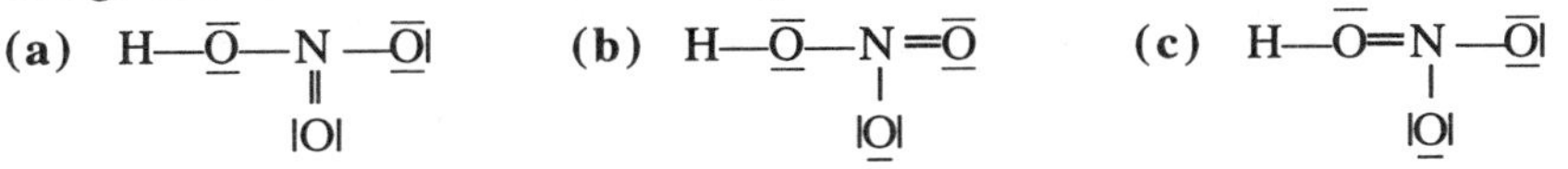

Odd-electron species

53. **(a)** CH_3 has a total of $(3 \times 1) + 4 = 7$ valence electrons, or 3 pairs and a lone electron. C is the central atom. A plausible Lewis structure is $H{-}\overset{\cdot}{\underset{|}{C}}{-}H\atop H$

  (b) ClO_2 has a total of $(2 \times 6) + 7 = 19$ valence electrons, or 9 pairs and a lone electron. Cl is the central atom. A plausible Lewis structure is $\overline{I}O{-}\overline{\underline{C}l}{-}\overline{O}{\cdot}$

  (c) NO_3 has a total of $(3 \times 6) + 5 = 23$ valence electrons, or 11 pairs, plus a lone electron. N is the central atom. A plausible Lewis structure is ${\cdot}\overline{O}{-}N{=}\overline{O}\atop{|}\atop{IOI}$ Resonance forms can also be drawn.

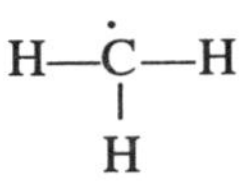

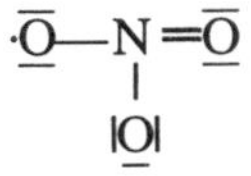

(c) NO_3 has a total of $(3 \times 6) + 5 = 23$ valence electrons, or 11 pairs, plus a lone electron. N is the central atom. A plausible Lewis structure is 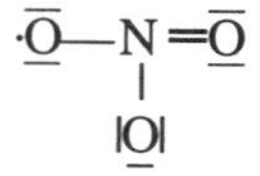 Resonance forms can also be drawn.

54. In NO_2, there are $5 + (2 \times 6) = 17$ valence electrons, 8 pairs and a lone electron. N is the central atom. A plausible Lewis structure is |O—Ṅ=O| which has a formal charge of +1 on N and one of −1 on the single-bonded O. Another Lewis structure, with zero formal charge on each atom, is |Ȯ—N=O| Because of the unpaired electron we expect NO_2 to be paramagnetic.

We would expect a bond—a pair of electrons—to form between two NO_2 molecules as a result of the pairing of the electrons than are unpaired in the NO_2 molecules. If the second structure for NO_2 is used, the one with zero formal charge on each atom, a plausible structure for N_2O_4 is O=N—O—O—N=O

If the first Lewis structure for NO_2 is used, a plausible Lewis structure contains a N—N bond. 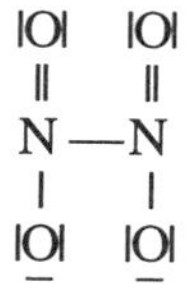

Resonance structures can be drawn for this second version of N_2O_4. A N—N bond is observed experimentally in N_2O_4. In either structure for N_2O_4, all electrons are paired; the molecule is expected to be (and is) diamagnetic.

Expanded octets

55. Lewis structures follow the discussion.

In PO_4^{3-} there are $5 + (4 \times 6) + 3 = 32$ valence electrons or 16 pairs. An expanded octet is not needed.
In PI_3 there are $5 + (3 \times 7) = 26$ valence electrons or 13 pairs. An expanded octet is not needed.
In ICl_3 there are $7 + (3 \times 7) = 28$ valence electrons or 14 pairs. An expanded octet is necessary
In $OSCl_2$ there are $6 + 6 + (2 \times 7) = 26$ valence electrons or 13 pairs. An expanded octet is not needed.
In SF_4 there are $6 + (4 \times 7) = 34$ valence electrons or 17 pairs. An expanded octet is necessary.
In ClO_4^{3-} there are $7 + (4 \times 6) + 1 = 32$ valence electrons or 16 pairs. An expanded octet is not needed.

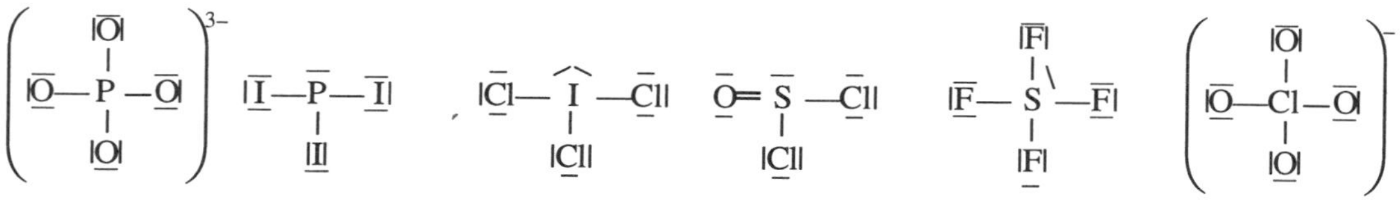

56. Let us draw the Lewis structure of H_2CSF_4. The molecule has $(2 \times 1) + 4 + 6 + (4 \times 7) = 40$ valence electrons, or 20 pairs. With only single bonds and all octets complete, there is a −1 formal charge on C, as in the structure below left. The structure below right avoids that difficulty by creating a carbon-to-sulfur double bond.

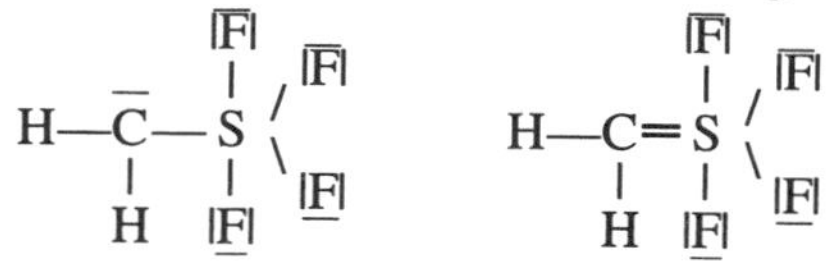

Molecular shapes

57. The AX_nE_m designations that are cited below are to be found in Table 11-2 of the text, along with a sketch and a picture of a model of each type of structure.

(a) |N≡N| — N_2 is linear; two points define a line.

(b) H—C≡N| — HCN is linear. The molecule belongs to the AX_2E_0 category, and these species are linear.

(c) 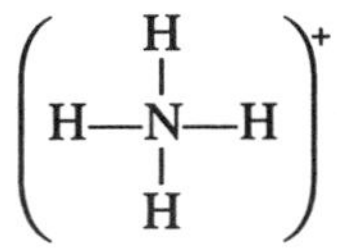

NH_4^+ is tetrahedral. The ion is of the AX_4 type, which has a tetrahedral electron-group geometry and a tetrahedral shape.

(d) 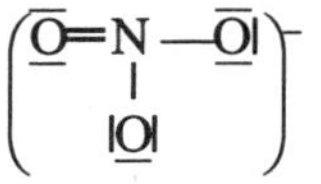

NO_3^- is trigonal planar. The ion is of the AX_3 type, which has a trigonal planar electron-group geometry and a trigonal planar shape. The other resonance forms are of the same type.

(e) 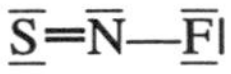

SNF is bent. The molecule is of the AX_2E type, which has a trigonal planar electron-group geometry and a bent shape.

58. The AX_nE_m designations that are cited below are to be found in Table 11-2 of the text, along with a sketch and a picture of a model of each type of structure.

(a) 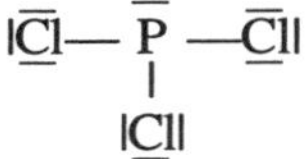

PCl_3 is a trigonal pyramid. The molecule is of the AX_3E_1 type, and has a tetrahedral electron-group geometry and a trigonal pyramid shape.

(b) 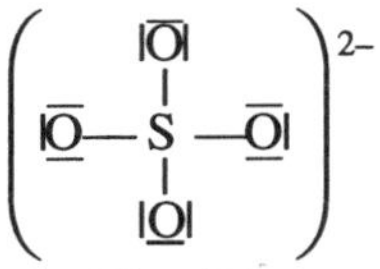

SO_4^{2-} has a tetrahedral shape. The ion is of the type AX_4, and has a tetrahedral electron-group geometry and a tetrahedral shape. The other resonance forms of the sulfate ion have the same shape.

(c) 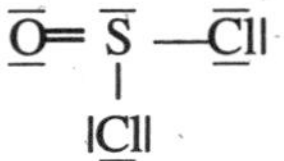

$SOCl_2$ has a trigonal pyramidal shape. This molecule is of the AX_3E_1 type and has a tetrahedral electron-group geometry and a trigonal pyramidal shape.

(d) 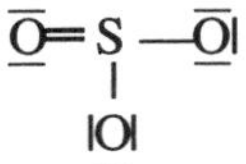

SO_3 has a trigonal planar shape. The molecule is of the AX_3 type,with a trigonal planar electron-group geometry and molecular shape. The other resonance forms have the same shape.

(e) 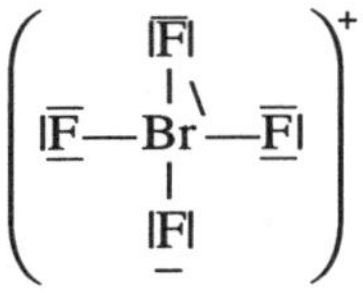

BrF_4^+ has a see-saw shape. The molecule is of the AX_4E type,with a trigonal bipyramid electron-group geometry and a see-saw molecular shape.

59. A trigonal planar shape requires that three ligands and no lone pairs be bonded to the central atom. Thus PF_6^- cannot have a trigonal planar shape, since six atoms are attached to the central atom. In addition, PO_4^{3-} cannot have a trigonal planar shape, since four O atoms are attached to the central P atom. We now draw the Lewis structure of each of the remaining ions, as a first step in predicting their shapes.

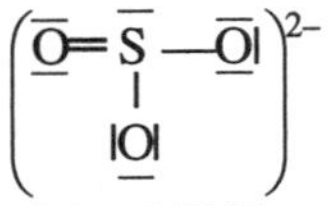 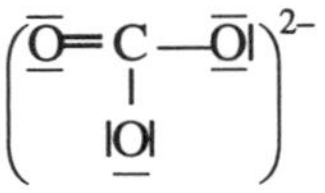

The SO_3^{2-} ion is of the AX_3E_1 type; it has a tetrahedral electron-group geometry and a trigonal pyramidal shape. The CO_3^{2-} ion is of the AX_3 type, and has a trigonal planar electron-group geometry and a trigonal planar shape.

60. We predict shapes by beginning with the Lewis structures of the species.

	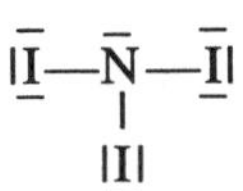	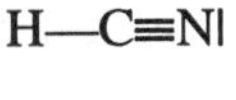	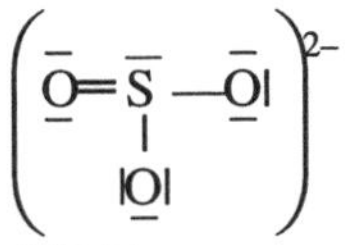	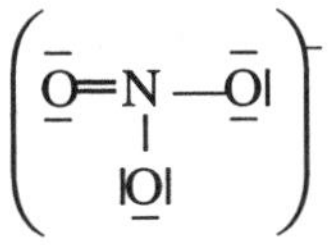
AX_nE_m	AX_3E_1	AX_2E_0	AX_3E_1	AX_3E_0
shape	trigonal pyramid	linear	trigonal pyramid	trigonal planar

Thus, SO_3^{2-} and NI_3 both have the same shape.

61. (a) In CO_2 there are a total of $4 + (2 \times 6) = 16$ valence electrons, or 8 pairs. The following Lewis structure is plausible. 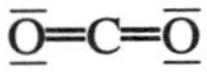 This is a molecule of type AX_2; CO_2 has a linear electron-group geometry and a linear shape.

(b) In Cl_2CO there are a total of $(2 \times 7) + 4 + 6 = 24$ valence electrons, or 12 pairs. The molecule can be represented by a Lewis structure with C as the central atom. $\overline{O}= C —\overline{Cl}|$ This molecule is of the AX_3 type; it has a trigonal planar electron-group geometry and molecular shape.

(c) In $ClNO_2$ there are a total of $7 + 5 + (2 \times 6) = 24$ valence electrons, or 12 pairs. N is the central atom. A plausible Lewis structure is $|\overline{Cl}—N=\overline{O}$ This molecule is of the AX_3 type; it has a trigonal planar electron-group geometry and a trigonal planar shape.

62. Lewis structures enable us to determine molecular shapes.

(a) N_2O_4 has $(2 \times 5) + (4 \times 6) = 34$ valence electrons, or 17 pairs.

(b) C_2N_2 has $(2 \times 4) + (2 \times 5) = 18$ valence electrons, or 9 pairs.

(c) C_2H_6 has $(2 \times 4) + (6 \times 1) = 14$ valence electrons, or 7 pairs.

(d) CH_3OCH_3 has 6 more valence electrons that $C_2H_6 = 20$ valence electrons or 10 pairs.

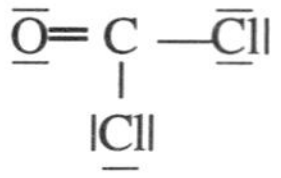

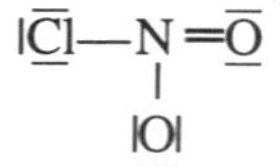

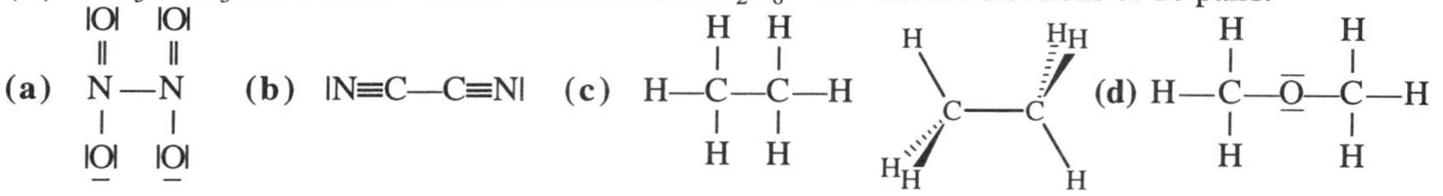

(a) There are three atoms bound to each N in N_2O_4, making the molecule triangular planar around each N. The entire molecule does not have to be planar, however, since there is free rotation around the N—N bond. The molecule is a bit more stable if it is planar, however.

(b) There are two atoms attached to each C, requiring a linear geometry. The entire molecule is linear.

(c) There are four atoms attached to each C. The molecular geometry around each C is tetrahedral, as shown in the sketch above.

(d) Around the central O, the electron-group geometry is tetrahedral. With two of the electron groups being lone pairs, the molecular geometry around the central atom is bent.

63. First we draw the Lewis structure of each species, then use these Lewis structures to predict the molecular shape.

(a) In ClO_4^- there are $7 + (4 \times 6) + 1 = 32$ valence electrons or 16 pairs. A plausible Lewis structure follows. Since there are four atoms and no lone pairs bonded to the central atom, the molecular shape and the electron-group geometry are the same: tetrahedral.

(b) In $S_2O_3^{2-}$ there are $(2 \times 6) + (3 \times 6) + 2 = 32$ valence electrons or 16 pairs. A plausible Lewis structure follows. Since there are four atoms and no lone pairs bonded to the central atom, the electron-group geometry and the molecular shape are the same: tetrahedral.

(c) In PF_6^- there are $5 + (6 \times 7) + 1 = 48$ valence electrons or 24 pairs. A plausible Lewis structure follows. Since there are six atoms and no lone pairs bonded to the central atom, the electron-group geometry and molecular shape are octahedral.

(d) In I_3^- there are $(3 \times 7) + 1 = 22$ valence electrons or 11 pairs. There are three lone pairs and two atoms bound to the central atom. The electron-group geometry is trigonal bipyramid; the molecular shape is linear.

(a) $\left(\begin{array}{c} |\overline{O}| \\ | \\ |\overline{O}—Cl—\overline{O}| \\ | \\ |\overline{O}| \end{array} \right)^-$ (b) $\left(\begin{array}{c} |S| \\ \| \\ |\overline{O}—S—\overline{O}| \\ | \\ |\overline{O}| \end{array} \right)^{2-}$ (c) $\left(\begin{array}{c} |\overline{F} \quad \overline{F}| \\ |\overline{F}—\overset{.}{S}—\overline{F}| \\ |\overline{F} \quad \overline{F}| \end{array} \right)^-$ (d) $[|\overline{I}—\overset{\frown}{I}—\overline{I}|]^-$

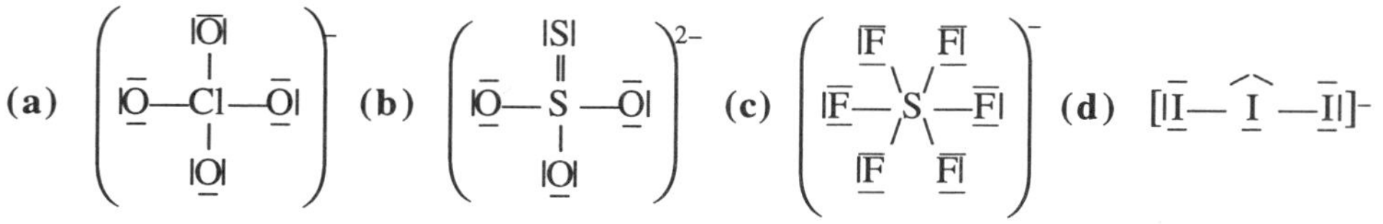

64. (a) In OSF_2, there are a total of $6 + 6 + (2 \times 7) = 26$ valence electrons, or 13 pairs. S is the central atom. A plausible Lewis structure is $|\overline{O}—\overline{S}—\overline{F}|$ This molecule is of the type AX_3E_1; it has a tetrahedral electron-group geometry and a trigonal pyramidal shape.

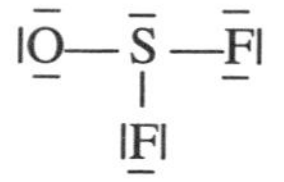

(b) In O_2SF_2, there are a total of $(2 \times 6) + 6 + (2 \times 7) = 32$ valence electrons, or 16 pairs. S is the central

atom. One plausible Lewis structure has zero formal charge on each atom. $\overline{O}{=}\,\overset{\displaystyle |\overline{F}|}{\underset{\displaystyle |\overline{F}|}{S}}\,{=}\overline{O}$ The molecule

is of the AX_4 type; it has a tetrahedral electron-group geometry and a tetrahedral shape. The structure with all single bonds is preferred because it does not have an expanded octet. It has the same shape.

(c) In SF_5^- there are $6 + (5 \times 7) + 1 = 42$ valence electrons, or 21 pairs. A plausible Lewis structure is

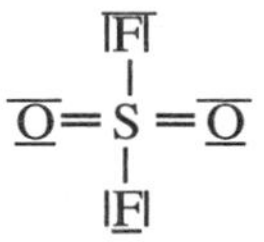

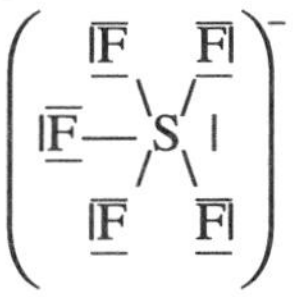

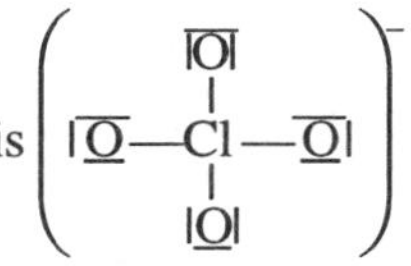

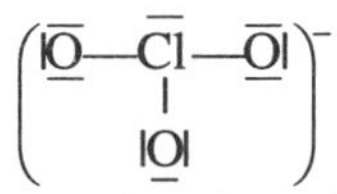

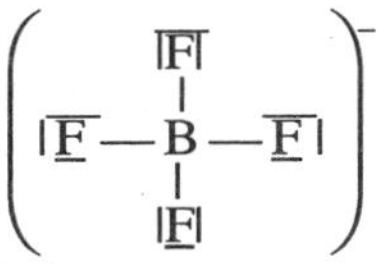

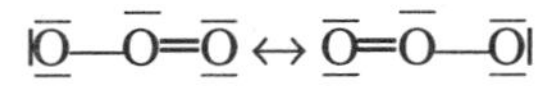

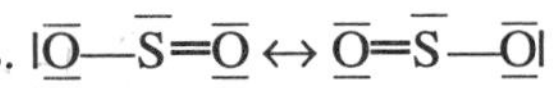

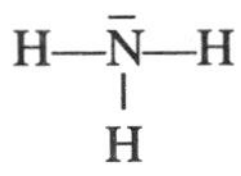

$$\left(\begin{array}{c} |\overline{F} \quad \overline{F}| \\ |\overline{F}{-}\overset{\diagdown}{\underset{\diagup}{S}}| \\ |\overline{F} \quad \overline{F}| \end{array} \right)^{-}$$ The ion is of the AX_5E type. It has an octahedral electron-group geometry and a

square pyramidal molecular shape.

(d) In ClO_4^-, there are a total of $1 + 7 + (4 \times 6) = 32$ valence electrons, or 16 pairs. Cl is the central atom.

A plausible Lewis structure is $$\left(\begin{array}{c} |\overline{O}| \\ | \\ |\overline{O}{-}Cl{-}\overline{O}| \\ | \\ |\overline{O}| \end{array} \right)^{-}$$ This ion is of the AX_4 type; it has a tetrahedral

electron-group geometry and a tetrahedral shape.

(e) In ClO_3^- the total number of valence electrons is $1 + (3 \times 6) + 7 = 26$ valence electrons, or 13 pairs. A

plausible Lewis structure is $$\left(\begin{array}{c} |\overline{O}{-}Cl{-}\overline{O}| \\ | \\ |\overline{O}| \end{array} \right)^{-}$$ The molecule is of the type AX_3E; it has a

tetrahedral electron-group geometry and a trigonal pyramidal molecular shape.

65. In BF_4^-, there are a total of $1 + 3 + (4 \times 7) = 32$ valence electrons, or 16 pairs. A plausible Lewis structure

has B as the central atom. $$\left(\begin{array}{c} |\overline{F}| \\ | \\ |\overline{F}{-}B{-}\overline{F}| \\ | \\ |\overline{F}| \end{array} \right)^{-}$$ This ion is of the type AX_4; it has a tetrahedral electron-group

geometry and a tetrahedral shape.

66. The O_3 molecule has $(3 \times 6) = 18$ valence electrons, 9 pairs. A plausible Lewis structure does indeed appear

to be linear; it has two resonance forms. $\quad |\overline{O}{-}\overline{O}{=}\overline{O} \leftrightarrow \overline{O}{=}\overline{O}{-}\overline{O}| \quad$ But in each resonance form, there

are two bonding pairs and one lone pair on the central O atom. This indicates a trigonal planar electron-group geometry and a bent molecular geometry.

Polar molecules

67. For each molecule, we first draw the Lewis structure, which we use to predict the shape.

(a) SO_2 has a total of $6 + (2 \times 6) = 18$ valence electrons, or 9 pairs. The molecule has two resonance

forms. $|\overline{O}{-}\overline{S}{=}\overline{O} \leftrightarrow \overline{O}{=}\overline{S}{-}\overline{O}|$ Each of these resonance forms is of the type AX_2E_1; it has a triangular

planar electron-group geometry and a bent shape. Since each S—O bond is polar toward O, and since the bond dipoles do not point in opposite directions, the molecule has a resultant dipole moment, pointing from S through a point midway between the two O atoms; SO_2 is polar.

(b) NH_3 has a total of $5 + (3 \times 1) = 8$ valence electrons, or 4 pairs. N is the central atom. A plausible

Lewis structure is $\;H{-}\overset{\displaystyle \,}{\underset{\displaystyle H}{\overline{N}}}{-}H\;$ The molecule is of the AX_3E_1 type; it has a tetrahedral electron-group

geometry and a trigonal pyramidal shape. Each N—H bond is polar toward N. Since the bonds do not symmetrically oppose each other, there is a resultant molecular dipole moment, pointing from the triangular base (formed by the three H atoms) through N. The molecule is polar.

(c) H_2S has a total of $6 + (2 \times 1) = 8$ valence electrons, or 4 pairs. S is the central atom and a plausible Lewis structure is H—$\overline{\underline{S}}$—H This molecule is of the AX_2E_2 type; it has a tetrahedral electron-group geometry and a bent shape. Each H—S bond is polar toward S. Since the bonds do not symmetrically oppose each other, the molecule has a net dipole moment, pointing through S from a point midway between the two H atoms. H_2S is polar.

(d) C_2H_4 consists of atoms that have all about the same electronegativities. Of course, the C—C bond is not polar, but neither are the C—H bonds. Thus, the entire molecule is nonpolar. The molecule is planar.

(e) SF_6 has a total of $6 + (6 \times 7) = 48$ valence electrons, or 24 pairs. S is the central atom. All atoms have zero formal charge in the Lewis structure. 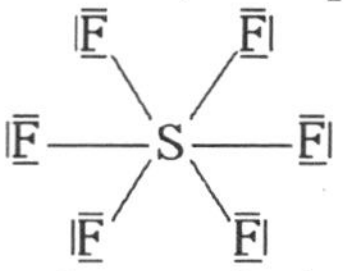 This molecule is of the AX_6E_0 type; it has an octahedral electron-group geometry and an octahedral shape. Even though each S—F bond is polar toward F, the bonds symmetrically oppose each other resulting in a molecule that is nonpolar.

(f) CH_2Cl_2 has a total of $4 + (2 \times 1) + (2 \times 7) = 20$ valence electrons, or 10 pairs. A plausible Lewis structure is drawn below. The molecule is tetrahedral and polar, since the two polar bonds (C—Cl) do not cancel the effect of each other. 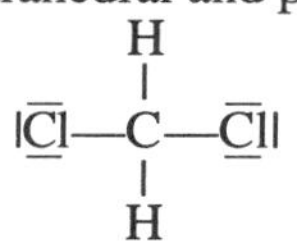

68. **(a)** HCN is a linear molecule, which can be derived from its Lewis structure. H—C≡N| The C≡N bond is strongly polar toward nitrogen, while the H—C bond is generally considered to be nonpolar. Thus, the molecule has a dipole moment, pointed toward N from C.

(b) SO_3 is a trigonal planar molecule, which can be derived from its Lewis structure. 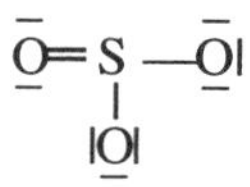

Each sulfur-oxygen bond is polar from S to O, but the three bonds are equally polar and are pointed in opposition so that they cancel. The SO_3 molecule has zero dipole moment.

(c) CS_2 is a linear molecule, which can be derived from its Lewis structure. $\overline{\underline{S}}$=C=$\overline{\underline{S}}$ Each carbon-sulfur bond is polar from C to S, but the two bonds are equally polar and are pointed in opposition to each other so that they cancel. The CS_2 molecule has zero dipole moment.

(d) OCS also is a linear molecule. Its Lewis structure is $\overline{\underline{O}}$=C=$\overline{\underline{S}}$ But the carbon-oxygen bond is more polar than the carbon-sulfur bond. Although both bond dipoles point from the central atom to the bonded atom, these two bond dipoles are unequal in strength. Thus, the molecule is polar in the direction from C to O.

(e) $SOCl_2$ is a trigonal pyramidal molecule. Its Lewis structure is 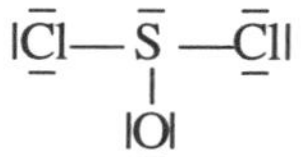 The lone pair is at one corner of the tetrahedron. Each bond in the molecule is polar, pointing away from the central atom, but the sulfur-chlorine bond is less polar than is the sulfur-oxygen bond, making the molecule polar. The dipole moment of the molecule points from the sulfur atom to the base of the trigonal pyramid, not toward the center of the base but slightly toward the O apex of that base.

(f) SiF_4 is a tetrahedral molecule, with the following Lewis structure 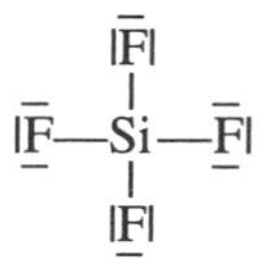 Each Si—F bond is polar, with its negative end away from the central atom toward F in each case. These four Si—F bond dipoles oppose each other in arrangement and thus cancel. SiF_4 has no dipole moment.

(g) POF_3 is a tetrahedral molecule, with the following Lewis structure

```
      _
     |F|
      |
 _    |    _
|F— P —O|
 ‾    |    ‾
     |F|
      ‾
```

All four bonds are polar, pointing away from the central atom, but the P—F bond polarity is stronger than that of the P—O bond. Thus, POF_3 is a polar molecule with its dipole moment pointing away from the P to the center of the triangle formed by the three F atoms.

69. In H_2O_2, there are a total of $(2 \times 1) + (2 \times 6) = 14$ valence electrons, 7 pairs. The two O atoms are central atoms. A plausible Lewis structure has zero formal charge on each atom. H—$\overline{\underline{O}}$—$\overline{\underline{O}}$—H In the hydrogen peroxide molecule, the O—O bond is non-polar, while the H—O bonds are polar, toward O. Since the molecule has a resultant dipole moment, it cannot be linear, for, if it were linear the two polar bonds would oppose each other and their polarities would cancel.

70. **(a)** FNO has a total of $7 + 5 + 6 = 18$ valence electrons, or 9 pairs. N is the central atom. A plausible Lewis structure is |$\overline{\underline{F}}$—$\overline{N}$=$\overline{\underline{O}}$ The formal charge on each atom in this structure is zero.

(b) FNO is of the AX_2E_1 type; it has a trigonal planar electron-group geometry and a bent shape.

(c) The N—F bond is polar toward F and the N—O bond is polar toward O. In FNO, these two bond dipoles point in the same general direction, producing a polar molecule. In FNO_2, however, the additional N—O bond dipole partially opposes the polarity of the other two bond dipoles, resulting in a smaller net dipole moment.

Bond Energies

71.

```
   H  H                      H  H
   |  |      _   _           |  |    _        _
H—C—C—H + |Cl—Cl| ⟶ H—C—C—Cl| + H—Cl|
   |  |      ‾   ‾           |  |    ‾        ‾
   H  H                      H  H
```

Analysis of the Lewis structures of products and reactants indicates that a C—H bond and a Cl—Cl bond are broken, and a C—Cl and a H—Cl bond are formed.

Energy required to break bonds = C—H + Cl—Cl = 414 kJ/mol + 243 kJ/mol = 657 kJ/mol

Energy realized by forming bonds = C—Cl + H—Cl = 339 kJ/mol + 431 kJ/mol = 770 kJ/mol

$\Delta H = 657 \text{ kJ/mol} - 770 \text{ kJ/mol} = -113 \text{ kJ/mol}$

72. The chemical equation in terms of Lewis structures is $\overline{\underline{O}}$=$\overline{O}$—$\overline{\underline{O}}$| + $\overline{\underline{O}}$=$\overline{N}$· ⟶ $\overline{\underline{O}}$=$\overline{N}$—$\overline{\underline{O}}$· + $\overline{\underline{O}}$=$\overline{\underline{O}}$

The net result is the breakage of an O—O bond (142 kJ/mol) and the formation of a N—O bond (230 kJ/mol). $\Delta H = 142 \text{ kJ/mol} - 230 \text{ kJ/mol} = -88 \text{ kJ/mol}$

73. In each case we write the formation reaction, but specify reactants and products with their Lewis structures. All species are assumed to be gases.

(a) $\frac{1}{2}$ $\overline{\underline{O}}$=$\overline{\underline{O}}$ + $\frac{1}{2}$ H—H ⟶ ·$\overline{\underline{O}}$—H

Bonds broken = $\frac{1}{2}$ O=O + $\frac{1}{2}$ H—H = 0.5 (498 kJ + 436 kJ) = 467 kJ

Bond formed = O—H = 464 kJ $\qquad$ $\Delta H_f^\circ = 467 \text{ kJ} - 464 \text{ kJ} = 3 \text{ kJ/mol}$

If the O—H bond dissociation energy of 428.0 kJ/mol from Figure 11-14 is used, $\Delta H_f^\circ = 39$ kJ/mol.

(b)

```
                            _   _
|N≡N| + 2 H—H ⟶ H—N—N—H
                            |   |
                            H   H
```

Bonds broken = N≡N + 2 H—H = 946 kJ + 2 × 436 kJ = 1818 kJ

Bonds formed = N—N + 4 N—H = 163 kJ + 4 × 389 kJ = 1719 kJ

$\Delta H_f^\circ = 1818 \text{ kJ} - 1719 \text{ kJ} = 99 \text{ kJ}$

74. The reaction of Example 11-13 is $CH_4(g) + Cl_2(g) \longrightarrow CH_3Cl(g) + HCl(g)$ $\quad \Delta H_{rxn} = -113$ kJ. In Appendix D are the following values: $\Delta H_f^\circ[CH_4(g)] = -74.81$ kJ, $\Delta H_f^\circ[HCl(g)] = -92.31$ kJ, $\Delta H_f^\circ[Cl_2(g)] = 0$

Thus, we have $\Delta H_{rxn} = \Delta H_f^\circ[CH_3Cl(g)] + \Delta H_f^\circ[HCl(g)] - \Delta H_f^\circ[CH_4(g)] - \Delta H_f^\circ[Cl_2(g)]$

$-113 \text{ kJ} = \Delta H_f^\circ[CH_3Cl(g)] - 92.31 \text{ kJ} - (-74.81 \text{ kJ}) - (0.00)$

$\Delta H_f^\circ[CH_3Cl(g)] = -113 \text{ kJ} + 92.31 \text{ kJ} - 74.81 \text{ kJ} = -96 \text{ kJ}$

75. $H_3C{-}CH_2{-}\overline{\underline{O}}{-}H + 3\ \overline{O}{=}\overline{O} \longrightarrow 3\ H{-}\overline{\underline{O}}{-}H + 2\ \overline{O}{=}C{=}\overline{O}$

(structure drawn: H–C(H)(H)–C(H)(H)–O–H)

Bonds broken = 5 C—H + C—C + C—O + O—H + 3 O=O

$= 5 \times 414 \text{ kJ} + 347 + 360 + 464 + 3 \times 498 \text{ kJ} = 4.74 \times 10^3 \text{ kJ}$

Bonds formed = 6 H—O + 4 C=O = $6 \times 464 \text{ kJ} + 4 \times 736 \text{ kJ} = 5.73 \times 10^3 \text{ kJ}$

$\Delta H^\circ = 4.74 \times 10^3 \text{ kJ} - 5.73 \times 10^3 \text{ kJ} = -9.9 \times 10^2 \text{ kJ}$

76. First determine the value of ΔH for reaction **(2)**.

(1) $C(s) \longrightarrow C(g)$ $\quad \Delta H = 717$ kJ

(2) $C(g) + 2\ H_2(g) \longrightarrow CH_4(g)$ $\quad \Delta H = ?$

Net: $C(s) + 2\ H_2(g) \longrightarrow CH_4(g)$ $\quad \Delta H_f = -75$ kJ

717 kJ + ? = −75 kJ or ? = −75 −717 kJ = −792 kJ

To determine the energy of a C—H bond, we need to analyze reaction **(2)** in some detail.

C + 2 H—H ⟶ H—C(H)(H)—H

In this reaction, 2 H—H bonds are broken and 4 C—H bonds are formed resulting in the production of 792 kJ/mol.

−792 kJ/mol = energy of broken bonds − energy of formed bonds

= (2 × 436 kJ/mol) − (4 × C—H)

4 × C—H = 792 kJ/mol + (2 × 436 kJ/mol) = 1664 kJ/mol

C—H = 1662 ÷ 4 = 416 kJ/mol

This value compares favorably with the value of 414 kJ/mol given in Table 11-3.

FEATURE PROBLEMS

A. **(1)** The average of the H—H and Cl—Cl bond energies is (436 + 243) ÷ 2 = 340 kJ/mol. The ionic resonance energy is the difference between this calculated value and the measured value of the H—Cl bond energy: IRE = 431 − 340 = 91 kJ/mol

(2) $\Delta EN = \sqrt{IRE / 96} = \sqrt{91 / 96} = 0.97$

(3) An electronegativity difference of 0.97 gives about a 23% ionic character, read from Figure 11-6. The result of Example 11-4 is that the H—Cl bond is 20% ionic. These values are in good agreement with each other.

B. **(1)** The two bond moments can be added geometrically, by placing the head of one at the tail of the other, as long as we do not change the direction or the length of the moved dipole. The resultant molecular dipole moment is represented by the arrow drawn from the tail of one bond dipole to the head of the other. This is shown in the figure at right. The 52° angle in the figure is one-half of the 104° bond angle in water. The length of the diagonal is computed as follows.

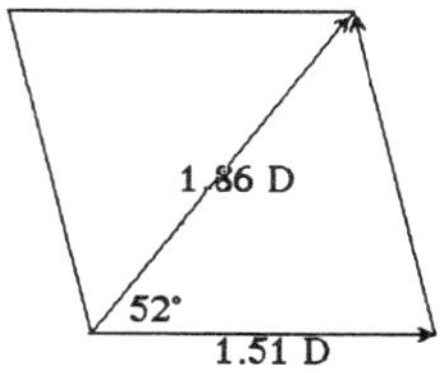

molecular dipole = 2 × 1.51 D × cos 52° = 1.86 D

(2) For H_2S, we do not know the bond angle. We shall represent this bond angle as α. Then, following the same procedure as above, we have the following.

$$\text{molecular dipole} = 2 \times \text{bond dipole} \times \cos\left(\frac{\alpha}{2}\right) \qquad 0.93 \text{ D} = 2 \times 0.67 \text{ D} \times \cos\left(\frac{\alpha}{2}\right)$$

$$\cos\left(\frac{\alpha}{2}\right) = \frac{0.93}{2 \times 0.67} = 0.694 \qquad \frac{\alpha}{2} = \cos^{-1}(0.694) = 46^\circ$$

Thus, the bond angle in $H_2S = 2 \times \alpha = 2 \times 46^\circ = 92^\circ$

12 CHEMICAL BONDING II:

ADDITIONAL ASPECTS

PRACTICE EXAMPLES

1A The valence-shell orbital diagrams of N and I are as follows.

N [He] $2s$[⇅] $2p$[↑][↑][↑]   I [Kr] $4d^{10}$ $5s$[⇅] $5p$[⇅][⇅][↑]

There are three half-filled $2p$ orbitals on N, and one half-filled $5p$ orbital on I. Each half-filled $2p$ orbital from N will overlap with one half-filled $5p$ orbital of a I. Thus, there will be three N—I bonds. The I atoms will be oriented in the same direction as the three $2p$ orbitals of N: toward the x-, y-, and z-directions of a Cartesian coordinate system. Thus, the I—N—I angles will be approximately 90° (probably larger because the I atoms will repel each other). The three I atoms will lie in the same plane at the points of a triangle, with the N atom centered above them. The molecule is trigonal pyramidal. (The same molecular shape is predicted if N is assumed to be sp^3 hybridized, but with 109.5° rather than 90° bond angles.)

1B The valence-shell orbital diagrams of N and H are as follows. N: [He] $2s$[⇅] $2p$[↑][↑][↑]   H: $1s$[↑]

There are three half-filled orbitals on N and one half-filled orbital on each H. There will be three N—H bonds, with bond angles of approximately 90°. The molecule is trigonal pyramid. (We obtain the same molecular shape if N is assumed to be sp^3 hybridized, but bond angles are closer to 109.5°, the tetrahedral bond angle.)

VSEPR theory begins with the Lewis structure and notes that there are three bond pairs and one lone pair attached to N. This produces a tetrahedral electron pair geometry and a trigonal pyramid molecular shape with bond angles a bit less than the tetrahedral angle of 109.5° because of the lone pair.

Since VSEPR theory makes a prediction closer to the experimental bond angle of 107°, it seems more appropriate in this case.

2A We begin by writing the Lewis structure. The H atoms are terminal atoms. There are three central atoms and $(3 \times 1) + 4 + 6 + 4 + (3 \times 1) = 20$ valence electrons, or 10 pairs. A plausible Lewis structure is drawn at right. Each central atom is surrounded by four electron pairs, requiring sp^3 hybridization. The valence-shell orbital diagrams for the atoms follow.

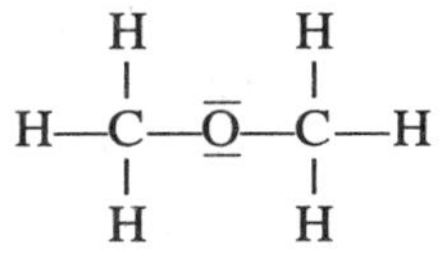

H $1s$[↑]   C [He] $2s$[⇅] $2p$[↑][ ][ ]   O [He] $2s$[⇅] $2p$[⇅][↑][↑]

The valence-shell orbital diagrams for the hybridized central atoms then are:

C [He] $_{sp^3}$[↑][↑][↑][↑]   O [He] $_{sp^3}$[⇅][⇅][↑][↑]

All bonds in the molecule are σ bonds. The H—C—H bond angles are 109.5°, as are the H—C—O bond angles. The C—O—C bond angle is possibly a bit smaller than 109.5° because of the repulsion of the two lone pairs of electrons on O. A wedge-and-dash sketch of the molecule is at right.

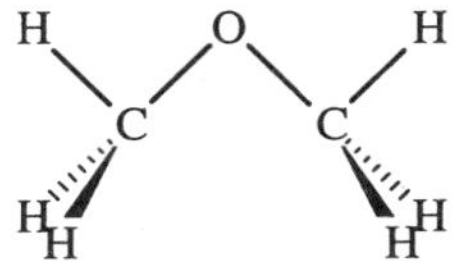

2B The H atoms and one O are terminal atoms in the Lewis structure, which has $3 \times 1 + 4 + 4 + 2 \times 6 + 1 = 24$ valence electrons, or 12 pairs. The left-most C and the right-most O are surrounded by four electron pairs, requiring sp^3 hybridization. The central carbon is surrounded by three electron groups and is sp^2 hybridized.

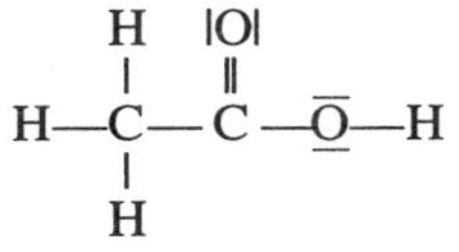

The orbital diagrams for the un hybridized atoms are H: $1s$[↑]   C: [He] $2s$[⇅] $2p$[↑][ ][ ]   O: [He] $2s$[⇅] $2p$[⇅][↑][↑]

Hybridized atom orbital diagrams: C: [He] $_{sp^3}$[↑][↑][↑][↑]   C: [He] $_{sp^2}$[↑][↑][↑] $2p$[↑]   O: [He] $_{sp^3}$[⇅][⇅][↑][↑]

There is one π bond in the molecule: between $2p$ on the central C and $2p$ on the terminal O. The remaining bonds are σ bonds. The H—C—H and H—C—C bond angles are 109.5°. The H—O—C angle is somewhat less, perhaps 105°. The C—C—O bond angles all are about 120°.

3A There are four bond pairs around the left-hand C, requiring sp^3 hybridization. Three of the bonds that form are C—H sigma bonds resulting from the overlap of a sp^3 hybrid orbital on C with a $1s$ orbital on H. The other C has two attached electron groups, producing sp hybridization. C: [He] sp [↑][↑] $2p$ [↑][↑] The N atom is not hybridized. N: [He] $2s$ [↑↓] $2p$ [↑][↑][↑] The two C atoms join with a sigma bond: overlap of sp^2 on the left-hand C with sp on the right-hand C. The three bonds between C and N consist of a sigma bond (sp on C with $2p$ on N), and two pi bonds ($2p$ on C with $2p$ on N).

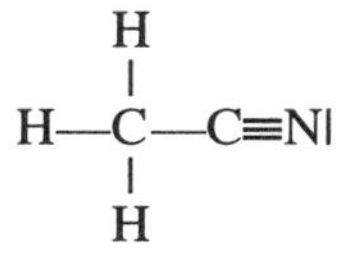

3B The bond lengths that are given indicate that N is the central atom. The molecule has $(2 \times 5) + 6 = 16$ valence electrons, or 8 pairs. Average bond lengths are: N—N = 145 pm, N═N = 123 pm, N≡N = 110 pm, N—O = 136 pm, N═O = 120 pm. Plausible resonance structures are (with subscripts for identification):

structure (1) $|N_a{\equiv}N_b{-}\bar{\underline{O}}|$ ↔ $|\bar{N}_a{=}N_b{=}\bar{\underline{O}}$ structure (2)

In both structures, the central N is attached to two other atoms, and has no attached lone pairs. The geometry of the molecule thus is linear and the hybridization on this central N is sp. N_b [He] sp [↑↓][↑] $2p$ [↑][↑]

We will assume that the terminal atoms are not hybridized in either structure. Their valence-shell orbital diagrams are: terminal N N_a [He] $2s$ [↑↓] $2p$ [↑][↑][↑] O [He] $2s$ [↑↓] $2p$ [↑↓][↑][↑]

In structure (1) the N≡N bond results from the overlap of three pairs of half-filled orbitals: (1) sp on N_b with $2p_x$ on N_a forming a σ bond, (2) $2p_y$ on N_b with $2p_y$ on N_a forming a π bond, and (3) $2p_z$ on N_b with $2p_z$ on N_a also forming a π bond. The N—O bond is a coordinate covalent bond, and requires that the electron configuration of O be written O [He] $2s$ [↑↓] $2p$ [↑↓][↑↓][ ] The N—O bond then forms by the overlap of the full sp orbital on N_b with the empty $2p$ orbital on O.

In structure (2) the N═O bond results from the overlap of two pairs of half-filled orbitals: (1) sp on N_b with $2p_y$ on O forming a σ bond and (2) $2p_z$ on N_b with $2p_z$ on O forming a π bond. The N═N σ bond is a coordinate covalent bond, and requires that the configuration of N_a be written N_a [He] $2s$ [↑↓] $2p$ [↑↓][↑][ ] The N═N bond is formed by two overlaps: (1) the overlap of the full sp orbital on N_b with the empty $2p_z$ orbital on N_a to form a σ bond, and (2) the overlap of the half-filled $2p_y$ orbital on N_b with the half-filled $2p_y$ orbital on N_a to form a π bond.

4A Removing an electron from Li_2 removes a bonding electron because the valence molecular orbital diagram for Li_2 is the same as that for H_2, just moved up a quantum level: $\sigma_{1s}{}^b$ [↑↓] $\sigma_{1s}{}^*$ [↑↓] $\sigma_{2s}{}^b$ [↑↓] $\sigma_{2s}{}^*$ [ ], So the molecular orbital diagram for $Li_2{}^+$ is $\sigma_{1s}{}^b$ [↑↓] $\sigma_{1s}{}^*$ [↑↓] $\sigma_{2s}{}^b$ [↑] $\sigma_{2s}{}^*$ [ ] The bond order is $Li_2{}^+$ is $\frac{1}{2}$, while that in Li_2 is one. Thus, the bond in $Li_2{}^+$ should be one half as strong as the bond in Li_2: 106 kJ/mol ÷ 2 = 53 kJ/mol $Li_2{}^+$

4B The H_2 ion contains 1 electron from each H plus 1 for the charge for a total of three electrons. Its molecular orbital diagram is σ_{1s} [↑↓] σ_{1s}^* [↑] There are 2 bonding electrons and there is 1 antibonding electron. For $H_2{}^-$ the bond order = (2 bonding electrons − 1 antibonding electron) ÷ 2 = $^1/_2$. Thus, we would expect the ion $H_2{}^-$ to be stable, with a bond strength about half that of a hydrogen molecule.

5A For each case, the empty molecular-orbital diagram has the following appearance. (KK indicates that the molecular orbitals formed from $1s$ atomic orbitals are full: KK = σ_{1s} [↑↓] σ_{1s}^* [↑↓])

KK σ_{2s} [ ] σ_{2s}^* [ ] π_{2p} [ ][ ] σ_{2p} [ ] π_{2p}^* [ ][ ] σ_{2p}^* [ ]

We need to simply count up the total number of electrons in each species, and place them in the valence-shell molecular-orbital diagram.

(a) $N_2{}^+$ has $(2 \times 5) - 1 = 9$ valence electrons. Its molecular orbital diagram is

$N_2{}^+$ KK σ_{2s} [↑↓] σ_{2s}^* [↑↓] π_{2p} [↑↓][↑↓] σ_{2p} [↑] π_{2p}^* [ ][ ] σ_{2p}^* [ ]

bond order = (7 bonding electrons − 2 antibonding electrons) ÷ 2 = 2.5

(b) $Ne_2{}^+$ has $(2 \times 8) - 1 = 15$ valence electrons. Its molecular orbital diagram is

$Ne_2{}^+$ KK σ_{2s} [↑↓] σ_{2s}^* [↑↓] π_{2p} [↑↓][↑↓] σ_{2p} [↑↓] π_{2p}^* [↑↓][↑↓] σ_{2p}^* [↑]

bond order = (8 bonding electrons – 7 antibonding electrons) ÷ 2 = 0.5

(c) C_2^{2-} has (2 × 4) + 2 = 10 valence electrons. Its molecular orbital diagram is

C_2^{2-} KK σ_{2s}[↑↓] σ^*_{2s}[↑↓] π_{2p}[↑↓][↑↓] σ_{2p}[↑↓] π^*_{2p}[ ][ ] σ^*_{2p}[ ]

bond order = (8 bonding electrons – 2 antibonding electrons) ÷ 2 = 3.0

5B For each case, the empty molecular-orbital diagram has the following appearance. (KK indicates that the molecular orbitals formed from 1*s* atomic orbitals are full: KK = σ_{1s}[↑↓] σ^*_{1s}[↑↓])

KK σ_{2s}[ ] σ^*_{2s}[ ] π_{2p}[ ][ ] σ_{2p}[ ] π^*_{2p}[ ][ ] σ^*_{2p}[ ]

All of these species are based on O_2, which has 2 × 6 = 12 valence electrons. We simply put the appropriate number of valence electrons in each diagram and determine the bond order.

O_2^+ 11 v.e. KK σ_{2s}[↑↓] σ^*_{2s}[↑↓] σ_{2p}[↑↓] π_{2p}[↑↓][↑↓] π^*_{2p}[↑][ ] σ^*_{2p}[ ]
bond order = (8 bonding electrons – 3 antibonding electrons) ÷ 2 = 2.5 bond length = 112 pm

O_2 12 v.e. KK σ_{2s}[↑↓] σ^*_{2s}[↑↓] σ_{2p}[↑↓] π_{2p}[↑↓][↑↓] π^*_{2p}[↑][↑] σ^*_{2p}[ ]
bond order = (8 bonding electrons – 4 antibonding electrons) ÷ 2 = 2.0 bond length = 121 pm

O_2^- 13 v.e. KK σ_{2s}[↑↓] σ^*_{2s}[↑↓] σ_{2p}[↑↓] π_{2p}[↑↓][↑↓] π^*_{2p}[↑↓][↑] σ^*_{2p}[ ]
bond order = (8 bonding electrons – 5 antibonding electrons) ÷ 2 = 1.5 bond length = 128 pm

O_2^{2-} 14 v.e. KK σ_{2s}[↑↓] σ^*_{2s}[↑↓] σ_{2p}[↑↓] π_{2p}[↑↓][↑↓] π^*_{2p}[↑↓][↑↓] σ^*_{2p}[ ]
bond order = (8 bonding electrons – 6 antibonding electrons) ÷ 2 = 1.0 bond length = 149 pm

We see that the bond length does indeed increase as the bond order decreases. Longer bonds are weaker bonds.

SUMMARIZING EXAMPLE CALCULATIONS

1. Average bond lengths are 136 pm for N—N, 123 pm for N=N, and 110 pm for N≡N. The bond lengths in the molecule are N_a—N_b = 124 pm and N_b—N_c = 113. Thus it is likely that N_a—N_b is a double bond, while N_b—N_c is close to a triple bond.

2. The HN_3 molecule has 1 + (3 × 5) = 16 valence electrons, or 8 pairs. N_a and N_b are central atoms. There are two plausible Lewis structures: (I) $\mathrm{H{-}\overline{N}_a{=}N_b{=}\overline{\underline{N}}_c}$ (II) $\mathrm{H{-}\overline{\underline{N}}_a{-}N_b{\equiv}\overline{N}_c}$

3. In both structures (I) and (II) there are no lone pairs on N_b, the central N, and there are two atoms bonded to it. According to VSEPR theory, the molecular geometry around N_b is linear. The hybridization on N_b is *sp*. Now consider N_a. In structure (I) there are two atoms and one lone pair attached to N_a. Its electron-pair geometry is trigonal planar, its hybridization is sp^2, its molecular geometry is bent, and the H—N_a—N_b bond angle approximates 120°. In structure (II) there are two atoms and two lone pairs attached to N_a. Its electron-pair geometry is tetrahedral, its hybridization is sp^3, its molecular geometry is bent, and the H—N_a—N_b bond angle approximates 109.5°. The orbital diagram for each atom and the orbitals that overlap to form each bond follow. (In showing the orbitals that overlap, the number of electrons in each orbital is indicated by a superscript outside the parenthetical designation of each orbital's type. To save space in the table, a coordinate covalent bond is indicated by its alternate name: dative bond.)

In both structures H $_{1s}$[↑] *and* N_b [He] $_{sp}$[↑↓][↑] $_{2p}$[↑][↑]

Structure (I) $\mathrm{H{-}\overline{N}_a{=}N_b{=}\overline{\underline{N}}_c}$

N_a [He] $_{sp^2}$[↑][↑][↑↓] $_{2p}$[↑] N_c [He] $_{2s}$[↑↓] $_{2p}$[ ][↑][↑↓]

σ bond between $H(1s)^1$ and $N_a(sp^2)^1$
σ bond between $N_a(sp^2)^1$ and $N_b(sp)^1$
lone pair in $N_a(sp^2)^2$
dative σ bond between $N_b(sp)^2$ and $N_c(2p_x)^0$
π bond between $N_a(2p_z)^1$ and $N_b(2p_z)^1$
π bond between $N_b(2p_y)^1$ and $N_c(2p_y)^1$
two lone pairs in $N_c(2p_y)^2$

Structure (II) $\mathrm{H{-}\overline{\underline{N}}_a{-}N_b{\equiv}\overline{N}_c}$

N_a [He] $_{sp^3}$[↑][ ][↑↓][↑↓] N_c [He] $_{2s}$[↑↓] $_{2p}$[↑][↑][↑]

σ bond between $H(1s)^1$ and $N_a(sp^3)^1$
dative σ bond between $N_a(sp^2)^0$ and $N_b(sp)^2$
two lone pairs in $N_a(sp^3)^2$
σ bond between $N_b(sp)^1$ and $N_c(2p_x)^1$
π bond between $N_b(2p_z)^1$ and $N_c(2p_z)^1$
π bond between $N_b(2p_y)^1$ and $N_c(2p_y)^1$
lone pair in $N_c(2p_y)^2$

In both structures, the three N's lie on a line, with a non-linear H—N—N bond angle: H \ N=N=$\underline{\mathrm{N}}$

REVIEW QUESTIONS

Note: In VSEPR theory the term "bond pair" is used for a single bond, a double bond, or a triple bond, even though a single bond consists of one pair of electrons, a double bond consists of two pairs, and a triple bond consists of three. To avoid any confusion between the number of electron pairs actually involved in the bonding to a central atom, and the number of atoms bonded to that central atom, we shall occasionally use the term "ligand" to indicate an atom or a group of atoms attached to the central atom. We also shall use the term "bond group" to refer to a single, a double, or a triple bond.

1. **(a)** An sp^2 hybrid orbital is one of a set of three orbitals that are identical in shape and energy and are formed by the combination of an s orbital and two p orbitals. The three orbitals are trigonal planar in orientation.

(b) A π_{2p} molecular orbital is a lower energy molecular orbital formed by the combination of two $2p$ atomic orbitals on adjacent atoms. The original atomic orbitals are oriented sidewise to each other.

(c) The bond order is an indication of the strength of a bond. In valence bond theory, it indicates the number of electron pairs shared by two atoms. In molecular orbital theory it is one half of the difference between the number of bonding and the number of antibonding electrons.

(d) A π bond is a bond that has regions of overlap above and below the internuclear axis.

2. **(a)** Hybridization of atomic orbitals is that process in which atomic orbitals are blended together to produce new atomic orbitals that point in directions other than those of pure atomic orbitals.

(b) The σ-bond framework consists of the atoms in the molecule and the single bonds that connect them, before the second (and third) bonds of multiple bonds are in place.

(c) The Kekulé structures of benzene are the two resonance forms of benzene in which the alternating single and double carbon-to-carbon bonds exchange positions.

(d) The band theory of metallic bonding describes the bonding between metal atoms in terms of molecular orbitals that extend throughout the crystal.

3. **(a)** A σ bond is located along the internuclear axis and is the first bond formed between two atoms. A π bond is located above and below the internuclear axis and only forms after a σ bond has formed.

(b) Localized electrons are those confined either to the region around one atom or to the region between two atoms. Delocalized electrons are free to move among three or more atoms.

(c) Valence-bond method attempts to explain why two atoms are joined in terms of the atomic orbitals that interact to produce a bond. VSEPR theory attempts to predict the shapes of molecules based on their electronic (Lewis) structures.

(d) A metal is a material that readily conducts electricity (due to a small or non-existent bond gap). A semiconductor conducts electricity only once it is heated or has a moderate voltage applied to it (due to an moderate band gap).

4. The Lewis structure of H_2Se, H—$\overline{\text{Se}}$—H, indicates that there are two atoms and two lone pairs attached to the central atom. The bond angle of 91° indicates that the $1s$ orbitals on H overlap the $4p$ orbitals on Se. This would give an undistorted bond angle of 90°. But repulsions between the H's open up the bond angle a slight bit.

5. Determining hybridization is made easier if we begin with Lewis structures. Only one resonance form is drawn for CO_3^{2-}, SO_2 and NO_2^-.

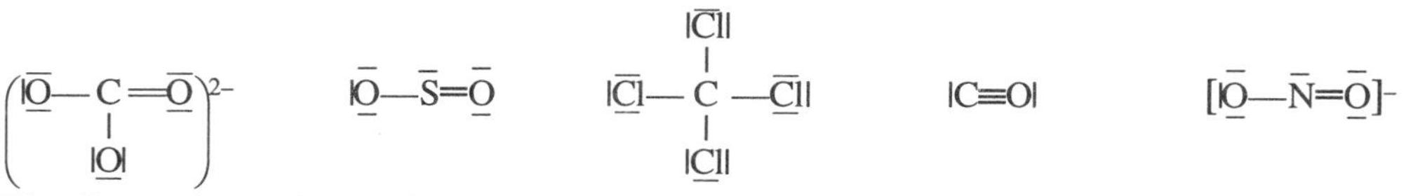

The C atom is attached to three ligands and no lone pairs and thus is sp^2 hybridized in CO_3^{2-}.
The S atom is attached to two ligands and one lone pair and thus is sp^2 hybridized in SO_2.
The C atom is attached to four ligands and no lone pairs and thus is sp^3 hybridized in CCl_4.
Neither atom needs to be hybridized in CO.
The N atom is attached to two ligands and one lone pair in NO_2^- and thus is sp^2 hybridized.

The central atom is sp^2 hybridized in SO_2, CO_3^{2-}, and NO_2^-.

6. Statement (b) is correct. The only consistently single bonds in C–H–O compounds are bonds to H atoms (a H atom can only form one bond), and single bonds must be σ bonds. Thus, both C—H and O—H bonds must be σ bonds, making (a) false and (b) true. Only C=C bonds (or C=O) bonds in these compounds, that is double bonds, consist of a σ bond and a π bond; (c) is false. A σ bond must form before a π bond forms to complete a double bond; (d) is false.

7. **(a)** HI: H $_{1s}$[↑] I [Kr]$4d^{10}$ $_{5s}$[↑↓] $_{5p}$[↑↓][↑↓][↑]
The $1s$ orbital of H overlaps the half-filled $5p$ orbital of I to produce a linear molecule.

(b) BrCl: Br [Ar]$3d^{10}$ $_{4s}$[↑↓] $_{4p}$[↑↓][↑↓][↑] Cl [Ne] $_{3s}$[↑↓] $_{3p}$[↑↓][↑↓][↑]
The half-filled $3p$ orbital of Cl overlaps the half-filled $4p$ orbital of Br to produce a linear molecule.

(c) H_2Se: H $_{1s}$[↑] Se [Ar]$3d^{10}$ $_{4s}$[↑↓] $_{4p}$[↑↓][↑][↑]
Each half-filled $4p$ orbital of Se overlaps with a half-filled $1s$ orbital of H to produce a bent molecule, with a bond angle of approximately 90°.

(d) OCl_2: O $_{1s}$[↑↓] $_{2s}$[↑↓] $_{2p}$[↑↓][↑][↑] Cl [Ne] $_{3s}$[↑↓] $_{3p}$[↑↓][↑↓][↑]
Each half-filled $2p$ orbital of O overlaps with a half-filled $3p$ orbital of Cl to produce a bent molecule, with a bond angle of approximately 90°.

8. The BF_3 molecule is planar with bond angles of 120°. The $2p$ orbitals are perpendicular to each other, with angles between them of 90°. Thus, three atoms bound to three $2p$ orbitals would not form a planar structure and would have the wrong bond angles. If two Fs bonded to two $2p$ orbitals on B and the third F bonded to the $2s$ orbital, the resulting molecule could indeed be planar, but the bond angles would not equal 120°, and two of the bonds would be different from the other one. By hybridizing the $2s$ and two $2p$ orbitals of B, we create three equivalent orbitals that produce the observed bond angles.

9. **(a)** The central C atom employs four sp^3 hybrid orbitals; these have a tetrahedral geometry. Each of the two H atoms employs a $1s$ orbital, and each of the two Cl atoms employs a half-filled $3p$ orbital.

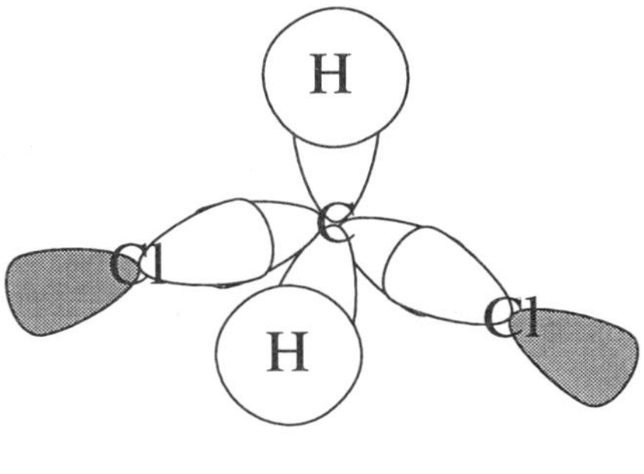

(b) The ground-state C atom has the electron configuration $_{1s}$[↑↓] $_{2s}$[↑↓] $_{2p}$[↑][↑][ ]. Based on this ground-state electron configuration, we would expect a bent molecule. To produce the two half-filled orbitals required to form two linear bonds an excited, and hybridized, electron configuration is required.

C* $_{1s}$[↑↓] $_{2s}$[↑] $_{2p}$[↑][↑][↑] ⟶ C*$_{\text{hyb}}$ $_{1s}$[↑↓] $_{sp}$[↑][↑] $_{2p}$[↑][↑]

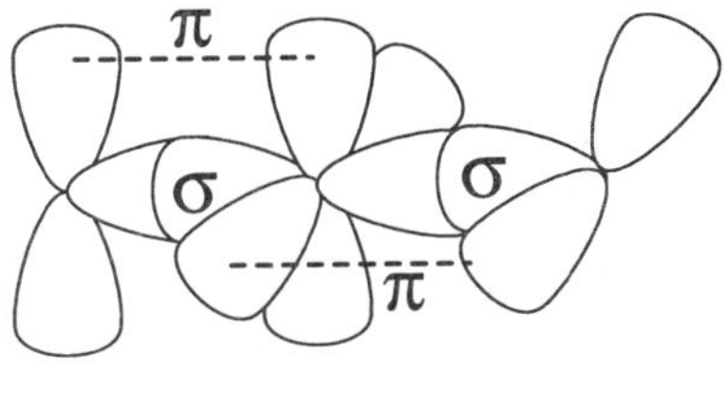

The overlap of the two half-filled sp orbitals of C, one with a half-filled $2p$ orbital on O and the other with a half-filled $2p$ orbital on N^-, produces the σ bond framework.

O $_{1s}$[↑↓] $_{2s}$[↑↓] $_{2p}$[↑↓][↑][↑] N^- $_{1s}$[↑↓] $_{2s}$[↑↓] $_{2p}$[↑↓][↑][↑]

The π bonds are produced by the sidewise overlap of the two half-filled $2p$ orbitals on C, one with a half-filled $2p$ orbital on O, the other with a half-filled $2p$ orbital on N^-.

(c) The ground-state electron configuration of B suggests an ability to form one B—F bond, rather than three.

$_{1s}$[↑↓] $_{2s}$[↑↓] $_{2p}$[↑][ ][ ]

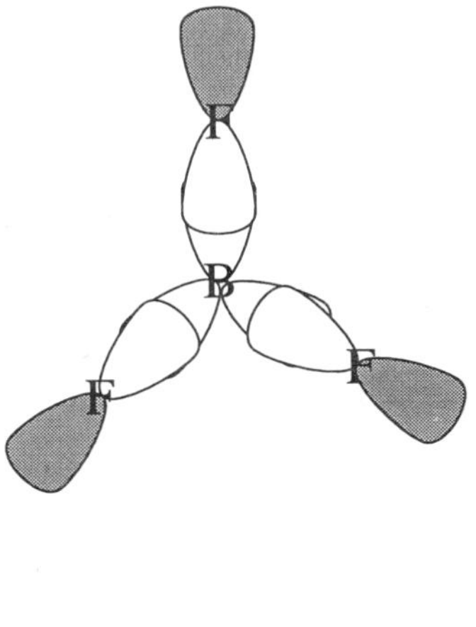

Again, excitation, or promotion of an electron to a higher energy orbital, followed by hybridization is required to form three equivalent half-filled B orbitals.

B* $_{1s}$[↑↓] $_{2s}$[↑] $_{2p}$[↑][↑][ ] ⟶ Be*$_{\text{hyb}}$ $_{1s}$[↑↓] $_{sp^2}$[↑][↑][↑] $_{2p}$[ ]

Recall that the ground-state electron configuration of F is [He] $_{2s}$[↑↓] $_{2p}$[↑↓][↑↓][↑]

The BF_3 molecule is trigonal planar, with F—B—F bond angles of 120°.

10. **(a)** In PF_6^- there are a total of 5 + (6 × 7) + 1 = 48 valence electrons, or 24 pairs. A plausible Lewis structure follows. In order to form the six P—F bonds, the hybridization on P must be sp^3d^2.

(b) In COS there are a total of 4 + 6 + 6 = 16 valence electrons, or 8 pairs. A plausible Lewis structure follows. In order to bond two atoms to the central C, the hybridization on that C atom must be *sp*.

(c) In $SiCl_4$, there are a total of 4 + (4 × 7) = 32 valence electrons, or 16 pairs. A plausible Lewis structure follows. In order to form four Si—Cl bonds, the hybridization on Si must be sp^3.

(d) In NO_3^-, there are a total of 5 + (3 × 6) + 1 = 24 valence electrons, or 12 pairs. A plausible Lewis structure follows. In order to bond three O atoms to the central N atom, with no lone pairs on that N atom, its hybridization must be sp^2.

(e) In AsF_5, there are a total of 5 + (5 × 7) = 40 valence electrons, or 20 pairs. A plausible Lewis structure follows. This is a molecule of the type AX_5E_0. To form five As—F bonds, the hybridization on As must be sp^3d.

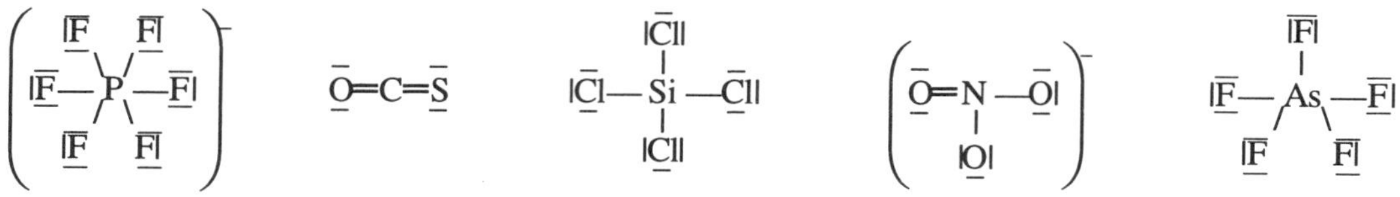

11. We base each hybridization scheme on the Lewis structure for the molecule.

(a) H—C≡N| There are two atoms bonded to C and no lone pairs in this molecule. Thus, the hybridization on C must be *sp*.

(b) 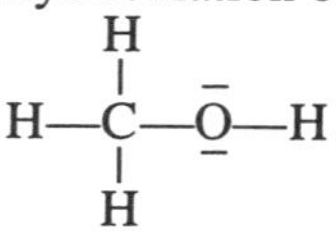 There are four atoms bonded to C and no lone pairs in this molecule. Thus, the hybridization on C must be sp^3.

(c) 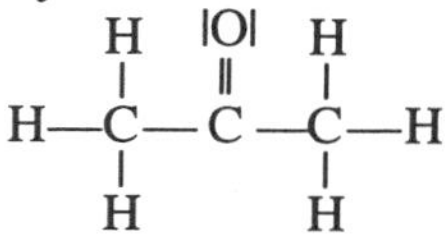 There are four atoms bonded to each terminal C and no lone pairs in this molecule. Thus, the hybridization on each terminal C must be sp^3. There are three atoms and no lone pairs bonded to the central C. Thus, the hybridization on the central C must be sp^2.

(d) 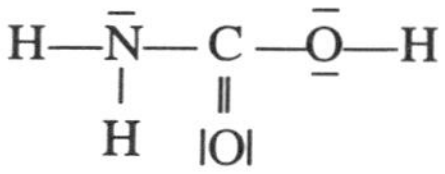 There are three atoms bonded to C and no lone pairs in this molecule. This requires sp^2 hybridization.

12. **(a)** 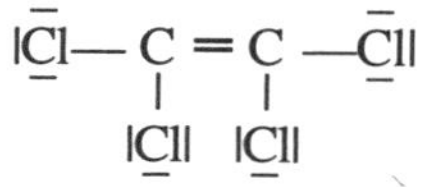 is a planar molecule. The hybridization on C is sp^2 (one bond to each of the three attached atoms).

(b) |N≡C—C≡N| is a linear molecule. The hybridization on each C is *sp* (one bond to each of the two ligands).

(c) 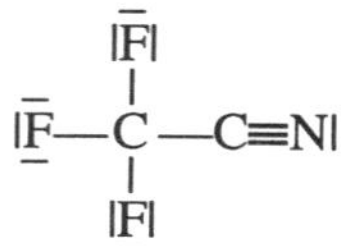 is neither linear nor planar. The shape around the left-hand C is tetrahedral and that C has sp^3 hybridization. The shape around the right-hand carbon is linear and that C has *sp* hybridization.

(d) [|S̄—C≡N|]$^-$ is a linear molecule. The hybridization on C is *sp* .

13. **(a)** In HCN, there are a total of 1 + 4 + 5 = 10 valence electrons, or 5 pairs. A plausible Lewis structure follows. H—C≡N| The H—C bond is a σ bond, and the C≡N bond is composed of 1 σ and 2 π bonds.

(b) In C_2N_2, there are a total of $(2 \times 4) + (2 \times 5) = 18$ valence electrons, or 9 pairs. A plausible Lewis structure follows. |N≡C—C≡N| The C—C bond is a σ bond, and each C≡N bond is composed of 1 σ and 2 π bonds.

(c) In $CH_3CHCHCCl_3$, there are a total of $(4 \times 4) + (5 \times 1) + (3 \times 7) = 42$ valence electrons, or 21 pairs. A plausible Lewis structure is 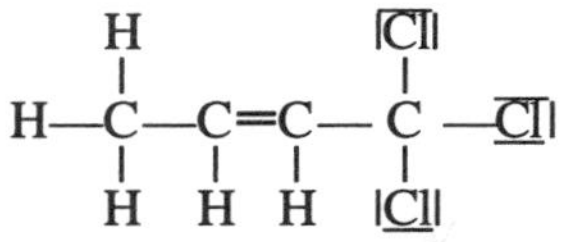 All bonds are σ bonds except one of the bonds that comprise the C═C bond. That bond is composed of one σ and one π bond.

(d) In HONO, there is a total of $1 + 5 + (2 \times 6) = 18$ valence electrons, or 9 pairs. A plausible Lewis structure is H—O—N═O All single bonds in this structure are σ bonds. The double bond is composed of one σ and one π bond.

14. **(a)** In CO_2, there are $4 + (2 \times 6) = 16$ valence electrons, or 8 pairs. A plausible Lewis structure is

O═C═O

(b) The molecule is linear and the hybridization on the C atom is *sp*. Each C═O bond is composed of a σ bond (end-to-end overlap of a half-filled *sp* orbital on C with a half-filled 2*p* orbital on O) and a π bond (overlap of a half-filled 2*p* orbital on O with a half-filled 2*p* orbital on C).

15. In CH_3NCO, there is a total of $(3 \times 1) + (2 \times 4) + 5 + 6 = 22$ valence electrons, or 11 pairs. A plausible Lewis structure is 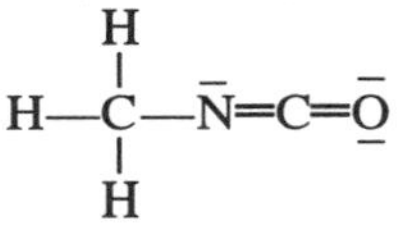

(a) Of course, all single bonds are σ bonds, and a double bond is composed of a σ bond and a π bond. Thus, each bond in the molecule consists of at least a σ bond. There are six (6) σ bonds in this molecule.

(b) Since each double bond consists of a σ bond and a π bond, there are two (2) π bonds in this molecule.

16. **(a)** A σ_{1s} orbital must be lower in energy than a σ^*_{1s} orbital. The bonding orbital always is lower than the antibonding orbital if they are derived from the same atomic orbitals.

(b) A σ_{2s} orbital should be lower in energy than a σ_{2p} orbital, since the 2*s* atomic orbital is lower in energy than the 2*p* atomic orbital, the orbitals from which the molecular orbitals, respectively, are derived.

(c) A σ^*_{1s} orbital should be lower than a σ_{2s} orbital, since the 1*s* atomic orbital is considerably lower in energy than is the 2*s* orbital.

(d) A σ_{2p} orbital should be lower in energy than a π^*_{2p} orbital. Both orbitals are from atomic orbitals in the same subshell but we expect a bonding orbital to be more stable than an antibonding orbital.

17. Based on Lewis theory, one would expect C_2 to have the greater bond energy due to the formation of quadruple bond, C≣C vs. Li—Li. The molecular orbital diagrams and bond orders are as follows.

C_2 KK σ_{2s} [⇅] σ^*_{2s} [⇅] π_{2p} [⇅][⇅] σ_{2p} [ ] bond order = (6 bonding e⁻ – 2 antibonding e⁻) ÷ 2 = 2

Li_2 KK σ_{2s} [⇅] σ^*_{2s} [ ] π_{2p} [ ][ ] σ_{2p} [ ] bond order = (2 bonding e⁻ – 0 antibonding e⁻) ÷ 2 = 1

Molecular orbital theory predicts that C_2 has a greater bond energy than does Li_2.

18. **(a)** In each case, the number of valence electrons in the species is determined first; this is followed by the molecular orbital diagram for each species.

		KK	σ_{2s}	σ^*_{2s}	π_{2p}	σ_{2p}	π^*_{2p}	σ^*_{2p}
C_2^+	no. valence elns = $(2 \times 4) - 1 = 7$	KK	[⇅]	[⇅]	[⇅][↑]	[ ]	[ ][ ]	[ ]
O_2^{2-}	no. valence elns = $(2 \times 6) + 2 = 14$	KK	[⇅]	[⇅]	[⇅]	[⇅][⇅]	[⇅][⇅]	[ ]
F_2^+	no. valence elns = $(2 \times 7) - 1 = 13$	KK	[⇅]	[⇅]	[⇅]	[⇅][⇅]	[⇅][↑]	[ ]

(b) Bond order = (no. bonding electrons – no. antibonding electrons) ÷ 2

C_2^+ bond order = (5 – 2)÷2 = $^3/_2$ This species is stable.

O_2^{2-} bond order = (8 – 6)÷2 = 1 This species is stable.

F_2^+ bond order = (8 – 5)÷2 = $^3/_2$ This species is stable.

(c) C_2^+ has an odd number of electrons and is paramagnetic, with one unpaired electron.

O_2^{2-} has an even number of electrons and is diamagnetic.

F_2^+ has an odd number of electrons and is paramagnetic, with one unpaired electron.

19. The electron sea model ascribes metallic bonding to a "glue" of electrons poured around atom kernels (nucleus plus nonvalence electrons). This electron "glue" can move in response to a potential (the metal carries a current) or a force (the metal is malleable and ductile). The electron sea theory does not explain lustre, the decrease in electrical conductivity with impurities or at high temperatures, or semiconductors.

20. The largest band gap between valence and conduction band is in an insulator. There is no energy gap between the valence and conduction bands in a metal. Metalloids are semiconductors; they have a small band gap between the valence and conduction bands.

EXERCISES

Valence-Bond Theory

21. There are several ways in which valence bond theory is superior to Lewis structures in describing covalent bonds. *First,* valence bond theory clearly distinguishes between sigma and pi bonds. In Lewis theory, a double bond appears to be just two bonds and it is not clear why a double bond is not simply twice as strong as a single bond. In valence bond theory, it is clear that a sigma bond must be stronger than a pi bond, for the orbitals overlap more efficiently in a sigma bond (end-to-end) than they do in a pi bond (side-to-side). *Second,* molecular geometries are more directly obtained in valence bond theory than in Lewis theory. Although valence bond theory requires the somewhat artificial introduction of hybridization to explain these geometries, Lewis theory does not predict geometries at all; it simply provides the basis from which VSEPR theory predicts geometries. *Third,* Lewis theory does not explain hindered rotation about double bonds. With valence bond theory, one realizes that any rotation about a double bond will produce poorer overlap of the *p* orbitals that produce the pi bond.

22. The overlap of pure atomic orbitals gives bond angles of 90°. These bond angles are suitable only for 3- and 4-atom compound in which the central atom is an atom of the third (or higher) period of the periodic table. For central atoms of the second period of the periodic table, 3- and 4-atom compounds have bond angles closer to 180°, 120°, and 109.5° than to 90°. These other bond angles can only be explained well through hybridization. *Secondly*, hybridization clearly distinguishes between the (hybrid) orbital that form σ bonds and the *p* orbitals that form π bonds. It places those *p* orbitals in their proper orientation so that they can overlap side-to-side to form π bonds. *Thirdly*, the bond angles of 90° and 120° that result when the octet of the central atom is expanded cannot be produced with pure atomic orbitals. Hybrid orbitals are necessary. *Finally*, the overlap of pure atomic orbitals does not usually result in all σ bonds being equivalent. Overlaps with hybrid orbitals produce equivalent bonds.

23. (a) Lewis theory really does not describe the shape of the water molecule. It does indicate that there is a single bond between each H atom and the O atom, and that there are two lone pairs attached to the O atom, but it says nothing about molecular shape.

(b) In valence bond theory using simple atomic orbitals, each H—O bond results from the overlap of a 1*s* orbital on H with a 2*p* orbital on O. Since the angle between 2p orbitals is 90°, this method initially predicts a 90° bond angle of 90°. The observed 104° bond angle is explained as being due to the repulsion between the two slightly positive H atoms.

(c) In VSEPR theory the H_2O molecule is categorized as being of the AX_2E_2 type, with two atoms and two lone pairs attached to the central oxygen atom. The lone pairs repel each other more than do the bond pairs, explaining the smaller than 109.5° bond angle.

(d) In valence bond theory using hybrid orbitals, each H—O bond results from the overlap of a 1*s* orbital on H with a sp^3 orbital on O. The angle between sp^3 orbitals is 109.5°. The observed bond angle of 104° is explained as the result of incomplete hybridization.

24. (a) Lewis theory really does not describe the shape of the molecule. It does indicate that there is a single bond between each Cl atom and the C atom, but it says nothing about molecular shape.

(b) In valence bond theory using simple atomic orbitals, each C—Cl bond results from the overlap of a $3p$ orbital on Cl with a $2p$ orbital on C. Since the angle between 2p orbitals is 90°, this method initially predicts a 90° bond angle of 90°. The observed 109.5° bond angle is explained as resulting from the repulsion between the two slightly negative Cl atoms. In addition, the molecule is predicted to have the formula CCl_2, since there are just two half-filled orbitals in the ground state of C.

(c) In VSEPR theory the CCl_4 molecule is categorized as being of the AX_4 type, with four atoms attached to the central carbon atom.

(d) In valence bond theory using hybrid orbitals, each C—Cl bond results from the overlap of a $3p$ orbital on Cl with a sp^3 orbital on C. The angle between sp^3 orbitals is 109.5°.

25. For each species, we first draw the Lewis structure, as an aid to explaining the bonding.

(a) In CO_2, there are a total of $4 + (2 \times 6) = 16$ valence electrons, or 8 pairs. C is the central atom. The

Lewis structure is $\bar{O}=C=\bar{O}$ The molecule is linear and C is sp hybridized.

(b) In $HONO_2$, there are a total of $1 + 5 + (3 \times 6) = 24$ valence electrons, or 12 pairs. N is the central

atom, and a plausible Lewis structure is H—$\bar{O}$—N$=\bar{O}$ The molecule is trigonal planar around

$|\bar{O}|$

N which is sp^2 hybridized. The left O is bent with sp^3 hybridization.

(c) In ClO_3^-, there are a total of $7 + (3 \times 6) + 1 = 26$ valence electrons, or 13 pairs. Cl is the central atom,

and a plausible Lewis structure is $\left(|\bar{O}—\bar{C}l—\bar{O}| \atop |\bar{O}| \right)^-$ The electron-group geometry around Cl

is tetrahedral, indicating that Cl is sp^3 hybridized.

(d) In BF_4^-, there are a total of $3 + (4 \times 7) + 1 = 32$ valence electrons, or 16 pairs. B is the central atom,

and a plausible Lewis structure is $\left({|\bar{F}| \atop \bar{F}—B—\bar{F}} \atop |\bar{F}| \right)^-$ The electron-group geometry is tetrahedral,

indicating that B is sp^3 hybridized.

26. In NSF there are $5 + 6 + 7 = 18$ valence electrons, or 9 pairs. The following Lewis structure has zero formal

charge on all atoms. $|N{\equiv}\bar{S}—\bar{F}|$ There are two atoms bonded to the central S, which has a lone pair. Thus,

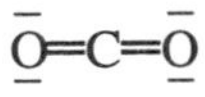

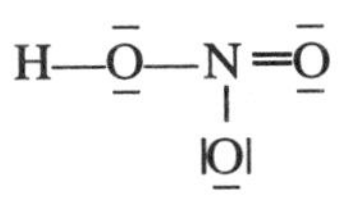

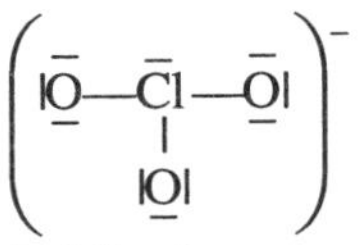

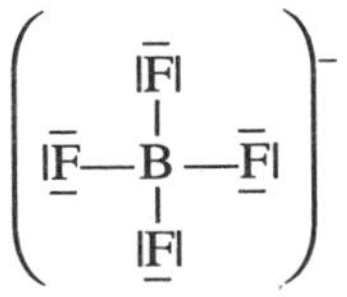

the electron group geometry around the central S is trigonal planar (with sp^2 hybridization) and the molecular shape is bent.

27. The Lewis structure of ClF_3 is $\bar{F}—\overset{\frown}{Cl}—\bar{F}$ There are three atoms and two lone pairs attached to the

$|\bar{F}|$

central atom, its hybridization is sp^3d, which is achieved as follows.

Cl_{unhyb} [Ne] $_{3s}\boxed{\uparrow\downarrow}$ $_{3p}\boxed{\uparrow\downarrow}\boxed{\uparrow\downarrow}\boxed{\uparrow}$ $_{3d}\boxed{||||}$ $\longrightarrow$ Cl_{hyb} [Ne] $_{dsp^3}\boxed{\uparrow\downarrow}\boxed{\uparrow\downarrow}\boxed{\uparrow}\boxed{\uparrow}\boxed{\uparrow}$ $_{3d}\boxed{|||}$

The orbital diagram of F is [He] $_{2s}\boxed{\uparrow\downarrow}$ $_{2p}\boxed{\uparrow\downarrow}\boxed{\uparrow\downarrow}\boxed{\uparrow}$ Each of the three sigma bonds are formed by the overlap of a $2p$ orbital on F with one of the half-filled dsp^3 orbitals on Cl. Since the dsp^3 hybridization has a trigonal bipyramidal shape, and the two lone pairs occupy the axial positions in the molecule, ClF_3 is T-shaped.

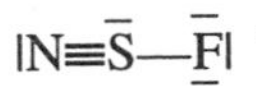

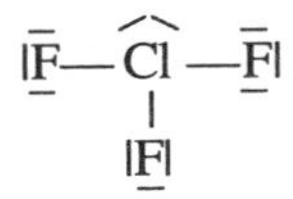

28. In SF_4 there are $6 + (4 \times 7) = 34$ valence electrons, or 17 pairs. A plausible Lewis structure has zero formal charge on each atom. $|\overline{F}— \overset{\displaystyle \overline{|F|}}{\underset{\displaystyle \overline{|F|}}{S}} —\overline{F}|$ This molecule has a see-saw shape. There are four atoms and one lone

pair attached to S, requiring sp^3d hybridization. Each sigma S—F bond results from the overlap of two half-filled orbitals: sp^3d on S and $2p$ on F. The orbital diagram of F is [He] $_{2s}\boxed{\uparrow\downarrow}$ $_{2p}\boxed{\uparrow\downarrow}\boxed{\uparrow}\boxed{\uparrow}$

S_{unhyb} [Ne] $_{3s}\boxed{\uparrow\downarrow}$ $_{3p}\boxed{\uparrow\downarrow}\boxed{\uparrow}\boxed{\uparrow}$ $_{3d}\boxed{}\boxed{}\boxed{}\boxed{}\boxed{}$ $\longrightarrow$ S_{hyb} [Ne] $_{dsp^3}\boxed{\uparrow\downarrow}\boxed{\uparrow}\boxed{\uparrow}\boxed{\uparrow}\boxed{\uparrow}$ $_{3d}\boxed{}\boxed{}\boxed{}\boxed{}$

29. **(a)** In CCl_4, there is a total of $4 + (4 \times 7) = 32$ valence electrons, or 8 pairs. C is the central atom. A

plausible Lewis structure is $|\overline{Cl}— \overset{\displaystyle |\overline{Cl}|}{\underset{\displaystyle |\overline{Cl}|}{C}} —\overline{Cl}|$ The geometry at C is tetrahedral; C is sp^3 hybridized.

Cl—C—Cl bond angles are 109.5°. Each C—Cl bond is represented by σ: $C(sp^3)^1$—$Cl(3p)^1$

(b) In ONCl, there is a total of $6 + 5 + 7 = 18$ valence electrons, or 9 pairs. N is the central atom. A

plausible Lewis structure is $\overline{O}\!=\!\overline{N}—\overline{Cl}|$ The geometry around N is triangular planar, and N is sp^2

hybridized. The O—N—Cl bond angle is about 120°. The three bonds are represented as follows.
σ: $O(2p_y)^1$—$N(sp^2)^1$ σ: $N(sp^2)^1$—$Cl(3p_z)^1$ π: $O(2p_z)^1$—$N(2p_z)^1$

(c) In HONO, there are a total of $1 + (2 \times 6) + 5 = 18$ valence electrons, or 9 pairs. A plausible Lewis

structure is $H—\overline{O}_a—\overline{N}\!=\!\overline{O}_b$ The geometry of O_a is tetrahedral, O_a is sp^3 hybridized, the H—O_a—N

bond angle is (at least close to) 109.5°. The geometry at N is C is trigonal, N is sp^2 hybridized, the
O_a—N—O_b bond angle is 120°. The four bonds are represented as follows.
σ: $H(1s)^1$—$O_a(sp^3)^1$ σ: $O_a(sp^3)^1$—$N(sp^2)^1$ σ: $N(sp^2)^1$—$O_b(2p_y)^1$ π: $N(2p_z)^1$—$O_b(2p_z)^1$

(d) In $COCl_2$, there are a total of $4 + 6 + (2 \times 7) = 24$ valence electrons, or 12 pairs. A plausible Lewis

structure is $|\overline{Cl}— \overset{\displaystyle |O|}{\overset{\displaystyle \|}{C}} —\overline{Cl}|$ The geometry around C is trigonal planar; all bond angles around C

are 120°, and the hybridization of C is sp^2. There are four bonds in the molecule:
2σ: $Cl(3p)^1$—$C(sp^2)^1$ σ: $O(2p)^1$—$C(sp^2)^1$ π: $O(2p)^1$—$C(2p)^1$

30. **(a)** In NO_2^-, there are a total of $5 + (2 \times 6) + 1 = 18$ valence electrons, or 9 pairs. A plausible Lewis

structure follows. $[\overline{O}_a\!=\!\overline{N}—\overline{O}_b|]^-$ The geometry around N is trigonal planar, all bond angles

around that atom are 120°, and the hybridization of N is sp^2. The three bonds in this molecule are:
σ: $O_a(2p)^1$—$N(sp^2)^1$ σ: $O_b(2p)^1$—$N(sp^2)^1$ π: $O_a(2p_z)^1$—$N(2p_z)^1$

(b) In I_3^-, there are a total of $(3 \times 7) + 1 = 22$ valence electrons, or 11 pairs. A plausible Lewis structure

follows. $[|\overline{I}— \overset{\frown}{\underline{I}} —\overline{I}|]^-$ Since there are three lone pairs and two ligands attached to the central I,

the electron-group geometry around that atom is trigonal bipyramidal; its hybridization is sp^3d. Since
the two ligand I atoms are located in the axial positions of the trigonal bipyramid, the I—I—I bond
angle is 180°. Each I—I sigma bond is the result of the overlap of a $5p$ orbital on a peripheral I with a
sp^3d^2 hybrid orbital on the central I.

(c) In $C_2O_4^{2-}$, there are a total of $(2 \times 4) + (4 \times 6) + 2 = 34$ valence electrons, or 17 pairs. A plausible

Lewis structure is $\left[\overline{O}— \overset{\displaystyle \|}{\underset{\displaystyle |O|}{C}} — \overset{\displaystyle \|}{\underset{\displaystyle |O_a|}{C}} —\overline{O}_b|\right]^{2-}$ As we see, the ion is symmetrical. There are three atoms

or groups of atoms attached to each C, the bond angles around each C are 120°, and the hybridization
on each C is sp^2. The bonding to one of these carbons (the right-hand one) is represented as follows.

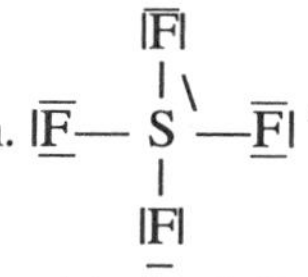

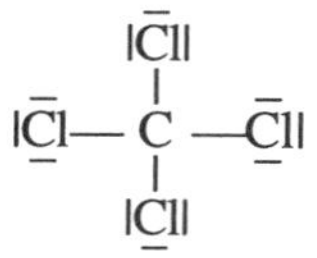

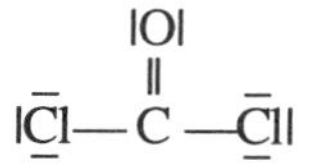

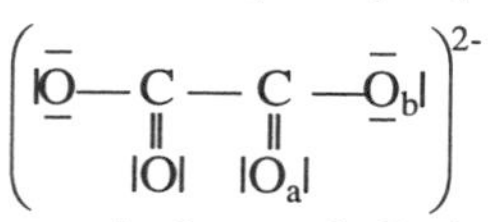

σ: $C(sp^2)^1$—$C(sp^2)^1$ σ: $C(sp^2)^1$—$O_a(2p_z)^1$ σ: $C(sp^2)^1$—$O_b(2p_z)^1$ π: $C(2p_y)^1$—$O_a(2p_y)^1$

(d) In HCO_3^-, there are $1 + 4 + (3 \times 6) + 1 = 24$ valence electrons, or 12 pairs. A plausible Lewis structure follows. 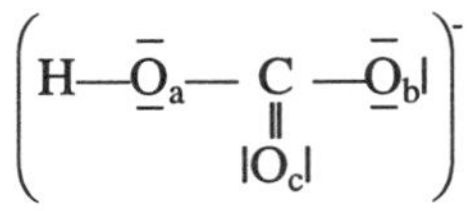 The geometry around C is trigonal planar, with a bond angle of 120°; the hybridization is sp^2. The geometry around O_a is bent with a bond angle less than 109.5°; the hybridization is sp^3. The five bonds in the molecule are represented as follows.

σ: $O_a(sp^3)^1$—$H(1s)^1$ σ: $C(sp^2)^1$—$O_a(sp^3)^1$ σ: $C(sp^2)^1$—$O_b(2p)^1$
σ: $C(sp^2)^1$—$O_c(2p_x)^1$ π: $C(2p_z)^1$—$O_c(2p_y)^1$

31. 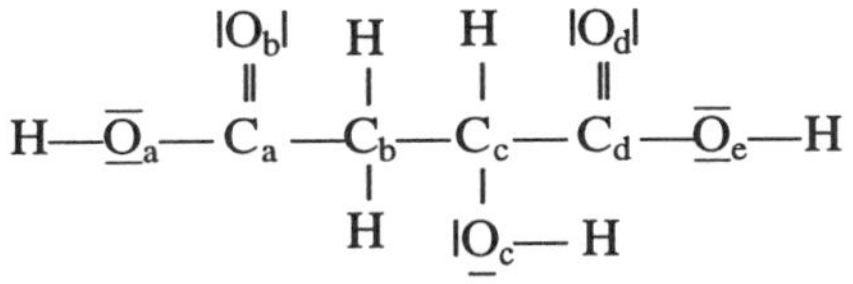 All interior atoms are sp^3 hybridized except C_a and C_d, which are sp^2 hybridized. The overlaps in the sixteen bonds of the molecule are represented as follows.

σ: $O_a(sp^3)^1$—$H(1s)^1$ σ: $C_a(sp^2)^1$—$O_a(sp^3)^1$ σ: $C_a(sp^2)^1$—$O_b(2p_x)^1$ π: $C_a(2p_z)^1$—$O_b(2p_y)^1$
σ: $O_e(sp^3)^1$—$H(1s)^1$ σ: $C_d(sp^2)^1$—$O_e(sp^3)^1$ σ: $C_d(sp^2)^1$—$O_d(2p_x)^1$ π: $C_d(2p_z)^1$—$O_d(2p_y)^1$
σ: $C_b(sp^3)^1$—$H(1s)^1$ σ: $C_b(sp^3)^1$—$H(1s)^1$ σ: $C_a(sp^2)^1$—$C_b(sp^3)^1$ σ: $C_b(sp^3)^1$—$C_c(sp^3)^1$
σ: $C_c(sp^3)^1$—$H(1s)^1$ σ: $C_c(sp^3)^1$—$O_c(sp^3)^1$ σ: $C_c(sp^3)^1$—$C_d(sp^2)^1$ σ: $O_c(sp^3)^1$—$H(1s)^1$

32. In glycine, there is a total of $(5 \times 1) + 5 + (2 \times 4) + (2 \times 6) = 30$ valence electrons, or 15 pairs. A plausible Lewis structure is 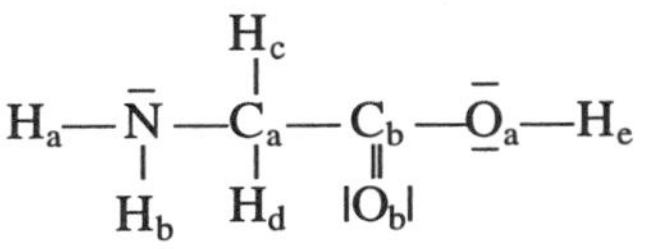 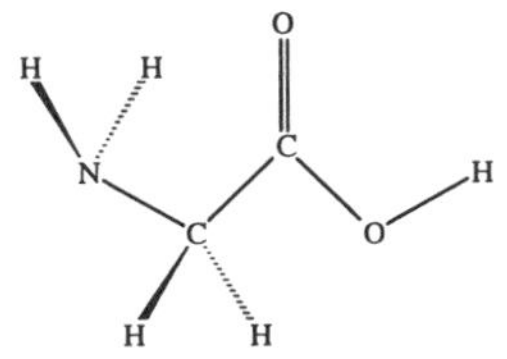

is a reasonable three-dimensional structure, drawn with wedges representing bonds projecting toward the viewer from the plane of the paper and dashes representing bonds projecting away from the viewer. C_b is trigonal planar, N is trigonal pyramidal, C_a is tetrahedral, and O_a is bent. C_a, O_a, and N each have sp^3 hybridization, while C_b has sp^2 hybridization. The overlap that produces each bond is as follows.

σ: $N(sp^3)^1$—$H_a(1s)^1$ σ: $N(sp^3)^1$—$H_b(1s)^1$ σ: $N(sp^3)^1$—$C_a(sp^3)^1$ σ: $C_a(sp^3)^1$—$H_c(1s)^1$
σ: $C_a(sp^3)^1$—$H_d(1s)^1$ σ: $C_a(sp^3)^1$—$C_b(sp^2)^1$ σ: $C_b(sp^2)^1$—$O_a(sp^3)^1$ σ: $O_a(sp^3)^1$—$H_e(1s)^1$
σ: $C_b(sp^2)^1$—$O_b(2p)^1$ π: $C_b(2p)^1$—$O_b(2p)^1$

33. The bond lengths are consistent with the left-hand C—C bond being a triple bond (120 pm), the remaining C—C bond being a single bond (154 pm) rather than a double bond (134 pm), and the C—O bond being a double bond (123 pm). Of course, the two C—H bonds are single bonds (110 pm). All of this is depicted in the following Lewis structure. 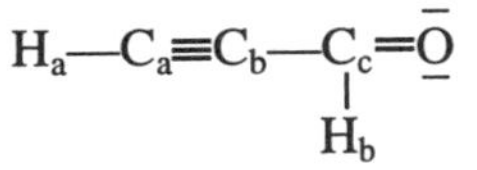 The overlap that produces each bond follows.

σ: $C_a(sp)^1$—$H_a(1s)^1$ σ: $C_a(sp)^1$—$C_b(sp)^1$ σ: $C_b(sp)^1$—$C_c(sp^2)^1$ σ: $C_c(sp^2)^1$—$H_b(1s)^1$
σ: $C_c(sp^2)^1$—$O(2p_y)^1$ π: $C_a(2p_y)^1$—$C_b(2p_y)^1$ π: $C_a(2p_z)^1$—$C_b(2p_z)^1$ π: $C_c(2p_z)^1$—$O(2p_z)^1$

34. All of the valence electron pairs in the molecule are indicated in the sketch in the problem statement; there are no lone pairs. The two end carbon atoms have three atoms attached to them; they are sp^2 hybridized. This means there is one half-filled $2p$ orbital available on each of these end carbon atoms for the formation of a π bond. The central carbon atom has two atoms attached to it; it is sp hybridized. Thus there are two (perpendicular) half-filled $2p$ orbitals (such as $2p_y$ and $2p_z$) on this central carbon atom available for the formation of π bonds. One π bond is formed from the central carbon atom to each of the terminal carbon atoms. Because the two $2p$ orbitals on the central carbon atom are perpendicular to each other, these two π bonds are also. Thus, the two H—C—H planes are at 90° from each other, because of the π bonding in the molecule.

Molecular-Orbital Theory

35. The valence-bond method describes a covalent bond as the result of the overlap of atomic orbitals. The more complete the overlap, the stronger the bond. Molecular orbital theory describes a bond as a region of space in a molecule where there is a good chance of finding the bonding electrons. The molecular orbital bond does not have to be created from atomic orbitals (although it often is) and the orientations of atomic orbitals do not have to be manipulated to obtain the correct geometric shape. There is little concept of the relative energies of bond in valence bond theory. In molecular orbital theory, bonds are ordered energetically. These energy orderings, in fact, provide a means of checking the predictions of the theory through spectroscopy of molecules.

36. C_2 has $(2 \times 4) = 8$ valence electrons. A plausible Lewis structure, that obeys the octet rule and has zero formal charge for each C atom, incorporates a quadruple bond C::::C *or* C≣C Molecular orbital theory, on the other hand, distributes those same 8 electrons as shown below, resulting in a bond order of $(6 - 2) \div 2 = 2$, that is, a double bond.
(ᵇ indicates a bonding orbital)

	σ_{2s}^{b}	σ_{2s}^{*}	π_{2p}^{b}	σ_{2p}^{b}	π_{2p}^{*}	σ_{2p}^{*}
C_2	[↑↓]	[↑↓]	[↑↓][↑↓]	[ ]	[ ][ ]	[ ]

37. The two molecular orbital diagrams follow.

		σ_{2s}	σ_{2s}^{*}	π_{2p}	σ_{2p}	π_{2p}^{*}	σ_{2p}^{*}
N_2^- no. valence elns = $(2 \times 5) + 1 = 11$	KK	[↑↓]	[↑↓]	[↑↓][↑↓]	[↑↓]	[↑][ ]	[ ]
N_2^{2-} no. valence elns = $(2 \times 5) + 2 = 12$	KK	[↑↓]	[↑↓]	[↑↓][↑↓]	[↑↓]	[↑][↑]	[ ]

N_2^- bond order $= (8 - 3) \div 2 = 2.5$ (stable) N_2^{2-} bond order $= (8 - 4) \div 2 = 2$ (stable)

38. B_2 has $(2 \times 5) = 10$ electrons. We compare the number of unpaired electrons predicted if the order is π_{2p}^{b} before σ_{2p}^{b} with the number predicted if the σ_{2p}^{b} orbital is before the π_{2p}^{b} orbital. (ᵇ indicates a bonding orbital)

		σ_{1s}^{b}	σ_{1s}^{*}	σ_{2s}^{b}	σ_{2s}^{*}	π_{2p}^{b}	σ_{2p}^{b}	π_{2p}^{*}	σ_{2p}^{*}
B_2	π_{2p}^{b} before σ_{2p}^{b}	[↑↓]	[↑↓]	[↑↓]	[↑↓]	[↑][↑]	[ ]	[ ][ ]	[ ]

		σ_{1s}^{b}	σ_{1s}^{*}	σ_{2s}^{b}	σ_{2s}^{*}	σ_{2p}^{b}	π_{2p}^{b}	π_{2p}^{*}	σ_{2p}^{*}
B_2	σ_{2p}^{b} before π_{2p}^{b}	[↑↓]	[↑↓]	[↑↓]	[↑↓]	[↑↓]	[ ][ ]	[ ][ ]	[ ]

The order of orbitals with π_{2p}^{b} below σ_{2p}^{b} in energy results in B_2 being paramagnetic, with two unpaired electrons, as observed experimentally. However, if σ_{2p}^{b} comes before the π_{2p}^{b} orbital in energy, a diamagnetic B_2 molecule is predicted, in contradiction to experimental evidence.

39. In order to have a bond order higher than triple, there would have to be a region in a molecular orbital diagram where four bonding orbitals occurred together in order of increasing energy, with no intervening antibonding orbitals. No such region exists in any of the molecular orbital diagrams in Figure 12-20. Alternatively, three bonding orbitals would have to occur together energetically, following an electron configuration which, when full, results in a bond order greater than zero. This arrangement does not occur either.

40. The statement is not true, for there are many instances when the bond is strengthened when a diatomic molecule loses an electron. Specifically, whenever the electron being lost is an antibonding electron (as is the case for $F_2 \longrightarrow F_2^+$ and $O_2 \longrightarrow O_2^+$), the bond will actually be stronger in the resulting cation than it was in the starting molecule.

41. (ᵇ indicates a bonding orbital)

				σ_{2s}^{b}	σ_{2s}^{*}	π_{2p}^{b}	σ_{2p}^{b}	π_{2p}^{*}	σ_{2p}^{*}
(a)	NO	$5 + 6 = 11$ valence electrons	KK	[↑↓]	[↑↓]	[↑↓][↑↓]	[↑↓]	[↑][ ]	[ ]
(b)	NO^+	$5 + 6 - 1 = 10$ valence electrons	KK	[↑↓]	[↑↓]	[↑↓][↑↓]	[↑↓]	[ ][ ]	[ ]
(c)	CO	$4 + 6 = 10$ valence electrons	KK	[↑↓]	[↑↓]	[↑↓][↑↓]	[↑↓]	[ ][ ]	[ ]
(d)	CN	$4 + 5 = 9$ valence electrons	KK	[↑↓]	[↑↓]	[↑↓][↑↓]	[↑]	[ ][ ]	[ ]
(e)	CN^-	$4 + 5 + 1 = 10$ valence electrons	KK	[↑↓]	[↑↓]	[↑↓][↑↓]	[↑↓]	[ ][ ]	[ ]
(f)	CN^+	$4 + 5 - 1 = 8$ valence electrons	KK	[↑↓]	[↑↓]	[↑↓][↑↓]	[ ]	[ ][ ]	[ ]
(g)	BN	$3 + 5 = 8$ valence electrons	KK	[↑↓]	[↑↓]	[↑↓][↑↓]	[ ]	[ ][ ]	[ ]

42. In Exercise 41, the species that are isoelectronic are:
with 8 electrons: CN^+ BN with 10 electrons: NO^+ CO CN^-

43. We first produce the molecular orbital diagram for each species.

(a) ([b] indicates a bonding orbital)

			σ_{2s}^{b}	σ_{2s}^{*}	π_{2p}^{b}	σ_{2p}^{b}	π_{2p}^{*}	σ_{2p}^{*}
NO^+	$5 + 6 - 1 = 10$ valence electrons	KK	↑↓	↑↓	↑↓ ↑↓	↑↓	☐ ☐	☐
N_2^+	$5 + 5 - 1 = 9$ valence electrons	KK	↑↓	↑↓	↑↓ ↑↓	↑	☐ ☐	☐

Since bond order = (no. bonding electrons – no. antibonding electrons)÷2

For NO^+ bond order $= (8 - 2)\div 2 = 3$ For N_2^+ bond order $= (7 - 2)\div 2 = 2.5$

(b) NO^+ is diamagnetic; it has no unpaired electrons. N_2^+ is paramagnetic; it has one unpaired electron.

(c) The molecule with the lower bond order will also have the greater bond length. Thus, N_2^+ should have the greater bond length.

44. We first produce the molecular orbital diagram for each species.

(a) ([b] indicates a bonding orbital)

			σ_{2s}^{b}	σ_{2s}^{*}	π_{2p}^{b}	σ_{2p}^{b}	π_{2p}^{*}	σ_{2p}^{*}
CO^+	$4 + 6 - 1 = 9$ valence electrons	KK	↑↓	↑↓	↑↓ ↑↓	↑	☐ ☐	☐
CN^-	$4 + 5 + 1 = 10$ valence electrons	KK	↑↓	↑↓	↑↓ ↑↓	↑↓	☐ ☐	☐

Since bond order = (no. bonding electrons – no. antibonding electrons)÷2

For CO^+ bond order $= (7 - 2)\div 2 = 2.5$ For CN^- bond order $= (8 - 2)\div 2 = 3.0$

(b) CN^- is diamagnetic; it has no unpaired electrons. CO^+ is paramagnetic; it has one unpaired electron.

(c) The molecule with the lower bond order will also have the greater bond length. Thus, CO^+ should have the greater bond length.

Delocalized Molecular Orbitals

45. With either Lewis structures or the valence bond method, two structures must be drawn (and "averaged") to explain the π bonding in C_6H_6. The σ bonding is well explained by assuming sp^2 hybridization on each C atom. But the π bonding requires that all six C—C π bonds must be equivalent. This can be achieved by creating six π molecular orbitals—three bonding and three antibonding—into which the 6 π electrons are placed. This creates one structure for the C_6H_6 molecule.

46. Resonance is replaced in molecular orbital theory with delocalized molecular orbitals. These orbitals extend over the entire molecule, rather than being localized between two atoms. Because Lewis theory represents all bonds as being localized, the only way to depict a situation in which a pair of electrons spreads over a larger region of the molecule is to draw several Lewis structures and envision the molecule as a combination, or average of these (localized) Lewis structures.

47. We expect to find delocalized orbitals in those species for which bonding cannot be represented thoroughly by one Lewis structure, that is, for compounds that require several resonance forms.

(a) In C_2H_4, there is a total of $(2 \times 4) + (4 \times 1) = 12$ valence electrons, or 6 pairs. C atoms are the central atoms.

```
H   H
|   |
C = C
|   |
H   H
```

The bonding description of C_2H_4 does not require the use of delocalized orbitals.

(b) In SO_2, there is a total of $6 + (2 \times 6) = 18$ valence electrons, or 9 pairs. N is the central atom. A plausible Lewis structure has two resonance forms. O=S—O| ↔ |O—S=O The bonding description of SO_2 will require the use of delocalized molecular orbitals.

(c) In H_2CO, there is a total of $(2 \times 1) + 4 + 6 = 12$ valence electrons, or 6 pairs. S is the central atom. A plausible Lewis structure is

```
H—C=O
  |
  H
```

Because one Lewis structure adequately represents the bonding in the molecule, the bonding description of H_2CO does not require the use of delocalized molecular orbitals.

48. We expect to find delocalized orbitals in those species for which bonding cannot be represented thoroughly by one Lewis structure, that is, for species that require several resonance forms.

(a) In HCO_2^-, there are $1 + 4 + (2 \times 6) + 1 = 18$ valence electrons, or 9 pairs. A plausible Lewis structure has two resonance forms.

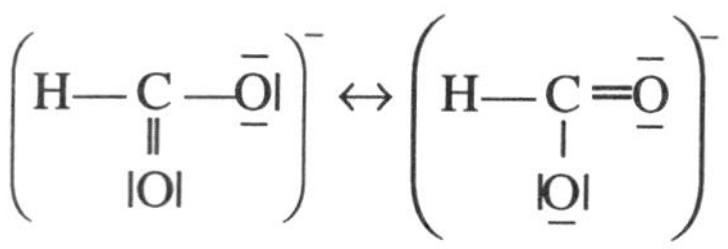

$$\left(\begin{matrix} H—C—\bar{\underline{O}}| \\ \| \\ |\underline{O}| \end{matrix}\right)^- \leftrightarrow \left(\begin{matrix} H—C=\bar{O} \\ | \\ |\underline{O}| \end{matrix}\right)^-$$

The bonding description of HCO_2^- requires the use of delocalized molecular orbitals.

(b) In CO_3^{2-}, there are $4 + (3 \times 6) + 2 = 24$ valence electrons, or 12 pairs. A plausible Lewis structure has three resonance forms.

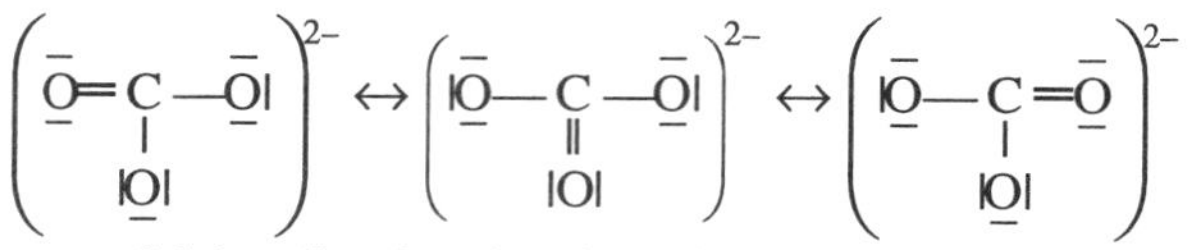

$$\left(\begin{matrix} \bar{O}=C—\bar{\underline{O}}| \\ | \\ |\underline{O}| \end{matrix}\right)^{2-} \leftrightarrow \left(\begin{matrix} |\bar{O}—C—\bar{\underline{O}}| \\ \| \\ |\underline{O}| \end{matrix}\right)^{2-} \leftrightarrow \left(\begin{matrix} |\bar{O}—C=\bar{O} \\ | \\ |\underline{O}| \end{matrix}\right)^{2-}$$

The bonding description of CO_3^{2-} requires the use of delocalized molecular orbitals.

(c) In CH_3^+, there are $4 + (3 \times 1) - 1 = 6$ valence electrons, or 3 pairs. The ion is adequately represented by one Lewis structure; no resonance forms are needed.

$$\left(\begin{matrix} H—C—H \\ | \\ H \end{matrix}\right)^+$$

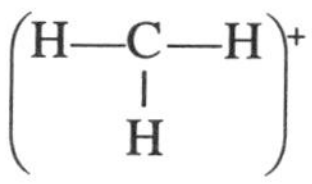

Metallic Bonding

49. **(a)** Atomic number, by itself, is not particularly important in determining whether a substance has metallic properties. However, atomic number determines where an element appears in the periodic table, and the location to the left and toward the bottom of the periodic table is a region where one finds atoms of high metallic character. Therefore atomic number—by locating an element in the periodic table—has some minimal predictive value in determining metallic character.

(b) The answer for this part is much the same as the answer to part (a), since atomic mass generally parallels atomic number for the elements.

(c) The number of valence electrons (electrons in the shell of highest principal quantum number) is important in predicting metallic character. Elements with few valence electrons—generally four or less—lose those electrons fairly readily to form cations, chemical behavior characteristic of metals. These elements also have a larger number of valence orbitals, which aids in metallic bond formation.

(d) The more vacant atomic orbitals there are in the valence shell, the more readily metallic bonding occurs.

(e) Because metals occur in every period of the periodic table—except the first—there is no particular relationship between number of electron shells and the metallic behavior of an element. (Remember that one shell is opened at the start of each period.)

50. We first determine the ground state electron configuration of Na, Fe, and Zn

$[Na] = [Ne]\ 3s^1$ $[Fe] = [Ar]\ 3d^6\ 4s^2$ $[Zn] = [Ar]\ 3d^{10}\ 4s^2$

Based on the number of electrons in the valence shell, we would expect Na to be the softest of these three metals. The one s electron per atom simply will not contribute as strongly to bonding as will two s electrons. Based on valence electron configurations, we expect Fe and Zn to be equally hard, equally strongly bonded together. However, the incomplete $3d$ shell in Fe contributes to bonding between Fe atoms, while the full $3d$ shell in Zn does not so contribute. Thus, Fe should be harder than Zn. Melting point should also reflect the strengths of bond between atoms, with higher melting points occurring in metals that are more strongly bonded. Thus in order, melting point: Fe (1530°C) > Zn (420°C) > Na (98°C)

and hardness: Fe (4.5) > Zn (2.5) > Na (0.4).

Parenthesized numbers are actual values, with hardness on a scale of 0 (talc) to 10 (diamond).

51. We first determine the number of Na atoms in the sample.

$$\text{no. Na atoms} = 26.8\ \text{mg Na} \times \frac{1\ \text{g Na}}{1000\ \text{mg Na}} \times \frac{1\ \text{mol Na}}{22.99\ \text{g Na}} \times \frac{6.022 \times 10^{23}\ \text{Na atoms}}{1\ \text{mol Na}} = 7.02 \times 10^{20}\ \text{Na atoms}$$

Because there is one $3s$ orbital per Na atom, and as many energy levels (molecular orbitals) are created as there are atomic orbitals initially present, there are 7.02×10^{20} energy levels present in the conduction band of this sample. Also, there is one $3s$ electron contributed by each Na atom, for a total of 7.02×10^{20} electrons. Because each energy level can hold two electrons, the conduction band is half full.

52. Although the band that results from the combination of $3s$ orbitals in magnesium metal is full, there is another empty band that overlaps this $3s$ band. This empty band results from the combination of $3p$ orbitals. Although it does not overlap the $3s$ band in the Mg_2 molecule or in atom clusters, once the metal crystal has grown to visible size, the overlap is complete.

13 Liquids, Solids, and Intermolecular Forces

Practice Examples

1A Values of ΔH_{vap} are in kJ/mol, so we first determine the amount in moles of diethyl ether.

$$\text{Heat} = 2.35\text{ g} \times \frac{1\text{ mol }(C_2H_5)_2O}{74.12\text{ g }(C_2H_5)_2O} \times \frac{29.1\text{ kJ}}{1\text{ mol }(C_2H_5)_2O} = 0.923\text{ kJ}$$

1B The molar mass of CH_3OH is 32.04 g/mol. The mass of CH_3OH needs to be converted to moles since ΔH_{vap} is given in kJ/mol.

total heat = heat needed to warm liquid + heat needed to vaporize liquid

$$= \left(215\text{ g} \times \frac{2.53\text{ J}}{\text{g °C}} \times (30.0 - 20.0)\text{ °C}\right) + \left(215\text{ g} \times \frac{1\text{ mol }CH_3OH}{32.04\text{ g}} \times \frac{38.0 \times 10^3\text{ J}}{1\text{ mol}}\right)$$

$$= 5.44 \times 10^3\text{ J} + 2.55 \times 10^5\text{ J} = 2.60 \times 10^5\text{ J} = 260\text{ kJ}$$

2A We first calculate pressure created by the water at 80.0 °C, assuming all 0.132 g H_2O vaporizes.

$$P_2 = \frac{nRT}{V} = \frac{\left(0.132\text{ g }H_2O \times \frac{1\text{ mol }H_2O}{18.02\text{ g }H_2}\right) \times 0.08206\frac{\text{L atm}}{\text{mol K}} \times 353.2\text{ K}}{0.525\text{ L}} \times \frac{760\text{ mmHg}}{1\text{ atm}}$$

$$= 307\text{ mmHg}$$

At 80.0 °C, the vapor pressure of water is 355.1 mmHg. Thus, all the water can exist as vapor at 80.0 °C.

2B The result of Example 13-2 is that 0.132 g H_2O would exert a pressure of 281 mmHg if it existed all as a vapor. Since that 281 mmHg is greater than the vapor pressure of water at this temperature, some of the water must exist as liquid. The calculation of the example is based on the equation $P = nRT/V$, which means that the pressure of water is proportional to its mass. Thus, the mass of water needed to produce a pressure of 92.5 mmHg under this situation is

$$\text{mass of water vapor} = 92.5\text{ mmHg} \times \frac{0.132\text{ g }H_2O}{281\text{ mmHg}} = 0.0435\text{ g }H_2O$$

mass of liquid water = 0.132 g H_2O total – 0.0435 g H_2O vapor = 0.089 g liquid water

3A From Table 13-1 we know that $\Delta H_{vap} = 38.0$ kJ/mol for methyl alcohol. We now can use the Clausius-Clapeyron equation to determine the vapor pressure at 25 °C = 298 K.

$$\ln\frac{P}{100\text{ mmHg}} = -\frac{38.0 \times 10^3\text{ J mol}^{-1}}{8.3145\text{ J mol}^{-1}\text{ K}^{-1}}\left(\frac{1}{298\text{ K}} - \frac{1}{(273.2 + 21.2)\text{ K}}\right) = +0.188$$

$$\frac{P}{100\text{ mmHg}} = e^{+0.188} = 1.21 \qquad P = 1.21 \times 100\text{ mmHg} = 121\text{ mmHg}$$

3B The vapor pressure at the normal boiling point (99.2 °C = 372.4 K) is 760 mmHg precisely. We can use the Clausius-Clapeyron equation to determine the vapor pressure at 25 °C = 298 K.

$$\ln\frac{P}{760\text{ mmHg}} = -\frac{35.76 \times 10^3\text{ J mol}^{-1}}{8.3145\text{ J mol}^{-1}\text{ K}^{-1}}\left(\frac{1}{298.2\text{ K}} - \frac{1}{372.4\text{ K}}\right) = -2.874$$

$$\frac{P}{760\text{ mmHg}} = e^{-2.874} = 0.0565 \qquad P = 0.0565 \times 760\text{ mmHg} = 42.9\text{ mmHg}$$

4A **(a)** At 1000 °C and 1 atm initially, the sample is graphite. Increasing the pressure at 1000 °C will yield diamond eventially, at about 2.5×10^4 atm.

(b) At 3800 K = 3500 °C and 1 atm initially the system is carbon vapor. As pressure is increased, at about 10 atm, solid graphite forms. At about 1.5×10^5 atm, diamond forms.

4B A heating curve is a plot of temperature versus time when heat is added to a sample at a constant rate. The specific heat of ice or of steam is about half the specific heat of liquid water, meaning that the temperature of ice or steam will rise about twice as rapidly as will the temperature of the liquid. The molar heat of vaporization of water is 44.0 kJ/mol, while its molar heat of fusion is 6.01 kJ/mol, meaning that the vaporization temperature will be maintained about seven times longer than the melting temperature. The heating curve below is not to scale in that the halts should take more time than the temperature rises.

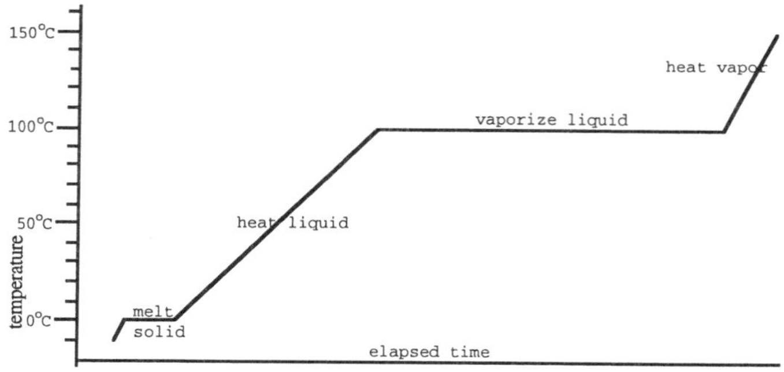

Heating curve for –10 °C ice to +150 °C water vapor at 1 atm

5A The substance with the highest boiling point will have the strongest intermolecular forces. The most important of van der Waals forces are London forces, which depend on molar mass: C_3H_8 is 44 g/mol, CO_2 is 44 g/mol, and CH_3CN is 41 g/mol. Thus the London forces are approximately equal for these three compounds. Next to consider are dipole-dipole forces. C_3H_8 is nonpolar; its bonds are not polar bonds. CO_2 is nonpolar; its two bond moments cancel each other. CH_3CN is polar and thus has the strongest intermolecular forces and should have the highest boiling point.

Boiling points are –78.44 °C for CO_2, –42.1 °C for C_3H_8, and 81.6 °C for CH_3CN.

5B Dispersion forces, which depend on molar mass, are the main determinants of boiling point. The molar masses are C_8H_{18} (114.2 g/mol), $CH_3CH_2CH_2CH_3$ (58.1 g/mol), $(CH_3)_3CH$ (58.1 g/mol), C_6H_5CHO (106.1 g/mol), and SO_3 (80.1 g/mol). We would expect $CH_3CH_2CH_2CH_3$ to have the lowest boiling point because it is not massive and it also is longer than $(CH_3)_3CH$, which should have the second highest boiling point. Then should follow SO_3. C_6H_5CHO should have a boiling point higher than the mre massive C_8H_{18} because benzaldehyde is polar while octane is not. Actual boiling points are given in parentheses in the following ranking. $(CH_3)_3CH$ (–11.6 °C) < $CH_3CH_2CH_2CH_3$ (–0.5 °C) < SO_3 (44.8 °C) < C_8H_{18} (125.7 °C) < C_6H_5CHO (178 °C)

6A Again we first look to molar masses: Ne (20.2 g/mol), He (4.0 g/mol), Cl_2 (70.9 g/mol), $(CH_3)_2CO$ (58.1 g/mol), O_2 (32.0 g/mol), and O_3 (48.0 g/mol). Both $(CH_3)_2CO$ and O_3 are polar, O_3 weakly so (because of its uneven distribution of electrons). We expect $(CH_3)_2CO$ to be the highest boiling, followed by Cl_2, O_3, O_2, Ne, and He. In the following ranking, actual boiling points are given in parentheses. $(CH_3)_2CO$ (56.2 °C), Cl_2 (–34.6 °C), O_3 (–111.9 °C), O_2 (–183.0 °C), Ne (–245.9 °C), and He (–268.9 °C).

6B The magnitude of the enthalpy of vaporization is strongly related to the strength of intermolecular forces: the stronger those forces are, the more endothermic will be the vaporization process. The first three substances all are nonpolar and therefore their only intermolecular forces are London forces, whose strength primarily depends on molar mass. The substances are arranged in order of increasing molar mass—H_2 = 2.0 g/mol, CH_4 = 16.0 g/mol, C_6H_6 = 78.1 g/mol—and also in order of increasing heat of vaporization. The last substance has a molar mass of 61.0 g/mol, which would produce intermolecular forces smaller than those of C_6H_6 if CH_3NO_2 were nonpolar. But the molecule is definitely polar. Dipole-dipole forces are strong enough that the intermolecular forces are stronger and the enthalpy of vaporization is larger for CH_3NO_2 than for C_6H_6.

7A Strong interionic forces correspond to high melting points. Strong interionic forces are created by ions with high charge and of small size. Thus, a compound with a lower melting point with KI would have ions of larger size, such as RbI or CsI. A compound with a melting point higher than CaO would have either smaller ions, such as MgO, or more highly charged ions, such as Ga_2O_3 or Ca_3N_2, or both, such as AlN or Mg_3N_2.

7B Mg^{2+} has a higher charge and a smaller size than does Na^+. In addition, Cl^- has a smaller size than I^-. Thus, interionic forces should be stronger in $MgCl_2$ than in NaI. We expect $MgCl_2$ to have lower solubility. In fact, 5.7 mol (543 g) $MgCl_2$ dissolves in a liter of water, while 12.3 (1840 g) mol NaI does, confirming our prediction.

8A The length (l) of a bcc unit cell and the radius (r) of the atom involved are related as $4r = l\sqrt{3}$. For potassium, $r = 227$ pm. Then $l = 4 \times 227 \text{ pm} / \sqrt{3} = 524$ pm

8B Consider just the face of Figure 13-34. Note that it is composed of one atom at each of four corners and one in the center. The four corner atoms touch the atom in the center, but not each other. Thus, the atoms are in contact across the diagonal of the face. If each atomic radius is designated r, then the length of the diagonal is $4r$ (= r for one corner atom + $2r$ for the center atom + r for the other corner atom). The diagonal also is related to the length of a side, l, by the Pythagorean theorem: $d^2 = l^2 + l^2 = 2l^2$ or $d = \sqrt{2}\, l$. We have two quantities equal to the diagonal, and thus to each other.

$$\sqrt{2}\, l = \text{diagonal} = 4r = 4 \times 143.1 \text{ pm} = 572.4 \text{ pm} \qquad l = \frac{572.4}{\sqrt{2}} = 404.7 \text{ pm}$$

The cubic unit cell volume, V, is equal to the cube of one side. $V = l^3 = (404.7 \text{ pm})^3 = 6.628 \times 10^7 \text{ pm}^3$

9A In a bcc unit cell, there are eight corner atoms, of which $\frac{1}{8}$ of each is apportioned to the unit cell. There also is one atom in the center. The total number of atoms per unit cell is 1 center + 8 corners $\times \frac{1}{8}$ = 2 atoms. Now we can compute the density, in g/cm³, of this cubic cell.

$$\text{density} = \frac{2 \text{ atoms}}{(524 \text{ pm})^3} \times \left(\frac{10^{12} \text{ pm}}{10^2 \text{ cm}}\right)^3 \times \frac{1 \text{ mol}}{6.022 \times 10^{23} \text{ atoms}} \times \frac{39.10 \text{ g K}}{1 \text{ mol K}} = 0.903 \text{ g/cm}^3$$

The tabulated density of potassium at 20 °C is 0.86 g/cm³.

9B In a fcc unit cell the number of atoms is computed as $^1/_8$ atom for each of the eight corner atoms (since each is shared among eight unit cells) plus $^1/_2$ atom for each of the six face atoms (since each is shared between two unit cells). This gives the total number of atoms per unit cell as:

atoms/unit cell = ($^1/_8$ corner atom × 8 corner atoms/unit cell) + ($^1/_2$ face atom × 6 face atoms/ unit cell)
= 4 atoms/unit cell

Now we can determine the mass per Al atom, and then the value the Avogadro constant.

$$\frac{\text{mass}}{\text{Al atom}} = \frac{2.6984 \text{ g Al}}{1 \text{ cm}^3} \times \left(\frac{100 \text{ cm}}{1 \text{ m}} \times \frac{1 \text{ m}}{10^{12} \text{ pm}}\right)^3 \times \frac{6.628 \times 10^7 \text{ pm}^3}{1 \text{ unit cell}} \times \frac{1 \text{ unit cell}}{4 \text{ Al atoms}}$$

$$= 4.471 \times 10^{-23} \text{ g/Al atom}$$

$$N_A = \frac{26.9815 \text{ g Al}}{1 \text{ mol Al}} \times \frac{1 \text{ Al atom}}{4.471 \times 10^{-23} \text{ g}} = 6.035 \times 10^{23} \frac{\text{atoms}}{\text{mol}}$$

10A Across the diagonal of a CsCl unit cell are Cs^+ and Cl^- ions, so that the body diagonal equals $2\,r(Cs^+) + 2\,r(Cl^-)$. This body diagonal equals $\sqrt{3}\, l$, where l is the length of the unit cell.

$$l = \frac{2\,r(Cs^+) + 2\,r(Cl^-)}{\sqrt{3}} = \frac{2(167 + 181) \text{ pm}}{\sqrt{3}} = 402 \text{ pm}$$

10B Since NaCl is fcc, the Na^+ ions are in the same locations as were the Al atoms in Practice Example 13-9, and there are 4 Na^+ ions per unit cell. For stoichiometric reasons there must also be 4 Cl^- ions. These are accounted for as follows. There is one Cl^- along each edge, and each of these edge Cl^- ions are shared among four unit cells. And there is one Cl^- precisely in the body center of the unit cell, not shared with any other unit cells. Thus, the number of Cl^- ions is given by:

Cl^- ions/unit cell = ($^1/_4$ Cl^- on edge × 12 edges per unit cell) + 1 Cl^- in body center = 4 Cl^-/unit cell.

The volume of this cubic unit cell is the cube of its length. Now we compute the density.

$$\text{NaCl density} = \frac{4 \text{ formula units}}{1 \text{ unit cell}} \times \frac{1 \text{ unit cell}}{(552 \text{ pm})^3} \times \left(\frac{10^{12} \text{ pm}}{1 \text{ m}} \times \frac{1 \text{ m}}{100 \text{ cm}}\right)^3 \times \frac{1 \text{ mole NaCl}}{6.022 \times 10^{23} \text{ f.u.}}$$

$$\times \frac{58.44 \text{ g NaCl}}{1 \text{ mol NaCl}} = 2.31 \text{ g/cm}^3$$

11A Sublimation of Cs(s):	$Cs(s) \longrightarrow Cs(g)$	$\Delta H_{sub} = +78.2$ kJ/mol
Ionization of Cs(g):	$Cs(g) \longrightarrow Cs^+(g) + e^-$	$\Delta I_1 = +375.7$ kJ/mol (Table 10.3)
$\frac{1}{2}$ Dissociation of $Cl_2(g)$:	$\frac{1}{2} Cl_2(g) \longrightarrow Cl(g)$	$DE = \frac{1}{2} \times 243 = 121.5$ kJ/mol (Table 11.3)
Cl(g) electron affinity:	$Cl(g) + e^- \longrightarrow Cl^-(g)$	$EA_1 = -349.0$ kJ/mol (Figure 10-11)
Lattice energy:	$Cs^+(g) + Cl^-(g) \longrightarrow CsCl(s)$	L.E.
Enthalpy of formation:	$Cs(s) + \frac{1}{2} Cl_2(s) \longrightarrow CsCl(s)$	$\Delta H_f^\circ = -442.8$ kJ/mol

-442.8 kJ/mol $= +78.2$ kJ/mol $+ 375.7$ kJ/mol $+ 121.5$ kJ/mol $- 349.0$ kJ/mol + L.E.
$= +226.4$ kJ/mol + L.E.
L.E. $= -442.8$ kJ $- 226.4$ kJ $= -669.2$ k/mol

11B Sublimation:	$Ca(s) \longrightarrow Ca(g)$	$\Delta H_{sub} = +178.2$ kJ/mol
First ionization energy:	$Ca(g) \longrightarrow Ca^+(g) + e^-$	$I_1 = +590$ kJ/mol
Second ionization energy:	$Ca^+(g) \longrightarrow Ca^{2+}(g) + e^-$	$I_2 = +1145$ kJ/mol
Dissociation energy:	$Cl_2(g) \longrightarrow 2\,Cl(g)$	D.E. $= (2 \times 122)$ kJ/mol
Electron Affinity:	$2\,Cl(g) + 2\,e^- \longrightarrow Cl_2(g)$	$2 \times$ E.A. $= 2\,(-349)$ kJ/mol
Lattice energy:	$Ca^{2+}(g) + 2\,Cl^-(g) \longrightarrow CaCl_2(s)$	L.E. $= -2255$ kJ/mol
Enthalpy of formation:	$Ca(s) + Cl_2(s) \longrightarrow CaCl_2(s)$	$\Delta H_f^\circ = ?$

$$\Delta H_f^\circ = \Delta H_{sub} + I_1 + I_2 + \text{D.E.} + (2 \times \text{E.A.}) + \text{L.E.}$$
$$= 178.2 \text{ kJ/mol} + 590 \text{ kJ/mol} + 1145 \text{ kJ/mol} + 244 \text{ kJ/mol} - 698 \text{ kJ/mol} - 2255 \text{ kJ/mol}$$
$$= -796 \text{ kJ/mol}$$

Summarizing Example Calculations

1. heat loss of water $= 100.0 \text{ g} \times \frac{4.18 \text{ J}}{\text{g °C}} \times (0.0 - 20.0)\text{°C} \times \frac{1 \text{ kJ}}{1000 \text{ J}} = -8.36 \text{ kJ}$
2. Remember that, because vaporization is the reverse of condensation, $\Delta H_{cond} = -\Delta H_{vap}$.
heat loss of steam = heat loss of condensation + heat loss for cooling
$$= \left(175 \text{ g } H_2O \times \frac{1 \text{ mol } H_2O}{18.015 \text{ g } H_2O} \times \frac{-40.7 \text{ kJ}}{1 \text{ mol } H_2O}\right)$$
$$+ \left(175 \text{ g} \times \frac{4.18 \text{ J}}{\text{g °C}} \times (0.0 - 100.0)\text{°C} \times \frac{1 \text{ kJ}}{1000 \text{ J}}\right)$$
$$= -395 \text{ kJ} - 73.2 \text{ kJ} = -468 \text{ kJ}$$
3. amount of melted ice $= (8.36 + 468) \text{ kJ} \times \frac{1 \text{ mol melted ice}}{6.01 \text{ kJ}} = 79.2$ mol melted ice
4. mass of melted ice $= 79.2 \text{ mol melted ice} \times \frac{18.02 \text{ g ice}}{1 \text{ mol ice}} \times \frac{1 \text{ kg}}{1000 \text{ g}} = 1.43$ kg melted ice
mass of unmelted ice = 1.65 kg ice initially – 1.43 kg melted ice = 0.22 kg unmelted ice

Review Questions

1. **(a)** ΔH_{vap} symbolizes the molar enthalpy of vaporization, the quantity of heat needed to convert one mole of a substance from liquid to vapor.
(b) T_c symbolizes the critical temperature, that temperature above which a gas cannot be condensed to a liquid, no matter how high the applied pressure.
(c) An instantaneous dipole is a momentary imbalance of the positive and negative charges in an atom or a molecule, caused when the electron cloud moves off center in a random manner.
(d) Coordination number refers to the number of atoms of one type surrounding another atom at the same distance, often those touching the central atom.
(e) A unit cell is the smallest portion of a crystal that will replicate the crystal through simple translations.
2. **(a)** Surface tension is the work required to create a unit (such as 1 cm^2) of liquid surface area.
(b) Viscosity refers to the resistance of a liquid to flowing.
(c) Sublimation is the physical change of a substance being converted directly from a solid to a vapor.

(d) Supercooling refers to the cooling of a liquid below its freezing point without the formation of any solid. The supercooled liquid is in a metastable state.
(e) The freezing point of a liquid is that place on the cooling curve of a liquid where the temperature remains constant while heat is withdrawn from the sample. During that period, solid forms as liquid freezes.

3. (a) Adhesive forces are those between different types of substances, usually forces between two condensed phases. Cohesive forces are those within a substance, again usually referring to liquids or solids.
(b) Vaporization is the physical process of transforming a liquid into a vapor. Condensentation is the reverse process: vapor to liquid.
(c) A triple point of a substance is the point where three phases coexist, often solid, liquid, and vapor. The critical point is that temperature and pressure above which liquid and vapor cannot be distinguished; just a fluid exists, called a supercritical fluid.
(d) Face-centered and body-centered cubic unit cells both have atoms at the vertices of a cube. In a face-centered unit cell there also is an atom on the center of each of the six faces of the cube. In contrast, in a body-centered cubic unit cell there is an atom exactly in the center of the three-dimensional cubic cell.
(e) A tetrahedral hole is one surrounded by four atoms at the vertices of a tetrahedron. An octahedral hole is surrounded by six atoms: above and below, in front and behind, and to the right and the left.

4. *Instantaneous dipole–induced dipole* forces occur after a chance distortion of the electron cloud of an atom or molecule creates an instantaneous imbalance of charge in that molecule: an instantaneous dipole. This imbalance creates a similar imbalance in an adjacent atom or molecule: an induced dipole. The two dipoles reinforce each other and can induce dipoles in other nearby atoms or molecules. The dipoles thus created attract each other and hold the particles together. *Dipole–dipole forces* exist between molecules that have permanent resultant dipole moments, which attract each other and hold these molecules together. *Hydrogen bonds* result from the attraction between a highly electronegative atom—usually N, O, or F—and a hydrogen atom bonded to another highly electronegative atom. *Relative strengths* of these forces are hydrogen bonds > dipole-dipole forces > instantaneous dipole-induced dipole forces. Each force in the preceding ordering is approximately ten times stronger than the one that follows it.

5. (a & b) It is possible for a substance not to have a normal melting point or a normal boiling point. Because normal means that the process occurs at 760 mmHg, if the solid-liquid-vapor triple point occurs at a pressure higher than 1 atm, for instance, then melting and vaporization must occur at pressures greater than 1 atm. In such a case, however, there will be a normal sublimation point.
(c) Every substance should have a critical point. There should always be a temperature above which molecules are moving so rapidly that forces of attraction between them are ineffective in producing a condensed phase.

6. One would expect the enthalpy of sublimation **(d)** to be the largest of the four quantities cited. Molar heat capacities are quite small, on the order of fractions of a kilojoule per mole-degree. (Remember that specific heats have values of joules per gram-degree.) All of the heats of transistion are positive numbers and on the order of kilojoules per mole. Since the heat of sublimation is the sum of the heat of fusion and the heat of vaporization, ΔH_{subl} must be the largest of the three.

7. (a) Intermolecular forces in a liquid *do* affect its vapor pressure. The stronger these forces are, the more difficult it is for molecules to vaporize and the lower the vapor pressure is at a given temperature.
(b) The volume of liquid present *does not* affect its vapor pressure. Realize that vaporization occurs at the surface of the liquid and thus volume of liquid should be of no consequence.
(c) The volume of vapor present also *does not* affect the vapor pressure of a liquid, for similar resons to those given in part (b).
(d) The size of the container also *does not* affect a liquid's vapor pressure, again following the reasoning in part (b). Of course, both liquid and vapor must be present at the point of equilibrium.
(e) The temperature of the liquid *does* affect that liquid's vapor pressure. Higher temperature means that more energy is present per molecule, each molecule has a larger fraction of the energy needed to overcome cohesive forces, and thus it is more likely that it will be able to overcome those cohesive forces and vaporize.

8. The product can lower the surface tension of water. Then the water can more easily wet a solid substance, because a greater surface area of water can be created with the same energy. (Surface tension equals the work needed to create a given quantity of surface area.) This greater water surface area means a greater area of contact with a solid object, such as a piece of fabric. More of the fabric being in contact with the water means that the water indeed is wetter.

9. (a) When the water vaporizes in the outer container, heat is required; vaporization is an endothermic process. When this vapor (steam) condenses on the outside walls of the inner container, that same heat is liberated; condensation is an exothermic process.
(b) Liquid water—condensed on the outside wall—is in equilibriuum with the water vapor that fills the space between the two containers. This equilibrium exists at the boiling point of water. We assume that the pressure is 1.000 atm, and thus, the temperature of the equilibrium must be 373.15 K or 100.00 °C. This is the maximum temperature that can be realized without pressurizing the apparatus.

10. (a) $\text{mass } CHCl_3 \text{ vaporized} = 6.62 \text{ kJ} \times \frac{1000 \text{ J}}{1 \text{ kJ}} \times \frac{1 \text{ g } CHCl_3}{247 \text{ J}} = 26.8 \text{ g } CHCl_3$

(b) $\Delta H_{vap} = \frac{247 \text{ J}}{1 \text{ g } CHCl_3} \times \frac{1 \text{ kJ}}{1000 \text{ J}} \times \frac{119.38 \text{ g } CHCl_3}{1 \text{ mol } CHCl_3} = 29.5 \text{ kJ/mol chloroform}$

(c) $\text{heat evolved} = 19.6 \text{ g } CHCl_3 \times \frac{247 \text{ J}}{1 \text{ g } CHCl_3} \times \frac{1 \text{ kJ}}{1000 \text{ J}} = 4.84 \text{ kJ}$

11. (a) We read up the 100 °C line until we arrive at the C_6H_{7N} curve (e). This occurs at about 45 mmHg.
(b) We read across the 760 mmHg line until we arrive at the C_7H_8 curve (d). This occurs at about 110 °C.

12. (a) The normal boiling point occurs where the vapor pressure is 760 mmHg, and thus $\ln P = 6.63$. For aniline, this occurs at about the uppermost data point (open circle) on the aniline line. This corresponds to $1/T = 2.18 \times 10^{-3} \text{ K}^{-1}$. Thus $T_{nbp} = 1 \div (2.18 \times 10^{-3} \text{ K}^{-1}) = 459 \text{ K}$.
(b) 25 °C = 298 K = T and thus $1/T = 3.36 \times 10^{-3} \text{ K}^{-1}$. This occurs at about $\ln P = 6.14$. Thus, $P = e^{6.14} =$ 464 mmHg.

13. We use the ideal gas equation, with $n = \text{moles } Br_2 = 0.486 \text{ g } Br_2 \times \frac{1 \text{ mol } Br_2}{159.8 \text{ g } Br_2} = 3.04 \times 10^{-3} \text{ mol } Br_2$.

$$P = \frac{nRT}{V} = \frac{3.04 \times 10^{-3} \text{ mol } Br_2 \times 0.08206 \text{ L atm mol}^{-1} \text{ K}^{-1} \times 298.2 \text{ K}}{0.2500 \text{ L}} \times \frac{760 \text{ mmHg}}{1 \text{ atm}}$$
$$= 226 \text{ mmHg}$$

14. We use the Clausius-Clapeyron equation (13.2).
$T_1 = (56.0 + 273.2) \text{ K} = 329.2 \text{ K}$ $\quad T_2 = (103.7 + 273.2) \text{ K} = 376.9 \text{ K}$

$$\ln \frac{10.0 \text{ mmHg}}{100.0 \text{ mmHg}} = -\frac{\Delta H_{vap}}{8.3145 \text{ J mol}^{-1} \text{ K}^{-1}} \left(\frac{1}{329.2 \text{ K}} - \frac{1}{376.9 \text{ K}}\right) = -2.30 = -4.624 \times 10^{-5} \Delta H_{vap}$$

$\Delta H_{vap} = 4.97 \times 10^4 \text{ J/mol} = 49.7 \text{ kJ/mol}$

15. We use the Clausius-Clapeyron equation (13.2). $T_1 = (5.0 + 273.2) \text{ K} = 278.2 \text{ K}$

$$\ln \frac{760.0 \text{ mmHg}}{40.0 \text{ mmHg}} = 2.944 = -\frac{38.0 \times 10^3 \text{ J/mol}}{8.3145 \text{ J mol}^{-1} \text{ K}^{-1}} \left(\frac{1}{T_{nbp}} - \frac{1}{278.2 \text{ K}}\right)$$

$$\left(\frac{1}{T_{nbp}} - \frac{1}{278.2 \text{ K}}\right) = -2.944 \times \frac{8.3145 \text{ K}^{-1}}{38.0 \times 10^3} = -6.44 \times 10^{-4} \text{ K}^{-1}$$

$$\frac{1}{T_{nbp}} = \frac{1}{278.2 \text{ K}} - 6.44 \times 10^{-4} \text{ K}^{-1} = 2.95 \times 10^{-3} \text{ K}^{-1} \qquad T_{nbp} = 339 \text{ K}$$

16. $\text{heat needed} = (5.08 \text{ cm})^3 \times \frac{0.92 \text{ g}}{1 \text{ cm}^3} \times \frac{1 \text{ mol}}{18.0 \text{ g}} \times \frac{6.01 \text{ kJ}}{1 \text{ mol}} = 40. \text{ kJ}$

17. (a) $\text{heat evolved} = 3.78 \text{ kg Cu} \times \frac{1000 \text{ g}}{1 \text{ kg}} \times \frac{1 \text{ mol Cu}}{63.55 \text{ g Cu}} \times \frac{13.05 \text{ kJ}}{1 \text{ mol Cu}} = 776 \text{ kJ}$

(b) $\text{heat absorbed} = (75 \text{ cm} \times 15 \text{ cm} \times 12 \text{ cm}) \times \frac{8.92 \text{ g}}{1 \text{ cm}^3} \times \frac{1 \text{ mol Cu}}{63.55 \text{ g Cu}} \times \frac{13.05 \text{ kJ}}{1 \text{ mol Cu}} = 2.5 \times 10^4 \text{ kJ}$

18. Let us use the ideal gas law to determine the final pressure in the container, assuming that all of the dry ice vaporizes. We then locate this pressure, at a temperature of 25 °C, on the phase diagram of Figure 13-18.

$$P = \frac{nRT}{V} = \frac{\left(80.0 \text{ g } CO_2 \times \dfrac{1 \text{ mol } CO_2}{44.0 \text{ g } CO_2}\right) \times 0.08206 \dfrac{\text{L atm}}{\text{mol K}} \times 298 \text{ K}}{0.500 \text{ L}} = 88.9 \text{ atm}$$

Although this point (25 °C and 88.9 atm) is most likely in the region labeled "liquid" in Figure 13-18, we computed its pressure assuming the CO_2 is a gas. Some of this gas should condense to a liquid. Thus, both liquid and gas are present in the container.

19. (After each formula is that substance's boiling point from a handbook.)

(a) $C_{10}H_{22}$ (174.1 °C) has a higher boiling point than C_7H_{16} (98.4 °C). All else being equal, the substance of the higher mole weight has the higher boiling point.

(b) CH_3OCH_3 ($\mathcal{M}$ = 46.0 g/mol) (–25 °C) has a higher boiling point than C_3H_8 ($\mathcal{M}$ = 44.0 g/mol) (–42.1 °C). For two substances with approximately equal mole weights, the polar substance has a higher boiling point than the nonpolar substance.

(c) CH_3CH_2OH ($\mathcal{M}$ = 46.0 g/mol) has a higher boiling point than CH_3CH_2SH ($\mathcal{M}$ = 62.1 g/mol). Even though CH_3CH_2SH (35 °C) is the heavier molecule, hydrogen bonds form between CH_3CH_2OH (78.5 °C) molecules, requiring more energy to vaporize them.

20. All of these substances are homonuclear molecules, composed of the same type of atoms. Thus boiling point should increase with increasing mole weight: N_2 ($\mathcal{M}$ = 28.0 g/mol), O_3 ($\mathcal{M}$ = 48.0 g/mol), F_2 ($\mathcal{M}$ = 38.0 g/mol), Ar ($\mathcal{M}$ = 39.9 g/mol), and Cl_2 ($\mathcal{M}$ = 70.9 g/mol). Thus, O_3 is seen to be out of order. The correct order is N_2 (–195.8 °C) < F_2 (–188.1 °C) < Ar (–185.7 °C) < O_3 (–111.9 °C) < Cl_2 (–34.6 °C). (Boiling points from a handbook.)

21. Both Ne and C_3H_8 are nonpolar; their solids are held together by relatively weak London forces, which are stronger for substances with high molecular weights. Ne has a lower melting point than C_3H_8. Next higher in melting are CH_3CH_2OH and $CH_2OHCHOHCH_2OH$, which are held together with hydrogen bonds in addition to London forces. There are more hydrogen bonds possible for a molecule of $CH_2OHCHOHCH_2OH$ than for one of CH_3CH_2OH, and $CH_2OHCHOHCH_2OH$ molecules are heavier besides; $CH_2OHCHOHCH_2OH$ has the higher melting point. KI, K_2SO_4, and MgO are ionic solids and have the highest melting points. For ionic solid the melting point is high for small, highly charged ions. Thus MgO has the highest melting point of these three, KI the lowest. Arranged in order of increasing melting point, the seven substances are as follows. (Melting points, in parentheses, are from a handbook.)

Ne < C_3H_8 < CH_3CH_2OH < $CH_2OHCHOHCH_2OH$ < KI < K_2SO_4 < MgO
(–248.6 °C < –187.7 °C < –97.8 °C < 18.2 °C < 681 °C < 1067 °C < 2825 °C)

22. The polarity of hydrazine (34.0 g/mol) might explain its high boiling point. But HCl (36.5 g/mol) also is polar, and it is a gas at room temperature. The distinction between the two substances is that hydrazine can form strong hydrogen bonds, while HCl cannot. This would raise the boiling point of hydrazine significantly above that of HCl (–114.18 °C).

23. **(a)** Si(s) is a network covalent solid. Its structure is similar to that of diamond; both elements are in the same family of the periodic table.

(b) $SiCl_4$(s) is a molecular solid. Each $SiCl_4$ formula unit is a discrete nonpolar molecule. The molecules are attracted to each other by London forces.

(c) $CaCl_2$(s) is an ionic solid. Calcium chloride consists of Ca^{2+} cations and Cl^- anions.

(d) Ag(s) is a metallic solid. Silver is one of the metals in the periodic table.

(e) SO_2(s) will form a molecular solid. Each SO_2 formula unit is a discrete polar molecule. The molecules are held to each other with London forces and dipole-dipole attractions.

24. Answer **(c)** is correct; atoms at the corners, along the edges, and on the faces of a unit cell are shared with adjacent unit cells. **(a)** is incorrect; often there are several formula units within a unit cell, as in the case for NaCl. **(b)** is incorrect; the unit cell need not be cubic; that of hexagonal close packing is not. **(d)** is incorrect; a unit cell will not contain the same number of cations as anions if their numbers are not equal in the formula of the compound.

25. In the unit cell of Figure 13-44, there is one Cs^+ ion totally contained within the unit cell. There are eight "corner" Cl^- ions, each shared by eight other unit cells. Thus, no. Cl^- ions = (8 × 1/8) = 1 Cl^- ions. Thus, the "formula of cesium chloride" is CsCl.

26. (a) The atoms touch along the face diagonal, d. The length of that diagonal thus is $4r$ (r from one corner atom + $2r$ from the center atom + r from the other corner atom). The Pythagorean theorem relates the length of the unit cell, l, to the face diagonal: $d^2 = l^2 + l^2$ (the cell is square; both sides are equal) or $d = \sqrt{2}\, l$. Two quantities equal to the face diagonal are equal to each other: $d = \sqrt{2}\, l = 4\, r = 4 \times 128$ pm

$$l = \frac{4 \times 128 \text{ pm}}{1.414} = 362 \text{ pm}$$

(b) The volume of the unit cell is that of a cube: $V = l^3 = (362 \text{ pm})^3 = 4.74 \times 10^7 \text{ pm}^3$.

(c) In a face-centered unit cell, there are eight "corner" Cu atoms, each shared by eight unit cells; there are six "face" Cu atoms, each shared by two unit cells. No. Cu atoms = $(8 \times \frac{1}{8}) + (6 \times \frac{1}{2}) = 4$ Cu atoms.

(d) The mass of 4 Cu atoms is determined from the atomic mass.

$$\frac{\text{mass}}{\text{unit cell}} = \frac{4 \text{ Cu atoms}}{\text{unit cell}} \times \frac{1 \text{ mol Cu atoms}}{6.022 \times 10^{23} \text{ Cu atoms}} \times \frac{63.55 \text{ g Cu}}{1 \text{ mol Cu atoms}} = \frac{4.221 \times 10^{-22} \text{ g Cu}}{1 \text{ unit cell}}$$

(e) $$\text{density} = \frac{\text{mass}}{\text{volume}} = \frac{4.221 \times 10^{-21} \text{ g}}{4.74 \times 10^7 \text{ pm}^3} \times \left(\frac{10^{12} \text{ pm}}{1 \text{ m}} \times \frac{1 \text{ m}}{100 \text{ cm}}\right)^3 = 8.91 \text{ g/cm}^3$$

EXERCISES

Surface Tension; Viscosity

27. Since both the silicone oil and the cloth or leather are composed of relatively nonpolar molecules, they attract each other. The oil thus adheres well to the material. Water, on the other hand is polar and adheres very poorly to the silicone oil (in fact, the water is repelled by the oil), much more poorly, in fact, than it adheres to the cloth or leather. This is because the oil is more nonpolar than is the cloth or the leather. Thus, water is repelled from the silicone-treated cloth or leather.

28. Both surface tension and viscosity deal with the work needed to overcome the attractions between molecules. Increasing the temperature of a liquid sample causes the molecules to move around somewhat. Some of the work has been done by adding thermal energy (heat) and less work needs to be done by the experimenter; both surface tension and viscosity decrease. The vapor pressure is a measure of the concentration of molecules that have come free of the surface. As thermal energy is added to the liquid sample, more and more molecules have enough energy to break free of the surface, and the vapor pressure increases.

Vaporization

29. The process of evaporation is an endothermic one; it requires energy. If evaporation occurs from an uninsulated container, this energy is obtained from the surroundings, through the walls of the container. However, if the evaporation occurs from an insulated container, the only source of the needed energy is the liquid that is evaporating. Therefore, the temperature of the liquid decreases.

30. We determine the total heat needed to vaporize 0.10 g $C_2H_5OH(l)$ and 0.90 g $H_2O(l)$.

$$\text{heat} = \left(0.10 \text{ g } C_2H_5OH \times \frac{1 \text{ mol } C_2H_5OH}{46.07 \text{ g } C_2H_5OH} \times \frac{42.6 \text{ kJ}}{1 \text{ mol } C_2H_5OH}\right)$$

$$+ \left(0.90 \text{ g } H_2O \times \frac{1 \text{ mol } H_2O}{18.02 \text{ g } H_2O} \times \frac{44.0 \text{ kJ}}{1 \text{ mol } H_2O}\right) = 0.092 \text{ kJ} + 2.2 \text{ kJ} = 2.3 \text{ kJ}$$

Let's now calculate the number of moles of air we could lower in temperature from 55 °C to 25 °C with each gram of solution. heat = amount × heat capacity × temperature change

$2.3 \times 10^3 \text{ J} = -n$ moles of air $\times 29 \text{ J mol}^{-1} \text{ K}^{-1} \times (25\text{ °C} - 55\text{ °C}) = 8.7 \times 10^2\, n$ J/mol

$$n = \frac{2.3 \times 10^3 \text{ J}}{8.7 \times 10^2 \text{ J/mol}} = 2.6 \text{ mol of air} \approx 71 \text{ L cooled by each gram of solution.}$$

If the volume of the car's interior is about 2 m³ (2000 L) only about 30 g of the solution is needed to cool the interior. Unfortuantely this does not take into account the heat stored in the seats, the dashboard, the steering wheel, etc. Because the heat capacity of a solid is considerably larger than that of a gas (in part because of its greater density) it is unlikely that this product would offer more than temporary relief. And it would produce a new problem: a very humid interior.

31. We use the quantity of heat to determine the number of moles of benzene that vaporize.

$$V = \frac{nRT}{P} = \frac{\left(1.54 \text{ kJ} \times \frac{1 \text{ mol}}{33.9 \text{ kJ}}\right) \times 0.08206 \frac{\text{L atm}}{\text{mol K}} \times 298 \text{ K}}{95.1 \text{ mmHg} \times \frac{1 \text{ atm}}{760 \text{ mmHg}}} = 8.88 \text{ L } C_6H_6(l)$$

32. $$n_{\text{acetonitrile}} = \frac{PV}{RT} = \frac{1.00 \text{ atm} \times 1.17 \text{ L}}{0.08206 \text{ L atm mol}^{-1} \text{K}^{-1} \times (273.2 + 81.6) \text{ K}} = 0.0402 \text{ mol acetonitrile}$$

$$\Delta H_{\text{vap}} = \frac{1.00 \text{ kJ}}{0.0402 \text{ mol}} = 24.9 \text{ kJ/mol acetonitrile}$$

33. $$\text{heat needed} = 3.78 \text{ L } H_2O \times \frac{1000 \text{ cm}^3}{1 \text{ L}} \times \frac{0.958 \text{ g}}{1 \text{ cm}^3} \times \frac{1 \text{ mol } H_2O}{18.02 \text{ g } H_2O} \times \frac{40.7 \text{ kJ}}{1 \text{ mol } H_2O} = 8.18 \times 10^3 \text{ kJ}$$

$$\text{amount } CH_4 \text{ needed} = 8.18 \times 10^3 \text{ kJ} \times \frac{1 \text{ mol } CH_4}{890 \text{ kJ}} = 9.19 \text{ mol } CH_4$$

$$V = \frac{nRT}{P} = \frac{9.19 \text{ mol} \times 0.08206 \text{ L atm mol}^{-1} \text{K}^{-1} \times 296.6 \text{ K}}{768 \text{ mmHg} \times \frac{1 \text{ atm}}{760 \text{ mmHg}}} = 221 \text{ L methane}$$

34. If not all of the water vaporizes, the final temperature of the system will be 100.00 °C. Let us proceed on that assumption and modify our final state if it is not true. First we determine the heat available from the iron in cooling down, then the heat needed to warm the water to boiling, and finally the mass of water that vaporizes.

$$\text{heat from Fe} = \text{mass} \times \text{sp.ht.} \times \Delta t = 50.0 \text{ g} \times \frac{0.45 \text{ J}}{\text{g °C}} \times (100.00 \text{ °C} - 152 \text{ °C}) = 1.17 \times 10^3 \text{ J}$$

$$\text{heat to warm water} = 20.0 \text{ g} \times \frac{4.21 \text{ J}}{\text{g °C}} \times (100.00 \text{ °C} - 89 \text{ °C}) = 9.3 \times 10^2 \text{ J}$$

$$\text{mass of water vaporized} = (11.7 \times 10^2 \text{ J available} - 9.3 \times 10^2 \text{ J used}) \times \frac{1 \text{ mol } H_2O \text{ vaporized}}{40.7 \times 10^3 \text{ J}} \times \frac{18.02 \text{ g } H_2O}{1 \text{ mol } H_2O}$$

$$= 0.11 \text{ g of water vaporize}$$

Clearly all of the water does not vaporize and our initial assumption was true.

Vapor Pressure and Boiling Point

35. With the Clausius-Clapeyron equation, we use the vapor pressure of water at 100.0 °C = 373.2 K and 120.0 °C = 393.2 K to determine ΔH_{vap} of water near its boiling point. We then use the equation again, to determine the temperature where water's vapor pressure is 2.00 atm.

$$\ln \frac{1489.1 \text{ mmHg}}{760.0 \text{ mmHg}} = -\frac{\Delta H_{\text{vap}}}{8.3145 \text{ J mol}^{-1} \text{K}^{-1}} \left(\frac{1}{393.2 \text{ K}} - \frac{1}{373.2 \text{ K}}\right) = 0.6726 = 1.639 \times 10^{-5} \Delta H_{\text{vap}}$$

$$\Delta H_{\text{vap}} = 4.104 \times 10^4 \text{ J/mol} = 41.04 \text{ kJ/mol}$$

$$\ln \frac{1520.0 \text{ mmHg}}{760.0 \text{ mmHg}} = 0.6931 = -\frac{41.04 \times 10^3 \text{ J/mol}}{8.3145 \text{ J mol}^{-1} \text{K}^{-1}} \left(\frac{1}{T} - \frac{1}{373.2 \text{ K}}\right)$$

$$\left(\frac{1}{T_{\text{bp}}} - \frac{1}{373.2 \text{ K}}\right) = -0.6931 \times \frac{8.1345 \text{ K}^{-1}}{41.03 \times 10^3} = -1.404 \times 10^{-4} \text{ K}^{-1}$$

$$\frac{1}{T_{\text{bp}}} = \frac{1}{373.2 \text{ K}} - 1.404 \times 10^{-4} \text{ K}^{-1} = 2.539 \times 10^{-3} \text{ K}^{-1} \qquad T_{\text{nbp}} = 393.9 \text{ K} = 120.7 \text{ °C}$$

36. **(a)** We need the temperature at which the vapor pressure of water is 640 mmHg. This is a temperature between 95.0 °C (633.9 mmHg) and 96.0 °C (657.6 mmHg), about one-third between, in fact. We estimate a boiling point of 95.3 °C.

(b) If the observed boiling point is 94 °C, the atmospheric pressure equals the vapor pressure of water at 94 °C, that is, 611 mmHg.

37. The 25.0 L of He becomes saturated with aniline vapor, at a pressure equal to the vapor pressure of aniline.

$$n_{\text{aniline}} = (6.220 \text{ g} - 6.108 \text{ g}) \times \frac{1 \text{ mol aniline}}{93.13 \text{ g aniline}} = 0.00120 \text{ mol aniline}$$

$$P = \frac{nRT}{V} = \frac{0.00120 \text{ mol} \times 0.08206 \text{ L atm mol}^{-1} \text{K}^{-1} \times 303.2 \text{ K}}{25.0 \text{ L}} = 0.00119 \text{ atm} = 0.907 \text{ mmHg}$$

38. This is essentially a Boyle's Law problem. The initial pressure is the pressure of the $N_2(g)$ and of the CCl_4 vapor: $P_1 = 742$ mmHg + 261 mmHg = 1003 mmHg, according to Dalton's law of partial pressures. The final pressure is 742 mmHg. The initial volume is 7.53 L.

$$V_f = V_i \times \frac{P_i}{P_f} = 7.53\text{ L} \times \frac{1003\text{ mmHg}}{742\text{ mmHg}} = 10.18\text{ L}$$

39.

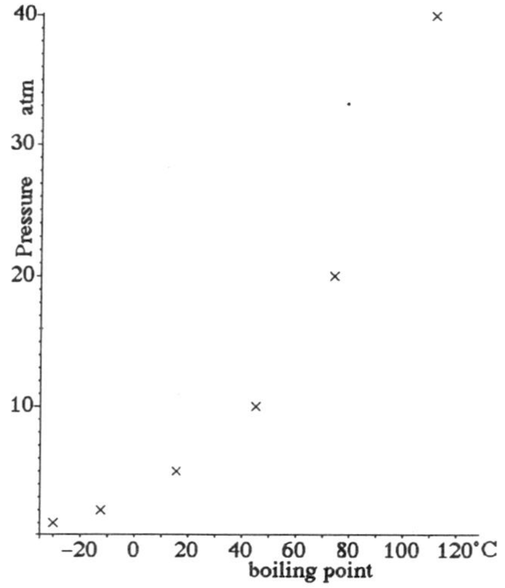

The graph of pressure *vs.* boiling point for Freon-12 is drawn at left.
At a temperature of 25 °C the vapor pressure is approximately 6.5 atm for Freon-12. Thus the compressor must be capable of producing pressure greater than 6.5 atm.

40. At 27 °C the vapor pressure of water is 26.7 mmHg. We use this value to determine the mass of water that could exist within the container as vapor.

$$\text{mass of water vapor} = \frac{\left(26.7\text{ mmHg} \times \frac{1\text{ atm}}{760\text{ mmHg}}\right) \times \left(1515\text{ mL} \times \frac{1\text{ L}}{1000\text{ mL}}\right)}{\frac{0.08206\text{ L atm}}{\text{mol K}} \times (27 + 273)\text{ K}} = 0.00216\text{ mol } H_2O(g)$$

$$\text{mass of water vapor} = 0.00216\text{ mol } H_2O(g) \times \frac{18.02\text{ g } H_2O}{1\text{ mol } H_2O} = 0.0389\text{ g } H_2O(g)$$

The Clausius-Clapeyron Equation

41. $t = 56.2$ °C is $T = 329.4$ K

$$\ln\frac{760\text{ mmHg}}{375\text{ mmHg}} = \frac{-25.5 \times 10^3\text{ J/mol}}{8.3145\text{ J mol}^{-1}\text{ K}^{-1}}\left(\frac{1}{329.4\text{ K}} - \frac{1}{T}\right) = 0.706$$

$$\left(\frac{1}{329.4\text{ K}} - \frac{1}{T}\right) = \frac{0.706 \times 8.3145}{-25.5 \times 10^3}\text{ K}^{-1} = -2.30 \times 10^{-4}\text{ K}^{-1} = 3.03_6 \times 10^{-3}\text{ K}^{-1} - 1/T$$

$1/T = (3.03_6 + 0.230) \times 10^{-3}\text{ K}^{-1} = 3.266 \times 10^{-3}\text{ K}^{-1}$ $\qquad T = 306\text{ K} = 33$ °C

42. First use the Clausius-Clapeyron equation and the data supplied to determine the value of ΔH_{vap}.
$T_1 = (113.5 + 273.2)\text{ K} = 386.7\text{ K}$ $\qquad T_2 = (380 + 273) = 653\text{ K}$

$$\ln\frac{145.4\text{ atm}}{1.00\text{ atm}} = -\frac{\Delta H_{vap}}{R}\left(\frac{1}{653\text{ K}} - \frac{1}{386.7\text{ K}}\right) = 0.00105_5\text{ K}^{-1}\frac{\Delta H_{vap}}{R} = 4.979$$

$$\Delta H_{vap} = \frac{4.979 \times \frac{8.3145\text{ J}}{\text{mol K}}}{0.00105_5\text{ K}^{-1}} \times \frac{1\text{ kJ}}{1000\text{ J}} = 39.2\text{ kJ/mol}$$

Then, at 100.0 °C = 373.2 K, we use the Clausius-Clapeyron equation to find the vapor pressure.

$$\ln\frac{P}{1.00\text{ atm}} = -\frac{39.2 \times 10^3\text{ J/mol}}{8.3145\text{ J mol}^{-1}\text{ K}^{-1}}\left(\frac{1}{373.2} - \frac{1}{386.7}\right) = -0.441 \qquad \frac{P}{1.00\text{ atm}} = e^{-0.441} = 0.643$$

$P = 0.643\text{ atm} = 489\text{ mmHg}$

Critical Point

43. A substance that can exist as a liquid at room temperature (about 20 °C) is one whose critical temperature is above 20 °C, 293 K. Of the substances listed in Table 13.3, this includes CO_2 (T_c = 304.2 K), HCl (T_c = 324.6 K), NH_3 (T_c = 405.7 K), SO_2 (T_c = 431.0 K), and H_2O (T_c = 647.3 K). In fact, CO_2 exists as a liquid in CO_2 fire extinguishers.

44. The critical temperature of SO_2 is 431.0 K, which is above the temperature of 0 °C, 273 K. The critical pressure of SO_2 is 77.7 atm, which is below the pressure of 100 atm. Thus, SO_2 can be maintained as a liquid at 0 °C and 100 atm.

The critical temperature of methane, CH_4, is 191.1 K, which is below the temperature of 0 °C, 273 K. Thus, CH_4 cannot exist as a liquid at 0 °C, no matter what pressure is applied.

States of Matter and Phase Diagrams

45. **(a)** As heat is added initially, the temperature of the ice rises from –20 °C to 0 °C. At (or just slightly below) 0 °C, ice begins to melt to liquid water. The temperature remains at 0 °C until all of the ice has melted. Adding heat then warms the liquid until a temperature of about 93.5 °C is reached, where the liquid begins to vaporize to steam. The temperature remains fixed until all the water is converted to steam. Adding heat then warms the steam to 200 °C. (Data for this part are taken from Figure 13-19 and Table 13.2.)

(b) As the pressure is raised, initially gaseous iodine is compressed. At about 91 mmHg, liquid iodine appears; the system remains at a fixed pressure with further compression until all vapor is converted to liquid. Increasing the pressure further simply compresses the liquid until a high pressure is reached, perhaps 50 atm, where solid iodine appears. Again the pressure remains fixed with further compression until all iodine is converted to solid. After this has occurred further compression raises the pressure of the system until 100 atm is reached. (Data for this part are from Figure 13-17 and the surrounding text.)

(c) Cooling of gaseous CO_2 simply lowers the temperature until a temperature of perhaps 20 °C is reached. At this point, liquid CO_2 appears. The temperature remains constant as more heat is removed until all the gas is converted to liquid. Further cooling then lowers the temperature of the liquid until a temperature of slightly higher than –56.7 °C is reached, where solid CO_2 appears. At this point, further cooling simply converts liquid to solid at constant temperature, until all solid has been converted to liquid. From this point, further cooling lowers the temperature of the solid. (Data for this part are taken from Figure 13-18 and Table 13.3.)

46. **(a)** The upper-right region of the phase diagram is the liquid region, while the lower-right region is the region of gas. One way to figure this out is to imagine moving from left to right (toward higher temperatures) at constant pressure. (Assume 45 atm for this case.) From experience, we known that the progression of states is solid (low temperature) $\longrightarrow$ liquid (intermediate temperature) $\longrightarrow$ gas (high temperature).

(b) Melting means that we convert the solid to a liquid. As the phase diagram shows, the lowest pressure at which liquid exists is at the triple point pressure, 43 atm. 1.00 atm is far below 43 atm; liquid cannot exist at this temperature.

(c) As we move from point *A* to point *B* by lowering the pressure, initially nothing happens. At a certain pressure, the solid liquefies. The pressure continues to drop, with the entire sample being liquid while it does, until another, lower pressure is reached. At this lower pressure the entire sample vaporizes. The pressure then continues to drop, with the gas becoming less dense as it does so, until point *B* is reached.

47. 0.240 g H_2O corresponds to 0.0133 mol H_2O. If the water does not vaporize completely, the pressure of the vapor in the flask equals the vapor pressure of water at the indicated temperature. However, if the water vaporizes completely, the pressure of the vapor is determined by the ideal gas law.

(a) 30.0 °C, vapor pressure of H_2O = 31.8 mmHg = 0.0418 atm

$$n = \frac{PV}{RT} = \frac{0.0418 \text{ atm} \times 3.20 \text{ L}}{0.08206 \text{ L atm mol}^{-1} \text{ K}^{-1} \times 303.2 \text{ K}}$$

= 0.00538 mol H_2O vapor < 0.0133 mol H_2O; not all the H_2O vaporizes. P = 0.0418 atm

(b) 50.0 °C, vapor pressure of H_2O = 92.5 mmHg = 0.122 atm

$$n = \frac{PV}{RT} = \frac{0.122 \text{ atm} \times 3.20 \text{ L}}{0.08206 \text{ L atm mol}^{-1} \text{ K}^{-1} \times 323.2 \text{ K}}$$

= 0.0147 mol H_2O vapor > 0.0133 mol H_2O; all the H_2O vaporizes. Thus,

$$P = \frac{nRT}{V} = \frac{0.0133 \text{ mol} \times 0.08206 \text{ L atm mol}^{-1} \text{ K}^{-1} \times 323.2 \text{ K}}{3.20 \text{ L}} = 0.110 \text{ atm} = 83.8 \text{ mmHg}$$

(c) 70.0 °C All the H_2O must vaporize, as this temperature is higher than that of part **(b)**. Thus,

$$P = \frac{nRT}{V} = \frac{0.0133 \text{ mol} \times 0.08206 \text{ L atm mol}^{-1} \text{ K}^{-1} \times 343.2 \text{ K}}{3.20 \text{ L}} = 0.117 \text{ atm} = 89.0 \text{ mmHg}$$

48. **(a)** If the pressure exerted by 2.50 g of H_2O(vapor) is less than the vapor pressure of water at 120. °C (393 K), then the water must exist entirely as a vapor.

$$P = \frac{nRT}{V} = \frac{\left(2.50 \text{ g} \times \frac{1 \text{ mol } H_2O}{18.02 \text{ g } H_2O}\right) \times 0.08206 \text{ L atm mol}^{-1} \text{ K}^{-1} \times 393 \text{ K}}{5.00 \text{ L}} = 0.895 \text{ atm}$$

Since this is less than the 1.00 atm vapor pressure at 100. °C, it must be less than the vapor pressure of water at 120 °C; the water exists entirely as a vapor.

(b) 0.895 atm = 680 mmHg. From Table 13-2, we see that this corresponds to a temperature of 97.0 °C (at which temperature the vapor pressure of water is 682.1 mmHg). Thus, at temperatures slightly less than 97.0 °C the water will begin to condense to liquid. But we have forgotten that, in this constant-volume container, the pressure (of water) will decrease as the temperature decreases. Calculating the precise decrease in both involves linking the Clausius-Clapeyron equation with the expression $P = k{\cdot}T$, but we can estimate the final temperature. First, we determine the pressure if we lower the temperature from 393 K to 97 °C (370 K). $P_f = \frac{370 \text{ K}}{393 \text{ K}} \times 680 \text{ mmHg} = 640 \text{ mmHg}$

From Table 13-2, this pressure occurs at a temperature of about 95 °C (368 K). We now determine the final pressure at this temperature. $P_f = \frac{368 \text{ K}}{393 \text{ K}} \times 680 \text{ mmHg} = 637 \text{ mmHg}$

This is just slightly above the vapor pressure of water (633.9 mmHg) at 95 °C, which we conclude must be the temperature at which liquid water appears.

49. **(a)** According to Figure 13-18, CO_2(s) exists at temperatures below –78.5 °C when the pressure is 1 atm or less. We do not expect to find temperatures this low and partial pressures of CO_2(g) of 1 atm on the surface of the earth.

(b) According to Table 13.3, the critical temperature of CH_4—the maximum temperature at which CH_4(l) can exist—is 191.1 K = –82.1 °C. We do not expect to find temperatures this low on the surface of the earth.

(c) Since, according to Table 13.3, the critical temperature of SO_2 is 431.0 K = 157.8 °C, SO_2(g) can be found on the surface of the earth.

(d) According to Figure 13-17, I_2(l) can exist at pressures less than 1.00 atm between the temperatures of 114 °C and 184 °C. There are very few places on the surface of the earth that reach temperatures this far above the boiling point of water, places such as volcano mouths. Essentially, I_2(l) is not found on the surface of the earth.

(e) According to Table 13.3, the critical temperature—the maximum temperature at which O_2(l) exists—is 154.8 K = –118.4 °C. Temperatures this low do not exist on the surface of the earth.

50. The final pressure specified (100. atm) is below the ice III-ice I-liquid water triple point, according to the text adjacent to Figure 13-19; no ice III is formed. The starting point is that of H_2O(g). When the pressure is increased to about 4.5 mmHg, the vapor condenses to ice I. As the pressure is raised, the melting point of ice I decreases—by 1 °C per 125 atm pressure increase. Thus, somewhere between 10 and 15 atm, the ice I melts to liquid water, in which state it remains until the final pressure is reached.

51. The mass of unmelted ice is 0.22 kg. First compute the heat needed to melt this ice.

$$\text{heat to melt ice} = 0.22 \text{ kg} \times \frac{1000 \text{ g ice}}{1 \text{ kg ice}} \times \frac{1 \text{ mol ice}}{18.02 \text{ g ice}} \times \frac{6.01 \text{ kJ}}{1 \text{ mol ice}} = 73._4 \text{ kJ}$$

Then compute the heat produced by 1 mole (18.02 g) of steam condensing and then lowering in temperature from 100.0 °C to 0.0 °C, the melting point of ice.

$$\text{heat evolved} = \left(1 \text{ mol} \times \frac{-40.7 \text{ kJ}}{1 \text{ mol}}\right) + \left(18.02 \text{ g} \times \frac{4.18 \text{ J}}{\text{g °C}} \times \frac{1 \text{ kJ}}{1000 \text{ J}} \times (0 - 100) \text{ °C}\right)$$

$= -40.7$ kJ/mol $- 7.53$ kJ/mol $= -48.2$ kJ/mol of steam.

Use the conversion factor just obtained to determine the mass of steam needed to melt the remaining ice.

$$\text{mass of steam} = 73.4 \text{ kJ} \times \frac{1 \text{ mol steam}}{48.2 \text{ kJ}} \times \frac{18.02 \text{ g steam}}{1 \text{ mol steam}} = 27 \text{ g steam.}$$

52. The heat gained by the ice equals the the negative of the heat lost by the water. Let us use this fact in a step-by-step approach. We first compute the heat needed to raise the temperature of the ice to 0.0 °C and then the heat given off when the temperature of the water is lowered to 0.0 °C.

$$\text{to heat the ice} = 54 \text{ cm}^3 \times \frac{0.917 \text{ g}}{1 \text{ cm}^3} \times 2.01 \text{ J g}^{-1} \text{ °C}^{-1} \times (0 \text{ °C} + 25.0 \text{ °C}) = 2.5 \times 10^3 \text{ J}$$

$$\text{to cool the water} = -400.0 \text{ cm}^3 \times \frac{0.998 \text{ g}}{1 \text{ cm}^3} \times 4.18 \text{ J g}^{-1} \text{ °C}^{-1} \times (0 \text{ °C} - 32.0 \text{ °C}) = 53.4 \times 10^3 \text{ J}$$

Thus, at 0 °C, we have 50. g ice, 399 g water, and $(53.4 - 2.5) \times 10^3$ J = 50.9 kJ of heat available

Since 50.0 g of ice is a bit less than 3 mol of ice and 50.9 kJ is enough heat to melt at least 8 mole of ice, all of the ice will melt. The heat needed to melt the ice is

$$50. \text{ g ice} \times \frac{1 \text{ mol } H_2O}{18.0 \text{ g } H_2O} \times \frac{6.01 \text{ kJ}}{1 \text{ mol ice}} = 17 \text{ kJ}$$

We now have 50.9 kJ – 17 kJ = 34 kJ of heat, and 399 g + 50. g = 449 g of water at 0 °C. We compute the temperature change that is produced by adding the heat to the water.

$$\Delta t = \frac{34 \times 10^3 \text{ J}}{449 \text{ g} \times 4.18 \text{ J g}^{-1} \text{ °C}^{-1}} = 18 \text{ °C}$$ The final temperature is 18 °C

53. The liquid in the can is supercooled. When the can is opened, gas bubbles released from the carbonated beverage serve as nuclei for the formation of ice crystals. The condition of supercooling is destroyed and the liquid reverts to the solid phase instantly.

An alternative explanation follows. The process of the gas coming out of solution is endothermic (heat is required). (We know this to be true because the reaction

$$\text{solution of gas in water} \longrightarrow \text{gas + liquid water}$$

proceeds to the right as the temperature is raised, a characteristic direction of an endothermic reaction.) The required heat is taken from the cooled liquid, causing it to freeze.

54. Both the melting point of ice and the boiling point of water are temperatures that vary as the pressure changes, the boiling point more substantially than the melting point. The triple point, however, does not vary with pressure. Solid, liquid, and vapor coexist only at one fixed temperature and pressure.

Van der Waals Forces

55. **(a)** HCl is not a very heavy diatomic molecule; London forces are expected to only moderately important. Hydrogen bonding is weak in the case of H—Cl bonds; Cl is not one of the three atoms (F, O, N) that form strong hydrogen bonds. Finally, because Cl is an electronegative atom, and H is only moderately electronegative, dipole-dipole interactions should be relatively strong.

(b) In Br_2 neither hydrogen bonds nor dipole-dipole attractions are important; there are no H atoms in the molecule, and homonuclear molecules are nonpolar. London forces are more important than in HCl since Br_2 is heavier.

(c) In ICl there are no hydrogen bonds since there are no H atoms in the molecule. The London forces are as strong as in Br_2 since the two molecules have the same number of electrons. However, dipole-dipole interactions are important in ICl; the molecule is polar toward Cl.

(d) In HF London forces are not very important; the molecule has only 10 electrons. Hydrogen bonding is quite important and definitely overshadows even the strong dipole-dipole interactions.

(e) In CH_4, H bonds are not important; the H atoms are not bonded to F, O, or N. In addition the molecule is not polar, so there are no dipole-dipole interactions. Finally, London forces are quite weak since the molecule contains only 10 electrons, and this is why CH_4 is a gas with a low critical temperature.

56. Substituting Cl for H both makes the molecule heavier and thus increases London forces, and makes it polar which increases dipole-dipole interactions. Both of these increases make it more difficult to disrupt the forces of attraction between molecules, increasing the boiling point. A substitution of Br for Cl increases London forces, but makes the molecule less polar. Since London forces are relatively more important than dipole-dipole interactions, the boiling point increases yet again. Finally, substituting OH for Br decreases London forces but both increases the dipole-dipole interactions and creates opportunities for hydrogen bonding. Since hydrogen bonds are much stronger than London forces, the boiling point increases yet once more.

Hydrogen Bonding

57. We expect CH_3OH to be a liquid from among the four substances listed. Of these four molecules, C_3H_8 has the most electrons and should have the strongest London forces. However, only CH_3OH satisfies the conditions for hydrogen bonding (H bonded to and attracted to N, O, or F) and its intermolecular attractions should be much stronger than those of the other substances.

58. **(a)** Intramolecular hydrogen bonding cannot occur in C_2H_6 since the conditions for hydrogen bonding (H bonded to and also attracted to N, O, or F) are not satisfied in this molecule. There is no N, O, or F in this molecule.

(b) Intramolecular hydrogen bonding will not occur in H_3CCH_2OH since there is not another F, O, or N in the molecule to which the H of —OH can hydrogen bond.

(c) Intramolecular hydrogen bonding is not important in H_3CCOOH. Although there is another O atom to which the H of —OH can hydrogen bond, the resulting configurations will create a four membered ring 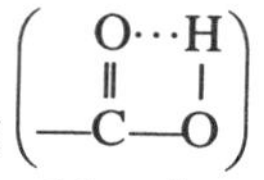 with bond angles of 90°, quite different, and therefore strained, compared to the normal bond angles of 109.5° and 120°.

(d) In orthophthalic acid, intramolecular hydrogen bonds can be important. The H of one —COOH group can be attracted to one of the O atoms of the other —COOH group. The resulting ring is seven atoms around and thus should not cause bond angle strain.

Network Covalent Solids

59. One would expect diamond to have a greater density than graphite. Although the bond distance in graphite—"one-and-a-half" bonds—would be expected to be shorter than the single bonds in diamond, the large spacing between the layers of C atoms in graphite makes its crystals much less dense than those of diamond.

60. Diamond works well in glass cutters because of its extreme hardness, a hardness that is due to the crystal being held together entirely by covalent bonds. Graphite will not function effectively in a glass cutter, since it is quite soft, soft enough to flake off in microscopic pieces when used in pencils. In fact, graphite is so soft that pure graphite is rarely used in common wooden pencils. Often clay or some other substance is mixed with the graphite to produce a mechanically strong pencil "lead".

61. **(a)** We expect Si and C atoms to alternate in the structure, as shown at right. The C atoms are on the corners ($8 \times 1/8 = 1$ C atom) and on the faces ($6 \times 1/2 = 3$ C atoms), a total of four C atoms/unit cell. The Si atoms are each totally within the cell, a total of four Si atoms/unit cell.

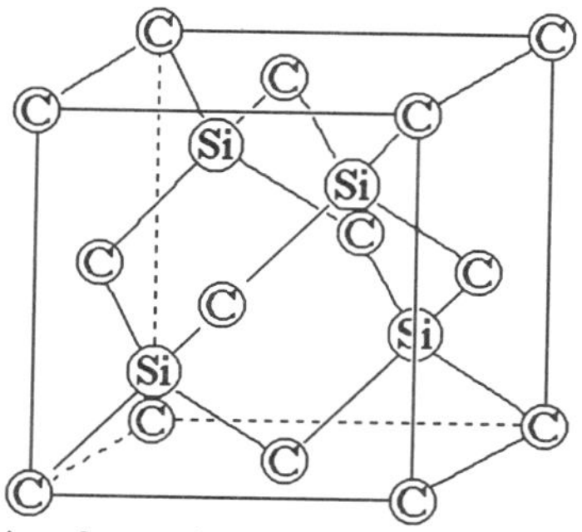

(b) To have a graphite structure, we expect sp^2 hybridization for each atom. The hybridization schemes for B and N atoms follow. B [He] sp^2 [1][1][1] $2p$ [ ] N [He] sp^2 [1][1][1] $2p$ [1l]

The half-filled sp^2 hybrid orbitals overlap to form the σ bonding structure, and a hexagonal array of atoms. The $2p_z$ orbitals then overlap to form the π bonding orbitals. There are as many π electrons in a sample of BN as there are in a sample of graphite, assuming both samples have the same number of atoms.

62. Buckminsterfullerene is composed of C_{60} spheres. These molecules of carbon are rather like the N_2, P_4 and S_8 molecules and produce a nonpolar molecular solid. The chain of alternating single and triple bonds could be a network covalent solid in the sense that graphite is. The long carbon chain should be linear and these rods of carbon (—C≡C—C≡C—C≡C—C≡C—C≡C—C≡C—) should fit together rather like spaghetti in a box, held together by the π electrons of the triple bonds from adjacent rods attracting each other.

Ionic Bonding and Properties

63. We expect forces in ionic compounds to increase as sizes of ions become smaller and as ionic charges become greater. As the forces between ions become stronger, a higher temperature is required to melt the crystal. In the series of compounds NaF, NaCl, NaBr, and NaI, the anions are progressively larger, and thus the ionic forces become weaker. We expect the melting points to decrease in this series from NaF to NaI. This is precisely what we see.

64. Coulomb's law (Appendix B) states that the force between two particles of charges Q_1 and Q_2 that are separated by a distance r is given by $F = Q_1Q_2/\varepsilon r^2$ where ε, the dielectric constant, equals 1 for a vacuum. Since we are comparing forces, we can use +1, –1, +2, and –2 for the charges on ions. We take r to equal the sum of the cation and anion radii, which are taken from Figure 12-35.

For NaCl, $r_+ = 99$ pm, $r_- = 181$ pm $\quad F = (+1)(-1)/(99 + 181)^2 = -1.3 \times 10^{-5}$

For MgO, $r_+ = 72$ pm, $r_- = 140$ pm $\quad F = (+2)(-2)/(72 + 140)^2 = -8.9 \times 10^{-5}$

Thus, it is clear that interionic forces are about seven times stronger in MgO than in NaCl.

Crystal Structures

65. In each layer of a closest packing arrangement of spheres, there are six spheres surrounding and touching any given sphere. A second similar layer then is placed on top of this first layer so that its spheres fit into the identations in the layer below. The two different closest packing arrangements arise from two different ways of placing the third layer on top of these two, with its spheres fitting into the indentations of the layer below. In one case, one can look down into these indentations and see a sphere of the bottom (first) layer. If these indentations are used, the closest packing arrangement *abab* results (hexagonal closest packing). In the other case, no first layer sphere is visible through the indentation; the closest packing arrangement *abcabc* results (cubic closest packing).

66. Physical properties are determined by the type of bonding between atoms in a crystal or the types of interactions between ions or molecules in the crystal. Different types of interactions can produce the same geometrical relationships of unit cell components. In both Ar and CO_2, London forces hold the particles in the crystal. In NaCl, the forces are interionic attractions, while metallic bonds hold Cu atoms in their crystals. It is the type of force, not the geometric arrangement of the components, that largely determines the physical properties of the crystalline material.

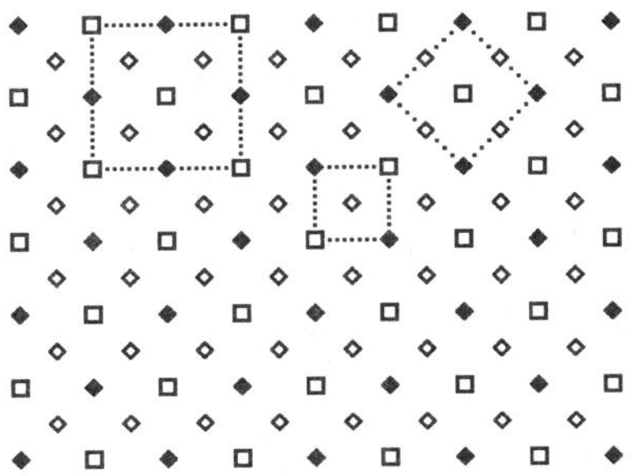

67. **(a)** We naturally tend to look at crystal structures in right-left, up-down terms. If we do that here, we might assign a unit cell as a square, with its corners at the centers of the light-colored squares. But the crystal does not "know" right and left or top and bottom. If we look at this crystal from the lower right corner, we see a unit cell that has its corners at the centers of dark-colored diamonds. These two types of unit cells are outlined at the top of the diagram at right.

(b) The unit cell has one light-colored square fully inside it. It has four light-colored "circles" (which the computer doesn't draw as very round) on the edges, each shared with one other unit cell. So a total of $4 \times {}^1/_2 = 2$ circles per unit cell. And the unit cell has a four dark-colored diamonds, one at each corner, and each shared with four other unit cells, for a total of $4 \times {}^1/_4 = 1$ diamond per unit cell.

(c) One example of an erroneous unit cell is the small square outlined near the center of the figure drawn in part (a). But notice that simply repeatedly translating this unit cell toward the right, so that its left edge sits where its right edge is now, will not generate the lattice.

68. This question reduces to asking what percentage of the area of a square is covered by a circle inscribed within it. The diameter of the circle equals the side of the square. Since diameter = 2 × radius, we have

$$\frac{\text{area of circle}}{\text{area of square}} = \frac{\pi r^2}{(2r)^2} = \frac{\pi}{4} = \frac{3.14159}{4} = 0.7854 \quad \textit{or} \quad 78.54\%$$

Area uncovered = 100.00% – 78.54% = 21.46%

69. In Figure 13-42 we see that the body diagonal of a cube has a length of $\sqrt{3}\, l$, where l is the length of one edge of the cube. The length of this body diagonal also equals $4r$, where r is the radius of the atom in the structure. Hence $4\,r = \sqrt{3}\,l$ or $l = 4\,r \div \sqrt{3}$ Recall that the volume of a cube is l^3, and $\sqrt{3} = 1.732$

$$\text{density} = \frac{\text{mass}}{\text{volume}} = \frac{\dfrac{2\text{ W atoms}}{1\text{ unit cell}} \times \dfrac{1\text{ mol W}}{6.022 \times 10^{23}\text{ W atoms}} \times \dfrac{183.85\text{ g W}}{1\text{ mol W}}}{\left(\dfrac{4 \times 139\text{ pm}}{1.732} \times \dfrac{1\text{ m}}{10^{12}\text{ pm}} \times \dfrac{100\text{ cm}}{1\text{ m}}\right)^3} = 18.5\text{ g/cm}^3$$

This compares well with a tabulated density of 19.25 g/cm^3.

70. One atom lies entirely within the hcp unit cell and there are eight corner atoms, each of which are shared among eight unit cells. Thus, there are a total of two atoms per unit cell. The volume of the unit cell is given by the product of its height times the area of its base. The base is a parallelogram, which can be subdivided along its shorter diagonal into two triangles. The height of either triangle, equivalent to the perpendicular distance between the parallel sides of the parallelogram, is given by 320. pm × sin(60°) = 277 pm.

Area of base $= 2 \times \text{area of triangle} = 2 \times (^1/_2 \times \text{base} \times \text{height})$

$= 277\text{ pm} \times 320.\text{ pm} = 8.86 \times 10^4\text{ pm}^2$

Volume of unit cell $= 8.86 \times 10^4\text{ pm}^2 \times 520.\text{ pm} = 4.61 \times 10^7\text{ pm}^3$

$$\text{Density} = \frac{\text{mass}}{\text{volume}} = \frac{2\text{ Mg atoms} \times \dfrac{1\text{ mol Mg}}{6.022 \times 10^{23}\text{ Mg atoms}} \times \dfrac{24.305\text{ g Mg}}{1\text{ mol Mg}}}{4.61 \times 10^7\text{ pm}^3 \times \left(\dfrac{1\text{ m}}{10^{12}\text{ pm}} \times \dfrac{100\text{ cm}}{1\text{ m}}\right)^3} = 1.75\text{ g/cm}^3$$

This is in reasonable agreement with the experimental value of 1.738 g/cm^3.

Ionic Crystal Structures

71. CaF_2 There are eight Ca^{2+} ions on the corners, each shared among eight unit cells, for a total of ($8 \times {}^1/_8$) one corner ion per unit cell. There are six Ca^{2+} ions on the faces, each shared between two unit cells, for a total of ($6 \times {}^1/_2$) three face ions per unit cell. This gives a total of four Ca^{2+} ions per unit cell. There are eight F^- ions, each wholly contained within the unit cell. The ratio of Ca^{2+} ions to F^- ions is 4 Ca^{2+} ions per 8 F^- ions: Ca_4F_8 or CaF_2.

TiO_2 There are eight Ti^{4+} ions on the corners, each shared among eight unit cells, for a total of ($8 \times {}^1/_8$) one Ti^{4+} corner ion per unit cell. There is one Ti^{4+} ion in the center, wholly contained within the unit cell. Thus, there are a total of two Ti^{4+} ions per unit cell. There are four O^{2-} ions on the faces of the unit cell, each shared between two unit cells, for a total of ($4 \times {}^1/_2$) two face atoms per unit cells. There are two O^{2-} ions totally contained within the unit cell. This gives a total of four O^{2-} ions per unit cell. The ratio of Ti^{4+} ions to O^{2-} ions is 2 Ti^{4+} ions per 4 O^{2-} ion: Ti_2O_4 or TiO_2.

72. The unit cell of CsCl is pictured in Figure 13-44. The length of the body diagonal equals $2\,r_+ + 2\,r_-$. From Figure 10-9, for Cl^-, $r_- = 181$ pm. Thus, the length of the body diagonal equals 2(169 pm + 181 pm) = 700. pm. From Figure 13-42, the length of this body diagonal is $\sqrt{3}\,l = 700.$ pm. The volume of the unit cell is $V = l^3$. $\sqrt{3} = 1.732$.

$$V = \left(\frac{700.\text{ pm}}{1.732} \times \frac{1\text{ m}}{10^{12}\text{ pm}} \times \frac{100\text{ cm}}{1\text{m}}\right)^3 = 6.60 \times 10^{-23}\text{ cm}^3$$

Per unit cell, there is one Cs^+ ion and one Cl^- ion (8 corner ions × $^1/_8$ corner ion per unit cell).

$$\text{density} = \frac{\text{mass}}{\text{volume}} = \frac{\left(1\text{ f.u. CsCl} \times \dfrac{1\text{ mol CsCl}}{6.022 \times 10^{23}\text{ CsCl f.u.'s}} \times \dfrac{168.4\text{ g CsCl}}{1\text{ mol CsCl}}\right)}{6.60 \times 10^{-23}\text{ cm}^3}$$

$= 4.24\text{ g/cm}^3$

73. **(a)** In a sodium chloride type of lattice, there are six cations around each anion and six anions around each cation. These oppositely charged ions are arranged as follows: one above, one below, one in front, one in back, one to the right, and one to the left. Thus the coordination number of Mg^{2+} is 6 and that of O^{2-} is 6 also.

(b) In the unit cell, there is an oxide ion at each of the eight corners; each of these is shared between eight unit cells. There also is an oxide ion at the center of each of the six faces; each of these oxide ions is shared between two unit cells. Thus, the total number of oxide ions is computed as follows.

$$\text{total number of oxide ions} = 8 \text{ corners} \times \frac{1 \text{ oxide ion}}{8 \text{ unit cells}} + 6 \text{ faces} \times \frac{1 \text{ oxide ion}}{2 \text{ unit cells}} = 4\ O^{2-} \text{ ions}$$

There is a magnesium ion on each of the twelve edges; each of these is shared between four unit cells. There also is a magnesium ion in the center which is not shared with another unit cell.

$$\text{total number of magnesium ions} = 12 \text{ edges} \times \frac{1 \text{ magnesium ion}}{4 \text{ unit cells}} + 1 \text{ central } Mg^{2+} \text{ ion} = 4\ Mg^{2+} \text{ ions}$$

Thus there are four formula units per unit cell of MgO.

(c) Along the edge of the unit cell, Mg^{2+} and O^{2-} ions are in contact. The length of the edge is equal to the radius of one O^{2-}, plus the diameter of Mg^{2+}, plus the radius of another O^{2-}.

edge length = $2 \times O^{2-}$ radius + $2 \times Mg^{2+}$ radius = 2×140 pm + 2×72 pm = 424 pm

The unit cell is a cube; its volume is the cube of its length.

$$\text{volume} = (424 \text{ pm})^3 = 7.62 \times 10^7 \text{ pm}\left(\frac{1 \text{ m}}{10^{12} \text{ pm}} \times \frac{100 \text{ cm}}{1 \text{ m}}\right)^3 = 7.62 \times 10^{-23} \text{ cm}^3$$

(d) $$\text{density} = \frac{\text{mass}}{\text{volume}} = \frac{4 \text{ MgO f.u}}{7.62 \times 10^{-23} \text{ cm}^3} \times \frac{1 \text{ mol MgO}}{6.022 \times 10^{23} \text{ f.u.}} \times \frac{40.30 \text{ g MgO}}{1 \text{ mol MgO}} = 3.51 \text{ g/cm}^3$$

74. From Figure 13-43 we conclude that there are four formula units of KCl in a unit cell and the length of the edge of that unit cell is twice the internuclear distance between K^+ and Cl^- ions. (The analysis is given in more detail in the answer to Exercise 73.) The unit cell is a cube; its volume is the cube of its length.

length = 2×314.54 pm = 629.08 pm

$$\text{volume} = (629.08 \text{ pm})^3 = 2.4895 \times 10^8 \text{ pm}\left(\frac{1 \text{ m}}{10^{12} \text{ pm}} \times \frac{100 \text{ cm}}{1 \text{ m}}\right)^3 = 2.4895 \times 10^{-22} \text{ cm}^3$$

unit cell mass = volume × density = 2.4895×10^{-22} cm^3 × 1.9893 g/cm^3 = 4.9524×10^{-22} g

= mass of 4 KCl formula units

$$N_A = \frac{4 \text{ KCl f.u.}}{4.9524 \times 10^{-22} \text{ g}} \times \frac{74.5510 \text{ g KCl}}{1 \text{ mol}} = 6.0214 \times 10^{23} \text{ f.u./mol}$$

Lattice Energy

75. Whether or not enthalpies of sublimation of the alkali metals are approximately the same, lattice energies of a series such as LiCl(s), NaCl(s), KCl(s), RbCl(s), and CsCl(s) will vary approximately with the size of the cation. A small cation will produce a more exothermic lattice energy. Thus, the lattice energy for LiCl(s) should be the most exothermic and CsCl(s) the least in this series.

76. The cycle of reactions is at right. Recall that Hess's law states that the enthalpy change is the same, whether a chemical change is produced by one reaction or several.

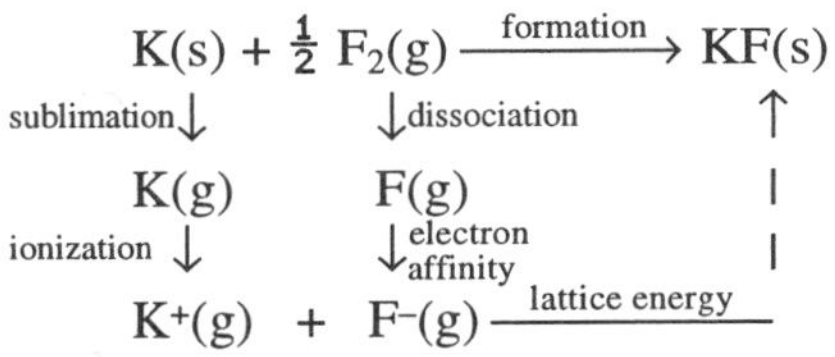

Formation reaction:	$K(s) + \frac{1}{2} F_2(g) \longrightarrow KF(s)$	$\Delta H_f^\circ = -567.3$ kJ/mol
Sublimation:	$K(s) \longrightarrow K(g)$	$\Delta H_{sub} = 89.24$ kJ/mol
Ionization:	$K(s) \longrightarrow K^+(g) + e^-$	$I_1 = 418.9$ kJ/mol
Dissociation:	$\frac{1}{2} F_2(g) \longrightarrow F(g)$	D.E. = (159/2) kJ/mol F
Electron Affinity:	$F(g) + e^- \longrightarrow F^-(g)$	E.A. = −328 kJ/mol

$\Delta H_f^\circ = \Delta H_{sub} + I_1$ + D.E. + E.A. + lattice energy (L.E.)

−567.3 kJ/mol = 89.24 kJ/mol + 418.9 kJ/mol + (159/2) kJ/mol − 328 kJ/mol + L.E.

L.E. = −827 kJ/mol

77. Second ionization energy: $Mg^+(g) \longrightarrow Mg^{2+}(g) + e^-$ $I_2 = 1451$ kJ/mol
Lattice energy: $Mg^{2+}(g) + 2\ Cl^-(g) \longrightarrow MgCl_2(s)$ L.E. = –2526 kJ/mol
Sublimation: $Mg(s) \longrightarrow Mg(g)$ $\Delta H_{sub} = 146$ kJ/mol
First ionization energy: $Mg(g) \longrightarrow Mg^+(g) + e^-$ $I_1 = 738$ kJ/mol
Dissociation energy: $Cl_2(g) \longrightarrow 2\ Cl(g)$ D.E. = (2 × 122) kJ/mol
Electron Affinity: $2\ Cl(g) + 2\ e^- \longrightarrow 2\ Cl^-(g)$ 2 × E.A. = 2 (–349) kJ/mol

$$\Delta H_f^\circ = \Delta H_{sub} + I_1 + I_2 + \text{D.E.} + (2 \times \text{E.A.}) + \text{L.E.}$$
$$= 146\text{ kJ/mol} + 738\text{ kJ/mol} + 1451\text{ kJ/mol} + 244\text{ kJ/mol} - 698\text{ kJ/mol} - 2526\text{ kJ/mol}$$
$$= -645\text{ kJ/mol}$$

In Example 13-11, the value of ΔH_f° for MgCl is calculated as –19 kJ/mol. Therefore, $MgCl_2$ is much more stable than MgCl, since considerably more energy is released when it forms. We expect $MgCl_2(s)$ to be more stable than MgCl(s) because a +2 cation more strongly attracts anions than does a +1 cation.

78. Formation reaction: $Na(s) + \frac{1}{2}\ H_2(g) \longrightarrow NaH(s)$ $\Delta H_f^\circ = -57$ kJ/mol
Heat of sublimation: $Na(s) \longrightarrow Na(g)$ $\Delta H_{sub} = +108$ kJ/mol
Ionization energy: $Na(g) \longrightarrow Na^+(g) + e^-$ $I_1 = +496$ kJ/mol
Dissociation energy: $\frac{1}{2}\ H_2(g) \longrightarrow H(g)$ D.E. = +218 kJ/mol H
Lattice energy: $Na^+(g) + H^-(g) \longrightarrow NaH(s)$ L.E. = –812 kJ/mol

$$\Delta H_f^\circ = \Delta H_{sub} + I_1 + \text{D.E.} + \text{E.A.} + \text{lattice energy (L.E.)}$$
$$-57\text{ kJ/mol} = 108\text{ kJ/mol} + 496\text{ kJ/mol} + \text{E.A.} + 218\text{ kJ/mol} - 812\text{ kJ/mol}$$
$$\text{E.A.} = -67\text{ kJ/mol}$$

FEATURE PROBLEMS

A. We obtain the surface tension by substituting the experimental values into the equation for surface tension.

$$h = \frac{2\gamma}{dgr} \qquad \gamma = \frac{hdgr}{2} = \frac{1.1\text{ cm} \times 0.789\text{ g cm}^{-3} \times 981\text{ cm s}^{-2} \times 0.050\text{ cm}}{2} = 21\text{ g/s}^2 = 0.021\text{ J/m}^2$$

B. As the can cools, the water vapor in the can condenses to liquid, and the pressure inside the can lowers to the vapor pressure of water at room temperature, around 20 to 30 mmHg. But the pressure on the outside of the can is, of course, atmospheric pressure, close to 760 mmHg. This huge difference in pressure, amounting to about 14 lb/in.^{2}, is sufficient to crush the can.

C. (1) A sketch of the phase diagram follows.

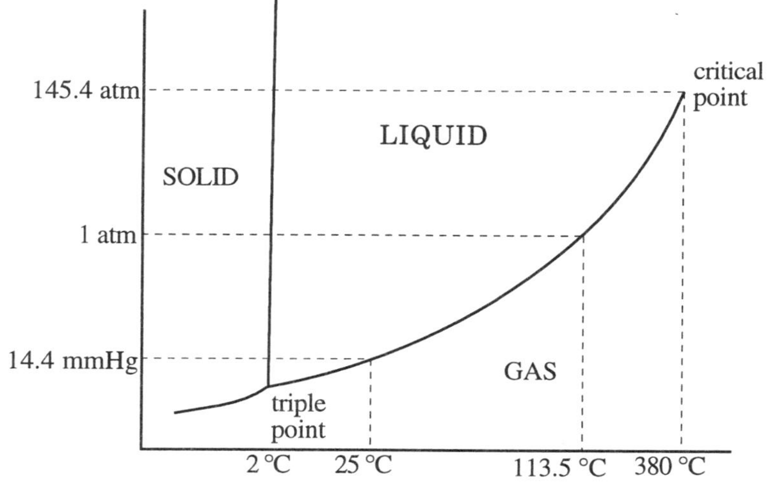

(2) We use the Clausius-Clapeyron equation to determine ΔH_{vap}

$$\ln \frac{760\text{ mmHg}}{14.4\text{ mmHg}} = 3.97 = -\frac{\Delta H_{vap}}{8.3145\text{ J mol}^{-1}\text{ K}^{-1}}\left(\frac{1}{(113.5 + 273.2)\text{ K}} - \frac{1}{298\text{ K}}\right)$$

$$3.97 = 9.26 \times 10^{-5}\, \Delta H_{vap} \qquad \Delta H_{vap} = \frac{3.97}{9.26 \times 10^{-5}} = 42.9 \text{ kJ/mol}$$

(3) $\Delta H_{subl} = \Delta H_{fus} + \Delta H_{vap} = 12.66 \text{ kJ/mol} + 42.9 \text{ kJ/mol} = 55.6 \text{ kJ/mol}$

(4) The sample is at a temperature of 0 °C. In these circumstances, hydrazine is a solid. Its pressure is somewhat less than 14.4 torr. First we determine the vapor pressure at the triple point, 2 °C = 275 K.

$$\ln \frac{P}{14.4 \text{ mmHg}} = -\frac{42.9 \times 10^3 \text{ J/mol}}{8.3145 \text{ J mol}^{-1} \text{ K}^{-1}} \left(\frac{1}{275 \text{ K}} - \frac{1}{298 \text{ K}}\right) = -1.45$$

$P = 14.4\, e^{-1.45} = 14.4 \times 0.235 = 3.38$ mmHg

Then we find the sublimation pressure at 273 K.

$$\ln \frac{P}{3.38 \text{ mmHg}} = -\frac{55.6 \times 10^3 \text{ J/mol}}{8.3145 \text{ J mol}^{-1} \text{ K}^{-1}} \left(\frac{1}{273 \text{ K}} - \frac{1}{275 \text{ K}}\right) = -0.178$$

$P = 3.38\, e^{-0.178} = 3.38 \times 0.837 = 2.83$ mmHg

D. **(1)** If we draw a dividing line from the "peak" of the S_β region toward the middle of its base (which meets the extensions of the rhombic sublimation line and the vaporization line), that line will represent the melting point line for S_α, which is rhombic sulfur.

(2) The liquid will freeze at 119 °C because it freezes into monoclinic sulfur, S_β. Monoclinic sulfur solidifies at a higher temperature than does rhombic sulfur.

E. We would expect to determine the melting point and boiling point by comparison with the corresponding values for other hydrodes of the group. Following are values taken from a handbook, with Z being the atomic number of the group 6A element.

Hydride	H_2O	H_2S	H_2Se	H_2Te
Z	8	16	34	52
m.p., °C	0.0	–85.5	–65.7	–49
b.p., °C	100.0	–60.3	–41.4	–2

If we assume that the H_2S value should be the average of the H_2O and H_2Se values, we obtain the following.

(a) melting point: $-85.5 = (-65.7 + H_2O)/2 \quad -85.5 \times 2 + 65.7 = -105$ °C = melting point of water.

(b) boiling point: $-60.3 = (-41.4 + H_2O)/2 \quad -60.3 \times 2 + 41.4 = -79$ °C = boiling point of water.

(c) The temperature of liquid water's maximum density should be just above the solid's melting point, at –105 °C.

(d) The solid should be more dense than the liquid; it should sink in the liquid.

The result of all of this is that there would be no liquid water at the temperatures that support the life we know. Thus the material of protoplasm would not exist and neither would life.

14 Solutions and Their Physical Properties

Practice Examples

1A To determine mass percent, we need both the mass of ethanol and the mass of solution. From volume percent, we know that 100.0 mL of solution contains 20.0 mL pure ethanol. The density of pure ethanol is 0.789 g/mL. We now can determine the mass of solute (ethanol) and solution. We perform the calculation in one set-up.

$$\text{mass percent ethanol} = \frac{20.0 \text{ mL ethanol} \times \dfrac{0.789 \text{ g}}{1 \text{ mL ethanol}}}{100.0 \text{ mL soln} \times \dfrac{0.977 \text{ g}}{1 \text{ mL soln}}} \times 100\% = 16.2\% \text{ ethanol by mass}$$

1B In each case, we use the definition of the concentration unit, making sure that numerator and denominator are converted to the correct units.

(a) We first determine the mass in grams of the solute and of the solution.

$$\text{mass } CH_3OH = 11.3 \text{ mL } CH_3OH \times \frac{0.793 \text{ g}}{1 \text{ mL}} = 8.96 \text{ g } CH_3OH$$

$$\text{mass soln} = 75.0 \text{ mL soln} \times \frac{0.980 \text{ g}}{1 \text{ mL}} = 73.5 \text{ g soln}$$

$$\text{amount of } H_2O = (73.5 \text{ g soln} - 8.96 \text{ g } CH_3OH) \times \frac{1 \text{ mol } H_2O}{18.02 \text{ g } H_2O} = 3.58 \text{ mol } H_2O \ (= 64.5 \text{ g } H_2O)$$

$$\text{amount of } CH_3OH = 8.96 \text{ g } CH_3OH \times \frac{1 \text{ mol } CH_3OH}{32.04 \text{ g } CH_3OH} = 0.280 \text{ mol } CH_3OH$$

$$H_2O \text{ mole fraction} = \frac{\text{amount of } CH_3OH \text{ in moles}}{\text{amount of soln in moles}} = \frac{3.58 \text{ mol } H_2O}{3.58 \text{ mol } H_2O + 0.280 \text{ mol } CH_3OH} = 0.927$$

(b) $$[CH_3OH] = \frac{\text{amount of } CH_3OH \text{ in moles}}{\text{volume of soln in L}} = \frac{0.280 \text{ mol } CH_3OH}{75.0 \text{ mL soln} \times \dfrac{1 \text{ L}}{1000 \text{ mL}}} = 3.73 \text{ M}$$

(c) $$\text{molality of } CH_3OH = \frac{\text{amount of } CH_3OH \text{ in moles}}{\text{mass of } H_2O \text{ in kg}} = \frac{0.280 \text{ mol } CH_3OH}{64.5 \text{ g } H_2O \times \dfrac{1 \text{ kg}}{1000 \text{ g}}} = 4.34 \, m$$

2A First we need the amount of each component in solution. We use a 100.00-g sample of solution, in which there are 16.00 g glycerol and 84.00 g water.

$$\text{amount glycerol} = 16.00 \text{ g } C_3H_5(OH)_3 \times \frac{1 \text{ mol } C_3H_5(OH)_3}{92.10 \text{ g } C_3H_5(OH)_3} = 0.1737 \text{ mol } C_3H_5(OH)_3$$

$$\text{amount water} = 84.00 \text{ g } H_2O \times \frac{1 \text{ mol } H_2O}{18.02 \text{ g } H_2O} = 4.661 \text{ mol } H_2O$$

$$\text{mole fraction of } C_3H_5(OH)_3 = \frac{\text{amount } C_3H_5(OH)_3}{\text{amount } [C_3H_5(OH)_3 \ \& \ H_2O]} = \frac{0.1737 \text{ mol } C_3H_5(OH)_3}{0.1737 \text{ mol } C_3H_5(OH)_3 + 4.661 \text{ mol } H_2O}$$

$= 0.03593$ OR 3.593 mole percent $C_3H_5(OH)_3$

2B First we need the amount of sucrose in solution. We use a 100.00-g sample of solution, in which there are 10.00 g sucrose and 90.00 g water.

$$\text{amount } C_{12}H_{22}O_{11} = 10.00 \text{ g } C_{12}H_{22}O_{11} \times \frac{1 \text{ mol } C_{12}H_{22}O_{11}}{342.30 \text{ g } C_{12}H_{22}O_{11}} = 0.02921 \text{ mol } C_{12}H_{22}O_{11}$$

(a) Molarity is amount of solute in moles per liter volume of solution. Convert the 100.00 g of solution to L with density as a conversion factor.

$$C_{12}H_{22}O_{11} \text{ molarity} = \frac{0.02921 \text{ mol } C_{12}H_{22}O_{11}}{100.0 \text{ g soln}} \times \frac{1.040 \text{ g soln}}{1 \text{ mL}} \times \frac{1000 \text{ mL}}{1 \text{ L}} = 0.3038 \text{ M}$$

(b) Molality is amount of solute in moles per kilogram mass of solvent. Convert 90.00 g of solute to kg.

$$C_{12}H_{22}O_{11}\ \text{molality} = \frac{0.02921\ \text{mol}\ C_{12}H_{22}O_{11}}{90.00\ \text{g}\ H_2O} \times \frac{1000\ \text{g}}{1\ \text{kg}} = 0.3246\ m$$

(c) Mole fraction is amount of solute per amount of solution. First compute the amount in 90.00 g H_2O.

$$\text{amount}\ H_2O = 90.00\ \text{g} \times \frac{1\ \text{mol}\ H_2O}{18.02\ \text{g}\ H_2O} = 4.994\ \text{mol}\ H_2O$$

$$\text{mole fraction}\ C_{12}H_{22}O_{11} = \frac{0.02921\ \text{mol}\ C_{12}H_{22}O_{11}}{0.02921\ \text{mol}\ C_{12}H_{22}O_{11} + 4.994\ \text{mol}\ H_2O} = 0.005815$$

OR 0.5815 mole %

3A Water is a polar, hydrogen bonding compound. It should mix well with other polar, hydrogen bonding compounds. (a) Toluene is nonpolar and should not be soluble in water. While (c) benzaldehyde is polar, it cannot form hydrogen bonds because none of its hydrogens are attached to N, O, or F. (c) Benzaldehyde should not be very soluble in water. (b) Oxalic acid is polar and can form hydrogen bonds. (b) Oxalic acid should be the most polar of these three compounds in water. Actual solubilities (w/w%) are: toluene (0.067%) < benzaldehyde (0.28%) < oxalic acid (14%).

3B Both I_2 and CCl_4 are nonpolar molecules. It does not take much energy to break the attractions among I_2 molecules, or among CCl_4 molecules. Further, there is not a strong I_2–CCl_4 attraction created when a solution forms. The result is that I_2 should dissolve well in CCl_4 by simple mixing. H_2O is extensively hydrogen bonded with strong intermolecular forces requiring significant energy to break, but there is not a strong I_2–H_2O attraction created when a solution forms. We expect I_2 to dissolve poorly in water. Actual solubilities are 2.603 g I_2/100 g CCl_4 soln and 0.033 g I_2/100 g H_2O soln.

4A The two suggestions are quoted first, followed by the means of achieving each one.

(1) Dissolve the 95 g NH_4Cl in just enough water to produce a saturated solution (55 g NH_4Cl/100 g H_2O) at 60 °C.

$$\text{mass of water needed} = 95\ \text{g}\ NH_4Cl \times \frac{100\ \text{g}\ H_2O}{55\ \text{g}\ NH_4Cl} = 173\ \text{g}\ H_2O$$

The mass of NH_4Cl in the saturated solution at 20 °C now is smaller.

$$\text{mass}\ NH_4Cl\ \text{dissolved} = 173\ \text{g}\ H_2O \times \frac{37\ \text{g}\ NH_4Cl}{100\ \text{g}\ H_2O} = 64\ \text{g}\ NH_4Cl\ \text{dissolved}$$

$$\text{crystallized mass}\ NH_4Cl = 95\ \text{g}\ NH_4Cl\ \text{total} - 64\ \text{g}\ NH_4Cl\ \text{dissolved at 20 °C} = 31\ \text{g}\ NH_4Cl\ \text{crystallized}$$

(2) Lower the final temperature to 0 °C, rather than 20 °C. From Figure 14-8, at 0 °C the solubility of NH_4Cl is 28.5 g NH_4Cl/100 g H_2O. From this (and knowing that there is 173 g H_2O present in the solution) we calculate the mass of NH_4Cl dissolved at this lower temperature.

$$\text{mass dissolved}\ NH_4Cl = 173\ \text{g}\ H_2O \times \frac{28.5\ \text{g}\ NH_4Cl}{100\ \text{g}\ H_2O} = 49.5\ \text{g}\ NH_4Cl\ \text{dissolved}$$

The mass of NH_4Cl recrystallized is 95 g – 49.5 g = 46 g   yield = (46/95) × 100% = 46%

4B Percent yield for the recrystallization can be defined as $\%\ \text{yield} = \dfrac{\text{mass crystallized}}{\text{mass dissolved (40 °C)}} \times 100\%$

$$\%\ \text{yield} = \frac{\text{mass dissolved (40 °C)} - \text{mass dissolved (20 °C)}}{\text{mass dissolved (40 °C)}} \times 100\%$$

Figure 14-8 solubilities per 100 g H_2O are followed by percent yield calculations.

$KClO_4$: 6.7 g at 40 °C, 3.5 g at 20 °C

$$KClO_4\ \%\ \text{yield} = \frac{6.7\ \text{g (40 °C)} - 3.5\ \text{g (20 °C)}}{6.7\ \text{g (40 °C)}} \times 100\% = 48\%\ KClO_4$$

KNO_3: 60.7 g at 40 °C, 32.3 g at 20 °C

$$KNO_3\ \%\ \text{yield} = \frac{60.7\ \text{g (40 °C)} - 32.3\ \text{g (20 °C)}}{60.7\ \text{g (40 °C)}} \times 100\% = 47\%\ KNO_3$$

K_2SO_4: 15.1 g at 40 °C, 11.9 g at 20 °C

$$K_2SO_4\ \%\ \text{yield} = \frac{15.1\ \text{g (40 °C)} - 11.9\ \text{g (20 °C)}}{15.1\ \text{g (40 °C)}} \times 100\% = 21\%\ K_2SO_4$$

Ranked in order from highest to lowest percent yield: $KClO_4$ (48%) > KNO_3 (47%) > K_2SO_4 (21%)

5A From Example 14-5, we know that the Henry's law constant for O_2 dissolved in water is $k = 2.18 \times 10^{-3}$ M atm^{-1}. First we solve Henry's law for the pressure of gas and then we substitute the values we have.

$$P_{gas} = \frac{C}{k} = \frac{8.34 \times 10^{-4}\ \text{M}}{2.18 \times 10^{-3}\ \text{M atm}^{-1}} = 0.383\ \text{atm}\ O_2\ \text{pressure}$$

5B The pressure required must be sufficient that 0.0100 mol CO will dissolve in each liter of solution, which we take as 1000 mL H_2O. We know that 1 atm pressure 35.4 mL CO dissolves in each liter of solution (0.0354 mL CO/mL solution. Thus, [CO] at 1 atm CO pressure is

$$[\text{CO}] = \frac{35.4 \text{ mL CO at STP} \times \dfrac{1 \text{ mol CO}}{22{,}414 \text{ mL CO at STP}}}{1 \text{ L soln}} = 0.00158 \text{ M}$$

Now, Henry's law: $k = \dfrac{c}{P_{gas}} = \dfrac{0.00158 \text{ M}}{1 \text{ atm}} = \dfrac{0.0100 \text{ M}}{P}$ $\quad P = \dfrac{0.0100 \text{ M} \times 1 \text{ atm}}{0.00158 \text{ M}} = 6.33 \text{ atm}$

6A Raoult's law enables us to determine the vapor pressure of each component.

$P_{hex} = \chi_{hex} P^{\circ}_{hex} = 0.750 \times 149.1 \text{ mmHg} = 112 \text{ mmHg}$

$P_{pen} = \chi_{pen} P^{\circ}_{pen} = 0.250 \times 508.5 \text{ mmHg} = 127 \text{ mmHg}$

We use Dalton's law to determine the total vapor pressure:

$P_{total} = P_{hex} + P_{pen} = 112 \text{ mmHg} + 127 \text{ mmHg} = 239 \text{ mmHg}$

6B Masses of solution components need to be converted to amounts in moles through the use of molar masses. Let us choose as our mass precisely 1.0000 mole of C_6H_6 = 78.11 g C_6H_6 and an equal mass of toluene.

$$\text{amount of toluene} = 78.11 \text{ g } C_7H_8 \times \frac{1 \text{ mol } C_7H_8}{92.14 \text{ g } C_7H_8} = 0.8477 \text{ mol } C_7H_8$$

$$\text{mole fraction toluene} = \chi_{tol} = \frac{0.8477 \text{ mol } C_7H_8}{0.8477 \text{ mol } C_7H_8 + 1.0000 \text{ mol } C_6H_6} = 0.4588$$

toluene vapor pressure = $\chi_{tol}P^{\circ}_{tol} = 0.4588 \times 28.4 \text{ mmHg} = 13.0 \text{ mmHg}$

benzene vapor pressure = $\chi_{benz}P^{\circ}_{benz} = (1.0000 - 0.4588) \times 95.1 \text{ mmHg} = 51.5 \text{ mmHg}$

total vapor pressure = 13.0 mmHg + 51.5 mmHg = 64.5 mmHg

7A The mole fraction composition of each component is that component's partial pressure divided by the total pressure. Again we note that the vapor is richer in the more volatile component.

$$y_{hexane} = \frac{P_{hexane}}{P_{total}} = \frac{112 \text{ mmHg hexane}}{239 \text{ mmHg total}} = 0.469 \qquad y_{pentane} = \frac{P_{pentane}}{P_{total}} = \frac{127 \text{ mmHg pentane}}{239 \text{ mmHg total}} = 0.531$$

7B The mole fraction composition of each component is that component's partial pressure divided by the total pressure. Again we note that the vapor is richer in the more volatile component.

$$y_t = \frac{P_{toluene}}{P_{total}} = \frac{13.0 \text{ mmHg toluene}}{64.5 \text{ mmHg total}} = 0.202 \qquad y_b = \frac{P_{benzene}}{P_{total}} = \frac{51.5 \text{ mmHg benzene}}{64.5 \text{ mmHg total}} = 0.798$$

8A We use the osmotic pressure equation, converting the mass of solute to amount in moles, the temperature to kelvins, and the solution volume to liters.

$$\pi = \frac{nRT}{V} = \frac{\left(1.50 \text{ g } C_{12}H_{22}O_{11} \times \dfrac{1 \text{ mol } C_{12}H_{22}O_{11}}{342.3 \text{ g } C_{12}H_{22}O_{11}}\right) \times 0.08206 \dfrac{\text{L atm}}{\text{mol K}} \times 298 \text{ K}}{125 \text{ mL} \times \dfrac{1 \text{ L}}{1000 \text{ mL}}} = 0.857 \text{ atm}$$

8B We use the osmotic pressure equation to determine the molarity of the solution.

$$c = \frac{\pi}{RT} = \frac{0.015 \text{ atm}}{0.08206 \text{ L atm mol}^{-1} \text{ K}^{-1} \times 298 \text{ K}} = 6.1 \times 10^{-4} \text{ M}$$

Then calculate the mass of urea.

$$\text{urea mass} = 225 \text{ mL soln} \times \frac{1 \text{ L}}{1000 \text{ mL}} \times \frac{6.1 \times 10^{-4} \text{ mol urea}}{1 \text{ L soln}} \times \frac{60.06 \text{ g CO(NH}_2)_2}{1 \text{ mol CO(NH}_2)_2} = 8.24 \times 10^{-3} \text{ g}$$

9A We could substitute directly into the equation for molar mass derived in Example 14-9, but let us rather think our way through each step of the process. First, we find the concentration of the solution, by rearranging $\pi = cRT$. We need to convert the osmotic pressure to atmospheres.

$$c = \frac{\pi}{RT} = \frac{8.73 \text{ mmHg} \times \dfrac{1 \text{ atm}}{760 \text{ mmHg}}}{0.08206 \dfrac{\text{L atm}}{\text{mol K}} \times 298 \text{ K}} = 4.70 \times 10^{-4} \text{ M}$$

Next, we determine the amount in moles of dissolved solute.

$$\text{solute amount} = 100.0 \text{ mL} \times \frac{1 \text{ L}}{1000 \text{ mL}} \times \frac{4.74 \times 10^{-4} \text{ mol solute}}{1 \text{ L solution}} = 4.74 \times 10^{-5} \text{ mol solute}$$

We use that amount—which weighs 4.04 g—to determine the molar mass. $\mathcal{M} = \dfrac{4.04\text{ g}}{4.74 \times 10^{-5}\text{ mol}} = 8.52 \times 10^4\text{ g/mol}$

9B We use the osmotic pressure equation along with the molarity of the solution.

$$\pi = cRT = \frac{2.12\text{ g} \times \dfrac{1\text{ mol}}{6.86 \times 10^4\text{ g}}}{75.00\text{ mL} \times \dfrac{1\text{ L}}{1000\text{ mL}}} \times \frac{0.08206\text{ L atm}}{\text{mol K}} \times (37.0 + 273.2)\text{ K} = 0.0105\text{ atm} = 7.97\text{ mmHg}$$

10A **(a)** The freezing point depression constant for water is $K_f = 1.86\ ^\circ\text{C}\ m^{-1}$.

$$\text{molality} = \frac{\Delta T_f}{-K_f} = \frac{-0.227\ ^\circ\text{C}}{-1.86\ ^\circ\text{C}\ m^{-1}} = 0.122\ m$$

(b) Use the definition of molality to determine the amount in moles of riboflavin dissolved. This amount weighs 0.833 g.

$$\text{riboflavin amount} = 18.1\text{ g solvent H}_2\text{O} \times \frac{1\text{ kg solvent}}{1000\text{ g}} \times \frac{0.122\text{ mol solute}}{1\text{ kg solvent}} = 2.21 \times 10^{-3}\text{ mol riboflavin}$$

$$\mathcal{M} = \frac{0.833\text{ g riboflavin}}{2.21 \times 10^{-3}\text{ mol}} = 377\text{ g/mol}$$

(c) Use the method of Chapter 3 to find riboflavin's empirical formula, starting with a 100.00-g sample.

$$54.25\text{ g C} \times \frac{1\text{ mol C}}{12.01\text{ gC}} \quad 5.36\text{ g H} \times \frac{1\text{ mol H}}{1.008\text{ g H}} \quad 25.51\text{ g O} \times \frac{1\text{ mol O}}{16.00\text{ g O}} \quad 14.89\text{ g N} \times \frac{1\text{ mol N}}{14.01\text{ g N}}$$

= 4.517 mol C (/1.063) = 5.32 mol H (/1.063) = 1.594 mol O (/1.063) = 1.063 mol N (/1.063)

= 4.249 mol C = 5.00 mol H = 1.500 mol O = 1.000 mol N

If we multiply each of these amounts by 4 (because 4.245 is almost equal to $4\frac{1}{4}$), the empirical formula is $C_{17}H_{20}O_6N_4$ with a molar mass of 376 g/mol. The molecular formula is $C_{17}H_{20}O_6N_4$.

10B The boiling point of pure water at 760.0 mmHg is 100.000 °C. For higher pressures, the boiling point occurs at a higher temperature; for lower pressures, a lower boiling point is observed. The boiling point elevation of this urea solution is calculated. $\Delta T_b = K_b \times m = 0.512\ ^\circ\text{C}\ m^{-1} \times 0.205\ m = 0.105\ ^\circ\text{C}$

We would expect this urea solution to boil at (100.000 + 0.105 =) 100.105 °C under 760.0 mmHg atmospheric pressure. Since it boils at a lower temperature, the atmospheric pressure must be lower than 760.0 mmHg.

11A We assume a van't Hoff factor of $i = 3.0$ and convert the temperature to kelvins, 298 K.

$$\pi = icRT = \frac{3.0\text{ mol ions}}{1\text{ mol MgCl}_2} \times \frac{0.0530\text{ mol MgCl}_2}{1\text{ L soln}} \times 0.08206\frac{\text{L atm}}{\text{mol K}} \times 298\text{ K} = 3.9\text{ atm}$$

11B We first determine the molality of the solution, and assume a van't Hoff factor of $i = 2.0$.

$$m = \frac{\Delta T_f}{-K_f \times i} = \frac{-0.100\ ^\circ\text{C}}{1.86\ ^\circ\text{C}\ m^{-1} \times 2.0} = 0.027\ m \approx 0.027\text{ M}$$

$$\text{volume of HCl(aq)} = 250.0\text{ mL final soln} \times \frac{0.0269\text{ mmol HCl}}{1\text{ mL soln}} \times \frac{1\text{ mL conc soln}}{12.0\text{ mmol HCl}} = 0.56\text{ mL conc soln}$$

SUMMARIZING EXAMPLE CALCULATIONS

1. Convert percents to masses in grams and then to amounts in moles. Divide all molar amounts by the smallest to obtain the simplest ratio of small whole numbers, and thus the empirical formula.

$$81.79\text{ g C} \times \frac{1\text{ mol C}}{12.011\text{ g C}} = 6.810\text{ mol C} \quad \div 0.7569 \longrightarrow 8.9976\text{ mol C}$$

$$6.10\text{ g H} \times \frac{1\text{ mol H}}{1.008\text{ g H}} = 6.05\text{ mol H} \quad \div 0.7569 \longrightarrow 7.99\text{ mol H} \qquad C_9H_8O\text{ empirical formula}$$

$$12.11\text{ g O} \times \frac{1\text{ mol O}}{15.999\text{ g O}} = 0.7569\text{ mol O} \quad \div 0.7569 \longrightarrow 1.000\text{ mol O}$$

2. Determine the molarity of the solution from the osmotic pressure equation.

$$c = \frac{\pi}{RT} = \frac{141\text{ mmHg} \times \dfrac{1\text{ atm}}{760\text{ mmHg}}}{\dfrac{0.08206\text{ L atm}}{\text{mol K}} \times 298\text{ K}} = 0.00759\text{ M} = \frac{0.00759\text{ mol cinnamaldehyde}}{1\text{ L soln}}$$

3. Determine the mass of cinnamaldehyde in 1 L soln.

solute mass = $1.00 \text{ L} \times \frac{1000 \text{ mL}}{1 \text{ L}} \times \frac{1.00 \text{ g}}{1 \text{ mL}} \times \frac{0.100 \text{ g solute}}{100.00 \text{ g soln}} = 1.00 \text{ g cinnamaldehyde}$

4. Determine the molar mass: $\mathcal{M} = \frac{\text{mass in grams}}{\text{amount in moles}} = \frac{1.00 \text{ g}}{0.00759 \text{ mol}} = 132 \text{ g/mol}$

5. The empirical mass of C_9H_8O is $(9 \times 12.0 \text{ g C}) + (8 \times 1.01 \text{ g H}) + 16.0 \text{ g O} = 132 \text{ g/mol}$. Since this is the same as the molar mass, the molecular formula is C_9H_8O.

REVIEW QUESTIONS

1. **(a)** χ_B is the symbol for the mole fraction of component B: the amount in moles of B divided by the total amount in moles in the solution.
(b) P_A° represents the vapor pressure of pure substance A.
(c) K_f is the freezing point depression constant, the distance that the freezing point of a solvent is lowered by having one mole of dissolved particles present in solution.
(d) i is the symbol for the van't Hoff factor, the measured value of a colligative property divided by the value calculated assuming that each mole of solute produces one mole of particles in solution.
(e) The activity of a solute is its effective concentration, the concentration it would have to have if solute particles did not interact with each other in solution.

2. **(a)** Henry's law states that the solubility of a gas in a solution is directly related to the pressure of that gas above the solution.
(b) Freezing point depression is the change in the freezing point of a liquid due to the presence of dissolved solute.
(c) Recrystallization describes a purification process. An impure solute is dissolved in a solvent. As solvent evaporates away, the less soluble pure solute crystallizes first. When this process is repeated with the crystals obtained they become purer still.
(d) A hydrated ion is an ion surrounded by and attracted to water molecules.
(e) Deliquescence is the absorbance by a solid of water vapor from the air to such an extent that a saturated solution is formed.

3. **(a)** Molarity is the amount of dissolved solute in moles per liter of solution, while molality is the amount of dissolved solute in moles per kilogram of solvent.
(b) In an ideal solution the attractions among particles of all types are approximately the same in strength. In a nonideal solution, there is a disparity in the strengths of these attractive forces.
(c) More solute can be dissolved in an unsaturated solution. The attempt to dissolve more solute in a supersaturated solution results in the precipitation of some solute.
(d) Fractional crystallization is the process of first producing a concentrated and warm solution and then chilling it to a point of lower solute solubility to precipitate out some of that solute. Repetition of this process produces quite pure solute. Fractional distillation is the repeated vaporization and condensation of a liquid mixture. This also produces quite pure components.
(e) Osmosis is the motion of solute through a membrane permeable only to it. This creates an imbalance of pressures, if the solvent and solution volumes are limited, known as osmotic pressure. Reverse osmisis describes obtaining pure solvent by pushing a solution through a semipermeable membrane. This, of course, leaves the solute behind.

4. $\% \text{ NaBr} = \frac{116 \text{ g NaBr}}{116 \text{ g NaBr} + 100 \text{ g H}_2\text{O}} \times 100\% = 53.7\% = 53.7 \text{ g NaBr/100 g solution}$

5. **(a)** $\% \text{ by volume} = \frac{12.8 \text{ mL CH}_3\text{CH}_2\text{CH}_2\text{OH}}{75.0 \text{ mL soln}} \times 100\% = 17.1\% \text{ CH}_3\text{CH}_2\text{CH}_2\text{OH by volume}$

(b) $\% \text{ by mass} = \frac{12.8 \text{ mL CH}_3\text{CH}_2\text{CH}_2\text{OH} \times \frac{0.803 \text{ g}}{1 \text{ mL}}}{75.0 \text{ mL} \times \frac{0.988 \text{ g}}{1 \text{ mL}}} \times 100\% = 13.9\% \text{ CH}_3\text{CH}_2\text{CH}_2\text{OH by mass}$

(c) $\% \text{ (mass/vol)} = \frac{12.8 \text{ mL CH}_3\text{CH}_2\text{CH}_2\text{OH} \times \frac{0.803 \text{ g}}{1 \text{ mL}}}{75.0 \text{ mL}} \times 100\% = 13.7 \frac{\text{mass}}{\text{vol}} \% \text{ CH}_3\text{CH}_2\text{CH}_2\text{OH}$

6. $\text{soln. volume} = 725 \text{ kg NaCl} \times \frac{1000 \text{ g NaCl}}{1 \text{ kg NaCl}} \times \frac{100.00 \text{ g soln}}{3.87 \text{ g NaCl}} \times \frac{75.0 \text{ mL soln}}{76.9 \text{ g soln}} \times \frac{1 \text{ L soln}}{1000 \text{ mL soln}}$

$$= 1.83 \times 10^4 \text{ L soln} = 18.3 \text{ kL soln}$$

7. $$\text{mass AgNO}_3 = 125.0 \text{ mL soln} \times \frac{1 \text{ L}}{1000 \text{ mL}} \times \frac{0.0321 \text{ mol AgNO}_3}{1 \text{ L soln}} \times \frac{169.9 \text{ g}}{1 \text{ mol AgNO}_3} \times \frac{100.00 \text{ g mixt.}}{99.81 \text{ g AgNO}_3}$$
$$= 0.683 \text{ g mixture}$$

8. $$\text{molality} = \frac{2.65 \text{ g C}_6\text{H}_4\text{Cl}_2 \times \frac{1 \text{ mol C}_6\text{H}_4\text{Cl}_2}{147.0 \text{ g}}}{50.0 \text{ mL} \times \frac{0.879 \text{ g}}{1 \text{ mL}} \times \frac{1 \text{ kg}}{1000 \text{ g}}} = 0.410\ m$$

9. $$\text{molarity} = \frac{6.00 \text{ g CH}_3\text{OH} \times \frac{1 \text{ mol CH}_3\text{OH}}{32.04 \text{ g CH}_3\text{OH}}}{100.00 \text{ g soln} \times \frac{1 \text{ mL}}{0.988 \text{ g}} \times \frac{1 \text{ L}}{1000 \text{ mL}}} = 1.85 \text{ M} = [\text{CH}_3\text{OH}]$$

10. total amount = 1.28 mol C_7H_{16} + 2.92 mol C_8H_{18} + 2.64 mol C_9H_{20} = 6.84 moles

(a) $\chi_7 = \frac{1.28 \text{ mol C}_7\text{H}_{16}}{6.84 \text{ moles total}} = 0.187$ **(b)** $\times\ 100\% = 18.7$ mol% C_7H_{16}

$\chi_8 = \frac{2.92 \text{ mol C}_8\text{H}_{18}}{6.84 \text{ moles total}} = 0.427$ $\times\ 100\% = 42.7$ mol% C_8H_{18}

$\chi_9 = \frac{2.64 \text{ mol C}_9\text{H}_{20}}{6.84 \text{ moles total}} = 0.386$ $\times\ 100\% = 38.6$ mol% C_9H_{20}

11. **(a)** $$[\text{C}_3\text{H}_8\text{O}_3] = \frac{62.0 \text{ g C}_3\text{H}_8\text{O}_3 \times \frac{1 \text{ mol C}_3\text{H}_8\text{O}_3}{92.09 \text{ g C}_3\text{H}_8\text{O}_3}}{100.0 \text{ g soln} \times \frac{1 \text{ mL soln}}{1.159 \text{ g soln}} \times \frac{1 \text{ L soln}}{1000 \text{ mL soln}}} = 7.80 \text{ M}$$

(b) $$[\text{H}_2\text{O}] = \frac{38.0 \text{ g H}_2\text{O} \times \frac{1 \text{ mol H}_2\text{O}}{18.02 \text{ g H}_2\text{O}}}{100.0 \text{ g soln} \times \frac{1 \text{ mL soln}}{1.159 \text{ g soln}} \times \frac{1 \text{ L soln}}{1000 \text{ mL soln}}} = 24.4 \text{ M}$$

(c) $$\text{H}_2\text{O molality} = \frac{38.0 \text{ g H}_2\text{O} \times \frac{1 \text{ mol H}_2\text{O}}{18.02 \text{ g}}}{62.0 \text{ g C}_3\text{H}_8\text{O}_3 \times \frac{1 \text{ kg}}{1000 \text{ g}}} = 34.0\ m$$

(d) $$\text{C}_3\text{H}_8\text{O}_3 \text{ mole fraction} = \frac{62.0 \text{ g C}_3\text{H}_8\text{O}_3 \times \frac{1 \text{ mol C}_3\text{H}_8\text{O}_3}{92.09 \text{ g C}_3\text{H}_8\text{O}_3}}{62.0 \text{ g C}_3\text{H}_8\text{O}_3 \times \frac{1 \text{ mol C}_3\text{H}_8\text{O}_3}{92.09 \text{ g C}_3\text{H}_8\text{O}_3} + 38.0 \text{ g H}_2\text{O} \times \frac{1 \text{ mol H}_2\text{O}}{18.02 \text{ g H}_2\text{O}}}$$
$$= \frac{0.674 \text{ mol C}_3\text{H}_8\text{O}_3}{0.674 \text{ mol C}_3\text{H}_8\text{O}_3 + 2.11 \text{ mol H}_2\text{O}} = 0.242$$

(e) $$\text{H}_2\text{O mole percent} = \frac{2.11 \text{ mol H}_2\text{O}}{0.674 \text{ mol C}_3\text{H}_8\text{O}_3 + 2.11 \text{ mol H}_2\text{O}} \times 100\% = 75.8 \text{ mol \% H}_2\text{O}$$

12. Solubilities in Figure 14-8 are given as grams of solute per 100 g of solvent.

$$\text{mass NH}_4\text{Cl} = 1.12 \text{ mol NH}_4\text{Cl} \times \frac{53.5 \text{ g NH}_4\text{Cl}}{1 \text{ mol NH}_4\text{Cl}} = 59.9 \text{ g NH}_4\text{Cl}$$

$$\text{NH}_4\text{Cl concentration} = \frac{59.9 \text{ g NH}_4\text{Cl}}{150.0 \text{ g H}_2\text{O}} \times 100 \text{ g H}_2\text{O} = 39.9 \text{ g NH}_4\text{Cl}$$

From Figure 14-8, the solubility of NH_4Cl in water at 30 °C is 42 g/100 g water. The solution described in the problem is unsaturated.

13. Answer **(d)** is correct. An ideal solution is most likely to form when the solute and the solvent are the same type of substance, and thus the forces between solvent and solute molecules are about the same strength as those among solute or among solvent molecules. In **(a)**, NaCl is an ionic solid, while H_2O is a polar liquid. In **(b)**, C_2H_5OH is a polar liquid, while C_6H_6(l) is a nonpolar one. In **(c)**, C_7H_{16} a nonpolar liquid, while H_2O is a highly polar one. However, in **(d)** both C_7H_{16} and C_8H_{18} are nonpolar liquids.

14. NH_2OH(s) should be the most water soluble. Both C_6H_6(l) and $C_{10}H_8$(s) are composed of nonpolar molecules, which are barely (if at all) soluble in water. Both NH_2OH(s) and $CaCO_3$(s) should be able to be attracted by water molecules. But $CaCO_3$(s) contains ions of high charge, difficult to dissolve because of the high lattice energy. (Recall the solubility rules of Chapter 5: most carbonates are insoluble in water.)

15. Butyl alcohol should be moderately soluble in both water and benzene. A solute that is moderately soluble in both solvents will have some properties that are common to each solvent. Both naphthalene and hexane are nonpolar molecules, like benzene, but have no properties in common with water molecules. Sodium chloride consists of charged ions, similar to the charges in the polar bonds of water. But butyl alcohol possesses both a nonpolar part (C_4H_9—) like benzene, and a polar bond (—O—H) like water. In fact, water and phenol can mutually hydrogen bond. Naphthalene and hexane both are nonpolar; they should be soluble in benzene but not in water. NaCl(s) is an ionic solid that should be soluble in water but not in benzene.

16. We first determine the number of moles of O_2 that have dissolved.

$$\text{amount } O_2 = \frac{PV}{RT} = \frac{1.00 \text{ atm} \times 0.02831 \text{ L}}{0.08206 \text{ L atm mol}^{-1} \text{ K}^{-1} \times 298 \text{ K}} = 1.16 \times 10^{-3} \text{ mol } O_2$$

$$[O_2] = \frac{1.16 \times 10^{-3} \text{ mol } O_2}{1.00 \text{ L soln}} = 1.16 \times 10^{-3} \text{ M}$$

The oxygen concentration now is computed at the higher pressure.

$$[O_2] = \frac{1.16 \times 10^{-3} \text{ M}}{1 \text{ atm } O_2} \times 3.86 \text{ atm } O_2 = 4.48 \times 10^{-3} \text{ M}$$

17. First determine the number of moles of each component, then its mole fraction in the solution, then the partial pressure due to that component above the solution, and finally the total pressure.

$$\text{amount benzene} = n_b = 35.8 \text{ g } C_6H_6 \times \frac{1 \text{ mol } C_6H_6}{78.11 \text{ g } C_6H_6} = 0.458 \text{ mol } C_6H_6$$

$$\text{amount toluene} = n_t = 56.7 \text{ g } C_7H_8 \times \frac{1 \text{ mol } C_7H_8}{92.14 \text{ g } C_7H_8} = 0.615 \text{ mol } C_7H_8$$

$$\chi_b = \frac{0.458 \text{ mol } C_6H_6}{(0.458 + 0.615) \text{ total moles}} = 0.427 \qquad \chi_t = \frac{0.615 \text{ mol } C_7H_8}{(0.458 + 0.615) \text{ total moles}} = 0.573$$

$$P_b = 0.427 \times 95.1 \text{ mmHg} = 40.6 \text{ mmHg} \qquad P_t = 0.573 \times 28.4 \text{ mmHg} = 16.3 \text{ mmHg}$$

total pressure = 40.6 mmHg + 16.3 mmHg = 56.9 mmHg

18. $$\text{vapor fraction of benzene} = f_b = \frac{40.6 \text{ mmHg}}{56.9 \text{ mmHg}} = 0.714$$

$$\text{vapor fraction of toluene} = f_t = \frac{16.3 \text{ mmHg}}{56.9 \text{ mmHg}} = 0.286$$

These are both pressure fractions, as calculated, and also are the mole fractions in the vapor phase.

19. The freezing point depression is proportional to the product of the solute's molality and the van't Hoff factor. For nonelectrolytes, such as C_2H_5OH (0.050 m), $i = 1$, and thus $i \cdot m = 0.050\ m$. For 1:1 electrolytes, such as $MgSO_4$ and NaCl, $i = 2$, and thus $i \cdot m = 0.020\ m$ for (0.010 m) $MgSO_4$ and $im = 0.022\ m$ for (0.011 m) NaCl. And for 1:2 electrolytes, such as MgI_2 (0.010 m), $i = 3$, and thus $i \cdot m = 0.030\ m$. The im product is largest for C_2H_5OH; thus its freezing point depression is greatest and its freezing point is lowest.

20. The solution with the largest freezing-point depression will have the lowest freezing point. Since C_2H_5OH is a nonelectrolyte, it will produce 1 mol of particles per mole of solute, and give the smallest freezing point depression, the highest freezing point. $HC_2H_3O_2$ is a weak electrolyte and produces slightly more than 1 mole of particles per mole of solute; its freezing point is next lower. NaCl, $MgBr_2$, and $Al_2(SO_4)_3$ all are strong electrolytes, producing 2, 3, and 5 moles of particles per mole of solute. They are listed in order of decreasing value of the freezing points of their solutions. Thus, the solutes, listed in order of decreasing freezing points of their 0.01 m aqueous solutions, are: $C_2H_5OH > HC_2H_3O_2 > NaCl > MgBr_2 > Al_2(SO_4)_3$

21. **(a)** $$[Na^+] = \frac{0.92 \text{ g NaCl}}{100 \text{ mL soln}} \times \frac{1 \text{ mol NaCl}}{58.44 \text{ g NaCl}} \times \frac{1 \text{ mol } Na^+}{1 \text{ mol NaCl}} \times \frac{1000 \text{ mL}}{1 \text{ L soln}} = 0.16 \text{ M } Na^+$$

(b) $[\text{ions}] = [Na^+] + [Cl^-] = 2 \times [Na^+] = 2 \times 0.16 \text{ M} = 0.32 \text{ M}$

(c) $$\pi = CRT = \frac{0.32 \text{ mol ions}}{1 \text{ L}} \times \frac{0.08206 \text{ L atm}}{\text{mol K}} \times 310 \text{ K} = 8.1 \text{ atm}$$

(d) To determine freezing point depression, we first need molality, which means we need mass of solvent.

$$\text{molality} = \frac{0.92 \text{ g NaCl} \times \frac{1 \text{ mol NaCl}}{58.44 \text{ g NaCl}} \times \frac{2 \text{ mol ions}}{1 \text{ mol NaCl}}}{\left(\left(100 \text{ mL soln} \times \frac{1.005 \text{ g}}{1 \text{ mL soln}}\right) - 0.92 \text{ g NaCl}\right) \times \frac{1 \text{ kg}}{1000\text{g}}} = 0.32\ m$$

$$\Delta T_f = -K_f \times m = -1.86\ °C/m \times 0.32\ m = -0.60\ °C \qquad T_f = -0.60\ °C$$

22. We compute the concentration of the solution. Then, assuming that the solution volume is the same as that of the solvent (0.2500 L), we determine the amount of solute dissolved, and finally the molar mass.

$$c = \frac{\pi}{RT} = \frac{1.67 \text{ mmHg} \times \dfrac{1 \text{ atm}}{760 \text{ mmHg}}}{0.0821 \text{ L atm mol}^{-1} \text{ K}^{-1} \times 298 \text{ K}} = 8.98 \times 10^{-5} \text{ M}$$

$$\text{solute amount} = 0.2500 \text{ L} \times \frac{8.98 \times 10^{-5} \text{ mol}}{1 \text{ L}} = 2.25 \times 10^{-5} \text{ mol} \qquad \mathcal{M} = \frac{0.61 \text{ g}}{2.25 \times 10^{-5} \text{ mol}} = 2.7 \times 10^{4} \text{ g/mol}$$

23. First compute the molality of the benzene solution, then the number of moles of solute dissolved, and finally the mole weight of the unknown compound.

$$m = \frac{\Delta T_f}{-K_f} = \frac{4.92 \text{ °C} - 5.53 \text{ °C}}{-5.12 \text{ °C}/m} = 0.12\, m$$

$$\text{amount solute} = 0.07522 \text{ kg benzene} \times \frac{0.12 \text{ mol solute}}{1 \text{ kg benzene}} = 9.0 \times 10^{-3} \text{ mol solute}$$

$$\mathcal{M} = \frac{1.10 \text{ g unknown compound}}{9.0 \times 10^{-3} \text{ mol}} = 1.2 \times 10^{2} \text{ g/mol}$$

24. We compute the value of the van't Hoff factor, then use this value to compute the boiling point elevation.

$$i = \frac{\Delta T_f}{-K_f\, m} = \frac{-0.072 \text{ °C}}{-1.86 \text{ °C}/m \times 0.010\, m} = 3.9 \qquad \Delta T_b = i\, K_b\, m = 3.9 \times 0.512 \text{ °C } m^{-1} \times 0.010\, m = 0.020 \text{ °C}$$

The solution begins to boil at 100.02 °C.

EXERCISES

Homogeneous and Heterogeneous Mixtures

25. (b) salicyl alcohol probably is moderately soluble in both benzene and water. The reason for this selection is that salicyl alcohol contains a benzene ring, which would make it soluble in benzene, and also can use its —OH groups to hydrogen bond to water molecules. On the other hand, (c) diphenyl contains only nonpolar benzene rings; it should be soluble in benzene but not in water. (a) *para*-dichlorobenzene contains a benzene ring, making it soluble in benzene, and two polar C—Cl bonds, which oppose each other, producing a nonpolar—and thus water-insoluble—molecule. (d) hydroxyacetic acid is a very polar molecule with many opportunities for hydrogen bonding. Its polar nature would make it insoluble in benzene, while the prospective hydrogen bonds will enhance aqueous solubility.

26. Vitamin C is a water-soluble vitamin. Its molecules contain a number of polar —OH groups, capable of hydrogen bonding with water, making these molecules soluble in water. Vitamin E is fat-soluble. Its molecules are composed of largely non-polar hydrocarbon chains, with few polar groups and thus should be soluble in nonpolar solvents.

27. **(c)** formic acid and **(f)** propylene glycol are soluble in water. They both can form hydrogen bonds with water, and they both have small nonpolar portions.
(b) benzoic acid and **(d)** butyl alcohol are only slightly soluble in water. Although they both can form hydrogen bonds with water, both molecules contain reasonably large nonpolar portions, which will not dissolve well in water.
(a) iodoform and **(e)** chlorobenzene are insoluble in water. Although both molecules are polar, this is not sufficient to enable them to successfully disrupt the hydrogen bonds in water and be dissolved by water.

28. Benzoic acid will react with NaOH to form a solution of sodium ions and benzoate ions. (In this equation, ϕ represents the benzene ring with one hydrogen omitted, C_6H_5—.)
$Na^+(aq) + OH^-(aq) + \phi\text{—COOH} \longrightarrow Na^+(aq) + \phi\text{—COO}^-(aq) + H_2O$
The result of this reaction is that the concentration of molecular benzoic acid decreases, and thus more of the acid can be dissolved before the solution is saturated.

29. We expect small, highly charged ions to form crystals with large lattice energies—a factor that decreases solubility. Based on this information, we would expect MgF_2 to be insoluble and KF to be soluble. It is probable that CaF_2 is insoluble due to a high lattice energy, but that NaF, with a smaller lattice energy, is soluble. KF is probably the most water soluble. Actual solubilities at 25 °C are: 0.00020 M CaF_2 < 0.0021 M MgF_2 < 0.95 M NaF < 16 M KF.

30. The nitrate ion is reasonably large and has a small negative charge. It thus should produce crystals with small lattice energies. In contrast, S^{2-} is highly charged and easily polarized. It should form crystals with relatively high lattice energies. In fact, partial covalent bonding occurs in many metal sulfides, as predicted from the polarizability of the anion. And these partially covalent solids are difficult to break apart. We would expect the sulfide ion to be most polarized by small, highly charged cations; these metal sulfides should be the least soluble. The most soluble sulfides thus should be those of large cations of low charge: K_2S, Rb_2S, Cs_2S, for example.

Percent Concentration

31. For water, the mass in grams and the volume in mL are about equal; the density of water is close to 1.0. For ethanol, on the other hand, the density is about 0.8. Thus, if the solution has about the same volume as its components, the percent by volume of ethanol will have to be larger than its percent by mass. This would not necessarily be true of other ethanol solutions. It would only be true in those cases where the density of the other component is greater than the density of ethanol.

32. Volume percent is not usually independent of temperature. As temperature increases, the volume of a liquid also usually increases. Unless the volume of the solution and the volume of the solute increase in the same proportion, the volume precent will be altered by these temperature increases. Mass, on the other hand, is unaffected by temperature, and thus mass percent is independent of temperature.

33. $$\text{mass } HC_2H_3O_2 = 355 \text{ mL vinegar} \times \frac{1.01 \text{ g vinegar}}{1 \text{ mL}} \times \frac{6.02 \text{ g } HC_2H_3O_2}{100.00 \text{ g vinegar}} = 21.6 \text{ g } HC_2H_3O_2$$

34. We start with the molarity and convert both solute amount (numerator) and solution volume (denominator) to mass in order to obtain the mass percent.

$$\text{mass \% } H_2SO_4 = \frac{6.00 \text{ mol } H_2SO_4 \times \dfrac{98.08 \text{ g } H_2SO_4}{1 \text{ mol } H_2SO_4}}{1 \text{ L soln} \times \dfrac{1000 \text{ mL}}{1 \text{ L}} \times \dfrac{1.338 \text{ g}}{1 \text{ mL}}} \times 100\% = 43.98\% \; H_2SO_4$$

Molarity

35. The solution of Example 14-1 is 1.71 M C_2H_5OH, or 1.71 mmol C_2H_5OH in each mL of solution.

$$\text{volume conc. soln} = 825 \text{ mL} \times \frac{0.285 \text{ mmol } C_2H_5OH}{1 \text{ mL soln}} \times \frac{1 \text{ mL conc. soln}}{1.71 \text{ mmol } C_2H_5OH} = 138 \text{ mL conc. soln}$$

36. $$HNO_3 \text{ molarity} = \frac{30.00 \text{ g } HNO_3 \times \dfrac{1 \text{ mol } HNO_3}{63.01 \text{ g } HNO_3}}{100.0 \text{ g soln} \times \dfrac{1 \text{ mL}}{1.18 \text{ g}} \times \dfrac{1 \text{ L}}{1000 \text{ mL}}} = 5.62 \text{ M at } 20\ °C$$

Molality

37. The mass of solvent in kg multiplied by the molality gives the amount in moles of the solute.

$$\text{mass } I_2 = \left(725.0 \text{ mL } CS_2 \times \frac{1.261 \text{ g}}{1 \text{ mL}} \times \frac{1 \text{ kg}}{1000 \text{ g}}\right) \times \frac{0.236 \text{ mol } I_2}{1 \text{ kg } CS_2} \times \frac{253.8 \text{ g } I_2}{1 \text{ mol } I_2} = 54.8 \text{ g } I_2$$

38. In the original solution the mass of CH_3OH associated with each 1.00 kg = 1000 g of water is

$$\text{mass } CH_3OH = 1.00 \text{ kg } H_2O \times \frac{1.38 \text{ mol } CH_3OH}{1 \text{ kg } H_2O} \times \frac{32.04 \text{ g } CH_3OH}{1 \text{ mol } CH_3OH} = 44.2 \text{ g } CH_3OH$$

Then we compute the mass of methanol in 1.0000 kg = 1000.0 g of the original solution.

$$\text{mass } CH_3OH = 1000.0 \text{ g soln} \times \frac{44.2 \text{ g } CH_3OH}{1044.2 \text{ g soln}} = 42.3 \text{ g } CH_3OH$$

Thus, 1.0000 kg of this original solution contains (1000.0 – 42.3) 957.7 g H_2O

Now, a 1.00 *m* CH_3OH solution contains 32.04 g CH_3OH (one mole) for every 1000.0 g H_2O. We can compute the mass of H_2O associated with 34.7 g CH_3OH in such a solution.

$$\text{mass } H_2O = 42.3 \text{ g } CH_3OH \times \frac{1000.0 \text{ g } H_2O}{32.04 \text{ g } CH_3OH} = 1320 \text{ g } H_2O$$

(We have temporarily retained an extra significant figure in the calculation.) Since we already have 957.7 g H_2O in the original solution, the mass of water that must be added is

mass of H_2O to be added = 1320 g – 957.7 g = 362 g H_2O = 3.6×10^2 g H_2O

39. $$H_3PO_4 \text{ molarity} = \frac{34.0 \text{ g } H_3PO_4 \times \frac{1 \text{ mol } H_3PO_4}{98.00 \text{ g } H_3PO_4}}{100.0 \text{ g soln} \times \frac{1 \text{ mL}}{1.209 \text{ g}} \times \frac{1 \text{ L}}{1000 \text{ mL}}} = 4.19 \text{ M}$$

$$\text{molality} = \frac{34.0 \text{ g } H_3PO_4 \times \frac{1 \text{ mol } HNO_3}{98.00 \text{ g } H_3PO_4}}{66.0 \text{ g solvent} \times \frac{1 \text{ kg}}{1000 \text{ g}}} = 5.25\ m$$

40. $$C_2H_5OH \text{ molality} = \frac{10.00 \text{ g } C_2H_5OH \times \frac{1 \text{ mol } C_2H_5OH}{46.069 \text{ g } C_2H_5OH}}{90.00 \text{ g } H_2O \times \frac{1 \text{ kg}}{1000 \text{ g}}} = 2.412\ m$$

The molality of this solution does not vary with temperature. We would expect the solution's molarity to vary with temperature, however, because the solution's density, and thus the volume of solution that contains one mole, varies with temperature. But, unlike the volume, the mass of the solution does not vary with temperature, and so the solution's molality does not vary either.

Mole Fraction, Mole Percent

41. **(a)** $$\text{amount } C_2H_5OH = 21.7 \text{ g } C_2H_5OH \times \frac{1 \text{ mol } C_2H_5OH}{46.07 \text{ g } C_2H_5OH} = 0.471 \text{ mol } C_2H_5OH$$

$$\text{amount } H_2O = 78.3 \text{ g } H_2O \times \frac{1 \text{ mol } H_2O}{18.02 \text{ g } H_2O} = 4.35 \text{ mol } H_2O$$

$$\chi_{\text{water}} = \frac{4.34 \text{ mol } H_2O}{(0.471 + 4.34) \text{ total moles}} = 0.902 \qquad \chi_{\text{ethanol}} = \frac{0.471 \text{ mol ethanol}}{4.81 \text{ total moles}} = 0.0979$$

(b) $$\text{amount } H_2O = 1000. \text{ g} \times \frac{1 \text{ mol } H_2O}{18.02 \text{ g } H_2O} = 55.49 \text{ mol } H_2O$$

$$\chi_{\text{water}} = \frac{55.49 \text{ mol } H_2O}{(55.49 + 0.684) \text{ total moles}} = 0.9879 \qquad \chi_{\text{urea}} = \frac{0.684 \text{ mol urea}}{56.17 \text{ total moles}} = 0.0122$$

42. **(a)** The amount of solvent is found after the solute's mass is subtracted from that of the solution.

$$\text{solvent amount} = \left(\left(1 \text{ L soln} \times \frac{1000 \text{ mL}}{1 \text{ L}} \times \frac{1.006 \text{ g}}{1 \text{ mL}}\right) - \left(0.112 \text{ mol } C_6H_{12}O_6 \times \frac{180.2 \text{ g } C_6H_{12}O_6}{1 \text{ mol } C_6H_{12}O_6}\right)\right)$$

$$= [1006 \text{ g soluntion} - 20.2 \text{ g } C_6H_{12}O_6] \times \frac{1 \text{ mol } H_2O}{18.02 \text{ g } H_2O} = 54.7 \text{ mol } H_2O$$

$$\chi_{\text{solute}} = \frac{0.112 \text{ mol } C_6H_{12}O_6}{0.112 \text{ mol } C_6H_{12}O_6 + 54.7 \text{ mol } H_2O} = 0.00204 \quad \text{OR} \quad 0.204 \text{ mol \% } C_6H_{12}O_6$$

(b) First determine the mass and the amount of ethanol. The amount of solvent is found after the ethanol's mass is subtracted from the solution's mass. Use a 100.00-mL sample of solution for computation.

$$3.20 \text{ mL } C_2H_5OH \times \frac{0.789 \text{ g}}{1 \text{ mL } C_2H_5OH} = 2.52 \text{ g } C_2H_5OH \times \frac{1 \text{ mol } C_2H_5OH}{46.07 \text{ g } C_2H_5OH} = 0.0547 \text{ mol } C_2H_5OH$$

$$\text{amount } H_2O = \left(\left(100.0 \text{ mL soln} \times \frac{0.993 \text{ g}}{1 \text{ mL}}\right) - 2.52 \text{ g } C_2H_5OH\right) \times \frac{1 \text{ mol } H_2O}{18.02 \text{ g } H_2O} = 5.37 \text{ mol } H_2O$$

$$\chi_{\text{solute}} = \frac{0.0547 \text{ mol } C_2H_5OH}{0.0547 \text{ mol } C_2H_5OH + 5.37 \text{ mol } H_2O} = 0.0101 \quad \text{OR} \quad 1.01 \text{ mol \% } C_2H_5OH$$

43. Solve the following relationship for n_{ethanol}, the number of moles of ethanol, C_2H_5OH. The number of moles of water is given in the Example is 5.01 moles, that of ethanol is 0.171 moles.

$$\chi_{\text{ethanol}} = \frac{n_{\text{ethanol}}}{n_{\text{ethanol}} + 5.01 \text{ mol } H_2O} = 0.0525 \qquad n_{\text{ethanol}} = 0.0525\ n_{\text{ethanol}} + 0.263$$

$$n_{\text{ethanol}} = \frac{0.263}{1.0000 - 0.0525} = 0.278 \text{ mol ethanol}$$

$$\text{mass added ethanol} = (0.278 - 0.171) \text{ mol } C_2H_5OH \text{ added} \times \frac{46.07 \text{ g } C_2H_5OH}{1 \text{ mol } C_2H_5OH} = 4.93 \text{ g } C_2H_5OH$$

44. The amount of water present in 1 kg is $n_{water} = 1000 \text{ g } H_2O \times \frac{1 \text{ mol } H_2O}{18.02 \text{ g } H_2O} = 55.49 \text{ mol } H_2O$

Now solve the following expression for n_{gly}, the amount of glycerol. 4.85% = 0.0485.

$\chi_{gly} = 0.0485 = \frac{n_{gly}}{n_{gly} + 55.49}$ $\quad n_{gly} = 0.0485\, n_{gly} + 2.69$

$n_{gly} = \frac{2.69}{(1.0000 - 0.0485)} = 2.83 \text{ mol glycerol}$

$\text{volume glycerol} = 2.83 \text{ mol } C_3H_8O_3 \times \frac{92.09 \text{ g } C_3H_8O_3}{1 \text{ mol } C_3H_8O_3} \times \frac{1 \text{ mL}}{1.26 \text{ g}} = 207 \text{ mL glycerol}$

Solubility Equilibrium

45. At 40 °C the solubility of NH_4Cl is 46.3 g per 100 g of H_2O. To determine molality, we calculate amount in moles of the solute and the solvent mass in kg.

$$\text{molality} = \frac{46.3 \text{ g} \times \frac{1 \text{ mol } NH_4Cl}{53.49 \text{ g } NH_4Cl}}{100 \text{ g } H_2O \times \frac{1 \text{ kg}}{1000 \text{ g}}} = 8.66\, m$$

46. The data in Figure 14-8 are given in mass of solute per 100 g H_2O. 0.200 m $KClO_4$ contains 0.200 mol $KClO_4$ associated with each 1 kg = 1000 g of H_2O. The mass of $KClO_4$ dissolved in 100 g H_2O is

$\text{mass } KClO_4 = 100 \text{ g } H_2O \times \frac{0.200 \text{ mol } KClO_4}{1000 \text{ g } H_2O} \times \frac{138.5 \text{ g } KClO_4}{1 \text{ mol } KClO_4} = 2.77 \text{ g } KClO_4$

This concentration is realized at a temperature of about 32 °C. At this temperature a saturated solution of $KClO_4$ is 0.200 m.

47. **(a)** The concentration of this mixture is calculated first.

$\frac{\text{mass solute}}{100 \text{ g } H_2O} = 100 \text{ g } H_2O \times \frac{20.0 \text{ g } KClO_4}{500.0 \text{ g water}} = 4.00 \text{ g } KClO_4$

At 40 °C a saturated $KClO_4$ solution has a concentration of about 4.6 g $KClO_4$ dissolved in 100 g water. Thus, the solution is unsaturated.

(b) The mass of $KClO_4$ that must be added is the difference between the mass now present in the mixture and the mass that is dissolved in 500 g H_2O to produce a saturated solution.

$\text{mass to be added} = \left(500.0 \text{ g } H_2O \times \frac{4.6 \text{ g } KClO_4}{100 \text{ g } H_2O}\right) - 20.0 \text{ g } KClO_4 = 3.0 \text{ g } KClO_4$

48. From Figure 14-8, the solubility of KNO_3 in water is 38 g KNO_3 per 100 g water at 25.0 °C, while at 0.0 °C its solubility is 15 g KNO_3 per 100 g water. At 25.0 °C, every 138 g solution contains 38 g KNO_3 and 100.0 g water; and at 0.0 °C every 115 g solution contains 15 g KNO_3 and 100.0 g water. We calculate the mass of water and of KNO_3 present at 25.0 °C.

$\text{mass water} = 335 \text{ g soln} \times \frac{100.0 \text{ g water}}{138 \text{ g soln}} = 243 \text{ g water}$

$\text{mass } KNO_3 = 335 \text{ g soln} - 243 \text{ g water} = 92 \text{ g } KNO_3$

At 0.0 °C the mass of water has been reduced 55 g by evaporation to (243 g – 55 g =) 188 g. The mass of KNO_3 soluble in this mass of water is

$\text{mass } KNO_3 \text{ dissolved} = 188 \text{ g } H_2O \times \frac{15 \text{ g } KNO_3}{100.0 \text{ g } H_2O} = 28 \text{ g } KNO_3$

Thus, of the 92 g KNO_3 originally present in solution, the mass that recrystallizes is

$\text{mass } KNO_3 \text{ recrystallized} = 92 \text{ g } KNO_3 \text{ originally dissolved} - 28 \text{ g } KNO_3 \text{ still dissolved} = 64 \text{ g } KNO_3$

Solubility of Gases

49. $\text{mass } CH_4 = 1.00 \times 10^3 \text{ kg } H_2O \times \frac{0.02 \text{ g } CH_4}{1 \text{ kg } H_2O\cdot\text{atm}} \times 20 \text{ atm} = 4 \times 10^2 \text{ g } CH_4$ (natural gas)

50. We first compute the amount of O_2 dissolved in 515 mL = 0.515 L at each temperature.

$\text{amount } O_2 = 0.515 \text{ L} \times \frac{2.18 \times 10^{-3} \text{ mol } O_2}{1 \text{ L}} = 1.12 \times 10^{-3} \text{ mol } O_2 \text{ at } 0\text{ °C}$

$\text{amount } O_2 = 0.515 \text{ L} \times \frac{1.26 \times 10^{-3} \text{ mol } O_2}{1 \text{ L}} = 6.49 \times 10^{-4} \text{ mol } O_2 \text{ at } 25\text{ °C}$

Determine the difference between these amounts, and the corresponding volume.

$$O_2 \text{ volume} = \frac{(11.2 - 6.49) \times 10^{-4} \text{ mol } O_2 \times \dfrac{0.08206 \text{ L atm}}{\text{mol K}} \times (25 + 273) \text{ K}}{1 \text{ atm}}$$

$$= 1.2 \times 10^{-2} \text{ L} = 12 \text{ mL expelled}$$

51. We use the STP molar volume (22.414 L = 22,414 mL) to determine the molarity of Ar under 1 atom pressure and then the Henry's law constant.

$$k_{Ar} = \frac{C}{P_{Ar}} = \frac{\dfrac{33.7 \text{ mL Ar}}{1 \text{ L soln}} \times \dfrac{1 \text{ mol Ar}}{22{,}414 \text{ mL at STP}}}{1 \text{ atm pressure}} = \frac{0.00150 \text{ M}}{\text{atm}}$$

In the atmosphere, the partial pressure of argon is $P_{Ar} = 0.00934$ atm. (Recall that pressure fractions equal volume fractions for ideal gases.) We now compute the concentration of argon in aqueous solution.

$$C = k_{Ar}P_{Ar} = \frac{0.00150 \text{ M}}{\text{atm}} \times 0.00934 \text{ atm} = 1.4 \times 10^{-5} \text{ M}$$

52. We use the STP molar volume (22.414 L = 22,414 mL) to determine the molarity of CO_2 under 1 atm pressure and then the Henry's law constant.

$$k_{gas} = \frac{C}{P_{gas}} = \frac{\dfrac{87.8 \text{ mL } CO_2}{1 \text{ L soln}} \times \dfrac{1 \text{ mol } CO_2}{22{,}414 \text{ mL at STP}}}{1 \text{ atm pressure}} = \frac{0.00392 \text{ M}}{\text{atm}}$$

In the atmosphere, the partial pressure of argon is $P_{gas} = 0.00036$ atm. (Recall that pressure fractions equal volume fractions for ideal gases.) We now compute the concentration of argon in aqueous solution.

$$C = k_{gas}P_{gas} = \frac{0.00392 \text{ M}}{\text{atm}} \times 0.00036 \text{ atm} = 1.4 \times 10^{-6} \text{ M}$$

Raoult's Law and Liquid-Vapor Equilibrium

53. We determine the mole fraction of water in this solution.

$$n_{urea} = 165 \text{ g } C_6H_{12}O_6 \times \frac{1 \text{ mol } C_6H_{12}O_6}{180.2 \text{ g } C_6H_{12}O_6} = 0.916 \text{ mol } C_6H_{12}O_6$$

$$n_{water} = 685 \text{ g } H_2O \times \frac{1 \text{ mol } H_2O}{18.02 \text{ g } H_2O} = 38.0 \text{ mol } H_2O \qquad \chi_{water} = \frac{38.0 \text{ mol } H_2O}{(38.0 + 0.916) \text{ total moles}} = 0.976$$

$$P_{water} = \chi_{water} P^*_{water} = 0.976 \times 23.8 \text{ mmHg} = 23.2 \text{ mmHg}$$

54. We determine the mole fraction of methanol in this solution.

$$n_{urea} = 17 \text{ g } CO(NH_2)_2 \times \frac{1 \text{ mol } CO(NH_2)_2}{60.06 \text{ g } CO(NH_2)_2} = 0.283 \text{ mol } CO(NH_2)_2$$

$$n_{methanol} = 100 \text{ mL } CH_3OH \times \frac{0.792 \text{ g}}{1 \text{ mL}} \times \frac{1 \text{ mol } CH_3OH}{32.04 \text{ g } CH_3OH} = 2.47 \text{ mol } CH_3OH$$

$$\chi_{methanol} = \frac{2.47 \text{ mol } CH_3OH}{(2.47 + 0.283) \text{ total moles}} = 0.897$$

$$P_{methanol} = \chi_{methanol} P^*_{methanol} = 0.897 \times 95.7 \text{ mmHg} = 85.8 \text{ mmHg}$$

55. We consider a sample of 100.0 g of the solution and determine the number of moles of each component in this sample. From this information and the given vapor pressures, we determine the vapor pressure of each component.

$$\text{amount styrene} = n_s = 38 \text{ g styrene} \times \frac{1 \text{ mol } C_8H_8}{104 \text{ g } C_8H_8} = 0.37 \text{ mol } C_8H_8$$

$$\text{amount ethylbenzene} = n_e = 62 \text{ g ethylbenzene} \times \frac{1 \text{ mol } C_8H_{10}}{106 \text{ g } C_8H_{10}} = 0.58 \text{ mol } C_8H_{10}$$

$$\chi_s = \frac{n_s}{n_s + n_e} = \frac{0.37 \text{ mol}}{(0.37 + 0.58) \text{ total moles}} = 0.39 \qquad P_s = 0.39 \times 134 \text{ mmHg} = 52 \text{ mmHg for } C_8H_8$$

$$\chi_e = \frac{n_e}{n_s + n_e} = \frac{0.58 \text{ mol}}{(0.37 + 0.58) \text{ total moles}} = 0.61 \qquad P_e = 0.61 \times 182 \text{ mmHg} = 111 \text{ mmHg for } C_8H_{10}$$

Then the mole fraction in the vapor can be determined.

$$y_e = \frac{P_e}{P_e + P_s} = \frac{111 \text{ mmHg}}{(111 + 52) \text{ mmHg}} = 0.68 \qquad y_s = 1.00 - 0.68 = 0.32$$

56. The vapor pressures of the pure liquids are: for benzene $P_b^\circ = 95.1$ mmHg and for toluene $P_t^\circ = 28.4$ mmHg. We know that $P_t = \chi_t P_t^\circ$ and $P_b = \chi_b P_b^\circ$. We also know that $\chi_t + \chi_b = 1.000$. We solve for χ_b in the following expression.

$$0.620 = \frac{P_b}{P_b + P_t} = \frac{\chi_b P_b^\circ}{\chi_b P_b^\circ + \chi_t P_t^\circ} = \frac{\chi_b P_b^\circ}{\chi_b P_b^\circ + (1 - \chi_b) P_t^\circ} = \frac{\chi_b 95.1}{\chi_b 95.1 + (1 - \chi_b) 28.4}$$

$$95.1\ \chi_b = 0.620[\chi_b 95.1 + (1 - \chi_b) 28.4] = 59.0\ \chi_b + 17.6 - 17.6\ \chi_b$$

$$17.6 = (95.1 - 59.0 + 17.6)\ \chi_b = 53.7\ \chi_b \qquad \chi_b = \frac{17.6}{53.7} = 0.328$$

Osmotic Pressure

57. Both the flowers and the cucumber contain solutions (plant sap), but both of these solutions are less concentrated than the salt solution. Thus, the solvent water in the plant material moves across the semipermeable membrane in an attempt to dilute the salt solution, leaving behind wilted flowers and shriveled pickles (less water in their tissues).

58. The main purpose of drinking water is to provide water for living tissues. In fresh water, water should move across the semipermeable membranes of a fish's organs as water moves past the gills for breathing. Thus, it should not be necessary for freshwater fish to drink. In contrast, the concentration of solutes in salt water should be about the same as their concentration in a fish's blood plasma. In this case, osmotic pressure will not be effective in moving water into a fish's tissues. The purpose of drinking for a salt water fish is to isolate the ingested water within the fish. Then mechanisms within the fish's body can move water across membranes into the fish's tissues. Or possibly, the fish has a means of desalinating the ingested water, in which case osmotic pressure would carry water into the fish's tissues.

59. We first determine the molarity of the solution. Let's work the problem with three significant figures.

$$c = \frac{\pi}{RT} = \frac{1.00 \text{ atm}}{0.08206 \text{ L atm mol}^{-1} \text{ K}^{-1} \times 273 \text{ K}} = 0.0446 \text{ M}$$

$$\text{volume} = 1 \text{ mol} \times \frac{1 \text{ L}}{0.0446 \text{ mol solute}} = 22.4 \text{ L solution} = 22.4 \text{ L solvent}$$

We have assumed that the solution is so dilute that its volume closely approximates the volume of the solvent constituting it. Note that this volume corresponds to the STP molar volume of an ideal gas. The osmotic pressure equation also resembles the idea gas equation.

60. First calculate the concentration of the solution, then the mass of solute in 100.0 mL of that solution.

$$[\text{solute}] = \frac{\pi}{RT} = \frac{7.25 \text{ mmHg} \times \frac{1 \text{ atm}}{760 \text{ mmHg}}}{0.08206 \text{ L atm mol}^{-1} \text{ K}^{-1} \times 298 \text{ K}} = 3.90 \times 10^{-4} \text{ M}$$

$$\text{mass of hemoglobin} = 100.0 \text{ mL} \times \frac{1 \text{ L}}{1000 \text{ mL}} \times \frac{3.90 \times 10^{-4} \text{ mol hemoglobin}}{1 \text{ L soln}} \times \frac{6.86 \times 10^4 \text{ g}}{1 \text{ mol hemoglobin}}$$
$$= 2.68 \text{ g hemoglobin}$$

61. First determine the concentration of the solution from the osmotic pressure, then the amount of solute dissolved, and finally the molar mass of that solute.

$$\pi = 5.1 \text{ mm soln} \times \frac{0.88 \text{ mmHg}}{13.6 \text{ mm soln}} \times \frac{1 \text{ atm}}{760 \text{ mmHg}} = 4.3 \times 10^{-4} \text{ atm}$$

$$c = \frac{\pi}{RT} = \frac{4.3 \times 10^{-4} \text{ atm}}{0.0821 \text{ L atm mol}^{-1} \text{ K}^{-1} \times 298 \text{ K}} = 1.8 \times 10^{-5} \text{ M}$$

$$\text{amount solute} = 100.0 \text{ mL} \times \frac{1 \text{ L}}{1000 \text{ mL}} \times 1.8 \times 10^{-5} \text{ M} = 1.8 \times 10^{-6} \text{ mol solute}$$

$$\mathcal{M} = \frac{0.50 \text{ g}}{1.8 \times 10^{-6} \text{ mol}} = 2.8 \times 10^5 \text{ g/mol}$$

62. We need the molar concentration of ions in the solution to determine its osmotic pressure. We assume that the solution has a density of 1.00 g/mL.

$$[\text{ions}] = \frac{0.92 \text{ g NaCl} \times \frac{1 \text{ mol NaCl}}{58.4 \text{ g NaCl}} \times \frac{2 \text{ mol ions}}{1 \text{ mol NaCl}}}{100.0 \text{ mL soln} \times \frac{1 \text{ L soln}}{1000 \text{ mL}}} = 0.32 \text{ M}$$

$$\pi = cRT = 0.32 \frac{\text{mol}}{\text{L}} \times 0.0821 \text{ L atm mol}^{-1} \text{ K}^{-1} \times (37.0 + 273.2)\text{K} = 8.1 \text{ atm}$$

63. The reverse osmosis process requires a pressure equal to or slightly greater than the osmotic pressure of the solution. We assume that this solution has a density of 1.00 g/mL. First, we determine the molar concentration of ions in the solution.

$$[\text{ions}] = \frac{2.5 \text{ g NaCl} \times \frac{1 \text{ mol NaCl}}{58.4 \text{ g NaCl}} \times \frac{2 \text{ mol ions}}{1 \text{ mol NaCl}}}{100.0 \text{ g soln} \times \frac{1 \text{ mL soln}}{1.00 \text{ g}} \times \frac{1 \text{ L soln}}{1000 \text{ mL}}} = 0.86 \text{ M}$$

$$\pi = cRT = 0.86 \frac{\text{mol}}{\text{L}} \times 0.0821 \text{ L atm mol}^{-1} \text{ K}^{-1} \times (25 + 273.2)\text{K} = 21 \text{ atm}$$

64. We need to determine the molarity of each solution. The solution with the higher molarity has the higher osmotic pressure, and water will flow from the other solution into that more concentrated solution, in an attempt to create two solutions of equal concentrations.

$$[C_3H_8O_3] = \frac{14.0 \text{ g } C_3H_8O_3 \times \frac{1 \text{ mol } C_3H_8O_3}{92.09 \text{ g } C_3H_8O_3}}{55.2 \text{ mL soln} \times \frac{1 \text{ L}}{1000 \text{ mL}}} = 2.75 \text{ M}$$

$$[C_{12}H_{22}O_{11}] = \frac{17.2 \text{ g } C_{12}H_{22}O_{11} \times \frac{1 \text{ mol } C_{12}H_{22}O_{11}}{342.3 \text{ g } C_{12}H_{22}O_{11}}}{62.5 \text{ mL soln} \times \frac{1 \text{ L}}{1000 \text{ mL}}} = 0.804 \text{ M}$$

Thus, water will move from the $C_{12}H_{22}O_{11}$ solution into the $C_3H_8O_3$ solution, that is, from right to left.

Freezing Point Depression and Boiling Point Elevation

65. **(a)** First determine the molality of the solution, then the value of the freezing-point depression constant.

$$m = \frac{1.00 \text{ g } C_6H_6 \times \frac{1 \text{ mol } C_6H_6}{78.11 \text{ g } C_6H_6}}{80.00 \text{ g solvent} \times \frac{1 \text{ kg}}{1000 \text{ g}}} = 0.160\ m \qquad K_f = \frac{\Delta T_f}{-m} = \frac{3.3\ °\text{C} - 6.5\ °\text{C}}{-0.160\ m} = 20.\ °\text{C}/m$$

(b) For benzene, $K_f = 5.12\ °\text{C}\ m^{-1}$. Cyclohexane is the better solvent for freezing point depression determinations of molar mass, since a less concentrated solution will still give a substantial freezing-point depression. For the same concentration, cyclohexane solutions will show a freezing point depression approximately four times that of benzene. (Another factor is that benzene is labeled as a carcinogen and should be avoided.)

66. The solution's boiling point is 0.40 °C above the boiling point of pure water. First compute the molality of the solution, then the mass of sucrose that must be added to 1 kg (1000 g) H_2O, and finally the mass% sucrose in the solution.

$$m = \frac{\Delta T_b}{K_b} = \frac{0.40\ °\text{C}}{0.512\ °\text{C}/m} = 0.78\ m \longrightarrow \frac{0.78 \text{ mol } C_{12}H_{22}O_{11}}{1 \text{ kg } H_2O} \times \frac{342.3 \text{ g } C_{12}H_{22}O_{11}}{1 \text{ mol } C_{12}H_{22}O_{11}} = 267 \text{ g } C_{12}H_{22}O_{11}$$

$$\text{mass\% } C_{12}H_{22}O_{11} = \frac{267 \text{ g } C_{12}H_{22}O_{11}}{(1000. + 267) \text{ g total}} \times 100\% = 21\% \ C_{12}H_{22}O_{11}$$

67. We determine the molality of the solution, then the number of moles of solute present, and then the molar mass of the solute. Then we determine the compound's empirical formula, and combine this with the molar mass to determine the molecular formula.

$$m = \frac{\Delta T_f}{-K_f} = \frac{1.37\ °\text{C} - 5.53\ °\text{C}}{-5.12\ °\text{C}/m} = 0.813\ m$$

$$\text{amount} = \left(50.0 \text{ mL } C_6H_6 \times \frac{0.879 \text{ g}}{1 \text{ mL}} \times \frac{1 \text{ kg}}{1000 \text{ g}}\right) \times \frac{0.813 \text{ mol solute}}{1 \text{ kg } C_6H_6} = 3.57 \times 10^{-2} \text{ mol}$$

$$\mathcal{M} = \frac{6.45 \text{ g}}{3.57 \times 10^{-2} \text{ mol}} = 181 \text{ g/mol}$$

$42.9 \text{ g C} \times \frac{1 \text{ mol C}}{12.01 \text{ g C}} = 3.57 \text{ mol C} \quad \div 1.19 \longrightarrow 3.00 \text{ mol C}$

$2.4 \text{ g H} \times \frac{1 \text{ mol H}}{1.01 \text{ g H}} = 2.4 \text{ mol H} \quad \div 1.19 \longrightarrow 2.0 \text{ mol H}$

$16.7 \text{ g N} \times \frac{1 \text{ mol N}}{14.01 \text{ g N}} = 1.19 \text{ mol N} \quad \div 1.19 \longrightarrow 1.00 \text{ mol N}$

$38.1 \text{ g O} \times \frac{1 \text{ mol O}}{16.00 \text{ g O}} = 2.38 \text{ mol O} \quad \div 1.19 \longrightarrow 2.00 \text{ mol O}$

The empirical formula is $C_3H_2NO_2$, with a formula mass of 84.0 g/mol. This is almost one-half the experimentally determined molar mass. Thus, the molecular formula is $C_6H_4N_2O_4$.

68. We determine the molality of the nitrobenzene ($K_f = 8.1$ °C/m) solution first, then the solute's molar mass.

$m = \frac{\Delta T_f}{-K_f} = \frac{-1.4 \text{ °C} - 5.7 \text{ °C}}{-8.1 \text{ °C}/m} = 0.88\ m$

$\text{amount solute} = 30.0 \text{ mL nitrobenzene} \times \frac{1.204 \text{ g}}{1 \text{ mL}} \times \frac{1 \text{ kg}}{1000 \text{ g}} \times \frac{0.88 \text{ mol solute}}{1 \text{ kg nitrobenzene}} = 0.032 \text{ mol solute}$

$\mathcal{M} = \frac{3.88 \text{ g nicotinamide}}{0.032 \text{ mol nicotinamide}} = 1.2 \times 10^2 \text{ g/mol}$

Then we determine the compound's empirical formula, and combine this with the molar mass to determine the molecular formula.

$59.0 \text{ g C} \times \frac{1 \text{ mol C}}{12.01 \text{ g C}} = 4.91 \text{ mol C} \quad \div 0.819 \longrightarrow 6.00 \text{ mol C}$

$5.0 \text{ g H} \times \frac{1 \text{ mol H}}{1.01 \text{ g H}} = 5.0 \text{ mol H} \quad \div 0.819 \longrightarrow 6.1 \text{ mol H}$

$22.9 \text{ g N} \times \frac{1 \text{ mol N}}{14.01 \text{ g N}} = 1.63 \text{ mol N} \quad \div 0.819 \longrightarrow 1.99 \text{ mol N}$

$13.1 \text{ g O} \times \frac{1 \text{ mol O}}{16.00 \text{ g O}} = 0.819 \text{ mol O} \quad \div 0.819 \longrightarrow 1.00 \text{ mol O}$

A reasonable empirical formula is $C_6H_6N_2O$, which has an empirical mass of 122. Since this is the same as the experimentally determined molar mass, the molecular formula of nicotinamide is $C_6H_6N_2O$.

69. We determine the molality of the benzene solution first, then the molar mass of the solute.

$m = \frac{\Delta T_f}{-K_f} = \frac{-1.183 \text{ °C}}{-5.12 \text{ °C}/m} = 0.231\ m$

$\text{amount solute} = 0.04456 \text{ kg benzene} \times \frac{0.231 \text{ mol solute}}{1 \text{ kg benzene}} = 0.0103 \text{ mol solute}$

$\mathcal{M} = \frac{0.867 \text{ g thiophene}}{0.0103 \text{ mol thiophene}} = 84.2 \text{ g/mol}$

Next, we determine the empirical formula from the masses of the combustion products.

$\text{amount C} = 4.913 \text{ g } CO_2 \times \frac{1 \text{ mol } CO_2}{44.010 \text{ g } CO_2} \times \frac{1 \text{ mol C}}{1 \text{ mol } CO_2} = 0.1116 \text{ mol C} \div 0.02791 \longrightarrow 3.999 \text{ mol C}$

$\text{amount H} = 1.005 \text{ g } H_2O \times \frac{1 \text{ mol } H_2O}{18.015 \text{ g } H_2O} \times \frac{2 \text{ mol H}}{1 \text{ mol } H_2O} = 0.1116 \text{ mol H} \div 0.02791 \longrightarrow 3.999 \text{ mol H}$

$\text{amount S} = 1.788 \text{ g } SO_2 \times \frac{1 \text{ mol } SO_2}{64.065 \text{ g } SO_2} \times \frac{1 \text{ mol S}}{1 \text{ mol } SO_2} = 0.02791 \text{ mol S} \div 0.02791 \longrightarrow 1.000 \text{ mol S}$

A reasonable empirical formula is C_4H_4S, which has an empirical mass of 84.1. Since this is the same as the experimentally determined molar mass, the molecular formula of thiophene if C_4H_4S.

70. First we determine the empirical formula of coniferin from the combustion data.

$0.698 \text{ g } H_2O \times \frac{1 \text{ mol } H_2O}{18.02 \text{ g } H_2O} \times \frac{2 \text{ mol H}}{1 \text{ mol } H_2O} = 0.0775 \text{ mol H} \times \frac{1.01 \text{ g H}}{1 \text{ mol H}} = 0.0783 \text{ g H}$

$2.479 \text{ g } CO_2 \times \frac{1 \text{ mol } CO_2}{44.01 \text{ g } CO_2} \times \frac{1 \text{ mol C}}{1 \text{ mol } CO_2} = 0.05633 \text{ mol C} \times \frac{12.011 \text{ g C}}{1 \text{ mol C}} = 0.6766 \text{ g C}$

$(1.205 \text{ g} - 0.0783 \text{ g H} - 0.6766 \text{ g C}) \times \frac{1 \text{ mol O}}{16.00 \text{ g O}} = 0.0281 \text{ mol O}$

$0.0775 \text{ mol H} \div 0.0281 = 2.76 \text{ mol H} \quad \times 4 \longrightarrow 11.04 \text{ mol H}$

$0.05633 \text{ mol C} \div 0.0281 = 2.00 \text{ mol C} \quad \times 4 \longrightarrow 8.00 \text{ mol C}$

$0.0281 \text{ mol O} \div 0.0281 = 1.00 \text{ mol O} \quad \times 4 \longrightarrow 4.00 \text{ mol O}$

Empirical formula = $C_8H_{11}O_4$ with a formula weight of 171 g/mol.

Now we determine the molality of the boiling solution, the number of moles of solute present in that solution, and the molar mass of the solute. $m = \frac{\Delta T_b}{K_b} = \frac{0.068\ °C}{0.512\ °C/m} = 0.13\ m$

$$\text{amount solute} = 48.68\ g\ H_2O \times \frac{1\ kg\ solvent}{1000\ g} \times \frac{0.13\ mol\ solute}{1\ kg\ solvent} = 6.3 \times 10^{-3}\ mol\ solute$$

$$\mathcal{M} = \frac{2.216\ g\ solute}{6.3 \times 10^{-3}\ mol\ solute} = 3.5 \times 10^2\ g/mol$$

This molar mass is twice the formula weight of the empirical formula. Thus, the molecular formula is twice the empirical formula. Molecular formula = $C_{16}H_{22}O_8$.

Strong Electrolytes, Weak Electrolytes, and Nonelectrolytes

71. The freezing point depression is given by $\Delta T_f = -i\ K_f\ m$. Since $K_f = 1.86$ °C/m for water, $\Delta T_f = -i\ 0.186$ °C for this group of 0.10 m solutions.

(a) $T_f = -0.19$ °C Urea is a nonelectrolyte, and $i = 1$.
(b) $T_f = -0.37$ °C NH_4NO_3 is a strong electrolyte, composed of two ions per formula unit; $i = 2$.
(c) $T_f = -0.37$ °C HCl also is a strong electrolyte, composed of two ions per formula unit; $i = 2$.
(d) $T_f = -0.56$ °C $CaCl_2$ is a strong electrolyte, composed of three ions per formula unit; $i = 3$.
(e) $T_f = -0.37$ °C $MgSO_4$ is a strong electrolyte, composed of two ions per formula unit; $i = 2$.
(f) $T_f = -0.19$ °C Ethanol is a nonelectrolyte; $i = 1$.
(g) $T_f < -0.19$ °C $HC_2H_3O_2$ is a weak electrolyte; i is somewhat larger than 1.

72. (a) First calculate the freezing point for a 0.050 m solution of a nonelectrolyte.

$\Delta T_{f,calc} = K_f\ m = 1.86\ °C/m \times 0.050\ m = 0.093\ °C$ $\qquad i = \frac{\Delta T_{f,obs}}{\Delta T_{f,calc}} = \frac{0.0986\ °C}{0.093\ °C} = 1.1$

(b) $[H^+] = [NO_2^-] = 6.91 \times 10^{-3}\ M = 0.00691\ M$ $\qquad [HNO_2] = 0.100\ M - 0.00691\ M = 0.093\ M$

$[\text{particles}] = 0.00691\ M + 0.00691\ M + 0.093\ M = 0.107\ M$ $\qquad i = \frac{0.107\ M}{0.100\ M} = 1.07$

73. The mixture of $NH_3(aq)$ and $HC_2H_3O_2(aq)$, results in the formation of $NH_4C_2H_3O_2(aq)$, a solution of an ionic substance and a strong electrolyte.

$NH_3(aq) + HC_2H_3O_2(aq) \longrightarrow NH_4C_2H_3O_2(aq) \longrightarrow NH_4^+(aq) + C_2H_3O_2^-(aq)$

This solution of strong electrolyte conducts a current very well.

74. For two solutions to be isotonic, they must contain the same concentration of particles. For them to also have the same % mass/volume would mean that each gram of each substance produces the same number of particles. This would be true if they had the same molar mass and the same van't Hoff factor. It would also be true if the quotient of molar mass and van't Hoff factor were the same. While this condition is not impossible to meet, it is pretty unlikely.

FEATURE PROBLEMS

A. A quick solution to this problem is to calculate the molarity of a 50% by volume solution, using the density of 0.789 g/mL for ethanol.

$$\text{ethanol molarity} = \frac{50\ mL\ C_2H_5OH \times \frac{0.789\ g}{1\ mL} \times \frac{1\ mol\ C_2H_5OH}{46.07\ g\ C_2H_5OH}}{100\ mL\ soln \times \frac{1\ L}{1000\ mL}} = 8.6\ M$$

Thus, solutions that are more than 8.6 M are more than 100 proof.

B. 1. A solution with $\chi_{HCl} = 0.50$ begins to boil at about 18 °C. At that temperature the composition of the vapor is about $\chi_{HCl} = 0.63$, reading directly across the tie line at 18 °C. The vapor has $\chi_{HCl} > 0.50$.

2. A solution with $\chi_{HCl} = 0.05$ begins to boil at about 102 °C. At that temperature, the composition of the vapor is about $\chi_{HCl} = 0.01$, reading directly across the tie line at 102 °C. The vapor has $\chi_{HCl} < 0.05$.

3. The composition of HCl(aq) changes as the solution boils in an open container because the vapor has a different composition than does the liquid. Thus, one component or another is depleated as the solution boils.

4. The azeotrope occurs at the maximum of the curve: at $\chi_{HCl} = 0.12$ and a temperature of 110 °C.

5. We first determine the amount of HCl in the sample.

$$\text{amount HCl} = 30.32 \text{ mL NaOH} \times \frac{1 \text{ L}}{1000 \text{ mL}} \times \frac{1.006 \text{ mol NaOH}}{1 \text{ L}} \times \frac{1 \text{ mol HCl}}{1 \text{ mol NaOH}} = 0.03050 \text{ mol}$$

The mass of water is the difference between the mass of solution and that of HCl.

$$\text{mass } H_2O = \left(5.00 \text{ mL soln} \times \frac{1.099 \text{ g}}{1 \text{ mL}}\right) - \left(0.03050 \text{ mol HCl} \times \frac{36.46 \text{ g HCl}}{1 \text{ mol HCl}}\right) = 4.38 \text{ g}$$

Now we determine the amount of H_2O and then the mole fraction of HCl.

$$\text{amount } H_2O = 4.38 \text{ g} \times \frac{1 \text{ mol } H_2O}{18.02 \text{ g } H_2O} = 0.243 \text{ mol } H_2O$$

$$\chi_{HCl} = \frac{0.03050 \text{ mol HCl}}{0.03050 \text{ mol HCl} + 0.243 \text{ mol } H_2O} = 0.111$$

C. 1. At 20 °C, the solubility of NaCl is 33.4 g NaCl / 100 g H_2O. We determine the mole fraction of H_2O in this solution

$$\text{amount } H_2O = 100 \text{ g } H_2O \times \frac{1 \text{ mol } H_2O}{18.02 \text{ g } H_2O} = 5.549 \text{ mol } H_2O$$

$$\text{amount NaCl} = 33.4 \text{ g NaCl} \times \frac{1 \text{ mol NaCl}}{58.44 \text{ g NaCl}} = 0.572 \text{ mol NaCl}$$

$$\chi_{water} = \frac{5.549 \text{ mol } H_2O}{0.572 \text{ mol NaCl} + 5.549 \text{ mol } H_2O} = 0.9066$$

The approximate relative humidity then will be 91% (90.66%), because the water vapor pressure above the NaCl saturated solution will be 90.66% of the vapor pressure of pure water at 20 °C.

2. $CaCl_2 \cdot 6H_2O$ deliquesces if the relative humidity is over 32%. Thus $CaCl_2 \cdot 6H_2O$ will deliquesce.

3. If the substance in the bottom of the dessicator has a high water solubility its saturated solution will have a low χ_{water}, which in turn will produce a low relative humidity. Relative humidity lower than 32% is needed to keep $CaCl_2 \cdot 6H_2O$ dry.

D. 1. We first compute the molality of a 0.92% mass/volume solution, assuming the solution's density is about 1.00 g/mL, meaning that 100.0 mL solution has a mass of 100.0 g.

$$\text{molality} = \frac{0.92 \text{ g NaCl} \times \frac{1 \text{ mol NaCl}}{58.44 \text{ g NaCl}}}{(100.0 \text{ g soln} - 0.92 \text{ g NaCl}) \times \frac{1 \text{ kg solvent}}{1000 \text{ g}}} = 0.16\ m$$

Then we compute the freezing point depression of this solution.

$$\Delta T_f = -i\, K_f\, m = -2.0 \frac{\text{mol ions}}{\text{mol NaCl}} \times \frac{1.86 \text{ °C}}{m} \times 0.16\ m = -0.60 \text{ °C}$$

The van't Hoff factor of NaCl most likely is no equal to 2.0, but a bit less and thus the two definitions are in fair agreement.

2. We calculate the amount of each solute, assume 1.00 L of solution weighs 1000 g, and subtract the mass of all solutes to determine the mass of solvent.

$$\text{amount NaCl ions} = 3.5 \text{ g NaCl} \times \frac{1 \text{ mol NaCl}}{58.44 \text{ g NaCl}} \times \frac{2 \text{ mol ions}}{1 \text{ mol NaCl}} = 0.12_0 \text{ mol ions}$$

$$\text{amount KCl ions} = 1.5 \text{ g} \times \frac{1 \text{ mol KCl}}{74.55 \text{ g KCl}} \times \frac{2 \text{ mol ions}}{1 \text{ mol KCl}} = 0.040 \text{ mol ions}$$

$$\text{amount } Na_3C_6H_5O_7 \text{ ions} = 2.9 \text{ g} \times \frac{1 \text{ mol } Na_3C_6H_5O_7}{258.07 \text{ g } Na_3C_6H_5O_7} \times \frac{4 \text{ mol ions}}{1 \text{ mol } Na_3C_6H_5O_7} = 0.045 \text{ mol ions}$$

$$\text{amount } C_6H_{12}O_6 = 20.0 \text{ g} \times \frac{1 \text{ mol } C_6H_{12}O_6}{180.2 \text{ g } C_6H_{12}O_6} = 0.111 \text{ mol}$$

$$\text{solvent mass} = 1000.0 \text{ g} - (3.5 \text{ g} + 1.5 \text{ g} + 2.9 \text{ g} + 20.0 \text{ g}) = 940.6 \text{ g } H_2O = 0.9406 \text{ kg } H_2O$$

$$\text{solution molality} = \frac{(0.120 + 0.040 + 0.045 + 0.111) \text{ mol}}{0.9406 \text{ kg } H_2O} = 0.336\ m$$

$$\Delta T_f = -K_f\, m = -1.86 \text{ °C}/m \times 0.336\ m = -0.62 \text{ °C}$$

This again is close to the defined freezing point of –0.52 °C, with the error most likely due to van't Hoff factors not being integral.

15 CHEMICAL KINETICS

PRACTICE EXAMPLES

1A Rate of raction of a reactant is expressed as the negative of the change in molarity divided by the time interval. rate of reaction of A $= \frac{-\Delta[A]}{\Delta t} = \frac{-(0.3187\ M - 0.3629\ M)}{8.25\ min} \times \frac{1\ min}{60\ sec} = 8.93 \times 10^{-5}$ M sec^{-1}

1B We use the rate of reaction of A to determine the rate of formation of B, noting from the balanced equation that 3 moles of B form (+3 moles B) when 2 moles of A react (–2 moles A). (Recall that "M" means "moles per liter.")

$$\text{rate of B formation} = \frac{0.5522\ M\ A - 0.5684\ M\ A}{2.50\ min \times \frac{60\ s}{1\ min}} \times \frac{+3\ moles\ B}{-2\ moles\ A} = 1.62 \times 10^{-4}\ M\ B\ s^{-1}$$

2A **(a)** The 2400-s tangent line intersects the 1200-s vertical line at 0.75 M and reaches 0 M at 3500 s. The slope of that tangent line is thus slope $= \frac{0\ M - 0.75\ M}{3500\ s - 1200\ s} = -3.3 \times 10^{-4}$ M s^{-1} = –rate of reaction

instantaneous rate of reaction = 3.3×10^{-4} M s^{-1}

(b) At 2400 s, $[H_2O_2] = 0.39$ M. At 2450 s, $[H_2O_2] = 0.39\ M + \text{rate} \times \Delta t$

At 2450 s, $[H_2O_2] = 0.39\ M + [-3.3 \times 10^{-4}\ mol\ H_2O_2\ L^{-1}\ s^{-1} \times 50\ s] = 0.39\ M - 0.017\ M = 0.37\ M$

2B With only the data of Table 15.2 we can use only the reaction rate during the first 400 seconds, $-\Delta[H_2O_2]/\Delta t = 15.0 \times 10^{-4}$ M s^{-1}, and the initial concentration, $[H_2O_2]_0 = 2.32$ M. We calculate the change in $[H_2O_2]$ and add it to $[H_2O_2]_0$ to determine $[H_2O_2]_{100}$.

$\Delta[H_2O_2]$ = rate of reaction of $H_2O_2 \times \Delta t = -15.0 \times 10^{-4}\ M\ s^{-1} \times 100\ s = -0.15\ M$

$[H_2O_2]_{100} = [H_2O_2]_0 + \Delta[H_2O_2] = 2.32\ M + (-0.15\ M) = 2.17\ M$

This value differs from the value of 2.15 M determined in *text* Example 15-2b because the *text* used the initial rate of reaction (17.4×10^{-4} M s^{-1}) which is a bit faster than the average rate over the first 400 seconds.

3A We write the equation for each rate, divide them into each other, and solve for n.

$R_1 = k \times [N_2O_5]_1^n = 5.45 \times 10^{-5}\ M\ s^{-1} = k\ (3.15\ M)^n \qquad R_2 = k \times [N_2O_5]_2^n = 1.35 \times 10^{-5}\ M\ s^{-1} = k\ (0.78\ M)^n$

$$\frac{R_1}{R_2} = \frac{5.45 \times 10^{-5}\ M\ s^{-1}}{1.35 \times 10^{-5}\ M\ s^{-1}} = 4.04 = \frac{k \times [N_2O_5]_1^n}{k \times [N_2O_5]_2^n} = \frac{k\ (3.15\ M)^n}{k\ (0.78\ M)^n} = \left(\frac{3.15}{0.78}\right)^n = (4.0_4)^n$$

We kept an extra significant figure ($_4$) to emphasize that the value of $n = 1$; the reaction is first-order in N_2O_5.

3B For the reaction, we know that $R = k[HgCl_2]^1[C_2O_4{}^{2-}]^2$. Thus, we can compare Expt. 4 to Expt. 1 as follows.

$$\frac{R_4}{R_1} = \frac{k[HgCl_2]_4^1[C_2O_4{}^{2-}]_4^2}{k[HgCl_2]_1^1[C_2O_4{}^{2-}]_1^2} = \frac{0.025\ M \times (0.045\ M)^2}{0.105\ M \times (0.150\ M)^2} = 0.0214 = \frac{R_4}{1.8 \times 10^{-5}\ M\ min^{-1}}$$

The desired rate is $R_4 = 0.0214 \times 1.8 \times 10^{-5}\ M\ min^{-1} = 3.8 \times 10^{-7}\ M\ min^{-1}$

4A We substitute into the rate law and solve for k.

rate $= k[A]^2[B] = 4.78 \times 10^{-2}\ M\ s^{-1} = k\ (1.12\ M)^2(0.87\ M) \qquad k = \frac{4.78 \times 10^{-2}\ M\ s^{-1}}{(1.12\ M)^2\ 0.87\ M} = 4.4 \times 10^{-2}\ M^{-2}\ s^{-1}$

4B We already know that $R = k[HgCl_2]^1[C_2O_4^{2-}]^2$ and $k = 7.6 \times 10^{-3}\ M^{-2}\ min^{-1}$ Thus, substitution yields:
Rate $= 7.6 \times 10^{-3}\ M^{-2}\ min^{-1}\ (0.050\ M)^1\ (0.025\ M)^2 = 2.4 \times 10^{-7}\ M\ s^{-1}$

5A We substitute directly into the integrated rate equation.
$\ln [A]_t = -kt + \ln [A]_0 = -3.02 \times 10^{-3}\ s^{-1} \times 325\ s + \ln (2.80) = -0.982 + 1.030 = 0.048$
$[A]_t = e^{0.048} = 1.0\ M$

5B We substitute the suggested values into *text* equation (15.12).
$\ln \dfrac{[H_2O_2]_t}{[H_2O_2]_0} = -kt = -k \times 600\ s = \ln \dfrac{1.49\ M}{2.32\ M} = -0.443 \qquad k = \dfrac{-0.443}{-600\ s} = 7.38 \times 10^{-4}\ s^{-1}$
Now we choose $[H_2O_2]_0 = 1.49$ M, $[H_2O_2]_t = 0.62$, $t = 1800\ s - 600\ s = 1200\ s$
$\ln \dfrac{[H_2O_2]_t}{[H_2O_2]_0} = -kt = -k \times 1200\ s = \ln \dfrac{0.62\ M}{1.49\ M} = -0.88 \qquad k = \dfrac{-0.88}{-1200\ s} = 7.3 \times 10^{-4}\ s^{-1}$
These two values agree within the limits of the experimental measurement.

6A We can solve the integrated rate equation for the ratio of the final and initial concentrations. This ratio equals the fraction of the initial concentration that remains.
$\ln \dfrac{[A]_t}{[A]_0} = -kt = -2.95 \times 10^{-3}\ s^{-1} \times 150\ s = -0.443 \qquad \dfrac{[A]_t}{[A]_0} = e^{-0.443} = 0.642 \qquad$ 64.2% [A] remains.

6B After two-thirds of the sample has decomposed, one-third of the sample remains. Thus $[H_2O_2]_t = [H_2O_2]_0 \div 3$, and we have $\ln \dfrac{[H_2O_2]_t}{[H_2O_2]_0} = -kt = \ln \dfrac{[H_2O_2]_0 \div 3}{[H_2O_2]_0} = \ln (1/3) = -1.099 = -7.30 \times 10^{-4}\ s^{-1}\ t$
$t = \dfrac{-1.099}{7.30 \times 10^{-4}\ s^{-1}} = 1.50 \times 10^3\ s = 25.2\ min$

7A We use partial pressures in place of concentrations in the integrated first-order rate equation. Notice first that more than 30 half-lives have elapsed, and thus the ethylene oxide pressure has declined to at most $(0.5)^{30} = 9 \times 10^{-10}$ of its initial value.
$\ln \dfrac{P_{30}}{P_0} = -kt = -2.05 \times 10^{-4}\ s^{-1} \times 30.0\ h \times \dfrac{3600\ s}{1\ h} = -22.1 \qquad \dfrac{P_{30}}{P_0} = e^{-22.1} = 2.4 \times 10^{-10}$
$P_{30} = 2.4 \times 10^{-10}\ P_0 = 2.4 \times 10^{-10} \times 782\ mmHg = 1.9 \times 10^{-7}\ mmHg$

7B At the end of one half-life the pressure of DTBP will have halved, to 400 mmHg. At the end of another half-life, at 160 min, the pressure of DTBP will have halved again, to 200 mmHg. Thus, the pressure of DTBP at 125 min will be intermediate between the pressure at 80.0 min (400 mmHg) and that at 160 min (200 mmHg). To obtain an exact answer, first we determine the value of the rate constant from the half-life.
$k = \dfrac{0.693}{t_{1/2}} = \dfrac{0.693}{80.0\ min} = 0.00866\ min^{-1} \qquad \ln \dfrac{(P_{DTBP})_t}{(P_{DTBP})_0} = -kt = -0.00866\ min^{-1} \times 125\ min = -1.08$
$\dfrac{(P_{DTBP})_t}{(P_{DTBP})_0} = e^{-1.08} = 0.340 \qquad (P_{DTBP})_t = 0.340 \times (P_{DTBP})_0 = 0.340 \times 800\ mmHg = 272\ mmHg$

8A We first begin by looking for a constant rate, indicative of a zero-order reaction. If the rate is constant, the concentration will decrease by the same quantity during the same time period. If we choose a 25-s time period, we note that the concentration decreases (0.88 M – 0.74 M =) 0.14 M during the first 25 s, (0.74 M – 0.62 M =) 0.12 M during the second 25 s, (0.62 M – 0.52 M =) 0.10 M during the third 25 s, and (0.52 M – 0.44 M =) 0.08 M during the fourth 25 s period. This is hardly a constant rate and we conclude that the reaction is not zero-order.

We next look for a constant half-life, indicative of a first-order reaction. The initial concentration of 0.88 M decreases to one half of that value, 0.44 M, during the first 100 s, indicating a 100-s half-life. The concentration halves again to 0.22 M in the second 100 s, another 100-s half-life. Finally, we note that the concentration halves also from 0.62 M at 50 s to 0.31 M at 150 s, yet another 100-s half-life. The rate is established as first order. The rate constant is $k = \dfrac{0.693}{t_{1/2}} = \dfrac{0.693}{100\ s} = 6.93 \times 10^{-3}\ s^{-1}$

8B We plot the data in three ways to determine the order. (1) A plot of [A] vs. time is linear if the reaction is zero order. (2) A plot of ln [A] vs. time is linear if the reaction is first order. (3) A plot of 1/[A] vs. time is linear if the reaction is second order. It is obvious from the plots below that the reaction is zero order. The negative of the slope of the line equals $k = -(0.083\ M - 0.250\ M) \div 18.00\ min = 9.28 \times 10^{-3}\ M/min$

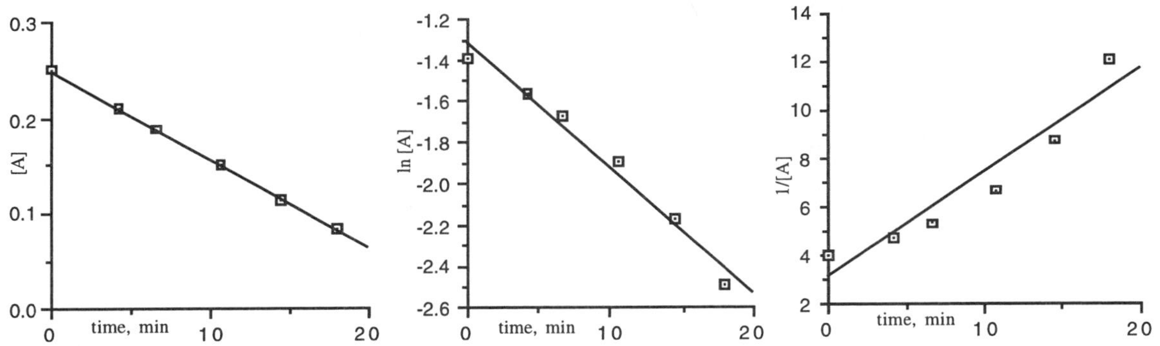

9A First we compute the value of the rate constant at 75.0 °C with the Arrhenius equation. We know that the activation energy is $E_a = 1.06 \times 10^5$ J/mol, and that $k = 3.46 \times 10^{-5}\ s^{-1}$ at 298 K. The temperature of 75.0 °C = 348.2 K.

$$\ln \frac{k_2}{k_1} = \ln \frac{k_2}{3.46 \times 10^{-5}\ s^{-1}} = \frac{E_a}{R}\left(\frac{1}{T_1} - \frac{1}{T_2}\right) = \frac{1.06 \times 10^5\ J/mol}{8.3145\ J\ mol^{-1}\ K^{-1}}\left(\frac{1}{298.2\ K} - \frac{1}{348.2\ K}\right) = 6.14$$

$$k_2 = 3.46 \times 10^{-5}\ s^{-1} \times e^{+6.14} = 3.46 \times 10^{-5}\ s^{-1} \times 4.6 \times 10^2 = 0.016\ s^{-1}$$

$$t_{1/2} = \frac{0.693}{k} = \frac{0.693}{0.016\ s^{-1}} = 43\ s \text{ at } 75.0\ °C$$

9B We use the integrated rate equation to determine the rate constant, realizing that one-third remains when two-thirds have decomposed.
$$\ln \frac{[N_2O_5]_t}{[N_2O_5]_0} = \ln \frac{[N_2O_5]_0 \div 3}{[N_2O_5]_0} = \ln \tfrac{1}{3} = -kt = -k\ (1.50\ h) = -1.099$$

$$k = \frac{1.099}{1.50\ h} \times \frac{1\ h}{3600\ s} = 2.04 \times 10^{-4}\ s^{-1}$$

Now use the Arrhenius equation to determine the temperature having the rate constant of this value.

$$\ln \frac{k_2}{k_1} = \ln \frac{2.04 \times 10^{-4}\ s^{-1}}{3.46 \times 10^{-5}\ s^{-1}} = 1.77 = \frac{E_a}{R}\left(\frac{1}{T_1} - \frac{1}{T_2}\right) = \frac{1.06 \times 10^5\ J/mol}{8.3145\ J\ mol^{-1}\ K^{-1}}\left(\frac{1}{298\ K} - \frac{1}{T_2}\right)$$

$$\frac{1}{T_2} = \frac{1}{298\ K} - \frac{1.77 \times 8.3145\ K^{-1}}{1.06 \times 10^5} = 3.22 \times 10^{-3}\ K^{-1} \qquad T_2 = 311\ K$$

10A The two steps of the mechanism must add, in a Hess's law fashion, to produce the overall reaction.

overall reaction:	$CO + NO_2 \longrightarrow CO_2 + NO$	*or*	$CO + NO_2 \longrightarrow CO_2 + NO$
– second step:	$-(NO_3 + CO \longrightarrow NO_2 + CO_2)$	*or*	$+(NO_2 + CO_2 \longrightarrow NO_3 + CO)$
first step:			$2\ NO_2 \longrightarrow NO + NO_3$

If the first step is the slow step, then it will be the rate-determining step, and the rate of that step will be the rate of the reaction: rate of reaction = $k_1\ [NO_2]^2$

10B (1) The steps of the mechanism must add, in a Hess's law fashion, to produce the overall reaction. This is done at right. The two intermediates, $NO_2F_2(g)$ and F(g), are each produced in one step and consumed in the next one.

Fast: $NO_2(g) + F_2(g) \rightleftharpoons NO_2F_2(g)$
Slow: $NO_2F_2(g) \longrightarrow NO_2F(g) + F(g)$
Fast: $F(g) + NO_2(g) \longrightarrow NO_2F(g)$
Net: $2\ NO_2(g) + F_2(g) \longrightarrow 2\ NO_2F(g)$

(2) The mechanism must agree with the rate law. We expect the rate-determining step to determine the reaction rate: Rate = $k_2[NO_2F_2]$. To eliminate $[NO_2F_2]$, we recognize that the first elementary reaction is very fast and will have the same rate forward as reverse: $R_f = k_f[NO_2][F_2] = k_r[NO_2F_2] = R_r$. We solve for the concentration of intermediate: $[NO_2F_2] = k_f[NO_2][F_2] / k_r$. We now substitute this expression for $[NO_2F_2]$ into the rate equation: Rate = $(k_2k_f / k_r)\ [NO_2F_2]$. This agrees with the experimental rate law.

SUMMARIZING EXAMPLE CALCULATIONS

1. Solve the rate equation for k, using number of molecules in place of concentration.

$$\text{rate} = k\ [PAN] \quad \text{becomes} \quad \frac{1.0 \times 10^{12}\ \text{molecules}}{L\ min} = k\ \frac{5.0 \times 10^{14}\ \text{molecules}}{L}$$

$$k = \frac{1.0 \times 10^{12} \text{ molecules L}^{-1} \text{ min}^{-1}}{5.0 \times 10^{14} \text{ molecules L}^{-1}} = 0.0020 \text{ min}^{-1}$$

2. Determine the value of the rate constant at 25 °C = 298 K $\quad k = \frac{0.693}{t_{1/2}} = \frac{0.693}{30.0 \text{ min}} = 0.0231 \text{ min}^{-1}$

3. Solve the Arrhenius equation for the unknown temperature. $E_a = 1.2 \times 10^2$ kJ/mol

$$\ln \frac{k_2}{k_1} = \ln \frac{0.0020 \text{ min}^{-1}}{0.0231 \text{ min}^{-1}} = -2.45 = \frac{E_a}{R}\left(\frac{1}{T_1} - \frac{1}{T_2}\right) = \frac{1.2 \times 10^5 \text{ J/mol}}{8.3145 \text{ J mol}^{-1} \text{ K}^{-1}}\left(\frac{1}{298 \text{ K}} - \frac{1}{T}\right)$$

$$\frac{-2.45 \times 8.3145}{1.2 \times 10^5} \text{K}^{-1} - \frac{1}{298 \text{ K}} = -\frac{1}{T} = -3.53 \times 10^{-3} \text{ K}^{-1} \qquad T = 283 \text{ K} = 10 \text{ °C}$$

REVIEW QUESTIONS

1. **(a)** $[A]_0$ symbolizes the concentration of some species A, usually a reactant, at time = 0.

(b) k is the symbol for the specific rate constant of a reaction, the speed of the reaction if all rate-determining species were present at unit molarity.

(c) $t_{1/2}$ is the symbol for the half-life of a reaction, the time during which the concentration of the reactant drops to one-half of its initial value.

(d) A zero-order reaction is one whose rate is independent of the concentration of reactant.

(e) A catalyst is a substance that alters the rate of a chemical reaction but that can be recovered unchanged at the end of the reaction.

2. **(a)** The method of initial rates is a means of determining the reaction order and the rate constant for a reaction by measuring the effect on the beginning reaction rate of different reactant concentrations.

(b) An activated complex is a high-energy species that exists as a result of the collision of two molecules.

(c) The reaction mechanism is the series of molecular processes that accounts for the overall reaction and its observed kinetics.

(d) Heterogeneous catalysis refers to catalytic activity that occurs at the interface between two phases, usually of a liquid or a gas in contact with a solid catalytic surface.

(e) The rate-determining step is the slowest step of a reaction mechanism.

3. **(a)** The rate of a first-order reaction depends on the first power of the concentration of a reactant; a second-order reaction's rate depends on the second power of the concentration of a reactant, or on the first power of each of two reactants.

(b) A rate equation is the relationship between the rate of a reaction and the powers of concentrations of the reactants. An integrated rate equation gives the dependence of concentration of reactant based upon elapsed time and initial concentration.

(c) The activation energy is the energy that must be supplied to energize reactants to the level of the activated complex; it always is endothermic. The enthalpy of reaction is the energy absorbed as the overall reaction occurs; it may be exothermic or endothermic.

(d) An elementary process is a description of how molecules interact with each other. The net reaction is the result of the several elementary processes that constitute it.

(e) An enzyme is a biological catalyst. A substrate is the reactant whose reaction the catalyst promotes.

4. **(a)** $\text{Rate} = \frac{\Delta[A]}{\Delta t} = \frac{0.2643 \text{ M} - 0.1832 \text{ M}}{35 \text{ min}} = 2.3 \times 10^{-3} \text{ M min}^{-1}$

(b) $\text{Rate} = 2.3 \times 10^{-3} \frac{\text{mol}}{\text{L·min}} \times \frac{1 \text{ min}}{60 \text{ s}} = 3.8 \times 10^{-5} \text{ M sec}^{-1}$

5. **(a)** $\frac{\Delta[B]}{\Delta t} = \frac{6.2 \times 10^{-4} \text{ mol A}}{\text{L s}} \times \frac{1 \text{ mol B}}{2 \text{ mol A}} = 3.1 \times 10^{-4} \text{ mol B L}^{-1} \text{ s}^{-1}$

(b) 3 mol D are produced (+3 mol D) for every 2 mol A that reacts (–2 mol A).

$$\frac{\Delta[D]}{\Delta t} = \frac{-6.2 \times 10^{-4} \text{ mol A}}{\text{L s}} \times \frac{+3 \text{ mol D}}{-2 \text{ mol A}} = 9.3 \times 10^{-4} \text{ mol D L}^{-1} \text{ s}^{-1}$$

6. In each case, we draw the tangent line to the plotted curve.

(a) The slope of the line is $\frac{\Delta[H_2O_2]}{\Delta t} = \frac{2.00 \text{ M} - 0.000 \text{ M}}{0 \text{ s} - 2300 \text{ s}} = -8.7 \times 10^{-4} \text{ M s}^{-1}$

Reaction rate $= -\dfrac{\Delta[H_2O_2]}{\Delta t} = 8.7 \times 10^{-4}$ M s^{-1}

(b) The slope of the line is $\dfrac{\Delta[H_2O_2]}{\Delta t} = \dfrac{1.35\text{ M} - 0.000\text{ M}}{0\text{ s} - 3400\text{ s}} = -4.0 \times 10^{-4}$ M s^{-1}

Reaction rate $= -\dfrac{\Delta[H_2O_2]}{\Delta t} = 4.0 \times 10^{-4}$ M s^{-1}

7. Statement **(b)** is correct. After each half-life—that is, after each 75 s—the amount of reactant remaining is half of the amount that was present at the beginning of that half-life. Statement **(a)** is incorrect; the quantity of A remaining after 150 s is half of what was present after 75 s. Statement **(c)** is incorrect because different quantities of A are consumed in each 75 s of the reaction: $^1/_2$ of the original amount in the first 75 s, $^1/_4$ of the original amount in the second 75 s, $^1/_8$ of the original amount in the third 75 s, and so on. Statement **(d)** is incorrect; it implies a constant rate during the first half life. The rate of a first-order reaction actually decreases as time passes and reactant is consumed.

8. Substitute the given values into the rate equation to obtain the rate of reaction.
Rate $= k\,[A]^2\,[B]^0 = (0.0103\text{ M}^{-1}\text{ min}^{-1})\,(0.116\text{ M})^2\,(3.83\text{ M})^0 = 1.39 \times 10^{-4}$ M/min
Recall that (any quantity)^0 = 1.

9. The rate constant is determined as $k = 0.693/t_{1/2} = 0.693/19.8\text{ min} = 0.0350\text{ min}^{-1}$ Then the rate can be determined. Rate $= k[A] = 0.0350\text{ min}^{-1} \times 0.632\text{ M} = 0.0221\text{ M min}^{-1}$

10. **(a)** A first-order reaction has a constant half-life. Thus, half of the initial concentration remains after 30.0 minutes, and at the end of another half-life—60.0 minutes total—half of the concentration present at 30.0 minutes is gone: the concentration has decreased to one-quarter of its initial value. Or, we could say that the reaction is 75% complete after two half-lives—60.0 minutes.

(b) A zero-order reaction proceeds at a constant rate. Thus, if the reaction is 50% complete in 30.0 minutes, in twice the time—60.0 minutes—the reaction will be 100% complete. (And in one-fifth the time—6.0 minutes—the reaction will be 10% complete. Alternatively, we can say that the rate of reaction is 10%/6.0 min.) time to be 75% complete $= 75\% \times \dfrac{60.0\text{ min}}{100\%} = 45$ min

11. **(a)** Although we might be tempted to apply equation (15.12) to the given data, and calculate k first, there is an easier method. During one half-life, [A] decreases from 2.00 M to 1.00 M. Then, during the second half-life, [A] decreases from 1.00 M to 0.500 M. And finally, during the third half-life, [A] decreases 0.500 M to 0.250 M. Consequently, three half-lives must have elapsed while [A] has decreased from 2.00 M to 0.250 M. $3\,t_{1/2} = 126$ min $\quad t_{1/2} = 126\text{ min} \div 3 = 42.0$ min

(b) $k = \dfrac{0.693}{t_{1/2}} = \dfrac{0.693}{42.0\text{ min}} = 0.0165\text{ min}^{-1}$

12. **(a)** The mass of A has decreased to one fourth of its original value, from 1.60 g to 0.40 g. Since $\frac{1}{4} = \frac{1}{2} \times \frac{1}{2}$, we see that two half-lives have elapsed. Thus, $2 \times t_{1/2} = 38$ min, or $t_{1/2} = 19$ min.

(b) $k = 0.693/t_{1/2} = 0.693/19\text{ min} = 0.036\text{ min}^{-1}$ $\quad \ln\dfrac{[A]_t}{[A]_0} = -kt = -0.036\text{ min}^{-1} \times 60\text{ min} = -2.2$

$\dfrac{[A]_t}{[A]_0} = e^{-2.2} = 0.11$ *or* $[A]_t = [A]_0 e^{-kt} = 1.60\text{ g A} \times 0.11 = 0.18\text{ g A}$

13. **(a)** $\ln\dfrac{[A]_t}{[A]_0} = -kt = \ln\dfrac{0.632\text{ M}}{0.816\text{ M}} = -0.256 \qquad k = -\dfrac{-0.256}{16.0\text{ min}} = 0.0160\text{ min}^{-1}$

(b) $t_{1/2} = \dfrac{0.693}{k} = \dfrac{0.693}{0.0160\text{ min}^{-1}} = 43.3$ min

(c) Solve the integrated rate equation for the elapsed time.
$\ln\dfrac{[A]_t}{[A]_0} = -kt = \ln\dfrac{0.235\text{ M}}{0.816\text{ M}} = -1.245 = -0.0160\text{ min}^{-1} \times t \qquad t = \dfrac{-1.245}{-0.0160\text{ min}^{-1}} = 77.8$ min

(d) $\ln\dfrac{[A]}{[A]_0} = -kt$ becomes $\dfrac{[A]}{[A]_0} = e^{-kt}$ which in turn becomes

$[A] = [A]_0\,e^{-kt} = 0.816\text{ M}\exp\left(-0.0160\text{ min}^{-1} \times 2.5\text{ h} \times \dfrac{60\text{ min}}{1\text{h}}\right) = 0.816 \times 0.0907 = 0.074\text{ M}$

14. (a) From Expt. 1 to Expt. 3, when [B] remains constant and [A] doubles, the rate increases by a factor of

$$\frac{6.75 \times 10^{-4}\ \text{M s}^{-1}}{3.35 \times 10^{-4}\ \text{M s}^{-1}} = 2.01 \approx 2$$ Thus, the reaction is first order with respect to A.

From Expt 1 to Expt. 2, when [A] remains constant and [B] doubles, the rate increases by a facor of

$$\frac{1.35 \times 10^{-3}\ \text{M s}^{-1}}{3.35 \times 10^{-4}\ \text{M s}^{-1}} = 4.03 \approx 4$$ Thus, the reaction is second order with respect to B.

(b) Overall rate = order with respect to A + order with respect to B = 1 + 2 = 3 The reaction is third order overall.

(c) Rate = 3.35×10^{-4} M s^{-1} = $k(0.185\ \text{M})(0.133\ \text{M})^2$

$$k = \frac{3.35 \times 10^{-4}\ \text{M s}^{-1}}{(0.185\ \text{M})(0.133\ \text{M})^2} = 0.102\ \text{M}^{-2}\ \text{s}^{-1}$$

15. In the first 500 s, [A] decreases from 2.00 M to 1.00 M; this is the first half-life. From 500 s to 1500 s, an elapsed period of 1000 s, [A] decreases by half again, from 1.00 M to 0.50 M; this is the second half-life. Since the half-life is not constant, the reaction is not first order.

During the first 500 s, $\text{Rate} = -\frac{1.00\ \text{M} - 2.00\ \text{M}}{500\ \text{s}} = 0.00200$ M/s. Then during the first 1500 s, the rate is computed as $\text{Rate} = -\frac{0.50\ \text{M} - 2.00\ \text{M}}{1500\ \text{s}} = 0.00100$ M/s. Since the rate is not constant, this reaction is not zero order.

We conclude that the reaction is second order, by the process of elimination. We confirm this conclusion by computing values of the second-order rate constant with the equation $1/[\text{A}]_t - 1/[\text{A}]_0 = kt$

$$\frac{1}{1.00\ \text{M}} - \frac{1}{2.00\ \text{M}} = 0.500\ \text{M}^{-1} = k \times 500\ \text{s} \qquad k = \frac{0.500\ \text{M}^{-1}}{500\ \text{s}} = 0.0010\ \text{M}^{-1}\ \text{s}^{-1}$$

$$\frac{1}{0.50\ \text{M}} - \frac{1}{2.00\ \text{M}} = 1.50\ \text{M}^{-1} = k \times 1500\ \text{s} \qquad k = \frac{1.50\ \text{M}^{-1}}{1500\ \text{s}} = 0.0010\ \text{M}^{-1}\ \text{s}^{-1}$$

$$\frac{1}{0.25\ \text{M}} - \frac{1}{2.00\ \text{M}} = 3.50\ \text{M}^{-1} = k \times 3500\ \text{s} \qquad k = \frac{3.50\ \text{M}^{-1}}{3500\ \text{s}} = 0.0010\ \text{M}^{-1}\ \text{s}^{-1}$$

The constant value of the second-order rate constant indicates that this reaction indeed is second order.

16. Statement (d) is correct; the activation energy of an endothermic reaction must at least equal the value of ΔH for that reaction. However, if $E_a = \Delta H_{rxn}$, the products will readily "slip back down the hill" to reactants; that is, no net reaction will occur. (There will be no activation energy for the reverse reaction, nothing to prevent that reaction from occuring regardless of the energy of its reactants.) Thus E_a should be at least a slight bit greater than ΔH_{rxn}.

17. (a) $$\ln\frac{k_1}{k_2} = \frac{E_a}{R}\left(\frac{1}{T_2} - \frac{1}{T_1}\right) = \ln\frac{5.4 \times 10^{-4}\ \text{L mol}^{-1}\ \text{s}^{-1}}{2.8 \times 10^{-2}\ \text{L mol}^{-1}\ \text{s}^{-1}} = \frac{E_a}{R}\left(\frac{1}{683\ \text{K}} - \frac{1}{599\ \text{K}}\right)$$

$$-3.95\ R = -E_a \times 2.05 \times 10^{-4}$$

$$E_a = \frac{3.95\ R}{2.05 \times 10^{-4}} = 1.93 \times 10^4\ \text{K}^{-1} \times 8.3145\ \text{J mol}^{-1}\ \text{K}^{-1} = 1.60 \times 10^5\ \text{J/mol} = 160\ \text{kJ/mol}$$

(b) $$\ln\frac{k_1}{k_2} = \frac{E_a}{R}\left(\frac{1}{T_2} - \frac{1}{T_1}\right) = \ln\frac{5.0 \times 10^{-3}\ \text{L mol}^{-1}\ \text{s}^{-1}}{2.8 \times 10^{-2}\ \text{L mol}^{-1}\ \text{s}^{-1}} = \frac{1.60 \times 10^5\ \text{J/mol}}{8.3145\ \text{J mol}^{-1}\ \text{K}^{-1}}\left(\frac{1}{683\ \text{K}} - \frac{1}{T}\right)$$

$$-1.72 = 1.92 \times 10^4\left(\frac{1}{683\ \text{K}} - \frac{1}{T}\right) \qquad \left(\frac{1}{683\ \text{K}} - \frac{1}{T}\right) = \frac{-1.72}{1.92 \times 10^4} = -8.96 \times 10^{-5}$$

$$\frac{1}{T} = 8.96 \times 10^{-5} + 1.46 \times 10^{-3} = 1.55 \times 10^{-3} \qquad T = 645\ \text{K}$$

18. (a) Set II is data from a zero-order reaction. We know this because the rate of set II is constant. 0.25 M/25 s = 0.010 M s^{-1}. A zero-order reaction has a constant rate.

(b) A first-order reaction has a constant half-life. In set I, the first half-life is slightly less than 75 sec, since the concentration decreases by slightly more than half (from 1.00 M to 0.47 M) in 75 s. Again, from 75 s to 150 s the concentration decreases from 0.47 M to 0.22 M, again by slightly more than half, in a time of 75 s. Finally, two half-lives should see the concentration decrease to one-fourth of its initial value. This is what we see: from 100 s to 250 sec, 150 s of elapsed time, the concentration decreases from 0.37 M to 0.08 M; that is, to slightly less than one-fourth of its initial value. Notice that we cannot make the same statement of constancy of half-life regarding set III. The first half-life is 100 s, but it takes more than 150 s (from 100 s to 250 s) for [A] to again decrease by half.

(c) For a second-order reaction, $1/[A]_t - 1/[A]_0 = kt$. For the initial 100 s, we have

$$\frac{1}{0.50\text{ M}} - \frac{1}{1.00\text{ M}} = 1.0\text{ L mol}^{-1} = k\ 100\text{ s} \qquad k = 0.010\text{ L mol}^{-1}\text{ s}^{-1}$$

For the initial 200 s, we have

$$\frac{1}{0.33\text{ M}} - \frac{1}{1.00\text{ M}} = 2.0\text{ L mol}^{-1} = k\ 200\text{ s} \qquad k = 0.010\text{ L mol}^{-1}\text{ s}^{-1}$$

Since we obtain the same value of the rate constant using the equation for second-order kinetics, set III must be second order.

19. For a zero-order reaction, the rate equals the rate constant: $k = \Delta[A]/\Delta t = 1.00\text{ M}/100\text{ s} = 0.0100\text{ M/s}$.

20. Set I is the data for a first-order reaction; we can analyze those items of data to determine the half-life. In the first 75 s, the concentration decreases by a bit more than half. This implies a half-life slightly less than 75 s, perhaps 70 s. This is consistent with the other time periods noted in the answer to Review Question 18 (b) and also to the fact that in the 150 s from 50 s to 200 s, the concentration decreases from 0.61 M to 0.14 M, a bit more than a factor-of-four decrease. The factor-of-four decrease, to one-fourth of the initial value, is what we would expect of two successive half-lives. We can determine the half-life more accurately, by obtaining a value of k from the relation $\ln([A]_t/[A]_0) = -kt$ followed by $t_{1/2} = 0.693/k$

21. We can determine an approximate initial rate by using data from the first 25 s.

$$\text{Rate} = -\frac{\Delta[A]}{\Delta t} = -\frac{0.80\text{ M} - 1.00\text{ M}}{25\text{ s} - 0\text{ s}} = 0.0080\text{ M s}^{-1}$$

22. The approximate rate at 75 s can be taken as the rate over the time period from 50 s to 100 s.

(a) $\text{Rate}_{II} = -\dfrac{\Delta[A]}{\Delta t} = -\dfrac{0.00\text{ M} - 0.50\text{ M}}{100\text{ s} - 50\text{ s}} = 0.010\text{ M s}^{-1}$

(b) $\text{Rate}_{I} = -\dfrac{\Delta[A]}{\Delta t} = -\dfrac{0.37\text{ M} - 0.61\text{ M}}{100\text{ s} - 50\text{ s}} = 0.0048\text{ M s}^{-1}$

(c) $\text{Rate}_{III} = -\dfrac{\Delta[A]}{\Delta t} = -\dfrac{0.50\text{ M} - 0.67\text{ M}}{100\text{ s} - 50\text{ s}} = 0.0034\text{ M s}^{-1}$

Alternatively we can use [A] at 75 s (the values given in the table) in the relationship $\text{Rate} = k[A]^m$, where $m = 0, 1,$ or 2.

(a) $\text{Rate} = 0.010\text{ M s}^{-1} \times (0.25\text{ mol/L})^0 = 0.010\text{ M s}^{-1}$

(b) Since $t_{1/2} = 70$ s, $k = 0.693/70\text{ s} = 0.0099\text{ s}^{-1}$ $\quad$ $\text{Rate} = 0.0099\text{ s}^{-1} \times (0.47\text{ mol/L})^1 = 0.0047\text{ M s}^{-1}$

(c) $\text{Rate} = 0.010\text{ L mol}^{-1}\text{ s}^{-1} \times (0.57\text{ mol/L})^2 = 0.0032\text{ M s}^{-1}$

23. We can combine the approximate rates from Review Question 22, with the fact that 10 s have elapsed, and the concentration at 100 s.

(a) $[A]_{II} = 0.00$ M $\quad$ There is no reactant left after 100 s.

(b) $[A]_I = [A]_{100} - (10\text{ s} \times \text{rate}) = 0.37\text{ M} - (10\text{ s} \times 0.0047\text{ M s}^{-1}) = 0.32\text{ M}$

(c) $[A]_{III} = [A]_{100} - (10\text{ s} \times \text{rate}) = 0.50\text{ M} - (10\text{ s} \times 0.0032\text{ M s}^{-1}) = 0.47\text{ M}$

24. (a) We can demonstrate consistency with the stoichiometry by adding the two reactions.

Sum of the reactions: $\quad A + B + I + B \longrightarrow I + C + D$ $\quad$ We then cancel the "I" that appears on both sides of the resultant equation. $\quad A + 2\,B \longrightarrow C + D$

We determine the rate of the reaction from the mechanism by assuming that the rate of the elementary slow step equals the reaction rate: Reaction rate $= k_{slow}[A][B]$. This is equivalent to the observed rate law.

(b) Again, we demonstrate consistency with the stoichiometry by adding the two reactions to obtain: $2\,B + A + B_2 \longrightarrow B_2 + C + D$ $\quad$ We then cancel the "B_2" that appears on both sides of the resultant equation, and finally have: $\quad 2\,B + A \longrightarrow C + D$

The reaction rate is the rate of the elementary slow step: $\quad$ Reaction rate $= k_{slow}[A][B_2]$

But B_2 is a reaction intermediate whose concentration is difficult to determine. We can determine that concentration by assuming that the fast initial step goes equally rapidly in the forward and reverse directions. $\quad k_f[B]^2 = k_r[B_2] \qquad [B_2] = (k_f/k_r)[B]^2$

The resulting expression for $[B_2]$ is substituted into the expression for reaction rate, and we see that the experimentally determined rate law is *not* recovered.
Reaction rate $= k_{slow}[A][B_2] = k_{slow}[A](k_f/k_r)[B]^2 = k'[A][B]^2$

EXERCISES

Rates of Reactions

25. $\text{Rate} = \dfrac{-\Delta[A]}{\Delta t} = \dfrac{-(0.474\text{ M} - 0.485\text{ M})}{82.4\text{ s} - 71.5\text{ s}} = 1.0 \times 10^{-3}\text{ M s}^{-1}$

26. (a) $\text{Rate} = -\dfrac{\Delta[A]}{\Delta t} = -\dfrac{0.1498\text{ M} - 0.1565\text{ M}}{1.00\text{ min} - 0.00\text{ min}} = 0.0067\text{ M min}^{-1}$

$\text{Rate} = -\dfrac{\Delta[A]}{\Delta t} = -\dfrac{0.1433\text{ M} - 0.1498\text{ M}}{2.00\text{ min} - 1.00\text{ min}} = 0.0065\text{ M min}^{-1}$

(b) The rates are not equal because, in all except zero-order reactions, the rate depends on the concentration of reactant. And, of course, as the reaction proceeds reactant is consumed and its concentration decreases, changing the rate of the reaction.

27. (a) $\Delta[A] = \dfrac{\Delta[A]}{\Delta t}\Delta t = -2.2 \times 10^{-2}\text{ M/min} \times (5.00\text{ min} - 4.40\text{ min}) = -0.013\text{ M}$

$[A] = [A]_i + \Delta[A] = 0.588\text{ M} - 0.013\text{ M} = 0.575\text{ M}$

(b) $\Delta[A] = 0.565\text{ M} - 0.588\text{ M} = -0.023\text{ M}$

$\Delta t = \Delta[A]\dfrac{\Delta t}{\Delta[A]} = \dfrac{-0.023\text{ M}}{-2.2 \times 10^{-2}\text{ M/min}} = 1.0\text{ min}$ time $= t + \Delta t = (4.40 + 1.0)\text{ min} = 5.4\text{ min}$

28. Initial concentrations are $[HgCl_2] = 0.105$ M and $[C_2O_4^{2-}] = 0.300$ M. The initial rate of the reaction is $7.1 \times 10^{-5}\text{ M min}^{-1}$. Of course, the reaction is $2\ HgCl_2 + C_2O_4^{2-} \longrightarrow 2\ Cl^- + 2\ CO_2 + Hg_2Cl_2$. The rate of reaction equals the rate of disappearance of $C_2O_4^{2-}$. Then, after 1 hour, assuming that the rate is the same as the initial rate,

(a) $[HgCl_2] = 0.105\text{ M} - \left(7.1 \times 10^{-5}\dfrac{\text{mol } C_2O_4^{2-}}{\text{L·s}} \times \dfrac{2\text{ mol } HgCl_2}{1\text{ mol } C_2O_4^{2-}} \times 1\text{ h} \times \dfrac{60\text{ min}}{1\text{ h}}\right) = 0.096\text{ M}$

(b) $[C_2O_4^{2-}] = 0.300\text{ M} - \left(7.1 \times 10^{-5}\dfrac{\text{mol}}{\text{L·min}} \times 1\text{ h} \times \dfrac{60\text{ min}}{1\text{ h}}\right) = 0.296\text{ M}$

29. (a) $\text{Rate} = \dfrac{-\Delta[A]}{\Delta t} = \dfrac{\Delta[C]}{2\ \Delta t} = 1.76 \times 10^{-5}\text{ M s}^{-1}$ $\dfrac{\Delta[C]}{\Delta t} = 2 \times 1.76 \times 10^{-5}\text{ M s}^{-1} = 3.52 \times 10^{-5}\text{ M/s}$

(b) $\dfrac{\Delta[A]}{\Delta t} = -\dfrac{\Delta[C]}{2\ \Delta t} = -1.76 \times 10^{-5}\text{ M s}^{-1}$ Assume this rate is constant.

$[A] = 0.3580\text{ M} + \left(-1.76 \times 10^{-5}\text{ M s}^{-1} \times 1.00\text{ min} \times \dfrac{60\text{ s}}{1\text{ min}}\right) = 0.3569\text{ M}$

(c) $\dfrac{\Delta[A]}{\Delta t} = -1.76 \times 10^{-5}\text{ M s}^{-1}$ $\Delta t = \dfrac{\Delta[A]}{-1.76 \times 10^{-5}\text{ M/s}} = \dfrac{0.3500\text{ M} - 0.3580\text{ M}}{-1.76 \times 10^{-5}\text{ M/s}} = 4.5 \times 10^2\text{ s}$

30. (a) $\dfrac{\Delta n[O_2]}{\Delta t} = 1.00\text{ L soln} \times \dfrac{5.7 \times 10^{-4}\text{ mol } H_2O_2}{1\text{ L soln·s}} \times \dfrac{1\text{ mol } O_2}{2\text{ mol } H_2O_2} = 2.9 \times 10^{-4}\text{ mol } O_2\text{/s}$

(b) $\dfrac{\Delta n[O_2]}{\Delta t} = 2.9 \times 10^{-4}\dfrac{\text{mol } O_2}{\text{s}} \times \dfrac{60\text{ s}}{1\text{ min}} = 1.7 \times 10^{-2}\text{ mol } O_2\text{/min}$

(c) $\dfrac{\Delta V[O_2]}{\Delta t} = 1.7 \times 10^{-2}\dfrac{\text{mol } O_2}{\text{min}} \times \dfrac{22{,}414\text{ mL } O_2\text{ at STP}}{1\text{ mol } O_2} = \dfrac{3.8 \times 10^2\text{ mL } O_2\text{ at STP}}{\text{min}}$

31. Notice that, for every 1000 mmHg drop in the pressure of A(g), there will be a corresponding 2000 mmHg rise in the pressure of B(g) plus a 1000 mmHg rise in the pressure of C(g).

(a) We set up the calculation with three lines of information below the balanced equation: (1) the initial conditions, (2) the changes that occur, which are related to each other by reaction stoichiometry, and (3) the final conditions, which simply are initial conditions + changes.

	A(g)	$\longrightarrow$	2 B(g)	+	C(g)
Initial	1000. mmHg		0. mmHg		0. mmHg

Changes	–1000. mmHg		+2000. mmHg	+1000. mmHg
Final	0. mmHg		2000. mmHg	1000. mmHg

Total final pressure = 0. mmHg + 2000. mmHg + 1000. mmHg = 3000. mmHg

(b)

	A(g)	$\longrightarrow$	2 B(g)	+ C(g)
Initial	1000. mmHg		0. mmHg	0. mmHg
Changes	–200. mmHg		+400. mmHg	+200. mmHg
Final	800. mmHg		400. mmHg	200. mmHg

Total pressure = 800. mmHg + 400. mmHg + 200. mmHg = 1400. mmHg

32. **(a)** We use the ideal gas law to determine N_2O_5 pressure

$$P\{N_2O_5\} = \frac{nRT}{V} = \frac{\left(1.00\text{ g} \times \frac{1\text{ mol }N_2O_5}{108.0\text{ g}}\right) \times 0.08206\,\frac{\text{L·atm}}{\text{mol·K}} \times (273+65)\text{ K}}{15\text{ L}} \times \frac{760\text{ mmHg}}{1\text{atm}}$$

$= 13$ mmHg

(b) After 2.38 min, one half-life has passed, so the initial pressure of N_2O_5 has decreased by half to 6.5 mmHg.

(c) From the balanced chemical equation, the reaction of 2 mol $N_2O_5(g)$ produces 4 mol $NO_2(g)$ and 1 mol $O_2(g)$. That is, the consumption of 2 mol of reactant gas produces 5 mol of product gas. When measured at the same temperature and confined to the same volume, pressures will behave as amounts: the reaction of 2 mmHg of reactant produces 5 mmHg of product.

$$P_{total} = 13\text{ mmHg }N_2O_5\text{ initially} - 6.5\text{ mmHg }N_2O_5\text{ react} + \left(6.5\text{ mmHg react} \times \frac{5\text{ mmHg product}}{2\text{ mmHg reactant}}\right)$$

$= (13 - 6.5 + 16)$ mmHg $= 23$ mmHg total

Method of Initial Rates

33. From Expt. 1 and Expt. 2 we see that [B] remains fixed while [A] triples. As a result, the initial rate increases from 4.2×10^{-3} M/min to 1.3×10^{-2} M/min, that is, the rate triples. Therefore the reaction is first order in [A]. Between Expt. 2 and Expt. 3, we see that [A] doubles, which would double the rate, and [B] doubles. As a consequence, the initial rate goes from 1.3×10^{-2} M/min to 5.2×10^{-2} M/min, that is, the rate quadruples. Since an additional doubling of the rate is due to the change in [B], the reaction is first order in [B]. Now we determine the value of the rate constant.

Rate $= k[A]^1[B]^1$ $\quad k = \frac{\text{Rate}}{[A][B]} = \frac{5.2 \times 10^{-2}\text{ M/min}}{3.00\text{ M} \times 3.00\text{ M}} = 5.8 \times 10^{-3}\text{ L mol}^{-1}\text{ min}^{-1}$

The rate law is Rate $= (5.8 \times 10^{-3}\text{ L mol}^{-1}\text{ min}^{-1})[A]^1[B]^1$

34. **(a)** From Expt. 1 to Expt. 2, [B] remains constant at 1.40 M and [C] remains constant at 1.00 M, but [A] is halved (× 0.50). At the same time the rate is halved (× 0.50). Thus the reaction is first order with respect to A, since $0.50^x = 0.50$ when $x = 1$.

From Expt. 2 to Expt. 3, [A] remains constant at 0.70 M and [C] remains constant at 1.00 M, but [B] is halved (× 0.50), from 1.40 M to 0.70 M. At the same time, the rate is quartered (× 0.25). Thus, the reaction is second order with respect to B, since $0.50^y = 0.25$ when $y = 2$.

From Expt. 1 to Expt. 4, [A] remains constant at 1.40 M and [B] remains constant at 1.40 M, but [C] is halved (× 0.5), from 1.00 M to 0.50 M. At the same time, the rate is increased by a factor of 2.0. [$R_4 = 16\,R_3 = 16 \times \frac{1}{4}R_2 = 4\,R_2 = 4 \times \frac{1}{2}R_1 = 2 \times R_1$.] Thus, the order of the reaction with respect to C is –1, since $0.5^z = 2.0$ when $z = -1$.

(b) $R_5 = k\,(0.70\text{ M})^1(0.70\text{ M})^2(0.50\text{ M})^{-1} = k\left(\frac{1.40\text{ M}}{2}\right)^1\left(\frac{1.40\text{ M}}{2}\right)^2\left(\frac{1.00\text{ M}}{2}\right)^{-1}$

$= k\,\frac{1}{2}^1(1.40\text{ M})^1\,\frac{1}{2}^2(1.40\text{ M})^2\,\frac{1}{2}^{-1}(1.00\text{ M})^{-1} = R_1\left(\frac{1}{2}\right)^{1+2-1} = R_1\left(\frac{1}{2}\right)^2 = \frac{1}{4}R_1$

This is based on $R_1 = k\,(1.40\text{ M})^1\,(1.40\text{ M})^2\,(1.00\text{ M})^{-1}$

First-Order Reactions

35. **(a)** TRUE The rate of the reaction does decrease as more and more of B and C are formed, but not because more and more of B and C are formed. Rather the rate decreases because the concentration of A must decrease to form more and more of B and C.

(b) FALSE The time required for one half of substance A to react—the half-life—is independent of the quantity of "A" present.

36. **(a)** FALSE A plot of ln [A] or log [A] vs. time yields a straight line. One of [A] vs. time yields a curved line.

(b) TRUE The rate of formation of "C" is related to the rate of disappearance of "A" by the stoichiometry of the reaction.

37. **(a)** Since the half-life is 180 s, after 900 s five half-lives have elapsed, and the original quantity of A has been cut in half five times.

final quantity of A = $(0.5)^5 \times$ initial quantity of A = $0.03125 \times$ initial quantity of A

3.125% of the original quantity of A remains unreacted after 900 s.

(b) For a first order reaction $k = \frac{0.693}{t_{1/2}} = \frac{0.693}{180\ \text{s}} = 0.00385\ \text{s}^{-1}$

Rate = $k\,[\text{A}] = 0.00385\ \text{s}^{-1} \times 0.50\ \text{M} = 0.00193\ \text{M/s}$

38. **(a)** We note that the final concentration is one-eighth of the initial concentration. Thus, three half-lives have elapsed, the first to reduce [A] from 0.800 M to 0.400 M, the second to reduce [A] from 0.400 M to 0.200 M, and the third half-life to reduce [A] from 0.200 M to 0.100 M.

$$54\ \text{min} = 3\,t_{1/2} \qquad t_{1/2} = \frac{54\ \text{min}}{3} = 18\ \text{min}$$

To reduce the concentration even further requires two additional half-lives: the first of these to lower [A] from 0.100 M to 0.050 M and the second of them to reduce [A] from 0.050 M to 0.025 M. These two half-lives equal 2×18 min = 36 min. Thus at 54 min + 36 min = 90 min from the start of the reaction [A] = 0.025 M.

(b) We determine the rate constant for this first-order reaction: $k = \frac{0.693}{t_{1/2}} = \frac{0.693}{18\ \text{min}} = 0.039\ \text{min}^{-1}$

Then we determine the rate: Rate = $k[\text{A}]^1 = 0.039\ \text{min}^{-1} \times 0.025\ \text{M} = 9.8 \times 10^{-4}\ \text{M/min}$

39. We determine the value of the first-order rate constant and from that we determine the half-life. If the reactant is 99% decomposed in 137 min, then only 1% (0.010) of the initial concentration remains.

$$\ln\frac{[\text{A}]_t}{[\text{A}]_0} = -kt = \ln\frac{0.010}{1.000} = -4.61 = -k \times 137\ \text{min} \qquad k = \frac{4.61}{137\ \text{min}} = 0.0336\ \text{min}^{-1}$$

$$t_{1/2} = \frac{0.0693}{k} = \frac{0.693}{0.0336\ \text{min}^{-1}} = 20.6\ \text{min}$$

40. If 99% of the radioactivity of ^{32}P is lost, 1% (0.010) of that radioactivity remains. First we compute the value of the rate constant from the half life. $k = \frac{0.693}{t_{1/2}} = \frac{0.693}{14.3\ \text{d}} = 0.0485\ \text{d}^{-1}$

Then we use the integrated rate equation to determine the elapsed time.

$$\ln\frac{[\text{A}]_t}{[\text{A}]_0} = -kt \qquad t = -\frac{1}{k}\ln\frac{[\text{A}]_t}{[\text{A}]_0} = -\frac{1}{0.0485\ \text{d}^{-1}}\ln\frac{0.010}{1.000} = 95\ \text{days}$$

41. **(a)** When acetoacetic acid is 65% decomposed, its concentration has decreased to 35% of its initial value, a fraction of 0.35. The value of the first-order rate constant is calculated from the given value of the half-life. $k = \frac{0.693}{t_{1/2}} = \frac{0.693}{144\ \text{min}} = 4.81 \times 10^{-3}\ \text{min}^{-1}$

$$\ln\frac{[\text{A}]_t}{[\text{A}]_0} = -k\,t = \ln\frac{0.35 \times 0.135\ \text{M}}{0.135\ \text{M}} = -1.05 = -4.81 \times 10^{-3}\ \text{min}^{-1}\,t$$

$$t = \frac{1.05}{4.81 \times 10^{-3}\ \text{min}^{-1}} = 218\ \text{min}$$

Notice that we really did not need to know the initial concentration of acetoacetic acid.

(b) We can use partial pressures for gases or numbers of atoms for radioisotopes in the integrated rate equation; we also can use masses. We determine the mass of acetoacetic acid remaining after 575 min.

$\ln\frac{m_t}{m_0} = -kt = -4.81\times10^{-3}\ \text{min}^{-1}\times 575\ \text{min} = -2.77$ $\qquad \frac{m_t}{m_0} = e^{-2.77} = 0.063 = \frac{m_t}{10.0\ \text{g}}$

$m_t = 0.063 \times 10.0\ \text{g} = 0.63\ \text{g}$ remain $\qquad 10.0\ \text{g} - 0.63\ \text{g} = 9.4\ \text{g}$ have decomposed.

In the balanced chemical equation, 1 mol CO_2 is produced for each mol of acetoacetic acid that decomposes. We use the STP molar volume to determine the volume of $CO_2(g)$ formed.

$$CO_2\ \text{volume} = 9.4\ \text{g acid}\times\frac{1\ \text{mol}\ CH_3COCH_2COOH}{102.1\ \text{g}\ CH_3COCH_2COOH}\times\frac{1\ \text{mol}\ CO_2}{1\ \text{mol}\ CH_3COCH_2COOH}\times\frac{22.414\ \text{L}\ CO_2}{1\ \text{mol}}$$
$$= 2.1\ \text{L}\ CO_2(g)\ \text{at STP}$$

42. **(a)** $\ln\frac{[A]_t}{[A]_0} = -kt = \ln\frac{2.5\ \text{g}}{80.0\ \text{g}} = -3.47 = -6.2\times10^{-4}\ \text{s}^{-1}\ t$ $\qquad t = \frac{3.47}{6.2\times10^{-4}\ \text{s}^{-1}} = 5.6\times10^{3}\ \text{s}$

We substituted masses for concentrations, because the same substance (with the same mole weight) is present initially, and at time t and because it is present in the same volume.

(b) amount $O_2 = 77.5\ \text{g}\ N_2O_5\times\frac{1\ \text{mol}\ N_2O_5}{108.0\ \text{g}\ N_2O_5}\times\frac{1\ \text{mol}\ O_2}{2\ \text{mol}\ N_2O_5} = 0.359\ \text{mol}\ O_2$

$$V = \frac{nRT}{P} = \frac{0.359\ \text{mol}\ O_2\times 0.08206\ \frac{\text{L·atm}}{\text{mol·K}}\times(45+273)\ \text{K}}{745\ \text{mmHg}\times\frac{1\ \text{atm}}{760\ \text{mmHg}}} = 9.56\ \text{L}\ O_2$$

43. **(a)** If the reaction is first order, we will obtain the same value of the rate constant from several sets of data.

$\ln\frac{[A]_t}{[A]_0} = -kt = \ln\frac{0.497\ \text{M}}{0.600\ \text{M}} = -k\times100\ \text{s} = -0.188$ $\qquad k = \frac{0.188}{100\ \text{s}} = 1.88\times10^{-3}\ \text{s}^{-1}$

$\ln\frac{[A]_t}{[A]_0} = -kt = \ln\frac{0.344\ \text{M}}{0.600\ \text{M}} = -k\times300\ \text{s} = -0.556$ $\qquad k = \frac{0.556}{300\ \text{s}} = 1.85\times10^{-3}\ \text{s}^{-1}$

$\ln\frac{[A]_t}{[A]_0} = -kt = \ln\frac{0.285\ \text{M}}{0.600\ \text{M}} = -k\times400\ \text{s} = -0.744$ $\qquad k = \frac{0.744}{400\ \text{s}} = 1.86\times10^{-3}\ \text{s}^{-1}$

$\ln\frac{[A]_t}{[A]_0} = -kt = \ln\frac{0.198\ \text{M}}{0.600\ \text{M}} = -k\times600\ \text{s} = -1.109$ $\qquad k = \frac{1.109}{600\ \text{s}} = 1.85\times10^{-3}\ \text{s}^{-1}$

The virtual constancy of the rate constant throughout the time of the reaction confirms that the reaction is first order.

(b) We assume that the rate constant equals the average of the values obtained in part (a).

$k = \frac{1.88+1.85+1.86+1.85}{4}\times10^{-3}\ \text{s}^{-1} = 1.86\times10^{-3}\ \text{s}^{-1}$

(c) We use the integrated first-order rate equation:

$[A]_{750} = [A]_0\exp(-kt) = 0.600\ \text{M}\exp(-1.86\times10^{-3}\ \text{s}^{-1}\times750\ \text{s}) = 0.600\ \text{M}\ e^{-1.40} = 0.148\ \text{M}$

44. **(a)** If the reaction is first order, we will obtain the same value of the rate constant from several sets of data.

$\ln\frac{P_t}{P_0} = -kt = \ln\frac{264\ \text{mmHg}}{312\ \text{mmHg}} = -k\times390\ \text{s} = -0.167$ $\qquad k = \frac{0.167}{390\ \text{s}} = 4.28\times10^{-4}\ \text{s}^{-1}$

$\ln\frac{P_t}{P_0} = -kt = \ln\frac{224\ \text{mmHg}}{312\ \text{mmHg}} = -k\times777\ \text{s} = -0.331$ $\qquad k = \frac{0.331}{777\ \text{s}} = 4.26\times10^{-4}\ \text{s}^{-1}$

$\ln\frac{P_t}{P_0} = -kt = \ln\frac{187\ \text{mmHg}}{312\ \text{mmHg}} = -k\times1195\ \text{s} = -0.512$ $\qquad k = \frac{0.512}{1195\ \text{s}} = 4.28\times10^{-4}\ \text{s}^{-1}$

$\ln\frac{P_t}{P_0} = -kt = \ln\frac{78.5\ \text{mmHg}}{312\ \text{mmHg}} = -k\times3155\ \text{s} = -1.38$ $\qquad k = \frac{1.38}{3155\ \text{s}} = 4.37\times10^{-4}\ \text{s}^{-1}$

The virtual constancy of the rate constant confirms that the reaction is first order.

(b) The value of the rate constant was determined in part (a) as $4.3\times10^{-4}\ \text{s}^{-1}$.

(c) At 390 s, the pressure of dimethyl ether has dropped to 264 mmHg. Thus, an amount of dimethyl ether equivalent to a pressure of (312 mmHg – 264 mmHg =) 48 mmHg has decomposed. For each 1 mmHg pressure of dimethyl ether that decomposes, 3 mmHg of pressure of the products is produced. Thus, the increase in the pressure of the products is 3 × 48 = 144 mmHg. The total pressure at this point is 264 mmHg + 144 mmHg = 408 mmHg. Done in a more systematic fashion:

	$(CH_3)_2O(g)$	$\longrightarrow$	$CH_4(g)$ +	$H_2(g)$ +	$CO(g)$
Initial	312 mmHg		0 mmHg	0 mmHg	0 mmHg
Changes	–48 mmHg		+48 mmHg	+48 mmHg	+48 mmHg
Final	264 mmHg		48 mmHg	48 mmHg	48 mmHg

$P_{total} = P_{DME} + P_{methane} + P_{hydrogen} + P_{CO}$
$= 264\ \text{mmHg} + 48\ \text{mmHg} + 48\ \text{mmHg} + 48\ \text{mmHg} = 408\ \text{mmHg}$

(d) In the same manner as we solved part (c) we have the following.

	$(CH_3)_2O(g)$	$\longrightarrow$ $CH_4(g)$ +	$H_2(g)$ +	$CO(g)$
Initial	312 mmHg	0 mmHg	0 mmHg	0 mmHg
Changes	–312 mmHg	+312 mmHg	+312 mmHg	+312 mmHg
Final	0 mmHg	312 mmHg	312 mmHg	312 mmHg

$P_{total} = P_{DME} + P_{methane} + P_{hydrogen} + P_{CO}$
$= 0 \text{ mmHg} + 312 \text{ mmHg} + 312 \text{ mmHg} + 312 \text{ mmHg} = 936 \text{ mmHg}$

(e) We first determine P_{DME} at 1000 s. $\ln \frac{P_{1000}}{P_0} = -kt = -4.3 \times 10^{-4}\ \text{s}^{-1} \times 1000\ \text{s} = -0.43$

$\frac{P_{1000}}{P_0} = e^{-0.43} = 0.65$ $\quad P_{1000} = 312 \text{ mmHg} \times 0.65 = 203 \text{ mmHg}$

Then we use the technique of parts (c) and (d)

	$(CH_3)_2O(g)$	$\longrightarrow$ $CH_4(g)$ +	$H_2(g)$ +	$CO(g)$
Initial	312 mmHg	0 mmHg	0 mmHg	0 mmHg
Changes	–109 mmHg	+109 mmHg	+109 mmHg	+109 mmHg
Final	203 mmHg	109 mmHg	109 mmHg	109 mmHg

$P_{total} = P_{DME} + P_{methane} + P_{hydrogen} + P_{CO}$
$= 203 \text{ mmHg} + 109 \text{ mmHg} + 109 \text{ mmHg} + 109 \text{ mmHg} = 530. \text{ mmHg}$

Second-Order Reactions

45. **(a)** We can graph 1/[ArSOOH] vs. time and obtain a straight line. We also graph [ArSOOH] *vs.* time and ln [ArSOOH] *vs.* time to demonstrate that they do not yield a straight line.

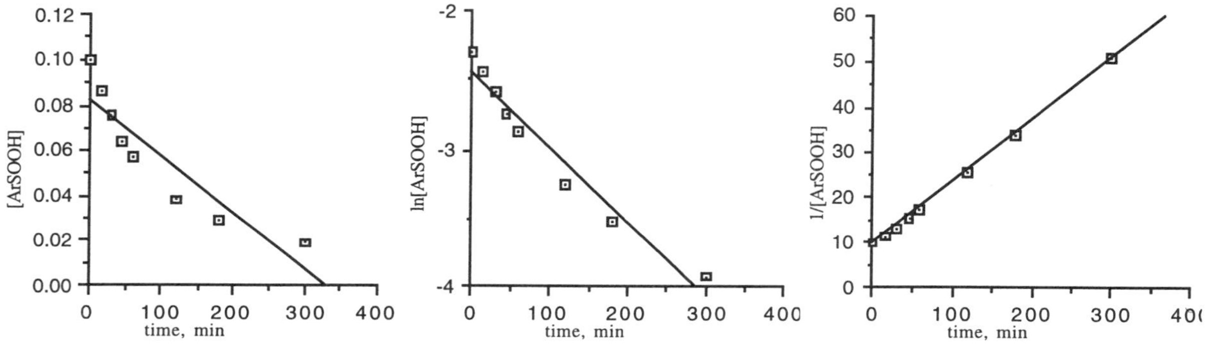

The straightness of the line indicates that the reaction is second order.

(b) We solve the rearranged integrated second-order rate law for the rate constant, using the longest time interval. $\frac{1}{[A]_t} - \frac{1}{[A]_0} = kt \qquad \frac{1}{t}\left(\frac{1}{[A]_t} - \frac{1}{[A]_0}\right) = k$

$$k = \frac{1}{300\ \text{min}}\left(\frac{1}{0.0196\ \text{M}} - \frac{1}{0.100\ \text{M}}\right) = 0.136\ \text{L mol}^{-1}\ \text{min}^{-1}$$

(c) We use the same equation as in part (b), but solved for *t*, rather than *k*.

$$t = \frac{1}{k}\left(\frac{1}{[A]_t} - \frac{1}{[A]_0}\right) = \frac{1}{0.136\ \text{L mol}^{-1}\ \text{min}^{-1}}\left(\frac{1}{0.0500\ \text{M}} - \frac{1}{0.100\ \text{M}}\right) = 73.5\ \text{min}$$

(d) We use the same equation as in part (b), but solved for *t*, rather than *k*.

$$t = \frac{1}{k}\left(\frac{1}{[A]_t} - \frac{1}{[A]_0}\right) = \frac{1}{0.136\ \text{L mol}^{-1}\ \text{min}^{-1}}\left(\frac{1}{0.0250\ \text{M}} - \frac{1}{0.100\ \text{M}}\right) = 221\ \text{min}$$

(e) We use the same equation as in part (b), but solved for *t*, rather than *k*.

$$t = \frac{1}{k}\left(\frac{1}{[A]_t} - \frac{1}{[A]_0}\right) = \frac{1}{0.136\ \text{L mol}^{-1}\ \text{min}^{-1}}\left(\frac{1}{0.0350\ \text{M}} - \frac{1}{0.100\ \text{M}}\right) = 137\ \text{min}$$

46. **(a)** We can either graph $1/[C_4H_6]$ vs. time and obtain a straight line, or we can determine the second-order rate constant from several data points. Then, if *k* indeed is a constant, the reaction is demonstrated to be second order. We shall use the second technique in this case. First we do a bit of algebra.

$$\frac{1}{[A]_t} - \frac{1}{[A]_0} = kt \qquad \frac{1}{t}\left(\frac{1}{[A]_t} - \frac{1}{[A]_0}\right) = k$$

$$k = \frac{1}{12.18 \text{ min}}\left(\frac{1}{0.0144 \text{ M}} - \frac{1}{0.0169 \text{ M}}\right) = 0.843 \text{ L mol}^{-1} \text{ min}^{-1}$$

$$k = \frac{1}{24.55 \text{ min}}\left(\frac{1}{0.0124 \text{ M}} - \frac{1}{0.0169 \text{ M}}\right) = 0.875 \text{ L mol}^{-1} \text{ min}^{-1}$$

$$k = \frac{1}{42.50 \text{ min}}\left(\frac{1}{0.0103 \text{ M}} - \frac{1}{0.0169 \text{ M}}\right) = 0.892 \text{ L mol}^{-1} \text{ min}^{-1}$$

$$k = \frac{1}{68.05 \text{ min}}\left(\frac{1}{0.00845 \text{ M}} - \frac{1}{0.0169 \text{ M}}\right) = 0.870 \text{ L mol}^{-1} \text{ min}^{-1}$$

The nearly constant value of the rate constant indicates that the reaction is second order.

(b) The rate constant is the average of the values obtained in part (a).

$$k = \frac{0.843 + 0.875 + 0.892 + 0.870}{4} \text{ L mol}^{-1} \text{ min}^{-1} = 0.870 \text{ L mol}^{-1} \text{ min}^{-1}$$

(c) We use the same equation as in part (a), but solved for t, rather than k.

$$t = \frac{1}{k}\left(\frac{1}{[\text{A}]_t} - \frac{1}{[\text{A}]_0}\right) = \frac{1}{0.870 \text{ L mol}^{-1} \text{ min}^{-1}}\left(\frac{1}{0.00423 \text{ M}} - \frac{1}{0.0169 \text{ M}}\right) = 204 \text{ min}$$

(d) We use the same equation as in part (a), but solved for t, rather than k.

$$t = \frac{1}{k}\left(\frac{1}{[\text{A}]_t} - \frac{1}{[\text{A}]_0}\right) = \frac{1}{0.870 \text{ L mol}^{-1} \text{ min}^{-1}}\left(\frac{1}{0.0050 \text{ M}} - \frac{1}{0.0169 \text{ M}}\right) = 162 \text{ min}$$

Establishing the Order of a Reaction

47. **(a)** $\text{Initial rate} = -\frac{\Delta[\text{A}]}{\Delta t} = -\frac{1.490 \text{ M} - 1.512 \text{ M}}{1.0 \text{ min} - 0.0 \text{ min}} = +0.022 \text{ M/min}$

$\text{Initial rate} = -\frac{\Delta[\text{A}]}{\Delta t} = -\frac{2.935 \text{ M} - 3.024 \text{ M}}{1.0 \text{ min} - 0.0 \text{ min}} = +0.089 \text{ M/min}$

(b) When the initial concentration is doubled (× 2.0), from 1.512 M to 3.024 M, the initial rate quadruples (× 4.0). Thus, the reaction is second order, since $2.0^x = 4.0$ when $x = 2$.

48. **(a)** Let us assess the possibilities. If the reaction is zero order, its rate will be constant. During the first 8 min, the rate is –(0.60 M – 0.80 M)/8 min = 0.025 M/min. Then, during the first 24 min, the rate is –(0.35 M – 0.80 M)/24 min = 0.019 M/min. Thus, the reaction is not zero order.

If the reaction is first order, it will have a constant half-life, that is consistent with its rate constant. The half-life can be assessed from the fact that 40 min elapse while the concentration drops from 0.80 M to 0.20 M, that is, to one-fourth of its initial value. Thus, 40 min equals two half-lives and $t_{1/2} = 20$ min. This gives $k = 0.693/t_{1/2} = 0.693/20 \text{ min} = 0.035 \text{ min}^{-1}$. Also

$$k\,t = -\ln\frac{[\text{A}]_t}{[\text{A}]_0} = -\ln\frac{0.35 \text{ M}}{0.80 \text{ M}} = 0.827 = k \times 24 \text{ min} \qquad k = \frac{0.827}{24 \text{ min}} = 0.034 \text{ min}^{-1}$$

The constancy of the value of k indicates that the reaction is first order.

(b) The value of rate constant is $k = 0.034 \text{ min}^{-1}$.

(c) Reaction rate $= \frac{1}{2}$(rate of formation of B) $= k[\text{A}]^1$ First we need [A] at t = 30. min

$$\ln\frac{[\text{A}]}{[\text{A}]_0} = -kt = -0.034 \text{ min}^{-1} \times 30. \text{ min} = -1.0_2 \qquad \frac{[\text{A}]}{[\text{A}]_0} = e^{-1.02} = 0.36$$

$[\text{A}] = 0.36 \times 0.80 \text{ M} = 0.29 \text{ M}$

rate of formation of B $= 2 \times 0.034 \text{ min}^{-1} \times 0.29 \text{ M} = 2.0 \times 10^{-2} \text{ M min}^{-1}$

49. The data appear to indicate a first-order reaction with a constant half-life. Notice that from $t = 0$ s with $[N_2O_5] = 1.46$ M to $t = 1116$ s with $[N_2O_5] = 0.72$ M, the $[N_2O_5]$ is approximately halved and the elapsed time is 1116 s. From $t = 1116$ s with $[N_2O_5] = 0.72$ M to $t = 2343$ s with $[N_2O_5] = 0.35$ M, the $[N_2O_5]$ is approximately halved and the elapsed time is 1227 s. And from $t = 423$ s with $[N_2O_5] = 1.09$ M to $t = 1582$ s with $[N_2O_5] = 0.54$ M, the $[N_2O_5]$ is approximately halved and the elapsed time is 1159 s. The half life is approximately constant. To make sure, we determine the value of k from several sets of data and see that it is indeed constant.

$$\ln\frac{[\text{A}]_t}{[\text{A}]_0} = -kt \qquad k = -\frac{1}{t}\ln\frac{[\text{A}]_t}{[\text{A}]_0} = -\frac{1}{423 \text{ s}}\ln\frac{1.09 \text{ M}}{1.46 \text{ M}} = 6.91 \times 10^{-4} \text{ s}^{-1}$$

$$k = -\frac{1}{1582 \text{ s}}\ln\frac{0.54 \text{ M}}{1.46 \text{ M}} = 6.29 \times 10^{-4} \text{ s}^{-1} \qquad k = -\frac{1}{1116 \text{ s}}\ln\frac{0.72 \text{ M}}{1.46 \text{ M}} = 6.33 \times 10^{-4} \text{ s}^{-1}$$

$$k = -\frac{1}{753 \text{ s}}\ln\frac{0.89 \text{ M}}{1.46 \text{ M}} = 6.57 \times 10^{-4} \text{ s}^{-1} \qquad k = -\frac{1}{1986 \text{ s}}\ln\frac{0.43 \text{ M}}{1.46 \text{ M}} = 6.16 \times 10^{-4} \text{ s}^{-1}$$

The approximate constancy of the rate constant over a wide range of elapsed times indicates that the reaction indeed is first order, with a rate constant of 6.5×10^{-4} s^{-1} (the average of the 5 values we have computed).

50. **(a)** In the first 22 s, [A] decreases from 0.715 M to 0.605 M, that is, $\Delta[A] = -0.110$ M. The rate in these 22 s is then determined. Rate $= \frac{-\Delta[A]}{\Delta t} = \frac{0.110 \text{ M}}{22 \text{ s}} = 5.0 \times 10^{-3}$ M/s In the first 74 s, $\Delta[A] =$ 0.345 M – 0.715 M, and the rate is determined. Rate $= \frac{-\Delta[A]}{\Delta t} = \frac{0.370 \text{ M}}{74 \text{ s}} = 5.0 \times 10^{-3}$ M/s

Finally, in the first 132 s, $\Delta[A] = 0.055$ M – 0.715 M, and the rate is determined as follows. Rate $= \frac{-\Delta[A]}{\Delta t} = \frac{0.660 \text{ M}}{132 \text{ s}} = 5.0 \times 10^{-3}$ M/s. Since the rate is constant for this reaction, it must be zero order.

(b) The half-life of this reaction is the time needed for one half of the initial [A] to react. Thus, $\Delta[A] =$ 0.715 M ÷ 2 = 0.358 M and $t_{1/2} = \frac{0.358 \text{ M}}{5.0 \times 10^{-3} \text{ M/s}} = 72$ s.

51. The half-life of the reaction depends on the concentration of "A" and thus this reaction cannot be first order. For a second-order reaction, the half-life varies inversely with the rate: $t_{1/2} = 1/(k[A]_0)$ or $k = 1/(t_{1/2}[A]_0)$. Let us attempt to verify the second-order nature of this reaction by seeing if the rate constant is fixed.

$k = \frac{1}{1.00 \text{ M} \times 50 \text{ min}} = 0.020$ L mol^{-1} min^{-1} $\quad k = \frac{1}{2.00 \text{ M} \times 25 \text{ min}} = 0.020$ L mol^{-1} min^{-1}

$k = \frac{1}{0.50 \text{ M} \times 100 \text{ min}} = 0.020$ L mol^{-1} min^{-1}

The constancy of the rate constant demonstrates that this reaction indeed is second order. The rate equation is Rate $= k[A]^2$ and $k = 0.020$ L mol^{-1} min^{-1}.

52. **(a)** The half-life depends on the initial [NH_3] and thus the reaction cannot be first order. Let us attempt to verify second-order kinetics.

$k = \frac{1}{[NH_3]_0 t_{1/2}}$ for a second-order reaction $\quad k = \frac{1}{0.0031 \text{ M} \times 7.6 \text{ min}} = 42$ L mol^{-1} min^{-1}

$k = \frac{1}{0.0015 \text{ M} \times 3.7 \text{ min}} = 180$ L mol^{-1} min^{-1} $\quad k = \frac{1}{0.00068 \text{ M} \times 1.7 \text{ min}} = 865$ L mol^{-1} min^{-1}

The reaction is not second order. But, if the reaction is zero order, its rate will be constant.

Rate $= \frac{[A]_0/2}{t_{1/2}} = \frac{0.0031 \text{ M} \div 2}{7.6 \text{ min}} = 2.0 \times 10^{-4}$ M/min $\quad$ Rate $= \frac{0.0015 \text{ M} \div 2}{3.7 \text{ min}} = 2.0 \times 10^{-4}$ M/min

Rate $= \frac{0.00068 \text{ M} \div 2}{1.7 \text{ min}} = 2.0 \times 10^{-4}$ M/min $\quad$ zero-order reaction

(b) The constancy of the rate indicates that the decomposition of ammonia under these conditions is zero order, and the rate constant is $k = 2.0 \times 10^{-4}$ M/min.

Collision Theory; Activation Energy

53. **(a)** The rate of a reaction depends on at least two factors other than the frequency of collisions. The first of these is whether each collision possesses sufficient energy to lead to a successful reaction. This depends on the activation energy of the reaction; the higher it is, the smaller will be the fraction of successful collisions. The second factor is whether the molecules in a given collision are oriented for a successful reaction. The more complicated the molecules are, the smaller the fraction of collisions that will be successfuly oriented.

(b) Although the collision frequency increases relatively slowly with temperature, the fraction of those collisions that have sufficient energy to overcome the activation energy increases much more rapidly. Therefore, the rate of reaction will increase dramatically with temperature.

(c) The addition of a catalyst has the net effect of decreasing the activation energy of the reaction, often by enabling another mechanism. The decreased activation energy means that a larger fraction of molecules has sufficient energy to lead to successful reaction. Thus the rate increases, even though the temperature does not.

54. **(a)** The activation energy for the reaction of hydrogen with oxygen is quite high, too high to be supplied by the energy ordinarily available in a mixture of the two gases. However, the spark supplies a suitably concentrated form of energy to initiate the reaction of at least a few molecules. Since the reaction is highly exothermic, the reaction of these first few molecules supplies sufficient energy for yet other molecules to react and the reaction proceeds to completion.

(b) A larger spark simply means that a larger number of molecules reacts initially. But the eventual course of the reaction remains the same, with the initial reaction producing enough energy to initiate still more molecules, and so on.

55. **(a)** The products are 21 kJ/mol closer in energy to the constant energy activated complex than are the reactants. Thus, the activation energy for the reverse reaction is

$$84 \text{ kJ/mol} - 21 \text{ kJ/mol} = 63 \text{ kJ/mol}.$$

(b) The reaction profile for this reaction is sketched at right.

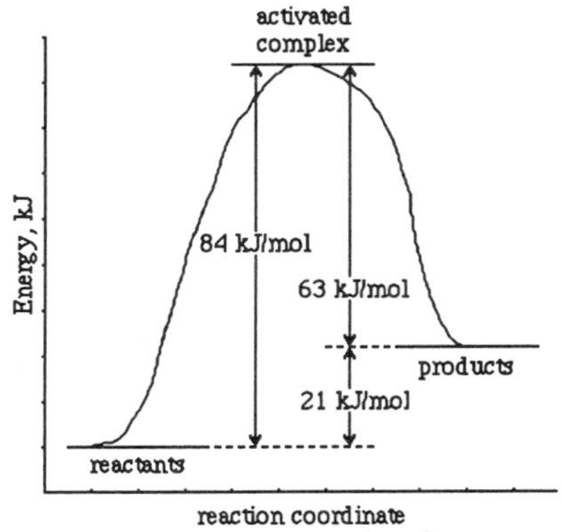

56. In an endothermic reaction, sketched at near right, we see that the value of the activation energy must be larger than the value of the enthalpy change for the reaction, ΔH. On the other hand, for an exothermic reaction (far right) there is no such requirement; we may have a small activation energy leading to a large value of ΔH.

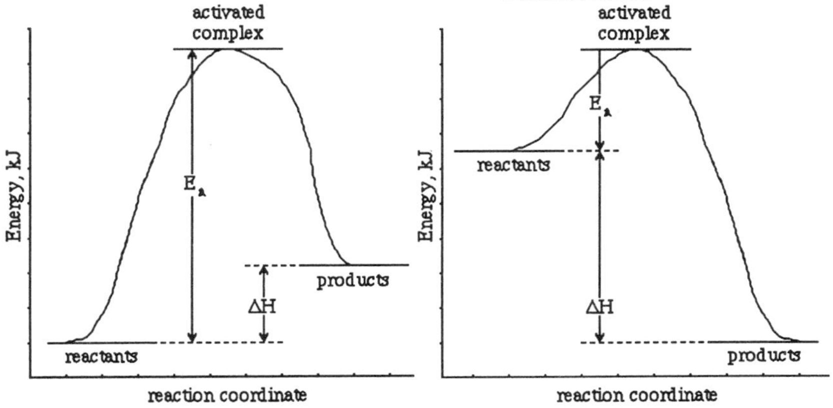

Effect of Temperature on Rates of Reaction

57. **(a)** First we need to compute values of ln *k* and $1/T$. Then we plot the graph.

t, °C	0 °C	10 °C	20 °C	30 °C
T, K	273 K	283 K	293 K	303 K
$1/T$, K^{-1}	0.00366	0.00353	0.00341	0.00330
k, s^{-1}	5.6×10^{-6}	3.2×10^{-5}	1.6×10^{-4}	7.6×10^{-4}
ln k	−12.09	−10.35	−8.74	−7.18

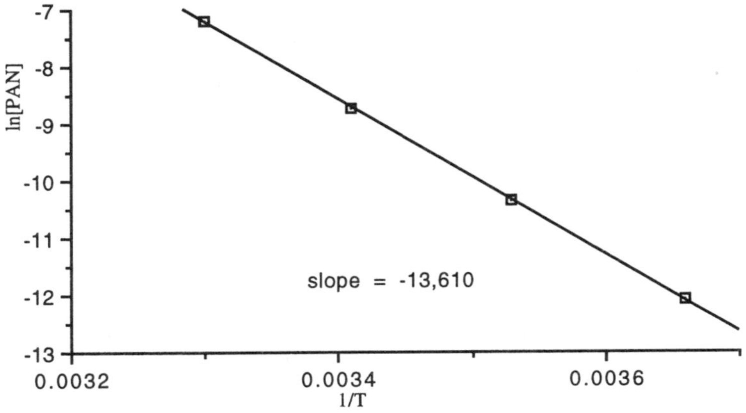

(b) The slope $= -E_a/R$.

$$E_a = -R \times \text{slope} = -8.3145 \frac{\text{J}}{\text{mol K}} \times -1.36 \times 10^4 \text{ K} = 113 \text{ kJ/mol}$$

(c) We apply the Arrhenius equation, with $k = 5.6 \times 10^{-6}\ s^{-1}$ at 0 °C (273 K), k = ? at 40 °C (313 K), and $E_a = 113 \times 10^3$ J/mol.

$$\ln \frac{k}{5.6 \times 10^{-6}\ s^{-1}} = \frac{E_a}{R}\left(\frac{1}{T_1} - \frac{1}{T_2}\right) = \frac{113 \times 10^3\ J/mol}{8.3145\ J\ mol^{-1}\ K^{-1}}\left(\frac{1}{273\ K} - \frac{1}{313\ K}\right) = 6.36$$

$$e^{6.36} = 578 = \frac{k}{5.6 \times 10^{-6}\ s^{-1}} \qquad k = 578 \times 5.6 \times 10^{-6}\ s^{-1} = 3.2 \times 10^{-3}\ s^{-1}$$

$$t_{1/2} = \frac{0.693}{k} = \frac{0.693}{3.2 \times 10^{-3}\ s^{-1}} = 2.2 \times 10^2\ s^{-1}$$

58. (a) We plot $\ln k$ vs. $1/T$. The slope of the straight line equals $-E_a/R$. First we tabulate the data to plot.

t	15.83	32.02	59.75	90.61	°C
T	288.98	305.17	332.90	363.76	K
$1/T$	0.0034604	0.0032769	0.0030039	0.0027491	K^{-1}
k	5.03×10^{-5}	3.68×10^{-4}	6.71×10^{-3}	0.119	$M^{-1}\ s^{-1}$
$\ln k$	−9.898	−7.907	−5.004	−2.129	

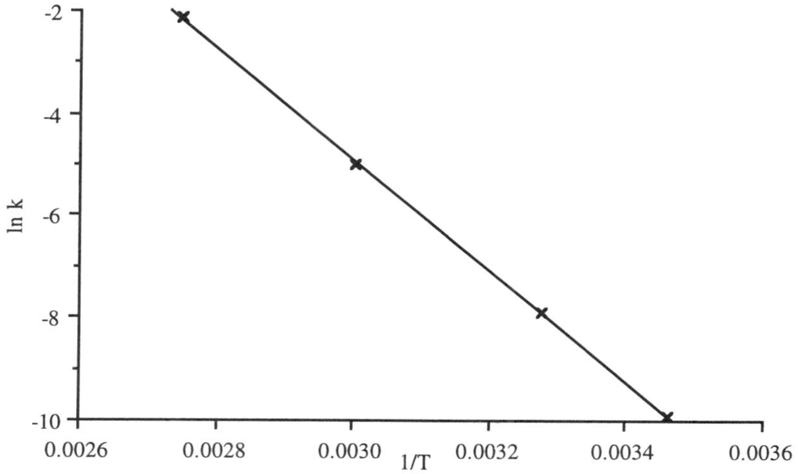

The slope of this graph = $-1.089 \times 10^4\ K = -E_a/R$

$E_a = -(-1.089 \times 10^4\ K) \times 8.3145\ J\ mol^{-1}\ K^{-1} = 9.055 \times 10^4$ J/mol = 90.55 kJ/mol

(b) We calculate the activation energy with the Arrhenius equation from the two extreme data points.

$$\ln \frac{k_2}{k_1} = \ln \frac{0.119}{5.03 \times 10^{-5}} = +7.769 = \frac{E_a}{R}\left(\frac{1}{T_1} - \frac{1}{T_2}\right) = \frac{E_a}{R}\left(\frac{1}{288.98\ K} - \frac{1}{363.76\ K}\right)$$

$$= 7.1138 \times 10^{-4}\ K^{-1}\ \frac{E_a}{R} \qquad E_a = \frac{7.769 \times 8.3145\ J\ mol^{-1}\ K^{-1}}{7.1138 \times 10^{-4}\ K^{-1}} = 9.080 \times 10^4\ J/mol$$

E_a = 90.80 kJ/mol The values are in quite good agreement, with a 0.3% deviation.

(c) We apply the Arrhenius equation, with $E_a = 9.080 \times 10^4$ J/mol, $k = 5.03 \times 10^{-5}\ M^{-1}\ s^{-1}$ at 15.83 °C (288.98 K), and k = ? at 100.0 °C (373.2 K).

$$\ln \frac{k}{5.03 \times 10^{-5}\ M^{-1}\ s^{-1}} = \frac{E_a}{R}\left(\frac{1}{T_1} - \frac{1}{T_2}\right) = \frac{90.80 \times 10^3\ J/mol}{8.3145\ J\ mol^{-1}\ K^{-1}}\left(\frac{1}{288.98K} - \frac{1}{373.2\ K}\right) = 8.528$$

$$e^{8.528} = 5.05 \times 10^3 = \frac{k}{5.03 \times 10^{-5}\ M^{-1}\ s^{-1}} \qquad k = 5.05 \times 10^3 \times 5.03 \times 10^{-5}\ M^{-1}\ s^{-1} = 0.254\ M^{-1}\ s^{-1}$$

59. The half-life of a first-order reaction is inversely proportional to its rate constant: $k = 0.693/t_{1/2}$. Thus we can apply a modified version of the Arrhenius equation.

(a) $$\ln \frac{k_2}{k_1} = \ln \frac{(t_{1/2})_1}{(t_{1/2})_2} = \frac{E_a}{R}\left(\frac{1}{T_1} - \frac{1}{T_2}\right) = \ln \frac{46.2\ min}{2.6\ min} = \frac{E_a}{R}\left(\frac{1}{298\ K} - \frac{1}{(102 + 273)\ K}\right)$$

$$2.88 = \frac{E_a}{R}\ 6.89 \times 10^{-4} \qquad E_a = \frac{2.88 \times 8.3145}{6.89 \times 10^{-4}} \times \frac{1\ kJ}{1000\ J} = 34.8\ kJ/mol$$

(b) $\ln\frac{10.0\ \text{min}}{46.2\ \text{min}} = \frac{34.8\times 10^3\ \text{J/mol}}{8.3145\ \text{J mol}^{-1}\ \text{K}^{-1}}\left(\frac{1}{T}-\frac{1}{298}\right) = -1.53 = 4.19\times 10^3\left(\frac{1}{T}-\frac{1}{298}\right)$

$\left(\frac{1}{T}-\frac{1}{298}\right) = \frac{-1.53}{4.19\times 10^3} = -3.65\times 10^{-4}$ $\qquad \frac{1}{T} = 2.99\times 10^{-3}$ $\qquad T = 334\ \text{K} = 61\ °\text{C}$

60. The half-life of a first-order reaction is inversely proportional to its rate constant: $k = 0.693/t_{1/2}$. Thus we can apply a modified version of the Arrhenius equation.

(a) $\ln\frac{k_2}{k_1} = \ln\frac{(t_{1/2})_1}{(t_{1/2})_2} = \frac{E_a}{R}\left(\frac{1}{T_1}-\frac{1}{T_2}\right) = \ln\frac{22.5\ \text{h}}{1.5} = \frac{E_a}{R}\left(\frac{1}{293\ \text{K}}-\frac{1}{(40+273)\ \text{K}}\right)$

$2.71 = \frac{E_a}{R}\ 2.18\times 10^{-4}$ $\qquad E_a = \frac{2.71\times 8.3145}{2.18\times 10^{-4}}\times\frac{1\ \text{kJ}}{1000\ \text{J}} = 103\ \text{kJ/mol}$

(b) The relationship is $k = A\exp(-E_a/RT)$

$k = 2.05\times 10^{13}\ \text{s}^{-1}\exp\left(\frac{-103\times 10^3\ \text{J mol}^{-1}}{8.3145\ \text{J mol}^{-1}\ \text{K}^{-1}\times(273+30)\ \text{K}}\right) = 2.05\times 10^{13}\ \text{s}^{-1}\ e^{-40.9}$

$= 3.5\times 10^{-5}\ \text{s}^{-1}$

61. **(a)** It is the change in the value of the rate constant that causes the reaction to go faster. Let k_1 be the rate constant at room temperature, 20 °C or 293 K. Then $k_2 = 2\,k_1$ is the rate constant ten degrees higher, at 30 °C or 303 K.

$\ln\frac{k_2}{k_1} = \ln\frac{2\,k_1}{k_1} = 0.693 = \frac{E_a}{R}\left(\frac{1}{T_1}-\frac{1}{T_2}\right) = \frac{E_a}{R}\left(\frac{1}{293}-\frac{1}{303\ \text{K}}\right) = 1.13\times 10^{-4}\ \text{K}^{-1}\frac{E_a}{R}$

$E_a = \frac{0.693\times 8.3145\ \text{J mol}^{-1}\ \text{K}^{-1}}{1.13\times 10^{-4}\ \text{K}^{-1}} = 5.1\times 10^4\ \text{J/mol} = 51\ \text{kJ/mol}$

(b) Since the activation energy for the formation of HI from its elements is 171 kJ/mol, we would not expect this reaction to follow the rule of thumb.

62. Under a pressure of 2.00 atm the boiling point of water is approximately 121 °C or 394 K. Under a pressure of 1 atm, the boiling point of water is 100 °C or 373 K. We assume an activation energy of 5.1×10^4 J/mol and compute the ratio of the two rates.

$\ln\frac{\text{Rate}_2}{\text{Rate}_1} = \frac{E_a}{R}\left(\frac{1}{T_1}-\frac{1}{T_2}\right) = \frac{5.1\times 10^4\ \text{J/mol}}{8.3145\ \text{J mol}^{-1}\ \text{K}^{-1}}\left(\frac{1}{373}-\frac{1}{394\ \text{K}}\right) = 0.88$

$\text{Rate}_2 = e^{0.88}\ \text{Rate}_1 = 2.4\ \text{Rate}_1$ $\qquad$ Cooking will occur 2.4 times faster in the pressure cooker.

Catalysis

63. **(a)** Although a catalyst is *recovered unchanged from a reaction*, it does "take part in the reaction." And some catalysis actually slow down the rate of a reaction. Usually, however, these negative catalysts are called inhibitors.

(b) The function of a catalyst is to *change the mechanism of a reaction*. The new mechanism is one that has a different (lower) activation energy than the original reaction.

64. If the reaction is first order, its half-life is 100 min, for in this time period [S] decreases from 1.00 M to 0.50 M, that is, by one half. This gives a rate constant of $k = 0.693/t_{1/2} = 0.693/100\ \text{min} = 0.00693\ \text{min}^{-1}$. The rate constant also can be determined from any two of the other sets of data.

$k\,t = \ln\frac{[A]_0}{[A]_t} = \ln\frac{1.00\ \text{M}}{0.70\ \text{M}} = 0.357 = k\times 60\ \text{min}$ $\qquad k = \frac{0.357}{60\ \text{min}} = 0.00595\ \text{min}^{-1}$

This is not a particularly constant rate constant. Let's try zero order, where the rate should be constant.

$\text{Rate} = -\frac{0.90\ \text{M} - 1.00\ \text{M}}{20\ \text{min}} = 0.0050\ \text{M/min}$ $\qquad \text{Rate} = -\frac{0.50\ \text{M} - 1.00\ \text{M}}{100\ \text{min}} = 0.0050\ \text{M/min}$

$\text{Rate} = -\frac{0.20\ \text{M} - 0.90\ \text{M}}{160\ \text{min} - 20\ \text{min}} = 0.0050\ \text{M/min}$ $\qquad \text{Rate} = -\frac{0.50\ \text{M} - 0.90\ \text{M}}{100\ \text{min} - 20\ \text{min}} = 0.0050\ \text{M/min}$

Thus, this reaction is zero order with respect to [S].

65. Both platinum and an enzyme can be considered heterogeneous catalysts in the sense that the catalytic activity occurs on their surfaces or near them. Even so, some would classify an enzyme as a homogeneous catalyst.

The most important difference, however, is one of specificity. Platinum is a rather nonspecific catalyst, catalyzing many different reactions. An enzyme, however, is quite specific, usually catalyzing only one reaction rather than all reactions of a given class.

66. In both the enzyme and the metal surface cases, the reaction occurs in a specialized location: either with the enzyme or on the surface of the catalyst. At high concentrations of reactant, the limiting factor in determining the rate is not the concentration of reactant present but how rapidly active sites become available for reaction to occur. Thus, the rate of the reaction depends on either the quantity of enzyme present or the surface area of the catalyst, rather than on how much reactant is present; the reaction is zero order. At low concentrations or gas pressures the reaction rate depends on how rapidly molecules can reach the available active sites. Thus, the rate depends on concentration or pressure of reactant and is first order.

Reaction Mechanisms

67. The molecularity of an elementary process is the number of reactant molecules in that process. This molecularity is equal to the order of the overall reaction only if the elementary process in question is the slowest and thus, the rate-determining, step of the overall reaction. In addition, the elementary process in question should be the only elementary step that influences the rate of the reaction.

68. If the type of molecule that is expressed in the rate law as being first order collides with other molecules that are present in much larger concentrations, the reaction will seem to depend only on the concentration of those types of molecules present in smaller concentration, since the larger concentration will be essentially unchanged during the course of the reaction. Such a situation is quite common, and has been given the name pseudo first order. Another situation is that molecules which do not participate directly in the reaction—including product molecules—strike the reactant molecules and impart to them sufficient energy to react.

69. The three elementary steps must sum to the overall reaction. All species in the equations below are gases.

overall:	$2\,NO + 2\,H_2 \longrightarrow N_2 + 2\,H_2O$		$2\,NO + 2\,H_2 \longrightarrow N_2 + 2\,H_2O$
–first:	$-(2\,NO \rightleftharpoons N_2O_2)$		$N_2O_2 \rightleftharpoons 2\,NO$
–third	$-(N_2O + H_2 \longrightarrow N_2 + H_2O)$	*or*	$N_2 + H_2O \longrightarrow N_2O + H_2$
the result is the second step, which is slow:			$H_2 + N_2O_2 \longrightarrow H_2O + N_2O$
The rate of this rate-determining step is:			Rate $= k_2\,[H_2]\,[N_2O_2]$

Since N_2O_2 does not appear in the overall reaction, we need to replace its concentration with the concentrations of species that do appear in the overall reaction. To do this, recall that the first step is rapid, with the forward reaction occurring at the same rate as the reverse reaction.

$k_1[NO]^2$ = forward rate = reverse rate = $k_{-1}[N_2O_2]$

This expression is solved for $[N_2O_2]$, which then is substituted into the rate equation for the overall reaction.

$$[N_2O_2] = \frac{k_1\,[NO]^2}{k_{-1}} \qquad \text{Rate} = \frac{k_2\,k_1}{k_{-1}}\,[H_2]\,[NO]^2$$

This reaction is first-order in $[H_2]$ and second-order in [NO], agreeing with the experimental determination.

70. A possible mechanism is: $O_3 \underset{k_2}{\overset{k_1}{\rightleftharpoons}} O_2 + O$ (fast) $\qquad O + O_3 \xrightarrow{k_3} 2\,O_2$ (slow)

The rate is that of the slow step: Rate $= k_3[O][O_3]$. But O is a reaction intermediate, whose concentration is difficult to determine. An expression for [O] can be found by assuming that the forward and reverse "fast" steps proceed equally rapidly.

$$\text{Rate}_1 = \text{Rate}_2 \qquad k_1[O_3] = k_2[O_2][O] \qquad [O] = \frac{k_1[O_3]}{k_2[O_2]}$$

Then substitute this expression into the rate law for the reaction. $\quad \text{Rate} = k_3\,\frac{k_1[O_3]}{k_2[O_2]}\,[O_3] = \frac{k_3\,k_1}{k_2}\,\frac{[O_3]^2}{[O_2]}$

This rate equation has the same form as the experimentally determined rate law.

FEATURE PROBLEMS

A. **1.** To determine the order of the reaction, we need $[C_6H_5N_2Cl]$ at each time. To determine this value, note that 58.3 mL $N_2(g)$ evolved corresponds to total depletion of $C_6H_5N_2Cl$, to $[C_6H_5N_2Cl]$ = 0.000 M.

Thus, at any point in time, $[C_6H_5N_2Cl] = 0.071\text{ M} - \left(\text{volume } N_2(g) \times \frac{0.071\text{ M } C_6H_5N_2Cl}{58.3\text{ mL } N_2(g)}\right)$

Consider 21 min: $[C_6H_5N_2Cl] = 0.071\ M - \left(44.3\ mL\ N_2 \times \dfrac{0.071\ M\ C_6H_5N_2Cl}{58.3\ mL\ N_2(g)}\right) = 0.017\ M$

The numbers in the following table are determined with this method.

time, min	0	3	6	9	12	15	18	21	24	27	30	∞
volume N_2, mL	0	10.8	19.3	26.3	32.4	37.3	41.3	44.3	46.5	48.4	50.4	58.3
$[C_6H_5N_2Cl]$, mM	71	58	47	39	32	26	21	17	14	12	10	0

[The concentration is given in thousandths of a mole per liter (mM).]

2.

time, min	0	3	6	9	12	15	18	21	24	27	30	∞
Δt, min		3	3	3	3	3	3	3	3	3	3	
$[C_6H_5N_2Cl]$, mM	71	58	47	39	32	26	21	17	14	12	10	0
$\Delta[C_6H_5N_2Cl]$, mM		−13	−11	−8	−7	−6	−5	−4	−3	−2	−2	
Reaction rate		4.3	3.7	2.7	2.3	2.0	1.7	1.3	1.0	0.7	0.7	

3. The two graphs are drawn on the same axes.

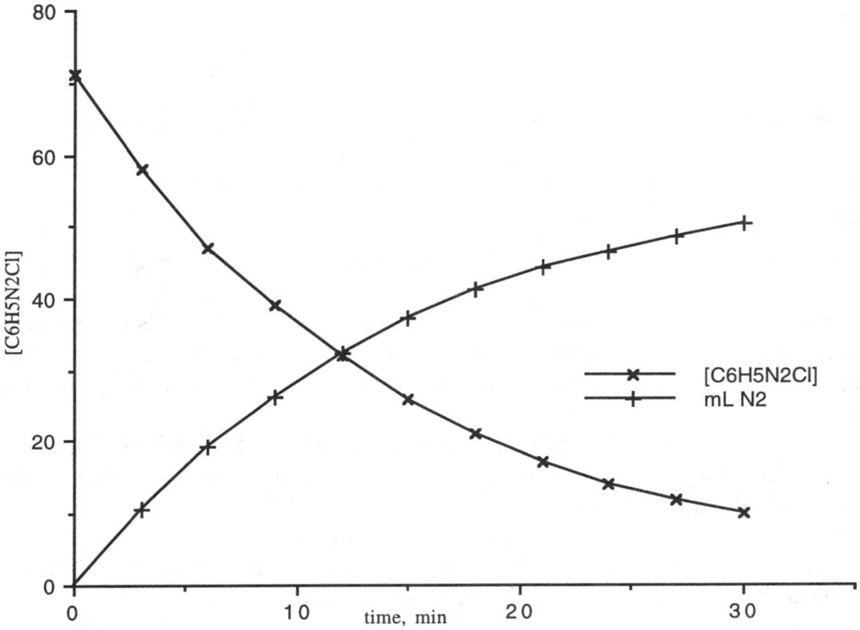

4. The rate of the reaction at $t = 21$ min is the slope of the tangent line to the $[C_6H_5N_2Cl]$ curve. The tangent line intercepts the vertical axis at about $[C_6H_5N_2Cl] = 39$ mN and the horizontal axis at about 37 min. $\text{Rate} = \dfrac{39 \times 10^{-3}\ M}{37\ min} = 1.0_5 \times 10^{-3}\ M\ min^{-1} = 1.1 \times 10^{-3}\ M\ min^{-1}$

The agreement with the reported value is very good.

5. The initial rate is the slope of the tangest line to the $[C_6H_5N_2Cl]$ curve at $t = 0$. The intercept with the vertical axis is 71 mM, of course. That with the horizontal axis is about 13 min.

$\text{Rate} = \dfrac{71 \times 10^{-3}\ M}{13\ min} = 5.5 \times 10^{-3}\ M\ min^{-1}$

6. The first-order rate law is Rate = $k\,[C_6H_5N_2Cl]$, which we solve for $k = \dfrac{\text{Rate}}{[C_6H_5N_2Cl]}$

$k_0 = \dfrac{5.5 \times 10^{-3}\ M\ min^{-1}}{71 \times 10^{-3}\ M} = 0.077\ min^{-1}$ $\qquad k_{21} = \dfrac{1.1 \times 10^{-3}\ M\ min^{-1}}{17 \times 10^{-3}\ M} = 0.065\ min^{-1}$

An average value would be a good estimate: $k_{avg} = 0.071\ min^{-1}$

7. The estimated rate constant gives one value of the half-life: $t_{1/2} = \dfrac{0.693}{k} = \dfrac{0.693}{0.071\ min^{-1}} = 9.8$ min

The first half-life occurs when $[C_6H_5N_2Cl]$ drops from 0.071 M to 0.0355 M. This seems to occur at about 10.5 min.

8. The reaction should be three-fourths complete in two half-lives, about 21 minutes.

9. The graph plots $[C_6H_5N_2Cl]$ (in millimoles/L) *vs.* time in minutes.

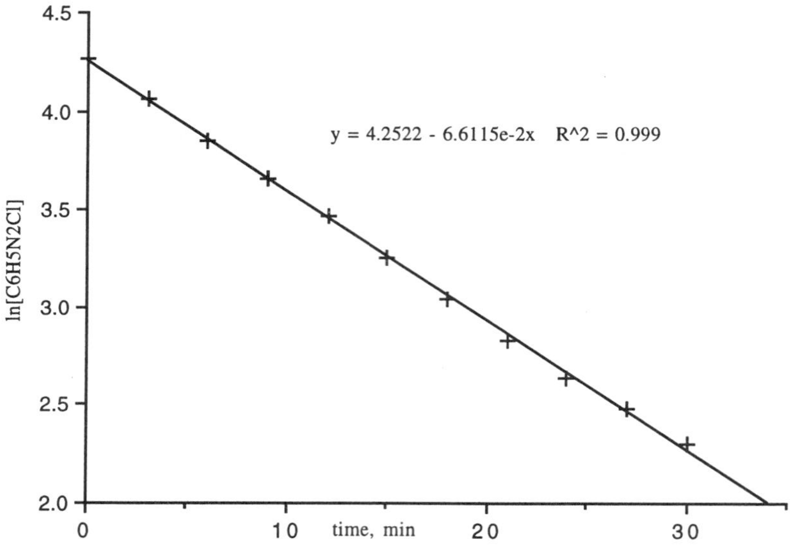

The linearity of the graph demonstrates that the reaction is first order.

10. $k = -\text{slope} = -(-6.61 \times 10^{-2})\ \text{min}^{-1} = 0.0661\ \text{min}^{-1}$

$t_{1/2} = \dfrac{0.693}{0.0661\ \text{min}^{-1}} = 10.5$ min in good agreement with our previously determined values.

B. 1. In Experiments 1 & 2, [KI] is the same (0.20 M), while $[(NH_4)_2S_2O_8]$ is halved, from 0.20 M to 0.10 M. As a consequence, the time to produce a color change doubles, that is, the rate is halved. This indicates that the reaction is first-order in $(NH_4)_2S_2O_8$. Experiments 2 and 3 produce a similar conclusion.

In Experiments 4 and 5, $[(NH_4)_2S_2O_8]$ is the same (0.20 M) while [KI] is halved, from 0.10 to 0.050 M. As a consequence, the time to produce a color change nearly doubles, that is, the rate is halved. This indicates that the reaction is also first-order in KI.

The reaction is (first + first) second order overall.

2. The blue color appears when all the $S_2O_3^{2-}$ has been consumed, for only then does reaction (b) cease. The same amount of $S_2O_3^{2-}$ is placed in each reaction mixture.

$$\text{amount } S_2O_3^{2-} = 10.0\ \text{mL} \times \frac{1\ \text{L}}{1000\ \text{mL}} \times \frac{0.010\ \text{mol}\ Na_2S_2O_3}{1\ \text{L}} \times \frac{1\ \text{mol}\ S_2O_3^{2-}}{1\ \text{mol}\ Na_2S_2O_3} = 1.0 \times 10^{-4}\ \text{mol}$$

Through stoichiometry, we determine the amount of each reactant that is reacts before this amount of $S_2O_3^{2-}$ will be consumed.

$$\text{amount } S_2O_8^{2-} = 1.0 \times 10^{-4}\ \text{mol}\ S_2O_3^{2-} \times \frac{1\ \text{mol}\ I_3^-}{2\ \text{mol}\ S_2O_3^{2-}} \times \frac{1\ \text{mol}\ S_2O_8^{2-}}{1\ \text{mol}\ I_3^-} = 5.0 \times 10^{-5}\ \text{mol}\ S_2O_8^{2-}$$

$$\text{amount } I^- = 5.0 \times 10^{-5}\ \text{mol}\ S_2O_8^{2-} \times \frac{2\ \text{mol}\ I^-}{1\ \text{mol}\ S_2O_8^{2-}} = 1.0 \times 10^{-4}\ \text{mol}\ I^-$$

Note that we do not use "3 mol I^-" from equation (a) since one mole has not been oxidized; it simply complexes with the product I_2.

The total volume of each solution is (25.0 mL + 25.0 mL + 10.0 mL + 5.0 mL =) 65.0 mL, or 0.0650 L. The amount of $S_2O_8^{2-}$ that reacts in each case is 5.0×10^{-5} mol and thus

$$\Delta[S_2O_8^{2-}] = \frac{-5.0 \times 10^{-5}\ \text{mol}}{0.0650\ \text{L}} = -7.7 \times 10^{-4}\ \text{M}$$

$$\text{Thus Rate}_1 = \frac{-\Delta[S_2O_8^{2-}]}{\Delta t} = \frac{+7.7 \times 10^{-4}\ \text{M}}{21\ \text{s}} = 3.7 \times 10^{-5}\ \text{M s}^{-1}$$

3. For Experiment 2, $\text{Rate}_2 = \dfrac{-\Delta[S_2O_8^{2-}]}{\Delta t} = \dfrac{+7.7 \times 10^{-4}\ \text{M}}{42\ \text{s}} = 1.8 \times 10^{-5}\ \text{M s}^{-1}$

To determine the value of k, we need initial concentrations, as altered by dilution.

$[S_2O_8^{2-}]_1 = 0.20\ \text{M} \times \dfrac{25.0\ \text{mL}}{65.0\ \text{mL total}} = 0.077\ \text{M}$ $[I^-]_1 = 0.20 \times \dfrac{25.0\ \text{mL}}{65.0\ \text{mL}} = 0.077\ \text{M}$

$\text{Rate}_1 = 3.7 \times 10^{-5}\ \text{M s}^{-1} = k[S_2O_8^{2-}]^1\ [I^-]^1 = k\ (0.077\ \text{M})^1\ (0.077\ \text{M})^1$

$$k = \frac{3.7 \times 10^{-5}\ \text{M s}^{-1}}{0.077\ \text{M} \times 0.077\ \text{M}} = 6.4 \times 10^{-3}\ \text{M}^{-1}\ \text{s}^{-1}$$

$$[S_2O_8^{2-}]_2 = 0.10\ \text{M} \times \frac{25.0\ \text{mL}}{65.0\ \text{mL total}} = 0.038\ \text{M} \quad [I^-]_1 = 0.20 \times \frac{25.0\ \text{mL}}{65.0\ \text{mL}} = 0.077\ \text{M}$$

$\text{Rate}_1 = 1.8 \times 10^{-5}\ \text{M s}^{-1} = k[S_2O_8^{2-}]^1\ [I^-]^1 = k\ (0.038\ \text{M})^1\ (0.077\ \text{M})^1$

$$k = \frac{1.8 \times 10^{-5}\ \text{M s}^{-1}}{0.038\ \text{M} \times 0.077\ \text{M}} = 6.2 \times 10^{-3}\ \text{M}^{-1}\ \text{s}^{-1}$$

4. First we determine concentrations for Experiment 4.

$$[S_2O_8^{2-}]_4 = 0.20\ \text{M} \times \frac{25.0\ \text{mL}}{65.0\ \text{mL total}} = 0.077\ \text{M} \quad [I^-]_4 = 0.050 \times \frac{25.0\ \text{mL}}{65.0\ \text{mL}} = 0.019\ \text{M}$$

We have two expressions for Rate; let us equate them and solve for the rate constant.

$$\text{Rate}_2 = \frac{-\Delta[S_2O_8^{2-}]}{\Delta t} = \frac{+7.7 \times 10^{-4}\ \text{M}}{\Delta t} = k[S_2O_8^{2-}]_4^1\ [I^-]_4^1 = k\ (0.077\ \text{M})\ (0.019\ \text{M})$$

$$k = \frac{7.7 \times 10^{-4}\ \text{M}}{\Delta t \times 0.077\ \text{M} \times 0.019\ \text{M}} = \frac{0.53\ \text{M}^{-1}}{\Delta t} \qquad k_3 = \frac{0.53\ \text{M}^{-1}}{189\ \text{s}} = 0.0028\ \text{M}^{-1}\ \text{s}^{-1}$$

$$k_{13} = \frac{0.53\ \text{M}^{-1}}{88\ \text{s}} = 0.0060\ \text{M}^{-1}\ \text{s}^{-1} \qquad k_{24} = \frac{0.53\ \text{M}^{-1}}{42\ \text{s}} = 0.013\ \text{M}^{-1}\ \text{s}^{-1}$$

$$k_{33} = \frac{0.53\ \text{M}^{-1}}{21\ \text{s}} = 0.025\ \text{M}^{-1}\ \text{s}^{-1}$$

5. We plot ln k *vs.* $1/T$ The slope of the line = $-E_a/R$.

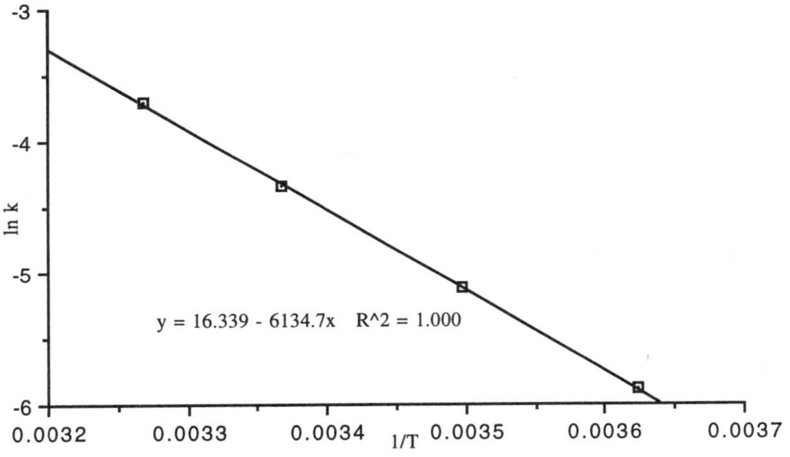

$E_a = +6135\ \text{K} \times 8.3145\ \text{J mol}^{-1}\ \text{K}^{-1} = 51.0 \times 10^3\ \text{J/mol} = 51.0\ \text{kJ/mol}$

The scatter of the data permits only a two significant figure result: 51 kJ/mol

6. For the mechanism to agree with the stoichiometry, the steps of the mechanism must add to the overall reaction, in the manner of Hess's law.

(slow) $I^- + S_2O_8^{2-} \longrightarrow IS_2O_8^{3-}$

(fast) $IS_2O_8^{3-} \longrightarrow 2\ SO_4^{2-} + I^+$

(fast) $I^+ + I^- \longrightarrow I_2$

(fast) $I_2 + I^- \longrightarrow I_3^-$

(net) $3\ I^- + S_2O_8^{2-} \longrightarrow 2\ SO_4^{2-} + I_3^-$

Each of the intermediates cancels: $IS_2O_8^{3-}$ is produced in the first step and consumed in the second, I^+ is produced in the second step and consumed in the third, I_2 is produced in the third step and consumed in the 4th. The mechanism is consistent with the stoichiometry.

The rate of the slow step of the mechanism is $\text{Rate}_1 = k_1\ [S_2O_8^{2-}]^1\ [I^-]^1$

This is exactly the same as the experimental rate law.

It is reasonable that the first step be slow since it involves two negatively charges species coming together. We know that like charges repel, and thus this should not be an easy or rapid process.

16 Principles of Chemical Equilibrium

Practice Examples

1A For the cited rection $CO(g) + 2\,H_2(g) \rightleftharpoons CH_3OH(g)$ $\qquad K = \dfrac{[CH_3OH]}{[CO][H_2]^2} = 14.5$

We know that $[CO] = [CH_3OH]$. Thus, one may be substituted for the other, and the resulting equation solved for $[H_2]$. $\quad K = \dfrac{[CO]}{[CO][H_2]^2} = 14.5 = \dfrac{1}{[H_2]^2} \qquad [H_2] = \sqrt{\dfrac{1}{14.5}} = 0.263\text{ M}$

1B We simply substitute the values we have into the equilibrium constant expression, and solve for $[H_2]$.

$$K_c = \frac{[NH_3]^2}{[N_2][H_2]^3} = 1.8 \times 10^4 = \frac{(2.00\text{ M})^2}{0.015\text{ M }[H_2]^3} \qquad [H_2] = \sqrt[3]{[H_2]^3} = \sqrt[3]{\frac{(2.00)^2}{1.8 \times 10^4 \times 0.015}} = 0.25\text{ M}$$

2A The Example gives $K_c = 3.6 \times 10^8$ for the reaction $N_2(g) + 3\,H_2(g) \rightleftharpoons 2\,NH_3(g)$. The reaction we are considering is one-third of this reaction. If we divide the reaction by 3, we should take the cube root of the equilibrium constant to obtain the value of the equilibrium constant for the "divided" reaction.

$$K_{c3} = \sqrt[3]{K_c} = \sqrt[3]{3.6 \times 10^8} = 7.1 \times 10^2$$

2B First we reverse the given reaction to put $NO_2(g)$ on the product side. The new equilibrium constant is the inverse of the given one. $\qquad NO_2(g) \rightleftharpoons NO(g) + \frac{1}{2}O_2(g) \qquad K_c' = 1/7.5 \times 10^2 = 0.0013$

Then we double the reaction to obtain 2 moles of $NO_2(g)$ as reactant. The equilibrium constant is then raised to the second power. $\qquad 2\,NO_2(g) \rightleftharpoons 2\,NO(g) + O_2(g) \qquad K_c = (0.0013)^2 = 1.7 \times 10^{-6}$

3A We use the expression $K_p = K_c(RT)^{\Delta n}$. In this case, $\Delta n_{gas} = 3 + 1 - 2 = 2$ and thus we have

$$K_p = K_c(RT)^2 = 2.8 \times 10^{-9}\left(0.08206\,\frac{\text{L·atm}}{\text{mol·K}} \times 298\text{K}\right)^2 = 1.7 \times 10^{-6}$$

3B We begin by writing the K_p expression. We then substitute $(n/V)RT = [\ ]RT$ for each pressure. We collect terms to obtain an expression relating K_c and K_p, into which we substitute to find the value of K_c.

$$K_p = \frac{\{P(H_2)\}^2\{P(S_2)\}}{\{P(H_2S)\}^2} = \frac{([H_2]RT)^2\,([S_2]RT)}{([H_2S]RT)^2} = \frac{[H_2]^2[S_2]}{[H_2S]^2}RT = K_cRT$$

We would have obtained the same result by using $K_p = K_c(RT)^{\Delta n_{gas}}$, since $\Delta n_{gas} = 2 + 1 - 2 = +1$.

$$K_c = \frac{K_p}{RT} = \frac{1.2 \times 10^{-2}}{\dfrac{0.08206\text{ L atm}}{\text{mol K}} \times (1065 + 273)\text{ K}} = 1.1 \times 10^{-4}$$

4A We remember that neither solids, such as $Ca_5(PO_4)_3OH(s)$, nor liquids, such as H_2O, appear in the equilibrium constant expression. Concentrations of products appear in the numerator, those of reactants in the denominator. $\qquad K_c = \dfrac{[Ca^{2+}]^5[HPO_4^{2-}]^3}{[H^+]^4}$

4B First we write the balanced chemical equation for the reaction. Then we write the equilibrium constant expressions, remembering that gases and solutes in aqueous solution appear in the K_c expression, but pure liquids and pure solids do not. $\qquad 3\,Fe(s) + 4\,H_2O(g) \rightleftharpoons Fe_3O_4(s) + 4\,H_2(g)$

$K_p = \dfrac{\{P(H_2)\}^4}{\{P(H_2O)\}^4} \qquad K_c = \dfrac{[H_2]^4}{[H_2O]^4} \qquad$ Because $\Delta n_{gas} = 4 - 4 = 0$, $K_p = K_c$

5A We compare the value of the reaction quotient, Q_c, with that of K_c.

$$Q_c = \frac{[PCl_3][Cl_2]}{[PCl_5]} = \frac{0.50\ M \times 0.20\ M}{4.50\ M} = 0.022 < 0.0454 = K_c$$

Because $Q_c < K_c$, the net reaction will proceed to the right, forming products.

5B We compute the value of Q_c. Each concentration equals the mass (m) of the substance divided by its molar mass (which quotient is the amount of the substance in moles) and further divided by the volume of the container.

$$Q_c = \frac{[CO_2][H_2]}{[CO][H_2O]} = \frac{\dfrac{m \times \dfrac{1\ \text{mol}}{44.0\ \text{g}\ CO_2}}{V} \times \dfrac{m \times \dfrac{1\ \text{mol}}{2.0\ \text{g}\ H_2}}{V}}{\dfrac{m \times \dfrac{1\ \text{mol}}{28.0\ \text{g}\ CO}}{V} \times \dfrac{m \times \dfrac{1\ \text{mol}}{18.0\ \text{g}\ H_2O}}{V}} = \frac{\dfrac{1}{44.0 \times 2.0}}{\dfrac{1}{28.0 \times 18.0}} = \frac{28.0 \times 18.0}{44.0 \times 2.0}$$

$$= 5.7 > 1.00 = K_c$$

(In evaluating the expression above, we cancelled the equal values of V, and we also cancelled the equal values of m.) Because the value of Q_c is larger than the value of K_c, reaction will proceed to the left to reach a state of equilibrium. Thus, at equilibrium there will be larger amounts of reactants, and smaller amounts of products than there are initially.

6A $O_2(g)$ is a reactant. The equilibrium system will shift right forming product in an attempt to consume some of the added reactant. Looked at in another way, $[O_2]$ is increased above its equilibrium value by the addition of oxygen. This makes Q_c smaller than K_c. (The $[O_2]$ factor is in the denominator of the expression.) And the system shifts right to compensate.

6B **(a)** The position of an equilibrium mixture is affected only by changing the concentration of substances that appear in the equilibrium constant expression, $K_c = [CO_2]$. Since CaO(s) is a pure solid, its concentration does not appear in the equilibrium constant expression and thus its addition will have no direct effect on the position of equilibrium.

(b) The addition of $CO_2(g)$ will increase $[CO_2]$ above its equilibrium value. The reaction will shift left to alleviate this increase; some $CaCO_3(s)$ will form.

7A We know that a decrease in volume or an increase in pressure on an equilibrium mixture of gases causes a net reaction in the direction producing the smaller number of moles of gas. In the reaction in question, that direction is to the left: one mole of $N_2O_4(g)$ is formed when two moles of $NO_2(g)$ combine. Thus, decreasing the cylinder volume would have the initial effect of doubling both $[N_2O_4]$ and $[NO_2]$. When the system regained equilibrium $[N_2O_4]$ would have increased still more and $[NO_2]$ would have decreased somewhat. It is unlikely, however, that $[NO_2]$ would decrease below its *starting* equilibrium value.

7B In the balanced chemical equation for the chemical reaction $\Delta n_{gas} = (1 + 1) - (1 + 1) = 0$. As a consequence, a change in overall volume or total gas pressure will have no effect on the position of equilibrium. In the equilibrium constant expression, the two partial pressures in the numerator will be affected to exactly the same degree as the two partial pressures in the denominator, and Q_p will continue to equal K_p.

8A The cited reaction is endothermic. Raising the temperature on an equilibrium mixture favors the endothermic reaction. Thus, $N_2O_4(g)$ should decompose more completely at higher temperatures and the amount of $NO_2(g)$ formed from a given amount of $N_2O_4(g)$ will be greater at high temperatures than at low ones.

8B The equation for the formation reaction is $\frac{1}{2}N_2(g) + \frac{3}{2}H_2(g) \longrightarrow NH_3(g) \quad \Delta H° = -46.11$ kJ/mol

This reaction is an exothermic reaction. Lowering temperature causes a shift in the direction of this exothermic reaction. The reaction will shift right at a lower temperature. $[NH_3(g)]$ will be greater at –100 °C.

9A We write the expression for K_c and then substitute expressions for molar concentrations.

$$K_c = \frac{[H_2]^2\ [S_2]}{[H_2S]^2} = \frac{\left(\dfrac{0.22\ \text{mol}\ H_2}{3.00\ \text{L}}\right)^2 \dfrac{0.11\ \text{mol}\ S_2}{3.00\ \text{L}}}{\left(\dfrac{2.78\ \text{mol}\ H_2S}{3.00\text{L}}\right)^2} = 2.3 \times 10^{-4}$$

9B We write the equilibrium constant expression and solve it for $[N_2O_4]$.

$K_c = 4.61 \times 10^{-3} = \frac{[NO_2]^2}{[N_2O_4]}$ $[N_2O_4] = \frac{[NO_2]^2}{4.61 \times 10^{-3}} = \frac{(0.0236\text{ M})^2}{4.61 \times 10^{-3}} = 0.121\text{ M}$

Then we determine the mass of N_2O_4 present in 2.26 L.

$$N_2O_4 \text{ mass} = 2.26\text{ L} \times \frac{0.121\text{ mol } N_2O_4}{1\text{ L}} \times \frac{92.01\text{ g } N_2O_4}{1\text{ mol } N_2O_4} = 25.2\text{ g } N_2O_4$$

10A We use the initial-change-equilibrium setup to establish the amount of each substance at equilibrium. We label each entry in the table in the order of its determination (1st, 2nd, 3rd, 4th, 5th), to better illustrate the technique. We know the inital amounts of all substances (1st). (There are zero moles of each product; none was put in.) Because "initial" + "change" = "equilibrium", the equilibrium amount (2nd) of $Br_2(g)$ enables us to determine "change" (3rd) for $Br_2(g)$. We then use stoichiometry to write other entries (4th) on the "change" line. And finally, we determine the remaining equilibrium amounts (5th).

reaction:	$2\ NOBr(g)$	$\rightleftharpoons$	$2\ NO(g)$	+	$Br_2(g)$
initial:	1.86 mol 1st		0.00 mol 1st		0.00 mol 1st
change:	–0.164 mol 4th		+0.164 mol 4th		+0.082 mol 3rd
equil.:	1.70 mol 5th		0.164 mol 5th		0.082 mol 2nd

$$K_c = \frac{[NO]^2\,[Br_2]}{[NOBr]^2} = \frac{\left(\frac{0.164\text{ mol NO}}{5.00\text{ L}}\right)^2 \frac{0.082\text{ mol } Br_2}{5.00\text{ L}}}{\left(\frac{1.70\text{ mol NOBr}}{5.00\text{L}}\right)^2} = 1.5 \times 10^{-4}$$

In this case, $\Delta n_{gas} = 2 + 1 - 2 = +1$. $K_p = K_c(RT)^{+1} = 1.5 \times 10^{-4} \times 0.08206 \frac{\text{L·atm}}{\text{mol·K}} \times 298\text{ K} = 3.7 \times 10^{-3}$

10B We use the amounts stated in the problem to determine the equilibrium amount of each substance.

reaction:	$2\ SO_3(g)$ $\rightleftharpoons$	$2SO_2(g)$ +	$O_2(g)$
initial:	0 mol	0.100 mol	0.100 mol
changes:	+0.0916 mol	–0.0916 mol	–0.0916/2 mol
equil.:	0.0916 mol	0.0084 mol	0.0542 mol
concns:	$\frac{0.916\text{ mol}}{1.52\text{ L}}$	$\frac{0.0084\text{ mol}}{1.52\text{ L}}$	$\frac{0.0542\text{ mol}}{1.52\text{ L}}$
concns:	0.0603 M	0.0055 M	0.0357 M

We use these values to compute K_c for the reaction and then the relationship $K_p = K_c(RT)^{\Delta n_{gas}}$ (with $\Delta n_{gas} = 1 + 1 - 1 = +1$) to determine the value of K_p.

$$K_c = \frac{[SO_2]^2[O_2]}{[SO_3]^2} = \frac{(0.0055)^2\,(0.0357)}{(0.0603)^2} = 3.0 \times 10^{-4} \quad K_p = 3.0 \times 10^{-4} \times \frac{0.08206\text{ L atm}}{\text{mol K}} \times 900\text{ K} = 0.022$$

11A The equilibrium constant expression is $K_p = P\{H_2O\}P\{CO_2\} = 0.231$ at 100 °C. From the balanced chemical equation, we see that one mole of $H_2O(g)$ is formed for each mole of $CO_2(g)$. Consequently, $P\{H_2O\} = P\{CO_2\}$ and $K_p = (P\{CO_2\})^2$. We solve this expression for $P\{CO_2\}$

$$P\{CO_2\} = \sqrt[2]{(P\{CO_2\})^2} = \sqrt[2]{K_c} = \sqrt[2]{0.231} = 0.481\text{ atm}$$

11B The equation for the reaction is $NH_4HS(s) \rightleftharpoons NH_3(g) + H_2S(g)$ $K_p = 0.108$ at 25 °C

The two partial pressures do not have to be equal at equilibrium. The only instance in which they must be equal is when all of the two gases come from the decomposition of $NH_4HS(s)$. In this case, some of the $NH_3(g)$ comes from another source. We can obtain the pressure of $H_2S(g)$ by substitution into the equilibrium constant expression, since we are given the equilibrium pressure of $NH_3(g)$.

$K_p = P\{H_2S\}\ P\{NH_3\} = 0.108 = P\{H_2S\} \times 0.500\text{ atm } NH_3$ $P\{H_2S\} = \frac{0.108}{0.500} = 0.216\text{ atm}$

12A We set up this problem in the same manner we have previously employed, designating the equilibrium amount of HI as $2x$. (Note that we use the same multipliers for x as the stoichiometric coefficients.)

Equation:	$H_2(g)$ +	$I_2(g)$ $\rightleftharpoons$	$2\ HI(g)$
Initial:	0.150 mol	0.200 mol	0 mol
Changes:	$-x$ mol	$-x$ mol	$+2x$ mol
Equil:	$(0.150 - x)$ mol	$(0.200 - x)$ mol	$2x$ mol

We substitute these expressions into the equilibrium constant expression, at right, and solve for x

$$K_c = \frac{\left(\frac{2x}{15.0}\right)^2}{\frac{0.150 - x}{15.0} \times \frac{0.200 - x}{15.0}} = \frac{(2x)^2}{(0.150 - x)(0.200 - x)} = 50.2$$

$4x^2 = (0.150 - x)(0.200 - x)\ 50.2 = 50.2\ (0.0300 - 0.350\ x + x^2) = 1.51 - 17.6\ x + 50.2\ x^2$

$0 = 46.2\,x^2 - 17.6\,x + 1.51$ Now we use the quadratic formula to obtain a value of x.

$$x = \frac{-b \pm \sqrt{b^2 - 4ac}}{2a} = \frac{17.6 \pm \sqrt{(17.6)^2 - 4 \times 46.2 \times 1.51}}{2 \times 46.2} = \frac{17.6 \pm 5.54}{92.4} = 0.250 \text{ or } 0.131$$

The first root cannot be used because we cannot have $(0.150 - 0.250 = -0.100)$ a negative amount of H_2. Thus, we have $2 \times 0.131 = 0.262$ mol HI at equilibrium. We check by substituting amounts into the K_c expression. (Notice that the volumes cancel.) The slight disagreement in the two values (52 compared to 50.2) is the result of rounding error.

$$K_c = \frac{(0.262)^2}{(0.150 - 0.131)(0.200 - 0.131)} = \frac{0.0686}{0.019 \times 0.069} = 52$$

12B **(a)** The equation for the reaction is $N_2O_4(g) \rightleftharpoons 2\,NO_2(g)$ $K_c = 4.61 \times 10^{-3}$ at 25 °C. In the example, this reaction is conducted in a 0.372-L flask. The effect of moving the mixture to the larger, 10.0-L container is that the reaction will shift to produce a greater number of moles of gas; $NO_2(g)$ will be produced and $N_2O_4(g)$ will dissociate. Consequently, the amount of N_2O_4 will decrease.

(b) The equilibrium constant expression from the Example, with 10.0 L substituted for 0.372 L, follows.

$$K_c = \frac{[NO_2]^2}{[N_2O_4]} = \frac{\left(\frac{2x}{10.0}\right)^2}{\frac{0.0240 - x}{10.0}} = \frac{4x^2}{10.0(0.0240 - x)} = 4.61 \times 10^{-3}$$

This can be solved with the quadratic equation, and the result is $x = 0.0118$ moles. We can attempt the method of successive approximations. *First*, assume that $x \ll 0.0240$. We obtain:

$$x = \sqrt{\frac{4.61 \times 10^{-3} \times 10.0\,(0.0240 - 0)}{4}} = \sqrt{4.61 \times 10^{-3} \times 2.50\,(0.0240 - 0)} = 0.0166$$

Clearly x is not considerably smaller than 0.0240. So, *second* assume $x \approx 0.0166$. We obtain:

$$x = \sqrt{4.61 \times 10^{-3} \times 2.50\,(0.0240 - 0.0166)} = 0.00925$$

This assumption is not valid either. So, *third*, assume $x \approx 0.00925$. We obtain:

$$x = \sqrt{4.61 \times 10^{-3} \times 2.50\,(0.0240 - 0.00925)} = 0.0130$$

Notice that each cycle the value we obtain for x gets closer to the value obtained from the quadratic equation. The values from the next several cycles follow.

cycle	fourth	fifth	sixth	seventh	eighth	ninth	tenth	eleventh
x value	0.0112	0.0121	0.0117	0.0119	0.0118_1	0.0118_6	0.0118_3	0.0118_4

The amount of N_2O_4 at equilibrium is 0.0118 mol, less than the 0.0210 mol N_2O_4 at equilibrium in the 0.372-L flask, as predicted.

13A Again we organize our solution on the balanced chemical equation.

equation:	$Ag^+(aq)$ +	$Fe^{2+}(aq)$	$\rightleftharpoons$	$Fe^{3+}(aq) + Ag(s)$	$K_c = 2.98$
initial:	0 M	0 M		1.20 M	
changes:	$+x$ M	$+x$ M		$-x$ M	
equil:	x M	x M		$(1.20 - x)$ M	

$$K_c = \frac{[Fe^{3+}]}{[Ag^+][Fe^{2+}]} = 2.98 = \frac{1.20 - x}{x^2} \qquad 2.98\,x^2 = 1.20 - x \qquad 0 = 2.98\,x^2 + x - 1.20$$

We use the quadratic formula to obtain a solution.

$$x = \frac{-b \pm \sqrt{b^2 - 4ac}}{2a} = \frac{-1.00 \pm \sqrt{(1.00)^2 + 4 \times 2.98 \times 1.20}}{2 \times 2.98} = \frac{-1.00 \pm 3.91}{5.96} = 0.489 \text{ or } -0.823$$

The negative root makes no physical sense. We obtain the equilibrium concentrations from x.

$[Ag^+] = [Fe^{2+}] = 0.489$ M $\qquad [Fe^{3+}] = 1.20 - 0.489 = 0.71$ M

13B We first calculate the value of Q_c to determine the direction of the reaction.

$$Q_c = \frac{[V^{2+}][Cr^{3+}]}{[V^{3+}][Cr^{2+}]} = \frac{0.150 \times 0.150}{0.0100 \times 0.0100} = 225 < 7.2 \times 10^2 = K_c$$

Because the reaction quotient has a smaller value than the equilibrium constant, a net reaction will occur to the right. We now set up this solution as we have others, based on the balanced chemical equation.

	$V^{3+}(aq)$	+	$Cr^{2+}(aq)$	$\rightleftharpoons$	$V^{2+}(aq)$	+	$Cr^{3+}(aq)$
initial:	0.0100 M		0.0100 M		0.150 M		0.150 M
changes:	$-x$ M		$-x$ M		$+x$ M		$+x$ M
equil:	$(0.0100 - x)$ M		$(0.0100 - x)$ M		$(0.150 + x)$ M		$(0.150 + x)$ M

$$K_c = \frac{[V^{2+}][Cr^{3+}]}{[V^{3+}][Cr^{2+}]} = \frac{(0.150 + x) \times (0.150 + x)}{(0.0100 - x) \times (0.0100 - x)} = 7.2 \times 10^2 = \left(\frac{0.150 + x}{0.0100 - x}\right)^2$$

If we take the square root of both sides of this expression, we have $\sqrt{7.2 \times 10^2} = \frac{0.150 + x}{0.0100 - x} = 27$

$0.150 + x = 0.27 - 27x$ which becomes $28x = 0.12$ and yields $x = 0.0043$ M. Then the equilibrium concentrations are: $[V^{3+}] = [Cr^{2+}] = 0.0100\text{ M} - 0.0043\text{ M} = 0.0057\text{ M}$

$[V^{2+}] = [Cr^{3+}] = 0.150\text{ M} + 0.0043\text{ M} = 0.154\text{ M}$

We should not over emphasize the importance of using Q_c to determine the direction of reaction. If we would have determined erroneously that net reaction occurs to the left, the final equation would be $0.150 - x = 0.270 + 27x$ from which we would have obtained $x = -0.0043$ M.

SUMMARIZING EXAMPLE CALCULATIONS

1. We use the balanced equation to determine the amount of $H_2(g)$ present at equilibrium and then convert that amount to the mass of $H_2(g)$. The 12.4 g CH_4 equals 0.773 mol CH_4 (16.04 g/mol).

	$CH_4(g)$	+	$2\ H_2O(g)$	$\rightleftharpoons$	$CO_2(g)$	+	$4\ H_2(g)$
initial:	1.00 mol		1.00 mol		0 mol		0 mol
changes:	−0.227 mol		−0.454 mol		+0.227 mol		+0.908 mol
equil.:	0.773		0.546 mol		0.227 mol		0.908 mol

$$\text{mass } H_2(g) = 0.908\text{ mol} \times \frac{2.016\text{ g } H_2}{1\text{ mol } H_2} = 1.83\text{ g } H_2(g)$$

2. Reverse reaction (2) and add the result to reaction (1) to obtain the desired reaction.

$-(2)\ H_2O(g) + CH_4(g) \rightleftharpoons CO(g) + 3\ H_2(g) \qquad \Delta H° = +230\text{ kJ}$

$+(1)\ CO(g) + H_2O(g) \rightleftharpoons CO_2(g) + H_2(g) \qquad \Delta H° = -40\text{ kJ}$

$CH_4(g) + 2\ H_2O(g) \rightleftharpoons CO_2(g) + 4\ H_2(g) \qquad \Delta H° = +\ 190\text{ kJ}$, an endothermic reaction

3. Predictions are based on the balanced chemical equation, and the enthalpy change of the reaction. The value of the equilibrium constant is not needed.

(a) add a catalyst: No change since a catalyst merely speeds up how rapidly a reaction reaches equilibrium, not the ultimate position of equilibrium.

(b) add CH_4: More $H_2(g)$ will form since adding a gas-phase reactant will cause the reaction to shift right.

(c) add CO_2: Less $H_2(g)$ will be present; adding a gas-phase product will cause the reaction to shift left.

(d) add He(g) at constant volume: No effect since none of the reagent's partial pressures will be altered.

(e) remove $H_2O(g)$: Less $H_2(g)$ is present; removing a gas-phase reactant causes the reaction to shift left.

(f) raise the temperature to 1200 K: More $H_2(g)$ will form; an endothermic reaction shifts right at high temperature.

(g) transfer the mixture to a 15.0-L flask: More $H_2(g)$ will form; the reaction will shift to the side (right) where it forms the greater number of moles of gas.

REVIEW QUESTIONS

1. **(a)** K_p is the symbol for the equilibrium constant in terms of partial pressures: the quotient of the equilibrium partial pressures of the products divided by those of the reactants, each pressure (in atmospheres) raised to a power equal to its stoichiometric coefficient.

(b) Q_c is the symbol for the reaction quotient in terms of molarities: the quotient of the molarities of the products divided by those of the reactants, each concentration raised to a power equal to its stoichiometric coefficient. These concentrations need not be those at equilibrium.

(c) Δn_{gas} is the difference between the sum of the stoichiometric coefficients of the gaseous products and the similar sum for gaseous reactants, using the coefficients from the balanced chemical equation.

2. **(a)** A dynamic equilibrium is a balanced situation that is created by opposing processes proceeding at such rates that the net result is no apparent change, although individual molecules may indeed be undergoing change, and probably are.

(b) The direction of net reaction describes whether a chemical reaction produces more product (to the right) or forms more reactant (to the left), generally as the result of the alteration of reaction conditions.

(c) Le Châtelier's principle states that if a stress is applied to a system that is at equilibrium, the system will react in such a way as to relieve the stress and attain a new equilibrium position.

(d) The effect of a catalyst on the position of equilibrium is nonexistent. A catalyst does, however, speed up the rate at which equilibrium is established.

3. **(a)** A reaction that goes to completion is one that reacts until the amount of one of the reactants is gone. A reversible reaction is one that comes to a point of balance before all reactants are depleted.

(b) The equilibrium constant K_c contains the equilibrium concentrations of products and reactants, while the K_p expression contains their partial pressures. The values of these two numbers are related by the expression $K_p = K_c(RT)^{\Delta n}$.

(c) The reaction quotient, Q_c, and the equilibrium constant expression, K_c, have the same functional form, with the concentrations of products over those of reactants, each raised to a power equal to the stoichiometric coefficient of that species. However, the concentrations that are used differ. In the K_c expression, these concentrations are equilibrium concentrations, while in the Q_c expression they are not necessarily the concentrations at equilibrium—often they are the initial concentrations. Similar statements can be made concerning Q_p and K_p, except that partial pressures replace concentrations.

(d) A homogeneous equilibrium is one that takes place in a solution, either entirely gas or entirely liquid. A heterogeneous equilibrium is one that is established when two phases of matter are present, generally a solid with either a liquid or a gaseous solution.

4. **(a)** $K_c = \dfrac{[NO_2]^2}{[NO]^2[O_2]}$ **(b)** $K_c = \dfrac{[Zn^{2+}]}{[Ag^+]^2}$ **(c)** $K_c = \dfrac{[OH^-]^2}{[CO_3^{2-}]}$

5. **(a)** $K_p = \dfrac{P\{CH_4\}P\{H_2S\}^2}{P\{CS_2\}P\{H_2\}^4}$ **(b)** $K_p = P\{O_2\}^{1/2}$ **(c)** $K_p = P\{CO_2\}P\{H_2O\}$

6. In each case we write the equation for the formation reaction and then the equilibrium constant expression for that reaction.

(a) $N_2(g) + \frac{1}{2}O_2(g) \rightleftharpoons N_2O(g)$ $\qquad K_c = \dfrac{[NO]}{[N_2][O_2]^{1/2}}$

(b) $\frac{1}{2}H_2(g) + \frac{1}{2}Br_2(g) \rightleftharpoons HBr(g)$ $\qquad K_c = \dfrac{[HBr]}{[H_2]^{1/2}[Br_2]^{1/2}}$

(c) $\frac{1}{2}N_2(g) + \frac{3}{2}H_2(g) \rightleftharpoons NH_3(g)$ $\qquad K_c = \dfrac{[NH_3]}{[N_2]^{1/2}[H_2]^{3/2}}$

(d) $\frac{1}{2}Cl_2(g) + \frac{3}{2}F_2(g) \rightleftharpoons ClF_3(g)$ $\qquad K_c = \dfrac{[ClF_3]}{[Cl_2]^{1/2}[F_2]^{3/2}}$

(e) $\frac{1}{2}N_2(g) + \frac{1}{2}O_2(g) + F_2(g) \rightleftharpoons NOF_2(g)$ $\qquad K_c = \dfrac{[NOF_2]}{[N_2]^{1/2}[O_2]^{1/2}[F_2]}$

7. **(a)** This is the reverse of the first reaction; K_c is the inverse value. $K_c = \dfrac{[CO][H_2O]}{[CO_2][H_2]} = \dfrac{1}{23.2} = 0.0431$

(b) This is twice the second reaction; K_c is its square. $K_c = \dfrac{[SO_3]^2}{[SO_2]^2[O_2]} = (56)^2 = 3.1 \times 10^3$

(c) This is half of the reverse of the third reaction; K_c is the the inverse of the square root. $K_c = \dfrac{[H_2S]}{[H_2][S_2]^{1/2}} = \dfrac{1}{\sqrt{2.3 \times 10^{-4}}} = 66$

8. We combine one-half of the reversed first reaction with the second reaction to obtain the desired reaction.

$\frac{1}{2}N_2(g) + \frac{1}{2}O_2(g) \rightleftharpoons NO(g)$ $\qquad K_c = \dfrac{1}{\sqrt{2.1 \times 10^{30}}}$

$NO(g) + \frac{1}{2}Br_2(g) \rightleftharpoons NOBr(g)$ $\qquad K_c = 1.4$

net: $\frac{1}{2}N_2(g) + \frac{1}{2}O_2(g) + \frac{1}{2}Br_2(g) \rightleftharpoons NOBr(g)$ $\qquad K_c = \dfrac{1.4}{\sqrt{2.1 \times 10^{30}}} = 9.7 \times 10^{-16}$

9. Answer (d) is correct. 1 mol I_2 is the limiting reagent in this reaction. If the reaction were to go to completion, 2 mol HI would be produced. Thus, the only way that 2 mol HI could be produced (answer b) would be if K_c were infinite. 1 mol HI could be produced (answer a), but without knowing the value of K_c for this reaction, we cannot be sure. The only certain answer is that something less than 2 mol HI are produced, answer (d).

10. Simply substitute into the K_c expression. $K_c = \frac{[C]^2}{[A]^2[B]} = \frac{(0.43\text{ M})^2}{(0.55\text{ M})^2 \times 0.33\text{ M}} = 1.9$

11. The formation of 2.0×10^{-11} mol S(g) does not significantly change the amount of $S_2(g)$ from its original 0.0040 mol. Thus, $K_c = \frac{[S]^2}{[S_2]} = \frac{\left(\frac{2.0 \times 10^{-11}\text{ mol}}{0.500\text{ L}}\right)^2}{\frac{0.0040\text{ mol}}{0.500\text{ L}}} = 2.0 \times 10^{-19}$

12. **(a)** If the amounts of SO_2 and SO_3 are equal, then $[SO_2] = [SO_3]$.

$K_c = 35.5 = \frac{[SO_3]^2}{[SO_2]^2[O_2]} = \frac{\cancel{[SO_2]^2}}{\cancel{[SO_2]^2}[O_2]} = \frac{1}{[O_2]}$ $\qquad [O_2] = \frac{1}{35.5} = 0.0282\text{ M}$

no. mol $O_2 = 0.0282\text{ M} \times 2.05\text{ L} = 0.0578$ mol O_2

(b) If there are twice as many moles of SO_3 in the flask as SO_2, then $[SO_3] = 2 \times [SO_2]$

$K_c = 35.5 = \frac{[SO_3]^2}{[SO_2]^2[O_2]} = \frac{2^2\cancel{[SO_3]^2}}{\cancel{[SO_2]^2}[O_2]} = \frac{4}{[O_2]}$ $\qquad [O_2] = \frac{4}{35.5} = 0.113\text{ M}$

no. mol $O_2 = 0.113\text{ M} \times 2.05\text{ L} = 0.232$ mol O_2

13. In each case we use the relationship $K_p = K_c(RT)^{\Delta n_{gas}}$

(a) For $CO_2(g) + H_2(g) \rightleftharpoons CO(g) + H_2O(g)$ $\qquad K_p = K_c = 0.0431$

(b) For $2\,SO_2(g) + O_2 \rightleftharpoons 2\,SO_3(g)$ $\qquad K_p = K_c(RT)^{-1} = \frac{3.1 \times 10^3}{0.0821 \times 900\text{K}} = 42$

(c) For $H_2(g) + \frac{1}{2}S_2(g) \rightleftharpoons H_2S(g)$ $\qquad K_p = K_c(RT)^{-1/2} = \frac{66}{\sqrt{0.0821 \times 1405\text{ K}}} = 6.1$

14. We first find the value of K_p for the reaction $2\,NO_2(g) \rightleftharpoons 2\,NO(g) + O_2(g)$ $\quad K_c = 1.8 \times 10^{-6}$ at 184 °C = 457 K. For this reaction $\Delta n_{gas} = 2 + 1 - 2 = +1$.

$K_p = K_c(RT)^{\Delta n_{gas}} = 1.8 \times 10^{-6}\,(0.08206\text{ L atm mol}^{-1}\text{ K}^{-1} \times 457\text{ K})^{+1} = 6.8 \times 10^{-5}$

To obtain the final reaction $NO(g) + \frac{1}{2}O_2(g) \rightleftharpoons NO_2(g)$ from the initial reaction, that initial reaction must be reversed and then divided by two. Thus, in order to determine the value of the equilibrium constant for the final reaction, the value of K_p for the initial reaction must be inverted, and the square root taken of the result. $\quad K_{p,final} = \sqrt{\frac{1}{6.8 \times 10^{-5}}} = 1.2 \times 10^2$

15. **(a)** $K_c = \frac{[CO][H_2O]}{[CO_2][H_2]} = \frac{\frac{n\{CO\}}{V}\,\frac{n\{H_2O\}}{V}}{\frac{n\{CO_2\}}{V}\,\frac{n\{H_2\}}{V}} = \frac{n\{CO\}n\{H_2O\}}{n\{CO_2\}n\{H_2\}}$ $\qquad$ Thus, K_c does not depend on the volume of the container for this reaction. Note also that the value of K_p does not depend on volume.

(b) Note that $K_p = K_c$ for this reaction, since $\Delta n_{gas} = 0$.

$K_c = K_p = \frac{0.224\text{ mol CO} \times 0.224\text{ mol }H_2O}{0.276\text{ mol }CO_2 \times 0.276\text{ mol }H_2} = 0.659$

16. **(a)** $K_c = \frac{[PCl_5]}{[PCl_3][Cl_2]} = \frac{\frac{0.105\text{ g }PCl_5}{2.50\text{ L}} \times \frac{1\text{ mol }PCl_5}{208.2\text{ g}}}{\left(\frac{0.220\text{ g }PCl_3}{2.50\text{ L}} \times \frac{1\text{ mol }PCl_3}{137.3\text{ g}}\right) \times \left(\frac{2.12\text{ g }Cl_2}{2.50\text{ L}} \times \frac{1\text{ mol }Cl_2}{70.9\text{ g}}\right)} = 26.3$

(b) $K_p = K_c(RT)^{\Delta n} = 26.3\,(0.08206\text{ L atm mol}^{-1}\text{ K}^{-1} \times 523\text{ K})^{-1} = 0.613$

17. **(a)** If 5% of the 1.00 mol I_2 is dissociated into atoms, then 0.05 mol I_2 has dissociated.

Equation:	$I_2(g)$	$\rightleftharpoons$	$2\,I(g)$
Initial:	1.00 mol		0.00 mol
Changes:	−0.05 mol		+0.10 mol
Equil:	0.95 mol		0.10 mol

$K_c = \frac{[I]^2}{[I_2]} = \frac{\left(\frac{0.10\text{ mol I}}{1.00\text{ L}}\right)^2}{\frac{0.95\text{ mol }I_2}{1.00\text{ L}}} = 0.011$

(b) $K_p = K_c(RT)^{+1} = 0.011\,(0.0821 \times 1473\text{ K}) = 1.3$

18. **(a)** We determine the concentration of each species in the gaseous solution, use these concentrations to determine the value of the reaction quotient, and compare this value of Q_c with the value of K_c.

$$[SO_2] = \frac{0.455 \text{ mol } SO_2}{1.90 \text{ L}} = 0.239 \text{ M} \qquad [O_2] = \frac{0.183 \text{ mol } O_2}{1.90 \text{ L}} = 0.0963 \text{ M}$$

$$[SO_3] = \frac{0.568 \text{ mol } SO_3}{1.90 \text{ L}} = 0.299 \text{ M} \qquad Q_c = \frac{[SO_3]^2}{[SO_2]^2[O_2]} = \frac{(0.299)^2}{(0.239)^2 0.0963} = 16.3$$

Since $Q_c = 16.3 \neq 2.8 \times 10^2 = K_c$, this mixture is not at equilibrium.

(b) Since the value of Q_c is smaller than that of K_c, the reaction will proceed to the right, forming products to reach equilibrium.

19. Increasing the volume of an equilibrium mixture causes that mixture to shift toward the side (reactants or products) where the sum of the stoichiometric coefficients of the gaseous species is the larger. That is: shifts to the right if $\Delta n_{gas} > 0$, shifts to the left if $\Delta n_{gas} < 0$, and does not shift if $\Delta n_{gas} = 0$

(a) $C(s) + H_2O(g) \rightleftharpoons CO(g) + H_2(g)$ $\quad \Delta n_{gas} > 0$, shift right, toward products

(b) $Ca(OH)_2(s) + CO_2(g) \rightleftharpoons CaCO_3(s) + H_2O(g)$ $\quad \Delta n_{gas} = 0$, no shift, no change in equilibrium position.

(c) $4\ NH_3(g) + 5\ O_2(g) \rightleftharpoons 4\ NO(g) + 6\ H_2O(g)$ $\quad \Delta n_{gas} > 0$, shift right, toward products

20. The equilibrium position for a reaction that is exothermic shifts to the left (favors reactants) when temperature is raised. For one that is endothermic, it shifts right (favors products) when temperature is raised.

(a) $NO(g) \rightleftharpoons \frac{1}{2} N_2(g) + \frac{1}{2} O_2(g)$ $\quad \Delta H° = -90.2$ kJ $\quad$ shifts left, % dissociation decreases

(b) $SO_3(g) \rightleftharpoons SO_2(g) + \frac{1}{2} O_2(g)$ $\quad \Delta H° = +98.9$ kJ $\quad$ shifts right, % dissociation increases

(c) $N_2H_4(g) \rightleftharpoons N_2(g) + 2\ H_2(g)$ $\quad \Delta H° = -95.4$ kJ $\quad$ shifts left, % dissociation decreases

(d) $COCl_2(g) \rightleftharpoons CO(g) + Cl_2(g)$ $\quad \Delta H° = +108.3$ kJ $\quad$ shifts right, % dissociation increases

21. $4\ HCl(g) + O_2(g) \rightleftharpoons 2\ H_2O(g) + 2\ Cl_2(g) \quad \Delta H° = -114$ kJ

(a) Adding $O_2(g)$ to the equilibrium mixture at constant volume will cause the position of equilibrium to shift to the right, increasing the equilibrium amount of $Cl_2(g)$.

(b) Removing HCl(g) from the equilibrium mixture at constant volume will cause the position of equilibrium to shift to the left, decreasing the amount of $Cl_2(g)$.

(c) Because $\Delta n_{gas} < 0$, transferring the equilibrium mixture to a container of twice the volume will cause the position of equilibrium to shift to the left, decreasing the amount of $Cl_2(g)$.

(d) Adding a catalyst to the equilibrium mixture will have no effect on the equilibrium position.

(e) Raising the temperature of this exothermic reaction will cause the equilibrium position to shift to the left, decreasing the amount of $Cl_2(g)$.

22. The information for the calculation is organized around the chemical equation. Let x = mol H_2 (or I_2) that reacts. Then use stoichiometry to determine the amount of HI formed, in terms of x, and finally solve for x.

Equation:	$H_2(g)$	+	$I_2(g)$	$\rightleftharpoons$	$2\ HI(g)$
Initial:	0.150 mol		0.150 mol		0.000 mol
Changes:	$-x$ mol		$-x$ mol		$+2x$ mol
Equil:	$0.150 - x$		$0.150 - x$		$2x$

$$K_c = \frac{[HI]^2}{[H_2][I_2]} = \frac{\left(\frac{2x}{3.25 \text{ L}}\right)^2}{\frac{0.150 - x}{3.25 \text{ L}} \times \frac{0.150 - x}{3.25 \text{ L}}}$$

Then take the square root of both sides: $\quad \sqrt{K_c} = \sqrt{50.2} = \dfrac{2x}{0.150 - x} = 7.09$

$2x = 1.06 - 7.09\ x \qquad x = \dfrac{1.06}{9.09} = 0.117$ mol $\qquad$ amount HI $= 2x = 2 \times 0.117$ mol $= 0.234$ mol HI

amount H_2 = amount $I_2 = (0.150 - x)$ mol $= (0.150 - 0.117)$ mol $= 0.033$ mol H_2 or I_2

23. This calculation is set up in a manner similar to the calculation of Review Question 22. In this case, we let x equal the amount of I_2 that is formed.

Equation:	$H_2(g)$	+	$I_2(g)$	$\rightleftharpoons$	$2\ HI(g)$
Initial:	0.000 mol		0.000 mol		0.150 mol
Changes:	$+x$ mol		$+x$ mol		$-2x$ mol
Equil:	x		x		$0.150 - 2x$

$$K_c = 50.2 = \frac{[HI]^2}{[H_2][I_2]} = \frac{\left(\frac{0.150 - 2x}{3.25\text{ L}}\right)^2}{\frac{x}{3.25\text{ L}} \times \frac{x}{3.25\text{ L}}} = \frac{(0.150 - 2x)^2}{x^2} \qquad \sqrt{50.2} = \frac{0.150 - 2x}{x} = 7.09$$

$$0.150 - 2x = 7.09x \qquad x = \frac{0.150}{2 + 7.09} = 0.0165 \qquad \text{amount } I_2 = x = 0.0165 \text{ mol } I_2$$

Note that we begin with half as much material as we had in the previous instance.

24. We first determine the initial pressure of NH_3.

$$P\{NH_3(g)\} = \frac{nRT}{V} = \frac{0.100 \text{ mol } NH_3 \times 0.08206 \text{ L atm mol}^{-1} \text{ K}^{-1} \times 298 \text{ K}}{2.58 \text{ L}} = 0.948 \text{ atm}$$

Equation:	$NH_4HS(s) \rightleftharpoons$	$NH_3(g)$ +	$H_2S(g)$
Initial:		0.948 atm	
Changes:		$+x$ atm	$+x$ atm
Equil:		$(0.948 + x)$atm	x atm

$K_p = P\{NH_3\}P\{H_2S\} = 0.108 = (0.948 + x)x = 0.948\,x + x^2 \qquad 0 = x^2 + 0.948\,x - 0.108$

$$x = \frac{-b \pm \sqrt{b^2 - 4ac}}{2a} = \frac{-0.948 \pm \sqrt{0.899 + 0.432}}{2} = 0.103 \text{ atm}, -1.05 \text{ atm}$$

The negative pressure makes no physical sense. Thus, the total gas pressure is obtained as follows.

$P_{tot} = P\{NH_3\} + P\{H_2S\} = (0.948 + x) + x = 0.948 + 2\,x = 0.948 + 2 \times 0.103 = 1.154$ atm

EXERCISES

25. **(a)** $2\,COF_2(g) \rightleftharpoons CO_2(g) + CF_4(g) \qquad K_c = \frac{[CO_2][CF_4]}{[COF_2]^2}$

(b) $Cu(s) + 2\,Ag^+(aq) \rightleftharpoons Cu^{2+}(aq) + 2\,Ag(s) \qquad K_c = \frac{[Cu^{2+}]}{[Ag^+]^2}$

(c) $S_2O_8^{2-}(aq) + 2\,Fe^{2+}(aq) \rightleftharpoons 2\,SO_4^{2-}(aq) + 2\,Fe^{3+}(aq) \qquad K_c = \frac{[SO_4^{2-}]^2\,[Fe^{3+}]^2}{[S_2O_8^{2-}]\,[Fe^{2+}]^2}$

26. **(a)** $4\,NH_3(g) + 3\,O_2(g) \rightleftharpoons 2\,N_2(g) + 6\,H_2O(g) \qquad K_c = \frac{[N_2]^2[H_2O]^6}{[NH_3]^4[O_2]^3}$

(b) $7\,H_2(g) + 2\,NO_2(g) \rightleftharpoons 2\,NH_3(g) + 4\,H_2O(g) \qquad K_c = \frac{[NH_3]^2[H_2O]^4}{[H_2]^7[NO_2]^2}$

(c) $N_2(g) + Na_2CO_3(s) + 4\,C(s) \rightleftharpoons 2\,NaCN(s) + 3\,CO(g) \qquad K_c = \frac{[CO]^3}{[N_2]}$

27. Since $K_p = K_c(RT)^{\Delta n}$, it is also true that $K_c = K_p(RT)^{-\Delta n}$.

(a) $K_c = \frac{[SO_2(g)][Cl_2(g)]}{[SO_2Cl_2(g)]} = K_p(RT)^{-(+1)} = 2.9 \times 10^{-2}\,(0.0821 \times 303\text{ K})^{-1} = 0.0012$

(b) $K_c = \frac{[NO_2]^2}{[NO]^2[O_2]} = K_p(RT)^{-(-1)} = (1.48 \times 10^4)(0.0821 \times 457\text{ K}) = 5.55 \times 10^5$

(c) $K_c = \frac{[H_2S]^3}{[H_2]^3} = K_p(RT)^0 = K_p = 0.429$

28. $K_p = K_c(RT)^{\Delta n}$, with $R = 0.0821$ L·atm mol^{-1} K^{-1}

(a) $K_p = \frac{P\{NO_2\}^2}{P\{N_2O_4\}} = K_c(RT)^{(+1)} = 4.61 \times 10^{-3}\,(0.0821 \times 298\text{ K})^1 = 0.113$

(b) $K_p = \frac{P\{C_2H_2\}\,P\{H_2\}^3}{P\{CH_4\}^2} = K_c(RT)^{(+2)} = (0.154)(0.0821 \times 2000\text{ K})^2 = 4.15 \times 10^3$

(c) $K_p = \frac{P\{H_2\}^4\,P\{CS_2\}}{P\{H_2S\}^2\,P\{CH_4\}} = K_c(RT)^{(+2)} = (5.27 \times 10^{-8})(0.0821 \times 973\text{ K})^2 = 3.36 \times 10^{-4}$

29. The equilibrium reaction is $H_2O(l) \rightleftharpoons H_2O(g)$ with $\Delta n_{gas} = +1$. $K_p = K_c(RT)^{\Delta n}$ gives $K_c = K_p(RT)^{-\Delta n}$.

$$K_p = P\{H_2O\} = 23.8 \text{ mmHg} \times \frac{1 \text{ atm}}{760 \text{ mmHg}} = 0.0313$$

$$K_c = K_p(RT)^{-(+1)} = \frac{K_p}{RT} = \frac{0.0313}{0.0821 \times 298 \text{ K}} = 1.28 \times 10^{-3}$$

30. The equilibrium reaction is $C_6H_6(l) \rightleftharpoons C_6H_6(g)$ with $\Delta n_{gas} = +1$. $K_p = K_c(RT)^{\Delta n}$ gives $K_c = K_p(RT)^{-\Delta n}$.

$K_p = K_c(RT) = 5.12 \times 10^{-3}\,(0.08206 \times 298\text{ K}) = 0.125 = P\{C_6H_6\}$

$$P\{C_6H_6\} = 0.125 \text{ atm} \times \frac{760 \text{ mmHg}}{1 \text{ atm}} = 95.0 \text{ mmHg}$$

31. We combine the several given reactions to obtain the net reaction.

$2\ N_2O(g) \rightleftharpoons 2\ N_2(g) + O_2(g)$ $\quad K_c = \dfrac{1}{(2.7 \times 10^{-18})^2}$

$4\ NO_2(g) \rightleftharpoons 2\ N_2O_4(g)$ $\quad K_c = \dfrac{1}{(4.6 \times 10^{-3})^2}$

$2\ N_2(g) + 4\ O_2(g) \rightleftharpoons 4\ NO_2(g)$ $\quad K_c = (4.1 \times 10^{-9})^4$

net: $2\ N_2O(g) + 3\ O_2(g) \rightleftharpoons 2\ N_2O_4(g)$ $\quad K_c = \dfrac{(4.1 \times 10^{-9})^4}{(2.7 \times 10^{-18})^2\,(4.6 \times 10^{-3})^2} = 1.8 \times 10^6$

32. We combine the K_c values to obtain the value of K_c for the overall reaction, and then convert this to a value for K_p.

$2\ CO_2(g) + 2\ H_2(g) \rightleftharpoons 2\ CO(g) + 2\ H_2O(g)$ $\quad K_c = (1.4)^2$

$2\ C(\text{graphite}) + O_2(g) \rightleftharpoons 2\ CO(g)$ $\quad K_c = (1 \times 10^8)^2$

$4\ CO(g) \rightleftharpoons 2\ C(\text{graphite}) + 2\ CO_2(g)$ $\quad K_c = \dfrac{1}{(0.64)^2}$

net: $2\ H_2(g) + O_2(g) \rightleftharpoons 2\ H_2O(g)$ $\quad K_c = \dfrac{(1.4)^2\,(1 \times 10^8)^2}{(0.64)^2} = 5 \times 10^{16}$

$$K_p = K_c(RT)^{\Delta n} = \frac{K_c}{RT} = \frac{5 \times 10^{16}}{0.08206 \times 1200\text{K}} = 5 \times 10^{14}$$

Experimental Determination of Equilibrium Constants

33. First, we determine the concentration of PCl_5 and of Cl_2 present initially and at equilibrium, respectively. Then we use the balanced equation to help us determine the concentration of each species present at equilibrium.

$$[PCl_5] = \frac{1.00 \times 10^{-3} \text{ mol } PCl_5}{0.250 \text{ L}} = 0.00400 \text{ M} \qquad [Cl_2] = \frac{9.65 \times 10^{-4} \text{ mol } Cl_2}{0.250 \text{ L}} = 0.00386 \text{ M}$$

Equation:	$PCl_5(g)$	$\rightleftharpoons$ PCl_3	+ $Cl_2(g)$
Initial:	0.00400 M	0.0 M	0.0 M
Changes:	–0.00386 M	+0.00386 M	+0.00386 M
Equil:	0.00014 M	0.00386 M	0.00386 M

$$K_c = \frac{[PCl_3][Cl_2]}{[PCl_5]} = \frac{(0.00386 \text{ M})(0.00386 \text{ M})}{0.00014 \text{ M}} = 0.106$$

34. First we determine the partial pressure of each gas.

$$P_{\text{initial}}\{H_2(g)\} = \frac{nRT}{V} = \frac{1.00 \text{ g } H_2 \times \dfrac{1 \text{ mol } H_2}{2.016 \text{ g } H_2} \times \dfrac{0.08206 \text{ L atm}}{\text{mol K}} \times 1670 \text{ K}}{0.500 \text{ L}} = 136 \text{ atm}$$

$$P_{\text{initial}}\{H_2S(g)\} = \frac{nRT}{V} = \frac{1.06 \text{ g } H_2S \times \dfrac{1 \text{ mol } H_2S}{34.08 \text{ g } H_2S} \times \dfrac{0.08206 \text{ L atm}}{\text{mol K}} \times 1670 \text{ K}}{0.500 \text{ L}} = 8.52 \text{ atm}$$

$$P_{\text{equil.}}\{S_2(g)\} = \frac{nRT}{V} = \frac{8.00 \times 10^{-6} \text{ mol } S_2 \times \dfrac{0.08206 \text{ L atm}}{\text{mol K}} \times 1670 \text{ K}}{0.500 \text{ L}} = 2.19 \times 10^{-3} \text{ atm}$$

Equation:	$2\ H_2(g)$ +	$S_2(g)$ $\rightleftharpoons$	$2\ H_2S(g)$
Initial:	136 atm	0 atm	8.52 atm
Changes:	+0.00438 atm	+0.00219 atm	–0.00438 atm
Equil:	136 atm	0.00219 atm	8.52 atm

$$K_p = \frac{P\{H_2S(g)\}^2}{P\{H_2(g)\}^2\,P\{S_2(g)\}} = \frac{(8.52)^2}{(136)^2\,0.00219} = 1.79$$

Equilibrium Relationships

35. $K_c = 281 = \dfrac{[SO_3]^2}{[SO_2]^2[O_2]} = \dfrac{[SO_3]^2}{[SO_2]^2} \times \dfrac{0.185 \text{ L}}{0.00247 \text{ mol}}$ $\qquad \dfrac{[SO_2]}{[SO_3]} = \sqrt{\dfrac{0.185}{0.00247 \times 281}} = 0.516$

36. $K_c = 0.011 = \dfrac{[I]^2}{[I_2]} = \dfrac{\left(\dfrac{0.37 \text{ mol I}}{V}\right)^2}{\dfrac{1.00 \text{ mol } I_2}{V}} = \dfrac{1}{V} \times 0.14 \qquad V = \dfrac{0.14}{0.011} = 13 \text{ L}$

37. **(a)** A possible equation for the oxidation of $NH_3(g)$ to $NO_2(g)$ follows.

$NH_3(g) + \frac{7}{4} O_2(g) \rightleftharpoons NO_2(g) + \frac{3}{2} H_2O(g)$

(b) We obtain K_p for the reaction in part (a) by appropriately combining the values of K_p given in the problem.

$NH_3(g) + \frac{5}{4} O_2(g) \rightleftharpoons NO(g) + \frac{3}{2} H_2O(g) \qquad K_p = 2.11 \times 10^{19}$

$NO(g) + \frac{1}{2} O_2(g) \rightleftharpoons NO_2(g) \qquad K_p = \dfrac{1}{0.524}$

net: $NH_3(g) + \frac{7}{4} O_2(g) \rightleftharpoons NO_2(g) + \frac{3}{2} H_2O(g) \qquad K_p = \dfrac{2.11 \times 10^{19}}{0.524} = 4.03 \times 10^{19}$

38. **(a)** We first determine $[H_2]$ and $[CH_4]$, and then $[C_2H_2]$. $\qquad [CH_4] = [H_2] = \dfrac{0.10 \text{ mol}}{1.00 \text{ L}} = 0.10 \text{ M}$

$K_c = \dfrac{[C_2H_2][H_2]^3}{[CH_4]^2} \qquad [C_2H_2] = \dfrac{K_c [CH_4]^2}{[H_2]^3} = \dfrac{0.154 \times 0.10^2}{0.10^3} = 1.5 \text{ M}$

In a 1.00-L container, each concentration numerically equals the amount of substance.

$\chi\{C_2H_2\} = \dfrac{1.5 \text{ mol } C_2H_2}{1.5 \text{ mol } C_2H_2 + 0.10 \text{ mol } CH_4 + 0.10 \text{ mol } H_2} = 0.88_2$

(b) The conversion of $CH_4(g)$ to $C_2H_2(g)$ is favored at low pressures, since the conversion reaction has a larger sum of the stoichiometric coefficients of gaseous products (4) than of reactants (2).

(c) Initially, all concentrations are halved when the mixture is transferred to a flask that is twice as large. In re-establishing equilibrium the system reacts to the right, forming more moles of gas. We base our solution on the balanced chemical equation, in the manner we have used before.

Equation:	$2\,CH_4(g)$	$\rightleftharpoons$	$C_2H_2(g)$	+	$3\,H_2(g)$
Initial:	0.10 M / 2		1.50 M / 2		0.10 M / 2
Changes:	$-2\,x$ M		$+x$ M		$+3\,x$ M
Equil:	$(0.050 - 2x)$ M		$(0.750 + x)$ M		$(0.050 + 3\,x)$ M

$K_c = \dfrac{[C_2H_2][H_2]^3}{[CH_4]^2} = \dfrac{(0.050 + 3x)^3\,(0.750 + x)}{(0.050 - 2x)^2} = 0.154$

We solve this fourth-order equation by successive approximations. First we guess that $x = 0.010$ M.

$x = 0.010 \quad Q_c = \dfrac{(0.050 + 3(0.010))^3\,(0.750 + 0.010)}{(0.050 - 2(0.010))^2} = \dfrac{(0.080)^3\,(0.760)}{(0.030)^2} = 0.433 > 0.154$

$x = 0.020 \quad Q_c = \dfrac{(0.050 + 3(0.020))^3\,(0.750 + 0.020)}{(0.050 - 2(0.020))^2} = \dfrac{(0.110)^3\,(0.770)}{(0.010)^2} = 10.2 > 0.154$

$x = 0.005 \quad Q_c = \dfrac{(0.050 + 3(0.005))^3\,(0.750 + 0.005)}{(0.050 - 2(0.005))^2} = \dfrac{(0.065)^3\,(0.755)}{(0.040)^2} = 0.129 < 0.154$

$x = 0.006 \quad Q_c = \dfrac{(0.050 + 3(0.006))^3\,(0.750 + 0.006)}{(0.050 - 2(0.006))^2} = \dfrac{(0.068)^3\,(0.756)}{(0.038)^2} = 0.165 > 0.154$

This is the maximum number of significant figures our system permits. We have $x = 0.006$ M.

$[CH_4] = 0.038$ M $\qquad [C_2H_2] = 0.756$ M $\qquad [H_2] = 0.068$ M

Because we have a 2.00-L container, molar amounts are double the numerical value of molarities.

$2.00 \text{ L} \times \dfrac{0.756 \text{ mol } C_2H_2}{1 \text{ L}} = 1.51 \text{ mol } C_2H_2 \qquad 2.00 \text{ L} \times \dfrac{0.038 \text{ mol } CH_4}{1 \text{ L}} = 0.076 \text{ mol } CH_4$

$2.00 \text{ L} \times \dfrac{0.068 \text{ mol } H_2}{1 \text{ L}} = 0.14 \text{ mol } H_2$

Direction and Extent of Chemical Change

39. We compute the value of Q_c for the given amounts of product and reactants.

$$Q_c = \frac{[SO_3]^2}{[SO_2]^2[O_2]} = \frac{\left(\dfrac{1.8 \text{ mol } SO_3}{7.2 \text{ L}}\right)^2}{\left(\dfrac{3.6 \text{ mol } SO_2}{7.2 \text{ L}}\right)^2 \dfrac{2.2 \text{ mol } O_2}{7.2 \text{ L}}} = 0.81 < K_c = 100$$

The mixture described cannot be maintained indefinitely. In fact, because $Q_c < K_c$, the reaction will proceed to the right, that is, toward products, until equilibrium is established.

40. We compute the value of Q_c for the given amounts of product and reactants.

$$Q_c = \frac{[NO_2]^2}{[N_2O_4]} = \frac{\left(\frac{0.0205 \text{ mol } NO_2}{5.25 \text{ L}}\right)^2}{\frac{0.750 \text{ mol } N_2O_4}{5.25 \text{ L}}} = 1.1 \times 10^{-4} < K_c = 4.61 \times 10^{-3}$$

The mixture described cannot be maintained indefinitely. In fact, because $Q_c < K_c$, the reaction will proceed to the right, that is, toward products, until equilibrium is established.

41. We use the balanced chemical equation as a basis to organize the information we have about the reaction.

Equation:	$SbCl_5(g)$	$\rightleftharpoons$ $SbCl_3(g)$ +	$Cl_2(g)$
Initial:	$\frac{0.00 \text{ mol}}{2.50 \text{ L}}$	$\frac{0.280 \text{ mol}}{2.50 \text{ L}}$	$\frac{0.160 \text{ mol}}{2.50 \text{ L}}$
Initial:	0.000 M	0.112 M	0.0640 M
Changes:	$+x$ M	$-x$ M	$-x$ M
Equil:	x M	$(0.112 - x)$M	$(0.0640 - x)$M

$$K_c = 0.025 = \frac{[SbCl_3][Cl_2]}{[SbCl_5]} = \frac{(0.112 - x)(0.0640 - x)}{x} = \frac{0.00717 - 0.176\,x + x^2}{x}$$

$0.025\,x = 0.00717 - 0.176\,x + x^2 \qquad x^2 - 0.201\,x + 0.00717 = 0$

$$x = \frac{-b \pm \sqrt{b^2 - 4ac}}{2a} = \frac{0.201 \pm \sqrt{0.0404 - 0.0287}}{2} = 0.0464 \text{ or } 0.155$$

The second of the two values for x gives a negative value of $[Cl_2]$ (= –0.091 M), and thus is physically meaningless. Thus, concentrations and amounts follow.

$[SbCl_5] = x = 0.0464$ M $\qquad$ amount $SbCl_5 = 2.50 \text{ L} \times 0.0464 \text{ M} = 0.116 \text{ mol } SbCl_5$

$[SbCl_3] = 0.112 - x = 0.066$ M $\qquad$ amount $SbCl_3 = 2.50 \text{ L} \times 0.066 \text{ M} = 0.17 \text{ mol } SbCl_3$

$[Cl_2] = 0.0640 - x = 0.0176$ M $\qquad$ amount $Cl_2 = 2.50 \text{ L} \times 0.0176 \text{ M} = 0.0440 \text{ mol } Cl_2$

Or, in the first line labeled "Initial," we could have set y = number of moles of $SbCl_5$ produced. Then our final answer would be $y = 0.116$ mol $SbCl_5$. That is, we would have obtained the answer more directly.

42. Use the chemical equation as a basis to organize the information we have about the reaction, and then determine the final number of moles of $Cl_2(g)$ present.

Equation:	$CO(g)$ +	$Cl_2(g)$	$\rightleftharpoons$	$COCl_2(g)$
Initial:	0.3500 mol			0.05500 mol
Changes:	$+x$ mol	$+x$ mol		$-x$ mol
Equil:	$(0.3500 + x)$mol	x mol		$(0.0550 - x)$mol

$$K_c = 1.2 \times 10^3 = \frac{[COCl_2]}{[CO][Cl_2]} = \frac{\frac{(0.05500 - x)\text{mol}}{3.050 \text{ L}}}{\frac{(0.3500 + x)\text{mol}}{3.050 \text{ L}} \times \frac{x \text{ mol}}{3.050 \text{ L}}} \qquad \frac{1.2 \times 10^3}{3.050} = \frac{0.0550 - x}{(0.3500 + x)\,x}$$

We assume that $x \ll 0.0550$ This produces the following expression.

$$\frac{1.2 \times 10^3}{3.050} = \frac{0.0550}{0.3500\,x} \qquad x = \frac{3.050 \times 0.0550}{0.3500 \times 1.2 \times 10^3} = 4.0 \times 10^{-4} \text{ mol } Cl_2$$

Our assumption that $x \ll 0.0550$ was not terribly good. We use the first value we obtained (4.0×10^{-4} = 0.0004) to arrive at a second value. $\quad x = \frac{3.050 \times (0.0550 - 0.0004)}{(0.3500 + 0.0004) \times 1.2 \times 10^3} = 4.0 \times 10^{-4} \text{ mol } Cl_2$

Because the value did not change on the second iteration, we have arrived at a solution.

43. We first compute the initial concentration of each species present. Then we determine the equilibrium concentrations of all species. And finally, we compute the mass of CO_2 present at equilibrium.

$$[CO]_i = \frac{1.00 \text{ g}}{1.41 \text{ L}} \times \frac{1 \text{ mol CO}}{28.01 \text{ g CO}} = 0.0253 \text{ M}$$

$$[H_2O]_i = \frac{1.00 \text{ g}}{1.41 \text{ L}} \times \frac{1 \text{ mol } H_2O}{18.02 \text{ g } H_2O} = 0.0394 \text{ M}$$

$$[H_2]_i = \frac{1.00 \text{ g}}{1.41 \text{ L}} \times \frac{1 \text{ mol } H_2}{2.016 \text{ g } H_2} = 0.352 \text{ M}$$

Equation:	$CO(g)$ +	$H_2O(g)$ $\rightleftharpoons$	$CO_2(g)$ +	$H_2(g)$
Initial:	0.0253 M	0.0394 M	0.0000 M	0.352 M
Changes:	$-x$ M	$-x$ M	$+x$ M	$+x$ M
Equil:	$(0.0253 - x)$M	$(0.0394 - x)$M	x M	$(0.352 + x)$M

$$K_c = \frac{[CO_2][H_2]}{[CO][H_2O]} = 23.2 = \frac{x(0.352 + x)}{(0.0253 - x)(0.0394 - x)} = \frac{0.352x + x^2}{0.000997 - 0.0647x + x^2}$$

$0.0231 - 1.50x + 23.2x^2 = 0.352x + x^2$ $\qquad$ $22.2x^2 - 1.852x + 0.0231 = 0$

$$x = \frac{-b \pm \sqrt{b^2 - 4ac}}{2a} = \frac{1.852 \pm \sqrt{3.430 - 2.051}}{44.4} = 0.0682 \text{ M}, 0.0153 \text{ M}$$

The first value of x gives a negative [CO] (= –0.0429 M) and a negative $[H_2O]$ (= –0.0288 M). Thus, x = 0.0153 M = $[CO_2]$. Now we find the mass of CO_2.

$$1.41 \text{ L} \times \frac{0.0153 \text{ mol } CO_2}{1 \text{ L mixture}} \times \frac{44.01 \text{ g } CO_2}{1 \text{ mol } CO_2} = 0.949 \text{ g } CO_2$$

44. We base each of our solutions on the balanced chemical equation.

(a)

Equation:	$PCl_5(g)$ $\rightleftharpoons$	$PCl_3(g)$ +	$Cl_2(g)$
Initial:	$\frac{0.550 \text{ mol}}{2.50 \text{ L}}$	$\frac{0.550 \text{ mol}}{2.50 \text{ L}}$	$\frac{0 \text{ mol}}{2.50 \text{ L}}$
Changes:	$\frac{-x \text{ mol}}{2.50 \text{ L}}$	$\frac{+x \text{ mol}}{2.50 \text{ L}}$	$\frac{+x \text{ mol}}{2.50 \text{ L}}$
Equil:	$\frac{(0.550-x) \text{ mol}}{2.50 \text{ L}}$	$\frac{(0.550+x) \text{ mol}}{2.50 \text{ L}}$	$\frac{x \text{ mol}}{2.50 \text{ L}}$

$$K_c = \frac{[PCl_3][Cl_2]}{[PCl_5]} = 3.8 \times 10^{-2} = \frac{\frac{(0.550+x)\text{mol}}{2.50 \text{ L}} \times \frac{x \text{ mol}}{2.50 \text{ L}}}{\frac{(0.550-x)\text{mol}}{2.50 \text{ L}}} \qquad \frac{x(0.550 + x)}{2.50(0.550-x)} = 3.8 \times 10^{-2}$$

$x^2 + 0.550\,x = 0.052 - 0.095\,x$ $\qquad$ $x^2 + 0.645\,x - 0.052 = 0$

$$x = \frac{-b \pm \sqrt{b^2 - 4ac}}{2a} = \frac{-0.645 \pm \sqrt{0.416 + 0.208}}{2} = 0.0725 \text{ mol}, -0.717 \text{ mol}$$

The second answer gives a negative amount of Cl_2, and that makes no physical sense.

amount PCl_5 = (0.550 – 0.0725) = 0.478 mol $\qquad$ amount PCl_3 = (0.550 + 0.0725) = 0.623 mol

amount $Cl_2 = x$ = 0.0725 mol

(b)

Equation:	$PCl_5(g)$ $\rightleftharpoons$	$PCl_3(g)$ +	$Cl_2(g)$
Initial:	$\frac{0.610 \text{ mol}}{2.50 \text{ L}}$		
Changes:	$\frac{-x \text{ mol}}{2.50 \text{ L}}$	$\frac{+x \text{ mol}}{2.50 \text{ L}}$	$\frac{+x \text{ mol}}{2.50 \text{ L}}$
Equil:	$\frac{0.610-x \text{ mol}}{2.50 \text{ L}}$	$\frac{(x \text{ mol})}{2.50 \text{ L}}$	$\frac{(x \text{ mol})}{2.50 \text{ L}}$

$$K_c = \frac{[PCl_3][Cl_2]}{[PCl_5]} = 3.8 \times 10^{-2} = \frac{\frac{(x \text{ mol})}{2.50 \text{ L}} \times \frac{(x \text{ mol})}{2.50 \text{ L}}}{\frac{0.610 - x \text{ mol}}{2.50 \text{ L}}} \qquad 2.50 \times 3.8 \times 10^{-2} = \frac{x^2}{0.610-x} = 0.095$$

$0.058 - 0.095\,x = x^2$ $\qquad$ $x^2 + 0.095\,x - 0.058 = 0$

$$x = \frac{-b \pm \sqrt{b^2 - 4ac}}{2a} = \frac{-0.095 \pm \sqrt{0.0090 + 0.23}}{2} = 0.20 \text{ mol}, -0.29 \text{ mol}$$

amount PCl_3 = 0.20 mol = amount Cl_2 $\qquad$ amount PCl_5 = 0.610 – 0.20 = 0.41 mol

45. **(a)** We use the balanced chemical equation as a basis to organize the information we have about the reactants and products.

Equation:	$2\,COF_2(g)$ $\rightleftharpoons$	$CO_2(g)$ +	$CF_4(g)$
Initial:	$\frac{0.145 \text{ mol}}{5.00 \text{ L}}$	$\frac{0.262 \text{ mol}}{5.00 \text{ L}}$	$\frac{0.074 \text{ mol}}{5.00 \text{ L}}$
Initial:	0.0290 M	0.0524 M	0.0148 M

And we now compute a value of Q_c to compare with the given value of K_c.

$$Q_c = \frac{[CO_2][CF_4]}{[COF_2]^2} = \frac{(0.0524)(0.0148)}{(0.0290)^2} = 0.922 < 2.00 = K_c$$

Because Q_c is not equal to K_c, the mixture is not at equilibrium.

(b) Because Q_c is smaller than K_c, the reaction will shift right, that is, products will be formed, in reaching a state of equilibrium.

(c) We continue the organization of information about reactants and products.

Equation:	$2\ COF_2(g)$	$\rightleftharpoons CO_2(g)$	+	$CF_4(g)$
Initial:	0.0290 M	0.0524 M		0.0148 M
Changes:	$-2x$ M	$+x$ M		$+x$ M
Equil:	$(0.0290 - 2x)$M	$(0.0524 + x)$M		$(0.0148 + x)$M

$$K_c = \frac{[CO_2][CF_4]}{[COF_2]^2} = \frac{(0.0524 + x)(0.0148 + x)}{(0.0290 - 2x)^2} = 2.00 = \frac{0.000776 + 0.0672x + x^2}{0.000841 - 0.1160\ x + 4x^2}$$

$$0.00168 - 0.232x + 8x^2 = 0.000776 + 0.0672x + x^2 \qquad 7x^2 - 0.299x + 0.000904 = 0$$

$$x = \frac{-b \pm \sqrt{b^2 - 4ac}}{2a} = \frac{0.299 \pm \sqrt{0.0894 - 0.0253}}{14} = 0.0033\text{ M},\ 0.0394\text{ M}$$

The second of these values for x (0.0394) gives a negative $[COF_2]$ (= –0.0498 M), clearly a nonsensical result. We now compute the concentration of each species at equilibrium, and check to ensure that the equilibrium constant is satisfied.

$[COF_2] = 0.0290 - 2x = 0.0290 - 2(0.0033) = 0.0224$ M

$[CO_2] = 0.0524 + x = 0.0524 + 0.0033 = 0.0557$ M

$[CF_4] = 0.0148 + x = 0.0148 + 0.0033 = 0.0181$ M

$$K_c = \frac{[CO_2][CF_4]}{[COF_2]^2} = \frac{0.0557\text{ M} \times 0.0181\text{ M}}{(0.0224\text{ M})^2} = 2.01$$

The agreement of this value of K_c with the cited value (2.00) indicates that this solution is correct. Now we determine the number of moles of each species at equilibrium.

no. mol $COF_2 = 5.00\text{ L} \times 0.0224\text{ M} = 0.112$ mol COF_2

no. mol $CO_2 = 5.00\text{ L} \times 0.0557\text{ M} = 0.279$ mol CO_2

no. mol $CF_4 = 5.00\text{ L} \times 0.0181\text{ M} = 0.0905$ mol CF_4

But suppose we had *incorrectly* concluded, in part (b), that reactants would be formed in reaching equilibrium. What result would we obtain? The set up follows.

Equation:	$2\ COF_2(g)$	$\rightleftharpoons CO_2(g)$	+	$CF_4(g)$
Initial:	0.0290 M	0.0524 M		0.0148 M
Changes:	$+2y$ M	$-y$ M		$-y$ M
Equil:	$(0.0290 + 2y)$M	$(0.0524 - y)$M		$(0.0148 - y)$M

$$K_c = \frac{[CO_2][CF_4]}{[COF_2]^2} = \frac{(0.0524 - y)(0.0148 - y)}{(0.0290 + 2y)^2} = 2.00 = \frac{0.000776 - 0.0672y + y^2}{0.000841 + 0.1160\ y + 4y^2}$$

$$0.00168 + 0.232y + 8y^2 = 0.000776 - 0.0672y + y^2 \qquad 7y^2 + 0.299y + 0.000904 = 0$$

$$y = \frac{-b \pm \sqrt{b^2 - 4ac}}{2a} = \frac{-0.299 \pm \sqrt{0.0894 - 0.0253}}{14} = -0.0033\text{ M},\ -0.0394\text{ M}$$

The second of these values for x (–0.0394) gives a negative $[COF_2]$ (= –0.0498 M), clearly a nonsensical result. We now compute the concentration of each species at equilibrium, and check to ensure that the equilibrium constant is satisfied.

$[COF_2] = 0.0290 + 2y = 0.0290 + 2(-0.0033) = 0.0224$ M

$[CO_2] = 0.0524 - y = 0.0524 + 0.0033 = 0.0557$ M

$[CF_4] = 0.0148 - y = 0.0148 + 0.0033 = 0.0181$ M

These are the same equilibrium concentrations that we obtained by making the correct decision regarding the direction that the reaction would take. Thus, you can be assured that, if you perform the algebra correctly, it will guide you even if you make the incorrect decision about the direction of the reaction.

46. (a) We calculate the initial amount of each substance.

$$n\{C_2H_5OH\} = 17.2\text{ g } C_2H_5OH \times \frac{1\text{ mol } C_2H_5OH}{46.07\text{ g } C_2H_5OH} = 0.373\text{ mol } C_2H_5OH$$

$$n\{CH_3CO_2H\} = 23.8\text{ g } CH_3CO_2H \times \frac{1\text{ mol } CH_3CO_2H}{60.05\text{ g } CH_3CO_2H} = 0.396\text{ mol } CH_3CO_2H$$

$$n\{CH_3CO_2C_2H_5\} = 48.6\text{ g } CH_3CO_2C_2H_5 \times \frac{1\text{ mol } CH_3CO_2C_2H_5}{88.11\text{ g } CH_3CO_2C_2H_5} = 0.552\text{ mol } CH_3CO_2C_2H_5$$

$$n\{H_2O\} = 71.2\text{ g } H_2O \times \frac{1\text{ mol } H_2O}{18.02\text{ g } H_2O} = 3.95\text{ mol } H_2O$$

Since we would divide each amount by the total volume, and since there are the same numbers of product and reactant stoichiometric coefficients, we can use amounts rather than concentrations in the Q_c expression. $Q_c = \dfrac{n\{CH_3CO_2C_2H_5\}\, n\{H_2O\}}{n\{C_2H_5OH\}\, n\{CH_3CO_2H\}} = \dfrac{0.552 \text{ mol} \times 3.95 \text{ mol}}{0.373 \text{ mol} \times 0.396 \text{ mol}} = 14.8 > K_c = 4.0$

Since $Q_c > K_c$ the reaction will shift to the left, forming reactants, as it attains equilibrium.

(b)

Equation:	C_2H_5OH +	CH_3CO_2H ⇌	$CH_3CO_2C_2H_5$ +	H_2O
Initial:	0.373 mol	0.396 mol	0.552 mol	3.95 mol
Changes:	$+x$ mol	$+x$ mol	$-x$ mol	$-x$ mol
Equil:	$(0.373 + x)$ mol	$(0.396 + x)$ mol	$(0.552 - x)$ mol	$(3.95 - x)$ mol

$$K_c = \frac{(0.552 - x)(3.95 - x)}{(0.373 + x)(0.396 + x)} = \frac{2.18 - 4.50\,x + x^2}{0.148 + 0.769\,x + x^2} = 4.0$$

$x^2 - 4.50\,x + 2.18 = 4\,x^2 + 3.08\,x + 0.59 \qquad 3\,x^2 + 7.58\,x - 1.59 = 0$

$$x = \frac{-b \pm \sqrt{b^2 - 4ac}}{2a} = \frac{-7.58 \pm \sqrt{57 + 19}}{6} = 0.19 \text{ moles}, -2.72 \text{ moles}$$

Negative amounts make no physical sense. We compute the equilibrium amount of each substance.

$n\{C_2H_5OH\} = 0.373 \text{ mol} + 0.19 \text{ mol} = 0.56 \text{ mol } C_2H_5OH$

$n\{CH_3CO_2H\} = 0.396 \text{ mol} + 0.19 \text{ mol} = 0.59 \text{ mol } CH_3CO_2H$

$n\{CH_3CO_2C_2H_5\} = 0.552 \text{ mol} - 0.19 \text{ mol} = 0.36\ CH_3CO_2C_2H_5$

$n\{H_2O\} = 3.95 \text{ mol} - 0.19 \text{ mol} = 3.76 \text{ mol } H_2O$

To check $K_c = \dfrac{n\{CH_3CO_2C_2H_5\}\, n\{H_2O\}}{n\{C_2H_5OH\}\, n\{CH_3CO_2H\}} = \dfrac{0.36 \text{ mol} \times 3.76 \text{ mol}}{0.56 \text{ mol} \times 0.59 \text{ mol}} = 4.10$

47. Again we organize the known data around the balanced chemical equation.

Equation:	$H_2(g)$ +	$I_2(g)$ ⇌	$2\ HI(g)$
Initial:	0.125 mol	0.125 mol	0.000 mol
Changes:	$-x$ mol	$-x$ mol	$+2x$ mol
Equil:	$(0.125 - x)$mol	$(0.125 - x)$mol	$2x$ mol

$$K_c = \frac{[HI]^2}{[H_2][I_2]} = \frac{\left(\frac{2x \text{ mol}}{6.14 \text{ L}}\right)^2}{\frac{(0.125 - x)\text{mol}}{6.14 \text{ L}} \times \frac{(0.125 - x)\text{mol}}{6.14 \text{ L}}} = \left(\frac{2x}{0.125 - x}\right)^2 = 50.2$$

$\sqrt{50.2} = \dfrac{2x}{0.125 - x} = 7.09 \qquad 2x = 0.886 - 7.09\,x \qquad 9.09\,x = 0.886 \qquad x = \dfrac{0.886}{9.09} = 0.0975 \text{ mol}$

amount HI $= 2x = 2 \times 0.0975 \text{ mol} = 0.195 \text{ mol}$

amount H_2 = amount $I_2 = 0.125 - x = 0.125 - 0.0975 = 0.0275$ mol

Total amount $= 0.195 \text{ mol HI} + 0.0275 \text{ mol } H_2 + 0.0275 \text{ mol } I_2 = 0.250 \text{ mol}$

mol% HI $= \dfrac{0.195 \text{ mol HI}}{0.250 \text{ mol total}} \times 100\% = 78.0\%$ HI

48. The final volume of the mixture is 0.750 L + 2.25 L = 3.00 L. Then use the balanced chemical equation to organize the data we have concerning the reaction. The reaction should shift to the right, that is form products, in reaching a new equilibrium, since the volume is greater.

Equation:	$N_2O_4(g)$ ⇌	$2\ NO_2(g)$
Initial:	$\dfrac{0.971 \text{ mol}}{3.00 \text{ L}}$	$\dfrac{0.0580 \text{ mol}}{3.00 \text{ L}}$
Initial:	0.324 M	0.0193 M
Changes:	$-x$ M	$+2x$ M
Equil:	$(0.324 - x)$M	$(0.0193 + 2x)$M

$$K_c = \frac{[NO_2]^2}{[N_2O_4]} = \frac{(0.0193 + 2x)^2}{0.324 - x} = \frac{0.000372 + 0.0772x + 4x^2}{0.324 - x} = 4.61 \times 10^{-3}$$

$0.000372 + 0.0772x + 4x^2 = 0.00149 - 0.00461x \qquad 4x^2 + 0.0818x - 0.00112 = 0$

$$x = \frac{-b \pm \sqrt{b^2 - 4ac}}{2a} = \frac{-0.0818 \pm \sqrt{0.00669 + 0.0179}}{8} = 0.00938 \text{ M}, -0.0298 \text{ M}$$

$[NO_2] = 0.0193 + (2 \times 0.00938) = 0.0381$ M $\qquad$ amount $NO_2 = 0.0381 \text{ M} \times 3.00 \text{ L} = 0.114 \text{ mol } NO_2$

$[N_2O_4] = 0.324 - 0.00938 = 0.314_6$ M $\qquad$ amount $N_2O_4 = 0.314_6 \text{ M} \times 3.00 \text{ L} = 0.944 \text{ mol } N_2O_4$

49. $[HCONH_2]_{init} = \dfrac{0.186 \text{ mol}}{2.16 \text{ L}} = 0.0861 \text{ M}$

Equation:	$HCONH_2(g)$ ⇌	$NH_3(g)$ +	$CO(g)$
Initial:	0.0861 M	0 M	0 M
Changes:	$-x$ M	$+x$ M	$+x$ M
Equil:	$(0.0861 - x)$ M	x M	x M

$K_c = \dfrac{[NH_3][CO]}{[HCONH_2]} = \dfrac{x \cdot x}{0.0861 - x} = 4.84 \qquad x^2 = 0.417 - 4.84\,x \qquad 0 = x^2 + 4.84\,x - 0.417$

$x = \dfrac{-b \pm \sqrt{b^2 - 4ac}}{2a} = \dfrac{-4.84 \pm \sqrt{23.4 + 1.67}}{2} = 0.084 \text{ M}, -4.92 \text{ M}$

The negative concentration obviously is physically meaningless. We determine the total concentration of all species, and then the total pressure.

$[\text{total}] = [NH_3] + [CO] + [HCONH_2] = x + x + 0.0861 - x = 0.0861 + 0.084 = 0.170 \text{ M}$

$P_{tot} = 0.170 \text{ mol L}^{-1} \times 0.08206 \text{ L atm mol}^{-1} \text{ K}^{-1} \times 400. \text{ K} = 5.58 \text{ atm}$

50. We determine the value of Q_p for this situation and compare it to K_p. We assume that the added solids are of negligible volume so that the initial partial pressures of $CO_2(g)$ and $H_2O(g)$ do not significantly change.

$P\{H_2O\} = \left(715 \text{ mmHg} \times \dfrac{1 \text{ atm}}{760 \text{ mmHg}}\right) = 0.941 \text{ atm } H_2O$

$Q_p = P\{CO_2\}P\{H_2O\} = 2.10 \text{ atm } CO_2 \times 0.941 \text{ atm } H_2O = 1.98 > 0.23 = K_p$

Because Q_p is larger than K_p, the reaction will proceed left toward reactants in establishing equilibrium; the partial pressures of the two gases will decrease.

51. We organize the given data around the balanced chemical equation.

Equation:	$Ag^+(aq)$ +	Fe^{2+} ⇌	$Fe^{3+}(aq)$ +	$Ag(s)$
Initial:	1.00 M	1.00 M		
Changes:	$-x$ M	$-x$ M	$+x$ M	
Equil:	$(1.00 - x)$M	$(1.00 - x)$M	x M	

$K_c = \dfrac{[Fe^{3+}]}{[Ag^+][Fe^{2+}]} = \dfrac{x}{(1.00 - x)^2} = 2.98 \qquad x = 2.98(1.00 - 2.00x + x^2) = 2.98 - 5.96x + 2.98x^2$

$2.98x^2 - 6.96x + 2.98 = 0 \qquad x = \dfrac{-b \pm \sqrt{b^2 - 4ac}}{2a} = \dfrac{6.96 \pm \sqrt{48.4 - 35.5}}{5.96} = 1.77 \text{ M}, 0.565 \text{ M}$

The first value of x (1.77 M) gives a negative $[Ag^+]$ and $[Fe^{2+}]$ (= –0.77 M). Thus, we choose $x = 0.565$ M.

$[Fe^{3+}] = 0.56 \text{ M} \qquad [Ag^+] = [Fe^{2+}] = 1.00 \text{ M} - 0.565 \text{ M} = 0.44 \text{ M}$

52. Again we base the set up of the problem around the balanced chemical equation.

	$Pb(s)$ +	$2\,Cr^{3+}(aq)$ ⇌	$Pb^{2+}(aq)$ +	$2\,Cr^{2+}(aq)$
initial:		0.100 M	0 M	0 M
changes:		$-2x$ M	$+x$ M	$+2x$ M
equil:		$(0.100 - 2x)$M	xM	$2x$ M

$K_c = 3.2 \times 10^{-10} \qquad K_c = \dfrac{[Pb^{2+}][Cr^{2+}]^2}{[Cr^{3+}]^2} = \dfrac{x\,(2x)^2}{(0.100 - 2x)^2}$

$K_c \approx \dfrac{x\,(2x)^2}{(0.100)^2} = 3.2 \times 10^{-10} \qquad 4x^3 = (0.100)^2 \times 3.2 \times 10^{-10} = 3.2 \times 10^{-12}$

$x = \sqrt[3]{\dfrac{3.2 \times 10^{-12}}{4}} = 9.3 \times 10^{-5} \text{ M}$

Our assumption, that $2x \ll 0.100$, is valid and thus also is our result: $[Pb^{2+}] = x = 9.3 \times 10^{-5}$ M

Partial Pressure Equilibrium Constant, K_p

53. We substitute the given equilibrium pressure into the equilibrium constant expression and solve for the other equilibrium pressure. $K_p = \dfrac{P\{O_2\}^3}{P\{CO_2\}^2} = 28.5 = \dfrac{P\{O_2\}^3}{(0.0721 \text{ atm } CO_2)^2}$

$P\{O_2\} = \sqrt[3]{P\{O_2\}^3} = \sqrt[3]{28.5\,(0.0712 \text{ atm})^2} = 0.525 \text{ atm } O_2$

$P_{total} = P\{CO_2\} + P\{O_2\} = 0.0721 \text{ atm } CO_2 + 0.525 \text{ atm } O_2 = 0.597 \text{ atm total}$

54. The composition of dry air is given in volume percent. Division of these percents by 100 gives the volume fraction, which equals the mole fraction and also the partial pressure in atmospheres, if the total pressure is 1.00 atm. Thus, we have $P\{O_2\} = 0.20946$ atm and $P\{CO_2\} = 0.00036$ atm. We substitute these two values into the expression for Q_p.

$$Q_p = \frac{P\{O_2\}^3}{P\{CO_2\}^2} = \frac{(0.20946 \text{ atm } O_2)^3}{(0.00036 \text{ atm } CO_2)^2} = 7.1 \times 10^4 > 28.5 = K_p$$

The value of Q_p in the atmosphere is much larger than the value of K_p. This reaction should proceed to the left. It should only occur to the right when the pressure of O_2 drops or that of CO_2 rises (as would be the case in self-contained breathing devices).

55. **(a)** We first determine the initial pressure of each gas.

$$P\{CO\} = P\{Cl_2\} = \frac{nRT}{V} = \frac{1.00 \text{ mol} \times 0.08206 \text{ L atm mol}^{-1} \text{ K}^{-1} \times 668 \text{ K}}{1.75 \text{ L}} = 31.3 \text{ atm}$$

Then we calculate equilibrium partial pressures, organizing our calculation around the balanced chemical equation. We see that the equilibrium constant is quite large, meaning that the equilibrium position favors the product. Thus, we begin by forming as much product as possible; then proceed to equilibrium. (Significant figure rules produce a somewhat erroneous answer if we do not initially form product.)

Equation:	$CO(g)$	+	$Cl_2(g)$	$\rightleftharpoons$	$COCl_2(g)$
Initial:	31.3 atm		31.3 atm		0 atm
To right:	0 atm		0 atm		31.3 atm
Changes:	$+x$ atm		$+x$ atm		$-x$ atm
Equil:	x atm		x atm		$(32.5 - x)$ atm

$$K_p = \frac{P\{COCl_2\}}{P\{CO\}\, P\{Cl_2\}} = 22.5 = \frac{31.3 - x}{x^2} \qquad 22.5\, x^2 = 31.3 - x \qquad 22.5x^2 + x - 31.3 = 0$$

$$x = \frac{-b \pm \sqrt{b^2 - 4ac}}{2a} = \frac{-1 \pm \sqrt{1 + 2817}}{45.0} = 1.16 \text{ atm}, -1.20 \text{ atm}$$

$P\{CO\} = P\{Cl_2\} = 1.16$ atm $\qquad P\{COCl_2\} = 31.3 \text{ atm} - 1.16 \text{ atm} = 30.1$ atm

(b) $P_{total} = P\{CO\} + P\{Cl_2\} + P\{COCl_2\} = 1.16 \text{ atm} + 1.16 \text{ atm} + 30.1 \text{ atm} = 32.4$ atm

56. The $I_2(s)$ maintains the presence of I_2 in the flask until it has all vaporized. Thus, if enough HI(g) is produced to deplete the $I_2(s)$, equilibrium will not be achieved.

$$P\{H_2S\} = 747.6 \text{ mmHg} \times \frac{1 \text{ atm}}{760 \text{ mmHg}} = 0.98376 \text{ atm}$$

Equation:	$H_2S(g) + I_2(s)$	$\rightleftharpoons$	$2\,HI(g) + S(s)$
Initial:	0.9837 atm		
Changes:	$-x$ atm		$+2x$ atm
Equil:	$(0.9837 - x)$atm		$2x$ atm

$$K_p = \frac{P\{HI\}^2}{P\{H_2S\}} = \frac{(2x)^2}{(0.9837 - x)} = 1.34 \times 10^{-5} \approx \frac{4x^2}{0.9837};\; x = \sqrt{\frac{1.34 \times 10^{-5} \times 0.9837}{4}} = 1.82 \times 10^{-3} \text{ atm}$$

Our assumption, that $0.9837 >> x$, clearly is valid. We now verify that sufficient $I_2(s)$ is present by computing the mass of I_2 needed to produce the predicted pressure of HI(g). There is 1.85 g I_2 available.

$$\text{mass } I_2 = \frac{1.82 \times 10^{-3} \text{ atm} \times 0.725 \text{ L}}{0.08206 \text{ L atm mol}^{-1} \text{ K}^{-1} \times 333 \text{ K}} \times \frac{1 \text{ mol } I_2}{2 \text{ mol HI}} \times \frac{253.8 \text{ g } I_2}{1 \text{ mol } I_2} = 0.00613 \text{ g } I_2$$

$$P_{tot} = P\{H_2S\} + P\{HI\} = (0.9837 - x) + 2x = 0.9837 + x = 0.9837 + 0.00182 = 0.9855 \text{ atm} = 749.0 \text{ mmHg}$$

Le Châtelier's Principle

57. Continuous removal of the product, of course, has the effect of decreasing the concentration of the products below their equilibrium values. Thus, the equilibrium system is disturbed by removing the products and the system will attempt (in vain, as it turns out) to re-establish the equilibrium by shifting toward the right, that is, by producing more products.

58. We notice that the density of the solid ice is smaller than is that of liquid water. This means that the same mass of liquid water is present in a smaller volume than an equal mass of ice. Thus, if pressure is placed on ice, attempting to force it into a smaller volume, the ice will be transformed into the less-space-occupying water at 0 °C. Thus, at 0 °C ice will melt to water under pressure. This behavior is *not* expected in most cases because generally a solid is *more* dense than its liquid phase.

59. **(a)** This reaction is exothermic with $\Delta H° = -150.$ kJ. Thus high temperatures favor the reverse reaction. The amount of $H_2(g)$ present at high temperatures will be less than that present at low temperatures.

(b) $H_2O(g)$ is one of the reactants involved. Introducing more will cause the equilibrium position to shift to the right, favoring products. The amount of $H_2(g)$ will increase.

(c) Doubling the volume of the container will favor the side of the reaction with the largest sum of gaseous stoichiometric coefficients. The sum of the stoichiometric coefficients of gaseous species is the same (4) on both sides of this reaction. Therefore, increasing the volume of the container will have no effect on the amount of $H_2(g)$ present at equilibrium.

(d) A catalyst merely speeds up the rate at which a reaction reaches the equilibrium position. The addition of a catalyst has no effect on the amount of $H_2(g)$ present at equilibrium.

60. **(a)** This reaction is endothermic, with $\Delta H°= +92.5$ kJ. Thus, high temperature will favor the forward reaction and increase the amount of HI(g) present at equilibrium.

(b) The introduction of more product will favor the reverse reaction and decrease the amount of HI(g) present at equilibrium.

(c) The sum of the stoichiometric coefficients of gaseous products is larger than that for gaseous reactants. Increasing the volume of the container will favor the forward reaction and increase the amount of HI(g) present at equilibrium.

(d) A catalyst merely speeds up the rate at which a reaction reaches the equilibrium position. The addition of a catalyst has no effect on the amount of HI(g) present at equilibrium.

(e) The addition of an inert gas to the constant-volume reaction mixture will not change any partial pressures. It will have no effect on the amount of HI(g) present at equilibrium.

61. **(a)** The formation of NO(g) from the elements is an endothermic reaction ($\Delta H° = +181$ kJ/mol). Since the equilibrium position of endothermic reactions is shifted toward products at high temperatures, we expect the formation of NO(g) from the elements to be enhanced at high temperatures.

(b) Reaction rates always are enhanced by high temperatures, since a larger fraction of the collisions will have an energy which surmounts the activation energy. This enhancement of rates affects both the forward and the reverse reactions. Thus, the position of equilibrium is reached more rapidly at high temperatures than at low temperatures.

62. If the reaction is endothermic ($\Delta H° > 0$), the forward reaction is favored at high temperatures. If the reaction is exothermic ($\Delta H° < 0$), the forward reaction is favored at low temperatures.

(a) $\Delta H° = \Delta H_f°[PCl_5(g)] - \Delta H_f°[PCl_3(g)] - \Delta H_f°[Cl_2(g)]$
$= -374.9 - (-287.0) - 0.00 = -87.9$ kJ/mol   low temperatures

(b) $\Delta H° = 2\,\Delta H_f°[H_2O(g)] + 3\,\Delta H_f°[S(\text{rhombic})] - \Delta H_f°[SO_2(g)] - 2\,\Delta H_f°[H_2S(g)]$
$= 2\,(-241.8) + 3\,(0.00) - (-296.8) - 2\,(-20.63) = -145.5$ kJ/mol   low temperatures

(c) $\Delta H° = 4\,\Delta H_f°[NOCl(g)] + 2\,\Delta H_f°[H_2O(g)] - 2\,\Delta H_f°[N_2(g)] - 3\,\Delta H_f°[O_2(g)] - 4\,\Delta H_f°[HCl(g)]$
$= 4\,(51.71) + 2\,(-241.8) - 2\,(0.00) - 3\,(0.00) - 4\,(-92.31) = +92.5$ kJ/mol   high temps

63. Although the yield of each of these reactions is greater at lower temperatures, both reactions are faster at higher temperatures (as are all reactions). The desire is to cause these reactions to proceed as rapidly as feasible. We need to be aware that the facility in which the reaction is being carried out costs money to build and to operate. The longer this facility is occupied in producing a given batch of product, the more costly that product will be. Besides, there are other ways to form large quantities of product from a given amount of reactant. Ways other than running the reaction under conditions where the equilibrium constant has a large value include constantly removing the product as it is formed, and feeding in reactant as it is consumed. Both of these strategies will enhance the formation of product.

64. **(a)** Since $\Delta H° = 0$, the position of the equilibrium for this reaction will not be affected by temperature. Because the position of equilibrium is expressed by the value of the equilibrium constant, we expect K_p to be unaffected by, or to remain constant with, temperature.

(b) From part (a), we know that the value of K_p will not change when the temperature is changed. The pressures of the gases, however, will change with temperature. (Recall the ideal gas law: $P = nRT/V$.) In fact, all pressures will increase. The stoichiometric coefficients in the reaction are such that higher pressure favor the formation of reactant in reattaining equilibrium. Thus, the amount of D(g) will be smaller when equilibrium is re-established at the higher temperature for the cited reaction.

$$A(s) \rightleftharpoons B(s) + 2\,C(g) + \tfrac{1}{2}\,D(g)$$

FEATURE PROBLEMS

A. We first determine the amount in moles of acetic acid in the equilibrium mixture.

$$\text{amount } CH_3CO_2H = 28.85 \text{ mL} \times \frac{1 \text{ L}}{1000 \text{ mL}} \times \frac{0.1000 \text{ mol Ba(OH)}_2}{1 \text{ L}} \times \frac{2 \text{ mol } CH_3CO_2H}{1 \text{ mol Ba(OH)}_2} \times \frac{\text{complete equilibrium mixture}}{0.01 \text{ of equilibrium mixture}} = 0.5770 \text{ mol } CH_3CO_2H$$

Equation:	C_2H_5OH +	CH_3CO_2H $\rightleftharpoons$	$CH_3CO_2C_2H_5$ +	H_2O
Initial:	0.500 mol	1.000 mol	0.000 mol	0.000 mol
Changes:	–0.423 mol	–0.423 mol	+0.423 mol	+0.423 mol
Equil:	0.077 mol	0.577 mol	0.423 mol	0.423 mol

$$K_c = \frac{[CH_3CO_2C_2H_5][H_2O]}{[C_2H_5OH][CH_3CO_2H]} = \frac{\frac{0.423 \text{ mol}}{V} \times \frac{0.423 \text{ mol}}{V}}{\frac{0.077 \text{ mol}}{V} \times \frac{0.577 \text{ mol}}{V}} = \frac{0.423 \times 0.423}{0.077 \times 0.577} = 4.0$$

B. From previous discussions, we know that the ammonia synthesis reaction as run industrially requires a catalyst. Thus, we surmise that the reaction is slow. Generally, we heat a reaction to provide molecules with activation energy and thus speed up the reaction. But we notice that this reaction is exothermic, meaning that we decrease the yield by heating it. Next we notice that the reaction produces two moles of gaseous $N_2(g)$ from four moles of gaseous reactants. This indicates that running the reaction at elevated pressure will increase the yield. Because the synthesis plant costs money to use, we want the reaction to run as fast as possible, consistent with a reasonable yield. So the reaction is run at high pressure and temperature with a catalyst.

We can separate the NH_3 from the equilibrium mixture by cooling it until it liquefies. This will leave gaseous N_2 and H_2 behind to be recycled into the reaction chamber.

The condition of high pressure conforms to LeChâtelier's principle, but that of high temperature does not. For this synthesis a rapid reaction is preferred over one displaced far to products. A rapid reaction means that expensive synthetic apparatus is not tied up waiting for a slow equilibrium to be achieved. It is practical in this case because it is relatively easy to separate product (by liquefying it) from the equilibrium mixture and to recycle the unused reactants.

C. **1.** $K_p = K_c\,(RT)^{\Delta n}$ For the reaction $CO(g) + 2\,H_2(g) \rightleftharpoons CH_3OH(g)$, $\Delta n_{gas} = 1 - (2 + 1) = -2$

$$K_p = 14.5\left(0.08206 \frac{\text{L·atm}}{\text{mol·K}} \times 483 \text{ K}\right)^{-2} = 9.23 \times 10^{-3}$$

2. We know that mole percents equal pressure percents for ideal gases.

$P\{CO\} = 0.350 \times 100.0 \text{ atm} = 35.0 \text{ atm}$ $\qquad$ $P\{H_2\} = 0.650 \times 100.0 \text{ atm} = 65.0 \text{ atm}$

3.

equation:	$CO(g)$ +	$2\,H_2(g)$ $\rightleftharpoons$	$CH_3OH(g)$
initial:	35.0 atm	65.0 atm	0 atm
changes:	$-P$ atm	$-2\,P$ atm	$+P$ atm
equil:	$35.0 - P$	$65.0 - 2P$	P

4. $$K_p = \frac{P\{CH_3OH\}}{P\{CO\} \times P\{H_2\}^2} = \frac{P}{(35.0 - P)\,(65.0 - 2P)^2} = 9.23 \times 10^{-3} = 0.00932$$

Guess $P = 30.0$ atm $\quad Q_p = \frac{30.0}{(35.0 - 30.0)\,(65.0 - 2(30.0))^2} = \frac{30.0}{5.0\,(5.0)^2} = 0.24 > 0.00923$

Guess $P = 20.0$ atm $\quad Q_p = \frac{20.0}{(35.0 - 20.0)\,(65.0 - 2(20.0))^2} = \frac{20.0}{15.0\,(25.0)^2} = 0.0021 < 0.00923$

Guess $P = 25.0$ atm $\quad Q_p = \frac{25.0}{(35.0 - 25.0)\,(65.0 - 2(25.0))^2} = \frac{25.0}{10.0\,(15.0)^2} = 0.0111 > 0.00923$

Guess $P = 24.0$ atm $\quad Q_p = \frac{24.0}{(35.0 - 24.0)\,(65.0 - 2(24.0))^2} = \frac{24.0}{11.0\,(17.0)^2} = 0.00755 < 0.00923$

Guess $P = 24.5$ atm $\quad Q_p = \frac{24.5}{(35.0 - 24.5)\,(65.0 - 2(24.5))^2} = \frac{24.5}{10.5\,(16.0)^2} = 0.00911 < 0.00923$

Guess $P = 24.6$ atm $\quad Q_p = \frac{24.6}{(35.0 - 24.6)\,(65.0 - 2(24.6))^2} = \frac{24.6}{10.4\,(15.8)^2} = 0.00948 > 0.00923$

To three significant figures, the answer is $P = 24.6$ atm $CH_3OH(g)$.

17 ACIDS AND BASES

PRACTICE EXAMPLES

1A **(a)** In the forward reaction, HF is a proton donor because F^- is a product; HF is an acid. The proton is donated to H_2O, which is a base. F^- is the proton acceptor (base) in the reverse reaction and H_3O^+ is the acid.

(b) HSO_4^- is the proton donor (acid) in the forward reaction because SO_4^{2-} is created. The proton is donated to NH_3, which is the base. In the reverse reaction SO_4^{2-} is the proton acceptor (base) and NH_4^+ is the acid.

(c) HCl is the proton donor (acid) in the forward reaction because Cl^- is created. The proton is donated to $C_2H_3O_2^-$, the base. In the reverse reaction, Cl^- accepts a proton and is a base. $HC_2H_3O_2$ is the acid.

1B We know that the formulas of most acids begin with H. Thus, we identify HNO_2 and HCO_3^- as acids.

$HNO_2(aq) + H_2O \rightleftharpoons NO_2^-(aq) + H_3O^+(aq)$ $\quad$ $HCO_3^-(aq) + H_2O \rightleftharpoons CO_3^{2-}(aq) + H_3O^+(aq)$

A negatively charged species will attract a positively charged proton and act as a base. Thus PO_4^{3-} and HCO_3^- can act as bases. We also know that PO_4^{3-} is a base because it cannot act as an acid—it has no protons to donate—and we know that all three species have acid–base properties.

$PO_4^{3-}(aq) + H_2O \rightleftharpoons HPO_4^{2-}(aq) + OH^-(aq)$ $\quad$ $HCO_3^-(aq) + H_2O \rightleftharpoons CO_2{\cdot}H_2O(aq) + OH^-(aq)$

Notice that HCO_3^- is the amphiprotic species, acting as both an acid and as a base.

2A $[H_3O^+]$ is readily computed from pH: $[H_3O^+] = 10^{-pH}$ $\quad$ $[H_3O^+] = 10^{-2.85} = 1.4 \times 10^{-3}$ M. $[OH^-]$ can be found in two ways: (1) from $K_w = [H_3O^+][OH^-]$, giving $[OH^-] = \dfrac{K_w}{[H_3O^+]} = \dfrac{1.0 \times 10^{-14}}{1.4 \times 10^{-3}} = 7.1 \times 10^{-12}$ M,

or (2) from pH + pOH = 14.00, giving pOH = 14.00 – pH = 14.00 – 2.85 = 11.15, and then $[OH^-] = 10^{-pOH} = 10^{-11.15} = 7.1 \times 10^{-12}$ M.

2B $[H_3O^+]$ is computed from pH in each case: $[H_3O^+] = 10^{-pH}$

$[H_3O^+]_{conc.} = 10^{-2.50} = 3.2 \times 10^{-3}$ M $\quad$ $[H_3O^+]_{dil.} = 10^{-3.10} = 7.9 \times 10^{-4}$ M

All of the H_3O^+ in the dilute solution comes from the concentrated solution.

$$\text{amount } H_3O^+ = 1.00 \text{ L conc. soln} \times \frac{3.2 \times 10^{-3} \text{ mol } H_3O^+}{1 \text{ L conc. soln}} = 3.2 \times 10^{-3} \text{ mol } H_3O^+$$

But it is diluted in the dilute solution.

$$\text{volume of dilute solution} = 3.2 \times 10^{-3} \text{ mol } H_3O^+ \times \frac{1 \text{ L dilute soln}}{7.9 \times 10^{-4} \text{ mol } H_3O^+} = 4.1 \text{ L dilute soln}$$

volume of water to be added = 3.1 L

3A pH is computed directly from the $[H_3O^+]$, pH = –log $[H_3O^+]$ = –log (0.0025) = 2.60. We know that HI is a strong acid and thus is completely dissociated into H_3O^+ and I^-. The consequence is that $[I^-] = [H_3O^+] =$ 0.0025 M. $[OH^-]$ is most readily computed from pH: pOH = 14.00 – pH = 14.00 – 2.60 = 11.40; $[OH^-] = 10^{-pOH} = 10^{-11.40} = 4.0 \times 10^{-12}$ M

3B The amount of HCl(g) is calculated from the ideal gas equation. Then $[H_3O^+]$ is calculated, based on the fact that HCl(aq) is a strong acid (1 mol H_3O^+ is produced from each moles of HCl).

$$\text{amount HCl(g)} = \frac{\left(747\ \text{mmHg} \times \dfrac{1\ \text{atm}}{760\ \text{mmHg}}\right) \times 0.535\ \text{L}}{\dfrac{0.08206\ \text{L atm}}{\text{mol K}} \times (26.5 + 273.2)\ \text{K}} = 0.0214\ \text{mol HCl(g)}$$

$$[H_3O^+] = \frac{0.0214\ \text{mol HCl}}{0.625\ \text{L soln}} \times \frac{1\ \text{mol}\ H_3O^+}{1\ \text{mol HCl}} = 0.0342\ \text{M} \qquad \text{pH} = -\log(0.0342) = 1.466$$

4A pH is most readily determined from pOH = $-\log[OH^-]$. We assume that $Mg(OH)_2$ is a strong base.

$$[OH^-] = \frac{9.63\ \text{mg}\ Mg(OH)_2}{100.0\ \text{mL soln}} \times \frac{1000\ \text{mL}}{1\ \text{L}} \times \frac{1\ \text{g}}{1000\ \text{mg}} \times \frac{1\ \text{mol}\ Mg(OH)_2}{58.32\ \text{g}\ Mg(OH)_2} \times \frac{2\ \text{mol}\ OH^-}{1\ \text{mol}\ Mg(OH)_2}$$

$= 0.00330$ M pOH $= -\log(0.00330) = 2.481$ pH $= 14.00 - \text{pOH} = 14.00 - 2.48 = 11.52$

4B Notice that KOH is a strong base, which means that each mole of KOH produces one mole of $OH^-(aq)$. First we calculate $[OH^-]$ and the pOH. We then use pH + pOH = 14.00 to determine pH.

$$[OH^-] = \frac{3.00\ \text{g KOH}}{100.00\ \text{g soln}} \times \frac{1\ \text{mol KOH}}{56.11\ \text{g KOH}} \times \frac{1\ \text{mol}\ OH^-}{1\ \text{mol KOH}} \times \frac{1.0242\ \text{g soln}}{1\ \text{mL soln}} \times \frac{1000\ \text{mL}}{1\ \text{L}} = 0.548\ \text{M}$$

pOH $= -\log(0.548) = 0.261$ pH $= 14.00 - \text{pOH} = 14.00 - 0.261 = 13.74$

5A First we determine $[H_3O^+] = 10^{-pH} = 10^{-4.18} = 6.6 \times 10^{-5}$ M. Then we organize the solution around the balanced chemical equation.

Equation:	$HOCl(aq) + H_2O \rightleftharpoons$	$H_3O^+(aq)$ +	$OCl^-(aq)$
Initial:	0.150 M		
Changes:	-6.6×10^{-5} M	$+6.6 \times 10^{-5}$ M	$+6.6 \times 10^{-5}$ M
Equil:	0.150 M	6.6×10^{-5} M	6.6×10^{-5} M

$$K_a = \frac{[H_3O^+][OCl^-]}{[HOCl]} = \frac{(6.6 \times 10^{-5})(6.6 \times 10^{-5})}{0.150} = 2.9 \times 10^{-8}$$

5B First, we use pH to determine $[OH^-]$. pOH $= 14.00 - \text{pH} = 14.00 - 10.08 = 3.92$. $[OH^-] = 10^{-pOH} = 10^{-3.92} = 1.2 \times 10^{-4}$ M. We determine the initial concentration of cocaine and then organize the solution around the balanced equation in the manner we have used before.

$$C_{17}H_{21}O_4N\ \text{molarity} = \frac{0.17\ \text{g}\ C_{17}H_{21}O_4N}{100\ \text{mL soln}} \times \frac{1000\ \text{mL}}{1\ \text{L}} \times \frac{1\ \text{mol}\ C_{17}H_{21}O_4N}{303.36\ \text{g}\ C_{17}H_{21}O_4N} = 5.6 \times 10^{-3}\ \text{M} = 0.0056\ \text{M}$$

Equation:	$C_{17}H_{21}O_4N + H_2O \rightleftharpoons$	$C_{17}H_{21}O_4NH^+(aq)$ +	$OH^-(aq)$
Initial:	0.0056 M		
Changes:	-1.2×10^{-4} M	$+1.2 \times 10^{-4}$ M	$+1.2 \times 10^{-4}$ M
Equil:	0.0055 M	1.2×10^{-4} M	1.2×10^{-4} M

$$K_b = \frac{[C_{17}H_{21}O_4NH^+][OH^-]}{[C_{17}H_{21}O_4N]} = \frac{(1.2 \times 10^{-4})(1.2 \times 10^{-4})}{0.0055} = 2.6 \times 10^{-6}$$

6A Again we organize our solution around the balanced chemical equation.

Equation:	$HC_2H_2FO_2 + H_2O \rightleftharpoons$	H_3O^+ +	$C_2H_2FO_2^-$
Initial:	0.100 M		
Changes:	$-x$ M	$+x$ M	$+x$ M
Equil:	$(0.100 - x)$ M	x M	x M

$$K_a = \frac{[H_3O^+][C_2H_2FO_2^-]}{[HC_2H_2FO_2^-]} = 2.6 \times 10^{-3} = \frac{(x)(x)}{0.100 - x} \approx \frac{x^2}{0.100}$$

$x = \sqrt{x^2} = \sqrt{0.100 \times 2.6 \times 10^{-3}} = 0.016\ \text{M} = [H_3O^+]$ pH $= -\log[H_3O^+] = -\log(0.016) = 1.80$

6B We first determine the concentration of undissociated acid. We then use this value in a set up that is based on the balanced chemical equation. $[\text{acid}] = \dfrac{1000\ \text{mg acid}}{325\ \text{mL soln}} \times \dfrac{1\ \text{mmol acid}}{180.2\ \text{mg acid}} = 0.0171\ \text{M}$

Equation:	$HC_9H_7O_4$ + $H_2O \rightleftharpoons$	H_3O^+ +	$C_9H_7O_4$
Initial:	0.0171 M	0 M	0 M
Changes:	$-x$ M	$+x$ M	$+x$M
Equil:	$(0.0171 - x)$ M	x M	x M

$$K_a = 3.3 \times 10^{-4} = \frac{[H_3O^+][C_9H_7O_4^-]}{[HC_9H_7O_4]} = \frac{x \cdot x}{0.0171 - x}$$

$x^2 + 3.3 \times 10^{-4} - 5.6 \times 10^{-6} = 0$

$$x = \frac{-3.3 \times 10^{-4} \pm \sqrt{1.1 \times 10^{-7} + 2.3 \times 10^{-5}}}{2} = 0.0022\ \text{M} \qquad \text{pH} = -\log(0.0022) = 2.66$$

7A Again we organize our solution around the balanced chemical equation.

Equation:	$HC_2H_2FO_2 + H_2O$ $\rightleftharpoons$	H_3O^+ +	$C_2H_2FO_2^-$
Initial:	0.015 M		
Changes:	$-x$ M	$+x$ M	$+x$ M
Equil:	$(0.015 - x)$ M	x M	x M

$$K_a = \frac{[H_3O^+][C_2H_2FO_2^-]}{[HC_2H_2FO_2^-]} = 2.6 \times 10^{-3} = \frac{(x)(x)}{0.015 - x} \approx \frac{x^2}{0.015}$$

$x = \sqrt{x^2} = \sqrt{0.015 \times 2.6 \times 10^{-3}} = 0.0062\text{ M} = [H_3O^+]$ Our assumption is invalid: 0.0062 is not quite small compared to 0.015. Thus we use another cycle of successive approximations.

$$K_a = \frac{(x)(x)}{0.015 - 0.0062} = 2.6 \times 10^{-3} \quad x = \sqrt{(0.015 - 0.0062) \times 2.6 \times 10^{-3}} = 0.0048\text{ M} = [H_3O^+]$$

$$K_a = \frac{(x)(x)}{0.015 - 0.0048} = 2.6 \times 10^{-3} \quad x = \sqrt{(0.015 - 0.0048) \times 2.6 \times 10^{-3}} = 0.0051\text{ M} = [H_3O^+]$$

$$K_a = \frac{(x)(x)}{0.015 - 0.0051} = 2.6 \times 10^{-3} \quad x = \sqrt{(0.015 - 0.0051) \times 2.6 \times 10^{-3}} = 0.0051\text{ M} = [H_3O^+]$$

Two successive identical results signal the solution. $pH = -\log [H_3O^+] = -\log(0.0051) = 2.29$
The quadratic equation gives the same result (0.0051 M) as this method of successive approximations.

7B We first determine the concentration of undissociated piperidine. We then use that value in a set up based on the balanced chemical equation. $[C_5H_{11}N] = \frac{114\text{ mg } C_5H_{11}N}{315\text{ mL soln}} \times \frac{1\text{ mmol } C_5H_{11}N}{85.15\text{ mg } C_5H_{11}N} = 0.00425\text{ M}$

Equation:	$C_5H_{11}N + H_2O$ $\rightleftharpoons$	$C_5H_{11}NH^+$ +	OH^-
Initial:	0.00425 M	0 M	0 M
Changes:	$-x$ M	$+x$ M	$+x$ M
Equil:	$(0.00425 - x)$ M	x M	x M

$$K_b = \frac{[C_5H_{11}NH^+][OH^-]}{[C_5H_{11}N]} = 1.6 \times 10^{-3} = \frac{x \cdot x}{0.00425 - x} \approx \frac{x \cdot x}{0.00425}$$

We assumed that $x << 0.00425$

$x = \sqrt{0.0016 \times 0.00425} = 0.0026\text{ M}$ The assumption is not valid. Let's assume $x \approx 0.0026$
$x = \sqrt{0.0016\,(0.00425 - 0.0026)} = 0.0016$ Let's try again, with $x \approx 0.0016$
$x = \sqrt{0.0016\,(0.00425 - 0.0016)} = 0.0021$ Yet another try, with $x \approx 0.0021$
$x = \sqrt{0.0016\,(0.00425 - 0.0021)} = 0.0019$ The last time, with $x \approx 0.0019$
$x = \sqrt{0.0016\,(0.00425 - 0.0019)} = 0.0019\text{ M} = [OH^-]$ $pOH = -\log(0.0019) = 2.72$
$pH = 14.00 - pOH = 14.00 - 2.72 = 11.28$ We could have solved the problem with the quadratic equation rather than by successive approximations. The same answer is obtained. In fact, if we substitute $x = 0.0019$ into the K_b expression, we obtain $(0.0019)^2 / (0.00425 - 0.0019) = 1.5 \times 10^{-3}$ compared to $K_b = 1.6 \times 10^{-3}$. The error is due to rounding, not to an incorrect value. Using $x = 0.0020$ gives a value of 1.8 $\times 10^{-3}$, while using $x = 0.0018$ gives 1.3×10^{-3}.

8A We organize the solution around the balanced chemical equation; a M is the initial concentration of HF.

Equation:	$HF + H_2O$ $\rightleftharpoons$	H_3O^+ +	F^-
Initial:	a M		
Changes:	$-x$ M	$+x$ M	$+x$ M
Equil:	$(a - x)$ M	x M	x M

$$K_a = \frac{[H_3O^+][F^-]}{[HF]} = \frac{(x)\,(x)}{a - x} \approx \frac{x^2}{a} = 6.6 \times 10^{-4}$$

$x = \sqrt{a \times 6.6 \times 10^{-4}}$

For 0.20 M HF, $x = \sqrt{0.20 \times 6.6 \times 10^{-4}} = 0.011\text{ M}$ $\quad \%\text{dissoc} = \frac{0.011\text{ M}}{0.20\text{ M}} \times 100\% = 5.5\%$

For 0.020 M HF, $x = \sqrt{0.020 \times 6.6 \times 10^{-4}} = 0.0036\text{ M}$ We need another cycle of approximation.

$x = \sqrt{(0.020 - 0.0036) \times 6.6 \times 10^{-4}} = 0.0033\text{ M}$ Yet another cycle with $x \approx 0.0033$ M

$x = \sqrt{(0.020 - 0.0033) \times 6.6 \times 10^{-4}} = 0.0033\text{ M}$ $\quad \%\text{dissoc} = \frac{0.0033\text{ M}}{0.020\text{ M}} \times 100\% = 17\%$

As expected, the weak acid is more dissociated in the less concentrated solution.

8B Since both H_3O^+ and $C_3H_5O_3^-$ come from the same source in equimolar amounts, their concentrations are equal. $[H_3O^+] = [C_3H_5O_3^-] = 0.067 \times 0.0284\text{ M} = 0.0019\text{ M}$

$$K_a = \frac{[H_3O^+][C_3H_5O_3^-]}{[HC_3H_5O_3]} = \frac{(0.0019)\,(0.0019)}{0.0284 - 0.0019} = 1.4 \times 10^{-4}$$

9A For an aqueous solution of a diprotic acid, the concentration of the divalent anion equals the second ionization constant: $[^-OOCCH_2COO^-] = K_{a2} = 2.0 \times 10^{-6}$ M. We organize around the chemical equation.

Equation:	$CH_2(COOH)_2 + H_2O \rightleftharpoons$	H_3O^+ +	$HCH_2(COO)_2^-$	
Initial:	1.0 M			
Changes:	$-x$ M	$+x$ M	$+x$ M	
Equil:	$(1.0 - x)$ M	x M	x M	

$$K_a = \frac{[H_3O^+][HCH_2(COO)_2^-]}{[CH_2(COOH)_2]} = \frac{(x)\,(x)}{1.0 - x} \approx \frac{x^2}{1.0} = 1.4 \times 10^{-3}$$

$x = \sqrt{1.0 \times 1.4 \times 10^{-3}} = 3.7 \times 10^{-2}$ M $= [H_3O^+] = [HOOCCH_2COO^-]$

The assumption is valid that $x \ll 1.0$ M.

9B We already know that K_{a2} = [doubly charged anion] for a polyprotic acid. Thus $K_{a2} = 5.3 \times 10^{-5} = [C_2O_4^{2-}]$. From the pH we can find the value of $[H_3O^+] = 10^{-pH} = 10^{-0.67} = 0.21$ M. We also recognize that $[HC_2O_4^-] = [H_3O^+]$, since the second ionization occurs to only a very small extent. We realize that $HC_2O_4^-$ is produced by the ionization of $H_2C_2O_4$. Each mole of $HC_2O_4^-$ present results from the ionization of 1 mole of $H_2C_2O_4$. Now we have sufficient information to determine the value of K_{a1}.

$$K_{a1} = \frac{[H_3O^+][HC_2O_4^-]}{[H_2C_2O_4]} = \frac{0.21 \times 0.21}{1.05 - 0.21} = 5.3 \times 10^{-2}$$

10A H_2SO_4 is a strong acid in its first ionization, and somewhat weak in its second, with $K_{a2} = 1.1 \times 10^{-2} = 0.011$. Because of the strong first ionization step, this problem is one of determining concentrations in a solution that initially is 0.20 M H_3O^+ and 0.20 M HSO_4^-. We base the setup on the balanced chemical equation.

Equation:	$HSO_4^-(aq) + H_2O \rightleftharpoons$	$H_3O^+(aq)$ +	$SO_4^{2-}(aq)$
Initial:	0.20 M	0.20 M	
Changes:	$-x$ M	$+x$ M	$+x$ M
Equil:	$(0.20 - x)$ M	$(0.20 + x)$ M	x M

$$K_{a2} = \frac{[H_3O^+][SO_4^{2-}]}{[HSO_4^-]} = \frac{(0.20 + x)\,x}{0.20 - x} = 0.011 \approx \frac{0.20 \times x}{0.20}$$, assuming that $x \ll 0.20$ M. $x = 0.011$ M

Try one cycle of approximation: $0.011 \approx \frac{(0.20 + 0.011)\,x}{(0.20 - 0.011)} = \frac{0.21x}{0.19}$ $\quad x = \frac{0.19 \times 0.011}{0.21} = 0.010$ M

This is the same answer as produced by the next cycle of approximations. $0.010\text{ M} = [SO_4^{2-}]$

$[H_3O^+] = 0.010 + 0.20\text{ M} = 0.21$ M $\qquad [HSO_4^-] = 0.20 - 0.010\text{ M} = 0.19$ M

10B We know that H_2SO_4 is a strong acid in its first ionization, and a somewhat weak acid in its second, with $K_{a2} = 1.1 \times 10^{-2} = 0.011$. Because of the strong first ionization step, the problem essentially reduces to determining concentrations in a solution that initially is 0.020 M H_3O^+ and 0.020 M HSO_4^-. We base the setup on the balanced chemical equation. The result is solved with the quadratic equation.

Equation:	$HSO_4^-(aq) + H_2O \rightleftharpoons$	$H_3O^+(aq)$ +	$SO_4^{2-}(aq)$
Initial:	0.020 M	0.020 M	
Changes:	$-x$ M	$+x$ M	$+x$ M
Equil:	$(0.020 - x)$ M	$(0.020 + x)$ M	x M

$$K_{a2} = \frac{[H_3O^+][SO_4^{2-}]}{[HSO_4^-]} = \frac{(0.020 + x)\,x}{0.020 - x} = 0.011 \qquad 0.020\,x + x^2 = 2.2 \times 10^{-4} - 0.011\,x$$

$$x^2 + 0.031\,x - 0.00022 = 0 \qquad x = \frac{-0.031 \pm \sqrt{0.00096 + 0.00088}}{2} = 0.0060\text{ M} = [SO_4^{2-}]$$

$[HSO_4^-] = 0.020 - 0.0060 = 0.014$ M $\qquad [H_3O^+] = 0.020 + 0.0060 = 0.026$ M

(The method of successive approximations converges to $x = 0.006$ M in 8 cycles.)

11A (a) $CH_3NH_3^+NO_3^-$ is the salt of the cation of a weak base that will hydrolyze to form an acidic solution ($CH_3NH_3^+ + H_2O \rightleftharpoons CH_3NH_2 + H_3O^+$), and the anion of a strong acid that will not hydrolyze. The aqueous solutions of this compounds will be acidic.

(b) NaI is the salt of the cation of a strong base and the anion of a strong acid, neither of which hydrolyzes in water. Solutions of this compound will be pH neutral.

(c) $NaNO_2$ is the salt of the cation of a strong base that will not hydrolyze in water and the cation of a weak acid that will hydrolyze to form an alkaline solution ($NO_2^- + H_2O \rightleftharpoons HNO_2 + OH^-$). Aqueous solutions of this compound will be basic (alkaline).

11B Without comparing values of K we can predict that the reaction that produces H_3O^+ occurs to the greater extent. The reason, of course, is that the solution produced is *acidic*, since its pH is less than 7.00. We write the two reactions of $H_2PO_4^-$ with water, along with the values of their equilibrium constants.

$H_2PO_4^-(aq) + H_2O \rightleftharpoons H_3O^+(aq) + HPO_4^-(aq) \qquad K_{a2} = 6.3 \times 10^{-8}$

$$H_2PO_4^-(aq) + H_2O \rightleftharpoons OH^-(aq) + H_3PO_4(aq) \qquad K_b = \frac{K_w}{K_{a1}} = \frac{1.0 \times 10^{-14}}{7.1 \times 10^{-3}} = 1.4 \times 10^{-12}$$

As predicted, the acid ionization occurs to the greater extent.

12A From the value of pK_b we determine the value of K_b and thenK_a for the cation.

$$\text{cocaine:} \quad K_b = 10^{-pK} = 10^{-8.41} = 3.9 \times 10^{-9} \qquad K_a = \frac{K_w}{K_b} = \frac{1.0 \times 10^{-14}}{3.9 \times 10^{-9}} = 2.6 \times 10^{-6}$$

$$\text{codeine:} \quad K_b = 10^{-pK} = 10^{-7.95} = 1.1 \times 10^{-8} \qquad K_a = \frac{K_w}{K_b} = \frac{1.0 \times 10^{-14}}{1.1 \times 10^{-8}} = 9.1 \times 10^{-7}$$

(This method may be a bit easier: $pK_a = 14.00 - pK_b = 14.00 - 8.41 = 5.59$, $K_a = 10^{-5.59} = 2.6 \times 10^{-6}$)
The acid with the larger K_a will produce the higher [H^+], and that solution will have the lower pH. Thus, the solution of codeine hydrochlorine will have the higher pH.

12B Both of the ions of $NH_4CN(aq)$ react with water in hydrolysis reactions.

$$NH_4^+(aq) + H_2O \rightleftharpoons NH_3(aq) + H_3O^+(aq) \quad K_a = \frac{K_w}{K_b} = \frac{1.0 \times 10^{-14}}{1.8 \times 10^{-5}} = 5.6 \times 10^{-10}$$

$$CN^-(aq) + H_2O \rightleftharpoons HCN(aq) + OH^-(aq) \quad K_b = \frac{K_w}{K_a} = \frac{1.0 \times 10^{-14}}{6.2 \times 10^{-10}} = 1.6 \times 10^{-5}$$

Since the value of the equilibrium constant for the hydrolysis reaction of cyanide ion is larger than that for the hydrolysis of ammonium ion, the cyanide ion hydrolysis reaction will proceed to a greater extent. The solution of $NH_4CN(aq)$ will be basic (alkaline).

13A NaF dissociates completely into sodium ion and fluoride ion. Fluoride ion hydrolyzes in aqueous solution to form hydroxide ion. The solution is organized around the balanced equation.

Equation:	$F^- + H_2O \rightleftharpoons$	HF	+	OH^-
Initial:	0.10 M			
Changes:	$-x$ M	$+x$ M		$+x$ M
Equil:	$(0.10 - x)$ M	x M		x M

$$K_b = \frac{K_w}{K_a} = \frac{1.0 \times 10^{-14}}{6.6 \times 10^{-4}} = 1.5 \times 10^{-11} = \frac{[HF][OH^-]}{[F^-]} = \frac{(x)(x)}{(0.10 - x)} \approx \frac{x^2}{0.10}$$

$x = \sqrt{0.10 \times 1.5 \times 10^{-11}} = 1.2 \times 10^{-6}\ M = [OH^-] \qquad pOH = -\log(1.2 \times 10^{-6}) = 5.92$

$pH = 14.00 - pOH = 14.00 - 5.92 = 8.08$

13B The cyanide ion hydrolyzes in solution, as shown in Practice Example 17-12. As a consequence of that hydrolysis, $[OH^-] = [HCN]$. $[OH^-]$ can be found from the pH of the solution and then values are substituted into the K_b expression for CN^-, which is solved for $[CN^-]$.

$pOH = 14.00 - pH = 14.00 - 10.38 = 3.62 \qquad [OH^-] = 10^{-pOH} = 10^{-3.62} = 2.4 \times 10^{-4}\ M = [HCN]$

$$K_b = \frac{[HCN][OH^-]}{[CN^-]} = 1.6 \times 10^{-5} = \frac{(2.4 \times 10^{-4})^2}{[CN^-]} \qquad [CN^-] = \frac{(2.4 \times 10^{-4})^2}{1.6 \times 10^{-5}} = 3.6 \times 10^{-3}\ M$$

14A First we draw the Lewis structures of the four acids. Lone pairs are not depicted since we are interested in the arrangements of atoms.

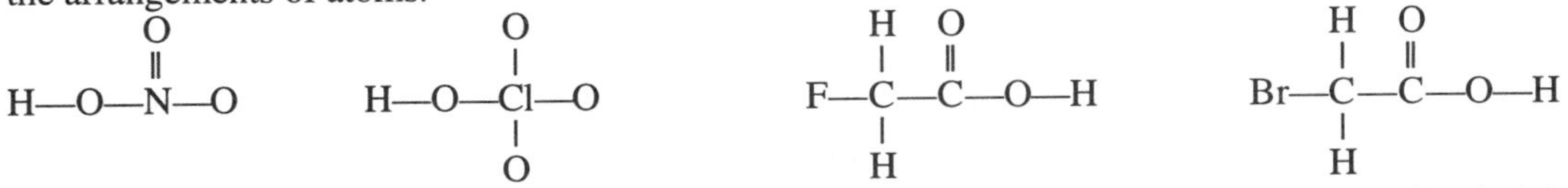

$HClO_4$ should be stronger than HNO_3 for two reasons. First, Cl is more electronegative than is N and will withdraw electrons to itself from the OH bond, making that bond more easily broken. Second, there are more terminal oxygens attached to Cl (three) than to N (two). This means that the charge on the anion produced by ionization can spread out through more of the molecule and a more stable anion will be produced. The more stable anion will be more easily formed.

CH_2FCOOH will be a stronger acid than $CH_2BrCOOH$ because F is a more electronegative atom than is Br. This will withdraw some electron density from the O—H bond, making that bond more easily broken.

14B First we draw the Lewis structures of the first two acids. Lone pairs are not depicted since we are interested in the arrangements of atoms.

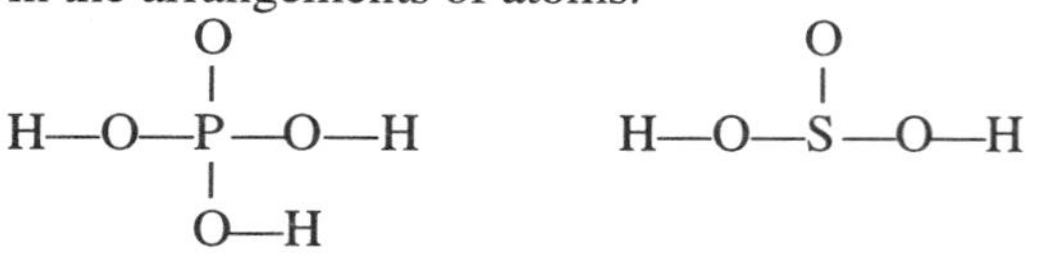

H_3PO_4 and H_2SO_3 both have one terminal oxygen atom, but S is more electronegative than is P. This indicates that H_2SO_3 ($K_{a1} = 1.3 \times 10^{-2}$) should be a stronger acid that H_3PO_4 ($K_{a1} = 7.1 \times 10^{-3}$), and it is. The only difference between CCl_3CH_2COOH and CCl_2FCH_2COOH is the replacement of Cl by F. Since F is more electronegative than Cl, CCl_3CH_2COOH should be a weaker acid than CCl_2FCH_2COOH.

15A We draw Lewis structures to help us decide.

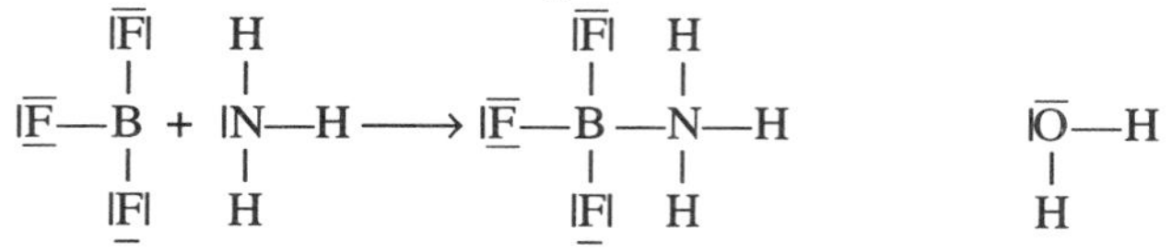

(a) Clearly, BF_3 is an electron pair acceptor, a Lewis acid, and NH_3 is an electron pair donor, a Lewis base.

(b) H_2O certainly has electron pairs (lone pairs) to donate and thus it can be a Lewis base. It is unlikely that the cation Cr^{3+} has any valence electrons to enable it to be a Lewis base; it is the Lewis acid.

15B The Lewis structures of the six species follow.

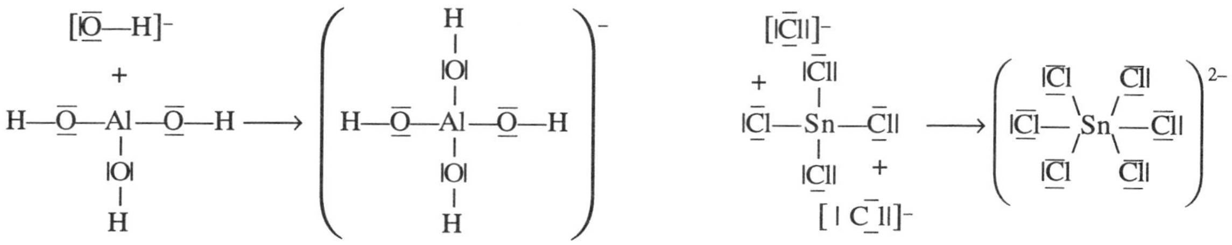

Both hydroxide ion and chloride ion have extra lone pairs of electrons; these two are the electron pair donors, the Lewis bases. $Al(OH)_3$ and $SnCl_4$ have additional spaces in their bonding schemes to accept pairs of electrons, which is what they do in forming the complex anions. $Al(OH)_3$ and $SnCl_4$ are the Lewis acids in these reactions.

SUMMARIZING EXAMPLE CALCULATIONS

1. We will arrange the solutions by increasing pH: from the most acidic to the most basic.

2. Each solution first is categorized as acidic, basic or neutral.

1.0 M NaBr	Neither ion hydrolyzes: pH neutral.
0.05 M $HC_2H_3O_2$	The solution of a weak acid: acidic.
0.05 M NH_3	The solution of a weak base: **basic**.
0.02 M $KC_2H_3O_2$	The acetate ion hydrolyzes, behaving as a very weak base: **basic**.
0.05 M $Ba(OH)_2$	The solution of a diprotic strong base: **basic**.
0.05 M H_2SO_4	The solution of a diprotic strong acid: acidic.
0.10 M HI	The solution of a monoprotic strong acid: acidic.
0.06 M NaOH	The solution of a monoprotic strong base: **basic**.
0.05 M NH_4Cl	The ammonium ion hydrolyzes, behaving as a very weak acid: acidic.
0.05 M $CH_2ClCOOH$	The solution of a moderately weak acid.

3. There are four basic solutions, with the word **basic** in boldface to make them easier to pick out. The most basic solution is that of 0.05 M $Ba(OH)_2$, it has $[OH^-] = 0.10$ M. Next comes the monoprotic strong base, 0.06 M NaOH. Then the weak base, since $K_b \approx 10^{-5}$ for NH_3. And finally the hydrolyzing anion, since $K_b \approx 10^{-10}$ for $C_2H_3O_2^-$ Result: 0.02 M $KC_2H_3O_2$ < 0.05 M NH_3 < 0.06 M NaOH < 0.05 M $Ba(OH)_2$

One of the two strong acids is the most acidic. But we remember that the second ionization of H_2SO_4 is not complete, and thus $[H_3O^+] < 2 \times 0.05$ M in 0.05 M H_2SO_4. Thus, 0.10 M HI has the lowest pH. The strongest of the weak acids is next, 0.05 M $CH_2ClCOOH$, followed by a slightly weaker acid, 0.05 M CH_3COOH, and then by the hydrolyzing cation, 0.05 M NH_4Cl.

4. Overall result: 0.10 M HI < 0.05 M H_2SO_4 < 0.05 M $CH_2ClCOOH$ < 0.05 M CH_3COOH < 0.05 M NH_4Cl < 1.0 M NaBr < 0.02 M $KC_2H_3O_2$ < 0.05 M NH_3 < 0.06 M NaOH < 0.05 M $Ba(OH)_2$

REVIEW QUESTIONS

1. **(a)** K_w is the symbol for the ion product of water: $K_w = [H_3O^+][OH^-] = 1.0 \times 10^{-14}$ at 25°C.
(b) pH symbolizes the negative of the (base 10) logarithm of the hydrogen ion (actually hydronium ion, H_3O^+) concentration: pH = $-\log [H_3O^+]$. High pH indicates an alkaline solution; low pH indicates an acidic one.
(c) pK_a symbolizes the negative of the logarithm of the ionization constant of an acid: $pK_a = -\log K_a$.
(d) Hydrolysis refers to the process where an ion reacts (as an acid or a base) with water to produce either hydrogen ion or hydroxide ion.
(e) A Lewis acid is an electron pair acceptor, the recipient of a coordinate covalent bond.

2. **(a)** A conjugate base is the species produced when another species (molecule or ion) donates, and thus loses, a proton, H^+.
(b) Percent ionization is the percent ratio of the concentration of the ion formed by a species divided by the original concentration of that species prior to ionization.
(c) An acid anhydride is a species, often a nonmetal oxide, whose aqueous solution is acidic.
(d) Amphoterism refers to the ability of some substances, often oxides, to behave as acids in the presence of strong bases, and behave as bases in the presence of strong acids.

3. **(a)** A Brønsted–Lowry acid is a proton (H^+) donor; a Brønsted–Lowry base is a proton acceptor.
(b) pH = $[H_3O^+]$; the one is a logarithmic expression of the other, necessary because of the extreme range of $[H_3O^+]$ values commonly encountered.
(c) K_a for NH_4^+ and K_b for NH_3 are inversely related to each other: $K_a \times K_b = K_w$.
(d) The leveling effect refers the the inability of a somewhat basic solvent to distinguish between the strengths of two strong acids; they both donate protons equally well to the solvent. The electron-withdrawing effect refers to the attraction of electrons to an electronegative atom or group within a molecule. If the electrons are drawn away from an O—H bond, that bond will become polar and more easily broken, producing a stronger acid.

4. **(a)** HNO_2 is an acid, a proton donor. The conjugate base is NO_2^-.
(b) OCl^- is a base, a proton acceptor. The conjugate acid is HOCl.
(c) NH_2^- is a base, a proton acceptor. The conjugate acid is NH_3.
(d) NH_4^+ is an acid, a proton donor. The conjugate base is NH_3.
(e) $CH_3NH_3^+$ is an acid, a proton donor. The conjugate base is CH_3NH_2.

5. We write the conjugate base as the first product of the equilibrium in which the acid is the first reactant.
(a) $HIO_3(aq) + H_2O \rightleftharpoons IO_3^-(aq) + H_3O^+(aq)$
(b) $C_6H_5COOH(aq) + H_2O \rightleftharpoons C_6H_5COO^-(aq) + H_3O^+(aq)$
(c) $HPO_4^{2-}(aq) + H_2O \rightleftharpoons PO_4^{3-}(aq) + H_3O^+(aq)$
(d) $C_2H_5NH_3^+(aq) + H_2O \rightleftharpoons C_2H_5NH_2(aq) + H_3O^+(aq)$

6. All of the solutes are strong acids or strong bases.

(a) $[H_3O^+] = 0.00165 \text{ M HCl} \times \dfrac{1 \text{ mol } H_3O^+}{1 \text{ mol } HNO_3} = 0.00165 \text{ M}$

$[OH^-] = \dfrac{K_w}{[H_3O^+]} = \dfrac{1.0 \times 10^{-14}}{0.00165 \text{ M}} = 6.1 \times 10^{-12} \text{ M}$

(b) $[OH^-] = 0.0087 \text{ M KOH} \times \dfrac{1 \text{ mol } OH^-}{1 \text{ mol KOH}} = 0.0087 \text{ M}$

$[H_3O^+] = \dfrac{K_w}{[OH^-]} = \dfrac{1.0 \times 10^{-14}}{0.0087 \text{ M}} = 1.1 \times 10^{-12} \text{ M}$

(c) $[OH^-] = 0.00213 \text{ M } Sr(OH)_2 \times \dfrac{2 \text{ mol } OH^-}{1 \text{ mol } Sr(OH)_2} = 0.00426 \text{ M}$

$[H_3O^+] = \dfrac{K_w}{[OH^-]} = \dfrac{1.0 \times 10^{-14}}{0.00426 \text{ M}} = 2.3 \times 10^{-12} \text{ M}$

(d) $[H_3O^+] = 5.8 \times 10^{-4}\ M \times \frac{1\ mol\ H_3O^+}{1\ mol\ HI} = 5.8 \times 10^{-4}\ M$ $\quad [OH^-] = \frac{K_w}{[H_3O^+]} = \frac{1.0 \times 10^{-14}}{5.8 \times 10^{-4}\ M} = 1.7 \times 10^{-11}\ M$

7. Again, all of the solutes are strong acids or strong bases.

(a) $[H_3O^+] = 0.0045\ M \times \frac{1\ mol\ H_3O^+}{1\ mol\ HCl} = 0.0045\ M$ $\quad pH = -\log(0.0045) = 2.35$

(b) $[H_3O^+] = 6.14 \times 10^{-4}\ M \times \frac{1\ mol\ H_3O^+}{1\ mol\ HNO_3} = 6.14 \times 10^{-4}\ M$ $\quad pH = -\log(6.14 \times 10^{-4}) = 3.212$

(c) $[OH^-] = 0.00683\ M \times \frac{1\ mol\ OH^-}{1\ mol\ NaOH} = 0.00683\ M$ $\quad pOH = -\log(0.00683) = 2.166$

$pH = 14.000 - pOH = 14.000 - 2.166 = 11.83$

(d) $[OH^-] = 4.8 \times 10^{-3}\ M \times \frac{2\ mol\ OH^-}{1\ mol\ Ba(OH)_2} = 9.6 \times 10^{-3}\ M$ $\quad pOH = -\log(9.6 \times 10^{-3}) = 2.02$

$pH = 14.00 - 2.02 = 11.98$

8. Most probably, 0.11 M = $[H_3O^+]$ in 0.10 M H_2SO_4. Although H_2SO_4 is completely ionized in dilute solution, this solution is not sufficiently dilute. (Recall the paragraph before Example 17-10 in the text.) Thus, 0.20 M is incorrect. However, the first ionization of H_2SO_4 is complete; 0.05 M cannot be correct. And the second ionization of H_2SO_4 is at least partially complete; 0.10 M incorrectly assumes that the only source of H_3O^+ is the first ionization.

9. We determine the amounts of H_3O^+ and OH^- and then the amount of the one that is in excess. We express molar concentration in millimoles/milliliter, equivalent to mol/L.

$$24.80\ mL \times \frac{0.248\ mmol\ HNO_3}{1\ mL\ soln} \times \frac{1\ mmol\ H_3O^+}{1\ mmol\ HNO_3} = 6.15\ mmol\ H_3O^+$$

$$15.40\ mL \times \frac{0.394\ mmol\ KOH}{1\ mL\ soln} \times \frac{1\ mmol\ OH^-}{1\ mmol\ KOH} = 6.07\ mmol\ OH^-$$

The net reaction is $H_3O^+(aq) + OH^-(aq) \longrightarrow 2\ H_2O$. There is an excess amount of H_3O^+ of (6.15 – 6.07 =) 0.08 mmol H_3O^+. This is an acidic solution. The total solution volume is (24.80 + 15.40 =) 40.20 mL.

$$[H_3O^+] = \frac{0.08\ mmol\ H_3O^+}{40.20\ mL} = 0.002\ M \qquad pH = -\log(0.002) = 2.7$$

10. Answer (4) is correct, pH < 13. If this were a strong base, its $[OH^-] = 0.10$, giving pOH = 1.0, and pH = 13.0. But since methylamine is a weak base, its $[OH^-] < 0.10$, giving pOH > 1.0, and pH < 13.0.

11. $[HC_3H_5O_2]_i = \frac{0.275\ mol}{625\ mL} \times \frac{1000\ mL}{1\ L} = 0.440\ M$ $\quad [HC_3H_5O_2]_e = 0.440\ M - 0.00239\ M = 0.438\ M$

$$K_a = \frac{[H_3O^+][C_3H_5O_2^-]}{[HC_3H_5O_2]} = \frac{(0.00239)^2}{0.438} = 1.30 \times 10^{-5}$$

12. (a) The set up is based on the balanced chemical equation.

Equation:	$HC_8H_7O_2(aq) + H_2O$	$\rightleftharpoons$	$H_3O^+(aq)$ +	$C_8H_7O_2^-(aq)$
Initial:	0.186 M		0 M	0 M
Changes:	$-x$ M		$+x$ M	$+x$ M
Equil:	$(0.186 - x)$ M		x M	x M

$$K_a = 4.9 \times 10^{-5} = \frac{[H_3O^+][C_8H_7O_2^-]}{[HC_8H_7O_2]} = \frac{x \cdot x}{0.186 - x} \approx \frac{x^2}{0.186}$$

$$x = \sqrt{0.186 \times 4.9 \times 10^{-5}} = 0.0030\ M = [H_3O^+] = [C_8H_7O_2^-]$$

(b)

Equation:	$HC_8H_7O_2 + H_2O$	$\rightleftharpoons$	H_3O^+ +	$C_8H_7O_2^-$
Initial:	0.121 M		0 M	0 M
Changes:	$-x$ M		$+x$ M	$+x$ M
Equil:	$(0.121 - x)$M		x M	x M

$$K_a = 4.9 \times 10^{-5} = \frac{[H_3O^+][C_8H_7O_2^-]}{[HC_8H_7O_2]} = \frac{x^2}{0.121 - x} \approx \frac{x^2}{0.121} \qquad x = 0.0024\ M = [H_3O^+]$$

We have assumed $x \ll 0.121$, an assumption that clearly is correct. $pH = -\log(0.0024) = 2.62$

13. We need first to determine the molarity S of the acid needed. $([H_3O^+] = 10^{-2.85} = 1.4 \times 10^{-3}\ M)$

Equation:	$HC_7H_5O_2 + H_2O$	$\rightleftharpoons$	H_3O^+	+	$C_7H_5O_2^-$
Initial:	S		0 M		0 M
Changes:	–0.0014 M		+0.0014 M		+0.0014 M
Equil:	S – 0.0014 M		0.0014 M		0.0014 M

$$K_a = \frac{[H_3O^+][C_7H_5O_2^-]}{[HC_7H_5O_2]} = \frac{(0.0014)^2}{S - 0.0014} = 6.3 \times 10^{-5} \qquad S - 0.0014 = \frac{(0.0014)^2}{6.3 \times 10^{-5}} = 0.031$$

$S = 0.031 + 0.0014 = 0.032\ M = [HC_7H_5O_2]$

$$350.0\ mL \times \frac{1\ L}{1000\ mL} \times \frac{0.032\ mol\ HC_7H_5O_2}{1\ L\ soln} \times \frac{122.1\ g\ HC_7H_5O_2}{1\ mol\ HC_7H_5O_2} = 1.4\ g\ HC_7H_5O_2$$

14. First determine $[OH^-]$, and thus $[(CH_3)_3NH^+]$, in the solution. Then use the K_b expression to find $[(CH_3)_3N]$. $pOH = 14.00 - pH = 14.00 - 11.12 = 2.88$ $[OH^-] = 10^{-pOH} = 10^{-2.88} = 0.0013\ M$

$$K_b = 6.3 \times 10^{-5} = \frac{[(CH_3)_3NH^+][OH^-]}{[(CH_3)_3N]_{equil}} = \frac{(0.0013)^2}{[(CH_3)_3N]_{equil}} \qquad [(CH_3)_3N]_{equil} = \frac{(0.0013)^2}{6.3 \times 10^{-5}} = 0.027\ M$$

Since some of the $(CH_3)_3N$ has ionized, $[(CH_3)_3N]_{initial} = 0.027\ M + 0.0013\ M = 0.028\ M$

15. For H_2CO_3, $K_1 = 4.4 \times 10^{-7}$ and $K_2 = 4.7 \times 10^{-11}$

(a) The first acid ionization proceeds to a far greater extent than does the second and determines the value of $[H_3O^+]$.

Equation:	$H_2CO_3 + H_2O \rightleftharpoons$	H_3O^+ +	HCO_3^-
Initial:	0.045 M	0 M	0 M
Changes:	$-x$ M	$+x$ M	$+x$ M
Equil:	$(0.045 - x)$M	x M	x M

$$K_1 = \frac{[H_3O^+][HCO_3^-]}{[H_2CO_3]} = \frac{x^2}{0.045 - x} = 4.4 \times 10^{-7} \approx \frac{x^2}{0.045} \qquad x = 1.4 \times 10^{-4}\ M = [H_3O^+]$$

(b) Since the second ionization occurs to only a limited extent, $[HCO_3^-] = 1.4 \times 10^{-4}\ M = 0.00014\ M$

(c) We use the second ionization to determine $[CO_3^{2-}]$.

Equation:	$HCO_3^- + H_2O \rightleftharpoons$	H_3O^+ +	CO_3^{2-}
Initial:	0.00014 M	0.00014 M	0 M
Changes:	$-x$ M	$+x$ M	$+x$ M
Equil:	$(0.00014 - x)$M	$(0.00014 + x)$M	x M

$$K_2 = \frac{[H_3O^+][CO_3^{2-}]}{[HCO_3^-]} = \frac{(0.00014+x)x}{0.00014 - x} = 4.7 \times 10^{-11} \approx \frac{(0.00014)\,x}{0.00014}$$

$x = 4.7 \times 10^{-11}\ M = [CO_3^{2-}]$

Again we assumed that $x << 0.00014$ M, which clearly is the case. We also note that our original assumption, that the second ionization is much less significant than the first, also is a valid one. As an alternative to all of this, we could have recognized that the concentration of the divalent anion equals the second ionization constant: $[CO_3^{2-}] = K_2$.

16. The species that hydrolyze are the cations of weak bases—NH_4^+ and $C_6H_5NH_3^+$— and the anions of weak acids—NO_2^- and $C_7H_5O_2^-$.

(a) $NH_4^+(aq) + NO_3^-(aq) + H_2O \rightleftharpoons NH_3(aq) + NO_3^-(aq) + H_3O^+(aq)$

(b) $Na^+(aq) + NO_2^-(aq) + H_2O \rightleftharpoons Na^+(aq) + HNO_2(aq) + OH^-(aq)$

(c) $K^+(aq) + C_7H_5O_2^-(aq) + H_2O \rightleftharpoons K^+(aq) + HC_7H_5O_2(aq) + OH^-(aq)$

(d) $K^+(aq) + Cl^-(aq) + Na^+(aq) + I^-(aq) + H_2O \longrightarrow$ no reaction

(e) $C_6H_5NH_3^+(aq) + Cl^-(aq) + H_2O \rightleftharpoons C_6H_5NH_2(aq) + Cl^-(aq) + H_3O^+(aq)$

17.

Equation:	$NH_4^+ + H_2O \rightleftharpoons$	H_3O^+ +	NH_3
Initial:	1.68 M	0 M	0 M
Changes:	$-x$ M	$+x$ M	$+x$ M
Equil:	$(1.68 - x)$M	x M	x M

$$K_a = \frac{[H_3O^+][NH_3]}{[NH_4^+]} = \frac{x^2}{1.68 - x} = 5.6 \times 10^{-10} \approx \frac{x^2}{1.68}$$

$x = 3.1 \times 10^{-5}\ M = [H_3O^+]$ $\qquad pH = -\log(3.1 \times 10^{-5}) = 4.51$

18. Recall that, for a conjugate weak acid-weak base pair, $K_a \times K_b = K_w$

(a) $K_a = \frac{K_w}{K_b} = \frac{1.0 \times 10^{-14}}{1.5 \times 10^{-9}} = 6.7 \times 10^{-6}$ for $C_5H_5NH^+$

(b) $K_b = \frac{K_w}{K_a} = \frac{1.0 \times 10^{-14}}{1.8 \times 10^{-4}} = 5.6 \times 10^{-11}$ for CHO_2^-

(c) $K_b = \frac{K_w}{K_a} = \frac{1.0 \times 10^{-14}}{1.0 \times 10^{-10}} = 1.0 \times 10^{-4}$ for $C_6H_5O^-$

19. We base the calculation on the balanced chemical equation for the hydrolysis of the anion.

Equation:	$C_2H_2ClO_2^-(aq) + H_2O \rightleftharpoons$	$HC_2H_2ClO_2(aq)$ +	$OH^-(aq)$
Initial:	2.05 M	0 M	0 M
Changes:	$-x$ M	$+x$ M	$+x$ M
Equil:	$(2.05 - x)$ M	x M	x M

$$K_b = \frac{K_w}{K_a} = \frac{1.0 \times 10^{-14}}{1.4 \times 10^{-3}} = 7.1 \times 10^{-12} = \frac{[HC_2H_2ClO_2][OH^-]}{[C_2H_2ClO_2^-]} = \frac{x \cdot x}{2.05 - x} \approx \frac{x^2}{2.05}$$

$x = \sqrt{2.05 \times 7.1 \times 10^{-12}} = 3.8 \times 10^{-6} = [OH^-]$ $\quad pOH = -\log(3.8 \times 10^{-6}) = 5.42$

$pH = 14.00 - pOH = 14.00 - 5.42 = 8.58$

20. CCl_3COOH is a stronger acid than CH_3COOH because in CCl_3COOH there are electronegative (electron withdrawing) Cl atoms bonded to the atom adjacent to the COOH group. The electron withdrawing power of these Cl atoms polarizes the O—H bond, creating a stronger acid.

21. **(a)** HI is the stronger acid because the H—I bond length is longer than the H—Br bond length and the H—I bond is weaker.
(b) HOClO is a stronger acid than HOI because there is a terminal O in HOClO but not in HOI and Cl is more electronegative than I.
(c) $H_3CCH_2CCl_2COOH$ is a stronger acid than $I_3CCH_2CH_2COOH$ both because Cl is more electronegative than is I and also because the Cl atoms are closer to the acidic hydrogen in the COOH group and thus can exert a stronger effect on it than can the more distant I atoms.

22. A Lewis base is an electron pair donor, while a Lewis acid is an electron pair acceptor. We draw Lewis structures to assist our interpretation.

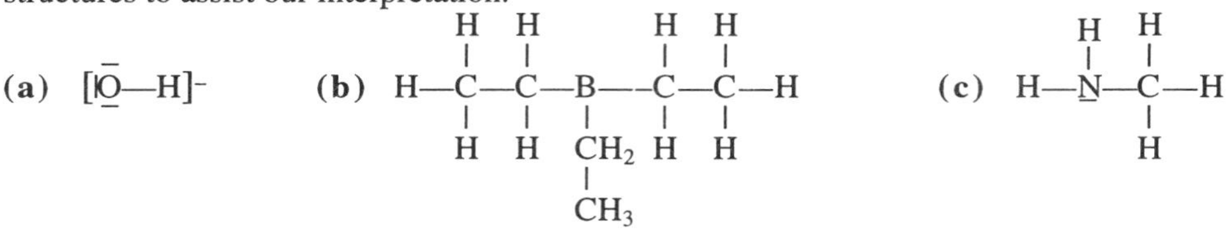

(a) The lone pairs on oxygen can readily be donated; this is a Lewis base.
(b) The incomplete octet of B provides a site for acceptance of an electron pair; this is a Lewis acid.
(c) The lone pair on nitrogen can be donated; this is a Lewis base.

23. A Lewis base is an electron pair donor, while a Lewis acid is an electron pair acceptor. We draw Lewis structures to assist our interpretation.
(a) According to the following Lewis structures, SO_3 appears to be the electron pair acceptor (the Lewis acid), and H_2O is the electron pair donor (the Lewis base). Note that an additional sulfur-to-oxygen bond is formed.

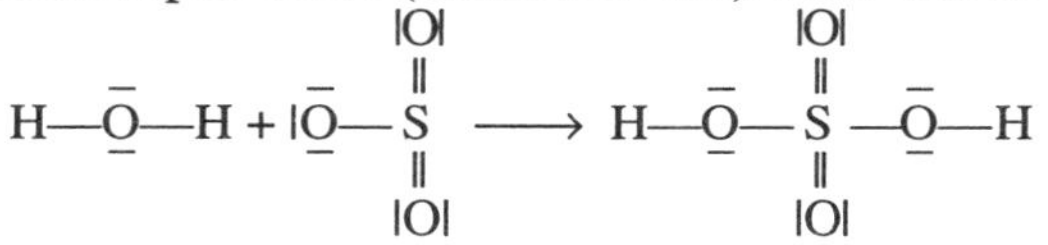

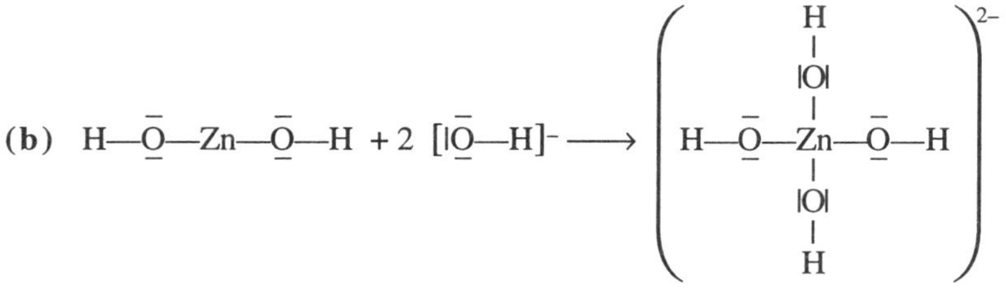

Zn in $Zn(OH)_2$ accepts a pair of electrons; $Zn(OH)_2$ is a Lewis acid. OH^- donates the pair of electrons that form the covalent bond; OH^- is a Lewis base. We have assumed that Zn has sufficient covalent character to form coordinate covalent bonds with hydroxide ions.

24. (a) $NH_4Cl(aq)$ contains a cation—$NH_4^+(aq)$—that is a weak acid; it should have a pH slightly below 7.00.
(b) $NH_3(aq)$ is a weak base; it should have a reasonably high pH.
(c) $NaC_2H_3O_2(aq)$ contains an anion—$C_2H_3O_2^-(aq)$—that is a weak base; it should have a pH slightly above 7.00.
(d) Neither $K^+(aq)$ nor $Cl^-(aq)$ is an acid or a base in aqueous solution; KCl(aq) should have a pH of 7.00.
In order of decreasing pH: $NH_3(aq) > NaC_2H_3O_2(aq) > KCl(aq) > NH_4Cl(aq)$

EXERCISES

Brønsted-Lowry Theory of Acids and Bases

25. The acids (proton donors) and bases (proton acceptors) are labeled below their formulas. Remember that a proton, in Brønsted-Lowry acid-base theory , is H^+.

(a) $HOBr + H_2O \rightleftharpoons H_3O^+ + OBr^-$
acid base acid base

(b) $HSO_4^- + H_2O \rightleftharpoons H_3O^+ + SO_4^{2-}$
acid base acid base

(c) $HS^- + H_2O \rightleftharpoons H_2S + OH^-$
base acid acid base

(d) $C_6H_5NH_3^+ + OH^- \rightleftharpoons C_6H_5NH_2 + H_2O$
acid base base acid

26. For each amphiprotic substance, we write its reactions with water: with the substance acting as an acid, and then with it acting as a base. Even for the substances that are not usually considered amphiprotic, both reactions are written, but one of them is labeled as unlikely. In some instances we have written an oxygen as Ø to keep track of it through the reaction.

(a) $ØH^- + H_2O \rightleftharpoons Ø^{2-} + H_3O^+$ (unlikely) $ØH^- + H_2O \rightleftharpoons H_2Ø + OH^-$
(b) $NH_4^+ + H_2O \rightleftharpoons NH_3 + H_3O^+$ NH_4^+ cannot add a proton without expanding octet
(c) $H_2Ø + H_2O \rightleftharpoons ØH^- + H_3O^+$ $H_2Ø + H_2O \rightleftharpoons H_3Ø^+ + OH^-$
(d) $HS^- + H_2O \rightleftharpoons S^{2-} + H_3O^+$ $HS^- + H_2O \rightleftharpoons H_2S + OH^-$
(e) NO_2^- cannot act as an acid, no protons $NO_2^- + H_2O \rightleftharpoons HNO_2 + OH^-$
(f) $HCO_3^- + H_2O \rightleftharpoons CO_3^{2-} + H_3O^+$ $HCO_3^- + H_2O \rightleftharpoons H_2CO_3 + OH^-$
(g) $HBr + H_2O \rightleftharpoons H_3O^+ + Br^-$ $HBr + H_2O \rightleftharpoons H_2Br^+ + OH^-$ (unlikely)

27. Answer (2), NH_3, is correct. $HC_2H_3O_2$ will react most completely with the strongest base. NO_3^- and Cl^- are very weak bases. H_2O is a weak base, but is amphiprotic, acting as an acid (donating protons), as in the presence of NH_3. Thus, NH_3 must be the strongest base and the most effective in causing $HC_2H_3O_2$ to react.

28. Lewis structures are given below each equation.

(a) $2\ NH_3(l) \rightleftharpoons NH_4^+ + NH_2^-$

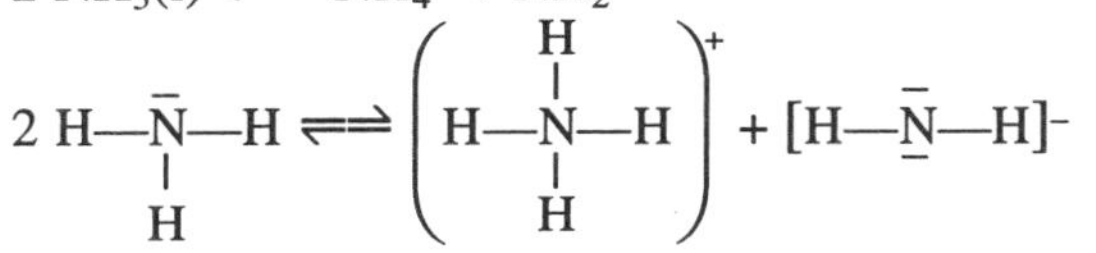

(b) $2\ HF(l) \rightleftharpoons H_2F^+ + F^-$

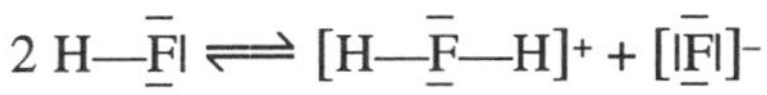

(c) $2\ CH_3OH(l) \rightleftharpoons CH_3OH_2^+ + CH_3O^-$

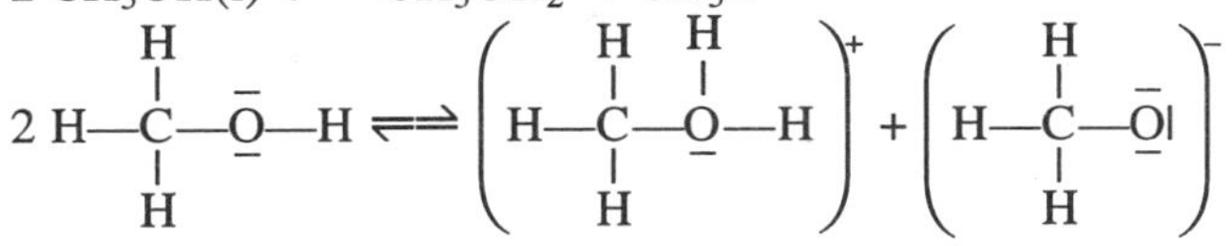

(d) $2\ HC_2H_3O_2(l) \rightleftharpoons H_2C_2H_3O_2^+ + C_2H_3O_2^-$

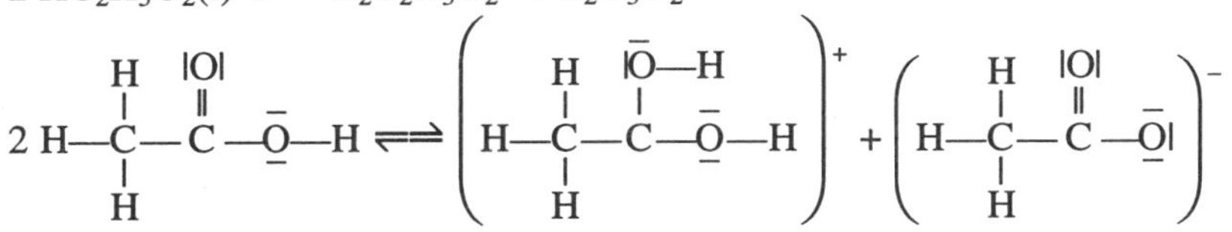

(e) $2\ H_2SO_4(l) \rightleftharpoons H_3SO_4^+ + HSO_4^-$

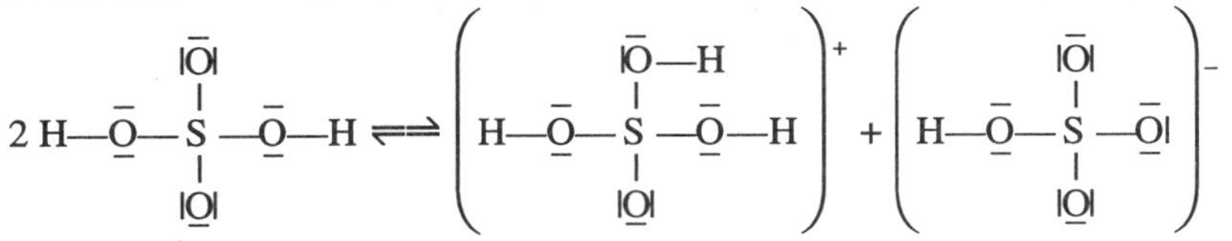

29. The principle here is that the weaker acid and the weaker base will predominate at equilibrium. The reason is that a strong acid will do a good job of donating its protons and, having done so, its conjugate base will be left behind. The preferred direction is: Strong acid + strong base $\longrightarrow$ weak (conjugate) base + weak (conjugate) acid.

(a) The reaction will favor the forward direction because OH^- (a strong base) > NH_3 as bases and NH_4^+ > H_2O as acids.
(b) The reaction will favor the reverse direction because HNO_3 > HSO_4^- (a weak acid in the second ionization) as acids, and SO_4^{2-} > NO_3^- as bases.
(c) The reaction will favor the the reverse direction because $HC_2H_3O_2$ > CH_3OH (not usually thought of as an acid) as acids, and CH_3O^- > $C_2H_3O_2^-$ as bases.

30. The principle is that the weaker acid and the weaker base predominate at equilibrium. This is because a strong acid will do a good job of donating its protons and, having done so, its conjugate base will remain. The preferred direction is: Strong acid + strong base ⟶ weak (conjugate) base + weak (conjugate) acid.
(a) The reaction will favor the forward direction because $HC_2H_3O_2$ (a moderate acid) > HCO_3^- (a quite weak acid) as acids and CO_3^{2-} > $C_2H_3O_2^-$ as bases.
(b) The reaction will favor the reverse direction because $HClO_4$ (a strong acid) > HNO_2 as acids, and NO_2^- > ClO_4^- as bases.
(c) The reaction will favor the forward direction, because H_2CO_3 > HCO_3^- as acids (because $K_1 > K_2$) and CO_3^{2-} > HCO_3^- as bases.

Strong Acids, Strong Bases, and pH

31. $$[OH^-] = \frac{3.9 \text{ g Ba(OH)}_2\cdot 8H_2O}{1000 \text{ mL soln}} \times \frac{1000 \text{ mL}}{1 \text{ L}} \times \frac{1 \text{ mol Ba(OH)}_2\cdot 8H_2O}{315.5 \text{ g Ba(OH)}_2\cdot 8 H_2O} \times \frac{2 \text{ mol OH}^-}{1 \text{ mol Ba(OH)}_2\cdot 8H_2O}$$

$$= 0.025 \text{ M} \qquad [H_3O^+] = \frac{K_w}{[OH^-]} = \frac{1.0 \times 10^{-14}}{0.025 \text{ M OH}^-} = 4.0 \times 10^{-13} \text{ M}$$

$pH = -\log(4.0 \times 10^{-13}) = 12.40$

32. The dissolved $Ca(OH)_2$ is completely dissociated into ions. $pOH = 14.00 - pH = 14.00 - 12.35 = 1.65$
$[OH^-] = 10^{-pOH} = 10^{-1.65} = 2.2 \times 10^{-2}$ M OH^- = 0.022 M OH^-

$$\text{solubility} = \frac{0.022 \text{ mol OH}^-}{1 \text{ L soln}} \times \frac{1 \text{ mol Ca(OH)}_2}{2 \text{ mol OH}^-} \times \frac{74.09 \text{ g Ca(OH)}_2}{1 \text{ mol Ca(OH)}_2} \times \frac{1000 \text{ mg}}{1 \text{ g}} = \frac{8.1 \times 10^2 \text{ mg Ca(OH)}_2}{1 \text{ Lsoln}}$$

In 100 mL the solubility is $100 \text{ mL} \times \frac{1 \text{ L}}{1000 \text{ mL}} \times \frac{8.1 \times 10^2 \text{ mg Ca(OH)}_2}{1 \text{ Lsoln}} = 81 \text{ mg Ca(OH)}_2$

33. First determine the amount of HCl, and then its concentration.

$$\text{amount HCl} = \frac{PV}{RT} = \frac{\left(751 \text{ mmHg} \times \frac{1 \text{ atm}}{760 \text{ mmHg}}\right) \times 0.205 \text{ L}}{0.08206 \text{ L atm mol}^{-1} \text{ K}^{-1} \times 295 \text{ K}} = 8.37 \times 10^{-3} \text{ mol HCl}$$

$$[H_3O^+] = \frac{8.37 \times 10^{-3} \text{ mol HCl}}{4.25 \text{ L soln}} \times \frac{1 \text{ mol H}_3O^+}{1 \text{ mol HCl}} = 1.97 \times 10^{-3} \text{ M}$$

34. First determine the concentration of NaOH, then of OH^-.

$$[OH^-] = \frac{0.125 \text{ L} \times \frac{0.606 \text{ mol NaOH}}{1 \text{ L}} \times \frac{1 \text{ mol OH}^-}{1 \text{ mol NaOH}}}{15.00 \text{ L final solution}} = 0.00505 \text{ M}$$

$pOH = -\log(0.00505 \text{ M}) = 2.297 \qquad pH = 14.00 - 2.297 = 11.70$

35. First determine the amount of HCl, and then the volume of the concentrated solution.

$$\text{amount HCl} = 12.5 \text{ L} \times \frac{10^{-2.10} \text{ mol H}_3O^+}{1 \text{ L soln}} \times \frac{1 \text{ mol HCl}}{1 \text{ mol H}_3O^+} = 0.099 \text{ mol HCl}$$

$$\text{concentrated solution volume} = 0.099 \text{ mol HCl} \times \frac{36.46 \text{ g HCl}}{1 \text{ mol HCl}} \times \frac{100.0 \text{ g soln}}{36.0 \text{ g HCl}} \times \frac{1 \text{ mL}}{1.18 \text{ g}} = 8.5 \text{ mL}$$

36. First determine the amount of KOH, and then the volume of the concentrated solution.
$pOH = 14.00 - pH = 14.00 - 11.55 = 2.45 \qquad [OH^-] = 10^{-pOH} = 10^{-2.45} = 0.0035 \text{ M}$

$$\text{amount KOH} = 25.0 \text{ L} \times \frac{0.0035 \text{ M mol OH}^-}{1 \text{ L soln}} \times \frac{1 \text{ mol KOH}}{1 \text{ mol OH}^-} = 0.088 \text{ mol KOH}$$

$$\text{concentrated solution volume} = 0.088 \text{ mol KOH} \times \frac{56.11 \text{ g KOH}}{1 \text{ mol KOH}} \times \frac{100.0 \text{ g soln}}{15.0 \text{ g KOH}} \times \frac{1 \text{ mL}}{1.14 \text{ g}} = 29 \text{ mL}$$

37. The volume of HCl(aq) needed is determined by first finding the amount of NH_3(aq) present, and then realizing that acid and base react in a 1:1 molar ratio.

$$\text{HCl(aq) volume} = 1.25 \text{ L base} \times \frac{0.265 \text{ mol NH}_3}{1 \text{ L base}} \times \frac{1 \text{ mol H}_3\text{O}^+}{1 \text{ mol NH}_3} \times \frac{1 \text{ mol HCl}}{1 \text{ mol H}_3\text{O}^+} \times \frac{1 \text{ L acid}}{6.15 \text{ mol HCl}} = 0.0539 \text{ L acid}$$

38. NH_3 and HCl react in a 1:1 molar ratio. Since, by Avogadro's hypothesis, equal volumes of gasses measured at the same temperature and pressure contain equal numbers of moles, we need to determine the volume at 762 mmHg and 21.0°C that is equivalent to 28.2 L at 742 mmHg and 25.0°C.

$$\text{volume NH}_3\text{(g)} = 28.2 \text{ L HCl(g)} \times \frac{742 \text{ mmHg}}{762 \text{ mmHg}} \times \frac{(273.2 + 21.0) \text{ K}}{(273.2 + 25.0) \text{ K}} \times \frac{1 \text{ L NH}_3\text{(g)}}{1 \text{ L HCl(g)}} = 27.1 \text{ L NH}_3\text{(g)}$$

Alternatively, we can make sure that we get the ratios correct with the approach of setting up and solving the ideal gas equation for the amount of each gas, and then equating these two expressions and solving for the volume of NH_3.

$$n\{\text{HCl}\} = \frac{742 \text{ mmHg} \cdot 28.2 \text{ L}}{R \cdot 298.2 \text{ K}} \qquad n\{\text{NH}_3\} = \frac{762 \text{ mmHg} \cdot V\{\text{NH}_3\}}{R \cdot 294.2 \text{ K}}$$

$$\frac{742 \text{ mmHg} \cdot 28.2 \text{ L}}{R \cdot 298.2 \text{ K}} = \frac{762 \text{ mmHg} \cdot V\{\text{NH}_3\}}{R \cdot 294.2 \text{ K}}$$

This yields $V\{NH_3\} = 27.1$ L NH_3(g)

39. We determine the amounts of H_3O^+ and OH^- and then the amount of the one that is in excess. We express molar concentration in millimoles/milliliter, equivalent to mol/L.

$$50.00 \text{ mL} \times \frac{0.0155 \text{ mmol HI}}{1 \text{ mL soln}} \times \frac{1 \text{ mmol H}_3\text{O}^+}{1 \text{ mmol HI}} = 0.775 \text{ mmol H}_3\text{O}^+$$

$$75.00 \text{ mL} \times \frac{0.0106 \text{ mmol KOH}}{1 \text{ mL soln}} \times \frac{1 \text{ mmol OH}^-}{1 \text{ mmol KOH}} = 0.795 \text{ mmol OH}^-$$

The net reaction is $H_3O^+(aq) + OH^-(aq) \longrightarrow 2\,H_2O$. There is an excess amount of OH^- of (0.795 – 0.775 =) 0.020 mmol OH^-. This is a basic solution. The total solution volume is (50.00 + 75.00 =) 125.00 mL.

$$[\text{OH}^-] = \frac{0.020 \text{ mmol OH}^-}{125.00 \text{ mL}} = 1.6 \times 10^{-4} \text{ M}, \quad \text{pOH} = -\log(1.6 \times 10^{-4}) = 3.80, \quad \text{pH} = 14.00 - 3.80 = 10.20$$

40. In each case, we need to determine the $[H_3O^+]$ or $[OH^-]$ of each solution being mixed, and then the amount of H_3O^+ or OH^-, so that we can determine the amount in excess. We express molar concentration in millimoles/milliliter, equivalent to mol/L.

$[H_3O^+] = 10^{-2.12} = 7.6 \times 10^{-3}$ M   amount $H_3O^+ = 25.00 \text{ mL} \times 7.6 \times 10^{-3} \text{ M} = 0.19$ mmol H_3O^+

pOH = 14.00 –12.65 = 1.35 $[OH^-] = 10^{-1.35} = 4.5 \times 10^{-2}$ M

amount $OH^- = 25.00 \text{ mL} \times 4.5 \times 10^{-2} \text{ M} = 1.13$ mmol OH^-

There is excess OH^- in the amount of 0.94 mmol OH^- (= 1.13 mmol OH^- – 0.19 mmol H_3O^+)

$$[\text{OH}^-] = \frac{0.94 \text{ mmol OH}^-}{25.00 \text{ mL} + 25.00 \text{ mL}} = 1.9 \times 10^{-2} \text{ M} \qquad \text{pOH} = -\log(1.9 \times 10^{-2}) = 1.72$$

pH = 14.00 – 1.72 = 12.28

Weak Acids, Weak Bases, and pH

41. We organize the solution around the balanced chemical equation.

Equation:	$HNO_2 + H_2O \rightleftharpoons$	H_3O^+	+ NO_2^-
Initial:	0.143 M	0 M	0 M
Changes:	$-x$ M	$+x$ M	$+x$ M
Equil:	$(0.143 - x)$M	x M	x M

$$K_a = \frac{[\text{H}_3\text{O}^+][\text{NO}_2^-]}{[\text{HNO}_2]} = \frac{x^2}{0.143 - x} = 7.2 \times 10^{-4} \approx \frac{x^2}{0.143} \quad \text{assuming } x << 0.143$$

$$x = \sqrt{0.143 \times 7.2 \times 10^{-4}} = 0.010 \text{ M}$$

We have assumed that $x << 0.143$ M, an almost-correct assumption. Another cycle of approximations:

$$x = \sqrt{(0.143 - 0.010) \times 7.2 \times 10^{-4}} = 0.0098 \text{ M} = [\text{H}_3\text{O}^+] \qquad \text{pH} = -\log(0.0098) = 2.01$$

This is the same result as is determined with the quadratic equation.

42. We organize the solution around the balanced chemical equation, and solve first for $[OH^-]$

Equation:	$C_2H_5NH_2 + H_2O \rightleftharpoons$	OH^-	+ $C_2H_5NH_3^+$
Initial:	0.085 M	0 M	0 M
Changes:	$-x$ M	$+x$ M	$+x$ M
Equil:	$(0.085 - x)$M	x M	x M

$K_b = \frac{[OH^-][C_2H_5NH_3^+]}{[C_2H_5NH_2]} = \frac{x^2}{0.085 - x} = 4.3 \times 10^{-4} \approx \frac{x^2}{0.085}$ assuming $x << 0.085$

$x = \sqrt{0.085 \times 4.3 \times 10^{-4}} = 0.0060$ M

We have assumed that $x << 0.085$ M, an almost-correct assumption. Another cycle of approximations:

$x = \sqrt{(0.085 - 0.0060) \times 4.3 \times 10^{-4}} = 0.0058$ M $= [OH^-]$ Yet another cycle produces:

$x = \sqrt{(0.085 - 0.0058) \times 4.3 \times 10^{-4}} = 0.0058$ M $= [OH^-]$ pOH $= -\log(0.0058) = 2.24$

pH $= 14.00 - 2.24 = 11.76$ $[H_3O^+] = 10^{-pH} = 10^{-11.76} = 1.7 \times 10^{-12}$ M

This is the same result as determined with the quadratic equation.

43. We base our set up on the balanced chemical equation. $[H_3O^+] = 10^{-1.56} = 2.8 \times 10^{-2}$ M

Equation:	$CH_2FCOOH(aq) + H_2O \rightleftharpoons$	$CH_2FCOO^-(aq)$ +	$H_3O^+(aq)$
Initial:	0.318 M	0 M	0 M
Changes:	−0.028 M	+0.028 M	+0.028 M
Equil:	0.290 M	0.028 M	0.028 M

$K_a = \frac{[H_3O^+][CH_2FCOO^-]}{[CH_2FCOOH]} = \frac{(0.028)(0.028)}{0.290} = 2.7 \times 10^{-3}$

44. $[HC_6H_{11}O_2] = \frac{11\text{ g}}{1\text{ L}} \times \frac{1\text{ mol }HC_6H_{11}O_2}{116.2\text{ g }HC_6H_{11}O_2} = 9.5 \times 10^{-2}$ M $= 0.095$ M

$[H_3O^+] = 10^{-2.94} = 1.1 \times 10^{-3}$ M $= 0.0011$ M

The stoichiometry of the reaction, written below, indicates that $[H_3O^+] = [C_6H_{11}O_2^-]$

Equation:	$HC_6H_{11}O_2 + H_2O \rightleftharpoons$	$C_6H_{11}O_2^-$ +	H_3O^+
Initial:	0.095 M	0 M	0 M
Changes:	−0.0011 M	+0.0011 M	+0.0011 M
Equil:	0.094 M	0.0011 M	0.0011 M

$K_a = \frac{[C_6H_{11}O_2^-][H_3O^+]}{[HC_6H_{11}O_2]} = \frac{(0.0011)^2}{0.094} = 1.3 \times 10^{-5}$

45. We organize the solution around the balanced chemical equation, then use the quadratic formula.

Equation:	$HClO_2 + H_2O \rightleftharpoons$	H_3O^+ +	ClO_2^-
Initial:	0.55 M	0 M	0 M
Changes:	$-x$ M	$+x$ M	$+x$ M
Equil:	$(0.55 - x)$M	x M	x M

$K_a = \frac{[H_3O^+][ClO_2^-]}{[HClO_2]} = \frac{x^2}{0.55 - x} = 1.1 \times 10^{-2} = 0.011$ $x^2 = 0.0061 - 0.011x$ $x^2 + 0.011x - 0.0061 = 0$

$x = \frac{-b \pm \sqrt{b^2 - 4ac}}{2a} = \frac{-0.011 \pm \sqrt{0.000121 + 0.0244}}{2} = 0.073\text{ M} = [H_3O^+]$

The method of successive approximations converges to the same answer in four cycles.

pH $= -\log[H_3O^+] = -\log(0.073) = 1.14$ pOH $= 14.00 -$ pH $= 14.00 - 1.14 = 12.86$

$[OH^-] = 10^{-pOH} = 10^{-12.86} = 1.4 \times 10^{-13}$ M

46. We organize the solution around the balanced chemical equation, and solve first for $[OH^-]$.

Equation:	$CH_3NH_2 + H_2O \rightleftharpoons$	OH^- +	$CH_3NH_3^+$
Initial:	0.386 M	0 M	0 M
Changes:	$-x$ M	$+x$ M	$+x$ M
Equil:	$(0.386 - x)$M	x M	x M

$K_b = \frac{[OH^-][CH_3NH_3^+]}{[CH_3NH_2]} = \frac{x^2}{0.386 - x} = 4.2 \times 10^{-4} \approx \frac{x^2}{0.386}$ assuming $x << 0.386$

$x = \sqrt{0.386 \times 4.2 \times 10^{-4}} = 0.013$ M

We have assumed that $x << 0.386$ M, an almost-correct assumption. Another cycle of approximations:

$x = \sqrt{(0.386 - 0.013) \times 4.2 \times 10^{-4}} = 0.013$ M $= [OH^-]$ pOH $= -\log(0.013) = 1.89$

pH $= 14.00 - 1.89 = 12.11$ $[H_3O^+] = 10^{-pH} = 10^{-12.11} = 7.8 \times 10^{-12}$ M

This is the same result as is determined with the quadratic equation.

47. $[C_{10}H_7NH_2] = \frac{1\text{ g}}{590\text{ g }H_2O} \times \frac{1.00\text{ g }H_2O}{1\text{ mL}} \times \frac{1000\text{ mL}}{1\text{ L}} \times \frac{1\text{ mol }C_{10}H_7NH_2}{143.2\text{ g }C_{10}H_7NH_2} = 1.2 \times 10^{-2}$ M $= 0.012$ M

$K_b = 10^{-pK} = 10^{-3.92} = 1.2 \times 10^{-4}$

Equation:	$C_{10}H_7NH_2 + H_2O \rightleftharpoons$	OH^- +	$C_{10}H_7NH_3^+$
Initial:	0.012 M	0 M	0 M
Changes:	$-x$ M	$+x$ M	$+x$ M
Equil:	$(0.012 - x)$M	x M	x M

$K_b = \frac{[OH^-][C_{10}H_7NH_3^+]}{[C_{10}H_7NH_2]} = \frac{x^2}{0.012 - x} = 1.2 \times 10^{-4} \approx \frac{x^2}{0.012}$ assuming $x << 0.012$

$x = \sqrt{0.012 \times 1.2 \times 10^{-4}} = 0.0012$ This is an almost-correct assumption. Another approximation cycle:

$x = \sqrt{(0.012 - 0.0012) \times 1.2 \times 10^{-4}} = 0.0011$ Yet another cycle seems necessary.

$x = \sqrt{(0.012 - 0.0011) \times 1.2 \times 10^{-4}} = 0.0011$ M$= [OH^-]$

$pOH = -\log [OH^-] = -\log(0.0011) = 2.96$ $pH = 14.00 - pOH = 11.04$

48. The stoichiometry of the ionization reaction indicates that $[H_3O^+] = [OC_6H_4NO_2^-]$. $K_a = 10^{-pK} = 10^{-7.23} = 5.9 \times 10^{-8}$ and $[H_3O^+] = 10^{-4.53} = 3.0 \times 10^{-5}$ M. We let S = the molar solubility of *o*-nitrophenol.

Equation:	$HOC_6H_4NO_2 + H_2O \rightleftharpoons$	$OC_6H_4NO_2^-$ +	H_3O^+
Initial:	S	0 M	0 M
Changes:	-3.0×10^{-5} M	$+3.0 \times 10^{-5}$ M	$+3.0 \times 10^{-5}$ M
Equil:	$S - 3.0 \times 10^{-5}$ M	3.0×10^{-5} M	3.0×10^{-5} M

$K_a = 5.9 \times 10^{-8} = \frac{[OC_6H_4NO_2^-][H_3O^+]}{[HOC_6H_5NO_2]} = \frac{(3.0 \times 10^{-5})^2}{S - 3.0 \times 10^{-5}}$

$S - 3.0 \times 10^{-5} = \frac{(3.0 \times 10^{-5})^2}{5.9 \times 10^{-8}} = 1.5 \times 10^{-2}$ M $\quad S = 1.5 \times 10^{-2}$ M $+ 3.0 \times 10^{-5}$ M $= 1.5 \times 10^{-2}$ M

Hence solubility $= \frac{1.5 \times 10^{-2} \text{ mol } HOC_6H_4NO_2}{1 \text{ L soln}} \times \frac{139.1 \text{ g } HOC_6H_4NO_2}{1 \text{ mol } HOC_6H_4NO_2} = 2.1$ g $HOC_6H_4NO_2$/L soln

49. We determine $[H_3O^+]$ which, because of the stoichiometry of the reaction, equals $[C_2H_3O_2^-]$.
$[H_3O^+] = 10^{-pH} = 10^{-4.52} = 3.0 \times 10^{-5}$ M $= [C_2H_3O_2^-]$
We solve for S, the concentration of $HC_2H_3O_2$ in the 0.750 L solution before it dissociates.

Equation:	$HC_2H_3O_2 + H_2O \rightleftharpoons$	$C_2H_3O_2^-$ +	H_3O^+
initial:	S M	0 M	0 M
changes:	-3.0×10^{-5} M	$+3.0 \times 10^{-5}$ M	$+3.0 \times 10^{-5}$ M
equil:	$(S - 3.0 \times 10^{-5})$M	3.0×10^{-5} M	3.0×10^{-5} M

$K_a = \frac{[H_3O^+][C_2H_3O_2^-]}{[HC_2H_3O_2]} = 1.8 \times 10^{-5} = \frac{(3.0 \times 10^{-5})^2}{(S - 3.0 \times 10^{-5})}$

$(3.0 \times 10^{-5})^2 = 1.8 \times 10^{-5}(S - 3.0 \times 10^{-5}) = 9.0 \times 10^{-10} = 1.8 \times 10^{-5}S - 5.4 \times 10^{-10}$

$S = \frac{9.0 \times 10^{-10} + 5.4 \times 10^{-10}}{1.8 \times 10^{-5}} = 8.0 \times 10^{-5}$ M

Now we determine the mass of vinegar needed.

mass vinegar $= 0.750 \text{ L} \times \frac{8.0 \times 10^{-5} \text{ mol } HC_2H_3O_2}{1 \text{ L soln}} \times \frac{60.05 \text{ g } HC_2H_3O_2}{1 \text{ mol } HC_2H_3O_2} \times \frac{100.0 \text{ g vinegar}}{5.7 \text{ g } HC_2H_3O_2}$

$= 0.063$ g vinegar

50. We determine $[OH^-]$ which, because of the stoichiometry of the reaction, equals $[NH_4^+]$.
$pOH = 14.00 - pH = 14.00 - 11.55 = 2.45$ $[OH^-] = 10^{-pOH} = 10^{-2.45} = 3.5 \times 10^{-3}$ M $= [NH_4^+]$
We solve for S, the concentration of NH_3 in the 0.625 L solution before it dissociates.

Equation:	$NH_3 + H_2O \rightleftharpoons$	NH_4^+ +	OH^-
initial:	S M	0 M	0 M
changes:	-0.0035 M	$+0.0035$ M	$+0.0035$ M
equil:	$(S - 0.0035)$M	0.0035 M	0.0035 M

$K_b = \frac{[NH_4^+][OH^-]}{[NH_3]} = 1.8 \times 10^{-5} = \frac{(0.0035)^2}{(S - 0.0035)}$

$(0.0035)^2 = 1.8 \times 10^{-5}(S - 0.0035) = 1.2 \times 10^{-5} = 1.8 \times 10^{-5}S - 6.3 \times 10^{-8}$

$$S = \frac{1.2 \times 10^{-5} + 6.3 \times 10^{-8}}{1.8 \times 10^{-5}} = 0.67 \text{ M}$$

Now we determine the volume of household ammonia needed.

$$\text{ammonia volume} = 0.625 \text{ L} \times \frac{0.67 \text{ mol NH}_3}{1 \text{ L soln}} \times \frac{17.03 \text{ g NH}_3}{1 \text{ mol NH}_3} \times \frac{100.0 \text{ g soln}}{6.8 \text{ g NH}_3} \times \frac{1 \text{ mL soln}}{0.97 \text{ g soln}}$$

$= 1.1 \times 10^2$ mL household ammonia solution

Percent Ionization

51. Let us first compute the $[H_3O^+]$ in this solution.

Equation:	$HC_3H_5O_2 + H_2O \rightleftharpoons$	H_3O^+ +	$C_3H_5O_2^-$
Initial:	0.45 M	0 M	0 M
Changes:	$-x$ M	$+x$ M	$+x$ M
Equil:	$(0.45 - x)$M	x M	x M

$$K_a = \frac{[H_3O^+][C_3H_5O_2^-]}{[HC_3H_5O_2]} = \frac{x^2}{0.45 - x} = 10^{-4.89} = 1.3 \times 10^{-5} \approx \frac{x^2}{0.45}$$

$x = 2.4 \times 10^{-3}$ M We have assumed that $x << 0.45$ M, an assumption that clearly is correct.

(a) $\alpha = \frac{[H_3O^+]_{eq}}{[HC_3H_5O_2]_i} = \frac{2.4 \times 10^{-3} \text{ M}}{0.45 \text{ M}} = 0.0053$ = degree of ionization

(b) % ionization = $\alpha \times 100\% = 0.0053 \times 100\% = 0.53\%$

52. We first determine $[H_3O^+]$ in this solution. $K_a = pK_a = 10^{-0.52} = 0.30$

Equation:	$HC_2Cl_3O_2 + H_2O \rightleftharpoons$	$C_2Cl_3O_2^-$ +	H_3O^+
Initial:	0.035 M	0 M	0 M
Changes:	$-x$ M	$+x$ M	$+x$ M
Equil:	$(0.035 - x)$M	x M	x M

$$K_a = \frac{[C_2Cl_3O_2^-][H_3O^+]}{[HC_2Cl_3O_2]} = 10^{-0.52} = 0.30 = \frac{x^2}{0.035 - x}$$

$x^2 = 0.011 - 0.30\,x$ $x^2 + 0.30\,x - 0.011 = 0$

$$x = \frac{-b \pm \sqrt{b^2 - 4ac}}{2a} = \frac{-0.30 \pm \sqrt{0.090 + 0.044}}{2} = 0.03 \text{ M} = [C_2Cl_3O_2^-]$$

(a) $\alpha = \frac{[C_2Cl_3O_2^-]_{eq}}{[HC_2Cl_3O_2]_i} = \frac{0.03 \text{ M}}{0.035 \text{ M}} = 0.9$ **(b)** % ionization = $9 \times 10^1\%$

53. The fact that the NH_3 is 4.2% ionized means that $[NH_4^+] = 0.042\,[NH_3]_{initial} = [OH^-]$. Expressed another way, this is $[NH_3]_{initial} = 23.8\,[OH^-]$. ($23.8 = 1 \div 0.042$.) Of course, we also know that $[NH_3]_{equil} = [NH_3]_{initial} - [OH^-]$. We use the base dissociation constant expression for NH_3 to organize our information.

$$K_b = 1.8 \times 10^{-5} = \frac{[NH_4^+][OH^-]}{[NH_3]_{equil}} = \frac{[OH^-]^2}{23.8 \times [OH^-] - [OH^-]} = \frac{[OH^-]}{22.8}$$

$[OH^-] = 22.8 \times 1.8 \times 10^{-5} = 4.1 \times 10^{-4}$ M $[NH_3]_{initial} = 23.8 \times 4.1 \times 10^{-4} = 0.0098$ M

54. No, the degree of ionization of $HC_2H_3O_2$ is not expected to be 13% in 0.0010 M $HC_2H_3O_2$ or 42% in 0.00010 M $HC_2H_3O_2$. The degrees of ionization at lower values of $[HC_2H_3O_2]$ are based on the assumption that $[HC_2H_3O_2]_i \approx [HC_2H_3O_2]_i - [HC_2H_3O_2]_{eq}$. As long as $[HC_2H_3O_2]_{eq}$ is less than 5% of $[HC_2H_3O_2]_i$ this is a reasonable assumption. Of course, it has broken down at 13% ionization. (13% is larger than 5%.)

Polyprotic Acids

55. Because H_3PO_4 is a weak acid, there is little HPO_4^{2-} (produced in the second ionization) compared to the H_3O^+ (produced in the first ionization). In turn, there is little PO_4^{3-} (produced in the third ionization) compared to the HPO_4^{2-}, and very little compared to the H_3O^+.

56. The main estimate involves assuming that the mass percents can be expressed as 0.057 g of 75% H_3PO_4 per 100. mL of solution and 0.084 g of 75% H_3PO_4 per 100. mL of solution. That is, that the density of the aqueous solution is essentially 1.00 g/mL. Based on this assumption, we can compute the initial concentrations of H_3PO_4.

$$[H_3PO_4] = \frac{0.057 \text{ g impure H}_3\text{PO}_4 \times \frac{75 \text{ g H}_3\text{PO}_4}{100 \text{ g impure H}_3\text{PO}_4} \times \frac{1 \text{ mol H}_3\text{PO}_4}{98.00 \text{ g H}_3\text{PO}_4}}{100. \text{ mL soln} \times \frac{1 \text{ L}}{1000 \text{ mL}}} = 0.0044 \text{ M}$$

$$[H_3PO_4] = \frac{0.084 \text{ g impure } H_3PO_4 \times \dfrac{75 \text{ g } H_3PO_4}{100 \text{ g impure } H_3PO_4} \times \dfrac{1 \text{ mol } H_3PO_4}{98.00 \text{ g } H_3PO_4}}{100. \text{ mL soln} \times \dfrac{1 \text{ L}}{1000 \text{ mL}}} = 0.0064 \text{ M}$$

Equation:	$H_3PO_4 + H_2O \rightleftharpoons$	$H_2PO_4^- +$	H_3O^+
Initial:	0.0044 M	0 M	0 M
Changes:	$-x$ M	$+x$ M	$+x$ M
Equil:	$(0.0044 - x)$M	x M	x M

$$K_1 = \frac{[H_2PO_4^-][H_3O^+]}{[H_2PO_4]} = \frac{x^2}{0.0044 - x} = 7.1 \times 10^{-3} \qquad x^2 + 0.0071\,x - 3.1 \times 10^{-5} = 0$$

$$x = \frac{-b \pm \sqrt{b^2 - 4ac}}{2a} = \frac{-0.0071 \pm \sqrt{5.0 \times 10^{-5} + 1.2 \times 10^{-4}}}{2} = 3 \times 10^{-3} \text{ M} = [H_3O^+]$$

The set up for the second concentration is the same as for the first, with the exception of substitution 0.0064 M for 0.0044 M.

$$K_1 = \frac{[H_2PO_4^-][H_3O^+]}{[H_2PO_4]} = \frac{x^2}{0.0064 - x} = 7.1 \times 10^{-3} \qquad x^2 + 0.0071\,x - 4.5 \times 10^{-5} = 0$$

$$x = \frac{-b \pm \sqrt{b^2 - 4ac}}{2a} = \frac{-0.0071 \pm \sqrt{5.0 \times 10^{-5} + 1.8 \times 10^{-4}}}{2} = 4 \times 10^{-3} \text{ M} = [H_3O^+]$$

The two values of pH now are determined, representing the pH range in a cola drink.

$pH = -\log(3 \times 10^{-3}) = 2.5$ $\qquad$ $pH = -\log(4 \times 10^{-3}) = 2.4$

57. **(a)**

Equation:	$H_2S + H_2O \rightleftharpoons$	$HS^- +$	H_3O^+
Initial:	0.075 M	0 M	0 M
Changes:	$-x$ M	$+x$ M	$+x$ M
Equil:	$(0.075 - x)$M	x M	$+x$ M

$$K_1 = \frac{[HS^-][H_3O^+]}{[H_2S]} = 1.0 \times 10^{-7} = \frac{x^2}{0.075 - x} \approx \frac{x^2}{0.075} \qquad x = 8.7 \times 10^{-5} \text{ M} = [H_3O^+]$$

$[HS^-] = 8.7 \times 10^{-5}$ M $\qquad [S^{2-}] = K_2 = 1 \times 10^{-19}$ M

(b) The set up for this problem is the same as for part (a), with the substitution of 0.0050 M for 0.075 M as the initial value of $[H_2S]$.

$$K_1 = \frac{[HS^-][H_3O^+]}{[H_2S]} = 1.0 \times 10^{-7} = \frac{x^2}{0.0050 - x} \approx \frac{x^2}{0.0050} \qquad x = 2.2 \times 10^{-5} \text{ M} = [H_3O^+]$$

$[HS^-] = 2.2 \times 10^{-5}$ M $\qquad [S^{2-}] = K_2 = 1 \times 10^{-19}$ M

(c) The set up for this part is the same as for part (a), with the substitution of 1.0×10^{-5} M for 0.075 M as the initial value of $[H_2S]$. The solution differs in that we cannot assume $x << 1.0 \times 10^{-5}$. Rather than solve the quadratic equation, however, we shall use the method of successive approximations. This involves making the unwarranted assumption referred to above to obtain a first approximate value of x. This approximate value of x then is used in the denominator of the K_1 expression (but not to substitute in x^2) to obtain a second approximate value of x. The cycle continues until two successive values of x agree.

$$K_1 = \frac{[HS^-][H_3O^+]}{[H_2S]} = 1.0 \times 10^{-7} = \frac{x^2}{1.0 \times 10^{-5} - x} \approx \frac{x^2}{1.0 \times 10^{-5}} \qquad x = 1.0 \times 10^{-6} \text{ M} = [H_3O^+]$$

$$K_1 = \frac{[HS^-][H_3O^+]}{[H_2S]} = 1.0 \times 10^{-7} = \frac{x^2}{1.0 \times 10^{-5} - 1.0 \times 10^{-6}} = \frac{x^2}{9.0 \times 10^{-6}} \qquad x = 9.5 \times 10^{-7} \text{ M}$$

$$K_1 = \frac{[HS^-][H_3O^+]}{[H_2S]} = 1.0 \times 10^{-7} = \frac{x^2}{1.0 \times 10^{-5} - 9.5 \times 10^{-7}} = \frac{x^2}{9.0 \times 10^{-6}} \qquad x = 9.5 \times 10^{-7} \text{ M}$$

$x = 9.5 \times 10^{-7}$ M $= [H_3O^+]$ $\qquad [HS^-] = 9.5 \times 10^{-7}$ M $\qquad [S^{2-}] = K_2 = 1 \times 10^{-19}$ M

58. In all cases, of course, the first ionization of H_2SO_4 is complete, and establishes the initial values of $[H_3O^+]$ and $[HSO_4^-]$. We deal with the second ionization in each case.

(a)

Equation:	$HSO_4^- + H_2O \rightleftharpoons$	$SO_4^{2-} +$	H_3O^+
Initial:	0.75 M	0 M	0.75 M
Changes:	$-x$ M	$+x$ M	$+x$ M
Equil:	$(0.75 - x)$M	x M	$(0.75 + x)$M

$K_2 = \frac{[SO_4^{2-}][H_3O^+]}{[HSO_4^-]} = 0.011 = \frac{x\,(0.75 + x)}{0.75 - x} \approx \frac{0.75\,x}{0.75}$ $x = 0.011\text{ M} = [SO_4^{2-}]$

We have assumed that $x << 0.75$ M, an assumption that clearly is correct.

$[HSO_4^-] = 0.75 - 0.011 = 0.74\text{ M}$ $[H_3O^+] = 0.75 + 0.011\text{ M} = 0.76\text{ M}$

(b) The set up for this part is similar to part (a), with the exception of substituting 0.075 M for 0.75 M.

$K_2 = \frac{[SO_4^{2-}][H_3O^+]}{[HSO_4^-]} = 0.011 = \frac{x\,(0.075 + x)}{0.075 - x}$ $0.011(0.075 - x) = 0.075x + x^2$

$x^2 + 0.086\,x - 8.3 \times 10^{-4} = 0$ $x = \frac{-b \pm \sqrt{b^2 - 4ac}}{2a} = \frac{-0.086 \pm \sqrt{0.0074 + 0.0033}}{2} = 0.0087\text{ M}$

$x = 0.0087\text{ M} = [SO_4^{2-}]$

$[HSO_4^-] = 0.075 - 0.0087 = 0.066\text{ M}$ $[H_3O^+] = 0.075 + 0.0088\text{ M} = 0.084\text{ M}$

(c) Again, the set up is the same as for part (a), with the exception of substituting 0.00075 M for 0.75 M.

$K_2 = \frac{[SO_4^{2-}][H_3O^+]}{[HSO_4^-]} = 0.011 = \frac{x\,(0.00075 + x)}{0.00075 - x}$ $0.011(0.00075 - x) = 0.00075\,x + x^2$

$x^2 + 0.0118x - 8.3 \times 10^{-6} = 0$

$x = \frac{-b \pm \sqrt{b^2 - 4ac}}{2a} = \frac{-0.0118 \pm \sqrt{1.39 \times 10^{-4} + 3.3 \times 10^{-5}}}{2} = 6.6 \times 10^{-4}$

$x = 6.6 \times 10^{-4}\text{ M} = [SO_4^{2-}]$ $[HSO_4^-] = 0.00075 - 0.00066 = 9 \times 10^{-5}\text{ M}$

$[H_3O^+] = 0.00075 + 0.00066\text{ M} = 1.41 \times 10^{-3}\text{ M}$ $[H_3O^+]$ is almost twice the initial value of $[H_2SO_4]$. The second ionization of H_2SO_4 is nearly complete in this dilute solution.

59. We first determine $[H_3O^+]$, with our calculation based, as usual, on the balanced chemical equation.

Equation:	$HOOC(CH_2)_4COOH(aq) + H_2O \rightleftharpoons$	$HOOC(CH_2)_4COO^-(aq)$ +	$H_3O^+(aq)$
Initial:	0.10 M	0 M	0 M
Changes:	$-x$ M	$+x$ M	$+x$ M
Equil:	$(0.10 - x)$ M	x M	x M

$K_{a_1} = \frac{[H_3O^+][HOOC(CH_2)_4COO^-]}{[HOOC(CH_2)_4COOH]} = 3.9 \times 10^{-5} = \frac{x \cdot x}{0.10 - x} \approx \frac{x^2}{0.10}$

$x = \sqrt{0.10 \times 3.9 \times 10^{-5}} = 2.0 \times 10^{-3}\text{ M} = [H_3O^+]$

We see that our simplifying assumption, that $x << 0.10$ M, is indeed valid. Now we consider the second ionization. We shall see that very little H_3O^+ is produced in this ionization because of the small size of two numbers: K_{a_2} and $[HOOC(CH_2)_4COO^-]$. Again we base our calculation on the balanced chemical equation.

Equation:	$HOOC(CH_2)_4COO^-(aq) + H_2O \rightleftharpoons$	$^-OOC(CH_2)_4COO^-(aq)$ +	$H_3O^+(aq)$
Initial:	2.0×10^{-3} M	0 M	2.0×10^{-3} M
Changes:	$-y$ M	$+y$ M	$+y$ M
Equil:	$(0.0020 - y)$ M	y M	$(0.0020 + y)$ M

$K_{a_2} = \frac{[H_3O^+][^-OOC(CH_2)_4COO^-]}{[HOOC(CH_2)_4COO^-]} = 3.9 \times 10^{-6} = \frac{y(0.0020 + y)}{0.0020 - y} \approx \frac{0.0020\,y}{0.0020}$ $y = 3.9 \times 10^{-6}\text{ M} = [^-OOC(CH_2)_4COO^-]$

Again, we see that our assumption, that $y << 0.0020$ M, is valid. In addition, we also see that virtually no H_3O^+ is created in this second ionization. The concentrations of all species have been calculated above, with the exception of $[OH^-]$.

$[OH^-] = \frac{K_w}{[H_3O^+]} = \frac{1.0 \times 10^{-14}}{2.0 \times 10^{-3}} = 5.0 \times 10^{-12}\text{ M}$ $[HOOC(CH_2)_4COOH] = 0.10\text{ M}$

$[H_3O^+] = [HOOC(CH_2)_4COO^-] = 2.0 \times 10^{-3}\text{ M}$ $[^-OOC(CH_2)_4COO^-] = 3.9 \times 10^{-6}\text{ M}$

60. (a) Recall that a base is a proton acceptor, in this case, accepting from H_2O.

First ionization: $C_{20}H_{24}O_2N_2 + H_2O \rightleftharpoons C_{20}H_{24}O_2N_2H^+ + OH^-$ $pK_1 = 6.0$

Second ionization: $C_{20}H_{24}O_2N_2H^+ + H_2O \rightleftharpoons C_{20}H_{24}O_2N_2H_2^{2+} + OH^-$ $pK_2 = 9.8$

(b) $[C_{20}H_{24}O_2N_2] = \frac{1.00\text{ g quinine} \times \frac{1\text{ mol quinine}}{324.4\text{ g quinine}}}{1900.\text{ mL} \times \frac{1\text{ L}}{1000\text{ mL}}} = 1.62 \times 10^{-3}\text{ M}$ $K_1 = 10^{-6.0} = 1 \times 10^{-6}$

Equation:	$C_{20}H_{24}O_2N_2 + H_2O \rightleftharpoons$	$C_{20}H_{24}O_2N_2H^+$ +	OH^-
Initial:	0.00162 M	0 M	0 M
Changes:	$-x$ M	$+x$ M	$+x$ M
Equil:	$(0.00162 - x)$M	x M	x M

$$K_1 = \frac{[C_{20}H_{24}O_2N_2H^+][OH^-]}{[C_{20}H_{24}O_2N_2]} = 1 \times 10^{-6} = \frac{x^2}{0.00162 - x} \approx \frac{x^2}{0.00162} \qquad x = 4 \times 10^{-5}\text{ M}$$

Our assumption, that $x << 0.00162$, clearly is valid.

$pOH = -\log(4 \times 10^{-5}) = 4.4$ $\qquad pH = 14.00 - 4.4 = 9.6$

Ions as Acids and Bases (Hydrolysis)

61. **(a)** KCl forms a neutral solution, being composed of the cation of a strong base, and the anion of a strong acid; no hydrolysis occurs.

(b) KF forms an alkaline (basic) solution, being composed of the cation of a strong base and the anion of a weak acid; the fluoride ion hydrolyzes. $F^- + H_2O \rightleftharpoons HF + OH^-$

(c) $NaNO_3$ forms a neutral solution, being composed of the cation of a strong base and the anion of a strong acid. No hydrolysis occurs.

(d) $Ca(OCl)_2$ forms an alkaline (basic) solution, being composed of the cation of a strong base, and the anion of a weak acid; the hypochlorite ion hydrolyzes. $OCl^- + H_2O \rightleftharpoons HOCl + OH^-$

(e) NH_4NO_2 forms an acidic solution. The salt is composed of the cation of a weak base and the anion of a weak acid; the ammonium ion hydrolyzes: $NH_4^+ + H_2O \rightleftharpoons NH_3(aq) + H_3O^+$ $\quad K_a = 5.6 \times 10^{-10}$ and so does the nitrite ion: $NO_2^- + H_2O \rightleftharpoons HNO_2 + OH^-$ $\quad K_b = 1.4 \times 10^{-11}$ Since NH_4^+ is a stronger acid than NO_2^- is a base, the solution will be acidic. The ionization constants were computed from data in Table 17-3 and with the relationship $K_w = K_a \times K_b$.

For NH_4^+, $K_a = \dfrac{1.0 \times 10^{-14}}{1.8 \times 10^{-5}} = 5.6 \times 10^{-10}$ $\qquad$ For NO_2^-, $K_b = \dfrac{1.0 \times 10^{-14}}{7.2 \times 10^{-4}} = 1.4 \times 10^{-11}$

Where $\quad 1.8 \times 10^{-5} = K_b$ of NH_3 $\qquad$ and $\qquad 7.2 \times 10^{-4} = K_a$ of HNO_2

62. Our list in order of increasing pH is in order of decreasing acidity. First we look for the strong acids; there is just HNO_3. Next we look for the weak acids; there is only one, $HC_2H_3O_2$. Next in order of decreasing acidity come salts with cations from weak bases and anions from strong acids; NH_4ClO_4 is in this category. Then come salts in which both ions hydrolyze to the same degree; $NH_4C_2H_3O_2$ is an example, forming a pH-neutral solution. Next come salts that have the cation of a strong base and the anion of a weak acid; $NaNO_2$ is in this category. Then come weak bases, of which $NH_3(aq)$ is an example. And finally, strong bases: NaOH is the only one. Thus, in order of increasing pH of their 0.010 M aqueous solutions, the solutes are:
$HNO_3 < HC_2H_3O_2 < NH_4ClO_4 < NH_4C_2H_3O_2 < NaNO_2 < NH_3 < NaOH$

63. NaOCl dissociates completely in aqueous solution into $Na^+(aq)$, which does not hydrolyze, and $OCl^-(aq)$, which does hydrolyze. We determine $[OH^-]$ in a 0.089 M solution of OCl^-, finding the value of the hydrolysis constant from the ionization constant of HOCl, $K_a = 2.9 \times 10^{-8}$.

Equation:	$OCl^- + H_2O$	$\rightleftharpoons$	$HOCl$	+	OH^-
Initial:	0.089 M		0 M		0 M
Changes:	$-x$ M		$+x$ M		$+x$ M
Equil:	$(0.089 - x)$M		x M		x M

$$K_b = \frac{K_w}{K_a} = \frac{1.0 \times 10^{-14}}{2.9 \times 10^{-8}} = 3.4 \times 10^{-7} = \frac{[HOCl][OH^-]}{[OCl^-]} = \frac{x^2}{0.089 - x} \approx \frac{x^2}{0.089}$$

$x = 1.7 \times 10^{-4}\text{ M} = [OH^-]$ $\qquad pOH = -\log(1.7 \times 10^{-4}) = 3.77$ $\qquad pH = 14.00 - 3.77 = 10.23$

64. NH_4Cl dissociates completely in aqueous solution into $NH_4^+(aq)$, which hydrolyzes, and $Cl^-(aq)$, which does not. We determine $[H_3O^+]$ in a 0.123 M solution of NH_4^+, finding the value of the hydrolysis constant from the ionization constant of NH_3, $K_b = 1.8 \times 10^{-5}$.

Equation:	$NH_4^+ + H_2O$	$\rightleftharpoons$	NH_3	+	H_3O^+
Initial:	0.123 M		0 M		0 M
Changes:	$-x$ M		$+x$ M		$+x$ M
Equil:	$(0.123 - x)$M		x M		x M

$$K_b = \frac{K_w}{K_a} = \frac{1.0 \times 10^{-14}}{1.8 \times 10^{-5}} = 5.6 \times 10^{-10} = \frac{[NH_3][H_3O^+]}{[NH_4^+]} = \frac{x^2}{0.123 - x} \approx \frac{x^2}{0.123}$$

$x = 8.3 \times 10^{-6}\text{ M} = [H_3O^+]$ $\qquad pH = -\log(8.3 \times 10^{-6}) = 5.08$

65. $KC_6H_7O_2$ dissociates completely in aqueous solution into $K^+(aq)$, which does not hydrolyze, and the ion $C_6H_7O_2^-(aq)$, which does hydrolyze. We determine $[OH^-]$ in 0.37 M $KC_6H_7O_2$ solution.

Equation:	$C_6H_7O_2^- + H_2O \rightleftharpoons$	$HC_6H_7O_2$ +	OH^-	$K_a = 10^{-pK} = 10^{-4.77} = 1.7 \times 10^{-5}$
Initial:	0.37 M	0 M	0 M	
Changes:	$-x$ M	$+x$ M	$+x$ M	
Equil:	$(0.37 - x)$M	x M	x M	

$$K_b = \frac{K_w}{K_a} = \frac{1.0 \times 10^{-14}}{1.7 \times 10^{-5}} = 5.9 \times 10^{-10} = \frac{[HC_6H_7O_2][OH^-]}{[C_6H_7O_2^-]} = \frac{x^2}{0.37 - x} \approx \frac{x^2}{0.37}$$

$x = 1.5 \times 10^{-5}$ M $= [OH^-]$ $\quad$ $pOH = -\log(1.5 \times 10^{-5}) = 4.82$ $\quad$ $pH = 14.00 - 4.82 = 9.18$

66.

Equation:	$C_5H_5NH^+ + H_2O \rightleftharpoons$	C_5H_5N +	H_3O^+
Initial:	0.0482 M	0 M	0 M
Changes:	$-x$ M	$+x$ M	$+x$ M
Equil:	$(0.0482 - x)$M	x M	x M

$$K_a = \frac{K_w}{K_b} = \frac{1.0 \times 10^{-14}}{1.5 \times 10^{-9}} = 6.7 \times 10^{-6} = \frac{[C_5H_5N][H_3O^+]}{[C_5H_5NH^+]} = \frac{x^2}{0.0482 - x} = \frac{x^2}{0.0482}$$

$x = 5.7 \times 10^{-4}$ M $= [H_3O^+]$ $\quad$ $pH = -\log(5.7 \times 10^{-4}) = 3.24$

67. **(a)** $HSO_3^- + H_2O \rightleftharpoons H_3O^+ + SO_3^{2-}$ $\quad$ $K_a = K_2 = 6.2 \times 10^{-8}$

$HSO_3^- + H_2O \rightleftharpoons OH^- + H_2SO_3$ $\quad$ $K_b = \dfrac{K_w}{K_1} = \dfrac{1.0 \times 10^{-14}}{1.3 \times 10^{-2}} = 7.7 \times 10^{-13}$

Since $K_a > K_b$, a solution of HSO_3^- is acidic.

(b) $HS^- + H_2O \rightleftharpoons H_3O^+ + S^{2-}$ $\quad$ $K_a = K_2 = 1 \times 10^{-19}$

$HS^- + H_2O \rightleftharpoons OH^- + H_2S$ $\quad$ $K_b = \dfrac{K_w}{K_1} = \dfrac{1.0 \times 10^{-14}}{1.0 \times 10^{-7}} = 1.0 \times 10^{-7}$

Since $K_a < K_b$, a solution of HS^- is alkaline, or basic.

(c) $HPO_4^{2-} + H_2O \rightleftharpoons H_3O^+ + PO_4^{3-}$ $\quad$ $K_a = K_3 = 4.2 \times 10^{-13}$

$HPO_4^{2-} + H_2O \rightleftharpoons OH^- + H_2PO_4^-$ $\quad$ $K_b = \dfrac{K_w}{K_2} = \dfrac{1.0 \times 10^{-14}}{6.3 \times 10^{-8}} = 1.6 \times 10^{-7}$

Since $K_a < K_b$, a solution of HPO_4^{2-} is alkaline, or basic.

68. pH = 8.65 is a basic solution. We need the salt of the cation of a strong base, and the anion of a weak acid, which will hydrolyze to produce this basic solution. This must be (c) KNO_2. (a) NH_4Cl is the salt of the cation of a weak base and the anion of a strong acid, and should form an acidic solution. (b) $KHSO_4$ and (d) $NaNO_3$ both are the salts of cations of strong bases with anions of strong acids; they form pH-neutral solutions. $\quad$ $pOH = 14.00 - 8.65 = 5.35$ $\quad$ $[OH^-] = 10^{-5.35} = 4.5 \times 10^{-6}$ M

Equation:	$NO_2^- + H_2O \rightleftharpoons$	HNO_2 +	OH^-
Initial:	S	0 M	0 M
Changes:	-4.5×10^{-6} M	$+4.5 \times 10^{-6}$ M	$+4.5 \times 10^{-6}$ M
Equil:	$(S - 4.5 \times 10^{-6}$ M)	4.5×10^{-6} M	4.5×10^{-6} M

$$K_b = \frac{K_w}{K_a} = \frac{1.0 \times 10^{-14}}{7.2 \times 10^{-4}} = 1.4 \times 10^{-11} = \frac{[HNO_2][OH^-]}{[NO_2^-]} = \frac{(4.5 \times 10^{-6})^2}{S - 4.5 \times 10^{-6}}$$

$$S - 4.5 \times 10^{-6} = \frac{(4.5 \times 10^{-6})^2}{1.4 \times 10^{-11}} = 1.4 \text{ M} \qquad S = 1.4 \text{ M} = [KNO_2]$$

Molecular Structure and Acid Strength

69. **(a)** $HClO_3$ should be a stronger acid than is $HClO_2$. In each acid there is an H—O—Cl grouping. The remaining oxygen atoms are attached directly to Cl as terminal O atoms. Thus, there are two terminal O atoms in $HClO_3$ and one terminal O atom in $HClO_2$. Of oxoacids of the same element, the one with the higher number of terminal oxygen atoms is the stronger. $K_a = 5 \times 10^2$ for $HClO_3$ and $K_a = 1.1 \times 10^{-2}$ for $HClO_2$

(b) HNO_2 and H_2CO_3 have the same number (one) of terminal oxygen atoms. They differ in N being more electronegative than C. HNO_2 ($K_a = 7.2 \times 10^{-4}$) is stronger than H_2CO_3 ($K_1 = 4.4 \times 10^{-7}$).

(c) H_3PO_4 and H_2SiO_3 have the same number (one) of terminal oxygen atoms. They differ in P being more electronegative than Si. H_3PO_4 ($K_1 = 7.1 \times 10^{-3}$) is stronger than H_2SiO_3 ($K_1 = 1.7 \times 10^{-10}$).

70. The weakest of the five acids is CH_3CH_2COOH. The reasoning is as follows. HBr is a strong acid, stronger than the carboxylic acids. A carboxylic acid—such as $CH_2ClCOOH$ and CH_2FCH_2COOH—with a strongly electronegative atom attached to the hydrocarbon chain will be stronger than one in which no such group is attached. But the I atom is so weakly electronegative that it influences acid strength hardly at all. Acid strengths of some of these acids (as values of pK_a) follow. (Larger values of pK_a indicate weaker acids.)
Strength: CH_3CH_2COOH (4.89) < $CH_2ClCOOH$ (2.85) < CH_2FCH_2COOH < CI_3COOH < HBr (–8.72)

Lewis Theory of Acids and Bases

71. **(a)**

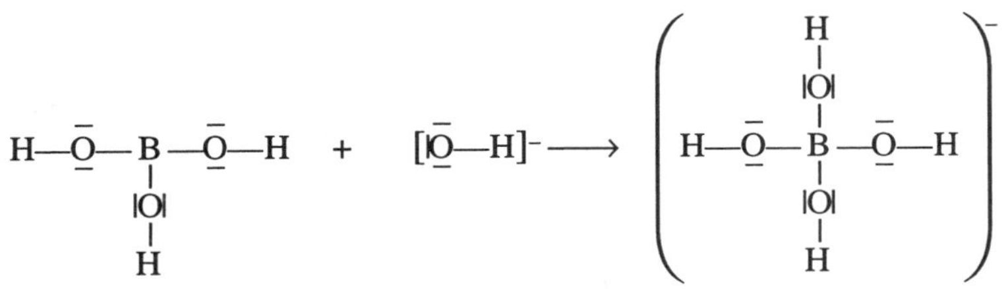

acid: eln pr acceptor base: eln pr donor

(b)

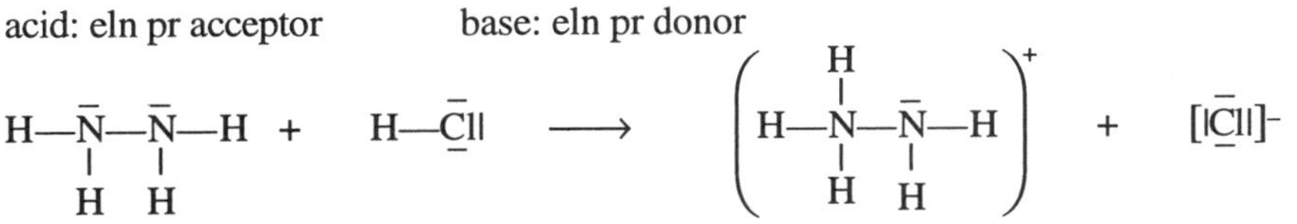

base secondary acid
The actual Lewis acid is H^+, which is supplied by HCl.

(c)

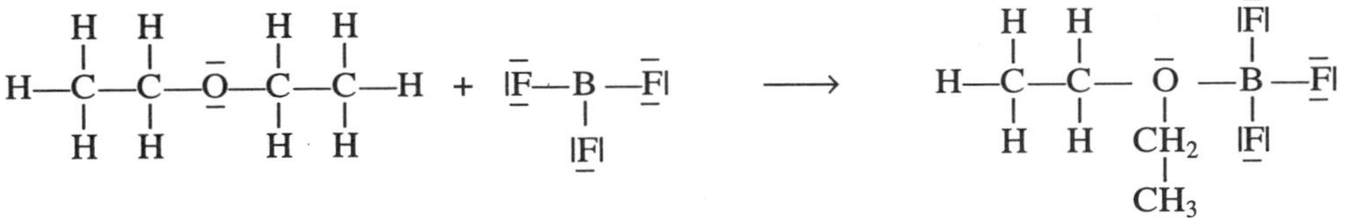

base: electron pair donor acid: eln pr acceptor

72. The C in a CO_2 molecule has location available for the acceptance of a pair of electrons; it is the Lewis acid. The hydroxide anion is a Lewis base, with pairs of electrons to donate.

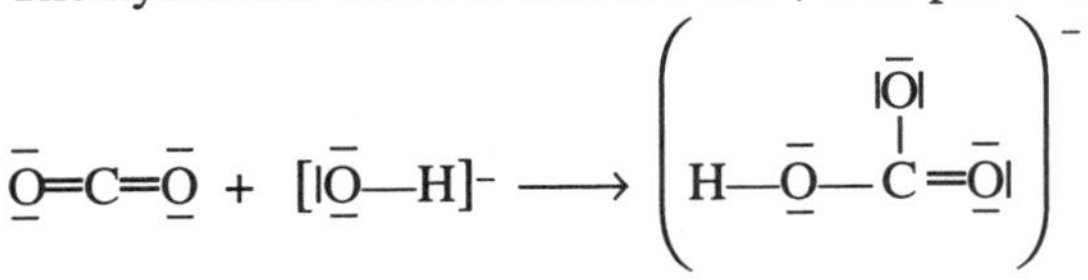

acid base

FEATURE PROBLEMS

A. **1.** From the combustion analysis we can determine the empirical formula. Note that the mass of oxygen is determined by difference.

$$\text{amount C} = 1.599 \text{ g } CO_2 \times \frac{1 \text{ mol } CO_2}{44.01 \text{ g } CO_2} \times \frac{1 \text{ mol C}}{1 \text{ mol } CO_2} = \frac{0.03633}{\text{mol C}} \times \frac{12.011 \text{ g C}}{1 \text{ mol C}} = 0.4364 \text{ g C}$$

$$\text{amount H} = 0.327 \text{ g } H_2O \times \frac{1 \text{ mol } H_2O}{18.02 \text{ g } H_2O} \times \frac{2 \text{ mol H}}{1 \text{ mol } H_2O} = \frac{0.03629}{\text{mol H}} \times \frac{1.008 \text{ g H}}{1 \text{ mol H}} = 0.03658 \text{ g H}$$

$$\text{amount O} = (1.054 \text{ g sample} - 0.4364 \text{ g C} - 0.03568 \text{ g H}) \times \frac{1 \text{ mol O}}{16.00 \text{ g O}} = 0.0364 \text{ mol O}$$

There are equal moles of the three elements. The empirical formula is CHO.
The freezing-point depression data are used to determine the molar mass.

$$\Delta T_f = -K_f\, m \qquad m = \frac{\Delta T_f}{-K_f} = \frac{-0.82\ °C}{-3.90\ °C/m} = 0.21\ m$$

$$\text{amount of solute} = 25.10 \text{ g H}_2\text{O solvent} \times \frac{1 \text{ kg}}{1000 \text{ g}} \times \frac{0.21 \text{ mol solute}}{1 \text{ kg solvent}} = 0.0053 \text{ mol}$$

$$\mathcal{M} = \frac{0.615 \text{ g solute}}{0.0053 \text{ mol solute}} = 1.2 \times 10^2 \text{ g/mol}$$

The formula weight of the empirical formula is: 12.0 g C + 1.0 g H + 16.0 g O = 29.0 g/mol

There thus are four empirical units in a molecule, the molecular formula is $C_4H_4O_4$, and the molar mass is 116.0 g/mol.

2. We determine the mass of maleic acid that reacts with one mol OH⁻.

$$\frac{\text{mass}}{\text{mol OH}^-} = \frac{0.4250 \text{ g maleic acid}}{34.03 \text{ mL base} \times \frac{1 \text{ L}}{1000 \text{ mL}} \times \frac{0.2152 \text{ mol KOH}}{1 \text{ L base}}} = 58.03 \text{ g/mol OH}^-$$

This means that one mole of maleic acid (116.0 g/mol) reacts with two moles of hydroxide ion. Maleic acid is a diprotic acid: $H_2C_4H_2O_4$.

3. Maleic acid will have two —COOH groups. This leaves C_2H_4 in the rest of the molecule. A plausible condensed structural formula is $HOOCCH_2CH_2COOH$. A full structural formula is:

$$
\begin{array}{ccccccc}
 & & O & H & H & O & \\
 & & \| & | & | & \| & \\
H & - O - & C & - C & - C & - C & - O - H \\
 & & & | & | & & \\
 & & & H & H & &
\end{array}
$$

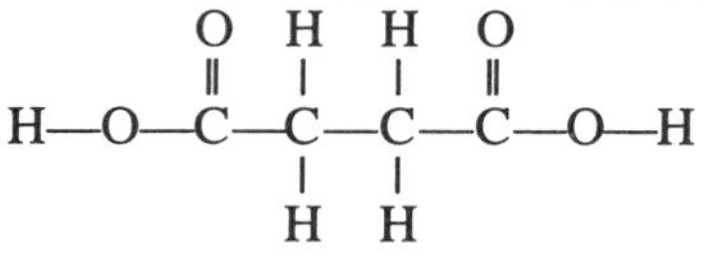

4. We first determine $[H_3O^+] = 10^{-pH} = 10^{-1.80} = 0.016 \text{ M}$ and then the initial concentration of acid.

$$[(CH_2COOH)_2]_i = \frac{0.215 \text{ g}}{50.00 \text{ mL}} \times \frac{1000 \text{ mL}}{1 \text{ L}} \times \frac{1 \text{ mol}}{116.0 \text{ g}} = 0.0371 \text{ M}$$

We use the first ionization to determine the value of K_{a_1}

Equation:	$(CH_2COOH)_2 + H_2O \rightleftharpoons$	$H(CH_2COO)_2^-$	+	H_3O^+
Initial:	0.0371 M			
Changes:	–0.016 M	+0.016 M		+0.016 M
Equil:	0.021 M	0.016 M		0.016 M

$$K_{a_1} = \frac{[H(CH_2COO)_2^-][H_3O^+]}{[(CH_2COOH)_2]} = \frac{(0.016)(0.016)}{0.021} = 1.2 \times 10^{-2}$$

The second ionization constant could be determined if we had some way to determine the total concentration of all ions in solution.

5. Practically all the $[H_3O^+]$ arises from the first ionization.

Equation:	$(CH_2COOH)_2 + H_2O \rightleftharpoons$	$H(CH_2COO)_2^-$	+	H_3O^+
Initial:	0.0500 M			
Changes:	–x M	+x M		+x M
Equil:	(0.0500 – x) M	x M		x M

$$K_{a_1} = \frac{[H(CH_2COO)_2^-][H_3O^+]}{[(CH_2COOH)_2]} = \frac{x^2}{0.0500 - x} = 1.2 \times 10^{-2} \qquad \begin{array}{l} x^2 = 0.00060 - 0.012\,x \\ 0 = x^2 + 0.012\,x - 0.00060 \end{array}$$

$$x = \frac{-b \pm \sqrt{b^2 - 4ac}}{2a} = \frac{-0.012 \pm \sqrt{0.00014 + 0.0024}}{2} = 0.019 \text{ M} = [H_3O^+]$$

$$pH = -\log(0.019) = 1.72$$

B. We organize the solution around the balanced chemical equation, as we have done before.

Equation:	$HClO_2 + H_2O \rightleftharpoons$	ClO_2^-	+	H_3O^+
Initial:	0.500 M			
Changes:	–x M	+x M		+x M
Equil:	x M	x M		x M

$$K_{a_1} = \frac{[ClO_2^-][H_3O^+]}{[HClO_2]} = \frac{x^2}{0.500 - x} = 1.2 \times 10^{-2} \approx \frac{x^2}{0.500} \qquad \text{Assuming } x \ll 0.500 \text{ M}$$

$x = \sqrt{x^2} = \sqrt{0.500 \times 1.2 \times 10^{-2}} = 0.077 \text{ M}$ This is not greatly smaller than 0.500 M.

Assume $x = 0.077$ $x = \sqrt{(0.500 - 0.077) \times 1.2 \times 10^{-2}} = 0.071 \text{ M}$ Try once more.

Assume $x = 0.071$ $x = \sqrt{(0.500 - 0.071) \times 1.2 \times 10^{-2}} = 0.072 \text{ M}$ One final time.

Assume $x = 0.072$ $x = \sqrt{(0.500 - 0.072) \times 1.2 \times 10^{-2}} = 0.072 \text{ M}$ Final result!

$[H_3O^+] = 0.072 \text{ M}$ $pH = -\log[H_3O^+] = -\log(0.072) = 1.14$

C. Here is the same problem attacked from a slightly different perspective, with the same result (except for rounding errors). $[H_3O^+] = 10^{-pH} = 10^{-2.15} = 7.1 \times 10^{-3}$ M

We assume initially that both ionization steps of H_2SO_4 go to completion. We then check this assumption.

$$H_2SO_4 \text{ concentration} = \frac{7.1 \times 10^{-3} \text{ mol } H_3O^+}{1 \text{ L soln}} \times \frac{1 \text{ mol } H_2SO_4}{2 \text{ mol } H_3O^+} = 3.6 \times 10^{-3} \text{ M } H_2SO_4$$

We see that $[H_2SO_4] < 0.001$ M is not true. We can see the extent of the second ionization by examining the second ionization constant.

$$K_{a2} = 0.011 = \frac{[H_3O^+][SO_4^{2-}]}{[HSO_4^-]} = \frac{7.1 \times 10^{-3}[SO_4^{2-}]}{[HSO_4^-]} \qquad \frac{[SO_4^{2-}]}{[HSO_4^-]} = \frac{0.011}{7.1 \times 10^{-3}} = 1.5$$

The ratio $[SO_4^{2-}]/[HSO_4^-]$ will be infinite for complete ionization of H_2SO_4. In the case of this solution, we can determine the concentrations of all ions with the concept of charge balance. That is, the sum of the positive ion concentrations must equal the sum of the negative ion concentrations.

$[H_3O^+] = 0.0071 \text{ M} = [HSO_4^-] + (2 \times [SO_4^{2-}]) + [OH^-]$

The "2" is present because the sulfate ion has twice the charge of the bisulfate ion. Now we know that $[OH^-]$ is quite small (1.4×10^{-12} M, actually) and can be safely neglected. And we also know that $[SO_4^{2-}] = 1.5 \times [HSO_4^-]$. We use these two relationships to obtain the following equation, and solve for $[HSO_4^-]$.

$$0.0071 \text{ M} = [HSO_4^-] + (2 \times 1.5)\,[HSO_4^-] = 4\,[HSO_4^-] \qquad [HSO_4^-] = \frac{0.0071 \text{ M}}{4} = 0.0018 \text{ M}$$

From this result we obtain $[SO_4^{2-}] = 1.5 \times [HSO_4^-] = 1.5 \times 0.0018 \text{ M} = 0.0027$ M

The total molarity of sulfuric acid then is the sum of the molarities of sulfate-ion containing species.

Sulfuric acid concentration $= [HSO_4^-] + [SO_4^{2-}] = 0.0018 \text{ M} + 0.0027 \text{ M} = 4.5 \times 10^{-3}$ M, about 25% more than our initial estimate.

C1. In this case, two equilibria must be satisfied simultaneously. The two have a common variable, $[H_3O^+] = z$.

Equation:	$HC_2H_3O_2 + H_2O \rightleftharpoons$	$C_2H_3O_2^-$ +	H_3O^+
Initial:	0.315 M	0 M	0 M
Changes:	$+x$ M	$+x$ M	$+z$ M
Equil:	$(0.315 - x)$M	x M	z M

$$K_a = \frac{[H_3O^+][C_2H_3O_2^-]}{[HC_2H_3O_2]} = \frac{xz}{0.315 - x} = 1.8 \times 10^{-5}$$

Equation:	$HCHO_2 + H_2O \rightleftharpoons$	CHO_2^- +	H_3O^+
Initial:	0.250 M	0 M	0 M
Changes:	$+y$ M	$+y$ M	$+z$ M
Equil:	$(0.250 - y)$M	y M	z M

$$K_a = \frac{[H_3O^+][CHO_2^-]}{[HCHO_2]} = \frac{yz}{0.250 - y} = 1.8 \times 10^{-4}$$

In this system, there are three variables, $x = [C_2H_3O_2^-]$, $y = [CHO_2^-]$, and $z = [H_3O^+]$. These three variables also represent the concentrations of the only charged species in the reaction, and thus $x + y = z$. We solve the two K_a expressions for x or y respectively. (Before we do so, however, we take advantages of the fact that x and y are quite small: $x << 0.315$ and $y << 0.250$.) Then we substitute these results into the expression for z, and solve to obtain a value of that variable.

$$xz = 1.8 \times 10^{-5}(0.315) = 5.7 \times 10^{-6} \qquad yz = 1.8 \times 10^{-4}(0.250) = 4.5 \times 10^{-5}$$

$$x = \frac{5.7 \times 10^{-6}}{z} \qquad y = \frac{4.5 \times 10^{-5}}{z}$$

$$z = x + y = \frac{5.7 \times 10^{-6}}{z} + \frac{4.5 \times 10^{-5}}{z} = \frac{5.07 \times 10^{-5}}{z} \qquad z = \sqrt{5.07 \times 10^{-5}} = 7.1 \times 10^{-3} \text{ M} = [H_3O^+]$$

$pH = -\log(7.1 \times 10^{-3}) = 2.15$ We see that our assumptions about the sizes of x and y must be valid, since each of them is smaller than z. (Remember that $x + y = z$.)

2. We first determine the initial concentration of each solute.

$$[NH_3] = \frac{12.5 \text{ g } NH_2}{0.375 \text{ L soln}} \times \frac{1 \text{ mol } NH_3}{17.03 \text{ g } NH_3} = 1.96 \text{ M} \qquad K_b = 1.8 \times 10^{-5}$$

$$[CH_3NH_2] = \frac{1.55 \text{ g } CH_3NH_2}{0.375 \text{ L soln}} \times \frac{1 \text{ mol } CH_3NH_2}{31.06 \text{ g } CH_3NH_2} = 0.113 \text{ M} \qquad K_b = 4.2 \times 10^{-4}$$

Now we solve simultaneously two equilibria, which have a common variable, $[OH^-] = z$.

Equation:	$NH_3(aq) + H_2O \rightleftharpoons$	$NH_4^+(aq)$ +	$OH^-(aq)$
Initial:	1.96 M		
Changes:	$-x$ M	$+x$ M	$+z$ M
Equil:	$(1.96 - x)$ M	x M	z M

$$K_b = \frac{[NH_4^+][OH^-]}{[NH_3]} = 1.8 \times 10^{-5} = \frac{xz}{1.96 - x} \approx \frac{xz}{1.96}$$

Equation:	$CH_3NH_2(aq) + H_2O \rightleftharpoons$	$CH_3NH_3^+(aq)$	$+ OH^-(aq)$		$K_b = \frac{[CH_3NH_3^+][OH^-]}{[CH_3NH_2]}$
Initial:	0.113 M				
Changes:	$-y$ M	$+y$ M	$+z$ M		$= 4.2 \times 10^{-4} = \frac{yz}{0.113 - y} \approx \frac{yz}{0.113}$
Equil:	$(0.113 - y)$ M	y M	z M		

In this system, there are three variables, $x = [NH_4^+]$, $y = [CH_3NH_3^+]$, and $z = [OH^-]$. These three variables also represent the concentrations of the only charged species in solution, and thus $x + y = z$. We solve the two K_b expressions for x and y respectively. Then we substitute these results into the expression for z, and solve to obtain the value of that variable.

$xz = 1.96 \times 1.8 \times 10^{-5} = 3.41 \times 10^{-5}$ $\qquad$ $yz = 0.113 \times 4.2 \times 10^{-5} = 4.7_5 \times 10^{-5}$

$$x = \frac{3.41 \times 10^{-5}}{z} \qquad y = \frac{4.7_5 \times 10^{-5}}{z}$$

$$z = x + y = \frac{3.41 \times 10^{-5}}{z} + \frac{4.7_5 \times 10^{-5}}{z} = \frac{8.16 \times 10^{-5}}{z} \qquad z^2 = 8.16 \times 10^{-5}$$

$z = 9.0 \times 10^{-3}$ M $= [OH^-]$ $\quad$ pOH $= -\log(9.0 \times 10^{-3}) = 2.05$ $\quad$ pH $= 14.00 - 2.05 = 11.95$

We see that our assumptions about x and y (that $x << 1.96$ M and $y << 0.113$ M) must be valid, since each of them is smaller than z. (Remember that $x + y = z$.)

3. In 1.0 M NH_4CN there are the following species: $NH_4^+(aq)$, $NH_3(aq)$, $CN^-(aq)$, $HCN(aq)$, $H_3O^+(aq)$, and $OH^-(aq)$. These species are related by the following six equations.

(1) $K_w = 1.0 \times 10^{-14} = [H_3O^+][OH^-]$ $\qquad$ $[OH^-] = \frac{1.0 \times 10^{-14}}{[H_3O^+]}$

(2) $K_a = 6.2 \times 10^{-10} = \frac{[H_3O^+][CN^-]}{[HCN]}$

(3) $K_b = 1.8 \times 10^{-5} = \frac{[NH_4^+][OH^-]}{[NH_3]}$

(4) $[NH_3] + [NH_4^+] = 1.0$ M $\qquad$ $[NH_3] = 1.0 - [NH_4^+]$

(5) $[HCN] + [CN^-] = 1.0$ M $\qquad$ $[HCN] = 1.0 - [CN^-]$

(6) $[NH_4^+] + [H_3O^+] = [CN^-] + [OH^-]$ or approximately $\qquad$ $[NH_4^+] = [CN^-]$

Equation (6) is the result of charge balance, that there must be the same quantity of positive and negative charge in the solution. The approximation is the result of remembering that not much H_3O^+ or OH^- will be formed as the result of hydrolysis of ions.

Substitute equation (4) into equation (3), and equation (5) into equation (2).

(3') $K_b = 1.8 \times 10^{-5} = \frac{[NH_4^+][OH^-]}{1.0 - [NH_4^+]}$ $\qquad$ (2') $K_a = 4.0 \times 10^{-10} = \frac{[H_3O^+][CN^-]}{1.0 - [CN^-]}$

Now substitute equation (6) into equation (2'), and equation (1) into equation (3').

(2") $K_a = 6.2 \times 10^{-10} = \frac{[H_3O^+][NH_4^+]}{1.0 - [NH_4^+]}$ $\qquad$ (3") $\frac{1.8 \times 10^{-5}}{1.0 \times 10^{-14}} = \frac{[NH_4^+]}{[H_3O^+](1.0 - [NH_4^+])}$

Now we solve both of these equations for $[NH_4^+]$.

('2): $6.2 \times 10^{-10} - 6.2 \times 10^{-10}[NH_4^+] = [H_3O^+][NH_4^+]$ $\qquad$ $[NH_4^+] = \frac{6.2 \times 10^{-10}}{6.2 \times 10^{-10} + [H_3O^+]}$

('3): $1.8 \times 10^9\,[H_3O^+] - 1.8 \times 10^9[H_3O^+][NH_4^+] = [NH_4^+]$ $\qquad$ $[NH_4^+] = \frac{1.8 \times 10^9\,[H_3O^+]}{1.00 + 1.8 \times 10^9[H_3O^+]}$

We equate the two results and solve for $[H_3O^+]$.

$$\frac{6.2 \times 10^{-10}}{6.2 \times 10^{-10} + [H_3O^+]} = \frac{1.8 \times 10^9\,[H_3O^+]}{1.00 + 1.8 \times 10^9[H_3O^+]}$$

$6.2 \times 10^{-10} + 1.1\,[H_3O^+] = 1.1\,[H_3O^+] + 1.8 \times 10^9\,[H_3O^+]^2$ $\qquad$ $6.2 \times 10^{-10} = 1.8 \times 10^9\,[H_3O^+]^2$

$[H_3O^+] = \sqrt{\frac{6.2 \times 10^{-10}}{1.8 \times 10^9}} = 5.9 \times 10^{-10}$ M $\qquad$ pH $= -\log(5.9 \times 10^{-10}) = 9.23$

Note that $[H_3O^+] = \sqrt{\frac{K_a \times K_w}{K_b}}$ *or* pH $= 0.500\,(pK_a + pK_w - pK_b)$

18 Additional Aspects of Acid–Base Equilibria

Practice Examples

1A We organize the solution around the balanced chemical equation, as we have done before.

Equation:	$HF(aq) + H_2O \rightleftharpoons$	$H_3O^+(aq)$ +	$F^-(aq)$
Initial:	0.500 M		
Changes:	$-x$ M	$+x$ M	$+x$ M
Equil:	$(0.500 - x)$ M	x M	x M

$K_a = \dfrac{[H_3O^+][F^-]}{[HF]} = \dfrac{(x)\,(x)}{0.500 - x} = 6.6 \times 10^{-4} \approx \dfrac{x^2}{0.500}$ assuming $x << 0.500$

$x = \sqrt{0.500 \times 6.6 \times 10^{-4}} = 0.018$ M One further cycle of approximations gives:

$x = \sqrt{(0.500 - 0.018) \times 6.6 \times 10^{-4}} = 0.018 \text{ M} = [H_3O^+]$ $[HF] = 0.500 \text{ M} - 0.018 \text{ M} = 0.482 \text{ M}$

We recognize that 0.100 M HCL means $[H_3O^+]_{initial} = 0.100$ M, because HCl is a strong acid.

Equation:	$HF(aq) + H_2O \rightleftharpoons$	$H_3O^+(aq)$ +	$F^-(aq)$
Initial:	0.100 M		0.100
Changes:	$-x$ M	$+x$ M	$+x$ M
Equil:	$(0.500 - x)$ M	x M	$(0.100 + x)$ M

$K_a = \dfrac{[H_3O^+][F^-]}{[HF]} = \dfrac{(x)\,(0.100 + x)}{0.500 - x} = 6.6 \times 10^{-4} \approx \dfrac{0.100x}{0.500}$ assuming $x << 0.100$

$x = \dfrac{6.6 \times 10^{-4} \times 0.500}{0.100} = 3.3 \times 10^{-3} \text{ M} = [H_3O^+]$ The assumption is valid. $[HF] = 0.500 \text{ M} - 0.003 \text{ M} = 0.497 \text{ M}$

1B From Example 17-6 *in the text*, we know that $[H_3O^+] = [C_2H_3O_2^-] = 1.3 \times 10^{-3}$ M in 0.100 M $HC_2H_3O_2$. We base our calculation, as usual, on the balanced chemical equation. The concentration of H_3O^+ from the added HCl is represented by x.

Equation:	$HC_2H_3O_2(aq) + H_2O \rightleftharpoons$	$H_3O^+(aq)$ +	$C_2H_3O_2^-(aq)$
Initial:	0.100 M		
Changes:	−0.00010 M	+0.00010 M	+0.00010 M
From HCl:		$+x$ M	
Equil:	0.100 M	$(0.00010 + x)$ M	+0.00010 M

$K_a = \dfrac{[H_3O^+][C_2H_3O_2^-]}{[HC_2H_3O_2]} = \dfrac{(0.00010 + x)\,0.00010}{0.100} = 1.8 \times 10^{-5}$

$0.0010 + x = \dfrac{1.8 \times 10^{-5} \times 0.100}{0.00010} = 0.018 \text{ M}$ $x = 0.018 \text{ M} - 0.00010 \text{ M} = 0.018 \text{ M}$

$\text{volume of 12 M HCl} = 1.00 \text{ L} \times \dfrac{0.018 \text{ mol } H_3O^+}{1 \text{ L}} \times \dfrac{1 \text{ mol HCl}}{1 \text{ mol } H_3O^+} \times \dfrac{1 \text{ L soln}}{12 \text{ mol HCl}} \times \dfrac{1000 \text{ mL}}{1 \text{ L}} \times \dfrac{1 \text{ drop}}{0.050 \text{ mL}}$

$= 30.$ drops

Since 30. drops corresponds to 1.5 mL of 12 M solution, we see that the volume of solution does indeed remain 1.00 L after dilution of the 12 M HCl.

2A We again organize the solution around the balanced chemical equation.

Equation:	$HCHO_2(aq) + H_2O \rightleftharpoons$	$CHO_2^-(aq)$ +	$H_3O^+(aq)$
Initial:	0.100 M	0.150 M	
Changes:	$-x$ M	$+x$ M	$+x$ M
Equil:	$(0.100 - x)$ M	$(0.150 + x)$ M	x M

$$K_a = \frac{[CHO_2^-][H_3O^+]}{[HCHO_2]} = \frac{(0.150 + x)(x)}{0.100 - x} = 1.8 \times 10^{-4} \approx \frac{0.150\,x}{0.100} \quad \text{assuming } x << 0.100$$

$$x = \frac{0.100 \times 1.8 \times 10^{-4}}{0.150} = 1.2 \times 10^{-4}\text{ M} = [H_3O^+] \qquad [CHO_2^-] = 0.150\text{ M} + 0.00012\text{ M} = 0.150\text{ M}$$

2B This problem is almost the same as the previous one, except that the anion now is being added to the weak acid solution rather than the cation, the hydronium ion. We let x represent the concentration of acetate ion from the added sodium acetate. Notice that sodium acetate is a strong electrolyte, completely dissociated in aqueous solution. $[H_3O^+] = 10^{-pH} = 10^{-5.00} = 1.0 \times 10^{-5}\text{ M} = 0.000010\text{ M}$

Equation:	$HC_2H_3O_2(aq) + H_2O \rightleftharpoons$	$C_2H_3O_2^-(aq)$ +	$H_3O^+(aq)$
Initial:	0.100 M		
Changes:	−0.000010 M	+0.000010 M	+0.000010 M
From NaAc:		$+x$ M	
Equil:	0.100 M	$(0.000010 + x)$ M	+0.000010 M

$$K_a = \frac{[H_3O^+][C_2H_3O_2^-]}{[HC_2H_3O_2]} = \frac{0.000010\,(0.000010 + x)}{0.100} = 1.8 \times 10^{-5}$$

$$0.00010 + x = \frac{1.8 \times 10^{-5} \times 0.100}{0.000010} = 0.18\text{ M} \qquad x = 0.18\text{ M} - 0.000010\text{ M} = 0.18\text{ M}$$

$$\text{mass of } NaC_2H_3O_2 = 1.00\text{ L} \times \frac{0.18\text{ mol } C_2H_3O_2^-}{1\text{ L}} \times \frac{1\text{ mol } NaC_2H_3O_2}{1\text{ mol } C_2H_3O_2^-} \times \frac{82.03\text{ g } NaC_2H_3O_2}{1\text{ mol } NaC_2H_3O_2}$$
$$= 15\text{ g } NaC_2H_3O_2$$

3A A strong acid dissociates completely, and essentially is a source of H_3O^+. $NaC_2H_3O_2$ also dissociates completely in solution. The hydronium ion and the acetate ion react to form acetic acid:
$H_3O^+(aq) + C_2H_3O_2^-(aq) \rightleftharpoons HC_2H_3O_2(aq) + H_2O$ All that is necessary for a buffer is that there be approximately equal amounts of a weak acid and its conjugate base in solution. This will be achieved if we add an amount of HCl equal to approximately half the original amount of acetate ion.

3B HCl dissociates completely and serves as a source of hydronium ion. This reacts with ammonia to form ammonium ion: $NH_3(aq) + H_3O^+(aq) \rightleftharpoons NH_4^+(aq) + H_2O$. Because a buffer contains approximately equal amounts of a weak base (NH_3) and its conjugate acid (NH_4^+), to prepare a buffer we simply add an amount of HCl equal to approximately half the amount of $NH_3(aq)$ initially present.

4A We first find the formate ion concentration, remembering that $NaCHO_2$ is a strong electrolyte, existing in solution as $Na^+(aq)$ and $CHO_2^-(aq)$.

$$[CHO_2^-] = \frac{23.1\text{ g } NaCHO_2}{500.0\text{ mL soln}} \times \frac{1000\text{ mL}}{1\text{ L}} \times \frac{1\text{ mol } NaCHO_2}{68.01\text{ g } NaCHO_2} \times \frac{1\text{ mol } CHO_2^-}{1\text{ mol } NaCHO_2} = 0.679\text{ M}$$

As usual, the solution to the problem is organized around the balanced chemical equation.

Equation:	$HCHO_2(aq) + H_2O \rightleftharpoons$	$CHO_2^-(aq)$ +	$H_3O^+(aq)$
Initial:	0.432 M	0.679 M	
Changes:	$-x$ M	$+x$ M	$+x$ M
Equil:	$(0.432 - x)$ M	$(0.679 + x)$ M	$+x$ M

$$K_a = \frac{[H_3O^+][CHO_2^-]}{[HCHO_2]} = \frac{x\,(0.679 + x)}{0.432 - x} = 1.8 \times 10^{-4} \approx \frac{0.679\,x}{0.432} \qquad x = \frac{0.432 \times 1.8 \times 10^{-4}}{0.679}$$

This gives $x = 1.1 \times 10^{-4}\text{ M} = [H_3O^+]$ The assumption, that $x << 0.432$, is clearly correct.
$pH = -\log[H_3O^+] = -\log(1.1 \times 10^{-4}) = 3.96$

4B The concentrations of the components in the 100.0 mL of buffer solution are found by dilution. Remember that $NaC_2H_3O_2$ is a strong electrolyte, existing in solution as $Na^+(aq)$ and $C_2H_3O_2^-(aq)$.

$$[HC_2H_3O_2] = 0.200\text{ M} \times \frac{63.0\text{ mL}}{100.0\text{ mL}} = 0.126\text{ M} \qquad [C_2H_3O_2^-] = 0.200\text{ M} \times \frac{37.0\text{ mL}}{100.0\text{ mL}} = 0.0740\text{ M}$$

As usual, the solution to the problem is organized around the balanced chemical equation.

Equation:	$HC_2H_3O_2(aq) + H_2O \rightleftharpoons$	$C_2H_3O_2^-(aq)$ +	$H_3O^+(aq)$
Initial:	0.126 M	0.0740 M	
Changes:	$-x$ M	$+x$ M	$+x$ M
Equil:	$(0.126 - x)$ M	$(0.0740 + x)$ M	$+x$ M

$$K_a = \frac{[H_3O^+][C_2H_3O_2^-]}{[HC_2H_3O_2]} = \frac{x\,(0.0740 + x)}{0.126 - x} = 1.8 \times 10^{-5} \approx \frac{0.0740\,x}{0.126}$$

$$x = \frac{1.8 \times 10^{-5} \times 0.126}{0.0740} = 3.1 \times 10^{-5} \text{ M} = [H_3O^+] \qquad pH = -\log [H_3O^+] = -\log 3.1 \times 10^{-5} = 4.51$$

Notice that our assumption is valid: $x << 0.0740 < 0.126$ and x can be neglected in adding to or subtracting from them.

5A We know the initial concentration of NH_3 in the buffer solution and can use the pH to find the equilibrium $[OH^-]$. The rest of the solution is organized around the balanced chemical equation. Our first goal is to determine the inital concentration of NH_4^+.

$pOH = 14.00 - pH = 14.00 - 9.00 = 5.00 \qquad [OH^-] = 10^{-pOH} = 10^{-5.00} = 1.0 \times 10^{-5}$ M

Equation:	$NH_3(aq) + H_2O \rightleftharpoons$	$NH_4^+(aq)$ +	$OH^-(aq)$
Initial:	0.35 M	x M	
Changes:	-1.0×10^{-5} M	$+1.0 \times 10^{-5}$ M	$+1.0 \times 10^{-5}$ M
Equil:	$(0.35 - 1.0 \times 10^{-5})$ M	$(x + 1.0 \times 10^{-5})$ M	1.0×10^{-5} M

$$K_b = \frac{[NH_4^+][OH^-]}{[NH_3]} = 1.8 \times 10^{-5} = \frac{(x + 1.0 \times 10^{-5})(1.0 \times 10^{-5})}{0.35 - 1.0 \times 10^{-5}} \approx \frac{1.0 \times 10^{-5} \times x}{0.35}$$

We assumed $x >> 1.0 \times 10^{-5}$ THEN $x = \frac{0.35 \times 1.8 \times 10^{-5}}{1.0 \times 10^{-5}} = 0.63$ M = initial NH_4^+ concentration

$$(NH_4)_2SO_4 \text{ mass} = 0.500 \text{ L} \times \frac{0.63 \text{ mol } NH_4^+}{1 \text{ L soln}} \times \frac{1 \text{ mol } (NH_4)_2SO_4}{2 \text{ mol } NH_4^+} \times \frac{132.1 \text{ g } (NH_4)_2SO_4}{1 \text{ mol } (NH_4)_2SO_4} = 20.8 \text{ g } (NH_4)_2SO_4$$

5B The solution in Figure 18-5(b) is composed of 300.0 mL of 0.250 M HCl—which thus has $[H_3O^+] = 0.250$ M—to which 33.05 g $NaC_2H_3O_2 \cdot 3H_2O$ has been added. The $NaC_2H_3O_2$ is a strong electrolyte that exists in solution as $Na^+(aq)$ and $C_2H_3O_2^-(aq)$ ions. First we calculate the initial $[C_2H_3O_2^-]$, assuming that the solution's volume remains at 300. mL.

$$[C_2H_3O_2^-] = \frac{33.05 \text{ g } NaC_2H_3O_2 \cdot 3H_2O}{300. \text{ mL soln}} \times \frac{1000 \text{ mL}}{1 \text{ L}} \times \frac{1 \text{ mol } NaC_2H_3O_2 \cdot 3H_2O}{136.08 \text{ g } NaC_2H_3O_2 \cdot 3H_2O} \times \frac{1 \text{ mol } C_2H_3O_2^-}{1 \text{ mol solute}}$$
$$= 0.8095 \text{ M}$$

We organize this information around the balanced chemical equation, as before. We could use the equation from Practice Example 18-3, and then consider the ionization of the weak acid, but there is really no benefit in doing so. We recognize that as much hydronium ion (a strong acid) as possible will react to produce the weak acid, acetic acid.

Equation:	$HC_2H_3O_2(aq) + H_2O \rightleftharpoons$	$C_2H_3O_2^-(aq)$ +	$H_3O^+(aq)$
Initial:	0 M	0.8095 M	0.250
Form HAc:	+0.250 M	−0.250 M	−0.250 M
	0.250 M	0.560 M	≈0 M
Changes:	$-x$ M	$+x$ M	$+x$ M
Equil:	$(0.250 - x)$ M	$(0.560 + x)$ M	$+x$ M

$$K_a = \frac{[H_3O^+][C_2H_3O_2^-]}{[HC_2H_3O_2]} = \frac{x\,(0.560 + x)}{0.250 - x} = 1.8 \times 10^{-5} \approx \frac{0.560\,x}{0.250}$$

$$x = \frac{1.8 \times 10^{-5} \times 0.250}{0.560} = 8.0 \times 10^{-6} \text{ M} = [H_3O^+] \qquad pH = -\log [H_3O^+] = -\log 8.0 \times 10^{-6} = 5.10$$

Notice that our assumption is valid: $x << 0.250 < 0.560$. The algebraic solution could be replaced with the Henderson-Hasselbalch equation, since the ratio $[C_2H_3O_2^-]/[HC_2H_3O_2]$ falls between 0.1 and 10.

$$pH = pK_a + \log \frac{[C_2H_3O_2^-]}{[HC_2H_3O_2]} = 4.74 + \log \frac{0.560}{0.250} = 5.09$$

6A **(a)** For formic acid, $pK_a = -\log(1.8 \times 10^{-4}) = 3.74$. The Henderson-Hasselbalch equation produces the pH of the original buffer solution: $\quad pH = pK_a + \log \frac{[CHO_2^-]}{[HCHO_2]} = 3.74 + \log \frac{0.350}{0.550} = 3.54$

(b) The added acid reacts with formate ion and produces formic acid. Each mole/L of added acid consumes one mole/L of formate ion and forms one mole/L of formic acid: $CHO_2^- + H_3O^+ \longrightarrow HCHO_2 + H_2O$. Thus, $[CHO_2^-] = 0.350 \text{ M} - 0.0050 \text{ M} = 0.345$ M and $[HCHO_2] = 0.550 \text{ M} + 0.0050 \text{ M} = 0.555$ M. We again use the Henderson-Hasselbalch equation.

$$pH = pK_a + \log \frac{[CHO_2^-]}{[HCHO_2]} = 3.74 + \log \frac{0.345}{0.555} = 3.53$$

(c) Added base reacts with formic acid and produces formate ion. Each mole/L of added base consumes one mole/L of formic acid and forms one mole/L of formate ion: $HCHO_2 + OH^- \longrightarrow CHO_2^- + H_2O$. Thus, $[CHO_2^-] = 0.350\text{ M} + 0.0050\text{ M} = 0.355\text{ M}$ and $[HCHO_2] = 0.550\text{ M} - 0.0050\text{ M} = 0.545\text{ M}$. We again use the Henderson-Hasselbalch equation.

$$pH = pK_a + \log\frac{[CHO_2^-]}{[HCHO_2]} = 3.74 + \log\frac{0.355}{0.545} = 3.55$$

6B The buffer cited has the same concentration of weak acid and its anion as does the buffer of Practice Example 18-5B. Our goal is pH = 5.03 or $[H_3O^+] = 10^{-pH} = 10^{-5.03} = 9.3 \times 10^{-6}$ M. Adding strong acid (H_3O^+), of course, produces $HC_2H_3O_2$ at the expense of $C_2H_3O_2^-$; it drives the reaction to the left. Again, we organize the data around the balanced chemical equation.

Equation:	$HC_2H_3O_2(aq) + H_2O \rightleftharpoons$	$C_2H_3O_2^-(aq)$	+	$H_3O^+(aq)$
Initial:	0.250 M	0.560 M		9.3×10^{-6} M
Add acid:				$+y$ M
Form HAc:	$+y$ M	$-y$ M		$-y$ M
	$(0.250 + y)$ M	$(0.560 - y)$ M		≈ 0 M
Changes:	$+x$ M	$-x$ M		$-x$ M
Equil:	$(0.250 + y - x)$ M	$(0.560 - y + x)$ M		9.3×10^{-6} M

$$K_a = \frac{[H_3O^+][C_2H_3O_2^-]}{[HC_2H_3O_2]} = \frac{9.3 \times 10^{-6}\,(0.560 - y + x)}{0.250 + y - x} = 1.8 \times 10^{-5} \approx \frac{9.3 \times 10^{-6}\,(0.560 - y)}{0.250 + y}$$

$$\frac{1.8 \times 10^{-5} \times (0.250 + y)}{9.3 \times 10^{-6}} = 0.484 + 1.94\,y = 0.560 - y \qquad y = \frac{0.560 - 0.484}{1.94 + 1.00} = 0.026\text{ M}$$

Notice that our assumption is valid: $x << 0.250 + y\ (= 0.276) < 0.560 - y\ (= 0.534)$.

$$\text{volume } HNO_3 = 300.0\text{ mL buffer} \times \frac{0.026\text{ mmol } H_3O^+}{1\text{ mL buffer}} \times \frac{1\text{ mL } HNO_3(aq)}{6.0\text{ mmol } H_3O^+} = 1.3\text{ mL of } 6.0\text{ M } HNO_3$$

The algebraic solution could be replaced with the Henderson-Hasselbalch equation, since the final pH falls within one pH unit of the pK_a of acetic acid. We let z indicate the increase in $[HC_2H_3O_2]$, and also the decrease in $[C_2H_3O_2^-]$

$$pH = pK_a + \log\frac{[C_2H_3O_2^-]}{[HC_2H_3O_2]} = 4.74 + \log\frac{0.560 - z}{0.250 + z} = 5.03 \qquad \frac{0.560 - z}{0.250 + z} = 10^{5.03 - 4.74} = 1.95$$

$$0.560 - z = 1.95\,(0.250 + z) = 0.488 + 1.95\,z \qquad z = \frac{0.560 - 0.488}{1.95 + 1.00} = 0.024\text{ M}$$

This is—and should be—almost exactly the value of y we obtained by the other method. The differences are due to imprecision: rounding errors.

7A **(a)** The initial pH is the pH of 0.150 M HCl, which we obtain from $[H_3O^+]$ of that strong acid solution.

$$[H_3O^+] = \frac{0.150\text{ mol HCl}}{1\text{ L soln}} \times \frac{1\text{ mol } H_3O^+}{1\text{ mol HCl}} = 0.150\text{ M}, \quad pH = -\log[H_3O^+] = -\log(0.150) = 0.824$$

(b) To determine $[H_3O^+]$ and then pH at the 50.0% point, we need the volume of the solution and the amount of H_3O^+ unreacted. First we calculate the amount of hydronium ion present and then the volume of base solution needed for its complete neutralization.

$$\text{amount } H_3O^+ = 25.00\text{ mL} \times \frac{0.150\text{ mmol HCl}}{1\text{ mL soln}} \times \frac{1\text{ mmol } H_3O^+}{1\text{ mmol HCl}} = 3.75\text{ mmol } H_3O^+$$

$$\text{acid volume} = 3.75\text{ mmol } H_3O^+ \times \frac{1\text{ mmol } OH^-}{1\text{ mmol } H_3O^+} \times \frac{1\text{ mmol NaOH}}{1\text{ mmol } OH^-} \times \frac{1\text{ mL titrant}}{0.250\text{ mmol NaOH}} = 15.0\text{ mL titrant}$$

At the 50.0% point, half (1.88 mmol H_3O^+) will remain unreacted and only half (7.50 mL titrant) of the titrant solution will be added. From this information, and the original 25.00-mL volume of the solution, we calculate $[H_3O^+]$ and then pH.

$$[H_3O^+] = \frac{1.88\text{ mmol } OH^-\text{ left}}{25.00\text{ mL original} + 7.50\text{ mL titrant}} = 0.0578\text{ M} \qquad pH = -\log(0.0578) = 1.238$$

(c) Since this is the titration of a strong base by a strong acid, at the neutralization point pH = 7.00. This is a solution of NaCl(aq), in which neither ion hydrolyzes.

(d) Beyond the end point, the added titrant contains the only species that determine solution pH. The volume of the solution is 25.00 mL + 1.00 mL = 26.00 mL. The amount of hydroxide ion in the excess titrant is calculated and used to determine $[OH^-]$, from which pH is computed.

$\text{amount } OH^- = 1.00 \text{ mL} \times \dfrac{0.250 \text{ mmol NaOH}}{1 \text{ mL}} = 0.250 \text{ mmol } OH^-$ $\quad [OH^-] = \dfrac{0.250 \text{ mmol } OH^-}{26.00 \text{ mL}} = 0.00962 \text{ M}$

$pOH = -\log(0.00962) = 2.017 \qquad pH = 14.00 - 2.017 = 11.98$

7B **(a)** The initial pH is simply the pH of 0.00812 M $Ba(OH)_2$, which we obtain from $[OH^-]$ of that solution, the solution of a strong base. $\quad [OH^-] = \dfrac{0.00812 \text{ mol } Ba(OH)_2}{1 \text{ L soln}} \times \dfrac{2 \text{ mol } OH^-}{1 \text{ mol } Ba(OH)_2} = 0.01624 \text{ M}$

$pOH = -\log[OH^-] = -\log(0.0162) = 1.790 \qquad pH = 14.00 - pOH = 14.00 - 1.790 = 12.21$

(b) To determine $[OH^-]$ and then pH at the 50.0% point, we need the volume of the solution and the amount of OH^- unreacted. First we calculate the amount of hydroxide ion present and then the volume of acid solution needed for its complete neutralization.

$$\text{amount } OH^- = 50.00 \text{ mL} \times \frac{0.00812 \text{ mmol } Ba(OH)_2}{1 \text{ mL soln}} \times \frac{2 \text{ mmol } OH^-}{1 \text{ mmol } Ba(OH)_2} = 0.812 \text{ mmol } OH^-$$

$$\text{acid volume} = 0.812 \text{ mmol } OH^- \times \frac{1 \text{ mmol } H_3O^+}{1 \text{ mmol } OH^-} \times \frac{1 \text{ mmol HCl}}{1 \text{ mmol } H_3O^+} \times \frac{1 \text{ mL titrant}}{0.0250 \text{ mmol HCl}}$$

$$= 32.48 \text{ mL titrant}$$

At the 50.0% point, half (0.406 mmol OH^-) will remain unreacted and only half (16.24 mL titrant) of the titrant solution will be added. From this information, and the original 50.00-mL volume of the solution, we calculate $[OH^-]$ and then pH.

$$[OH^-] = \frac{0.406 \text{ mmol } OH^- \text{ left}}{50.00 \text{ mL original} + 16.24 \text{ mL titrant}} = 0.00613 \text{ M}$$

$pOH = -\log(0.00613) = 2.213 \qquad pH = 14.00 - pOH = 11.79$

(c) Since this is the titration of a strong base by a strong acid, at the neutralization point pH = 7.00. This is a solution of $BaCl_2(aq)$, in which neither ion hydrolyzes.

8A **(a)** The initial pH is just that of 0.150 M HF. [The initial solution contains $20.00 \text{ mL} \times \dfrac{0.150 \text{ mmol HF}}{1 \text{ mL}}$ = 3.00 mmol HF. Additionally, $pK_a = -\log(6.6 \times 10^{-4}) = 3.18$.]

Equation:	$HF(aq) + H_2O \rightleftharpoons$	$H_3O^+(aq) +$	$F^-(aq)$
Initial:	0.150 M		
Changes:	$-x$ M	$+x$ M	$+x$ M
Equil:	$(0.150 - x)$ M	x M	x M

$$K_b = \frac{[H_3O^+][F^-]}{[HF]} = \frac{x \cdot x}{0.150 - x} \approx \frac{x^2}{0.150} = 6.6 \times 10^{-4}$$

$$x = \sqrt{0.150 \times 6.6 \times 10^{-4}} = 9.9 \times 10^{-3} \text{ M}$$

$[H_3O^+] = 9.9 \times 10^{-3}$ M $\qquad$ The assumption is valid $\qquad pH = -\log(9.9 \times 10^{-3}) = 2.00$

(b) When the titration is 25.0% complete, there are $(0.25 \times 3.00 =)$ 0.75 mmol F^- for every 3.00 mmol HF that were present initially. $(3.00 - 0.75 =)$ 2.25 mmol HF remain untitrated. We designate the solution volume (the volume holding these 3.00 mmol total) as V and use the Henderson-Hasselbalch equation.

$$pH = pK_a + \log\frac{[F^-]}{[HF]} = 3.18 + \log\frac{0.75 \text{ mmol}/V}{2.25 \text{ mmol}/V} = 2.70$$

(c) At the midpoint of the titration of a weak base, $pH = pK_a = 3.18$

(d) At the endpoint of the titration, the anion of the weak acid hydrolyzes. We calculate the amount of that anion and the volume of the solution in order to calculate its initial concentration.

$$\text{amount } F^- = 25.00 \text{ mL} \times \frac{0.150 \text{ mmol HF}}{1 \text{ mL soln}} \times \frac{1 \text{ mmol } F^-}{1 \text{ mmol HF}} = 3.75 \text{ mmol } F^-$$

$$\text{volume titrant} = 3.75 \text{ mmol HF} \times \frac{1 \text{ mmol } OH^-}{1 \text{ mmol HF}} \times \frac{1 \text{ mL titrant}}{0.250 \text{ mmol } OH^-} = 15.0 \text{ mL titrant}$$

$$[F^-] = \frac{3.75 \text{ mmol } NH_4^+}{25.00 \text{ mL original volume} + 15.0 \text{ mL titrant}} = 0.0938 \text{ M}$$

We organize the solution of the hydrolysis problem around its balanced chemical equation.

Equation:	$F^-(aq) + H_2O \rightleftharpoons$	$HF(aq) +$	$OH^-(aq)$
Initial:	0.0938 M		
Changes:	$-x$ M	$+x$ M	$+x$ M
Equil:	$(0.0938 - x)$ M	x M	x M

$$K_a = \frac{[HF][OH^-]}{[F^-]} = \frac{K_w}{K_a} = \frac{1.0 \times 10^{-14}}{6.6 \times 10^{-4}} = 1.5 \times 10^{-11} = \frac{x \cdot x}{0.0938 - x} \approx \frac{x^2}{0.0938}$$

$x = \sqrt{0.0938 \times 1.5 \times 10^{-11}} = 1.2 \times 10^{-6} \text{ M} = [OH^-]$ $\qquad$ The assumption is valid.

$pOH = -\log(1.2 \times 10^{-6}) = 5.92 \qquad pH = 14.00 - pOH = 14.00 - 5.92 = 8.08$

8B **(a)** The initial pH is simply that of 0.106 M NH_3.

Equation:	$NH_3(aq) + H_2O$ $\rightleftharpoons$	$NH_4^+(aq)$ +	$OH^-(aq)$
Initial:	0.106 M		
Changes:	$-x$ M	$+x$ M	$+x$ M
Equil:	$(0.106 - x)$ M	x M	x M

$$K_b = \frac{[NH_4^+][OH^-]}{[NH_3]} = \frac{x \cdot x}{0.106 - x} \approx \frac{x^2}{0.106} = 1.8 \times 10^{-5}$$

$$x = \sqrt{0.106 \times 1.8 \times 10^{-5}} = 1.4 \times 10^{-3}\ M$$

$[OH^-] = 1.4 \times 10^{-3}$ M The assumption is valid. pOH = $-\log(0.0014) = 2.86$ pH = 11.14

(b) When the titration is 25.0% complete, there are 25.0 mmol NH_4^+ for every 100.0 mmol NH_3 that were present initially. 75.0 mmol NH_3 remain untitrated. We designate the solution volume (the volume holding these 100.0 mmol total) as V and use a version of the Henderson-Hasselbalch equation.

$$pOH = pK_b + \log\frac{[NH_4^+]}{[NH_3]} = 4.74 + \log\frac{25.0\ mmol/V}{75.0\ mmol/V} = 4.26 \qquad pH = 14.00 - 4.26 = 9.74$$

(c) At the midpoint of the titration of a weak base, pOH = $pK_b = 4.74$ pH = 14.00 – 4.74 = 9.26

(d) At the endpoint of the titration, the cation of the weak base hydrolyzes. We calculate the amount of that cation and the volume of the solution in order to calculate its initial concentration.

$$\text{amount } NH_4^+ = 50.00\ mL \times \frac{0.106\ mmol\ NH_3}{1\ mL\ soln} \times \frac{1\ mmol\ NH_4^+}{1\ mmol\ NH_3} = 5.30\ mmol\ NH_4^+$$

$$\text{volume titrant} = 5.30\ mmol\ NH_3 \times \frac{1\ mmol\ H_3O^+}{1\ mmol\ NH_3} \times \frac{1\ mL\ titrant}{0.225\ mmol\ H_3O^+} = 23.6\ mL\ titrant$$

$$[NH_4^+] = \frac{5.30\ mmol\ NH_4^+}{50.00\ mL\ original\ volume + 23.6\ mL\ titrant} = 0.0720\ M$$

We organize the solution of the hydrolysis problem around its balanced chemical equation.

Equation:	$NH_4^+(aq) + H_2O$ $\rightleftharpoons$	$NH_3(aq)$ +	$H_3O^+(aq)$
Initial:	0.0720 M		
Changes:	$-x$ M	$+x$ M	$+x$ M
Equil:	$(0.0720 - x)$ M	x M	x M

$$K_b = \frac{[NH_3][H_3O^+]}{[NH_4^+]} = \frac{K_w}{K_b} = \frac{1.0 \times 10^{-14}}{1.8 \times 10^{-5}} = 5.6 \times 10^{-10} = \frac{x \cdot x}{0.0720 - x} \approx \frac{x^2}{0.0720}$$

$x = \sqrt{0.0720 \times 5.6 \times 10^{-10}} = 6.3 \times 10^{-6}\ M = [H_3O^+]$ The assumption is valid.

$pH = -\log(6.3 \times 10^{-6}) = 5.20$

9A The acidity of the solution is principally due to the hydrolysis of the carbonate ion, considered first.

Equation:	$CO_3^{2-}(aq) + H_2O$ $\rightleftharpoons$	$HCO_3^-(aq)$ +	$OH^-(aq)$
Initial:	1.0 M		
Changes:	$-x$ M	$+x$ M	$+x$ M
Equil:	$(1.0 - x)$ M	x M	x M

$$K_b = \frac{K_w}{K_a(HCO_3^-)} = \frac{1.0 \times 10^{-14}}{4.7 \times 10^{-11}} = 2.1 \times 10^{-4} = \frac{[HCO_3^-][OH^-]}{[CO_3^{2-}]} = \frac{x \cdot x}{1.0 - x} \approx \frac{x^2}{1.0}$$

$x = \sqrt{1.0 \times 2.1 \times 10^{-4}} = 1.5 \times 10^{-2}\ M = 0.015\ M = [OH^-]$ The assumption is valid.

Now we consider the hydrolysis of the bicarbonate ion.

Equation:	$HCO_3^-(aq) + H_2O$ $\rightleftharpoons$	$H_2CO_3(aq)$ +	$OH^-(aq)$
Initial:	0.015 M		0.015 M
Changes:	$-y$ M	$+y$ M	$+y$ M
Equil:	$(0.015 - y)$ M	y M	$(0.015 + y)$ M

$$K_b = \frac{K_w}{K_a(H_2CO_3)} = \frac{1.0 \times 10^{-14}}{4.4 \times 10^{-7}} = 2.3 \times 10^{-8} = \frac{[H_2CO_3][OH^-]}{[HCO_3^-]} = \frac{y\,(0.015 + y)}{0.015 - x} \approx \frac{0.015\,y}{0.015} = y$$

The assumption is valid and $y = [H_2CO_3] = 2.3 \times 10^{-8}$ M. The second hydrolysis makes a negligible contribution to the acidity of the solution. For the entire solution, then

$pOH = -\log[OH^-] = -\log(0.015) = 1.82$ pH = 14.00 – 1.82 = 12.18

9B The acidity of the solution is principally due to the hydrolysis of the carbonate ion, considered first.

Equation:	$SO_3^{2-}(aq) + H_2O$ $\rightleftharpoons$	$HSO_3^-(aq)$ +	$OH^-(aq)$
Initial:	0.500 M		
Changes:	$-x$ M	$+x$ M	$+x$ M
Equil:	$(0.500 - x)$ M	x M	x M

$$K_b = \frac{K_w}{K_a(HSO_3^-)} = \frac{1.0 \times 10^{-14}}{6.2 \times 10^{-8}} = 1.6 \times 10^{-7} = \frac{[HSO_3^-][OH^-]}{[SO_3^{2-}]} = \frac{x \cdot x}{0.500 - x} \approx \frac{x^2}{1.0}$$

$x = \sqrt{0.500 \times 1.6 \times 10^{-7}} = 2.8 \times 10^{-4}$ M = 0.00028 M = $[OH^-]$ The assumption is valid.

Now we consider the hydrolysis of the bicarbonate ion.

Equation:	$HSO_3^-(aq) + H_2O \rightleftharpoons$	$H_2SO_3(aq)$ +	$OH^-(aq)$
Initial:	0.00028 M		0.00028 M
Changes:	$-y$ M	$+y$ M	$+y$ M
Equil:	$(0.00028 - y)$ M	y M	$(0.00028 + y)$ M

$$K_b = \frac{K_w}{K_a(H_2SO_3)} = \frac{1.0 \times 10^{-14}}{1.3 \times 10^{-2}} = 7.7 \times 10^{-13} = \frac{[H_2SO_3][OH^-]}{[HSO_3^-]} = \frac{y\,(0.00028 + y)}{0.00028 - y} \approx \frac{0.00028\,y}{0.00028} = y$$

The assumption is valid and $y = [H_2CO_3] = 7.7 \times 10^{-13}$ M. The second hydrolysis makes a negligible contribution to the acidity of the solution. For the entire solution, then

$pOH = -\log [OH^-] = -\log (0.00028) = 3.55$ $\quad$ $pH = 14.00 - 3.55 = 10.45$

SUMMARIZING EXAMPLE CALCULATIONS

1. For bromcresol green $K_{HIn} = 2.1 \times 10^{-5}$ and $pK_{HIn} = -\log (2.1 \times 10^{-5}) = 4.68$. The color change range for an indicator extends one pH unit on either side of pK_{HIn}: $3.68 < pH < 5.68$.

2. At the equivalence point of the titration of 10.00 mL of 0.100 M NH_3 with 0.350 M HCl, the solution is one of $NH_4Cl(aq)$, whose pH is determined by the hydrolysis of the ammonium ion. In order to determine the ion's initial concentration, we calculate the amount of that ion and the volume of titrant needed to reach the equivalence point. $\quad NH_3(aq) + H_3O^+(aq, \text{from HCl}) \longrightarrow NH_4^+(aq) + H_2O$

$$\text{amount } NH_4 = 10.00 \text{ mL} \times \frac{1.00 \text{ mmol } NH_3}{1 \text{ mL soln}} \times \frac{1 \text{ mmol } NH_4^+}{1 \text{ mmol } NH_3} = 10.0 \text{ mmol } NH_4^+$$

$$\text{titrant volume} = 10.0 \text{ mmol } NH_4^+ \times \frac{1 \text{ mmol } H_3O^+}{1 \text{ mmol } NH_4^+} \times \frac{1 \text{ mL titrant}}{0.350 \text{ mmol } H_3O^+} = 28.6 \text{ mL titrant}$$

$$[NH_4^+] = \frac{10.0 \text{ mmol } NH_4^+}{28.6 \text{ mL titrant} + 10.0 \text{ mL original}} = 0.259 \text{ M}$$

3. We organize the solution of the hydrolysis problem around its balanced chemical equation.

Equation:	$NH_4^+(aq) + H_2O \rightleftharpoons$	$NH_3(aq)$ +	$H_3O^+(aq)$
Initial:	0.259 M		
Changes:	$-x$ M	$+x$ M	$+x$ M
Equil:	$(0.259 - x)$ M	x M	x M

$$K_b = \frac{[NH_3][H_3O^+]}{[NH_4^+]} = \frac{K_w}{K_b} = \frac{1.0 \times 10^{-14}}{1.8 \times 10^{-5}} = 5.6 \times 10^{-10} = \frac{x \cdot x}{0.259 - x} \approx \frac{x^2}{0.259}$$

$x = \sqrt{0.259 \times 5.6 \times 10^{-10}} = 1.2 \times 10^{-5}$ M = $[H_3O^+]$ $\quad$ The assumption is valid.

$pH = -\log (1.2 \times 10^{-5}) = 4.92$

4. pH = 4.92 falls within the range for bromcresol green: $3.68 < pH < 5.68$.

REVIEW QUESTIONS

1. **(a)** The abbreviation "mmol" stands for millimole, one thousandth of a mole. This amount of material often is a convenient one to use in calculations involving solutions where volumes are measured in mL.

(b) HIn is a generalized symbol for the formula of the acid form of an indicator.

(c) The equivalence point of a titration is that point when sufficient added titrant is present to react with all of the substance being titrated, with no titrant left over.

(d) A titration curve is a plot of the pH of solution versus the volume of added titrant (or sometimes versus the percentage of substance that has been titrated).

2. **(a)** The common ion effect refers to the suppression of an ionization equilibrium caused by the presence (or addition) of some of the ions produced by the ionization.

(b) A buffer solution maintains constant pH by consuming added strong acid or added strong base. The buffer solution contains components that react with each.

(c) The value of pK_a can be determined from the titration curve of a monoprotic weak acid since the pH at the half equivalence point is equal to the pK_a of the weak acid.

(d) pH is measured with a series of acid-base indicators that change color at different pH values. By determining the color of each indicator in a sample of the solution being tested, the pH of that solution can be placed within a narrow range.

3. **(a)** Buffer capacity is the amount of strong acid or base that can be added to a specified volume of a buffer before a significant change in pH occurs. Buffer range is the range of pH values within which a buffer solution will resist changes in pH.

(b) Hydrolysis refers to the reaction of an ion with water to produce either $H_3O^+(aq)$ or $OH^-(aq)$. Neutralization refers to the reaction of an acid with a base to produce a situation with pH closer to 7.0.

(c) The first and second equivalence points of a weak acid are the points where, respectively, the first proton has been completely ionized, and the second proton has been completely ionized.

(d) The equivalence point of a titration is the point where stoichiometric amounts of acid and base have been combined. The end point is when the indicator changes color. Careful selectron of the indicator can ensure that these two points coincide.

4. **(a)** Note that HI is a strong acid and the initial $[H_3O^+] = [HI] = 0.0892$ M

Equation:	$HC_3H_5O_2 + H_2O \rightleftharpoons$	$C_3H_5O_2^- \;+$	H_3O^+
Initial:	0.275 M		0.0892 M
Changes:	$-x$ M	$+x$ M	$+x$ M
Equil:	$(0.275 - x)$M	x M	$(0.0892 + x)$M

$$K_a = \frac{[C_3H_5O_2^-][H_3O^+]}{[HC_3H_5O_2]} = 1.3 \times 10^{-5} = \frac{x(0.0892 + x)}{0.275 - x} \approx \frac{0.0892\,x}{0.275} \qquad x = 4.0 \times 10^{-5}\text{ M}$$

We have assumed that $x << 0.0892$ M, an assumption that clearly is correct. $[H_3O^+] = 0.0892$ M

(b) $[OH^-] = \dfrac{K_w}{[H_3O^+]} = \dfrac{1.0 \times 10^{-14}}{0.0892} = 1.1 \times 10^{-13}$ M

(c) $[C_3H_5O_2^-] = x = 4.0 \times 10^{-5}$ M

(d) $[I^-] = [HI]_i = 0.0892$ M

5. **(a)** The NH_4Cl dissociates completely and thus $[NH_4^+]_i = [Cl^-]_i = 0.102$ M

Equation:	$NH_3 + H_2O \rightleftharpoons$	$NH_4^+ \;+$	OH^-
Initial:	0.164 M	0.102 M	
Changes:	$-x$ M	$+x$ M	$+x$ M
Equil:	$(0.164 - x)$M	$(0.102 + x)$M	x M

$$K_b = \frac{[NH_4^+][OH^-]}{[NH_3]} = \frac{(0.102 + x)\,x}{0.164 - x} = 1.8 \times 10^{-5} \approx \frac{0.102\,x}{0.164} \qquad x = 2.9 \times 10^{-5}\text{ M}$$

In solving this problem, we have assumed $x << 0.102$ M, a clearly valid assumption.

$[OH^-] = x = 2.9 \times 10^{-5}$ M

(b) $[NH_4^+] = 0.102 + x = 0.102$ M

(c) $[Cl^-] = 0.102$ M

(d) $[H_3O^+] = \dfrac{1.0 \times 10^{-14}}{2.9 \times 10^{-5}} = 3.4 \times 10^{-10}$ M

6. Adding $H_3O^+(aq)$ represents reaction with strong acid; adding $OH^-(aq)$ represents reaction with strong base.

(a) $CHO_2^-(aq) + H_3O^+(aq) \longrightarrow HCHO_2(aq) + H_2O$

$HCHO_2(aq) + OH^-(aq) \longrightarrow CHO_2^-(aq) + H_2O$

(b) $C_6H_5NH_2(aq) + H_3O^+(aq) \longrightarrow C_6H_5NH_3^+(aq) + H_2O$

$C_6H_5NH_3^+(aq) + OH^-(aq) \longrightarrow C_6H_5NH_2(aq) + H_2O$

(c) $HPO_4^{2-}(aq) + H_3O^+(aq) \longrightarrow H_2PO_4^-(aq) + H_2O$

$H_2PO_4^-(aq) + OH^-(aq) \longrightarrow HPO_4^{2-}(aq) + H_2O$

7. **(a)**

Equation:	$HC_7H_5O_2(aq) + H_2O \rightleftharpoons$	$C_7H_5O_2^- \;+$	H_3O^+
Initial:	0.012 M	0.033 M	
Changes:	$-x$ M	$+x$ M	$+x$ M
Equil:	$(0.013 - x)$M	$(0.033 + x)$M	x M

$$K_a = \frac{[H_3O^+][C_7H_5O_2^-]}{[HC_7H_5O_2]} = 6.3 \times 10^{-5} = \frac{x(0.033 + x)}{0.012 - x} \approx \frac{0.033\,x}{0.012} \qquad x = 2.3 \times 10^{-5}\text{ M}$$

To determine the value of x, we assumed $x << 0.012$ M, an assumption that clearly is correct.

$[H_3O^+] = 2.3 \times 10^{-5}$ M $\qquad$ $pH = -\log(2.3 \times 10^{-5}) = 4.64$

(b)

Equation:	$NH_3(aq) + H_2O \rightleftharpoons$	$NH_4^+(aq)$ +	$OH^-(aq)$
Initial:	0.408 M	0.153 M	
Changes:	$-x$ M	$+x$ M	$+x$ M
Equil:	$(0.408 - x)$M	$(0.153 + x)$M	x M

$$K_b = \frac{[NH_4^+][OH^-]}{[NH_3]} = 1.8 \times 10^{-5} = \frac{x(0.153 + x)}{0.408 - x} \approx \frac{0.153\,x}{0.408} \qquad x = 4.8 \times 10^{-5}\text{ M}$$

To determine the value of x, we assumed $x << 0.153$, a clearly valid assumption.

$[OH^-] = 4.8 \times 10^{-5}$ M $\quad$ pOH $= -\log(4.8 \times 10^{-5}) = 4.32$ $\quad$ pH $= 14.00 - 4.32 = 9.68$

8. $[H_3O^+] = 10^{-4.06} = 8.7 \times 10^{-5}$ M. We let S = initial $[CHO_2^-]$

Equation:	$HCHO_2(aq) + H_2O \rightleftharpoons$	$CHO_2^-(aq)$ +	$H_3O^+(aq)$
Initial:	0.366 M	S M	
Changes:	-8.7×10^{-5} M	$+8.7 \times 10^{-5}$ M	$+8.7 \times 10^{-5}$M
Equil:	0.366 M	$(S + 8.7 \times 10^{-5})$	8.7×10^{-5} M

$$K_a = \frac{[H_3O^+][CHO_2^-]}{[HCHO_2]} = 1.8 \times 10^{-4} = \frac{(S + 8.7 \times 10^{-5})\,8.7 \times 10^{-5}}{0.366} \approx \frac{8.7 \times 10^{-5}\,S}{0.366}$$

$S = 0.76$ M $\quad$ To determine S, we assumed $S >> 8.7 \times 10^{-5}$ M, clearly a valid assumption.

Or, we could have used the Henderson-Hasselbalch equation. $pK_a = -\log(1.8 \times 10^{-4}) = 3.74$

$$4.06 = 3.74 + \log\frac{[CHO_2^-]}{[HCHO_2]} \qquad \frac{[CHO_2^-]}{[HCHO_2]} = 2.1 \qquad [CHO_2^-] = 2.1 \times 0.366 = 0.77\text{ M}$$

The difference in the two answers is due simply to rounding.

9. We use the Henderson-Hasselbalch equation. $pK_b = -\log(1.8 \times 10^{-5}) = 4.74$

$$pK_a = 14.00 - pK_b = 14.00 - 4.74 = 9.26 \qquad pH = 9.12 = 9.26 + \log\frac{[NH_3]}{[NH_4^+]}$$

$$\frac{[NH_3]}{[NH_4^+]} = 10^{-0.14} = 0.72 \qquad [NH_3] = 0.72 \times [NH_4^+] = 0.72 \times 0.732\text{ M} = 0.53\text{ M}$$

10. 0.60 mol $NaC_2H_3O_2$ will raise the pH of 1.00 L of 0.50 M HCl to the greatest extent. Both OH^- and $C_2H_3O_2^-$ are bases that will react with the strong acid (H_3O^+) in HCl(aq), resulting in an increase in the pH of the solution. Since there is more acetate ion available than hydroxide ion, 0.60 mol $NaC_2H_3O_2$ will consume more H_3O^+ than will 0.40 mol NaOH, causing a greater increase in the pH of the solution. In fact, 0.60 mol $NaC_2H_3O_2$ will consume all of the H_3O^+ in 1.00 L of 0.50 M HCl, leaving only a solution of acetic acid and sodium acetate, while 0.40 mol NaOH will leave a solution of HCl (with a concentration of 0.10 M) and NaCl. With regard to the other two possibilities, 0.70 mol NaCl will not affect the pH and 0.50 mol $HC_2H_3O_2$ will lower the pH because $HC_2H_3O_2$ is an acid.

11. We use the Henderson-Hasselbalch equation to determine K_a of lactic acid.

$$[C_3H_5O_3^-] = \frac{1.00\text{ g }NaC_3H_5O_3}{100.0\text{ mL soln}} \times \frac{1000\text{ mL}}{1\text{ L soln}} \times \frac{1\text{ mol }NaC_3H_5O_3}{112.0\text{ g }NaC_3H_5O_3} \times \frac{1\text{ mol }C_3H_5O_3^-}{1\text{ mol }NaC_3H_5O_3} = 0.0893\text{ M}$$

$$pH = 4.11 = pK_a + \log\frac{[C_3H_5O_3^-]}{[HC_3H_5O_3]} = pK_a + \log\frac{0.0893\text{ M}}{0.0500\text{ M}} = pK_a + 0.252$$

$$pK_a = 4.11 - 0.252 = 3.86 \qquad K_a = 10^{-3.86} = 1.4 \times 10^{-4}$$

12. **(a)** Use the Henderson-Hasselbalch equation to determine $[CHO_2^-]$ in the buffer solution.

$$pH = pK_a + \log\frac{[CHO_2^-]}{[HCHO_2]} = 3.82 = 3.74 + \log\frac{[CHO_2^-]}{[HCHO_2]} \qquad \log\frac{[CHO_2^-]}{[HCHO_2]} = 3.82 - 3.74 = 0.08$$

$$\frac{[CHO_2^-]}{[HCHO_2]} = 1.2 \qquad [CHO_2^-] = 1.2\,[HCHO_2] = 1.2 \times 0.465\text{ M} = 0.56\text{ M}$$

$$\text{mass }NaCHO_2 = 0.250\text{ L} \times \frac{0.56\text{ mol }CHO_2^-}{1\text{ L soln}} \times \frac{1\text{ mol }NaCHO_2}{1\text{ mol }CHO_2^-} \times \frac{68.0\text{ g }NaCHO_2}{1\text{ mol }NaCHO_2}$$

$= 9.5$ g $NaCHO_2$ $\quad$ [Note that $pK_a = -\log K_a = -\log(1.8 \times 10^{-4}) = 3.74$]

(b) $$[OH^-]_i = \frac{0.20\text{ g NaOH}}{0.250\text{ L}} \times \frac{1\text{ mol NaOH}}{40.0\text{ g NaOH}} \times \frac{1\text{ mol }OH^-}{1\text{ mol NaOH}} = 0.020\text{ M }OH^-$$

Thus, $[HCHO_2]$ will decrease by 0.020 M and $[CHO_2^-]$ will increase by 0.020 M because of the reaction with the added hydroxide ion, as a result of: $\quad HCHO_2 + OH^- \longrightarrow CHO_2^- + H_2O$

$$pH = 3.74 + \log\frac{0.56 + 0.02}{0.465 - 0.02} = 3.86 \qquad \text{The pH has increased 0.04 unit.}$$

13. **(a)** The pH color change range is 1.00 pH unit on either side of pK_{HIn}. If the pH color change range is below pH = 7.00, the indicator changes color in acidic solution. If it is above pH = 7.00, the indicator changes color in alkaline solution. If pH = 7.00 falls within the pH color change ranges, the indicator changes color near the neutral point.

indicator	K_{HIn}	pK_{HIn}	pH color change range	changes color in?
bromophenol blue	1.4×10^{-4}	3.85	2.9 (yellow) to 4.9 (blue)	acidic solution
bromocresol green	2.1×10^{-5}	4.68	3.7 (yellow) to 5.7 (blue)	acidic solution
bromothymol blue	7.9×10^{-8}	7.10	6.1 (yellow) to 8.1 (blue)	neutral soln
2,4-dinitrophenol	1.3×10^{-4}	3.89	2.9 (clrless) to 4.9 (yellow)	acidic solution
chlorophenol red	1.0×10^{-6}	6.00	5.0 (yellow) to 7.0 (red)	sl. acidic soln
thymolphthalein	1.0×10^{-10}	10.00	9.0 (clrless) to 11.0 (blue)	basic solution

(b) If bromcresol green is green, the pH is between 3.7 and 5.7, probably about pH = 4.7.
If chlorophenol red is orange, the pH is between 5.0 and 7.0, probably about pH = 6.0.

14. We need to determine the pH of each solution, and then use the table developed in Exercise 13(a) to predict the color of the indicator. (The weakly acidic or basic character of the indicator does not affect the pH of the solution, since the indicator is added in very small quantity.)

(a) $[H_3O^+] = 0.100\text{ M HCl} \times \dfrac{1\text{ mol }H_3O^+}{1\text{ mol HCl}} = 0.100\text{ M} \qquad pH = -\log(0.100\text{ M}) = 1.000$

2,4-dinitrophenol assumes its acid color in a solution with pH = 1.000. The solution is colorless.

(b) Solutions of NaCl(aq) are pH neutral, with pH = 7.000. Chlorophenol red assumes its basic color in such a solution; the solution is red.

(c)

Equation:	$NH_3 + H_2O$	$\rightleftharpoons$	NH_4^+ +	OH^-
Initial:	1.00 M			
Changes:	$-x$ M		$+x$ M	$+x$ M
Equil:	$(1.00 - x)$M		x M	x M

$$K_b = \frac{[NH_4^+][OH^-]}{[NH_3]} = 1.8 \times 10^{-5} = \frac{x^2}{1.00 - x} \approx \frac{x^2}{1.00} \qquad x = 4.2 \times 10^{-3}\text{ M} = [OH^-]$$

$pOH = -\log(4.2 \times 10^{-3}) = 2.38 \qquad pH = 14.00 - 2.38 = 11.62$

Thymolphthalein assumes its basic color in a solution with pH = 11.62; the solution is blue.

(d) From Figure 17-3, seawater has pH = 8.50 to 10.0. Bromcresol green assumes its basic color in this solution; the solution is blue.

15. **(a)** The titration reaction is $KOH(aq) + HI(aq) \longrightarrow KI(aq) + H_2O$

$$\text{vol. KOH soln} = 25.00\text{ mL} \times \frac{0.212\text{ mmol HI}}{1\text{ mL soln}} \times \frac{1\text{ mmol KOH}}{1\text{ mmol HI}} \times \frac{1\text{ mL soln}}{0.146\text{ mmol KOH}}$$
$$= 36.3\text{ mL KOH soln}$$

(b) The titration reaction is $2\,KOH(aq) + H_2SO_4(aq) \longrightarrow K_2SO_4(aq) + 2\,H_2O$

$$\text{vol. KOH soln} = 20.00\text{ mL} \times \frac{0.0942\text{ mol }H_2SO_4}{1\text{ L soln}} \times \frac{2\text{ mol KOH}}{1\text{ mol }H_2SO_4} \times \frac{1\text{ mL soln}}{0.146\text{ mmol KOH}}$$
$$= 25.8\text{ mL KOH soln}$$

16. We can just sketch approximate titration curves of pH vs. percent of titration, since we do not have the concentration of the acid or the base, or the volume of solution being titrated. We can, however, precisely determine the pH at the half-equivalence point [mid-way between untitrated and completely titrated for a weak acid (or base)]; this is equal to the pK_a of the weak acid (or $pOH = pK_b$ of the weak base). Further, if we assume all solutions are 1.00 M, we can determine the pH at each equivalence point. It equals 14.00 + $\log(\sqrt{K_w/K_a})$ for the titration of a weak acid. [Indicator choices are given in square brackets.]

(a) In this titration (assuming 1.00 M KOH and 1.00 M HNO_3), the initial pH = 14.00 and at the equivalence point pH = 7.00. The pH drops rapidly after the equivalence point, rapidly reaching pH = 1.00. [Bromothymol blue changes from blue at pH = 8 to yellow at pH = 6.]

(b) The initial pH is that of 1.00 M NH_3, pH = 11.6. That of the half-equivalence point is $pOH = pK_b$ = 4.76 and thus pH = 9.24. That of the equivalence point is $-\log(\sqrt{K_w/K_b})$ = pH = 4.62. The pH then drops rapidly with added strong acid to pH = 1.00. [Methyl red changes from yellow at pH = 6.2 to red at pH = 4.5.]

(c) The initial pH is that of 1.00 M $HC_2H_3O_2$, pH = 2.38. That of the half-equivalence point is pH = pK_a = 4.76. The pH at the equivalence point is determined from $pOH = -\log(\sqrt{K_w/K_a}) = 4.62$ or pH = 9.38. The pH then rises rapidly with the addition of strong base, rapidly reaching pH = 13.00. [Phenolphthalein changes from colorless at pH = 8 to red at pH = 10.]

(d) The initial pH is that of 1.00 M $H_2PO_4^-$, pH = 3.60. That of the first half-equivalence point is pH = $pK_2 = 7.20$. That of the first equivalence point results from the hydrolysis of HPO_4^{2-} ion, about pH = 10.6. That of the second half equivalence point would be pH = $pK_3 = 12.37$, but this will be somewhat hard to reach; the solution is becoming pretty dilute. Finally, the pH of the second equivalence point will be that of the hydrolysis of the phosphate ion, about pH = 13.2, which again will be hard to reach without adding excessively concentrated base. [Alizarin yellow R changes from yellow at pH = 10 to violet at pH = 12. These is no suitable indicator given in Figure 18-6 for the second equivalence point.]

These curves are sketched below.

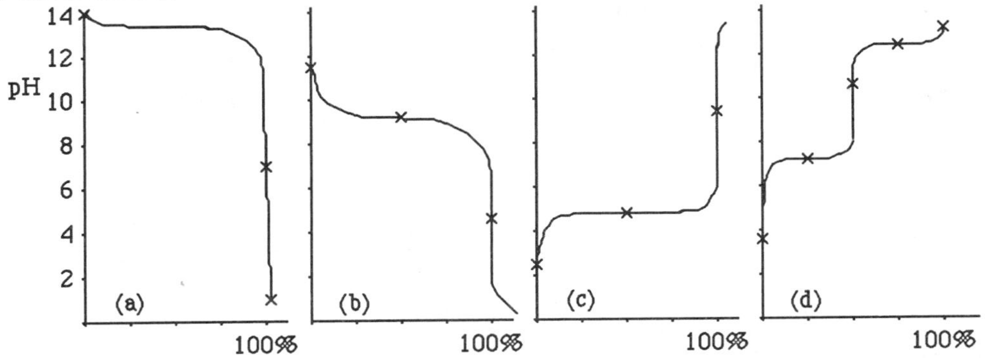

17. First we calculate the amount of HCl. The titration reaction is $HCl(aq) + KOH(aq) \longrightarrow KCl(aq) + H_2O$

$$\text{amount HCl} = 25.00 \text{ mL} \times \frac{0.160 \text{ mmol HCl}}{1 \text{ mL soln}} = 4.00 \text{ mmol HCl} = 4.00 \text{ mmol } H_3O^+ \text{ present}$$

Then, in each case, we calculate the amount of OH^- that has been added, determine which ion—$OH^-(aq)$ or $H_3O^+(aq)$—is in excess, compute the concentration of that ion, and determine the pH.

(a) $\text{amount } OH^- = 10.00 \text{ mL} \times \frac{0.242 \text{ mmol } OH^-}{1 \text{ mL soln}} = 2.42 \text{ mmol } OH^-$ H_3O^+ is in excess.

$$[H_3O^+] = \frac{4.00 \text{ mmol } H_3O^+ - \left(2.42 \text{ mmol } OH^- \times \frac{1 \text{ mmol } H_3O^+}{1 \text{ mmol } OH^-}\right)}{25.00 \text{ mL originally} + 10.00 \text{ mL titrant}} = 0.0451 \text{ M}$$

$pH = -\log(0.0451) = 1.346$

(b) $\text{amount } OH^- = 15.00 \text{ mL} \times \frac{0.242 \text{ mmol } OH^-}{1 \text{ mL soln}} = 3.63 \text{ mmol } OH^-$ H_3O^+ is in excess.

$$[H_3O^+] = \frac{4.00 \text{ mmol } H_3O^+ - \left(3.63 \text{ mmol } OH^- \times \frac{1 \text{ mmol } H_3O^+}{1 \text{ mmol } OH^-}\right)}{25.00 \text{ mL originally} + 15.00 \text{ mL titrant}} = 0.00925 \text{ M}$$

$pH = -\log(0.00925) = 2.034$

18. The titration reaction is $KOH(aq) + HCl(aq) \longrightarrow KCl(aq) + H_2O$

$$\text{no. mmol KOH} = 20.00 \text{ mL} \times \frac{0.275 \text{ mmol KOH}}{1 \text{ mL soln}} = 5.50 \text{ mmol KOH}$$

(a) The total volume of the solution is $V = 20.00 \text{ mL} + 15.00 \text{ mL} = 35.00 \text{ mL}$

$$\text{no. mmol HCl} = 15.00 \text{ mL} \times \frac{0.350 \text{ mmol HCl}}{1 \text{ mL soln}} = 5.25 \text{ mmol HCl}$$

$$\text{no. mmol excess } OH^- = (5.50 \text{ mmol KOH} - 5.25 \text{ mmol HCl}) \times \frac{1 \text{ mmol } OH^-}{1 \text{ mmol KOH}} = 0.25 \text{ mmol } OH^-$$

$[OH^-] = \frac{0.25 \text{ mmol } OH^-}{35.00 \text{ mL soln}} = 0.0071 \text{ M}$ $pOH = -\log(0.0071) = 2.15$ $pH = 14.00 - 2.15 = 11.85$

(b) The total volume of solution is $V = 20.00 \text{ mL} + 20.00 \text{ mL} = 40.00 \text{ mL}$

$$\text{no. mmol HCl} = 20.00 \text{ mL} \times \frac{0.350 \text{ mmol HCl}}{1 \text{ mL soln}} = 7.00 \text{ mmol HCl}$$

$$\text{no. mmol excess } H_3O^+ = (7.00 \text{ mmol HCl} - 5.50 \text{ mmol KOH}) \times \frac{1 \text{ mmol } H_3O^+}{1 \text{ mmol HCl}} = 1.50 \text{ mmol } H_3O^+$$

$[H_3O^+] = \dfrac{1.50 \text{ mmol } H_3O^+}{40.00 \text{ mL}} = 0.0375 \text{ M}$ $\text{pH} = -\log(0.0375) = 1.426$

19. The titration reaction is $HNO_2(aq) + NaOH(aq) \longrightarrow NaNO_2(aq) + H_2O$

amount $HNO_2 = 25.00 \text{ mL} \times \dfrac{0.132 \text{ mmol } HNO_2}{1 \text{ mL soln}} = 3.30 \text{ mmol } HNO_2$

(a) The volume of the solution is 25.00 mL + 10.00 mL = 35.00 mL

amount NaOH $= 10.00 \text{ mL} \times \dfrac{0.116 \text{ mmol NaOH}}{1 \text{ mL soln}} = 1.16 \text{ mmol NaOH}$

1.16 mmol $NaNO_2$ are formed in this reaction and there is an excess of (3.30 mmol HNO_2 – 1.16 mmol NaOH =) 2.14 mmol HNO_2. We can use the Henderson-Hasselbalch equation to determine the pH of the solution. $pK_a = -\log(7.2 \times 10^{-4}) = 3.14$

$$\text{pH} = pK_a + \log\frac{[NO_2^-]}{[HNO_2]} = 3.14 + \log\frac{1.16 \text{ mmol } NO_2^-/35.00 \text{ mL}}{2.14 \text{ mmol } HNO_2/35.00 \text{ mL}} = 2.87$$

(b) The volume of the solution is 25.00 mL + 20.00 mL = 45.00 mL

amount NaOH $= 20.00 \text{ mL} \times \dfrac{0.116 \text{ mmol NaOH}}{1 \text{ mL soln}} = 2.32 \text{ mmol NaOH}$

2.32 mmol $NaNO_2$ are formed in this reaction and there is an excess of (3.30 mmol HNO_2 – 2.32 mmol NaOH =) 0.98 mmol HNO_2.

$$\text{pH} = pK_a + \log\frac{[NO_2^-]}{[HNO_2]} = 3.14 + \log\frac{2.32 \text{ mmol } NO_2^-/45.00 \text{ mL}}{0.98 \text{ mmol } HNO_2/45.00 \text{ mL}} = 3.51$$

20. The calculation is very similar to that of Review Question 19. In this case, however, the titration reaction is $NH_3(aq) + HCl(aq) \longrightarrow NH_4Cl(aq) + H_2O$

amount $NH_3 = 20.00 \text{ mL} \times \dfrac{0.318 \text{ mmol } NH_3}{1 \text{ mL soln}} = 6.36 \text{ mmol } NH_3$

(a) The volume of the solution is 20.00 mL + 10.00 mL = 30.00 mL

amount HCl $= 10.00 \text{ mL} \times \dfrac{0.475 \text{ mmol NaOH}}{1 \text{ mL soln}} = 4.75 \text{ mmol HCl}$

4.75 mmol NH_4Cl is formed in this reaction and there is an excess of (6.36 mmol NH_3 – 4.75 mmol HCl =) 1.61 mmol NH_3. We can use the Henderson-Hasselbalch equation to determine the pH of the solution. $pK_b = -\log(1.8 \times 10^{-5}) = 4.74$ $pK_a = 14.00 - pK_b = 14.00 - 4.74 = 9.26$

$$\text{pH} = pK_a + \log\frac{[NH_3]}{[NH_4^+]} = 9.26 + \log\frac{1.61 \text{ mmol } NH_3/30.00 \text{ mL}}{4.75 \text{ mmol } NH_4^+/30.00 \text{ mL}} = 8.79$$

(b) The volume of the solution is 20.00 mL + 15.00 mL = 35.00 mL

amount HCl $= 15.00 \text{ mL} \times \dfrac{0.475 \text{ mmol NaOH}}{1 \text{ mL soln}} = 7.13 \text{ mmol HCl}$

6.36 mmol NH_4Cl is formed in this reaction and there is an excess of (7.13 mmol HCl – 6.36 mmol NH_3 =) 0.77 mmol HCl; this excess HCl determines the pH of the solution.

$[H_3O^+] = \dfrac{0.77 \text{ mmol HCl}}{35.00 \text{ mL soln}} \times \dfrac{1 \text{ mmol } H_3O^+}{1 \text{ mmol HCl}} = 0.022 \text{ M}$ $\text{pH} = -\log(0.022) = 1.66$

21. **(a)** This is the pH of a 0.01000 M $HC_7H_5O_2$ solution.

Equation:	$HC_7H_5O_2 + H_2O \rightleftharpoons$	$C_7H_5O_2^-$ +	H_3O^+	
Initial:	0.01000 M			
Changes:	$-x$ M	$+x$ M	$+x$ M	
Equil:	$(0.01000 - x)$M	x M	x M	

$$K_a = \frac{[C_7H_5O_2^-][H_3O^+]}{[HC_7H_5O_2]} = 6.3 \times 10^{-5} = \frac{x^2}{0.0100 - x} = \frac{x^2}{0.0100}$$

$x = 7.9 \times 10^{-4} \text{ M} = [H_3O^+]$ $\text{pH} = -\log(7.9 \times 10^{-4}) = 3.10$

(b) amount of $HC_7H_5O_2 = 25.0 \text{ mL} \times 0.01000 \text{ M} = 0.250 \text{ mmol } HC_7H_5O_2$

We determine the volume of 0.01000 M $Ba(OH)_2$ to reach the equivalence point.

The titration reaction is $Ba(OH)_2(aq) + 2\, HC_7H_5O_2(aq) \longrightarrow Ba(C_7H_5O_2)_2 + 2\, H_2O$

$$\text{vol. base} = 0.250 \text{ mmol } HC_7H_5O_2 \times \frac{1 \text{ mmol } Ba(OH)_2}{2 \text{ mmol } HC_7H_5O_2} \times \frac{1 \text{ mL soln}}{0.01000 \text{ mmol } Ba(OH)_2} = 12.5 \text{ mL}$$

Thus, the addition of 6.25 mL of 0.01000 M $Ba(OH)_2$ brings us to the half-equivalence point, where $\text{pH} = pK_a = 4.20$.

(c) At the equivalence point, there is 0.250 mmol $C_7H_5O_2^-$ in 25.00 + 12.50 = 37.50 mL solution. It is the hydrolysis of this anion that determines the pH of the solution.

$$[C_7H_5O_2^-] = \frac{0.250 \text{ mmol } C_7H_5O_2^-}{37.50 \text{ mL soln}} = 6.67 \times 10^{-3} \text{ M}$$

Equation:	$C_7H_5O_2^- + H_2O \rightleftharpoons$	$HC_7H_5O_2$ +	OH^-
Initial:	0.00667 M		
Changes:	$-x$ M	$+x$ M	$+x$ M
Equil:	$(0.00667 - x)$M	x M	x M

$$K_b = \frac{K_w}{K_a} = \frac{1.0 \times 10^{-14}}{6.3 \times 10^{-5}} = \frac{x^2}{0.00667 - x} \approx \frac{x^2}{0.00667} \qquad x = 1.0 \times 10^{-6} \text{ M} = [OH^-]$$

$pOH = -\log(1.0 \times 10^{-6}) = 6.00 \qquad pH = 14.00 - 6.00 = 8.00$

(d) The excess added base determines the pH of the solution.

$$\text{excess amount OH}^- = 2.5 \text{ mL} \times \frac{0.0100 \text{ mmol Ba(OH)}_2}{1 \text{ mL Ba(OH)}_2} \times \frac{2 \text{ mmol OH}^-}{1 \text{ mmol Ba(OH)}_2} = 0.050 \text{ mmol OH}^-$$

$$[OH^-] = \frac{0.050 \text{ mmol OH}^-}{25.0 \text{ mL} + 15.0 \text{ mL}} = 1.3 \times 10^{-3} \text{ M} \qquad pOH = -\log(1.3 \times 10^{-3}) = 2.89$$

$pH = 14.00 - 2.89 = 11.11$

22. 0.10 M solutions of $NaHSO_4$ are the most acidic. The Na^+ cation does not contribute to the acidity of the solution; it does not hydrolyze. Of the anions involved, S^{2-} hydrolyzes to produce an alkaline solution. Each of the other anions (HSO_4^-, HCO_3^-, HPO_4^{2-}) can ionize further to yield H_3O^+, but HSO_4^- does so to a far greater extent than do the other two; it is quite strong for a weak acid. In fact, HPO_4^{2-} is expected to hydrolyze to form a basic solution.

EXERCISES

The Common Ion Effect

23. (a) We first determine the pH of 0.100 M HNO_2.

Equation:	$HNO_2 + H_2O \rightleftharpoons$	NO_2^- +	H_3O^+
Initial:	0.100 M		
Changes:	$-x$ M	$+x$ M	$+x$ M
Equil:	$(0.100 - x)$M	x M	x M

$$K_a = \frac{[NO_2^-][H_3O^+]}{[HNO_2]} = 7.2 \times 10^{-4} = \frac{x^2}{0.100 - x} = \frac{x^2}{0.100} \qquad x = 8.5 \times 10^{-3} \text{ M} = [H_3O^+]$$

$pH = -\log(8.5 \times 10^{-3}) = 2.07$

When 0.100 mol $NaNO_2$ is added to a 1.00 L of a 0.100 M HNO_2 solution, we have a solution with $[NO_2^-] = 0.100 \text{ M} = [HNO_2]$. By the Henderson-Hasselbalch equation, this solution has $pH = pK_a = -\log(7.2 \times 10^{-4}) = 3.14$ The addition has caused a pH change of 1.07 units.

(b) Although $NaNO_3$ contributes to the solution an ion—nitrate ion, NO_3^-—that is produced by the ionization of HNO_3, HNO_3 is a strong acid. There is no HNO_3(aq) in equilibrium with hydrogen and nitrate ions. Thus, there is no equilibrium to be shifted by the addition of one of its products. The $[H_3O^+]$ and the pH are unaffected by the addition of $NaNO_3$ to a solution of nitric acid.

The difference occurs because there is an equilibrium system to be shifted in the first solution, whereas there is no equilibrium—just total ionization—in the second solution.

24. The explanation for the different result is that each of these solutions has present an ion—acetate ion, $C_2H_3O_2^-$—that is produced in the ionization of acetic acid. The presence of this ion suppresses the ionization of acetic acid, thus minimizing the increase in $[H_3O^+]$. All three solutions are buffer solutions and their pH can be found with the aid of the Henderson-Hasselbalch equation.

(a) $$pH = pK_a + \log\frac{[C_2H_3O_2^-]}{[HC_2H_3O_2]} = 4.74 + \log\frac{0.10}{1.0} = 3.74 \qquad [H_3O^+] = 10^{-3.74} = 1.8 \times 10^{-4} \text{ M}$$

$$\% \text{ ionization} = \frac{[H_3O^+]}{[HC_2H_3O_2]_i} \times 100\% = \frac{1.8 \times 10^{-4} \text{ M}}{1.0 \text{ M}} \times 100\% = 0.018\%$$

(b) $$pH = pK_a + \log\frac{[C_2H_3O_2^-]}{[HC_2H_3O_2]} = 4.74 + \log\frac{0.10}{0.10} = 4.74 \qquad [H_3O^+] = 10^{-4.78} = 1.8 \times 10^{-5} \text{ M}$$

$$\% \text{ ionization} = \frac{[H_3O^+]}{[HC_2H_3O_2]_i} \times 100\% = \frac{1.8 \times 10^{-5} \text{ M}}{0.10 \text{ M}} \times 100\% = 0.018\%$$

(c) $pH = pK_a + \log \frac{[C_2H_3O_2^-]}{[HC_2H_3O_2]} = 4.74 + \log \frac{0.10}{0.010} = 5.74$ $[H_3O^+] = 10^{-5.74} = 1.8 \times 10^{-6}$ M

$\% \text{ ionization} = \frac{[H_3O^+]}{[HC_2H_3O_2]_i} \times 100\% = \frac{1.8 \times 10^{-6} \text{ M}}{0.010 \text{ M}} \times 100\% = 0.018\%$

25. **(a)** The strong acid HCl suppresses the ionization of the weak acid HOCl so much that a negligible concentration of H_3O^+ is contributed to the solution by HOCl. Thus $[H_3O^+] = [HCl] = 0.035$ M

(b) This is a buffer solution. We can use the Henderson-Hasselbalch equation to determine its pH.

$pK_a = -\log(7.2 \times 10^{-4}) = 3.14$ $pH = pK_a + \log \frac{[NO_2^-]}{[HNO_2]} = 3.14 + \log \frac{0.100 \text{ M}}{0.0550 \text{ M}} = 3.40$

$[H_3O^+] = 10^{-3.40} = 4.0 \times 10^{-4}$ M

(c) This also is a buffer solution, as we see by an analysis of the reaction between the components.

Equation:	H_3O^+(aq, from HCl) +	$C_2H_3O_2^-$(aq, from $NaC_2H_3O_2$) $\longrightarrow$	$HC_2H_3O_2$(aq) + H_2O
In soln:	0.0525 M	0.0768 M	
Produce HAc:	−0.0525 M	−0.0525 M	+0.0525 M
Initial:	≈0 M	0.0243 M	0.0525 M

Now the Henderson-Hasselbalch equation can be used. $pK_a = -\log(1.8 \times 10^{-5}) = 4.74$

$pH = pK_a + \log \frac{[C_2H_3O_2^-]}{[HC_2H_3O_2]} = 4.74 + \log \frac{0.0243 \text{ M}}{0.0525 \text{ M}} = 4.41$ $[H_3O^+] = 10^{-4.41} = 3.9 \times 10^{-5}$ M

26. **(a)** Neither Ba^{2+}(aq) nor Cl^-(aq) hydrolyzes or ionizes to affect the acidity of the solution. $[OH^-]$ is determined entirely by the $Ba(OH)_2$ solute.

$[OH^-] = \frac{0.0062 \text{ mol } Ba(OH)_2}{1 \text{ L soln}} \times \frac{2 \text{ mol } OH^-}{1 \text{ mol } Ba(OH)_2} = 0.012_4 \text{ M}$

(b) We use the Henderson-Hasselbalch equation for this buffer solution.

$[NH_4^+] = 0.315 \text{ M } (NH_4)_2SO_4 \times \frac{2 \text{ mol } NH_4^+}{1 \text{ mol } (NH_4)_2SO_4} = 0.630 \text{ M}$ $pK_a = 9.26$ for NH_4^+.

$pH = pK_a + \log \frac{[NH_3]}{[NH_4^+]} = 9.26 + \log \frac{0.486 \text{ M}}{0.630 \text{ M}} = 9.15$ $pOH = 14.00 - 9.15 = 4.85$

$[OH^-] = 10^{-4.85} = 1.4 \times 10^{-5}$ M

(c) This solution also is a buffer solution, as analysis of the reaction between its components shows.

Equation:	NH_4^+(aq, from NH_4Cl) +	OH^-(aq, from NaOH) $\longrightarrow$	NH_3(aq) + H_2O
In soln:	0.264 M	0.196 M	
Form NH_3:	−0.196 M	−0.196 M	+0.196 M
Initial:	0.068 M	≈0 M	0.196 M

$pH = pK_a + \log \frac{[NH_3]}{[NH_4^+]} = 9.26 + \log \frac{0.196 \text{ M}}{0.068 \text{ M}} = 9.72$ $pOH = 14.00 - 9.72 = 4.28$

$[OH^-] = 10^{-4.28} = 5.2 \times 10^{-5}$ M

Buffer Solutions

27. **(a)** 0.100 M NaCl is not a buffer solution. There is neither a weak acid nor a weak base present.

(b) 0.100 M NaCl—0.100 M NH_4Cl is not a buffer solution. Although a weak acid, NH_4^+, is present, its conjugate base is not.

(c) 0.100 M CH_3NH_2—0.150 M $CH_3NH_3^+Cl^-$ is a buffer solution. Both a weak base, CH_3NH_2, and its conjugate acid, $CH_3NH_3^+$, are present in approximately equal concentrations.

(d) 0.100 M HCl—0.050 M $NaNO_2$ is not a buffer solution. All the NO_2^- is converted to HNO_2 and thus the solution is a mixture of a strong acid and a weak acid.

(e) 0.100 M HCl—0.200 M $NaC_2H_3O_2$ is a buffer solution. All of the HCl reacts with half of the $C_2H_3O_2^-$ to form a solution with 0.100 M $HC_2H_3O_2$, a weak acid, and 0.100 M $C_2H_3O_2^-$, its conjugate base.

(f) 0.100 M $HC_2H_3O_2$—0.125 M $NaC_3H_5O_2$ is not a buffer in the strict sense because it does not contain a weak acid and its conjugate base, but rather the conjugate base of another weak acid. These two weak acids (acetic, with $K_a = 1.8 \times 10^{-5}$ and propionic, with $K_a = 1.35 \times 10^{-5}$) have approximately the same strength, however, so this solution would resist changes in its pH on the addition of strong acid or strong base, although its pH would not be as narrowly controlled as in a "classic" buffer solution.

28. **(a)** Reaction with added acid: $HPO_4^{2-} + H_3O^+ \longrightarrow H_2PO_4^- + H_2O$

Reaction with added base: $H_2PO_4^- + OH^- \longrightarrow HPO_4^{2-} + H_2O$

(b) We assume initially that the buffer has equal concentrations of the two ions, $[H_2PO_4^-] = [HPO_4^{2-}]$

$$pH = pK_2 + \log \frac{[HPO_4^{2-}]}{[H_2PO_4^-]} = 7.20 + 0.00 = 7.20$$ is the pH where the buffer is most effective.

(c) $$pH = 7.20 + \log \frac{[HPO_4^{2-}]}{[H_2PO_4^-]} = 7.20 + \log \frac{0.150\ M}{0.050\ M} = 7.20 + 0.48 = 7.68$$

29. $$\text{amount of solute} = 1.15\ \text{mg} \times \frac{1\ \text{g}}{1000\ \text{mg}} \times \frac{1\ \text{mol}\ C_6H_5NH_3^+Cl^-}{129.6\ \text{g}} \times \frac{1\ \text{mol}\ C_6H_5NH_3^+}{1\ \text{mol}\ C_6H_5NH_3^+Cl^-}$$

$$= 8.87 \times 10^{-6}\ \text{mol}\ C_6H_5NH_3^+$$

$$[C_6H_5NH_3^+] = \frac{8.87 \times 10^{-6}\ \text{mol}\ C_6H_5NH_3^+}{3.18\ \text{L soln}} = 2.79 \times 10^{-6}\ M$$

Equation:	$C_6H_5NH_2(aq) + H_2O \rightleftharpoons$	$C_6H_5NH_3^+(aq)$	+	$OH^-(aq)$
Initial:	0.105 M	2.79×10^{-6} M		
Changes:	$-x$ M	$+x$ M		$+x$ M
Equil:	$(0.105 - x)$M	$(2.79 \times 10^{-6} + x)$M		x M

$$K_b = \frac{[C_6H_5NH_3^+][OH^-]}{[C_6H_5NH_2]} = 7.4 \times 10^{-10} = \frac{(2.79 \times 10^{-6} + x)\,x}{0.105 - x}$$

$$7.4 \times 10^{-10}(0.105 - x) = (2.79 \times 10^{-6} + x)x = 7.8 \times 10^{-11} - 7.4 \times 10^{-10}x = 2.79 \times 10^{-6}\,x + x^2$$

$$x^2 + (2.79 \times 10^{-6} + 7.4 \times 10^{-10})x - 7.8 \times 10^{-11} = 0 = x^2 + 2.79 \times 10^{-6}x - 7.8 \times 10^{-11}$$

$$x = \frac{-b \pm \sqrt{b^2 - 4ac}}{2a} = \frac{-2.79 \times 10^{-6} \pm \sqrt{7.78 \times 10^{-12} + 3.1 \times 10^{-10}}}{2} = 7.5 \times 10^{-6} M = [OH^-]$$

$pOH = -\log(7.5 \times 10^{-6}) = 5.12$ $\qquad pH = 14.00 - 5.12 = 8.88$

30. We determine the concentration of the cation of the weak base.

$$[C_6H_5NH_3^+] = \frac{8.50\ \text{g} \times \frac{1\ \text{mol}\ C_6H_5NH_3^+Cl^-}{129.6\ \text{g}} \times \frac{1\ \text{mol}\ C_6H_5NH_3^+}{1\ \text{mol}\ C_6H_5NH_3^+Cl^-}}{750\ \text{mL} \times \frac{1\ \text{L}}{1000\ \text{mL}}} = 0.0874\ M$$

In order to be an effective buffer, each concentration must exceed the ionization constant ($K_b = 7.4 \times 10^{-10}$) by a factor of 100, which clearly is true. Also, the ratio of the two concentrations must fall between 0.1 and 10: $\frac{[C_6H_5NH_3^+]}{[C_6H_5NH_2]} = \frac{0.0874\ M}{0.215\ M} = 0.407$. This solution will be an effective buffer.

31. **(a)** First use the Henderson-Hasselbalch equation [$pK_b = -\log(1.8 \times 10^{-5}) = 4.74$, $pK_a = 14.00 - 4.74 = 9.26$] to determine $[NH_4^+]$ in the buffer solution.

$$pH = 9.45 = pK_a + \log \frac{[NH_3]}{[NH_4^+]} = 9.26 + \log \frac{[NH_3]}{[NH_4^+]} \qquad \log \frac{[NH_3]}{[NH_4^+]} = 9.45 - 9.26 = +0.19$$

$$\frac{[NH_4^+]}{[NH_3]} = 10^{-0.19} = 0.65 \qquad [NH_4^+] = 0.65 \times [NH_3] = 0.65 \times 0.258\ M = 0.17\ M$$

We now assume that the volume of the solution does not change significantly when the solid is added.

$$\text{mass}\ (NH_4)_2SO_4 = 425\ \text{mL} \times \frac{1\ \text{L soln}}{1000\ \text{mL}} \times \frac{0.17\ \text{mol}\ NH_4^+}{1\ \text{L soln}} \times \frac{1\ \text{mol}\ (NH_4)_2SO_4}{2\ \text{mol}\ NH_4^+}$$
$$\times \frac{132.2\ \text{g}\ (NH_4)_2SO_4}{1\ \text{mol}\ (NH_4)_2SO_4} = 4.8\ \text{g}\ (NH_4)_2SO_4$$

(b) Let's use the Henderson-Hasselbalch equation to determine the ratio of concentrations of cation and weak base in the altered solution.

$$pH = 9.30 = pK_a + \log \frac{[NH_3]}{[NH_4^+]} = 9.26 + \log \frac{[NH_3]}{[NH_4^+]} \qquad \log \frac{[NH_3]}{[NH_4^+]} = 9.30 - 9.26 = +0.04$$

$$\frac{[NH_4^+]}{[NH_3]} = 10^{-0.04} = 0.91 = \frac{0.17\ M + x\ M}{0.258} \qquad 0.17 + x = 0.91 \times 0.258 = 0.234$$

The reason we decided to add x to the numerator follows. (Notice we cannot remove a component.) A pH of 9.30 is more acidic than a pH of 9.45 and therefore the conjugate acid's (NH_4^+) concentration must increase. Additionally, mathematics tells us that for the concentration ratio to increase from 0.65 to 0.91, its numerator must increase. We solve this expression for x. $\quad x = 0.234 - 0.17 = 0.06$. We need to add ammonium ion to increase its concentration by 0.07 M in 750 mL of solution.

$(NH_4)_2SO_4 \text{ mass} = 0.100 \text{ L} \times \frac{0.06 \text{ mol } NH_4^+}{1 \text{ L}} \times \frac{1 \text{ mol } (NH_4)_2SO_4}{2 \text{ mol } NH_4^+} \times \frac{132.2 \text{ g } (NH_4)_2SO_4}{1 \text{ mol } (NH_4)_2SO_4}$

$= 0.40 \text{ g } (NH_4)_2SO_4$ to add

32. **(a)** amount $HC_7H_5O_2 = 2.00 \text{ g } HC_7H_5O_2 \times \frac{1 \text{ mol } HC_7H_5O_2}{122.1 \text{ g } HC_7H_5O_2} = 0.0164 \text{ mol } HC_7H_5O_2$

amount $C_7H_5O_2^- = 2.00 \text{ g } NaC_7H_5O_2 \times \frac{1 \text{ mol } NaC_7H_5O_2}{144.1 \text{ g } NaC_7H_5O_2} \times \frac{1 \text{ mol } C_7H_5O_2^-}{1 \text{ mol } NaC_7H_5O_2}$

$= 0.0139 \text{ mol } C_7H_5O_2^-$

$pH = pK_a + \log \frac{[C_7H_5O_2^-]}{[HC_7H_5O_2]} = -\log(6.3 \times 10^{-5}) + \log \frac{0.0139 \text{ mol } C_7H_5O_2^-/0.7500 \text{ L}}{0.0164 \text{ mol } HC_7H_5O_2/0.7500 \text{ L}}$

$= 4.20 - 0.0718 = 4.13$

(b) To lower the pH of this buffer solution, that is, to make it more acidic, benzoic acid must be added. The quantity is determined as follows. We use amounts rather than concentrations because all components are present in the same volume of solution.

$4.00 = 4.20 + \log \frac{0.0139 \text{ mol } C_7H_5O_2^-}{x \text{ mol } HC_7H_5O_2}$ $\qquad \log \frac{0.0139 \text{ mol } C_7H_5O_2^-}{x \text{ mol } HC_7H_5O_2} = -0.20$

$\frac{0.0139 \text{ mol } C_7H_5O_2^-}{x \text{ mol } HC_7H_5O_2} = 10^{-0.20} = 0.63$ $\qquad x = \frac{0.0139}{0.63} = 0.022 \text{ mol } HC_7H_5O_2$ total

added $HC_7H_5O_2 = 0.022 \text{ mol } HC_7H_5O_2 - 0.0164 \text{ mol } HC_7H_5O_2 = 0.006 \text{ mol } HC_7H_5O_2$

added mass $HC_7H_5O_2 = 0.006 \text{ mol } HC_7H_5O_2 \times \frac{122.1 \text{ g } HC_7H_5O_2}{1 \text{ mol } HC_7H_5O_2} = 0.7 \text{ g } HC_7H_5O_2$

33. The added HCl will react with the ammonia, and the pH of the buffer solution will decrease. The original buffer solution has $[NH_3] = 0.258$ M and $[NH_4^+] = 0.17$ M. We first calculate the [HCl] in solution, reduced from 12 M because of dilution. $\qquad [HCl] \text{ added} = 12 \text{ M} \times \frac{0.55 \text{ mL}}{100.6 \text{ mL}} = 0.066 \text{ M}$

We determine pK_a for ammonium ion: $pK_b = -\log(1.8 \times 10^{-5}) = 4.74 \qquad pK_a = 14.00 - 4.74 = 9.26$

Equation:	$NH_3(aq)$ +	$HCl(aq)$ ⟶	$NH_4^+(aq)$ +	$Cl^-(aq)$
Buffer:	0.258	0 M	0.17	
Added:		0.066 M		
Changes:	−0.066 M	−0.066 M	+0.066 M	
Final:	0.192 M	0 M	0.24 M	

$pH = pK_a + \log \frac{[NH_3]}{[NH_4^+]} = 9.26 + \log \frac{0.192}{0.24} = 9.16$

34. The added NH_3 will react with the benzoic acid, and the pH of the buffer solution will increase. The original buffer solution has $[C_7H_5O_2^-] = 0.0139 \text{ mol } C_7H_5O_2^-/0.7500 \text{ L} = 0.0185$ M and $[HC_7H_5O_2] = 0.0164 \text{ mol } HC_7H_5O_2/0.7500 \text{ L} = 0.0219$ M. We first calculate the $[NH_3]$ in solution, reduced from 15 M because of dilution. $\quad [NH_3] \text{ added} = 15 \text{ M} \times \frac{0.35 \text{ mL}}{750.4 \text{ mL}} = 0.0070 \text{ M}$

For benzoic acid, $pK_a = -\log(6.3 \times 10^{-5}) = 4.20$

Equation:	$NH_3(aq)$ +	$HC_7H_5O_2(aq)$ ⟶	$NH_4^+(aq)$ +	$C_7H_5O_2^-(aq)$
Buffer:		0.0219 M		0.0185 M
Added:	0.0070 M			
Changes:	−0.0070 M	−0.0070 M		+0.0070 M
Final:	0.000 M	0.0149 M		0.0255 M

$pH = pK_a + \log \frac{[C_7H_5O_2^-]}{[HC_7H_5O_2]} = 4.20 + \log \frac{0.0255}{0.0149} = 4.43$

35. The pK_a's of the three acids help us chose the one to be used in the buffer. It is the acid with pK_a within 1.00 pH unit of 3.50. $pK_a = 3.74$ for $HCHO_2$, $pK_a = 4.74$ for $HC_2H_3O_2$, and $pK_1 = 2.15$ for H_3PO_4. Thus we choose $HCHO_2$ and $NaCHO_2$ to prepare a buffer with pH = 3.50. The Henderson-Hasselbalch equation is used to determine the relative amounts of each component present in the buffer solution.

$pH = 3.50 = 3.74 + \log \frac{[CHO_2^-]}{[HCHO_2]} \qquad \log \frac{[CHO_2^-]}{[HCHO_2]} = 3.50 - 3.74 = -0.24 \qquad \frac{[CHO_2^-]}{[HCHO_2]} = 10^{-0.24} = 0.58$

This ratio of concentrations is also the ratio of the number of moles of each component in the buffer solution, since both concentrations are a number of moles in a certain volume, and the volumes are the same

(the two solutes are in the same solution). This ratio also is the ratio of the volumes of the two solutions, since both solutions being mixed contain the same concentration of solute.
If we assume 100. mL of acid solution, V_{acid} = 100. mL. Then the volume of salt solution is
$V_{salt} = 0.58 \times 100.$ mL = 58 mL 0.100 M $NaCHO_2$

36. a. We can lower the pH of the 0.250 M $HC_2H_3O_2$—0.560 M $C_2H_3O_2^-$ buffer solution by increasing $[HC_2H_3O_2]$ or lowering $[C_2H_3O_2^-]$. NaCl solutions will have no effect, and the addition of NaOH(aq) or $NaC_2H_3O_2$(aq) will raise the pH. A solution of $HC_2H_3O_2$ will lower the pH; the addition of 0.100 M $HC_2H_3O_2$ to a 0.250 M $HC_2H_3O_2$ solution actually will lower $[HC_2H_3O_2]$, but $[C_2H_3O_2^-]$ will be lowered even more by dilution, and thus the ratio $[C_2H_3O_2^-]/[HC_2H_3O_2]$ decreases, lowering pH by the Henderson-Hasselbalch equation. The addition of 0.150 M HCl will raise $[HC_2H_3O_2]$ and lower $[C_2H_3O_2^-]$ through the reaction $H_3O^+ + C_2H_3O_2^- \longrightarrow HC_2H_3O_2 + H_2O$

b. We first use the Henderson-Hasselbalch equation to determine the ratio of the concentration of acetate ion and acetic acid. $pH = 5.00 = 4.74 + \log \dfrac{[C_2H_3O_2^-]}{[HC_2H_3O_2]}$

$\log \dfrac{[C_2H_3O_2^-]}{[HC_2H_3O_2]} = 5.00 - 4.74 = 0.26$ $\qquad \dfrac{[C_2H_3O_2^-]}{[HC_2H_3O_2]} = 10^{0.26} = 1.8$

Now we compute the amount of each component in the original buffer solution.

$$\text{amount of } C_2H_3O_2^- = 300.\text{ mL} \times \frac{0.560 \text{ mmol } C_2H_3O_2^-}{1 \text{ mL soln}} = 168 \text{ mmol } C_2H_3O_2^-$$

$$\text{amount of } HC_2H_3O_2 = 300.\text{ mL} \times \frac{0.250 \text{ mmol } HC_2H_3O_2}{1 \text{ mL soln}} = 75.0 \text{ mmol } HC_2H_3O_2$$

⟶ Now let x represent the amount of H_3O^+ added in mmol.

$1.8 = \dfrac{168 - x}{75.0 + x}$ $\qquad 168 - x = 1.8(75 + x) = 13_5 + 1.8x$ $\qquad 168 - 13_5 = 2.7\,x$

$x = \dfrac{168 - 13_5}{2.7} = 12 \text{ mmol } H_3O^+$

$$\text{no. mL 0.150 M HCl} = 12 \text{ mmol } H_3O^+ \times \frac{1 \text{ mmol HCl}}{1 \text{ mmol } H_3O^+} \times \frac{1 \text{ mL soln}}{0.150 \text{ mmol HCl}}$$
$$= 80 \text{ mL 0.150 M HCl solution}$$

⟶ For adding 0.100 M $HC_2H_3O_2$, we let x = volume in mL of added weak acid. Then, the number of mmols of added $HC_2H_3O_2$ is 0.100 x.

$[C_2H_3O_2^-] = \dfrac{168}{300 + x}$ $\qquad [HC_2H_3O_2] = \dfrac{75.0 + 0.100x}{300 + x}$

$1.8 = \dfrac{[C_2H_3O_2^-]}{[HC_2H_3O_2]} = \dfrac{168}{75.0 + 0.100\,x}$ $\qquad 13_5 + 0.180\,x = 168$ $\qquad x = \dfrac{168 - 13_5}{0.180} = 1.8 \times 10^2$ mL

37. (a) The pH of the buffer is determined from the Henderson-Hasselbalch equation.

$pH = pK_a + \log \dfrac{[C_3H_5O_2^-]}{[HC_3H_5O_2]} = 4.89 + \log \dfrac{0.100 \text{ M}}{0.100 \text{ M}} = 4.89$

The effective pH range is the same for every propionate buffer: from pH = 3.89 to pH = 5.89, one pH unit on either side of pK_a for propionic acid.

(b) To each liter of 0.100 M $HC_3H_5O_2$—0.100 M $NaC_3H_5O_2$ we can add 0.100 mol OH^- before all of the $HC_3H_5O_2$ is consumed, and we can add 0.100 mol H_3O^+ before all of the $C_3H_5O_2^-$ is consumed. The buffer capacity thus is 100. millimoles (0.100 mol) of acid or base per liter of buffer solution.

38. (a) The solution will be an effective buffer one pH unit on either sides of the pK_a of methylammonium ion, $CH_3NH_3^+$, $K_b = 4.2 \times 10^{-4}$ for methylamine, $pK_b = -\log(4.2 \times 10^{-4}) = 3.38$. For methylammonium cation, $pK_a = 14.00 - 3.38 = 10.62$. Thus, this buffer will be effective from a pH of 9.62 to a pH of 11.62.

(b) The capacity of the buffer is reached when all of the weak base or all of the conjugate acid has been neutralized by added strong acid or strong base. Because their concentrations are the same in the same volume of solution, there is the same amount of weak base and its cation.

$$\text{amount of weak base or its cation} = 125 \text{ mL} \times \frac{0.0500 \text{ mmol}}{1 \text{ mL}} = 6.25 \text{ mmol } CH_3NH_2 \text{ or } CH_3NH_3^+$$

The buffer capacity thus is 6.25 millimoles of acid or base per 125 mL of buffer solution.

39. **(a)** The pH of this buffer solution can be determined with the Henderson-Hasselbalch equation.

$$pH = pK_a + \log\frac{[CHO_2^-]}{[HCHO_2]} = -\log(1.8 \times 10^{-4}) + \log\frac{8.5 \text{ mmol}/75.0 \text{ mL}}{15.5 \text{ mmol}/75.0 \text{ mL}} = 3.74 - 0.26 = 3.48$$

This solution is almost not a buffer, because $[CHO_2^-] = 1.1 \times 10^{-1}$ which is about 600 times K_a

(b) The amount of added $OH^- = 0.25 \text{ mmol Ba(OH)}_2 \times \frac{2 \text{ mmol OH}^-}{1 \text{ mmol Ba(OH)}_2} = 0.50 \text{ mmol OH}^-$

The OH^- added reacts with the formic acid and produces formate ion.

Equation:	$HCHO_2$ +	OH^- ⇌	$CHO_2^- + H_2O$
Buffer:	15.5 mmol		8.5 mmol
Add base:		0.50 mmol	
React:	–0.50 mmol	–0.50 mmol	+0.50 mmol
Final:	15.0 mmol		9.0 mmol

$$pH = pK_a + \log\frac{[CHO_2^-]}{[HCHO_2]} = -\log(1.8 \times 10^{-4}) + \log\frac{9.0 \text{ mmol}/75.0 \text{ mL}}{15.0 \text{ mmol}/75.0 \text{ mL}} = 3.74 - 0.22 = 3.52$$

(c) The amount of added $H_3O^+ = 1.05 \text{ mL acid} \times \frac{12 \text{ mmol HCl}}{1 \text{ mL acid}} \times \frac{1 \text{ mmol } H_3O^+}{1 \text{ mmol HCl}} = 13 \text{ mmol } H_3O^+$

The OH^- added reacts with the formate ion and produces formic acid.

Equation:	CHO_2^- +	H_3O^+ ⇌	$HCHO_2 + H_2O$
Buffer:	8.5 mmol		15.5 mmol
Add acid:		13 mmol	
React:	–8.5 mmol	– 8.5 mmol	+ 8.5 mmol
Final:	0 mmol	4.5 mmol	24.0 mmol

The buffer's capacity has been exceeded. The pH of the solution is determined by the strong acid present. $[H_3O^+] = \frac{4.5 \text{ mmol}}{75.0 \text{ mL} + 1.05 \text{ mL}} = 0.059 \text{ M}$ $\quad pH = -\log(0.059) = 1.23$

40. For NH_3, $pK_b = -\log(1.8 \times 10^{-5})$ $\quad$ For NH_4^+, $pK_a = 14.00 - pK_b = 14.00 - 4.74 = 9.26$

(a) $[NH_3] = \frac{1.68 \text{ g NH}_3}{0.500 \text{ L}} \times \frac{1 \text{ mol NH}_3}{17.03 \text{ g NH}_3} = 0.197 \text{ M}$

$[NH_4^+] = \frac{4.05 \text{ g (NH}_4)_2SO_4}{0.500 \text{ L}} \times \frac{1 \text{ mol (NH}_4)_2SO_4}{132.1 \text{ g (NH}_4)_2SO_4} \times \frac{2 \text{ mol NH}_4^+}{1 \text{ mol (NH}_4)_2SO_4} = 0.123 \text{ M}$

$pH = pK_a + \log\frac{[NH_3]}{[NH_4^+]} = 9.26 + \log\frac{0.197 \text{ M}}{0.123 \text{ M}} = 9.46$

(b) The $OH^-(aq)$ reacts with the $NH_4^+(aq)$ to produce an equivalent amount of $NH_3(aq)$.

$[OH^-]_i = \frac{0.88 \text{ g NaOH}}{0.500 \text{ L}} \times \frac{1 \text{ mol NaOH}}{40.00 \text{ g NaOH}} \times \frac{1 \text{ mol OH}^-}{1 \text{ mol NaOH}} = 0.044 \text{ M}$

Equation:	$NH_4^+(aq)$ +	$OH^-(aq)$ ⇌	$NH_3(aq) + H_2O$
Initial:	0.123 M		0.197 M
Add NaOH:		0.044 M	
React:	–0.044 M	–0.044 M	+0.044 M
Final:	0.079 M	0.0000 M	0.241 M

$pH = pK_a + \log\frac{[NH_3]}{[NH_4^+]} = 9.26 + \log\frac{0.241 \text{ M}}{0.079 \text{ M}} = 9.74$

(c)

Equation:	$NH_3(aq)$ +	$H_3O^+(aq)$ ⇌	$NH_4^+(aq) + H_2O$
Initial:	0.197 M		0.123 M
Add HCl:		$+x$ M	
React:	$-x$ M	$-x$ M	$+x$ M
Final:	$(0.197 - x)$ M	10^{-9} M	$(0.123 + x)$M

$pH = 9.00 = pK_a + \log\frac{[NH_3]}{[NH_4^+]} = 9.26 + \log\frac{(0.197 - x) \text{ M}}{(0.123 + x) \text{ M}}$

$\log\frac{(0.197 - x) \text{ M}}{(0.123 + x) \text{ M}} = 9.00 - 9.26 = -0.26$ $\qquad \frac{(0.197 - x) \text{ M}}{(0.123 + x) \text{ M}} = 10^{-0.26} = 0.55$

$0.197 - x = 0.55(0.123 + x) = 0.068 + 0.55\,x$ $\qquad 1.55\,x = 0.197 - 0.068 = 0.129$

$x = \frac{0.129}{1.55} = 0.0832 \text{ M}$

$\text{volume HCl} = 0.500 \text{ L} \times \frac{0.0832 \text{ mol } H_3O^+}{1 \text{ L soln}} \times \frac{1 \text{ mol HCl}}{1 \text{ mol } H_3O^+} \times \frac{1000 \text{ mL}}{12 \text{ mol HCl}} = 3.5 \text{ mL}$

41. **(a)** We use the Henderson-Hasselbalch equation to determine the pH of the solution. The total solution volume is 36.00 mL + 64.00 mL = 100.00 mL. $pK_a = 14.00 - pK_b = 14.00 + \log(1.8 \times 10^{-5}) = 9.26$

$$[NH_3] = \frac{36.00 \text{ mL} \times 0.200 \text{ M } NH_3}{100.00 \text{ mL}} = \frac{7.20 \text{ mmol } NH_3}{100.0 \text{ mL}} = 0.0720 \text{ M}$$

$$[NH_4^+] = \frac{64.00 \text{ mL} \times 0.200 \text{ M } NH_4^+}{100.00 \text{ mL}} = \frac{12.8 \text{ mmol } NH_4^+}{100.0 \text{ mL}} = 0.128 \text{ M}$$

$$pH = pK_a + \log\frac{[NH_3]}{[NH_4^+]} = 9.26 + \log\frac{0.0720}{0.128 \text{ M}} = 9.01$$

(b) The solution has $[OH^-] = 10^{-4.99} = 1.0 \times 10^{-5}$ M

The Henderson-Hasselbalch equation depends on the assumption: $[NH_3] >> 1.8 \times 10^{-5} \text{ M} << [NH_4^+]$

If the solution is diluted to 1.00 L, $[NH_3] = 7.20 \times 10^{-3}$ M, and $[NH_4^+] = 1.28 \times 10^{-2}$ M. These two concentrations are consistent with the assumption.

However, if the solution is diluted to 1000. L, $[NH_3] = 7.2 \times 10^{-6}$ M, and $[NH_3] = 1.28 \times 10^{-5}$ M, and these two concentrations are not consistent with the assumption. Thus, in 1000. L of solution, the given quantities of NH_3 and NH_4^+ will not produce a solution with pH = 9.00. With sufficient dilution, the solution will become indistinguishable from pure water; its pH will equal 7.00.

(c) The 0.20 mL of added 1.00 M HCl does not significantly affect the volume of the solution, but it does add 0.20 mL × 1.00 M HCl = 0.20 mmol H_3O^+. This added H_3O^+ reacts with NH_3, decreasing its amount from 7.20 mmol NH_3 to 7.00 mmol NH_3, and increasing the amount of NH_4^+ from 12.8 mmol NH_4^+ to 13.0 mmol NH_4^+, through the reaction: $NH_3 + H_3O^+ \longrightarrow NH_4^+ + H_2O$

$$pH = 9.26 + \log\frac{7.00 \text{ mmol } NH_3/100.20 \text{ mL}}{13.0 \text{ mmol } NH_4^+/100.20 \text{ mL}} = 8.99$$

(d) We see in the calculation of part (c) that the total volume of the solution does not affect the pOH of the solution, at least as long as the Henderson-Hasselbalch equation is obeyed. We let x represent the number of millimoles of H_3O^+ added, through 1.00 M HCl. This increases the amount of NH_4^+ and decreases the amount of NH_3, through the reaction $NH_3 + H_3O^+ \longrightarrow NH_4^+ + H_2O$

$$pH = 8.90 = 9.26 + \log\frac{7.20 - x}{12.8 + x} \qquad \log\frac{7.20 - x}{12.8 + x} = 8.90 - 9.26 = -0.36$$

Inverting, we have: $\dfrac{12.8 + x}{7.20 - x} = 10^{0.36} = 2.29 \qquad 12.8 + x = 2.29(7.20 - x) = 16.5 - 2.29\,x$

$$x = \frac{16.5 - 12.8}{1.00 + 2.29} = 1.1 \text{ mmol } H_3O^+$$

$$\text{vol } 1.00 \text{ M HCl} = 1.1 \text{ mmol } H_3O^+ \times \frac{1 \text{ mmol HCl}}{1 \text{ mmol } H_3O^+} \times \frac{1 \text{ mL soln}}{1.00 \text{ mmol HCl}} = 1.1 \text{ mL } 1.00 \text{ M HCl}$$

42. **(a)** $$[C_2H_3O_2^-] = \frac{12.0 \text{ g } NaC_2H_3O_2}{0.300 \text{ L soln}} \times \frac{1 \text{ mol } NaC_2H_3O_2}{82.03 \text{ g } NaC_2H_3O_2} \times \frac{1 \text{ mol } C_2H_3O_2^-}{1 \text{ mol } NaC_2H_3O_2} = 0.488 \text{ M } C_2H_3O_2^-$$

Equation:	$HC_2H_3O_2 + H_2O$	$\rightleftharpoons$	$C_2H_3O_2^-$ +	H_3O^+
Initial:				0.200 M
Add $NaC_2H_3O_2$			0.488 M	
Consume H_3O^+	+0.200 M		−0.200 M	−0.200 M
Buffer:	0.200 M		0.288 M	

Then use the Henderson-Hasselbalch equation.

$$pH = pK_a + \log\frac{[C_2H_3O_2^-]}{[HC_2H_3O_2]} = 4.74 + \log\frac{0.288 \text{ M}}{0.200 \text{ M}} = 4.74 + 0.16 = 4.90$$

(b) We can calculate the initial $[OH^-]$ due to the $Ba(OH)_2$.

$$[OH^-] = \frac{1.00 \text{ g } Ba(OH)_2}{0.300 \text{ L}} \times \frac{1 \text{ mol } Ba(OH)_2}{171.3 \text{ g } Ba(OH)_2} \times \frac{2 \text{ mol } OH^-}{1 \text{ mol } Ba(OH)_2} = 0.0389 \text{ M}$$

Then $HC_2H_3O_2$ is consumed.

Equation:	$HC_2H_3O_2$ +	OH^-	$\rightleftharpoons$	$C_2H_3O_2^-$ +	H_2O
Initial:	0.200 M	0.0389 M		0.288 M	
Consume OH^-	0.161 M			0.327 M	

Then use the Henderson-Hasselbalch equation.

$$pH = pK_a + \log\frac{[C_2H_3O_2^-]}{[HC_2H_3O_2]} = 4.74 + \log\frac{0.327 \text{ M}}{0.161 \text{ M}} = 4.74 + 0.31 = 5.05$$

(c) $Ba(OH)_2$ can be added until all of the $HC_2H_3O_2$ is consumed.

$Ba(OH)_2 + 2\ HC_2H_3O_2 \longrightarrow Ba(C_2H_3O_2)_2 + 2\ H_2O$

$$\text{quantity of } Ba(OH)_2 = 0.300\ \text{L} \times \frac{0.200\ \text{mol}\ HC_2H_3O_2}{1\ \text{L soln}} \times \frac{1\ \text{mol}\ Ba(OH)_2}{2\ \text{mol}\ HC_2H_3O_2} = 0.0300\ \text{mol}\ Ba(OH)_2$$
$$\times \frac{171.3\ \text{g}\ Ba(OH)_2}{1\ \text{mol}\ Ba(OH)_2} = 5.14\ \text{g}\ Ba(OH)_2$$

(d) This is an excess of 0.36 g $Ba(OH)_2$ and it is this excess that determines the pOH of the solution.

$$[OH^-] = \frac{0.36\ \text{g}\ Ba(OH)_2}{0.300\ \text{L soln}} \times \frac{1\ \text{mol}\ Ba(OH)_2}{171.3\ \text{g}\ Ba(OH)_2} \times \frac{2\ \text{mol}\ OH^-}{1\ \text{mol}\ Ba(OH)_2} = 1.4 \times 10^{-2}\ \text{M}\ OH^-$$

$\text{pOH} = -\log(1.4 \times 10^{-2}) = 1.85$ $\qquad$ $\text{pH} = 14.00 - 1.85 = 12.15$

Acid-Base Indicators

43. (a) In an acid-base titration, the pH of the solution changes sharply at a definite pH that is known prior to titration. (This pH change occurs during the addition of a very small volume of titrant.) In determining the pH of a solution, on the other hand, that pH is not known in advance. Since each indicator only serves to fix the pH over a quite small region, often less than 2.0 pH units, several indicators—carefully chosen to span the entire range of 14 pH units—must be employed to even narrow the pH to ±1 pH unit.

(b) An indicator is, after all, a weak acid. Its addition to a solution will affect the acidity of that solution. Thus, one adds only enough indicator to show a color change and not enough to affect solution acidity.

44. (a) We use an equation similar to the Henderson-Hasselbalch equation to determine the relative concentrations of indicator, HIn, and its anion, In^-, in this solution.

$$\text{pH} = pK_{HIn} + \log\frac{[In^-]}{[HIn]} \qquad 4.55 = 4.95 + \log\frac{[In^-]}{[HIn]} \qquad \log\frac{[In^-]}{[HIn]} = 4.55 - 4.95 = -0.40$$

$$\frac{[In^-]}{[HIn]} = 10^{-0.40} = 0.40 = \frac{x}{100 - x} \qquad x = 40 - 0.40\,x \qquad x = \frac{40}{1.40} = 29\%\ In^- \text{ and } 71\%\ HIn$$

(b) When the indicator is in a solution whose pH equals its pK_a (4.95), the ratio $[In^-]/[HIn] = 1.00$. And yet, at the midpoint of its color change range (about pH = 5.3), the ratio $[In^-]/[HIn]$ is greater than 1.00. Even though $[HIn] < [In^-]$ at this midpoint, the contribution of HIn to establishing the color of the solution is about the same as the contribution of In^-. This must mean that HIn (red) is more strongly colored than In^- (yellow).

45. (a) 0.10 M KOH is an alkaline solution and phenol red will display its basic color in such a solution; the solution will be red.

(b) 0.10 M $HC_2H_3O_2$ is an acidic solution—although that of a weak acid—and phenol red will display its acidic color in such a solution; the solution will be yellow.

(c) 0.10 M NH_4NO_3 is an acidic solution due to the hydrolysis of the ammonium ion. Phenol red will display its acidic solor—that is, yellow—in this solution.

(d) 0.10 M HBr is an acidic solution, the aqueous solution of a strong acid. Phenol red will display its acidic color in this solution; the solution will be yellow.

(e) 0.10 M NaCN is an alkaline solution because of the hydrolysis of the cyanide ion. Phenol red will display its basic color—red—in this solution.

(f) An equimolar acetic acid–potassium acetate buffer has $\text{pH} = pK_a = 4.74$ for acetic acid. In this solution phenol red will display its acid color, yellow.

46. (a) $\text{pH} = -\log(0.205) = 0.688$ The indicator is red in this solution.

(b) The total volume of the solution is 600.0 mL. We compute the amount of each solute.

amount $H_3O^+ = 350.0\ \text{mL} \times 0.205\ \text{M} = 71.8\ \text{mmol}\ H_3O^+$

amount $NO_2^- = 250.0\ \text{mL} \times 0.500\ \text{M} = 125\ \text{mmol}\ NO_2^-$

$$[H_3O^+] = \frac{71.8\ \text{mmol}}{600.0\ \text{mL}} = 0.120\ \text{M} \qquad [NO_2^-] = \frac{125\ \text{mmol}}{600.0\ \text{mL}} = 0.208\ \text{M}$$

The H_3O^+ and NO_2^- react to produce a buffer solution in which $[HNO_2] = 0.120$ M and $[NO_2^-] = 0.208 - 0.120 = 0.088$ M. We use the Henderson-Hasselbalch equation to determine the pH of this solution. $pK_a = -\log(7.2 \times 10^{-4}) = 3.14$

$$\text{pH} = pK_a + \log\frac{[NO_2^-]}{[HNO_2]} = 3.14 + \log\frac{0.088\ \text{M}}{0.120\ \text{M}} = 3.01$$ The indicator is yellow in this solution.

(c) The total volume of the solution is 750. mL. We compute the amount and then the concentration of each solute. amount $OH^- = 150 \text{ mL} \times 0.100 \text{ M} = 15.0 \text{ mmol } OH^-$

This OH^- reacts with HNO_2 in the buffer solution to neutralize some of it and leave 56.8 mmol (= 71.8 – 15.0) unneutralized.

$$[HNO_2] = \frac{56.8 \text{ mmol}}{750. \text{ mL}} = 0.0757 \text{ M} \qquad [NO_2^-] = \frac{(125 + 15) \text{ mmol}}{750. \text{ mL}} = 0.187 \text{ M}$$

We use the Henderson Hasselbalch equation to determine the pH of this solution.

$$pH = pK_a + \log \frac{[NO_2^-]}{[HNO_2]} = 3.14 + \log \frac{0.187 \text{ M}}{0.0757 \text{ M}} = 3.53$$ The indicator is yellow in this solution.

(d) We determine the $[OH^-]$ due to the added $Ba(OH)_2$.

$$[OH^-] = \frac{5.00 \text{ g } Ba(OH)_2}{0.750 \text{ L}} \times \frac{1 \text{ mol } Ba(OH)_2}{171.34 \text{ g } Ba(OH)_2} \times \frac{2 \text{ mol } OH^-}{1 \text{ mol } Ba(OH)_2} = 0.0778 \text{ M}$$

This is sufficient $[OH^-]$ to react with the existing $[HNO_2]$ and leave an excess $[OH^-] = 0.0778 \text{ M} - 0.0757 \text{ M} = 0.0021 \text{ M}$. [$pOH = -\log(0.0021) = 2.68$. $pH = 14.00 - 2.68 = 11.32$.] The indicator is blue in this solution.

Neutralization Reactions

47. To the second equivalence point the reaction is: $H_3PO_4(aq) + 2\,KOH(aq) \longrightarrow K_2HPO_4(aq) + 2\,H_2O$
The molarity of the H_3PO_4 solution is determined in the following manner.

$$H_3PO_4 \text{ molarity} = \frac{31.15 \text{ mL KOH soln} \times \frac{0.2420 \text{ mmol KOH}}{1 \text{ mL KOH soln}} \times \frac{1 \text{ mmol } H_3PO_4}{2 \text{ mmol KOH}}}{25.00 \text{ mL } H_3PO_4 \text{ soln}} = 0.1508 \text{ M}$$

48. From first to second equivalence points the reaction is: $NaH_2PO_4(aq) + NaOH(aq) \longrightarrow Na_2HPO_4(aq) + H_2O$
The molarity of the H_3PO_4 solution is determined in the following manner.

$$H_3PO_4 \text{ molarity} = \frac{18.67 \text{ mL NaOH soln} \times \frac{0.1885 \text{ mmol NaOH}}{1 \text{ mL NaOH soln}} \times \frac{1 \text{ mmol } H_3PO_4}{1 \text{ mmol NaOH}}}{20.00 \text{ mL } H_3PO_4 \text{ soln}} = 0.1760 \text{ M}$$

49. We determine the amount of H_3O^+ or OH^- in each solution, followed by the amount of the reagent in excess.

A. $$\text{amount } H_3O^+ = 50.00 \text{ mL} \times \frac{0.0150 \text{ mmol } H_2SO_4}{1 \text{ mL soln}} \times \frac{2 \text{ mmol } H_3O^+}{1 \text{ mmol } H_2SO_4} = 1.50 \text{ mmol } H_3O^+$$

B. $$\text{amount } OH^- = 50.00 \text{ mL} \times \frac{0.0385 \text{ mmol NaOH}}{1 \text{ mL soln}} \times \frac{1 \text{ mmol } OH^-}{1 \text{ mmol NaOH}} = 1.93 \text{ mmol } OH^-$$

Result:

Titration reaction:	$OH^-(aq)$ +	$H_3O^+(aq)$	$\rightleftharpoons 2\,H_2O$
Initial amounts:	1.93 mmol	1.50 mmol	
After reaction:	0.43 mmol	0 mmol	

$$[OH^-] = \frac{0.43 \text{ mmol } OH^-}{100.0 \text{ mL soln}} = 4.3 \times 10^{-3} \text{ M}$$

$$pOH = -\log(4.3 \times 10^{-3}) = 2.37$$
$$pH = 14.00 - 2.37 = 11.63$$

50. We determine the amount of solute in each solution, followed by the amount of the reagent in excess.

A. $[H_3O^+] = 10^{-2.50} = 0.0032 \text{ M}$

$$\text{mmol HCl} = 100.0 \text{ mL} \times \frac{0.0032 \text{ mmol } H_3O^+}{1 \text{ mL soln}} \times \frac{1 \text{ mmol HCl}}{1 \text{ mmol } H_3O^+} = 0.32 \text{ mmol HCl}$$

B. $pOH = 14.00 - 11.00 = 3.00$ $\qquad [OH^-] = 10^{-3.00} = 1.0 \times 10^{-3} \text{ M}$

$$\text{mmol NaOH} = 100.0 \text{ mL} \times \frac{0.0010 \text{ mmol } OH^-}{1 \text{ mL soln}} \times \frac{1 \text{ mmol NaOH}}{1 \text{ mmol } OH^-} = 0.10 \text{ mmol NaOH}$$

Result:

Titration reaction:	NaOH(aq) +	HCl(aq)	$\rightleftharpoons$ NaCl(aq) + H_2O
Initial amounts:	0.10 mmol	0.32 mmol	
After reaction:	0.00 mmol	0.22 mmol	

$$[H_3O^+] = \frac{0.22 \text{ mmol HCl}}{200.0 \text{ mL soln}} \times \frac{1 \text{ mmol } H_3O^+}{1 \text{ mmol HCl}} = 1.1 \times 10^{-3} \text{ M} \qquad pH = -\log(1.1 \times 10^{-3}) = 2.96$$

Titration Curves

51. In each case, the volume of acid and its molarity are the same. Thus, also the amount of acid is the same in each case. The volume of titrant needed to reach the equivalence point will also be the same in both cases, since the titrant has the same concentration in each case, and it is the same amount of base that reacts with a given amount (in moles) of acid. Realize that, as the titration of a weak acid proceeds, the weak acid will ionize—replenishing the H_3O^+ in solution. This will occur until all of the weak acid has ionized, and has reacted with the strong base. At the equivalence point in the titration of a strong acid with a strong base is an aqueous solution of ions that do not hydrolyze. But the equivalence point solution of the titration of a weak acid with a strong base contains the anion of a weak acid which will hydrolyze to produce a basic (alkaline) solution. (There also is the cation of a strong base that will not hydrolyze.)

52. **(a)** This equivalence point is the result of the titration of a weak acid with a strong base. The species present in solution is CO_3^{2-} which, through its hydrolysis, will form an alkaline, or basic solution. The other ionic species in solution—Na^+—will not hydrolyze. Thus, pH > 7.0

(b) This is the titration of a strong acid with a weak base. The species present in solution is NH_4^+ which hydrolyzes to form an acidic solution. Cl^- does not hydrolyze. Thus pH < 7.0

(c) This is the titration of a strong acid with a strong base. Two ions are present in the solution at the equivalence point—K^+ and Cl^-—and neither of these hydrolyze. The solution will have a pH of 7.00.

53. **(a)** Initial $[OH^-] = 0.100$ M OH^- pOH = $-\log(0.100) = 1.000$ pH = 13.00

Since this is the titration of a strong base with a strong acid, KI is the solute present at the equivalence point and pH = 7.00. The titration reaction is $KOH(aq) + HI(aq) \longrightarrow KI(aq) + H_2O$

$$\text{no. mL HI} = 25.0 \text{ mL KOH soln} \times \frac{0.100 \text{ mmol KOH soln}}{1 \text{ mL soln}} \times \frac{1 \text{ mmol HI}}{1 \text{ mmol KOH}} \times \frac{1 \text{ mL HI soln}}{0.200 \text{ mmol HI}}$$

$$= 12.5 \text{ mL HI soln}$$

Initial amount of KOH present = 25.0 mL KOH soln × 0.100 M = 2.50 mmol KOH

At the 40% titration point: 5.00 mL HI soln × 0.200 M HI = 1.00 mmol HI

excess KOH = 2.50 mmol KOH – 1.00 mmol HI = 1.50 mmol KOH

$$[OH^-] = \frac{1.50 \text{ mmol KOH}}{30.0 \text{ mL total}} \times \frac{1 \text{ mmol OH}^-}{1 \text{ mmol KOH}} = 0.0500 \text{ M} \qquad \text{pOH} = -\log(0.0500) = 1.30 \quad \text{pH} = 14.00 - 1.30 = 12.70$$

At the 80% titration point: 10.00 mL HI soln × 0.200 M HI = 2.00 mmol HI

excess KOH = 2.50 mmol KOH – 2.00 mmol HI = 0.50 mmol KOH

$$[OH^-] = \frac{0.50 \text{ mmol KOH}}{35.0 \text{ mL total}} \times \frac{1 \text{ mmol OH}^-}{1 \text{ mmol KOH}} = 0.0143 \text{ M} \qquad \text{pOH} = -\log(0.0143) = 1.84 \quad \text{pH} = 14.00 - 1.84 = 12.16$$

At the 110% titration point: 13.75 mL HI soln × 0.200 M HI = 2.75 mmol HI

excess HI = 2.75 mmol KOH – 2.50 mmol HI = 0.25 mmol KOH

$$[H_3O^+] = \frac{0.25 \text{ mmol KOH}}{38.8 \text{ mL total}} \times \frac{1 \text{ mmol OH}^-}{1 \text{ mmol KOH}} = 0.0064 \text{ M} \qquad \text{pH} = -\log(0.0064) = 2.19$$

Since the pH changes very rapidly at the equivalence point, from about pH = 10 to about pH = 4, most of the indicators in Figure 18-6 can be used. The main exceptions are alizarin yellow R, bromophenol blue, thymol blue (in its acid range), and methyl violet.

(b) *Initial pH:*

Equation:	$NH_3(aq) + H_2O \rightleftharpoons$	$NH_4^+(aq)$ +	$OH^-(aq)$
Initial:	1.00 M		
Changes:	$-x$ M	$+x$ M	$+x$ M
Equil:	$(1.00 - x)$M	x M	x M

$$K_b = \frac{[NH_4^+][OH^-]}{[NH_3]} = 1.8 \times 10^{-5} = \frac{x^2}{1.00 - x} \approx \frac{x^2}{1.00}$$

$x = 4.2 \times 10^{-3}$ M = $[OH^-]$, pOH = $-\log(4.2 \times 10^{-3}) = 2.38$, pH = 14.00 – 2.38 = 11.62 = initial pH

Volume of titrant: $NH_3 + HCl \longrightarrow NH_4Cl + H_2O$

$$\text{no. mL HCl} = 10.0 \text{ mL} \times \frac{1.00 \text{ mmol NH}_3}{1 \text{ mL soln}} \times \frac{1 \text{ mmol HCl}}{1 \text{ mmol NH}_3} \times \frac{1 \text{ mL HCl soln}}{0.250 \text{ mmol HCl}} = 40.0 \text{ mL soln}$$

pH at equivalence point: The total solution volume at the equivalence point is 10.0 + 40.0 = 50.0 mL Also at the equivalence point, all of the NH_3 has reacted to form NH_4^+. It is this NH_4^+ that hydrolyzes to determine the pH of the solution.

$$[NH_4^+] = \frac{10.0 \text{ mL} \times \frac{1.00 \text{ mmol NH}_3}{1 \text{ mL soln}} \times \frac{1 \text{ mmol NH}_4^+}{1 \text{ mmol NH}_3}}{50.0 \text{ mL total solution}} = 0.200 \text{ M}$$

Equation:	$NH_4^+(aq) + H_2O \rightleftharpoons$	$NH_3(aq)$ +	$H_3O^+(aq)$
Initial:	0.200 M		
Changes:	$-x$ M	$+x$ M	$+x$ M
Equil:	$(0.200 - x)$M	x M	x M

$$K_a = \frac{K_w}{K_b} = \frac{1.0 \times 10^{-14}}{1.8 \times 10^{-5}} = \frac{[NH_3][H_3O^+]}{[NH_3]} = \frac{x^2}{0.200 - x} \approx \frac{x^2}{0.200}$$

$x = 1.1 \times 10^{-5}$ M $\quad [H_3O^+] = 1.1 \times 10^{-5}$ M $\quad$ pH $= -\log(1.1 \times 10^{-5}) = 4.96$

Of the indicators in Figure 18-6, the only one that has the pH of the equivalence point within its pH color change range is methyl red (yellow at pH = 6.2 and red at pH = 4.5).

At the 50% titration point, $[NH_3] = [NH_4^+]$ and pOH = pK_b = 4.74 $\quad$ pH = 14.00 – 4.74 = 9.26

The titration curves for parts (a) and (b) follow.

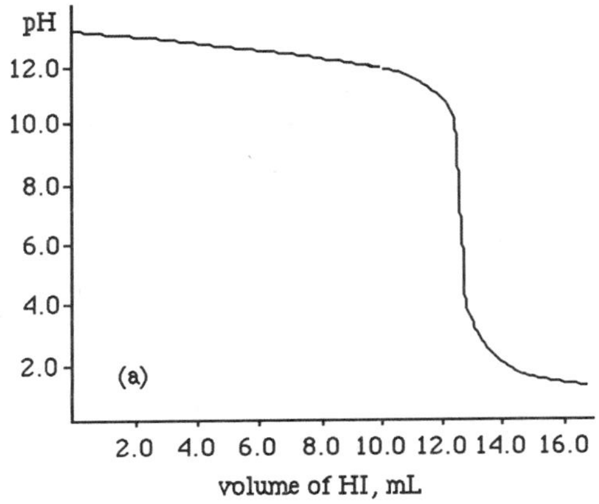

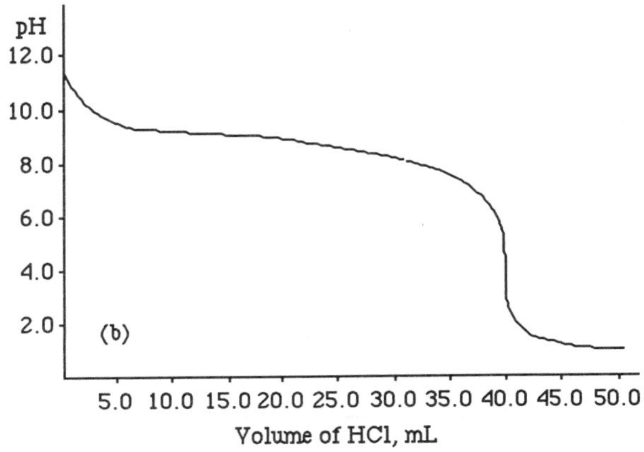

54. **(a)** This is the pH of 0.275 M NH_3.

Equation:	$NH_3(aq) + H_2O \rightleftharpoons$	$NH_4^+(aq)$ +	$OH^-(aq)$
Initial:	0.275 M		
Changes:	$-x$ M	$+x$ M	$+x$ M
Equil:	$(0.275 - x)$M	x M	x M

$$K_b = \frac{[NH_4^+][OH^-]}{[NH_3]} = 1.8 \times 10^{-5} = \frac{x^2}{0.275 - x} \approx \frac{x^2}{0.275} \qquad x = 2.2 \times 10^{-3}\ \text{M} = [OH^-]$$

pOH $= -\log(2.2 \times 10^{-3}) = 2.66$ $\qquad$ pH = 11.34 M

(b) This is the volume of titrant needed to reach the equivalence point.

The titration reaction is $\quad NH_3(aq) + HI(aq) \longrightarrow NH_4I(aq)$

$$\text{vol. HI} = 20.00\ \text{mL}\ NH_3(aq) \times \frac{0.275\ \text{mmol}\ NH_3}{1\ \text{mL}\ NH_3\ \text{soln}} \times \frac{1\ \text{mmol HI}}{1\ \text{mmol}\ NH_3} \times \frac{1\ \text{mL HI soln}}{0.325\ \text{mmol HI}} = \begin{matrix}16.9\ \text{mL}\\ \text{HI soln}\end{matrix}$$

(c) The pOH at the half-equivalence point of the titration of a weak base with a strong acid is equal to the pK_b of the weak base. $\quad$ pOH = pK_b = 4.74 $\quad$ pH = 14.00 – 4.74 = 9.26

(d) NH_4^+ is formed during the titration, and its hydrolysis determines the pH of the solution.

total volume of solution = 20.00 mL + 16.9 mL = 36.9 mL

$$\text{mmol}\ NH_4^+ = 20.00\ \text{mL}\ NH_3(aq) \times \frac{0.275\ \text{mmol}\ NH_3}{1\ \text{mL}\ NH_3\ \text{soln}} \times \frac{1\ \text{mmol}\ NH_4^+}{1\ \text{mmol}\ NH_3} = 5.50\ \text{mmol}\ NH_4^+$$

$$[NH_4^+] = \frac{5.50\ \text{mmol}\ NH_4^+}{36.9\ \text{mL soln}} = 0.149\ \text{M}$$

Equation:	$NH_4^+(aq) + H_2O \rightleftharpoons$	$NH_3(aq)$ +	$H_3O^+(aq)$
Initial:	0.149 M		
Changes:	$-x$ M	$+x$ M	$+x$ M
Equil:	$(0.149 - x)$M	x M	x M

$$K_a = \frac{K_w}{K_b} = \frac{1.0 \times 10^{-14}}{1.8 \times 10^{-5}} = \frac{[NH_3][H_3O^+]}{[NH_4^+]} = \frac{x^2}{0.149 - x} \approx \frac{x^2}{0.149} \qquad x = 9.1 \times 10^{-6}\ \text{M} = [H_3O^+]$$

pH $= -\log(9.1 \times 10^{-6}) = 5.04$

55. A pH greater than 7.00 in the titration of a strong base with a strong acid means that the base is not completely titrated. A pH less than 7.00 means that excess acid has been added.

(a) We can determine $[OH^-]$ of the solution from the pH. $[OH^-]$ is also the quotient of the amount of hydroxide ion in excess divided by the volume of the solution: 20.00 mL base + x mL added acid.

pOH = 14.00 – pH = 14.00 – 12.55 = 1.45 $\qquad [OH^-] = 10^{-pOH} = 10^{-1.45} = 0.035$ M

$$[OH^-] = \frac{\left(20.00 \text{ mL base} \times \frac{0.175 \text{ mmol OH}^-}{1 \text{ mL base}}\right) - \left(x \text{ mL acid} \times \frac{0.200 \text{ mmol H}_3\text{O}^+}{1 \text{ mL acid}}\right)}{20.00 \text{ mL} + x \text{ mL}} = 0.035 \text{ M}$$

$3.50 - 0.200\,x = 0.70 + 0.035\,x$ $\quad 3.50 - 0.70 = 0.035\,x + 0.200\,x = 2.80 = 0.235\,x$

$x = \frac{2.80}{0.235} = 11.9$ mL acid added.

(b) The set-up here is the same as for part **(a)**.
pOH = 14.00 – pH = 14.00 – 10.80 = 3.20 $\quad [OH^-] = 10^{-pOH} = 10^{-3.20} = 0.00063$ M

$$[OH^-] = \frac{\left(20.00 \text{ mL base} \times \frac{0.175 \text{ mmol OH}^-}{1 \text{ mL base}}\right) - \left(x \text{ mL acid} \times \frac{0.200 \text{ mmol H}_3\text{O}^+}{1 \text{ mL acid}}\right)}{20.00 \text{ mL} + x \text{ mL}} = 0.00063$$

$3.50 - 0.200\,x = 0.0126 + 0.00063\,x$ $\quad 3.50 - 0.0126 = 0.00063\,x + 0.200\,x = 3.49 = 0.201\,x$

$x = \frac{3.49}{0.201} = 17.4$ mL acid added. Almost at the equivalence point, 17.5 mL

(c) Here the acid is in excess, so we reverse the set-up of part **(a)**. We know that we are just slightly beyond the equivalence point. This is close to the "mirror image" of part **(b)**.
$[H_3O^+] = 10^{-pH} = 10^{-4.25} = 0.000056$ M

$$[H_3O^+] = \frac{\left(x \text{ mL acid} \times \frac{0.200 \text{ mmol H}_3\text{O}^+}{1 \text{ mL acid}}\right) - \left(20.00 \text{ mL base} \times \frac{0.175 \text{ mmol OH}^-}{1 \text{ mL base}}\right)}{20.00 \text{ mL} + x \text{ mL}}$$

$= 5.6 \times 10^{-5}$ M

$0.200\,x - 3.50 = 0.0011 + 5.6 \times 10^{-5}\,x$ $\quad 3.50 + 0.0011 = -5.6 \times 10^{-5}x + 0.200\,x = 3.50 = 0.200\,x$

$x = \frac{3.50_1}{0.200} = 17.5_1$ mL acid added. Just beyond the equivalence point, 17.5 mL

56. In the titration of a weak acid with a strong base, the middle range of the titration, with the pH within one unit of pK_a (= 4.74 for acetic acid), is known as the buffer region. The Henderson-Hasselbalch equation can be used to determine the ratio of weak acid and anion concentrations. The amount of weak acid then is used in these calculations to determine the amount of base to be added.

(a) $pH = pK_a + \log \frac{[C_2H_3O_2^-]}{[HC_2H_3O_2]} = 3.85 = 4.74 + \log \frac{[C_2H_3O_2^-]}{[HC_2H_3O_2]}$ $\quad \log \frac{[C_2H_3O_2^-]}{[HC_2H_3O_2]} = 3.85 - 4.74$

$\frac{[C_2H_3O_2^-]}{[HC_2H_3O_2]} = 10^{-0.89} = 0.13$ $\quad$ amount of $HC_2H_3O_2 = 25.00 \text{ mL} \times \frac{0.100 \text{ mmol HC}_2\text{H}_3\text{O}_2}{1 \text{ mL acid}} = 2.50$ mmol

Since acetate ion and acetic acid are in the same solution, we can use their amounts in millimoles in place of their concentrations. The amount of acetate ion is the amount created by the addition of strong base, one millimole of acetate ion for each millimole of strong based added. The amount of acetic acid is reduced by the same number of millimoles. $\quad HC_2H_3O_2 + OH^- \longrightarrow C_2H_3O_2^- + H_2O$

$$0.13 = \frac{x \text{ mL base} \times \frac{0.200 \text{ mmol OH}^-}{\text{mL base}} \times \frac{1 \text{ mmol C}_2\text{H}_3\text{O}_2^-}{1 \text{ mmol OH}^-}}{2.50 \text{ mmol HC}_2\text{H}_3\text{O}_2 - \left(x \text{ mL base} \times \frac{0.200 \text{ mmol OH}^-}{\text{mL base}}\right)} = \frac{0.200\,x}{2.50 - 0.200\,x}$$

$0.200\,x = 0.13(2.50 - 0.200\,x) = 0.33 - 0.026\,x$ $\quad 0.33 = 0.200\,x + 0.026\,x = 0.226\,x$

$x = \frac{0.33}{0.226} = 1.5$ mL base

(b) This is the same set-up as part **(a)**, except for a different ratio of concentrations.

$pH = 5.25 = 4.74 + \log \frac{[C_2H_3O_2^-]}{[HC_2H_3O_2]}$ $\quad \log \frac{[C_2H_3O_2^-]}{[HC_2H_3O_2]} = 5.25 - 4.74 = 0.51$

$\frac{[C_2H_3O_2^-]}{[HC_2H_3O_2]} = 10^{0.51} = 3.2$ $\quad 3.2 = \frac{0.200\,x}{2.50 - 0.200\,x}$

$0.200\,x = 3.2(2.50 - 0.200\,x) = 8.0 - 0.64\,x$ $\quad 8.0 = 0.200\,x + 0.64\,x = 0.84\,x$

$x = \frac{8.0}{0.84} = 9.5$ mL base $\quad$ This is close to the equivalence point, 12.5 mL base.

(c) This is after the equivalence point, where the pH is determined by the excess added base.
pOH = 14.00 – pH = 14.00 – 11.10 = 2.90 $\quad [OH^-] = 10^{-pOH} = 10^{-2.90} = 0.0013$ M

$$[OH^-] = 0.0013 \text{ M} = \frac{x \text{ mL} \times \frac{0.200 \text{ mmol OH}^-}{1 \text{ mL base}}}{x \text{ mL} + (12.50 \text{ mL} + 25.00 \text{ mL})} = \frac{0.200\,x}{37.50 + x}$$

$0.200\,x = 0.0013(37.50 + x) = 0.049 + 0.0013\,x \qquad x = \frac{0.049}{0.200 - 0.0013} = 0.25 \text{ mL excess}$

Total base added = 12.5 mL to equivalence point + 0.25 mL excess = 12.8 mL

57. For each of the titrations, the pH at the half-equivalence point equals the pK_a of the acid.

The initial pH is that of 0.1000 M weak acid: $\qquad K_a = \frac{x^2}{0.1000} \qquad x = \sqrt{0.1000 \times K_a} = [H_3O^+]$

The pH at the equivalence point is that of 0.05000 M anion of the weak acid, for which the $[OH^-]$ is determined as follows. $\qquad K_b = \frac{K_w}{K_a} = \frac{x^2}{0.05000} \qquad x = \sqrt{\frac{K_w}{K_a} 0.0500} = [OH^-]$

And finally when 0.100 mL of base has been added beyond the equivalence point, the pH is determined by the excess added base, as follows (for all three titrations).

$$[OH^-] = \frac{0.100 \text{ mL} \times \frac{0.1000 \text{ mmol NaOH}}{1 \text{ mL NaOH soln}} \times \frac{1 \text{ mmol OH}^-}{1 \text{ mmol NaOH}}}{20.1 \text{ mL soln total}} = 4.98 \times 10^{-4} \text{ M}$$

$pOH = -\log(4.98 \times 10^{-4}) = 3.303 \qquad pH = 14.000 - 3.303 = 10.697$

(a) Initial: $[H_3O^+] = \sqrt{0.1000 \times 7.0 \times 10^{-3}} = 0.026 \text{ M} \qquad pH = 1.59$

Equiv: $[OH^-] = \sqrt{\frac{1.0 \times 10^{-14}}{7.0 \times 10^{-3}} \times 0.0500} = 2.7 \times 10^{-7} \qquad pOH = 6.57 \quad pH = 14.00 - 6.57 = 7.43$

Indicator: bromothymol blue, yellow at pH = 6.2 and blue at pH = 7.8

(b) Initial: $[H_3O^+] = \sqrt{0.1000 \times 3.0 \times 10^{-4}} = 0.0055 \text{ M} \qquad pH = 2.26$

Equiv: $[OH^-] = \sqrt{\frac{1.0 \times 10^{-14}}{3.0 \times 10^{-4}} \times 0.0500} = 1.3 \times 10^{-6} \qquad pOH = 5.89 \quad pH = 14.00 - 5.89 = 8.11$

Indicator: thymol blue, yellow at pH = 8.0 and blue at pH =10.0

(c) Initial: $[H_3O^+] = \sqrt{0.1000 \times 2.0 \times 10^{-8}} = 0.000045 \text{ M} \qquad pH = 4.35$

Equiv: $[OH^-] = \sqrt{\frac{1.0 \times 10^{-14}}{2.0 \times 10^{-8}} \times 0.0500} = 1.6 \times 10^{-4} \qquad pOH = 3.80 \quad pH = 14.00 - 3.80 = 10.20$

Indicator: alizarin yellow R, yellow at pH = 10.0 and violet at pH = 12.0

The three titration curves are sketched on the same axes below.

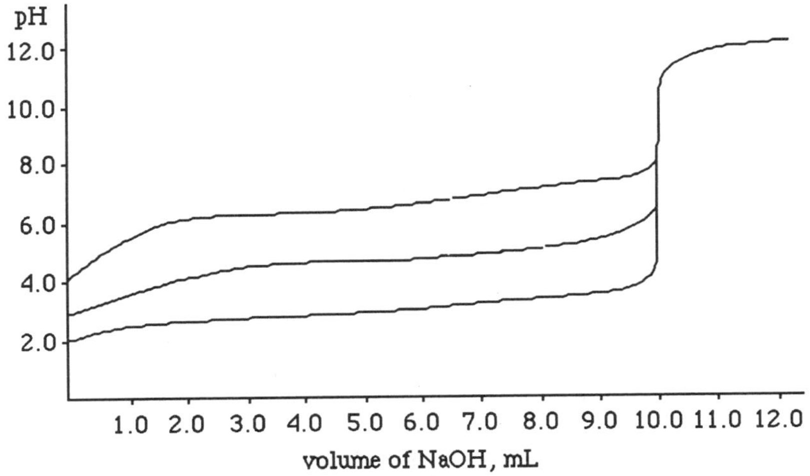

58. For each of the titrations, the pOH at the half-equivalence point equals the pK_b of the base.

The initial pOH is that of 0.1000 M weak base, determined as follows.

$K_b = \frac{x^2}{0.1000} \qquad x = \sqrt{0.1000 \times K_a} = [OH^-]$

The pH at the equivalence point is that of 0.05000 M cation of the weak base, for which the $[H_3O^+]$ is determined as follows. $\qquad K_a = \frac{K_w}{K_b} = \frac{x^2}{0.05000} \qquad x = \sqrt{\frac{K_w}{K_b} 0.0500} = [H_3O^+]$

And finally when 0.100 mL of acid has been added beyond the equivalence point, the pH is determined by the excess added acid, as follows.

$$[H_3O^+] = \frac{0.100 \text{ mL} \times \dfrac{0.1000 \text{ mmol HCl}}{1 \text{ mL HCl soln}} \times \dfrac{1 \text{ mmol } H_3O^+}{1 \text{ mmol HCl}}}{20.1 \text{ mL soln total}} = 4.98 \times 10^{-4} \text{ M}$$

$\text{pH} = -\log(4.98 \times 10^{-4}) = 3.303$

(a) Initial: $[OH^-] = \sqrt{0.1000 \times 1 \times 10^{-3}} = 0.01$ M pOH = 2.0 pH = 12.0

Half-equiv: $\text{pOH} = -\log(1 \times 10^{-3}) = 3.0$ pH = 11.0

Equiv: $[H_3O^+] = \sqrt{\dfrac{1.0 \times 10^{-14}}{1 \times 10^{-3}} \times 0.0500} = 7 \times 10^{-7}$ pH = 6.2

Indicator: methyl red, yellow at pH = 6.3 and red at pH = 4.5

(b) Initial: $[OH^-] = \sqrt{0.1000 \times 3 \times 10^{-6}} = 5 \times 10^{-4}$ M pOH = 3.3 pH = 10.7

Half-equiv: $\text{pOH} = -\log(3 \times 10^{-6}) = 5.5$ pH = 8.5

Equiv: $[H_3O^+] = \sqrt{\dfrac{1.0 \times 10^{-14}}{3 \times 10^{-6}} \times 0.0500} = 1 \times 10^{-5}$ pH = 5.0

Indicator: methyl red, yellow at pH = 6.3 and red at pH = 4.5

(c) Initial: $[OH^-] = \sqrt{0.1000 \times 7 \times 10^{-8}} = 8 \times 10^{-5}$ M pOH = 4.1 pH = 9.9

Half-equiv: $\text{pOH} = -\log(7 \times 10^{-8}) = 7.2$ pH =6.8

Equiv: $[H_3O^+] = \sqrt{\dfrac{1.0 \times 10^{-14}}{7 \times 10^{-8}} \times 0.0500} = 8 \times 10^{-5}$ pH = 4.1

Indicator: bromophenol blue, blue at pH = 4.5 and yellow at pH = 3.0

The three titration curves are sketched on the same axes below.

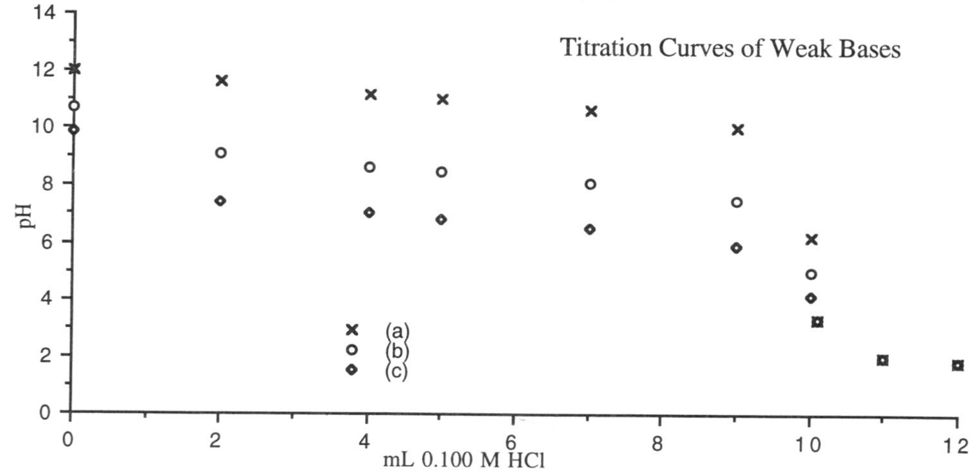

pH of Salts of Polyprotic Acids

59. We expect a solution of Na_2S to be alkaline, or basic. This alkalinity is created by the hydrolysis of the sulfide ion, the anion of a very weak acid ($K_2 = 1 \times 10^{-19}$ for H_2S). $S^{2-}(aq) + H_2O \rightleftharpoons HS^-(aq) + OH^-(aq)$

60. We expect the pH of a solution of sodium dihydrogen citrate, NaH_2Cit, to be acidic because pK_a values of first and second ionization constants of polyprotic acids are reasonably large. The pH of a solution of the salt is the average of pK_1 and pK_2. For citric acid, in fact, this average is $(3.13 + 4.76) \div 2 = 3.95$. This is an acidic solution.

61. **(a)** $H_3PO_4(aq) + CO_3^{2-}(aq) \longrightarrow H_2PO_4^-(aq) + HCO_3^-(aq)$

$H_2PO_4^-(aq) + CO_3^{2-}(aq) \longrightarrow HPO_4^{2-}(aq) + HCO_3^-(aq)$

$HPO_4^{2-}(aq) + OH^-(aq) \longrightarrow PO_4^{3-}(aq) + H_2O$

(b) The pH values of 1.00 M solutions of the three ions are;
1.0 M OH^- has pH = 14.00   1.0 M CO_3^{2-} has pH = 12.11   1.0 M PO_4^{3-} has pH = 13.19
Thus, we see that CO_3^{2-} is not a strong enough base to remove the third proton from H_3PO_4.
As an alternative method of solving this problem, we can compute the equilibrium constant for each of the reactions of carbonate ion with the various protonated species of H_3PO_4.

$$H_3PO_4 + CO_3^{2-} \longrightarrow H_2PO_4^- + HCO_3^- \qquad K = \frac{K_1\{H_3PO_4\}}{K_2\{H_2CO_3\}} = \frac{7.1 \times 10^{-3}}{4.7 \times 10^{-11}} = 1.5 \times 10^8$$

$$H_2PO_4^- + CO_3^{2-} \longrightarrow HPO_4^{2-} + HCO_3^- \qquad K = \frac{K_2\{H_3PO_4\}}{K_2\{H_2CO_3\}} = \frac{6.3 \times 10^{-8}}{4.7 \times 10^{-11}} = 1.3 \times 10^3$$

$$HPO_4^{2-} + CO_3^{2-} \longrightarrow PO_4^{3-} + HCO_3^- \qquad K = \frac{K_3\{H_3PO_4\}}{K_2\{H_2CO_3\}} = \frac{4.2 \times 10^{-13}}{4.7 \times 10^{-11}} = 8.9 \times 10^{-3}$$

Since the equilibrium constant for the third reaction is much smaller than 1.00, we conclude that it proceeds to the right to only a negligible extent and thus is not a practical method of producing PO_4^{3-}. The other two reactions have reasonably large equilibrium constants, and would be expected to form product. They have the advantage of using an inexpensive base and, even if they do not go to completion, they will be drawn to completion by reaction with OH^- in the last step of the process.

62. We expect the bicarbonate ion to hydrolyze and that to establish the pH of the solution.

Equation:	$HCO_3^- + H_2O \rightleftharpoons$	"H_2CO_3" +	$OH^-(aq)$
Initial:	1.00 M		
Changes:	$-x$ M	$+x$ M	$+x$ M
Equil:	$(1.00 - x)$ M	x M	x M

$$K_b = \frac{K_w}{K_1} = \frac{1.00 \times 10^{-14}}{4.4 \times 10^{-7}} = 2.3 \times 10^{-8} = \frac{[H_2CO_3][OH^-]}{[HCO_3^-]} = \frac{(x)\,(x)}{1.00 - x} \approx \frac{x^2}{1.00}$$

$x = \sqrt{1.00 \times 2.3 \times 10^{-8}} = 1.5 \times 10^{-4}$ M = $[OH^-]$ pOH = $-\log(1.5 \times 10^{-4}) = 3.82$ pH = 10.18
For 1.00 M NaOH, $[OH^-] = 1.00$ pOH = $-\log(1.00) = 0.00$ pH = 14.00
Both 1.00 M $NaHCO_3$ and 1.00 M NaOH have an equal capacity to neutralize acids since one mole of each neutralizes one mole of strong acid. $NaOH(aq) + H_3O^+(aq) \longrightarrow Na^+(aq) + 2\,H_2O$
$NaHCO_3(aq) + H_3O^+(aq) \longrightarrow Na^+(aq) + CO_2(g) + 2\,H_2O$ But on a per gram basis, the one with the smaller molar mass is the more effective. Because the molar mass of NaOH is 40.0 g/mol, while that of $NaHCO_3$ is 84.0 g/mol, NaOH(s) is more than twice as effective than $NaHCO_3$(s) on a per gram basis.

$NaHCO_3$ is preferred in laboratories for safety and expense reasons. If NaOH were used, one would have to continually worry about the hazards of the NaOH, a strong base. But $NaHCO_3$, baking soda, is relatively nonhazardous. It also is much cheaper than NaOH.

63. (a) $Ba(OH)_2$ is a strong base.   pOH = 14.00 − 11.88 = 2.12 $[OH^-] = 10^{-2.12} = 0.0076$ M

$$[Ba(OH)_2] = \frac{0.0076 \text{ mol } OH^-}{1 \text{ L}} \times \frac{1 \text{ mol } Ba(OH)_2}{2 \text{ mol } OH^-} = 0.0038 \text{ M}$$

(b) $pH = 4.52 = pK_a + \log\frac{[C_2H_3O_2^-]}{[HC_2H_3O_2]} = 4.74 + \log\frac{0.294 \text{ M}}{[HC_2H_3O_2]}$ $\log\frac{0.294 \text{ M}}{[HC_2H_3O_2]} = 4.52 - 4.74$

$$\frac{0.294 \text{ M}}{[HC_2H_3O_2]} = 10^{-0.22} = 0.60 \qquad [HC_2H_3O_2] = \frac{0.294 \text{ M}}{0.60} = 0.49 \text{ M}$$

64. (a) pOH = 14.00 − 8.95 = 5.05 $[OH^-] = 10^{-5.05} = 8.9 \times 10^{-6}$ M

Equation:	$C_6H_5NH_2(aq) + H_2O \rightleftharpoons$	$C_6H_5NH_3^+(aq)$ +	$OH^-(aq)$
Initial:	x M		
Changes:	-8.9×10^{-6} M	$+8.9 \times 10^{-6}$ M	$+8.9 \times 10^{-6}$ M
Equil:	$(x - 8.9 \times 10^{-6})$M	8.9×10^{-6} M	8.9×10^{-6} M

$$K_b = \frac{[C_6H_5NH_3^+][OH^-]}{[C_6H_5NH_2]} = 7.4 \times 10^{-10} = \frac{(8.9 \times 10^{-6})^2}{x - 8.9 \times 10^{-6}}$$

$$x - 8.9 \times 10^{-6} = \frac{(8.9 \times 10^{-6})^2}{7.4 \times 10^{-10}} = 0.11 \text{ M} \qquad x = 0.11 \text{ M} = \text{molarity of aniline}$$

(b) $[H_3O^+] = 10^{-5.12} = 7.6 \times 10^{-6}$ M

Equation:	$NH_4^+(aq) + H_2O \rightleftharpoons$	$NH_3(aq)$ +	$H_3O^+(aq)$
Initial:	x M		
Changes:	-7.6×10^{-6} M	$+7.6 \times 10^{-6}$ M	$+7.6 \times 10^{-6}$ M
Equil:	$(x - 7.6 \times 10^{-6})$ M	7.6×10^{-6} M	7.6×10^{-6} M

$$K_a = \frac{[NH_3][H_3O^+]}{[NH_4^+]} = \frac{K_w}{K_b \text{ for } NH_3} = \frac{1.0 \times 10^{-14}}{1.8 \times 10^{-5}} = 5.6 \times 10^{-10} = \frac{(7.6 \times 10^{-6})^2}{x - 7.6 \times 10^{-6}}$$

$$x - 7.6 \times 10^{-6} = \frac{(7.6 \times 10^{-6})^2}{5.6 \times 10^{-10}} = 0.10 \text{ M} \qquad x = [NH_4^+] = [NH_4Cl] = 0.10 \text{ M}$$

65. **(a)** pOH = 14.00 – 6.07 = 7.93 $\qquad [OH^-] = 10^{-7.93} = 1.2 \times 10^{-8}$ M

$$Q_b = \frac{[NH_4^+][OH^-]}{[NH_3]} = \frac{(0.10 \text{ M})(1.2 \times 10^{-8} \text{ M})}{0.10} = 1.2 \times 10^{-8}$$

Since the value of Q_b does not equal the tabulated value of $K_b = 1.8 \times 10^{-5}$, the solution described cannot exist.

(b) These solutes can be added to the same solution, but the final solution will have an appreciable $[HC_2H_3O_2]$ because of the reaction of $H_3O^+(aq)$ with $C_2H_3O_2^-(aq)$.

Equation:	$H_3O^+(aq)$ +	$C_2H_3O_2^-(aq) \longrightarrow$	$HC_2H_3O_2(aq) + H_2O$
Initial:	0.058 M	0.10 M	
Changes:	–0.058 M	–0.058 M	+0.058 M
Final:	0.000 M	0.04 M	0.058 M

Of course, some H_3O^+ will exist in the final solution, but not equivalent to 0.058 M HI.

(c) Both 0.10 M KNO_2 and 0.25 M KNO_3 can exist together. Some hydrolysis of the $NO_2^-(aq)$ ion will occur, forming $HNO_2(aq)$.

(d) $Ba(OH)_2$ is a strong base and will react as much as possible with the weak conjugate acid NH_4^+, to form $NH_3(aq)$. We will end up with a solution of $BaCl_2(aq)$, $NH_3(aq)$, and unreacted $NH_4Cl(aq)$.

(e) This will be a benzoic acid–benzoate ion buffer solutions. Since the two components have the same concentration, the buffer solution will have pH = pK_a = $-\log(6.3 \times 10^{-5})$ = 4.20. This solution indeed can exist.

(f) The first three components contain no ions that will hydrolyze. But $C_2H_3O_2^-$ is the anion of a weak acid and will hydrolyze to form a slightly basic solution. Since pH = 6.4 is an acidic solution, the solution described cannot exist.

66. **(a)** When $[H_3O^+]$ and $[HC_2H_3O_2]$ are high and $[C_2H_3O_2^-]$ is very low, a common ion—H_3O^+—has been added to a solution of acetic acid, suppressing its ionization.

(b) When $[C_2H_3O_2^-]$ is high and $[H_3O^+]$ and $[HC_2H_3O_2]$ are very low, we are dealing with a solution of acetate ion, which hydrolyzes to produce a limited concentration of $HC_2H_3O_2$.

(c) When $[HC_2H_3O_2]$ is high and both $[H_3O^+]$ and $[C_2H_3O_2^-]$ are low, the solution is an acetic acid solution, in which the solute is partially ionized.

(d) When both $[HC_2H_3O_2]$ and $[C_2H_3O_2^-]$ are high while $[H_3O^+]$ is low, the solution is a buffer solution, in which the presence of acetate ion suppresses the ionization of acetic acid.

FEATURE PROBLEMS

A1. The two curves cross that the point where half the acetate is present as acetic acid and half is present as acetate ion. This is the half equivalence point in a titration, where pH = pK_a = 4.74.

A2. For carbonic acid, there are three carbonate containing species: "H_2CO_3" which predominates at low pH, HCO_3^-, and CO_3^{2-} which prediminates in alkaline solution. The points of intersection should occur at the half-equivalence points in each step-wise titration: at pH = pK_{a_1} = $-\log(4.4 \times 10^{-7})$ = 6.36 and at pH = pK_{a_2} = $-\log(4.7 \times 10^{-11})$ = 10.33. The following graph was computer-calculated (and then drawn) from these equations. f in each instance represents the fraction of the species whose formula is in parentheses.

$$\frac{1}{f(H_2A)} = 1 + \frac{K_1}{[H^+]} + \frac{K_1 K_2}{[H^+]^2} \qquad \frac{1}{f(HA^-)} = \frac{[H^+]}{K_1} + 1 + \frac{K_2}{[H^+]} \qquad \frac{1}{f(A^{2-})} = \frac{[H^+]^2}{K_1 K_2} + \frac{[H^+]}{K_2} + 1$$

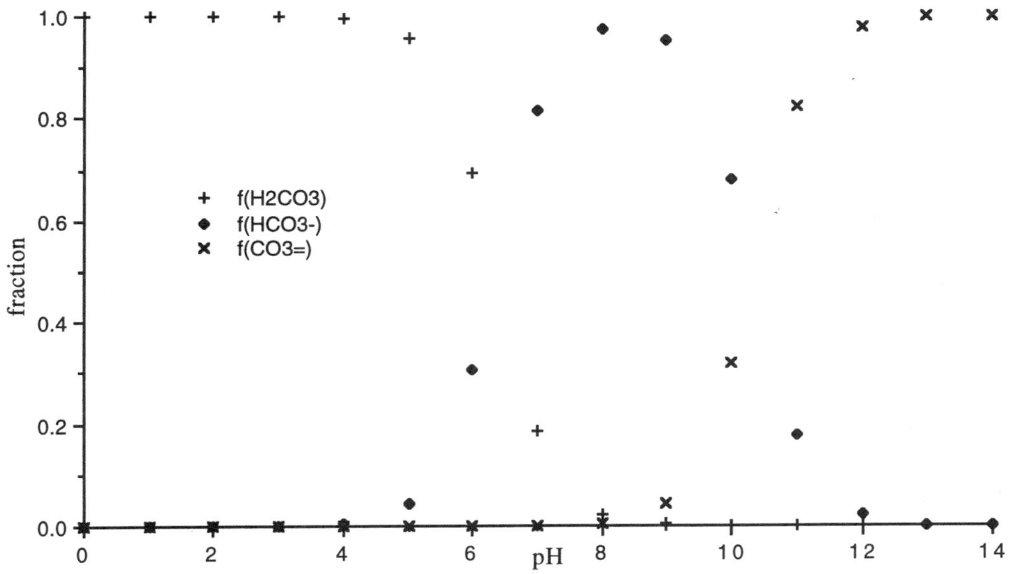

A3. For phosphoric acid, there are four phosphate containing species: H_3PO_4 under acidic conditions, $H_2PO_4^-$, HPO_4^{2-}, and PO_4^{3-} which predominates in alkaline solution. The points of intersection should occur at pH = $pK_{a_1} = -\log(7.1 \times 10^{-3}) = 2.15$, pH = $pK_{a_2} = -\log(6.3 \times 10^{-8}) = 7.20$, and pH = $pK_{a_3} = -\log(4.2 \times 10^{-13}) = 12.38$, a quite alkaline solution. The graph that follows was computer-calculated and drawn.

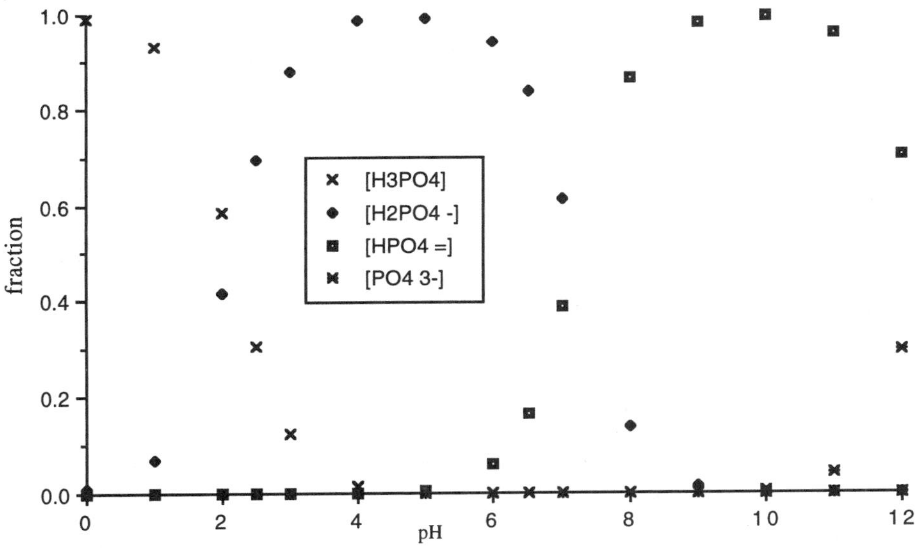

B1. In the titration of 25.00 mL of 0.100 M NaOH with 0.100 M HCl, the initial pOH is that of 0.100 M NaOH, pOH = –log(0.100) = 1.000.
The pH at the equivalence point of strong acid titrating strong base is 7.000.
In all calculations, we take advantage of HCl being a strong acid and realize that there is 0.100 mmol H_3O^+ in each mL of 0.100 M HCl, and similarly that there is 0.100 mmol OH^- in each mL of 0.100 M NaOH, a strong base.
After the addition of 2.50 mL acid (10.0% titrated),

$$[OH^-] = \frac{\left(25.00 \text{ mL base} \times \dfrac{0.100 \text{ mmol OH}^-}{1 \text{ mL base}}\right) - \left(2.50 \text{ mL acid} \times \dfrac{0.100 \text{ mmol H}_3\text{O}^+}{1 \text{ mL acid}}\right)}{25.00 \text{ mL} + 2.50 \text{ mL}}$$

$$= \frac{2.500 - 0.250}{27.50} = 0.0818 \text{ M} \qquad \text{pOH} = -\log(0.0818) = 1.087$$

Notice that there is a general formula, if we let f = fraction titrated = $\frac{\%\text{ titrated}}{100}$

$$[OH^-] = \frac{2.500 \text{ mmol } OH^- - (f \times 2.500) \text{ mmol } H_3O^+}{25.00 \text{ mL acid} + (f \times 25.00) \text{ mL base}}$$

At 25% titrated $[OH^-] = \frac{2.500 - (0.25 \times 2.500)}{25.00 + (25.00 \times 0.250)} = 0.0600$ M pOH = –log(0.0600) = 1.222

We let a spread sheet on a computer calculate these points again (to check) as well as the remaining points.

% titr	0	5	10	20	30	40	50	60	70	80	90	95	100
pH	1	1.04	1.09	1.18	1.27	1.37	1.48	1.60	1.75	1.95	2.28	2.59	7.00
mL HCl	0	1.25	2.50	5.00	7.50	10.00	12.50	15.00	17.50	20.00	22.50	23.75	25.00

After the endpoint of the titration, the solution is acidic, with its pH determined by the excess added acid.

At 110% titrated
2.50 mL in excess:
$$[H_3O^+] = \frac{2.50 \text{ mL acid} \times \frac{0.100 \text{ mmol } H_3O^+}{1 \text{ mL acid}}}{50.00 \text{ mL at endpoint} + 2.50 \text{ mL excess acid}} = 0.00476 \text{ M}$$

pH = –log(0.00476) = 2.32 pOH = 14.00 – pH = 14.00 – 2.32 = 11.68

Again we notice a general formula; we let e = excess fraction titrated = $\frac{\%\text{overtitrated}}{100}$

$$[H_3O^+] = \frac{(2.500 \times e) \text{ mmol } H_3O^+}{50.00 \text{ mL at end point} + (25.00 \times e) \text{ mL excess acid}}$$

Again, we let the computer calculate the necessary points, and also plot the result.

% titr	100	105	110	120	130	140
pH	7.00	11.39	11.68	11.96	12.12	12.22
mL HCl	25.00	26.25	27.50	30.00	32.50	35.00

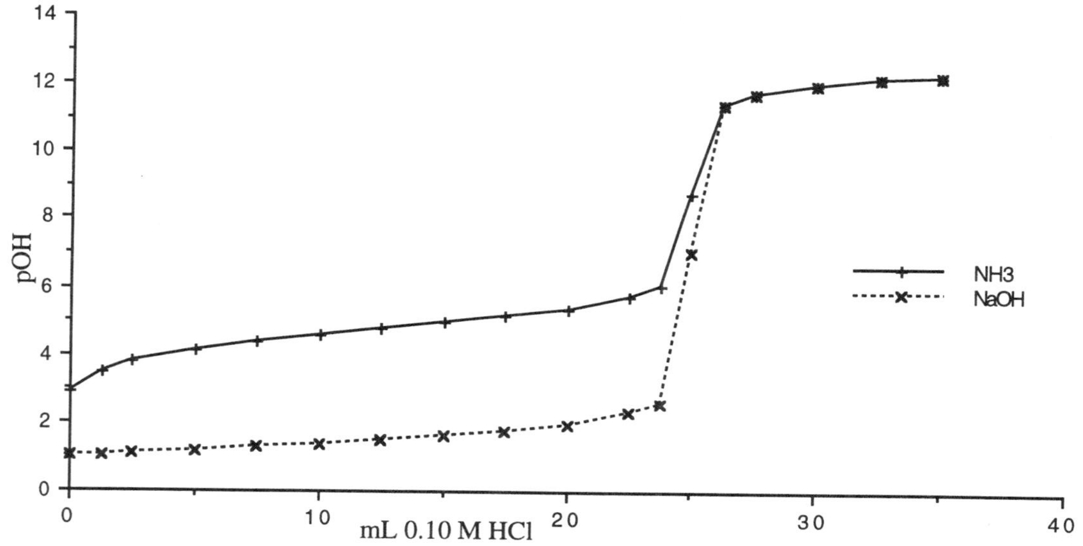

For the titration of 0.100 M NH_3, the initial pOH is that of 0.100 M NH_3, computed as follows.

$$K_b = \frac{[NH_4^+][OH^-]}{[NH_3]} = 1.8 \times 10^{-5} = \frac{(x)(x)}{0.100 - x} \approx \frac{x^2}{0.100} \qquad x = \sqrt{0.100 \times 1.8 \times 10^{-5}} = 0.0013_4 \text{ M}$$

pOH = $-\log(0.0013_4)$ = 2.87

The pH at the equivalence point is that of 0.0500 M NH_4Cl, computed as follows.

$$K_a = \frac{[NH_3][H_3O^+]}{[NH_4^+]} = \frac{K_w}{K_b} = \frac{1.0 \times 10^{-14}}{1.8 \times 10^{-5}} = 5.6 \times 10^{-10} = \frac{(x)(x)}{0.0500 - x} \approx \frac{x^2}{0.0500}$$

$x = \sqrt{0.0500 \times 5.6 \times 10^{-10}} = 5.3 \times 10^{-6}$ M pH = $-\log(5.3 \times 10^{-6})$ = 5.28

pOH = 14.00 – 5.28 = 8.72

The pOH of points between 0% and 100% titrated is computed with the Henderson–Hasselbalch equation, slightly modified, as follows.

$$pH = pK_a + \log\frac{[NH_3]}{[NH_4^+]} \qquad pOH = 14.00 - pH = 14.00 - pK_a - \log\frac{[NH_3]}{[NH_4^+]} = pK_w - pK_a + \log\frac{[NH_4^+]}{[NH_3]}$$

Now $pK_w - pK_a = -\log \frac{K_w}{K_a} = -\log(K_b) = pK_b$ Thus, $pOH = pK_b + \log \frac{[NH_4^+]}{[NH_3]}$

From calculations we have performed in this chapter, we know that we can use amounts (in millimoles) in place of molar concentrations in this expression. Thus, at 10% titrated (0.10 titrated), we have:

$$pOH = pK_b + \log \frac{[NH_4^+]}{[NH_3]} = 4.74 + \log \frac{0.10 \times 2.50 \text{ mmol } NH_4^+}{0.90 \times 2.50 \text{ mmol } NH_3} = 4.74 - 0.95 = 3.79$$

This works because 10% of the original 2.50 mmol NH_3 has reacted with HCl and formed $NH_4^+Cl^-$, and 90% of the NH_3 remains unreacted. $NH_3(aq) + HCl(aq) \longrightarrow NH_4^+(aq) + Cl^-(aq)$

Notice also that the fraction $[NH_4^+]/[NH_3\}$ can be represented by (10%) ÷ (90%), or (percent titrated) ÷ (percent untitrated). We allow the computer to perform these calculations. Notice that the pOH at each point after the equivalence point is the same as in the strong acid–strong base titration.

pH	2.87	3.47	3.79	4.14	4.38	4.57	4.74	4.92	5.11	5.35	5.70	6.02	8.72
% titr	0	5	10	20	30	40	50	60	70	80	90	95	100
mL HCl	0	1.25	2.50	5.00	7.50	10.00	12.50	15.00	17.50	20.00	22.50	23.75	25.00

To obtain pH all we need to do is calculate pH = 14.00 – pOH. That has been done for both sets of data, and the plotted results follow.

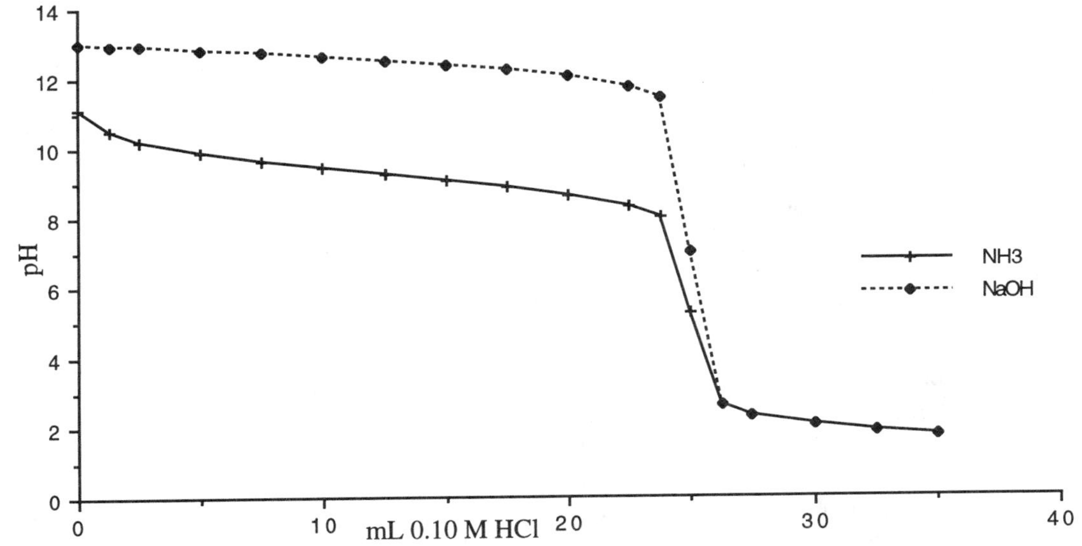

C1. This is exactly the same titration curve we would obtain for the titration of 25.00 mL of 0.200 M HCl with 0.200 M NaOH, because the acid species being titrated is H_3O^+. Both acids are strong acids and have ionized completely before titration begins. The initial pH is that of 0.200 M NaOH = 0.200 M OH^-, pOH = –log(0.200) = 0.699, pH = 14.00 – 0.699 = 13.301. At the equivalence point, pH = 7.000. We use the expressions we derived in Feature Problem B to determine the pOH during the titration and after excess titrant is added, modified for the different molarity of the solutions, and the fact we are determining pH, rather than pOH.

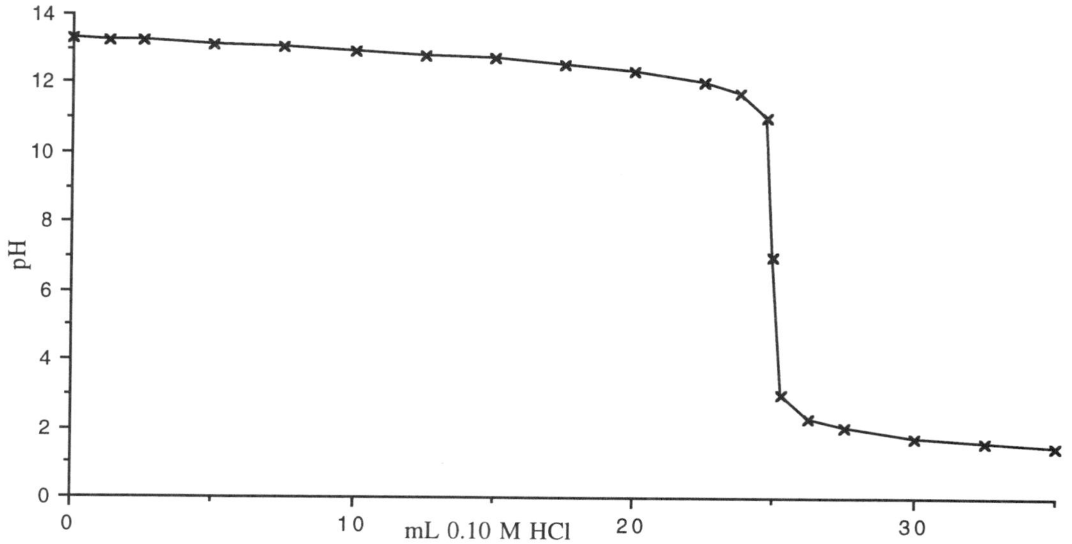

C2. In Figure 18-7, we note that the equivalence point of the titration of a strong acid occurs at pH = 7.00, but that the strong acid is essentially completely neutralized at pH = 4. In Figure 18-10, we see that the first equivalence point of H_3PO_4 occurs at about pH = 4.6. Thus, the first equivalence point represents the complete neutralization of HCl and the neutralization of H_3PO_4 to $H_2PO_4^-$. Then, the second equivalence point represents the neutralization of $H_2PO_4^-$ to HPO_4^{2-}. To reach the first equivalence point requires about 20.0 mL of 0.216 M NaOH, while to reach the second one requires a total of 30.0 mL of 0.216 M NaOH, or an additional 10.0 mL of base beyond the first equivalence point. The equations for the two titration reactions are as follows.

To the first equivalence point: $NaOH + H_3PO_4 \longrightarrow NaH_2PO_4 + H_2O$

$NaOH + HCl \longrightarrow NaCl + H_2O$

To the second equivalence point: $NaOH + NaH_2PO_4 \longrightarrow Na_2HPO_4 + H_2O$

There is a third equivalence point, not shown in the figure, that would require an additional 10.0 mL of base to reach. Its titration reaction is represented by the following equation.

To the third equivalence point: $NaOH + Na_2HPO_4 \longrightarrow Na_3PO_4 + H_2O$

We determine the molar concentration of H_3PO_4 and then of HCl. Notice that only 10.0 mL of the NaOH needed to reach the first equivalence point reacts with the HCl(aq); the rest reacts with H_3PO_4.

$$\frac{(30.0 - 20.0)\text{ mL NaOH(aq)} \times \dfrac{0.216\text{ mmol NaOH}}{1\text{ mL NaOH soln}} \times \dfrac{1\text{ mmol } H_3PO_4}{1\text{ mmol NaOH}}}{10.00\text{ mL acid soln}} = 0.216\text{ M } H_3PO_4$$

$$\frac{(20.0 - 10.0)\text{ mL NaOH(aq)} \times \dfrac{0.216\text{ mmol NaOH}}{1\text{ mL NaOH soln}} \times \dfrac{1\text{ mmol HCl}}{1\text{ mmol NaOH}}}{10.00\text{ mL acid soln}} = 0.216\text{ M HCl}$$

C3. We start with a phosphoric acid–dihydrogen phosphate buffer solution and titrate until all H_3PO_4 is consumed. We begin with $10.00\text{ mL} \times \dfrac{0.0400\text{ mmol } H_3PO_4}{1\text{ mL}} = 0.400\text{ mmol } H_3PO_4$ and the diprotic anion, $10.00\text{ mL} \times \dfrac{0.0150\text{ mmol } H_2PO_4^-}{1\text{ mL}} = 0.150\text{ mmol } H_2PO_4^{2-}$. The volume of 0.200 M NaOH needed is $0.400\text{ mmol } H_3PO_4 \times \dfrac{1\text{ mL base}}{0.0200\text{ mmol NaOH}} = 20.00\text{ mL}$ to reach the first equivalence point.

The pH of points during this titration are computed with the Henderson-Hasselbalch equation.

Initially: $pH = pK_1 + \log\dfrac{[H_2PO_4^-]}{[H_3PO_4]} = -\log(7.1 \times 10^{-3}) + \log\dfrac{0.150}{0.200} = 2.15 - 0.12 = 2.03$

At 5.00 mL: $pH = 2.15 + \log\dfrac{(0.150 + 0.050)}{(0.200 - 0.050)} = 2.15 + 0.12 = 2.27$

At 10.0 mL, pH = 2.54 At 15.0 mL, pH = 2.93

This is the first equivalence point, a solution of 30.00 mL (= 10.00 mL originally + 20.00 mL titrant), containing 0.400 mmol $H_2PO_4^-$ from the titration and the 0.150 mmol $H_2PO_4^{2-}$ originally present. This is a solution with $[H_2PO_4^-] = \dfrac{(0.400 + 0.150)\text{ mmol } H_2PO_4^-}{12.00\text{ mL}} = 0.0458$ M, which has $pH = \frac{1}{2}(pK_1 + pK_2) =$ $0.50\,(2.15 - \log(6.3 \times 10^{-8})) = 0.50\,(2.15 + 7.20) = 4.68$

To reach the second equivalence point means titrating 0.550 mmol $H_2PO_4^-$, which requires an additional volume of titrant given by $0.550\text{ mmol } H_2PO_4^- \times \dfrac{1\text{ mL base}}{0.0200\text{ mmol NaOH}} = 27.5$ mL. To determine pH during this titration, we divide the region into five equal portions of 5.5 mL and use the Henderson–Hasselbalch equation.

At (20.0 + 5.5) mL, $pH = pK_2 + \log \dfrac{[HPO_4^{2-}]}{[H_2PO_4^-]} = 7.20 + \log \dfrac{(0.20 \times 0.550)\text{ mmol } HPO_4^{2-}\text{ formed}}{(0.80 \times 0.550)\text{ mmol } H_2PO_4^-\text{ remaining}}$

$= 7.20 - 0.60 = 6.60$

At (20.0 + 11.0) mL = 33.0 mL, $pH = 7.20 + \log \dfrac{0.40 \times 0.550}{0.60 \times 0.550} = 7.02$

At 36.5 mL, pH = 7.37 At 42.0 mL, pH = 7.80

The pH at the second equivalence point is given by $pH = \frac{1}{2}(pK_2 + pK_3) = 0.50\,(7.20 - \log(4.2 \times 10^{-13})) =$ $0.50\,(7.20 + 12.38) = 9.79$. Another 27.50 mL of 0.020 M NaOH would be required to reach the third equivalence point. pH values at each of four equally spaced volumes of 5.50 mL additional 0.0200 M NaOH are computed as before, assuming the Henderson-Hasselbalch equation is valid.

At (47.50 + 5.50) mL = 53.00 mL, $pH = pK_3 + \log \dfrac{[PO_4^{3-}]}{[HPO_4^{2-}]} = 12.38 + \log \dfrac{0.20 \times 0.550}{0.80 \times 0.550}$

$= 12.38 - 0.60 = 11.78$

At 58.50 mL, pH = 12.20 At 64.50 mL, pH = 12.56 At 89.50 mL, pH = 13.33

At the last equivalence point, the solution will contain 0.550 mmol PO_4^{3-} in a total of 10.00 + 20.00 + 27.50 + 27.50 mL = 85.00 mL of solution, with $[PO_4^{3-}] = \dfrac{0.550\text{ mmol}}{85.00\text{ mL}} = 0.00647$ M. But we can never reach this point, because the pH of the 0.0200 M NaOH titrant is 12.30. And the titrant is diluted by its addition to the solution. Thus, our titration will cease sometime shortly after the second equivalence point. We never will see the third equivalence point, largely because the titrant is too dilute. Our results are plotted below.

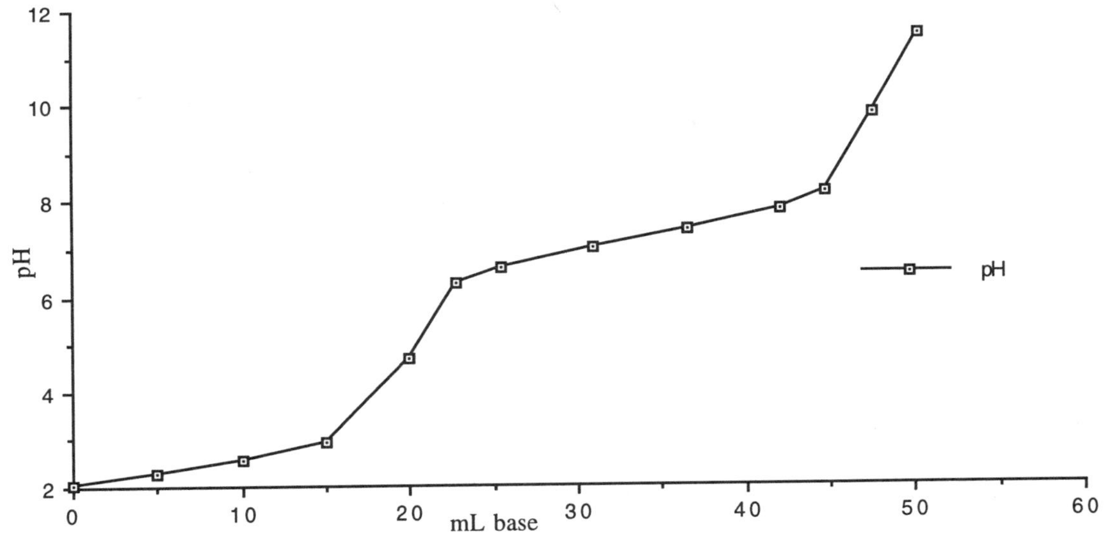

19 Solubility and Complex-Ion Equilibria

Practice Examples

1A In each case, we first write the balanced equation for the solubility equilibrium and then the equilibrium constant expression for that equilibrium, the K_{sp} expression.

(a) $MgCO_3(s) \rightleftharpoons Mg^{2+}(aq) + CO_3^{2-}(aq)$ $\quad K_{sp} = [Mg^{2+}][CO_3^{2-}]$

(b) $Ag_3PO_4(s) \rightleftharpoons 3\ Ag^+(aq) + PO_4^{3-}(aq)$ $\quad K_{sp} = [Ag^+]^3[PO_4^{3-}]$

1B **(a)** The hydrogen phosphate ion is not expected to ionize in aqueous solution to a significant extent because of the quite small values of the second and third ionization constants of phosphoric acid.
$CaHPO_4(s) \rightleftharpoons Ca^{2+}(aq) + HPO_4^{2-}(aq)$

(b) The ionization constant is written in the manner of a K_c expression: $K_{sp} = [Ca^{2+}][HPO_4^{2-}]$

2A We calculate the solubility of silver cyanate, s, as a molarity. We then use the solubility equation to (1) relate the concentrations of the ions and (2) write the K_{sp} expression.

$$s = \frac{7\text{ mg AgOCN}}{100\text{ mL}} \times \frac{1000\text{ mL}}{1\text{ L}} \times \frac{1\text{ g}}{1000\text{ mg}} \times \frac{1\text{ mol AgOCN}}{149.9\text{ g AgOCN}} = 5 \times 10^{-4}\text{ M}$$

Equation:	$AgOCN(s) \rightleftharpoons$	$Ag^+(aq)$	$+ OCN^-(aq)$
Solubility:		s	s

$K_{sp} = [Ag^+][OCN^-] = (s)\ (s) = s^2 = (5 \times 10^{-4})^2 = 3 \times 10^{-7}$

2B We calculate the solubility of lithium phosphate, s, as a molarity. We then use the solubility equation to (1) relate the concentrations of the ions and (2) write the K_{sp} expression.

$$s = \frac{0.034\text{ g }Li_3PO_4}{100\text{ mL soln}} \times \frac{1000\text{ mL}}{1\text{ L}} \times \frac{1\text{ mol }Li_3PO_4}{115.79\text{ g }Li_3PO_4} = 0.0029\text{ M}$$

Equation:	$Li_3PO_4(s) \rightleftharpoons$	$3\ Li^+(aq)$	$+ PO_4^{3-}(aq)$
Solubility:		$3s$	s

$K_{sp} = [Li^+]^3[PO_4^{3-}] = (3s)^3\ (s) = 27\ s^4 = 27\ (0.0029)^4 = 1.9 \times 10^{-9}$

3A We use the solubility equilibrium to write the K_{sp} expression, which we then solve to obtain the molar solubility, s, of $Fe(OH)_3$.

	$Fe(OH)_3(s) \rightleftharpoons$	$Fe^{3+}(aq)$	$+ 3\ OH^-(aq)$
Solubility:		s	$3s$

$K_{sp} = [Fe^{3+}][OH^-]^3 = (s)\ (3s)^3 = 27\ s^4 = 4 \times 10^{-38}$

$$s = \sqrt[4]{\frac{4 \times 10^{-38}}{27}} = 2 \times 10^{-10}\text{ M}$$

3B First we determine the solubility of $BaSO_4$, and then find the mass dissolved.

$BaSO_4(aq) \rightleftharpoons Ba^{2+}(aq) + SO_4^{2-}(aq)$ $\quad K_{sp} = [Ba^{2+}][SO_4^{2-}] = s^2$

The last relationship is true because $[Ba^{2+}] = [SO_4^{2-}]$ in a solution produced by dissolving $BaSO_4$ in pure water. Thus, $s = \sqrt{K_{sp}} = \sqrt{1.1 \times 10^{-10}} = 1.0_5 \times 10^{-5}$ M

$$\text{mass }BaSO_4 = 225\text{ mL} \times \frac{1.0_5 \times 10^{-5}\text{ mmol }BaSO_4}{1\text{ mL sat'd soln}} \times \frac{233.39\text{ mg }BaSO_4}{1\text{ mmol }BaSO_4} = 0.55\text{ mg }BaSO_4$$

4A For PbI_2, $K_{sp} = [Pb^{2+}][I^-]^2 = 7.1 \times 10^{-9}$ The solubility equilibrium is the basis of the calculation.

Equation:	$PbI_2(s) \rightleftharpoons$	$Pb^{2+}(aq)$ +	$2\ I^-(aq)$
Initial:		0.10 M	
Changes:		$+s$ M	$+2s$ M
Equil:		$(0.10 + s)$ M	$2s$ M

$K_{sp} = [Pb^{2+}][I^-]^2 = 7.1 \times 10^{-9} = (0.10 + s)(2s)^2 \approx 0.40\, s^2 \qquad s = \sqrt{\dfrac{7.1 \times 10^{-9}}{0.40}} = 1.3 \times 10^{-4}\ M$

This value of s is the solubility of PbI_2 in 0.10 M $Pb(NO_3)_2(aq)$.

4B We find pOH from the given pH: pOH = 14.00 – 8.20 = 5.80; $[OH^-] = 10^{-pOH} = 10^{-5.80} = 1.6 \times 10^{-6}$ M
We assume that pOH remains constant. and use the K_{sp} expression for $Fe(OH_3)$.

$K_{sp} = [Fe^{3+}][OH^-]^3 = 4 \times 10^{-38} = [Fe^{3+}](1.6 \times 10^{-6})^3 \qquad [Fe^{3+}] = \dfrac{4 \times 10^{-38}}{(1.6 \times 10^{-6})^3} = 1 \times 10^{-20}\ M$

The dissolved $Fe(OH)_3$ does not significantly affect $[OH^-]$.

5A First we determine $[I^-]$ as altered by dilution. We then compute Q_{sp} and compare it with K_{sp}.

$$[I^-] = \frac{3\ \text{drops} \times \dfrac{0.05\ \text{mL}}{1\ \text{drop}} \times \dfrac{0.20\ \text{mmol KI}}{1\ \text{mL}} \times \dfrac{1\ \text{mmol } I^-}{1\ \text{mmol KI}}}{100.0\ \text{mL soln}} = 3 \times 10^{-4}\ M$$

$Q_{sp} = [Ag^+][I^-] = (0.010\ M)(3 \times 10^{-4}) = 3 \times 10^{-6} > 8.5 \times 10^{-17} = K_{sp}$ Precipitation should occur.

5B We first use the solubility product constant expression for PbI_2 to determine the $[I^-]$ needed in solution to just form a precipitate when $[Pb^{2+}] = 0.010$ M. We assume that the volume of solution added is small and that $[Pb^{2+}]$ remains at 0.010 M throughout.

$K_{sp} = [Pb^{2+}][I^-]^2 = 7.1 \times 10^{-9} = (0.010)\,[I^-]^2 \qquad [I^-] = \sqrt{\dfrac{7.1 \times 10^{-9}}{0.010}} = 8.4 \times 10^{-4}\ M$

We determine the volume of 0.20 M KI needed.

$$\text{volume of KI(aq)} = 100.0\ \text{mL} \times \frac{8.4 \times 10^{-4}\ \text{mmol } I^-}{1\ \text{mL}} \times \frac{1\ \text{mmol KI}}{1\ \text{mmol } I^-} \times \frac{1\ \text{mL KI(aq)}}{0.20\ \text{mmol KI}} \times \frac{1\ \text{drop}}{0.050\ \text{mL}}$$
$$= 8.4\ \text{drops} = 9\ \text{drops}$$

Since one additional drop is needed, 10 drops will be required. This is an insufficient volume to dilute the original solution, so we are justified in assuming that $[Pb^{2+}]$ remains constant.

6A We find $[Ca^{2+}]$ that can coexist with $[OH^-] = 0.040$ M. $K_{sp} = 5.5 \times 10^{-6}$ for $Ca(OH)_2$.

$K_{sp} = 5.5 \times 10^{-6} = [Ca^{2+}][OH^-]^2 = [Ca^{2+}](0.040)^2 \quad [Ca^{2+}] = \dfrac{5.5 \times 10^{-6}}{(0.040)^2} = 3.4 \times 10^{-3}\ M$

For precipitation to be considered complete, $[Ca^{2+}]$ should be less than 0.1% of its original value. 3.4×10^{-3} M is 34% of 0.010 M and therefore precipitation of $Ca(OH)_2$ is not complete under these conditions.

6B We begin by finding $[Mg^{2+}]$ that corresponds to 1 μg Mg^{2+}/L. $\quad [Mg^{2+}] = \dfrac{1\ \mu g\ Mg^{2+}}{1\ \text{L soln}} \times \dfrac{1\ g\ Mg^{2+}}{10^6\ \mu g} \times \dfrac{1\ \text{mol } Mg^{2+}}{24.3\ g\ Mg^{2+}} = 4 \times 10^{-8}\ M$

Now we use the K_{sp} expression for $Mg(OH)_2$ to determine $[OH^-]$.

$K_{sp} = 1.8 \times 10^{-11} = [Mg^{2+}][OH^-]^2 = (4 \times 10^{-8})[OH^-]^2 \qquad [OH^-] = \sqrt{\dfrac{1.8 \times 10^{-11}}{4 \times 10^{-8}}} = 0.02\ M$

7A Let us first determine $[Ag^+]$ when AgCl(s) just begins to precipitate. At that point, Q_{sp} and K_{sp} are equal.

$K_{sp} = 1.8 \times 10^{-10} = [Ag^+][Cl^-] = Q_{sp} = [Ag^+] \times 0.115\ M \qquad [Ag^+] = \dfrac{1.8 \times 10^{-10}}{0.115} = 1.6 \times 10^{-9}\ M$

Now let us determine $[Br^-]$ that can coexist with this $[Ag^+]$.

$K_{sp} = 5.0 \times 10^{-13} = [Ag^+][Br^-] = 1.6 \times 10^{-9}\ M \times [Br^-] \qquad [Br^-] = \dfrac{5.0 \times 10^{-13}}{1.6 \times 10^{-9}} = 3.1 \times 10^{-4}\ M$

The remaining bromide ion has precipitated as AgBr(s) with the addition of $AgNO_3(aq)$.

$$\text{Percent of remaining } Br^- = \frac{[Br^-]_{final}}{[Br^-]_{initial}} \times 100\% = \frac{3.1 \times 10^{-4}\ M}{0.264\ M} \times 100\% = 0.12\%$$

7B Since the ions have the same charge and the same concentrations, we look for two K_{sp} values with the same anion that are as far apart as possible. The K_{sp} values for the hydroxides are similar, while those for the carbonates are quite different. Because K_{sp} for $CoCO_3$ (1.4×10^{-13}) is so much smaller than K_{sp} for $NiCO_3$ (6.6×10^{-9}), $NiCO_3$ will precipitate second. $NiCO_3$ will begin to precipitate when $[CO_3^{2-}]$ has the value:

$[CO_3^{2-}] = \frac{K_{sp}}{[Ni^{2+}]} = \frac{6.6 \times 10^{-9}}{0.10} = 6.6 \times 10^{-8}$ M At this point $[Co^{2+}]$ is found as follows.

$[Co^{2+}] = \frac{K_{sp}}{[CO_3^{2-}]} = \frac{1.4 \times 10^{-13}}{6.6 \times 10^{-8}} = 2.1 \times 10^{-5}$ M $[Co^{2+}]$ has dropped to 0.021% or its initial value and therefore is considered to be completely precipitated, before $NiCO_3$ begins to precipitate. The two ions are effectively separated as carbonates.

8A First determine $[OH^-]$ resulting from the hydrolysis of acetate ion.

Equation:	$C_2H_3O_2^-(aq) + H_2O \rightleftharpoons$	$HC_2H_3O_2(aq)$ +	$OH^-(aq)$
Initial:	0.10 M		
Changes:	$-x$ M	$+x$ M	$+x$ M
Equil:	$(0.10 - x)$ M	x M	x M

$K_b = \frac{1.0 \times 10^{-14}}{1.8 \times 10^{-5}} = 5.6 \times 10^{-10} = \frac{[HC_2H_3O_2][OH^-]}{[C_2H_3O_2^-]} = \frac{x \cdot x}{0.10-x} \approx \frac{x^2}{0.10}$

$x = [OH^-] = \sqrt{0.10 \times 5.6 \times 10^{-10}} = 7.5 \times 10^{-6}$ M

Now compute the value of the ion product in this solution and compare it with the value of K_{sp} for $Mg(OH)_2$.

$Q_{sp} = [Mg^{2+}][OH^-]^2 = (0.010 \text{ M}) (7.5 \times 10^{-6} \text{ M})^2 = 5.6 \times 10^{-13} < 1.8 \times 10^{-11} = K_{sp}[Mg(OH)_2]$

Because Q_{sp} is smaller than K_{sp}, this solution is unsaturated and precipitation of $Mg(OH)_2(s)$ will not occur.

8B With the Henderson–Hasselbalch equation we can determine the pH of the acetic acid–acetate ion buffer.

$pH = pK_a + \log \frac{[C_2H_3O_2^-]}{[HC_2H_3O_2]} = -\log (1.8 \times 10^{-5}) + \log \frac{0.250 \text{ M}}{0.150 \text{ M}} = 4.74 + 0.22 = 4.96$

$pOH = 14.00 - pH = 14.00 - 4.96 = 9.04$ $[OH^-] = 10^{-pOH} = 10^{-9.04} = 9.1 \times 10^{-10}$ M

Now we determine Q_{sp} to see if precipitation will occur.

$Q_{sp} = [Fe^{3+}][OH^-]^3 = (0.013 \text{ M})(9.1 \times 10^{-10})^3 = 9.8 \times 10^{-30} > 4 \times 10^{-38} = K_{sp}$ Precipitation should occur.

9A Let us determine $[OH^-]$, and then pH necessary to prevent the precipitation of $Mn(OH)_2$.

$K_{sp} = 1.9 \times 10^{-13} = [Ni^{2+}][OH^-]^2 = (0.0050 \text{ M})[OH^-]^2$ $[OH^-] = \sqrt{\frac{1.9 \times 10^{-13}}{0.0050}} = 2.0 \times 10^{-6}$ M

$pOH = -\log(6.2 \times 10^{-6}) = 5.21$ $pH = 14.00 - 5.21 = 8.79$

We use this pH in the Henderson–Hasselbalch equation to determine $[NH_4^+]$. $pK_b = 4.74$ for NH_3.

$pH = pK_a + \log \frac{[NH_3]}{[NH_4^+]} = 8.79 = (14.00 - 4.74) + \log \frac{0.0250 \text{ M}}{[NH_4^+]}$

$\log \frac{0.0250 \text{ M}}{[NH_4^+]} = 8.79 - (14.00 - 4.74) = -0.47$ $\frac{0.0250}{[NH_4^+]} = 10^{-0.47} = 0.34$

$[NH_4^+] = \frac{0.0250}{0.34} = 0.074$ M

9B (a) The solution has $[Al^{3+}] = 0.010$ M. We determine $[OH^-]$ that can be maintained before precipitation occurs. This also is the minumum $[OH^-]$ needed to form a precipitate.

$K_{sp} = 1.3 \times 10^{-33} = [Al^{3+}][OH^-]^3 = 0.010 \text{ M } [OH^-]^3$ $[OH^-] = \sqrt[3]{\frac{1.3 \times 10^{-33}}{0.010}} = 5.1 \times 10^{-11}$ M

$pOH = -\log [OH^-] = -\log (5.1 \times 10^{-11}) = 10.29$ $pH = 14.00 - pOH = 14.00 - 10.29 = 3.71$

(b) We use the Henderson-Hasselbalch equation to find the desired ratio. $pK_a = -\log (1.8 \times 10^{-4}) = 4.74$

$pH = pK_a + \log \frac{[C_2H_3O_2^-]}{[HC_2H_3O_2]} = 3.71 = 4.74 + \log \frac{[C_2H_3O_2^-]}{[HC_2H_3O_2]}$ $\log \frac{[C_2H_3O_2^-]}{[HC_2H_3O_2]} = 3.71 - 4.74 = -1.03$

$\frac{[C_2H_3O_2^-]}{[HC_2H_3O_2]} = 10^{-1.03} = 0.093$ Even though this ratio is just slightly below the guideline of 0.10 for the use of the Henderson-Hasselbalch equation, it still is a valid result, because of its closeness to that guideline.

10A (a) In solution are $Cu^{2+}(aq)$, $SO_4^{2-}(aq)$, $Na^+(aq)$, and $OH^-(aq)$. $Cu^{2+}(aq) + 2\ OH^-(aq) \longrightarrow Cu(OH)_2(s)$

(b) In solution above $Cu(OH)_2(s)$ is $Cu^{2+}(aq)$: $Cu(OH)_2(s) \rightleftharpoons Cu^{2+}(aq) + 2\ OH^-(aq)$

This $Cu^{2+}(aq)$ reacts with the added $NH_3(aq)$: $Cu^{2+}(aq) + 4\ NH_3(aq) \rightleftharpoons [Cu(NH_3)_4]^{2+}(aq)$

The overall result therefore is: $Cu(OH)_2(s) + 4\ NH_3(aq) \rightleftharpoons [Cu(NH_3)_4]^{2+}(aq) + 2\ OH^-(aq)$

(c) $HNO_3(aq)$ as a strong acid, produces $H_3O^+(aq)$, which reacts with both $OH^-(aq)$ and $NH_3(aq)$.

$OH^-(aq) + H_3O^+(aq) \longrightarrow 2\ H_2O$ $NH_3(aq) + H_3O^+(aq) \longrightarrow NH_4^+(aq) + H_2O$

As $NH_3(aq)$ is consumed, this reaction shifts to the left.

$$Cu(OH)_2(s) + 4\ NH_3(aq) \rightleftharpoons [Cu(NH_3)_4]^{2+}(aq) + 2\ OH^-(aq)$$

But as $OH^-(aq)$ is consumed, this reaction shifts to the right: $Cu(OH)_2(s) \rightleftharpoons Cu^{2+}(aq) + 2\ OH^-(aq)$

The species in solution at the end of all this are

$$Cu^{2+}(aq),\ NO_3^-(aq),\ NH_4^+(aq),\ \text{excess}\ H_3O^+(aq),\ Na^+(aq),\ \text{and}\ SO_4^{2-}(aq)$$

10B **(a)** In solution are $Zn^{2+}(aq)$, $SO_4^{2-}(aq)$, and $NH_3(aq)$. $Zn^{2+}(aq) + 4\ NH_3(aq) \rightleftharpoons [Zn(NH_3)_4]^{2+}(aq)$

(b) $HNO_3(aq)$, a strong acid, produces $H_3O^+(aq)$, which reacts with $NH_3(aq)$.

$NH_3(aq) + H_3O^+(aq) \longrightarrow NH_4^+(aq) + H_2O$ As $NH_3(aq)$ is consumed, complex ion is destroyed.

$[Zn(NH_3)_4]^{2+}(aq) + 4\ H_3O^+(aq) \rightleftharpoons [Zn(H_2O)_4]^{2+}(aq) + 4\ NH_4^+(aq)$

(c) Another possibility is reversing the effect of the reaction of part **(b)**.

$[Zn(H_2O)_4]^{2+}(aq) + 4\ NH_4^+(aq) + 4\ OH^-(aq) \rightleftharpoons [Zn(NH_3)_4]^{2+}(aq) + 8\ H_2O(l)$

(c) $NaOH(aq)$ is a strong base that produces $OH^-(aq)$, forming of a hydroxide precipitate.

$[Zn(H_2O)_4]^{2+}(aq) + 2\ OH^-(aq) \rightleftharpoons Zn(OH)_2(s) + 4\ H_2O(l)$

Another possibility is reversing the effect of the reaction of part **(b)**.

$[Zn(H_2O)_4]^{2+}(aq) + 4\ NH_4^+(aq) + 4\ OH^-(aq) \rightleftharpoons [Zn(NH_3)_4]^{2+}(aq) + 8\ H_2O(l)$

(d) The precipitate dissolves in excess base. $Zn(OH)_2(s) + 2\ OH^-(aq) \rightleftharpoons [Zn(OH)_4]^{2-}(aq)$

11A We first determine $[Ag^+]$ in a solution that is 0.100 M Ag^+(aq from $AgNO_3$) and 0.225 M $NH_3(aq)$. Because of the large value of $K_f = 1.6 \times 10^7$, we have the reagents form as much complex ion as possible, and approach equilibrium from that point.

Equation:	$Ag^+(aq)$ +	$2\ NH_3(aq)$ $\rightleftharpoons$	$[Ag(NH_3)_2]^+(aq)$
In soln:	0.100 M	0.225 M	
Form complex:	–0.100 M	–0.200 M	+0.100 M
Initial:	0 M	0.025 M	0.100 M
Changes:	$+x$ M	$+2x$ M	$-x$ M
Equil:	x M	$(0.025 + x)$ M	$(0.100 - x)$ M

$$K_f = 1.6 \times 10^7 = \frac{[[Ag(NH_3)_2]^+]}{[Ag^+][NH_3]^2} = \frac{0.100 - x}{x\ (0.025 + x)^2} \approx \frac{0.100}{x\ (0.025)^2}$$

$$x = \frac{0.100}{(0.025)^2\ 1.6 \times 10^7} = 1.0 \times 10^{-5}\ M = [Ag^+] = \text{concentration of free silver ion}$$

The $[Cl^-]$ is diluted: $[Cl^-]_{final} = [Cl^-]_{initial} \times \dfrac{1.00\ mL_{initial}}{1{,}500\ mL_{final}} = 3.50\ M \div 1500 = 0.00233\ M$

Finally we compare Q_{sp} with K_{sp} to determine if precipitation of AgCl(s) will occur.

$Q_{sp} = [Ag^+][Cl^-] = (1.0 \times 10^{-5}\ M)\ (0.00233\ M) = 2.3 \times 10^{-8} > 1.8 \times 10^{-10} = K_{sp}$

Because the value of the ion product is larger than the value of the solubility product constant, precipitation should occur.

11B We organize the solution around the balanced equation of the formation reaction.

Equation:	$Pb^{2+}(aq)$ +	$EDTA^{4-}(aq)$ $\rightleftharpoons$	$[PbEDTA]^{2-}(aq)$
Initial:	0.100 M	0.250 M	
Form Complex:	0 M	(0.250 – 0.100) M	0.100 M
Equil:	x M	$(0.150 + x)$ M	$(0.100 - x)$ M

$$K_f = \frac{[[PbEDTA]^{2-}]}{[Pb^{2+}][EDTA^{4-}]} = 2 \times 10^{18} = \frac{0.100 - x}{x\ (0.150 + x)} \approx \frac{0.100}{0.150\ x} \qquad x = \frac{0.100}{0.150 \times 2 \times 10^{18}} = 3 \times 10^{-19}\ M$$

We calculate Q_{sp} to determine if precipitation should occur.

$Q_{sp} = [Pb^{2+}][I^-]^2 = (3 \times 10^{-19}\ M)(0.10\ M)^2 = 3 \times 10^{-21} < 7.1 \times 10^{-9} = K_{sp}$ Precipitation will not occur.

12A We first determine the maximum concentration of free Ag^+. $K_{sp} = [Ag^+][Cl^-] = 1.8 \times 10^{-10}$

$[Ag^+] = \dfrac{1.8 \times 10^{-10}}{0.0075} = 2.4 \times 10^{-8}\ M$ This is so small that we assume that all the silver(I) in solution is present as complex ion: $[[Ag(NH_3)_2]^+] = 0.13$ M. We use the formation constant expression to determine the concentration of free NH_3. $K_f = \dfrac{[[Ag(NH_3)_2]^+]}{[Ag^+][NH_3]^2} = 1.6 \times 10^7 = \dfrac{0.13\ M}{2.4 \times 10^{-8}\ [NH_3]^2}$

$[NH_3] = \sqrt{\dfrac{0.13}{2.4 \times 10^{-8} \times 1.6 \times 10^7}} = 0.58\ M.$ If we combine this with the ammonia present in the complex ion, the total ammonia concentration is $0.58\ M + (2 \times 0.13\ M) = 0.84\ M$.

12B We use the solubility product constant expression to determine the maximum $[Ag^+]$ that can be present with 0.010 M Cl^- without precipitation occurring.

$$K_{sp} = 1.8 \times 10^{-10} = [Ag^+][Cl^-] = [Ag^+](0.010\text{ M}) \qquad [Ag^+] = \frac{1.8 \times 10^{-10}}{0.010} = 1.0 \times 10^{-8}\text{ M}$$

This is also the concentration of free silver ion in the K_f expression. Because of the large value of K_f, practically all of the silver ion in solution is present as the complex ion, and $[[Ag(S_2O_3)_2]^{3-}] = 0.10$ M. We solve the expression for $[S_2O_3^{2-}]$ and then add the $[S_2O_3^{2-}]$ "tied up" in the complex ion.

$$K_f = 1.7 \times 10^{13} = \frac{[[Ag(S_2O_3)_2]^{3-}]}{[Ag^+][S_2O_3^{2-}]^2} = \frac{0.10\text{ M}}{1.0 \times 10^{-8}\text{ M }[S_2O_3^{2-}]^2}$$

$$[S_2O_3^{2-}] = \sqrt{\frac{0.10}{1.0 \times 10^{-8} \times 1.7 \times 10^{13}}} = 7.7 \times 10^{-4}\text{ M} = \text{concentration of free } S_2O_3^{2-}$$

$$\text{total } [S_2O_3^{2-}] = 7.7 \times 10^{-4}\text{ M} + 0.10\text{ M } [Ag(S_2O_3)_2]^{3-} \times \frac{2\text{ mol } S_2O_3^{2-}}{1\text{ mol } [Ag(S_2O_3)_2]^{3-}}$$

$$= 0.20 + 0.00077\text{ M} = 0.20\text{ M}$$

13A We combine the two equilibrium expressions, for K_f and for K_{sp}.

$$Fe(OH)_3(s) \rightleftharpoons Fe^{3+}(aq) + 3\,OH^-(aq) \qquad K_{sp} = 4 \times 10^{-38}$$

$$Fe^{3+}(aq) + 3\,C_2O_4^{2-}(aq) \rightleftharpoons [Fe(C_2O_4)_3]^{3-}(aq) \qquad K_f = 2 \times 10^{20}$$

$$Fe(OH)_3(s) + 3\,C_2O_4^{2-}(aq) \rightleftharpoons [Fe(C_2O_4)_3]^{3-}(aq) + 3\,OH^-(aq) \qquad K = 8 \times 10^{-18}$$

	$3\,C_2O_4^{2-}$	$[Fe(C_2O_4)_3]^{3-}$	$3\,OH^-$
initial:	0.100 M		
changes:	$-3x$ M	$+x$ M	$+3x$ M
equil:	$(0.100 - 3x)$ M	x M	$3x$ M

$$K = \frac{[[Fe(C_2O_4)_3]^{3-}][OH^-]^3}{[C_2O_4^{2-}]^3} = \frac{(x)\,(3x)^3}{(0.100 - 3x)^3} = 8 \times 10^{-18} \approx \frac{27\,x^4}{(0.100)^3} \qquad \text{assuming } 3x << 0.100$$

$$x = \sqrt[4]{\frac{(0.100)^3\; 8 \times 10^{-18}}{27}} = 4 \times 10^{-6}\text{ M}$$

The assumption is valid. The solubility of $Fe(OH)_3$ in 0.100 M $C_2O_4^{2-}$ is 4×10^{-6} M.

13B In Example 19-3 we saw that the expression for the solubility, s, of a silver halide in an aqueous ammonia solution, where $[NH_3]$ is the concentration of aqueous ammonia, is given by:

$$K_{sp} \times K_f = \left(\frac{s}{[NH_3] - 2s}\right)^2 \quad or \quad \sqrt{K_{sp} \times K_f} = \frac{s}{[NH_3] - 2s}$$

We are comparing situations of constant $[NH_3] = 0.100$ M and $K_f = 1.6 \times 10^7$. We see that s will decrease as does K_{sp}. The values are: $K_{sp}(AgCl) = 1.8 \times 10^{-10}$, $K_{sp}(AgBr) = 5.0 \times 10^{-13}$, $K_{sp}(AgI) = 8.5 \times 10^{-17}$.

14A For FeS, we know that $K_{spa} = 6 \times 10^2$, For Ag_2S, $K_{spa} = 6 \times 10^{-30}$. We compute the value of Q_{spa} in each case, with $[H_2S] = 0.10$ M and $[H_3O^+] = 0.30$ M.

$$\text{For FeS, } Q_{spa} = \frac{0.020 \times 0.10}{(0.30)^2} = 0.022 < 6 \times 10^2 = K_{spa}$$ Precipitation of FeS should not occur.

$$\text{For } Ag_2S,\; Q_{spa} = \frac{(0.010)^2 \times 0.10}{(0.30)^2} = 1.1 \times 10^{-4} > 6 \times 10^{-30} = K_{spa}$$ Precipitation of Ag_2S should occur.

14B The $[H_3O^+]$ needed to just form a precipitate can be obtained by direct substitution into the K_{spa} expression. When that expression is satisfied, precipitate will just form.

$$K_{spa} = \frac{[Fe^{2+}][H_2S]}{[H_3O^+]^2} = 6 \times 10^2 = \frac{(0.015\text{ M } Fe^{2+})(0.10\text{ M } H_2S)}{[H_3O^+]^2},\; [H_3O^+] = \sqrt{\frac{0.015 \times 0.10}{6 \times 10^2}} = 0.002\text{ M}$$

$$pH = -\log [H_3O^+] = -\log (0.002) = 2.7$$

SUMMARIZING EXAMPLE CALCULATIONS

1. We determine the pH of saturated $Ca(OH)_2(aq)$ from the K_{sp} expression.

$$K_{sp} = 5.5 \times 10^{-6} = [Ca^{2+}][OH^-]^2 = (s)(2s)^2 = 4s^3 \qquad s = \sqrt[3]{\frac{5.5 \times 10^{-6}}{4}} = 0.011\text{ M} = [Ca^{2+}]$$

$[OH^-] = 2 \times [Ca^{2+}] = 0.022$ M $\qquad$ pOH $= -\log (0.022) = 1.66$ $\qquad$ pH $= 14.00 - 1.66 = 12.34$

2. The two reactions are added to each other after the one is reversed. Thus, the resultant equilibrium constant is the quotient of the other to equilibrium constants.

$Ca(OH)_2(s) \rightleftharpoons Ca^{2+}(aq) + 2\,OH^-(aq)$ $\qquad K_{sp} = 5.5 \times 10^{-6}$

$Ca^{2+}(aq) + CO_3^{2-}(aq) \rightleftharpoons CaCO_3(s)$ $\qquad 1/K_{sp} = 1 \div 2.8 \times 10^{-9}$

$$Ca(OH)_2(s) + CO_3^{2-}(aq) \rightleftharpoons CaCO_3(s) + 2\,OH^-(aq) \qquad K_c = \frac{[OH^-]^2}{[CO_3^{2-}]} = \frac{5.5 \times 10^{-6}}{2.8 \times 10^{-9}} = 2.0 \times 10^3$$

3. The set up is based on the balanced chemical equation, but recognizing that the equilibrium position lies far to the right. With that insight, it is not necessary to solve a quadratic equation.

Equation:	$Ca(OH)_2(s)$ +	$CO_3^{2-}(aq)$ $\rightleftharpoons$	$CaCO_3(s)$ +	$2\,OH^-(aq)$
In soln:		1.0 M		
Form products:		−1.0 M		+2.0 M
Initial:		0 M		2.0 M
Changes:		$+x$ M		$-2x$ M
Equil:		x M		$(2.0 - 2x)$ M

$$K_c = 2.0 \times 10^3 = \frac{[OH^-]^2}{[CO_3^{2-}]} = \frac{(2.0-2x)^2}{x} \approx \frac{(2.0)^2}{x} \qquad x = \frac{(2.0)^2}{2.0 \times 10^3} = 2.0 \times 10^{-3}\text{ M}$$

$[OH^-] = 2.0 - 2x = 2.0$ M; pOH = −log (2.0) = −0.3 $\qquad$ pH = 14.00 − pOH = 14.00 −(−0.3) = 14.3

REVIEW QUESTIONS

1. **(a)** K_{sp} is the symbol for the solubility product constant, representing the equilibrium constant for the dissolving of a sparingly soluble (ionic) compound.

(b) K_f is the symbol for the formation constant of a complex ion: one complex ion is formed (is the product) from the reactants: a simple cation and the appropriate ligands.

(c) A ligand is a species that attaches to a central metal ion by forming a (coordinate) covalent bond; it is a Lewis base.

(d) The coordination number in a complex ion is the number of locations where ligands have attached.

2. **(a)** The common ion effect in the dissolving of a sparingly soluble ionic compound means that less of that compound will dissolve because of the presence of one of its constituent ions in solution.

(b) Fractional precipitation refers to adding a precipitating agent (such as an anion) to a solution containing two other ions (such as cations) gradually so that only one compound precipitates.

(c) Ion pair formation refers to the result of the attraction between solvated ions of unlike charges in solution. These solvated ions cluster together and behave to some extent as one particle.

(d) Qualitative cation analysis is a scheme of alternating selective precipitation and dissolving, with the goal of isolating each type of cation into its own sample, where a definitive test for it can be performed.

3. **(a)** The solubility of a compound refers to the concentration of that compound in solution, either as a molarity or as a mass per unit volume. The solubility product constant is the equilibrium constant in terms of concentrations of ions, for the dissolving equilibrium.

(b) The common ion effect describes the lowering of the solubility of a compound in a solution due to the presence of one of the ions of that compound in the solution. The salt effect describes the enhancement of the solubility of a compound in a solution due to the presence of a different type of ion in solution.

(c) A complex ion is a cation with its attached ligands. A coordination compound is a compound that contains one or more types of complex ions.

4. **(a)** $Ag_2SO_4(s) \rightleftharpoons 2\,Ag^+(aq) + SO_4^{2-}(aq)$ $\qquad K_{sp} = [Ag^+]^2[SO_4^{2-}]$

(b) $Ra(IO_3)_2(s) \rightleftharpoons Ra^{2+}(aq) + 2\,IO_3^-(aq)$ $\qquad K_{sp} = [Ra^{2+}][IO_3^-]^2$

(c) $Ni_3(PO_4)_2(s) \rightleftharpoons 3\,Ni^{2+}(aq) + 2\,PO_4^{3-}(aq)$ $\qquad K_{sp} = [Ni^{2+}]^3[PO_4^{3-}]^2$

(d) $PuO_2CO_3(s) \rightleftharpoons PuO_2^{2+}(aq) + CO_3^{2-}(aq)$ $\qquad K_{sp} = [PuO_2^{2+}][CO_3^{2-}]$

5. **(a)** $K_{sp} = [Fe^{3+}][OH^-]^3$ $\qquad Fe(OH)_3(s) \rightleftharpoons Fe^{3+}(aq) + 3\,OH^-(aq)$

(b) $K_{sp} = [BiO^+][OH^-]$ $\qquad BiOOH(s) \rightleftharpoons BiO^+(aq) + OH^-(aq)$

(c) $K_{sp} = [Hg_2^{2+}][I^-]^2$ $\qquad Hg_2I_2(s) \rightleftharpoons Hg_2^{2+}(aq) + 2\,I^-(aq)$

(d) $K_{sp} = [Pb^{2+}]^3[AsO_4^{3-}]^2$ $\qquad Pb_3(AsO_4)_2(s) \rightleftharpoons 3\,Pb^{2+}(aq) + 2\,AsO_4^{3-}(aq)$

6. **(a)** $CrF_3(s) \rightleftharpoons Cr^{3+}(aq) + 3\,F^-(aq)$ $\qquad K_{sp} = [Cr^{3+}][F^-]^3 = 6.6 \times 10^{-11}$

(b) $Au_2(C_2O_4)_3(s) \rightleftharpoons 2\,Au^{3+}(aq) + 3\,C_2O_4^{2-}(aq)$ $\qquad K_{sp} = [Au^{3+}]^2[C_2O_4^{2-}]^3 = 1 \times 10^{-10}$

(c) $Cd_3(PO_4)_2(s) \rightleftharpoons 3\ Cd^{2+}(aq) + 2\ PO_4^{3-}(aq)$ $K_{sp} = [Cd^{2+}]^3[PO_4^{3-}]^2 = 2.1 \times 10^{-33}$

(d) $SrF_2(s) \rightleftharpoons Sr^{2+}(aq) + 2\ F^-(aq)$ $K_{sp} = [Sr^{2+}][F^-]^2 = 2.5 \times 10^{-9}$

7. Let s = solubility of each compound in moles of compound per liter of solution.

(a) $K_{sp} = [Ba^{2+}][CrO_4^{2-}] = (s)(s) = s^2 = 1.2 \times 10^{-10}$ $s = 1.1 \times 10^{-5}$ M

(b) $K_{sp} = [Pb^{2+}][Br^-]^2 = (s)(2s)^2 = 4s^3 = 4.0 \times 10^{-5}$ $s = 2.2 \times 10^{-2}$ M

(c) $K_{sp} = [Ce^{3+}][F^-]^3 = (s)(3s)^3 = 27s^4 = 8 \times 10^{-16}$ $s = 7 \times 10^{-5}$ M

(d) $K_{sp} = [Mg^{2+}]^3[AsO_4^{3-}]^2 = (3s)^3(2s)^2 = 108s^5 = 2.1 \times 10^{-20}$ $s = 4.5 \times 10^{-5}$ M

8. Again, let s = solubility of each compound in moles of solute per liter of solution.

(a) $K_{sp} = [Cs^+][MnO_4^-] = (s)(s) = s^2 = (3.8 \times 10^{-3})^2 = 1.4 \times 10^{-5}$

(b) $K_{sp} = [Pb^{2+}][ClO_2^-]^2 = (s)(2s)^2 = 4s^3 = 4(2.8 \times 10^{-3})^3 = 8.8 \times 10^{-8}$

(c) $K_{sp} = [Li^+]^3[PO_4^{3-}] = (3s)^3(s) = 27s^4 = 27(2.9 \times 10^{-3})^4 = 1.9 \times 10^{-9}$

9. Statement **(d)** is correct. Consider the solubility equation : $PbI_2(s) \rightleftharpoons Pb^{2+}(aq) + 2\ I^-(aq)$
The stoichiometry of the dissolving reaction indicates that two I^- ions are formed for each Pb^{2+} ion. Thus $[Pb^{2+}] = 0.5\ [I^-]$. The relationship between K_{sp} and $[Pb^{2+}]$ is $[Pb^{2+}] = \sqrt[3]{K_{sp} / 4}$.

10. We let s = molar solubility of $Mg(OH)_2$ in moles solute per liter of solution.

(a) $K_{sp} = [Mg^{2+}][OH^-]^2 = (s)(2s)^2 = 4s^3 = 1.8 \times 10^{-11}$ $s = 1.7 \times 10^{-4}$ M

(b)

Equation:	$Mg(OH)_2(s)$	$\rightleftharpoons$	$Mg^{2+}(aq)$ +	$2\ OH^-(aq)$
Initial:			0.0862 M	
Changes:			$+s$ M	$+2s$ M
Equil:			$(0.0862 + s)$M	$2s$ M

$K_{sp} = (0.0862 + s)(2s)^2 = 1.8 \times 10^{-11} \approx (0.0862)(2s)^2 = 0.34\ s^2$ $s = 7.3 \times 10^{-6}$ M

(c) $[OH^-] = [KOH] = 0.0355$ M

Equation:	$Mg(OH)_2(s)$	$\rightleftharpoons$	$Mg^{2+}(aq)$ +	$2\ OH^-(aq)$
Initial:				0.0355 M
Changes:			$+s$ M	$+2s$ M
Equil:			s M	$(0.0355 + 2s)$M

$K_{sp} = (s)(0.0355 + 2s)^2 = 1.8 \times 10^{-11} \approx (s)(0.0355)^2 = 0.0013\ s$ $s = 1.4 \times 10^{-8}$ M

11. The solubility equilibrium is $CaCO_3(s) \rightleftharpoons Ca^{2+}(aq) + CO_3^{2-}(aq)$

(a) The addition of $Na_2CO_3(aq)$ produces $CO_3^{2-}(aq)$ in solution. This common ion suppresses the solubility of $CaCO_3(s)$.

(b) HCl(aq) is a strong acid that reacts with carbonate ion: $CO_3^{2-}(aq) + 2\ H_3O^+(aq) \longrightarrow CO_2(g) + 3\ H_2O$. This decreases $[CO_3^{2-}]$ in the solution and more $CaCO_3(s)$ will dissolve.

(c) $HSO_4^-(aq)$ is a moderately weak acid. It is strong enough to protonate carbonate ion, decreasing $[CO_3^{2-}]$ and enhancing the solubility of $CaCO_3(s)$, as the value of K_c indicates.

$HSO_4^-(aq) + CO_3^{2-}(aq) \rightleftharpoons SO_4^{2-}(aq) + HCO_3^-(aq)$ $K_c = \dfrac{K(HSO_4^-)}{K(HCO_3^-)} = \dfrac{0.011}{4.7 \times 10^{-11}} = 2.3 \times 10^8$

12. In each case, compute Q and compare its value with the value of K_{sp}. If $Q > K_{sp}$, a precipitate should form.

(a) $Q = [Mg^{2+}][CO_3^{2-}] = (0.0037)(0.0068) = 2.5 \times 10^{-5} > 3.5 \times 10^{-8} = K_{sp}$
Precipitation should occur.

(b) $Q = [Ag^+]^2[SO_4^{2-}] = (0.018)^2(0.0062) = 2.0 \times 10^{-6} < 1.4 \times 10^{-5} = K_{sp}$
Precipitation is not expected to occur.

(c) pOH = 14.00 – 3.20 = 10.80 $[OH^-] = 10^{-10.80} = 1.6 \times 10^{-11}$ M
$Q = [Cr^{3+}][OH^-]^3 = (0.038)(1.6 \times 10^{-11})^3 = 1.6 \times 10^{-34} < 6.3 \times 10^{-31} = K_{sp}$
Precipitation is not expected to occur.

13. We use the solubility product expression to determine $[Ca^{2+}]$ that can coexist with $[SO_4^{2-}] = 0.750$ M.

$K_{sp} = 9.1 \times 10^{-6} = [Ca^{2+}][SO_4^{2-}] = [Ca^{2+}](0.750\text{ M})$ $[Ca^{2+}] = \dfrac{9.1 \times 10^{-6}}{0.750\text{ M}} = 1.2 \times 10^{-5}$ M

$$\% \text{ unprecipitated} = \frac{1.2 \times 10^{-5}\text{ M}}{0.103\text{ M}} \times 100\% = 0.012\% \text{ unprecipitated, essentially complete precipitation.}$$

Of course, $[SO_4^{2-}]$ actually does decrease because it is consumed in the precipitation. One mole of sulfate ion is removed from solution for each mole of calcium ion that precipitates, which amounts to 0.103 moles from each liter of solution. So, in this case $[SO_4^{2-}] = 0.750\text{ M} - 0.103\text{ M} = 0.647\text{ M}$. But this just changes the final $[Ca^{2+}]$ slightly, to 1.4×10^{-5} M.

14. **(a)** Determine $[I^-]$ when AgI just begins to precipitate, and $[I^-]$ when PbI_2 just begins to precipitate.

$K_{sp} = [Ag^+][I^-] = 8.5 \times 10^{-17} = (0.10)[I^-]$ $\qquad [I^-] = 8.5 \times 10^{-16}$ M

$K_{sp} = [Pb^{2+}][I^-]^2 = 7.1 \times 10^{-9} = (0.10)[I^-]^2$ $\qquad [I^-] = \sqrt{\frac{7.1 \times 10^{-9}}{0.10}} = 2.7 \times 10^{-4}$ M

Since 8.5×10^{-16} M is less than 2.7×10^{-4} M, AgI will precipitate before PbI_2.

(b) $[I^-] = 2.7 \times 10^{-4}$ M before the second cation—Pb^{2+}—begins to precipitate.

(c) $K_{sp} = [Ag^+][I^-] = 8.5 \times 10^{-17} = [Ag^+](2.7 \times 10^{-4})$ $\qquad [Ag^+] = 3.1 \times 10^{-13}$ M

(d) Since $[Ag^+]$ has decreased to less than 0.1% of its initial value before PbI_2 begins to precipitate, we conclude that Ag^+ and Pb^{2+} can be separated by precipitation with iodide ion.

15. $Mg(OH)_2(s)$ will be the most soluble in a solution of $NaHSO_4(aq)$. $NaHSO_4$ will form an acidic solution, via the reaction $HSO_4^-(aq) + H_2O \rightleftharpoons H_3O^+(aq) + SO_4^{2-}(aq)$ and this H_3O^+ will react with OH^- ion: $OH^-(aq) + H_3O^+(aq) \rightleftharpoons 2\ H_2O$ causing the solubility equilibrium to shift right $Mg(OH)_2(s) \rightleftharpoons Mg^{2+}(aq) + 2\ OH^-(aq)$. Since NaOH has an ion in common with $Mg(OH)_2$, its use will actually decrease the solubility of $Mg(OH)_2$. Addition of Na_2CO_3 will also decrease $Mg(OH)_2$ solubility, through hydrolysis of the carbonate ion: $H_2O + CO_3^{2-} \rightleftharpoons HCO_3^- + OH^-$ followed by the common ion effect of OH^-.

16. **(a)** $Ag^+(aq) + NO_3^-(aq) + Na^+(aq) + Br^-(aq) \longrightarrow AgBr(s) + Na^+(aq) + NO_3^-(aq)$

(b) $Cu^{2+}(aq) + NO_3^-(aq) + H_3O^+(aq) + Cl^-(aq) \longrightarrow$ no reaction

(c) $Fe^{2+}(aq) + H_2S(aq, \text{ in } 0.3\text{ M HCl}) \longrightarrow$ no reaction

(d) $Cu(OH)_2(s) + 4\ NH_3(aq) \longrightarrow [Cu(NH_3)_4]^{2+}(aq) + 2\ OH^-(aq)$

(e) $Fe^{3+}(aq) + 3\ NH_3(aq) + 3\ H_2O(l) \longrightarrow Fe(OH)_3(s) + 3\ NH_4^+(aq)$

(f) $Ag_2SO_4(s) + 4\ NH_3(aq) \longrightarrow 2\ [Ag(NH_3)_2]^+(aq) + SO_4^{2-}(aq)$

(g) $CaSO_3(s) + 2\ H_3O^+(aq) \longrightarrow Ca^{2+}(aq) + 3\ H_2O + SO_2(g)$

17. $Cu(OH)_2$ dissolves readily in acidic solutions [HCl(aq) and HNO_3(aq)] and in ammoniacal solutions [NH_3(aq)]. The net ionic equations for these two reactions are, respectively:

$Cu(OH)_2(s) + 2\ H_3O^+(aq) \longrightarrow Cu^{2+}(aq) + 4\ H_2O$

$Cu(OH)_2(s) + 4\ NH_3(aq) \longrightarrow [Cu(NH_3)_4]^{2+}(aq) + 2\ OH^-(aq)$

18. The best choice is HCl(aq). AgCl is insoluble in water, while $CuCl_2$ is water soluble. On the other hand, both Cu^{2+} and Ag^+ form insoluble hydroxides and carbonates, so NaOH(aq) and $(NH_4)_2CO_3$(aq) would not serve to separate these two cations.

19. Both Cu^{2+} and Mg^{2+} form hydroxides, although $Cu(OH)_2$ ($K_{sp} = 2.2 \times 10^{-20}$) is much less soluble than $Mg(OH)_2$ ($K_{sp} = 1.8 \times 10^{-11}$). Thus, NaOH(aq) might be one choice to produce a precipitate of $Cu(OH)_2(s)$ while leaving $Mg^{2+}(aq)$ in solution. A better choice, however, seems to be NH_3(aq) in which $[Cu(NH_3)_4]^{2+}$ forms ($K_f = 1.1 \times 10^{13}$) while simultaneously creating an alkaline solution from which $Mg(OH)_2(s)$ precipitates.

20. First we determine $[OH^-]$ in this buffer solution.

$pH = pK_a + \log \frac{[C_2H_3O_2^-]}{[HC_2H_3O_2]} = 4.74 + \log \frac{0.35\text{ M}}{0.45\text{ M}} = 4.63$ $\qquad pOH = 14.00 - 4.63 = 9.37$

$[OH^-] = 10^{-9.37} = 4.3 \times 10^{-10}$ M $\quad$ Now, we compute the value of Q_{sp} for $Al(OH)_3$.

$Q = [Al^{3+}][OH^-]^3 = (0.275)(4.3 \times 10^{-10})^3 = 2.2 \times 10^{-29} > 1.3 \times 10^{-33} = K_{sp}$

Precipitation should occur from this solution.

21. We first find the concentration of free metal ion. Then we determine the value of Q_{sp} for the precipitation reaction, and compare that value with the value of K_{sp} to determine whether precipitation will occur.

Equation:	$Ag^+(aq)$ +	$2\ CN^-(aq)$ $\rightleftharpoons$	$[Ag(CN)_2]^-(aq)$
Initial:		1.05 M	0.012 M
Changes:	$+x$ M	$+2x$ M	$-x$ M
Equil:	x M	$(1.05 + 2x)$M	$(0.012 - x)$M

$$K_f = \frac{[[Ag(CN)_2]^-]}{[Ag^+][CN^-]^2} = 5.6 \times 10^{18} = \frac{0.012 - x}{x\,(1.05 + 2x)^2} \approx \frac{0.012}{1.05^2\,x} \qquad x = 1.9 \times 10^{-21}\text{ M} = [Ag^+]$$

$Q = [Ag^+][I^-] = (1.9 \times 10^{-21})(2.0) = 3.8 \times 10^{-21} < 8.5 \times 10^{-17} = K_{sp}$ Precipitation should not occur.

22. We use the formation constant expression to determine the concentration of free silver ion, $[Ag^+]$.

$$K_f = 1.6 \times 10^7 = \frac{[[Ag(NH_3)_2]^+]}{[Ag^+][NH_3]^2} = \frac{1.6\text{ M}}{[Ag^+](1.25\text{ M})^2} \qquad [Ag^+] = \frac{1.6}{1.6 \times 10^7\,(1.25)^2} = 6.4 \times 10^{-8}\text{ M}$$

The K_{sp} expression is used to determine the $[Cl^-]$ that can coexist with this $[Ag^+]$.

$$K_{sp} = 1.8 \times 10^{-10} = [Ag^+][Cl^-] = (6.4 \times 10^{-8}\text{ M})[Cl^-] \qquad [Cl^-] = \frac{1.8 \times 10^{-10}}{6.4 \times 10^{-8}} = 2.8 \times 10^{-3}\text{ M}$$

EXERCISES

K_{sp} and Solubility

23. We use the value of K_{sp} for each compound in determining the molar solubility saturated solution. In each case, s represents the molar solubility of the compound.

AgCN	$K_{sp} = [Ag^+][CN^-] = (s)(s) = s^2 = 1.2 \times 10^{-16}$	$s = 1.1 \times 10^{-8}$ M
$AgIO_3$	$K_{sp} = [Ag^+][IO_3^-] = (s)(s) = s^2 = 3.0 \times 10^{-8}$	$s = 1.7 \times 10^{-4}$ M
AgI	$K_{sp} = [Ag^+][I^-] = (s)(s) = s^2 = 8.5 \times 10^{-17}$	$s = 9.2 \times 10^{-9}$ M
$AgNO_2$	$K_{sp} = [Ag^+][NO_2^-] = (s)(s) = s^2 = 6.0 \times 10^{-4}$	$s = 2.4 \times 10^{-2}$ M
Ag_2SO_4	$K_{sp} = [Ag^+]^2[SO_4^{2-}] = (2s)^2(s) = 4s^3 = 1.4 \times 10^{-5}$	$s = 1.5 \times 10^{-2}$ M

In order of increasing molar solubility, smallest to largest: AgI < AgCN < $AgIO_3$ < Ag_2SO_4 < $AgNO_2$

24. We use the value of K_{sp} for each compound in determining $[Mg^{2+}]$ in its saturated solution. In each case, s represents the molar solubility of the compound.

(a) $MgCO_3$ $\quad K_{sp} = [Mg^{2+}][CO_3^{2-}] = (s)(s) = s^2 = 3.5 \times 10^{-8}$
$s = 1.9 \times 10^{-4}$ M $\quad [Mg^{2+}] = 1.9 \times 10^{-4}$ M

(b) MgF_2 $\quad K_{sp} = [Mg^{2+}][F^-]^2 = (s)(2s)^2 = 4s^3 = 3.7 \times 10^{-8}$
$s = 2.1 \times 10^{-3}$ M $\quad [Mg^{2+}] = 2.1 \times 10^{-3}$ M

(c) $Mg_3(PO_4)_2$ $\quad K_{sp} = [Mg^{2+}]^3[PO_4^{3-}]^2 = (3s)^3(2s)^2 = 108s^5 = 1 \times 10^{-25}$
$s = 4 \times 10^{-6}$ M $\quad [Mg^{2+}] = 1 \times 10^{-5}$ M

A saturated solution of MgF_2 has the highest $[Mg^{2+}]$.

25. We determine $[F^-]$ in saturated CaF_2, and from that value the concentration of fluoride ion in ppm.
For CaF_2 $\quad K_{sp} = [Ca^{2+}][F^-]^2 = (s)(2s)^2 = 4s^3 = 5.3 \times 10^{-9}$ $\quad s = 1.1 \times 10^{-3}$ M
The solubility in ppm is the number of grams of CaF_2 in 10^6 g solution. We assume a solution density of 1.00 g/mL.

$$\text{mass of F}^- = 10^6\text{ g soln} \times \frac{1\text{ mL}}{1.00\text{ g soln}} \times \frac{1\text{ L soln}}{1000\text{ mL}} \times \frac{1.1 \times 10^{-3}\text{ mol CaF}_2}{1\text{ L soln}} \times \frac{2\text{ mol F}^-}{1\text{ mol CaF}_2} \times \frac{19.0\text{ g F}^-}{1\text{ mol F}^-} = 42\text{ g F}^-$$

This is 42 times more concentrated than the optimum concentration of fluoride ion. CaF_2 is, in fact, more soluble than is necessary. Its uncontrolled use might lead to excessive F^- in solution.

26. We determine $[OH^-]$ in a saturated solution. From this $[OH^-]$ we determine pH.
$K_{sp} = [BiO^+][OH^-] = 4 \times 10^{-10} = s^2$ $\quad s = 2 \times 10^{-5}\text{ M} = [OH^-] = [BiO^+]$
$pOH = -\log(2 \times 10^{-5}) = 4.7$ $\quad pH = 9.3$

27. We first assume that the volume of the solution does not change appreciably when its temperature is lowered. Then we determine the mass of $Mg(C_{16}H_{31}O_2)_2$ dissolved in each solution, recognizing that the molar solubility of $Mg(C_{16}H_{31}O_2)_2$ equals the cube root of one fourth of its solubility product constant, since it is the only solute in the solution. $K_{sp} = 4s^3$ $\quad s = \sqrt[3]{K_{sp} / 4}$

At 50°C: $s = \sqrt[3]{4.8 \times 10^{-12}/4} = 1.1 \times 10^{-4}$ M $\qquad$ At 25°C: $s = \sqrt[3]{3.3 \times 10^{-12}/4} = 9.4 \times 10^{-5}$ M

$$\text{amount of } Mg(C_{16}H_{31}O_2)_2 \text{ at } 50°C = 0.965 \text{ L} \times \frac{1.1 \times 10^{-4} \text{ mol } Mg(C_{16}H_{31}O_2)_2}{1 \text{ L soln}} = 1.1 \times 10^{-4} \text{ mol}$$

$$\text{mass of } Mg(C_{16}H_{31}O_2)_2 \text{ at } 25°C = 0.965 \text{ L} \times \frac{9.4 \times 10^{-5} \text{ mol } Mg(C_{16}H_{31}O_2)_2}{1 \text{ L soln}} = 0.91 \times 10^{-4} \text{ mol}$$

$$\text{mass } Mg(C_{16}H_{31}O_2)_2 \text{ precipitated} = (1.1 - 0.91) \times 10^{-4} \text{ mol} \times \frac{535.15 \text{ g } Mg(C_{16}H_{31}O_2)_2}{1 \text{ mol } Mg(C_{16}H_{31}O_2)_2} \times \frac{1000 \text{ mg}}{1 \text{ g}} = 11 \text{ mg}$$

28. We first assume that the volume of the solution does not change appreciably when its temperature is lowered. Then we determine the mass of CaC_2O_4 dissolved in each solution, recognizing that the molar solubility of CaC_2O_4 equals the square root of its solubility product constant, since it is the only solute in the solution.

At 95 °C: $s = \sqrt{1.2 \times 10^{-8}} = 1.1 \times 10^{-4}$ M $\qquad$ At 13 °C: $s = \sqrt{2.7 \times 10^{-9}} = 5.2 \times 10^{-5}$ M

$$\text{mass of } CaC_2O_4 \text{ at } 95 °C = 0.725 \text{ L} \times \frac{1.1 \times 10^{-4} \text{ mol } CaC_2O_4}{1 \text{ L soln}} \times \frac{128.1 \text{ g } CaC_2O_4}{1 \text{ mol } CaC_2O_4} = 0.010 \text{ g } CaC_2O_4$$

$$\text{mass of } CaC_2O_4 \text{ at } 13°C = 0.725 \text{ L} \times \frac{5.2 \times 10^{-5} \text{ mol } PbSO_4}{1 \text{ L soln}} \times \frac{128.1 \text{ g } CaC_2O_4}{1 \text{ mol } CaC_2O_4} = 0.0048 \text{ g } CaC_2O_4$$

$$\text{mass } CaC_2O_4 \text{ precipitated} = (0.010 \text{ g} - 0.0048 \text{ g}) \times \frac{1000 \text{ mg}}{1 \text{ g}} = 5 \text{ mg}$$

29. First we determine $[I^-]$ in the saturated solution.

$K_{sp} = [Pb^{2+}][I^-]^2 = 7.1 \times 10^{-9} = (s)(2s)^2 = 4s^3$ $\qquad s = 1.2 \times 10^{-3}$ M

The $AgNO_3$ reacts with the I^- in this saturated solution in the titration. $\quad Ag^+(aq) + I^-(aq) \longrightarrow AgI(s)$

We determine the amount of Ag^+ needed for this titration, and then $[AgNO_3]$ in the titrant.

$$\text{amount } Ag^+ = 0.02500 \text{ L} \times \frac{1.2 \times 10^{-3} \text{ mol } PbI_2}{1 \text{ L soln}} \times \frac{2 \text{ mol } I^-}{1 \text{ mol } PbI_2} \times \frac{1 \text{ mol } Ag^+}{1 \text{ mol } I^-} = 6.0 \times 10^{-5} \text{ mol } Ag^+$$

$$AgNO_3 \text{ molarity} = \frac{6.0 \times 10^{-5} \text{ mol } Ag^+}{0.0133 \text{ L soln}} \times \frac{1 \text{ mol } AgNO_3}{1 \text{ mol } Ag^+} = 4.5 \times 10^{-3} \text{ M}$$

30. We determine $[C_2O_4{}^{2-}] = s$, the solubility of the saturated solution.

$$[C_2O_4{}^{2-}] = \frac{4.8 \text{ mL} \times \dfrac{0.00134 \text{ mmol } KMnO_4}{1 \text{ mL soln}} \times \dfrac{5 \text{ mmol } C_2O_4{}^{2-}}{2 \text{ mmol } MnO_4{}^-}}{250.0 \text{ mL}} = 6.4 \times 10^{-5} \text{ M} = s = [Ca^{2+}]$$

$K_{sp} = [Ca^{2+}][C_2O_4{}^{2-}] = (s)(s) = s^2 = (6.4 \times 10^{-5})^2 = 4.1 \times 10^{-9}$

31. We use the ideal gas law to determine the amount in moles of H_2S gas used.

$$n = \frac{PV}{RT} = \frac{\left(748 \text{ mmHg} \times \dfrac{1 \text{ atm}}{760 \text{ mmHg}}\right) \times \left(30.4 \text{ mL} \times \dfrac{1 \text{ L}}{1000 \text{ mL}}\right)}{0.08206 \text{ L atm mol}^{-1} \text{ K}^{-1} \times (23 + 273)\text{K}} = 1.23 \times 10^{-3} \text{ moles}$$

If we assume that all the H_2S is consumed in forming Ag_2S, we can compute the $[Ag^+]$ in the $AgBrO_3$ solution. This assumption is valid if the equilibrium constant for the cited reaction is large, which we see that it is, as follows.

$2\,Ag^+(aq) + HS^-(aq) + OH^-(aq) \rightleftharpoons Ag_2S(s) + H_2O$ $\qquad 1/K_{sp} = 1/6 \times 10^{-51}$

$H_2S(aq) + H_2O \rightleftharpoons HS^-(aq) + H_3O^+(aq)$ $\qquad K_1 = 1.0 \times 10^{-7}$

$2\,H_2O \rightleftharpoons H_3O^+(aq) + OH^-(aq)$ $\qquad K_w = 1.0 \times 10^{-14}$

$$2\,Ag^+(aq) + H_2S(aq) + 2\,H_2O \rightleftharpoons Ag_2S(s) + 2\,H_3O^+(aq) \quad K = \frac{1.0 \times 10^{-7} \times 1.0 \times 10^{-14}}{6 \times 10^{-51}} = 1.7 \times 10^{29}$$

$$[Ag^+] = \frac{1.23 \times 10^{-3} \text{ mol } H_2S}{338 \text{ mL soln}} \times \frac{1000 \text{ mL}}{1 \text{ L soln}} \times \frac{2 \text{ mol } Ag^+}{1 \text{ mol } H_2S} = 7.28 \times 10^{-3} \text{ M}$$

Then, for $AgBrO_3$ $K_{sp} = [Ag^+][BrO_3^-] = (7.28 \times 10^{-3})^2 = 5.30 \times 10^{-5}$

32. The titration reaction is $Ca(OH)_2(aq) + 2\,HCl(aq) \longrightarrow CaCl_2(aq) + 2\,H_2O$

$$[OH^-] = \frac{10.7\text{ mL HCl} \times \frac{1\text{ L}}{1000\text{ mL}} \times \frac{0.1032\text{ mol HCl}}{1\text{ L}} \times \frac{1\text{ mol OH}^-}{1\text{ mol HCl}}}{50.00\text{ mL Ca(OH)}_2\text{ soln} \times \frac{1\text{ L}}{1000\text{ mL}}} = 0.022\text{ M}$$

In a saturated solution of $Ca(OH)_2$, $[Ca^{2+}] = [OH^-] \div 2$

$K_{sp} = [Ca^{2+}][OH^-]^2 = (0.022 \div 2)(0.022)^2 = 5.3 \times 10^{-6}$ Compare with 5.5×10^{-6} in Appendix D.

The Common Ion Effect

33. The presence of KI in a solution produces a significant $[I^-]$ in the solution. Through the solubility product principle, not as much AgI can dissolve in such a solution as in pure water, since the ion product, $[Ag^+][I^-]$, cannot exceed the value of K_{sp}. In similar fashion, $AgNO_3$ produces a significant $[Ag^+]$ in solution, again influencing the value of the ion product; not as much AgI can dissolve as in pure water.

34. If the solution contains KNO_3, more AgI can dissolve than in pure water, because the activity of each ion will be less than its molarity. On an ionic level, the reason is that ion pairs—such as $Ag^+NO_3^-(aq)$ and $K^+I^-(aq)$—form in the solution, preventing Ag^+ and I^- ions from coming together and precipitating.

35.

Equation:	$Ag_2SO_4(s) \rightleftharpoons$	$2\,Ag^+(aq)$	+	$SO_4^{2-}(aq)$
Original:				0.150 M
Add solid:		$+x$ M		$+x/2$ M
Equil:		x M		$(0.150 + x/2)$M

$x = [Ag^+] = 9.7 \times 10^{-3}$ M $K_{sp} = [Ag^+]^2[SO_4^{2-}] = (9.7 \times 10^{-3})^2(0.150 + 0.0049) = 1.5 \times 10^{-5}$

36.

Equation:	$CaSO_4(s) \rightleftharpoons$	$Ca^{2+}(aq)$	+	$SO_4^{2-}(aq)$
Soln:				0.0025 M
+ $CaSO_4(s)$		$+x$ M		$+x$ M
Equil:		x M		$(0.0025 + x)$M

$K_{sp} = [Ca^{2+}][SO_4^{2-}] = 9.1 \times 10^{-6} = x(0.0025 + x) = 0.0025\,x + x^2$ $x^2 + 0.0025\,x - 9.1 \times 10^{-6} = 0$

$$x = \frac{-b \pm \sqrt{b^2 - 4ac}}{2a} = \frac{-0.0025 \pm \sqrt{6.3 \times 10^{-6} + 3.6 \times 10^{-5}}}{2} = 2.0 \times 10^{-3}\text{ M} = [CaSO_4]$$

$$\text{mass CaSO}_4 = 100.0\text{ mL} \times \frac{1\text{ L}}{1000\text{ mL}} \times \frac{2.0 \times 10^{-3}\text{ mol CaSO}_4}{1\text{ L soln}} \times \frac{136.1\text{ g CaSO}_4}{1\text{ mol CaSO}_4} = 0.027\text{ g CaSO}_4$$

37. For PbI_2, $K_{sp} = 7.1 \times 10^{-9} = [Pb^{2+}][I^-]^2$

In a solution with 1.5×10^{-4} mol PbI_2/L, $[Pb^{2+}] = 1.5 \times 10^{-4}$ M, and $[I^-] = 2\,[Pb^{2+}] = 3.0 \times 10^{-4}$ M

Equation:	$PbI_2(s) \rightleftharpoons$	$Pb^{2+}(aq)$	+	$2\,I^-(aq)$
Initial:		1.5×10^{-4} M		3.0×10^{-4} M
Add lead(II):		$+x$ M		
Equil:		$(0.00015 + x)$M		0.00030 M

$K_{sp} = 7.1 \times 10^{-9} = (0.00015 + x)(0.00030)^2$ $(0.00010 + x) = 0.079$ $x = 0.079\text{ M} = [Pb^{2+}]$

38. For PbI_2, $K_{sp} = 7.1 \times 10^{-9} = [Pb^{2+}][I^-]^2$

In a solution with 1.5×10^{-5} mol PbI_2/L, $[Pb^{2+}] = 1.5 \times 10^{-5}$ M, and $[I^-] = 2\,[Pb^{2+}] = 3.0 \times 10^{-5}$ M

Equation:	$PbI_2(s) \rightleftharpoons$	$Pb^{2+}(aq)$	+	$2\,I^-(aq)$
Initial:		1.5×10^{-5} M		3.0×10^{-5} M
Add KI:				$+x$ M
Equil:		1.5×10^{-5} M		$(3.0 \times 10^{-5} + x)$M

$K_{sp} = 7.1 \times 10^{-9} = (1.5 \times 10^{-5})(3.0 \times 10^{-5} + x)^2$ $(3.0 \times 10^{-5} + x) = \sqrt{\dfrac{7.1 \times 10^{-9}}{1.5 \times 10^{-5}}} = 2.2 \times 10^{-2}$

$x = 2.2 \times 10^{-2} - 3.0 \times 10^{-5} = 2.2 \times 10^{-2}\text{ M} = [I^-]$

39. For Ag_2CrO_4, $K_{sp} = 1.1 \times 10^{-12} = [Ag^+]^2[CrO_4^{2-}]$

In a solution with 5.0×10^{-8} mol Ag_2CrO_4/L, $[CrO_4^{2-}] = 5.0 \times 10^{-8}$ M and $[Ag^+] = 1.0 \times 10^{-7}$ M

Equation:	$Ag_2CrO_4(s) \rightleftharpoons$	$2\,Ag^+(aq)$ +	$CrO_4^{2-}(aq)$
Initial:		1.0×10^{-7} M	5.0×10^{-8} M
Add chromate:			$+x$ M
Equil:		1.0×10^{-7} M	$(5.0 \times 10^{-8} + x)$M

$K_{sp} = 1.1 \times 10^{-12} = (1.0 \times 10^{-7})^2(5.0 \times 10^{-8} + x)$ $\qquad (5.0 \times 10^{-8} + x) = 1.1 \times 10^2$

$x = 1.1 \times 10^2\text{ M} = [CrO_4^{2-}]$ This is an impossibly high concentration to reach; we cannot lower the solubility of Ag_2CrO_4 to 5.0×10^{-8} mol Ag_2CrO_4/L with CrO_4^{2-} as the common ion. What about Ag^+?

Equation:	$Ag_2CrO_4(s) \rightleftharpoons$	$2\,Ag^+(aq)$ +	$CrO_4^{2-}(aq)$
Initial:		1.0×10^{-7} M	5.0×10^{-8} M
Add silver(I):		$+x$ M	
Equil:		$(1.0 \times 10^{-7} + x)$M	5.0×10^{-8} M

$K_{sp} = 1.1 \times 10^{-12} = (1.0 \times 10^{-7} + x)^2(5.0 \times 10^{-8})$ $\qquad (1.0 \times 10^{-7} + x) = \sqrt{\dfrac{1.1 \times 10^{-12}}{5.0 \times 10^{-8}}} = 4.7 \times 10^{-3}$

$x = 4.7 \times 10^{-3} - 1.0 \times 10^{-7} = 4.7 \times 10^{-3}\text{ M} = [I^-]$ This is an easy-to-reach concentration.

40. Even though $BaCO_3$ is more soluble than $BaSO_4$, it will still precipitate when 0.50 M $Na_2CO_3(aq)$ is added to a saturated solution of $BaSO_4$, because there is a sufficient $[Ba^{2+}]$ in such a solution for the product $[Ba^{2+}][CO_3^{2-}]$ to exceed the value of K_{sp} for the compound. An example will demonstrate this phenomenon. Let us assume that the two solutions being mixed are of equal volume and—to make the situation even more unfavorable—that the saturated $BaSO_4$ solution is not in contact with solid $BaSO_4$, meaning that it does not maintain its saturation when it is diluted. First we determine $[Ba^{2+}]$ in saturated $BaSO_4(aq)$.

$K_{sp} = [Ba^{2+}][SO_4^{2-}] = 1.1 \times 10^{-10} = s^2$ $\qquad s = \sqrt{1.1 \times 10^{-10}} = 1.0 \times 10^{-5}$ M

Mixing solutions of equal volumes means that the concentrations of solutes not common to the two solutions are halved by dilution.

$$[Ba^{2+}] = \tfrac{1}{2} \times \frac{1.0 \times 10^{-5}\text{ mol BaSO}_4}{1\text{ L}} \times \frac{1\text{ mol Ba}^{2+}}{1\text{ mol BaSO}_4} = 5.0 \times 10^{-6}\text{ M}$$

$$[CO_3^{2-}] = \tfrac{1}{2} \times \frac{0.50\text{ mol Na}_2\text{CO}_3}{1\text{ L}} \times \frac{1\text{ mol CO}_3^{2-}}{1\text{ mol Na}_2\text{CO}_3} = 0.25\text{ M}$$

$Q\{BaCO_3\} = [Ba^{2+}][CO_3^{2-}] = (5.0 \times 10^{-6})(0.25) = 1.3 \times 10^{-6} > 5.0 \times 10^{-9} = K_{sp}\{BaCO_3\}$

Thus, precipitation of $BaCO_3$ indeed should occur under the conditions described.

41. $$[Ca^{2+}] = \frac{115\text{ g Ca}^{2+}}{10^6\text{ g soln}} \times \frac{1\text{ mol Ca}^{2+}}{40.08\text{ g Ca}^{2+}} \times \frac{1000\text{ g soln}}{1\text{ L soln}} = 2.87 \times 10^{-3}\text{ M}$$

$[Ca^{2+}][F^-]^2 = K_{sp} = 5.3 \times 10^{-9} = (2.87 \times 10^{-3})[F^-]^2$ $\qquad [F^-] = 1.4 \times 10^{-3}$ M

$$\text{ppm F}^- = \frac{1.4 \times 10^{-3}\text{ mol F}^-}{1\text{ L soln}} \times \frac{19.00\text{ g F}^-}{1\text{ mol F}^-} \times \frac{1\text{ L soln}}{1000\text{ g}} \times 10^6\text{ g soln} = 27\text{ ppm}$$

42. We first calculate the $[Ag^+]$ and the $[Cl^-]$ in the saturated solution.

$K_{sp} = [Ag^+][Cl^-] = 1.8 \times 10^{-10} = (s)(s) = s^2$ $\qquad s = 1.3 \times 10^{-5}\text{ M} = [Ag^+] = [Cl^-]$

Both of these concentrations are marginally diluted by the addition of 1 mL of NaCl(aq)

$$[Ag^+] = [Cl^-] = 1.3 \times 10^{-5}\text{ M} \times \frac{100.0\text{ mL}}{100.0\text{ mL} + 1.0\text{ mL}} = 1.3 \times 10^{-5}\text{ M}$$

The $[Cl^-]$ in the NaCl(aq) also is diluted. $\qquad [Cl^-] = 1.0\text{ M} \times \dfrac{1.0\text{ mL}}{100.0\text{ mL} + 1.0\text{ mL}} = 9.9 \times 10^{-3}$ M

Let us use this $[Cl^-]$ to determine the $[Ag^+]$ that can exist in this solution.

$[Ag^+][Cl^-] = 1.8 \times 10^{-10} = [Ag^+](9.9 \times 10^{-3}\text{ M})$ $\qquad [Ag^+] = \dfrac{1.8 \times 10^{-10}}{9.9 \times 10^{-3}} = 1.8 \times 10^{-8}$ M

We compute the amount of AgCl in this final solution, and in the initial solution.

$$\text{mmol AgCl final} = 101.0\text{ mL} \times \frac{1.8 \times 10^{-8}\text{ mol Ag}^+}{1\text{ L soln}} \times \frac{1\text{ mmol AgCl}}{1\text{ mmol Ag}^+} = 1.8 \times 10^{-6}\text{ mmol AgCl}$$

$$\text{mmol AgCl final} = 101.0\text{ mL} \times \frac{1.3\times10^{-5}\text{ mol Ag}^+}{1\text{ L soln}} \times \frac{1\text{ mmol AgCl}}{1\text{ mmol Ag}^+} = 1.3\times10^{-3}\text{ mmol AgCl}$$

The difference between these two amounts is the amount of AgCl that precipitates. We compute its mass.

$$\text{mass AgCl} = (1.3\times10^{-3} - 1.8\times10^{-6})\text{ mmol AgCl} \times \frac{143.3\text{ mg AgCl}}{1\text{ mmol AgCl}} = 0.19\text{ mg}$$

We conclude that the precipitate will not be visible to the unaided eye, since it weighs less than 1 mg.

Criterion for Precipitation from Solution

43. We first determine $[Mg^{2+}]$, and then determine the value of Q_{sp} in order to compare it to the value of K_{sp}. We express molarity in millimoles per milliliter, entirely equivalent to moles per liter.

$$[Mg^{2+}] = \frac{22.5\text{ mg MgCl}_2}{325\text{ mL soln}} \times \frac{1\text{ mmol MgCl}_2\cdot6H_2O}{203.3\text{ mg MgCl}_2\cdot6H_2O} \times \frac{1\text{ mmol Mg}^{2+}}{1\text{ mmol MgCl}_2} = 3.41\times10^{-4}\text{ M}$$

$Q_{sp} = [Mg^{2+}][F^-]^2 = (3.41\times10^{-4})(0.035)^2 = 4.1\times10^{-7} > 3.7\times10^{-8} = K_{sp}$

Thus, precipitation of $MgF_2(s)$ should occur from this solution.

44. The solutions mutually dilute each other.

$$[Cl^-] = 0.016\text{ M} \times \frac{155\text{ mL}}{155\text{ mL} + 245\text{ mL}} = 6.2\times10^{-3}\text{ M}$$

$$[Pb^{2+}] = 0.175\text{ M} \times \frac{245\text{ mL}}{245\text{ mL} + 155\text{ mL}} = 0.107\text{ M}$$

Then we compute the value of the ion product and compare it to the solubility product constant value.

$Q_{sp} = [Pb^{2+}][Cl^-]^2 = (0.107)(6.2\times10^{-3})^2 = 4.2\times10^{-6} < 1.6\times10^{-5} = K_{sp}$

Thus, precipitation of $PbCl_2(s)$ will not occur from these mixed solutions.

45. We determine the $[OH^-]$ needed to just initiate precipitation of $Cd(OH)_2$

$K_{sp} = [Cd^{2+}][OH^-]^2 = 2.5\times10^{-14} = (0.0055\text{ M})[OH^-]^2$ $\qquad [OH^-] = \sqrt{\frac{2.5\times10^{-14}}{0.0055}} = 2.1\times10^{-6}\text{ M}$

$pOH = -\log(2.1\times10^{-6}) = 5.68$ $\qquad pH = 14.00 - 5.68 = 8.32$

$Cd(OH)_2$ will precipitate from a solution with pH >8.32.

46. We determine the $[OH^-]$ needed to just initiate precipitation of $Cr(OH)_3$

$K_{sp} = [Cr^{3+}][OH^-]^3 = 6.3\times10^{-31} = (0.086\text{ M})[OH^-]^3$ $\qquad [OH^-] = \sqrt[3]{\frac{6.3\times10^{-31}}{0.086}} = 1.9\times10^{-10}\text{ M}$

$pOH = -\log(1.9\times10^{-10}) = 9.72$ $\qquad pH = 14.00 - 9.72 = 4.28$

$Cr(OH)_3$ will precipitate from a solution with pH > 4.28.

47. **(a)** First we determine $[Cl^-]$ due to the added NaCl.

$$[Cl^-] = \frac{0.10\text{ mg NaCl}}{1.0\text{ L soln}} \times \frac{1\text{ g}}{1000\text{ mg}} \times \frac{1\text{ mol NaCl}}{58.4\text{ g NaCl}} \times \frac{1\text{ mol Cl}^-}{1\text{ mol NaCl}} = 1.7\times10^{-6}\text{ M}$$

Then we determine the value of the ion product and compare it to the solubility product constant value.

$Q = [Ag^+][Cl^-] = (0.10)(1.7\times10^{-6}) = 1.7\times10^{-7} > 1.8\times10^{-10} = K_{sp}$ for AgCl

Precipitation of AgCl(s) should occur.

(b) The KBr(aq) is diluted on mixing, but the $[Ag^+]$ and $[Cl^-]$ are barely affected by dilution.

$$[Br^-] = 0.10\text{ M} \times \frac{0.05\text{ mL}}{0.05\text{ mL} + 250\text{ mL}} = 2\times10^{-5}\text{ M}$$

Now we determine $[Ag^+]$ in a saturated AgCl solution.

$K_{sp} = [Ag^+][Cl^-] = (s)(s) = s^2 = 1.8\times10^{-10}$ $\qquad s = 1.3\times10^{-5}\text{ M}$

Then we determine the value of the ion product and compare it to the solubility product constant value.

$Q = [Ag^+][Br^-] = (1.3\times10^{-5})(2\times10^{-5}) = 3\times10^{-10} > 5.0\times10^{-13} = K_{sp}$ for AgBr

Precipitation of AgBr(s) should occur.

(c) The hydroxide ion is diluted by mixing the two solutions.

$$[OH^-] = 0.0150\text{ M} \times \frac{0.05\text{ mL}}{0.05\text{ mL} + 3000\text{ mL}} = 3\times10^{-7}\text{ M}$$

But the $[Mg^{2+}]$ does not change significantly.

$$[Mg^{2+}] = \frac{2.0\text{ mg Mg}^{2+}}{1.0\text{ L soln}} \times \frac{1\text{ g}}{1000\text{ mg}} \times \frac{1\text{ mol Mg}^{2+}}{24.3\text{ g Mg}} = 8.2\times10^{-5}\text{ M}$$

Then we determine the value of the ion product and compare it to the solubility product constant value.

$Q = [Mg^{2+}][OH^-]^2 = (3 \times 10^{-7})(8.2 \times 10^{-5})^2 = 2 \times 10^{-15} < 1.8 \times 10^{-11} = K_{sp}$ for $Mg(OH)_2$

Thus, precipitation of $Mg(OH)_2(s)$ will not occur.

48. We determine the amount of H_2 produced during the electrolysis, and then determine $[OH^-]$.

$$\text{amount } H_2 = \frac{PV}{RT} = \frac{752 \text{ mmHg} \times \dfrac{1 \text{ atm}}{760 \text{ mmHg}} \times 0.652 \text{ L}}{0.08206 \text{ L atm mol}^{-1} \text{ K}^{-1} \times 295 \text{ K}} = 0.0267 \text{ mol } H_2$$

$$[OH^-] = \frac{0.0267 \text{ mol } H_2 \times \dfrac{2 \text{ mol } OH^-}{1 \text{ mol } H_2}}{0.315 \text{ L sample}} = 0.170 \text{ M}$$

$Q_{sp} = [Mg^{2+}][OH^-]^2 = (0.185)(0.170)^2 = 5.35 \times 10^{-3} > 1.8 \times 10^{-11} = K_{sp}$

Yes, precipitation of $Mg(OH)_2(s)$ should occur during the electrolysis.

49. We determine $[C_2O_4^{2-}]$ in this solution. From key idea 3 in Section 17-6, $[C_2O_4^{2-}] = K_{a_2} = 5.4 \times 10^{-5}$

$Q_{sp} = [Ca^{2+}][C_2O_4^{2-}] = (0.150)(5.4 \times 10^{-5}) = 8.1 \times 10^{-6} > 4 \times 10^{-9} = K_{sp}$

Thus CaC_2O_4 should precipitate from this solution. (The quantity of $H_2C_2O_4$ used is immaterial because $[C_2O_4^{2-}]$ is reasonably independent of the oxalic acid concentration.)

50. The solutions mutually dilute each other. We first determine the solubility of each compound in its saturated solution and then its concentration after dilution.

$K_{sp} = [Ag^+]^2[SO_4^{2-}] = 1.4 \times 10^{-5} = (2s)^2 s = 4s^3$ $\qquad s = \sqrt[3]{\dfrac{1.4 \times 10^{-5}}{4}} = 1.5 \times 10^{-2}$ M

$[SO_4^{2-}] = 0.015 \text{ M} \times \dfrac{100.0 \text{ mL}}{100.0 \text{ mL} + 250.0 \text{ mL}} = 0.0043$ M $\qquad [Ag^+] = 0.0086$ M

$K_{sp} = [Pb^{2+}][CrO_4^{2-}] = 2.8 \times 10^{-13} = (s)(s) = s^2$ $\qquad s = \sqrt{2.8 \times 10^{-13}} = 5.3 \times 10^{-7}$ M

$[Pb^{2+}] = [CrO_4^{2-}] = 5.3 \times 10^{-7} \times \dfrac{250.0 \text{ mL}}{250.0 \text{ mL} + 100.0 \text{ mL}} = 3.8 \times 10^{-7}$ M

From the balanced chemical equation, we see that the two possible precipitates are $PbSO_4$ and Ag_2CrO_4. (Neither $PbCrO_4$ nor Ag_2SO_4 can precipitate because they have been diluted below their saturated concentrations.) $\qquad PbCrO_4 + Ag_2SO_4 \longrightarrow PbSO_4 + Ag_2CrO_4$ $\qquad$ Thus, we compute the value of Q for each of these compounds and compare those values with the solubity constant product value.

$Q = [Pb^{2+}][SO_4^{2-}] = (3.8 \times 10^{-7})(0.0043) = 1.6 \times 10^{-9} < 1.6 \times 10^{-8} = K_{sp}$ for $PbSO_4$

Thus, $PbSO_4(s)$ will not precipitate.

$Q = [Ag^+]^2[CrO_4^{2-}] = (0.0086)^2(3.8 \times 10^{-7}) = 2.8 \times 10^{-11} > 1.1 \times 10^{-12} = K_{sp}$ for Ag_2CrO_4

Thus, $Ag_2CrO_4(s)$ should precipitate.

Completeness of Precipitation

51. First determine that a precipitate forms. The solutions mutually dilute each other.

$[CrO_4^{2-}] = 0.350 \text{ M} \times \dfrac{200.0 \text{ mL}}{200.0 \text{ mL} + 200.0 \text{ mL}} = 0.175$ M

$[Ag^+] = 0.0100 \text{ M} \times \dfrac{200.0 \text{ mL}}{200.0 \text{ mL} + 200.0 \text{ mL}} = 0.00500$ M

We determine the value of the ion product and compare it to the solubility product constant value.

$Q = [Ag^+]^2[CrO_4^{2-}] = (0.00500)^2(0.175) = 4.4 \times 10^{-6} > 1.1 \times 10^{-12} = K_{sp}$ for Ag_2CrO_4

Ag_2CrO_4 should precipitate.

Now, we assume that as much solid forms as possible, and then we approach equilibrium by dissolving that solid in a solution that contains the ion in excess.

Equation:	$Ag_2CrO_4(s)$ $\rightleftharpoons$	$2\,Ag^+(aq)$	+	$CrO_4^{2-}(aq)$
Orig. soln:		0.00500		0.175 M
Form solid:		0 M		0.170 M
Changes:		$+2x$ M		$+x$ M
Equil:		$2x$ M		$(0.170 + x)$M

$K_{sp} = [Ag^+]^2[CrO_4^{2-}] = 1.1 \times 10^{-12} = (2x)^2(0.170 + x) \approx (4x^2)(0.170)$

$$x = \sqrt{\frac{1.1 \times 10^{-12}}{4 \times 0.170}} = 1.3 \times 10^{-6}\text{ M} \qquad [\text{Ag}^+] = 2x = 2.6 \times 10^{-6}\text{ M}$$

$$\%\ \text{Ag}^+ \text{ unprecipitated} = \frac{2.6 \times 10^{-6}\text{ M final}}{0.00500\text{ M initial}} \times 100\% = 0.052\% \text{ unprecipitated}$$

52. We first use the solubility product constant expression to determine $[Pb^{2+}]$ in a solution with 0.100 M Cl^-.

$$K_{sp} = [\text{Pb}^{2+}][\text{Cl}^-]^2 = 1.6 \times 10^{-5} = [\text{Pb}^{2+}](0.100)^2 \qquad [\text{Pb}^{2+}] = \frac{1.6 \times 10^{-5}}{(0.100)^2} = 1.6 \times 10^{-3}\text{ M}$$

$$\%\ \text{unprecipitated} = \frac{1.6 \times 10^{-3}\text{ M}}{0.065\text{ M}} \times 100\% = 2.5\%$$

Now, we want $[\text{Pb}^{2+}]_f = 1\%\ [\text{Pb}^{2+}]_i = 0.010 \times 0.065\text{ M} = 6.5 \times 10^{-4}\text{ M}$

$$K_{sp} = [\text{Pb}^{2+}][\text{Cl}^-]^2 = 1.6 \times 10^{-5} = (6.5 \times 10^{-4})[\text{Cl}^-]^2 \qquad [\text{Cl}^-] = \sqrt{\frac{1.6 \times 10^{-5}}{6.5 \times 10^{-4}}} = 0.16\text{ M}$$

Fractional Precipitation

53. The concentrations of silver ion that are cited in Example 19-7 range from 5.0×10^{-11} M to 1.0×10^{-5} M. These are incredibly small concentrations, especially the first. Virtually any $AgNO_3$(aq) solution that we would prepare by usual means would have at least these concentrations. However, there is the matter of dilution to be considered. If one drop (0.05 mL) of $AgNO_3$(aq) is added to 500.0 mL of solution, the $[Ag^+]$ will decrease by a factor of 10^4. Thus we would have to begin with 0.15 M $AgNO_3$ for this dilution to produce 1.5×10^{-5} M. So we cannot be too careless and use extremely dilute $AgNO_3$(aq).

54. (a) 0.10 M NaCl will not work at all, since both $BaCl_2$ and $CaCl_2$ are soluble in water.

(b) $K_{sp} = 1.1 \times 10^{-10}$ for $BaSO_4$ and $K_{sp} = 9.1 \times 10^{-6}$ for $CaSO_4$. Since these values differ by more than 1000, 0.05 M Na_2SO_4 would effectively separate Ba^{2+} from Ca^{2+}.
We first compute $[SO_4^{2-}]$ when $BaSO_4$ begins to precipitate.

$$[\text{Ba}^{2+}][\text{SO}_4^{2-}] = (0.050)[\text{SO}_4^{2-}] = 1.1 \times 10^{-10} \qquad [\text{SO}_4^{2-}] = \frac{1.1 \times 10^{-10}}{0.050} = 2.2 \times 10^{-9}\text{ M}$$

And then $[SO_4^{2-}]$ when $[Ba^{2+}]$ has decreased to 0.1% of its initial value, that is, to 5.0×10^{-5} M

$$[\text{Ba}^{2+}][\text{SO}_4^{2-}] = (5.0 \times 10^{-5})[\text{SO}_4^{2-}] = 1.1 \times 10^{-10} \qquad [\text{SO}_4^{2-}] = \frac{1.1 \times 10^{-10}}{5.0 \times 10^{-5}} = 2.2 \times 10^{-6}\text{ M}$$

And finally $[SO_4^{2-}]$ when $CaSO_4$ begins to precipitate.

$$[\text{Ca}^{2+}][\text{SO}_4^{2-}] = (0.050)[\text{SO}_4^{2-}] = 9.1 \times 10^{-6} \qquad [\text{SO}_4^{2-}] = \frac{9.1 \times 10^{-6}}{0.050} = 1.8 \times 10^{-4}\text{ M}$$

(c) Now, $K_{sp} = 5 \times 10^{-3}$ for $Ba(OH)_2$ and $K_{sp} = 5.5 \times 10^{-6}$ for $Ca(OH)_2$. The fact that these two K_{sp} values differ by almost a factor of 1000 does not tell the entire story, because $[OH^-]$ appears squared in both K_{sp} expressions. We compute $[OH^-]$ when $Ca(OH)_2$ begins to precipitate.

$$[\text{Ca}^{2+}][\text{OH}^-]^2 = 5.5 \times 10^{-6} = (0.050)[\text{OH}^-]^2 \qquad [\text{OH}^-] = \sqrt{\frac{5.5 \times 10^{-6}}{0.050}} = 1.0 \times 10^{-2}\text{ M}$$

Precipitation will not proceed this far; we only have 0.001 M NaOH, which has $[OH^-] = 1 \times 10^{-3}$ M

(d) Finally, $K_{sp} = 5.1 \times 10^{-9}$ for $BaCO_3$ and $K_{sp} = 2.8 \times 10^{-9}$ for $CaCO_3$. Since these two values differ by less than a factor of 2, 0.50 M Na_2CO_3 would not effectively separate Ba^{2+} from Ca^{2+}.

55. Normally we would worry about the mutual dilution of the two solutions, but the values of the solubility product constants are so small that only a very small volume of 0.50 M $Pb(NO_3)_2$ solution needs to be added, as we shall see.

(a) Since the two anions are present at the same concentration and they have the same type of formula (one anion per cation), the one forming the compound with the smallest K_{sp} value will precipitate first. Thus, CrO_4^{2-} is the first ion to precipitate.

(b) At the point where SO_4^{2-} begins to precipitate, we have

$K_{sp} = [Pb^{2+}][SO_4^{2-}] = 1.6 \times 10^{-8} = [Pb^{2+}](0.010\text{ M})$ $\quad [Pb^{2+}] = \dfrac{1.6 \times 10^{-8}}{0.010} = 1.6 \times 10^{-6}\text{ M}$

Now we can test our original assumption, that only a very small volume of 0.50 M $Pb(NO_3)_2$ solution has been added. We assume that we have 1.00 L of the original solution, the one with the two anions dissolved in it, and compute the volume of 0.50 M $Pb(NO_3)_2$ that has to be added to achieve $[Pb^{2+}] = 1.6 \times 10^{-6}$ M

$$\text{added soln volume} = 1.00\text{ L} \times \frac{1.6 \times 10^{-6}\text{ mol }Pb^{2+}}{1\text{ L soln}} \times \frac{1\text{ mol }Pb(NO_3)_2}{1\text{ mol }Pb^{2+}} \times \frac{1\text{ L }Pb^{2+}\text{soln}}{0.50\text{ mol }Pb(NO_3)_2} \times \frac{1000\text{ mL}}{1\text{ L}} = 3.2 \times 10^{-3}\text{ mL }Pb^{2+}\text{ soln} = 0.0032\text{ mL }Pb^{2+}\text{ soln}$$

This is less than one drop (0.05 mL) of the Pb^{2+} solution, clearly a very small volume.

(c) The two anions are effectively separated if $[Pb^{2+}]$ has not reached 1.6×10^{-6} M when $[CrO_4^{2-}]$ is reduced to 0.1% of its original value, that is, to $0.010 \times 10^{-3}\text{ M} = 1.0 \times 10^{-5}\text{ M} = [CrO_4^{2-}]$

$K_{sp} = [Pb^{2+}][CrO_4^{2-}] = 2.8 \times 10^{-13} = [Pb^{2+}](1.0 \times 10^{-5})$

$[Pb^{2+}] = \dfrac{2.8 \times 10^{-13}}{1.0 \times 10^{-5}} = 2.8 \times 10^{-8}\text{ M}$

Thus, the two anions can be effectively separated by fractional precipitation.

56. (a) We answer this question by determining the $[Ag^+]$ needed to initiate precipitation of each compound.

AgCl: $K_{sp} = [Ag^+][Cl^-] = 1.8 \times 10^{-10} = [Ag^+](0.250)$ $\quad [Ag^+] = \dfrac{1.8 \times 10^{-10}}{0.250} = 7.2 \times 10^{-10}\text{ M}$

AgBr: $K_{sp} = [Ag^+][Br^-] = 5.0 \times 10^{-13} = [Ag^+](0.0022)$ $\quad [Ag^+] = \dfrac{5.0 \times 10^{-13}}{0.0022} = 2.3 \times 10^{-10}\text{ M}$

Br^- precipitates first, as AgBr, because it requires a lower $[Ag^+]$.

(b) $[Ag^+] = 7.2 \times 10^{-10}$ M when chloride ion, the second ion, begins to precipitate.

(c) Cl^- and Br^- cannot be separated by this fractional precipitation. $[Ag^+]$ will have to rise to 1000 times its initial value, to 2.3×10^{-7} M, before AgBr is completely precipitated. But as soon as $[Ag^+]$ reaches 7.2×10^{-10} M, AgCl will begin to precipitate.

Solubility and pH

57. In each case we indicate whether the compound is more soluble in acid than in water. We write the net ionic equation for the reaction in which the solid dissolves in acid. Substances are soluble in acid if either (1) an acid-base reaction occurs [as in (d)] or (2) a gas is produced, since escape of the gas from the reaction mixture causes the reaction to shift to the right.

(a) Same: KCl

(b) Acid: $MgCO_3(s) + 2\,H^+(aq) \longrightarrow Mg^{2+}(aq) + H_2O + CO_2(g)$

(c) Acid: $FeS(s) + 2\,H^+(aq) \longrightarrow Fe^{2+}(aq) + H_2S(g)$

(d) Acid: $Ca(OH)_2(s) + 2\,H^+(aq) \longrightarrow Ca^{2+}(aq) + 2\,H_2O$

(e) Water: C_6H_5COOH should be less soluble in acid, because of the H_3O^+ common ion.

58. In each case we indicate whether the compound is more soluble in base than in water. We write the net ionic equation for the reaction in which the solid dissolves in base. Substances are soluble in base if either (1) an acid-base reaction occurs [as in (b)] or (2) a gas is produced, since escape of the gas from the reaction mixture causes the reaction to shift to the right.

(a) Water: $BaSO_4$ should be less soluble in base because hydrolysis of SO_4^{2-} will be repressed.

(b) Base: $H_2C_2O_4(s) + 2\,OH^-(aq) \longrightarrow C_2O_4^{2-}(aq) + 2\,H_2O$

(c) Water: $Fe(OH)_3$ should be less soluble in base because of the OH^- common ion.

(d) Same: $NaNO_3$

(e) Water: MnS should be less soluble in base because hydrolysis of S^{2-} will be repressed.

59. We determine $[Mg^{2+}]$ in the solution.

$$[Mg^{2+}] = \frac{0.65\text{ g }Mg(OH)_2}{1\text{ L soln}} \times \frac{1\text{ mol }Mg(OH)_2}{58.3\text{ g }Mg(OH)_2} \times \frac{1\text{ mol }Mg^{2+}}{1\text{ mol }Mg(OH)_2} = 0.011\text{ M}$$

Then we determine $[OH^-]$ in the solution, and its pH.

$K_{sp} = [Mg^{2+}][OH^-]^2 = 1.8 \times 10^{-11} = (0.011)[OH^-]^2$ $\qquad [OH^-] = \sqrt{\dfrac{1.8 \times 10^{-11}}{0.011}} = 4.0 \times 10^{-5}$ M

$pOH = -\log(4.0 \times 10^{-5}) = 4.40$ $\qquad pH = 14.00 - 4.39 = 9.60$

60. First determine the $[Mg^{2+}]$ and $[NH_3]$ that result from dilution to a total volume of 0.500 L.

$[Mg^{2+}] = 0.100 \text{ M} \times \dfrac{0.150 \text{ L}_{initial}}{0.500 \text{ L}_{final}} = 0.0300 \text{ M}$ $\qquad [NH_3] = 0.150 \text{ M} \times \dfrac{0.350 \text{ L}_{initial}}{0.500 \text{ L}_{final}} = 0.105 \text{ M}$

Then determine the $[OH^-]$ that will allow $[Mg^{2+}] = 0.0300$ M in this solution.

$K_{sp} = [Mg^{2+}][OH^-]^2 = (0.0300)[OH^-]^2$ $\qquad [OH^-] = \sqrt{\dfrac{1.8 \times 10^{-11}}{0.0300}} = 2.4 \times 10^{-5}$ M

This $[OH^-]$ is maintained by the NH_3–NH_4^+ buffer, for which we use the Henderson-Hasselbalch equation.

$pH = 14.00 - pOH = 14.00 + \log(2.4 \times 10^{-5}) = 9.38 = pK_a + \log\dfrac{[NH_3]}{[NH_4^+]} = 9.26 + \dfrac{[NH_3]}{[NH_4^+]}$

$\log\dfrac{[NH_3]}{[NH_4^+]} = 9.38 - 9.26 = +0.12$ $\qquad \dfrac{[NH_3]}{[NH_4^+]} = 10^{+0.12} = 1.3$ $\qquad [NH_4^+] = \dfrac{0.105 \text{ M } NH_3}{1.3} = 0.081$ M

$\text{mass } (NH_4)_2SO_4 = 0.500 \text{ L} \times \dfrac{0.081 \text{ mol } NH_4^+}{\text{L soln}} \times \dfrac{1 \text{ mol } (NH_4)_2SO_4}{2 \text{ mol } NH_4^+} \times \dfrac{132.1 \text{ g } (NH_4)_2SO_4}{1 \text{ mol } (NH_4)_2SO_4} = 2.7 \text{ g}$

61. **(a)** We calculate $[OH^-]$ needed for precipitation.

$K_{sp} = [Al^{3+}][OH^-]^3 = 1.3 \times 10^{-33} = (0.075 \text{ M})[OH^-]^3$ $\qquad [OH^-] = \sqrt[3]{\dfrac{1.3 \times 10^{-33}}{0.075}} = 2.6 \times 10^{-11}$

$pOH = -\log(2.6 \times 10^{-11}) = 10.59$ $\qquad pH = 14.00 - 10.59 = 3.41$

(b) We use the Henderson-Hasselbalch equation to determine $[C_2H_3O_2^-]$.

$pH = 3.41 = pK_a + \log\dfrac{[C_2H_3O_2^-]}{[HC_2H_3O_2]} = 4.74 + \log\dfrac{[C_2H_3O_2^-]}{1.00 \text{ M}}$

$\log\dfrac{[C_2H_3O_2^-]}{1.00 \text{ M}} = 3.41 - 4.74 = -1.33$ $\qquad \dfrac{[C_2H_3O_2^-]}{1.00 \text{ M}} = 10^{-1.33} = 0.047$ $\qquad [C_2H_3O_2^-] = 0.047$ M

This situation does not quite obey the guideline that the ratio of concentrations fall in the range 0.10 to 10.0, but the resulting error is a small one in this circumstance.

$\text{mass } NaC_2H_3O_2 = 0.2500 \text{ L} \times \dfrac{0.047 \text{ mol } C_2H_3O_2^-}{1 \text{ L soln}} \times \dfrac{1 \text{ mol } NaC_2H_3O_2}{1 \text{ mol } C_2H_3O_2^-} \times \dfrac{82.03 \text{ g } NaC_2H_3O_2}{1 \text{ mol } NaC_2H_3O_2}$

$= 0.96 \text{ g } NaC_2H_3O_2$

62. **(a)** Since HI is a strong acid, $[I^-] = 1.05 \times 10^{-3} \text{ M} + 1.05 \times 10^{-3} \text{ M} = 2.10 \times 10^{-3}$ M

We determine the value of the ion product and compare it to the solubility product constant value.

$Q_{sp} = [Pb^{2+}][I^-]^2 = (1.1 \times 10^{-3})(2.10 \times 10^{-3})^2 = 4.9 \times 10^{-9} < 7.1 \times 10^{-9} = K_{sp}$ for PbI_2

A precipitate of PbI_2 will not form under these conditions.

(b) We compute the $[OH^-]$ needed for precipitation.

$K_{sp} = [Mg^{2+}][OH^-]^2 = 1.8 \times 10^{-11} = (0.0150)[OH^-]^2$ $\qquad [OH^-] = \sqrt{\dfrac{1.8 \times 10^{-11}}{0.0150}} = 3.5 \times 10^{-5}$ M

Then we compute $[OH^-]$ in this solution, resulting from the ionization of NH_3.

$[NH_3] = 6.00 \text{ M} \times \dfrac{0.05 \times 10^{-3} \text{ L}}{2.50 \text{ L}} = 1 \times 10^{-4}$ M

Even though NH_3 is a weak base, $[OH^-]$ produced from NH_3 will approximate 4×10^{-5} M in this very dilute solution. (Recall that degree of ionization is high in dilute solution.) And since $[OH^-] = 3.5 \times 10^{-5}$ M is needed for precipitation to occur, we conclude that $Mg(OH)_2$ will just barely precipitate from this solution.

(c) 0.010 M $HC_2H_3O_2$ and 0.010 M $NaC_2H_3O_2$ is a buffer solution with $pH = pK_a$ of acetic acid, since the acid and its anion are present in equal concentrations. (If they were not, we would have used the Henderson-Hasselbalch equation to determine the pH.) From this, we determine the $[OH^-]$.

$pH = 4.74$ $\qquad pOH = 14.00 - 4.74 = 9.26$ $\qquad [OH^-] = 10^{-9.26} = 5.5 \times 10^{-10}$

$Q = [Al^{3+}][OH^-]^3 = (0.010)(5.5 \times 10^{-10})^3 = 1.7 \times 10^{-30} > 1.3 \times 10^{-33} = K_{sp}$ of $Al(OH)_3$

Thus, $Al(OH)_3(s)$ should precipitate from this solution.

Complex-Ion Equilibria

63. Lead(II) ion forms a complex ion with chloride ion. It forms no such complex ion with nitrate ion. The formation of this complex ion decreases the concentrations of free $Pb^{2+}(aq)$ and free $Cl^-(aq)$. Thus, more $PbCl_2$ can dissolve before the value of the solubility product is satisfied.
$Pb^{2+}(aq) + 3\,Cl^-(aq) \rightleftharpoons [PbCl_3]^-(aq)$

64. $Zn^{2+}(aq) + 4\,NH_3(aq) \rightleftharpoons [Zn(NH_3)_4]^{2+}(aq) \qquad K_f = 4.1 \times 10^8$
$NH_3(aq)$ will be least effective in reducing the concentration of the complex ion. In fact, the addition of $NH_3(aq)$ will increase the concentration of the complex ion by favoring a shift of the equilibrium to the right. $NH_4^+(aq)$ will have a similar effect, but not as direct. $NH_3(aq)$ is formed by the hydrolysis of $NH_4^+(aq)$ and thus increasing $[NH_4^+]$ will eventually increase $NH_3(aq)$: $\quad NH_4^+(aq) + H_2O \rightleftharpoons NH_3(aq) + H_3O^+(aq)$.
The addition of HCl(aq) will decrease the concentration of complex ion the most. HCl(aq) will react with $NH_3(aq)$ to decrease its concentration (by forming NH_4^+) and that will cause the complex ion equilibrium reaction to shift left.

65. We substitute the given concentrations directly into the K_f expression.

$$K_f = \frac{[[Cu(CN)_4^{3-}]}{[Cu^+][CN^-]^4} = \frac{0.0500}{(6.1 \times 10^{-32})(0.80)^4} = 2.0 \times 10^{30}$$

66. The solution is organized around the balanced chemical equation. Free $[NH_3]$ is 6.0 M at equilibrium. The size of the equilibrium constant indicates that most copper(II) is present as the complex ion.

Equation:	$Cu^{2+}(aq)$ +	$4\,NH_3(aq)$ ⇌	$[Cu(NH_3)_4]^{2+}(aq)$
Initial:	0.10 M		
Form Complex:	0 M		0.10 M
Changes:	$+x$ M		$-x$ M
Equil:	x M	6.0 M	$(0.10 - x)$ M

$$K_f = \frac{[[Cu(NH_3)_4]^{2+}]}{[Cu^{2+}][NH_3]^4} = 1.1 \times 10^{13} = \frac{0.10 - x}{x\,6.0^4} \approx \frac{0.10}{1.3 \times 10^3\,x}$$

$$x = \frac{0.10}{1.3 \times 10^3 \times 1.1 \times 10^{13}} = 7.0 \times 10^{-18}\text{ M} = [Cu^{2+}]$$

67. We first find the concentration of free metal ion. Then we determine the value of Q_{sp} for the precipitation reaction, and compare its value with the value of K_{sp} to determine whether precipitation should occur.

Equation:	$Ag^+(aq)$ +	$2\,S_2O_3^{2-}(aq)$ ⇌	$[Ag(S_2O_3)_2]^{3-}(aq)$
Initial:		0.76 M	0.048 M
Changes:	$+x$ M	$+2x$ M	$-x$ M
Equil:	x M	$(0.76 + 2x)$M	$(0.048 - x)$M

$$K_f = \frac{[[Ag(S_2O_3)_2]^{3-}]}{[Ag^+][S_2O_3^{2-}]^2} = 1.7 \times 10^{13} = \frac{0.048 - x}{x\,(0.76 + 2x)^2} \approx \frac{0.048}{0.76^2\,x}$$

$x = 4.9 \times 10^{-15}\text{ M} = [Ag^+] \qquad Q = [Ag^+][I^-] = (4.9 \times 10^{-15})(2.0) = 9.8 \times 10^{-15} > 8.5 \times 10^{-17} = K_{sp}$
Because $Q > K_{sp}$, precipitation of AgI(s) should occur.

68. We determine $[OH^-]$ in this solution, and also the free $[Cu^{2+}]$.

$$pH = pK_a + \log\frac{[NH_3]}{[NH_4^+]} = 9.26 + \log\frac{0.10\text{ M}}{0.10\text{ M}} = 9.26$$

$pH = 14.00 - 9.26 = 4.74$
$[OH^-] = 10^{-4.74} = 1.8 \times 10^{-5}$ M

$Cu^{2+}(aq) + 4\,NH_3(aq) \rightleftharpoons [Cu(NH_3)_4]^{2+}(aq)$

$$K_f = \frac{[[Cu(NH_3)_4]^{2+}]}{[Cu^{2+}][NH_3]^4} = 1.1 \times 10^{13} = \frac{0.015}{[Cu^{2+}]\,0.10^4} \qquad [Cu^{2+}] = \frac{0.015}{1.1 \times 10^{13} \times 0.10^4} = 1.4 \times 10^{-11}\text{ M}$$

Now we determine the value of Q and compare it with the value of K_{sp} for $Cu(OH)_2$.
$Q = [Cu^{2+}][OH^-]^2 = (1.4 \times 10^{-11})(1.8 \times 10^{-5})^2 = 4.5 \times 10^{-21} < 2.2 \times 10^{-20} = K_{sp}$ for $Cu(OH)_2$
Precipitation of $Cu(OH)_2(s)$ from this solution should not occur.

69. We first compute the free $[Ag^+]$ in the original solution. The size of the equilibrium constant indicates that the reaction lies to the right, so we form as much complex ion as possible.

Equation:	$Ag^+(aq)$ +	$2\,NH_3(aq)$ ⇌	$[Ag(NH_3)_2]^+(aq)$
In soln:	0.10 M	1.00 M	
Form complex:	−0.10 M	−0.20 M	+0.10 M
	0 M	0.80 M	0.10 M
Changes:	$+x$ M	$+2x$ M	$-x$ M
Equil:	x M	$(0.80 + 2x)$ M	$(0.10 - x)$ M

$$K_f = 1.6 \times 10^7 = \frac{[[Ag(NH_3)_2]^+]}{[Ag^+][NH_3]^2} = \frac{0.10 - x}{x\,(0.80 + 2x)^2} \approx \frac{0.10}{x\,(0.80)^2} \quad x = \frac{0.10}{1.6 \times 10^7\,(0.80)^2} = 9.8 \times 10^{-9}\text{ M}$$

Thus, free $[Ag^+] = 9.8 \times 10^{-9}$. We determine the $[I^-]$ that can coexist in this solution without precipitation.

$$K_{sp} = [Ag^+][I^-] = 8.5 \times 10^{-17} = (9.8 \times 10^{-9})[I^-] \qquad [I^-] = \frac{8.5 \times 10^{-17}}{9.8 \times 10^{-9}} = 8.7 \times 10^{-9}\text{ M}$$

And now we determine the mass of KI needed to produce this $[I^-]$

$$\text{mass KI} = 1.00\text{ L soln} \times \frac{8.7 \times 10^{-9}\text{ mol I}^-}{1\text{ L soln}} \times \frac{1\text{ mol KI}}{1\text{ mol I}^-} \times \frac{166.0\text{ g KI}}{1\text{ mol KI}} = 1.4 \times 10^{-6}\text{ g KI}$$

70. First we determine $[Ag^+]$ that can exist with this $[Cl^-]$. We know that $[Cl^-]$ will be unchanged because precipitation will not be allowed to occur.

$$K_{sp} = [Ag^+][Cl^-] = 1.8 \times 10^{-10} = [Ag^+]\,0.100\text{ M} \qquad [Ag^+] = \frac{1.8 \times 10^{-10}}{0.100} = 1.8 \times 10^{-9}\text{ M}$$

We now consider the complex ion equilibrium. If the complex ion's final concentration is x, then the decrease in $[NH_3]$ is $2x$, because 2 mol NH_3 react to form each mol of complex ion, as follows.

$Ag^+(aq) + 2\,NH_3(aq) \rightleftharpoons [Ag(NH_3)_2]^+(aq)$ We can solve the K_f expression for x.

$$K_f = 1.6 \times 10^7 = \frac{[[Ag(NH_3)_2]^+]}{[Ag^+][NH_3]^2} = \frac{x}{1.8 \times 10^{-9}\,(1.00 - 2\,x)^2}$$

$$x = (1.6 \times 10^7)(1.8 \times 10^{-9})(1.00 - 2\,x)^2 = 0.029\,(1.00 - 4.00\,x + 4.00\,x^2) = 0.029 - 0.12x + 0.12x^2$$

$0 = 0.029 - 1.12x + 0.12x^2$ We use the quadratic formula to solve for x.

$$x = \frac{-b \pm \sqrt{b^2 - 4ac}}{2a} = \frac{1.12 \pm \sqrt{(1.12)^2 - 4 \times 0.029 \times 0.12}}{2 \times 0.012} = \frac{1.12 \pm 1.114}{0.024} = 93,\ 0.25$$

Thus, we can add 0.25 mol $AgNO_3$ to this solution. Since the molar mass of $AgNO_3$ is 169.9 g/mol, this means we can add about 42 g $AgNO_3$ to this solution before we see precipitate of AgCl form.

Precipitation and Solubilities of Metal Sulfides

71. We know that $K_{spa} = 3 \times 10^7$ for MnS and $K_{spa} = 6 \times 10^2$ for FeS. The metal sulfide will begin to precipitate when $Q_{spa} = K_{spa}$. Let us determine $[H_3O^+]$ just necessary to form each precipitate. We assume that the solution is saturated with H_2S, $[H_2S] = 0.10$ M.

$$K_{spa} = \frac{[M^{2+}][H_2S]}{[H_3O^+]^2} \qquad [H_3O^+] = \sqrt{\frac{[M^{2+}][H_2S]}{K_{spa}}} = \sqrt{\frac{(0.10\text{ M})(0.10\text{ M})}{3 \times 10^7}} = 1.8 \times 10^{-5}\text{ M for MnS}$$

$$[H_3O^+] = \sqrt{\frac{(0.10\text{ M})(0.10\text{ M})}{6 \times 10^2}} = 4.1 \times 10^{-3}\text{ M for FeS}$$

Thus, if the solution is maintained at an acidity just a bit higher than $1.8 \times 10^{-5}\text{ M} = [H_3O^+]$, FeS will precipitate and $Mn^{2+}(aq)$ will remain in solution. To determine if the separation is complete, we see whether $[Fe^{2+}]$ has decreased to 0.1% or less of its original value when the solution is held at the aforementioned acidity. Let $[H_3O^+] = 2.0 \times 10^{-5}$ M and calculate $[Fe^{2+}]$.

$$K_{spa} = \frac{[Fe^{2+}][H_2S]}{[H_3O^+]^2} = 6 \times 10^2 = \frac{[Fe^{2+}](0.10\text{ M})}{(2.0 \times 10^{-5}\text{ M})^2} \qquad [Fe^{2+}] = \frac{(6 \times 10^2)(2.0 \times 10^{-5})^2}{0.10} = 2.4 \times 10^{-6}\text{ M}$$

$$\%\ Fe^{2+}(aq)\text{ remaining} = \frac{2.4 \times 10^{-6}\text{ M}}{0.10\text{ M}} \times 100\% = 0.0024\% \qquad \text{Separation is complete.}$$

72. Since the cation concentrations are identical, the value of Q_{spa} is the same for each one. It is this value of Q_{spa} that we compare with K_{spa} to determine if precipitation occurs.

$$Q_{spa} = \frac{[M^{2+}][H_2S]}{[H_3O^+]^2} = \frac{0.05\text{ M} \times 0.10\text{ M}}{(0.010\text{ M})^2} = 5 \times 10^1$$

If $Q_{spa} > K_{spa}$ precipitation of the metal sulfide should occur. But if $Q_{spa} < K_{spa}$, precipitation will not occur.

For CuS, $K_{spa} = 6 \times 10^{-16} < Q_{spa} = 5 \times 10^1$ Precipitation of CuS(s) should occur.

For HgS, $K_{spa} = 2 \times 10^{-32} < Q_{spa} = 5 \times 10^1$ Precipitation of HgS(s) should occur.

For MnS, $K_{spa} = 3 \times 10^7 > Q_{spa} = 5 \times 10^1$ Precipitation of MnS(s) will not occur.

73. **(a)** First we calculate $[H_3O^+]$ in the buffer solution with the Henderson-Hasselbalch equation.

$$pH = pK_a + \log \frac{[C_2H_3O_2^-]}{[HC_2H_3O_2]} = 4.74 + \log \frac{0.15\ M}{0.25\ M} = 4.52 \qquad [H_3O^+] = 10^{-4.52} = 3.0 \times 10^{-5}\ M$$

We use this information to calculate a value of Q_{spa} for MnS in this solution and compare that with K_{spa}

$$Q_{spa} = \frac{[Mn^{2+}][H_2S]}{[H_3O^+]^2} = \frac{(0.15)(0.10)}{(3.0 \times 10^{-5})^2} = 1.7 \times 10^7 < 3 \times 10^7 = K_{spa} \text{ for MnS}$$

Precipitation of MnS(s) will not occur.

(b) We need to change $[H_3O^+]$ so that $Q_{spa} = 3 \times 10^7 = \frac{(0.15)(0.10)}{[H_3O^+]^2}$ $\qquad [H_3O^+] = \sqrt{\frac{(0.15)(0.10)}{3 \times 10^7}}$

$[H_3O^+] = 2.2 \times 10^{-5}\ M$ $\qquad$ pH = 4.66 $\qquad$ This is a more basic solution, which we can achieve by increasing the basic component of the buffer solution, the acetate ion. We find out the new acetate ion concentration with the Henderson-Hasselbalch equation.

$$pH = pK_a + \log \frac{[C_2H_3O_2^-]}{[HC_2H_3O_2]} = 4.66 = 4.74 + \log \frac{[C_2H_3O_2^-]}{0.25\ M} \qquad \log \frac{[C_2H_3O_2^-]}{0.25\ M} = 4.66 - 4.74 = -0.08$$

$$\frac{[C_2H_3O_2^-]}{0.25\ M} = 10^{-0.08} = 0.83 \qquad [C_2H_3O_2^-] = 0.83 \times 0.25\ M = 0.21\ M$$

74. **(a)** CuS is in the hydrogen sulfide group of qualitative analysis. Its precipitation occurs when 0.3 M HCl is saturated with H_2S. It will certainly precipitate from a (non-acidic) saturated solution of H_2S, which has a much higher $[S^{2-}]$.

$$Cu^{2+}(aq) + H_2S(\text{satd aq}) \longrightarrow CuS(s) + 2\ H^+(aq)$$

This reaction proceeds to a significant extent in the forward direction.

(b) MgS is soluble, according to the solubility rules listed in Chapter 5.

$$Mg^{2+}(aq) + H_2S(\text{satd aq}) \xrightarrow{0.3\ M\ HCl} \text{no reaction}$$

(c) As in part (a), PbS is in the qualitative analysis hydrogen sulfide group, which precipitates from a 0.3 M HCl solution saturated with H_2S. Therefore, PbS does not dissolve appreciably in 0.3 M HCl.

$$PbS(s) + HCl\ (0.3\ M) \longrightarrow \text{no reaction}$$

(d) Since ZnS(s) does not precipitate in the Hydrogen Sulfide group, we conclude that it is soluble in acidic solution. $\qquad ZnS(s) + 2\ HNO_3(aq) \longrightarrow Zn(NO_3)_2(aq) + H_2S(g)$

Qualitative Analysis

75. The purpose of adding hot water is to separate Pb^{2+} from AgCl and Hg_2Cl_2. Thus, the most important consequence of the omission is that there not longer is a valid test for the presence or absence of Pb^{2+}. In addition, if we add NH_3 first, $PbCl_2$ may form $Pb(OH)_2$. If $Pb(OH)_2$ does form, it will be present with Hg_2Cl_2 in the solid, although $Pb(OH)_2$ will not darken with added NH_3. Thus, we might falsely conclude that Ag^+ is present, but not falsely conclude that Hg_2^{2+} is present.

76. For $PbCl_2(aq)$, $2\,[Pb^{2+}] = [Cl^-]$ and we let s = molar solubility of $PbCl_2$. Thus $s = [Pb^{2+}]$.

$$K_{sp} = [Pb^{2+}][Cl^-]^2 = (s)(2s)^2 = 4s^3 = 1.6 \times 10^{-5} \qquad s = \sqrt[3]{1.6 \times 10^{-5} \div 4} = 1.6 \times 10^{-2}\ M = [Pb^{2+}]$$

Both $[Pb^{2+}]$ and $[CrO_4^{2-}]$ are diluted by mixing the two solutions.

$$[Pb^{2+}] = 0.016\ M \times \frac{1.00\ mL}{1.05\ mL} = 0.015\ M \qquad [CrO_4^{2-}] = 1.0\ M \times \frac{0.05\ mL}{1.05\ mL} = 0.048\ M$$

$$Q = [Pb^{2+}][CrO_4^{2-}] = (0.015\ M)(0.048\ M) = 7.2 \times 10^{-4} > 2.8 \times 10^{-13} = K_{sp}$$

Thus, precipitation should occur from the solution described.

77. **(a)** Ag^+ and/or Hg_2^{2+} are probably present. Both of these cations form precipitates from an acidic solution of chloride ion.

(b) We cannot tell whether Mg^{2+} is present or not. Both MgS and $MgCl_2$ are water soluble.

(c) Pb^{2+} possibly is absent; it is the only cation of those given which forms a precipitate in an acidic solution that is treated with H_2S, and no sulfide precipitate was formed.

(d) We cannot tell whether Fe^{2+} is present or not. FeS will not precipitate from an acidic solution that is treated with H_2S; the solution must be alkaline for a FeS precipitate to form.

(a) and **(c)** are the valid conclusions.

78. **(a)** $Pb^{2+}(aq) + 2\ Cl^-(aq) \longrightarrow PbCl_2(s)$

(b) $Zn(OH)_2(s) + 2\ OH^-(aq) \longrightarrow [Zn(OH)_4]^{2-}(aq)$

(c) $Fe(OH)_3(s) + 3\ H_3O^+(aq) \longrightarrow Fe^{3+}(aq) + 6\ H_2O$ *or* $[Fe(H_2O)_6]^{3+}(aq)$

(d) $Cu^{2+}(aq) + H_2S(aq) \longrightarrow CuS(s) + 2\ H^+(aq)$

FEATURE PROBLEMS

A. We know that $[Ca^{2+}] = [SO_4^{2-}]$ in the saturated solution. Let us first determine the amount of H_3O^+ in the 100.0-mL diluted effluent. $H_3O^+(aq) + NaOH(aq) \longrightarrow 2\ H_2O + Na^+(aq)$

$$\text{amount } H_3O^+ = 100.0\text{ mL} \times \frac{8.25\text{ mL base}}{10.00\text{ mL sample}} \times \frac{0.0105\text{ mmol NaOH}}{1\text{ mL base}} \times \frac{1\text{ mmol } H_3O^+}{1\text{ mmol NaOH}}$$
$$= 0.866\text{ mmol } H_3O^+(aq)$$

Now we determine $[Ca^{2+}]$ in the original 25.00 mL sample, remembering that 2 H_3O^+ were produced for each Ca^{2+}.

$$[Ca^{2+}] = \frac{0.866\text{ mmol } H_3O^+(aq) \times \dfrac{1\text{ mmol } Ca^{2+}}{2\text{ mmol } H_3O^+}}{25.00\text{ mL}} = 0.0173\text{ M}$$

$K_{sp} = [Ca^{2+}][SO_4^{2-}] = (0.0173)^2 = 3.0 \times 10^{-4}$ Compares with 9.1×10^{-6} in Appendix D.

B. **1.** We assume that there is little of each ion present in solution at equilibrium; that this is a simple stoichiometric calculation. This is true because we are titrating: we stop when just enough silver ion has been added. $Ag^+(aq) + Cl^-(aq) \longrightarrow AgCl(s)$

$$V = 100.0\text{ mL} \times \frac{29.5\text{ mg } Cl^-}{1000\text{ mL}} \times \frac{1\text{ mmol } Cl^-}{35.45\text{ mg } Cl^-} \times \frac{1\text{ mmol } Ag^+}{1\text{ mmol } Cl^-} \times \frac{1\text{ mL}}{0.01000\text{ mmol } AgNO_3}$$
$$= 8.32\text{ mL}$$

2. We first calculate the concentration of each ion as the consequence of dilution. Then we determine the $[Ag^+]$ from the value of K_{sp}

$$\text{initial } [Ag^+] = 0.0100\text{ M} \times \frac{8.32\text{ mL added}}{108.3\text{ mL final volume}} = 7.68 \times 10^{-4}\text{ M}$$

$$\text{initital}[Cl^-] = \frac{29.5\text{ mg } Cl^- \times \dfrac{1\text{ mmol } Cl^-}{35.45\text{ mg } Cl^-}}{1000\text{ mL}} \times \frac{100.0\text{ mL taken}}{108.3\text{ mL final volume}} = 7.68 \times 10^{-4}\text{ M}$$

The slight excess of each ion will precipitate until the solubility constant is satisfied.

$$[Ag^+] = [Cl^-] = \sqrt{K_{sp}} = \sqrt{1.8 \times 10^{-10}} = 1.3 \times 10^{-5}\text{ M}$$

3. If we want Ag_2CrO_4 to appear just when AgCl has completed precipitation, $[Ag^+] = 1.3 \times 10^{-5}$ M. We determine $[CrO_4^{2-}]$ from the K_{sp} expression.

$$K_{sp} = [Ag^+]^2[CrO_4^{2-}] = 1.1 \times 10^{-12} = (1.3 \times 10^{-5})^2[CrO_4^{2-}] \qquad [CrO_4^{2-}] = \frac{1.1 \times 10^{-12}}{(1.3 \times 10^{-5})^2} = 0.0065\text{ M}$$

4. **(a)** If $[CrO_4^{2-}]$ were greater than just computed, Ag_2CrO_4 would appear before all Cl^- had precipitated, leading to a false early endpoint. We would calculate a falsely low $[Cl^-]$ of the original solution.

(b) If $[CrO_4^{2-}]$ were less than computed in part 3, Ag_2CrO_4 would appear somewhat after all Cl^- had precipitated, leading one to conclude there was more Cl^- in solution than actually was the case.

5. If Ag^+ were in the sample being titrated, it would react immediately with the CrO_4^{2-} in that sample, forming a red-orange precipitate. This precipitate would not be likely to dissolve so that AgCl could form. There would be no visual indication of the endpoint.

C. **1.** We need $[Mg^{2+}]$ in saturated $Mg(OH)_2$, $K_{sp} = 1.8 \times 10^{-11} = [Mg^{2+}][OH^-]^2 = (s)(2s)^2 = 4s^3$

$$s = \sqrt[3]{\frac{1.8 \times 10^{-11}}{4}} = 1.7 \times 10^{-4}\text{ M} = [Mg^{2+}]$$

2. Even though water has been added to the original solution, it still is saturated because it is in equilibrium with the undissolved solid $Mg(OH)_2$. $[Mg^{2+}] = 1.7 \times 10^{-4}$ M.

3. Although HCl(aq) reacts with OH^-, it will not react with Mg^{2+}. The solution simply is diluted.

$$[Mg^{2+}] = 1.7 \times 10^{-4}\text{ M} \times \frac{100.0\text{ mL initial volume}}{(100.0 + 500.)\text{ mL final volume}} = 2.8 \times 10^{-5}\text{ M}$$

4. In this instance, we have a dual dilution to a 275.0 mL total volume, followed by a common ion problem.

$$\text{initial } [Mg^{2+}] = \frac{\left(25.00 \text{ mL} \times \dfrac{1.7 \times 10^{-4} \text{ mmol } Mg^{2+}}{1 \text{ mL}}\right) + \left(250.0 \text{ mL} \times \dfrac{0.065 \text{ mmol } Mg^{2+}}{1 \text{ mL}}\right)}{275.0 \text{ mL total volume}}$$

$$= 0.059 \text{ M}$$

$$\text{initial } [OH^-] = \frac{25.00 \text{ mL} \times \dfrac{1.7 \times 10^{-4} \text{ mmol } Mg^{2+}}{1 \text{ mL}} \times \dfrac{2 \text{ mmol } OH^-}{1 \text{ mmol } Mg^{2+}}}{275.0 \text{ mL total volume}} = 3.1 \times 10^{-5} \text{ M}$$

Let's see if precipitation occurs.

$Q_{sp} = [Mg^{2+}][OH^-]^2 = (0.059)(3.1 \times 10^{-5})^2 = 5.7 \times 10^{-11} > 1.8 \times 10^{-11} = K_{sp}$

Precipitation barely occurs. If $[OH^-]$ goes down by 1.4×10^{-5} M (which means that $[Mg^{2+}]$ drops by 0.7×10^{-5} M) then $[OH^-] = 1.7 \times 10^{-5}$ M and $[Mg^{2+}] = (0.059 \text{ M} - 0.7 \times 10^{-5} \text{ M} =)$ 0.059 M, then $Q_{sp} < K_{sp}$ and precipitation will stop. Thus, $[Mg^{2+}] = 0.059$ M.

5. Again we have a dual dilution, now to a 200.0 mL final volume, followed by a common ion problem.

$$\text{initial } [Mg^{2+}] = \frac{50.00 \text{ mL} \times \dfrac{1.7 \times 10^{-4} \text{ mmol } Mg^{2+}}{1 \text{ mL}}}{200.0 \text{ mL total volume}} = 4.3 \times 10^{-5} \text{ M}$$

$$\text{initial } [OH^-] = 0.150 \text{ M} \times \frac{150.0 \text{ mL initial volume}}{200.0 \text{ mL final volume}} = 0.0113 \text{ M}$$

Now it is evident that precipitation will occur. We determine $[Mg^{2+}]$ that can exist in solution with 0.0375 M OH^-. It is clear that $[Mg^{2+}]$ will drop dramatically to satisfy the K_{sp} expression but the larger value of $[OH^-]$ will scarcely be affected.

$K_{sp} = [Mg^{2+}][OH^-]^2 = 1.8 \times 10^{-11} = [Mg^{2+}](0.0113 \text{ M})^2$ $\qquad$ $[Mg^{2+}] = \dfrac{1.8 \times 10^{-11}}{(0.0113)^2} = 1.4 \times 10^{-9}$ M

20 SPONTANEOUS CHANGE: ENTROPY AND FREE ENERGY

PRACTICE EXAMPLES

1A The "souring" of cream and the formation of a green patina both are spontaneous changes. They occur despite the measures that we take to present them: keeping the cream cold and well covered, making "corrosion resistant" bronze alloys, and coating the bronze with a clear finish. But obtaining gold nuggets by "panning" requires active intervention and the input not only of work but of knowledge (the process must be performed correctly or it will not work). Obtaining gold nuggets by panning is *not* a spontaneous change.

1B The pairing of socks in a laundry dryer is a nonspontaneous process. Although there is plenty of energy supplied, by the rotting drum, it is highly unlikely that the socks will pair unless we intervene. Both combustion of gasoline and dissolving sugar in hot coffee are spontaneous: once they have started, they proceed on their own with no further intervention. This is not true of riding on a swing.We need to push hard and often to get the swinging process started.

2A In general, $\Delta S > 0$ if $\Delta n_{gas} > 0$ also. This is because gases are highly disordered; they have a high entropy. Recall that Δn_{gas} is the difference between the sum of the stoichiometric coefficients of the gaseous products and a similar sum for the reactants. **(a)** $\Delta n_{gas} = 2 + 0 - (2 + 1) = -1$. One mole of high-entropy gas is consumed here. We predict $\Delta S < 0$. **(b)** $\Delta n_{gas} = 1 + 0 - 0 = +1$. Since a mole of high-entropy gas is produced, we predict $\Delta S > 0$.

2B **(a)** The outcome is uncertain in the reaction of ZnS(s) and Ag_2O(s). We have used Δn_{gas} to estimate the sign of entropy change. There is no gas involved in this reaction and thus our prediction is uncertain. **(b)** In the chlor-alkali process we are confident that entropy increases because two moles of gas have formed were none were originally present.

3A **(a)** Because $\Delta n_{gas} = 2 - (1 + 3) = -2$ for the synthesis of ammonia, we would predict $\Delta S < 0$ for the reaction. We already know that $\Delta H < 0$. Thus, the reaction falls into case 2, spontaneous at low temperatures and nonspontaneous at high temperatures.

(b) For the formation of ethylene $\Delta n_{gas} = 1 - (2 + 0) = -1$ and thus $\Delta S < 0$. We are given that $\Delta H > 0$ and thus this reaction corresponds to case 4, a reaction that is nonspontaneous at all temperatures.

3B **(a)** Because $\Delta n_{gas} = +1$ for the synthesis of ammonia, we would predict $\Delta S > 0$ for the reaction, favoring the reaction at high temperatures. High temperatures also favor this endothermic ($\Delta H^\circ > 0$) reaction.

(b) Roasting ZnS(s) has $\Delta n_{gas} = 2 - 3 = -1$ and thus $\Delta S < 0$. We are given that $\Delta H < 0$; thus this reaction corresponds to case 2, a reaction spontaneous at low temperatures, and nonspontaneous at high ones.

4A For a vaporization, $\Delta G^\circ_{vap} = 0 = \Delta H^\circ_{vap} - T\Delta S^\circ_{vap}$. Thus, $\Delta S^\circ_{vap} = \Delta H^\circ_{vap}/T_{vap}$. We substitute the given values. $\Delta S^\circ_{vap} = \dfrac{\Delta H^\circ_{vap}}{T_{vap}} = \dfrac{20.2 \text{ kJ mol}^{-1}}{(-29.79 + 273.15) \text{ K}} = 83.0 \text{ J mol}^{-1} \text{ K}^{-1}$

4B For a phase change, $\Delta G^\circ_{tr} = 0 = \Delta H^\circ_{tr} - T\Delta S^\circ_{tr}$. Thus, $\Delta H^\circ_{tr} = T\Delta S^\circ_{tr}$. We substitute the given values.
$\Delta H^\circ_{tr} = T\Delta S^\circ_{tr} = (95.5 + 273.2) \text{ K} \times 1.09 \text{ J mol}^{-1} \text{ K}^{-1} = 402 \text{ J/mol}$

5A The entropy change for the reaction is expressed in terms of the standard entropies of the reagents.
$\Delta S^\circ = 2S^\circ[NH_3(g)] - S^\circ[N_2(g)] - 3\,S^\circ[H_2(g)]$
$= 2 \times 192.3 \text{ J mol}^{-1} \text{ K}^{-1} - 191.5 \text{ J mol}^{-1} \text{ K}^{-1} - 3 \times 130.6 \text{ J mol}^{-1} \text{ K}^{-1} = -198.7 \text{ J mol}^{-1} \text{ K}^{-1}$
To form *one* mole of NH_3(g), the standard entropy change is $-99.4 \text{ J mol}^{-1} \text{ K}^{-1}$

5B The entropy change for the reaction is expressed in terms of the standard entropies of the reagents.
$\Delta S° = S°[NO(g)] + S°[NO_2(g)] - S°[N_2O_3(g)]$
$138.5 \text{ J mol}^{-1} \text{ K}^{-1} = 210.7 \text{ J mol}^{-1} \text{ K}^{-1} + 240.0 \text{ J mol}^{-1} \text{ K}^{-1} - S°[N_2O_3(g)]$
$= 450.7 \text{ J mol}^{-1} \text{ K}^{-1} - S°[N_2O_3(g)]$
$S°[N_2O_3(g)] = 450.7 \text{ J mol}^{-1} \text{ K}^{-1} - 138.5 \text{ J mol}^{-1} \text{ K}^{-1} = 312.2 \text{ J mol}^{-1} \text{ K}$

6A The expression $\Delta G° = \Delta H° - T\Delta S°$ is used with $T = 298.15$ K.
$\Delta G° = \Delta H° - T\Delta S° = -1648 \text{ kJ} - 298.15 \text{ K} \times (-549.3 \text{ J K}^{-1}) \times (1 \text{ kJ}/1000 \text{ J})$
$= -1648 \text{ kJ} + 163.8 \text{ kJ} = -1484 \text{ kJ}$

6B We just need to substitute values from Appendix D into the supplied expression.
$\Delta G° = 2\,\Delta G_f°[NO_2(g)] - 2\,\Delta G_f°[NO(g)] - \Delta G_f°[O_2(g)]$
$= 2 \times 51.29 \text{ kJ mol}^{-1} - 2 \times 86.55 \text{ kJ mol}^{-1} - 0.00 \text{ kJ mol}^{-1} = -70.52 \text{ kJ mol}^{-1}$

7A Pressures of gases and molarities of solutes in aqueous solution appear in thermodynamic equilibrium constant expressions. Pure solids and liquids (including solvents) do not appear.
(a) $K_{eq} = \dfrac{P\{SiCl_4\}}{P\{Cl_2\}^2} = K_p$ **(b)** $K_{eq} = \dfrac{[HOCl][H^+][Cl^-]}{P\{Cl_2\}}$
$K_{eq} = K_p$ for **(a)** because all factors in the K_{eq} expression are gas pressures.

7B We need the balanced chemical equation in order to write the equilibrium constant expression. We start by translating names into formulas. $PbS(s) + HNO_3(aq) \longrightarrow Pb(NO_3)_2(aq) + S(s) + NO(g)$
The equation then is balanced with the ion-electron method.
oxidation: $PbS(s) \longrightarrow Pb^{2+}(aq) + S(s) + 2\,e^-$ $\times 3$
reduction: $NO_3^-(aq) + 4\,H^+(aq) + 3\,e^- \longrightarrow NO(g) + 2\,H_2O$ $\times 2$

net ionic: $3\,PbS(s) + 2\,NO_3^-(aq) + 8\,H^+(aq) \longrightarrow 3\,Pb^{2+}(aq) + 3\,S(s) + 2\,NO(g) + 4\,H_2O$
In writing the thermodynamic equilibrium constant, recall that neither pure solids [PbS(s) and S(s)] nor pure liquids (H_2O) appear in the thermodynamic equilibrium constant expression. [Even though the water is not "pure", since it contains dissolved nitric acid and lead(II) nitrate, the activity of water in dilute solutions is quite close to unity.] Note also that we have written $H^+(aq)$ here for brevity even though we understand that $H_3O^+(aq)$ is the acidic species in aqueous solution. $K = \dfrac{[Pb^{2+}]^3\,P\{NO\}^2}{[NO_3^-]^2[H^+]^8}$

8A We first determine the value of $\Delta G°$ and then set $\Delta G° = -RT\ln K_{eq}$ to determine K_{eq}.
$\Delta G° = \Delta G°_f[Ag^+(aq)] + \Delta G°_f[I^-(aq)] - \Delta G°_f[AgI(s)] = [(-51.57 + 77.11) - (-66.19)] \text{ kJ/mol} = +91.73 \text{ kJ/mol}$
$\ln K_{eq} = \dfrac{-\Delta G°}{RT} = -\dfrac{+91.73 \text{ kJ/mol}}{8.3145 \text{ J mol}^{-1} \text{ K}^{-1} \times 298.15 \text{ K}} \times \dfrac{1000 \text{ J}}{1 \text{ kJ}} = -37.00$ $K_{eq} = e^{-37.00} = 8.5 \times 10^{-17}$
This is precisely equal to the value of K_{sp} listed in Appendix D.

8B We begin by translating names into formulas. $MnO_2(s) + HCl(aq) \longrightarrow Mn^{2+}(aq) + Cl_2(aq)$
Then we produce a balanced net ionic equation with the ion-electron method.
oxidation: $2\,Cl^-(aq) \longrightarrow Cl_2(g) + 2\,e^-$
reduction: $MnO_2(s) + 4\,H^+(aq) + 2\,e^- \longrightarrow Mn^{2+}(aq) + 2\,H_2O$

net ionic: $MnO_2(s) + 4\,H^+(aq) + 2\,Cl^-(aq) \longrightarrow Mn^{2+}(aq) + Cl_2(g) + 2\,H_2O$
Next we determine the value of $\Delta G°$ for the reaction and then the value of the equilibrium constant.
$\Delta G° = \Delta G_f°[Mn^{2+}(aq)] + \Delta G_f°[Cl_2(g)] + 2\,\Delta G_f°[H_2O(l)] - \Delta G_f°[MnO_2(s)] - 4\,\Delta G_f°[H^+(aq)] - 2\,\Delta G_f°[Cl^-(aq)]$
$= -228.1 \text{ kJ} + 0.0 \text{ kJ} + 2 \times (-237.2 \text{ kJ}) - (-465.1 \text{ kJ}) - 4 \times 0.0 \text{ kJ} - 2 \times (-131.2 \text{ kJ}) = +25.0 \text{ kJ}$
$\ln K_{eq} = \dfrac{-\Delta G°}{RT} = \dfrac{-(+25.0 \times 10^3 \text{ J mol}^{-1})}{8.3145 \text{ J mol}^{-1} \text{ K}^{-1} \times 298.15 \text{ K}} = -10.0_8$ $K_{eq} = e^{-10.08} = 4 \times 10^{-5}$
Because the value of K_{eq} is so much smaller than unity, we do not expect an appreciable reaction.

9A We set equal the two expressions for $\Delta G°$ and solve for the absolute temperature.
$\Delta G° = \Delta H° - T\Delta S° = -RT\ln K_{eq}$ $\Delta H° = T\Delta S° - RT\ln K_{eq} = T(\Delta S° - R\ln K_{eq})$

$$T = \frac{\Delta H^\circ}{\Delta S^\circ - R\ln K_{eq}} = \frac{-114.1 \times 10^3 \text{ J/mol}}{[-146.4 - 8.3145 \ln(150)] \text{ J mol}^{-1} \text{ K}^{-1}} = 607 \text{ K}$$

9B We expect the value of the equilibrium constant to increase as temperature decreases since this is an exothermic reaction and it should become more spontaneous (shift right) at lower temperatures. Thus, we expect K_{eq} to be larger than 1000, its value at 4.3×10^2 K.

(a) The value of the equilibrium constant at 25 °C is obtained directly from the value of ΔG°, since that value is also for 25 °C. Note: $\Delta G^\circ = \Delta H^\circ - T\Delta S^\circ = -77.1$ kJ/mol – 298.15 K (–0.1212 kJ/mol·K) = –41.0 kJ/mol

$$\ln K_{eq} = \frac{-\Delta G^\circ}{RT} = \frac{-(-41.0 \times 10^3 \text{ J mol}^{-1})}{8.3145 \text{ J mol}^{-1} \text{ K}^{-1} \times 298.15 \text{ K}} = +16.5 \qquad K_{eq} = e^{+16.5} = 1.5 \times 10^7$$

(b) First, we solve for ΔG° at 75 °C = 348 K

$$\Delta G^\circ = \Delta H^\circ - T\Delta S^\circ = -77.1 \frac{\text{kJ}}{\text{mol}} \times \frac{1000 \text{ J}}{1 \text{ kJ}} - \left(348.15 \text{ K} \times \left(-121.2 \frac{\text{J}}{\text{mol K}}\right)\right) = -34.9 \times 10^3 \text{ J/mol}$$

Then we use this value to obtain the value of the equilibrium constant, as in part (a).

$$\ln K_{eq} = \frac{-\Delta G^\circ}{RT} = \frac{-(-34.9 \times 10^3 \text{ J mol}^{-1})}{8.3145 \text{ J mol}^{-1} \text{ K}^{-1} \times 348.15 \text{ K}} = +12.0 \qquad K_{eq} = e^{+12.0} = 1.8 \times 10^5$$

10A We use the value of $K_p = 9.1 \times 10^2$ at 800 K and $\Delta H^\circ = -1.8 \times 10^5$ J/mol in the van't Hoff equation.

$$\ln \frac{5.8 \times 10^{-2}}{9.1 \times 10^2} = \frac{-1.8 \times 10^5 \text{ J/mol}}{8.3145 \text{ J mol}^{-1} \text{ K}^{-1}} \left(\frac{1}{800 \text{ K}} - \frac{1}{T \text{ K}}\right) = -9.66 \qquad \frac{1}{T} = \frac{1}{800} + \frac{9.66 \times 8.3145}{1.8 \times 10^5}$$

$1/T = 1.25 \times 10^{-3} + 4.5 \times 10^{-4} = 1.7 \times 10^{-3}$ $\qquad T = 588 \text{ K} = 315$ °C

This temperature is an estimate because it extrapolates beyond the range of the data supplied.

10B The temperature we are considering is 235 °C = 508 K. We use the value of $K_p = 9.1 \times 10^2$ at 800 K and $\Delta H^\circ = -1.8 \times 10^5$ J/mol in the van't Hoff equation.

$$\ln \frac{K_p}{9.1 \times 10^2} = \frac{-1.8 \times 10^5 \text{ J/mol}}{8.3145 \text{ J mol}^{-1} \text{ K}^{-1}} \left(\frac{1}{800 \text{ K}} - \frac{1}{508 \text{ K}}\right) = +15._6 \qquad \frac{K_p}{9.1 \times 10^2} = e^{+15.6} = 6 \times 10^6$$

$K_p = 6 \times 10^6 \times 9.1 \times 10^2 = 5 \times 10^9$ $\qquad$ (If we use 16 rather than $15._6$, we obtain $K_p = 8 \times 10^9$.)

SUMMARIZING EXAMPLE CALCULATIONS

1. For the reaction $CO(g) + 2\,H_2(g) \rightleftharpoons CH_3OH(l)$, we compute the value of ΔG°

$\Delta G^\circ = \Delta G^\circ_f[CH_3OH(g)] - \Delta G^\circ_f[CO(g)] - 2\,\Delta G^\circ_f[H_2(g)] = -162.0$ kJ/mol – (–137.2 kJ/mol) – 2 × 0.0 kJ/mol = –24.8 kJ/mol = -24.8×10^3 J/mol

2. We use this value of ΔG° to calculate a value of K_{eq} at 298 K.

$$\ln K_{eq} = \frac{-\Delta G^\circ}{RT} = \frac{-(-24.8 \times 10^3 \text{ J mol}^{-1})}{8.3145 \text{ J mol}^{-1} \text{ K}^{-1} \times 298.15 \text{ K}} = +10.0 \qquad K_{eq} = e^{+10.0} = 2.2 \times 10^4$$

3. ΔH° is computed in a manner similar to that used for ΔG°

$\Delta H^\circ = \Delta H^\circ_f[CH_3OH(g)] - \Delta H^\circ_f[CO(g)] - 2\,\Delta H^\circ_f[H_2(g)] = -200.7$ kJ/mol – (–110.5 kJ/mol) – 2 × 0.0 kJ/mol = –90.2 kJ/mol = -90.2×10^3 J/mol

4. The van't Hoff equation then is used to determine the value of K_p at 500 K.

$$\ln \frac{K_p}{2.2 \times 10^4} = \frac{-9.02 \times 10^4 \text{ J/mol}}{8.3145 \text{ J mol}^{-1} \text{ K}^{-1}} \left(\frac{1}{298 \text{ K}} - \frac{1}{500 \text{ K}}\right) = -14.7 \qquad \frac{K_p}{2.2 \times 10^4} = e^{-14.7} = 4.1 \times 10^{-7}$$

$K_p = 4.1 \times 10^{-7} \times 2.2 \times 10^4 = 9.0 \times 10^{-3}$

REVIEW QUESTIONS

1. **(a)** ΔS_{univ} is the symbol for the entropy change of the universe. If this quantity is positive, the accompanying process is spontaneous.

(b) ΔG°_f is the symbol for the standard free energy of formation, the free energy change of the standard state reaction when one mole of substance is produced from stable forms of its elements.

(c) K_{eq} is the symbol for the thermodynamic equilibrium constant; in this expression gases are represented by pressures and solutes in aqueous solution by molarities.

2. **(a)** *Absolute* molar entropy refers to the entropy of a substance referred to pure perfect crystals at 0 K, which have zero entropy.

(b) A reversible process is one occurring in a state of constant equilibrium; a small opposing stress would reverse the direction of the reaction.

(c) Trouton's rule states that the entropy change of vaporization approximates 88 J mol^{-1} K^{-1}. It works well for nonpolar liquids.

(d) An equilibrium constant is evaluated from tabulated thermodynamic data by combining values of ΔG_f° to obtain ΔG_{rxn}° and then using $\Delta G_{rxn}^\circ = -RT \ln K_{eq}$.

3. **(a)** A spontaneous process is one that occurs without (or sometimes in spite of) external intervention. A nonspontaneous process will only occur when some external agency operates on it.

(b) The second law of thermodynamics places limitations on converting heat into work and on the spontaneous directions of processes. The third law of thermodynamics establishes a zero point for entropy.

(c) ΔG refers to the free energy change for any process, while ΔG° requires that both the initial and final states of the process are standard ones: gases at 1 atm pressure, aqueous solutes at 1 molar concentration, and pure solids and liquids.

4. **(a)** Increase in entropy because a gas has been created from a liquid, a condensed phase.

(b) Decrease in entropy because a condensed phase, a solid, is created from a solid and a gas.

(c) No change in entropy that can be easily detected, since the number of moles of gas produced is the same as the number that reacted.

(d) $2\ H_2S(g) + 3\ O_2(g) \longrightarrow 2\ H_2O(g) + 2\ SO_2(g)$ Decrease in entropy since five moles of gas with high entropy become only four moles of gas, with about the same quantity of entropy per mole.

5. **(a)** At 75 °C, 1 mol H_2O (g, 1 atm) has a greater entropy than 1 mol H_2O (l, 1 atm) since a gas is much more disordered than a liquid.

(b) $50.0\text{ g Fe} \times \frac{1\text{ mol Fe}}{55.8\text{ g Fe}} = 0.896$ mol Fe has a higher entropy than 0.80 mol Fe, both (s) at 1 atm and 5 °C, because entropy is an extensive property that depends on the amount of substance present.

(c) 1 mol Br_2(l, 1 atm, 8 °C) has a higher entropy than 1 mol Br_2(s, 1atm, –8 °C) because solids are more ordered substances than are liquids, and the temperature is higher.

(d) 0.312 mol SO_2 (g, 0.110 atm, 32.5 °C) has a higher entropy than 0.284 mol O_2 (g, 15.0 atm, 22.3 °C) for at least three reasons. First, entropy is an extensive property that depends on the amount of substance present. Second, entropy increases with temperature. Third, entropy is greater at lower pressures. Furthermore, entropy generally is higher per mole for more complicated molecules.

6. We predict the sign of ΔS based on the number of moles of gas in the balanced chemical equation; if $\Delta n_{gas} < 0$, ΔS will be smaller than zero. If $\Delta n_{gas} > 0$, ΔS will be greater than zero.

(a) $\Delta S > 0$ and $\Delta H > 0$ This is case 3 in Table 20-1.

(b) $\Delta S < 0$ and $\Delta H < 0$ This is case 2 in Table 20-1.

(c) $\Delta S > 0$ and $\Delta H < 0$ This is case 1 in Table 20-1.

(d) $\Delta S < 0$ and $\Delta H > 0$ This is case 4 in Table 20-1.

7. Answer **(b)** is correct. Br—Br bonds are broken in this reaction, meaning that it is endothermic, with $\Delta H > 0$. Since the number of moles of gas increases during the reaction, $\Delta S > 0$. And, because $\Delta G = \Delta H - T\,\Delta S$, this reaction is nonspontaneous at low temperatures where the ΔH term predominates, with $\Delta G > 0$, and spontaneous at high temperatures where the $T\,\Delta S$ term predominates, with $\Delta G < 0$.

8. Answer **(d)** is correct. A reaction that proceeds only through electrolysis is a reaction that is nonspontaneous. Such a reaction has $\Delta G > 0$.

9. Answer **(d)** is correct. Because $\Delta G^\circ = -RT \ln K_{eq}$, if $\Delta G^\circ = 0$, $\ln K_{eq} = 0$, which means that $K_{eq} = 1.00$.

10. $\Delta H^\circ = \Delta H_f^\circ[NH_4Cl(s)] - \Delta H_f^\circ[NH_3(g)] - \Delta H_f^\circ[HCl(g)]$

$= -314.4$ kJ/mol $- (-46.11$ kJ/mol $- 92.31$ kJ/mol$) = -176.0$ kJ/mol

$\Delta G^\circ = \Delta G_f^\circ[NH_4Cl(s)] - \Delta G_f^\circ[NH_3(g)] - \Delta G_f^\circ[HCl(g)]$

$= -202.9$ kJ/mol $- (-16.48$ kJ/mol $- 95.31$ kJ/mol$) = -91.1$ kJ/mol

$\Delta G^\circ = \Delta H^\circ - T\Delta S^\circ$ $\quad \Delta S^\circ = \frac{\Delta H^\circ - \Delta G^\circ}{T} = \frac{-176.0\text{ kJ/mol} + 91.1\text{ kJ/mol}}{298.15\text{ K}} \times \frac{1000\text{ J}}{1\text{ kJ}} = -285\text{ J mol}^{-1}\text{ K}^{-1}$

11. (a) $\Delta G^\circ = \Delta G^\circ_f[C_2H_6(g)] - \Delta G^\circ_f[C_2H_2(g)] - 2\,\Delta G^\circ_f[H_2(g)]$
$= -32.89$ kJ/mol $- 209.2$ kJ/mol $- 2(0.00$ kJ/mol$) = -242.1$ kJ/mol

(b) $\Delta G^\circ = 2\,\Delta G^\circ_f[SO_2(g)] + \Delta G^\circ_f[O_2(g)] - 2\,\Delta G^\circ_f[SO_3(g)]$
$= 2(-300.2$ kJ/mol$) + 0.00$ kJ/mol $- 2(-371.1$ kJ/mol$) = +141.8$ kJ/mol

(c) $\Delta G^\circ = 3\,\Delta G^\circ_f[Fe(s)] + 4\,\Delta G^\circ_f[H_2O(g)] - \Delta G^\circ_f[Fe_3O_4(s)] - 4\,\Delta G^\circ_f[H_2(g)]$
$= 3\,(0.00$ kJ/mol$) + 4\,(-228.6$ kJ/mol$) - (-1015$ kJ/mol$) - 4\,(0.00$ kJ/mol$) = 101$ kJ/mol

(d) $\Delta G^\circ = 2\,\Delta G^\circ_f[Al^{3+}(aq)] + 3\,\Delta G^\circ_f[H_2(g)] - 2\,\Delta G^\circ_f[Al(s)] - 6\,\Delta G^\circ_f[H^+(aq)]$
$= 2\,(-485$ kJ/mol$) + 3\,(0.00$ kJ/mol$) - 2\,(0.00$ kJ/mol$) - 6\,(0.00$ kJ/mol$) = -970$ kJ/mol

12. (a) This temperature is the melting point of $I_2(s)$ at 1 atm pressure. From the phase diagram for iodine, we see that this process occurs at 114 °C.

(b) $\Delta G^\circ = 0$ for $I_2(s, 1\text{ atm}) \rightleftharpoons I_2(l, 1\text{ atm})$. The point of the crossing of the lines in Figure 20-9 is the point where $\Delta G^\circ = 0$. In addition the melting point of a substance is the temperature where liquid and solid have the same free energy.

13. (a) $$\Delta S^\circ_{vap} = \frac{\Delta H^\circ_{vap}}{T_{vap}} = \frac{3.86\text{ kcal/mol}}{-85.05\text{ °C} + 273.15\text{ K}} \times \frac{1000\text{ cal}}{1\text{ kcal}} \times \frac{4.184\text{ J}}{1\text{ cal}} = 85.9\text{ J mol}^{-1}\text{ K}^{-1}$$

(b) $$\Delta S^\circ_{fus} = \frac{\Delta H^\circ_{fus}}{T_{fus}} = \frac{27.05\text{ cal/g}}{97.82\text{ °C} + 273.15\text{ K}} \times \frac{22.99\text{ g Na}}{1\text{ mol Na}} \times \frac{4.184\text{ J}}{1\text{ cal}} = 7.014\text{ J mol}^{-1}\text{ K}^{-1}$$

14. (a) $K_{eq} = \dfrac{P\{NO_2(g)\}^2}{P\{NO(g)\}^2\,P\{O_2\}} = K_p$

(c) $K_{eq} = \dfrac{[H_3O^+][C_2H_3O_2^-]}{[HC_2H_3O_2]} = K_c$

(b) $K_{eq} = P\{SO_2(g)\} = K_p$

(d) $K_{eq} = P\{H_2O(g)\}\,P\{CO_2(g)\} = K_p$

(e) $K_{eq} = \dfrac{[Mn^{2+}(aq)]\,P\{Cl_2(g)\}}{[H^+]^4\,[Cl^-]^2}$ neither K_p nor K_c

15. $$\Delta G^\circ = -RT\ln K_p = -(8.3145\text{ J mol}^{-1}\text{ K}^{-1})(1000.\text{ K})\left(\frac{1\text{ kJ}}{1000\text{ J}}\right)\ln(2.45 \times 10^{-7}) = 126.6\text{ kJ/mol}$$

16. $\Delta G^\circ = 2\,\Delta G^\circ_f[NO(g)] - \Delta G^\circ_f[N_2O(g)] - 0.5\,\Delta G^\circ_f[O_2(g)]$
$= 2\,(86.55$ kJ/mol$) - (104.2$ kJ/mol$) - 0.5\,(0.00$ kJ/mol$) = 68.9$ kJ/mol
$= -RT\ln K_p = -(8.3145 \times 10^{-3}\text{ kJ mol}^{-1}\text{ K}^{-1})(298\text{ K})\ln K_p$

$$\ln K_p = -\frac{68.9\text{ kJ/mol}}{8.3145 \times 10^{-3}\text{ kJ mol}^{-1}\text{ K}^{-1} \times 298.15\text{ K}} = -27.8$$

$K_p = e^{-27.8} = 8.4 \times 10^{-13}$

17. We first balance the chemical equation in each case, and then calculate the value of ΔG° with data from Appendix D, and finally calculate the value of K_{eq} with the use of $\Delta G^\circ = -RT\ln K$.

(a) $4\,HCl(g) + O_2(g) \rightleftharpoons 2\,H_2O(g) + 2\,Cl_2(s)$
$\Delta G^\circ = 2\,\Delta G^\circ_f[H_2O(g)] + 2\,\Delta G^\circ_f[Cl_2(g)] - 4\,\Delta G^\circ_f[HCl(g)] - \Delta G^\circ_f[O_2(g)]$
$= 2 \times (-228.6$ kJ/mol$) + 2 \times 0.00$ kJ/mol $- 4 \times (-95.30$ kJ/mol$) - 0.00$ kJ/mol $= -76.0$ kJ/mol

$$\ln K_{eq} = \frac{-\Delta G^\circ}{RT} = \frac{+76.0 \times 10^3\text{ J/mol}}{8.3145\text{ J mol}^{-1}\text{ K}^{-1} \times 298\text{ K}} = +30.7 \qquad K_{eq} = e^{+30.7} = 2.2 \times 10^{13}$$

(b) $3\,Fe_2O_3(s) + H_2(g) \rightleftharpoons 2\,Fe_3O_4(s) + H_2O(g)$
$\Delta G^\circ = 2\,\Delta G^\circ_f[Fe_3O_4(s)] + \Delta G^\circ_f[H_2O(g)] - 3\,\Delta G^\circ_f[Fe_2O_3(s)] - \Delta G^\circ_f[H_2(g)]$
$= 2 \times (-1015$ kJ/mol$) - 228.6$ kJ/mol $- 3 \times (-742.2$ kJ/mol$) - 0.00$ kJ/mol $= -32$ kJ/mol

$$\ln K_{eq} = \frac{-\Delta G^\circ}{RT} = \frac{32 \times 10^3\text{ J/mol}}{8.3145\text{ J mol}^{-1}\text{ K}^{-1} \times 298\text{ K}} = 13 \qquad K_{eq} = e^{+13} = 4 \times 10^5$$

(c) $2\,Ag^+(aq) + SO_4^{2-}(aq) \rightleftharpoons Ag_2SO_4(s)$
$\Delta G^\circ = \Delta G^\circ_f[Ag_2SO_4(s)] - 2\,\Delta G^\circ_f[Ag^+(aq)] - \Delta G^\circ_f[SO_4^{2-}(aq)]$
$= -618.4$ kJ/mol $- 2 \times 77.11$ kJ/mol $- (-744.5$ kJ/mol$) = -28.1$ kJ/mol

$$\ln K_{eq} = \frac{-\Delta G^\circ}{RT} = \frac{28.1 \times 10^3\text{ J/mol}}{8.3145\text{ J mol}^{-1}\text{ K}^{-1} \times 298\text{ K}} = 11.3 \qquad K_{eq} = e^{+11.3} = 8.1 \times 10^4$$

18. $\Delta S^\circ = S^\circ\{CO_2(g)\} + S^\circ\{H_2(g)\} - S^\circ\{CO(g)\} - S^\circ\{H_2O(g)\}$
$= 213.6\text{ J mol}^{-1}\text{ K}^{-1} + 130.6\text{ J mol}^{-1}\text{ K}^{-1} - 197.6\text{ J mol}^{-1}\text{ K}^{-1} - 188.7\text{ J mol}^{-1}\text{ K}^{-1}$
$= -42.1\text{ J mol}^{-1}\text{ K}^{-1}$

19. We only compute a value of $\Delta G°$ for each reaction. If this value is significantly less than zero, we conclude that the reaction is expected to occur to some extent at 298.15 K.

(a) $\Delta G° = 2\,\Delta G^\circ_f[O_3(g)] - 3\,\Delta G^\circ_f[O_2(g)] = 2\,(163.2 \text{ kJ/mol}) - 3\,(0.00 \text{ kJ/mol})$
$= 326.4$ kJ/mol Does not occur to a significant extent.

(b) $\Delta G° = 2\,\Delta G^\circ_f[NO_2(g)] - \Delta G^\circ_f[N_2O_4(g)] = 2\,(51.29 \text{ kJ/mol}) - 97.82 \text{ kJ/mol}$
$= 4.76$ kJ/mol Occurs only to a very small extent.

(c) $\Delta G° = 2\,\Delta G^\circ_f[BrCl(g)] - \Delta G^\circ_f[Cl_2(g)] - \Delta G^\circ_f[Br_2(l)]$
$= 2\,(-0.98 \text{ kJ/mol}) - 0.00 \text{ kJ/mol} - 0.00 \text{ kJ/mol}$
$= -2.0$ kJ/mol Occurs to some small extent.

We could have predicted the first and last result by simply looking at the free energy of formation of product. The small size of the result in the other case indicates that some calculation is necessary.

20. $\Delta G = \Delta H - T\,\Delta S \qquad T\,\Delta S = \Delta H - \Delta G \qquad T = \dfrac{\Delta H - \Delta G}{\Delta S}$

$$T = \frac{-24.8 \times 10^3 \text{ J} - (-45.5 \times 10^3 \text{ J})}{15.2 \text{ J/K}} = 1.36 \times 10^3 \text{ K}$$

21. **(a)** $\Delta G° = 2\,\Delta G^\circ_f[N_2O_5(g)] - 2\,\Delta G^\circ_f[N_2O_4(g)] - \Delta G^\circ_f[O_2(g)]$
$= 2\,(115.0 \text{ kJ/mol}) - 2\,(97.82 \text{ kJ/mol}) - (0.00 \text{ kJ/mol}) = 34.4$ kJ/mol

(b) $\Delta G° = -R\,T \ln K_p \qquad \ln K_p = -\dfrac{\Delta G°}{R\,T} = -\dfrac{34.4 \times 10^3 \text{ J/mol}}{8.3145 \text{ J mol}^{-1} \text{ K}^{-1} \times 298.15 \text{ K}} = -13.9$

$K_p = e^{-13.9} = 9 \times 10^{-7}$

22. **(a)** $\Delta S° = S°[Na_2CO_3(s)] + S°[H_2O(l)] + S°[CO_2(g)] - 2\,S°[NaHCO_3(s)]$
$= 135.0 \text{ J mol}^{-1} \text{ K}^{-1} + 69.91 \text{ J mol}^{-1} \text{ K}^{-1} + 213.6 \text{ J mol}^{-1} \text{ K}^{-1} - 2\,(101.7 \text{ J mol}^{-1} \text{ K}^{-1})$
$= 215.1 \text{ J mol}^{-1} \text{ K}^{-1}$

(b) $\Delta H° = \Delta H^\circ_f[Na_2CO_3(s)] + \Delta H^\circ_f[H_2O(l)] + \Delta H^\circ_f[CO_2(g)] - 2\,\Delta H^\circ_f[NaHCO_3(s)]$
$= -1131 \text{ kJ/mol} - 285.8 \text{ kJ/mol} - 393.5 \text{ kJ/mol} - 2\,(-950.8 \text{ kJ/mol})$
$= +91$ kJ/mol

(c) $\Delta G° = \Delta H° - T\,\Delta S° = 91 \text{ kJ/mol} - (298.15 \text{ K})(215.1 \times 10^{-3} \text{ kJ mol}^{-1} \text{ K}^{-1})$
$= 91 \text{ kJ/mol} - 64.13 \text{ kJ/mol} = 27$ kJ/mol

(d) $\Delta G° = -R\,T \ln K_{eq} \qquad \ln K_{eq} = -\dfrac{\Delta G°}{R\,T} = -\dfrac{27 \times 10^3 \text{ J/mol}}{8.3145 \text{ J mol}^{-1} \text{ K}^{-1} \times 298 \text{ K}} = -10.9$

$K_{eq} = e^{-10.9} = 2 \times 10^{-5}$

23. **(a)** $\Delta S° = S°[CH_3CH_2OH(g)] + S°[H_2O(g)] - S°[CO(g)] - 2\,S°[H_2(g)] - S°[CH_3OH(g)]$
$= 282.6 \text{ J mol}^{-1} \text{ K}^{-1} + 188.7 \text{ J mol}^{-1} \text{ K}^{-1}$
$- 197.6 \text{ J mol}^{-1} \text{ K}^{-1} - 2\,(130.6 \text{ J mol}^{-1} \text{ K}^{-1}) - 239.7 \text{ J mol}^{-1} \text{ K}^{-1}$
$= -227.2 \text{ J mol}^{-1} \text{ K}^{-1}$

$\Delta H° = \Delta H^\circ_f[CH_3CH_2OH(g)] + \Delta H^\circ_f[H_2O(g)] - \Delta H^\circ_f[CO(g)] - 2\,\Delta H^\circ_f[H_2(g)] - \Delta H^\circ_f[CH_3OH(g)]$
$= -235.1 \text{ kJ/mol} - 241.8 \text{ kJ/mol} - (-110.5 \text{ kJ/mol}) - 2\,(0.00 \text{ kJ/mol}) - (-200.7 \text{ kJ/mol})$
$= -165.7$ kJ/mol

$\Delta G° = -165.7 \text{ kJ/mol} - (298.15 \text{ K})(-227.2 \times 10^{-3} \text{ kJ mol}^{-1} \text{ K}^{-1}) = -165.7 \text{ kJ/mol} + 67.7 \text{ kJ/mol}$
$= -98.0$ kJ/mol

(b) $\Delta H° < 0$ for this reaction; it is favored at low temperatures. And because $\Delta n_{gas} = +2 - 4 = -2$, which is less than zero, the reaction is favored at high pressures.

(c) We assume that neither $\Delta S°$ nor $\Delta H°$ varies significantly with temperature. Then we compute a value of $\Delta G°$ at 750. K From this value of $\Delta G°$, we compute a value of K_p.
$\Delta G° = \Delta H° - T\,\Delta S° = -165.7 \text{ kJ/mol} - (750. \text{ K})(-227.2 \times 10^{-3} \text{ kJ mol}^{-1} \text{ K}^{-1})$
$= -165.7 \text{ kJ/mol} + 170. \text{ kJ/mol} = +4 \text{ kJ/mol} = -R\,T \ln K_p$

$\ln K_p = -\dfrac{\Delta G°}{R\,T} = -\dfrac{4 \times 10^3 \text{ J/mol}}{8.3145 \text{ J mol}^{-1} \text{ K}^{-1} \times 750. \text{ K}} = -0.6 \qquad K_p = e^{-0.6} = 0.55$

24. We use the van't Hoff equation with $\Delta H° = -1.8 \times 10^5$ J/mol, $T_1 = 800.$ K, $T_2 = 100.$ °C = 373 K, and $K_1 = 9.1 \times 10^2$.

$$\ln\frac{K_2}{K_1} = \frac{\Delta H°}{R}\left(\frac{1}{T_1} - \frac{1}{T_2}\right) = \frac{-1.8 \times 10^5 \text{ J/mol}}{8.3145 \text{ J mol}^{-1}\text{ K}^{-1}}\left(\frac{1}{800 \text{ K}} - \frac{1}{373 \text{ K}}\right) = 31$$

$$\frac{K_2}{K_1} = e^{31} = 3 \times 10^{13} = \frac{K_2}{9.1 \times 10^2} \qquad K_2 = (3 \times 10^{13})(9.1 \times 10^2) = 3 \times 10^{16}$$

25. Answer **(b)** is incorrect; the negative of this expression is correct: $\Delta S° = (\Delta H° - \Delta G°)/T$. **(a)** $K_{eq} = K_c$ for a reaction involving gases. **(c)** For a reaction, such as this one, that involves only gases, $K_{eq} = K_p$: and this equation combined with $\Delta G° = -RT \ln K_{eq}$ produces $K_p = \text{antiln}(-\Delta G°/RT) = e^{-\Delta G/RT}$. **(d)** $\Delta G = \Delta G° + RT \ln Q$ is a general equation, valid for all reactions.

26. $\Delta G°$ is directly related to the equilibrium constant of a chemical reaction through $\Delta G° = -RT \ln K$. And the equilibrium constant can, of course, be used to determine the relative proportions of reactants and products at equilibrium. At equilibrium ΔG (without the °) is equal to zero (that is one definition of equilibrium). The size of $\Delta G°$ is an indication of how far the equilibrium conditions differ from standard conditions.

EXERCISES

Spontaneous Change, Entropy, and Disorder

27. **(a)** The freezing of ethanol involves a *decrease* in the entropy of the system, since solids are more ordered than liquids, in general.
(b) The sublimation of dry ice involves converting a quite ordered solid into a very disordered vapor. Thus the entropy of the system *increases* substantially.
(c) The burning of rocket fuel involves converting a somewhat ordered liquid fuel into the highly disordered mixture of the gaseous combustion products. The entropy of the system *increases* substantially.

28. Although there is a substantial change in entropy involved in **(a)** changing H_2O(l, 1 atm) to H_2O(g, 1 atm), it is not as large as **(c)** converting the liquid to a gas at 10 mmHg. The gas is less ordered at lower pressures. In turn, **(b)** if we start with a solid and convert it to a gas at the lower pressure, the entropy change should be even larger, since a solid is more ordered than a liquid. Thus, in order of increasing ΔS, the processes are: **(a) < (c) < (b)**.

29. The first law of thermodynamics states that energy is neither created nor destroyed (thus, "The energy of the world is constant"). A consequence of the second law of thermodynamics is that entropy increases for all spontaneous—that is, naturally occurring—processes (and therefore, "the entropy of the world increases toward a maximum").

30. When environmental pollutants are produced they are dispersed throughout the environment. These pollutants thus start in a relatively compact form and end up spread throughout a large volume. In this large volume, because they are mixed with many other substances, they are highly disordered and thus have a high entropy. Returning them to their original compact form requires reducing this entropy, a highly nonspontaneous process. If we have had enough foresight to retain these pollutants in a reasonably compact form—such as disposing of them in a *secure* landfill, rather than dispersing them in the atmosphere or in rivers and seas—the task of permanently removing them from the environment, and perhaps even converting them to useful forms, will be considerably easier.

31. **(a)** Negative A liquid (of moderate entropy) combines with a solid to form another solid.
(b) Positive One mole of high entropy gas forms where no gas was present before.
(c) Positive One mole of high entropy gas forms where no gas was present before.
(d) Uncertain There are the same number of moles of gaseous products as of gaseous reactants.
(e) Negative Two moles of gas (along with a solid) combine to form just one mole of gas.

32. The entropy of formation of a compound would be the difference between the absolute entropy of one mole of the compound and the sum of the absolute entropies of the appropriate amounts of the elements constituting the compound, each in its most stable form. It seems as though CS_2(l) would have the highest molar entropy of formation of the compounds listed, since it is the only substance whose formation does not involve the consumption of high entropy gaseous reactants.

(a) $C(graphite) + 2\,H_2(g) \rightleftharpoons CH_4(g)$

$\Delta S^\circ_f[CH_4(g)] = S^\circ[CH_4(g)] - S^\circ[C(graphite)] - 2\,S^\circ[H_2(g)]$

$= 186.2\text{ J mol}^{-1}\text{ K}^{-1} - 5.74\text{ J mol}^{-1}\text{ K}^{-1} - 2 \times 130.6\text{ J mol}^{-1}\text{ K}^{-1} = -80.7\text{ J mol}^{-1}\text{ K}^{-1}$

(b) $2\,C(graphite) + 3\,H_2(g) + \frac{1}{2}O_2(g) \rightleftharpoons CH_3CH_2OH(l)$

$\Delta S^\circ_f[CH_3CH_2OH(l)] = S^\circ[CH_3CH_2OH(l)] - 2\,S^\circ[C(graphite)] - 3\,S^\circ[H_2(g)] - \frac{1}{2}S^\circ[O_2(g)]$

$= 160.7\text{ J mol}^{-1}\text{ K}^{-1} - 2 \times 5.74\text{ J mol}^{-1}\text{ K}^{-1} - 3 \times 130.6\text{ J mol}^{-1}\text{ K}^{-1} - \frac{1}{2} \times 205.0\text{ J mol}^{-1}\text{ K}^{-1}$

$= -345.1\text{ J mol}^{-1}\text{ K}^{-1}$

(c) $C(graphite) + 2\,S(rhombic) \rightleftharpoons CS_2(l)$

$\Delta S^\circ_f[CS_2(g)] = S^\circ[CS_2(l)] - S^\circ[C(graphite)] - 2\,S^\circ[S(rhombic)]$

$= 151.3\text{ J mol}^{-1}\text{ K}^{-1} - 5.74\text{ J mol}^{-1}\text{ K}^{-1} - 2 \times 31.80\text{ J mol}^{-1}\text{ K}^{-1} = 82.0\text{ J mol}^{-1}\text{ K}^{-1}$

Phase Transitions

33. **(a)** $\Delta H^\circ_{vap} = \Delta H^\circ_f[H_2O(g)] - \Delta H^\circ_f[H_2O(l)] = -241.8\text{ kJ/mol} - (-285.8\text{ kJ/mol}) = +44.0\text{ kJ/mol}$

$\Delta S^\circ_{vap} = S^\circ[H_2O(g)] - S^\circ[H_2O(l)] = 188.7\text{ J mol}^{-1}\text{ K}^{-1} - 69.91\text{ J mol}^{-1}\text{ K}^{-1} = 118.8\text{ J mol}^{-1}\text{ K}^{-1}$

There is an alternate, but incorrect, method of obtaining ΔS°_{vap}.

$$\Delta S^\circ_{vap} = \frac{\Delta H^\circ_{vap}}{T} = \frac{44.0 \times 10^3\text{ J/mol}}{298.15\text{ K}} = 148\text{ J mol}^{-1}\text{ K}^{-1}$$

This method is invalid because the temperature in the denominator of the equation must be the temperature at which the transition is at equilibrium. Liquid water and water vapor at 1 atm pressure (standard state, indicated by °) are in equilibrium only at 100° C = 373 K.

(b) The reason why ΔH°_{vap} is different from its value at 100 °C has to do with the heat required to bring the reactants and products down to 298 K from 373 K. The specific heat of liquid water is higher than the heat capacity of steam. Thus, more heat is given off (this is negative heat, an exothermic process) by lowering the temperature of the liquid water from 100 °C to 25 °C than is given off by lowering the temperature of the same amount of steam. Another way to think of this is that hydrogen bonding is more disrupted in water at 100° than at 25° (because the molecules are in rapid—thermal—motion), and hence there is not as much energy needed to convert liquid to vapor.

The reason why ΔS°_{vap} has a larger value at 25 °C than at 100 °C has to do with disorder. A vapor at 1 atm pressure (the case at both temperatures) has about the same entropy. On the other hand, liquid water is more disordered at higher temperatures since more of the hydrogen bonds are disrupted by thermal motion. (The hydrogen bonds are totally disrupted in the two vapors).

34. **(a)** We use Trouton's rule ($\Delta S^\circ_{vap} \approx 88\text{ J mol}^{-1}\text{ K}^{-1}$) to estimate the normal boiling point.

$$T_{nbp} = \frac{\Delta H^\circ_{vap}}{\Delta S^\circ_{vap}} = \frac{[-146.9\text{ kJ/mol} - (-173.5)] \times \frac{1000\text{ J}}{1\text{ kJ}}}{88\text{ J mol}^{-1}\text{ K}^{-1}} = 302\text{ K}$$

(b) $\Delta G^\circ_{vap} = \Delta H^\circ_{vap} - T\,\Delta S^\circ_{vap}$

$$= [-146.9 - (-173.5)]\text{ kJ/mol} - \frac{88\text{ J mol}^{-1}\text{ K}^{-1} \times 298\text{ K}}{\frac{1000\text{ J}}{1\text{ kJ}}} = 26.6\text{ kJ/mol} - 26.2\text{ kJ/mol}$$

$= +0.4\text{ kJ/mol}$

(c) A positive value of ΔG°_{vap} indicates that normal boiling (having a vapor pressure of 1.00 atm) should be nonspontaneous (will not occur) at 298 K for pentane. The vapor pressure of pentane at 298 K should be less than 1.00 atm.

35. Trouton's rule is obeyed most closely for liquids that do not have a high degree of order within the liquid. In both HF and CH_3OH hydrogen bonds create considerable order within the liquid. In $C_6H_5CH_3$ the only attractive forces are non-directional London forces, which cause the molecules to attract each other, but have no preferred orientation as hydrogen bonds do. Thus, liquid $C_6H_5CH_3$ is not a particularly ordered liquid.

36. $\Delta H^\circ_{vap} = \Delta H^\circ_f[Br_2(g)] - \Delta H^\circ_f[Br_2(l)] = 30.91\text{ kJ/mol} - 0.00\text{ kJ/mol} = 30.91\text{ kJ/mol}$

$$\Delta S^\circ_{vap} = \frac{\Delta H^\circ_{vap}}{T_{vap}} \approx 88\text{ J mol}^{-1}\text{ K}^{-1} \quad\text{or}\quad T_{vap} = \frac{\Delta H^\circ_{vap}}{\Delta S^\circ_{vap}} \approx \frac{30.91 \times 10^3\text{ J/mol}}{88\text{ J mol}^{-1}\text{ K}^{-1}} = 3.5 \times 10^2\text{ K}$$

The accepted value of the boiling point of bromine is 58.8 °C = 332 K = 3.32×10^2 K. Thus, our estimate is in reasonable agreement with the measured value.

Free Energy and Spontaneous Change

37. **(a)** $\Delta H° < 0$ and $\Delta S° < 0$ (since $\Delta n_{gas} < 0$) for this reaction. Thus, this reaction is case 2 of Table 20-1. It is spontaneous at low temperatures and nonspontaneous at high temperatures.

(b) We are unable to predict the sign of $\Delta S°$ for this reaction, since $\Delta n_{gas} = 0$. Thus, no prediction as to the temperature behavior of this reaction can be made.

(c) $\Delta H° > 0$ and $\Delta S° > 0$ (since $\Delta n_{gas} > 0$) for this reaction. This is case 3 of Table 20-1. It is nonspontaneous at low temperatures, but spontaneous at high temperatures.

38. **(d)** $\Delta H° > 0$ and $\Delta S° < 0$ (since $\Delta n_{gas} < 0$) for this reaction. This is case 4 of Table 20-1. It is nonspontaneous at all temperatures.

(e) $\Delta H° < 0$ and $\Delta S° > 0$ (since $\Delta n_{gas} > 0$) for this reaction. This is case 1 of Table 20-1. It is spontaneous at all temperatures.

(f) $\Delta H° < 0$ and $\Delta S° < 0$ (since $\Delta n_{gas} > 0$) for this reaction. This is case 2 of Table 20-1. It is spontaneous at low temperatures and nonspontaneous at high temperatures.

39. First of all, the process is clearly spontaneous, and therefore $\Delta G < 0$. In addition, the gases are more disordered when they are at a lower pressure and therefore $\Delta S > 0$. We also conclude that $\Delta H = 0$, because the gases are ideal and thus there are no forces of attraction or repulsion between them, producing no energy of interaction.

40. Because an ideal solution forms spontaneously, $\Delta G < 0$. Also, the molecules of solvent and solute are mixed together in the solution, a more disordered state than the separated solvent and solute. Therefore, $\Delta S > 0$. However, in an ideal solution, the attractive forces between solvent and solute molecules equals those forces between solvent molecules and those between solute molecules. Thus, $\Delta H = 0$, there is no energy of interaction.

41. **(a)** An exothermic reaction (one that gives off heat) may not occur spontaneously if, at the same time, the system becomes more ordered, that is , $\Delta S° < 0$. This is particularly true at a high temperature, where the $T\Delta S$ term dominates the ΔG expression. An example of such a process is freezing water (clearly exothermic because the reverse process, melting ice, is endothermic) at temperatures above 0 °C.

(b) A reaction in which $\Delta S > 0$ need not be spontaneous if that process also is endothermic. This is particularly true at low temperatures, where the ΔH term dominates the ΔG expression. An example is the vaporization of water (clearly an endothermic process, one that requires heat, and one that produces a gas, so $\Delta S > 0$) at low temperatures, that is, below 100 °C.

42. Because this reaction produces more moles of gas than it consumes, $\Delta S > 0$. The reaction also is endothermic, since energy is required to break the A—B bond. Hence $\Delta H > 0$. Therefore, this reaction is of case 3 in Table 20-1. It is nonspontaneous at low temperatures, but eventually becomes spontaneous as the temperature is raised.

Standard Free Energy Change

43. **(a)** $\Delta S° = 2\,S°[POCl_3(l)] - 2\,S°[PCl_3(g)] - S°[O_2(g)]$
$= 2\,(222.4\text{ J/K}) - 2\,(311.7\text{ J/K}) - 205.0\text{ J/K} = -383.6\text{ J/K}$

$\Delta G° = \Delta H° - T\,\Delta S° = -620.2 \times 10^3\text{ J} - (298\text{ K})(-383.6\text{ J/K}) = -506 \times 10^3\text{ J} = -506\text{ kJ}$

(b) The reaction proceeds spontaneously in the forward direction when reactants and products are in their standard states, because the value of $\Delta G°$ is less than zero.

44. **(a)** $\Delta S° = S°[Br_2(l)] + 2\,S°[HNO_2(aq)] - 2\,S°[H^+(aq)] - 2\,S°[Br^-(aq)] - 2\,S°[NO_2(g)]$
$= 152.2\text{ J/K} + 2\,(135.6\text{ J/K}) - 2\,(0\text{ J/K}) - 2\,(82.4\text{ J/K}) - 2\,(240.0\text{ J/K}) = -221.4\text{ J/K}$

$\Delta G° = \Delta H° - T\,\Delta S° = -61.6 \times 10^3\text{ J} - (298\text{ K})(-221.4\text{ J/K}) = +4.4 \times 10^3\text{ J} = +4.4\text{ kJ}$

(b) The reaction does not proceed spontaneously in the forward direction when reactants and products are in their standard states, because the value of $\Delta G°$ is greater than zero.

45. We combine the reactions in the same way as for Hess's law calculations.

(a) $N_2O(g) \longrightarrow N_2(g) + \frac{1}{2} O_2(g)$ $\quad \Delta G° = -\frac{1}{2}(+208.4 \text{ kJ}) = -104.2 \text{ kJ}$

$N_2(g) + 2\, O_2(g) \longrightarrow 2\, NO_2(g)$ $\quad \Delta G° = +102.6 \text{ kJ}$

Net: $N_2O(g) + \frac{3}{2} O_2(g) \longrightarrow 2\, NO_2(g)$ $\quad \Delta G° = -104.2 + 102.6 = -1.6 \text{ kJ}$

This reaction reaches an equilibrium condition, a conclusion we reach based on the relatively small absolute value of $\Delta G°$.

(b) $2\, N_2(g) + 6\, H_2(g) \longrightarrow 4\, NH_3(g)$ $\quad \Delta G° = 2\,(-33.0 \text{ kJ}) = -66.0 \text{ kJ}$

$4\, NH_3(g) + 5\, O_2(g) \longrightarrow 4\, NO(g) + 6\, H_2O(l)$ $\quad \Delta G° = -1010.5 \text{ kJ}$

$4\, NO(g) \longrightarrow 2\, N_2(g) + 2\, O_2(g)$ $\quad \Delta G° = -2\,(+173.1 \text{ kJ}) = -346.2 \text{ kJ}$

Net: $6\, H_2(g) + 3\, O_2(g) \longrightarrow 6\, H_2O(l)$ $\quad \Delta G° = -66.0 \text{ kJ} - 1010.5 \text{ kJ} - 346.2 \text{ kJ} = -1422.7 \text{ kJ}$

This reaction is three times the desired reaction, which therefore has $\Delta G° = -1422.7 \text{ kJ} \div 3 = -474.33$ kJ. The quite large negative value of this value of $\Delta G°$ indicates that this reaction would tend to go to completion at 25 °C.

(c) $4\, NH_3(g) + 5\, O_2(g) \longrightarrow 4\, NO(g) + 6\, H_2O(l)$ $\quad \Delta G° = -1010.5 \text{ kJ}$

$4\, NO(g) \longrightarrow 2\, N_2(g) + 2\, O_2(g)$ $\quad \Delta G° = -2\,(+173.1 \text{ kJ}) = -346.2 \text{ kJ}$

$2\, N_2(g) + O_2(g) \longrightarrow 2\, N_2O(g)$ $\quad \Delta G° = +208.4 \text{ kJ}$

Net: $4\, NH_3(g) + 4\, O_2(g) \rightarrow 2\, N_2O(g) + 6\, H_2O(l)$ $\quad \Delta G° = -1010.5 \text{ kJ} - 346.2 \text{ kJ} + 208.4 \text{ kJ} = -1148.3 \text{ kJ}$

This reaction is twice the desired reaction, which therefore has $\Delta G° = -11498.3 \text{ kJ} \div 2 = -574.15 \text{ kJ}$. The very large negative value of $\Delta G°$ for this reaction indicates that it will go to completion.

46. We combine the reactions in the same way as for Hess's law calculations.

(a) $COS(g) + 2\, CO_2(g) \longrightarrow SO_2(g) + 3\, CO(g)$ $\quad \Delta G° = -(-246.4 \text{ kJ}) = +246.6 \text{ kJ}$

$2\, CO(g) + 2\, H_2O(g) \longrightarrow 2\, CO_2(g) + 2\, H_2(g)$ $\quad \Delta G° = 2\,(-28.6 \text{ kJ}) = -57.2 \text{ kJ}$

Net: $COS(g) + 2\, H_2O(g) \longrightarrow SO_2(g) + CO(g) + 2\, H_2(g)$ $\quad \Delta G° = +246.6 - 57.2 = +189.4 \text{ kJ}$

This is spontaneous in the reverse direction, because of the large positive value of $\Delta G°$.

(b) $COS(g) + 2\, CO_2(g) \longrightarrow SO_2(g) + 3\, CO(g)$ $\quad \Delta G° = -(-246.4 \text{ kJ}) = +246.6 \text{ kJ}$

$3\, CO(g) + 3\, H_2O(g) \longrightarrow 3\, CO_2(g) + 3\, H_2(g)$ $\quad \Delta G° = 3\,(-28.6 \text{ kJ}) = -85.8 \text{ kJ}$

Net: $COS(g) + 3\, H_2O(g) \longrightarrow CO_2(g) + SO_2(g) + 3\, H_2(g)$ $\quad \Delta G° = +246.6 - 85.8 = +160.8 \text{ kJ}$

This is spontaneous in the reverse direction, because of the large positive value of $\Delta G°$.

(c) $COS(g) + H_2(g) \longrightarrow CO(g) + H_2S(g)$ $\quad \Delta G° = -(+1.4) \text{ kJ}$

$CO(g) + H_2O(g) \longrightarrow CO_2(g) + H_2(g)$ $\quad \Delta G° = -28.6 \text{ kJ} = -28.6 \text{ kJ}$

Net: $COS(g) + H_2O(g) \rightarrow CO_2(g) + H_2S(g)$ $\quad \Delta G° = -1.4 \text{ kJ} - 28.6 \text{ kJ} = -30.0 \text{ kJ}$

The negative value of $\Delta G°$ for this reaction indicates that it is spontaneous in the forward direction.

47. The combustion reaction is $C_6H_6(l) + \frac{15}{2} O_2(g) \longrightarrow 6\, CO_2(g) + 3\, H_2O(g \text{ or } l)$

(a) $\Delta G° = 6\, \Delta G°_f[CO_2(g)] + 3\, \Delta G°_f[H_2O(g)] - \Delta G°_f[C_6H_6(l)] - \frac{15}{2} \Delta G°_f[O_2(g)]$

$= 6\,(-394.4 \text{ kJ}) + 3\,(-228.6 \text{ kJ}) - (+124.4 \text{ kJ}) - \frac{15}{2}(0.00 \text{ kJ}) = -3176 \text{ kJ}$

(b) $\Delta G° = 6\, \Delta G°_f[CO_2(g)] + 3\, \Delta G°_f[H_2O(l)] - \Delta G°_f[C_6H_6(l)] - \frac{15}{2} \Delta G°_f[O_2(g)]$

$= 6\,(-394.4 \text{ kJ}) + 3\,(-237.2 \text{ kJ}) - (+124.4 \text{ kJ}) - \frac{15}{2}(0.00 \text{ kJ}) = -3202 \text{ kJ}$

(c) We could determine the difference between the two values of $\Delta G°$ by noting the difference between the two products: $3\, H_2O(l) \longrightarrow 3\, H_2O(g)$ and determining the value of $\Delta G°$ for this difference:

$\Delta G° = 3\, \Delta G°_f[H_2O(g)] - 3\, \Delta G°_f[H_2O(l)] = 3\,[-228.6 - (-237.2)] \text{ kJ} = 25.8 \text{ kJ}$

48. We wish a value of $\Delta H°$ for the given reaction: $F_2(g) \longrightarrow 2\, F(g)$

$\Delta S° = 2\, S°[F(g)] - S°[F_2(g)] = 2\,(158.7 \text{ J K}^{-1}) - (202.7 \text{ J K}^{-1}) = +114.7 \text{ J K}^{-1}$

$\Delta H° = \Delta G° + T\Delta S° = 123.8 \times 10^3 \text{ J} + (298 \text{ K} \times 114.7 \text{ J/K}) = 158.0 \text{ kJ/mole of bonds}$

The value in Table 11.3 is 159 kJ/mol, quite good agreement.

Free Energy Change and Equilibrium

49. The state point of 110 °C and 1 atm pressure is in the region labeled "solid" in the phase diagram of iodine. Thus the state of I_2 that exists at 110 °C and 1 atm pressure is solid iodine. Therefore, $I_2(s)$ has a lower free energy under these conditions than does $I_2(l)$, since the reaction $I_2(s) \longrightarrow I_2(l)$ is nonspontaneous; $\Delta G < 0$.

50. The state point of 1 atm and –60 °C is in the region labeled "gas" in the phase diagram of carbon dioxide. Thus the state of CO_2 that exists at 1 atm and –60 °C is gaseous CO_2. Of the three states of carbon dioxide $CO_2(g)$ has the lowest free energy under these conditions.

The Thermodynamic Equilibrium Constant

51. In all three cases, $K_{eq} = K_p$ because only gases, solids, and liquids are present in the chemical equations. There are no factors for solids and liquids in K_{eq} expressions, and gases appear as partial pressures in atmospheres. That makes K_{eq} the same as K_p for these three reactions. We now recall that $K_p = K_c(RT)^{\Delta n}$. Therefore, in these three cases we have:

(a) $2\,SO_2(g) + O_2(g) \rightleftharpoons 2\,SO_3(g)$ $\quad \Delta n_{gas} = 2 - (2 + 1) = -1 \quad K_{eq} = K_p = K_c(RT)^{-1}$

(b) $HI(g) \rightleftharpoons \frac{1}{2}H_2(g) + \frac{1}{2}I_2(g)$ $\quad \Delta n_{gas} = 1 - (\frac{1}{2} + \frac{1}{2}) = 0 \quad K_{eq} = K_p = K_c$

(c) $NH_4HCO_3(s) \rightleftharpoons NH_3(g) + CO_2(g) + H_2O(l)$ $\quad \Delta n_{gas} = 2 - (0) = +2 \quad K_{eq} = K_p = K_c(RT)^2$

52. **(a)** $K_{eq} = \dfrac{P\{H_2(g)\}^4}{P\{H_2O(g)\}^4}$

(b) Factors for both solids—Fe(s) and $Fe_3O_4(s)$—are properly excluded from the thermodynamic equilibrium constant expression. (Actually, each solid has an activity of 1.00.) Thus the equilibrium partial pressures of both $H_2(g)$ and $H_2O(g)$ do not depend on the amounts of the two solids present, as long as some of each solid is present. One way to understand this is that any chemical reaction occurs on the surface of the solids, and thus is unaffected by their amount present.

(c) We can produce $H_2(g)$ from $H_2O(g)$ without regard to the proportions of Fe(s) and $Fe_3O_4(s)$ with one qualification, of course. There must always be some Fe(s) present for the production of $H_2(g)$ to continue.

Relationships Involving ΔG, ΔG°, Q, and K

53. **(a)** To determine K_p we need the equilibrium partial pressures. For the ideal gas law, for each partial pressure, $P = nRT/V$. Because R, T, and V are the same for each gas, and because there are the same number of partial pressure factors in the numerator as in the denominator of the K_p expression, we can use the ratio of amounts to determine K_p.

$$K_p = \frac{P[CO(g)]\,P[H_2O(g)]}{P[CO_2(g)]\,P[H_2(g)]} = \frac{n[CO(g)]\,n[H_2O(g)]}{n[CO_2(g)]\,n[H_2(g)]} = \frac{0.224 \text{ mol CO} \times 0.224 \text{ mol } H_2O}{0.276 \text{ mol } CO_2 \times 0.276 \text{ mol } H_2} = 0.659$$

(b) $\Delta G° = -RT \ln K_p = -\,8.3145 \text{ J mol}^{-1} \text{ K}^{-1} \times 1000.\text{ K} \times \ln(0.659) = 3.47 \times 10^3 \text{ J/mol} = 3.47 \text{ kJ/mol}$

(c) $Q_p = \dfrac{0.034 \text{ mol CO} \times 0.065 \text{ mol } H_2O}{0.075 \text{ mol } CO_2 \times 0.095 \text{ mol } H_2} = 0.310 < 0.659 = K_p$

Since Q_p is smaller than K_p, the reaction will proceed to the right, forming products, in attaining equilibrium.

54. **(a)** We know that $K_p = K_c\,(RT)^{\Delta n}$. For the reaction $2\,SO_2(g) + O_2(g) \rightleftharpoons 2\,SO_3(g)$, $\Delta n_{gas} = 2 - (2 + 1) = -1$, and therefore a value of K_p can be obtained.

$$K_p = K_c\,(RT)^{-1} = \frac{2.8 \times 10^2}{\dfrac{0.08206 \text{ L atm}}{\text{mol K}} \times 1000 \text{ K}} = 3.41 = K_{eq}$$

We recognize that $K_{eq} = K_p$ since all of the substances involved in the reaction are gases. We can now evaluate $\Delta G°$.

$$\Delta G° = -RT \ln K_{eq} = -\frac{8.3145 \text{ J}}{\text{mol K}} \times 1000 \text{ K} \times \ln(3.41) = -1.02 \times 10^4 \text{ J/mol} -10.2 \text{ kJ/mol}$$

(b) We can evaluate Q_c for this situation and compare the value with that of $K_c = 2.8 \times 10^2$.

$$Q_c = \frac{[SO_3]^2}{[SO_2]^2[O_2]} = \frac{\left(\frac{0.72 \text{ mol } SO_3}{2.50 \text{ L}}\right)^2}{\left(\frac{0.40 \text{ mol } SO_2}{2.50 \text{ L}}\right)^2 \times \frac{0.18 \text{ mol } O_2}{2.50 \text{ L}}} = 45 < 2.8 \times 10^2 = K_c$$

Since Q_c is smaller than K_c the reaction will shift right, products will form, until the two values are equal.

55. (a) $K_{eq} = K_c \quad \Delta G° = -RT \ln K_{eq} = -(8.3145 \times 10^{-3} \text{ kJ mol}^{-1} \text{ K}^{-1})(445 + 273)\text{K} \ln 50.2 = -23.4 \text{ kJ}$

(b) $K_{eq} = K_p = K_c(RT)^{\Delta n} = 1.7 \times 10^{-13}\ (0.0821 \times 298)^{1/2} = 8.4 \times 10^{-13}$
$\Delta G° = -RT \ln K_p = -(8.3145 \times 10^{-3} \text{ kJ mol}^{-1} \text{ K}^{-1})(298 \text{ K}) \ln (8.4 \times 10^{-13}) = +68.89 \text{ kJ/mol}$

(c) $K_{eq} = K_p = K_c(RT)^{\Delta n} = 4.61 \times 10^{-3}\ (0.08206 \times 298)^{+1} = 0.113$
$\Delta G° = -RT \ln K_p = -(8.3145 \times 10^{-3} \text{ kJ mol}^{-1} \text{ K}^{-1})(298 \text{ K}) \ln (0.113) = +5.40 \text{ kJ/mol}$

(d) $K_{eq} = K_c = 9.14 \times 10^{-6}$
$\Delta G° = -RT \ln K_c = -(8.3145 \times 10^{-3} \text{ kJ mol}^{-1} \text{ K}^{-1})(298 \text{ K}) \ln (9.14 \times 10^{-6}) = +28.75 \text{ kJ/mol}$

56. (a) The first equation involves the formation of one mole of $Mg^{2+}(aq)$ from $Mg(OH)_2(s)$ and 2 $H^+(aq)$, while the second equation involves the formation of only one-half mole of $Mg^{2+}(aq)$. We would expect only half as large a free energy change if only half the product is formed.

(b) The value of K_{eq} for the first reaction is the square of the value of K_{eq} for the second reaction. This is the same way the equilibrium constant expressions are related. $K_1 = \frac{[Mg^{2+}]}{[H^+]^2} = \left(\frac{[Mg^{2+}]^{1/2}}{[H^+]}\right)^2 = (K_2)^2$

(c) The equilibrium solubilities will be the same no matter which of the two expressions are used. The equilibrium conditions (solubilities in this instance) are the same no matter how we choose to express them in an equilibrium constant expression.

57. $\Delta G° = -RT \ln K_p = -(8.3145 \times 10^{-3} \text{ kJ mol}^{-1} \text{ K}^{-1})(298 \text{ K}) \ln (6.5 \times 10^{11}) = -63.9 \text{ kJ/mol}$

$CO(g) + Cl_2(g) \longrightarrow COCl_2(g) \qquad \Delta G° = -67.4 \text{ kJ/mol}$

$\frac{1}{2} C(\text{graphite}) + \frac{1}{2} O_2(g) \longrightarrow CO(g) \qquad \Delta G_f° = -137.2 \text{ kJ/mol}$

Net: $\frac{1}{2} C(\text{graphite}) + \frac{1}{2} O_2(g) + Cl_2(g) \longrightarrow COCl_2(g) \qquad \Delta G_f° = -204.6 \text{ kJ/mol}$

$\Delta G_f°$ of $COCl_2(g)$ given in Appendix D is –204.6 kJ/mol, excellent agreement.

58. In each case, we determine the value of $\Delta G°$ for the solubility reaction. From that, we determine the value of the equilibrium constant, K_{sp}, for that solubility reaction.

(a) $AgBr(s) \rightleftharpoons Ag^+(aq) + Br^-(aq)$
$\Delta G° = \Delta G_f°[Ag^+(aq)] + \Delta G_f°[Br^-(aq)] - \Delta G_f°[AgBr(s)]$
$= 77.11 \text{ kJ/mol} - 104.0 \text{ kJ/mol} - (-109.8 \text{ kJ/mol}) = +82.4 \text{ kJ/mol}$

$$\ln K_{eq} = \frac{-\Delta G°}{RT} = \frac{-82.4 \times 10^3 \text{ J/mol}}{8.3145 \text{ J mol}^{-1} \text{ K}^{-1} \times 298.15 \text{ K}} = -33.2 \qquad K_{sp} = e^{-33.2} = 3.8 \times 10^{-15}$$

(b) $CaSO_4(s) \rightleftharpoons Ca^{2+}(aq) + SO_4^{2-}(aq)$
$\Delta G° = \Delta G_f°[Ca^{2+}(aq)] + \Delta G_f°[SO_4^{2-}(aq)] - \Delta G_f°[CaSO_4(s)]$
$= -553.6 \text{ kJ/mol} - 744.5 \text{ kJ/mol} - (-1332 \text{ kJ/mol}) = +24 \text{ kJ/mol}$

$$\ln K_{eq} = \frac{-\Delta G°}{RT} = \frac{-24 \times 10^3 \text{ J/mol}}{8.3145 \text{ J mol}^{-1} \text{ K}^{-1} \times 298.15 \text{ K}} = -9.7 \qquad K_{sp} = e^{-9.7} = 6 \times 10^{-5}$$

(c) $Fe(OH)_3(s) \rightleftharpoons Fe^{3+}(aq) + 3\ OH^-(aq)$
$\Delta G° = \Delta G_f°[Fe^{3+}(aq)] + 3\ \Delta G_f°[OH^-(aq)] - \Delta G_f°[Fe(OH)_3(s)]$
$= -4.7 \text{ kJ/mol} + 3 \times (-157.2 \text{ kJ/mol}) - (-696.5 \text{ kJ/mol}) = +220.2 \text{ kJ/mol}$

$$\ln K_{eq} = \frac{-\Delta G°}{RT} = \frac{-220.2 \times 10^3 \text{ J/mol}}{8.3145 \text{ J mol}^{-1} \text{ K}^{-1} \times 298.15 \text{ K}} = -88.83 \qquad K_{sp} = e^{-88.83} = 2.6 \times 10^{-39}$$

59. (a) We can determine the equilibrium partial pressure from the value of the equilibrium constant.

$$\Delta G° = -RT \ln K_p \qquad \ln K_p = -\frac{\Delta G°}{RT} = -\frac{58.54 \times 10^3 \text{ J/mol}}{8.3145 \text{ J mol}^{-1} \text{ K}^{-1} \times 298 \text{ K}} = -23.63$$

$K_p = P\{O_2(g)\}^{1/2} = e^{-23.63} = 5.5 \times 10^{-11}$ $\qquad P\{O_2(g)\} = (5.5 \times 10^{-11})^2 = 3.0 \times 10^{-21}$ atm

(b) Lavoisier did two things to increase the quantity of oxygen that he obtained. First, he ran the reaction at a high temperature, which shifts the equilibrium to the right for this endothermic reaction which also has $\Delta S > 0$. Second, the oxygen was continuously removed from the vicinity of the Hg(l) as it was formed, promoting formation of product (Le Châtelier's principle).

60. **(a)** We determine the values of $\Delta H°$ and $\Delta S°$ from the data in Appendix D, and then the value of $\Delta G°$ at 25 °C = 298 K.

$\Delta H° = \Delta H_f°[CH_3OH(g)] + \Delta H_f°[H_2O(g)] - \Delta H_f°[CO_2(g)] - 3\,\Delta H_f°[H_2(g)]$

$= 200.7$ kJ/mol + (–241.8 kJ/mol) – (–393.5 kJ/mol) – 3 (0.00 kJ/mol) = –49.0 kJ/mol

$\Delta S° = S°[CH_3OH(g)] + S°[H_2O(g)] - S°[CO_2(g)] - 3\,S°[H_2(g)]$

$= 239.7 + 188.7 - 213.6 - 3 \times 130.6 = -177.0$ J mol⁻¹ K⁻¹

$\Delta G° = \Delta H° - T\Delta S° = +49.0 \text{ kJ/mol} - 298 \text{ K} (-0.1770 \text{ kJ mol}^{-1} \text{K}^{-1}) = +3.75 \text{ kJ/mol}$

Because the value of $\Delta G°$ is positive, this reaction is not favored at 25 °C.

(b) Because the value of $\Delta H°$ is negative and that of $\Delta S°$ is positive, the reaction is *nonspontaneous* at high temperatures, if reactants and products are *in their standard states.* The reaction will proceed slightly in the forward direction, however, to produce an equilibrium mixture with small quantities of $C_2H_2(g)$ and $H_2O(g)$. And, because the forward reaction is exothermic, this reaction is favored by lowering the temperature. That is, the value of K_{eq} decreases with temperature.

(c) $\Delta G°_{1000} = \Delta H° - T\Delta S° = +49.0 \text{ kJ/mol} - 1000. \text{ K} (-0.1770 \text{ kJ mol}^{-1} \text{K}^{-1}) = 128.0 \text{ kJ/mol}$

$= 128.0 \times 10^3 \text{ J/mol} = -RT \ln K_p$

$$\ln K_p = \frac{-\Delta G°}{RT} = \frac{-128.0 \times 10^3 \text{ J/mol}}{8.3145 \text{ J mol}^{-1} \text{K}^{-1} \times 1000. \text{ K}} = -15.39 \qquad K_p = e^{-15.39} = 2.07 \times 10^{-7}$$

(d)

Reaction:	$CO_2(g)$ +	$3\,H_2(g)$ ⇌	$CH_3OH(g)$	+ $H_2O(g)$
Initial:	1.00 atm	1.00 atm		
Changes:	$-x$ atm	$-3x$ atm	$+x$ atm	$+x$ atm
Equil:	$(1.00 - x)$ atm	$(1.00 - 3x)$ atm	x atm	x atm

$$K_p = 2.07 \times 10^{-7} = \frac{P[CH_3OH]\,P[H_2O]}{P[CO_2]\,P[H_2]^3} = \frac{x \cdot x}{(1.00 - x)(1.00 - 3x)^3} \approx x^2$$

$x = \sqrt{2.07 \times 10^{-9}} = 4.55 \times 10^{-4} \text{ atm} = P[CH_3OH]$

Our assumption, that $5x \ll 1.00$ atm, is valid.

$\Delta G°$ as a Function of Temperature

61. We first determine the value of $\Delta G°$ at 100° C, from the values of $\Delta H°$ and $\Delta S°$, which are determined from information listed in Appendix D.

$\Delta H° = 2\,\Delta H_f°[NH_3(g)] - \Delta H_f°[N_2(g)] - 3\,\Delta H_f°[H_2(g)]$

$= 2(-46.11 \text{ kJ/mol}) - (0.00) - 3(0.00) = -92.22$ kJ/mol

$\Delta S° = 2\,S°[NH_3(g)] - S°[N_2(g)] - 3\,S°[H_2(g)]$

$= 2(192.3 \text{ J mol}^{-1} \text{K}^{-1}) - (191.5 \text{ J mol}^{-1} \text{K}^{-1}) - 3(130.6 \text{ J mol}^{-1} \text{K}^{-1}) = -198.7 \text{ J mol}^{-1} \text{K}^{-1}$

$\Delta G° = \Delta H° - T\Delta S° = -92.22 \text{ kJ/mol} - 673 \text{ K} \times (-0.1987 \text{ kJ mol}^{-1} \text{K}^{-1}) = +41.51 \text{ kJ/mol} = -RT \ln K_p$

$$\ln K_p = \frac{-\Delta G°}{RT} = \frac{-41.51 \times 10^3 \text{ J/mol}}{8.3145 \text{ J mol}^{-1} \text{K}^{-1} \times 673 \text{ K}} = -7.42 \qquad K_p = e^{-7.42} = 6.0 \times 10^{-4}$$

62. **(a)** $\Delta H° = \Delta H_f°[CO_2(g)] + \Delta H_f°[H_2(g)] - \Delta H_f°[CO(g)] - \Delta H_f°[H_2O(g)]$

$= -393.5 \text{ kJ/mol} - 0.00 \text{ kJ/mol} - (-110.5 \text{ kJ/mol}) - (-241.8 \text{ kJ/mol}) = -41.2$ kJ/mol

$\Delta S° = S°[CO_2(g)] + S°[H_2(g)] - S°[CO(g)] - S°[H_2O(g)]$

$= 213.6 \text{ J mol}^{-1} \text{K}^{-1} + 130.6 \text{ J mol}^{-1} \text{K}^{-1} - 197.6 \text{ J mol}^{-1} \text{K}^{-1} - 188.7 \text{ J mol}^{-1} \text{K}^{-1}$

$= -42.1 \text{ J mol}^{-1} \text{K}^{-1}$

$\Delta G° = \Delta H° - T\Delta S° = -41.2 \text{ kJ/mol} - (298.15 \text{ K})(-42.1 \times 10^{-3} \text{ kJ mol}^{-1} \text{K}^{-1})$

$= -41.2 \text{ kJ/mol} + 12.6 \text{ kJ/mol} = -28.6$ kJ/mol

(b) $\Delta G° = \Delta H° - T\Delta S° = -41.2 \text{ kJ/mol} - (875 \text{ K})(-42.1 \times 10^{-3} \text{ kJ mol}^{-1} \text{K}^{-1})$

$= -41.2 \text{ kJ/mol} + 36.8 \text{ kJ/mol} = -4.4 \text{ kJ/mol} = -RT \ln K_p$

$$\ln K_p = -\frac{\Delta G^\circ}{R\,T} = -\frac{-4.4 \times 10^3 \text{ J/mol}}{8.3145 \text{ J mol}^{-1}\text{ K}^{-1} \times 875 \text{ K}} = +0.60 \qquad K_p = e^{+0.60} = 1.8$$

63. We assume that both ΔH° and ΔS° are constant with temperature.

$\Delta H^\circ = 2\,\Delta H_f^\circ[SO_3(g)] - 2\,\Delta H_f^\circ[SO_2(g)] - \Delta H_f^\circ[O_2(g)]$
$= 2\,(-395.7 \text{ kJ/mol}) - 2\,(-296.8 \text{ kJ/mol}) - (0.00 \text{ kJ/mol}) = -197.8 \text{ kJ/mol}$

$\Delta S^\circ = 2\,S^\circ[SO_3(g)] - 2\,S^\circ[SO_2(g)] - S^\circ[O_2(g)]$
$= 2\,(256.7 \text{ J mol}^{-1}\text{ K}^{-1}) - 2\,(248.1 \text{ J mol}^{-1}\text{ K}^{-1}) - (205.0 \text{ J mol}^{-1}\text{ K}^{-1}) = -187.8 \text{ J mol}^{-1}\text{ K}^{-1}$

$$\Delta G^\circ = \Delta H^\circ - T\,\Delta S^\circ = -RT \ln K_{eq} \qquad \Delta H^\circ = T\,\Delta S^\circ - RT \ln K_{eq} \qquad T = \frac{\Delta H^\circ}{\Delta S^\circ - R \ln K_{eq}}$$

$$T = \frac{-197.8 \times 10^3 \text{ J/mol}}{-187.8 \text{ J mol}^{-1}\text{ K}^{-1} - 8.3145 \text{ J mol}^{-1}\text{ K}^{-1} \ln(1.0 \times 10^6)} = 653.5 \text{ K}$$

This value compares very favorably with the value of $T = 6.37 \times 10^2$ that was obtained in Example 20-10.

64. We use the van't Hoff equation to determine the value of ΔH°. 448 °C = 721 K and 350 °C = 623 K.

$$\ln\frac{K_1}{K_2} = \frac{\Delta H^\circ}{R}\left(\frac{1}{T_2} - \frac{1}{T_1}\right) = \ln\frac{50.0}{66.9} = -0.291 = \frac{\Delta H^\circ}{R}\left(\frac{1}{623} - \frac{1}{721}\right) = \frac{\Delta H^\circ}{R}\,(2.2 \times 10^{-4})$$

$$\frac{\Delta H^\circ}{R} = \frac{-0.291}{2.2 \times 10^{-4}} = -1.3 \times 10^3 \qquad \Delta H^\circ = -1.3 \times 10^3 \times 8.3145 = -11 \times 10^3 \text{ J/mol} = -11 \text{ kJ/mol}$$

ΔG° as a Function of Temperature; The van't Hoff Equation

65. **(a)**
$$\ln\frac{K_2}{K_1} = \frac{\Delta H^\circ}{R}\left(\frac{1}{T_1} - \frac{1}{T_2}\right) = \frac{57.2 \times 10^3 \text{J/mol}}{8.3145 \text{ J mol}^{-1}\text{ K}^{-1}}\left(\frac{1}{298 \text{ K}} - \frac{1}{273 \text{ K}}\right) = -2.1$$

$$\frac{K_2}{K_1} = e^{-2.1} = 0.12 \qquad K_2 = 0.12 \times 0.113 = 0.014 \text{ at } 273 \text{ K}$$

(b)
$$\ln\frac{K_2}{K_1} = \frac{\Delta H^\circ}{R}\left(\frac{1}{T_1} - \frac{1}{T_2}\right) = \frac{57.2 \times 10^3 \text{J/mol}}{8.3145 \text{ J mol}^{-1}\text{ K}^{-1}}\left(\frac{1}{T_1} - \frac{1}{298 \text{ K}}\right) = \ln\frac{0.113}{1.00} = -2.180$$

$$\left(\frac{1}{T_1} - \frac{1}{298 \text{ K}}\right) = \frac{-2.180 \times 8.3145}{57.2 \times 10^3} \text{ K}^{-1} = -3.17 \times 10^{-4} \text{ K}^{-1}$$

$$\frac{1}{T_1} = \frac{1}{298} - 3.17 \times 10^{-4} = 3.36 \times 10^{-3} - 3.17 \times 10^{-4} = 3.04 \times 10^{-3} \text{ K}^{-1} \qquad T_1 = 329 \text{ K}$$

66. First we calculate ΔG° at 298 K to obtain a value of K_{eq} at that temperature.

$\Delta G^\circ = 2\,\Delta G_f^\circ[NO_2(g)] - 2\,\Delta G_f^\circ[NO(g)] - \Delta G_f^\circ[O_2(g)]$
$= 2\,(51.29 \text{ kJ/mol}) - 2\,(86.55 \text{ kJ/mol}) - 0.00 \text{ kJ/mol} = -70.52 \text{ kJ/mol}$

$$\ln K_{eq} = \frac{-\Delta G^\circ}{RT} = -\frac{-70.52 \times 10^3 \text{ J/mol K}}{\frac{8.3145 \text{ J}}{\text{mol K}} \times 298.15 \text{ K}} = 28.45 \qquad K_{eq} = e^{28.45} = 2.3 \times 10^{12}$$

Now we calculate ΔH° for the reaction so that we can apply the van't Hoff equation.

$\Delta H^\circ = 2\,\Delta H_f^\circ[NO_2(g)] - 2\,\Delta H_f^\circ[NO(g)] - \Delta H_f^\circ[O_2(g)]$
$= 2\,(33.18 \text{ kJ/mol}) - 2\,(90.25 \text{ kJ/mol}) - 0.00 \text{ kJ/mol} = -114.14 \text{ kJ/mol}$

$$\ln\frac{K_2}{K_1} = \frac{\Delta H^\circ}{R}\left(\frac{1}{T_1} - \frac{1}{T_2}\right) = \frac{-114.14 \times 10^3 \text{J/mol}}{8.3145 \text{ J mol}^{-1}\text{ K}^{-1}}\left(\frac{1}{298 \text{ K}} - \frac{1}{373 \text{ K}}\right) = -9.3$$

$$\frac{K_2}{K_1} = e^{-9.3} = 9 \times 10^{-5} \qquad K_2 = 9 \times 10^{-5} \times 2.3 \times 10^{12} = 2 \times 10^8$$

Another way, without using the van't Hoff equation is to compute ΔH° (–114.14 kJ/mol) from ΔH_f° values and ΔS° (–146.4 J mol^{-1} K^{-1}) from S° values. Then determine ΔG° (–59.5 kJ/mol), and find K_p with the expression $\Delta G^\circ = -RT\ln K_p$. We obtain the same result, $K_p = 2 \times 10^8$.

67. First, the van't Hoff equation is used to obtain a value of ΔH°. 200 °C = 473 K and 260 °C = 533 K.

$$\ln\frac{K_2}{K_1} = \frac{\Delta H^\circ}{R}\left(\frac{1}{T_1} - \frac{1}{T_2}\right) = \ln\frac{2.15 \times 10^{11}}{4.56 \times 10^8} = 6.156 = \frac{\Delta H^\circ}{8.3145 \text{ J mol}^{-1}\text{ K}^{-1}}\left(\frac{1}{533} - \frac{1}{473 \text{ K}}\right)$$

$$6.156 = -2.9 \times 10^{-5}\,\Delta H^\circ \qquad \Delta H^\circ = \frac{6.156}{-2.9 \times 10^{-5}} = -2.1 \times 10^5 \text{ J/mol} = -2.1 \times 10^2 \text{ kJ/mol}$$

Another route to $\Delta H°$ is the combination of standard enthalpies of formation.

$CO(g) + 3\,H_2(g) \rightleftharpoons CH_4(g) + H_2O(g)$

$\Delta H° = \Delta H_f°[CH_4(g)] + \Delta H_f°[H_2O(g)] - \Delta H_f°[CO(g)] - 3\,\Delta H_f°[H_2(g)]$

$= -74.81\text{ kJ/mol} - 241.8\text{ kJ/mol} - (-110.5) - 3 \times 0.00\text{ kJ/mol} = -206.1\text{ kJ/mol}$

Within the precision of the data supplied, the results are in agreement.

68. **(a)**

t, °C	T, K	$1/T$, K^{-1}	K_p	$\ln K_p$
30.	303	3.30×10^{-3}	1.66×10^{-5}	−11.006
50.	323	3.10×10^{-3}	3.90×10^{-4}	−7.849
70.	343	2.92×10^{-3}	6.27×10^{-3}	−5.072
100.	373	2.68×10^{-3}	2.31×10^{-1}	−1.465

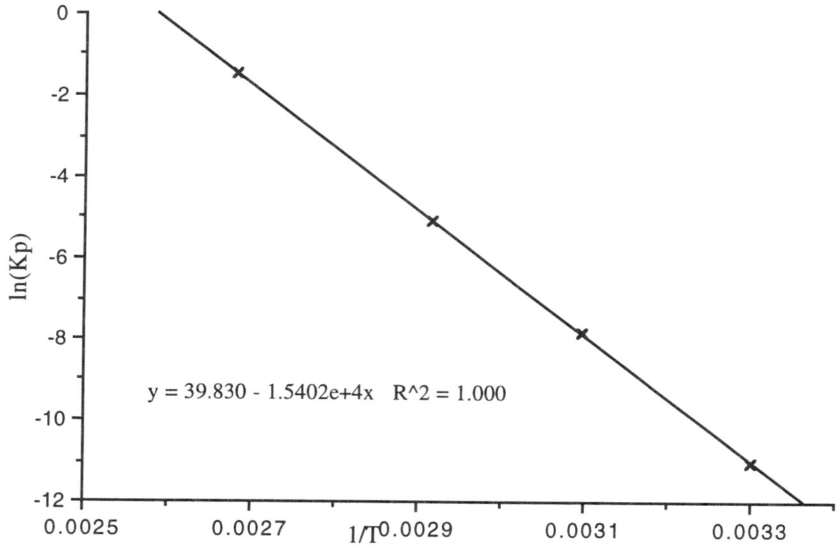

The slope of this graph is $-\Delta H°/R = -1.54 \times 10^4$ K

$\Delta H° = -(8.3145\text{ J mol}^{-1}\text{ K}^{-1})(-1.54 \times 10^4\text{ K}) = 128 \times 10^3\text{ J/mol} = 128\text{ kJ/mol}$

(b) When the total pressure is 2.00 atm, and both gases have been produced from $NaHCO_3(s)$,

$P\{H_2O(g)\} = P\{CO_2(g)\} = 1.00$ atm $\qquad K_p = P\{H_2O(g)\}P\{CO_2(g)\} = (1.00)(1.00) = 1.00$

Thus, $\ln K_p = \ln(1.00) = 0.000$ The point corresponds to $1/T = 2.59 \times 10^{-3}\text{ K}^{-1}$ $\qquad T = 386$ K

We can compute the same temperature from the van't Hoff equation.

$$\ln\frac{K_2}{K_1} = \frac{\Delta H°}{R}\left(\frac{1}{T_1} - \frac{1}{T_2}\right) = \frac{128 \times 10^3\text{J/mol}}{8.3145\text{ J mol}^{-1}\text{ K}^{-1}}\left(\frac{1}{T_1} - \frac{1}{303\text{ K}}\right) = \ln\frac{1.66 \times 10^{-5}}{1.00} = -11.006$$

$$\left(\frac{1}{T_1} - \frac{1}{303\text{ K}}\right) = \frac{-11.006 \times 8.3145}{128 \times 10^3}\text{ K}^{-1} = -7.15 \times 10^{-4}\text{ K}^{-1}$$

$$\frac{1}{T_1} = \frac{1}{303} - 7.15 \times 10^{-4} = 3.30 \times 10^{-3} - 7.15 \times 10^{-4} = 2.59 \times 10^{-3}\text{ K}^{-1} \qquad T_1 = 386\text{ K}$$

This result agrees well with the result obtained from the graph.

69. **(a)** We compute $\Delta G°$ for the given reaction.

$\Delta H° = \Delta H_f°[TiCl_4(l)] + \Delta H_f°[O_2(g)] - \Delta H_f°[TiO_2(s)] - 2\,\Delta H_f°[Cl_2(g)]$

$= -804.2\text{ kJ/mol} + 0.00\text{ kJ/mol} - (-944.7\text{ kJ/mol}) - 2(0.00\text{ kJ/mol}) = +140.5\text{ kJ/mol}$

$\Delta S° = S°[TiCl_4(l)] + S°[O_2(g)] - S°[TiO_2(s)] - 2\,S°[Cl_2(g)]$

$= 252.3\text{ J mol}^{-1}\text{ K}^{-1} + 205.0\text{ J mol}^{-1}\text{ K}^{-1} - (50.33\text{ J mol}^{-1}\text{ K}^{-1}) - 2\,(223.0\text{ J mol}^{-1}\text{ K}^{-1})$

$= -38.9\text{ J mol}^{-1}\text{ K}^{-1}$

$\Delta G° = \Delta H° - T\Delta S° = +140.5\text{ kJ/mol} - (298.2\text{ K})(-39.0 \times 10^{-3}\text{ kJ mol}^{-1}\text{ K}^{-1})$

$= +140.5\text{ kJ/mol} + 11.6\text{ kJ/mol} = +152.1\text{ kJ/mol}$

This reaction is nonspontaneous at 25 °C. (We also could have used values of $\Delta G_f°$ to calculate $\Delta G°$.)

(b) For the cited reaction $\qquad \Delta G° = 2\,\Delta G_f°[CO_2(g)] - 2\,\Delta G_f°[CO(g)] - \Delta G_f°[O_2(g)]$

$\Delta G° = 2\,(-394.4\text{ kJ/mol}) - 2\,(-137.2\text{ kJ/mol}) - 0.00\text{ kJ/mol} = -514.4\text{ kJ/mol}$

Then we couple the two reactions.

$TiO_2(s) + 2\,Cl_2(g) \longrightarrow TiCl_4(l) + O_2(g)$ $\Delta G^\circ = +152.1$ kJ/mol

$2\,CO(g) + O_2(g) \longrightarrow 2\,CO_2(g)$ $\Delta G^\circ = -514.4$ kJ/mol

Net: $TiO_2(s) + 2\,Cl_2(g) + 2\,CO(g) \longrightarrow TiCl_4(l) + 2\,CO_2(g)$ $\Delta G^\circ = +152.1 \text{ kJ/mol} - 514.4 \text{ kJ/mol}$
$= -362.3$ kJ/mol

The coupled reaction has $\Delta G^\circ < 0$ and therefore is spontaneous.

70. If $\Delta G^\circ < 0$ for the sum of the coupled reactions, reduction of the oxide with carbon is spontaneous.

(a) $NiO(s) \longrightarrow Ni(s) + \frac{1}{2}O_2(g)$ $\Delta G^\circ = +115$ kJ

$C(s) + \frac{1}{2}O_2(g) \longrightarrow CO(g)$ $\Delta G^\circ = -250$ kJ

Net: $NiO(s) + C(s) \longrightarrow Ni(s) + CO(g)$ $\Delta G^\circ = +115 \text{ kJ} - 250 \text{ kJ} = -135$ kJ Spontaneous

(b) $MnO(s) \longrightarrow Mn(s) + \frac{1}{2}O_2(g)$ $\Delta G^\circ = +280$ kJ

$C(s) + \frac{1}{2}O_2(g) \longrightarrow CO(g)$ $\Delta G^\circ = -250$ kJ

Net: $MnO(s) + C(s) \longrightarrow Mn(s) + CO(g)$ $\Delta G^\circ = +280 \text{ kJ} - 250 \text{ kJ} = +30$ kJ Nonspontaneous

(c) $TiO_2(s) \longrightarrow Ti(s) + O_2(g)$ $\Delta G^\circ = +630$ kJ

$2\,C(s) + O_2(g) \longrightarrow 2\,CO(g)$ $\Delta G^\circ = 2\,(-250 \text{ kJ}) = -500$ kJ

Net: $TiO_2(s) + 2\,C(s) \longrightarrow Ti(s) + 2\,CO(g)$ $\Delta G^\circ = +630 \text{ kJ} - 500 \text{ kJ} = +130$ kJ Nonspontaneous

FEATURE PROBLEMS

A. **1.** The first method involves combining the values of ΔG°_f. The second is using $\Delta G^\circ = \Delta H^\circ - T\Delta S^\circ$

$\Delta G^\circ = \Delta G^\circ_f[H_2O(g)] - \Delta G^\circ_f[H_2O(l)] = -228.572 \text{ kJ/mol} - (-237.129 \text{ kJ/mol}) = +8.557 \text{ kJ/mol}$

$\Delta H^\circ = \Delta H^\circ_f[H_2O(g)] - \Delta H^\circ_f[H_2O(l)] = -241.818 \text{ kJ/mol} - (-285.830 \text{ kJ/mol}) = +44.012 \text{ kJ/mol}$

$\Delta S^\circ = S^\circ[H_2O(g)] - S^\circ[H_2O(l)] = 188.825 \text{ J mol}^{-1}\text{ K}^{-1} - 69.91 \text{ J mol}^{-1}\text{ K}^{-1} = +118.92 \text{ J mol}^{-1}\text{ K}^{-1}$

$\Delta G^\circ = \Delta H^\circ - T\Delta S^\circ = 44.015 \text{ kJ/mol} - 298.15 \text{ K} \times 118.92 \times 10^{-3} \text{ kJ mol}^{-1}\text{ K}^{-1} = +8.559 \text{ kJ/mol}$

2. We use the average value: $\Delta G^\circ = +8.558 \times 10^3 \text{ J/mol} = -RT\ln K_{eq}$

$$\ln K_{eq} = -\frac{8558 \text{ J/mol}}{8.3145 \text{ J mol}^{-1}\text{ K}^{-1} \times 298.15 \text{ K}} = -3.452 \qquad K_{eq} = e^{-3.452} = 0.03168 \text{ bar}$$

3. $$P\{H_2O\} = 0.0317 \text{ bar} \times \frac{1 \text{ atm}}{1.01325 \text{ bar}} \times \frac{760 \text{ mmHg}}{1 \text{ atm}} = 23.78 \text{ mmHg}$$

4. $$\ln K_{eq} = -\frac{8590 \text{ J/mol}}{8.3145 \text{ J mol}^{-1}\text{ K}^{-1} \times 298.15 \text{ K}} = -3.465 \qquad K_{eq} = e^{-3.465} = 0.03127 \text{ atm}$$

$$P\{H_2O\} = 0.0313 \text{ atm} \times \frac{760 \text{ mmHg}}{1 \text{ atm}} = 23.77 \text{ mmHg}$$

B. **1.** When we combine two reactions and obtain the overall value of ΔG°, we subtract the value on the plot of the reaction that becomes a reduction from the value on the plot of the reaction that is an oxidation. Thus, to reduce ZnO with elemental Mg, we subtract the values on the line labeled "$2\,Zn + O_2 \longrightarrow 2\,ZnO$" from those on the line labeled "$2\,Mg + O_2 \longrightarrow 2\,MgO$." The result—for the overall ΔG°—always will be negative because every point on the "zinc" line is above the corresponding point on the "magnesium" line. There is no way that the value of ΔG° can be negative for the reduction of MgO by Zn; the "zinc" line is everywhere above the "magnesium" line.

2. In contrast, the "carbon" line is only below the "zinc" line at temperatures above about 1000 °C. At only these elevated temperatures can ZnO be reduced by carbon.

3. The decomposition of zinc oxide to its elements is the reverse of the plotted reaction. the value of ΔG° for the decomposition becomes negative, and the reaction becomes spontaneous, where the value of ΔG° for the plotted reaction becomes positive. This occurs above about 1850 °C.

4. The "carbon" line has a negative slope, indicating that carbon monoxide becomes more stable as temperature rises. The point where CO(g) would become less stable looks to be below −1000 °C (by extrapolating the line to lower temperatures). Based on this plot, it is not possible to decompose CO(g) to C(s) and $O_2(g)$ in a spontaneous reaction.

21 ELECTROCHEMISTRY

PRACTICE EXAMPLES

1A The conventions state that the anode material is written first, and the cathode material is written last.

Anode, oxidation: $Sc(s) \longrightarrow Sc^{3+}(aq) + 3\ e^-$

Cathode, reduction: $Ag^+(aq) + e^- \longrightarrow Ag(s) \quad \times 3$

Net: $Sc(s) + 3\ Ag^+(aq) \longrightarrow Sc^{3+}(aq) + 3\ Ag(s)$

1B Oxidation of Al(s) at the anode: $Al(s) \longrightarrow Al^{3+}(aq) + 3\ e^-$

Reduction of $Ag^+(aq)$ at the cathode: $Ag^+(aq) + e^- \longrightarrow Ag(s)$

Overall reaction: $Al(s) + 3\ Ag^+(aq) \longrightarrow Al^{3+}(aq) + 3\ Ag(s)$

Diagram: $Al|Al^{3+}(aq)||Ag^+(aq)|Ag$

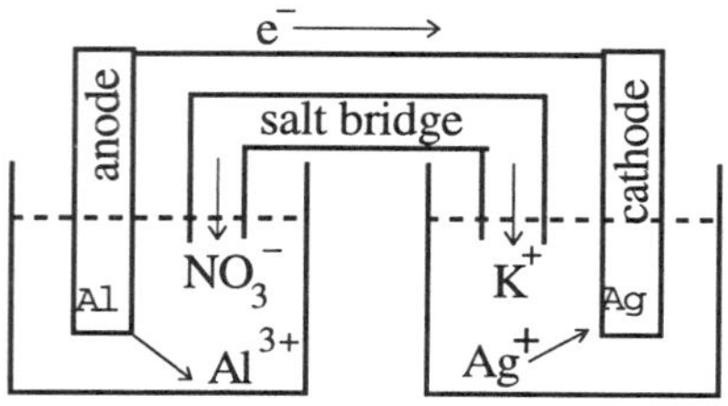

2A We obtain the two balanced half-equations and the half-cell potentials from Table 21-1.

oxidation: $\{Fe^{2+}(aq) \longrightarrow Fe^{3+}(aq) + e^-\} \quad \times 2 \quad -E° = -0.771\ V$

reduction: $Cl_2(g) + 2\ e^- \longrightarrow 2\ Cl^-(aq) \quad E° = +1.358\ V$

net: $2\ Fe^{2+}(aq) + Cl_2(g) \longrightarrow 2\ Fe^{3+}(aq) + 2\ Cl^-(aq) \quad E°_{cell} = +1.358\ V - 0.771\ V = +0.587\ V$

2B Since we need to refer to Table 21-1 in any event, it is perhaps a bit easier to locate the two balanced half-equations in the table. There is only one half-equation involving both $Fe^{2+}(aq)$ and $Fe^{3+}(aq)$ ions. It is reversed and written as an oxidation below. The half-equation involving $MnO_4^-(aq)$ is also written below. [Actually, we need to know that in acidic solution $Mn^{2+}(aq)$ is the principal reduction product of $MnO_4^-(aq)$.]

oxidation: $Fe^{2+}(aq) \longrightarrow Fe^{3+}(aq) + e^- \quad \times 5 \quad -E° = -0.771\ V$

reduction: $MnO_4^-(aq) + 8\ H^+(aq) + 5\ e^- \longrightarrow Mn^{2+}(aq) + 4\ H_2O \quad E° = +1.51\ V$

net: $MnO_4^-(aq) + 5\ Fe^{2+}(aq) + 8\ H^+(aq) \longrightarrow Mn^{2+}(aq) + 5\ Fe^{3+}(aq) + 4\ H_2O$

$E°_{cell} = +1.51\ V - 0.771\ V = +0.74\ V$

3A We write the oxidation half-equation with the method of Chapter 5, and obtain the reduction half-equation from Table 21-1, along with the reduction half-cell potential.

oxidation: $[H_2C_2O_4(aq) \longrightarrow 2\ CO_2(aq) + 2\ H^+(aq) + 2\ e^-] \quad \times 3 \quad -E°\{CO_2/H_2C_2O_4\}$

reduction: $Cr_2O_7^{2-}(aq) + 14\ H^+(aq) + 6\ e^- \longrightarrow 2\ Cr^{3+}(aq) + 7\ H_2O \quad E° = +1.33\ V$

net: $Cr_2O_7^{2-}(aq) + 8\ H^+(aq) + 3\ H_2C_2O_4(aq) \longrightarrow 2\ Cr^{3+}(aq) + 7\ H_2O + 6\ CO_2(aq)$

$E°_{cell} = +1.81\ V = +1.33\ V - E°\{CO_2/H_2C_2O_4\} \quad E°\{CO_2/H_2C_2O_4\} = 1.33\ V - 1.81\ V = -0.48\ V$

3B The second half-equation was found by looking for one with $O_2(g)$ as reactant and H_2O as product.

oxidation: $Cr^{2+}(aq) \longrightarrow Cr^{3+}(aq) + e^- \quad \times 4 \quad -E°\{Cr^{3+}|Cr^{2+}\}$

reduction: $O_2(g) + 4\ H^+(aq) + 4\ e^- \longrightarrow 2\ H_2O \quad E° = +1.229\ V$

net: $O_2(g) + 4\ H^+(aq) + 4\ Cr^{2+}(aq) \longrightarrow 2\ H_2O + 4\ Cr^{3+}(aq)$

$E°_{cell} = +1.653\ V = +1.229\ V - E°\{Cr^{3+}|Cr^{2+}\} \quad E°\{Cr^{3+}|Cr^{2+}\} = 1.229\ V - 1.653\ V = -0.424\ V$

4A First we write the two half-equations, obtain the half-cell potential for each, and then calculate E°_{cell}. From that value, we determine ΔG°

oxidation: $\{Al(s) \longrightarrow Al^{3+}(aq) + 3\,e^-\}$ $\times 2$ $E^\circ = +1.676$ V

reduction: $\{Br_2(l) + 2\,e^- \longrightarrow 2\,Br^-(aq)\}$ $\times 3$ $E^\circ = +1.065$ V

net: $2\,Al(s) + 3\,Br_2(l) \longrightarrow 2\,Al^{3+}(aq) + 6\,Br^-(aq)$ $E^\circ_{cell} = 1.676\text{ V} + 1.065\text{ V} = 2.741\text{ V}$

$$\Delta G^\circ = -n\mathcal{F}E^\circ_{cell} = -\,6\text{ mol e}^- \times \frac{96{,}485\text{ C}}{1\text{ mol e}^-} \times 2.741\text{ V} = -1.587 \times 10^6\text{ J} = 1587\text{ kJ}$$

4B First we write the two half-equations, one of which is the reduction equation from the previous example. The other is the oxidation that occurs in the standard hydrogen electrode.

oxidation: $2\,H_2(g) \longrightarrow 4\,H^+(aq) + 4\,e^-$

reduction: $O_2(g) + 4\,H^+(aq) + 4\,e^- \longrightarrow 2\,H_2O$

net: $2\,H_2(g) + O_2(g) \longrightarrow 2\,H_2O(l)$ $n = 4$ in this reaction.

This net reaction is simply twice the formation reaction for H_2O and therefore

$\Delta G^\circ = 2\,\Delta G^\circ_f[H_2O(l)] = 2 \times (-237.2\text{ kJ}) = -474.4 \times 10^3\text{ J} = -n\mathcal{F}E^\circ_{cell}$

$$E^\circ_{cell} = \frac{-\Delta G^\circ}{n\mathcal{F}} = -\frac{-474.4 \times 10^3\text{ J}}{4\text{ mol elns} \times \dfrac{96{,}485\text{ C}}{\text{mol e}^-}} = +1.229\text{ V} = E^\circ\{O_2(g)|H_2O(l)\}$$, as we might expect.

5A Cu(s) will displace from solution metal ions of a metal less active than copper. Silver ion is one example.

oxidation: $Cu(s) \longrightarrow Cu^{2+}(aq) + 2\,e^-$ $-E^\circ = -0.337$ V

reduction: $Ag^+(aq) + e^- \longrightarrow Ag(s)$ $\times 2$ $E^\circ = +0.800$ V

net: $2\,Ag^+(aq) + Cu(s) \longrightarrow Cu^{2+}(aq) + 2\,Ag(s)$ $E^\circ_{cell} = -0.337\text{ V} + 0.800\text{ V} = +0.463\text{ V}$

5B We determine the value for the hypothetical reaction's cell potential.

oxidation: $\{Na(s) \longrightarrow Na^+(aq) + e^-\}$ $\times 2$ $-E^\circ = +2.713$ V

reduction: $Mg^{2+}(aq) + 2e^- \longrightarrow Mg(s)$ $E^\circ = -2.356$ V

net: $2\,Na(s) + Mg^{2+}(aq) \longrightarrow 2\,Na^+(aq) + Mg(s)$ $E^\circ_{cell} = 2.713\text{ V} - 2.356\text{V} = +0.357\text{ V}$

The method is not feasible because another reaction occurs that has an even larger cell potential, the reaction of Na(s) to produce $H_2(g)$ and NaOH(aq), which has $E^\circ_{cell} = 2.713\text{ V} - 0.828\text{ V} = +1.885\text{ V}$.

6A The oxidation is that of sulfate ion to peroxodisulfate ion, the reduction is that of oxygen to water. The two half equations follow.

oxidation: $2\,SO_4^{2-}(aq) \longrightarrow S_2O_8^{2-}(aq) + 2\,e^-$ $\times 2$ $-E^\circ = -2.01$ V

reduction: $O_2(g) + 4\,H^+(aq) + 4\,e^- \longrightarrow 2\,H_2O$ $E^\circ = +1.229$ V

net: $O_2(g) + 4\,H^+(aq) + 2\,SO_4^{2-}(aq) \longrightarrow S_2O_8^{2-}(aq) + 2\,H_2O$ $E^\circ_{cell} = -2.01\text{ V} + 1.229\text{ V} = -0.78\text{ V}$

Because the standard cell potential is negative, we conclude that this cell reaction is nonspontaneous. This would not be a feasible method of producing peroxodisulfate ion.

6B (1) The oxidation is that of $Sn^{2+}(aq)$ to $Sn^{4+}(aq)$, the reduction is that of oxygen to water.

oxidation: $\{Sn^{2+}(aq) \longrightarrow Sn^{4+}(aq) + 2\,e^-\}$ $\times 2$ $-E^\circ = -0.154$ V

reduction: $O_2(g) + 4\,H^+(aq) + 4\,e^- \longrightarrow 2\,H_2O$ $E^\circ = +1.229$ V

net: $O_2(g) + 4\,H^+(aq) + 2\,Sn^{2+}(aq) \longrightarrow 2\,Sn^{4+}(aq) + 2\,H_2O$ $E^\circ_{cell} = -0.154\text{ V} + 1.229\text{ V} = +1.075\text{ V}$

Because the standard cell potential is positive, we conclude that this cell reaction is spontaneous.

(2) Now, the oxidation is that of Sn(s) to $Sn^{2+}(aq)$, the reduction is still that of oxygen to water.

oxidation: $\{Sn(s) \longrightarrow Sn^{2+}(aq) + 2\,e^-\}$ $\times 2$ $-E^\circ = +0.137$ V

reduction: $O_2(g) + 4\,H^+(aq) + 4\,e^- \longrightarrow 2\,H_2O$ $E^\circ = +1.229$ V

net: $O_2(g) + 4\,H^+(aq) + 2\,Sn(s) \longrightarrow 2\,Sn^{2+}(aq) + 2\,H_2O$ $E^\circ_{cell} = 0.137\text{ V} + 1.229\text{ V} = +1.336\text{ V}$

The standard cell potential for this reaction is more positive than that of situation (1); reaction (2) should occur preferentially. Also, if $Sn^{4+}(aq)$ is formed, it should react with Sn(s) to form $Sn^{2+}(aq)$.

oxidation: $Sn(s) \longrightarrow Sn^{2+}(aq) + 2\,e^-$ $\quad -E° = +0.137$ V

reduction: $Sn^{4+}(aq) + 2\,e^- \longrightarrow Sn^{2+}(aq)$ $\quad E° = +0.154$ V

net: $Sn^{4+}(aq) + Sn(s) \longrightarrow 2\,Sn^{2+}(aq)$ $\quad E°_{cell} = +0.137\text{ V} + 0.154\text{ V} = +0.291\text{ V}$

7A For the reaction $2\,Al(s) + 3\,Cu^{2+}(aq) \longrightarrow 2\,Al^{3+}(aq) + 3\,Cu(s)$ we know $n = 6$ and $E°_{cell} = +2.013$ V. We calculate the value of K_{eq}.

$$E°_{cell} = \frac{0.0257}{n}\ln K_{eq} \qquad \ln K_{eq} = \frac{n\,E°_{cell}}{0.0257} = \frac{6 \times +2.013}{0.0257} = 470 \qquad K_{eq} = e^{470} = 10^{204}$$

The huge size of the equilibrium constant indicates that this reaction indeed will go to completion.

7B We first determine the value of $E°_{cell}$ from the half-cell potentials.

oxidation: $Sn(s) \longrightarrow Sn^{2+}(aq) + 2\,e^-$ $\quad -E° = +0.137$ V

reduction: $Pb^{2+}(aq) + 2\,e^- \longrightarrow Pb(s)$ $\quad E° = -0.125$ V

net: $Pb^{2+}(aq) + Sn(s) \longrightarrow Pb(s) + Sn^{2+}(aq)$ $\quad E°_{cell} = +0.137\text{ V} - 0.125\text{ V} = +0.012\text{ V}$

$$E°_{cell} = \frac{0.0257}{n}\ln K_{eq} \qquad \ln K_{eq} = \frac{n\,E°_{cell}}{0.0257} = \frac{2 \times +0.012}{0.0257} = 0.93 \qquad K_{eq} = e^{0.93} = 2.5$$

The equilibrium constant's small size indicates that this reaction will reach equilibrium, not go to completion.

8A We first need to determine the standard cell voltage and the cell reaction.

oxidation: $Al(s) \longrightarrow Al^{3+}(aq) + 3\,e^-$ $\quad \times 2 \quad -E° = +1.676$ V

reduction: $Sn^{4+}(aq) + 2\,e^- \longrightarrow Sn^{2+}(aq)$ $\quad \times 3 \quad E° = +0.154$ V

net: $2\,Al(s) + 3\,Sn^{4+}(aq) \longrightarrow 2\,Al^{3+}(aq) + 3\,Sn^{2+}(aq)$ $\quad E°_{cell} = +1.676\text{ V} + 0.154\text{ V} = +1.830\text{ V}$

Note that $n = 6$. We now set up and substitute into the Nernst equation.

$$E_{cell} = E°_{cell} - \frac{0.0592}{n}\log\frac{[Al^{3+}]^2[Sn^{2+}]^3}{[Sn^{4+}]^3} = +1.830 - \frac{0.0592}{6}\log\frac{(0.36\text{ M})^2\,(0.54\text{ M})^3}{(0.086\text{ M})^3}$$

$$= +1.830\text{ V} - 0.0149\text{ V} = +1.815\text{ V}$$

8B We first need to determine the standard cell voltage and the cell reaction.

oxidation: $2\,Cl^-(1.0\text{ M}) \longrightarrow Cl_2(1\text{ atm}) + 2\,e^-$ $\quad -E° = -1.358$ V

reduction: $PbO_2(s) + 4\,H^+(aq) + 2\,e^- \longrightarrow Pb^{2+}(aq) + 2\,H_2O$ $\quad E° = +1.455$ V

net: $PbO_2(s) + 4\,H^+(0.10\text{ M}) + 2\,Cl^-(1.0\text{ M}) \longrightarrow Cl_2(1\text{ atm}) + Pb^{2+}(0.050) + 2\,H_2O$

$E°_{cell} = -1.358\text{ V} + 1.455\text{ V} = +0.097\text{ V}.$

Note that $n = 2$. We now set up and substitute into the Nernst equation.

$$E_{cell} = E°_{cell} - \frac{0.0592}{n}\log\frac{P\{Cl_2\}[Pb^{2+}]}{[H^+]^4[Cl^-]^2} = +0.097 - \frac{0.0592}{2}\log\frac{(1.0\text{ atm})\,(0.050\text{ M})}{(0.10\text{ M})^4\,(1.0\text{ M})^2}$$

$$= +0.097\text{ V} - 0.080\text{ V} = +0.017\text{ V}$$

9A The cell reaction is $2\,Fe^{3+}(0.35\text{ M}) + Cu(s) \longrightarrow 2\,Fe^{2+}(0.25) + Cu^{2+}(0.15\text{ M})$ with $n = 2$, and $E°_{cell} = -0.337\text{ V} + 0.771\text{ V} = 0.434\text{ V}$

We set up and substitute into the Nernst equation for this situation.

$$E_{cell} = E°_{cell} - \frac{0.0592}{n}\log\frac{[Fe^{2+}]^2[Cu^{2+}]}{[Fe^{3+}]^2} = 0.434 - \frac{0.0592}{2}\log\frac{(0.25)^2\,0.15}{(0.35)^2} = 0.434 - 0.033 = +0.401\text{ V}$$

9B The reaction is not spontaneous in either direction when $E_{cell} = 0.000$ V. We use the standard cell potential from Example 21-9.

$$E_{cell} = E°_{cell} - \frac{0.0592}{n}\log\frac{[Sn^{2+}]}{[Pb^{2+}]} = 0.000\text{ V} = 0.012\text{ V} - \frac{0.0592}{2}\log\frac{[Sn^{2+}]}{[Pb^{2+}]}$$

$$\log\frac{[Sn^{2+}]}{[Pb^{2+}]} = \frac{0.012 \times 2}{0.0592} = 0.41 \qquad \frac{[Sn^{2+}]}{[Pb^{2+}]} = 10^{0.41} = 2.6$$

10A In this concentration cell $E°_{cell} = 0.000$ V because the same reaction occurs at anode and cathode, only the concentrations of the ions differ. $[Ag^+] = 0.100$ M in the cathode compartment. The anode compartment contains a saturated solution of AgCl(aq).

$$K_{sp} = 1.8 \times 10^{-10} = [Ag^+][Cl^-] = s^2 \qquad s = \sqrt{1.8 \times 10^{-10}} = 1.3 \times 10^{-5}\text{ M}$$

Now we apply the Nernst equation. The cell reaction is $Ag^+(0.100\text{ M}) \longrightarrow Ag^+(1.3 \times 10^{-5}\text{ M})$

$$E_{cell} = 0.000 - \frac{0.0592}{1}\log\frac{1.3 \times 10^{-5}\text{ M}}{0.100\text{ M}} = +0.23\text{ V}$$

10B In this concentration cell $E^\circ_{cell} = 0.000$ V because the same reaction occurs at anode and cathode, only the concentrations of the ions differ. $[Pb^{2+}] = 0.100$ M in the cathode compartment. The anode compartment contains a saturated solution of PbI_2. We use the Nernst equation (with $n = 2$) to determine $[Pb^{2+}]$ in the saturated solution.

$$E_{cell} = +0.0567\text{ V} = 0.000 - \frac{0.0592}{2}\log\frac{x\text{ M}}{0.100\text{ M}} \qquad \log\frac{x\text{ M}}{0.100\text{ M}} = \frac{2 \times 0.0567}{-0.0592} = -1.92$$

$$\frac{x\text{ M}}{0.100\text{ M}} = 10^{-1.92} = 0.012 \qquad [Pb^{2+}] = x\text{ M} = 0.012 \times 0.100\text{ M} = 0.0012\text{ M} \qquad [I^-] = 2 \times 0.0012\text{ M}$$

$K_{sp} = [Pb^{2+}][I^-]^2 = (0.0012)(0.0024)^2 = 6.9 \times 10^{-9}$ compared with 7.1×10^{-9} in Appendix D

11A We obtain from Table 21-1 all the possible oxidations and reductions and choose one of each to get the least negative cell voltage. That pair is the most likely pair of half-reactions to occur.

oxidation: $2\,I^-(aq) \longrightarrow I_2(s) + 2\,e^-$ $\quad -E^\circ = -0.535$ V

$2\,H_2O \longrightarrow O_2(g) + 4\,H^+(aq) + 4\,e^-$ $\quad -E^\circ = -1.229$ V

reduction: $K^+(aq) + e^- \longrightarrow K(s)$ $\quad E^\circ = -2.924$ V

$2\,H_2O + 2\,e^- \longrightarrow H_2(g) + 2\,OH^-(aq)$ $\quad E^\circ = -0.828$ V

The least negative standard cell potential (–0.535 V – 0.828 V = –1.363 V) occurs when $I_2(s)$ is produced by oxidation at the anode, and $H_2(g)$ is produced by reduction at the cathode.

11B We obtain from Table 21-1 all the possible oxidations and reductions and choose one of each to get the least negative cell voltage. That pair is the most likely pair of half-reactions to occur.

oxidation: $2\,H_2O \longrightarrow O_2(g) + 4\,H^+(aq) + 4\,e^-$ $\quad -E^\circ = -1.229$ V

$Ag(s) \longrightarrow Ag^+(aq) + e^-$ $\quad -E^\circ = -0.800$ V

[We cannot further oxidize $NO_3^-(aq)$ or $Ag^+(aq)$.]

reduction: $Ag^+(aq) + e^- \longrightarrow Ag(s)$ $\quad E^\circ = +0.800$ V

$2\,H_2O + 2\,e^- \longrightarrow H_2(g) + 2\,OH^-(aq)$ $\quad E^\circ = -0.828$ V

Thus, we expect to form silver metal at the cathode and $Ag^+(aq)$ at the anode.

12A The half-cell equation is $Cu^{2+}(aq) + 2\,e^- \longrightarrow Cu(s)$, indicating that two moles of electrons are required for each mole of copper deposited. Current is measured in amperes, or coulombs per second. We convert the mass of copper to coulombs of electrons needed for the reduction and the time in hours to seconds.

$$\text{current} = \frac{12.3\text{ g Cu} \times \frac{1\text{ mol Cu}}{63.55\text{ g Cu}} \times \frac{2\text{ mol e}^-}{1\text{ mol Cu}} \times \frac{96{,}485\text{ C}}{1\text{ mol e}^-}}{5.50\text{ h} \times \frac{60\text{ min}}{1\text{ h}} \times \frac{60\text{ s}}{1\text{ min}}} = \frac{3.73_5 \times 10^4\text{ C}}{1.98 \times 10^4\text{ s}} = 1.89\text{ amperes}$$

12B We first determine the amount of $O_2(g)$ produced with the ideal gas equation.

$$\text{amount } O_2(g) = \frac{\left(738\text{ mmHg} \times \frac{1\text{ atm}}{760\text{ mmHg}}\right) \times 2.62\text{ L}}{\frac{0.08206\text{ L atm}}{\text{mol K}} \times (26.2 + 273.2)\text{ K}} = 0.104\text{ mol } O_2$$

Then we determine the time needed to produce this amount.

$$\text{elapsed time} = 0.104\text{ mol } O_2 \times \frac{4\text{ mol e}^-}{1\text{ mol } O_2} \times \frac{96{,}485\text{ C}}{1\text{ mol e}^-} \times \frac{1\text{ s}}{2.13\text{ C}} \times \frac{1\text{ h}}{3600\text{ s}} = 5.23\text{ h}$$

SUMMARIZING EXAMPLE CALCULATIONS

1. The two half-equations are given in the text.

oxidation: $C_3H_8(g) + 6\,H_2O(l) \longrightarrow 3\,CO_2(g) + 20\,H^+(aq) + 20\,e^-$

reduction: $O_2(g) + 4\,H^+(aq) + 4\,e^- \longrightarrow 2\,H_2O(l)$ $\quad \times 5$ $\quad E^\circ = 1.229$ V

net: $C_3H_8 + 5\,O_2(g) \longrightarrow 3\,CO_2(g) + 4\,H_2O(l)$ $\quad$ Note that $n = 20$

2. ΔG° is evaluated for the net reaction.

$\Delta G^\circ = 3\,\Delta G^\circ_f[CO_2(g)] + 4\,\Delta G^\circ_f[H_2O(l)] - \Delta G^\circ_f[C_3H_8(g)] - 5\,\Delta G^\circ_f[O_2(g)]$

$= 3 \times (-394.4 \text{ kJ}) + 4 \times (-237.2 \text{ kJ}) - (-23.56 \text{ kJ}) - 5 \times 0.00 \text{ kJ} = -2108 \text{ kJ}$

3. We now evaluate the standard cell potential from the standard free energy change.

$$\Delta G^\circ = -n\mathcal{F}E^\circ_{cell} \quad E^\circ_{cell} = \frac{-\Delta G^\circ}{n\mathcal{F}} = -\frac{-2108 \times 10^3 \text{ kJ}}{20 \text{ mol e}^- \times \dfrac{96{,}485 \text{ C}}{1 \text{ mol e}^-}} = +1.092 \text{ V}$$

4. And now determine the half-cell potential.

$E^\circ_{cell} = +1.092 \text{ V} = -E^\circ\{CO_2(g)|C_3H_8(g)\} + E^\circ\{O_2(g)|H_2O(l)\} = -E^\circ\{CO_2(g)|C_3H_8(g)\} + 1.229 \text{ V}$

$E^\circ\{CO_2(g)|C_3H_8(g)\} = -1.092 \text{ V} + 1.229 \text{ V} = +0.137 \text{ V}$

REVIEW QUESTIONS

1. (a) E° is the symbol for the standard cell potential, the voltage measured when no current is flowing and all cell reagents are in their standard states.
   (b) $\mathcal{F}$ is the symbol for Faraday's constant, the charge on one mole of electrons, 96,485 coulombs.
   (c) The anode is the electrode where oxidation occurs and toward which anions move.
   (d) The cathode is the electrode where reduction occurs and toward which cations move.

2. (a) A salt bridge is a tube filled with electrolyte that is used to join two half-cells in such a way that the electrochemical circuit is completed but the contents of the half-cells do not mix.
   (b) The standard hydrogen electrode is based on the reduction of hydrogen ion, at 1.000 M concentration, to hydrogen gas at 1.000 atm pressure. It is assigned a half-cell potential of 0.000 V.
   (c) Cathodic protection is achieved by electrically joining a more active metal to a less active one that is to be protected. The more active metal will be oxidized, protecting the less active one from corrosion.
   (d) A fuel cell is a voltaic cell in which a reaction that normally occurs as a combustion reaction serves as the cell reaction. Reactants are continually supplied and products removed from such a cell.

3. (a) A half-reaction is either the oxidation reaction or the reduction reaction. On the other hand, the net reaction is an oxidation-reduction reaction, the combination of two half-reactions.
   (b) In a voltaic or galvanic cell, a chemical change produces electricity. This type of cell has $E > 0$. In an electrolytic cell, the passage of electric current produces a chemical change. This type of cell has $E < 0$.
   (c) A primary battery is nonreversible, like a dry cell. A secondary battery can be recharged and reused.
   (d) E°_{cell} is the cell potential when all reactants and all products are in their standard states. E_{cell} is the cell potential when reactants and products are not necessarily in their standard states.

4. The correct statement is **(d)**. Electrons are produced at the anode and move toward the cathode, regardless of the electrode material. The electrons do not move through the salt bridge; ions do. Reduction occurs at the cathode in both galvanic and electrolytic cells—in all types of electrochemical cells.

5. (a)
   Oxidation: $Fe(s) \longrightarrow Fe^{2+}(aq) + 2\,e^-$
   Reduction: $Cu^{2+}(aq) + 2\,e^- \longrightarrow Cu(s)$
   Net: $Fe(s) + Cu^{2+}(aq) \longrightarrow Fe^{2+}(aq) + Cu(s)$

   (b)
   Oxidation: $2\,Br^-(aq) \longrightarrow Br_2(aq) + 2\,e^-$
   Reduction: $Cl_2(aq) + 2\,e^- \longrightarrow 2\,Cl^-(aq)$
   Net: $2\,Br^-(aq) + Cl_2(aq) \longrightarrow Br_2(aq) + 2\,Cl^-(aq)$

   (c)
   Oxidation: $Al(s) \longrightarrow Al^{3+}(aq) + 3\,e^-$
   Reduction: $\{Fe^{3+}(aq) + e^- \longrightarrow Fe^{2+}(aq)\} \quad \times 3$
   Net: $Al(s) + 3\,Fe^{3+}(aq) \longrightarrow Al^{3+}(aq) + 3\,Fe^{2+}(aq)$

   (d)
   Oxidation: $\{Cl^-(aq) + 3\,H_2O \longrightarrow ClO_3^-(aq) + 6\,H^+(aq) + 6\,e^-\} \quad \times 5$
   Reduction: $\{MnO_4^-(aq) + 8\,H^+(aq) + 5\,e^- \longrightarrow Mn^{2+}(aq) + 4\,H_2O\} \quad \times 6$
   Net: $5\,Cl^-(aq) + 6\,MnO_4^-(aq) + 18\,H^+(aq) \longrightarrow 5\,ClO_3^-(aq) + 6\,Mn^{2+}(aq) + 9\,H_2O$

   (e)
   Oxidation: $S^{2-}(aq) + 8\,OH^-(aq) \longrightarrow SO_4^{2-}(aq) + 4\,H_2O + 8\,e^-$
   Reduction: $\{O_2(g) + 2\,H_2O + 4\,e^- \longrightarrow 4\,OH^-(aq)\} \quad \times 2$

Net: $S^{2-}(aq) + 2\ O_2(g) \longrightarrow SO_4^{2-}(aq)$

6. Since $Zn^{2+}(aq)$ undergoes reduction, and the other metal (M) undergoes oxidation,

$E^\circ_{cell} = E^\circ_{Zn} - E^\circ_{M}$ or $E^\circ_{M} = E^\circ_{Zn} - E^\circ_{cell}$

(a) $E^\circ_{Mn} = E^\circ_{Zn} - E^\circ_{cell} = -0.763\ V - 0.417\ V = -1.180\ V$

(b) $E^\circ_{Po} = E^\circ_{Zn} - E^\circ_{cell} = -0.763\ V + 1.13\ V = +0.37\ V$

(c) $E^\circ_{Ti} = E^\circ_{Zn} - E^\circ_{cell} = -0.763\ V - 0.87\ V = -1.63\ V$

(d) $E^\circ_{V} = E^\circ_{Zn} - E^\circ_{cell} = -0.763\ V - 0.37\ V = -1.13\ V$

7. **(a)** Oxidation: $\{Al(s) \longrightarrow Al^{3+}(aq) + 3\ e^-\} \times 2$ $\qquad -E^\circ = +1.676\ V$

Reduction: $\{Sn^{2+}(aq) + 2\ e^- \longrightarrow Sn(s)\} \times 3$ $\qquad E^\circ = -0.137\ V$

Net: $2\ Al(s) + 3\ Sn^{2+}(aq) \longrightarrow 2\ Al^{3+}(aq) + 3\ Sn(s)$ $\qquad E^\circ_{cell} = +1.539\ V$

(b) Oxidation: $Fe^{2+}(aq) \longrightarrow Fe^{3+}(aq) + e^-$ $\qquad -E^\circ = -0.771\ V$

Reduction: $Ag^+(aq) + e^- \longrightarrow Ag(s)$ $\qquad E^\circ = +0.800\ V$

Net: $Fe^{2+}(aq) + Ag^+(aq) \longrightarrow Fe^{3+}(aq) + Ag(s)$ $\qquad E^\circ_{cell} = +0.029\ V$

(c) Oxidation: $Cu(s) \longrightarrow Cu^{2+}(aq) + 2\ e^-$ $\qquad -E^\circ = -0.337\ V$

Reduction: $Cl_2(g) + 2\ e^- \longrightarrow 2\ Cl^-(aq)$ $\qquad E^\circ = +1.358\ V$

Net: $Cu(s) + Cl_2(g) \longrightarrow Cu^{2+}(aq) + 2\ Cl^-(aq)$ $\qquad E^\circ_{cell} = +1.021\ V$

8. **(a)** Oxidation: $2\ Cl^-(aq) \longrightarrow Cl_2(g) + 2\ e^-$ $\qquad -E^\circ = -1.358\ V$

Reduction: $PbO_2(s) + 4\ H^+(aq) + 2\ e^- \longrightarrow Pb^{2+}(aq) + 2\ H_2O$ $\qquad E^\circ = +1.455\ V$

Net: $2\ Cl^-(aq) + PbO_2(s) + 4\ H^+(aq) \longrightarrow Cl_2(g) + Pb^{2+}(aq) + H_2O$ $\qquad E^\circ_{cell} = 0.097\ V$

(b) Oxidation: $\{Mg(s) \longrightarrow Mg^{2+}(aq) + 2\ e^-\} \times 3$ $\qquad -E^\circ = +2.356\ V$

Reduction: $\{Sc^{3+}(aq) + 3\ e^- \longrightarrow Sc(s)\} \times 2$ $\qquad E^\circ\{Sc^{3+}/Sc\}$

Net: $3\ Mg(s) + 2\ Sc^{3+}(aq) \longrightarrow 3\ Mg^{2+}(aq) + 2\ Sc(s)$ $\qquad E^\circ_{cell} = +0.33\ V$

$E^\circ\{Sc^{3+}/Sc\} = +0.33\ V - 2.356\ V = -2.03\ V$

(c) Oxidation: $Cu^+(aq) \longrightarrow Cu^{2+}(aq) + e^-$ $\qquad -E^\circ\{Cu^{2+}/Cu^+\}$

Reduction: $Ag^+(aq) + e^- \longrightarrow Ag(s)$ $\qquad E^\circ = +0.800\ V$

Net: $Cu^+(aq) + Ag^+(aq) \longrightarrow Cu^{2+}(aq) + Ag(s)$ $\qquad E^\circ_{cell} = +0.641\ V$

$-E^\circ\{Cu^{2+}/Cu^+\} = +0.641 - 0.800\ V = -0.159\ V$ $\qquad E^\circ\{Cu^{2+}/Cu^+\} = +0.159\ V$

9. **(a)** Oxidation: $Sn(s) \longrightarrow Sn^{2+}(aq) + 2\ e^-$ $\qquad -E^\circ = +0.137\ V$

Reduction: $Pb^{2+}(aq) + 2\ e^- \longrightarrow Pb(s)$ $\qquad E^\circ = -0.125\ V$

Net: $Sn(s) + Pb^{2+}(aq) \longrightarrow Sn^{2+}(aq) + Pb(s)$ $\qquad E^\circ_{cell} = +0.012\ V$ $\qquad$ Spontaneous

(b) Oxidation: $2\ I^-(aq) \longrightarrow I_2(s) + 2\ e^-$ $\qquad -E^\circ = -0.535\ V$

Reduction: $Cu^{2+}(aq) + 2\ e^- \longrightarrow Cu(s)$ $\qquad E^\circ = +0.337\ V$

Net: $2\ I^-(aq) + Cu^{2+}(aq) \longrightarrow Cu(s) + I_2(s)$ $\qquad E^\circ_{cell} = -0.198\ V$ $\qquad$ Nonspontaneous

(c) Oxidation: $\{2\ H_2O \longrightarrow O_2(g) + 4\ H^+(aq) + 4\ e^-\} \times 3$ $\qquad -E^\circ = -1.229\ V$

Reduction: $\{NO_3^- + 4\ H^+(aq) + 3\ e^- \longrightarrow NO(g) + 2\ H_2O\} \times 4$ $\qquad E^\circ = +0.956\ V$

Net: $4\ NO_3^-(aq) + 4\ H^+(aq) \longrightarrow 3\ O_2(g) + 4\ NO(g) + 2\ H_2O$ $\qquad E^\circ_{cell} = -0.273\ V$

This is a nonspontaneous cell reaction.

(d) Oxidation: $Cl^-(aq) + 2\ OH^-(aq) \longrightarrow OCl^-(aq) + H_2O + 2\ e^-$ $\qquad -E^\circ = -0.890\ V$

Reduction: $O_3(g) + H_2O + 2\ e^- \longrightarrow O_2(g) + 2\ OH^-(aq)$ $\qquad E^\circ = +1.246\ V$

Net: $Cl^-(aq) + O_3(g) \longrightarrow OCl^-(aq) + O_2(g)$ $\qquad E^\circ_{cell} = +0.356\ V$ $\qquad$ Spontaneous

10. Hg(l) is more difficult to oxidize to Hg_2^{2+} (–0.797 V) than H^+ is to reduce to H_2 (0.000 V); Hg(l) will not dissolve in 1 M HCl. The reduction of nitrate ion to NO(g) in acidic solution is quite spontaneous (+0.956 V). This can overcome the reluctance of Hg to be oxidized. Hg(l) will react with and dissolve in $HNO_3(aq)$.

11. **(a)** Oxidation: $Mg(s) \longrightarrow Mg^{2+}(aq) + 2\,e^-$ $-E° = +2.356$ V
Reduction: $Pb^{2+}(aq) + 2\,e^- \longrightarrow Pb(s)$ $E° = -0.125$ V
Net: $Mg(s) + Pb^{2+}(aq) \longrightarrow Mg^{2+}(aq) + Pb(s)$ $E°_{cell} = +2.231$ V This reaction occurs to a significant extent.

(b) Oxidation: $Sn(s) \longrightarrow Sn^{2+}(aq) + 2\,e^-$ $-E° = +0.137$ V
Reduction: $2\,H^+(aq) \longrightarrow H_2(g)$ $E° = 0.000$ V
Net: $Sn(s) + 2\,H^+(aq) \longrightarrow Sn^{2+}(aq) + H_2(g)$ $E°_{cell} = +0.137$ V
This reaction will occur to a significant extent.

(c) Oxidation: $Sn^{2+}(aq) \longrightarrow Sn^{4+}(aq) + 2\,e^-$ $-E° = -0.154$ V
Reduction: $SO_4^{2-}(aq) + 4\,H^+(aq) + 2\,e^- \longrightarrow SO_2(g) + 2\,H_2O$ $E° = +0.17$ V
Net: $Sn^{2+}(aq) + SO_4^{2-}(aq) + 4\,H^+(aq) \longrightarrow Sn^{4+}(aq) + SO_2(g) + 2\,H_2O$ $E°_{cell} = +0.02$ V
This reaction will barely occur, but not to a large extent.

(d) Oxidation: $\{H_2O_2(aq) \longrightarrow O_2(g) + 2\,H^+(aq) + 2\,e^-\}$ $\times 5$ $-E° = -0.695$ V
Reduction: $\{MnO_4^-(aq) + 8\,H^+(aq) + 5\,e^- \longrightarrow Mn^{2+}(aq) + 4\,H_2O\}$ $\times 2$ $E° = +1.51$ V
Net: $5\,H_2O_2(aq) + 2\,MnO_4^-(aq) + 6\,H^+(aq) \longrightarrow 5\,O_2(g) + 2\,Mn^{2+}(aq) + 8\,H_2O$ $E°_{cell} = +0.82$ V
This reaction will occur to a significant extent.

(e) Oxidation: $2\,Br^-(aq) \longrightarrow Br_2(aq) + 2\,e^-$ $-E° = -1.065$ V
Reduction: $I_2(s) + 2\,e^- \longrightarrow 2\,I^-(aq)$ $E° = +0.535$ V
Net: $2\,Cl^-(aq) + I_2(s) \longrightarrow Cl_2(aq) + 2\,I^-(aq)$ $E°_{cell} = -0.530$ V
This reaction will not occur to a significant extent.

12. The relatively small value of $E°_{cell}$ for a reaction indicates that the reaction will proceed in the forward reaction, but will stop short of completion. A much larger value of $E°_{cell}$ would be necessary before we would conclude that the reaction goes to completion. For example, we can compute the value of the equilibrium constant for this reaction. A value of 1000 or more is needed for a reaction that we would describe as going to completion.

$$E°_{cell} = \frac{0.0257}{n} \ln K_{eq} \qquad \ln K_{eq} = \frac{n \times E°_{cell}}{0.0257} = \frac{2 \times 0.02}{0.0257} = 2 \qquad K_{eq} = e^2 = 7$$

13. If $E°_{cell}$ is positive, the reaction will occur. For the reduction of dicromate ion to $Cr^{3+}(aq)$ we have the following. $Cr_2O_7^{2-}(aq) + 14\,H^+(aq) + 6\,e^- \longrightarrow 2\,Cr^{3+}(aq) + 7\,H_2O$ $E° = +1.33$ V
Thus, if an oxidation has $-E°$ that is smaller (more negative) than –1.33 V, the oxidation will not occur.

(a) $Sn^{2+}(aq) \longrightarrow Sn^{4+}(aq) + 2\,e^-$ $-E° = -0.154$ V
$Sn^{2+}(aq)$ can be oxidized to $Sn^{4+}(aq)$ by $Cr_2O_7^{2-}(aq)$.

(b) $I_2(s) + 6\,H_2O \longrightarrow 2\,IO_3^-(aq) + 12\,H^+(aq) + 10\,e^-$ $-E° = -1.20$ V
$I_2(s)$ can be oxidized to $IO_3^-(aq)$ by $Cr_2O_7^{2-}(aq)$.

(c) $Mn^{2+}(aq) + 4\,H_2O \longrightarrow MnO_4^-(aq) + 8\,H^+(aq) + 5\,e^-$ $-E° = -1.51$ V
$Mn^{2+}(aq)$ cannot be oxidized to $MnO_4^-(aq)$ by $Cr_2O_7^{2-}(aq)$.

14. **(a)** Oxidation: $\{Al(s) \longrightarrow Al^{3+}(aq) + 3\,e^-\}$ $\times 2$ $-E° = +1.676$ V
Reduction: $\{Cu^{2+}(aq) + 2\,e^- \longrightarrow Cu(s)\}$ $\times 3$ $E° = +0.337$ V
Net: $2\,Al(s) + 3\,Cu^{2+}(aq) \longrightarrow 2\,Al^{3+}(aq) + 3\,Cu(s)$ $E°_{cell} = +2.013$ V
$\Delta G° = -n\,\mathcal{F} E°_{cell} = -$ (6 mol elns)(96,485 C/mol elns)(2.013 V) $= -1.165 \times 10^6$ J $= -1.165 \times 10^3$ kJ

(b) Oxidation: $\{2\,I^-(aq) \longrightarrow I_2(s) + 2\,e^-\} \times 2$ $-E° = -0.535$ V
Reduction: $O_2(g) + 4\,H^+(aq) + 4\,e^- \longrightarrow 2\,H_2O$ $E° = +1.229$ V
Net: $4\,I^-(aq) + O_2(g) + 4\,H^+(aq) \longrightarrow 2\,I_2(s) + 2\,H_2O$ $E°_{cell} = +0.694$ V

$\Delta G° = -n\,\mathcal{F}E°_{cell} = -(4 \text{ mol elns})(96{,}485 \text{ C/mol elns})(0.694 \text{ V}) = -2.68 \times 10^5 \text{ J} = -268 \text{ kJ}$

(c) Oxidation: $\{Ag(s) \longrightarrow Ag^+(aq) + e^-\} \times 6$ $\quad -E° = -0.800$ V

Reduction: $Cr_2O_7^{2-}(aq) + 14\,H^+(aq) + 6\,e^- \longrightarrow 2\,Cr^{3+}(aq) + 7\,H_2O$ $\quad E° = +1.33$ V

Net: $6\,Ag(s) + Cr_2O_7^{2-}(aq) + 14\,H^+(aq) \longrightarrow 6\,Ag^+(aq) + 2\,Cr^{3+}(aq) + 7\,H_2O$

$E°_{cell} = -0.800 \text{ V} + 1.33 \text{ V} = +0.53 \text{ V}$

$\Delta G° = -n\,\mathcal{F}E°_{cell} = -(6 \text{ mol elns})(96{,}485 \text{ C/mol elns})(0.53 \text{ V}) = -3.1 \times 10^5 \text{ J} = -3.1 \times 10^2 \text{ kJ}$

15. $\Delta G° = -n\,\mathcal{F}E°_{cell} = -RT \ln K \qquad \ln K = \dfrac{n\,\mathcal{F}E°_{cell}}{R\,T}$ This becomes $\ln K = \dfrac{n}{0.0257} E°_{cell}$

(a) Oxidation: $\{Ag(s) \longrightarrow Ag^+(aq) + e^-\} \times 2$ $\quad -E° = -0.800$ V

Reduction: $Sn^{4+}(aq) + e^- \longrightarrow Sn^{2+}(aq)$ $\quad E° = +0.154$ V

Net: $2\,Ag(s) + Sn^{4+}(aq) \longrightarrow 2\,Ag^+(aq) + Sn^{2+}(aq)$ $\quad E°_{cell} = -0.646$ V

$$\ln K_{eq} = \frac{n}{0.0257} E°_{cell} = \frac{2 \text{ mol e}^- \times (-0.646 \text{ V})}{0.0257} = -50.3; \quad K_{eq} = e^{-50.3} = 1 \times 10^{-22} = \frac{[Sn^{2+}][Ag^+]^2}{[Sn^{4+}]}$$

(b) Oxidation: $2\,Cl^-(aq) \longrightarrow Cl_2(g) + 2\,e^-$ $\quad -E° = -1.358$ V

Reduction: $MnO_2(s) + 4\,H^+(aq) + 2\,e^- \longrightarrow Mn^{2+}(aq) + 2\,H_2O$ $\quad E° = +1.23$ V

Net: $2\,Cl^-(aq) + MnO_2(s) + 4\,H^+(aq) \longrightarrow Mn^{2+}(aq) + Cl_2(g) + 2\,H_2O$ $\quad E°_{cell} = -0.13$ V

$$\ln K_{eq} = \frac{2 \text{ mol elns} \times (-0.13 \text{ V})}{0.0257} = -10._1 \qquad K_{eq} = e^{-10.1} = 4 \times 10^{-5} = \frac{[Mn^{2+}]\,P\{Cl_2(g)\}}{[Cl^-]^2[H^+]^4}$$

(c) Oxidation: $4\,OH^-(aq) \longrightarrow O_2(g) + 2\,H_2O + 4\,e^-$ $\quad -E° = -0.401$ V

Reduction: $\{OCl^-(aq) + H_2O + e^- \longrightarrow Cl^-(aq) + 2\,OH^-\} \times 2$ $\quad E° = +0.890$ V

Net: $2\,OCl^-(aq) \longrightarrow 2\,Cl^-(aq) + O_2(g)$ $\quad E°_{cell} = +0.489$ V

$$\ln K_{eq} = \frac{4 \text{ mol e}^- \times (0.489 \text{ V})}{0.0257} = 76.1 \qquad K_{eq} = e^{76.1} = 1 \times 10^{33} = \frac{[Cl^-]^2\,P\{O_2(g)\}}{[OCl^-]^2}$$

16. (a) Oxidation: $Fe(s) \longrightarrow Fe^{2+}(aq) + 2\,e^-$ $\quad -E° = +0.440$ V

Reduction: $Cu^{2+}(aq) + 2\,e^- \longrightarrow Cu(s)$ $\quad E° = +0.337$ V

Net: $Fe(s) + Cu^{2+}(aq) \longrightarrow Fe^{2+}(aq) + Cu(s)$ $\quad E°_{cell} = +0.777$ V

Electrons flow from the Fe electrode (electrode B) to the Cu electrode (electrode A) with a potential of +0.777 V.

(b) Oxidation: $Sn^{2+}(aq) \longrightarrow Sn^{4+}(aq) + 2\,e^-$ $\quad -E° = -0.154$ V

Reduction: $\{Ag^+(aq) + e^- \longrightarrow Ag(s)\} \times 2$ $\quad E° = +0.800$ V

Net: $Sn^{2+}(aq) + 2\,Ag^+(aq) \longrightarrow Sn^{4+}(aq) + 2\,Ag(s)$ $\quad E°_{cell} = +0.646$ V

Electrons flow from the Pt electrode (electrode A) to the Ag electrode (electrode B) with a potential of +0.646 V.

(c) Oxidation: $Zn(s) \longrightarrow Zn^{2+}(aq) + 2\,e^-$ $\quad -E° = +0.763$ V

Reduction: $Fe^{2+}(aq) + 2\,e^- \longrightarrow Fe(s)$ $\quad E° = -0.440$ V

Net: $Zn(s) + Fe^{2+}(aq) \longrightarrow Zn^{2+}(aq) + Fe(s)$ $\quad E°_{cell} = +0.323$ V

$$E = E°_{cell} - \frac{0.0592}{n} \log \frac{[Zn^{2+}]}{[Fe^{2+}]} = +0.323 \text{ V} - \frac{0.0592}{2} \log \frac{0.10 \text{ M}}{1.0 \times 10^{-3} \text{ M}}$$

$= +0.323 \text{ V} - 0.0592 \text{ V} = +0.264\text{V}$

Electrons flow from the Zn electrode (electrode A) to the Fe electrode (electrode B) with a potential of +0.264 V.

17. Oxidation: $Zn(s) \longrightarrow Zn^{2+}(aq) + 2\,e^-$ $\quad -E° = +0.763$ V

Reduction: $\{Ag^+(aq) + e^- \longrightarrow Ag(s)\} \times 2$ $\quad E° = +0.800$ V

Net: $Zn(s) + 2\,Ag^+(aq) \longrightarrow Zn^{2+}(aq) + 2\,Ag(s)$ $\quad E°_{cell} = +1.563$ V

$$E = E^\circ_{cell} - \frac{0.0592}{n}\log\frac{[Zn^{2+}]}{[Ag^+]^2} = +1.563\text{ V} - \frac{0.0592}{2}\log\frac{1.00\text{ M}}{x^2} = +1.250\text{ V}$$

$$\log\frac{1.00\text{ M}}{x^2} = \frac{-2\times(1.250-1.563)}{0.0592} = 10.6 \qquad x = \sqrt{2.5\times10^{-11}} = 5.0\times10^{-6}\text{ M}$$

18. In each case, we employ the equation $E_{cell} = 0.0592$ pH.

(a) $E_{cell} = 0.0592\text{ pH} = 0.0592 \times 5.25 = 0.311\text{ V}$

(b) $\text{pH} = -\log(0.0103) = 1.987$ $\qquad E_{cell} = 0.0592\text{ pH} = 0.0592\times1.987 = 0.118\text{ V}$

(c) $K_a = \frac{[H^+][C_2H_3O_2^-]}{[HC_2H_3O_2]} = 1.8\times10^{-5} = \frac{x^2}{0.158-x} \approx \frac{x^2}{0.158}$

$x = \sqrt{0.158\times1.8\times10^{-5}} = 1.7\times10^{-3}\text{ M}$ $\qquad \text{pH} = -\log(1.7\times10^{-3}) = 2.77$

$E_{cell} = 0.0592\text{ pH} = 0.0592\times2.77 = 0.164\text{ V}$

19. We predict the possible products at the anode and at the cathode. Then we choose the oxidation and the reduction which, when combined, yield the least negative cell potential.

(a) Possible products (of oxidation) at the anode are the following.

$2\,H_2O \longrightarrow O_2(g) + 4\,H^+(aq) + 4\,e^-$ $\quad -1.229$ V

$2\,Cl^-(aq) \longrightarrow Cl_2(g) + 2\,e^-$ $\quad -1.358$ V

Possible products (of reduction) at the cathode are the following.

$2\,H^+(aq) + 2\,e^- \longrightarrow H_2(g)$ $\quad 0.000$ V

$Cu^{2+}(aq) + 2\,e^- \longrightarrow Cu(s)$ $\quad +0.337$ V

Because of the high overpotential for the production of $O_2(g)$, the products of electrolysis of $CuCl_2(aq)$ will be $Cl_2(g)$ at the anode and Cu(s) at the cathode.

(b) Possible products (of oxidation) at the anode are the following.

$2\,H_2O \longrightarrow O_2(g) + 4\,H^+(aq) + 4\,e^-$ $\quad -1.229$ V

$2\,SO_4^{2-}(aq) \longrightarrow S_2O_8^{2-}(aq) + 2\,e^-$ $\quad -2.01$ V

Possible products (of reduction) at the cathode are the following.

$2\,H^+(aq) + 2\,e^- \longrightarrow H_2(g)$ $\quad 0.000$ V

$Na^+(aq) + e^- \longrightarrow Na(s)$ $\quad -2.713$ V

The products of electrolysis of $Na_2SO_4(aq)$ will be $O_2(g)$ at the anode and $H_2(g)$ at the cathode.

(c) The only possible product (of oxidation) at the anode is the following.

$2\,Cl^-(l) \longrightarrow Cl_2(g) + 2\,e^-$

The only possible product (of reduction) at the cathode is the following.

$Ba^{2+}(l) + 2\,e^- \longrightarrow Ba(l)$

The products of electrolysis of $BaCl_2(l)$ will be $Cl_2(g)$ at the anode and Ba(l) at the cathode.

(d) The only possible product (of oxidation) at the anode is the following.

$4\,OH^- \longrightarrow O_2(g) + 2\,H_2O + 4\,e^-$ $\quad -0.401$ V

Possible products (of reduction) at the cathode are the following.

$2\,H_2O + 2\,e^- \longrightarrow H_2(g) + 2\,OH^-(aq)$ $\quad -0.828$ V

$K^+(aq) + e^- \longrightarrow K(s)$ $\quad -2.924$ V

The products of electrolysis of KOH(aq) will be $O_2(g)$ at the anode and $H_2(g)$ at the cathode.

20. When $MgCl_2(l)$ is electrolyzed, Mg(l) is produced at the cathode, with a half-cell voltage of -2.356 V. On the other hand, when $MgCl_2(aq)$ is electrolyzed, $H_2(g)$ is produced at the cathode, with a standard half-cell voltage of 0.000 V. If we wish to produce elemental magnesium, we must electrolyze molten $MgCl_2$, not its aqueous solution.

21. We calculate the total amount of charge passed and the number of moles of electrons.

$$\text{amount } e^- = 75\text{ min}\times\frac{60\text{ s}}{1\text{ min}}\times\frac{2.15\text{ C}}{1\text{ s}}\times\frac{1\text{ mol }e^-}{96485\text{ C}} = 0.10\text{ mol }e^-$$

(a) $\text{mass Zn} = 0.10\text{ mol }e^-\times\frac{1\text{ mol }Zn^{2+}}{2\text{ mol }e^-}\times\frac{1\text{ mol Zn}}{1\text{ mol }Zn^{2+}}\times\frac{65.39\text{ g Zn}}{1\text{ mol Zn}} = 3.3\text{ g Zn}$

(b) $\text{mass Al} = 0.10\text{ mol }e^-\times\frac{1\text{ mol }Al^{3+}}{3\text{ mol }e^-}\times\frac{1\text{ mol Al}}{1\text{ mol }Al^{3+}}\times\frac{26.98\text{ g Al}}{1\text{ mol Al}} = 0.90\text{ g Al}$

(c) $\text{mass Ag} = 0.10 \text{ mol e}^- \times \frac{1 \text{ mol Ag}^+}{1 \text{ mol e}^-} \times \frac{1 \text{ mol Ag}}{1 \text{ mol Ag}^+} \times \frac{107.9 \text{ g Ag}}{1 \text{ mol Ag}} = 11 \text{ g Ag}$

(d) $\text{mass Ni} = 0.10 \text{ mol e}^- \times \frac{1 \text{ mol Ni}^{2+}}{2 \text{ mol e}^-} \times \frac{1 \text{ mol Ni}}{1 \text{ mol Ni}^{2+}} \times \frac{58.69 \text{ g Ni}}{1 \text{ mol Ni}} = 2.9 \text{ g Ni}$

22. The two half reactions follow: $Cu^{2+}(aq) + 2\,e^- \longrightarrow Cu(s)$ and $2\,H^+(aq) + 2\,e^- \longrightarrow H_2(g)$ Thus, two moles of electrons are needed to produce each mole of Cu(s) and two moles of electrons are needed to produce each mole of $H_2(g)$. With this information, we compute the amount of $H_2(g)$ that will be produced.

$$\text{amount } H_2(g) = 3.28 \text{ g Cu} \times \frac{1 \text{ mol Cu}}{63.55 \text{ g Cu}} \times \frac{2 \text{ mol e}^-}{1 \text{ mol Cu}} \times \frac{1 \text{ mol } H_2(g)}{2 \text{ mol e}^-} = 0.0516 \text{ mol } H_2(g)$$

Then we use the ideal gas equation to find the volume of $H_2(g)$.

$$\text{volume of } H_2(g) = \frac{0.0516 \text{ mol } H_2 \times \frac{0.08206 \text{ L atm}}{\text{mol K}} \times (273.2 + 28.2) \text{ K}}{763 \text{ mmHg} \times \frac{1 \text{ atm}}{760 \text{ mmHg}}} = 1.27 \text{ L}$$

This assumes the $H_2(g)$ is not collected over water; we do not have to subtract the vapor pressure of water.

EXERCISES

23. **(a)** If the metal dissolves in HNO_3, it has a reduction potential that is smaller than $E°[NO_3^-(aq)/NO(g)] = 0.956$ V. If it also does not dissolve in HCl, it has a reduction potential that is larger than $E°[H^+(aq)/H_2(g)] = 0.000$ V. If it displaces $Ag^+(aq)$ from solution, then it has a reduction potential that is smaller than $E°[Ag^+(aq)/Ag(s)] = 0.800$ V. But if it does not displace $Cu^{2+}(aq)$ from solution, then its reduction potential is larger than $E°[Cu^{2+}(aq)/Cu(s)] = 0.337$ V $\quad 0.337 \text{ V} < E° < 0.800 \text{ V}$

(b) If the metal dissolves in HCl, it has a reduction potential that is smaller than $E°[H^+(aq)/H_2(g)] = 0.000$ V. If it does not displace $Zn^{2+}(aq)$ from solution, its reduction potential is larger than $E°[Zn^{2+}(aq)/Zn(s)] = -0.763$ V. If it also does not displace $Fe^{2+}(aq)$ from solution, its reduction potential is larger than $E°[Fe^{2+}(aq)/Fe(s)] = -0.440$ V. $\quad -0.440 \text{ V} < E° < 0.000 \text{ V}$

24. We place a strip of solid indium in each of the metal ion solutions and see if that metal plates out on the indium strip. Eventually, we find a pair of metals which, first, are adjacent to each other in Table 21-1 and, second, for one of which indium displaces that metal from solution and for the other no such displacement occurs. (In fact, for the second metal, the solid metal displaces indium metal from a solution of In^{3+}.) The standard electrode potential lies between the standard electrode potentials of these two metals. This technique will work only if indium metal does not react with water, that is, if the standard reduction potential of $In^{3+}(aq)/In(s)$ is greater than about -1.8 V. The inaccuracy inherent in this technique is due to overpotentials, which can be as much as 0.200 V. Its imprecision is limited by the closeness of the reduction potentials of the other two metals. Probably the metals furthest apart are U and Mg, which are separated by 0.7 V.

25. We separate the given equation into its two half-equations. One of them is the reduction of nitrate ion in acidic solution, whose standard half-cell potential we retrieve from Table 21-1 and use to solve the problem.

Oxidation: $\{Pt(s) + 4\,Cl^-(aq) \longrightarrow [PtCl_4]^{2-}(aq) + 2\,e^-\}$ $\times 3$ $\quad -E°\{Pt(s)|[PtCl_4]^{2-}(aq)]\}$

Reduction: $\{NO_3^-(aq) + 4\,H^+(aq) + 3\,e^- \longrightarrow NO(g) + 2\,H_2O\}$ $\times 2$ $\quad E° = +0.956$ V

Net: $3\,Pt(s) + 2\,NO_3^-(aq) + 8\,H^+(aq) + 12\,Cl^-(aq) \longrightarrow 3\,[PtCl_4]^{2-}(aq) + 2\,NO(g) + 6\,H_2O(l)$

$E°_{cell} = 0.201 \text{ V} = +0.956 \text{ V} - E°\{Pt(s)|[PtCl_4]^{2-}(aq)]\}$

$E°\{Pt(s)|[PtCl_4]^{2-}(aq)]\} = 0.956 \text{ V} - 0.201 \text{ V} = +0.755 \text{ V}$

26. We separate the given equation into its two half-equations. One of them is the reduction of $Cl_2(g)$ to $Cl^-(aq)$ whose standard half-cell potential we obtain from Table 21-1 and use to solve the problem.

Oxidation: $Na(\text{in Hg}) \longrightarrow Na^+(aq) + e^-$ $\times 2$ $\quad -E°[Na^+(aq)|Na(\text{in Hg})]$

Reduction: $Cl_2(g) + 2\,e^- \longrightarrow 2\,Cl^-(aq)$ $\quad E° = +1.358$ V

Net: $2\,Na(\text{in Hg}) + Cl_2(g) \longrightarrow 2\,Na^+(aq) + 2\,Cl^-(aq)$ $\quad E°_{cell} = 3.20$ V

$E°_{cell} = 3.20 \text{ V} = +1.358 \text{ V} - E°[Na^+(aq)|Na(\text{in Hg})]$

$E°[Na^+(aq)|Na(\text{in Hg})] = 1.358 \text{ V} - 3.20 \text{ V} = -1.84 \text{ V}$

27. We divide the net cell equation into two half-equations.

Oxidation: $Al(s) + 4\ OH^-(aq) \longrightarrow [Al(OH)_4]^-(aq) + 3\ e^-$ $\times 4$ $-E^\circ\{Al(s)|[Al(OH)_4]^-(aq)\}$

Reduction: $O_2(g) + 2\ H_2O + 4\ e^- \longrightarrow 4\ OH^-(aq)$ $\times 3$ $E^\circ = +0.401$ V

Net: $4\ Al(s) + 3\ O_2(g) + 6\ H_2O + 4\ OH^-(aq) \longrightarrow 4\ [Al(OH)_4]^-(aq)$ $E^\circ_{cell} = 2.71$ V

$E^\circ_{cell} = 2.71\ V = +0.401\ V - E^\circ\{Al(s)|[Al(OH)_4]^-(aq)\}$

$E^\circ\{Al(s)|[Al(OH)_4]^-(aq)\} = 0.401\ V - 2.71\ V = -2.31\ V$

28. We divide the net cell equation into two half-equations.

Oxidation: $CH_4(g) + 2\ H_2O \longrightarrow CO_2(g) + 8\ H^+(aq) + 8\ e^-$ $-E^\circ[CH_4(g)|CO_2(g)]$

Reduction: $O_2(g) + 4\ H^+(aq) + 4\ e^- \longrightarrow 2\ H_2O$ $\times 2$ $E^\circ = +1.229$ V

Net: $CH_4(g) + 2\ O_2(g) \longrightarrow CO_2(g) + 2\ H_2O$ $E^\circ_{cell} = 1.06$ V

$E^\circ_{cell} = 1.06\ V = +1.229\ V - E^\circ[CH_4(g)|CO_2(g)]$ $E^\circ[CH_4(g)|CO_2(g)] = 1.229\ V - 1.06\ V = +0.17\ V$

Predicting Oxidation-Reduction Reactions

29. Na(s) does not displace $Zn^{2+}(aq)$ from aqueous solution; solid sodium reacts with the solvent instead.

$2\ Na(s) + Zn^{2+}(aq) \longrightarrow 2\ Na^+(aq) + Zn(s)$ $E^\circ_{cell} = -(-2.713\ V) + (-0.763\ V) = +1.950\ V$

$2\ Na(s) + H_2O \longrightarrow 2\ Na^+(aq) + H_2(g) + 2\ OH^-(aq)$ $E^\circ_{cell} = -(-2.713\ V) + (-0.828\ V) = +1.885\ V$

In contrast, aluminum metal reacts with $Zn^{2+}(aq)$ to dispace it from solution.

$2\ Al(s) + 3\ H_2O \longrightarrow 2\ Al^{3+}(aq) + 3\ H_2(g) + 6\ OH^-(aq)$ $E^\circ_{cell} = -(-1.676\ V) + (-0.828\ V) = +0.840\ V$

$2\ Al(s) + 3\ Zn^{2+}(aq) \longrightarrow 2\ Al^{3+}(aq) + 3\ Zn(s)$ $E^\circ_{cell} = -(-1.676\ V) + (-0.763\ V) = +0.913\ V$

The standard cell potentials do not tell the whole story, for we might think that the reaction with the more positive value of E°_{cell} would occur. In truth, Al(s) is coated with a thin layer of tightly adhering $Al_2O_3(s)$, which protects the metal from attack by water. Na(s) has no such protective coating.

30. The reaction that occurs is $Zn(s) \longrightarrow Zn^{2+}(aq) + 2\ e^-$ Some of the electrons produced in this reaction move to the copper surface, where the following reduction occurs. $2\ H^+(aq) + 2\ e^- \longrightarrow H_2(g)$. The copper does not react with the HCl(aq) even under these circumstances.The copper surface is one on which bubbles of hydrogen form more readily than they do on a zinc metal surface. We say that the copper surface has a lower overpotential. Not as high a voltage is required to produce $H_2(g)$ because of the arrangement of atoms on the copper surface.

31. **(a)** Oxidation: $\{Ag(s) \longrightarrow Ag^+(aq) + e^-\} \times 3$ $-E^\circ = -0.800$ V

Reduction: $NO_3^-(aq) + 4\ H^+(aq) + 3\ e^- \longrightarrow NO(g) + 2\ H_2O$ $E^\circ = +0.956$ V

Net: $3\ Ag(s) + NO_3^-(aq) + 4\ H^+(aq) \longrightarrow 3\ Ag^+(aq) + NO(g) + 2\ H_2O$ $E^\circ_{cell} = +0.16$ V

Ag(s) reacts with $HNO_3(aq)$ to form a solution of $AgNO_3(aq)$.

(b) Oxidation: $Zn(s) \longrightarrow Zn^{2+}(aq) + 2\ e^-$ $-E^\circ = +0.763$ V

Reduction: $2\ H^+(aq) + 2\ e^- \longrightarrow H_2(g)$ $E^\circ = 0.000$ V

Net: $Zn(s) + 2\ H^+(aq) \longrightarrow Zn^{2+}(aq) + H_2(g)$ $E^\circ_{cell} = +0.763$ V

Zn(s) reacts with HCl(aq) to form a solution of $ZnCl_2(aq)$.

(c) Oxidation: $Au(s) \longrightarrow Au^{3+}(aq) + 3\ e^-$ $-E^\circ = -1.52$ V

Reduction: $NO_3^-(aq) + 4\ H^+(aq) + 3\ e^- \longrightarrow NO(g) + 2\ H_2O$ $E^\circ = +0.956$ V

Net: $Au(s) + NO_3^-(aq) + 4\ H^+(aq) \longrightarrow Au^{3+}(aq) + NO(g) + 2\ H_2O$ $E^\circ_{cell} = -0.52$ V

Au(s) does not react with 1.00 M $HNO_3(aq)$.

32. In each case, we determine whether E°_{cell} is greater than zero; if so, the reaction will occur.

(a) Oxidation: $Zn(s) \longrightarrow Zn^{2+}(aq) + 2\ e^-$ $-E^\circ = +0.763$ V

Reduction: $Fe^{2+}(aq) + 3\ e^- \longrightarrow Fe(s)$ $E^\circ = -0.440$ V

Net: $Zn(s) + Fe^{2+}(aq) \longrightarrow Zn^{2+}(aq) + Fe(s)$ $E^\circ_{cell} = +0.323$ V

This reaction occurs as written.

(b) Oxidation: $\{2\ Cl^-(aq) \longrightarrow Cl_2(g) + 2\ e^-\}$ $\times 5$ $-E° = -1.358$ V

Reduction: $\{MnO_4^-(aq) + 8\ H^+(aq) + 5\ e^- \longrightarrow Mn^{2+}(aq) + 4\ H_2O\} \times 2$ $E° = 1.51$ V

Net: $10\ Cl^-(aq) + 2\ MnO_4^-(aq) + 16\ H^+(aq) \longrightarrow 5\ Cl_2(g) + 2\ Mn^{2+}(aq) + 8\ H_2O$ $E°_{cell} = +0.15$ V

This reaction occurs as written.

(c) Oxidation: $\{Ag(s) \longrightarrow Ag^+(aq) + e^-\}$ $\times 2$ $-E° = -0.800$ V

Reduction: $2\ H^+(aq) + 2\ e^- \longrightarrow H_2(g)$ $E° = +0.000$ V

Net: $2\ Ag(s) + 2\ H^+(aq) \longrightarrow 2\ Ag^+(aq) + H_2(g)$ $E°_{cell} = -0.800$ V

This reaction will not occur as written; Ag(s) does not react with HCl(aq).

(d) Oxidation: $2\ Cl^-(aq) \longrightarrow Cl_2(g) + 2\ e^-$ $\times 2$ $-E° = -1.358$ V

Reduction: $O_2(g) + 4\ H^+(aq) + 4\ e^- \longrightarrow 2\ H_2O$ $E° = +1.229$ V

Net: $4\ Cl^-(aq) + 4\ H^+(aq) + O_2(g) \longrightarrow 2\ Cl_2(g) + 2\ H_2O$ $E°_{cell} = -0.129$ V

This reaction will not occur as written.

Voltaic Cells

33. **(a)** Oxidation: $Cu(s) \longrightarrow Cu^{2+}(aq) + 2\ e^-$ $-E° = -0.337$ V

Reduction: $\{Fe^{3+}(aq) + e^- \longrightarrow Fe^{2+}(aq)\}$ $\times 2$ $E° = +0.771$ V

Net: $Cu(s) + 2\ Fe^{3+}(aq) \longrightarrow Cu^{2+}(aq) + 2\ Fe^{2+}(aq)$ $E°_{cell} = +0.434$ V

(b) Oxidation: $\{Al(s) \longrightarrow Al^{3+}(aq) + 3\ e^-\}$ $\times 2$ $-E° = +1.676$ V

Reduction: $\{Pb^{2+}(aq) + 2\ e^- \longrightarrow Pb(s)\}$ $\times 3$ $E° = -0.125$ V

Net: $2\ Al(s) + 3\ Pb^{2+}(aq) \longrightarrow 2\ Al^{3+}(aq) + 3\ Pb(s)$ $E°_{cell} = +1.551$ V

(c) Oxidation: $2\ H_2O \longrightarrow O_2(g) + 4\ H^+(aq) + 4\ e^-$ $-E° = -1.229$ V

Reduction: $\{Cl_2(g) + 2\ e^- \longrightarrow 2\ Cl^-(aq)\}$ $\times 2$ $E° = +1.358$ V

Net: $2\ H_2O + 2\ Cl_2(g) \longrightarrow O_2(g) + 4\ H^+(aq) + 4\ Cl^-(aq)$ $E°_{cell} = +0.129$ V

The cells for parts (a) and (c) follow. The anode is on the left in each case.

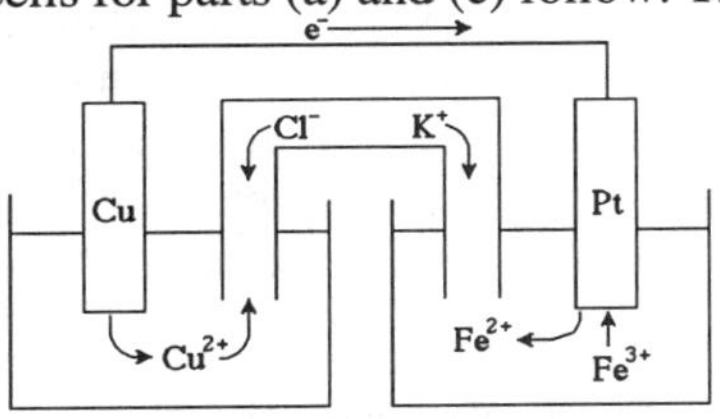

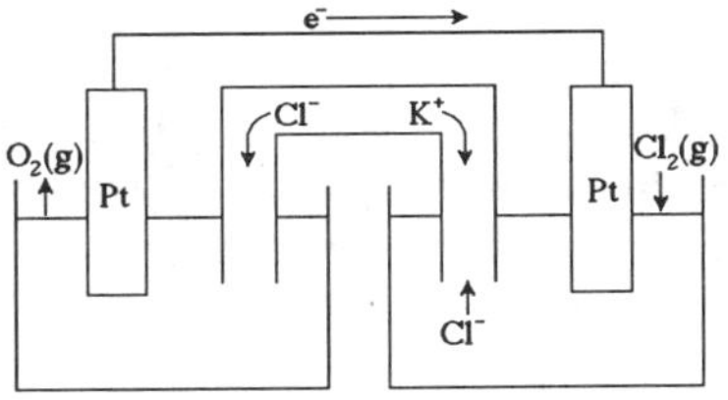

(d) Anode, Oxdn: $\{Zn(s) \longrightarrow Zn^{2+}(aq) + 2\ e^-\}$ $\times 3$ $-E° = +0.763$ V

Cathode, Redn: $\{NO_3^-(aq) + 4\ H^+(aq) + 3\ e^- \longrightarrow NO(g) + 2\ H_2O\}$ $\times 2$ $E° = +0.956$ V

Net rxn: $3\ Zn(s) + 2\ NO_3^-(aq) + 8\ H^+(aq) \longrightarrow 3\ Zn^{2+}(aq) + 2\ NO(g) + 4\ H_2O$ $E°_{cell} = +1.719$ V

Cell diagram: $Zn(s)|Zn^{2+}(aq)||H^+(aq),NO_3^-(aq)|NO(g)|Pt(s)$

34. **(a)** $Fe(s)|Fe^{2+}(aq)||Cl^-(aq)|Cl_2(g)|Pt(s)$

Oxidation: $Fe(s) \longrightarrow Fe^{2+}(aq) + 2\ e^-$ $-E° = +0.440$ V

Reduction: $Cl_2(g) + 2\ e^- \longrightarrow 2\ Cl^-(aq)$ $E° = +1.358$ V

Net: $Fe(s) + Cl_2(g) \longrightarrow Fe^{2+}(aq) + 2\ Cl^-(aq)$ $E°_{cell} = +1.798$ V

(b) $Zn(s)|Zn^{2+}(aq)||Ag^+(aq)|Ag(s)$

Oxidation: $Zn(s) \longrightarrow Zn^{2+}(aq) + 2\ e^-$ $-E° = -0.763$ V

Reduction: $\{Ag^+(aq) + e^- \longrightarrow Ag(s)\} \times 2$ $E° = +0.800$ V

Net: $Zn(s) + 2\ Ag^+(aq) \longrightarrow Zn^{2+}(aq) + 2\ Ag(s)$ $E°_{cell} = +0.037$ V

(c) $Pt(s)|Cu^+(aq),Cu^{2+}(aq)||Cu^+(aq)|Cu(s)$

Oxidation: $Cu^+(aq) \longrightarrow Cu^{2+}(aq) + e^-$ $-E° = -0.159$ V

Reduction: $Cu^+(aq) + e^- \longrightarrow Cu(s)$ $E° = +0.520$ V

Net: $2\ Cu^+(aq) \longrightarrow Cu^{2+}(aq) + Cu(s)$ $E°_{cell} = -0.361$ V

(d) $Mg(s)|Mg^{2+}(aq)||Br^-(aq)|Br_2(l)|Pt(s)$

Oxidation:	$Mg(s) \longrightarrow Mg^{2+}(aq) + 3\ e^-$	$-E° = +2.356$ V
Reduction:	$Br_2(l) + 2\ e^- \longrightarrow 2\ Br^-(aq)$	$E° = +1.065$ V
Net:	$Mg(s) + Br_2(l) \longrightarrow Mg^{2+}(aq) + 2\ Br^-(aq)$	$E°_{cell} = +3.421$ V

$\Delta G°$, $E°_{cell}$, and K

35. **(a)**

Oxidation:	$\{Mn^{2+} + 4\ H_2O \longrightarrow MnO_4^- + 8\ H^+(aq) + 5\ e^-\}$	$\times 2$	$-E° = -1.51$ V
Reduction:	$\{H_2O_2(aq) + 2\ H^+(aq) + 2\ e^- \longrightarrow 2\ H_2O\}$	$\times 5$	$E° = +1.763$ V

Net: $2\ Mn^{2+} + 5\ H_2O_2(aq) \longrightarrow 2\ MnO_4^- + 6\ H^+(aq) + 2\ H_2O$ $\qquad E°_{cell} = +0.25$ V

(b) $\Delta G° = -n\mathcal{F}E°_{cell} = -(10 \text{ mol e}^-)(96{,}485 \text{ C/mol e}^-)(0.25 \text{ V}) = -2.4 \times 10^5 \text{ J} = -2.4 \times 10^2 \text{ kJ}$

(c) $\ln K_{eq} = -\dfrac{\Delta G°}{RT} = -\dfrac{-2.4 \times 10^5}{8.3145 \text{ J mol}^{-1} \text{ K}^{-1} \times 298 \text{ K}} = 97 \qquad K = e^{97} = 1 \times 10^{42}$

(d) Based on the extremely large value of K_{eq}, we conclude that this reaction should go to completion.

36. **(a)**

Oxidation:	$Fe(s) \longrightarrow Fe^{2+}(aq) + 2\ e^-$		$-E° = +0.440$ V
Reduction:	$\{Cr^{3+}(aq) + e^- \longrightarrow Cr^{2+}(aq)\}$	$\times 2$	$E° = -0.424$V

Net: $2\ Cr^{3+}(aq) + Fe(s) \longrightarrow 2\ Fe^{2+}(aq) + 2\ Cr^{2+}(aq)$ **(b)** $E°_{cell} = +0.016$ V

(c) $\Delta G° = -n\mathcal{F}E°_{cell} = -(2 \text{ mol e}^-)(96{,}485 \text{ C/mol e}^-)(0.016 \text{ V}) = -3.1 \times 10^3 \text{ J} = -3.1 \text{ kJ}$

(d) $\ln K_{eq} = -\dfrac{\Delta G°}{RT} = -\dfrac{-3.1 \times 10^3}{8.3145 \text{ J mol}^{-1} \text{ K}^{-1} \times 298 \text{ K}} = 1.3 \qquad K = e^{1.3} = 3.7$

(e) Based on the very modest value of K_{eq}, we conclude that this reaction will not go to completion.

37. **(a)** A negative value of $E°_{cell}$ (-0.0050 V) indicates that $\Delta G° = -n\mathcal{F}E°_{cell}$ is positive which in turn indicates that K_{eq} is less than one ($K_{eq} < 1.00$); $\Delta G° = -RT \ln K_{eq}$. Now $K_{eq} = \dfrac{[Cu^{2+}]^2[Sn^{2+}]}{[Cu^+]^2[Sn^{4+}]}$ Thus, when all concentrations are the same, the ion product, Q, equals 1.00, and thus cannot equal the value of K_{eq} ($K_{eq} < 1.00$). Therefore, all the concentrations cannot be 0.500 M at the same time.

(b) In order to establish equilibrium, that is, to have the ion product become less than 1.00, and equal the equilibrium constant, the concentrations of the products must decrease and those of the reactants must increase. A net reaction to the left will occur.

38. **(a)** We calculate the value of the equilibrium constant from the standard cell potential.

$$E°_{cell} = \frac{0.0257}{n} \ln K_{eq} \qquad \ln K_{eq} = \frac{nE°_{cell}}{0.0257} = \frac{2 \text{ mol e}^- \times 0.0020 \text{ V}}{0.0257} = 0.16 \qquad K_{eq} = e^{0.16} = 1.2$$

To determine if the described solution is possible, we compare K_{eq} with Q. Now $K_{eq} = \dfrac{[Ni^{2+}]}{[V^{3+}]^2}$ Thus, when $[V^{2+}] = 0.600$ M and $[V^{3+}] = [Ni^{2+}] = 0.675$ M, the ion product, $Q = \dfrac{(0.600)^2\ 0.675}{(0.675)^2} = 0.533 < 1.2 = K_{eq}$. Therefore, the described situation cannot occur.

(b) In order to establish equilibrium, that is, to have the ion product (0.533) become equal to 1.2, the equilibrium constant, the concentrations of the products must increase and those of the reactants must decrease. A slight net reaction to the right should occur.

39. Cell reaction: $Zn(s) + Ag_2O(s) \longrightarrow ZnO(s) + 2\ Ag(s)$ We assume that the cell operates at 298 K.

$$\Delta G° = \Delta G°_f[ZnO(s)] + 2\ \Delta G°_f[Ag(s)] - \Delta G°_f[Zn(s)] - \Delta G°_f[Ag_2O(s)]$$
$$= -318.3 \text{ kJ/mol} + 2\ (0.00 \text{ kJ/mol}) - 0.00 \text{ kJ/mol} - (-11.20 \text{ kJ/mol}) = -307.1 \text{ kJ/mol} = -n\mathcal{F}E°_{cell}$$

$$E°_{cell} = -\frac{\Delta G°}{n\mathcal{F}} = -\frac{-307.1 \times 10^3 \text{ J/mol}}{2 \text{ mol e}^-/\text{mol rxn} \times 96{,}485 \text{ C/mol e}^-} = 1.591 \text{ V}$$

40. From equation (21.28) we know that $n = 12$ and the cell equation. First we compute the value of $\Delta G°$.

$$\Delta G° = -n\mathcal{F}E°_{cell} = -12 \text{ mol e}^- \times \frac{96485 \text{ C}}{1 \text{ mol e}^-} \times 2.71 \text{ V} = -3.14 \times 10^6 \text{ J} = -3.14 \text{ MJ}$$

Then we use this value with the balanced equation and values of $\Delta G°_f$ to calculate $\Delta G°_f[[Al(OH)_4]^-]$.

$4\,Al(s) + 3\,O_2(g) + 6\,H_2O + 4\,OH^-(aq) \longrightarrow 4\,[Al(OH)_4]^-(aq)$

$\Delta G° = 4\,\Delta G_f°[[Al(OH)_4]^-] - 4\,\Delta G_f°[Al(s)] - 3\,\Delta G_f°[O_2(g)] - 6\,\Delta G_f°[H_2O(l)] - 4\,\Delta G_f°[OH^-(aq)]$

$-3.14 \times 10^3\text{ kJ} = 4\,\Delta G_f°[[Al(OH)_4]^-] - 4 \times 0.00\text{ kJ} - 3 \times 0.00\text{ kJ} - 6 \times (-237.1\text{ kJ}) - 4 \times (-157.2)$

$= 4\,\Delta G_f°[[Al(OH)_4]^-] + 2051.4\text{ kJ}$

$\Delta G_f°[[Al(OH)_4]^-] = (-3.14 \times 10^3\text{ kJ} - 2051.4\text{ kJ}) \div 4 = -1.30 \times 10^3\text{ kJ} = -1.30\text{ MJ}$

Concentration Dependence of E_{cell}—the Nernst Equation

41. We first calculate $E°_{cell}$ for each reaction and then use the Nernst equation to calculate E_{cell}..

(a) Oxidation: $\{Al(s) \longrightarrow Al^{3+}(0.18\text{ M}) + 3\,e^-\} \times 2$ $\quad -E° = +1.676\text{ V}$

Reduction: $\{Fe^{2+}(0.85\text{ M}) + 2\,e^- \longrightarrow Fe(s)\} \times 3$ $\quad E° = -0.440\text{ V}$

Net: $2\,Al(s) + 3\,Fe^{2+}(0.85\text{ M}) \longrightarrow 2\,Al^{3+}(0.18\text{ M}) + Fe(s)$ $\quad E°_{cell} = +1.236\text{ V}$

$$E_{cell} = E°_{cell} - \frac{0.0592}{n}\log\frac{[Al^{3+}]^2}{[Fe^{2+}]^3} = 1.236\text{ V} - \frac{0.0592}{6}\log\frac{(0.18\text{ M})^2}{(0.85\text{ M})^3} = 1.249\text{ V}$$

(b) Oxidation: $\{Ag(s) \longrightarrow Ag^+(0.34\text{ M}) + e^-\} \times 2$ $\quad -E° = -0.800\text{ V}$

Reduction: $Cl_2(0.55\text{ atm}) + 2\,e^- \longrightarrow 2\,Cl^-(0.098\text{ M})$ $\quad E° = +1.358\text{ V}$

Net: $Cl_2(0.55\text{ atm}) + 2\,Ag(s) \longrightarrow 2\,Cl^-(0.098\text{ M}) + 2\,Ag^+(0.34\text{ M})$ $\quad E°_{cell} = +0.558\text{ V}$

$$E_{cell} = E°_{cell} - \frac{0.0592}{n}\log\frac{[Cl^-]^2[Ag^+]^2}{P\{Cl_2(g)\}} = +0.558 - \frac{0.0592}{2}\log\frac{(0.34)^2(0.098)^2}{0.55} = +0.638\text{ V}$$

42. **(a)** Oxidation: $Mn(s) \longrightarrow Mn^{2+}(0.40\text{ M}) + 2\,e^-$ $\quad -E° = +1.18\text{ V}$

Reduction: $\{Cr^{3+}(0.35\text{ M}) + 1\,e^- \longrightarrow Cr^{2+}(0.25\text{ M})\} \times 2$ $\quad E° = -0.424\text{ V}$

Net: $2\,Cr^{3+}(0.35\text{ M}) + Mn(s) \longrightarrow 2\,Cr^{2+}(0.25\text{ M}) + Mn^{2+}(0.40\text{ M})$ $\quad E°_{cell} = +0.76\text{ V}$

$$E_{cell} = E°_{cell} - \frac{0.0592}{n}\log\frac{[Cr^{2+}]^2[Mn^{2+}]}{[Cr^{3+}]^2} = +0.76\text{ V} - \frac{0.0592}{2}\log\frac{(0.25\text{ M})^2\,(0.40)}{(0.35\text{ M})^2} = +0.78\text{ V}$$

(b) Oxidation: $\{Mg(s) \longrightarrow Mg^{2+}(0.016\text{ M}) + 2\,e^-\} \times 3$ $\quad -E° = +2.356\text{ V}$

Reduction: $\{[Al(OH)_4]^-(0.25\text{ M}) + 3\,e^- \longrightarrow 4\,OH^-(0.042\text{ M}) + Al(s)\} \times 2$ $\quad E° = -2.310\text{ V}$

Net: $3\,Mg(s) + 2\,[Al(OH)_4]^-(0.25\text{ M}) \longrightarrow 3\,Mg^{2+}(0.016\text{ M}) + 8\,OH^-(0.042\text{ M}) + 2\,Al(s)$ $\quad E°_{cell} = +0.046\text{ V}$

$$E_{cell} = E°_{cell} - \frac{0.0592}{6}\log\frac{[Mg^{2+}]^3[OH^-]^8}{[[Al(OH)_4]^-]^2} = +0.046 - \frac{0.0592}{6}\log\frac{(0.016)^3(0.042)^8}{(0.25)^2}$$

$$= 0.046\text{ V} + 0.150\text{ V} = 0.196\text{ V}$$

43. All these observations can be understood in terms of the means we use to balance half-equations: the ion–electron method.

(a) The reactions for which E depends on pH are those that contain either $H^+(aq)$ or $OH^-(aq)$ in the balanced half equation. These reactions are those that involve oxoacids and oxoanions in which the central atom changes oxidation state.

(b) $H^+(aq)$ will inevitably be on the left side of the reduction of an oxoanion because reduction is accompanied by not only a decrease in oxidation state, but also by the loss of oxygen atoms, as in $ClO_3^-|ClO_2^-$, $SO_4^{2-}|SO_2$, and $NO_3^-|NO$. These oxygen atoms appear on the right-hand side as H_2O molecules. The hydrogens that are added to the right-hand side with the water molecules are then balanced with $H^+(aq)$ on the left-hand side.

(c) If a half-equation with $H^+(aq)$ ions present is transferred to basic solution, it may be re-balanced by adding to each side $OH^-(aq)$ ions equal in number to the $H^+(aq)$ originally present. This results in H_2O on the side that had $H^+(aq)$ ions (the left side in this case) and $OH^-(aq)$ ions on the other side (the right side.)

44. Oxidation: $\{2\,Cl^-(aq) \longrightarrow Cl_2(g) + 2\,e^-\} \times 2$ $\quad -E° = -1.358\text{ V}$

Reduction: $PbO_2(s) + 4\,H^+(aq) + 4\,e^- \longrightarrow Pb^{2+}(aq) + 2\,H_2O$ $\quad E° = +1.455\text{ V}$

Net: $PbO_2(s) + 4\,H^+(aq) + 4\,Cl^-(aq) \longrightarrow Pb^{2+}(aq) + 2\,H_2O + 2\,Cl_2(g)$ $\quad E°_{cell} = +0.097\text{ V}$

We derive an expression for E_{cell} that depends on just the changing $[H^+]$.

$$E_{cell} = E^\circ_{cell} - \frac{0.0592}{4} \log \frac{P\{Cl_2\}[Pb^{2+}]}{[H^+]^4[Cl^-]^4} = +0.097 - 0.0148 \log \frac{(1.00 \text{ atm})(1.00 \text{ M})}{[H^+]^4(1.00)^4}$$

$$= +0.097 + 4 \times 0.0148 \log[H^+] = +0.097 + 0.0592 \log [H^+] = +0.097 - 0.0592 \text{ pH}$$

(a) $E_{cell} = +0.097 + 0.0592 \log(6.0) = +0.143$ V Some reaction occurs.

(b) $E_{cell} = +0.097 + 0.0592 \log(1.2) = +0.102$ V Yes, some reaction occurs.

(c) $E_{cell} = +0.097 - 0.0592 \times 4.25 = -0.155$ V Very little reaction occurs.

Reaction occurs for all these conditions, but becomes more spontaneous as the solution becomes more acidic, that is, as pH decreases.

45. Oxidation: $Zn(s) \longrightarrow Zn^{2+}(aq) + 2\,e^-$ $-E^\circ = +0.763$ V

Reduction: $Cu^{2+}(aq) + 2\,e^- \longrightarrow Cu(s)$ $E^\circ = +0.337$ V

Net: $Zn(s) + Cu^{2+}(aq) \longrightarrow Cu(s) + Zn^{2+}(aq)$ $E^\circ_{cell} = +1.100$ V

(a) We set $E = 0.000$ V, $[Zn^{2+}] = 1.00$ M, and solve for $[Cu^{2+}]$ in the Nernst equation.

$$E_{cell} = E^\circ_{cell} - \frac{0.0592}{2} \log \frac{[Zn^{2+}]}{[Cu^{2+}]} = 0.000 = 1.100 - 0.0296 \log \frac{1.0 \text{ M}}{[Cu^{2+}]}$$

$$\log \frac{1.0 \text{ M}}{[Cu^{2+}]} = \frac{0.000 - 1.100}{-0.0296} = 37.2 \qquad [Cu^{2+}] = 10^{-37.2} = 6 \times 10^{-38} \text{ M}$$

(b) If we work the problem the other way—by assuming initial concentrations of $[Cu^{2+}]_i = 1.0$ M and $[Zn^{2+}]_i = 0.0$ M—we obtain $[Cu^{2+}]_f = 6 \times 10^{-38}$ M and $[Zn^{2+}]_f = 1.0$ M. We would say that this reaction goes to completion.

46. Oxidation: $Sn(s) \longrightarrow Sn^{2+}(aq) + 2\,e^-$ $-E^\circ = +0.137$ V

Reduction: $Pb^{2+}(aq) + 2\,e^- \longrightarrow Pb(s)$ $E^\circ = -0.125$ V

Net: $Sn(s) + Pb^{2+}(aq) \longrightarrow Sn^{2+}(aq) + Pb(s)$ $E^\circ_{cell} = +0.012$ V

Now we wish to find out if $Pb^{2+}(aq)$ will be completely displaced—that is, will $[Pb^{2+}]$ reach 0.0010 M—if $[Sn^{2+}]$ is fixed at 1.00 M? We use the Nernst equation to determine if the cell voltage still is positive under these conditions.

$$E_{cell} = E^\circ_{cell} - \frac{0.0592}{2} \log \frac{[Sn^{2+}]}{[Pb^{2+}]} = +0.012 - \frac{0.0592}{2} \log \frac{1.00}{0.0010} = +0.012 - 0.089 = -0.077 \text{ V}$$

No, this reaction will not go to completion under the conditions stated. The reaction stops being spontaneous when $E_{cell} = 0$. We can work this the other way as well: assume that $[Pb^{2+}] = (1.0 - x)$ M and calculate $[Sn^{2+}] = x$ M at equilibrium, that is, where $E_{cell} = 0$.

$$E_{cell} = 0.00 = E^\circ_{cell} - \frac{0.0592}{2} \log \frac{[Sn^{2+}]}{[Pb^{2+}]} = +0.012 - \frac{0.0592}{2} \log \frac{x}{1.0 - x}$$

$$\log \frac{x}{1.0 - x} = \frac{2 \times 0.012}{0.0592} = 0.41 \qquad x = 10^{0.41}(1.0 - x) = 2.6 - 2.6\,x \qquad x = \frac{2.6}{3.6} = 0.72 \text{ M}$$

We would expect the final $[Sn^{2+}]$ to equal 1.0 M (or at least 0.999 M) if the reaction went to completion. Instead it equals 0.72 M.

47. **(a)** The two half-equations and the cell equation are given below. $E^\circ_{cell} = 0.000$ V

Oxidation: $H_2(g) \longrightarrow 2\,H^+(0.65 \text{ M KOH}) + 2\,e^-$

Reduction: $2\,H^+(1.0 \text{ M}) + 2\,e^- \longrightarrow H_2(g)$

Net: $2\,H^+(1.0 \text{ M}) \longrightarrow 2\,H^+(0.65 \text{ M KOH})$

$$[H^+]_{base} = \frac{K_w}{[OH^-]} = \frac{1.00 \times 10^{-14}}{0.65} = 1.5 \times 10^{-14} \text{ M}$$

$$E_{cell} = E^\circ_{cell} - \frac{0.0592}{2} \log \frac{[H^+]^2_{base}}{[H^+]^2_{acid}} = 0.000 - \frac{0.0592}{2} \log \frac{(1.5 \times 10^{-14})^2}{(1.0)^2} = +0.818 \text{ V}$$

(b) For the reduction of H_2O to $H_2(g)$ in basic solution, $2\,H_2O + 2\,e^- \longrightarrow 2\,H_2(g) + 2\,OH^-(aq)$, $E^\circ = -0.828$ V. This reduction is the reverse of the reaction that occurs in the anode of the cell described, with one small difference: in the standard half-cell $[OH^-] = 1.00$ M, while in the anode half-cell in the case $[OH^-] = 0.65$ M. Or, viewed in another way, in 1.00 M KOH, $[H^+]$ is smaller still than in 0.65 M KOH. The forward reaction (dilution of H^+) should occur to an even greater extent with 1.00 M KOH than with 0.65 M KOH. This means that E°_{cell} (0.828 V) should be slightly larger than E_{cell} (0.818 V), which in fact is the case.

48. **(a)** Because $NH_3(aq)$ is a weaker base than KOH(aq), $[OH^-]$ will be smaller than it is in the previous problem. Therefore the $[H^+]$ will be higher. Its logarithm will be less negative, and the cell voltage will be smaller. Or, viewed as in Exercise 47(b), the difference in $[H^+]$ between 1.0 M H^+ and 0.65 M KOH is greater than the difference in $[H^+]$ between 1.0 M H^+ and 0.65 M NH_3. The forward reaction is "less spontaneous" and E°_{cell} is smaller.

(b)

Reaction:	$NH_3(aq) + H_2O \rightleftharpoons$	$NH_4^+(aq)$	+	$OH^-(aq)$
Initial:	0.65 M			
Changes:	$-x$ M	$+x$ M		$+x$ M
Equil:	$(0.65 - x)$ M	x M		x M

$$K_b = \frac{[NH_4^+][OH^-]}{[NH_3]} = 1.8 \times 10^{-5} = \frac{x \cdot x}{0.65 - x} \approx \frac{x^2}{0.65}$$

$$x = [OH^-] = \sqrt{0.65 \times 1.8 \times 10^{-5}} = 3.4 \times 10^{-3}\ \text{M} \qquad [H_3O^+] = \frac{1.00 \times 10^{-14}}{3.4 \times 10^{-3}} = 2.9 \times 10^{-12}\ \text{M}$$

$$E_{cell} = E^\circ_{cell} - \frac{0.0592}{2} \log \frac{[H^+]^2_{base}}{[H^+]^2_{acid}} = 0.000 - \frac{0.0592}{2} \log \frac{(2.9 \times 10^{-12})^2}{(1.0)^2} = +0.683\ \text{V}$$

49. First we need $[Ag^+]$ in a saturated solution of Ag_2CrO_4.

$$K_{sp} = [Ag^+]^2[CrO_4^{2-}] = (2s)^2(s) = 4s^3 = 1.1 \times 10^{-12} \qquad s = \sqrt[3]{\frac{1.1 \times 10^{-12}}{4}} = 6.5 \times 10^{-5}\ \text{M}$$

The cell diagrammed is a concentration cell, for which $E^\circ_{cell} = 0.000$ V, $n = 1$, $[Ag^+] = 2s = 1.3 \times 10^{-4}$ M

Cell reaction: $Ag(s) + Ag^+(0.125\ M) \longrightarrow Ag(s) + Ag^+(1.3 \times 10^{-4}\ M)$

$$E_{cell} = E^\circ_{cell} - \frac{0.0592}{1} \log \frac{1.3 \times 10^{-4}\ \text{M}}{0.125\ \text{M}} = 0.000 + 0.177\ \text{V} = 0.177\ \text{V}$$

50. We need to determine $[Ag^+]$ in the saturated solution of Ag_3PO_4.
The cell diagrammed is a concentration cell, for which $E^\circ_{cell} = 0.000$ V. $n = 1$.
Cell reaction: $Ag(s) + Ag^+(0.140\ M) \longrightarrow Ag(s) + Ag^+(x\ M)$

$$E_{cell} = 0.180\ \text{V} = E^\circ_{cell} - \frac{0.0592}{1} \log \frac{x\ \text{M}}{0.140\ \text{M}} \qquad \log \frac{x\ \text{M}}{0.140\ \text{M}} = \frac{0.180}{-0.0592} = -3.04$$

$$x\ \text{M} = 0.140\ \text{M} \times 10^{-3.04} = 0.140\ \text{M} \times 9.1 \times 10^{-4} = 1.3 \times 10^{-4}\ \text{M} = [Ag^+]$$

$$K_{sp} = [Ag^+]^3[PO_4^{3-}] = (3s)^3(s) = (1.3 \times 10^{-4})^3(1.3 \times 10^{-4} \div 3) = 9.5 \times 10^{-17}$$

51. (a)

Oxidation:	$Sn(s) \longrightarrow Sn^{2+}(0.075\ M) + 2\ e^-$	$-E^\circ = +0.137$ V
Reduction:	$Pb^{2+}(0.600\ M) + 2\ e^- \longrightarrow Pb(s)$	$E^\circ = -0.125$ V
Net:	$Sn(s) + Pb^{2+}(0.600\ M) \longrightarrow Pb(s) + Sn^{2+}(0.075\ M)$	$E^\circ_{cell} = +0.012$ V

$$E_{cell} = E^\circ_{cell} - \frac{0.0592}{2} \log \frac{[Sn^{2+}]}{[Pb^{2+}]} = 0.012 - 0.0296 \log \frac{0.075}{0.600} = 0.012 + 0.027 = 0.039\ \text{V}$$

(b) E_{cell} will decrease with time because, as the reaction proceeds, $[Sn^{2+}]$ will increase and $[Pb^{2+}]$ decrease.

(c) When $[Pb^{2+}] = 0.500\ M = 0.600\ M - 0.100\ M$, $[Sn^{2+}] = 0.075\ M + 0.100\ M$, because, by the stoichiometry of the reaction, a mole of Sn^{2+} is produced for every mole of Pb^{2+} that reacts.

$$E_{cell} = E^\circ_{cell} - \frac{0.0592}{2} \log \frac{[Sn^{2+}]}{[Pb^{2+}]} = 0.012 - 0.0296 \log \frac{0.175}{0.500} = 0.012 + 0.013 = 0.025\ \text{V}$$

(d)

Reaction:	$Sn(s)$ +	$Pb^{2+}(aq)$	$\longrightarrow$	$Pb(s)$ +	$Sn^{2+}(aq)$
Initial:		0.600 M			0.075 M
Changes:		$-x$ M			$+x$ M
Final:		$(0.600 - x)$M			$(0.075 + x)$M

$$E_{cell} = E^\circ_{cell} - \frac{0.0592}{2} \log \frac{[Sn^{2+}]}{[Pb^{2+}]} = 0.020 = 0.012 - 0.0296 \log \frac{0.075 + x}{0.600 - x}$$

$$\log \frac{0.075 + x}{0.600 - x} = \frac{E_{cell} - 0.012}{-0.0296} = \frac{0.020 - 0.012}{-0.0296} = -0.27 \qquad \frac{0.075 + x}{0.600 - x} = 10^{-0.27} = 0.54$$

$$0.075 + x = 0.54(0.600 - x) = 0.324 - 0.54\,x \qquad x = \frac{0.324 - 0.075}{1.54} = 0.162\ \text{M}$$

$[Sn^{2+}] = 0.075 + 0.162 = 0.237$ M

(e) Use the expression developed in part (d).

$$\log \frac{0.075 + x}{0.600 - x} = \frac{E_{cell} - 0.012}{-0.0296} = \frac{0.000 - 0.012}{-0.0296} = +0.41$$

$$\frac{0.075 + x}{0.600 - x} = 10^{+0.41} = 2.6; \quad 0.075 + x = 2.6(0.600 - x) = 1.6 - 2.6\,x \qquad x = \frac{1.6 - 0.075}{3.6} = 0.42\ \text{M}$$

$[Sn^{2+}] = 0.075 + 0.42 = 0.50$ M $\qquad [Pb^{2+}] = 0.600 - 0.42 = 0.18$ M

52. (a) Oxidation: $Ag(s) \longrightarrow Ag^+(0.015\ M) + e^-$ $\quad -E° = -0.800\ V$

Reduction: $Fe^{3+}(0.055\ M) + e^- \longrightarrow Fe^{2+}(0.045\ M)$ $\quad E° = +0.771\ V$

Net: $Ag(s) + Fe^{3+}(0.055\ M) \longrightarrow Ag^+(0.015\ M) + Fe^{2+}(0.045\ M)$ $\quad E°_{cell} = -0.029\ V$

$$E_{cell} = E°_{cell} - \frac{0.0592}{1}\log\frac{[Ag^+][Fe^{2+}]}{[Fe^{3+}]} = -0.029 - 0.0592\log\frac{0.015\times 0.045}{0.055}$$
$$= -0.029\ V + 0.113\ V = +0.084\ V$$

(b) E_{cell} will decrease with time because, as the reaction proceeds, $[Ag^+]$ and $[Fe^{2+}]$ will increase and $[Fe^{3+}]$ decrease.

(c) When $[Ag^+] = 0.020\ M = 0.015\ M + 0.005\ M$, $[Fe^{2+}] = 0.045\ M + 0.005\ M = 0.050\ M$ and $[Fe^{3+}] = 0.055\ M - 0.005\ M = 0.500\ M$, because, by the stoichiometry of the reaction, a mole of Fe^{2+} is produced and a mole of Fe^{3+} is consumed for every mole of Ag^+ produced.

$$E_{cell} = E°_{cell} - \frac{0.0592}{1}\log\frac{[Ag^+][Fe^{2+}]}{[Fe^{3+}]} = -0.029 - 0.0592\log\frac{0.020\times 0.050}{0.050}$$
$$= -0.029\ V + 0.101\ V = +0.072\ V$$

(d)

Reaction:	$Ag(s)$ +	$Fe^{3+}(aq)$ $\longrightarrow$	$Ag^+(0.015\ M)$ +	$Fe^{2+}(0.045\ M)$
Initial:		0.055 M	0.015 M	0.045 M
Changes:		$-x$ M	$+x$ M	$+x$ M
Final:		$(0.055 - x)$M	$(0.015 + x)$M	$(0.045 + x)$M

$$E_{cell} = E°_{cell} - \frac{0.0592}{1}\log\frac{[Ag^+][Fe^{2+}]}{[Fe^{3+}]} = -0.029 - 0.0592\log\frac{(0.015 + x)\ (0.045 + x)}{(0.055 - x)}$$
$$\log\frac{(0.015 + x)\ (0.045 + x)}{(0.055 - x)} = \frac{E_{cell} + 0.029}{-0.0592} = \frac{0.010 + 0.029}{-0.0592} = -0.66$$
$$\frac{(0.015 + x)\ (0.045 + x)}{(0.055 - x)} = 10^{-0.66} = 0.22$$
$$0.00068 + 0.060\,x + x^2 = 0.22(0.055 - x) = 0.012 - 0.22\,x \qquad x^2 + 0.28\,x - 0.011 = 0$$
$$x = \frac{-b \pm \sqrt{b^2 - 4ac}}{2a} = \frac{-0.28 \pm \sqrt{(0.28)^2 + 4\times 0.011}}{2} = 0.035\ M$$

$[Ag^+] = 0.015\ M + 0.035\ M = 0.050\ M$ $\qquad [Fe^{2+}] = 0.045\ M + 0.035\ M = 0.080\ M$

$[Fe^{3+}] = 0.055\ M - 0.035\ M = 0.020\ M$

(e) We use the expression that we developed in part (d).

$$\log\frac{(0.015 + x)\ (0.045 + x)}{(0.055 - x)} = \frac{E_{cell} + 0.029}{-0.0592} = \frac{0.000 + 0.029}{-0.0592} = -0.49$$
$$\frac{(0.015 + x)\ (0.045 + x)}{(0.055 - x)} = 10^{-0.49} = 0.32$$
$$0.00068 + 0.060\,x + x^2 = 0.32(0.055 - x) = 0.018 - 0.32\,x \qquad x^2 + 0.38\,x - 0.017 = 0$$
$$x = \frac{-b \pm \sqrt{b^2 - 4ac}}{2a} = \frac{-0.38 \pm \sqrt{(0.38)^2 + 4\times 0.017}}{2} = 0.040\ M$$

$[Ag^+] = 0.015\ M + 0.040\ M = 0.055\ M$ $\qquad [Fe^{2+}] = 0.045\ M + 0.040\ M = 0.085\ M$

$[Fe^{3+}] = 0.055\ M - 0.040\ M = 0.015\ M$

Batteries and Fuel Cells

53. (a) The cell diagram begins with the anode and ends with the cathode.

Cell diagram: $Cr(s) \mid Cr^{2+}(aq), Cr^{3+}(aq) \parallel Fe^{2+}(aq), Fe^{3+}(aq) \mid Fe(s)$

(b) Oxidation: $Cr^{2+}(aq) \longrightarrow Cr^{3+}(aq) + e^-$ $\quad -E° = +0.424\ V$

Reduction: $Fe^{3+}(aq) + e^- \longrightarrow Fe^{2+}(aq)$ $\quad E° = +0.771\ V$

Net: $Cr^{2+}(aq) + Fe^{3+}(aq) \longrightarrow Cr^{3+}(aq) + Fe^{2+}(aq)$ $\quad E°_{cell} = +1.195\ V$

54. (a) Oxidation: $Zn(s) \longrightarrow Zn^{2+}(aq) + 2\ e^-$ $\quad -E° = +0.763\ V$

Reduction: $2\ MnO_2(s) + H_2O + 2\ e^- \longrightarrow Mn_2O_3(s) + 2\ OH^-(aq)$

Acid-base: $\{NH_4^+(aq) + OH^-(aq) \longrightarrow NH_3(g) + H_2O(l)\} \quad \times 2$

Complex: $Zn^{2+}(aq) + 2\ NH_3(aq) + 2\ Cl^-(aq) \longrightarrow [Zn(NH_3)_2]Cl_2(s)$

Net: $Zn(s) + 2\ MnO_2(s) + 2\ NH_4^+(aq) + 2\ Cl^-(aq) \longrightarrow Mn_2O_3(s) + H_2O(l) + [Zn(NH_3)_2]Cl_2(s)$

(b) $\Delta G° = -n\mathcal{F}E°_{cell} = -(2\text{ mol e}^-)(96485\text{ C/mol e}^-)(1.55\text{ V}) = -2.99 \times 10^5\text{ J/mol}$
This is the standard free energy change for the entire reaction, which is composed of the four reactions in part (a). We can determine the values of $\Delta G°$ for the acid-base and complex formation reactions with data from Table 17-2 and $pK_f = -4.81$.
$\Delta G°_{a\text{-}b} = -RT\ln K_b^2 = -(8.3145\text{ J mol}^{-1}\text{ K}^{-1})(298.15\text{ K})\ln(1.8 \times 10^{-5})^2 = 5.417 \times 10^4\text{ J/mol}$
$\Delta G°_{cmplx} = -RT\ln K_f = -(8.3145\text{ J mol}^{-1}\text{ K}^{-1})(298.15\text{ K})\ln(10^{4.81}) = -2.746 \times 10^4\text{ J/mol}$
Then $\Delta G°_{total} = \Delta G°_{redox} + \Delta G°_{a\text{-}b} + \Delta G°_{cmplx}$
$\Delta G°_{redox} = \Delta G°_{total} - \Delta G°_{a\text{-}b} - \Delta G°_{cmplx}$
$= -2.99 \times 10^5\text{ J/mol} - 5.417 \times 10^4 + 2.746 \times 10^4\text{ J/mol} = -3.26 \times 10^5\text{ J/mol}$

Thus, the voltage of the redox reactions alone is $E° = \dfrac{-3.26 \times 10^5\text{ J}}{-2\text{ mol e}^- \times 96485\text{ C/mol e}^-} = 1.69\text{ V}$

$1.69\text{ V} = +0.763\text{ V} + E°(MnO_2/Mn_2O_3)$
$E°(MnO_2/Mn_2O_3) = 1.69\text{ V} - 0.763\text{ V} = +0.93\text{ V}$

55. (a) Cell reaction: $2\,H_2(g) + O_2(g) \longrightarrow 2\,H_2O(l)$
$\Delta G°_{rxn} = 2\,\Delta G°_f[H_2O(l)] = 2\,(-237.2\text{ kJ/mol}) = -474.4\text{ kJ/mol}$
$E°_{cell} = -\dfrac{\Delta G°}{n\mathcal{F}} = -\dfrac{-474.4 \times 10^3\text{ J/mol}}{4\text{ mol e}^- \times 96485\text{ C/mol e}^-} = 1.229\text{ V}$

(b)
Anode, Oxdn: $\{Zn(s) \longrightarrow Zn^{2+}(aq) + 2\,e^-\} \times 2$ $\quad -E° = +0.763\text{ V}$
Cathode, Redn: $O_2(g) + 4\,H^+(aq) + 4\,e^- \longrightarrow 2\,H_2O$ $\quad E° = +1.229\text{ V}$
Net: $2\,Zn(s) + O_2(g) + 4\,H^+(aq) \longrightarrow 2\,Zn^{2+}(aq) + 2\,H_2O$ $\quad E°_{cell} = +1.992\text{ V}$

(c)
Anode, Oxdn: $Mg(s) \longrightarrow Mg^{2+}(aq) + 2\,e^-$ $\quad -E° = +2.356\text{ V}$
Cathode, Redn: $I_2(s) + 2\,e^- \longrightarrow 2\,I^-(aq)$ $\quad E° = +0.535\text{ V}$
Net: $Mg(s) + I_2(s) \longrightarrow Mg^{2+}(aq) + 2\,I^-(aq)$ $\quad E°_{cell} = +2.891\text{ V}$

56. (a) A voltaic cell with a voltage of 0.1000 V would be possible by using two half-cells whose standard reduction potentials differ by approximately 0.10 V, such as the following pair.
Oxidation: $2\,Cr^{3+}(aq) + 7\,H_2O \longrightarrow Cr_2O_7^{2-}(aq) + 14\,H^+(aq) + 6\,e^-$ $\quad -E° = -1.33\text{ V}$
Reduction: $\{PbO_2(s) + 4\,H^+(aq) + 2\,e^- \longrightarrow Pb^{2+}(aq) + 2\,H_2O\} \times 3$ $\quad E° = +1.455\text{ V}$
Net: $2\,Cr^{3+}(aq) + 3\,PbO_2(s) + H_2O \longrightarrow Cr_2O_7^{2-}(aq) + 3\,Pb^{2+}(aq) + 2\,H^+(aq)$ $\quad E°_{cell} = 0.125\text{ V}$
The voltage can be adjusted to 0.1000 V by suitable alteration of the concentrations. $[Pb^{2+}]$ or $[H^+]$ could be increased or $[Cr^{3+}]$ could be decreased, or any combination of the three of these.

(b) To produce a cell with a voltage of 2.500 V requires that one start with two half-cells whose reduction potentials differ by about that much. An interesting pair follows.
Oxidation: $Al(s) \longrightarrow Al^{3+}(aq) + 3\,e^-$ $\quad -E° = +1.676\text{ V}$
Reduction: $\{Ag^+(aq) + e^- \longrightarrow Ag(s)\} \times 3$ $\quad E° = +0.800\text{ V}$
Net: $Al(s) + 3\,Ag^+(aq) \longrightarrow Al^{3+}(aq) + 3\,Ag(s)$ $\quad E°_{cell} = +2.476\text{ V}$
Again, the desired voltage can be obtained by adjusting the concentrations: in this case increasing $[Ag^+]$ and/or decreasing $[Al^+]$.

(c) Since no pair of half-cells has a potential difference larger than about 6 volts, we conclude that producing a single cell with a potential of 10.00 V is impossible. It is possible, however, to join several cells together into a battery that delivers a voltage of 10.00 V. Four of the cells from part (b) would do quite nicely.

Electrochemical Mechanism of Corrosion

57. (a) Because copper is a less active metal than is iron, this situation would be similar to that of an iron or steel can plated with tin in which the coating has been scratched. Oxidation of iron metal to $Fe^{2+}(aq)$ should be enhanced in the body of the nail (blue precipitate), and hydroxide ion should be produced in the vicinity of the copper wire (pink color), which serves as the cathode.

(b) Because a scratch stresses the iron, it is more susceptible to corrosion. We expect enhanced blue precipitate in the vicinity of the scratch.

(c) Zinc should protect the iron nail from corrosion. There should be almost no blue precipitate; the zinc corrodes instead. The pink color of hydroxide ion continues to form as before.

58. The anode reaction—that of oxidation—is the formation of $Fe^{2+}(aq)$. This occurs far below the water line. The cathode reaction—that of reduction—is the formation of $OH^-(aq)$ from $O_2(g)$. It is logical that this reaction would occur at or near the water line. This reduction reaction requires $O_2(g)$ from the atmosphere and H_2O from the water. The oxidation reaction, on the other hand simply requires iron from the pipe and also an aqueous solution into which the $Fe^{2+}(aq)$ can disperse and not build up to such a high concentration that corrosion is inhibited.

Anode, Oxidation: $Fe(s) \longrightarrow Fe^{2+}(aq) + 2\,e^-$

Cathode, Reduction: $O_2(g) + 2\,H_2O + 4\,e^- \longrightarrow 4\,OH^-(aq)$

Electrolysis Reactions

59. We determine the standard cell voltage of each chemical reaction. Those voltages that are negative are those of chemical reactions that require electrolysis.

(a) Oxidation: $2\,H_2O \longrightarrow 4\,H^+(aq) + O_2(g) + 4\,e^-$ $\quad -E° = -1.229$ V

Reduction: $\{2\,H^+(aq) + 2\,e^- \longrightarrow H_2(g)\} \times 2$ $\quad E° = 0.000$ V

Net: $2\,H_2O \longrightarrow 2\,H_2(g) + O_2(g)$ $\quad E°_{cell} = -1.229$ V

This reaction requires electrolysis, with an applied voltage of at least +1.229 V.

(b) Oxidation: $Zn(s) \longrightarrow Zn^{2+}(aq) + 2\,e^-$ $\quad -E° = +0.763$ V

Reduction: $Fe^{2+}(aq) + 2\,e^- \longrightarrow Fe(s)$ $\quad E° = -0.440$ V

Net: $Zn(s) + Fe^{2+}(aq) \longrightarrow Fe(s) + Zn^{2+}(aq)$ $\quad E°_{cell} = +0.323$ V

This is a spontaneous reaction.

(c) Oxidation: $\{Fe^{2+}(aq) \longrightarrow Fe^{3+}(aq) + e^-\} \times 2$ $\quad -E° = -0.771$ V

Reduction: $I_2(s) + 2\,e^- \longrightarrow 2\,I^-(aq)$ $\quad E° = +0.535$ V

Net: $2\,Fe^{2+}(aq) + I_2(s) \longrightarrow 2\,Fe^{3+}(aq) + 2\,I^-(aq)$ $\quad E°_{cell} = -0.236$ V

This reaction requires electrolysis, with an applied voltage of at least +0.236 V.

(d) Oxidation: $Cu(s) \longrightarrow Cu^{2+}(aq) + 2\,e^-$ $\quad -E° = -0.337$ V

Reduction: $Sn^{4+}(aq) + 2\,e^- \longrightarrow Sn^{2+}(aq) \times 2$ $\quad E° = +0.154$ V

Net: $Cu(s) + Sn^{4+}(aq) \longrightarrow Cu^{2+}(aq) + Sn^{2+}(aq)$ $\quad E°_{cell} = -0.183$ V

This reaction requires electrolysis, with an applied voltage of at least +0.183 V.

60. (a) Because oxidation occurs at the anode, we know that the product cannot be H_2 (since it is produced from H_2O in a reduction reaction), SO_2 (which is a reduction product of SO_4^{2-}), or SO_3 (which is produced from SO_4^{2-} without a change of oxidation state; it is the dehydration product of H_2SO_4). But O_2 indeed is the result of the oxidation of H_2O.

(b) Reduction should occur at the cathode. The possible species that can be reduced are H_2O to $H_2(g)$, $K^+(aq)$ to $K(s)$, and $SO_4^{2-}(aq)$ to perhaps $SO_2(g)$. Because potassium is a highly active metal, it will not be produced in aqueous solution. In order for $SO_4^{2-}(aq)$ to be reduced, a negatively charged ion would have to be attracted to the negatively charged cathode, which should not occur. Thus, $H_2(g)$ is produced at the cathode.

(c) At the anode: $2\,H_2O \longrightarrow 4\,H^+(aq) + O_2(g) + 4\,e^-$ $\quad -E° = -1.229$ V

At the cathode: $\{2\,H^+(aq) + 2\,e^- \longrightarrow H_2(g)\} \times 2$ $\quad E° = 0.000$ V

Net cell reaction: $2\,H_2O \longrightarrow 2\,H_2(g) + O_2(g)$ $\quad E°_{cell} = -1.229$ V

The minimum voltage required is +1.229 volts. Because of the high overpotential required for the formation of gases, we expect that a higher voltage will be necessary.

61. (a) The two gases that are produced are $H_2(g)$ and $O_2(g)$.

(b) At the anode: $2\,H_2O \longrightarrow 4\,H^+(aq) + O_2(g) + 4\,e^-$ $\quad -E° = -1.229$ V

At the cathode: $\{2\,H^+(aq) + 2\,e^- \longrightarrow H_2(g)\} \times 2$ $\quad E° = 0.000$ V

Net cell reaction: $2\,H_2O \longrightarrow 2\,H_2(g) + O_2(g)$ $\quad E°_{cell} = -1.229$ V

62. The product of the electrolysis of Na_2SO_4(aq) at the anode is oxygen, $-E°[O_2(g)|H_2O] = -1.229$ V. The other possible product is $S_2O_8^{2-}$(aq), which will not form, since it has a considerably less favorable half-cell potential, $-E°[S_2O_8^{2-}(aq)|SO_4^{2-}(aq)] = -2.01$ V. $H_2(g)$ is formed at the cathode.

$$\text{mol } O_2 = 3.75\text{ h} \times \frac{3600\text{ s}}{1\text{ h}} \times \frac{2.83\text{ C}}{1\text{ s}} \times \frac{1\text{ mol e}^-}{96485\text{ C}} \times \frac{1\text{ mol } O_2}{4\text{ mol e}^-} = 0.0990\text{ mol } O_2$$

The vapor pressure of water at 25°C, from Table 12-2, is 23.8 mmHg.

$$V = \frac{nRT}{P} = \frac{0.0990\text{ mol} \times 0.08206\text{ L atm mol}^{-1}\text{ K}^{-1} \times 298\text{ K}}{(742 - 23.8)\text{ mmHg} \times \dfrac{1\text{ atm}}{760\text{ mmHg}}} = 2.56\text{ L } O_2(g)$$

63. **(a)** $Zn^{2+}(aq) + 2\,e^- \longrightarrow Zn(s)$

$$\text{mass of Zn} = 42.5\text{ min} \times \frac{60\text{ s}}{1\text{ min}} \times \frac{1.87\text{ C}}{1\text{ s}} \times \frac{1\text{ mol e}^-}{96485\text{ C}} \times \frac{1\text{ mol Zn}}{2\text{ mol e}^-} \times \frac{65.39\text{ g Zn}}{1\text{ mol Zn}} = 1.62\text{ g Zn}$$

(b) $2\,I^-(aq) \longrightarrow I_2(s) + 2\,e^-$

$$\text{time needed} = 2.79\text{ g } I_2 \times \frac{1\text{ mol } I_2}{253.8\text{ g } I_2} \times \frac{2\text{ mol e}^-}{1\text{ mol } I_2} \times \frac{96485\text{ C}}{1\text{ mol e}^-} \times \frac{1\text{ s}}{1.75\text{ C}} \times \frac{1\text{ min}}{60\text{ s}} = 20.2\text{ min}$$

64. **(a)** $Cu^{2+}(aq) + 2\,e^- \longrightarrow Cu(s)$

$$\text{mmol } Cu^{2+}\text{ removed} = 282\text{ s} \times \frac{2.68\text{ C}}{1\text{ s}} \times \frac{1\text{ mol e}^-}{96485\text{ C}} \times \frac{1\text{ mol } Cu^{2+}}{2\text{ mol e}^-} \times \frac{1000\text{ mmol}}{1\text{ mol}}$$

$$= 3.92\text{ mmol } Cu^{2+}$$

$$\text{decrease in } [Cu^{2+}] = \frac{3.92\text{ mmol } Cu^{2+}}{425\text{ mL}} = 0.00922\text{ M}$$

final $[Cu^{2+}] = 0.366\text{ M} - 0.00922\text{ M} = 0.357$ M

(b) mmol Ag^+ removed = 255 mL (0.196 M − 0.175 M) = 5.36 mmol Ag^+

$$\text{time needed} = 5.36\text{ mmol } Ag^+ \times \frac{1\text{ mol } Ag^+}{1000\text{ mmol } Ag^+} \times \frac{1\text{ mol e}^-}{1\text{ mol } Ag^+} \times \frac{96485\text{ C}}{1\text{ mol e}^-} \times \frac{1\text{ s}}{1.84\text{ C}} = 281\text{ s}$$

65. **(a)** $$\text{charge} = 1.206\text{ g Ag} \times \frac{1\text{ mol Ag}}{107.87\text{ g Ag}} \times \frac{1\text{ mol e}^-}{1\text{ mol Ag}} \times \frac{96{,}485\text{C}}{1\text{ mol e}^-} = 1079\text{ C}$$

(b) $$\text{current} = \frac{1079\text{ C}}{1412\text{ s}} = 0.7642\text{ A}$$

66. **(a)** Anode, Oxdn: $2\,H_2O \longrightarrow 4\,H^+(aq) + 4\,e^- + O_2(g)$ $\quad -E° = -1.229$ V

Cathode, Redn: $\{Ag^+(aq) + e^- \longrightarrow Ag(s)\} \times 4$ $\quad E° = +0.800$ V

Net: $2\,H_2O + 4\,Ag^+(aq) \longrightarrow 4\,H^+(aq) + O_2(g) + 4\,Ag(s)$ $\quad E°_{cell} = -0.429$ V

(b) $$\text{charge} = (25.8639 - 25.0782)\text{g Ag} \times \frac{1\text{ mol Ag}}{107.87\text{ g Ag}} \times \frac{1\text{ mol e}^-}{1\text{ mol Ag}} \times \frac{96485\text{ C}}{1\text{ mol e}^-} = 702.8\text{ C}$$

$$\text{current} = \frac{702.8\text{ C}}{2.00\text{ h}} \times \frac{1\text{ h}}{3600\text{ s}} = 0.0976\text{ A}$$

(c) The gas is oxygen.

$$V = \frac{nRT}{P} = \frac{\left(702.8\text{ C} \times \dfrac{1\text{ mol e}^-}{96485\text{ C}} \times \dfrac{1\text{ mol } O_2}{4\text{ mol e}^-}\right) 0.08206\,\dfrac{\text{L atm}}{\text{mol K}}\,(23°\text{C} + 273)\text{K}}{755\text{ mmHg} \times \dfrac{1\text{ atm}}{760\text{ mmHg}}}$$

$$= 0.0445\text{ L } O_2 \times \frac{1000\text{ mL}}{1\text{ L}} = 44.5\text{ mL}$$

FEATURE PROBLEMS

A. **1.** **(a)**

$Cr^{3+}(aq) + e^- \longrightarrow Cr^{2+}(aq)$ $\quad \Delta G°_A = -1 \times \mathcal{F} \times (-0.424\text{ V})$

$Cr^{2+}(aq) + 2\,e^- \longrightarrow Cr(s)$ $\quad \Delta G°_B = -2 \times \mathcal{F} \times (-0.90\text{ V})$

Result: $Cr^{3+}(aq) + 3\,e^- \longrightarrow Cr(s)$ $\quad \Delta G°_C = \Delta G°_A + \Delta G°_B = -3 \times \mathcal{F} \times E°$

$$-1 \times \mathcal{F} \times (-0.424\ \text{V}) - 2 \times \mathcal{F} \times (-0.90\ \text{V}) = -3 \times \mathcal{F} \times E°$$

$$(-0.424\ \text{V}) + 2 \times (-0.90\ \text{V}) = 3\ E° \qquad E° = \frac{(-0.424\ \text{V}) + 2 \times (-0.90\ \text{V})}{3} = -0.74\ \text{V}$$

(b) $MnO_2(s) + 4\ H^+(aq) + 2\ e^- \longrightarrow Mn^{2+}(aq) + 2\ H_2O \qquad \Delta G°_A = -2 \times \mathcal{F} \times (+1.23\ \text{V})$

$Mn^{2+}(aq) + 2\ e^- \longrightarrow Mn(s) \qquad \Delta G°_B = -2 \times \mathcal{F} \times (-1.18\ \text{V})$

Result: $MnO_2(s) + 4\ H^+(aq) + 2\ e^- \longrightarrow Mn(s) + 2\ H_2O \qquad \Delta G°_C = \Delta G°_A + \Delta G°_B = -4 \times \mathcal{F} \times E°$

$$-2 \times \mathcal{F} \times (-0.424\ \text{V}) - 2 \times \mathcal{F} \times (-0.90\ \text{V}) = -3 \times \mathcal{F} \times E°$$

$$2 \times (+1.23\ \text{V}) + 2 \times (-1.18\ \text{V}) = 4\ E° \qquad E° = \frac{2 \times (-1.18\ \text{V}) + 2 \times (+1.23\ \text{V})}{4} = +0.025\ \text{V}$$

2. Anode: $\{Al(s) \longrightarrow Al^{3+}(aq) + 3\ e^-\} \times 2 \qquad \Delta G°_A = -6 \times \mathcal{F} \times (+1.676\ \text{V})$

Cathode: $\{Cu^{2+}(aq) + 2\ e^- \longrightarrow Cu(s)\} \times 3 \qquad \Delta G°_B = -6 \times \mathcal{F} \times (+0.337\ \text{V})$

Net: $2\ Al(s) + 3\ Cu^{2+}(aq) \longrightarrow 2\ Al^{3+}(aq) + 3\ Cu(s) \qquad \Delta G°_C = \Delta G°_A + \Delta G°_B = -6 \times \mathcal{F} \times E°$

$6 \times (+1.676\ \text{V}) + 6 \times (+0.337\ \text{V}) = 6\ E° \qquad E° = +1.676\ \text{V} + 0.337\ \text{V} = 2.013\ \text{V}$

B. 1. Anode: $H_2(g,\ 1\ \text{atm}) \longrightarrow 2\ H^+(1\ M) + 2\ e^- \qquad -E = -0.0000\ \text{V}$

Cathode: $\{Ag^+(x\ M) + e^- \longrightarrow Ag(s)\} \times 2 \qquad E° = 0.800\ \text{V}$

Net: $H_2(g,\ 1\ \text{atm}) + 2\ Ag^+(aq) \longrightarrow 2\ H^+(1\ M) + 2\ Ag(s) \qquad E°_{cell} = 0.800$

2. Since the voltage in the anode half-cell remains constant, we use the Nernst equation to calculate the half-cell voltage in the cathode half-cell, with two moles of electrons. This is then added to $-E$ for the anode half-cell. Because $-E° = 0.000$ for the anode half cell, $E_{cell} = E_{cathode}$

$$E = E° - \frac{0.0592}{2} \log \frac{1}{[Ag^+]^2} = 0.800 + 0.0592 \log [Ag^+]$$

3. *Initially*: $[Ag^+] = 0.0100 \qquad E = 0.800 + 0.0592 \log 0.0100 = 0.682\ \text{V} = E_{cell}$

At the equivalence point, we have a saturated solution of AgI, for which

$$[Ag^+] = \sqrt{K_{sp}[AgI]} = \sqrt{8.5 \times 10^{-17}} = 9.2 \times 10^{-9}$$

$$E = 0.800 + 0.0592 \log (9.2 \times 10^{-9}) = 0.324\ \text{V} = E_{cell}$$

Note that 50.0 mL of titrant is required for the titration, since both $AgNO_3$ and KI have the same concentrations and they react in equimolar ratios.

After 10.0 mL of titrant is added, the total volume of solution is 60.0 mL and the unreacted Ag^+ is that in the untitrated 40.0 mL of 0.0100 M $AgNO_3(aq)$.

$$[Ag^+] = \frac{40.0\ \text{mL} \times 0.0100\ \text{M}\ Ag^+}{60.0\ \text{mL}} = 0.00667\ \text{M} \qquad E = 0.800 + 0.0592 \log (0.00667) = 0.671\ \text{V} = E_{cell}$$

After 20.0 mL of titrant is added, the total volume of solution is 70.0 mL and the unreacted Ag^+ is that in the untitrated 30.0 mL of 0.0100 M $AgNO_3(aq)$.

$$[Ag^+] = \frac{30.0\ \text{mL} \times 0.0100\ \text{M}\ Ag^+}{70.0\ \text{mL}} = 0.00429\ \text{M} \qquad E = 0.800 + 0.0592 \log (0.00429) = 0.660\ \text{V} = E_{cell}$$

After 30.0 mL of titrant is added, the total volume of solution is 80.0 mL and the unreacted Ag^+ is that in the untitrated 20.0 mL of 0.0100 M $AgNO_3(aq)$.

$$[Ag^+] = \frac{20.0\ \text{mL} \times 0.0100\ \text{M}\ Ag^+}{80.0\ \text{mL}} = 0.00250\ \text{M} \qquad E = 0.800 + 0.0592 \log (0.00250) = 0.646\ \text{V} = E_{cell}$$

After 40.0 mL of titrant is added, the total volume of solution is 90.0 mL and the unreacted Ag^+ is that in the untitrated 10.0 mL of 0.0100 M $AgNO_3(aq)$.

$$[Ag^+] = \frac{10.0\ \text{mL} \times 0.0100\ \text{M}\ Ag^+}{90.0\ \text{mL}} = 0.00111\ \text{M} \qquad E = 0.800 + 0.0592 \log (0.00111) = 0.625\ \text{V} = E_{cell}$$

After 45.0 mL of titrant is added, the total volume of solution is 95.0 mL and the unreacted Ag^+ is that in the untitrated 5.0 mL of 0.0100 M $AgNO_3(aq)$.

$$[Ag^+] = \frac{5.0\ \text{mL} \times 0.0100\ \text{M}\ Ag^+}{95.0\ \text{mL}} = 0.0053\ \text{M} \qquad E = 0.800 + 0.0592 \log (0.00053) = 0.605\ \text{V} = E_{cell}$$

After 49.0 mL of titrant is added, the total volume of solution is 99.0 mL and the unreacted Ag^+ is that in the untitrated 1.0 mL of 0.0100 M $AgNO_3(aq)$.

$$[Ag^+] = \frac{1.0 \text{ mL} \times 0.0100 \text{ M } Ag^+}{99.0 \text{ mL}} = 0.00010 \text{ M} \qquad E = 0.800 + 0.0592 \log (0.00010) = 0.563 \text{ V} = E_{cell}$$

After the equivalence point, the $[Ag^+]$ is determined by the $[I^-]$ resulting from the excess KI(aq). *When 51.0 mL of titrant is added*, the total volume of solution is 101.0 mL and the excess I^- is that in 1.0 mL of 0.0100 M KI(aq).

$$[I^-] = \frac{1.0 \text{ mL} \times 0.0100 \text{ M } I^-}{101.0 \text{ mL}} = 0.000099 \text{ M} \qquad [Ag^+] = \frac{K_{sp}}{[I^-]} = \frac{8.5 \times 10^{-17}}{0.000099} = 8.6 \times 10^{-13}$$

$E = 0.800 + 0.0592 \log (8.6 \times 10^{-13}) = 0.086 \text{ V} = E_{cell}$

When 55.0 mL of titrant is added, the total volume of solution is 105.0 mL and the excess I^- is that in 5.0 mL of 0.0100 M KI(aq).

$$[I^-] = \frac{5.0 \text{ mL} \times 0.0100 \text{ M } I^-}{105.0 \text{ mL}} = 0.000048 \text{ M} \qquad [Ag^+] = \frac{K_{sp}}{[I^-]} = \frac{8.5 \times 10^{-17}}{0.000048} = 1.8 \times 10^{-13}$$

$E = 0.800 + 0.0592 \log (8.5 \times 10^{-13}) = 0.045 \text{ V} = E_{cell}$

When 60.0 mL of titrant is added, the total volume of solution is 110.0 mL and the excess I^- is that in 10.0 mL of 0.0100 M KI(aq).

$$[I^-] = \frac{10.0 \text{ mL} \times 0.0100 \text{ M } I^-}{110.0 \text{ mL}} = 0.00091 \text{ M} \qquad [Ag^+] = \frac{K_{sp}}{[I^-]} = \frac{8.5 \times 10^{-17}}{0.00091} = 9.3 \times 10^{-14}$$

$E = 0.800 + 0.0592 \log (9.3 \times 10^{-14}) = 0.029 \text{ V} = E_{cell}$

(c) The titration curve is presented below.

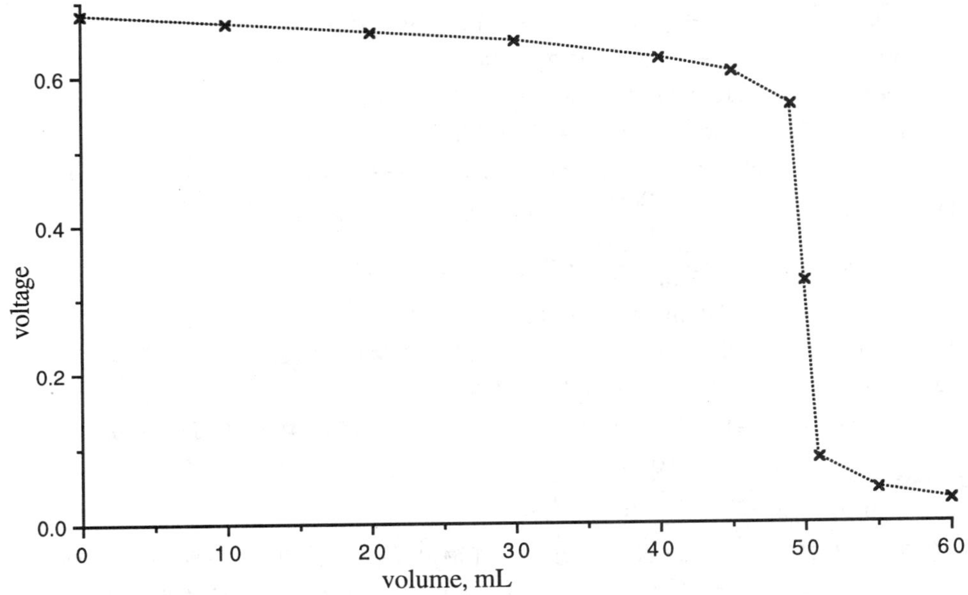

C. As soon as the iron and the copper came into contact, and electrochemical cell was created, in which the more electrochemically active metal (Fe) oxidized. In this way the iron behaved as a sacrificial anode, protecting the copper from corrosion. The two half-reactions and the net cell reaction follow.

Anode, Oxidation:	$Fe(s) \longrightarrow Fe^{2+} + 2\,e^-$	$-E° = +0.440$ V
Cathode, Reduction:	$Cu^{2+}(aq) + 2\,e^- \longrightarrow Cu(s)$	$E° = +0.337$ V
Net cell reaction:	$Fe(s) + Cu^{2+}(aq) \longrightarrow Cu(s) + Fe^{2+}(aq)$	$E°_{cell} = +0.777$ V

Note that, because of the presence of the iron, and its electrical contact with the copper, any copper that does corrode will be reduced back to the metal.

D. During corrosion, the metal that corrodes produces electrons. Hence, this metal is negatively charged. Thus, we wish to make the pipe the cathode, where reduction occurs, in an electrochemical cell, since corrosion proceeds by oxidation. Thus, the inert electrode will be the anode, and any possible oxidation will occur there. This means that the inert electode will be the positive electrode and the pipe will be the negative electrode.

E. 1. The metal has to have a reduction potential more negative than –0.691 V, so that its oxidation can reverse the tarnishing reaction's –0.691 V reduction potential. A metal that is inexpensive and readily available, and yet will not react with water is aluminum, with $E° = -1.676$ V. We expect that there might be an overpotential associated with the tarnishing reaction and thuse we do not choose zinc, $E° = -0.763$ V.

2. Oxidation: $\{Al(s) \longrightarrow Al^{3+}(aq) + 3\,e^-\} \times 2$

Reduction: $\{Ag_2S(s) + 2\,e^- \longrightarrow 2\,Ag(s) + S^{2-}(aq)\} \quad \times 3$

Net: $2\,Al(s) + 3\,Ag_2S(s) \longrightarrow 6\,Ag(s) + 3\,S^{2-}(aq) + 2\,Al^{3+}(aq)$

3. The dissolved $NaHCO_3(s)$ serves as an electrolyte. It would also enhance the electrical contact between the two objects.

4. There are, of course, several chemicals involved: Al, H_2O, and $NaHCO_3$ at a minimum. Although the aluminum plate will wear away very slowly because silver tarnish is quite thin, it will eventually erode away. We should be able to detect weight loss after many uses.

F. $Li^+(aq) + e^- \longrightarrow Li(s)$ We calculate $\Delta G°$ for the reaction as follows.

$\Delta G° = \Delta G_f^\circ[Li(s)] - \Delta G_f^\circ[Li^+(aq)] = 0\text{ kJ/mol} - (-293.3\text{ kJ/mol}) = +293.3\text{ kJ/mol} = -n\mathcal{F}E°$

$$E° = -\frac{293.3 \times 10^3\text{ J}}{1\text{ mol e}^- \times \dfrac{96{,}485\text{ C}}{1\text{ mol e}^-} \times \dfrac{1\text{ J}}{1\text{ V·C}}} = -3.040\text{ V}$$

This is precisely the value in Appendix D.

G. 1. (1) anode: $Na(s) \longrightarrow Na^+(\text{in ethylamine}) + e^-$

cathode: $Na^+(\text{in ethylamine}) \longrightarrow Na(\text{amalg, } 0.206\ \%)$

net: $Na(s) \longrightarrow Na(\text{amalg, } 0.206\%)$

(2) anode: $2\,Na(\text{amalg, } 0.206\%) \longrightarrow 2\,Na^+(1\text{ M}) + 2\,e^-$

cathode: $2\,H^+(aq) + 2\,e^- \longrightarrow H_2(g, 1\text{ atm})$

net: $2\,Na(\text{amalg, } 0.206\%) + 2\,H^+(aq) \longrightarrow 2\,Na^+(1\text{ M}) + H_2(g, 1\text{ atm})$

2. (1) $\Delta G° = -1\text{ mol e}^- \times \dfrac{96{,}485\text{ C}}{1\text{ mol e}^-} \times 0.8453\text{ V} = -8.156 \times 10^4\text{ J}$

(2) $\Delta G° = -2\text{ mol e}^- \times \dfrac{96{,}485\text{ C}}{1\text{ mol e}^-} \times 1.8673\text{ V} = -36.033 \times 10^4\text{ J}$

3. (1) $2\,Na(s) \longrightarrow 2\,Na(\text{amalg, } 0.206\%)$ $\quad \Delta G_1^\circ = -2 \times 8.156 \times 10^4\text{ J}$

(2) $2\,Na(\text{amalg, } 0.206\%) + 2\,H^+(aq) \longrightarrow 2\,Na^+(1\text{ M}) + H_2(g, 1\text{ atm})$ $\quad \Delta G_2^\circ = -36.033 \times 10^4\text{ J}$

overall: $2\,Na(s) + 2\,H^+(aq) \longrightarrow 2\,Na^+(1\text{ M}) + H_2(g, 1\text{ atm})$

$\Delta G° = \Delta G_1^\circ + \Delta G_2^\circ = -16.312 \times 10^4\text{ J} - 36.033 \times 10^4\text{ J} = -52.345 \times 10^4\text{ J} = -n\mathcal{F}E°$

4. $$E° = -\frac{-52.345 \times 10^4\text{ J}}{2\text{ mol e}^- \times \dfrac{96{,}485\text{ C}}{1\text{ mol e}^-} \times \dfrac{1\text{ J}}{1\text{ V·C}}} = +2.713\text{ V}$$

$+2.713\text{ V} = E°[H^+(1\text{ M})|H_2(1\text{ atm})] - E°[Na^+(1\text{ M})|Na(s)] = 0.000\text{ V} - E°[Na^+(1\text{ M})|Na(s)]$

$E°[Na^+(1\text{ M})|Na(s)] = -2.713\text{ V}$ This is precisely the value in Appendix D.

22 Chemistry of the Representative (Main Group) Elements I: Metals

Practice Examples

1A From Figure 22-2, the route from sodium chloride to sodium nitrate begins with electrolysis of $NaCl(aq)$ to form $NaOH(aq)$ $2\,NaCl(aq) + 2\,H_2O(l) \xrightarrow{\text{electrolysis}} 2\,NaOH(aq) + H_2(g) + Cl_2(g)$
followed by addition of $NO_2(g)$ to $NaOH(aq)$. $2\,NaOH(aq) + 3\,NO_2(g) \longrightarrow 2\,NaNO_3(aq) + NO(g) + H_2O$

1B From Figure 22-2, we see that the route from sodium chloride to sodium thiosulfate (1) begins with the electrolysis of $NaCl(aq)$ to produce $NaOH(aq)$,
$2\,NaCl(aq) + 2\,H_2O(l) \xrightarrow{\text{electrolysis}} 2\,NaOH(aq) + H_2(g) + Cl_2(g)$
(2) continues through the reaction of $SO_2(g)$ with the $NaOH(aq)$ in an acid-base reaction [$SO_2(g)$ is an acid anhydride] to produce $Na_2SO_3(aq)$: $2\,NaOH(aq) + SO_2(g) \longrightarrow Na_2SO_3(aq) + H_2O$
and (3) concludes with the addition of S to the boiling solution: $Na_2SO_3(aq) + S \xrightarrow{\text{boil}} Na_2S_2O_3(aq)$

2A The solution with the lower freezing point will have the higher molality of ions (moles of ions per kilogram of solvent), which in this case means the most ions in 1.00 gram of solute. With compounds of the same formula type (such as NaCl and KF, or as $MgCl_2$ and $CaCl_2$), the one of lower molar mass has more ions per gram. This means that $MgCl_2$ (95.21 g/mol) has more ions per gram than $CaCl_2$ (110.99 g/mol), and that KF (58.10 g/mol) has more ions per gram than NaCl (58.44 g/mol). We assume that the van't Hoff factor is integral in all cases.

$$\frac{\text{mol ions}}{\text{gram KF}} = \frac{2\text{ mol ions}}{1\text{ mol KF}} \times \frac{1\text{ mol KF}}{58.10\text{ g KF}} = 0.0344\text{ mol ions/g KF}$$

$$\frac{\text{mol ions}}{\text{gram }MgCl_2} = \frac{3\text{ mol ions}}{1\text{ mol }MgCl_2} \times \frac{1\text{ mol }MgCl_2}{95.21\text{ g }MgCl_2} = 0.0315\text{ mol ions/g }MgCl_2$$

The KF(aq) solution should begin to freeze at a lower temperature.

2B The solutions will have the same freezing points if they have the same number of particles per kg of water. We assume that these two solutions are dilute enough that the van't Hoff factors will be integral: 2 for NaCl (1 Na^+ ion and 1 Cl^- ion) and 3 for $CaCl_2$ (1 Ca^{2+} ion and 2 Cl^- ions).

$$\text{mass }CaCl_2 = 5.00\text{ g NaCl} \times \frac{1\text{ mol NaCl}}{58.44\text{ g NaCl}} \times \frac{2\text{ mol ions}}{1\text{ mol NaCl}}\ (= 0.171\text{ mol ions})$$
$$\times \frac{1\text{ mol }CaCl_2}{3\text{ mol ions}} \times \frac{110.98\text{ g }CaCl_2}{1\text{ mol }CaCl_2} = 6.33\text{ g }CaCl_2$$

We use the amount of moles of ions to determine the molality of the solution.

$$\Delta T_f = K_f \times m = 1.86\ °C/m \times \frac{0.171\text{ mol ions}}{1.000\text{ kg water}} = 0.318°C \qquad t_f = -0.318°C$$

3A The equilibrium that involves the two species mentioned is the following.
$CaCO_3(s) + H_2O + CO_2 \rightleftharpoons Ca(HCO_3)_2(aq) \quad K = 2.2 \times 10^{-5}$

The question essentially asks whether the addition of CaO will cause this reaction to shift left. The means by which this happens is that, first, CaO(s) forms the strong base $Ca(OH)_2(aq)$. When CaO(s) reacts with sufficient water to dissolve completely, it is present in solution as $Ca^{2+}(aq)$ and $OH^-(aq)$ ions.
$CaO(s) + H_2O \rightleftharpoons Ca^{2+}(aq) + 2\,OH^-(aq)$

This strong base, $OH^-(aq)$ then reacts with the weak acid, H_2CO_3(aq, or $CO_2 \cdot H_2O$), to produce the very weak base $CO_3^{2-}(aq)$ and the very weak acid H_2O. [Recall that acid-base reactions tend to produce weak acids and bases from stronger ones.]
$H_2CO_3(aq, CO_2 \cdot H_2O) + 2\,OH^-(aq) \longrightarrow 2\,H_2O + CO_3^{2-}(aq)$

3B The added $NH_3(aq)$ is a base that reacts with bicarbonate ion.

$HCO_3^-(aq) + NH_3(aq) \longrightarrow NH_4^+(aq) + CO_3^{2-}(aq)$ $\quad K_A = K_1\{HCO_3^-\} \times K_b\{NH_3\}$

$= 4.7 \times 10^{-11} \times 1.8 \times 10^{-5} = 8.5 \times 10^{-16}$

And the carbonate ion produced reacts with free calcium ion.

$Ca^{2+}(aq) + CO_3^{2-}(aq) \longrightarrow CaCO_3(s)$ $\quad K_S = 1/K_{sp} = 1/2.8 \times 10^{-9} = 3.6 \times 10^8$

The overall reaction is predicted to proceed to a slight extent.

$Ca(HCO_3)_2(aq) + NH_3(aq) \longrightarrow CaCO_3(s) + HCO_3^-(aq) + NH_4^+(aq)$ $\quad K = K_A \times K_S = 3.1 \times 10^{-7}$

SUMMARIZING EXAMPLE CALCULATIONS

1. The amount of H^+ present in the sample equals the amount of OH^- used for its titration.

$$\text{amount } H^+ = 7.59 \text{ mL} \times \frac{0.0133 \text{ mmol NaOH}}{1 \text{ mL}} \times \frac{1 \text{ mmol } OH^-}{1 \text{ mmol NaOH}} \times \frac{1 \text{ mmol } H^+}{1 \text{ mmol } OH^-} = 0.101 \text{ mmol } H^+$$

2. This H^+ was produced by 2 H^+ ions being exchanged for every Ca^{2+} ion in the original sample. We determine the mass of Ca^{2+} in that sample.

$$\text{mass } Ca^{2+} = 0.101 \text{ mmol } H^+ \times \frac{1 \text{ mmol } Ca^{2+}}{2 \text{ mmol } H^+} \times \frac{1 \text{ mol } Ca^{2+}}{1000 \text{ mmol } Ca^{2+}} \times \frac{40.078 \text{ g } Ca^{2+}}{1 \text{ mol } Ca^{2+}} = 2.02 \times 10^{-3} \text{ g } Ca^{2+}$$

3. If we assume a density of 1.00 g/mL for the water sample, we can calculate the quantity of Ca^{2+} as ppm.

$$\text{ppm } Ca^{2+} = \frac{2.02 \times 10^{-3} \text{ g } Ca^{2+}}{25.00 \text{ mL sample}} \times \frac{1 \text{ mL}}{1.00 \text{ g sample}} \times \frac{1 \text{ ppm}}{10^{-6} \text{ g } Ca^{2+}} = 80.8 \text{ ppm } Ca^{2+}$$

REVIEW QUESTIONS

1. **(a)** A dimer is a molecule that is formed by the joining together of two identical simpler molecules (called monomers). For instance, N_2O_4 is a dimer of NO_2.
(b) An adduct is formed when two simple molecules come together and a covalent bond, often a coordinate covalent bond, forms between them. $NH_3 \cdot BF_3$ is an adduct of NH_3 and BF_3.
(c) Calcination is the process of heating a carbonate strongly to drive off $CO_2(g)$ and form an oxide.
(d) An amphoteric oxide is one that will react with either strong acid or with strong base.

2. **(a)** A diagonal relationship refers to the similarity between two elements that are diagonally related to each other in the periodic table, such as Li and Mg, or Be and Al. Even though these elements are in different periodic families, they have some similarities in behavior.
(b) Deionized water is prepared by ion exchange by passing it through material in which the ions H^+ or OH^- are present. These ions substitute or exchange for the ions in the water, in a two-step process: first the cations in the water are replaced by H^+, then the anions by OH^-. The result is water virtually free of ionic contaminants.
(c) The thermite reaction refers to the reduction of a metal oxide with another, more active, metal in a highly exothermic reaction, for instance: $Fe_2O_3(s) + 2\,Al(s) \longrightarrow 2\,Fe(\text{liquid!}) + Al_2O_3(s)$
(d) The "inert pair" effect refers to the tendency of heavier representative metals to have oxidation states in which they have lost their *np* electrons, but not their ns^2 electrons. Thus, they have an oxidation state two units less than their periodic table family number. Examples include Pb^{2+}, Sn^{2+}, Bi^{3+}, Sb^{3+}, Tl^+.

3. **(a)** The peroxide ion is O_2^{2-}; the superoxide ion is O_2^-.
(b) Quicklime is the common name for $CaO(s)$; slaked lime is the common name for $Ca(OH)_2(s)$.
(c) Temporary hard water contains divalent cations, such as Ca^{2+}, Mg^{2+}, and Fe^{2+} and the bicarbonate anion, HCO_3^-. Heating produces, H_2O, $CO_2(g)$, and a carbonate precipitate. Permanent hard water does not form a precipitate upon heating, since the anion is one such as SO_4^{2-} that is thermally stable.
(d) A soap is a potassium or sodium salt of a natural or slightly altered carboxylic acid (—COOH) that has a long hydrocarbon chain. A detergent is a synthetic sodium or potassium salt of a long hydrocarbon chain sulfonic acid (—OSO_3H).

4. **(a)** PbO lead(II) oxide **(b)** SnF_2 tin(II) fluoride
(c) $CaSO_4 \cdot \frac{1}{2}H_2O$ calcium sulfate hemihydrate **(d)** Li_3N lithium nitride
(e) $Ca(OH)_2$ calcium hydroxide **(f)** KO_2 potassium superoxide
(f) $Mg(HCO_3)_2$ magnesium hydrogen carbonate

5. **(a)** $Li_2CO_3(s) \xrightarrow{\Delta} Li_2O(s) + CO_2(g)$ **(b)** $CaCO_3(s) + 2\,HCl(aq) \longrightarrow CaCl_2(aq) + CO_2(g) + H_2O$

(c) $2\ Al(s) + 2\ Na^+(aq) + 2\ OH^-(aq) + 6\ H_2O \longrightarrow 2\ Na^+(aq) + 2\ [Al(OH)_4]^-(aq) + 3\ H_2(g)$
(d) $BaO(s) + H_2O \longrightarrow Ba(OH)_2$(s, in limited water)
(e) $2\ Na_2O_2(s) + 2\ CO_2(g) \longrightarrow 2\ Na_2CO_3(s) + O_2(g)$

6. (a) $MgCO_3(s) + 2\ HCl(aq) \longrightarrow MgCl_2(aq) + CO_2(g) + H_2O$
(b) $2\ Na(s) + 2\ H_2O \longrightarrow 2\ NaOH(aq) + H_2(g)$
$2\ Al(s) + 2\ NaOH(aq) + 6\ H_2O \longrightarrow 2\ Na[Al(OH)]_4(aq) + 3\ H_2(g)$
(c) $2\ NaCl(s) + H_2SO_4$(conc., aq) $\longrightarrow 2\ HCl(g) + Na_2SO_4(s)$

7. (a) $K_2CO_3(aq) + Ba(OH)_2(aq) \longrightarrow BaCO_3(s) + 2\ KOH(aq)$
(b) $Mg(HCO_3)_2(aq) \xrightarrow{\Delta} MgCO_3(s) + CO_2(g) + H_2O(g)$ (c) $SnO(s) + C(s) \xrightarrow{\Delta} Sn(l) + CO(g)$
(d) $CaF_2(s) + H_2SO_4$(conc., aq) $\longrightarrow 2\ HF(g) + CaSO_4(s)$
(e) $NaHCO_3(s) + HCl(aq) \longrightarrow NaCl(aq) + H_2O + CO_2(g)$
(f) $PbO_2(s) + 4\ HBr(aq) \longrightarrow PbBr_2(s) + Br_2(l) + 2\ H_2O$

8. Replace the names with chemical formulas and balance the result.
$CaSO_4{\cdot}2H_2O(s) + (NH_4)_2CO_3(aq) \longrightarrow (NH_4)_2SO_4(aq) + CaCO_3(s) + 2\ H_2O$

9. Temporary hard water is softened by the addition of an alkaline (basic) material. All the substances listed form alkaline solutions except NH_4Cl. NH_4^+ hydrolyzes to form an acidic solution, and could not be used to soften temporary hard water.

10. (a) $SrCO_3(s) \xrightarrow{\Delta} SrO(s) + CO_2(g)$ (b) $Al_2O_3(s) \xrightarrow{\Delta}$ no reaction
(c) $Li_2CO_3(g) \xrightarrow{\Delta} Li_2O(s) + CO_2(g)$

11. (a) $Pb(NO_3)_2(aq) + 2\ NaHCO_3(aq) \longrightarrow PbCO_3(s) + H_2O + CO_2(g) + 2\ NaNO_3(aq)$
(b) $Li_2O(s) + (NH_4)_2CO_3(aq) \longrightarrow Li_2CO_3(s) + 2\ NH_3(g) + H_2O$ $NH_4^+(aq)$ is present in very limited amount because the solution is strongly basic; Li_2O is the anhydride of a strong base.
(c) $H_2SO_4(aq) + BaO_2(aq) \longrightarrow H_2O_2(aq) + BaSO_4(s)$
(d) $PbO(s) + OCl^-(aq) \longrightarrow PbO_2(s) + Cl^-(aq)$

12. To answer this one, you really do not need to formally study chemistry. You just have to observe the world around you. Neither aluminum foil nor aluminum cookware reacts with water. But to explain that answer, you have to know that of these four active metals, only aluminum forms an oxide that adheres tightly to its surface and prevents further reaction.

13. The correct answer is **(a)**. Balanced equations for the three pairs follow.

$Ca(s) + 2\ H_2O \longrightarrow Ca(OH)_2(s) + H_2(g)$ $CaH_2(s) + 2\ H_2O \longrightarrow Ca(OH)_2(s) + 2\ H_2(g)$
$2\ Na(s) + 2\ H_2O \longrightarrow 2\ NaOH(aq) + H_2(g)$ $2\ Na_2O_2(s) + 2\ H_2O \longrightarrow 4\ NaOH(aq) + O_2(g)$
$2\ K(s) + 2\ H_2O \longrightarrow 2\ KOH(aq) + H_2(g)$ $4\ KO_2(s) + 2\ H_2O \longrightarrow 4\ KOH(aq) + 3\ O_2(g)$

14. (a) Stalactites are $CaCO_3(s)$ (b) Gypsum is $CaSO_4{\cdot}2H_2O$
(c) "bathtub ring" is a salt of Ca^{2+} and a long-carbon chain carboxylate anion. An example would be calcium palmitate: $Ca[CH_3(CH_2)_{14}COO]_2(s)$
(d) barium "milkshake" is an aqueous temporary suspension of $BaSO_4(s)$
(e) blue sapphires are Al_2O_3 with Fe^{3+} and Ti^{4+} ions replacing some Al^{3+} ions.

EXERCISES

Alkali (Group 1A) Metals

15. (a) $2\ Cs(s) + Cl_2(g) \longrightarrow 2\ CsCl(s)$ (b) $2\ Na(s) + O_2(g) \longrightarrow Na_2O_2(s)$
(c) $Li_2CO_3(s) \xrightarrow{\Delta} Li_2O(s) + CO_2(g)$ (d) $Na_2SO_4(s) + 4\ C(s) \longrightarrow Na_2S(s) + 4\ CO(g)$
(e) $K(s) + O_2(g) \longrightarrow KO_2(s)$

16. **(a)** $2\ Rb(s) + 2\ H_2O(l) \longrightarrow RbOH_2(aq) + H_2(g)$
(b) $2\ KHCO_3(aq) \xrightarrow{\Delta} K_2CO_3(aq) + H_2O + CO_2(g)$
(c) $2\ Li(s) + O_2(g) \longrightarrow Li_2O_2(s)$ **(d)** $2\ KCl(s) + H_2SO_4(aq) \longrightarrow K_2SO_4(aq)(aq) + 2\ HCl(g)$
(f) $2\ LiH(s) + 2\ H_2O(l) \longrightarrow 2\ LiOH(aq) + H_2(g)$

17. Both LiCl and KCl are soluble in water, but Li_3PO_4 is not very soluble. Hence the addition of $K_3PO_4(aq)$ to a solution of the white solid will produce a precipitate if the white solid is LiCl, but no precipitate if the white solid is KCl. Another possibility is a flame test; lithium gives a red color to a flame, while the potassium flame test is violet.

18. When heated, Li_2CO_3 decomposes to $CO_2(g)$, $H_2O(g)$, and $Li_2O(l)$. $Na_2CO_3(s)$ simply melts when heated. The evolution of $CO_2(g)$ bubbles should be sufficient indication of the difference in behavior. The sodium flame test is yellow-orange, while that of potassium is violet.

19. In addition to $OH^-(aq)$, the other expected product is $O_2(g)$ in the case of peroxide and superoxide. The resulting equations are readily balanced by inspection if one pays attention to charge balance.
Oxide: $O^{2-} + H_2O \longrightarrow 2\ OH^-(aq)$ Peroxide: $2\ O_2^{2-} + 2\ H_2O \longrightarrow 4\ OH^-(aq) + O_2(g)$
Superoxide: $4\ O_2^- + 2\ H_2O \longrightarrow 4\ OH^-(aq) + 3\ O_2$

20. We know that sodium metal was produced at the cathode from the reduction of sodium ion, Na^+. Thus, hydroxide must have been involved in oxidation at the anode. The hydrogen in hydroxide ion already is in its highest oxidation state and thus could not be oxidized. This leaves oxidation of the hydroxide ion to elementary oxygen as the remaining reaction.
Cathode, reduction: $\{Na^+ + e^- \longrightarrow Na(l)\} \times 4$
Anode, oxidation: $4\ OH^- \longrightarrow O_2(g) + 2\ H_2O(g) + 4\ e^-$
Net: $4\ Na^+ + 4\ OH^- \longrightarrow 4\ Na(l) + O_2(g) + 2\ H_2O(g)$

21. **(a)** $H_2(g)$ and $Cl_2(g)$ are produced during the electrolysis of NaCl(aq), summarized in equation (22.7). The electrode reactions are: Anode, Oxdn: $2\ Cl^-(aq) \longrightarrow Cl_2(g) + 2\ e^-$
Cathode, Redn: $2\ H_2O + 2\ e^- \longrightarrow H_2(g) + 2\ OH^-(aq)$
We can compute the amount of OH^- produced at the cathode.

$$\text{mol } OH^- = 2.50 \text{ min} \times \frac{60 \text{ s}}{1 \text{ min}} \times \frac{0.810 \text{ C}}{1 \text{ s}} \times \frac{1 \text{ mol } e^-}{96500 \text{ C}} \times \frac{2 \text{ mol } OH^-}{2 \text{ mol } e^-} = 1.26 \times 10^{-3} \text{ mol } OH^-$$

Then we compute the $[OH^-]$ and, from that, the pH of the solution.

$$[OH^-] = \frac{1.26 \times 10^{-3} \text{ mol } OH^-}{0.872 \text{ L soln}} = 1.45 \times 10^{-3} \text{ M} \qquad pOH = -\log(1.45 \times 10^{-3}) = 2.839$$

$pH = 14.00 - 2.839 = 11.16$

(b) The final pH of the solution might depend on the initial [NaCl] of the solution. Once the $[Cl^-]$ drops to a quite low value, another reaction can occur at the anode.
Anode, Oxdn: $2\ H_2O \longrightarrow 4\ H^+(aq) + 4\ e^- + O_2(g)$
Note that this reaction produces 1 mol H^+ for every mole of electrons. Thus it precisely counters the cathode reaction and, once $Cl_2(g)$ is no longer produced at the anode, the pH of the solution will cease changing. Because the number of moles of OH^- produced at the cathode (1.26×10^{-3} mol OH^-) equals the number of moles of Cl^- consumed at the anode, and this is quite a small number, it is unlikely that the second anode reaction will occur.

22. **(a)**

$$\text{total energy} = 3.0 \text{ V} \times 0.50 \text{ A h} \times \frac{3600 \text{ s}}{1 \text{ hr}} \times \frac{1 \text{ C/s}}{1 \text{ A}} \times \frac{1 \text{ J}}{1 \text{ V}\cdot\text{C}} = 5.4 \times 10^3 \text{ J}$$

$$\text{time} = 5.4 \times 10^3 \text{ J} \times \frac{1 \text{ s}}{5 \times 10^{-6} \text{ J}} = 1._1 \times 10^9 \text{ s} \times \frac{1 \text{ hr}}{3600 \text{ s}} \times \frac{1 \text{ day}}{24 \text{ hr}} \times \frac{1 \text{ y}}{365 \text{ day}} = 34 \text{ y}$$

We obtained the first conversion factor for time as follows.

$$5.0\ \mu\text{W} \times \frac{1 \times 10^{-6} \text{ W}}{1\ \mu\text{W}} \times \frac{1 \text{ J/s}}{1 \text{ W}} = \frac{5.0 \times 10^{-6} \text{ J}}{1 \text{ s}}$$

(b) The capacity of the battery is determined by the mass of Li present.

$$\text{mass Li} = 0.50 \text{ A h} \times \frac{1 \text{ C/s}}{1 \text{ A}} \times \frac{3600 \text{ s}}{1 \text{ h}} \times \frac{1 \text{ mol e}^-}{96500 \text{ C}} \times \frac{1 \text{ mol Li}}{1 \text{ mol e}^-} \times \frac{6.941 \text{ g Li}}{1 \text{ mol Li}} = 0.13 \text{ g Li}$$

23. **(a)** We first compute the mass of $NaHCO_3$ that should be produced from 1.00 ton NaCl, assuming that all of the Na in the NaCl ends up in the $NaHCO_3$. We use the unit, ton-mole, to simplify the calculations.

$$\text{mass } NaHCO_3 = 1.00 \text{ ton NaCl} \times \frac{1 \text{ ton-mol NaCl}}{58.4 \text{ ton NaCl}} \times \frac{1 \text{ ton-mol Na}}{1 \text{ ton-mol NaCl}} \times \frac{1 \text{ ton-mol } NaHCO_3}{1 \text{ ton-mol Na}} \times \frac{84.0 \text{ ton } NaHCO_3}{1 \text{ ton-mol } NaHCO_3} = 1.44 \text{ ton } NaHCO_3$$

$$\% \text{ yield} = \frac{1.03 \text{ ton } NaHCO_3 \text{ produced}}{1.44 \text{ ton } NaHCO_3 \text{ expected}} \times 100\% = 71.5\% \text{ yield}$$

(b) NH_3 is used in the principal step of the Solvay proces to produce a solution in which $NaHCO_3$ is formed and from which it will precipitate. The solution remaining from the precipitation contains NH_4Cl, from which NH_3 is recovered by treatment with $Ca(OH)_2$. Thus NH_3 is simply used during the Solvay process to produce the proper conditions for the desired reactions. Any net consumption of NH_3 is the result of unavoidable losses during production.

24. **(a)** $Ca(OH)_2(s) + SO_4^{2-}(aq) \rightleftharpoons CaSO_4(s) + 2\,OH^-(aq)$

(b) We sum two solubility reactions and combine their values of K_{sp}.

$Ca(OH)_2(s) \rightleftharpoons Ca^{2+}(aq) + 2\,OH^-(aq)$ $\quad K_{sp} = 5.5 \times 10^{-6}$

$Ca^{2+}(aq) + SO_4^{2-}(aq) \rightleftharpoons CaSO_4(s)$ $\quad 1/K_{sp} = 1/9.1 \times 10^{-6}$

$Ca(OH)_2(s) + SO_4^{2-}(aq) \rightleftharpoons CaSO_4(s) + 2\,OH^-(aq)$ $\quad K_{eq} = \dfrac{5.5 \times 10^{-6}}{9.1 \times 10^{-6}} = 0.60$

Because this value of K_{eq} is not significantly different from 1.00, we conclude that the reaction lies neither very far to the right (it does not go to completion) nor to the left.

(c)

Reaction:	$Ca(OH)_2(s)$ +	$SO_4^{2-}(aq)$	$\rightleftharpoons CaSO_4(s)$ +	$2\,OH^-(aq)$
Initial:		1.00 M		
Changes:		$-x$ M		$+2x$ M
Equil:		$(1.00 - x)$M		$2x$ M

$K = \dfrac{[OH^-]^2}{[SO_4^{2-}]} = 0.60 = \dfrac{4x^2}{1.00 - x}$ $\qquad 4x^2 = 0.60 - 0.60\,x$ $\qquad 4x^2 + 0.60\,x - 0.60 = 0$

$x = \dfrac{-b \pm \sqrt{b^2 - 4ac}}{2a} = \dfrac{-0.60 \pm \sqrt{0.36 + 9.60}}{8} = 0.32 \text{ M}$

$[SO_4^{2-}] = 1.00 - x = 0.68 \text{ M}$ $\qquad [OH^-] = 2x = 0.64 \text{ M}$

Alkaline earth (group 2A) metals

25. $CaO \xleftarrow{\Delta} CaCO_3 \xleftarrow{CO_2} Ca(OH)_2 \xrightarrow{HCl} CaCl_2 \xrightarrow{\text{electrolysis}} Ca$

$CaHPO_4 \xleftarrow{H_3PO_4} Ca(OH)_2 \xrightarrow{H_2SO_4} CaSO_4$

The reactions are as follows.

$Ca(OH)_2(s) + 2\,HCl(aq) \longrightarrow CaCl_2(aq) + 2\,H_2O$

$CaCl_2(l) \xrightarrow{\Delta,\ \text{electrolysis}} Ca(l) + Cl_2(g)$

$Ca(OH)_2(s) + CO_2(g) \longrightarrow CaCO_3(s) + H_2O(g)$

$CaCO_3(s) \xrightarrow{\Delta} CaO(s) + CO_2(g)$

$Ca(OH)_2(s) + H_2SO_4(aq) \longrightarrow CaSO_4(s) + 2\,H_2O$

$Ca(OH)_2(s) + H_3PO_4(aq) \longrightarrow CaHPO_4(aq) + 2\,H_2O$

Actually $CaCO_3(s)$ is the industrial starting material from which $Ca(OH)_2$ is made by slaking lime. $\quad CaO(s) + H_2O \longrightarrow Ca(OH)_2(s)$

26. $MgO \xleftarrow{\Delta} MgCO_3 \xleftarrow{CO_2} Mg(OH)_2 \xrightarrow{HCl} MgCl_2 \xrightarrow{\text{electrolysis}} Mg \xrightarrow{N_2} Mg_3N_2$

$MgHPO_4 \xleftarrow{H_3PO_4} Mg(OH)_2 \xrightarrow{H_2SO_4} MgSO_4$

Once we return to $Mg(OH)_2$ from $MgSO_4$, other substances can be made by the indicated pathways. The return reaction is a precipitation: $MgSO_4(aq) + 2\,NaOH(aq) \longrightarrow Mg(OH)_2(s) + Na_2SO_4(aq)$. Then the other reactions are: $3\,Mg(s) + N_2(g) \longrightarrow Mg_3N_2(s)$ $\quad Mg(OH)_2(s) + 2\,HCl(aq) \longrightarrow MgCl_2(aq) + 2\,H_2O$

$MgCl_2(l) \xrightarrow{\Delta,\ \text{electrolysis}} Mg(l) + Cl_2(g)$ $\quad Mg(OH)_2(s) + CO_2(g) \longrightarrow MgCO_3(s) + H_2O(g)$

$MgCO_3(s) \xrightarrow{\Delta} MgO(s) + CO_2(g)$ $\quad$ $Mg(OH)_2(s) + H_2SO_4(aq) \longrightarrow MgSO_4(s) + 2\,H_2O$

$Mg(OH)_2(s) + H_3PO_4(aq) \longrightarrow MgHPO_4(aq) + 2\,H_2O$ $\quad$ $MgO(s) + H_2O \longrightarrow Mg(OH)_2(s)$

27. (a) $BeF_2 + Mg \longrightarrow Be + MgF_2(s)$ $\quad$ (c) $UO_2(s) + 2\,Ca(s) \longrightarrow U(s) + 2\,CaO(s)$

(b) $Ba(s) + Br_2(l) \longrightarrow BaBr_2(s)$ $\quad$ (d) $MgCO_3{\cdot}CaCO_3(s) \xrightarrow{\Delta} MgO(s) + CaO(s) + 2\,CO_2(g)$

(e) $2\,H_3PO_4(aq) + 3\,CaO(s) \longrightarrow Ca_3(PO_4)_2(s) + 3\,H_2O(l)$

28. (a) $Mg(HCO_3)_2(s) \xrightarrow{\text{heat}} MgO(s) + 2\,CO_2(g) + H_2O$

(b) $BaCl_2(l) \xrightarrow{\text{electricity}} Ba(l) + Cl_2(g)$ $\quad$ (c) $Sr(s) + 2\,HBr(aq) \longrightarrow SrBr_2(aq) + H_2(g)$

(d) $H_2SO_4(aq) + Ca(OH)_2(s) \longrightarrow CaSO_4(s) + 2\,H_2O(l)$

(e) $CaSO_4{\cdot}2H_2O(s) \xrightarrow{\Delta} CaSO_4{\cdot}\frac{1}{2}H_2O(s) + \frac{3}{2}\,H_2O(g)$

29. Let us compute the value of the equilibrium constant for each reaction by combining two solubility product constant values. Large values of equilibrium constants indicate that the reaction is displaced far to the right. Values of K that are quite a bit smaller than 1 indicate that the reaction is displaced far to the left.

(a) $BaSO_4(s) \rightleftharpoons Ba^{2+}(aq) + SO_4^{2-}(aq)$ $\quad$ $K_{sp} = 1.1 \times 10^{-10}$

$Ba^{2+}(aq) + CO_3^{2-}(aq) \rightleftharpoons BaCO_3(s)$ $\quad$ $1/K_{sp} = 1/5.1 \times 10^{-9}$

$BaSO_4(s) + CO_3^{2-}(aq) \rightleftharpoons BaCO_3(s) + SO_4^{2-}(aq)$ $\quad$ $K = \dfrac{1.1 \times 10^{-10}}{5.1 \times 10^{-9}} = 2.2 \times 10^{-2}$

Equilibrium lies slightly to the left.

(b) $Mg_3(PO_4)_2(s) \rightleftharpoons 3\,Mg^{2+}(aq) + 2\,PO_4^{3-}(aq)$ $\quad$ $K_{sp} = 1 \times 10^{-25}$

$3\,\{Mg^{2+}(aq) + CO_3^{2-}(aq) \rightleftharpoons MgCO_3(s)\}$ $\quad$ $1/K_{sp} = 1/3.5 \times 10^{-8}$

$Mg_3(PO_4)_2(s) + 3\,CO_3^{2-}(aq) \rightleftharpoons 3\,MgCO_3(s) + 2\,PO_4^{3-}(aq)$ $\quad$ $K = \dfrac{1 \times 10^{-25}}{(3.5 \times 10^{-8})^3} = 2.3 \times 10^{-3}$

Equilibrium lies to the left.

(c) $Ca(OH)_2(s) \rightleftharpoons Ca^{2+}(aq) + 2\,OH^-(aq)$ $\quad$ $K_{sp} = 5.5 \times 10^{-6}$

$Ca^{2+}(aq) + 2\,F^-(aq) \rightleftharpoons CaCO_3(s)$ $\quad$ $1/K_{sp} = 1/5.3 \times 10^{-9}$

$CaF_2(s) + CO_3^{2-}(aq) \rightleftharpoons CaCO_3(s) + 2\,F^-(aq)$ $\quad$ $K = \dfrac{5.5 \times 10^{-6}}{5.3 \times 10^{-9}} = 1.0 \times 10^3$ $\quad$ Equilibrium lies far to the right.

30. We expect the reaction to occur to a significant extent if its equilibrium constant has a large value.

(a) $BaCO_3(s) \rightleftharpoons Ba^{2+}(aq) + CO_3^{2-}(aq)$ $\quad$ $K_{sp} = 5.1 \times 10^{-9}$

$2\,HC_2H_3O_2(aq) \rightleftharpoons 2\,H^+(aq) + 2\,C_2H_3O_2^-(aq)$ $\quad$ $1/K_a^2 = 1/(1.8 \times 10^{-5})^2$

$H^+(aq) + CO_3^{2-}(aq) \rightleftharpoons HCO_3^-(aq)$ $\quad$ $1/K_{a_2} = 1/4.7 \times 10^{-11}$

$H^+(aq) + HCO_3^-(aq) \rightleftharpoons H_2O + CO_2(g)$ $\quad$ $1/K_{a_1} = 4.2 \times 10^{-7}$

$BaCO_3(s) + 2\,HC_2H_3O_2(aq) \rightleftharpoons Ba(C_2H_3O_2)_2(aq) + H_2O + CO_2(g)$

$$K = \frac{5.1 \times 10^{-9}}{(1.8 \times 10^{-5})^2\ 4.7 \times 10^{-11} \times 4.2 \times 10^{-7}} = 7.6 \times 10^{17}$$

(b) $Ca(OH)_2(s) \rightleftharpoons Ca^{2+}(aq) + 2\,OH^-(aq)$ $\quad$ $K_{sp} = 5.5 \times 10^{-6}$

$2\,NH_4^+(aq) + 2\,OH^-(aq) \rightleftharpoons 2\,NH_3(aq) + 2\,H_2O$ $\quad$ $1/K_b^2 = 1/(1.8 \times 10^{-5})^2$

$Ca(OH)_2(s) + 2\,NH_4^+(aq) \rightleftharpoons Ca^{2+}(aq) + 2\,NH_3(aq) + 2\,H_2O$ $\quad$ $K = \dfrac{5.5 \times 10^{-6}}{(1.8 \times 10^{-5})^2} = 1.7 \times 10^4$

(c) $BaF_2(s) \rightleftharpoons Ba^{2+}(aq) + 2\,F^-(aq)$ $\quad$ $K_{sp} = 1.0 \times 10^{-6}$

$2\,H_3O^+(aq) + 2\,F^-(aq) \rightleftharpoons 2\,HF(aq) + 2\,H_2O$ $\quad$ $1/K_a^2 = 1/(6.6 \times 10^{-4})^2$

$Ba(OH)_2(s) + 2\,F^-(aq) \rightleftharpoons Ba^{2+}(aq) + 2\,HF(aq) + 2\,H_2O$ $\quad$ $K = \dfrac{1.0 \times 10^{-6}}{(6.6 \times 10^{-4})^2} = 2.3$

This reaction would occur to some extent, certainly not to completion.

Hard Water

31. **(a)** There will be two HCO_3^- ions for each Ca^{2+} ion.

$$\frac{\text{g } Ca^{2+}}{10^6 \text{ g soln}} = \frac{185.0 \text{ g } HCO_3^-}{10^6 \text{ g soln}} \times \frac{1 \text{ mol } HCO_3^-}{61.02 \text{ g } HCO_3^-} \times \frac{1 \text{ mol } Ca^{2+}}{2 \text{ mol } HCO_3^-} \times \frac{40.08 \text{ g } Ca^{2+}}{1 \text{ mol } Ca^{2+}}$$

$$= 60.76 \text{ ppm } Ca^{2+}$$

(b) $$\frac{\text{mg } Ca^{2+}}{1 \text{ L}} = \frac{60.76 \text{ g } Ca^{2+}}{10^6 \text{ g soln}} \times \frac{1.00 \text{ g soln}}{1 \text{ mL}} \times \frac{1000 \text{ mL}}{1 \text{ L}} \times \frac{1000 \text{ mg}}{1 \text{ g}} = 60.76 \text{ mg } Ca^{2+}/\text{L}$$

(c) First we combine equations representing the neutralization of bicarbonate ion with hydroxide ion and the formation of a generalized carbonate precipitate [$MCO_3(s)$] to determine the overall stoichiometry of the reaction. We also add to the combination the slaking of lime, CaO(s). Remember that every two HCO_3^- ions are associated with an ion of $M^{2+}(aq)$.

$$CaO(s) + H_2O \longrightarrow Ca(OH)_2(s) \xrightarrow{H_2O} Ca^{2+}(aq) + 2\ OH^-(aq)$$

$$2\ \{HCO_3^-(aq) + OH^-(aq) \longrightarrow H_2O + CO_3^{2-}(aq)\}$$

$$CO_3^{2-}(aq) + M^{2+}(aq) \longrightarrow MCO_3(s)$$

$$CO_3^{2-}(aq) + Ca^{2+}(aq) \longrightarrow CaCO_3(s)$$

$$Ca(OH)_2(s) + M^{2+}(aq) + 2\ HCO_3^-(aq) \longrightarrow CaCO_3(s) + MCO_3(s) + 2\ H_2O$$

Then we determine the mass of $Ca(OH)_2$ needed.

$$\text{mass } Ca(OH)_2 = 1.00 \times 10^6 \text{ g water} \times \frac{185.0 \text{ g } HCO_3^-}{10^6 \text{ g water}} \times \frac{1 \text{ mol } HCO_3^-}{61.02 \text{ g } HCO_3^-} \times \frac{1 \text{ mol } Ca(OH)_2}{2 \text{ mol } HCO_3^-}$$

$$\times \frac{74.09 \text{ g CaO}}{1 \text{ mol CaO}} = 112.3 \text{ g } Ca(OH)_2$$

32. **(a)** $$\text{mass } CaCO_3 = 4.48 \times 10^5 \text{ g } Ca(OH)_2 \times \frac{1 \text{ mol } Ca(OH)_2}{78.09 \text{ g } Ca(OH)_2} \times \frac{2 \text{ mol } CaCO_3}{1 \text{ mol } Ca(OH)_2} \times \frac{100.1 \text{ g } CaCO_3}{1 \text{ mol } CaCO_3}$$

$$= 1.15 \times 10^6 \text{ g } CaCO_3$$

(b) It is clear from the equations that are summed in the answer to Exercise 31, and especially the net equation (that follows), that half the Ca^{2+} is derived from the CaO and half from the hard water itself.

$$Ca(OH)_2(s) + M^{2+}(aq) + 2\ HCO_3^-(aq) \longrightarrow CaCO_3(s) + MCO_3(s) + 2\ H_2O$$

33. The $[H^+]$ is computed from the data supplied in the problem.

$$[H^+] = \frac{86.4 \text{ g } Ca^{2+} \times \dfrac{1 \text{ mol } Ca^{2+}}{40.08 \text{ g } Ca^{2+}} \times \dfrac{2 \text{ mol } H^+}{1 \text{ mol } Ca^{2+}}}{1000 \text{ L water}} = 4.31 \times 10^{-3} \text{ M}$$

$$pH = -\log(4.31 \times 10^{-3}) = 2.366$$

34. Let us determine the $[OH^-]$ in this solution first. $2\ OH^-(aq) + H_2SO_4(aq) \longrightarrow 2\ H_2O + SO_4^{2-}(aq)$

$$[OH^-] = \frac{22.42 \text{ mL acid} \times \dfrac{1.00 \times 10^{-3} \text{ mmol } H_2SO_4}{1 \text{ mL acid}} \times \dfrac{2 \text{ mmol } OH^-}{1 \text{ mmol } H_2SO_4}}{25.00 \text{ mL base}} = 1.794 \times 10^{-3} \text{ M}$$

Now we compute the number of grams of $CaSO_4$ in 10^6 g of hard water. This is the ppm hardness. The anion exchange reaction is $R(OH)_2 + SO_4^{2-}(aq) \longrightarrow RSO_4 + 2\ OH^-(aq)$

$$\text{mass } CaSO_4 = 10^6 \text{ g water} \times \frac{1 \text{ mL}}{1.00 \text{ g}} \times \frac{1 \text{ L}}{1000 \text{ mL}} \times \frac{1.794 \times 10^{-3} \text{ mol } OH^-}{1 \text{ L}} \times \frac{1 \text{ mol } SO_4^{2-}}{2 \text{ mol } OH^-}$$

$$\times \frac{1 \text{ mol } CaSO_4}{1 \text{ mol } SO_4^{2-}} \times \frac{136.1 \text{ g } CaSO_4}{1 \text{ mol } CaSO_4} = 122 \text{ g } CaSO_4 \qquad (122 \text{ ppm } CaSO_4)$$

35. We first write the balanced equation for the formation of bathtub ring.

$$Ca^{2+}(aq) + 2\ CH_3(CH_2)_{16}COO^-K^+(aq) \longrightarrow Ca^{2+}[CH_3(CH_2)_{16}COO^-]_2(s, \text{"ring"}) + 2\ K^+(aq)$$

$$\text{“ring” mass} = 32.1 \text{ L} \times \frac{82.6 \text{ g } Ca^{2+}}{1000 \text{ L water}} \times \frac{1 \text{ mol } Ca^{2+}}{40.08 \text{ g } Ca^{2+}} \times \frac{1 \text{ mol “ring”}}{1 \text{ mol } Ca^{2+}} \times \frac{607.0 \text{ g “ring”}}{1 \text{ mol “ring”}} = 40.2 \text{ g bathtub ring}$$

36. We assume that the density of the solution is 1.00 g/mL. The solubility of magnesium palmitate equals the concentration of Mg^{2+}, $[Mg^{2+}]$, because there is one mole Mg^{2+} in each mole of compound.

$$[Mg^{2+}] = \frac{80 \text{ g } Mg^{2+}}{10^6 \text{ g soln}} \times \frac{1 \text{ mol } Mg^{2+}}{24.3 \text{ g } Mg^{2+}} \times \frac{1.00 \text{ g soln}}{1 \text{ mL soln}} \times \frac{10^3 \text{ mL soln}}{1 \text{ L soln}} = 3.3 \times 10^{-3} \text{ M} = s$$

From this, we calculate the value of the solubility product. Stoichiometry gives $[Palm^-] = 2\,[Mg^{2+}]$
$K_{sp} = [Mg^{2+}][Palm^-]^2 = (s)(2s)^2 = 4s^3 = 4\,(3.3 \times 10^{-3})^3 = 1.4 \times 10^{-7}$

Aluminum

37. (a) $2\,Al(s) + 6\,HCl(aq) \longrightarrow 2\,AlCl_3(aq) + 3\,H_2(g)$ (b) $2\,Al(s) + 3\,Br_2(g) \longrightarrow 2\,AlBr_3(s)$

(c) $2\,NaOH(aq) + 2\,Al(s) + 6\,H_2O(l) \longrightarrow 2\,Na^+(aq) + 2\,[Al(OH)_4]^-(aq) + 3\,H_2(g)$

(d) Oxidation: $\{Al(s) \longrightarrow Al^{3+}(aq) + 3\,e^-\}$ $\times 2$

Reduction: $\{SO_4^{2-}(aq) + 4\,H^+(aq) + 2\,e^- \longrightarrow SO_2(g) + 2\,H_2O\}$ $\times 3$

Net: $2\,Al(s) + 3\,SO_4^{2-}(aq) + 12\,H^+(aq) \longrightarrow 2\,Al^{3+}(aq) + 3\,SO_2(aq) + 6\,H_2O$

38. (a) Oersted: $2\,Al_2O_3(s) + 3\,C(s) + 6\,Cl_2(g) \xrightarrow{\Delta} 4\,AlCl_3(s) + 3\,CO_2(g)$

(b) Wöhler: $AlCl_3(s) + 3\,K(s) \xrightarrow{\Delta} Al(s) + 3\,KCl(s)$

(c) $2\,Al(s) + Cr_2O_3(s) \xrightarrow{heat} 2\,Cr(l) + Al_2O_3(s)$

(d) $Fe_2O_3(s) + OH^-(aq) \longrightarrow$ no reaction $Al_2O_3(s) + 2\,OH^-(aq) + 3\,H_2O \longrightarrow 2\,[Al(OH)_4]^-(aq)$

39. One method of analyzing this reaction is to envision the HCO_3^- ion as a combination of CO_2 and OH^-. Then the OH^- reacts with Al^{3+} and forms $Al(OH)_3$. [This method of envisioning HCO_3^- is somewhat of a trick, but it does have its basis in reality. After all, $H_2CO_3\,(= H_2O + CO_2) + OH^- \longrightarrow HCO_3^- + H_2O$.]
$Al^{3+}(aq) + 3\,HCO_3^-(aq) \longrightarrow Al(OH)_3(s) + 3\,CO_2(g)$
Another method is to consider the reaction as, first, the hydrolysis of hydrated aluminum ion to produce $Al(OH)_3(s)$ and an acidic solution, followed by the reaction of the acid with bicarbonate ion.
$[Al(H_2O)_6]^{3+}(aq) + 3\,H_2O \longrightarrow Al(OH)_3(s) + 3\,H_3O^+(aq) + 3\,H_2O$
$3\,H_3O^+(aq) + 3\,HCO_3^-(aq) \longrightarrow 6\,H_2O + 3\,CO_2(g)$
This gives the same net reaction: $[Al(H_2O)_6]^{3+}(aq) + 3\,HCO_3^-(aq) \longrightarrow Al(OH)_3(s) + 3\,CO_2(g) + 6\,H_2O$

40. The $Al^{3+}(aq)$ ion hydrolyzes. $Al^{3+}(aq) + 3\,H_2O \longrightarrow Al(OH)_3(s) + 3\,H^+(aq)$
And the hydrogen ion that is produced reacts with bicarbonate ion to liberate $CO_2(g)$.
$H^+(aq) + HCO_3^-(aq) \longrightarrow H_2O + CO_2(g)$

41. Aluminum and its oxide are soluble in both acid and in base.
$2\,Al(s) + 6\,H^+(aq) \longrightarrow 2\,Al^{3+}(aq) + 3\,H_2(g)$ $Al_2O_3(s) + 6\,H^+(aq) \longrightarrow 2\,Al^{3+}(aq) + 3\,H_2O$
$2\,Al(s) + 2\,OH^-(aq) + 6\,H_2O \longrightarrow 2\,[Al(OH)_4]^-(aq) + 3\,H_2(g)$
$Al_2O_3(s) + 2\,OH^-(aq) + 3\,H_2O \longrightarrow 2\,[Al(OH)_4]^-(aq)$
The resistance of Al(s) to corrosion in the pH range 4.5 to 8.5 (and its susceptibility to corrosion outside that range) is a reflection of the ability of either acid or base to attack aluminum or its protective oxide coating. Thus, aluminum is nonreactive only when the medium to which it is exposed is not highly acidic or highly basic.

42. Both Al and Mg are attacked by acid and their ions both are precipitated by hydroxide ion.
$2\,Al(s) + 6\,H^+(aq) \longrightarrow 2\,Al^{3+}(aq) + 3\,H_2(g)$ $Mg(s) + 2\,H^+(aq) \longrightarrow Mg^{2+}(aq) + H_2(g)$
$Al^{3+}(aq) + 3\,OH^-(aq) \longrightarrow Al(OH)_3(s)$ $Mg^{2+}(aq) + 2\,OH^- \longrightarrow Mg(OH)_2(s)$.
But, of these two solid hydroxides, only $Al(OH)_3(s)$ redissolves in excess $OH^-(aq)$.
$Al(OH)_3(s) + OH^-(aq) \longrightarrow [Al(OH)_4]^-(aq)$
Thus, the procedure for analysis consists of dissolving the sample in HCl(aq) and then treating the resulting solution with NaOH(aq) until a precipitate forms. If this precipitate dissolves completely when excess NaOH(aq) is added, the sample was Aluminum 2S. If at least some of the precipitate does not dissolve, the sample was magnalium.

43. $CO_2(g)$ is, of course, the anhydride of an acid. The reaction here is an acid-base reaction.
$[Al(OH)_4]^-(aq) + CO_2(aq) \longrightarrow Al(OH)_3(s) + HCO_3^-(aq)$

44. The reaction described in Exercise 43 is: $[Al(OH)_4]^-(aq) + CO_2(aq) \longrightarrow Al(OH)_3(s) + HCO_3^-(aq)$
Substituting HCl(aq) for $CO_2(aq)$ gives: $[Al(OH)_4]^-(aq) + HCl(aq) \longrightarrow Al(OH)_3(s) + H_2O + Cl^-(aq)$

Thus HCl(aq) could be used in place of CO_2(aq), but it has two disadvantages. First, HCl(aq) is a strong acid; and the addition of too much HCl(aq) will dissolve the $Al(OH)_3$(s); this will not happen if excess CO_2(aq) is added. Second, CO_2(aq) is cheaper and more readily available (also safer to use) than HCl(aq). Recall that CO_2(g) is produced during the electrolytic production of aluminum.

45. **(a)** $Fe_2O_3(s) + 2\,Al(s) \longrightarrow Al_2O_3(s) + 2\,Fe(s)$

$\Delta H^\circ_{rxn} = \Delta H^\circ_f[Al_2O_3(s)] + 2\,\Delta H^\circ_f[Fe(s)] - \Delta H^\circ_f[Fe_2O_3(s)] - 2\,\Delta H^\circ_f[Al(s)]$

$= -1676\text{ kJ/mol} + 2\,(0.00\text{ kJ/mol}) - (-824.2\text{ kJ/mol}) - 2\,(0.00\text{ kJ/mol}) = -852\text{ kJ/mol}$

(b) $3\,MnO_2(s) + 4\,Al(s) \longrightarrow 2\,Al_2O_3(s) + 4\,Mn(s)$

$\Delta H^\circ_{rxn} = 2\,\Delta H^\circ_f[Al_2O_3(s)] + 4\,\Delta H^\circ_f[Mn(s)] - 3\,\Delta H^\circ_f[MnO_2(s)] - 4\,\Delta H^\circ_f[Al(s)]$

$= 2\,(-1676\text{ kJ/mol}) + 4\,(0.00\text{ kJ/mol}) - 3\,(-520.0\text{ kJ/mol}) - 4\,(0.00\text{ kJ/mol}) = -1792\text{ kJ/mol}$

(c) $3\,MgO(s) + 2\,Al(s) \longrightarrow 3\,Mg(s) + Al_2O_3(s)$

$\Delta H^\circ_{rxn} = \Delta H^\circ_f[Al_2O_3(s)] + 3\,\Delta H^\circ_f[Mg(s)] - 3\,\Delta H^\circ_f[MgO(s)] - 2\,\Delta H^\circ_f[Al(s)]$

$= -1676\text{ kJ/mol} + 2\,(0.00\text{ kJ/mol}) - 3\,(-601.7\text{ kJ/mol}) - 2\,(0.00\text{ kJ/mol})$

$= +129\text{ kJ/mol}$ An endothermic reaction

46. We rewrite (slightly) equation (22.23) for the electrolysis of Al_2O_3(s). It has $n = 12$.

$3\,C(s) + 2\,Al_2O_3(s) \longrightarrow 4\,Al(s) + 3\,CO_2(g)$

$\Delta G^\circ = 4\,\Delta G^\circ_f[Al(s)] + 3\,\Delta G^\circ_f[CO_2(g)] - 3\,\Delta G^\circ_f[C(s)] - 2\,\Delta G^\circ_f[Al_2O_3(s)]$

$= 4\,(0.00\text{ kJ/mol}) + 3\,(-394.4\text{ kJ/mol}) - 3\,(0.00\text{ kJ/mol}) - 2\,(-1582\text{ kJ/mol})$

$= +1982\text{ kJ/mol} = 1.982 \times 10^6\text{ J/mol} = -nFE^\circ$

$$E^\circ = -\frac{\Delta G^\circ}{nF} = -\frac{1.982 \times 10^6\text{ J/mol}}{12\text{ mol e}^-\text{/mol rxn} \times 96{,}485\text{ C/mol e}^-} = -1.712\text{ V}$$

Our result is an estimate because we have used values of ΔG° at 298 K and the reaction occurs at a much higher temperature. If the oxidation of C(s) to CO_2(g) did not occur, then the cell reaction would be the reverse of the formation reaction of Al_2O_3(s), with $n = 6$.

$$E^\circ = -\frac{\Delta G^\circ}{nF} = -\frac{1.582 \times 10^6\text{ J/mol}}{6\text{ mol e}^-\text{/mol rxn} \times 96{,}485\text{ C/mol e}^-} = -2.733\text{ V}$$

Tin and Lead

47. **(a)** $PbO(s) + 2\,HNO_3(aq) \longrightarrow Pb(NO_3)_2(s) + H_2O$ **(b)** $SnCO_3(s) \xrightarrow{\Delta} SnO(s) + CO_2(g)$

(c) $PbO(s) + C(s) \xrightarrow{\Delta} Pb(l) + CO(g)$

(d) $2\,Fe^{3+}(aq) + Sn^{2+}(aq) \longrightarrow 2\,Fe^{2+}(aq) + Sn^{4+}(aq)$

(e) $2\,PbS(s) + 3\,O_2(g) \xrightarrow{\Delta} 2\,PbO(s) + 2\,SO_2(g)$ $2\,SO_2(g) + O_2(g) \longrightarrow 2\,SO_3(g)$

$SO_3(g) + PbO(s) \longrightarrow PbSO_4(s)$ Or perhaps simply: $PbS(s) + 2\,O_2(g) \longrightarrow PbSO_4(s)$

Yet a third possibility: $PbO(s) + SO_2(s) \longrightarrow PbSO_3(s)$ and $2\,PbSO_3(s) + O_2(s) \longrightarrow PbSO_4(s)$

48. **(a)** Treat tin(II) oxide with hydrochloric acid. $SnO(s) + 2\,HCl(aq) \longrightarrow SnCl_2(aq) + H_2O$

(b) Attack tin with chlorine. $Sn(s) + 2\,Cl_2(g) \longrightarrow SnCl_4(s)$

(c) First we dissolve PbO_2(s) in HNO_3(aq) and then treat the resulting solution with K_2CrO_4(aq)

$2\,PbO_2(s) + 4\,HNO_3(aq) \longrightarrow 2\,Pb(NO_3)_2(aq) + O_2(g) + 2\,H_2O$

$Pb(NO_3)_2(aq) + K_2CrO_4(aq) \longrightarrow PbCrO_4(aq) + 2\,KNO_3(aq)$

49. We use the Nernst equation to determine whether the cell voltage still is positive when the reaction has gone to completion.

(a)

Oxidation:	$\{Fe^{2+}(aq) \longrightarrow Fe^{3+}(aq) + e^-\}$	$\times 2$	$-E^\circ = -0.771\text{ V}$
Reduction:	$PbO_2(s) + 4\,H^+(aq) + 2\,e^- \longrightarrow Pb^{2+}(aq) + 2\,H_2O$		$E^\circ = +1.455\text{ V}$
Net:	$2\,Fe^{2+}(aq) + PbO_2(s) + 4\,H^+ \longrightarrow 2\,Fe^{3+}(aq) + Pb^{2+}(aq) + 2\,H_2O$		

$E^\circ_{cell} = -0.771\text{ V} + 1.455\text{ V} = +0.684\text{ V}$

In this case, when the reaction has gone to completion, $[Fe^{2+}] = 0.001$ M, $[Fe^{3+}] = 0.999$ M, and $[Pb^{2+}] = 0.500$ M.

$$E_{cell} = E^\circ_{cell} - \frac{0.0592}{2}\log\frac{[Fe^{3+}]^2\,[Pb^{2+}]}{[Fe^{2+}]^2} = 0.684 - \frac{0.0592}{2}\log\frac{(0.999)^2\,(0.500)}{(0.001)^2}$$

$= 0.684 - 0.169 = +0.515$ V Yes, this reaction will go to completion.

(b) Oxidation: $2\ SO_4^{2-}(aq) \longrightarrow S_2O_8^{2-}(aq) + 2\ e^-$ $-E° = -2.01$ V

Reduction: $PbO_2(s) + 4\ H^+(aq) + 2\ e^- \longrightarrow Pb^{2+}(aq) + 2\ H_2O$ $E° = +1.455$ V

Net: $2\ SO_4^{2-}(aq) + PbO_2(s) + 4\ H^+ \longrightarrow S_2O_8^{2-}(aq) + Pb^{2+}(aq) + 2\ H_2O$

$E°_{cell} = -2.01 + 1.455 = -0.56$ V This reaction is not even spontaneous initially.

(c) Oxdn: $\{Mn^{2+}(0.0001\ M) + 4\ H_2O \longrightarrow MnO_4^-(aq) + 8\ H^+(aq) + 5\ e^-\} \times 2$ $-E° = -1.51$ V

Redn: $\{PbO_2(s) + 4\ H^+(aq) + 2\ e^- \longrightarrow Pb^{2+}(aq) + 2\ H_2O\}$ $\times 5$ $E° = +1.455$ V

Net: $2\ Mn^{2+}(0.0001\ M) + 5\ PbO_2(s) + 4\ H^+ \longrightarrow 2\ MnO_4^-(aq) + 5\ Pb^{2+}(aq) + 2\ H_2O$

$E°_{cell} = -1.51 + 1.455 = -0.06$ V The standard cell potential indicates that this reaction is not spontaneous when all concentrations are 1 M. Since the concentration of a reactant (Mn^{2+}) is lower than 1.00 M, this reaction is even less spontaneous than the standard cell potential indicates.

50. A positive value of $E°_{cell}$ indicates that a reaction should occur.

(a) Oxidation: $Sn^{2+}(aq) \longrightarrow Sn^{4+}(aq) + 2\ e^-$ $-E° = -0.154$ V

Reduction: $I_2(s) + 2\ e^- \longrightarrow 2\ I^-(aq)$ $E° = +0.535$ V

Net: $Sn^{2+}(aq) + I_2(s) \longrightarrow Sn^{4+}(aq) + 2\ I^-(aq)$ $E°_{cell} = +0.381$ V

Yes, $Sn^{2+}(aq)$ will reduce I_2 to I^-.

(b) Oxidation: $Sn^{2+}(aq) \longrightarrow Sn^{4+}(aq) + 2\ e^-$ $-E° = -0.154$ V

Reduction: $Fe^{2+}(aq) + 2\ e^- \longrightarrow Fe(s)$ $E° = -0.440$ V

Net: $Sn^{2+}(aq) + Fe^{2+}(aq) \longrightarrow Sn^{4+}(aq) + Fe(s)$ $E°_{cell} = -0.594$ V

No, $Sn^{2+}(aq)$ will not reduce $Fe^{2+}(aq)$ to Fe(s).

(c) Oxidation: $Sn^{2+}(aq) \longrightarrow Sn^{4+}(aq) + 2\ e^-$ $-E° = -0.154$ V

Reduction: $Cu^{2+}(aq) + 2\ e^- \longrightarrow Cu(s)$ $E° = +0.337$ V

Net: $Sn^{2+}(aq) + Cu^{2+}(aq) \longrightarrow Sn^{4+}(aq) + Cu(s)$ $E°_{cell} = +0.183$ V

Yes, $Sn^{2+}(aq)$ will reduce $Cu^{2+}(aq)$ to Cu(s).

(d) Oxidation: $Sn^{2+}(aq) \longrightarrow Sn^{4+}(aq) + 2\ e^-$ $-E° = -0.154$ V

Reduction: $\{Fe^{3+}(aq) + e^- \longrightarrow 2\ Fe^{2+}(aq)$ $E° = +0.771$ V

Net: $Sn^{2+}(aq) + 2\ Fe^{3+}(aq) \longrightarrow Sn^{4+}(aq) + 2\ Fe^{2+}(aq)$ $E°_{cell} = +0.617$ V

Yes, $Sn^{2+}(aq)$ will reduce $Fe^{3+}(aq)$ to $Fe^{2+}(aq)$.

23 Chemistry of the Representative (Main Group) Elements II: Nonmetals

Practice Examples

1A The dissociation reaction is the reverse of the formation reaction and thus $\Delta G°$ for the dissociation reaction is the negative of ΔG_f° $\quad HF(g) \longrightarrow \frac{1}{2} H_2(g) + \frac{1}{2} F_2 \quad \Delta G° = -\Delta G_f^\circ = -(-273.2 \text{ kJ/mol})$

We know that $\Delta G° = -RT \ln K_P \qquad +273.2 \times 10^3 \text{ J/mol} = -8.3145 \text{ J mol}^{-1} \text{ K}^{-1} \times 298.15 \text{ K} \times \ln K_P$

$$\ln K_p = \frac{273.2 \times 10^3 \text{ J mol}^{-1}}{-8.3145 \text{ J mol}^{-1} \text{ K}^{-1} \times 298.15 \text{ K}} = -110 \qquad K_p = e^{-110} = 2 \times 10^{-48}$$

Virtually no dissociation of HF(g) into its elements occurs.

1B The dissociation reaction with all integer coefficients is twice the reverse of the formation reaction.

$2\,HCl(g) \rightleftharpoons H_2(g) + Cl_2(g) \qquad \Delta G° = -2 \times (-95.30 \text{ kJ/mol}) = +\ 190.6 \text{ kJ/mol} = -RT \ln K_p$

$$\ln K_p = \frac{-\Delta G°}{RT} = \frac{-190.6 \times 10^3 \text{ J/mol}}{8.3145 \text{ J mol}^{-1} \text{ K}^{-1} \times 298 \text{ K}} = -76.9 \qquad K_p = e^{-76.9} = 4 \times 10^{-34}$$

We assume an initial HCl(g) pressure of P atm, and calculate the final pressure of $Cl_2(g)$ and $H_2(g)$, x atm.

Reaction:	$2\,HCl(g)$	$\rightleftharpoons$	$H_2(g)$	+	$Cl_2(g)$
Initial:	P atm		0 atm		0 atm
Changes:	$-2x$ atm		$+x$ atm		$+x$ atm
Equil:	$(P - 2x)$ atm		x atm		x atm

$$K_p = \frac{P\{H_2(g)\}P\{Cl_2(g)\}}{P\{HCl(g)\}^2} = \frac{x \cdot x}{(P - 2x)^2} = \left(\frac{x}{P - 2x}\right)^2 \qquad \frac{x}{P - 2x} = \sqrt{4 \times 10^{-34}} = 2 \times 10^{-17}$$

$x = 2 \times 10^{-17} (P - 2x) \approx 2 \times 10^{-17} P$

$$\% \text{ decomposition} = \frac{2x}{P} \times 100\% = 2 \times 2 \times 10^{-17} \times 100\% = 4 \times 10^{-15}\% \text{ decomposed}$$

2A The first two reactions are those of Example 23-1, to get to B_2O_3.

$Na_2B_4O_7 \cdot 10H_2O + H_2SO_4 \longrightarrow 4\,B(OH)_3 + Na_2SO_4 + 5\,H_2O \qquad 2\,B(OH)_3 \xrightarrow{\Delta} B_2O_3 + H_2O$

The next reaction is conversion to BCl_3 with heat, carbon, and chlorine.

$2\,B_2O_3(s) + 3\,C(s) + 6\,Cl_2(g) \xrightarrow{\Delta} 4\,BCl_3(l) + 3\,CO_2(g)$

$LiAlH_4$ is used as a reducing agent to produce diborane.

$4\,BCl_3(l) + 3\,LiAlH_4 \longrightarrow 2\,B_2H_6(g) + 3\,LiCl + 3\,AlCl_3$

2B We first roast ZnS(s) to produce $SO_2(g)$. $\quad 2\,ZnS(s) + 3\,O_2(g) \longrightarrow 2\,ZnO(s) + 2\,SO_2(g)$

The $SO_2(g)$ is oxidized to $SO_3(g)$. $\quad 2\,SO_2(g) + O_2(g) \longrightarrow 2\,SO_3(g) \qquad V_2O_5$ catalyst.

The $SO_3(g)$ is hydrated in two steps. $\quad H_2SO_4(l) + SO_3(g) \longrightarrow H_2S_2O_7(l)$

$H_2S_2O_7(l) + H_2O(l) \longrightarrow 2\,H_2SO_4(l)$

Summarizing Example Calculations

1. The balancing of the disproportionation by inspection is possible if one realizes that the ratio of thiosulfate to sulfur dioxide to sulfur must be 1:1:1. Oxygen then is balanced with H_2O, hydrogen with H^+.

$S_2O_3^{2-}(aq) \longrightarrow SO_2(g) + S(s)$ becomes $S_2O_3^{2-}(aq) \longrightarrow SO_2(g) + S(s) + H_2O$ and finally

$S_2O_3^{2-}(aq) + 2\,H^+(aq) \longrightarrow SO_2(g) + S(s) + H_2O$

2. The half-equations begin from the couples: $S_2O_3^{2-}(aq) \longrightarrow SO_2(g)$ and $S_2O_3^{2-}(aq) \longrightarrow S(s)$

3. We begin with the two couples, write the two half-equations, and combine them.

oxidation: $H_2O + S_2O_3^{2-}(aq) \longrightarrow 2\ SO_2(g) + 2\ H^+(aq) + 4\ e^-$ $\quad -E° = -0.40$ V

reduction: $S_2O_3^{2-}(aq) + 6\ H^+(aq) + 4\ e^- \longrightarrow 2\ S(s) + 3\ H_2O$ $\quad E° = +0.60$ V

4. net: $2\ S_2O_3^{2-}(aq) + 4\ H^+(aq) \longrightarrow 2\ SO_2(g) + 2\ S(s) + 2\ H_2O$ $\quad E°_{cell} = +0.20$ V

The positive value of the cell voltage signifies a spontaneous reaction with all species in standard states.

simplified: $S_2O_3^{2-}(aq) + 2\ H^+(aq) \longrightarrow SO_2(g) + S(s) + H_2O$ $\quad \Delta G°_1 = -2\mathcal{F}E°_{cell}$

5. Another way to predict voltage in basic solution is combining chemical equations and their $\Delta G°$ values.

combine: $2\ H_2O \longrightarrow 2\ H^+(aq) + 2\ OH^-(aq)$ $\quad \Delta G°_2 = -RT\ln K_w^2$

overall: $S_2O_3^{2-}(aq) + H_2O \longrightarrow SO_2(g) + S(s) + 2\ OH^-(aq)$ $\quad \Delta G° = \Delta G°_1 + \Delta G°_2$

$$\Delta G° = -2\mathcal{F}E°_{cell} - RT\ln K_w^2 = -2 \text{ mol e}^- \times \frac{96{,}485 \text{ C}}{1 \text{ mol e}^-} \times (+0.20 \text{ V}) - 8.3145 \frac{\text{J}}{\text{mol·K}}\ 298 \text{ K} \times \ln(1.0 \times 10^{-14})^2$$

$$= -3.86 \times 10^4 \text{ J} + 1.597 \times 10^5 \text{ J} = -2\mathcal{F}E° = +1.211 \times 10^5 \text{ J}$$

$$E° = -\frac{+1.211 \times 10^5 \text{ J}}{2 \text{ mol e}^- \times 96485 \text{ C/mol e}^-} = -0.63 \text{ V}$$

The reaction clearly is non-spontaneous in alkaline solution.

REVIEW QUESTIONS

1. **(a)** A polyhalide ion is an anion consisting of two or more halogen atoms, such as I_3^-.

(b) A polyphosphate is a condensation product of several ortho phosphate ions. $P_2O_7^{4-}$ is diphosphate, $P_3O_{10}^{5-}$ is triphosphate, etc.

(c) Allotropy refers to an element existing in several forms in the same state (usually the solid state) of matter. Examples include diamond and graphite, red and white phosphorus, and the metallic and nonmetallic forms of arsenic and antimony.

(d) Disproportionation refers to a substance being both oxidized and reduced in the same reaction.

2. **(a)** The Frasch process is used to mine underground deposits of sulfur. Superheated water and compressed air are pumped down, melting the sulfur and forcing it to the surface, where it cools and dries to form solid sulfur.

(b) The contact process refers to oxidizing $SO_2(g)$ with $O_2(g)$ to $SO_3(g)$ using a $V_2O_5(s)$ catalyst. It is an essential step in the modern industrial production of sulfuric acid.

(c) Eutrophication is the result of an excess of nutrients in bodies of water, such as ponds, which causes rapid and excessive plant growth in the water. This depletes the oxygen content of the water and marine life dies off. The decaying plant and animal life fill the pond from the bottom, eventually turning it into a marsh.

(d) A three-center bond is a pair of electrons spanning three atoms, such as occurs in boron hydrides.

3. **(a)** An acid salt's anion is a partly ionized polyprotic acid: HSO_4^- or HPO_4^{2-}. An acid anhydride is a compound that forms an acid on the addition of water. Typically nonmetal oxides: SO_3 and CO_2.

(b) The azide ion is N_3^-; its salts typically are unstable, often explosively so. The nitride ion is N^{3-}; it often forms when active metals, such as magnesium, react with $N_2(g)$.

(c) A silane is a compound of silicon and hydrogen. A silicone has a —Si—O—Si— backbone terminated by —OH groups, with hydrocarbon groups attached to each of the silicon atoms.

(d) A colloidal disperson of a solid in a liquid that flows readily is known as a sol. One that has had sufficient liquid removed that it does not flow readily is called a gel.

4.

(a)	$KBrO_3$	potassium bromate	**(b)**	I_3^-	triiodide ion
(c)	NaIO	sodium hypoiodite	**(d)**	NaH_2PO_4	sodium dihydrogen phosphate
(e)	LiN_3	lithium azide	**(f)**	$Na_2S_2O_3$	sodium thiosulfate

5. **(a)** $CaCl_2(s) + H_2SO_4(\text{conc. aq}) \xrightarrow{\Delta} CaSO_4(s) + 2\ HCl(g)$

(b) $I_2(s) + Cl^-(aq) \longrightarrow$ no reaction

(c) $NH_3(aq) + HClO_4(aq) \longrightarrow NH_4ClO_4(aq)$

6. **(a)** $Cl_2(aq) + 2\ NaOH(aq) \longrightarrow NaCl(aq) + NaOCl(aq) + H_2O$

(b) Oxidation: $2\ I^-(aq) \longrightarrow I_2(s) + 2\ e^-$

Reduction: $SO_4^{2-}(aq) + 2\ e^- + 4\ H^+(aq) \longrightarrow SO_2(g) + 2\ H_2O$

Net: $2\ I^-(aq) + SO_4^{2-}(aq) + 4\ H^+(aq) \longrightarrow I_2(g) + SO_2(g) + 2\ H_2O$

$2\ NaI(s) + 2\ H_2SO_4(aq) \longrightarrow I_2(g) + 2\ SO_2(g) + H_2O + Na_2SO_4(aq)$

(c) $Cl_2(g) + 2\ Br^-(aq) \longrightarrow 2\ Cl^-(aq) + Br_2(l)$

7. **(a)** $KI(s) + H_3PO_4(l) \longrightarrow HI(g) + KH_2PO_4(s)$ Because H_3PO_4 is not a terribly strong acid, this reaction probably does not proceed beyond the first ionization.

(b) $Na_2SiF_6 + 4\ Na(s) \longrightarrow Si + 6\ NaF(s)$

8. **(a)** $As_4O_6(s) + 6\ CO(g) \longrightarrow 4\ As(s) + 6\ CO_2(g)$

(b) $NH_3(g) + H_3PO_4(aq) \longrightarrow NH_4H_2PO_4(aq)$ monoammonium phosphate

$2\ NH_3(g) + H_3PO_4(aq) \longrightarrow (NH_4)_2HPO_4(aq)$ diammonium phosphate

9. First compute the theoretical yield of P from 8.00 ton phosphate rock, then the percent yield in this case.

$$\text{no. ton P} = 8.00 \text{ ton rock} \times \frac{31 \text{ ton } P_4O_{10}}{100.0 \text{ ton rock}} \times \frac{1 \text{ ton-mol } P_4O_{10}}{283.8 \text{ ton } P_4O_{10}} \times \frac{4 \text{ ton-mol P}}{1 \text{ ton-mol } P_4O_{10}} \times \frac{31.0 \text{ ton P}}{1 \text{ ton-mol } P_2O_5} = 1.1 \text{ ton P}$$

The mass of a ton-mole in tons is numerically equal to the mass of a (gram-)mole in grams.

$$\%\text{yield} = \frac{1.00 \text{ ton P produced}}{1.1 \text{ ton P calculated}} \times 100\% = 91\% \text{ yield}$$

10. $3\ Cl_2(g) + I^-(aq) + 3\ H_2O \longrightarrow 6\ Cl^-(aq) + IO_3^-(aq) + 6\ H^+(aq)$

$Cl_2(g) + 2\ Br^-(aq) \longrightarrow 2\ Cl^-(aq) + Br_2(aq)$

11.
$$\text{seawater volume} = 38 \times 10^6 \text{ ton } H_2SO_4 \times \frac{2000 \text{ lb}}{1 \text{ ton}} \times \frac{454 \text{ g}}{1 \text{ lb}} \times \frac{1 \text{ mol } H_2SO_4}{98.1 \text{ g } H_2SO_4} \times \frac{1 \text{ mol } SO_4^{2-}}{1 \text{ mol } H_2SO_4} \times \frac{96.1 \text{ g } SO_4^{2-}}{1 \text{ mol } SO_4^{2-}} \times \frac{1000 \text{ mg}}{1 \text{ g}} \times \frac{1 \text{ L seawater}}{2650 \text{ mg } SO_4^{2-}} \times \frac{1 \text{ m}^3}{1000 \text{ L}} \times \frac{1 \text{ km}^3}{10^9 \text{ m}^3} = 13 \text{ km}^3 \text{ seawater}$$

12. **(a)**

Reduction:	$Cl_2(g) + 2\ e^- \longrightarrow 2\ Cl^-(aq)$	$E° = +1.358$ V
Oxidations:	$2\ I^-(aq) \longrightarrow I_2(s) + 2\ e^-$	$-E° = -0.535$ V
	$2\ H_2O \longrightarrow O_2(g) + 4\ H^+(aq) + 4\ e^-$	$-E° = -1.229$ V

The production of $I_2(s)$ is more likely; it results in the larger standard cell potential.

(b) NH_4^+ has nitrogen in its lowest common oxidation state. NH_4^+ cannot be reduced further. Thus, NH_4^+ is oxidized and H_2O_2 must be reduced to H_2O.

13. We draw the Lewis structures of the products and reactants of each reaction.

(a) $\frac{1}{2}$ Cl—Cl + $\frac{1}{2}$ F—F $\longrightarrow$ Cl—F

Break: 0.5 Cl—Cl + 0.5 F—F = 0.5(243 + 159) = 201 kJ/mol
Form: Cl—F = 251 kJ/mol
$\Delta H° = 201$ kJ/mol – 251 kJ/mol = – 50. kJ/mol

(b) F—F + $\frac{1}{2}$ O=O $\longrightarrow$ F—O—F

Break: 0.5 O=O + F—F = (0.5 × 498) + 159 = 408 kJ/mol
Form: 2 O—F = 2 × 213 = 426 kJ/mol
$\Delta H° = 408$ kJ/mol – 426 kJ/mol = –18 kJ/mol

(c) Cl—Cl + $\frac{1}{2}$ O=O $\longrightarrow$ Cl—O—Cl

Break: 0.5 O=O + Cl—Cl = (0.5 × 498) + 243 = 492 kJ/mol
Form: 2 Cl—O = 2 × 205 = 410 kJ/mol
$\Delta H° = 492$ kJ/mol – 410 kJ/mol = +82 kJ/mol

(d) $\frac{3}{2}$ F—F + $\frac{1}{2}$ N≡N $\longrightarrow$ F—N(—F)—F

Break: 0.5 N≡N + 1.5 F—F = (0.5 × 946) + (1.5 × 159) = 712 kJ/mol

Form: 3 N—F = 3 × 280 = 840 kJ/mol
$\Delta H° = 712$ kJ/mol – 840 kJ/mol = –128 kJ/mol

14. **(a)** AgAt silver astatide **(b)** CSe_2 carbon diselenide **(c)** MgPo magnesium polonide
(d) H_2TeO_3 tellurous acid **(e)** K_2SeSO_3 potassium thioselenate **(f)** $KAtO_4$ perastatic acid

EXERCISES

Noble Gas Compounds

15. **(a)** 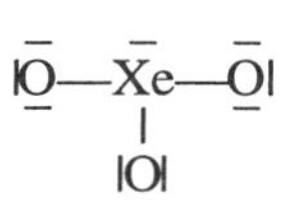 **(b)** 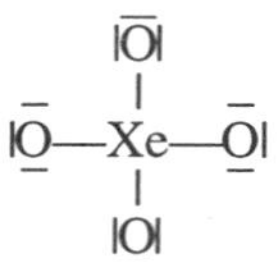 **(c)** 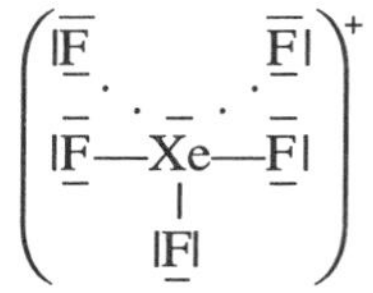

(a) The Lewis structure has three ligands and one lone pair on Xe. XeO_3 has a trigonal pyramidal shape.
(b) The Lewis structure has four ligands on Xe. XeO_4 has a tetrahedral shape.
(c) There are five ligands and one lone pair on Xe in XeF_5^+. Its shape is square pyramidal.

16. We use VSEPR theory to predict the shapes of the species involved.
(a) O_2XeF_2 has a total of $2 \times 6 + 8 + 2 \times 7 = 34$ valence electrons = 17 pairs.
(b) O_3XeF_2 has a total of $3 \times 6 + 8 + 2 \times 7 = 40$ valence electrons = 20 pairs.
(c) $OXeF_4$ has a total of $6 + 8 + 4 \times 7 = 42$ valence electrons = 21 pairs.
We draw a plausible Lewis structure for each species below.

(a) 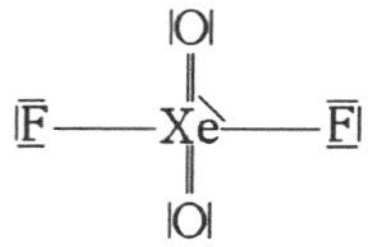 **(b)** 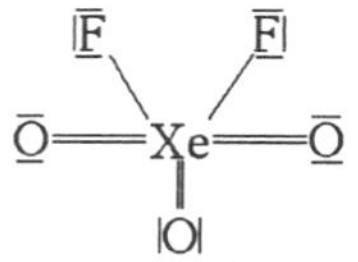 **(c)** 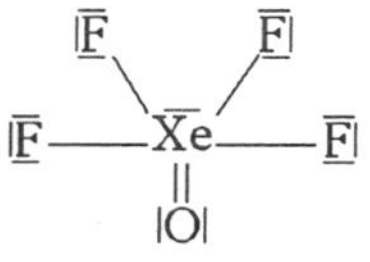

(a) There are four ligands and one lone pair on Xe in O_2XeF_2. Its shape is an irregular tetrahedron.
(b) There are five ligands and no lone pairs on Xe in O_3XeF_2. Its shape is trigonal bipyramidal.
(c) The Lewis structure has five ligands and one lone pair on Xe. $OXeF_4$ has a square pyramidal shape.

The Halogens

17. Iodide ion is slowly oxidized to iodine, which is yellow-brown in aqueous solution, by oxygen in the air.

Oxidation:	$2\,I^-(aq) \longrightarrow I_2(aq) + 2\,e^-$	× 2	$-E° = -0.535$ V
Reduction:	$O_2(g) + 4\,H^+(aq) + 4\,e^- \longrightarrow 2\,H_2O$		$E° = +1.229$ V
Net:	$4\,I^-(aq) + O_2(g) + 4\,H^+(aq) \longrightarrow 2\,I_2(aq) + 2\,H_2O$		$E°_{cell} = +0.694$ V

Possibly followed by: $I_2(aq) + I^-(aq) \longrightarrow I_3^-(aq)$

18. $MnF_6^{2-} + 2\,SbF_5 \longrightarrow MnF_4 + 2\,SbF_6^-$ $\qquad 2\,MnF_4 \longrightarrow 2\,MnF_3 + F_2(g)$

19. In general, elemental forms of the lighter elements in the halogen family will displace the ions of heavier ions in the family from their aqueous solutions. That is, Cl_2 will displace Br^- and I^-; and Br_2 will displace I^-.

20. We first list values of halogen properties.

	F	Cl	Br	I
(a) Covalent radius:	64	99.5	114	133 pm
(b) Ionic radius:	133	181	195	220 pm
(c) First ionization energy:	1681	1251	1140	1008 kJ/mol
(d) electron affinity:	–328	–349	–325	–295 kJ/mol
(e) electronegativity:	4.0	3.0	2.8	2.5
(f) standard reduction potential:	+2.886 V	+1.358 V	+1.065 V	+0.535 V

We can do a reasonable job of predicting the properties of astatine by simply looking at the difference between Br and I and assuming that the same difference exists between I and At.

(a) Covalent radius: ≈152 pm **(b)** Ionic radius: ≈247 pm
(c) 1st ionization energy: ≈880 kJ/mol **(d)** Eln. affinity: ≈–265 kJ/mol
(e) electronegativity: ≈2.3 **(e)** Std. redn pot'l: ≈0.20 V to 0.30 V

Depending on the technique that you use, you may arrive at different answers. For example, –259 kJ/mol is a reasonable estimation of the electron affinity by the following reasoning. The difference between the value for Cl and Br is 24 kJ/mol, the difference between the values for Br and I is 30 kJ/mol, so the difference between I and At should be about 36 kJ/mol.

21. **(a)** $\text{mass } F_2 = 1 \text{ km}^3 \times \left(\frac{1000 \text{ m}}{1 \text{ km}} \times \frac{100 \text{ cm}}{1 \text{ m}}\right)^3 \times \frac{1.03 \text{ g}}{1 \text{ cm}^3} \times \frac{1 \text{ lb}}{454 \text{ g}} \times \frac{1 \text{ ton}}{2000 \text{ lb}} \times \frac{1 \text{ g F}^-}{1 \text{ ton}}$

$= 1 \times 10^9 \text{ g } F_2 = 1 \times 10^6 \text{ kg } F_2$

(b) Bromine is extracted by displacing it from solution with $Cl_2(g)$. Since there is no chemical oxidizing agent that is stronger than F_2, this method of displacement would not work for F_2. Even if there were a chemical oxidizing agent stronger than F_2, it would displace O_2 before it displaced F_2. Obtaining F_2 would require electrolysis of its molten salts, obtained by evaporating the seawater.

22. $\text{mass } F_2 = 1.00 \times 10^3 \text{ kg rock} \times \frac{1000 \text{ g}}{1 \text{ kg}} \times \frac{1 \text{ mol } 3Ca_3(PO_4)_2 \cdot CaF_2}{1009 \text{ g}} \times \frac{2 \text{ mol F}}{1 \text{ mol } 3Ca_3(PO_4)_2 \cdot CaF_2}$

$\times \frac{19.00 \text{ g F}}{1 \text{ mol F}} \times \frac{1 \text{ kg}}{1000 \text{ g}} = 37.66 \text{ kg fluorine}$

Oxygen

23.

Reaction:	$H_2O_2(aq) + H_2O \rightleftharpoons$	$HO_2^-(aq)$ +	$H_3O^+(aq)$
Initial:	1.0 M		
Changes:	$-x$ M	$+x$ M	$+x$ M
Equil:	$(1.0 - x)$M	x M	x M

$K_a = \frac{[H_3O^+][HO_2^-]}{[H_2O_2]} = 10^{-11.75} = 1.8 \times 10^{-12} = \frac{x \cdot x}{1.0 - x} \approx \frac{x^2}{1.0}$

$x = \sqrt{1.8 \times 10^{-12}} = 1.3 \times 10^{-6} \text{ M} = [H_3O^+]$ $\quad pH = -\log(1.3 \times 10^{-6}) = 5.89$

24. Oxide ion reacts with water to form hydroxide ion. $Li_2O(s) + H_2O(l) \longrightarrow 2\,Li^+(aq) + 2\,OH^-(aq)$
We first calculate $[OH^-]$ and then the solution's pH.

$[OH^-] = \frac{0.050 \text{ g } Li_2O}{750.0 \text{ mL}} \times \frac{1000 \text{ mL}}{1 \text{ L}} \times \frac{1 \text{ mol } Li_2O}{29.88 \text{ g } Li_2O} \times \frac{2 \text{ mol } OH^-}{1 \text{ mol } Li_2O} = 0.0045 \text{ M}$

$pOH = -\log(0.0045) = 2.35$ $\quad pH = 14.00 - pOH = 14.00 - 2.35 = 11.65$

25. 3 |O=O| ⟶ 2 O···O···O

Bonds broken: 3 O=O = 3 × 498 kJ/mol = 1494 kJ/mol Bonds formed: 4 O···O
$\Delta H° = +285$ kJ/mol = bonds broken – bonds formed = 1494 kJ/mol – 4 O···O
4 O···O = 1494 kJ/mol – 285 kJ/mol = 1209 kJ/mol O···O = 1209 kJ/mol ÷ 4 = 302 kJ/mol

26. 3 O=O ⟶ 2 |O—O=O Average bond energy = $\frac{\text{O=O} + \text{O—O}}{2} = \frac{498 + 142}{2} = 320$ kJ/mol

Since bond energies actually are average values taken from many sources, the result of 320 kJ/mol is in reasonably good agreement with the specific value of 302 kJ/mol obtained in Exercise 25.

27. **(a)** H_2S, while polar, forms only weak hydrogen bonds. H_2O forms much stronger hydrogen bonds, which are much more difficult to disrupt, leading to a higher boiling point.

(b) All electrons are paired in O_3, producing a diamagnetic molecule. |O—O=O

28. **(a)** O=O |O—O=O H—O—O—H

The O—O bond in O_2 is a double bond, which should be short (121 pm). That in O_3 is a "one-and-a-half" bond, of intermediate length (128 pm). And that of H_2O_2 is a longer (148 pm) single bond.

(b) The O_2^+ ion has one fewer electron than does the O_2 molecule. Based on Lewis structures,

O=O O·–O

one would predict that O_2^+ would have a bond order of 1.5, weaker than the double bond of O_2, and therefore longer, in contradiction to the experimental evidence. But the molecular orbital picture of the two species is different, because the highest energy electrons are in antibonding orbitals. The bond order of O_2 is 2, while that of O_2^+ is 2.5, producing a stronger bond for O_2^+ than for O_2.

	σ_{1s}	σ_{1s}^*	σ_{2s}	σ_{2s}^*	π_{2p}	σ_{2p}	π_{2p}^*	σ_{2p}^*
O_2	[↑↓]	[↑↓]	[↑↓]	[↑↓]	[↑↓][↑↓]	[↑↓]	[↑][↑]	[ ]

bond order = (no. bonding electrons – no. antibonding electrons) ÷ 2 = (10 – 6) ÷ 2 = 2.0

O_2^+	[↑↓]	[↑↓]	[↑↓]	[↑↓]	[↑↓][↑↓]	[↑↓]	[↑][ ]	[ ]

bond order = (no. bonding electrons – no. antibonding electrons) ÷ 2 = (10 – 5) ÷ 2 = 1.5

Sulfur

29. **(a)** ZnS zinc sulfide **(b)** $KHSO_3$ potassium hydrogen sulfite
(c) $K_2S_2O_3$ potassium thiosulfate **(d)** SF_4 sulfur tetrafluoride

30. **(a)** $CaSO_4 \cdot 2H_2O$ calcium sulfate dihydrate **(b)** $H_2S(aq)$ hydrosulfuric acid
(c) $NaHSO_4$ sodium hydrogen sulfate **(d)** $H_2S_2O_7(aq)$ disulfuric acid

31. **(a)** $FeS(s) + 2\,HCl(aq) \longrightarrow FeCl_2(aq) + H_2S(g)$ $MnS(s)$, $ZnS(s)$, etc. also are possible.
(b) $CaSO_3(s) + 2\,HCl(aq) \longrightarrow CaCl_2(aq) + H_2O + SO_2(g)$
(c) Oxidation: $SO_2(aq) + 2\,H_2O \longrightarrow SO_4^{2-}(aq) + 4\,H^+(aq) + 2\,e^-$
Reduction: $MnO_2(s) + 4\,H^+(aq) + 2\,e^- \longrightarrow Mn^{2+}(aq) + 2\,H_2O$
Net: $SO_2(aq) + MnO_2(s) \longrightarrow Mn^{2+}(aq) + SO_4^{2-}(aq)$
(d) Oxidation: $S_2O_3^{2-}(aq) + H_2O \longrightarrow 2\,SO_2(g) + 2\,H^+(aq) + 4\,e^-$
Reduction: $S_2O_3^{2-}(aq) + 6\,H^+(aq) + 4\,e^- \longrightarrow 2\,S(s) + 3\,H_2O$
Net: $S_2O_3^{2-}(aq) + 2\,H^+(aq) \longrightarrow S(s) + SO_2(g) + H_2O$

32. **(a)** $S(s) + O_2(g) \longrightarrow SO_2(g)$ $2\,Na(s) + 2\,H_2O \longrightarrow 2\,NaOH(aq) + H_2(g)$
$2\,NaOH(aq) + SO_2(g) \longrightarrow Na_2SO_3(aq) + H_2O$
(b) $S(s) + O_2(g) \longrightarrow SO_2(g)$
$SO_2(g) + 2\,H_2O + Cl_2(g) \longrightarrow SO_4^{2-}(aq) + 2\,Cl^-(aq) + 4\,H^+(aq)$ ($E^\circ_{cell} = +1.19$ V)
$2\,Na(s) + 2\,H_2O \longrightarrow 2\,NaOH(aq) + H_2(g)$
$2\,NaOH(aq) + 2\,H^+(aq) + SO_4^{2-}(aq) \longrightarrow Na_2SO_4(aq) + 2\,H_2O$
(c) $S(s) + O_2(g) \longrightarrow SO_2(g)$ $2\,Na(s) + 2\,H_2O \longrightarrow 2\,NaOH(aq) + H_2(g)$
$2\,NaOH(aq) + SO_2(g) \longrightarrow Na_2SO_3(aq) + H_2O$
$Na_2SO_3(aq) + S(s) \longrightarrow NaS_2O_3(aq)$ (Boil the reactants in an alkline solution.)

33. The decomposition of thiosulfate ion is promoted in an acidic solution. Thus, simply adding a strong acid, such as HCl(aq), to the solid will result in no reaction for Na_2SO_4. But HCl(aq) added to $Na_2S_2O_3$ results in the production of $SO_2(g)$ and the formation of a S(s) precipitate.

$$S_2O_3^{2-}(aq) + 2\,H^+(aq) \longrightarrow S(s) + SO_2(g) + H_2O$$

34. Sulfites are easily oxidized to sulfates by, for example, O_2 in the atmosphere. On the other hand, there are no oxidizing agents naturally available in reasonable concentrations that can oxidize SO_4^{2-} to a higher oxidation state, such as in $S_2O_8^{2-}$. Similarly, the atmosphere of Earth is an oxidizing one, reducing agents are not present to reduce SO_4^{2-} to a species with a lower oxidation state, except in localized areas.

35. Neither $Na^+(aq)$ nor $HSO_4^-(aq)$ will hydrolyze, being respectively the cation of a strong base and the anion of a strong acid. But $HSO_4^-(aq)$ will ionize further, $K_2 = 1.1 \times 10^{-2}$ for $HSO_4^-(aq)$. We set up the situation, and solve the quadratic equation to obtain $[H_3O^+]$.

$$[HSO_4^-] = \frac{12.5\text{ g NaHSO}_4}{250.0\text{ mL soln}} \times \frac{1000\text{ mL}}{1\text{ L soln}} \times \frac{1\text{ mol NaHSO}_4}{120.1\text{ g NaHSO}_4} \times \frac{1\text{ mol HSO}_4^-}{1\text{ mol NaHSO}_4} = 0.416\text{ M}$$

Reaction: $HSO_4^-(aq) + H_2O \rightleftharpoons SO_4^{2-}(aq) + H_3O^+(aq)$

Initial: 0.416 M

Changes: $-x$ M $+x$ M $+x$ M

Equil: $(0.416 - x)$M x M x M

$$K_2 = \frac{[H_3O^+][SO_4^{2-}]}{[HSO_4^-]} = 0.011 = \frac{x^2}{0.416 - x} \qquad x^2 = 0.0046 - 0.011\,x \qquad 0 = x^2 + 0.011\,x - 0.0046$$

$$x = \frac{-b \pm \sqrt{b^2 - 4ac}}{2a} = \frac{-0.011 \pm \sqrt{1.2 \times 10^{-4} + 1.8 \times 10^{-2}}}{2} = 0.062 = [H_3O^+]$$

$$pH = -\log(0.062) = 1.21$$

36. Oxidation: $\{SO_3^{2-}(aq) + H_2O \longrightarrow SO_4^{2-}(aq) + 2\,H^+(aq) + 2\,e^-\} \qquad \times 5$

Reduction: $\{MnO_4^-(aq) + 8\,H^+(aq) + 5\,e^- \longrightarrow Mn^{2+}(aq) + 4\,H_2O\} \qquad \times 2$

Net: $5\,SO_3^{2-}(aq) + 2\,MnO_4^-(aq) + 6\,H^+(aq) \longrightarrow 5\,SO_4^{2-}(aq) + 2\,Mn^{2+}(aq) + 3\,H_2O$

$$mass\ Na_2SO_3 = 26.50\ mL \times \frac{1\ L}{1000\ mL} \times \frac{0.0510\ mol\ MnO_4^-}{1\ L\ soln} \times \frac{5\ mol\ SO_3^{2-}}{2\ mol\ MnO_4^-} \times \frac{1\ mol\ Na_2SO_3}{1\ mol\ SO_3^{2-}}$$

$$\times \frac{126.0\ g\ Na_2SO_3}{1\ mol\ Na_2SO_3} = 0.426\ g\ Na_2SO_3$$

Nitrogen

37. **(a)** $2\,NO_2(g) \rightleftharpoons N_2O_4(g)$ **(b)** $NH_4NO_3(s) \xrightarrow{\Delta} N_2O(g) + 2\,H_2O(g)$

     (c) $2\,NH_3(g) + H_2SO_4(aq) \longrightarrow (NH_4)_2SO_4(aq)$

38. **(a)** $3\,Ag(s) + 4\,H^+(aq) + 4\,NO_3^-(aq) \longrightarrow 3\,AgNO_3(aq) + NO(g) + 2\,H_2O$

     (b) $(CH_3)_2NNH_2(l) + 4\,O_2(g) \longrightarrow 2\,CO_2(g) + 4\,H_2O(l) + N_2(g)$

     (c)   We would expect to obtain the most stable species from this situation.

        $2\,NO(g) + 2\,CO(g) \longrightarrow N_2(g) + 2\,CO_2(g)$

39. **(a)**   In hydrazine, N_2H_4, there are $2 \times 5 + 4 \times 1 = 14$ valence electrons = 7 pairs.

     (b)   In hydrogen azide, HN_3, there are $1 + 3 \times 5 = 16$ valence electrons = 8 pairs.

     (c)   In dinitrogen tetroxide, N_2O_4, there are $2 \times 5 + 4 \times 6 = 34$ valence electrons = 17 pairs

(a) $H\!-\!\bar{N}\!-\!\bar{N}\!-\!H$ **(b)** $H\!-\!\bar{N}\!=\!N\!=\!\bar{N} \leftrightarrow H\!-\!\bar{N}\!\equiv\!N|$ **(c)** $\bar{O}\!=\!N\!-\!N\!=\!\bar{O}$
    | | |Ō| |Ō|
    H H

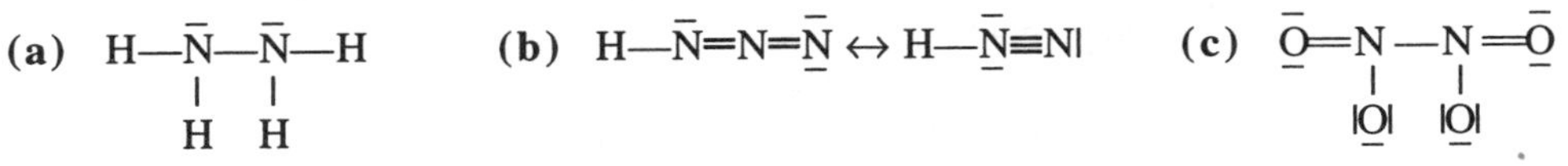

40. hyponitrous acid, a weak diprotic acid   nitramide, contains the amide group, $-NH_2$

$H\!-\!\bar{O}\!-\!N\!=\!\bar{N}\!-\!\bar{O}\!-\!H$ $H\!-\!\bar{N}\!-\!\bar{O}\!-\!\bar{N}\!=\!\bar{O}$ or $H\!-\!\bar{N}\!-\!N\!=\!\bar{O}$
                                        | | |
                                      H   H |Ō|

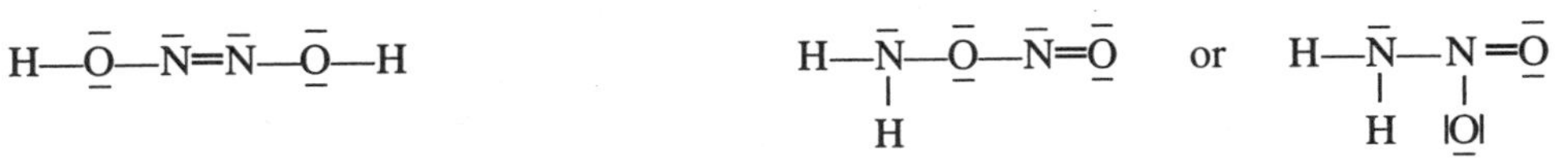

Both Lewis structures for nitramide are plausible based on the information supplied. To choose between them would require further information, such as whether the molecule contains a nitrogen-nitrogen bond. Experimental evidence indicates the structure to be the one with the nitrogen-nitrogen bond.

41. **(a)** HPO_4^{2-} hydrogen phosphate ion   **(c)** $H_6P_4O_{13}$ tetrapolyphosphoric acid

     (b) $Ca_2P_2O_7$ calcium pyrophosphate or calcium diphosphate

42. **(a)** $HONH_2$ hydroxylamine   **(c)** Li_3N lithium nitride

     (b) $CaHPO_4$ calcium hydrogen phosphate

Carbon and Silicon

43. In the sense that diamonds react imperceptibly slowly at room temperature (either with oxygen to form carbon dioxide, or in its transformation to the more stable graphite), it is true that "diamonds last forever." However, at elevated temperatures, diamond will burn to form $CO_2(g)$ and thus the statement is false. Also, the transformation C(diamond) $\longrightarrow$ C(graphite) might occur more rapidly under other conditions. Eventually, of course, the transition occurs.

44. The graphite in the pencil "lead" is a good dry lubricant that will lubricate the stickiness (or reluctance) in the lock and enable it to work smoothly. The key carries the graphite to the site that needs to be lubricated within the lock mechanism.

45. A silane is a silicon-hydrogen compound, with the general formula Si_nH_{2n+2}. A silanol is a compound in which one or more of the hydrogens of silane is replaced by an —OH group. Then, the general formula becomes $Si_nH_{2n+1}(OH)$. In both of these classes of compounds the number of silicon atoms, n, ranges from 1 to 6. Silicones are produced when silanols condense into chains, with the elimination of a water molecule between every two silanol molecules.

HO—Si_nH_{2n}—OH + HO—Si_nH_{2n}—OH $\longrightarrow$ HO—Si_nH_{2n}—O—Si_nH_{2n}—OH + H_2O

46. Both the alkali metal carbonates and the alkali metal silicates are soluble in water. Hence, they also are soluble in acids. However, the carbonates will produce gaseous carbon dioxide—CO_2—on reaction with acid ($CO_3^{2-} + 2 H^+ \longrightarrow$ "H_2CO_3" $\longrightarrow H_2O + CO_2$), while the silicates will produce silica—SiO_2— in an analogous reaction ($SiO_4^{4-} + 4 H^+ \longrightarrow$ "H_4SiO_4" $\longrightarrow 2 H_2O + SiO_2$). The silica is produced in many forms depending on reaction conditions: a colloidal dispersion, a gelatinous precipitate, or a semisolid gel. Silicates of cations other than those of alkali metals are insoluble in water, as are the analogous carbonates. However, the carbonates will dissolve in acids (witness the reaction of acid rain on limesone carvings and marble statues, both forms of $CaCO_3$), while the silicates will not. (Silicate rocks are not significantly affected by acid rain.)

47. **(1)** $2 CH_4(s) + S_8(g) \longrightarrow 2 CS_2(g) + 4 H_2S(g)$

  (2) $CS_2(g) + 3 Cl_2(g) \longrightarrow CCl_4(l) + S_2Cl_2(l)$

  (3) $4 CS_2(g) + 8 S_2Cl_2(g) \longrightarrow 4 CCl_4(l) + 3 S_8$

48. **(a)** $(CH_3)_3SiCl + H_2O \longrightarrow (CH_3)_3Si$—OH + HCl

   $2 (CH_3)_3Si$—OH $\longrightarrow (CH_3)_3Si$—O—$Si(CH_3)_3 + H_2O$

  (b) A silicone polymer does not form from $(CH_3)_3Si$—Cl. Only a dimer is produced.

  (c) The product that results from the treatment of CH_3SiCl_3 is two long Si—O—Si chains, with CH_3 groups on the outside, and joined by Si—O—Si bridges. Part of one of these chains and the beginnings of the bridges are shown below.

$$\begin{array}{ccccccccccccc}
& | & & | & & | & & | & & | & & | & \\
& O & & O & & O & & O & & O & & O & \\
& \downarrow & & | & & | & & | & & | & & | & \\
—O— & Si & —O— & Si & —O— & Si & —O— & Si & —O— & Si & —O— & Si & —O— \\
& | & & | & & | & & | & & | & & | & \\
& CH_3 & & CH_3 & & CH_3 & & CH_3 & & CH_3 & & CH_3 &
\end{array}$$

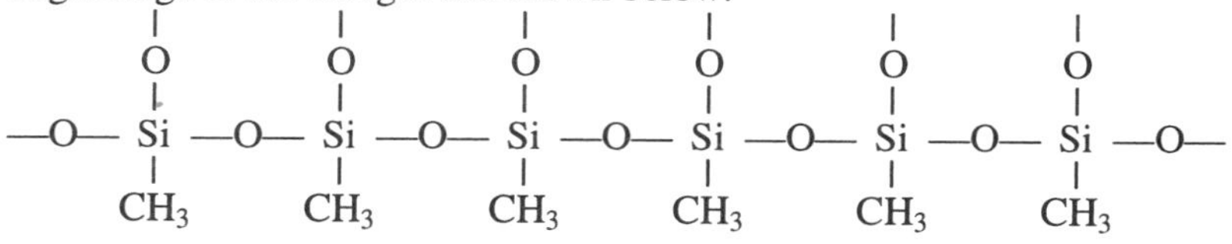

Boron

49. **(a)** B_4H_{10} contains a total of $4 \times 3 + 10 \times 1 = 22$ valence electrons or 11 pairs. Ten of these pairs could be allocated to form 10 B—H bonds, leaving but one pair to bond the four B atoms together, clearly an electron deficient situation.

  (b) In our analysis in the first part, we noted that the four B atoms had but one electron pair to bond them together. To bond these four atoms into a chain requires three electron pairs. Since each electron pair in a bridging bond replaces two "normal" bonds, there must be at least two bridging bonds in the B_4H_{10} molecules. By analogy with B_2H_6, we might write the structure below left. But this structure uses only a total of 20 electrons. (The bridge bonds are shown as dots, normal bonds—electron pairs—as dashes.) In the structure at right below, we have retained some of the form of B_2H_6, and produced a compound with the formula B_4H_{10} and 11 electron pairs. (The experimentally determined structure of B_4H_{10} consists of a four–membered ring of alternating B and H atoms, held together by bridging bonds. Two of the B atoms have two H atoms bonds to each of them by normal covalent bonds. The other two B atoms have one H atom covalently bonded to each. One final B—B bond joins these last two B atoms, across the diameter of the ring.)

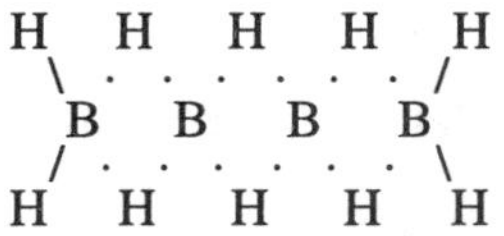

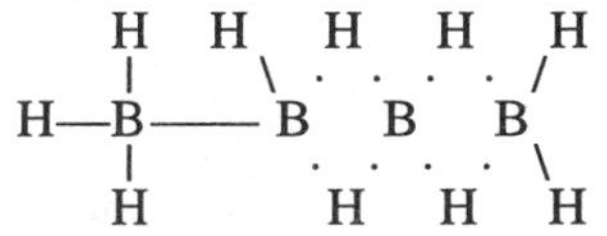

(c) C_4H_{10} contains a total of $4 \times 4 + 10 \times 1 = 26$ valence electrons or 13 pairs. A plausible Lewis structure

follows.
$$\begin{array}{ccccc} & H & H & H & H \\ & | & | & | & | \\ H- & C- & C- & C- & C-H \\ & | & | & | & | \\ & H & H & H & H \end{array}$$

50. (a)
$$\left[\begin{array}{c} \overline{|F|} \\ | \\ \overline{|F}-B-\overline{F|} \\ | \\ \overline{|F|} \end{array} \right]^{-}$$

(b)
$$\begin{array}{ccccc} \overline{|F|} & H & H & H \\ | & | & | & | \\ \overline{|F}-B\leftarrow N-C-C-H \\ | & | & | & | \\ \overline{|F|} & H & H & H \end{array}$$

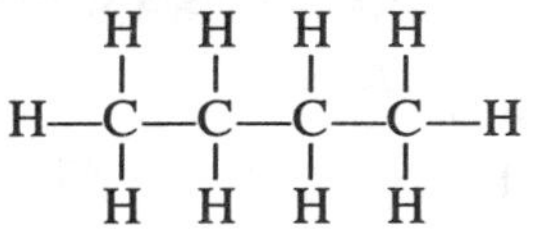

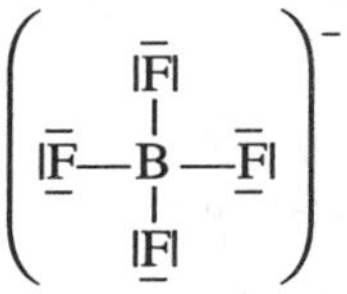

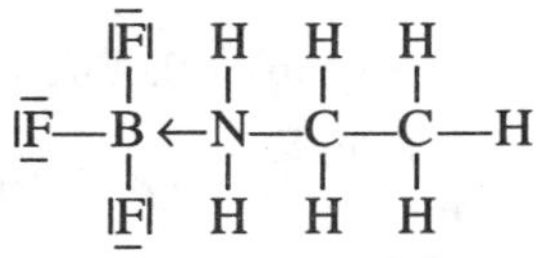

FEATURE PROBLEMS

A. 1. $E°[ClO_4^-|ClO_3^-] = +1.19$ V $E°[ClO_3^-|ClO_2] = +1.175$ V $E°[Cl_2|Cl^-] = +1.358$ V
$E°[ClO_3^-|Cl^-] = +1.450$ V   all in acidic solution
$E°[ClO_3^-|Cl^-] = +0.622$ V $E°[ClO^-|Cl^-] = +0.890$ V   both in basic solution

2. $ClO_3^-(aq) + 2\ H^+(aq) + e^- \longrightarrow ClO_2(aq) + H_2O$ $\Delta G_1° = -1\ \mathscr{F}(+1.175\ V)$

$ClO_2(aq) + H^+(aq) + e^- \longrightarrow HClO_2(aq)$ $\Delta G_2° = -1\ \mathscr{F}(+1.188\ V)$

$ClO_3^-(aq) + 3\ H^+(aq) + 2\ e^- \longrightarrow HClO_2(aq) + H_2O$ $\Delta G° = \Delta G_1° + \Delta G_2° = -2\ \mathscr{F}E°$

$E° = \dfrac{1.175\ V + 1.188\ V}{2} = 1.182\ V$ in acidic solution

$ClO_3^-(aq) + H_2O + e^- \longrightarrow ClO_2(aq) + 2\ OH^-(aq)$ $\Delta G_1° = -1\ \mathscr{F}(-0.481\ V)$

$ClO_2(aq) + H_2O + e^- \longrightarrow HClO_2(aq) + OH^-(aq)$ $\Delta G_2° = -1\ \mathscr{F}(+1.071\ V)$

$ClO_3^-(aq) + 2\ H_2O + 2\ e^- \longrightarrow HClO_2(aq) + 3\ OH^-(aq)$ $\Delta G° = \Delta G_1° + \Delta G_2° = -2\ \mathscr{F}E°$

$E° = \dfrac{1.071\ V - 0.481\ V}{2} = 0.295\ V$ in basic solution

$ClO_3^-(aq) + 2\ H_2O + 4\ e^- \longrightarrow ClO^-(aq) + 4\ OH^-(aq)$ $\Delta G_1° = -4\ \mathscr{F}E_1°$

$ClO^-(aq) + H_2O + 2\ e^- \longrightarrow Cl^-(aq) + 2\ OH^-(aq)$ $\Delta G_2° = -2\ \mathscr{F}(+0.890\ V)$

$ClO_3^-(aq) + 3\ H_2O + 6\ e^- \longrightarrow Cl^-(aq) + 6\ OH^-(aq)$ $\Delta G° = \Delta G_1° + \Delta G_2° = -6\ \mathscr{F}(+0.622\ V)$

$E_1° = \dfrac{6\ (+0.622\ V) - 2\ (0.890\ V)}{4} = 0.488\ V$

3. $E°[SO_3^{2-}|S] = -0.66$ V in basic solution, from Appendix D.

$4\ SO_3^{2-}(aq) + 6\ H_2O + 6\ e^- \longrightarrow S_4O_6^{2-}(aq) + 12\ OH^-(aq)$ $\Delta G_1° = -6\ \mathscr{F}E_1°$

$S_4O_6^{2-}(aq) + 2\ e^- \longrightarrow 2\ S_2O_3^{2-}(aq)$ $\Delta G_2° = -2\ \mathscr{F}(+0.080\ V)$

$4\ SO_3^{2-}(aq) + 6\ H_2O + 8\ e^- \longrightarrow 2\ S_2O_3^{2-}(aq) + 12\ OH^-(aq)$ $\Delta G° = -8\ \mathscr{F}(-0.576\ V) = \Delta G_1° + \Delta G_2°$

$E_1°° = \dfrac{8\ (-0.576\ V) - 2\ (-0.080\ V)}{6} = -0.795\ V$ $E°[SO_3^{2-}|S_4O_6^{2-}] = -0.795\ V$

4. Reduction: $S_2O_3^{2-}(aq) + 3\ H_2O + 4\ e^- \longrightarrow 2\ S(s) + 6\ OH^-(aq)$ $E° = -0.74\ V$

Oxidation: $S_2O_3^{2-}(aq) + 6\ OH^-(aq) \longrightarrow 2\ SO_3^{2-}(aq) + 3\ H_2O + 4\ e^-$ $-E° = +0.576\ V$

Overall: $2\ S_2O_3^{2-}(aq) \longrightarrow 2\ SO_3^{2-}(aq) + 2\ S(s)$ $E° = -0.16\ V$

This reaction clearly is non-spontaneous, although clearly a bit different from the result in the Summarizing Example. Can you see the difference?

B. 1. As before, we can organize a solution around the balanced chemical equation.

Equation:	$I_2(aq)$	$\rightleftharpoons$	$I_2(CCl_4)$
Initial:	1.33×10^{-3} M		0 M
Initial:	0.0133 mmol		
Changes:	$-x$ mmol		$+x$ mmol
Equil:	$(0.0133 - x)$ mmol		x mmol

$$K_c = 85.5 = \frac{[I_2(CCl_4)]}{[I_2(aq)]} = \frac{\dfrac{x}{10.0\text{ mL}}}{\dfrac{0.0133 - x}{10.0\text{ mL}}} \qquad x = 1.13_7 - 85.5\,x \qquad 86.5x = 1.13_7$$

$$x = \frac{1.13_7}{86.5} = 0.01315 \text{ mmol } I_2 \text{ in } CCl_4 \qquad I_2(aq) = (0.0133 - 0.01315)\text{ mmol} = 0.00015\text{ mmol}$$

$$\text{mass } I_2 = 0.00015\text{ mmol} \times \frac{253.8\text{ mg } I_2}{1\text{ mol } I_2} = 5 \times 10^{-3}\text{ mg } I_2 = 0.038\text{ mg } I_2$$

2. We have the same set-up, except that the initial amount $I_2(aq) = 0.00015$ mmol.

$$K_c = 85.5 = \frac{[I_2(CCl_4)]}{[I_2(aq)]} = \frac{x}{0.00015 - x} \qquad x = 0.0127 - 85.5\,x \qquad 86.5x = 0.0127$$

$$x = \frac{0.0127}{86.5} = 0.0001488\text{ mmol} = I_2(CCl_4) \quad I_2(aq) = 0.00015 - 0.0001488 = 1.2 \times 10^{-6}\text{ mmol } I_2$$

$$\text{mass } I_2 = 1.2 \times 10^{-6}\text{ mmol in } H_2O \times \frac{253.8\text{ mg } I_2}{1\text{ mol } I_2} = 3 \times 10^{-4}\text{ mg } I_2 = 0.0003\text{ mg } I_2$$

3. If twice the volume of CCl_4 were used for the initial extraction, the equilibrium concentrations would have been different from part **1.**

Equation:	$I_2(aq)$	$\rightleftharpoons$	$I_2(CCl_4)$
Initial:	1.33×10^{-3} M		0 M
Initial:	0.0133 mmol		
Changes:	$-x$ mmol		$+x$ mmol
Equil:	$(0.0133 - x)$ mmol		x mmol

$$K_c = 85.5 = \frac{[I_2(CCl_4)]}{[I_2(aq)]} = \frac{\dfrac{x}{20.0\text{ mL}}}{\dfrac{0.0133 - x}{10.0\text{ mL}}} \qquad x \div 2 = 1.13_7 - 85.5\,x \qquad 86.0x = 1.13_7$$

$$x = \frac{1.13_7}{86.0} = 1.32_2 \times 10^{-2}\text{ mmol} = \text{amount } I_2(CCl_4)$$

$$\text{total mass } I_2 = \frac{1.33 \times 10^{-3}\text{ mmol } I_2}{1\text{ mL}} \times 10.0\text{ mL soln} \times \frac{253.8\text{ mg } I_2}{1\text{ mol } I_2} = 3.376\text{ mg } I_2$$

$$\text{mass } I_2 \text{ in } CCl_4 = 1.32_2 \times 10^{-2}\text{ mmol} \times \frac{253.8\text{ mg } I_2}{1\text{ mol } I_2} = 3.356\text{ mg } I_2$$

mass I_2 remaining in water = 3.376 mg – 3.356 mg = 0.020 mg

Two extractions are much more efficient than one.

24 THE TRANSITION ELEMENTS

PRACTICE EXAMPLES

1A **(a)** Cu_2O should form. $2\ Cu_2S(s) + 3\ O_2(g) \longrightarrow 2\ Cu_2O(s) + 2\ SO_2(g)$
(b) W(s) is the reduction product. $WO_3(s) + 3\ H_2(g) \longrightarrow W(s) + 3\ H_2O(g)$
(c) Hg(l) forms. $2\ HgO(s) \xrightarrow{\Delta} 2\ Hg(l) + O_2(g)$

1B **(a)** $SiO_2(s)$ should be the oxidation product of Si. $3\ Si(s) + 2\ Cr_2O_3(s) \xrightarrow{\Delta} 3\ SiO_2(s) + 4\ Cr(l)$
(b) Roasting is simple heating in air. $2\ Co(OH)_3(s) \xrightarrow{\Delta} Co_2O_3(s) + 3\ H_2O(g)$
(c) $MnO_2(s)$ forms; assume an acidic solution. $Mn^{2+}(aq) + 2\ H_2O \longrightarrow MnO_2(s) + 4\ H^+(aq) + 2\ e^-$

2A In each case, chemical formulas are substituted for the names given in the problem and the result is balanced by inspection. The physical states—(s) or (l)—are those that seem reasonable.
$2\ V_2O_5(s) + 5\ Si(s) \xrightarrow{\Delta} 4\ V(l) + 5\ SiO_2(s)$ $CaO(s) + SiO_2(s) \xrightarrow{\Delta} CaSiO_3(l)$

2B In each case, chemical formulas are substituted for the names given in the problem and the result is balanced by inspection. The physical states—(s) or (l)—are those that seem reasonable.
$4\ FeWO_4(s) + 4\ Na_2CO_3(s) + O_2(g) \xrightarrow{\Delta} 2\ Fe_2O_3(s) + 4\ Na_2WO_4(aq) + 4\ CO_2(g)$
$Na_2WO_4(s) + H_2SO_4(aq) \longrightarrow Na_2SO_4(aq) + H_2O(l) + WO_3(s)$
$WO_3(s) + 3\ H_2(g) \longrightarrow W(s) + 3\ H_2O(g)$

3A We write and combine the half-equations for oxidation and reduction. If $E° > 0$, the reaction is spontaneous.
oxidation: $\{V^{3+}(aq) + H_2O \longrightarrow VO^{2+}(aq) + 2\ H^+(aq) + e^-\} \times 3$ $E° = -0.337$ V
reduction: $NO_3^-(aq) + 4\ H^+(aq) + 3\ e^- \longrightarrow NO(g) + 2\ H_2O$ $E° = +0.956$ V
net: $NO_3^-(aq) + 3\ V^{3+}(aq) + H_2O \longrightarrow NO(g) + 3\ VO^{2+}(aq) + 2\ H^+(aq)$ $E°_{cell} = +0.619$ V
Nitric acid can be used to oxidize $V^{3+}(aq)$ to $VO^{2+}(aq)$.

3B The couple that we seek must have a half-cell potential of such a size that a positive sum results when this half-cell potential (1) is combined with $E°[VO^{2+}(aq)|V^{3+}(aq)] = +0.337$ V, or (2) when it is combined with $E°[V^{3+}(aq)|V^{2+}(aq)] = -0.255$ V, but (3) a negative sum must be produced when this half-cell potential is combined with $E°[V^{2+}(aq)|V(s)] = -1.18$ V. Now recall that we are speaking of an *oxidation* reaction for this half-cell. Thus, our three relationships are, in terms of the *reduction* potential, $E°$,
(1) $-E° + 0.337\ V > 0$ (2) $-E° - 0.255 > 0$ (3) $-E° - 1.18\ V < 0$
Multiply each relationship by –1, which changes the direction of the inequality, and all signs.
(1) $+E° - 0.337\ V < 0$ (2) $+E° + 0.255 < 0$ (3) $+E° + 1.18\ V > 0$
Solve each inequality: (1) $E° < +0.337$ V (2) $E° < -0.255$ V (3) $E° > -1.18$ V
Thus, possible reducing couples from Table 21-1 have $+0.255\ V < -E° < +1.18$ V. Some are:
$-E°[Cr^{2+}(aq)|Cr^{3+}(aq)] = +0.424$ V $-E°[Fe(s)|Fe^{2+}(aq)] = +0.440$ V $-E°[Zn(s)|Zn^{2+}(aq)] = +0.763$ V

SUMMARIZING EXAMPLE CALCULATIONS

3. Combine the half-equations to determine $E°_{cell}$.
Oxidation: $Cu^+(aq) \longrightarrow Cu^{2+}(aq) + e^-$ $-E° = -0.154$ V
Reduction: $Cu^+(aq) + e^- \longrightarrow Cu(s)$ $E° = +0.520$ V
Cell: $2\ Cu^+(aq) \longrightarrow Cu^{2+}(aq) + Cu(s)$ $E°_{cell} = +0.366$ V

4. The disproportionation reaction is spontaneous. A solution with a large $[Cu^+]$ cannot be prepared.

REVIEW QUESTIONS

1. (a) A domain is a region within a metal within which the magnetic moments of the metal atoms align. In some ways, it behaves as a small magnet.
   (b) Flotation is a process of separating metal ore from the extra rock, called *gangue*. It involves agitating the crushed rock and ore with water and a detergent. The ore particles stick to the bubbles and are skimmed off.
   (c) Leaching refers to treating ore-bearing material with a chemical solution to dissolve the desired metallic elements. The solution generally is trickled through the material, and the resulting metal-bearing solution collected at the bottom.
   (d) An amalgam is an alloy (or solution) of another metal in mercury.

2. (a) The lanthanide contraction refers to the steady decrease in the size of atoms and (particularly) +3 ions as one proceeds across the lanthanide series from La through Lu.
   (b) Zone refining uses a ring oven to melt a portion of a cylindrical metal bar. Impurities are more soluble in the molten metal and tend to remain in the molten portion. The molten region then is swept down the bar to the end, carrying the impurities with it. The process can be repeated several times, concentrating impurities at one end of the bar, which is cut off and returned to be refined.
   (c) The basic oxygen process is a relatively recently (1950s) developed technique of steel making in which molten pig iron is treated with oxygen gas (at about 10 atm pressure) and powdered limestone to reduce the carbon content and remove the undesirable impurities.
   (d) Slag formation occurs when a flux, such as limestone, is added to an ore being reduced or a metal being refined. CaO(s) forms from thermal decomposition and this unites with SiO_2 to form $CaSiO_3(l)$. In this molten slag, impurities dissolve. The slag also floats on top of the reduced metal, protecting it from oxidation. Other nonmetal oxides, such as P_4O_{10}, can take the place of SiO_2 giving slags of other formulas: $Ca_3(PO_4)_2$. In this way low-level impurities are removed from the metal.

3. (a) Paramagnetism indicates that atoms, molecules, or ions have one or more unpaired electrons. In Ferromagnetism, the unpaired electrons on several adjacent atoms or ions align with each other, producing a stronger magnetic field.
   (b) Roasting refers to heating a concentrated ore, usually to convert hydroxides, carbonates, or sulfides to oxides. Refining is the process of purifying the crude metal that results from the reduction of the ore.
   (c) In hydrometallurgy, reactions in aqueous solution are used in concentration, reduction, and/or refining. In pyrometallurgy, the process of reduction and/or refining are carried out at high temperatures.
   (d) Chromate ion is CrO_4^{2-}, a good precipitating agent; dichromate ion is $Cr_2O_7^{2-}$, a good oxidizing agent.

4. (a) $Sc(OH)_3$ scandium hydroxide (b) Cu_2O copper(I) oxide
   (c) $TiCl_4$ titanium(IV) chloride *or* titanium tetrachloride (d) V_2O_5 vanadium(V) oxide
   (e) K_2CrO_4 potassium chromate (f) K_2MnO_4 potassium manganate

5. (a) CrO_3 chromium(VI) oxide (b) $FeSiO_3$ iron(II) silicate
   (c) $BaCr_2O_7$ barium dichromate (d) CuCN copper(I) cyanide
   (e) $CoCl_2·6H_2O$ cobalt(II) chloride hexahydrate

6. (a) Pig iron is the iron obtained from the blast furnace. Once poured into molds it is called cast iron. "It contains about 95% Fe, 3% to 4% C, and various other impurities."
   (b) Ferromanganese is an alloy of iron and manganese, produced by reducing the mixed oxides of the two metals with carbon.
   (c) Chromite ore is the principal ore of chromium, $Fe(CrO_2)_2$
   (d) Brass is a mixture of copper and zinc, with small quantities of Sn, Pb, and Fe.
   (e) Aqua regia is a mixture of three parts HCl(aq) and one part $HNO_3(aq)$. It both is an oxidizing acid and can form complex ions with cations in solution, making it the only acid that can dissolve difficultly soluble substances, such as Au and HgS.
   (f) Blister copper is the impure copper that results from the reduction process. The blisters are frozen bubbles of the $SO_2(g)$ that forms during that process.

(g) Stainless steel contains iron, and a significant proportion (about 10% each) of chromium and nickel. It is not ferromagnetic.

7. **(a)** $TiCl_4(g) + 4\,Na(l) \xrightarrow{\Delta} Ti(s) + 4\,NaCl(l)$ **(b)** $Cr_2O_3(s) + 2\,Al(s) \xrightarrow{\Delta} 2\,Cr(l) + Al_2O_3(s)$

(c) $Ag(s) + HCl(aq) \longrightarrow$ no reaction **(d)** $K_2Cr_2O_7(aq) + 2\,KOH(aq) \longrightarrow 2\,K_2CrO_4(aq) + H_2O$

(e) $MnO_2(s) + 2\,C(s) \xrightarrow{\Delta} Mn(l) + 2\,CO(g)$

8. **(a)** Oxidation: $\{Fe_2S_3(s) + 6\,OH^-(aq) \longrightarrow 2\,Fe(OH)_3(s) + 3\,S(s) + 6\,e^-\}$ $\times 2$

Reduction: $\{O_2(g) + 2\,H_2O + 4\,e^- \longrightarrow 4\,OH^-(aq)\}$ $\times 3$

Net: $2\,Fe_2S_3(s) + 3\,O_2(g) + 6\,H_2O \longrightarrow 4\,Fe(OH)_3(s) + 6\,S(s)$

(b) Oxidation: $\{Mn^{2+}(aq) + 4\,H_2O \longrightarrow MnO_4^-(aq) + 8\,H^+(aq) + 5\,e^-\} \times 2$

Reduction: $\{S_2O_8^{2-}(aq) + 2\,e^- \longrightarrow 2\,SO_4^{2-}(aq)\}$ $\times 5$

Net: $2\,Mn^{2+}(aq) + 8\,H_2O + 5\,S_2O_8^{2-}(aq) \longrightarrow 2\,MnO_4^-(aq) + 16\,H^+(aq) + 10\,SO_4^{2-}(aq)$

(c) Oxidation: $\{Ag(s) + 2\,CN^-(aq) \longrightarrow [Ag(CN)_2]^-(aq) + e^-\}$ $\times 4$

Reduction: $O_2(g) + 2\,H_2O + 4\,e^- \longrightarrow 4\,OH^-(aq)$

Net: $4\,Ag(s) + 8\,CN^-(aq) + O_2(g) + 2\,H_2O \longrightarrow 4\,[Ag(CN)_2]^-(aq) + 4\,OH^-(aq)$

9. **(a)** $Cr(s) + 2\,HCl(aq) \longrightarrow CrCl_2(aq) + H_2(g)$ Virtually any first period transition metal except Cu can be substituted for Cr, with a different metallic oxidation state, as well.

(b) $Cr_2O_3(s) + 2\,OH^-(aq) + 3\,H_2O \longrightarrow 2\,Cr(OH)_4^-(aq)$
The oxide must be amphoteric. Thus, Sc_2O_3, TiO_2, ZrO_2, and ZnO could be substituted for Cr_2O_3.

(c) $2\,La(s) + 6\,HCl(aq) \longrightarrow 2\,LaCl_3(aq) + 3\,H_2(g)$
Any lanthanide or actinide element can be substituted for lanthanum.

10. **(a)** Ti [Ar] 3d [↑][↑][ ][ ][ ] 4s [↑↓] **(b)** V^{3+} [Ar] 3d [↑][↑][ ][ ][ ] 4s [ ]

(c) Cr^{2+} [Ar] 3d [↑][↑][↑][↑][ ] 4s [ ] **(d)** Mn^{4+} [Ar] 3d [↑][↑][↑][ ][ ] 4s [ ]

(e) Mn^{2+} [Ar] 3d [↑][↑][↑][↑][↑] 4s [ ] **(f)** Fe^{3+} [Ar] 3d [↑][↑][↑][↑][↑] 4s [ ]

11. We first give the orbital diagram for each of the species, and then count the number of unpaired electrons.

Fe	[Ar] 3d [↑↓][↑][↑][↑][↑]	4s [↑↓]	4 unpaired electrons
Sc^{3+}	[Ar] 3d [ ][ ][ ][ ][ ]	4s [ ]	0 unpaired electrons
Ti^{2+}	[Ar] 3d [↑][↑][ ][ ][ ]	4s [ ]	2 unpaired electrons
Mn^{4+}	[Ar] 3d [↑][↑][↑][ ][ ]	4s [ ]	3 unpaired electrons
Cr	[Ar] 3d [↑][↑][↑][↑][↑]	4s [↑]	6 unpaired electrons
Cu^{2+}	[Ar] 3d [↑↓][↑↓][↑↓][↑↓][↑]	4s [ ]	1 unpaired electron

Finally, arrange in order of decreasing number of unpaired electrons: $Cr > Fe > Mn^{4+} > Ti^{2+} > Cu^{2+} > Sc^{3+}$

12. (1) Transition metals tend to have higher melting points than representative metals. (2) Because they are metals, transition elements have fairly low ionization energies. (3) The ions of transition metals often are colored in aqueous solution. (4) Because they are metals and thus readily form cations, they have negative standard reduction potentials. Their compounds often have unpaired electrons because of the diversity of *d*-electron configurations, and thus (5) they often are paramagnetic.

13. Sc^{3+} has the electron configuration of the preceding noble gas (Ar). This is obviously an electron configuration in which all electrons are paired, a diamagnetic species. All of the other species have at least one unpaired electron, as orbital diagrams show.

Cr^{2+} [Ar] 3d [↑][↑][↑][↑][ ] 4s [ ] Fe^{3+} [Ar] 3d [↑][↑][↑][↑][↑] 4s [ ]

Sc^{3+} [Ar] 3d [ ][ ][ ][ ][ ] 4s [ ] Cu^{2+} [Ar] 3d [↑↓][↑↓][↑↓][↑↓][↑] 4s [ ]

14. Ti should not display a +6 oxidation state, because in order for Ti to display a +6 oxidation state, two electrons would have to be removed from the noble gas electron configuration of Ar. This is quite unlikely.

15. The ion that is the best oxidizing agent in aqueous solution is the ion that can most readily be reduced. And we can assess that property either by standard reduction potentials—$E°[Ag^+(aq)|Ag(s)] = +0.800$ V,

$E°[Cu^{2+}(aq)|Cu(s)] = +0.337$ V, $E°[Zn^{2+}(aq)|Zn(s)] = -0.763$ V, $E°[Na^{+}(aq)|Na(s)] = -2.714$ V—with the most positive value indicating the best oxidizing agent, or from our own experience with the reverse reaction. We know that silver is the least likely of all the metals listed to be oxidized, and thus its ion is the most readily reduced.

16. The +3 state probably is the most stable oxidation state of iron because in this oxidation state iron has a d^5 electron configuration. As we have seen before, an electron configuration containing a half-filled subshell is unexpectedly stable energetically, compared to other configurations with partially filled subshells. In the cases of cobalt and nickel, the ions with d^5 configurations are Co^{3+} and Ni^{4+}. The slight advantage of a d^5 configuration is not sufficient to stabilize ions with high charges, although Co^{3+} is stable under certain conditions, as in complex ions.

EXERCISES

Properties of the Transition Elements

17. A given main group metal typically displays one oxidation state, usually equal to its family number in the periodic table. Exceptions are elements such as Tl (+1 and +3), Pb (+2 and +4), and Sn (+2 and +4) in which the lower oxidation state represents a pair of *s* electrons not being ionized (a so-called "inert pair").

Main group metals do not form a wide variety of complex ions, with Al^{3+}, Sn^{2+}, Sn^{4+}, and Pb^{2+} being major exceptions. On the other hand, most transition metal ions form an extensive variety of complex ions; it is, in fact, the unusual transition metal cation that does not form complex ions.

Most compounds of main group metals are colorless; exceptions occur when the anion is colored. On the other hand, many of the compounds of transition metal cations are colored.

Virtually every main group metal cation has no unpaired electrons and hence is diamagnetic. On the other hand, many transition metals cations have one or more unpaired electrons and therefore are paramagnetic.

18. As we proceed from Sc to Cr the valence electron configuration has an increasing number of unpaired electrons, which are capable of forming bonds to adjacent atoms. As we continue beyond Cr, however, these electrons become paired, and the resulting atoms are less able to form bonds with their neighbors.

19. When an electron is added to a main group element to create the element of next highest atomic number, this electron is added to the outer shell of the atom, far from the nucleus. Thus, it has a major influence on the size of the atom. However, when an electron is added to a transition metal atom to create the atom of next highest atomic number, it is added to the electronic shell inside the outermost. The electron thus has been added to a position close to the nucleus to which it is attracted quite strongly and thus it has small effect on the size of the atom.

20. The reason why the radii of Pd (138 pm) and Pt (139 pm) are so similar, and so different from the radius of Ni (125 pm) is because the lanthanide series intrudes between Pd and Pt. Because of the lanthanide contraction, elements in the second transition row are almost identical in size to their cogeners (family members) in the third transition row.

21. Of the first transition series, manganese exhibits the greatest number of different oxidation states in its compounds, every state from +1 to +7. One possible explanation might be the $3d^54s^2$ electron configuration. Removing one electron produces an electron configuration ($3d^54s^1$) with two half-filled subshells, removing two produces one with a half-filled and an empty subshell. Then there is no point of semistability until the remaining five *d* electrons are removed. These higher oxidation states all are stabilized by being present in oxides (MnO_2) or oxoanions (MnO_4^-). In contrast, Fe quickly attains an electron configuration ($3d^54s^0$) that has more stable features than its original electron configuration.

22. At the beginning of the series, there are few electrons beyond the last noble gas than can be ionized and thus the maximum oxidation state is limited. Toward the end of the series, many of the electrons are paired up or the *d* subshell is filled, and thus a somewhat stable situation would be disrupted by ionization.

23. The greater ease of forming lanthanide cations compared to forming transition metal cations, is due to the larger size of lanthanide atoms. The valence (outer shell) electrons of these larger atoms are much further

from the nucleus, much less strongly attracted to the positive charge of the nucleus, and thus are removed much more readily.

24. As we proceed across the transition series, electrons are being added to a shell next to the valence shell. Thus, the electronic character of the outside of each atom is changing somewhat, leading to changes in their chemical properties. (Recall that this valence shell consists of only two 4*s* electrons.) As we proceed across the lanthanides, on the other hand, the electrons are being added to a shell two removed from the outermost shell; there is less of an effect on the electronic character of the valence shell. (Recall also that there are 10 electrons in subshells outside the one being filled: $5s^2$, $5p^6$, and $6s^2$.)

Reactions of Transition Metals and Their Compounds

25. **(a)** $Sc(OH)_3(s) + 3\ H^+(aq) \longrightarrow Sc^{3+}(aq) + 3\ H_2O$

(b) $3\ Fe^{2+}(aq) + MnO_4^-(aq) + 2\ H_2O(aq) \longrightarrow 3\ Fe^{3+}(aq) + MnO_2(s) + 4\ OH^-$

(c) $2\ KOH(l) + TiO_2(s) \xrightarrow{\Delta} K_2TiO_3(s) + H_2O(g)$

(d) $Cu(s) + 2\ H_2SO_4(conc, aq) \longrightarrow CuSO_4(aq) + SO_2(g) + 2\ H_2O$

26. **(a)** $2\ Sc_2O_3(l, \text{in } Na_3ScF_6) + 3\ C(s) \xrightarrow{\text{electrolysis}} 4\ Sc(l) + 3\ CO_2(g)$
[By analogy with the equation for the electrolytic production of Al]

(b) $Cr(s) + 2\ HCl(aq) \longrightarrow Cr^{2+}(aq) + 2\ Cl^-(aq) + H_2(g)$

(c) $4\ Cr^{2+}(aq) + O_2(g) + 4\ H^+(aq) \longrightarrow 4\ Cr^{3+}(aq) + 2\ H_2O$

(d) $Ag(s) + 2\ HNO_3(aq) \longrightarrow AgNO_3(aq) + NO_2(g) + H_2O$

27. We write some of the following reactions as total equations rather than as net ionic equations, so that the reagents used are indicated.

(a) $FeS(s) + 2\ HCl(aq) \longrightarrow FeCl_2(aq) + H_2S(g)$
$4\ Fe^{2+}(aq) + O_2(g) + 4\ H^+(aq) \longrightarrow 4\ Fe^{3+}(aq) + 2\ H_2O$ $\quad Fe^{3+}(aq) + 3\ OH^-(aq) \longrightarrow Fe(OH)_3(s)$

(b) $BaCO_3(s) + 2\ HCl(aq) \longrightarrow BaCl_2(aq) + H_2O + CO_2(g)$
$2\ BaCl_2(aq) + K_2Cr_2O_7(aq) + 2\ NaOH(aq) \longrightarrow 2\ BaCrO_4(s) + 2\ KCl(aq) + 2\ NaCl(aq) + H_2O$

28. **(a)** Metal hydroxides thermally decompose to oxides. $Cu(OH)_2(s) \xrightarrow{\Delta} CuO(s) + H_2O(g)$

(b) $(NH_4)_2Cr_2O_7(s) \xrightarrow{\Delta} N_2(g) + 4\ H_2O(g) + Cr_2O_3(s)$
$Cr_2O_3(s) + 6\ HCl(aq) \longrightarrow CrCl_3(aq) + 3\ H_2O$

Oxidation-reduction

29. **(a)** Reduction: $VO^{2+}(aq) + 2\ H^+(aq) + e^- \longrightarrow V^{3+}(aq) + H_2O$

(b) Oxidation: $Cr^{2+}(aq) \longrightarrow Cr^{3+}(aq) + e^-$

30. **(a)** Oxidation: $Fe(OH)_3(s) + 5\ OH^-(aq) \longrightarrow FeO_4^{2-}(aq) + 4\ H_2O + 3\ e^-$

(b) Reduction: $[Ag(CN)_2]^-(aq) + e^- \longrightarrow Ag(s) + 2\ CN^-(aq)$

31. **(a)** First we need the reduction potential for the couple $VO_2^+(aq)|V^{2+}(aq)$. We use a method learned in a previous chapter's Feature Problem.

$VO_2^+(aq) + 2\ H^+(aq) + e^- \longrightarrow VO^{2+}(aq) + H_2O$ $\quad \Delta G° = -1\ \mathcal{F}(+1.000\ V)$

$VO^{2+}(aq) + 2\ H^+(aq) + e^- \longrightarrow V^{3+}(aq) + H_2O$ $\quad \Delta G° = -1\ \mathcal{F}(+0.337\ V)$

$V^{3+}(aq) + e^- \longrightarrow V^{2+}(aq)$ $\quad \Delta G° = -1\ \mathcal{F}(-0.255\ V)$

$VO_2^+(aq) + 4\ H^+(aq) + 3\ e^- \longrightarrow V^{2+}(aq) + 2\ H_2O$ $\quad \Delta G° = -3\ \mathcal{F}E°$

$E° = \dfrac{1.000\ V + 0.337\ V - 0.255\ V}{3} = +0.361\ V$ We analyze the oxidation-reduction reaction.

Oxidation: $\{2\ Br^-(aq) \longrightarrow Br_2(l) + 2\ e^-\}$ $\quad \times 3$ $\quad -E° = -1.065\ V$

Reduction: $\{VO_2^+(aq) + 4\ H^+(aq) + 3\ e^- \longrightarrow V^{2+}(aq) + 2\ H_2O\}$ $\quad \times 2$ $\quad E° = +0.361\ V$

Net: $6\ Br^-(aq) + 2\ VO_2^+(aq) + 8\ H^+(aq) \longrightarrow 3\ Br_2(l) + 2\ V^{2+}(aq) + 4\ H_2O$ $\quad E°_{cell} = -0.704\ V$

This reaction does not occur to a significant extent as written.

(b) Oxidation: $Fe^{2+}(aq) \longrightarrow Fe^{3+}(aq) + e^-$ $\quad -E° = -0.771$ V

Reduction: $VO_2^+(aq) + 2\,H^+(aq) + e^- \longrightarrow VO^{2+}(aq) + H_2O$ $\quad E° = +1.000$ V

Net: $Fe^{2+}(aq) + VO_2^+(aq) + 2\,H^+(aq) \longrightarrow Fe^{3+}(aq) + VO^{2+}(aq) + H_2O$ $\quad E°_{cell} = +0.229$ V

This reaction does occur to a significant extent under standard conditions.

(c) Oxidation: $H_2O_2 \longrightarrow 2\,H^+(aq) + 2\,e^- + O_2(g)$ $\quad -E° = -0.695$ V

Reduction: $MnO_2(s) + 4\,H^+(aq) + 2\,e^- \longrightarrow Mn^{2+}(aq) + 2\,H_2O$ $\quad E° = +1.23$ V

Net: $H_2O_2 + MnO_2(s) + 2\,H^+(aq) \longrightarrow O_2(g) + Mn^{2+}(aq) + 2\,H_2O$ $\quad E°_{cell} = +0.54$ V

This reaction does occur to a significant extent under standard conditions.

32. When a species acts as a reducing agent, it is oxidized. From Appendix D we obtain the potentials for each of the following couples.

$-E°[Zn^{2+}(aq)/Zn(s)] = +0.763$ V $\quad -E°[Sn^{4+}(aq)/Sn^{2+}(aq)] = -0.154$ V $\quad -E°[I_2(s)/I^-(aq)] = -0.535$ V

Each of these potentials is combined with the reduction potential cited in each part. If the resulting value of $E°_{cell}$ is positive, then the reducing agent will be effective in accomplishing the desired reduction, which we indicate with "yes"; if not, we write "no".

(a) $E°[Cr_2O_7^{2-}(aq)/Cr^{3+}(aq)] = +1.33$ V

$E°_{cell} = E°[Cr_2O_7^{2-}(aq)/Cr^{3+}(aq)] - E°[Zn^{2+}(aq)/Zn(s)] = +1.33\text{ V} + 0.763\text{ V} = +2.09$ V $\quad$ yes

$E°_{cell} = E°[Cr_2O_7^{2-}(aq)/Cr^{3+}(aq)] - E°[Sn^{4+}(aq)/Sn^{2+}(aq)] = +1.33\text{ V} - 0.154\text{ V} = +1.18$ V $\quad$ yes

$E°_{cell} = E°[Cr_2O_7^{2-}(aq)/Cr^{3+}(aq)] - E°[I_2(s)/I^-(aq)] = +1.33\text{ V} - 0.535\text{ V} = +0.80$ V $\quad$ yes

(b) $E°[Cr^{3+}(aq)/Cr^{2+}(aq)] = -0.424$ V

$E°_{cell} = E°[Cr^{3+}(aq)/Cr^{2+}(aq)] - E°[Zn^{2+}(aq)/Zn(s)] = -0.424\text{ V} + 0.763\text{ V} = +0.339$ V $\quad$ yes

$E°_{cell} = E°[Cr^{3+}(aq)/Cr^{2+}(aq)] - E°[Sn^{4+}(aq)/Sn^{2+}(aq)] = -0.424\text{ V} - 0.154\text{ V} = -0.578$ V $\quad$ no

$E°_{cell} = E°[Cr^{3+}(aq)/Cr^{2+}(aq)] - E°[I_2(s)/I^-(aq)] = -0.424\text{ V} - 0.535\text{ V} = -0.959$ V $\quad$ no

(c) $E°[SO_4^{2-}(aq)/SO_2(g)] = +0.17$ V

$E°_{cell} = E°[SO_4^{2-}(aq)/SO_2(g)] - E°[Zn^{2+}(aq)/Zn(s)] = +0.17\text{ V} + 0.763\text{ V} = +0.93$ V $\quad$ yes

$E°_{cell} = E°[SO_4^{2-}(aq)/SO_2(g)] - E°[Sn^{4+}(aq)/Sn^{2+}(aq)] = +0.17\text{ V} - 0.154\text{ V} = +0.02$ V $\quad$ yes

$E°_{cell} = E°[SO_4^{2-}(aq)/SO_2(g)] - E°[I_2(s)/I^-(aq)] = +0.17\text{ V} - 0.535\text{ V} = -0.37$ V $\quad$ no

33. We would expect $Fe^{2+}(aq)$ to be oxidized to the +3 cation by $O_2(g)$ in acidic solution. One reason for this expectation is the fact that Fe^{3+} has a d^5 electron configuration, and we have previously seen that a half-filled subshell is unusually stable. We combine half-cell potentials to demonstrate this.

Oxidation: $\{Fe^{2+}(aq) \longrightarrow Fe^{3+}(aq) + e^-\} \quad \times 4$ $\quad -E° = -0.771$ V

Reduction: $O_2(g) + 4\,H^+(aq) + 4\,e^- \longrightarrow 2\,H_2O$ $\quad E° = +1.229$ V

Net: $4\,Fe^{2+}(aq) + O_2(g) + 4\,H^+(aq) \longrightarrow 4\,Fe^{3+}(aq) + 2\,H_2O$ $\quad E°_{cell} = +0.458$ V

34. The couple that we seek must have a half-cell potential of such a size that a positive sum results when this half-cell potential (1) is combined with $E°[VO^{2+}(aq)|V^{3+}(aq)] = +0.337$ V, but (2) a negative sum must be produced when this half-cell potential is combined with $E°[V^{3+}(aq)|V^{2+}(aq)] = -0.255$ V, or (3) when it is combined with $E°[V^{2+}(aq)|V(s)] = -1.18$ V. Now recall that we are speaking of an *oxidation* reaction for this half-cell. Thus, our three relationships are, in terms of the *reduction* potential, $E°$,

(1) $-E° + 0.337\text{ V} > 0$ $\quad$ (2) $-E° - 0.255 < 0$ $\quad$ (3) $-E° - 1.18\text{ V} < 0$

Multiply each relationship by –1, which changes the direction of the inequality, and all signs.

(1) $+E° - 0.337\text{ V} < 0$ $\quad$ (2) $+E° + 0.255 > 0$ $\quad$ (3) $+E° + 1.18\text{ V} > 0$

Solve each inequality: (1) $E° < +0.337$ V $\quad$ (2) $E° > -0.255$ V $\quad$ (3) $E° > -1.18$ V

Thus, possible reducing couples from Table 21-1 have $+0.255\text{ V} > -E° > -0.337$ V and include:

$-E°[Sn(s)|Sn^{2+}(aq)] = +0.137$ V $\quad -E°[Pb(s)|Pb^{2+}(aq)] = +0.125$ V $\quad -E°[H_2(g)|H^+(aq)] = +0.000$ V

$-E°[H_2S(g)|S(s)] = -0.14$ V $\quad -E°[Sn^{2+}(aq)|Sn^{4+}(aq)] = -0.154$ V

$-E°[SO_2(g)|SO_4^{2-}(aq)] = -0.17$ V $\quad -E°[Cu(s)|Cu^{2+}(aq)] = -0.337$ V

Chromium and Chromium Compounds

35. Orange dichromate ion is in equilibrium with yellow chromate ion in aqueous solution.

$Cr_2O_7^{2-}(aq) + H_2O \rightleftharpoons 2\,CrO_4^{2-}(aq) + 2\,H^+(aq)$

The chromate ion in solution then reacts with lead(II) ion to form a precipitate of yellow lead(II) dichromate.
$Pb^{2+}(aq) + CrO_4^{2-}(aq) \rightleftharpoons PbCrO_4(s)$
$PbCrO_4(s)$ will form until $[H^+]$ from the first equilibrium increases to a significant value and both equilibria are simultaneously satisfied.

36. The initial dissolving reaction forms orange dichromate ion.
$2\ BaCrO_4(s) + 2\ HCl(aq) \rightleftharpoons 2\ Ba^{2+}(aq) + Cr_2O_7^{2-}(aq) + 2\ Cl^-(aq) + H_2O$
Dichromate ion is a good oxidizing agent, sufficiently strong to oxidize $Cl^-(aq)$ to $Cl_2(g)$ if the concentrations of reactants are high, the solution is acidic, and the product $Cl_2(g)$ is allowed to escape.
$6\ Cl^-(aq) + Cr_2O_7^{2-}(aq) + 14\ H^+(aq) \rightleftharpoons 3\ Cl_2(g) + 2\ Cr^{3+}(aq) + 7\ H_2O$
Chromium(III) can hydrolyze in solution to produce green $[Cr(OH)_4]^-(aq)$, but the solution needs to be alkaline for this to occur. A more likely source of the green color is a complex ion such as $[Cr(H_2O)_4Cl_2]^+$.

37. Oxidation: $\{Zn(s) \longrightarrow Zn^{2+}(aq) + 2\ e^-\}\quad \times 3$
Reduction: $Cr_2O_7^{2-}(aq, \text{orange}) + 14\ H^+(aq) + 6\ e^- \longrightarrow 2\ Cr^{3+}(aq, \text{green}) + 7\ H_2O$
Net: $3\ Zn(s) + Cr_2O_7^{2-}(aq) + 14\ H^+(aq) \longrightarrow 3\ Zn^{2+}(aq) + 2\ Cr^{3+}(aq) + 7\ H_2O$

Oxidation: $Zn(s) \longrightarrow Zn^{2+}(aq) + 2\ e^-$
Reduction: $\{Cr^{3+}(aq, \text{green}) + e^- \longrightarrow Cr^{2+}(aq, \text{blue})\}\quad \times 2$
Net: $Zn(s) + 2\ Cr^{3+}(aq) \longrightarrow Zn^{2+}(aq) + 2\ Cr^{2+}(aq)$
The green color is most likely due to a chloro complex of $Cr^{3+}(aq)$, such as $[Cr(H_2O)_4Cl_2]^+(aq)$.

Oxidation: $\{Cr^{2+}(aq, \text{blue}) \longrightarrow Cr^{3+}(aq, \text{green}) + e^-\}\quad \times 4$
Reduction: $O_2(g) + 4\ H^+(aq) + 4\ e^- \longrightarrow 2\ H_2O$
Net: $4\ Cr^{2+}(aq) + O_2(g) + 4\ H^+(aq) \longrightarrow 4\ Cr^{3+}(aq) + 2\ H_2O$

38. $CO_2(g)$, as the oxide of a nonmetal, is an acid anhydride. Its function is to make the solution acidic. A resonable equation for the reaction that occurs follows.
$2\ CrO_4^{2-}(aq) + 2\ H^+(aq) \rightleftharpoons Cr_2O_7^{2-}(aq) + H_2O$
$2\ H_2O + 2\ CO_2(aq) \rightleftharpoons 2\ H^+(aq) + 2\ HCO_3^-(aq)$
$2\ CrO_4^{2-}(aq) + 2\ CO_2(aq) + H_2O \rightleftharpoons Cr_2O_7^{2-}(aq) + 2\ HCO_3^-(aq)$

39. Simple substitution into expression (24.18) yields $[Cr_2O_7^{2-}]$ in each case. In fact, the expression is readily solved for the desired concentration: $[Cr_2O_7^{2-}] = 3.2 \times 10^{14}\ [H^+]^2\ [CrO_4^{2-}]^2$ In each case, we use the value of pH to determine $[H^+] = 10^{-pH}$.
(a) $[Cr_2O_7^{2-}] = 3.2 \times 10^{14}\ (10^{-6.62})^2\ (0.20)^2 = 0.74\ M$
(b) $[Cr_2O_7^{2-}] = 3.2 \times 10^{14}\ (10^{-8.85})^2\ (0.20)^2 = 2.6 \times 10^{-5}\ M$

40. We use expression (24.18).
$$\frac{[Cr_2O_7^{2-}]}{[CrO_4^{2-}]^2} = 3.2 \times 10^{14}\ [H^+]^2 = 3.2 \times 10^{14}\ (10^{-7.55})^2 = 3.2 \times 10^{14}\ (2.8 \times 10^{-8})^2 = 0.254$$
$$[CrO_4^{2-}]_{initial} = \frac{1.505\ g\ Na_2CrO_4}{0.345\ L\ soln} \times \frac{1\ mol\ Na_2CrO_4}{161.97\ g\ Na_2CrO_4} \times \frac{1\ mol\ CrO_4^{2-}}{1\ mol\ Na_2CrO_4} = 0.0269\ M$$

Reaction:	$2\ CrO_4^{2-}(aq) + 2\ H^+(aq)$	$\rightleftharpoons$	$Cr_2O_7^{2-}(aq) + H_2O$
Initial:	0.0269 M		
Changes:	$-2x$ M		$+x$ M
Equil:	$(0.0269 - 2x)$M		x M

$$\frac{[Cr_2O_7^{2-}]}{[CrO_4^{2-}]^2} = 0.254 = \frac{x}{(0.0269 - 2x)^2} \qquad x = 0.254\ (0.000724 - 0.108x + 4x^2)$$
$$x = 0.000184 - 0.0274\ x + 1.016\ x^2 \qquad 1.016\ x^2 - 1.0274\ x + 0.000184 = 0$$
$$x = \frac{-b \pm \sqrt{b^2 - 4ac}}{2a} = \frac{1.0274 \pm \sqrt{1.0556 - 0.000736}}{2.032} = 1.0111\ M,\ 0.00016\ M$$
We choose the second root because the first root gives a negative $[CrO_4^{2-}]$.
$[Cr_2O_7^{2-}] = x = 0.00016\ M \qquad [CrO_4^{2-}] = 0.0269 - 2x = 0.0266\ M$
We carry extra significant figures to avoid a significant rounding error in this problem.

41. Each mole of chromium metal plated out from a chromate ion solution requires six moles of electrons.

$$\text{mass Cr} = 1.00\text{ h} \times \frac{3600\text{ s}}{1\text{ hr}} \times \frac{3.4\text{ C}}{1\text{ s}} \times \frac{1\text{ mol e}^-}{96485\text{ C}} \times \frac{1\text{ mol Cr}}{6\text{ mol e}^-} \times \frac{52.00\text{ g Cr}}{1\text{ mol Cr}} = 1.10\text{ g Cr}$$

42. First we compute the amount of Cr(s) deposited.

$$\text{amount Cr} = \left(0.0010\text{ mm} \times \frac{1\text{ cm}}{10\text{ mm}} \times 0.375\text{ m}^2 \times \frac{10^4\text{ cm}^2}{1\text{ m}^2}\right) \times \frac{7.14\text{ g}}{1\text{ cm}^3} \times \frac{1\text{ mol Cr}}{52.0\text{ g Cr}}$$

$$= 5.1 \times 10^{-2}\text{ mol Cr}$$

Recall that the deposition of each mole of chromium requires six moles of electrons. We now compute the time required to deposit the 3.1×10^{-4} mol Cr.

$$\text{time} = 5.1 \times 10^{-2}\text{ mol Cr} \times \frac{6\text{ mol e}^-}{1\text{ mol Cr}} \times \frac{96485\text{ C}}{1\text{ mol e}^-} \times \frac{1\text{ s}}{3.5\text{ C}} \times \frac{1\text{ h}}{3600\text{ s}} = 2.3\text{ h}$$

43. Dichromate ion is the prevalent species in acidic solution. Oxoanions are better oxidizing agents in acidic solution because increasing the concentration of hydrogen ion favors formation of product. The half-equation is $Cr_2O_7^{2-}(aq) + 14\,H^+(aq) + 6\,e^- \longrightarrow 2\,Cr^{3+}(aq) + 7\,H_2O$. Note that most precipitation occurs effectively in alkaline solution. In fact, adding an acid to a compound is often an effective way of dissolving a water-insoluble compound. Thus, we expect to see the form that predominates in alkaline solution be the most effective precipitating agent. Notice also that CrO_4^{2-} is smaller than is $Cr_2O_7^{2-}$, giving it a higher lattice energy in its compounds and making those compounds harder to dissolve.

44. Both metal ions precipitate as hydroxides when $[OH^-]$ is moderate.

$Mg^{2+}(aq) + 2\,OH^-(aq) \longrightarrow Mg(OH)_2(s)$ $\qquad$ $Cr^{3+}(aq) + 3\,OH^-(aq) \longrightarrow Cr(OH)_3(s)$

Because chromium(III) oxides and hydroxides are amphoteric (and those of magnesium ion are not), $Cr(OH)_3(s)$ will dissolve in excess base. $\quad Cr(OH)_3(s) + NaOH(aq) \longrightarrow Na^+(aq) + [Cr(OH)_4]^-(aq)$

The Coinage Metals

45. **(a)** $Cu^{2+}(aq) + H_2(g) \longrightarrow Cu(s) + 2\,H^+(aq)$ $\qquad$ **(b)** $Au^+(aq) + Fe^{2+}(aq) \longrightarrow Au(s) + Fe^{3+}(aq)$

(c) $2\,Cu^{2+}(aq) + SO_2(g) + 2\,H_2O \longrightarrow 2\,Cu^+(aq) + SO_4^{2-}(aq) + 4\,H^+(aq)$

46. In either case, the acid acts as an oxidizing agent and dissolves silver; the gold is unaffected.

Oxidation: $\{Ag(s) \longrightarrow Ag^+(aq) + e^-\}$ $\times 3$

Reduction: $NO_3^-(aq) + 4\,H^+(aq) + 3\,e^- \longrightarrow NO(g) + 2\,H_2O$

Net: $3\,Ag(s) + NO_3^-(aq) + 4\,H^+(aq) \longrightarrow 3\,Ag^+(aq) + NO(g) + 2\,H_2O$

Oxidation: $\{Ag(s) \longrightarrow Ag^+(aq) + e^-\}$ $\times 2$

Reduction: $SO_4^{2-}(aq) + 4\,H^+(aq) + 2\,e^- \longrightarrow SO_2(g) + 2\,H_2O$

Net: $2\,Ag(s) + SO_4^{2-}(aq) + 4\,H^+(aq) \longrightarrow 2\,Ag^+(aq) + SO_2(g) + 2\,H_2O$

47. We know that $E^\circ_{cell} = +0.366$ V for the reaction $2\,Cu^+(aq) \longrightarrow Cu(s) + Cu^{2+}(aq)$. We wish to evaluate the equilibrium constant for this reaction. $\quad \Delta G^\circ = -RT \ln K_{eq} = -nFE^\circ_{cell}$

$$\ln K_{eq} = \frac{nFE^\circ_{cell}}{RT} = \frac{1\text{ mol e}^- \times \dfrac{96485\text{ C}}{1\text{ mol e}^-} \times +0.366\text{ V}}{\dfrac{8.3145\text{ J}}{\text{mol K}} \times 298.15\text{ K}} = 14.2 \qquad K_{eq} = e^{14.2} = 1.5 \times 10^6$$

48. In Exercise 47, we determined that $K_c = 1.4 \times 10^6 = \dfrac{[Cu^{2+}]}{[Cu^+]^2}$ or $[Cu^{2+}] = 1.4 \times 10^6\,[Cu^+]^2$

(a) When $[Cu^+] = 0.20$ M, $[Cu^{2+}] = 1.4 \times 10^6\,(0.20)^2 = 5.6 \times 10^4$ M. This is an impossibly high concentration. Thus $[Cu^+] = 0.20$ M can never be achieved.

(b) When $[Cu^+] = 1.0 \times 10^{-10}$ M, $[Cu^{2+}] = 1.4 \times 10^6\,(1.0 \times 10^{-10})^2 = 1.4 \times 10^{-14}$ M. This is an entirely reasonable (even though small) concentration; $[Cu^+] = 1.0 \times 10^{-10}$ M can be maintained in solution.

Group 2B Metals

49. We calculate the wavelength of light absorbed in order to promote an electron across each band gap. First a few relationships. $\quad E_{mole} = N_A E_{photon} \qquad E_{photon} = h\nu \qquad c = \nu\lambda$ or $\nu = c/\lambda$

Then, some algebra. $E_{mole} = N_A E_{photon} = N_A h\nu = N_A hc/\lambda$ or $\lambda = N_A hc/E_{mole}$

For ZnO, $\lambda = \dfrac{6.022 \times 10^{23}\ \text{mol}^{-1} \times 6.626 \times 10^{-34}\ \text{J s} \times 2.998 \times 10^{8}\ \text{m s}^{-1}}{290 \times 10^{3}\ \text{J mol}^{-1}} \times \dfrac{10^{9}\ \text{nm}}{1\ \text{m}}$

$= \dfrac{1.196 \times 10^{8}\ \text{J mol}^{-1}\ \text{nm}}{290 \times 10^{3}\ \text{J mol}^{-1}} = 413\ \text{nm}$ violet light

For CdS, $\lambda = \dfrac{1.196 \times 10^{8}\ \text{J mol}^{-1}\ \text{nm}}{250 \times 10^{3}\ \text{J mol}^{-1}} = 479\ \text{nm}$ blue light

The blue light absorbed by CdS is subtracted from the white light incident on the surface of the solid. The remaining reflected light is colored, yellow in this case. When the violet light is subtracted from the white light incident on the ZnO surface, the reflected light appears white.

50. The color that we see is the complementary color to the color absorbed. The color absorbed, in turn, is determined by the energy separation of the band gap. A short wavelength absorbed color indicates a band gap of large energy.

The yellow color of CdS means that the complementary color, violet, is absorbed. Since violet light has a quite short wavelength (about 410 nm), CdS must have a very large band gap energetically.

HgS is red, meaning that the color absorbed is green, which has a moderate wavelength (about 520 nm). HgS must have a band gap of intermediate energy.

CdSe is black, meaning that light of all wavelengths and thus all energies is absorbed. This would occur if CdSe has a very small band gap, smaller than that of least energetic red light (about 650 nm).

FEATURE PROBLEMS

A. **1.** A temperature-independent plot of $\Delta G°$ means that $\Delta H°$ is independent of temperature and $\Delta S°$ is close to zero. We expect that a reaction in which $\Delta n_{gas} = 0$ will fulfill the condition of $\Delta S° \approx 0$, and that is the case in this instance. In like fashion, a reaction that has $\Delta S° > 0$ (due to $\Delta n_{gas} > 0$) should have $\Delta G°$ decrease as temperature rises; this is the case with reaction (a). And finally, a reaction that has $\Delta S° < 0$ (due to $\Delta n_{gas} < 0$) should have $\Delta G°$ increase as temperature rises; this is the case with reaction (c).

2. When $K_{eq} = 1$, $\Delta G° = -RT\ln K_{eq} = 0$. We are looking for a temperature where the value of $\Delta G°$ for the reaction (d): $2\ H_2(g) + O_2(g) \longrightarrow 2\ H_2O(g)$ will be the same as $\Delta G°$ for reaction (a): $2\ C(s) + O_2(g) \longrightarrow 2\ CO(g)$. The water gas reaction is half of the sum that results when reaction (d) is subtracted from reaction (a). This occurs at a temperature of about 690 °C.

(a)	$2\ C(s) + O_2(g) \longrightarrow 2\ CO(g)$
–(d)	$2\ H_2O(g) \longrightarrow 2\ H_2(g) + O_2(g)$
(a) – (d)	$2\ C(s) + 2\ H_2O(g) \longrightarrow 2\ CO(g) + 2\ H_2(g)$

3. For the reaction given, $\Delta H° = \Delta H_f°[TiO_2(s)] = -944.7$ kJ/mol

$\Delta S° = S°[TiO_2(s)] - S°[Ti(s)] - 2\ S°[O_2(g)]$

$= 50.33\ \text{J mol}^{-1}\ \text{K}^{-1} - 30.63\ \text{J mol}^{-1}\ \text{K}^{-1} - 2 \times 205.0\ \text{J mol}^{-1}\ \text{K}^{-1}$

$= -390.3\ \text{J mol}^{-1}\ \text{K}^{-1} = -0.3903\ \text{kJ mol}^{-1}\ \text{K}^{-1}$

We calculate $\Delta G°$ at several temperatures:

0 °C = 273 K	$\Delta G° = -944.7$ kJ/mol + (0.3903×273) kJ/mol = -838.1 kJ/mol
1000 °C = 1273 K	$\Delta G° = -944.7$ kJ/mol + (0.3903×1273) kJ/mol = -447.8 kJ/mol
2000 °C = 2273 K	$\Delta G° = -944.7$ kJ/mol + (0.3903×2273) kJ/mol = -57.5 kJ/mol

Because of the possible formation of titanium carbides we combine the reduction of titanium dioxide $[TiO_2(s) \longrightarrow Ti(s) + 2\ O_2(g)]$ with reaction (c): $2\ CO(g) + O_2(g) \longrightarrow 2\ CO_2(g)$ The reduction of $TiO_2(s)$ with CO(g) becomes a spontaneous process where the plot of $\Delta G°$ of the formation reaction for $TiO_2(s)$ (the three data points calculated above) crosses the plot of $\Delta G°$ for reaction (c). This occurs at about 1310 °C. Above 1310 °C the reduction of $TiO_2(s)$ to Ti(s) with CO(g) is spontaneous.

B. **1.** Amphoteric cations are Al^{3+}, Cr^{3+}, and Zn^{2+}. Later indications are that Mn^{2+} precipitates as $MnO_2(s)$.

$Fe^{3+}(aq) + 3\ OH^-(aq) \longrightarrow Fe(OH)_3(s)$ $\quad$ $Al^{3+}(aq) + 4\ OH^-(aq) \longrightarrow [Al(OH)_4]^-(aq)$

$Co^{2+}(aq) + 2\ OH^-(aq) \longrightarrow Co(OH)_2(s)$ $\quad$ $Cr^{3+}(aq) + 4\ OH^-(aq) \longrightarrow [Cr(OH)_4]^-(aq)$

$Ni^{2+}(aq) + 2\,OH^-(aq) \longrightarrow Ni(OH)_2(aq)$ $\quad Zn^{2+}(aq) + 4\,OH^-(aq) \longrightarrow [Zn(OH)_4]^{2-}(aq)$

2. Of the three hydroxide precipitates, only Co^{2+} is easily oxidized.

$2\,Co(OH)_2(s) + H_2O_2(aq) \longrightarrow 2\,Co(OH)_3(s)$

3. We know CrO_4^{2-}, chromate ion, to be yellow.

$2\,[Cr(OH)_4]^-(aq) + 3\,H_2O_2(aq) + 2\,OH^-(aq) \longrightarrow 2\,CrO_4^{2-}(aq, \text{yellow}) + 8\,H_2O$

4. MnO_2 oxidizes HCl(aq) to form $Cl_2(g)$. The other hydroxides dissolve in strong acid.

$Fe(OH)_3(s) + 3\,H^+(aq) \longrightarrow Fe^{3+}(aq) + 3\,H_2O$ $\quad Co(OH)_3(s) + 3\,H^+(aq) \longrightarrow Co^{3+}(aq) + 3\,H_2O$

$Ni(OH)_2(s) + 2\,H^+(aq) \longrightarrow Ni^{2+}(aq) + 2\,H_2O$

$MnO_2(s) + 4\,HCl(aq) \longrightarrow MnCl_2(aq) + 2\,H_2O + Cl_2(g)$

5. $Fe(OH)_3$, $K_{sp} = 4 \times 10^{-38}$, is much less soluble than $Mn(OH)_2$, $K_{sp} = 1.9 \times 10^{-13}$. Both Co^{3+} and Ni^{2+} form ammine complex ions. $\quad Fe^{3+}(aq) + 3\,NH_3(aq) + 3\,H_2O \longrightarrow Fe(OH)_3(s) + 3\,NH_4^+(aq)$

C. 1. $E°[VO_2^+|VO^{2+}] = +1.000\text{ V}$ $\quad E°[VO^{2+}|V^{3+}] = +0.337\text{ V}$ $\quad E°[V^{3+}|V^{2+}] = -0.255\text{ V}$ $\quad E°[V^{2+}|V] = -1.13\text{ V}$

$VO_2^+(aq) + 2\,H^+(aq) + e^- \longrightarrow VO^{2+}(aq) + H_2O$ $\quad \Delta G_5° = -1\,\mathcal{F}(+1.000\text{ V})$

$VO^{2+}(aq) + 2\,H^+(aq) + e^- \longrightarrow V^{3+}(aq) + H_2O$ $\quad \Delta G_4° = -1\,\mathcal{F}(+0.337\text{ V})$

$VO_2^+(aq) + 4\,H^+(aq) + 2\,e^- \longrightarrow V^{3+}(aq) + 2\,H_2O$ $\quad \Delta G_a° = \Delta G_5° + \Delta G_4° = -2\,\mathcal{F}E_a°$

$$E_a° = \frac{1.000\text{ V} + 0.337\text{ V}}{2} = +0.669\text{ V} = E°[VO_2^+|V^{3+}]$$

$VO_2^+(aq) + 4\,H^+(aq) + 2\,e^- \longrightarrow V^{3+}(aq) + 2\,H_2O$ $\quad \Delta G_a° = -2\,\mathcal{F}(+0.669\text{ V})$

$V^{3+}(aq) + e^- \longrightarrow V^{2+}(aq)$ $\quad \Delta G_3° = -1\,\mathcal{F}(-0.255\text{ V})$

$VO_2^+(aq) + 4\,H^+(aq) + 3\,e^- \longrightarrow V^{2+}(aq) + 2\,H_2O$ $\quad \Delta G_b° = \Delta G_a° + \Delta G_3° = -3\,\mathcal{F}E_b°$

$$E_b° = \frac{1.337\text{ V} - 0.255\text{ V}}{3} = +0.361\text{ V} = E°[VO_2^+|V^{2+}]$$

$VO_2^+(aq) + 4\,H^+(aq) + 3\,e^- \longrightarrow V^{2+}(aq) + 2\,H_2O$ $\quad \Delta G_b° = -3\,\mathcal{F}(+0.361\text{ V})$

$V^{2+}(aq) + e^- \longrightarrow V^0(s)$ $\quad \Delta G_2° = -2\,\mathcal{F}(-1.13\text{ V})$

$VO_2^+(aq) + 4\,H^+(aq) + 5\,e^- \longrightarrow V^0(s) + 2\,H_2O$ $\quad \Delta G_c° = \Delta G_b° + \Delta G_2° = -5\,\mathcal{F}E_c°$

$$E_c° = \frac{1.083\text{ V} - 2.26\text{ V}}{5} = -0.24\text{ V} = E°[VO_2^+|V^0]$$

2(a) The reduction half-equation is significantly different in alkaline solution, as is its half-cell potential.

Oxidation: $MnO_4^{2-}(aq) \longrightarrow MnO_4^-(aq) + e^-$ $\quad -E° = -0.56\text{ V}$

Reduction: $MnO_4^{2-}(aq) + H_2O + e^- \longrightarrow MnO_3^-(aq) + 2\,OH^-(aq)$ $\quad E° = +0.3\text{ V}$

Net: $2\,MnO_4^{2-}(aq) + H_2O \longrightarrow MnO_4^-(aq) + MnO_3^-(aq) + 2\,OH^-(aq)$ $\quad E°_{cell} = -0.3\text{ V}$

The negative value of the cell voltage indicates that $MnO_4^{2-}(aq)$ is stable in alkaline solution, possibly because of the availability of the Mn(V) species in alkaline solution.

(b) From the electrode potential diagrams,

in acidic solution $\quad E°[MnO_4^-|MnO_2] = +1.70\text{ V}$ and $E°[MnO_2|Mn^{2+}] = +1.23\text{ V}$

in basic solution $\quad E°[MnO_4^-|MnO_2] = +0.60\text{ V}$ and $E°[MnO_2|Mn(OH)_2] = -0.05\text{ V}$

Therefore, in acidic solution, reduction does not stop with Mn(IV) because further reduction to Mn(II) is spontaneous. In alkaline (basic) solution, further reduction to Mn(II) is not spontaneous, and reduction stops with Mn(IV).

(c) We combine two half-equations: one in acidic solution, and the other in alkaline (basic) solution.

Reduction: $Mn(OH)_2(s) + 2\,e^- \longrightarrow Mn(s) + 2\,OH^-(aq)$ $\quad E° = -1.56\text{ V}$

Oxidation: $Mn(s) \longrightarrow Mn^{2+}(aq) + 2\,e^-$ $\quad -E° = +1.18\text{ V}$

Net: $Mn(OH)_2(s) \longrightarrow Mn^{2+}(aq) + 2\,OH^-(aq)$ $\quad E°_{cell} = -0.38\text{ V}$

$\Delta G° = -RT\ln K_{sp} = -n\mathcal{F}E°_{cell}$

$$\ln K_{sp} = \frac{n\mathcal{F}E°_{cell}}{RT} = \frac{2\text{ mol e}^- \times 96{,}485\text{ C/mol e}^- \times (-0.38\text{ V})}{8.3145\,\frac{\text{J}}{\text{mol K}} \times 298\text{ K}} = -29.6$$

$K_{sp} = e^{-29.6} = 1.4 \times 10^{-13}$ $\quad$ compare with 1.9×10^{-13} in Appendix D.

25 Complex Ions and Coordination Compounds

Practice Examples

1A There are two different kinds of ligands in this complex ion, iodo and cyano. Both are monodentate ligands; they form but one bond to the central atom. Since there are five ligands total in the complex ion, the coordination number is 5: C.N. = 5. Each cyano ligand has a charge of –1, as does the iodo ligand. Thus, the O.S. must be such that: O.S. + $[(4 + 1) \times (-1)] = -3 =$ O.S. – 5. Therefore, O.S. = +2.

1B The ligands are CN^-. Fe^{3+} is the central metal ion. The complex ion then is $[Fe(CN)_6]^{3-}$.

2A There are six "Cl" ligands (chloro), each with a charge of –1. Platinum is the metal ion with an oxidation state of +4. Thus, the complex ion is $[PtCl_6]^{2-}$, and we need two K^+ to balance charge: $K_2[PtCl_6]$

2B The "SCN" ligand is thiocyanato, with a charge of –1. The "NH_3" ligand is ammine and is not charged. There are five (penta) ammine ligands. The oxidation state of cobalt is +3. The complex ion is not negatively charged, so its name does not end with "ate". The name of the compound is

pentaamminethiocyanatocobalt(III) chloride

3A The oxalato ligand must occupy two *cis* positions. Either the two ammine or the two chloro ligands can be coplanar with the oxalate ligand, leaving the other two ligands axial. The other isomer has one ammine and one chloro ligand coplanar with the oxalate ligand. The structures are sketched below.

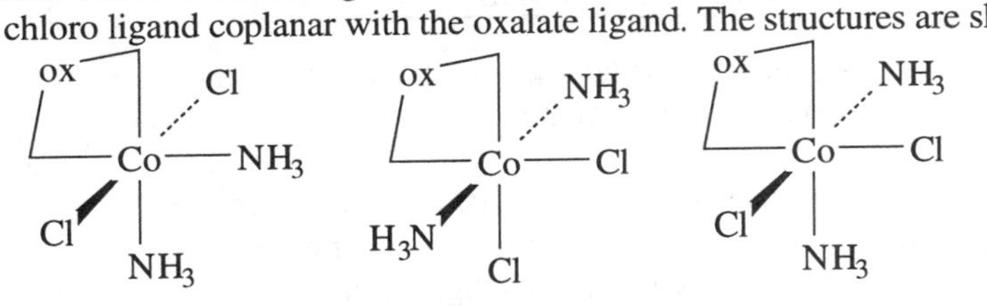

3B We start out with the two pyridines, C_5H_5N, located cis to each other. With that imposed, we can have the two chloros trans and the two carbonyls cis, the two carbonyls trans and the two chloros cis, or both the chloros and the carbonyls cis. If we now consider the two pyridines trans, we can either have both other pairs trans, or both other pairs cis. There are five geometric isomers. They follow, in the order described.

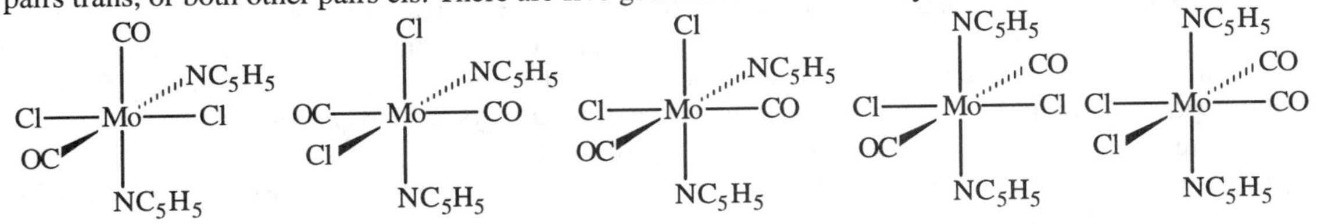

4A Fluoro is a weak field ligand. $[MnF_6]^{4-}$ is an octahedral complex. Mn^{2+} has five 3*d* electrons. The ligand field splitting diagram is sketched at right. There are five unpaired electrons.

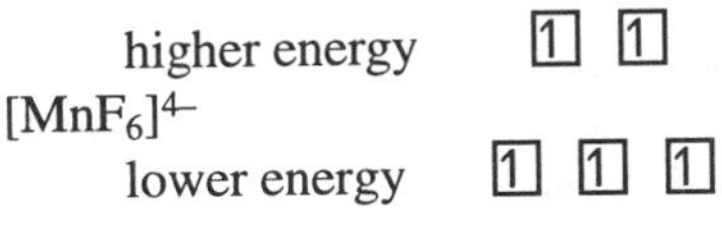

4B Co^{2+} has seven 3*d* electrons.
Chloro is a weak field ligand.
Aqua is a moderate field ligand.

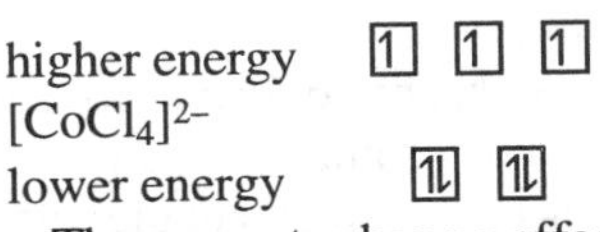

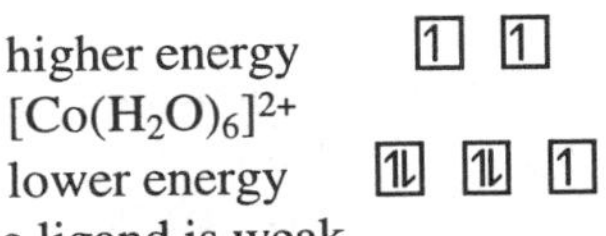

There are three unpaired electrons in each case. The geometry has no effect if the ligand is weak.

5A Cyano is a strong field ligand. Co^{2+} has seven 3*d* electrons. In the absence of a crystal field—all five *d* orbitals of the same energy—seven *d* electrons will produce three unpaired electrons. Thus, we need an orbital splitting diagram in which there are three orbitals of the same energy at the highest energy. This is the case with a tetrahedral orbital diagram. Thus, $[Co(CN)_4]^{2-}$ must be tetrahedral.

5B Ammine is a strong field ligand. Cu^{2+} has nine 3*d* electrons. There is no other way to arrange nine electrons in five orbitals than to have four of them filled, with two paired electrons in each, and one half-filled. Thus, the complex ion must be paramagnetic to the extent of one unpaired electron, no matter what the geometry of the ligands.

6A We are certain that $[Co(H_2O)_6]^{2-}$ is octahedral with a moderate field ligand. Tetrahedral $[CoCl_4]^{2-}$ has a weak field ligand. The relative values of ligand field splitting for the same ligand are $\Delta_t = 0.44\ \Delta_o$. Thus, $[Co(H_2O)_6]^{2+}$ absorbs light of higher energy, blue or green light, leaving a light pink as the complementary color we observe. $[CoCl_4]^{2-}$ absorbs lower energy red light, leaving blue light to pass through and be seen.

6B In order, the two complex ions are $[Fe(H_2O)_6]^{2+}$ and $[Fe(CN)_6]^{4-}$. We know that cyano is a strong field ligand; it should create the larger value of Δ and result in the absorption of light of the shorter wavelength. We would expect the cyano complex to absorb blue or violet light and thus $K_4[Fe(CN)_6]\cdot 3H_2O$ should appear yellow. The compound $[Fe(H_2O)_6](NO_3)_2$, containing the weak field complex thus should be green, because the weak field would result in the absorption of light of long wavelength, red light.

Summarizing Example Calculations

1. The maximum in the absorbance spectrum occurs at about 500 nm. In the electromagnetic spectrum in Chapter 9, we see that this corresponds to green light.
2. The light transmitted is what remains of white light after the green component has been removed. From looking at the dispersed spectrum of visible light, we see that the colors remaining after green is removed are red, blue, violet, and a little yellow. (A little because yellow is close to green in wavelength; much of it will be absorbed since the absorbance spectrum shows a broad peak.) These combine to give a magenta (nearly violet) color.

Review Questions

1. (a) The coordination number of a complex refers to the number of points of attachment for ligands on the central atom. If all ligands are monodentate, the coordination number equals the number of ligands attached to the central atom.
   (b) Δ_o is the octahedral crystal field splitting energy: the energy difference between the lower- and higher-energy *d* orbitals in a crystal field splitting diagram for an octahedral complex.
   (c) An ammine complex is one with ammonia molecule(s) as the ligand(s).
   (d) An enantiomer is one of two optical isomers: molecules of identical formula and bonding which are nonsuperimposable mirror images of each other.

2. (a) The spectrochemical series is a listing of the common ligands, ordered from strongest to weakest Lewis base, that is, from those creating the strongest to those creating the weakest ligand field.
   (b) Crystal field theory explains magnetism and colors of complex ions in terms of the difference in energy of *d* electrons created by the electrostatic field imposed by point charges at the locations of the ligands.
   (c) Optical isomerism refers to two compounds that differ in physical and chemical properties only in the direction in which each rotates the plane of polarized light.
   (d) Structural isomerism refers to differences in the ligands that are bonded to the central atom, and in the ligand atoms that are directly attached to the central atom.

3. (a) The coordination number of a central atom is the number of sites where ligands are bonded to that central atom. It is thus also the number of monodentate ligands bonded to a central atom. It also is known as the secondary valence. The oxidation number, or primary valence, is equal to the charge of the isolated central metal ion.
   (b) A monodentate ligand is one that has only one point of attachment to the central metal atom. A polydentate ligand bonds to the central metal atom at two or more sites.
   (c) In a *cis* isomer, two ligands have a small angular distance between their bonds to the central atom. In a *trans* isomer, two ligands are on opposite sides of the central atom.

(d) dextrorotatory and levorotatory isomers differ in the direction in which they rotate the plane of polarized light. A dextrorotatory isomer rotates this plane to the right (clockwise), while an levorotatory isomer rotates it to the left (counterclockwise).

(e) In a high-spin complex, the *d* electrons of the central atom are unpaired as much as possible; spread out in all five of the *d* orbitals. In a low-spin complex, the *d* electrons are grouped into two or three orbitals, thus pairing their spins and making the total net spin as small as possible.

4. **(a)** $[CrCl_4(NH_3)_2]^-$ diamminetetrachlorochromate(III)
(b) $[Fe(CN)_6]^{3-}$ hexacyanoferrate(III)
(c) $[Cr(en)_3][Ni(CN)_4]$ tris(ethylenediamine)chromium(III) tetracyanonickelate(II)

5. **(a)** $[Co(NH_3)_6]^{2+}$ The coordination number of Co is 6; there are six monodentate NH_3 ligands attached to Co. Since the NH_3 ligand does not have a charge, the oxidation state of cobalt is +2, the same as that of the complex ion; hexaamminecobalt(III) ion
(b) $[AlF_6]^{3-}$ The coordination number of Al is six; F^- is monodentate. Each F^- has a –1 charge; thus the oxidation state of Al is +3; hexafluoroaluminate(III) ion
(c) $[Cu(CN)_4]^{2-}$ The coordination number of Cu is 4; CN^- is monodendate. CN^- has a –1 charge; thus the oxidation state of Cu is +2; tetracyanocuprate(II) ion
(d) $[CrBr_2(NH_3)_4]^+$ The coordination number of Cr is 6; NH_3 and Br^- are monodentate. NH_3 has no charge; Br^- has a –1 charge. The oxidation state of chromium is +3; tetraamminediibromochromium(III)
(e) $[Co(ox)_3]^{4-}$ The coordination number of Co is 6; oxalate is bidentate. $C_2O_4^{2-}$ (ox) has a –2 charge; thus the oxidation state of cobalt is +2; trioxalatocobaltate(II)
(f) $[Ag(S_2O_3)_2]^{3-}$ The coordination number of Ag is 2; $S_2O_3^{2-}$ is monodentate. $S_2O_3^{2-}$ has a –2 charge; thus the oxidation state of silver is +1; dithiosulfatoargentate(I)

6. **(a)** $[AgI_2]^-$ diiodoargentate(I) ion
(b) $[Al(OH)(H_2O)_5]^{2+}$ pentaaquahydroxoaluminum(III) ion
(c) $[Zn(CN)_4]^{2-}$ tetracyanozincate(II) ion
(d) $[Pt(en)_2]^{2+}$ bis(ethylenediamine)platinum(II) ion
(e) $[CoCl(NO_2)(NH_3)_4]^+$ tetraamminechloronitrocobalt(III) ion

7. **(a)** $[CoBr(NH_3)_5]SO_4$ pentaamminebromocobalt(III) sulfate
(b) $[CoSO_4(NH_3)_5]Br$ pentaamminesulfatocobalt(III) bromide
(c) $[Cr(NH_3)_6][Co(CN)_6]$ hexaamminechromium(III) hexacyanocobaltate(III)
(d) $Na_3[Co(NO_2)_6]$ sodium hexanitrocobaltate(III)
(e) $[Co(en)_3]Cl_3$ tris(ethylenediamine)cobalt(III) chloride

8. **(a)** $[Ag(CN)_2]^-$ dicyanoargentate(I) ion
(b) $[Pt(NO_2)(NH_3)_3]^+$ triamminenitroplatinum(II) ion
(c) $[CoCl(en)_2(H_2O)]^{2+}$ aquachlorobis(ethylenediamine)cobalt(III) ion
(d) $K_4[Cr(CN)_6]$ potassium hexacyanochromate(II)

9. The Lewis structures are grouped together at the end.
(a) H_2O has $2 \times 1 + 6 = 8$ valence electrons, or 8 pairs.
(b) CH_3NH_2 has $4 + 3 \times 1 + 5 + 2 \times 1 = 14$ valence electrons, or 7 pairs.
(c) ONO^- has $2 \times 6 + 5 + 1 = 18$ valence electrons, or 9 pairs.
The structure has a –1 formal charge on the single-bonded oxygen.
(d) SCN^- has $6 + 4 + 5 + 1 = 16$ valence electrons, or 8 pairs.
This structure, appropriately, gives a –1 formal charge to N.

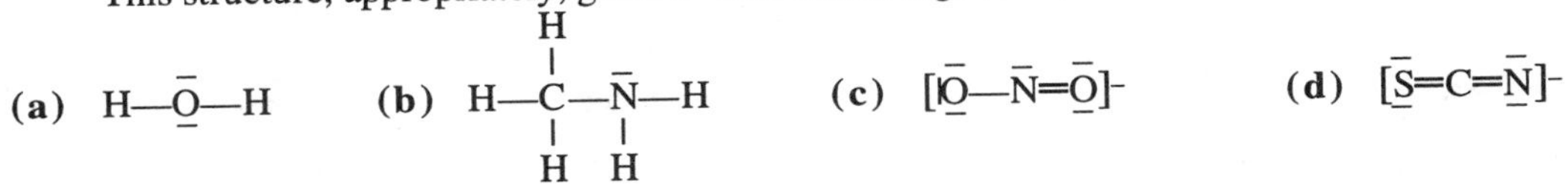

10. **(a)** manganese(II) sulfate hexahydrate $MnSO_4 \cdot 6H_2O$ $[Mn(H_2O)_6]SO_4$
(b) potassium hexacyanochromate(II) trihydrate

11. **(a)** $[PtCl_4]^{2-}$ tetrachloroplatinate(IV)
(b) $[FeCl_4(en)]^-$ tetrachloro(ethylenediamine)ferrate(III)
(c) *cis*-$[FeCl_2(ox)(en)]^-$ *cis*-dichloro(ethylenediamine)(oxalato)ferrate(III)
(d) *trans*-$[CrCl(OH)(NH_3)_4]^+$ *trans*–tetraamminechlorohydroxochromium(III) ion

(a) 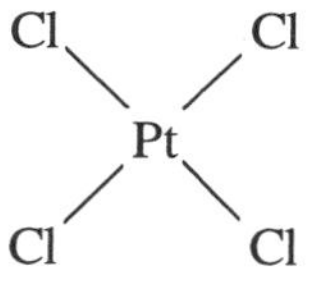 **(b)** 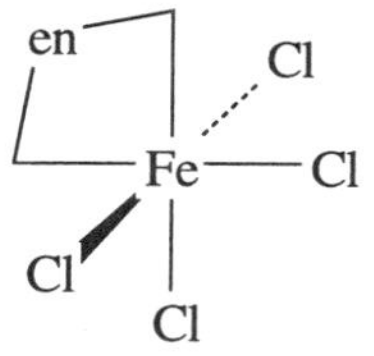 **(c)** 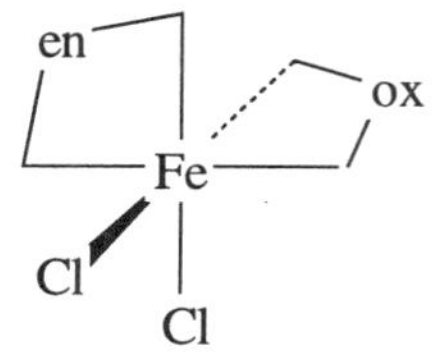 **(d)** 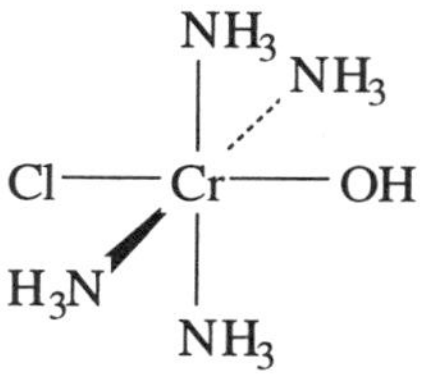

12. We assume that all of these complexes are octahedral in shape.
(a) $[Co(H_2O)(NH_3)_5]^{3+}$ has just one isomer. Other supposed isomers are rotations of that one.
(b) $[Co(H_2O)_2(NH_3)_4]^{3+}$ has two isomers, a *cis*-isomer (drawn on the left following) and a *trans*-isomer (drawn on the right). The two H_2O ligands are 90° from each other (*cis*-) or 180° from each other (*trans*-).

(a) 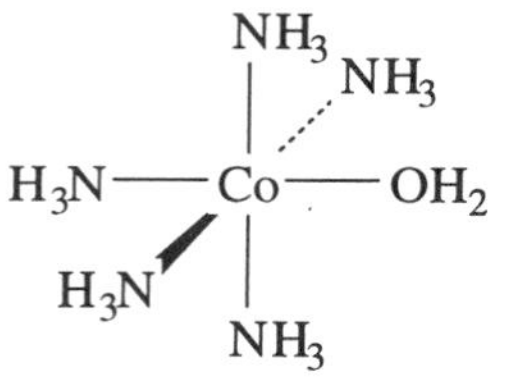
(b) 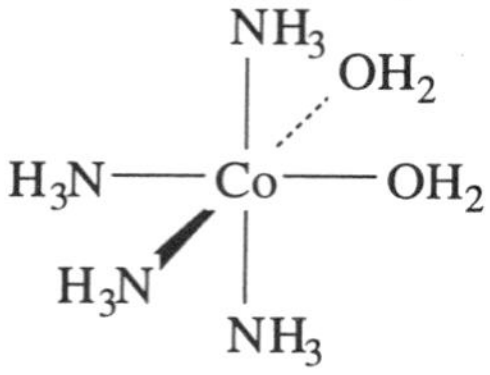 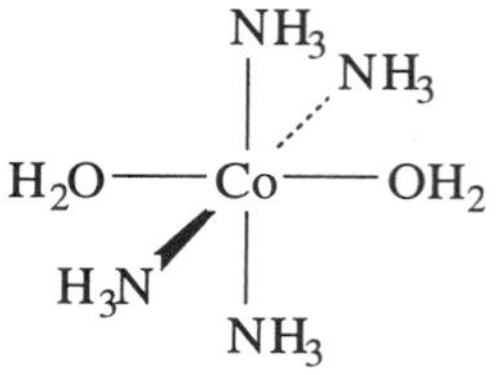

cis-isomer *trans*-isomer

(c) $[Co(H_2O)_3(NH_3)_3]^{3+}$ has two isomers, a *fac*-isomer (in which the three NH_3 ligands are 90° from each other, drawn at left following; also denoted a *cis*- isomer) and a *mer*-isomer (in which two of the NH_3 ligands are 180° from each other, on opposite sides of the central Co atom, drawn at right following; also denoted a *trans*-isomer).
(d) $[Co(H_2O)_4(NH_3)_2]^{3+}$ has two isomers, a *cis*-isomer (drawn at left following) and a *trans*-isomer drawn on the right). In this case it is the two NH_3 groups that are 90° from each other (*cis*-) or 180° from each other (*trans*-).

(c) 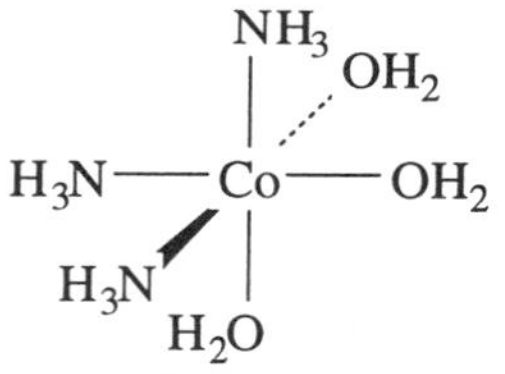
fac-isomer
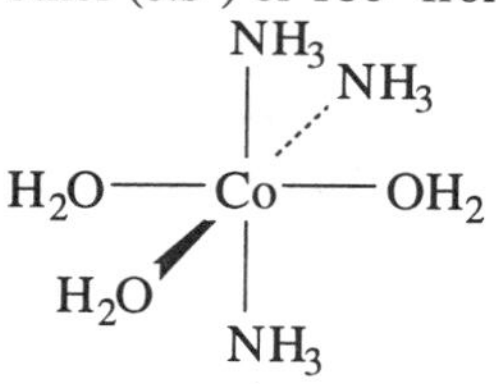
mer-isomer

(d) 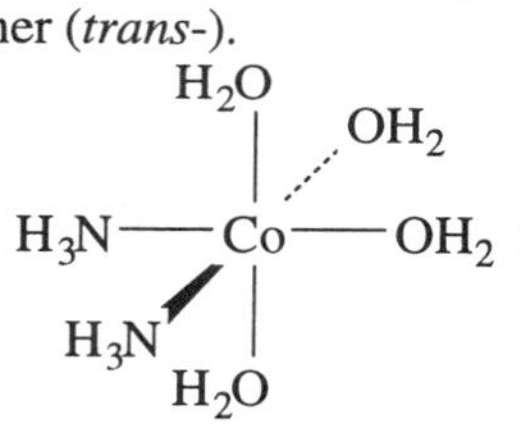
cis-isomer
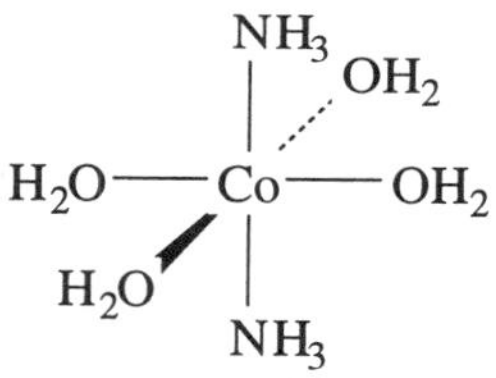
trans-isomer

13. **(a)** $[Zn(NH_3)_4][CuCl_4]$ can display coordination isomerism. Another isomer is $[Cu(NH_3)_4][ZnCl_4]$.
(b) $[Fe(CN)_5SCN]^{4-}$ displays linkage isomerism. The other isomer is $[Fe(CN)_5NCS]^{4-}$
(c) $[NiCl(NH_3)_5]^+$ does not display isomerism.
(d) $[PtBrCl_2(py)]^-$ displays geometric isomerism, because the complex is square planar.
(e) $[Cr(OH)_3(NH_3)_3]^-$ displays geometric isomerism. There is a *fac*-isomer (at left) and a *mer*-isomer (on the right). These could also be labeled *cis*- and *trans*-, respectively.

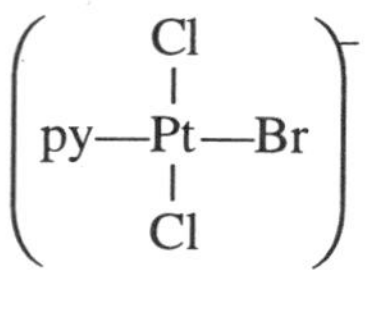
trans-isomer
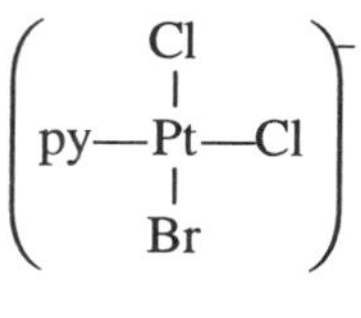
cis-isomer
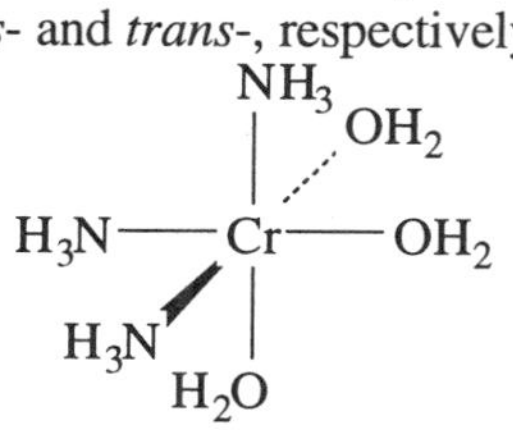
fac-isomer
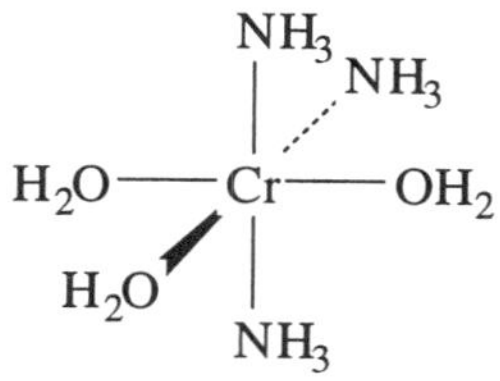
mer-isomer

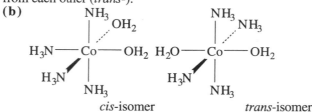

14. The yellow color indicates that the complex absorbs blue light, while the blue color indicates that the complex absorbs red and yellow light. Blue light has more energy per photon than does red and yellow light. The complex with the larger value of Δ will absorb the higher energy light. The ethylenediamine ligand produces a larger crystal field splitting energy than does the aqua ligand, according to the spectrochemical series. Thus, the yellow complex is $[Co(en)_3]^{3+}$ and the blue complex is $[Co(H_2O)_6]^{3+}$

EXERCISES

Nomenclature

15. **(a)** $[Co(OH)(H_2O)_4(NH_3)]^{2+}$ amminetetraaquahydroxocobalt(III) ion
(b) $[Co(ONO)_3(NH_3)_3]$ triamminetrinitritocobalt(III)
(c) $[Pt(H_2O)_4][PtCl_6]$ tetraaquaplatinum(II) hexachloroplatinate(IV)
(d) $[Fe(ox)_2(H_2O)_2]^-$ diaquadioxalatoferrate(III) ion
(e) $Ag_2[HgI_4]$ silver(I) tetraiodomercurate(II)

16. **(a)** $K_3[Fe(CN)_6]$ potassium hexacyanoferrate(III)
(b) $[Cu(en)_2]^{2+}$ bis(ethylenediamine)copper(II) ion
(c) $[Al(OH)(H_2O)_5]Cl_2$ pentaaquahydroxoaluminum(III) chloride
(d) $[CrCl(en)_2NH_3]SO_4$ amminechlorobis(ethylenediammine)chromium(III) sulfate
(e) $[Fe(en)_3]_4[Fe(CN)_6]_3$ tris(ethylenediamine)iron(III) hexacyanoferrate(II)

Bonding and Structure in Complex Ions

17. We assume that $[PtCl_4]^{2-}$ is square planar by analogy with $[Ni(CN)_4]^{2-}$ or $[PtCl_2(NH_3)_2]$ in Figure 25-3. The other two complex ions are octahedral.

(a)

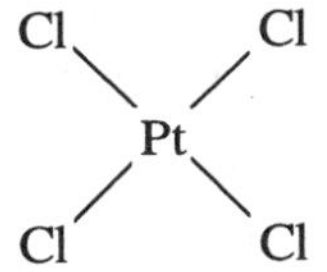

(b)

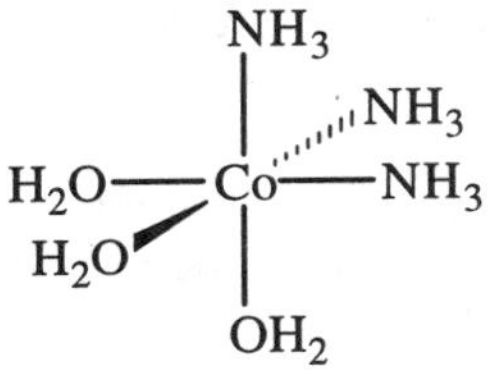

(c)

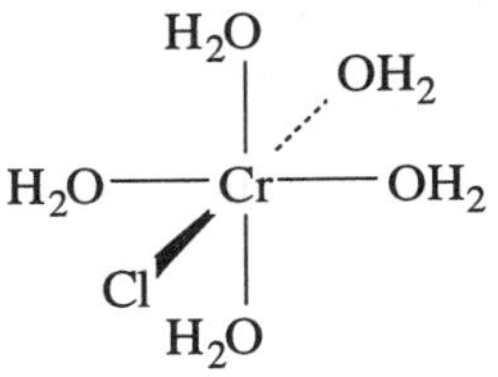

18. Structures of $[Pt(ox)_2]^{2-}$ and $[Cr(ox)_3]^{3-}$ are drawn below. The structure of $[Fe(EDTA)]^{2-}$ is the same as the structure of $[Pb(EDTA)]^{2-}$ drawn in Figure 25-18, with the substitution of Fe^{2+} for M^{n+}.

(a)

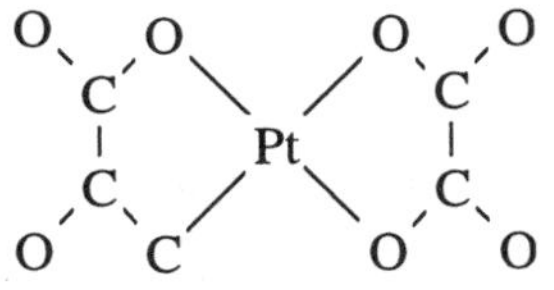

(b)

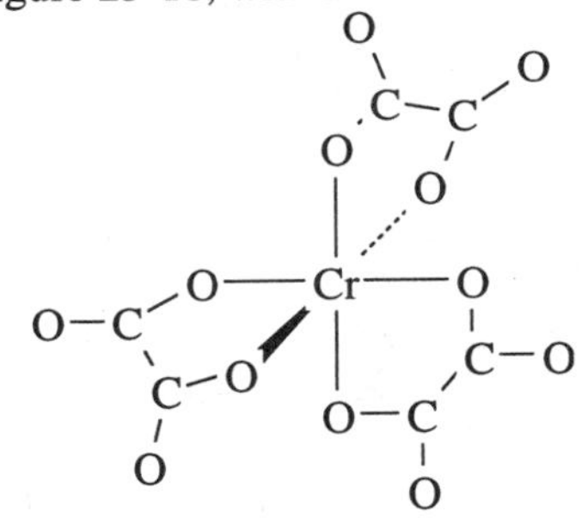

19. **(a)**

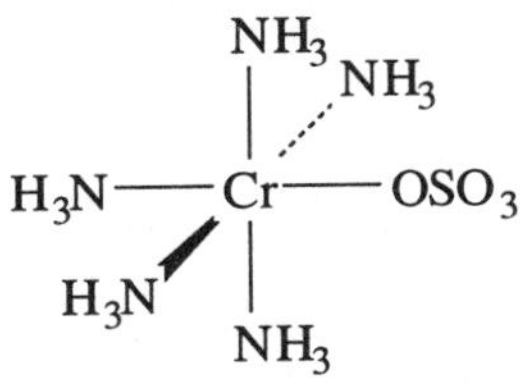

pentaamminesulfatochromium(III)

(b)

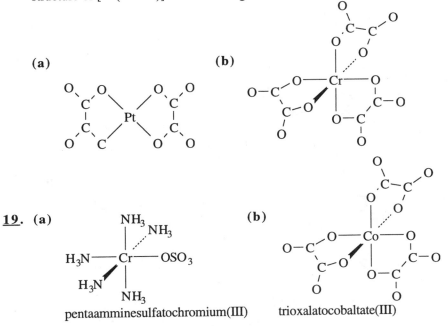

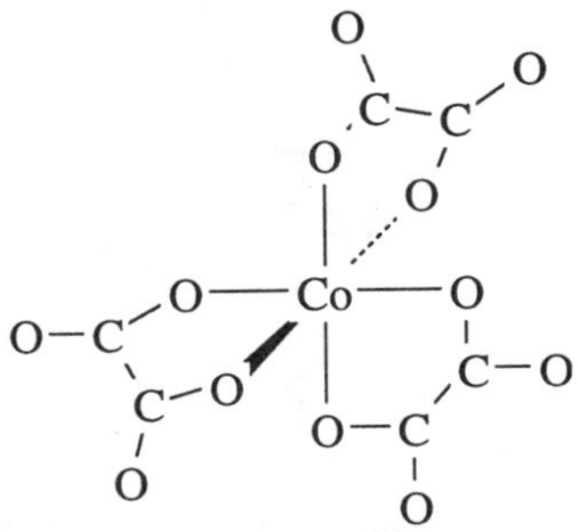

trioxalatocobaltate(III)

(c)

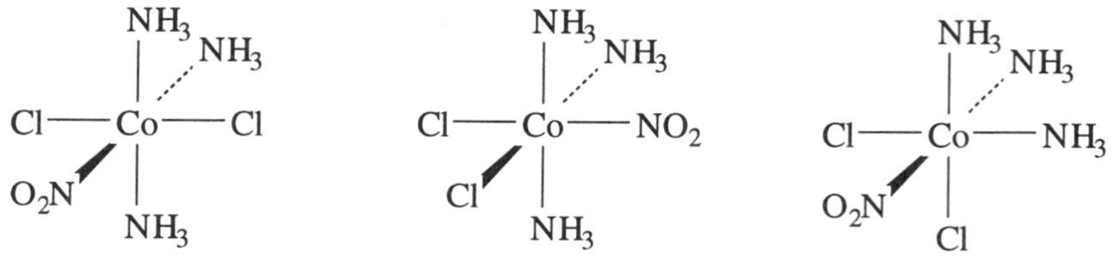

mer-triammine-*trans*-dichloronitrocobalt(III) ↑ *fac*-triamminedichloronitritocobalt(III)

mer-triammine-*cis*-dichloronitrocobalt(III)

20. **(a)** pentaamminenitrocobalt(III) ion

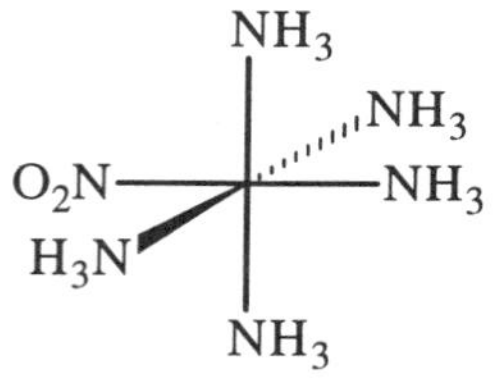

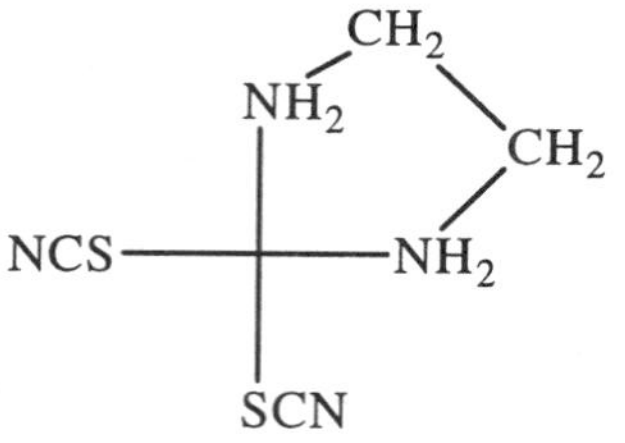

(c) hexaaquanickel(II) ion

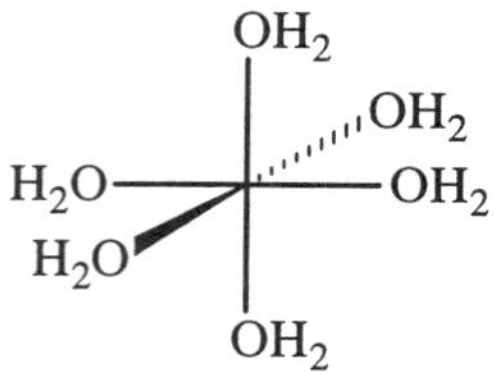

(b) ethylenediaminedithiocyanatocopper(II)

Isomerism

21. **(a)** *cis-trans* isomerism cannot occur with tetrahedral structures because all of the ligands are separated by the same angular distance from each other. One ligand cannot be on the other side of the central atom from another.

(b) Square planar structures can show *cis-trans* isomerism. Examples are drawn following, with the *cis*-isomer drawn on the left, and the *trans*-isomer drawn on the right.

cis-isomer *trans*-isomer

(c) Linear structures do not display *cis-trans* isomerism; there is only one way to bond the two ligands to the central atom.

22. All of the isomers are drawn together after the answer to this question.

(a) $[CrOH(NH_3)_5]^{2+}$ has one isomer.

(b) $[CrCl_2(H_2O)(NH_3)_3]^+$ has three isomers.

(c) $[CrCl_2(en)_2]^+$ has two geometric isomers, *cis*- and *trans*-.

(d) $[CrCl_4(en)]^-$ has only one isomer since the en ligand cannot bond *trans* to the central atom.

(e) $[Cr(en)_3]^{3+}$ has only one geometric isomer; it has two optical isomers.

(a)

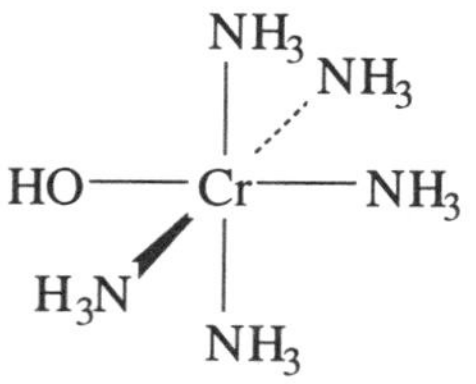

(b)

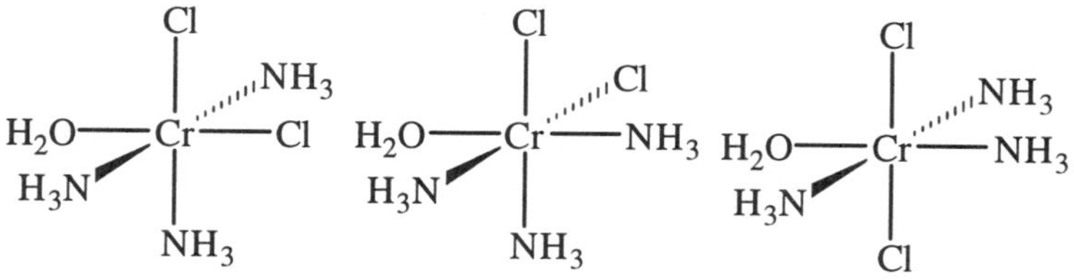

(c)

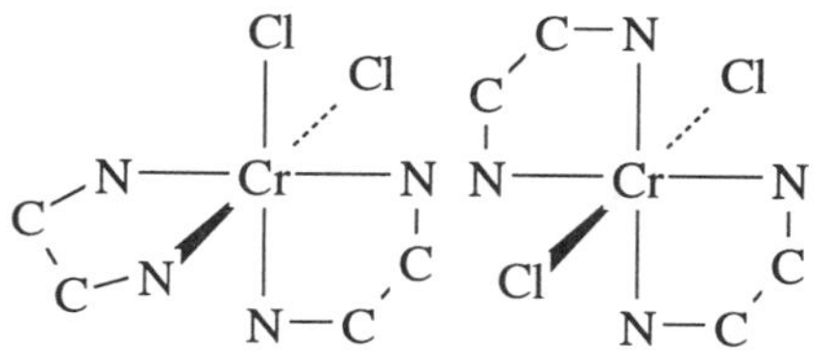

(d)

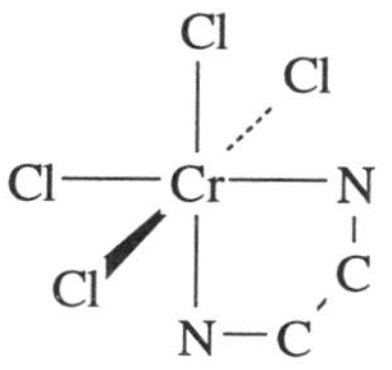

(e)

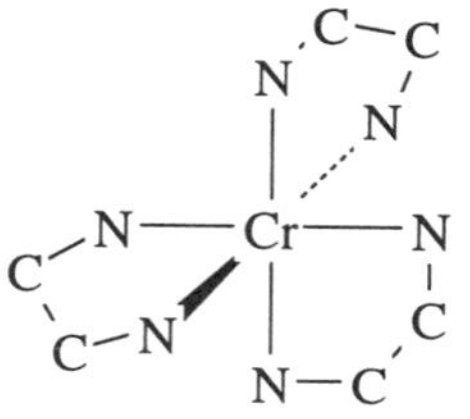

23. (a) There are three different square planar isomers, with D, C, and B, respectively, *trans* to the A ligand. They are drawn below.

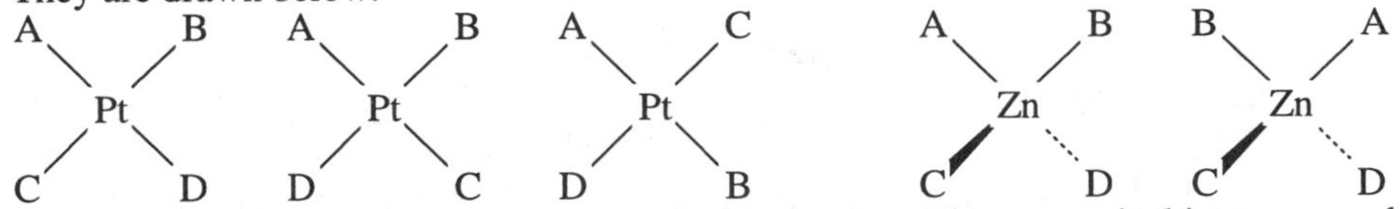

(b) Tetrahedral $[ZnABCD]^{2+}$ does display optical isomerism. The two optical isomers are drawn above.

24. (a) The three geometric isomers of $[CoCl_2(NH_3)_4]^+$ are sketched below. For clarity, we have shown only the chloro ligands.

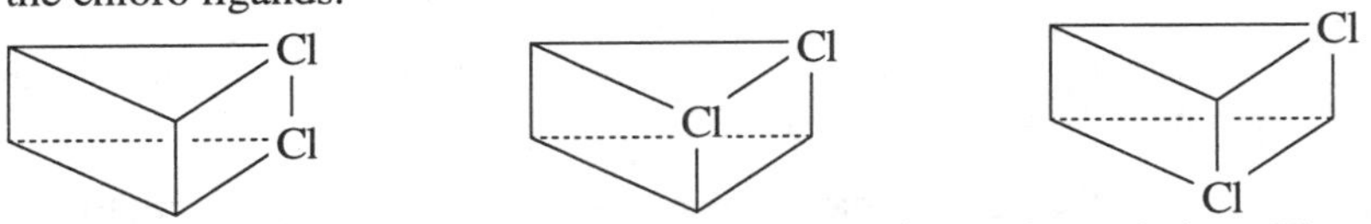

(b) Attempts to produce optical isomers of $[Co(en)_3]^{3+}$ are shown below. The ethylenediamine ligand is shown as an arc in each structure. The only successful attempt occurs when the ligand connects the diagonal corners of a face, which may be too long a distance to span.

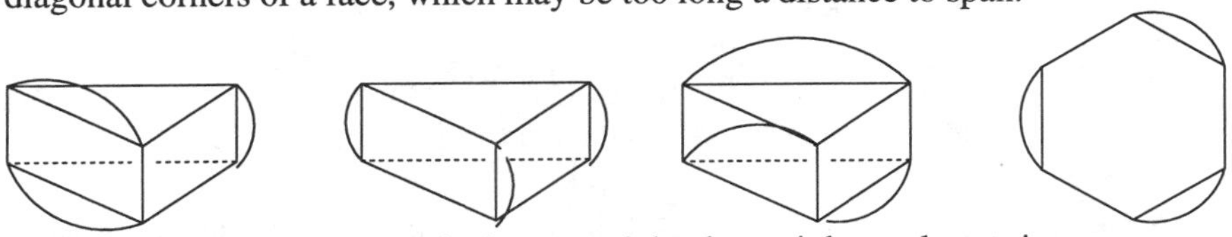

A planar hexagonal structure is drawn at right above; it has only one isomer.

25. There are a total of four coordination isomers. They are listed below. We assume that the oxidation state of each central metal ion is +3.

$[Co(en)_3][Cr(ox)_3]$	tris(ethylenediamine)cobalt(III) trioxalatochromate(III)
$[Co(ox)(en)_2][Cr(ox)_2(en)]$	bis(ethylenediamine)oxalatocobalt(III) (ethylenediamine)dioxalatochromate(III)
$[Cr(ox)(en)_2][Co(ox)_2(en)]$	bis(ethylenediamine)oxalatochromium(III) (ethylenediamine)dioxalatocobaltate(III)
$[Cr(en)_3][Co(ox)_3]$	tris(ethylenediamine)chromium(III) trioxalatocobaltate(III)

26. The *cis*-dichlorobis(ethylenediamine)cobalt(III) ion is optically active. The two optical isomers are drawn below. But the *trans*-isomer is not optically active. We can tell that in advance by noting that the ion and its mirror image are superimposable.

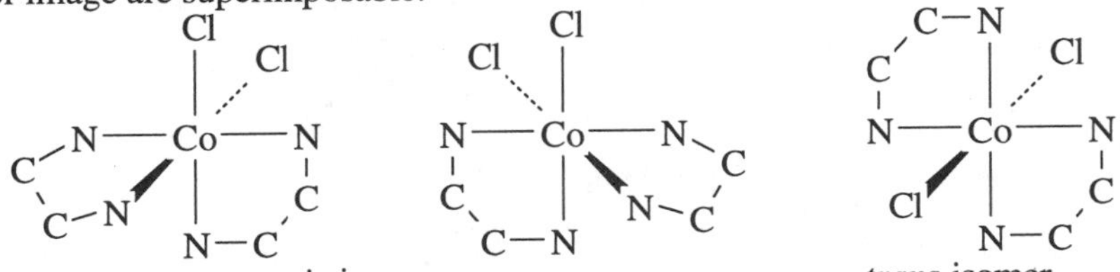

cis-isomers *trans*-isomer

Crystal Field Theory

27. In crystal field theory, the five *d* orbitals of a central transition metal ion are split into two (or more) groups of different energies. The energy spacing between these groups often corresponds to the energy of a photon of visible light. Thus, the complex ion will absorb light with energy corresponding to this spacing. If white light is incident on the complex ion, the light remaining after absorption will be missing some of its components. That is light of certain wavelengths (corresponding to the energies absorbed) will no longer be present in the formerly white light. The resulting light is colored. For example, if blue light is absorbed from white light, the remaining light will be yellow in color.

28. The difference in color is due to the difference in the value of Δ, the ligand field splitting energy. When the value of Δ is high, short wavelength light, which has a blue color, is absorbed, and the substance or its solution appears yellow. On the other hand, when the value of Δ is low, light of long wavelength, which has a red or yellow color, is absorbed, and the substance or its solution appears blue. The cyano ligand is a

strong field ligand, producing a large value of Δ, and thus yellow complexes. On the other hand, the aqua and chloro ligands are weak field ligands, producing a small value of Δ, and blue complexes.

29. We begin with the orbital diagram for Cr^{2+} [Ar] $_{3d}$[↑][↑][↑][↑][ ] The strong field and weak field diagrams for octahedral complexes follow, along with the number of unpaired electrons in each case.

strong field [ ][ ]
[↑↓][↑][↑] 2 unpaired electrons

weak field [↑][ ]
[↑][↑][↑] 4 unpaired electrons

30. We begin with the orbital diagram for Cr^{3+} [Ar] $_{3d}$[↑][↑][↑][ ][ ] The strong field and weak field diagrams for octahedral complexes follow, along with the number of unpaired electrons in each case.

strong field [ ][ ]
[↑][↑][↑] 3 unpaired electrons

weak field [ ][ ]
[↑][↑][↑] 3 unpaired electrons

The number of unpaired electrons is the same in each case.

31. **(a)** Both of the central atoms have the same oxidation state, +3. We give the electron configuration of the central atom to the left, then the completed crystal field diagram in the center, and finally the number of unpaired electrons.

The chloro ligand is a weak field ligand in the spectrochemical series.

Mo^{3+} [Kr] $4d^3$ [ ][ ] weak field
[Kr] $_{4d}$[↑][↑][↑][ ][ ] [↑][↑][↑] 3 unpaired electrons; paramagnetic

The ethylenediamine ligand is a strong field ligand in the spectrochemical series.

Co^{3+} [Ar] $3d^6$ [ ][ ] strong field
[Ar] $_{3d}$[↑↓][↑][↑][↑][↑] [↑↓][↑↓][↑↓] no unpaired electrons; diamagnetic

(b) In $[CoCl_4]^{2-}$ the oxidation state of cobalt is +2. Chloro is a weak field ligand. The electron configuration of Co^{2+} is [Ar] $3d^7$ or [Ar] $_{3d}$[↑↓][↑↓][↑][↑][↑] The tetrahedral ligand field diagram follows. weak field [↑][↑][↑]
[↑↓][↑↓] 3 unpaired electrons

32. **(a)** In $[Cu(py)_4]^{2+}$ the oxidation state of copper is +2. Pyridine is a strong field ligand. The electron configuration of Cu^{2+} is [Ar] $3d^9$ or [Ar] $_{3d}$[↑↓][↑↓][↑↓][↑↓][↑] There is no possible way that an odd number of electrons can be paired up, without at least one electron being unpaired. $[Cu(py)_4]^{2+}$ is paramagnetic.

(b) In $[Mn(CN)_6]^{3-}$ the oxidation state of manganese is +3. Cyano is a strong field ligand. The electron configuration of Mn^{3+} is [Ar] $3d^4$ or [Ar] $_{3d}$[↑][↑][↑][↑][ ] The ligand field diagram follows, at left. In $[FeCl_4]^-$ the oxidation state of iron is +3. Chloro is a weak field ligand. The electron configuration of Fe^{3+} is [Ar] $3d^5$ or [Ar] $_{3d}$[↑][↑][↑][↑][↑] The ligand field diagram follows, at right.

strong field [ ][ ]
[↑↓][↑][↑] 2 unpaired electrons

weak field [↑][↑][↑]
[↑][↑] 5 unpaired electrons

There are more unpaired electrons in $[FeCl_4]^-$ than in $[Mn(CN)_6]^{3-}$.

33. The electron configuration of Ni^{2+} is [Ar] $3d^8$ or [Ar] $_{3d}$[↑↓][↑↓][↑↓][↑][↑] Ammine is a strong field ligand. The ligand field diagrams follow, octahedral at left, tetrahedral in the center and square planar at right.

octahedral [↑][↑]
[↑↓][↑↓][↑↓]

tetrahedral [↑↓][↑][↑]
[↑↓][↑↓]

square planar [ ]
[↑↓]
[↑↓]
[↑↓][↑↓]

Since each configuration of the first two has the same number of unpaired electrons (that is, 2 unpaired electrons), we cannot use magnetic properties to determine whether the ammine complex of nickel(II) is octahedral or tetrahedral. But we can determine if the complex is square planar, since the square planar complex is diamagnetic with zero unpaired electrons.

34. The difference is due to the fact that $[Fe(CN_6)]^{4-}$ is a stong field complex ion, while $[Fe(H_2O)_6]^{2+}$ is a weak field complex ion. The electron configurations for an iron atom and an iron(II) ion, and the ligand field diagrams for the two complex ions follow.

iron atom [Ar] $_{3d}$[↑↓][↑][↑][↑][↑] $_{4s}$[↑↓]
iron(II) ion [Ar] $_{3d}$[↑↓][↑][↑][↑][↑] $_{4s}$[ ]

$[Fe(CN)_6]^{4-}$
[ ][ ]
[↑↓][↑↓][↑↓]

$Fe(H_2O)_6]^{2+}$
[↑][↑]
[↑↓][↑][↑]

Complex Ion Equilibria

35. **(a)** $Zn(OH)_2(s) + 4\ NH_3(aq) \rightleftharpoons [Zn(NH_3)_4]^{2+}(aq) + 2\ OH^-(aq)$

(b) $Cu^{2+}(aq) + 2\ OH^-(aq) \rightleftharpoons Cu(OH)_2(s)$

The blue color is most likely due to some unreacted $[Cu(H_2O)_4]^{2+}$(aq, pale blue)

$Cu(OH)_2(s) + 4\ NH_3(aq) \rightleftharpoons [Cu(NH_3)_4]^{2+}$(aq, dark blue) $+ 2\ OH^-(aq)$

$[Cu(NH_3)_4]^{2+}(aq) + 4\ H_3O^+(aq) \rightleftharpoons [Cu(H_2O)_4]^{2+}(aq) + 4\ NH_4^+(aq)$

36. **(a)** $CuCl_2(s) + 2\ Cl^-(aq) \rightleftharpoons [CuCl_4]^{2-}$(aq, yellow)

$2\ [CuCl_4]^{2-}(aq) + 4\ H_2O \rightleftharpoons [CuCl_4]^{2-}$(aq, yellow) $+ [Cu(H_2O)_4]^{2+}$(aq, pale blue) $+ 4\ Cl^-(aq)$

or $\rightleftharpoons 2\ [Cu(H_2O)_2Cl_2]$(aq, green) $+ 4\ Cl^-(aq)$

Either $[CuCl_4]^{2-}(aq) + 4\ H_2O \rightleftharpoons [Cu(H_2O)_4]^{2+}(aq) + 4\ Cl^-(aq)$

Or $[Cu(H_2O)_2Cl_2](aq) + 2\ H_2O \rightleftharpoons [Cu(H_2O)_4]^{2+}(aq) + 2\ Cl^-(aq)$

(b) The blue solution is that of $[Cr(H_2O_6]^{2+}$. This is quickly oxidized to Cr^{3+} by $O_2(g)$ from the atmosphere. The green color is due to $[CrCl_2(H_2O)_4]^{3+}$.

$4\ [Cr(H_2O)_6]^{2+}$(aq, blue) $+ 4\ H^+(aq) + 8\ Cl^-(aq) + O_2 \longrightarrow 4\ [CrCl_2(H_2O)_4]^+$(aq, green) $+ 10\ H_2O$

Over a period of time, we might expect volatile HCl(g) to escape, leading to complex ions with more H_2O and less Cl^-. $H^+(aq) + Cl^-(aq) \rightleftharpoons HCl(g)$

$[CrCl_2(H_2O)_4]^+$(aq, green) $+ H_2O \rightleftharpoons [CrCl(H_2O)_5]^{2+}$(aq, blue-green) $+ Cl^-(aq)$

$[CrCl(H_2O)_5]^{2+}$(aq, blue-green) $+ H_2O \rightleftharpoons [Cr(H_2O)_6]^{3+}$(aq, violet) $+ Cl^-(aq)$

Actually, to ensure the these final two reactions occur, and more rapidly, it would be helpful to dilute the solution with water after chromium metal has dissolved.

37. $[Co(en)_3]^{3+}$ should have the largest overall K_f value. We expect a complex ion with polydentate ligands to have a larger value for its formation constant than complexes that contain only monodentate ligands. This is an expression of the chelate effect. Once one end of a polydentate ligand becomes attached to the central metal, the attachment of the remaining electron pairs is relatively easy because they already are close to the central metal (and do not have to migrate in from a distant point in the solution).

38. **(a)** $[Zn(NH_3)_4]^{2+}$ $\beta_4 = K_1 \times K_2 \times K_3 \times K_4 = 3.9 \times 10^2 \times 2.1 \times 10^2 \times 1.0 \times 10^2 \times 50. = 4.1 \times 10^8$

(b) $[Ni(H_2O)_2(NH_3)_4]^{2+}$ $\beta_4 = K_1 \times K_2 \times K_3 \times K_4 = 6.3 \times 10^2 \times 1.7 \times 10^2 \times 54 \times 15 = 8.7 \times 10^7$

39.

First:	$[Fe(H_2O)_6]^{3+}(aq) + en(aq) \rightleftharpoons [Fe(H_2O)_4(en)]^{3+}(aq) + 2\ H_2O$	$K_1 = 10^{4.34}$
Second:	$[Fe(H_2O)_4(en)]^{3+}(aq) + en(aq) \rightleftharpoons [Fe(H_2O)_2(en)_2]^{3+}(aq) + 2\ H_2O$	$K_2 = 10^{3.31}$
Third:	$[Fe(H_2O)_2(en)_2]^{3+}(aq) + en(aq) \rightleftharpoons [Fe(en)_3]^{3+}(aq) + 2\ H_2O$	$K_3 = 10^{2.05}$
Net:	$[Fe(H_2O)_6]^{3+}(aq) + 3\ en(aq) \rightleftharpoons [Fe(en)_3]^{3+}(aq) + 6\ H_2O$	$K_f = K_1 \times K_2 \times K_3$

$\log K_f = 4.34 + 3.31 + 2.05 = 9.70$ $K_f = 10^{9.70} = 5.0 \times 10^9 = \beta_3$

40. Since the overall formation constant is the product of the individual stepwise formation constants, the logarithm of the overall formation constant is the sum of the logarithms of the stepwise formation constants.

$\log K_f = \log K_1 + \log K_2 + \log K_3 + \log K_4 = 2.80 + 1.60 + 0.49 + 0.73 = 5.62$

$K_f = 10^{5.62} = 4.2 \times 10^5$

41. **(a)** Aluminum(III) forms a stable (and soluble) hydroxo complex but not a stable ammine complex.

$[Al(H_2O)_3(OH)_3](s) + OH^-(aq) \rightleftharpoons [Al(H_2O)_2(OH)_4]^-(aq) + H_2O$

(b) Although zinc(II) forms a soluble stable ammine complex ion, its formation constant is not sufficiently large to dissolve highly insoluble ZnS. However, it is sufficiently large to dissolve the moderately insoluble $ZnCO_3$. Said another way, ZnS does not produce sufficient $[Zn^{2+}]$ to permit the complex ion to form. $ZnCO_3(s) + 4\ NH_3(aq) \rightleftharpoons [Zn(NH_3)_4]^{2+}(aq) + CO_3^{2-}(aq)$

(c) Chloride ion forms a stable complex ion with silver(I) ion, that dissolves the AgCl(s) that formed when $[Cl^-]$ is low. $AgCl(s) \rightleftharpoons Ag^+(aq) + Cl^-(aq)$ $Ag^+(aq) + 2\ Cl^-(aq) \rightleftharpoons [AgCl_2]^-(aq)$

overall: $AgCl(s) + Cl^-(aq) \rightleftharpoons [AgCl_2]^-(aq)$

42. **(a)** Because of the large value of the formation constant for the complex ion, $[Co(NH_3)_6]^{3+}(aq)$, the concentration of free $Co^{3+}(aq)$ is too small to enable it to oxidize water to $O_2(g)$. Since there is not a complex ion present—except, of course, $[Co(H_2O)_6]^{3+}(aq)$—when $CoCl_3$ is dissolved in water, the $[Co^{3+}]$ is sufficiently high for the oxidation-reduction reaction to be spontaneous.

$4\,Co^{3+}(aq) + 2\,H_2O \longrightarrow 4\,Co^{2+}(aq) + 4\,H^+(aq) + O_2(g)$

(b) Although AgI(s) is insoluble, there is a small concentration of $Ag^+(aq)$ present because of the solubility equilibrium: $AgI(s) \rightleftharpoons Ag^+(aq) + I^-(aq)$ These silver ions react with thiosulfate ion to form the stable dithiosulfatoargentate(I) complex ion: $Ag^+(aq) + 2\,S_2O_3^{2-}(aq) \rightleftharpoons [Ag(S_2O_3)_2]^-(aq)$

Acid-Base Properties

43. $[Al(H_2O)_6]^{3+}(aq)$ is capable of ionizing: $[Al(H_2O)_6]^{3+}(aq) + H_2O \rightleftharpoons [AlOH(H_2O)_5]^{2+}(aq) + H_3O^+(aq)$ The value of its ionization constant ($pK_a = 5.01$, from a handbook) approximates that of acetic acid.

44. **(a)** $[CrOH(H_2O)_5]^{2+}(aq) + OH^-(aq) \longrightarrow [Cr(OH)_2(H_2O)_4]^+(aq) + H_2O$

(b) $[CrOH(H_2O)_5]^{2+}(aq) + H_3O^+(aq) \longrightarrow [Cr(H_2O)_6]^{3+}(aq) + H_2O$

45. **(a)**

Reaction:	$[Fe(H_2O)_6]^{3+}(aq) + H_2O$	$\rightleftharpoons$	$[FeOH(H_2O)_5]^{2+}(aq)$	+	$H_3O^+(aq)$
Initial:	0.100 M				
Changes:	$-x$ M		$+x$ M		$+x$ M
Equil:	$(0.100 - x)$M		x M		x M

$$K_a = \frac{[[FeOH(H_2O)_5]^{2+}][H_3O^+]}{[[Fe(H_2O)_6]^{3+}]} = 9 \times 10^{-4} = \frac{x \cdot x}{0.100 - x} \approx \frac{x^2}{0.100}$$

$x = \sqrt{0.100 \times 9 \times 10^{-4}} = 9 \times 10^{-3}\text{ M} = [H_3O^+]$ $\qquad pH = -\log(9 \times 10^{-3}) = 2.0$

(b)

Reaction:	$[Fe(H_2O)_6]^{3+}(aq) + H_2O$	$\rightleftharpoons$	$[FeOH(H_2O)_5]^{2+}(aq)$	+	$H_3O^+(aq)$
Initial:	0.100 M				0.100
Changes:	$-x$ M		$+x$ M		$+x$ M
Equil:	$(0.100 - x)$M		x M		$(0.100 + x)$M

$$K_a = \frac{[[FeOH(H_2O)_5]^{2+}][H_3O^+]}{[[Fe(H_2O)_6]^{3+}]} = 9 \times 10^{-4} = \frac{x(0.100 + x)}{0.100 - x} \approx \frac{0.100x}{0.100}$$

$x = 9 \times 10^{-4}\text{ M} = [[Fe(H_2O)_5OH]^{2+}]$

(c) We simply substitute $[[FeOH(H_2O)_5]^{2+}] = 1 \times 10^{-6}$ M into the K_a expression, with $[[Fe(H_2O)_6]^{3+}] =$ 0.100 M, and determine $[H_3O^+]$.

$$K_a = \frac{[[FeOH(H_2O)_5]^{2+}][H_3O^+]}{[[Fe(H_2O)_6]^{3+}]} = 9 \times 10^{-4} = \frac{1 \times 10^{-6}\,[H_3O^+]}{0.100}$$

$$[H_3O^+] = \frac{0.100 \times 9 \times 10^{-4}}{1 \times 10^{-6}} = 90\text{ M}$$

This is an impossibly high $[H_3O^+]$. It is impossible to maintain $[[FeOH(H_2O)_5]^{2+}] = 1 \times 10^{-6}$ M.

46. Let us first determine the concentration of uncomplexed $Pb^{2+}(aq)$. Because of the large value of the equilibrium constant, we assume that most of the lead(II) is present as $[Pb(EDTA)]^{2-}(aq)$. We attain equilibrium from that point. Remember that $[EDTA^{4-}]$ remains constant at 0.20 M

Reaction:	$Pb^{2+}(aq)$	+	$EDTA^{4-}(aq)$	$\rightleftharpoons$	$[Pb(EDTA)]^{2-}(aq)$
Initial:	0 M		0.20 M		0.010 M
Changes:	$+x$ M				$-x$ M
Equil:	x M		0.20 M		$(0.010 - x)$M

$$K_f = \frac{[[Pb(EDTA)]^{2-}]}{[Pb^{2+}][EDTA^{4-}]} = 2 \times 10^{18} = \frac{0.010 - x}{0.20\,x} \approx \frac{0.010}{0.20\,x} \qquad x = \frac{0.010}{0.20 \times 2 \times 10^{18}} = 3 \times 10^{-20}\text{ M}$$

From Chapter 19, for PbS $K_{spa} = 3 \times 10^{-7}$. We substitute in the values that we have. Recall $[H_2S] = 0.10$ M.

$$K_{spa} = 3 \times 10^{-7} = \frac{[Pb^{2+}][H_2S]}{[H_3O^+]^2} = \frac{(3 \times 10^{-20})(0.10)}{[H_3O^+]^2} \qquad [H_3O^+] = \sqrt{\frac{(3 \times 10^{-20})(0.10)}{3 \times 10^{-7}}} = 1 \times 10^{-7}$$

PbS(s) should precipitate. This situation is a bit unusual. We would expect that the solution would have to be somewhat alkaline (basic) in order to maintain $EDTA^{4-}$ in its anion form. If the solution were not alkaline, however, then we would expect precipitation to occur.

Applications

47. **(a)** Solubility: $AgBr(s) \rightleftharpoons Ag^+(aq) + Br^-(aq)$ $\quad K_{sp} = 5.0 \times 10^{-13}$

Formation: $Ag^+(aq) + 2\,S_2O_3^{2-}(aq) \rightleftharpoons [Ag(S_2O_3)_2]^{3-}(aq)$ $\quad K_f = 1.7 \times 10^{13}$

Net: $AgBr(s) + 2\,S_2O_3^{2-}(aq) \rightleftharpoons [Ag(S_2O_3)_2]^{3-}(aq)$ $\quad K = K_{sp} \times K_f$

$K = 5.0 \times 10^{-13} \times 1.7 \times 10^{13} = 8.5$

With a reasonably high $[S_2O_3^{2-}]$, this reaction will go essentially to completion.

(b) $NH_3(aq)$ cannot be used in the fixing of photographic film because of the relatively small value of K_f for $[Ag(NH_3)_2]^+(aq)$, $K_f = 1.6 \times 10^7$. This would produce a value of $K = 8.0 \times 10^{-6}$ in the expression above, too small to indicate a reaction that goes to completion.

48. Oxidation: $HO_2^-(aq) + OH^-(aq) \longrightarrow H_2O + O_2(g) + 2\,e^-$ $\quad -E° = +0.076\text{ V}$

Reduction: $\{[Co(NH_3)_6]^{3+}(aq) + e^- \longrightarrow [Co(NH_3)_6]^{2+}(aq)\} \times 2$ $\quad E° = +0.10\text{ V}$

Net: $HO_2^-(aq) + OH^-(aq) + 2\,[Co(NH_3)_6]^{3+}(aq) \longrightarrow 2\,[Co(NH_3)_6]^{2+}(aq) + H_2O + O_2(g)$

$E°_{cell} = +0.076\text{ V} + 0.10\text{ V} = +0.18\text{ V}$

The positive value of the standard cell potential indicates that this is a spontaneous reaction.

26 NUCLEAR CHEMISTRY

PRACTICE EXAMPLES

1A A β^- has a mass number of zero and an "atomic number" of –1. Emission of this electron has the effect of transforming a neutron into a proton. $^{241}_{94}Pu \longrightarrow ^{241}_{95}Am + ^{0}_{-1}\beta$

1B ^{58}Ni has a mass number of 58 and an atomic number of 28. A positron has a mass number of 0 and an effective atomic number of +1. Emission of a positron has the seeming effect of transforming a proton into a neutron. The parent nuclide must be copper-58. $^{58}_{29}Cu \longrightarrow ^{58}_{28}Ni + ^{0}_{+1}\beta$

2A The sum of the mass numbers (139 + 12 = ? + 147) tells us that the other product species has $A = 4$. The atomic number of La is 57, that of C is 6, and that of Eu is 63. The atomic number sum (57 + 6 = ? + 63) indicates that the atomic number of this product species is zero. Therefore, four neutrons must have been emitted. $^{139}_{57}La + ^{12}_{6}C \longrightarrow ^{147}_{63}Eu + 4\,^{0}_{1}n$

2B An alpha particle is $^{4}_{2}He$ and a positron is $^{0}_{+1}\beta$. We note that the total mass number in the first equation is 125; the mass number of the product is 1. The total atomic number is 53; the atomic number of the product is 0. It is a neutron. In the second equation, the positron has a mass number of 0, meaning that the mass number of the product is 124. Because the atomic number of the positron is +1, that of the product is 52; it is $^{124}_{52}Te$.

$^{121}_{51}Sb + ^{4}_{2}He \longrightarrow ^{124}_{53}I + ^{1}_{0}n \qquad ^{124}_{53}I \longrightarrow ^{0}_{+1}\beta + ^{124}_{52}Te$

3A **(a)** The decay constant is found from the 8.040-day half-life.

$$\lambda = \frac{0.693}{8.040 \text{ d}} = 0.0862 \text{ d}^{-1} \times \frac{1 \text{ d}}{24 \text{ h}} \times \frac{1 \text{ h}}{60 \text{ min}} \times \frac{1 \text{ min}}{60 \text{ s}} = 9.98 \times 10^{-7} \text{ s}^{-1}$$

(b) The number of ^{131}I atoms is used to find the activity.

$$\text{no. } ^{131}I \text{ atoms} = 2.05 \text{ mg} \times \frac{1 \text{ g}}{1000 \text{ mg}} \times \frac{1 \text{ mol } ^{131}I}{131 \text{ g } ^{131}I} \times \frac{6.022 \times 10^{23} \text{ atoms}}{1 \text{ mol } ^{131}I} = 9.42 \times 10^{18} \text{ atoms } ^{131}I$$

Activity = $\lambda N = 9.98 \times 10^{-7} \text{ s}^{-1} \times 9.42 \times 10^{18}$ atoms = 9.40×10^{12} disintegrations/second

(c) We now determine the number of atoms remaining after 16 days. Because two half-lives elapse in 16 days, the number of atoms has halved twice, to one-forth (25%) the original number of atoms.

$N_t = 0.25 \times N_0 = 0.25 \times 9.42 \times 10^{18}$ atoms $= 2.36 \times 10^{18}$ atoms

(d) The rate after 14 days is determined by the number of atoms present.

rate = $\lambda N_t = 9.98 \times 10^{-7} \text{ s}^{-1} \times 2.36 \times 10^{18}$ atoms $= 2.36 \times 10^{12}$ dis/s

3B First we determine the value of λ: $\lambda = \dfrac{0.693}{t_{1/2}} = \dfrac{0.693}{11.4 \text{ d}} = 0.0608 \text{ d}^{-1}$

Then we allow $N_t = 1\%\, N_0 = 0.010\, N_0$ in equation (26.13).

$$\ln\frac{N_t}{N_0} = -\lambda t = \ln\frac{0.010\, N_0}{N_0} = \ln(0.010) = -4.61 = -(0.0608 \text{ d}^{-1})t \qquad t = \frac{-4.61}{-0.0608 \text{ d}^{-1}} = 75.8 \text{ d}$$

4A The half-life of ^{14}C is 5730 y and $\lambda = 1.21 \times 10^{-4} \text{ y}^{-1}$. The activity of ^{14}C when the object supposedly stopped growing was 15 dis/min per g C. We use equation (26.13) with activities (λN) in place of numbers of atoms (N).

$$\ln\frac{A_t}{A_0} = -\lambda t = \ln\frac{8.5 \text{ dis/min}}{15 \text{ dis/min}} = -(1.21 \times 10^{-4} \text{ y}^{-1})\, t = -0.568 \qquad t = \frac{0.568}{1.21 \times 10^{-4} \text{ y}^{-1}} = 4.69 \times 10^{3} \text{ y}$$

4B The half-life of ^{14}C is 5730 y and $\lambda = 1.21 \times 10^{-4}\ y^{-1}$. The activity of ^{14}C when the object supposedly stopped growing was 15 dis/min per g C. We use equation (26.13) with activities (λN) in place of numbers of atoms (N). $\ln \frac{A_t}{A_0} = -\lambda t = \ln \frac{A_t}{15\ \text{dis/min}} = -(1.21 \times 10^{-4}\ y^{-1})(1100\ y) = -0.133$

$\frac{A_t}{15\ \text{dis/min}} = e^{-0.133} = 0.875$ $\qquad A_t = 0.875 \times 15\ \text{dis/min} = 13\ \text{dis/min}$

5A Mass defect. = 145.913053 u (^{146}Sm) – 141.907719 u (^{142}Nd) – 4.002603 u (^{4}He) = 0.002731 u

Then, from the text, we have 935.1 MeV = 1 u $\qquad E = 0.002731\ \text{u} \times \frac{931.5\ \text{MeV}}{1\ \text{u}} = 2.544\ \text{MeV}$

5B Unfortunately, we cannot use the result of Example 26-5 (0.0045 u = 4.2 MeV) because it is expressed to only two significant figures, and we begin with four significant figures. But, we essentially work backwards through that calculation. The last conversion factor is from Table 2-1.

$$E = 5.590\ \text{MeV} \times \frac{1.602 \times 10^{-13}\ \text{J}}{1\ \text{MeV}} = 8.955 \times 10^{-13}\ \text{J} = mc^2 = m(2.9979 \times 10^8\ \text{m/s})^2$$

$$m = \frac{8.955 \times 10^{-13}\ \text{J}}{(2.9979 \times 10^8\ \text{m/s})^2} \times \frac{1000\ \text{g}}{1\ \text{kg}} \times \frac{1.0073\ \text{u}}{1.673 \times 10^{-24}\ \text{g}} = 0.005999\ \text{u}$$

Or we could use $\qquad m = 5.590\ \text{MeV} \times \frac{1\ \text{u}}{931.5\ \text{MeV}} = 0.006001\ \text{u} = 6.001\ \text{mu}$

6A **(a)** ^{88}Sr has an even atomic number (38) and an even neutron number (50); its mass number (88) is not too far from the average mass (87.6) of Sr. It should be stable.

(b) ^{118}Cs has an odd atomic number (55) and a mass number (118) that is pretty far from the average mass of Cs (132.9). It should be radioactive.

(c) ^{30}S has an even atomic number (16) and an even neutron number (14); but its mass number (30) is too far from the average mass of S (32.1). It should be radioactive.

6B We know that ^{19}F is stable, with approximately the same number of neutrons and protons: 9 protons, and 10 neutrons. Thus, nuclides of light elements with approximately the same number of neutrons and protons will be stable. In Practice Example 26-1 we saw that positron emission has the seeming effect of transforming a proton into a neutron. β^- emission has the opposite effect: seemingly transforming a neutron into a proton. The mass number does not change in either case. Now let us analyze our two nuclides.

^{17}F has 9 protons and 8 neutrons. Replacing a proton with a neutron would produce a more stable nuclide. Thus, we predict positron emission by ^{17}F to produce ^{17}O.

^{22}F has 9 protons and 14 neutrons. Replacing a neutron with a proton would produce a more stable nuclide. Thus, we predict β^- emission by ^{22}F to produce ^{22}Ne.

SUMMARIZING EXAMPLE CALCULATIONS

1. We calculate the mass of ^{131}I remaining after 1 month = 30 days.

$\lambda = \frac{0.693}{8.04\ \text{d}} = 0.0862\ \text{d}^{-1}$ $\qquad \ln \frac{N_t}{N_0} = -\lambda t = \ln \frac{N_t}{250.\ \text{g}} = -(0.0862\ \text{d}^{-1})(30\ \text{d}) = -2.59$

$\frac{N_t}{250.\ \text{g}} = e^{-2.59} = 0.0750$ $\qquad N_t = 0.0750 \times 250.\ \text{g} = 19\ \text{g}$

2. The number of ^{131}I atoms remaining after 1 month is

$$^{131}\text{I atoms} = 19\ \text{g}\ ^{131}\text{I} \times \frac{1\ \text{mol}\ ^{131}\text{I}}{131\ \text{g}\ ^{131}\text{I}} \times \frac{6.022 \times 10^{23}\ ^{131}\text{I atoms}}{1\ \text{mol}\ ^{131}\text{I}} = 8.7 \times 10^{22}\ \text{atoms}\ ^{131}\text{I}$$

3. $\lambda(\text{in s}^{-1}) = 0.0862\ \text{d}^{-1} \times \frac{1\ \text{day}}{24\ \text{h}} \times \frac{1\ \text{h}}{3600\ \text{s}} = 9.98 \times 10^{-7}\ \text{s}^{-1}$

4. Activity = $N\lambda = (8.64 \times 10^{22}\ ^{131}\text{I atoms})(9.98 \times 10^{-7}\ \text{s}^{-1}) = 8.7 \times 10^{16}$ dis/s

5. Activity (curies) = $8.7 \times 10^{16}\ \text{dis/s} \times \frac{1\ \text{curie}}{3.7 \times 10^{10}\ \text{dis/s}} = 2.4 \times 10^6$ curies = 2.4 megacuries

REVIEW QUESTIONS

1. **(a)** α refers to an alpha particle, a helium-4 nucleus: $^{4}_{2}He^{2+}$.

(b) β^- refers to a beta particle, that is, an electron.

(c) β^+ refers to a positron, identical in all respects to an electron but with opposite (positive) charge.
(d) γ refers to gamma radiation, electromagnetic radiation with energy measured in MeV per photon. In contrast, visible light has energies measured in eV per photon.
(e) $t_{1/2}$ indicates half-life, the time needed for half of a sample to decay radioactively.

2. (a) A radioactive decay series is the series of successive products resulting from one nuclide, which decays to a second nuclide, which in turn decays to yet a third, and so forth, until a stable nuclide is produced.
(b) A charged-particle accelerator is a device that produces particles of very high speeds. It imparts energy and thus speed to these particles by using the influence of either an electric field or a magnetic field on the charge of the particle.
(c) The neutron-to-proton ratio is the quotient of the number of neutrons divided by the number of protons. It is a general indication of the stability of a nuclide (when combined with atomic number). Certainly, for light elements (up to about $Z = 20$), a neutron-to-proton ration of 1-to-1 is that of a stable nuclide.
(d) The mass–energy relationship is that first promulgated by Einstein: $E = mc^2$.
(e) Background radiation is that radiation consistently present on earth from natural sources: cosmic rays, radioactive elements in the soil and the air, etc.

3. (a) Both electrons and positrons have the same mass, about 0.00055 u. However, the electron is negatively charged, while the positron has the same magnitude of charge but a positive one.
(b) The half-life, $t_{1/2}$, of a radioactive nuclide is the time necessary for half of a sample to decay. The decay constant is the rate constant for that first-order decay process: $\lambda = 0.693/t_{1/2}$.
(c) The mass defect is the difference between the mass sum of the protons, neutrons, and electrons that constitute a nuclide and the nuclidic mass. The binding energy is that defect expressed as an energy.
(d) Nuclear fission refers to the process in which the nucleus of an atom is split into smaller fragments, usually with the release of energy. Nuclear fusion refers to the process in which two lighter nuclei are combined into a nucleus of higher atomic number.
(e) Primary ionization refers to those ions created by collision of the radiation with the atoms of the sample. However, when these atoms ionize they give off electrons, which often are energetic enough to produce further ionization, called secondary ionization, in the sample.

<u>4</u>. (a) γ rays penetrate through matter the greatest distance, largely because they are uncharged and thus do not interact with matter extensively.
(b) α particles have the greatest ionizing power, principally because they have the largest charge and mass.
(c) β particles are deflected the most in a magnetic field, because of their small mass and relatively large charge, that is, because they have the largest charge-to-mass ratio.

<u>5</u>. (a) ${}^{160}_{74}W \longrightarrow {}^{156}_{72}Hf + {}^{4}_{2}He$ (b) ${}^{38}_{17}Cl \longrightarrow {}^{38}_{18}Ar + {}^{0}_{-1}\beta$
(c) ${}^{214}_{83}Bi \longrightarrow {}^{214}_{84}Po + {}^{0}_{-1}\beta$ (d) ${}^{32}_{17}Cl \longrightarrow {}^{32}_{16}S + {}^{0}_{+1}\beta$

<u>6</u>. (a) ${}^{23}_{11}Na + {}^{2}_{1}H \longrightarrow {}^{24}_{11}Na + {}^{1}_{1}H$ (b) ${}^{59}_{27}Co + {}^{1}_{0}n \longrightarrow {}^{56}_{25}Mn + {}^{4}_{2}He$
(c) ${}^{238}_{92}U + {}^{2}_{1}H \longrightarrow {}^{240}_{94}Pu + {}^{0}_{-1}\beta$ (d) ${}^{246}_{96}Cm + {}^{13}_{6}C \longrightarrow {}^{254}_{102}No + 5\,{}^{1}_{0}n$
(e) ${}^{238}_{92}U + {}^{14}_{7}N \longrightarrow {}^{246}_{99}Es + 6\,{}^{1}_{0}n$

<u>7</u>. (a) ${}^{214}Ra \longrightarrow {}^{210}Rn + {}^{4}_{2}He$ (b) ${}^{205}At \longrightarrow {}^{205}Po + \beta^+$
(c) ${}^{212}Fr + {}^{0}e^{-1} \longrightarrow {}^{212}Rn$ (d) ${}^{2}H + {}^{2}H \longrightarrow {}^{3}He + {}^{1}n$
(e) ${}^{241}_{95}Am + {}^{4}_{2}He \longrightarrow {}^{243}_{97}Bk + 2\,{}^{1}_{0}n$

<u>8</u>. (a) Since the decay constant is inversely related to the half-life, the nuclide with the smallest half-life also has the largest value of its decay constant. This is the nuclide ${}^{214}_{84}Po$ with a half-life of 1.64×10^{-4} s.
(b) The nuclide that displays a 75% reduction in its radioactivity has passed through two half-lives in a period of one month. Thus, this is the nuclide with a half-life of approximately two weeks. This is the nuclide ${}^{32}_{15}P$, with a half-life of 14.3 days.

(c) If more than 99% of the radioactivity is lost, less than 1% remains. Thus $(\frac{1}{2})^n < 0.010$. Now, when $n = 7$, $(\frac{1}{2})^n = 0.0078$. Thus, seven half-lives have elapsed in one month, and each half-life approximates 4.3 days. The longest lived nuclide that fits this description is $^{222}_{86}Rn$, which has a half-life of 3.823 days. Of course, all other nuclides with shorter half-lives also meet this criterion, specifically the following nuclides. $^{13}_{8}O$ (8.7×10^{-3} s), $^{28}_{12}Mg$ (21 h), $^{80}_{35}Br$ (17.6 min), and $^{214}_{84}Po$ (1.64×10^{-4} s).

9. Since $16 = 2^4$, four half-lives have elapsed in 18.0 h, and each half-life equals 4.50 h. The half-life of isotope B thus is 2.5×4.50 h = 11.25 h. Now, since $32 = 2^5$, five half-lives must elapse before the decay rate of isotope B falls to $\frac{1}{32}$ of its original value. Thus, the time elapsed for this amount of decay is:

$$\text{time elapsed} = 5 \text{ half-lives} \times \frac{11.25 \text{ h}}{1 \text{ half-life}} = 56.3 \text{ h}.$$

10. We use expression (26.13), substituting the disintegration rate for the number of atoms, since we recognize that in this first-order reaction the rate is directly proportional to the amount of reactant, that is, the number of atoms. (All radioactive decay processes are first-order reactions.) We also use equation (26.14), rearranged to $\lambda = \frac{0.693}{t_{1/2}}$. Thus $\ln \frac{N_t}{N_0} = -\frac{0.693\,t}{t_{1/2}}$.

(a) $\ln \frac{253 \text{ dis/min}}{1.00 \times 10^3 \text{ dis/min}} = -\frac{0.693\,t}{87.9 \text{ d}} = -1.374$ $\qquad t = \frac{1.374 \times 87.9 \text{ d}}{0.693} = 174 \text{ d}$

(b) $\ln \frac{104 \text{ dis/min}}{1.00 \times 10^3 \text{ dis/min}} = -\frac{0.693\,t}{87.9 \text{ d}} = -2.26$ $\qquad t = \frac{2.26 \times 87.9 \text{ d}}{0.693} = 287 \text{ d}$

(c) $\ln \frac{52 \text{ dis/min}}{1.00 \times 10^3 \text{ dis/min}} = -\frac{0.693\,t}{87.9 \text{ d}} = -2.96$ $\qquad t = \frac{2.96 \times 87.9 \text{ d}}{0.693} = 375 \text{ d}$

11. The principal equation that we shall employ is $E = mc^2$, along with conversion factors.

(a) $E = 6.02 \times 10^{-23} \text{ g} \times \frac{1 \text{ kg}}{1000 \text{ g}} \times (3.00 \times 10^8 \text{ m/s})^2 = 5.42 \times 10^{-9} \text{ kg m}^2 \text{ s}^{-2} = 5.42 \times 10^{-9} \text{ J}$

(b) $E = 4.0015 \text{ u} \times \frac{931.5016 \text{ MeV}}{1 \text{ u}} = 3727.4 \text{ MeV}$

(c) $\text{no. neutrons} = 6.75 \times 10^6 \text{ MeV} \times \frac{1 \text{ u}}{931.5016 \text{ MeV}} \times \frac{1 \text{ neutron}}{1.0087 \text{ u}} = 7.18 \times 10^3 \text{ neutrons.}$

12. $\text{Mass of individual particles} = \left(47 \text{ p} \times \frac{1.0073 \text{ u}}{1 \text{ p}}\right) \times \left(60 \text{ n} \times \frac{1.0087 \text{ u}}{1 \text{ n}}\right)$

$$= 47.343 \text{ u} + 60.522 \text{ u} = 107.865 \text{ u}$$

$$\frac{\text{Binding energy}}{\text{nucleon}} = \frac{107.865 \text{ u} - 106.905092 \text{ u}}{107 \text{ nucleons}} \times \frac{931.5016 \text{ MeV}}{1 \text{ u}} = 8.36 \text{ MeV/nucleon}$$

13. (c) ^{80}Br does not occur naturally; it has an odd number of protons (35), an odd number of neutrons (45).
(d) ^{132}Cs also does not occur naturally, for the same reason.

14. (a) A radioisotope with a long half-life is giving off few disintegrations per second, and thus is not exceedingly hazardous unless a large quantity of it is present. Radioisotopes with short half-lives, however, give off large quantities of radiation in a short time, but they also "burn themselves out" quickly and thus are hazardous for only a short time. On the other hand, radioisotopes with an intermediate half-life are both giving off reasonably large quantities of radiation and also are in existence for relatively long times.

(b) The radioisotopes that are hazardous from a distance are those that give off γ or high energy β radiation, radiation with large or moderate penetrating power. On the other hand, α particles are not hazardous at long distances because of their short penetrating power, but they are highly ionizing, and thus do significant damage when they are encountered at short distances.

(c) Potassium-40 is a radioisotope that decays by β emission to argon-40. This accounts for a large majority of the argon in the atmosphere. Helium-4 is also produced by radioactive decay, but it is so light that it escapes from the atmosphere at a much faster rate than does argon-40.

(d) Francium is produced by radioactive decay, and thus we should not expect it to occur where we find the other alkali metal compounds. It also is quite short-lived and thus there is not much present on the earth at any one time.

(e) Fusion involves joining positively charged nuclei together. Such a process requires quite energetic nuclei. This, in turn, means that the nuclei must be at very high temperatures.

EXERCISES

Radioactive Processes

15. (a) ${}^{234}_{94}\text{Pu} \longrightarrow {}^{230}_{92}\text{U} + {}^{4}_{2}\text{He}$

(b) ${}^{248}_{97}\text{Bk} \longrightarrow {}^{248}_{98}\text{Cf} + {}^{0}_{-1}\text{e}$

(c) ${}^{196}_{82}\text{Pb} + {}^{0}_{-1}\text{e} \longrightarrow {}^{196}_{81}\text{Tl}$ ${}^{196}_{81}\text{Tl} + {}^{0}_{-1}\text{e} \longrightarrow {}^{196}_{80}\text{Hg}$

16. (a) ${}^{214}_{82}\text{Pb} \longrightarrow {}^{214}_{83}\text{Bi} + {}^{0}_{-1}\text{e}$ ${}^{214}_{83}\text{Bi} \longrightarrow {}^{214}_{84}\text{Po} + {}^{0}_{-1}\text{e}$

(b) ${}^{226}_{88}\text{Ra} \longrightarrow {}^{222}_{86}\text{Rn} + {}^{4}_{2}\text{He}$ ${}^{222}_{86}\text{Rn} \longrightarrow {}^{218}_{84}\text{Po} + {}^{4}_{2}\text{He}$ ${}^{218}_{84}\text{Po} \longrightarrow {}^{214}_{82}\text{Pb} + {}^{4}_{2}\text{He}$

(c) ${}^{69}_{33}\text{As} \longrightarrow {}^{69}_{32}\text{Ge} + {}^{0}_{+1}\text{e}$

Radioactive Decay Series

17. We first write conventional nuclear reactions for each step in the decay series.

${}^{232}_{90}\text{Th} \longrightarrow {}^{228}_{88}\text{Ra} + {}^{4}_{2}\text{He}$ ${}^{228}_{88}\text{Ra} \longrightarrow {}^{228}_{89}\text{Ac} + {}^{0}_{-1}\text{e}$ ${}^{228}_{89}\text{Ac} \longrightarrow {}^{228}_{90}\text{Th} + {}^{0}_{-1}\text{e}$

${}^{228}_{90}\text{Th} \longrightarrow {}^{224}_{88}\text{Ra} + {}^{4}_{2}\text{He}$ ${}^{224}_{88}\text{Ra} \longrightarrow {}^{220}_{86}\text{Rn} + {}^{4}_{2}\text{He}$ ${}^{220}_{86}\text{Rn} \longrightarrow {}^{216}_{84}\text{Po} + {}^{4}_{2}\text{He}$

Now for a branch in the series

these two ${}^{216}_{84}\text{Po} \longrightarrow {}^{212}_{82}\text{Pb} + {}^{4}_{2}\text{He}$ ${}^{212}_{82}\text{Pb} \longrightarrow {}^{212}_{83}\text{Bi} + {}^{0}_{-1}\text{e}$

or these ${}^{216}_{84}\text{Po} \longrightarrow {}^{216}_{85}\text{At} + {}^{0}_{-1}\text{e}$ ${}^{216}_{85}\text{At} \longrightarrow {}^{212}_{83}\text{Bi} + {}^{4}_{2}\text{He}$

And now a second branch

these two ${}^{212}_{83}\text{Bi} \longrightarrow {}^{208}_{81}\text{Tl} + {}^{4}_{2}\text{He}$ ${}^{208}_{81}\text{Tl} \longrightarrow {}^{208}_{82}\text{Pb} + {}^{0}_{-1}\text{e}$

or these ${}^{212}_{83}\text{Bi} \longrightarrow {}^{212}_{84}\text{Po} + {}^{0}_{-1}\text{e}$ ${}^{216}_{84}\text{Po} \longrightarrow {}^{208}_{82}\text{Pb} + {}^{4}_{2}\text{He}$

Both branches end with the isotope ${}^{208}_{82}\text{Pb}$. The graph, similar to Figure 26-2, is drawn below.

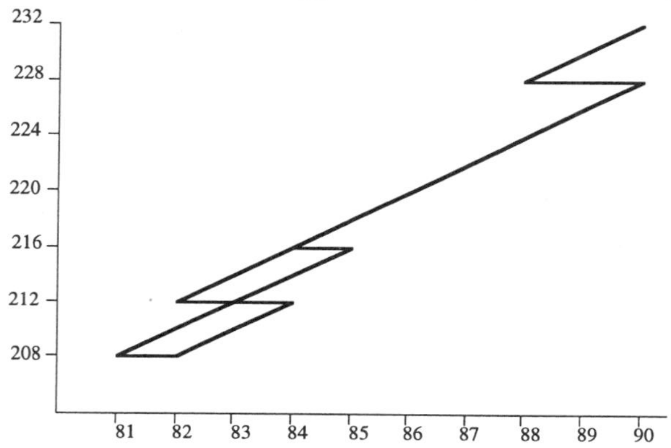

18. The series begins with uranium-235, and ends with lead-207.

${}^{235}_{92}\text{U} \longrightarrow {}^{231}_{90}\text{Th} + {}^{4}_{2}\text{He}$ ${}^{231}_{90}\text{Th} \longrightarrow {}^{231}_{91}\text{Pa} + {}^{0}_{-1}\text{e}$ ${}^{231}_{91}\text{Pa} \longrightarrow {}^{227}_{89}\text{Ac} + {}^{4}_{2}\text{He}$

Now the series branches

these two ${}^{227}_{89}\text{Ac} \longrightarrow {}^{223}_{87}\text{Fr} + {}^{4}_{2}\text{He}$ ${}^{223}_{87}\text{Fr} \longrightarrow {}^{223}_{88}\text{Ra} + {}^{0}_{-1}\text{e}$

or these ${}^{227}_{89}\text{Ac} \longrightarrow {}^{227}_{90}\text{Th} + {}^{0}_{-1}\text{e}$ ${}^{227}_{90}\text{Th} \longrightarrow {}^{223}_{88}\text{Ra} + {}^{4}_{2}\text{He}$

${}^{223}_{88}\text{Ra} \longrightarrow {}^{219}_{86}\text{Rn} + {}^{4}_{2}\text{He}$ ${}^{219}_{86}\text{Rn} \longrightarrow {}^{215}_{84}\text{Po} + {}^{4}_{2}\text{He}$ ${}^{215}_{84}\text{Po} \longrightarrow {}^{211}_{82}\text{Pb} + {}^{4}_{2}\text{He}$

${}^{211}_{82}\text{Pb} \longrightarrow {}^{211}_{83}\text{Bi} + {}^{0}_{-1}\text{e}$

The series branches again

these two ${}^{211}_{83}\text{Bi} \longrightarrow {}^{207}_{81}\text{Tl} + {}^{4}_{2}\text{He}$ ${}^{207}_{81}\text{Tl} \longrightarrow {}^{207}_{82}\text{Pb} + {}^{0}_{-1}\text{e}$

or these ${}^{211}_{83}\text{Bi} \longrightarrow {}^{211}_{84}\text{Po} + {}^{0}_{-1}\text{e}$ ${}^{211}_{84}\text{Po} \longrightarrow {}^{207}_{82}\text{Pb} + {}^{4}_{2}\text{He}$

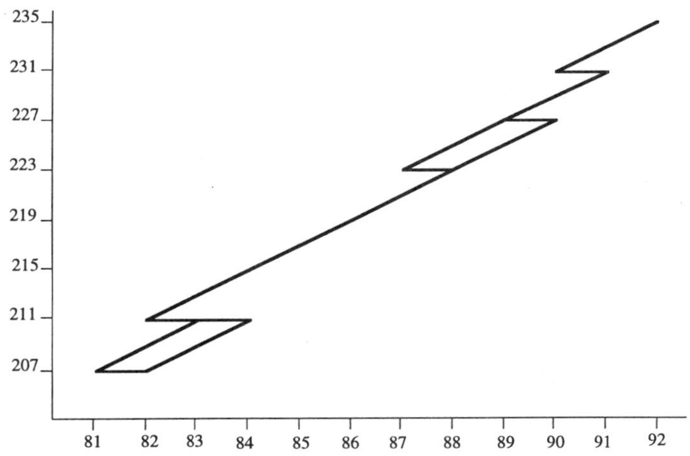

19. In Figure 26-2, only the following mass numbers are represented: 206, 210, 214, 218, 222, 226, 230, 234, and 238. We see that these mass numbers are separated from each other by 4 units. The first of them, 206, equals $(4 \times 51) + 2$, that is $4n + 2$, where $n = 51$.

20. The series to which each nuclide belongs is determined by dividing its mass number by 4 and obtaining the remainder.

(a) The mass number of $^{214}_{83}Bi$ is 214, and the remainder of its division by 4 is 2. This nuclide is a member of the $4n + 2$ series.

(b) The mass number of $^{216}_{84}Po$ is 216, and the remainder of its division by 4 is 0. This nuclide is a member of the $4n$ series.

(c) The mass number of $^{215}_{85}At$ is 215, and the remainder of its division by 4 is 3. This nuclide is a member of the $4n + 3$ series.

(d) The mass number of $^{235}_{92}U$ is 235, and the remainder of its division by 4 is 3. This nuclide is a member of the $4n + 3$ series.

21. We would expect a neutron:proton ratio that is closer to 1:1 than that of ^{14}C. This would be achieved if the product were ^{14}N, which will be the result of β^- decay: $^{14}_{6}C \longrightarrow ^{14}_{7}N + ^{0}_{-1}e$.

22. A nuclide with a closer to 1:1 neutron:proton ratio (than that of tritium) is helium-3, arrived at by beta emission: $^{3}_{1}H \longrightarrow ^{3}_{2}He + ^{0}_{-1}e$. Another possible product is deuterium, which is arrived at by neutron emission: $^{3}_{1}H \longrightarrow ^{2}_{1}H + ^{1}_{0}n$

Nuclear Reactions

23. **(a)** $^{7}_{3}Li + ^{1}_{1}H \longrightarrow ^{8}_{4}Be + \gamma$ **(b)** $^{9}_{4}Be + ^{2}_{1}H \longrightarrow 10\ 5B + ^{1}_{0}n$ **(c)** $^{14}_{7}N + ^{1}_{0}n \longrightarrow ^{14}_{6}C + ^{1}_{1}H$

24. **(a)** $^{238}_{92}U + ^{4}_{2}He \longrightarrow ^{239}_{94}Pu + ^{1}_{0}n$ **(b)** $^{3}_{1}H + ^{2}_{1}H \longrightarrow ^{4}_{2}He + ^{1}_{0}n$ **(c)** $^{33}_{16}S + ^{1}_{0}n \longrightarrow ^{33}_{15}P + ^{1}_{1}H$

Rate of Radioactive Decay

25. We use expression (26.13) to determine λ and then expression (26.11) to determine the number of atoms.

$$\lambda = \frac{0.693}{5.2\ y} \times \frac{1\ y}{365.25\ d} \times \frac{1\ d}{24\ h} = 1.5 \times 10^{-5}\ h^{-1}$$

$$N = \frac{\text{rate of decay}}{\lambda} = \frac{6740\ \text{atoms/h}}{1.5 \times 10^{-5}\ h^{-1}} = 4.4 \times 10^{8}\ ^{60}_{27}\text{Co atoms}$$

26. This is a first order reaction (as are all radioactive decay processes) with a rate of decay directly proportional to the number of atoms. We therefore use expression (26.12), with rates substituted for numbers of atoms.

$$6740\ \frac{\text{dis}}{\text{h}} \times \frac{1\ \text{h}}{60\ \text{min}} = 112\ \frac{\text{dis}}{\text{min}} \qquad \ln \frac{R_t}{R_o} = -\lambda t = \ln \frac{75\ \text{dis/min}}{112\ \text{dis/min}} = -1.5 \times 10^{-5}\ h^{-1}\ t = -0.401$$

$$t = \frac{0.401}{1.5 \times 10^{-5}\ \text{h}^{-1}} = 2.7 \times 10^4\ \text{h} \times \frac{1\ \text{d}}{24\ \text{h}} \times \frac{1\ \text{y}}{365.25\ \text{d}} = 3.0\ \text{y}$$

27. Let us use the first and the last values to determine the decay constant.

$$\ln\frac{R_t}{R_o} = -\lambda t = \ln\frac{138\ \text{cpm}}{1000\ \text{cpm}} = -\lambda\ 250\ \text{h} = -1.981 \qquad \lambda = \frac{1.981}{250\ \text{h}} = 0.00792\ \text{h}^{-1}$$

$$t_{1/2} = \frac{0.693}{\lambda} = \frac{0.693}{0.00792\ \text{h}^{-1}} = 87.5\ \text{h}$$

A slightly different value of $t_{1/2}$ may result from other combinations of R_o and R_t.

28. First calculate the decay constant. $\lambda = \frac{0.693}{1.7 \times 10^7\ \text{y}} \times \frac{1\ \text{y}}{365.25\ \text{d}} \times \frac{1\ \text{d}}{24\ \text{h}} \times \frac{1\ \text{h}}{3600\ \text{s}} = 1.3 \times 10^{-15}\ \text{s}^{-1}$

$$N = 1.00\ \text{mg}\ {}^{129}\text{I} \times \frac{1\ \text{g}}{1000\ \text{mg}} \times \frac{1\ \text{mol}\ {}^{129}\text{I}}{129\ \text{g}} \times \frac{6.022 \times 10^{23}\ \text{atoms}}{1\ \text{mol}\ {}^{129}\text{I}} = 4.67 \times 10^{18}\ {}^{129}\text{I atoms}$$

decay rate $= \lambda N = 1.3 \times 10^{-15}\ \text{s}^{-1} \times 4.67 \times 10^{18}\ \text{atoms} = 6.1 \times 10^3\ \text{dis/s}$

29. First we determine the decay constant. $\lambda = \frac{0.693}{1.39 \times 10^{10}\ \text{y}} = 4.99 \times 10^{-11}\ \text{y}^{-1}$

Then we can determine the ratio of the number of thorium atoms after (N_t) 2.7×10^9 y to the number (N_0) initially. $\ln\frac{N_t}{N_0} = -kt = -(4.99 \times 10^{-11}\ \text{y}^{-1})(2.7 \times 10^9\ \text{y}) = -0.13 \qquad \frac{N_t}{N_0} = 0.87$

Thus, for every mole of ^{232}Th present initially, there is present after 2.7×10^9 y, 0.87 mol ^{232}Th and 0.13 mol ^{208}Pb. From this information, we can compute the mass ratio.

$$\frac{0.13\ \text{mol}\ {}^{208}\text{Pb}}{0.87\ \text{mol}\ {}^{232}\text{Th}} \times \frac{1\ \text{mol}\ {}^{232}\text{Th}}{232\ \text{g}\ {}^{232}\text{Th}} \times \frac{208\ \text{g}\ {}^{208}\text{Pb}}{1\ \text{mol}\ {}^{208}\text{Pb}} = \frac{0.13\ \text{g}\ {}^{208}\text{Pb}}{1\ \text{g}\ {}^{232}\text{Th}}$$

30. First we determine the decay constant. $\lambda = \frac{0.693}{1.4 \times 10^{10}\ \text{y}} = 4.99 \times 10^{-11}\ \text{y}^{-1}$

The rock currently contains 1.00 g ^{232}Th and 0.25 g ^{208}Pb. We can calculate the mass of ^{232}Th that must have been present to produce this 0.25 g ^{208}Pb, and from that find the original mass of ^{232}Th.

original mass $^{232}\text{Th} = 1.00\ \text{g}\ {}^{232}\text{Th now} + \left(0.25\ \text{g}\ {}^{208}\text{Pb} \times \frac{232\ \text{g}\ {}^{232}\text{Th}}{208\ \text{g}\ {}^{208}\text{Pb}}\right) = (1.00 + 0.28)\ \text{g} = 1.28\ \text{g}$

$$\ln\frac{N_t}{N_0} = -\lambda t = \ln\frac{1.00\ \text{g}\ {}^{232}\text{Th now}}{1.28\ \text{g originally}} = -0.247 = -4.99 \times 10^{-11}\ \text{y}^{-1}\ t; \quad t = \frac{0.247}{4.99 \times 10^{-11}\ \text{y}^{-1}} = 4.95 \times 10^9\ \text{y}$$

Radiocarbon Dating

31. Again we use expression (26.12) and (26.13) to determine the time elapsed. The initial rate of decay is about 15 dis/min. First we compute the decay constant. $\lambda = \frac{0.693}{5730\ \text{y}} = 1.21 \times 10^{-4}\ \text{y}^{-1}$

$$\ln\frac{10\ \text{dis/min}}{15\ \text{dis/min}} = -0.405 = -\lambda t \qquad t = \frac{0.405}{1.21 \times 10^{-4}\ \text{y}^{-1}} = 3.35 \times 10^3\ \text{y}$$

The object is a bit more than 3000 years old, probably not dating from the pyramid era, about 3000 BC.

32. We use the value of λ from the previous exercise.

$$\ln\frac{R_t}{R_o} = -\lambda t = -(1.21 \times 10^{-4}\ \text{y}^{-1})t = \ln\frac{0.03\ \text{dis min}^{-1}\ \text{g}^{-1}}{15\ \text{dis min}^{-1}\ \text{g}^{-1}} = -6.2$$

$$t = \frac{6.2}{1.21 \times 10^{-4}\ \text{y}^{-1}} = 5.2 \times 10^4\ \text{y}$$

Energetics of Nuclear Reactions

33. The mass defect is the difference between the mass of the nuclide and the sum of the masses of its constituent particles. The binding energy is this mass defect expressed as an energy.

particle mass $= 9\,p + 10\,n + 9\,e = 9\,(p + n + e) + n$

$= 9\,(1.0073 + 1.0087 + 0.0005486)\ \text{u} + 1.0087\ \text{u} = 19.1576\ \text{u}$

mass defect $= 19.1576\ \text{u} - 18.998403\ \text{u} = 0.1592\ \text{u}$

$$\text{binding energy per nucleon} = \frac{0.1592 \text{ u} \times \dfrac{931.5 \text{ MeV}}{1 \text{ u}}}{19 \text{ nucleons}} = 7.805 \text{ MeV/nucleon}$$

34. The mass defect is the difference between the mass of the nuclide and the sum of the masses of its constituent particles. The binding energy is this mass defect expressed as an energy.
particle mass $= 26\,p + 30\,n + 26\,e = 26\,(p + n + e) + 4\,n$
$= 26\,(1.0073 + 1.0087 + 0.0005486)\text{ u} + 4 \times 1.0087\text{ u} = 56.4651\text{ u}$
mass defect $= 56.4651\text{ u} - 55.934939\text{ u} = 0.5302\text{ u}$

$$\text{binding energy per nucleon} = \frac{0.5302 \text{ u} \times \dfrac{931.5 \text{ MeV}}{1 \text{ u}}}{56 \text{ nucleons}} = 8.819 \text{ MeV/nucleon}$$

35. mass defect $= (10.01294\text{ u} + 4.00260\text{ u}) - (13.00335\text{ u} + 1.00783\text{ u}) = 0.00436\text{ u}$

$$\text{energy} = 0.00436\text{ u} \times \frac{931.5016\text{ MeV}}{1\text{ u}} = 4.06\text{ MeV}$$

36. mass defect $= (6.01513\text{ u} + 1.008665\text{ u}) - (4.00260\text{ u} + 3.01604\text{ u}) = 0.00516\text{ u}$

$$\text{energy} = 0.00516\text{ u} \times \frac{931.5016\text{ MeV}}{1\text{ u}} = 4.81\text{ MeV}$$

Nuclear Stability

37. **(a)** We expect ^{20}Ne to be more stable than ^{22}Ne. A neutron-to-proton ratio of 1-to-1 is associated with stability for elements of low atomic number (with $Z \leq 20$).
(b) We expect ^{18}O to be more stable than ^{17}O. An even number of protons and an even number of neutrons is associated with a stable istope.
(c) We expect ^{7}Li to be more stable than ^{6}Li. Both isotopes have an odd number of protons, but only ^{7}Li has an even number of neutrons.

38. **(a)** We expect ^{40}Ca to be more stable than ^{42}Ca. A neutron-to-proton ratio of 1-to-1 is associated with stability for elements of low atomic number (with $Z \leq 20$).
(b) We expect ^{31}P to be more stable than ^{32}P. Both isotopes have an odd number of protons, but only ^{31}P has an even number of neutrons.
(c) We expect ^{64}Zn to be more stable than ^{63}Zn. An even number of protons and an even number of neutrons is associated with a stable istope.

39. β^- emission has the effect of "converting" a neutron to a proton. β^+ emission, on the other hand, has the effect of "converting" a proton to a neutron.
(a) The most stable isotope of phosphorus is ^{31}P, with a neutron-to-proton ratio of close to 1-to-1 and an even number of neutrons. Thus, ^{29}P has "too few" neutrons, or too many protons. It should decay by β^+ emission. In contrast, ^{33}P has "too many" neutrons, or "too few" protons. Therefore, ^{33}P should decay by β^- emission.
(b) Based on the atomic weight of I (126.90447), we expect the isotopes of iodine to have mass numbers close to 127. This means that ^{120}I has "too few" neutrons and therefore should decay by β^+ emission, whereas ^{134}I has "too many" neutrons (or "too few" protons) and therefore should decay by β^- emission.

40. β^- emission has the effect of converting a neutron to a proton, while β^+ emission has the effect of converting a proton to a neutron.
(a) Based on the fact that elements of low atomic number have about the same number of protons as neutrons, $^{28}_{15}$P—with 15 protons and 13 neutrons—has too few neutrons. Therefore, it should decay by β^+ emission.
(b) Again based on the fact that elements of low atomic number have about the same number of protons as neutrons, $^{45}_{19}$K—with 19 protons and 26 neutrons—has too many neutrons. Therefore, it should decay by β^- emission.

(c) Based on the atomic weight of zinc (65.39) we expect most of its isotopes to have about 36 neutrons. There are 42 neutrons in ${}^{72}_{30}Zn$, more than we expect. Thus we expect this nuclide to decay by β^- emission.

41. A "doubly magic" nuclide is one in which the atomic number is a magic number (2, 8, 20, 28, 50, 82, 114) and the number of neutrons also is a magic number (2, 8, 20, 28, 50, 82, 126, 184, 196). The nuclides that fit this description are the following.

nuclide	${}^{4}He$	${}^{16}O$	${}^{40}Ca$	${}^{56}Ni$	${}^{208}Pb$
no. of protons	2	8	20	28	82
no. of neutrons	2	8	20	28	126

42. In isotopes of high atomic number, stable nuclides are characterized by a neutron-to-proton ratio that is greater than 1 and that increases with increasing atomic number. Naturally occurring isotopes of high atomic number decrease their atomic number by losing an alpha particle, which has a neutron-to-proton ratio of 1. This leaves the neutron-to-proton ratio of the daughter higher than that of the parent, when it should be slightly lower. In order to redress this, the number of neutrons needs to be decreased, the number of protons increases. Beta emission accomplishes this.

In contrast, artificially produced isotopes have no definite neutron-to-proton ratio. Thus, sometimes, the number of neutrons needs to be decreased, which is accomplished by beta emission, while at other times the number of protons needs to be decreased, which is accomplished by positron emission.

Fission and Fusion

43. We use the conversion factor between number of curies and mass of ${}^{131}I$ which was developed in the Summarizing Example.

$$\text{no. g I-131} = 170 \text{ curies} \times \frac{19 \text{ g I-131}}{2.4 \times 10^{6} \text{ curie}} = 1.35 \times 10^{-3} \text{ g} = 1.35 \text{ mg}$$

44. We use $\Delta H_f^\circ[CO_2(g)] = -393.51$ kJ/mol as the heat of combustion of 1 mole of carbon. In the text, the energy produced by the fusion of 1.00 g ${}^{235}U$ is determined as 8.20×10^{7} kJ.

$$\text{no. metric tons coal} = 1.00 \text{ kg } {}^{235}U \times \frac{1000 \text{ g}}{1 \text{ kg}} \times \frac{8.20 \times 10^{7} \text{ kJ}}{1.00 \text{ g } {}^{235}U} \times \frac{1 \text{ mol C}}{393.5 \text{ kJ}} \times \frac{12.01 \text{ g C}}{1 \text{ mol C}}$$
$$\times \frac{1.00 \text{ g coal}}{0.85 \text{ g C}} \times \frac{1 \text{ kg}}{1000 \text{ g}} \times \frac{1 \text{ metric ton}}{1000 \text{ kg}} = 2.9 \times 10^{3} \text{ metric tons (Mg)}$$

Effect of Radiation on Matter

45. The term "rem" is an acronym for "radiation equivalent–man," and takes into account the quantity of biological damage done by a given dosage of radiation. On the other hand, the rad is the dosage of radiation that will place 0.010 J of energy into each kg of irradiated matter. Thus, for living tissue, the rem gives us a good idea of how much tissue damage a certain kind and quantity of radiation damage will do. But, for nonliving materials, the rad is often preferred, and indeed is often the only unit of utility.

46. Low-level radiation is very close in its dosage to background radiation and one problem is to separate out the effects of the two sources (low-level and background). The other problem is that low-level radiation does not produce severe damage in a short period of time. Thus the effects of low-level radiation will only accumulate over a long time period. Of course other effects, such as chemical and biological toxins, will also be effective over these time periods, and we have to try to separate these two types of effects. (There also is the genetic heritage of the organism to consider, of course.)

47. One reason why ${}^{90}Sr$ is hazardous is because strontium is in the same family of the periodic table as calcium, and hence often reacts in a similar fashion to calcium. One site where calcium is incorporated into the body is in bones, where it resides for a long time. Strontium is expected to behave in a similar fashion. Thus, it will be retained in the body for a long time. Bone is an especially dangerous place for a radioisotope to be present—even if it has low penetrating power, as do β rays—because blood cells are produced in bone marrow.

48. It is not particularly hazardous to be near a flask of ^{222}Rn, because it is unlikely that the alpha particles can get through the walls of the flask. (Note that since radon is a gas, the flask must be sealed.) The decay products of ^{222}Rnmay produce other forms of radiation that are more penetrating, such as β particles and γ rays, so being near the flask could still be unhealthy. ^{222}Rn can be potentially hazardous if one breathes the gas.

Application of Radioisotopes

49. If a small amount of tritium, as $T_2(g)$ or $^3H_2(g)$, were mixed with the $H_2(g)$ supply, it would behave chemically in much the same way as $H_2(g)$. But the small leak would be readily detected with a radiation detector.

50. In neutron activation analysis, the sample is bombarded with neutrons. Radioisotopes are produced by this process. These radioisotopes can be easily detected even in very small quantities, much smaller than the quantities that can be detected by conventional means of quantitative analysis. These radioisotopes are produced in quantities that are proportional to the quantity of each element originally present in the sample. And each radioisotope is characteristic of the element from which it was produced by neutron bombardment. Even microscopic samples can be analyzed by this technique. Finally, neutron activation analysis is a nondestructive technique, while the conventional techniques of precipitation or titration require that the sample, or at least part of it, be destroyed.

51. The recovered sample will be radioactive. When NaCl(s) and $NaNO_3(s)$ are dissolved in solution, the ions (Na^+, Cl^-, and NO_3^-) are free to move throughout the solution. A given anion does not remain associated with a particular cation. Thus, all the anions and cations are shuffled and some of the radioactive ^{24}Na will end up in the crystallized $NaNO_3$.

52. We would expect the tritium label to appear in both the $NH_3(g)$ and the H_2O. When $NH_4^+(aq)$ is formed, one of the four chemically and spatially equivalent H atoms is a tritium atom, In the subsequent reaction with NaOH to form $NH_3(g)$ and H_2O there are three chances in four that a tritium atom will remain attached to N in NH_3 and one chance in four that a tritium ion will unite with a hydroxide ion to form H_2O.

27 Organic Chemistry

Practice Examples

1A We show only the C atoms and the bonds between them. Remember that there are four bonds to each C atom; the remaining bonds not shown are to H atoms. First we realize there is only one isomer with all six C atoms in one line. Then we draw the isomers with one 1-C branch. These are the center two structures below. The isomers with two 1-C branches can have them both on the same atom or on different atoms, as in the rightmost two structures. This accounts for all five isomers.

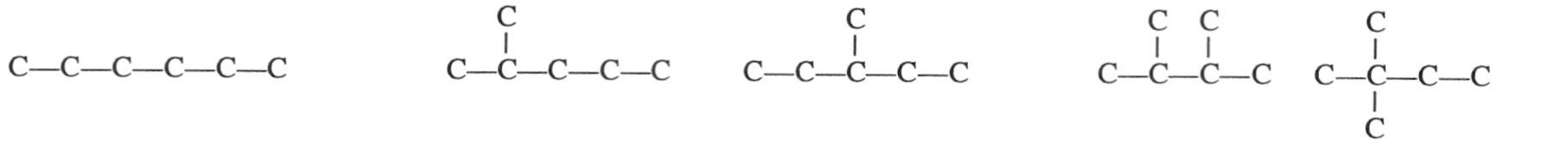

$CH_3CH_2CH_2CH_2CH_2CH_3$ $(CH_3)_2CHCH_2CH_2CH_3$ $CH_3CH_2CH(CH_3)CH_2CH_3$ $(CH_3)_2CHCH(CH_3)_2$ $(CH_3)_3CCH_2CH_3$

1B We show only the C atoms and the bonds between them. Remember that there are four bonds to each C atom; the remaining bonds not shown are to H atoms.

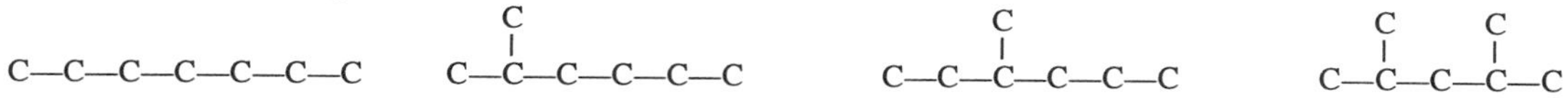

$CH_3CH_2CH_2CH_2CH_2CH_2CH_3$ $(CH_3)_2CCH_2CH_2CH_2CH_3$ $CH_3CH_2CH(CH_3)CH_2CH_2CH_3$ $(CH_3)_2CHCH_2CH(CH_3)_2$

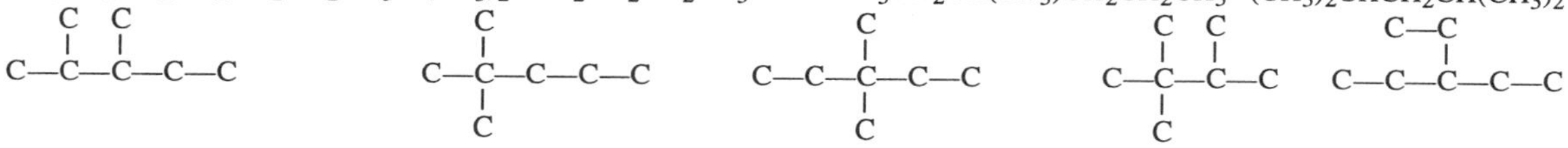

$(CH_3)_2CHCH(CH_3)CH_2CH_3$ $(CH_3)_3CCH_2CH_2CH_3$ $CH_3CH_2C(CH_3)_2CH_2CH_3$ $(CH_3)_3CCH(CH_3)_2$ $C(CH_2CH_3)_3$

2A

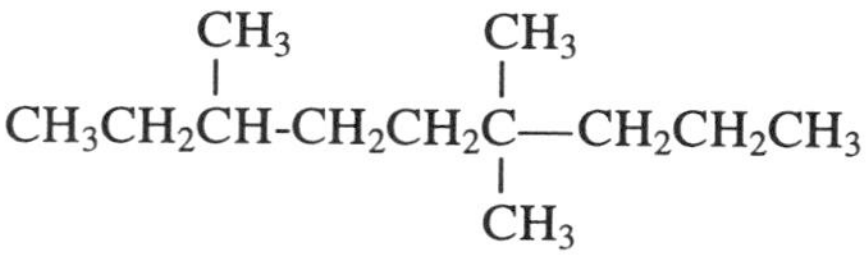

Numbering starts from the left and goes right. 3,6,6-trimethylnonane.

2B In the structural formula we show only the C atoms and the bonds between them. Remember that there are four bonds to each C atom; the remaining bonds not shown are to H atoms. The longest chain has 8 C atoms, making this an octane.

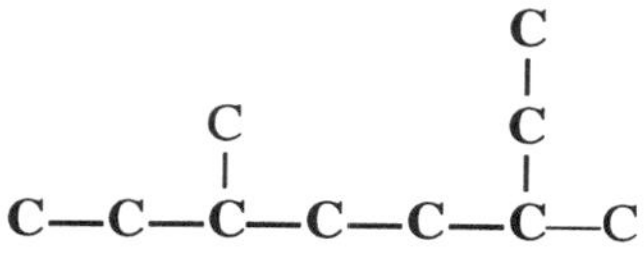

There are two methyl groups on the chain. The IUPAC name is 2,5-dimethyloctane.

3A We write the structural formula as before, showing only the C skeleton. The main chain is 7 C's long; we number it from left to right. Methyl groups are 1 carbon chains, an ethyl group is two.

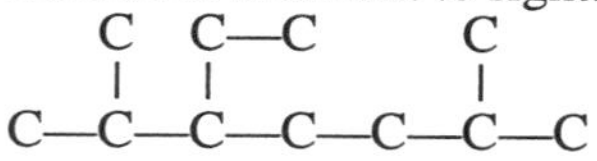

condensed structural formula

$CH_3CH(CH_3)CH(CH_2CH_3)CH_2CH_2CH(CH_3)_2$

3B Pentane is a five-carbon chain. There is a —CH_3 group on carbon 2, a $CH_3\overset{|}{C}HCH_3$ group on carbon 3.

$CH_3CH(CH_3)CH[CH(CH_3)_2]CH_2CH_3$

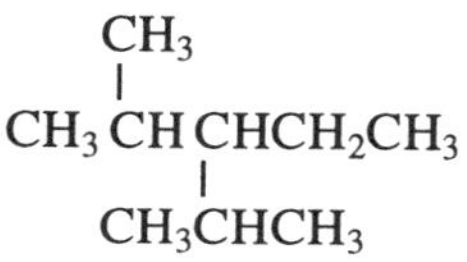

4A The aldehyde group is a meta director. Thus, the product of the mononitration of benzaldehyde should be the compound whose structure is drawn at right: *m*-nitrobenzaldehyde or 3-nitrobenzaldehyde.

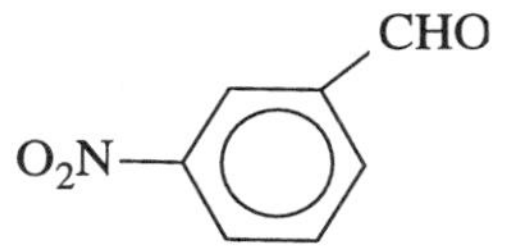

4B —Cl is an ortho,para director. The possible products are 1,3-dichloro-2-nitrobenzene and 2,4-dichloro-1-nitrobenzene.

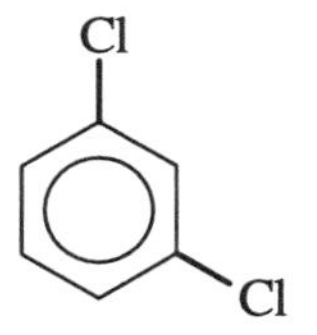

1,3-dichlorobenzene

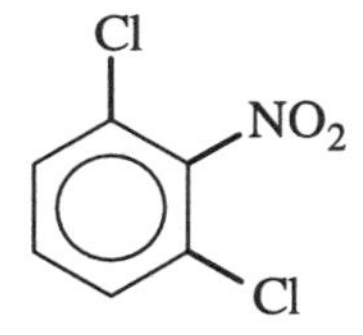

1,3-dichloro-2 -nitrobenzene

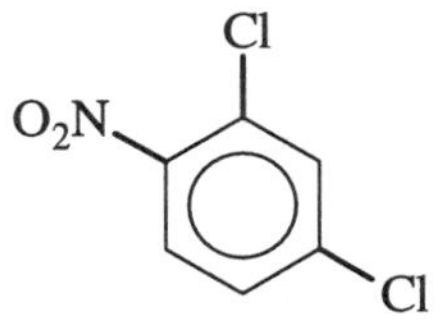

2,4-dichloro-1 -nitrobenzene

SUMMARIZING EXAMPLE CALCULATIONS

1. $CaCO_3(s) \xrightarrow{\text{heat}} CaO(s) + CO_2(g)$
   $CaO(s) + 3\,C(s) \xrightarrow[\text{2000°C}]{\text{electric furnace}} CaC_2(s) + CO_2(g)$
   $CaC_2(s) + 2\,H_2O(l) \longrightarrow HC{\equiv}CH(g) + Ca(OH)_2(s)$
2. $HC{\equiv}CH(g) + H_2(g) \xrightarrow[\text{heat/pressure}]{\text{Pt or Pd}} H_2C{=}CH_2(g)$
3. $H_2C{=}CH_2(g) + H_2O(l) \xrightarrow{\text{H2SO4}} CH_3CH_2OH(l)$
4. $CH_3CH_2OH(l) \xrightarrow[\text{acid}]{\text{K2Cr2O7}} CH_3COOH(l)$
5. $CH_3COOH(l) + HOCH_2CH_3(l) \longrightarrow CH_3COOCH_2CH_3(l) + H_2O$

REVIEW QUESTIONS

1. **(a)** *t* is the symbol for tertiary; it refers to a carbon atom to which three other carbon atoms are singly bonded.
   (b) R— is the symbol for an alkyl group, a carbon-hydrogen chain with no multiple bonds.
   (c) A hexagon with an inscribed circle is the symbol for a benzene ring, the molecule with formula C_6H_6.
   (d) A carbonyl group is a carbon atom with an oxygen double bonded to it, $\rangle C{=}O$.
   (e) A primary amine is an $—NH_2$ group with an attached hydrocarbon chain.

2. **(a)** In a substitution reaction, one atom or group of atoms replaces another atom or group of atoms.
   (b) The octane rating of gasoline is a measure of its quality. Gasoline's octane number is equal to the percent of isooctane in an isooctane–*n*-heptane mixture that has the same combustion characteristics as the gasoline.
   (c) Stereoisomerism refers to isomers that have similar bonding, but for which the substituent groups are oriented differently. It includes geometric isomerism and optical isomerism.
   (d) An ortho, para director is a group attached to an aromatic ring that causes substitution reactions to occur preferentially at positions ortho- and para- to it.
   (e) Step reaction (condensation) polymerization occurs when the monomers join the polymer chain by the reaction of two functional groups and the elimination of a small molecule such as water.

3. **(a)** An alkane is a hydrocarbon in which there are no multiple bonds; in an alkene there is at least one double bond.
   (b) An aliphatic compound is an alkane, a compound of carbon and hydrogen in which all bonds are single bonds and in which the carbon skeleton consists of straight or branched chains. In an aromatic compound there is at least one benzene ring.
   (c) An alcohol is an aliphatic compound with an attached —OH group, a phenol has the —OH group attached to a benzene ring.
   (d) An ether has a C—O—C linkage. An ester has a $C{—}O{—}\overset{\overset{\displaystyle O}{\|}}{C}{—}$ linkage.
   (e) An amine contains N attached to a C atom; ammonia is NH_3.

4. **(a)** Structural (or skeletal) isomers differ in how the atoms are joined together, the carbon chains present.

(b) Positional isomers differ in the location of a functional group along the carbon chain of the molecule.

(c) *cis-*, *trans-* isomers differ in the arrangement of atoms around a double bond.

(d) ortho-, meta-, and para- isomers differ in the location of substituent groups on an aromatic ring.

5. Just in terms of straight and branched chains, C_4H_8 has more isomers than C_4H_{10}, as shown by the structures below, in which the C_4H_8 structures are at right. In addition, only C_4H_8 can form rings: one four-membered ring isomer and one three-membered ring isomer.

$$
\begin{array}{ccc}
& \text{C} & \\
& | & \\
\text{C—C—C—C} & \text{C—C—C} &
\end{array}
\qquad
\begin{array}{cccc}
& & \text{C} & \text{C—C} \\
& & | & | \\
\text{C=C—C—C} & \text{C—C=C—C} & \text{C—C=C} & \text{C—C}
\end{array}
\qquad
\begin{array}{c}
\text{C} \\
| \\
\text{C}{-}\text{C}
\end{array}
$$

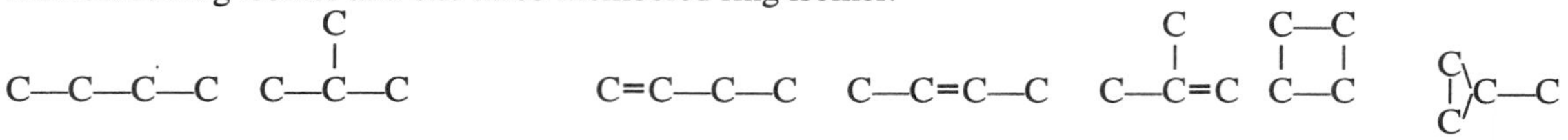

6. Cyclobutane has the formula C_4H_8; there is one carbon for every two hydrogens. $CH_3CH{=}CHCH_3$ has the formula C_4H_8, and $CH_3C{\equiv}CCH_3$ has the formula C_4H_6. Only in $CH_3CH{=}CHCH_3$ is there the same carbon-to-hydrogen ratio as in cyclobutane.

7. **(a)** $\overset{\displaystyle \text{Br}}{\overset{|}{\text{CH}_3\,\text{C}\,\text{HCH}_3}}$ alkyl halide (bromide)   **(b)** $CH_3CH_2\overset{\displaystyle O}{\overset{\|}{C}}OH$ carboxylic acid

2-bromo butane   butanoic acid

(c) $C_6H_5CH_2\overset{\displaystyle O}{\overset{\|}{C}}H$ aldehyde   **(d)** $(CH_3)_2CH{-}O{-}CH_3$ ether

benzaldehyde   isopropylmethyl ether

(e) $CH_3\overset{\displaystyle O}{\overset{\|}{C}}CH_2CH_3$ ketone   **(f)** $CH_3CH(NH_2)CH_2CH_3$ amine

methyl ethyl ketone   2-aminobutane

(g) $HO{-}C_6H_4{-}OH$ alcohol phenyl group   **(h)** $CH_3\overset{\displaystyle O}{\overset{\|}{C}}OCH_3$ ester

4-hydroxyphenol   methyl acetate

8. **(a)**
$$
\begin{array}{ccccc}
\text{H} & & |\overset{\_}{\text{Cl}}| & & \text{H} \\
| & & | & & | \\
\text{H—C} & {-} & \text{C} & {-} & \text{C—H} \\
| & & | & & | \\
\text{H} & & \text{H} & & \text{H}
\end{array}
$$
(b)
$$
\begin{array}{ccccc}
& \text{H} & & \text{H} & \\
& | & & | & \\
\text{H—}\overset{\_}{\text{O}} & {-} & \text{C} & {-}\text{C}{-}\text{}\overset{\_}{\text{O}}\text{—H} \\
& & | & | & \\
& & \text{H} & \text{H} &
\end{array}
$$
(c)
$$
\begin{array}{ccc}
& \text{H} & \\
& | & \\
\text{H—C—C}{=}\overset{\_}{\text{O}} \\
& | & | \\
& \text{H} & \text{H}
\end{array}
$$

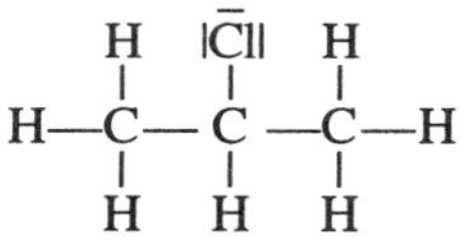

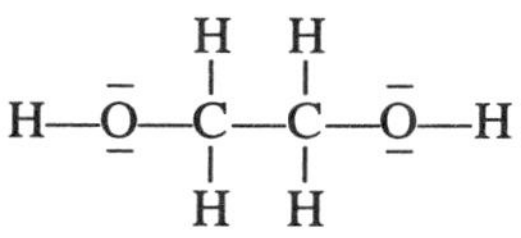

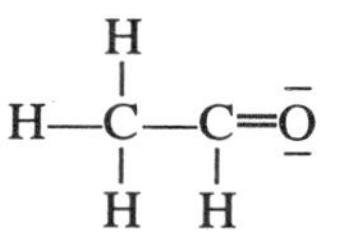

9. We draw only the carbon and chlorine atoms in each structure. Remember that there are four bonds to each carbon atom; the bonds not shown in these structures are C—H bonds.

$$
\begin{array}{c}
\text{Cl} \\
| \\
\text{Cl—C—C}{-}\text{C}
\end{array}
\qquad
\begin{array}{c}
\text{Cl} \\
| \\
\text{Cl—C}{-}\text{C—C}
\end{array}
\qquad
\text{Cl—C—C—C—Cl}
\qquad
\begin{array}{c}
\text{Cl} \\
| \\
\text{C—C}{-}\text{C} \\
| \\
\text{Cl}
\end{array}
$$

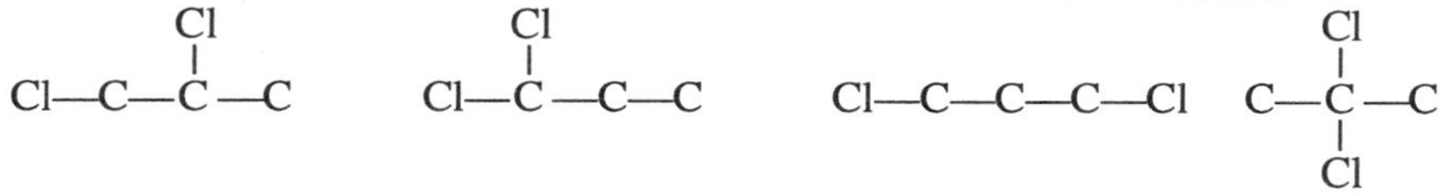

10. **(a)** These are not isomers, because they have different formulas: C_4H_{10} and C_4H_8.

(b) These two compounds are skeletal isomers. (Only the carbon skeleton is shown below).

$$
\begin{array}{c}
\text{C} \\
| \\
\text{C—C—C—C—C—C—C—C}
\end{array}
\qquad
\begin{array}{c}
\text{C} \\
| \\
\text{C—C—C—C—C—C—C—C}
\end{array}
$$

(c) These two compounds are not isomers; they have different formulas: C_4H_9Cl and $C_5H_{11}Cl$.

(d) These two compounds are identical; simply rotated.

(e) These two compounds are identical; simply rotated.

(f) These are two ortho-para isomers, ortho-nitrophenol on the left, and para-nitrophenol on the right.

11. **(a)** The longest chain is eight carbons long, the two substituent groups are methyl groups, and they are attached to the number 3 and number 5 carbon atom. This is 3,5-dimethyloctane.

(b) The longest carbon chain is three carbons long, the two substituent groups are methyl groups, and they are both attached to the number 2 carbon atom. This is 2,2-dimethylpropane.

(c) The longest carbon chain is 7 carbon atoms long, there are two chloro groups attached to carbon atom 3, and an ethyl group attached to carbon 5. This is 3,3-dichloro-5-ethylheptane.

12. (a) There are 2 chloro groups at the 1 and 3 positions on a benzene ring. This is 1,3-dichlorobenzene or, more appropriately, *meta*-dichlorobenzene.

(b) There is a methyl group at position 1 on a benzene ring, and a nitro group at position 3. This is 3-nitrotoluene or *meta*-nitrotoluene.

(c) There is a —COOH group at position 1 on the benzene ring, and a —NH$_2$ group at position 4, or para to the —COOH group. This is 4-aminobenzoic acid or *p*-aminobenzoic acid.

13. In the structural formulas drawn below, we omit the hydrogen atoms. Remember that there are four bonds to each C atom. The bonds that are not shown are C—H bonds.

(a) 3-isopropyloctane
$$\begin{array}{ccc} & C & C-C \\ & | & | \\ C-C-C-C-C-C-C \end{array}$$
(b) 2-chloro-3-methylpentane
$$\begin{array}{ccc} Cl & C \\ | & | \\ C-C-C-C-C \end{array}$$

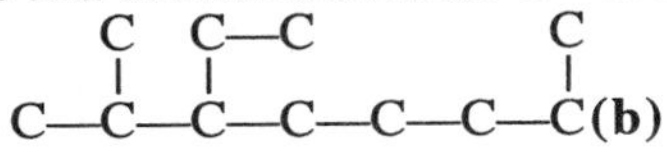

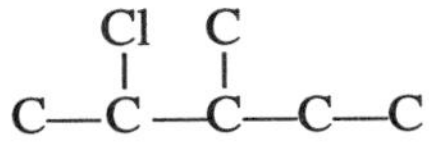

(c) 2-pentene   C—C=C—C—C

(d) dipropyl ether   C—C—C—O—C—C—C

(e) *p*-bromophenol   Br—⟨benzene ring⟩—OH

14. (a) isopropyl alcohol
CH$_3$CH(OH)CH$_3$
(b) 1,1,1-chlorodifluoroethane
CClF$_2$CH$_3$
(c) 2-methyl-1,3-butadiene
CH$_2$=C(CH$_3$)CH=CH$_2$

15. (a)

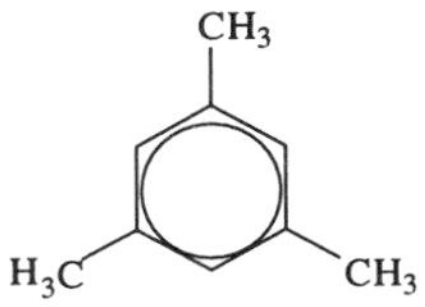

(b)

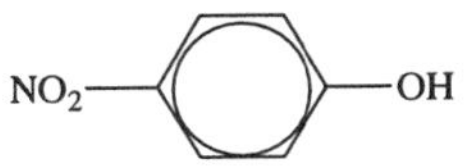

(c)

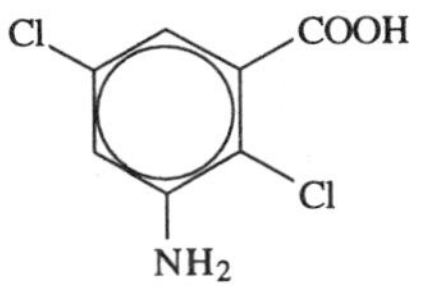

16. Methyl ethyl ketone has the abbreviated structure
$$\begin{array}{c} O \\ \| \\ C-C-C-C \end{array}$$
Thus, the alcohol from which it is produced must have an —OH group on carbon 2 and it must be a four-carbon-atom straight-chain alcohol. Of the alcohols listed, 2-butanol satisfies these criteria.

17. (a) CH$_3$CH$_2$CH=CH$_2$ $\xrightarrow{\text{H}_2\text{O, H}_2\text{SO}_4}$ CH$_3$CH$_2$CH(OH)CH$_3$

(b) CH$_3$CH$_2$CH$_3$ + Cl$_2$ $\xrightarrow{h\nu}$ CH$_3$CH$_2$CH$_2$Cl + CH$_3$CHClCH$_3$ + other chlorinated propanes

(c) ⟨benzene ring⟩—COOH + (CH$_3$)$_2$OH $\xrightarrow{\Delta}$ ⟨benzene ring⟩—COOCH(CH$_3$)$_2$

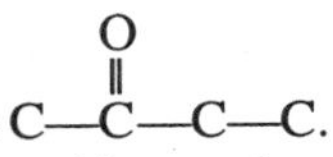

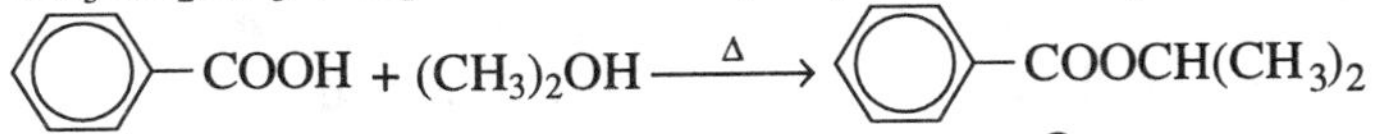

(d) CH$_3$CH(OH)CH$_2$CH$_3$ $\xrightarrow{\text{Cr}_2\text{O}_7^{2-},\ \text{H}^+}$ $\begin{array}{c} O \\ \| \\ CH_3-C-CH_2CH_3 \end{array}$

18. (a) C$_6$H$_6$ would have a higher boiling point that C$_6$H$_{12}$ because its London forces are stronger because C$_6$H$_6$ has many π electrons which interact with each other. There are no dipole-dipole forces or hydrogen bonds between these molecules.

(b) Both C$_3$H$_7$OH and C$_7$H$_{15}$OH are alcohols that can hydrogen bond, but C$_7$H$_{15}$OH has a much longer nonpolar hydrocarbon chain, interfering with its aqueous solubility. C$_3$H$_7$OH is more water soluble.

(c) Functional groups determine acidity of these two otherwise alike molecules. The aldehyde hydrogen is not notably acidic while the carboxylic hydrogen is. C$_6$H$_5$COOH is more acidic in aqueous solution.

EXERCISES

Organic Structures

19. In the following structural formulas, the hydrogen atoms are omitted for simplicity. Remember that there are four bonds to each carbon atom. The missing bonds are C—H bonds.

(a) CH$_3$CH$_2$CHBrCHBrCH$_3$

(c) (C$_2$H$_5$)$_2$CHCH=CHCH$_2$CH$_3$

(b) (CH$_3$)$_3$CCH$_2$C(CH$_3$)$_2$CH$_2$CH$_2$CH$_3$

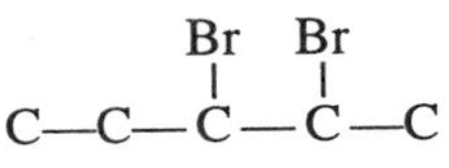

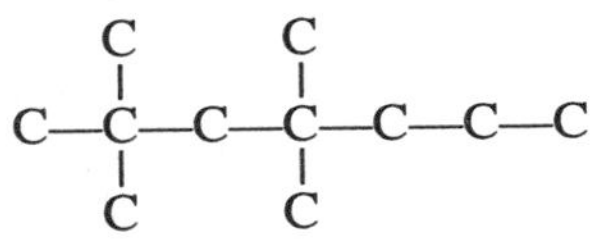

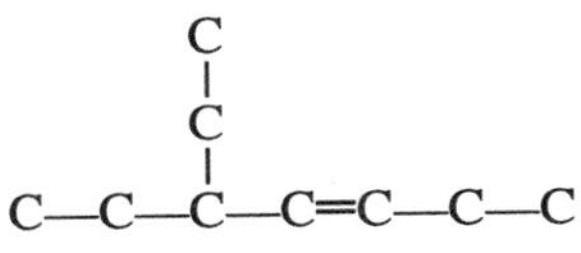

20. In the following structural formulas, the hydrogen atoms are omitted for simplicity. Remember that there are four bonds to each carbon atom. The missing bonds are C—H bonds.

(a) $(CH_3)_3CCH_2CH(CH_3)CH_2CH_2CH_3$ **(b)** $(CH_3)_2CHCH_2C(CH_3)_2CH_2Br$ **(c)** $Cl_3CCH_2CH(CH_3)CH_2Cl$

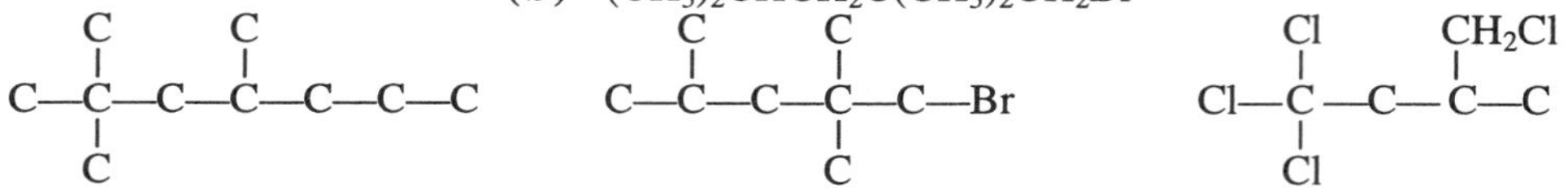

21. (a) Each carbon atom is sp^3 hybridized. All of the C—H bonds in the structure (drawn below) are sigma bonds, between the $1s$ orbital of H and the sp^3 orbital of C. The C—C bond is between sp^3 orbitals on each C atom.

(b) Each carbon atom is sp^2 hybridized. All of the C—H bonds in the structure (drawn below) are sigma bonds between the $1s$ orbital of H and the sp^2 orbital of C. The C—Cl bond is between the sp^2 orbital on C and the $3p$ orbital on Cl. The C=C double bond is composed of a sigma bond between the sp^2 orbitals on each C atom and a pi bond between the $2p_z$ orbitals on those C atoms.

(c) The left-most C atom (in the structure drawn below) is sp^3 hybridized, and the C—H bonds to that C atom are between the sp^3 orbitals on C and the $1s$ orbital on H. The other two C atoms are sp hybridized. The right-hand C—H bond is between the sp orbital on C and the $1s$ orbital on H. The C≡C triple bond is composed of one sigma bond formed by overlap of sp orbitals, one from each C atom, and two pi bonds, each formed by the overlap of two $2p$ orbitals, one from each C atom (that is a $2p_y$—$2p_y$ overlap and a $2p_z$—$2p_z$ overlap).

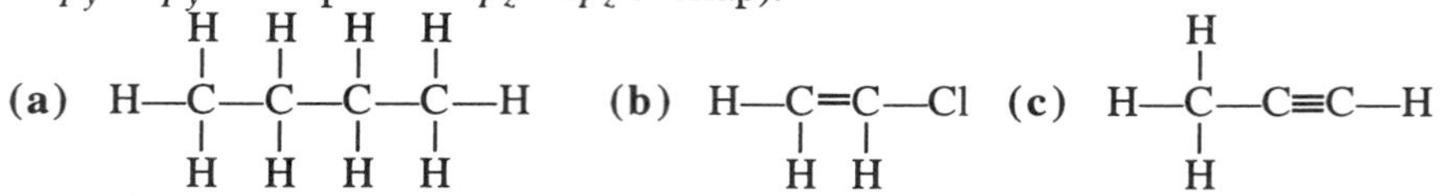

22. (a) The left- and right-most C atoms in the structure (drawn below) are sp^3 hybridized. All C—H bonds are sigma bonds formed by the overlap of an sp^3 orbital on C with a $1s$ orbital on H. The central C atom is sp^2 hybridized; both C—C bonds are sigma bonds, formed by the overlap between the sp^3 orbital on the terminal C atom and an sp^2 orbital on the central C atom. The C=O double bond is composed of a σ bond between the sp^2 orbital on the central C atom and a $2p_y$ orbital on the O atom, and a π bond between the $2p_z$ orbital on the central C atom and the $2p_z$ orbital on the O atom.

(b) The left C atom in the structure (drawn below) is sp^3 hybridized. All C—H bonds are sigma bonds formed by the overlap of an sp^3 orbital on C with a $1s$ orbital on H. The central C atom is sp^2 hybridized; both C—C bonds are sigma bonds, formed by the overlap between the sp^3 orbital on the terminal C atom and an sp^2 orbital on the central C atom. The C=O double bond is composed of a σ bond between the sp^2 orbital on the central C atom and a $2p_y$ orbital on the O atom, and a π bond between the $2p_z$ orbital on the central C atom and the $2p_z$ orbital on the O atom. The right O atom is sp^3 hybridized; the C—O sigma bond forms by the overlap of C(sp^2) with O(sp^3) and O(sp^3) with H($1s$).

(c) The two end C's are sp^2 hybridized; all C—H bonds form by the overlap of C(sp^2) with H($1s$). The central C is sp hybridized. Both C=C bonds consist of a σ bond formed by $C_{central}(sp)$ with $C_{end}(sp^2)$ overlap and a π bond formed by $C_{central}(2p)$ with $C_{end}(2p)$ overlap.

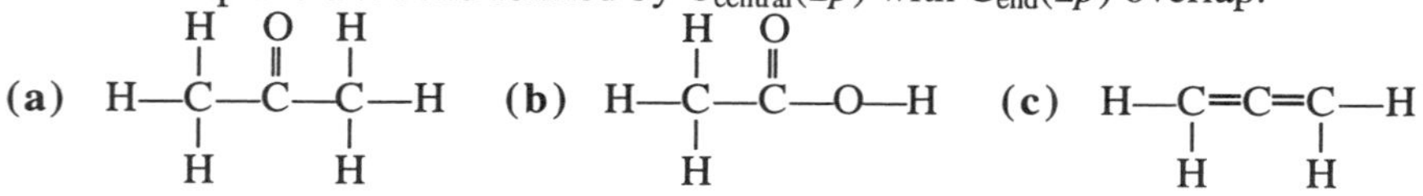

Isomers

23. Structural (skeletal) isomers differ from each other in the length of their carbon atom chains and in the length of the side chains. The carbon skeleton differs between these isomers. Positional isomers differ in the location or position where functional groups are attached to the carbon skeleton. Geometric isomers differ in whether two substituents are on the same side of the molecule or on opposite sides of the molecule from each other; usually they are on opposite sides or the same side of a double bond.

(a) The chloro group is attached to the terminal carbon in $CH_3CH_2CH_2Cl$ and to the central carbon atom in $CH_3CHClCH_3$. These are positional isomers.

(b) $CH_3CH(CH_3)CH_2CH_3$, methylbutane, and $CH_3(CH_2)_3CH_3$, pentane, are structural isomers.
(c) Ortho-aminotoluene and meta-aminotoluene differ in the position of the —NH_2 group on the benzene ring. They are positional isomers.

24. (a) The two chloro groups in CHCl═CHCl are on different carbon atoms. In CH_2═CCl_2 they are on the same atoms. These are positional isomers.
(b) CH_3CH═$CHCH_3$, 2-butene, and CH_2═$C(CH_3)_2$, methylpropene, are structural isomers. They might be termed positional isomers, differing as they do in the position of a CH_3 group. But since that group's location changes the structure of the molecule, they are structural isomers.
(c) These two compounds differ in the location of the two —COOH groups; the left-hand one has the two groups on the same side of the double bond (the *cis* isomer) and the right-hand one has the two groups on opposite sides of the double bond (the *trans* isomer). These are geometric isomers.

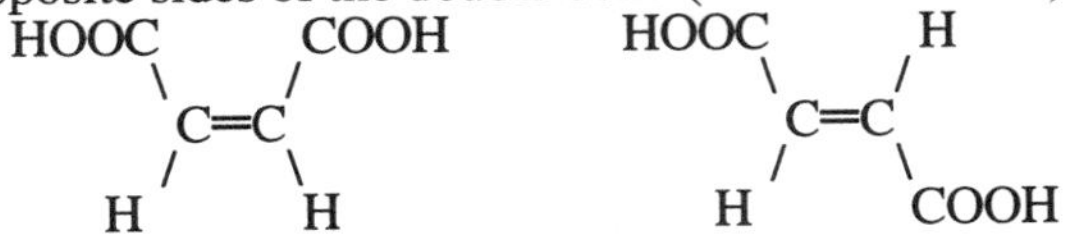

25. We show only the carbon atom skeleton in each case. Remember that there are four bonds to carbon. The bonds that are not indicated in these structures are C—H bonds.

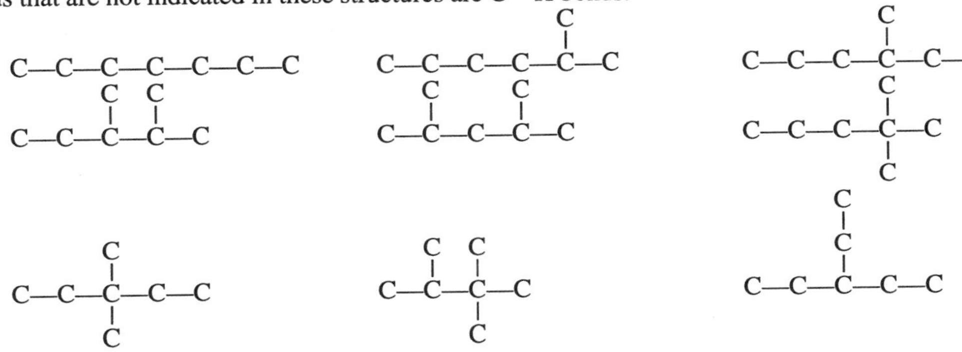

26. In each case, we draw only the carbon skeleton. It is understood that there are four bonds to each carbon atom. The bonds that are not shown are C—H bonds.

(a)
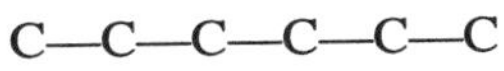
n-hexane

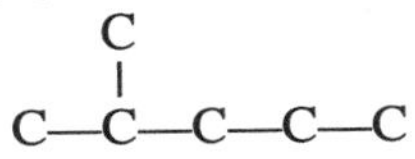
2-methylpentane

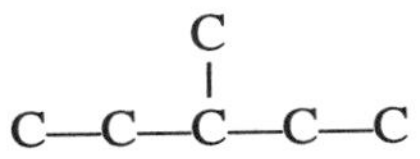
3-methylpentane

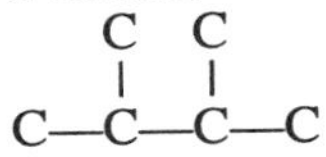
2,3-dimethylbutane

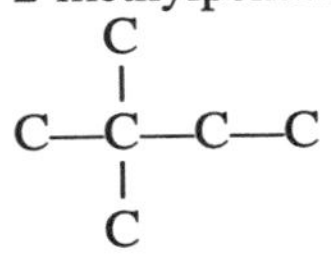
2,2-dimethylbutane

(b)
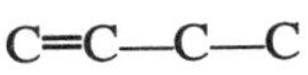
1-butene

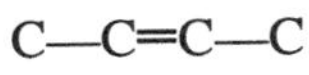
2-butene

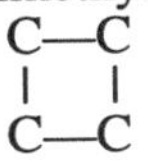
cyclobutane

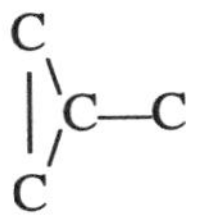
methylcyclopropane

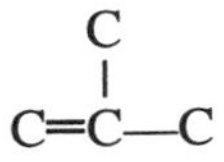
2-methyl-1-propene

(c)
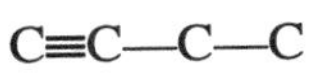
1-butyne

2-butyne

cyclobutene

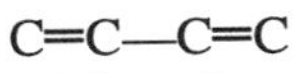
1,3-butadiene

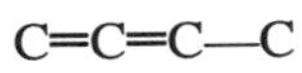
1,2-butadiene

Functional Groups

27. (a) A carbonyl group is $-\overset{\overset{\displaystyle O}{\|}}{C}-$, while a carboxyl group is $-\overset{\overset{\displaystyle O}{\|}}{C}-OH$. The essential difference between them is the hydroxyl group, —OH.

(**b**) An aldehyde has a carbonyl group at the end of a molecule, $R-\overset{\overset{O}{\|}}{C}-H$. In a ketone, $R-\overset{\overset{O}{\|}}{C}-R'$, the carbonyl group is in the center of the molecule.

(**c**) An acid (that is, carboxylic acid) group is $-\overset{\overset{O}{\|}}{C}-OH$, while an acyl group is $R-\overset{\overset{O}{\|}}{C}-$. The essential difference is the presence of the alkyl group, $-R$, and the removal of the $-OH$ group.

28. (**a**) aromatic nitro compound $\quad$ $C_6H_5NO_2$ $\quad$ nitrobenzene
(**b**) aliphatic amine $\quad$ $C_2H_5NH_2$ $\quad$ ethylamine
(**c**) chlorophenol $\quad$ p-ClC_6H_4OH $\quad$ *para*-chlorophenol $\quad$ (drawn below)
(**d**) aliphatic diol $\quad$ $HOCH_2CH_2OH$ $\quad$ 1,2-ethanediol
(**e**) unsaturated aliphatic alcohol $\quad$ $CH_2{=}CH(CH_2)_2OH$ $\quad$ 3-butene-1-ol
(**f**) alicyclic ketone $\quad$ $C_6H_{11}O$ $\quad$ cyclohexanone $\quad$ (drawn below)
(**g**) halogenated alkane $\quad$ $CH_3CHICH_2CH_3$ $\quad$ 2-iodobutane
(**h**) aromatic dicarboxylic acid $\quad$ o-$C_6H_4(COOH)_2$ $\quad$ *ortho*-phthalic acid $\quad$ (drawn below)
(**c**) (**f**) (**h**)

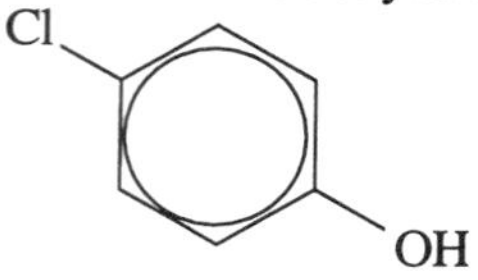

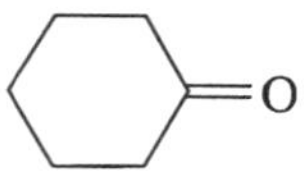

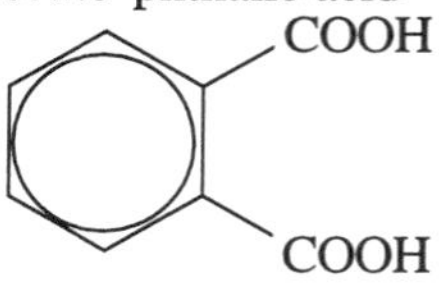

Nomenclature and Formulas

29. (**a**) The longest carbon chain has four carbon atoms and there are 2 methyl groups attached to carbon 2. This is 2,2-dimethylbutane.
(**b**) The longest chain is three carbons long, there is a double bond between carbons 1 and 2, and a methyl group attached to carbon 2. This is 2-methylpropene.
(**c**) Two methyl groups are attached to a three-carbon ring. This is 1,2-dimethylcyclopropane.
(**d**) The longest chain is 5 carbons long, there is a triple bond between carbons 2 and 3 and a methyl group attached to carbon 4. This is 4-methyl-2-pentyne.
(**e**) The longest chain is 6 carbons long. There are two methyl groups attached to carbons 3 and 4. This compound is 3,4-dimethylhexane.
(**f**) The longest carbon chain that contains the double bond is 5 carbons long. The double bond is between carbons 1 and 2. There is a propyl group attached to carbon 2, and two methyl groups attached to carbons 3 and 4. This is 3,4-dimethyl-2-propyl-1-pentene.

30. Again we do not show the hydrogen atoms in the structures below. But we realize that there are four bonds to each C atom and the missing bonds are C—H bonds.

(**a**) isopentane $\quad$ $C-\overset{\overset{C}{|}}{C}-C-C$
(**b**) cyclohexene
(**c**) 2-methyl-3-hexyne $\quad$ $C-\overset{\overset{C}{|}}{C}-C{\equiv}C-C-C$
(**d**) 2-butanol $\quad$ $C-\overset{\overset{OH}{|}}{C}-C-C-C$
(**e**) isopropyl ethyl ether $\quad$ $C-\overset{\overset{C}{|}}{C}-O-C-C$
(**f**) propionaldehyde $\quad$ $C-C-\overset{\overset{O}{\|}}{C}-H$

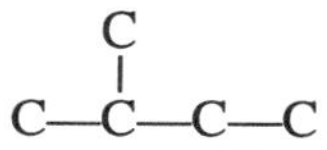

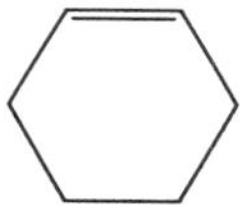

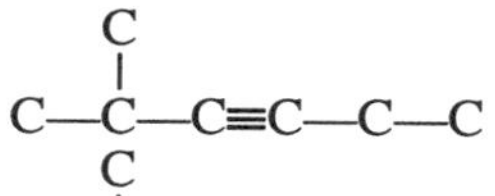

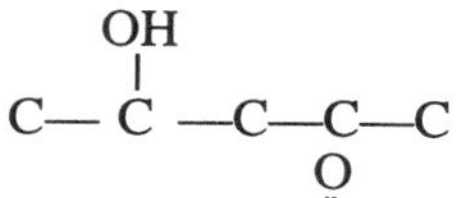

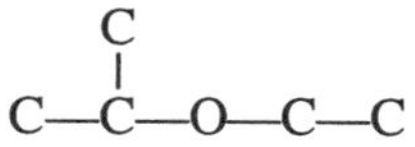

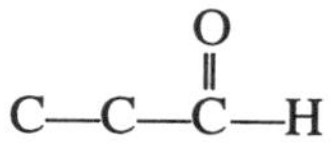

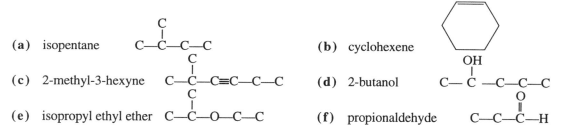

31. (**a**) The name pentene is insufficient. 1-pentene is $CH_2{=}CHCH_2CH_2CH_3$ and 2-pentene is $CH_3CH{=}CHCH_2CH_3$.
(**b**) The name butanone is sufficiently precise. There is only one four-carbon ketone.
(**c**) The name butyl alcohol is insufficiently precise. There are numerous butanols. 1-butanol is $HOCH_2CH_2CH_2CH_3$, 2-butanol is $CH_3CH(OH)CH_2CH_3$, isobutanol is $HOCH_2CH(CH_3)_2$, and *t*-butanol is $(CH_3)_3COH$.
(**d**) The name methylaniline is not sufficiently precise. It specifies $CH_3-C_6H_4-NH_2$, with no relative locations of the $-NH_2$ and $-CH_3$ substituents.
(**e**) The name methylcyclopentane is sufficiently precise. It does not matter where on a five-carbon ring with only C-C single bonds a methyl group is placed.

(f) Dibromobenzene is not sufficient. It specifies Br—C_6H_4—Br, with no indication of the relative locations of the two bromo groups on the ring.

32. **(a)** 3-pentene specifies $CH_3CH_2CH{=}CHCH_3$. The proper name for this compound is 2-pentene. The number for the functional group should be as small as possible.

(b) Pentadiene is insufficiently precise. Possible pentadienes are 1,2 pentadiene, $CH_2{=}C{=}CHCH_2CH_3$, 1,3-pentadiene, $CH_2{=}CHCH{=}CHCH_3$, and 2,3 pentadiene, $CH_3CH{=}C{=}CHCH_3$.

(c) 1-propanone is incorrect. There cannot be a ketone on the first carbon of a chain. The compound specified is either propanaldehyde, CH_3CH_2CHO, or propanone (2-propanone), CH_3COCH_3 (acetone).

(d) Bromopropane is insufficiently precise. It could be either 1-bromopropane, $BrCH_2CH_2CH_3$, or 2-bromopropane, $CH_3CHBrCH_3$.

(e) Although 2,6-dichlorobenzene conveys enough information, the proper name for the compound is meta-dichlorobenzene, or 1,3-dichlorobenzene. Substituent numbers should be as small as possible.

(f) 2-methyl-3-pentyne is $(CH_3)_2CHC{\equiv}CCH_3$. The proper name for this compound is 4-methyl-2-pentyne. The number of the triple bond should be as small as possible.

33. **(a)** 2,4,6-trinitrotoluene

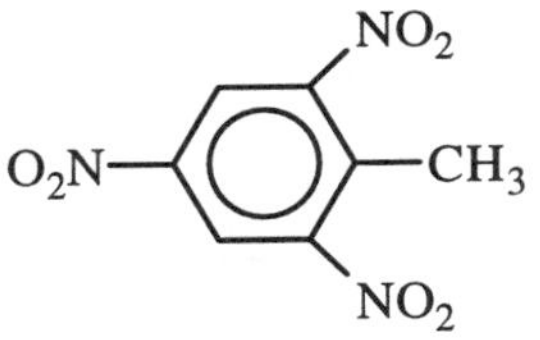

(b) methylsalicylate

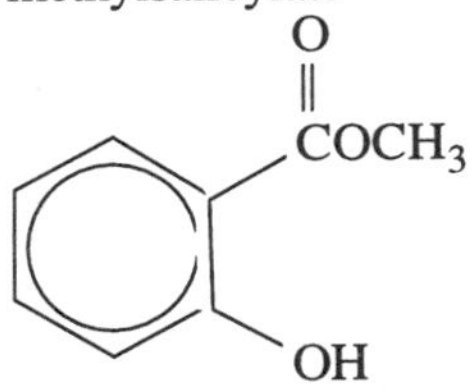

(c) 2-hydroxy-1,2,3-propanetricarboxylic acid $HOOCCH_2C(OH)(COOH)CH_2COOH$

34. **(a)** *o*-*t*-butylphenol

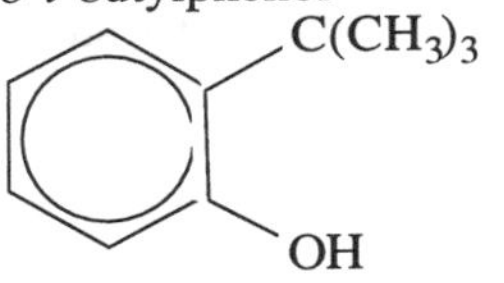

(b) 1-phenyl-2-aminopropane
$C_6H_5CH_2CH(NH_2)CH_3$

(c) 2-methylheptadecane
$(CH_3)_2CH(CH_2)_{14}CH_3$

Alkanes

35. The general formula of an alkane is C_nH_{2n+2}. Thus an alkane with a molecular mass of 72 u has the molecular formula C_5H_{12}.

(a) If the compound forms four different monochlorination products, there must be four different kinds of carbons in the molecule, each with H atoms attached. This compound is 2-methylbutane, on the left.

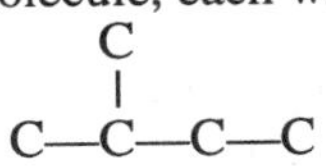

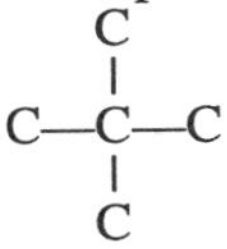

(b) If the compound forms but one monochlorination product, every carbon atom in the molecule with H atoms attached must be the same. This compound is 2,2-dimethylpropane, shown at right.

36. **(a)** An alkane with a molecular mass of 44 u must be a propane, since 3 carbons have a molecular mass of 36 u. There is only one propane, *n*-propane, and it has the condensed formula $CH_3CH_2CH_3$. Its two monochlorination products are $CH_3CH_2CH_2Cl$ and $CH_3CHClCH_3$

(b) An alkane with a molecular mass of 58 u cannot be a pentane, since five carbons have a mass of 60 u, so it must be a butane. Both normal butane and isobutane have only two monobromination products.

From *n*-butane, $CH_3CH_2CH_2CH_3$ $CH_3CH_2CH_2CH_2Br$ and $CH_3CH_2CHBrCH_3$
From iso-butane, $CH_3CH(CH_3)_2$ $BrCH_2CH(CH_3)_2$ and $CH_3CBr(CH_3)_2$

Alkenes

37. In the case of ethene there are only two carbon atoms between which there can be a double bond. Thus, specifying the compound as 1-ethene is unnecessary. In the case of propene, there can be a double bond only between the central carbon atom and a terminal carbon atom. Thus here also, specifying the compound as 1-propene is unnecessary. The case of butene is different, however, since 1-butene, $CH_2{=}CHCH_2CH_3$, is distinct from 2-butene, $CH_3CH{=}CHCH_3$.

38. In an alkene there is a C=C double bond. On the other hand, in an cyclic alkane there are no double bonds, but rather a chain of carbon atoms joined at the ends into a ring.

39. These reactions either saturate or create double bonds.

(a) $CH_2{=}CHCH_3 + H_2 \xrightarrow{\text{Pt, heat}} CH_3CH_2CH_3$

(b) $CH_3CHOHCH_2CH_3 \xrightarrow{H_2SO_4,\ \text{heat}} CH_3CH{=}CHCH_3$

40. (a) $CH_3\overset{\overset{\displaystyle Cl}{|}}{C}{=}CH_2 + HCl \longrightarrow CH_3CCl_2CH_3$

(b) $CH_3C{\equiv}CH + HCN \longrightarrow CH_3C(CN){=}CH_2$

(c) $CH_3CH{=}C(CH_3)_2 + HCl \longrightarrow CH_3CH_2CCl(CH_3)_2$

Aromatic Compounds

41. (a) phenylacetylene

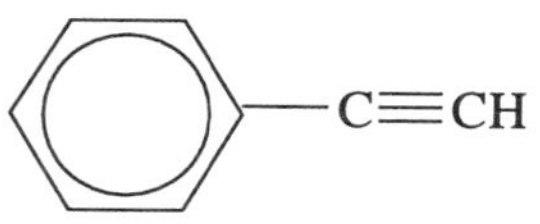

(b) meta-dichlorobenzene

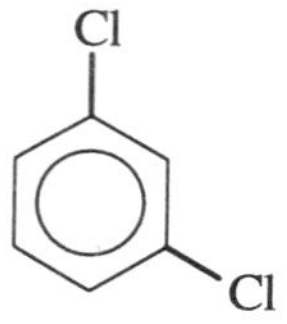

(c) 3,5-dihydroxyphenol

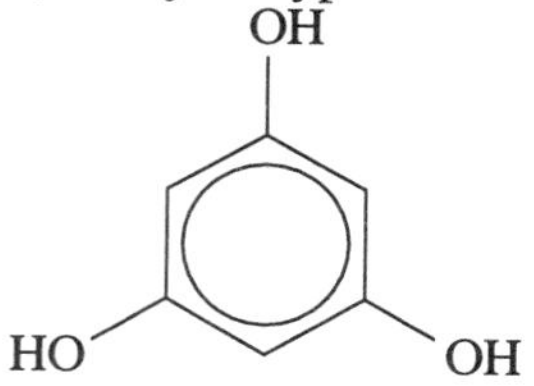

42. (a) *p*-phenylphenol

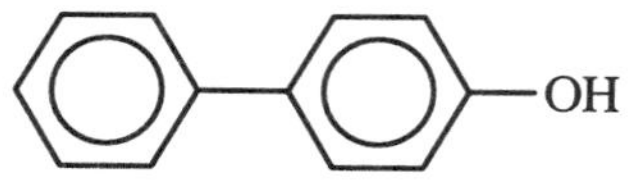

(b) 2-hydroxy-4-isopropyltoluene

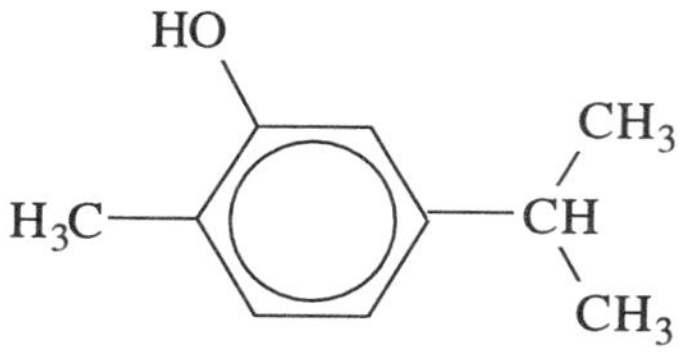

(c) meta-dinitrobenzene

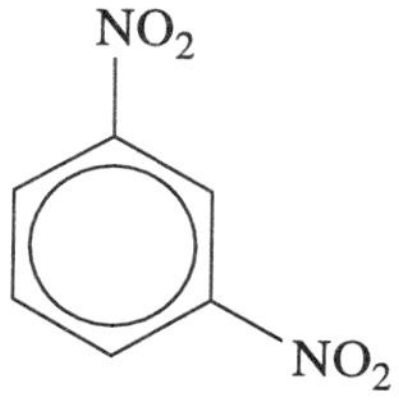

Organic Reactions

43. (a) In an aliphatic substitution reaction, an atom (usually H) of an alkane molecule is replaced by another atom or group of atoms. An example is equation (27.1).

(b) In an aromatic substitution reaction, an atom (usually H) of a phenyl group is replaced by another atom or group of atoms. Examples are in equations (27.7) and (27.8).

(c) In an addition reaction, the result is that a small molecule breaks into two parts and the fragments add, one to each of the carbon atoms of a double bond. An example is equation (27.9).

(d) In an elimination reaction, two groups from within the same molecule join to form a small molecule—such as H_2 or H_2O— and are eliminated from the larger molecule. Often a multiple bond results as well. An example is equation (27.12).

44. (a) $HC{\equiv}CH \xrightarrow{Br_2} CHBr{=}CHBr \xrightarrow{Br_2} CHBr_2CHBr_2$

(b) $HC{\equiv}CH \xrightarrow{\text{Pt, } H_2} CH_2{=}CH_2 \xrightarrow{H_2O,\ H_2SO_4} CH_3CH_2OH \xrightarrow{Cr_2O_7^{2-},\ H^+} CH_3CHO$

45. (a) $CH_3CH_2CH_2CH_2CH_2OH \xrightarrow{Cr_2O_7^{2-},\ H^+} CH_3CH_2CH_2CH_2COOH$

(b) $CH_3CH_2CH_2COOH + HOCH_2CH_3 \xrightarrow{H^+} CH_3CH_2CH_2\overset{\overset{\displaystyle O}{\|}}{C}OCH_2CH_3 + H_2O$

(c) $CH_3CH_2C(CH_3){=}CH_2 + H_2O \xrightarrow{H_2SO_4} CH_3CH_2C(CH_3)(OH){-}CH_3$ 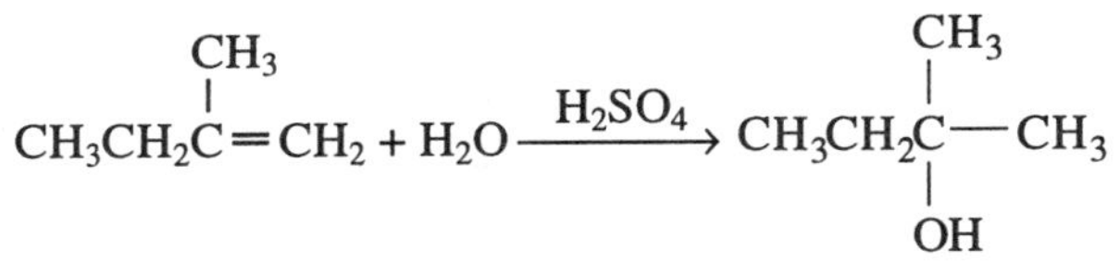

46. **(a)** Nitro groups are meta directors. Thus the only product formed is 5-bromo-1,3-dinitrobenzene, drawn below.

(b) Amine is an ortho, para director. There are two products formed: *ortho*-bromoaniline and *para*-bromoaniline, both of which are drawn below.

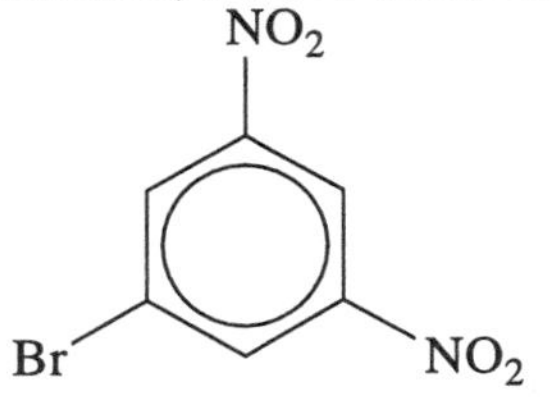

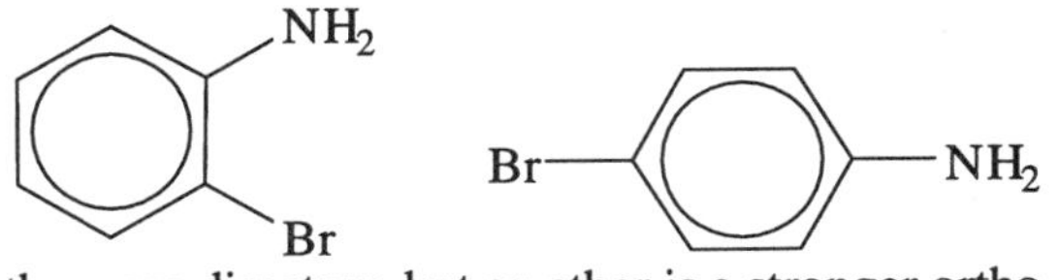

(c) Both an ether and a bromo group are ortho, para directors, but an ether is a stronger ortho, para director. The principal product we expect is 2,4-dibromoanisole, drawn below.

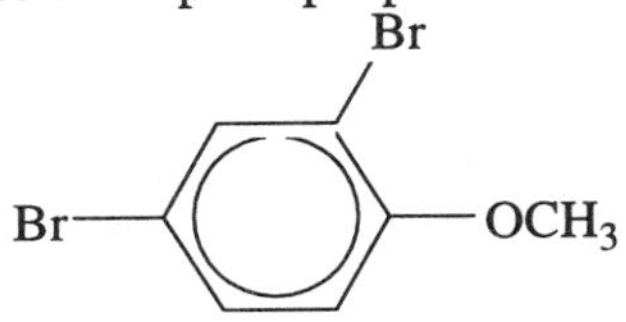

Polymerization Reactions

47. In a simple molecular substance like benzene, all molecules are identical (C_6H_6). No matter how many of these molecules are present in one sample, any other sample with the same number of molecules has the same mass. The mass in grams of one mole of molecules—the molar mass—is a unique quantity. In a polymer the situation is quite different. The number of monomer units in a polymer chain is not a constant number but is widely variable. Thus individual polymer molecules differ very much in mass. Thus, a mass of a mole of their molecules also is quite variable. However, when we take a sample for analysis, we obtain many molecules of each chain length or molecular mass. The resulting determination of molar mass obtains the average mass of all of these different sized polymer molecules.

48. An ester linkage is formed by the condensation of a carboxylic acid and an alcohol. Accompanying this condensation is the elimination of a water molecule between the ester and the carboxylic acid. Dacron is formed by the condensation of a *di*carboxylic acid with a *diol*. Thus, Dacron contains ester linkages. Because there are a large number of these ester linkages in Dacron—joining many subunits together— it is appropriate to call the polymer a polyester.

To determine the percent of oxygen in Dacron, we refer to the basic unit of the polymer, which is shown in Table 27-5. This unit has the formula $C_{10}H_8O_4$. $\%O = \frac{(4 \times 16.00)\text{ g O}}{192.2\text{ g polymer}} \times 100\% = 33.30\%O$

49. The polymerization of 1,6-hexanediamine with sebacyl chloride proceeds in the following manner. The italicized *H* and *Cl* atoms are removed in this condensation reaction.

$$H{-}NH{-}(CH_2)_6{-}NH{-}\mathit{H} + \mathit{Cl}{-}C(=O){-}(CH_2)_8{-}C(=O){-}\mathit{Cl} + \mathit{H}{-}NH{-}(CH_2)_6{-}NH{-}\mathit{H} + \mathit{Cl}{-}C(=O){-}(CH_2)_8{-}C(=O){-}Cl$$
$$\longrightarrow \left({-}C(=O){-}(CH_2)_8{-}C(=O){-}NH{-}(CH_2)_6{-}NH{-} \right)_x$$

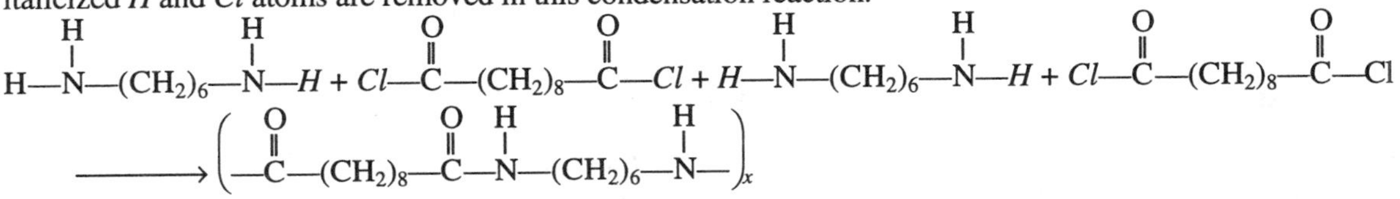

50. In order to form long-chain molecules, every monomer must have at least two functional groups, one on each end of the molecule. Ethyl alcohol has only one functional group, a hydroxy group (—OH). It cannot participate in a polymerization reaction with dimethyl terephthalate. But glycerol, with three hydroxy functional groups, can participate in this polymerization reaction.

28 CHEMISTRY OF THE LIVING STATE

PRACTICE EXAMPLES

1A $CH_2OOC(CH_2)_{10}CH_3$ lauro
|
$CHOOC(CH_2)_{10}CH_3$ lauro
|
$CH_2OOC(CH_2)_{10}CH_3$ laurate

Lauric acid is $C_{11}H_{23}COOH$.

This is glyceryl trilaurate, or trilaurin.

1B Lauric acid is $C_{11}H_{23}COOH$, myristic acid is $C_{13}H_{27}COOH$, and linoleic acid is $C_{17}H_{31}COOH$. The first two are saturated fatty acids; the last is doubly unsaturated, at carbons 9 and 12. The structural formula is:

$CH_2OOC(CH_2)_{10}CH_3$ lauro
|
$CHOOC(CH_2)_{12}CH_3$ myristo
|
$CH_2OOC(CH_2)_7CH{=}CHCH_2CH{=}CH(CH_2)_4CH_3$ linoleate

2A
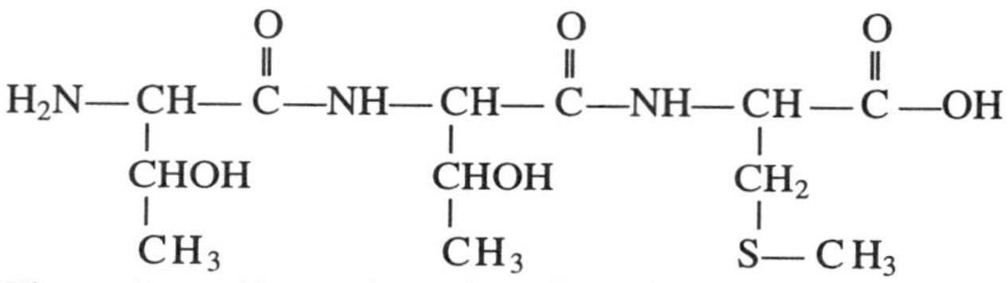

O O O
H2N—CH—C—NH—CH—C—NH—CH—C—OH
CHOH CHOH CH2
CH3 CH3 S—CH3

The amino acids are threonine, threonine, and methionine. This tripeptide is dithreonylmethionine.

2B The amino acids are: serine, glycine, and valine. The N terminus is first.

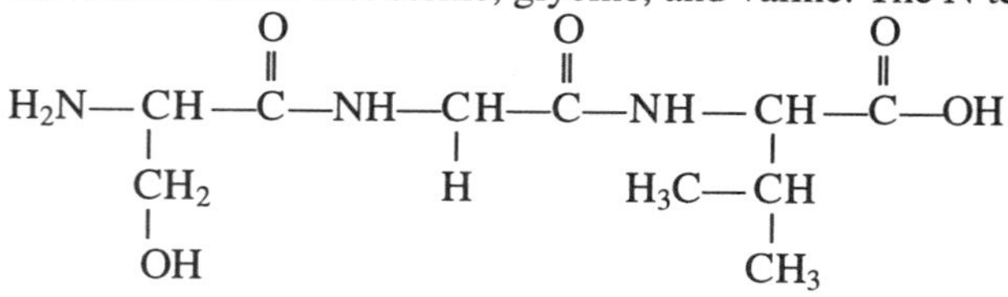

O O O
H2N—CH—C—NH—CH—C—NH—CH—C—OH
CH2 H H3C—CH
OH CH3

A. Because it is a pentapeptide and five amino acids have been identified, no amino acid is repeated. The sequences fall into place, as follows.

		Gly	Cys			second fragment
		Cys	Val	Phe		third fragment
			Val	Phe		first fragment
				Phe	Tyr	fourth fragment
pentapeptide sequence	Gly	Cys	Val	Phe	Tyr	

B. Because it is a hexapeptide and there are five distinct amino acids, one amino acid must appear twice. The fragmentation pattern indicates that the doubled amino acid is glycine. The sequences fall into place if we begin with the N-terminal end.

	Ser	Gly	Gly				third fragment
		Gly	Gly	Ala			second fragment
				Ala	Val	Trp	fourth fragment
					Val	Trp	first fragment
hexapeptide sequence	Ser	Gly	Gly	Ala	Val	Trp	

SUMMARIZING EXAMPLE CALCULATIONS

1. The formula of triolein is $CH_2CHCH_2(OOC(CH_2)_7CH{=}CH(CH_2)_7CH_3)_3$.

2. For simplicity in determining the molar mass, we group like atoms together in the formula: $C_3H_5(O_2C_{18}H_{33})_3$ or $C_{57}H_{104}O_6$. Molar mass = 57 × 12.011 g C + 104 × 1.00794 g H + 6 × 15.9994 g O = 684.63 g C + 104.826 g H + 95.9964 g O = 885.45 g triolein/mol.

3. Since there are three C=C bonds in triolein, the H_2 molecules react with each triolein molecule and three moles of H_2 react with each mole of triolein.

4. $$\text{amount } H_2(g) = 15.5 \text{ kg triolein} \times \frac{1000 \text{ g triolein}}{1 \text{ kg triolein}} \times \frac{1 \text{ mol triolein}}{885.45 \text{ g triolein}} \times \frac{3 \text{ mol } H_2(g)}{1 \text{ mol triolein}}$$
$$= 52.5 \text{ mol } H_2(g)$$

5. $$\text{volume } H_2 = \frac{52.5 \text{ mol } H_2(g) \times \dfrac{0.08206 \text{ L atm}}{\text{mol K}} \times (25.5 + 273.15) \text{ K}}{756 \text{ mmHg} \times \dfrac{1 \text{ atm}}{760 \text{ mmHg}}} = 1.29 \times 10^3 \text{ L } H_2$$

REVIEW QUESTIONS

1. **(a)** (+) is another way of designating a dextrorotatory compound.
(b) L indicates that, in the Fisher projection of the compound, the —OH group on the penultimate carbon atom is to the left and the —H group is to the right.
(c) A sugar is a carbohydrate that is either a monosaccharide or an oligosaccharide.
(d) An α-amino acid is a carboxylic acid that has an amine group on the carbon next to the —COOH group, that is, on the α carbon atom. Glycine (H_2NCH_2COOH) is the simplest α-amino acid.
(e) The pH at which the zwitterion form of an amino acid predominates in solution is known as the isoelectric point. The isoelectric point of glycine is $pI = 5.97$.

2. **(a)** Saponification is the hydrolysis of a glyceride in alkaline solution to produce glycerine and soaps: salts of fatty acids.
(b) A chiral carbon atom is one that exhibits optical isomerism, usually one to which four different groups are attached.
(c) A racemic mixture is one composed of equal amounts of an optically active compound and its enantiomer. Since these two compounds rotate polarized light by the same amount but in opposite directions, such a mixture does not rotate the plane of polarized light.
(d) The denaturation of a protein is the process in which the protein is treated with somewhat harsh conditions and loses at least some of its secondary or tertiary structure, temporarily or permanently, leading to a loss of biological activity.

3. **(a)** A fat is a triglyceride in which the fatty acid chains are saturated hydrocarbon chains. Fats are solids at room temperature. An oil is a triglyceride in which the fatty acid chains are to some degree unsaturated. Oils are liquids at room temperature.
(b) Enantiomers are optically active isomers of a compound that are mirror images of each other. Diastereomers are optically active isomers that are not mirror images.
(c) The primary structure of a protein refers to the sequence of amino acids. The secondary structure describes the shape of that polypeptide chain.
(d) ADP is adenosine diphosphate, ATP is adenosine triphosphate. They differ by a phosphate group.
(e) DNA is deoxyribonucleic acid, RNA is ribonucleic acid. They differ by an O atom. DNA is found in the cell nucleus, RNA in the cytoplasm.

<u>4</u>. **(a)** $C_{15}H_{31}COOH$ is palmitic acid. $C_{17}H_{29}COOH$ is linolenic acid or eleosteric acid. $C_{11}H_{23}COOH$ is lauric acid. Thus, the given compound is glyceryl palmitolinolenolaurate or glyceryl palmitoeleosterolaurate.
(b) $C_{17}H_{33}COOH$ is oleic acid. Thus, the compound is glyceryl trioleate or triolein.
(c) $C_{13}H_{27}COOH$ is myristicic acid. Thus, the compound is sodium myristate.

5. (a) glyceryl palmitolauroeleosterate

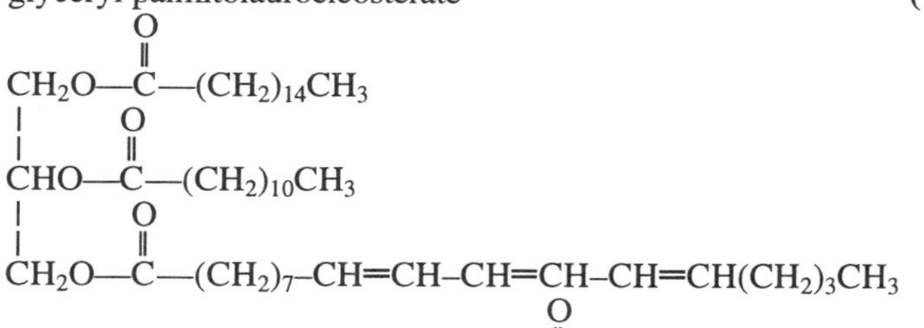

$$CH_2O-\overset{\overset{\displaystyle O}{\|}}{C}-(CH_2)_{14}CH_3$$
$$CHO-\overset{\overset{\displaystyle O}{\|}}{C}-(CH_2)_{10}CH_3$$
$$CH_2O-\overset{\overset{\displaystyle O}{\|}}{C}-(CH_2)_7-CH=CH-CH=CH-CH=CH(CH_2)_3CH_3$$

(b) tripalmitin

$$CH_2O-\overset{\overset{\displaystyle O}{\|}}{C}-(CH_2)_{14}CH_3$$
$$CHO-\overset{\overset{\displaystyle O}{\|}}{C}-(CH_2)_{14}CH_3$$
$$CH_2O-\overset{\overset{\displaystyle O}{\|}}{C}-(CH_2)_{14}CH_3$$

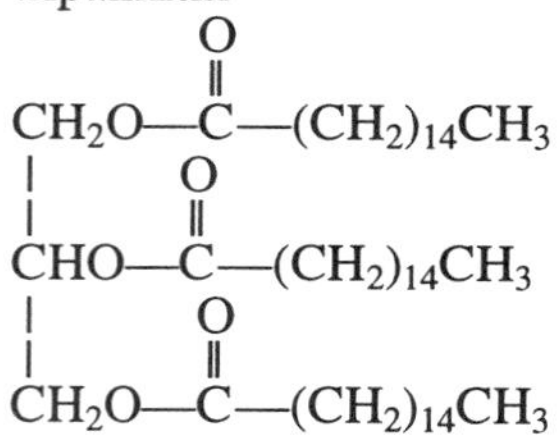

(c) potassium myristate $CH_3(CH_2)_{12}-\overset{\overset{\displaystyle O}{\|}}{C}-O^- K^+$

(d) butyl oleate $CH_3(CH_2)_3-O-\overset{\overset{\displaystyle O}{\|}}{C}-(CH_2)_7-CH=CH-(CH_2)_7-CH_3$

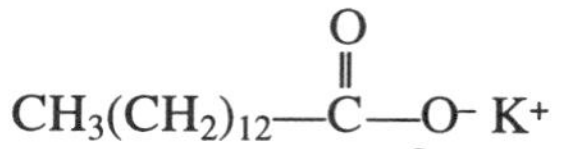

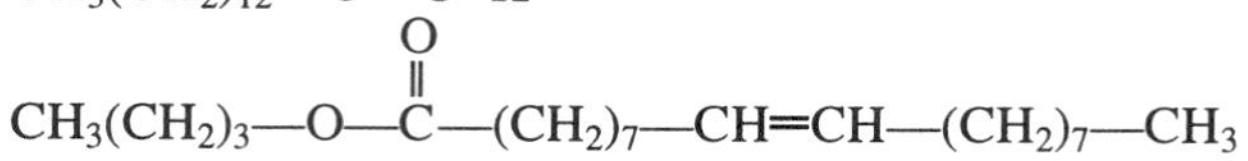

6. A DL mixture contains equal amounts of both enantiomers. This is also known as a racemic mixture. The optical activities of the two enantiomers cancel each other out, and the mixture rotates the plane of polarized light neither to the right nor to the left.

7. (a) D-(–)-arabinose is the optical isomer of L-(+)-arabinose. Its structure follows, at left.
(b) A diastereomer of L-(+)-arabinose is a molecule that is its optical isomer, but not its mirror image. There are several such diastereomers, one of which follows, at right.

(a)
$$H-C=O$$
$$HO-C-H$$
$$H-C-OH$$
$$H-C-OH$$
$$CH_2OH$$

(b)
$$H-C=O$$
$$HO-C-H$$
$$HO-C-H$$
$$H-C-OH$$
$$CH_2OH$$

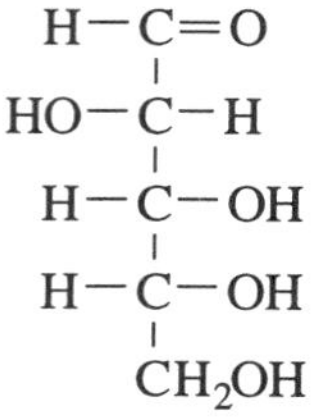

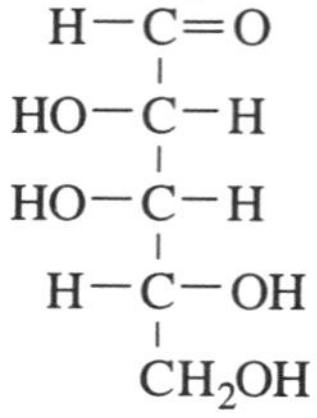

8. The pI of phenylalanine is 5.48. Thus, phenylalanine is in the form of a cation in 1.0 M HCl (pH = 0.0), in the form of an anion in 1.0 M NaOH (pH = 14.0), and in the form of a zwitterion at pH = 5.7. These three structures follow.

(a) $\bigcirc-CH_2-\underset{\underset{\displaystyle NH_3^+ Cl^-}{|}}{CH}\ COOH$

(b) $\bigcirc-CH_2-\underset{\underset{\displaystyle NH_2}{|}}{CH}\ COO^- Na^+$

(c) $\bigcirc-CH_2-\underset{\underset{\displaystyle NH_3^+}{|}}{CH}\ COO^-$

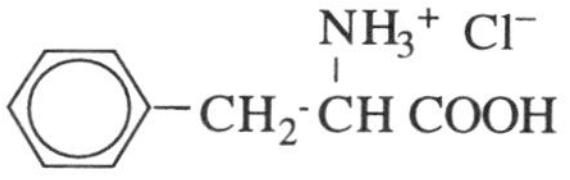

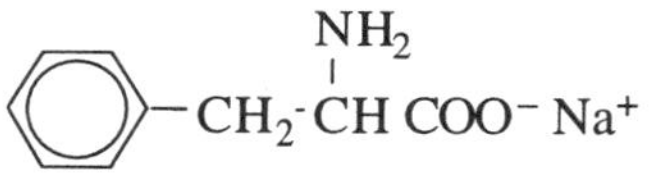

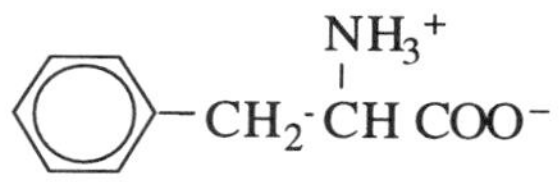

9. (a) alanylcysteine

$$H_2N-\underset{\underset{\displaystyle CH_3}{|}}{CH}-\overset{\overset{\displaystyle O}{\|}}{C}-NH-\underset{\underset{\underset{\displaystyle SH}{|}}{CH_2}}{CH}-\overset{\overset{\displaystyle O}{\|}}{C}-OH$$

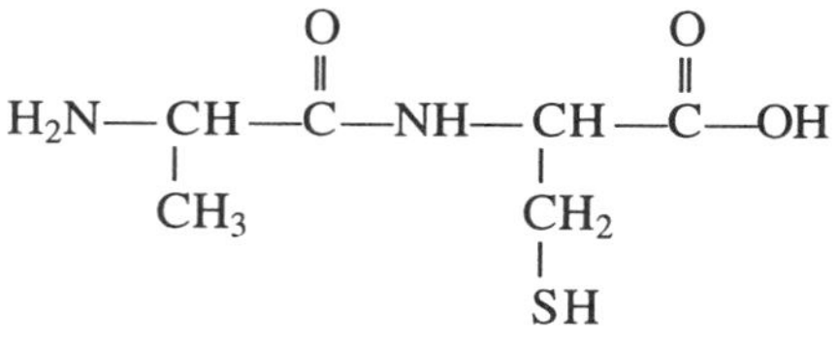

(b) threonylvalylglycine

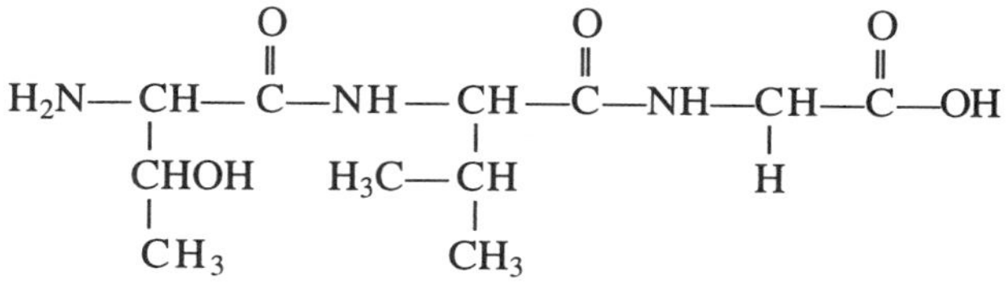

$$H_2N-\underset{\underset{\underset{\displaystyle CH_3}{|}}{CHOH}}{CH}-\overset{\overset{\displaystyle O}{\|}}{C}-NH-\underset{\underset{\underset{\displaystyle CH_3}{|}}{H_3C-CH}}{CH}-\overset{\overset{\displaystyle O}{\|}}{C}-NH-\underset{\underset{\displaystyle H}{|}}{CH}-\overset{\overset{\displaystyle O}{\|}}{C}-OH$$

10. (a) $H_2N-\underset{\underset{\underset{\displaystyle S-CH_3}{|}}{CH_2}}{CH}-\overset{\overset{\displaystyle O}{\|}}{C}-NH-\underset{\underset{\underset{\displaystyle CH_3}{|}}{H_3C-CH}}{CH}-\overset{\overset{\displaystyle O}{\|}}{C}-NH-\underset{\underset{\underset{\displaystyle CH_3}{|}}{CHOH}}{CH}-\overset{\overset{\displaystyle O}{\|}}{C}NH-\underset{\underset{\underset{\displaystyle SH}{|}}{CH_2}}{CH}-\overset{\overset{\displaystyle O}{\|}}{C}-OH$

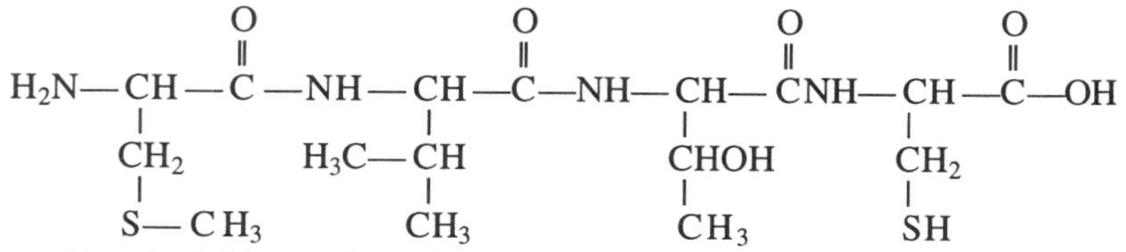

(b) methionylvalylthreonylcysteine

11. Heat can cause some of the very weak bonds that hold proteins together into their secondary and tertiary structure to break and re-form, altering that structure. pH can alter the ionic structure of the amino acids in a protein (protonating —NH₂ or deprotonating —COOH, for example) and these different ionic forms also can have an effect on the secondary and tertiary structure of a protein. Finally, metal ions can bond to the sites responsible for secondary and tertiary structure or they can bond strongly to the active site of the enzyme. Alteration of the structure of the protein or blockage of the active site will affect the activity of an enzyme.

Enzymes are so specific in the reactions they catalyze because the active site is a three dimensional pocket into which the substrate (the reactant) fits precisely. The surface of platinum, on the other hand, is simply a two-dimensional surface demanding little of the shape of the reactant, other than requiring its atoms to be spaced appropriately so that they can bond to that surface, usually two at a time. (And this spacing requirement can often be accomodated in a number of ways.)

12. The bases are on the right side of the Figure. From the top down, they are: adenine (purine), uracil (pyrimidine), guanine (purine), and cytosine (pyrimidine). The pentose sugars have the five-membered rings as their carbon skeletons; they are ribose sugars, and hence this is a chain of RNA. The phosphate groups are the five-membered groups centered on P atoms and surrounded by O atoms.

EXERCISES

Structure and Composition of the Cell

13. The volume of a cylinder is given by $V = \pi r^2 h = \pi d^2 h/4$.

$$V = [3.14159\ (1 \times 10^{-6}\ \text{m})^2\ (2 \times 10^{-6}\ \text{m}) \div 4] \times \frac{1000\ \text{L}}{1\ \text{m}^3} = 1.6 \times 10^{-15}\ \text{L}$$

The volume of the solution in the cell is $V_{soln} = 0.80 \times 1.6 \times 10^{-15}\ \text{L} = 1.3 \times 10^{-15}\ \text{L}$.

(a) $[H^+] = 10^{-6.4} = 4 \times 10^{-7}$ M

$$\text{no. } H_3O^+ \text{ ions} = 1.3 \times 10^{-15}\ \text{L} \times \frac{4 \times 10^{-7}\ \text{mol } H^+ \text{ ions}}{1\ \text{L soln}} \times \frac{6.022 \times 10^{23}\ \text{ions}}{1\ \text{mol ions}} = 3 \times 10^2\ H_3O^+ \text{ ions}$$

(b) $$\text{no. } K^+ \text{ ions} = 1.3 \times 10^{-15}\ \text{L} \times \frac{1.5 \times 10^{-4}\ \text{mol } K^+ \text{ ions}}{1\ \text{L soln}} \times \frac{6.022 \times 10^{23}\ \text{ions}}{1\ \text{mol ions}} = 1.2 \times 10^5\ K^+ \text{ ions}$$

14. Mass of all lipid molecules $= 0.02 \times 2 \times 10^{-12}\ \text{g} = 4 \times 10^{-14}$ g.

$$\text{no. of lipid molecules} = 4 \times 10^{-14}\ \text{g} \times \frac{1\ \text{lipid molecule}}{700\ \text{u}} \times \frac{1\ \text{u}}{1.66 \times 10^{-24}\ \text{g}} = 3 \times 10^7 \text{ lipid molecules}$$

15. mass of protein in cytoplasm $= 0.15 \times 0.90 \times 2 \times 10^{-12}\ \text{g} = 2._7 \times 10^{-13}$ g.

$$\text{no. of protein molecules} = 2._7 \times 10^{-13}\ \text{g} \times \frac{1\ \text{mol protein}}{3 \times 10^4\ \text{g}} \times \frac{6.022 \times 10^{23}\ \text{molecules}}{1\ \text{mol protein}}$$

$$= 5 \times 10^6 \text{ protein molecules}$$

16. $$\text{DNA length} = 4.5 \times 10^6 \text{ mononucleotides} \times \frac{450\ \text{pm}}{1\ \text{mononucleotide}} \times \frac{10^{-12}\ \text{m}}{1\ \text{pm}} = 2 \times 10^{-3}\ \text{m} = 2\ \text{mm}$$

2 mm $= 2 \times 10^3$ μm, the length of the stretched out DNA is one thousand times 2 μm, the length of the cell. Thus, the DNA must be wrapped up, or coiled, within the cell.

Lipids

17. (a) A triglyceride is an ester of the trihydroxy alcohol glycerol, $CH_2OHCHOHCH_2OH$, and three long-chain carboxylic acids. Tristearin is an example of a triglyceride, as is glyceryl laurooleostearate.

(b) A simple glyceride is a triglyceride (see the answer to part c) in which all three fatty acids are the same. Tristearin in an example of a simple glyceride.

(c) In a mixed glyceride, the three fatty acids are not all the same. Glyceryl laurooleostearate is an example of a mixed glyceride.

18. (a) Lipids are those compounds in living tissue that are soluble in nonpolar solvents, such as hydrocarbon liquids. Fats, oils, waxes, and triglycerides all are lipids.

(b) A fatty acid is a carboxylic acid with a long hydrocarbon chain. Stearic acid, $CH_3(CH_2)_{16}COOH$, is an example of a fatty acid.

(c) A soap is a compound of a fatty acid anion and a cation such as Na^+ or K^+. Sodium stearate is an example of a soap, $Na^+\ {}^-OOC(CH_2)_{16}CH_3$.

19. Polyunsaturated fatty acids are characterized by a large number of C=C double bonds in their hydrocarbon chain. Stearic acid has no C=C double bonds and therefore is not unsaturated, let alone polyunsaturated. But eleostearic acid has three C=C double bonds and thus is polyunsaturated. Polyunsaturated fatty acids are recommended in dietary programs since saturated fats are linked to a high incidence of heart disease. Of the lipids listed in Table 28-2, safflower oil has the highest percentage of unsaturated fatty acids, predominately linoleic acid, a diunsaturated fatty acid.

20. Safflower oil contains a larger percentage of the di-unsaturated fatty acid, linoleic acid (75-80%) than does corn oil (34-62%). It also contains a smaller percentage of saturated fatty acids, particularly palmitic acid (6-7%) than does corn oil (8-12%). And the two oils contain about the same proportion of the mono-unsaturated fatty acid, oleic acid. Consequently, safflower oil should consume the greater amount of $H_2(g)$ in its complete hydrogenation to a solid fat.

21. tripalmitin

$CH_2OOC(CH_2)_{14}CH_3$
|
$CHOOC(CH_2)_{14}CH_3$
|
$CH_2OOC(CH_2)_{14}CH_3$

saponification products of trilaurin: palmitic acid and glycerol

$CH_2OHCHOHCH_2OH$

$Na^+ {}^-OOC(CH_2)_{14}CH_3$

22. First we write the equation for the saponification reaction.

$CH_2OOCC_{13}H_{27}$
|
$CHOOCC_{13}H_{27} + 3\ NaOH \longrightarrow CH_2OHCHOHCH_2OH + 3\ Na^+ {}^-OOCC_{13}H_{27}$
|
$CH_2OOCC_{13}H_{27}$

$$\text{mass of soap} = 105 \text{ g triglyceride} \times \frac{1 \text{ mol triglyceride}}{723.2 \text{ g triglyceride}} \times \frac{3 \text{ mol soap}}{1 \text{ mol triglyceride}} \times \frac{250.4 \text{ g soap}}{1 \text{ mol soap}}$$
$$= 109 \text{ g soap}$$

Carbohydrates

23. **(a)** A monosaccharide is the simplest carbohydrate. Glucose, galactose, fructose, and mannose are examples of monosaccharides.
(b) A disaccharide is two monosaccharides joined together. The most familiar disaccharide is sucrose, common table sugar, a disaccharide composed of glucose and fructose.
(c) An oligosaccharide is several monosaccharide units (from two to ten) joined together.
(d) A polysaccharide is many monosaccharide units joined together. Two common polysaccharides are starch and cellulose.
(e) Sugar is a common name given to monosaccharides and oligosaccharides.

24. **(a)** The term glycose is another term for carbohydrate.
(b) An aldose is a monosaccharide in which the C=O group is at the terminal carbon atom; glucose is an aldose.
(c) A ketose is a monosaccharide in which the C=O group is on one of the central atoms of the molecule.
(d) A pentose is a five-carbon monosaccharide; fructose is a pentose.
(e) A hexose is a six-carbon monosaccharide; glucose is a hexose.

25. **(a)** A dextrorotatory compound is one that rotates the plane of polarized light to the right: clockwise.
(b) A levorotatory compound is one that rotates the plane of polarized light to the left: counterclockwise.
(c) A racemic mixture is one composed of equal amounts of an optically active compound and its enantiomer. Since these two compounds rotate polarized light by the same amount but in opposite directions, such a mixture does not rotate the plane of polarized light.
(d) (+) is another way of designating a dextrorotatory compound.

26. **(a)** Two compounds that are optical isomers of each other—they have different locations of the substituent groups around their chiral carbons—but which are not mirror images of each other are diastereomers.
(b) Two isomers that are nonsuperimposable mirror images of each other are enantiomers.
(c) (–) is another way of designating a levorotatory compound.
(d) D indicates that, in the Fisher projection of the compound, the —OH group on the penultimate carbon atom is to the right and the —H group is to the left.

27. A reducing sugar has a sufficient amount of the straight-chain form present at equilibrium that the sugar will reduce $Cu^{2+}(aq)$ to insoluble, red $Cu_2O(s)$. The free aldehyde group must be available to reduce the copper(II) ion.

28. The structure of L-glucose is as follows

Since D-glucose is dextrorotatory,

L-glucose is levorotatory

$$
\begin{array}{c}
\text{H—C}\!\!=\!\!\text{O} \\
| \\
\text{HO—C—H} \\
| \\
\text{H—C—OH} \\
| \\
\text{HO—C—H} \\
| \\
\text{HO—C—H} \\
| \\
\text{CH}_2\text{OH}
\end{array}
$$

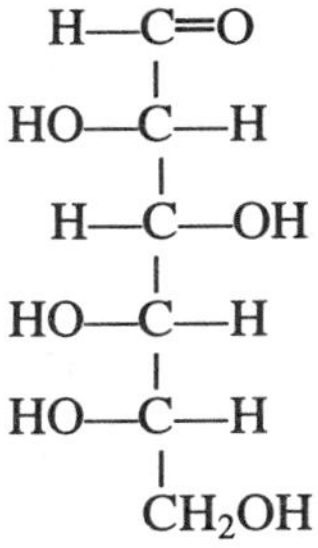

29. Enantiomers are alike in all respects, including in the degreee to which they rotate polarized light. They differ only in the direction in which this rotation occurs. Since α-glucose and β-glucose rotate the plane of polarized light by different degrees, they are diastereomers, not enantiomers.

30. We let x represent the fraction of α-D-glucose. Then $1 - x$ is the fraction of β-D-glucose.
$$+52.7° = x\,(112°) + (1 - x)(18.7°) = 112°x + 18.7° - 18.7°x = 93°x + 18.7°$$
$$x = \frac{52.7° - 18.7°}{93°} = 0.37 \qquad \text{The solution is 37\% } \alpha\text{-D-glucose, and thus 63\% } \beta\text{-D-glucose.}$$

Amino acids, Polypeptides, and Proteins

31. **(a)** An α-amino acid is a carboxylic acid that has an amine group on the carbon next to the —COOH group, that is, on the α carbon atom. Glycine (H_2NCH_2COOH) is the simplest α-amino acid.

   (b) A zwitterion is a form of an amino acid where the amine group is protonated (—NH_3^+) and the carboxyl group is deprotonated (—COO^-). The zwitterion form of glycine is $^+H_3NCH_2COO^-$.

   (c) The pH at which the zwitterion form of an amino acid predominates in solution is known as the isoelectric point. The isoelectric point of glycine is $pI = 5.97$.

   (d) The peptide bond is the bond that forms between the carbonyl group of one amino acid and the amine group of another, with the elimination of a water molecule between them. The peptide bond between two glycine molecules is shown as an outlined bold dash (═) in the structure below.

$$
\begin{array}{c}
\qquad\qquad\;\; \text{O} \qquad\qquad\qquad \text{O} \\
\qquad\qquad\;\; \| \qquad\qquad\qquad\; \| \\
\text{H}_2\text{N—CH}_2\text{—C—O═NH—CH}_2\text{—C—OH}
\end{array}
$$

   (e) Tertiary structure describes how a coiled protein chain further interacts with itself to wrap into a cluster.

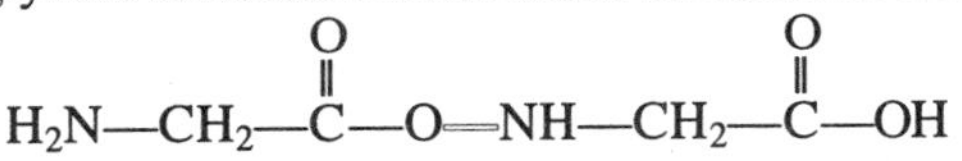

32. **(a)** A polypeptide is a long chain of amino acids, joined together by peptide bonds.

   (b) A protein is another name for a polypeptide, but there is a distinction that often is drawn. Proteins are longer chains than are polypeptides, and proteins are biologically active.

   (c) The N-terminal amino acid in a polypeptide is the one at the end of a molecule where there is a free NH_2 group at the end of the polypeptide chain. The N-terminal amino acid is at the left end of the structure of diglycine in the answer to the previous exercise.

   (d) An α helix is a natural secondary structure adopted by many proteins. It is rather like a spiral rising upward to the right (that is, clockwise as viewed from the bottom). This is a right-handed screw. The alpha helix is shown in Figure **28-9**.

   (e) Denaturation is that process in which at least some of the structure of a protein is disrupted, either thermally (with heat), mechanically, or by changing the pH or changing the ionic strength of the medium in which the protein is enveloped. Denaturation is accompanied by a decrease in the biological activity. Denaturation can be temporary or permanent.

33. $pH = 6.3$ is the isoelectric point of proline. Thus proline will not migrate. But $pH = 6.3$ is more acidic than the isoelectric point of lysine ($pI = 9.94$). Thus, lysine is positively charged in this solution and will migrate toward the negatively charged cathode. And $pH = 6.3$ is less acidic than the isoelectric point of aspartic acid ($pI = 2.77$). Aspartic acid, therefore, is negatively charged in this solution and will migrate toward the positively charged anode.

34. $pH = 5.5$ is the isoelectric point of phenylalanine. Thus phenylalanine will not migrate. But $pH = 5.5$ is more acidic than the isoelectric point of histidine ($pI = 7.65$). Thus, histidine is positively charged in this solution and will migrate toward the negatively charged cathode. And pH = 5.5 is less acidic than the isoelectric point of glutamic acid (pI = 3.22). Glutamic acid, therefore, is negatively charged in this solution and will migrate toward the positively charged anode.

35. (a) in strongly acidic solution

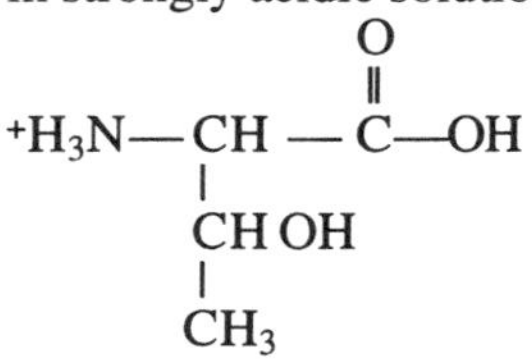

$^{+}H_3NCH(CHOHCH_3)COOH$

(b) at the isoelectric point

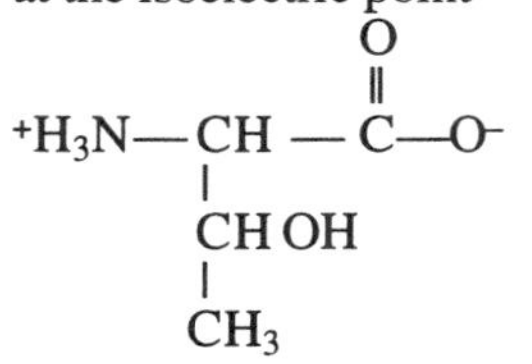

$^{+}H_3NCH(CHOHCH_3)COO^-$

(c) in strongly basic solution

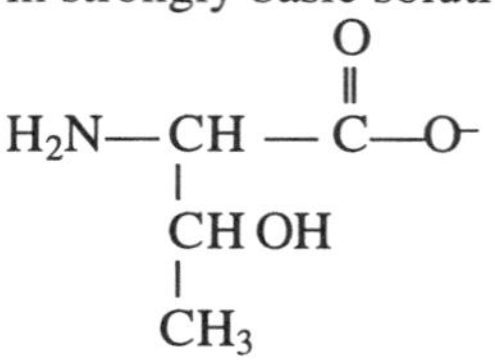

$H_2NCH(CHOHCH_3)COO^-$

36. (a) aspartic acid

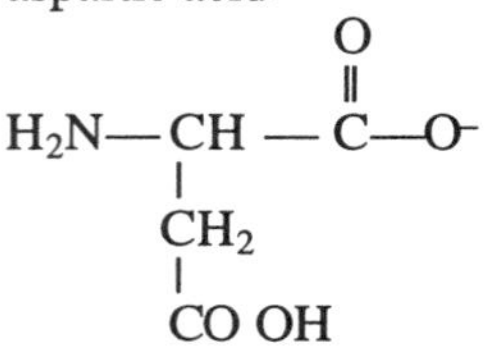

$H_2NCH(CH_2COOH)COO^-$

(b) lysine

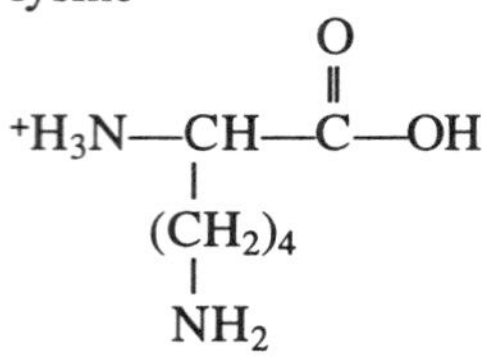

$^{+}H_3NCH[(CH_2)_4NH_2)COOH$

(c) alanine

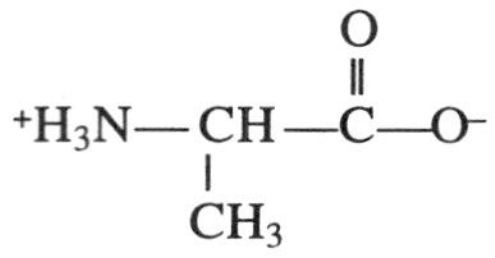

$^{+}H_3NCH(CH_3)COO^-$

37. (a) The structures of the tripeptides that contain one each of the amino acids alanine, serine, and lysine are drawn below.

Lys-Ser-Ala

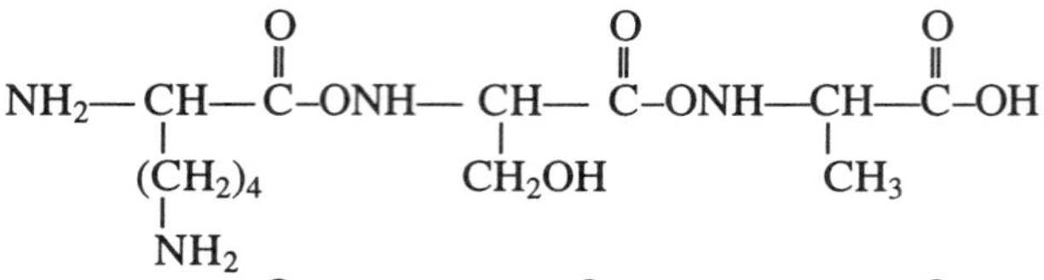

Lys-Ala-Ser

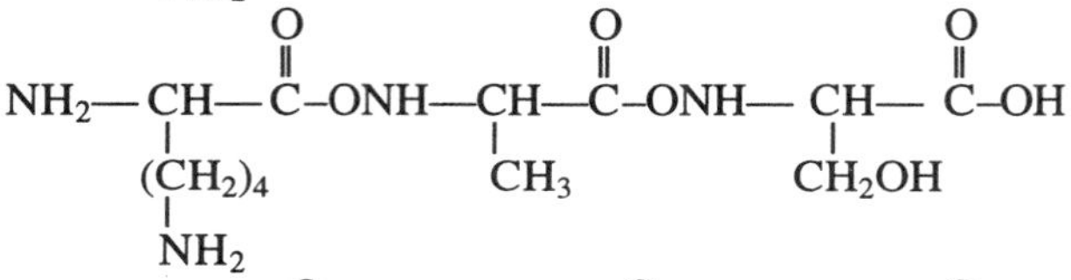

Ser-Lys-Ala

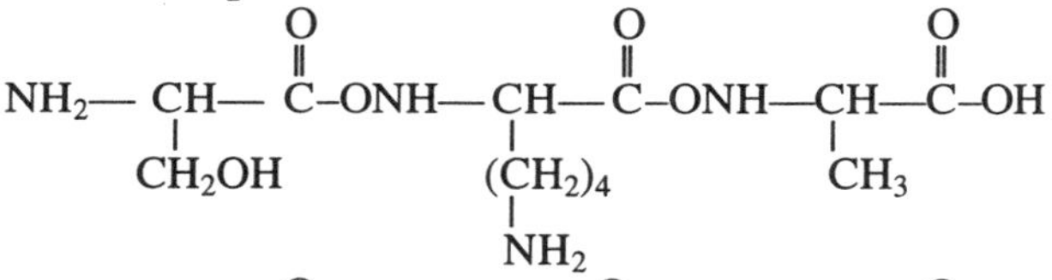

Ser-Ala-Lys

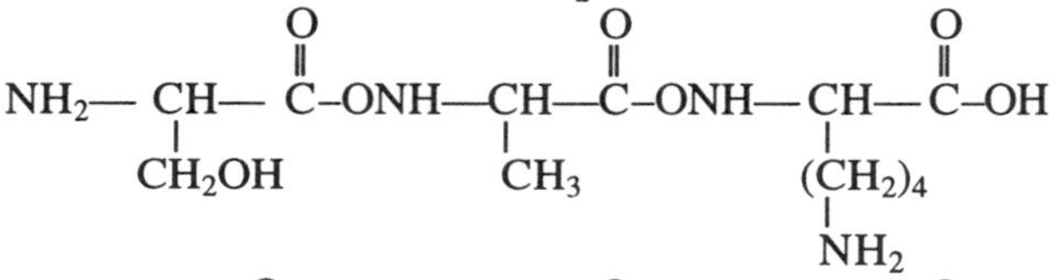

Ala-Ser-Lys

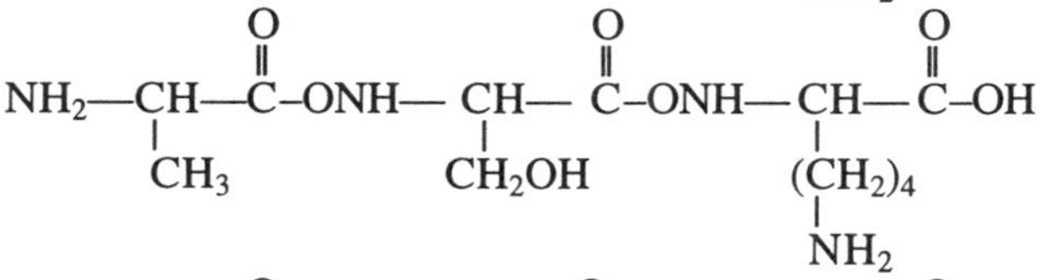

Ala-Lys-Ser

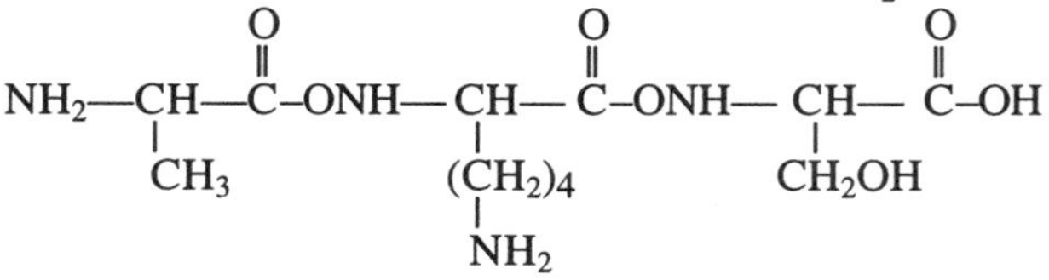

(b) The structures of the tetrapeptides that contain two serine and two alanine amino acids each follow.

Ala-Ser-Ala-Ser

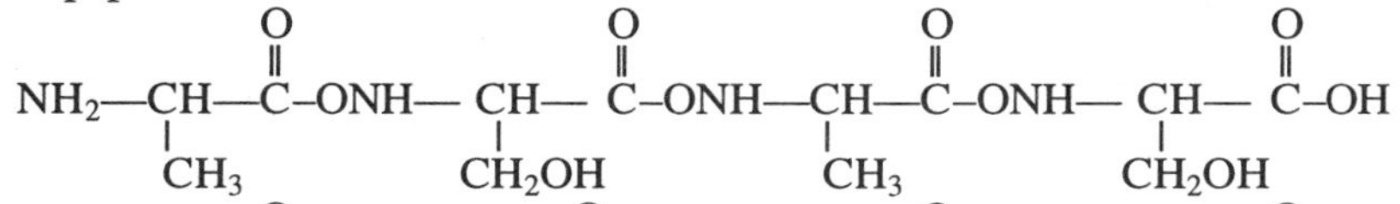

$$NH_2-CH-\overset{\overset{O}{\|}}{C}-ONH-CH-\overset{\overset{O}{\|}}{C}-ONH-CH-\overset{\overset{O}{\|}}{C}-ONH-CH-\overset{\overset{O}{\|}}{C}-OH$$
$$\quad\quad\;\;CH_3\quad\quad\quad CH_2OH\quad\quad\quad CH_3\quad\quad\quad CH_2OH$$

Ala-Ala-Ser-Ser

$$NH_2-CH-\overset{\overset{O}{\|}}{C}-ONH-CH-\overset{\overset{O}{\|}}{C}-ONH-CH-\overset{\overset{O}{\|}}{C}-ONH-CH-\overset{\overset{O}{\|}}{C}-OH$$
$$\quad\quad\;\;CH_3\quad\quad\quad\;\; CH_3\quad\quad\quad CH_2OH\quad\quad CH_2OH$$

Ala-Ser-Ser-Ala

$$NH_2-CH-\overset{\overset{O}{\|}}{C}-ONH-CH-\overset{\overset{O}{\|}}{C}-ONH-CH-\overset{\overset{O}{\|}}{C}-ONH-CH-\overset{\overset{O}{\|}}{C}-OH$$
$$\quad\quad\;\;CH_3\quad\quad\quad CH_2OH\quad\quad CH_2OH\quad\quad\;\; CH_3$$

Ser-Ser-Ala-Ala

$$NH_2-CH-\overset{\overset{O}{\|}}{C}-ONH-CH-\overset{\overset{O}{\|}}{C}-ONH-CH-\overset{\overset{O}{\|}}{C}-ONH-CH-\overset{\overset{O}{\|}}{C}-OH$$
$$\quad\quad\;\;CH_2OH\quad\quad CH_2OH\quad\quad\;\; CH_3\quad\quad\quad\;\; CH_3$$

Ser-Ala-Ser-Ala

$$NH_2-CH-\overset{\overset{O}{\|}}{C}-ONH-CH-\overset{\overset{O}{\|}}{C}-ONH-CH-\overset{\overset{O}{\|}}{C}-ONH-CH-\overset{\overset{O}{\|}}{C}-OH$$
$$\quad\quad\;\;CH_2OH\quad\quad\;\; CH_3\quad\quad\quad CH_2OH\quad\quad\;\; CH_3$$

Ser-Ala-Ala-Ser

$$NH_2-CH-\overset{\overset{O}{\|}}{C}-ONH-CH-\overset{\overset{O}{\|}}{C}-ONH-CH-\overset{\overset{O}{\|}}{C}-ONH-CH-\overset{\overset{O}{\|}}{C}-OH$$
$$\quad\quad\;\;CH_2OH\quad\quad\;\; CH_3\quad\quad\quad\;\; CH_3\quad\quad\quad CH_2OH$$

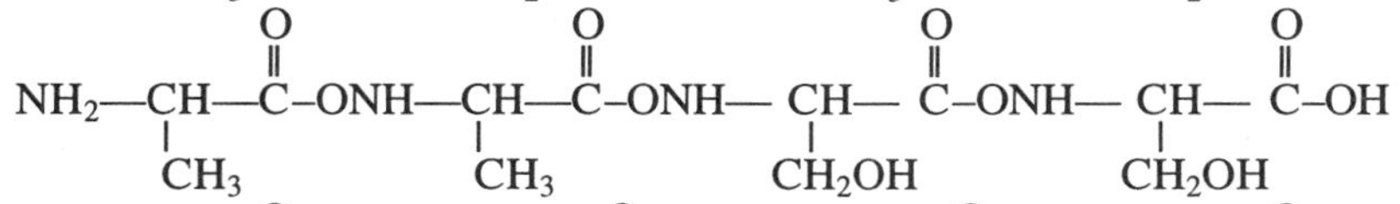

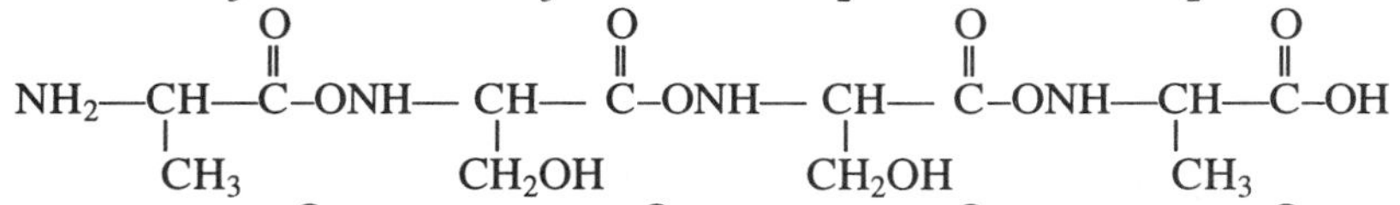

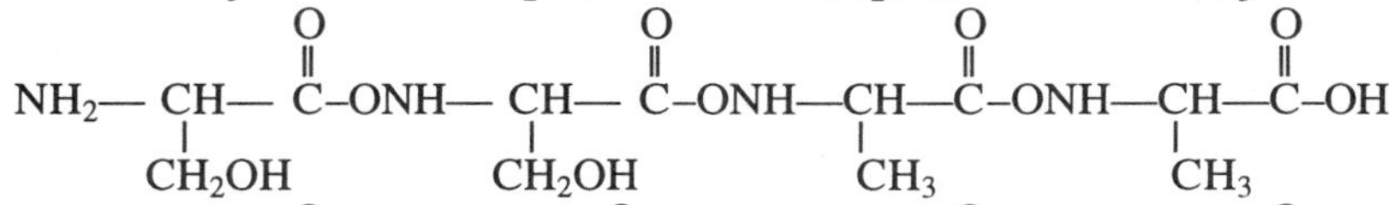

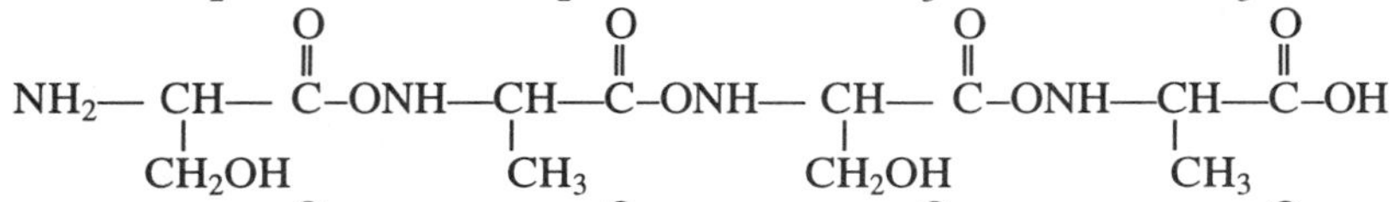

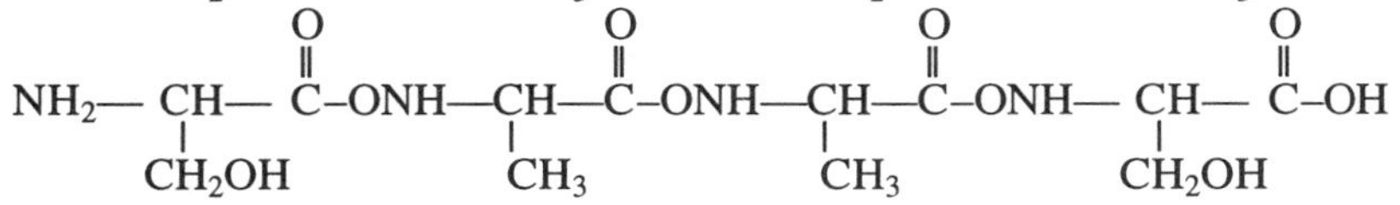

38. There are sixteen possibilities. They are listed here.

Ala-Lys-Ser-Phe	Lys-Ala-Ser-Phe	Ser-Ala-Phe-Lys	Phe-Ala-Ser-Lys
Ala-Lys-Phe-Ser	Lys-Ala-Phe-Ser	Ser-Ala-Lys-Phe	Phe-Ala-Lys-Ser
Ala-Ser-Lys-Phe	Lys-Ser-Phe-Ala	Ser-Lys-Phe-Ala	Phe-Lys-Ser-Ala
Ala-Ser-Phe-Lys	Lys-Ser-Ala-Phe	Ser-Lys-Ala-Phe	Phe-Lys-Ala-Ser
Ala-Phe-Ser-Lys	Lys-Phe-Ser-Ala	Ser-Phe-Ala-Lys	Phe-Ser-Ala-Lys
Ala-Phe-Lys-Ser	Lys-Phe-Ala-Ser	Ser-Phe-Lys-Ala	Phe-Ser-Lys-Ala

39. (a) We put the fragments together as follows, starting from the Ala end, simply placing them down in a matching pattern. We do not assume that the fragments are given with the N-terminal end first.

Fragments: Ala Ser   3rd fragment
                 Ser Gly Val   1st fragment
                      Gly Val Thr   5th fragment
                           Val Thr   2nd fragment, reversed
                           Val Thr Leu 4th fragment, reversed

Result: Ala Ser Gly Val Thr Leu

(b) alanyl-seryl-glycyl-valyl-threonyl-leucine, or alanylserylglycylvalylthreonylleucine

40. (a) We put the fragments together as follows, starting from the Ala end, simply placing them down in a matching pattern. We do not assume that the fragments are given with the N-terminal end first.

Fragments: Ala Lys Ser   1st fragment
                 Lys Ser Gly   5th fragment
                      Ser Gly   3rd fragment
                           Gly Phe   4th fragment
                           Gly Phe Gly 2nd fragment

Result: Ala Lys Ser Gly Phe Gly

(b) alanyl-lysyl-seryl-glycyl-phenylalanyl-glycine, or alanyllysylserylglycylphenylalanylglycine

41. The *primary* structure of an amino acid is the sequence of amino acids in the chain of the polypeptide. The *secondary* structure describes how the protein chain is folded, coiled, or convoluted. Possible methods include α-helices and β-pleated sheets. This secondary structure is held together principally by hydrogen bonds. The *tertiary* structure of a protein refers to how different parts of the molecules, often quite distant from each other, are held together into crudely spherical shapes. Although hydrogen bonding is involved here as well, disulfide linkages, hydrophobic interactions, and hydrophilic interactions (salt linkages) are responsible as well for tertiary structure. Finally, *quaternary* structure refers to how two or more protein molecules pack together into a larger protein complex. Not all proteins have a quaternary structure since many proteins have but one polypeptide chain.

42. The sole difference between sickle cell hemoglobin and normal hemoglobin is due to the substitution of one amino acid for another (valine for glutamic acid) at one position in the entire protein. This changes the quaternary structure of the hemoglobin. We can think of the fault as being due to the mistaken incorporation of one molecule for another during protein synthesis. This is why the name "molecular disease" is apt.

Nucleic Acids

43. The two major types of amino acids are DNA, deoxyribonucleic acid, and RNA, ribonucleic acid. Both of them contain phosphate groups. These phosphate groups alternate with sugars in forming the backbone of the molecule. These sugars are deoxyribose in the case of DNA and ribose in the case of RNA. Attached to each sugar is a purine or a pyrimidine base. The purine bases are adenine and guanine. One pyrimidine base is cytosine. In the case of RNA the other pyrimidine base is uracil, while in the case of DNA the other pyrimidine base is thymine.

44. The "thread of life" is an apt name for DNA, being both a literal and a figurative description. It is literal in that it is thread-like—long and narrow—in shape and is an essential molecule for life. It is figurative in that DNA is essential for the continuance of life and runs like a thread through all stages in the life of the organism, from origin through growth and reproduction to final death.

45. The complementary sequence to AGC is TCG. The hydrogen bonding is shown in the drawing below. One polynucleotide chain is completely shown, as is the hydrogen bonding to the bases in the other polynucleotide chain. Because of the distortions that result from depicting a three-dimensional structure in two dimensions, the hydrogen bonds themselves are distorted (they are all of approximately equal length) and the second sugar phosphate chain has been omitted.

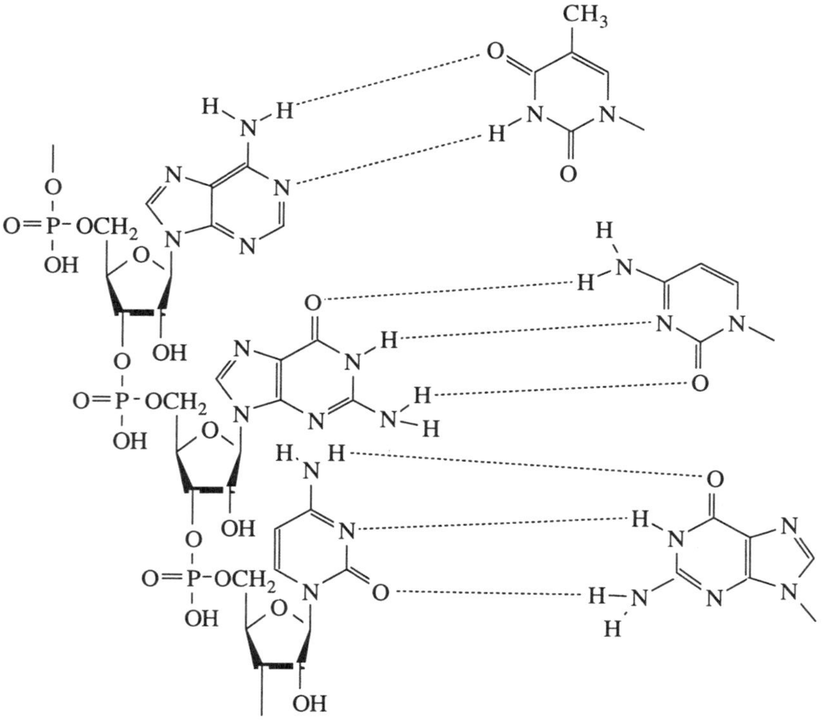

46. The DNA sequence CGCGGGCCC is "translated" to the complementary mRNA sequence in with each G is paired with C, and each C is paired with G. The code is then transcribed by grouping the bases into codons, groups of three, and reading the genetic code. The mRNA sequence is GCG CCC GGG. GCG codes for alanine, CCC codes for proline, and GGG codes for glycine. The tripeptide is Ala-Pro-Gly or alanylprolylglycine.